AF775162

Advanced Millimeter-wave Technologies

Advanced Millimeter-wave Technologies

Antennas, Packaging and Circuits

Dr Duixian Liu
IBM, USA

Mr Brian Gaucher
IBM, USA

Dr Ulrich Pfeiffer
University of Wuppertal, Germany

Dr Janusz Grzyb
Huber & Suhner AG, Switzerland

WILEY
A John Wiley and Sons, Ltd, Publication

This edition first published 2009
© 2009 John Wiley & Sons Ltd.

Registered office
John Wiley & Sons Ltd, The Atrium, Southern Gate, Chichester, West Sussex, PO19 8SQ, United Kingdom

For details of our global editorial offices, for customer services and for information about how to apply for permission to reuse the copyright material in this book please see our website at www.wiley.com.

The right of the author to be identified as the author of this work has been asserted in accordance with the Copyright, Designs and Patents Act 1988.

All rights reserved. No part of this publication may be reproduced, stored in a retrieval system, or transmitted, in any form or by any means, electronic, mechanical, photocopying, recording or otherwise, except as permitted by the UK Copyright, Designs and Patents Act 1988, without the prior permission of the publisher.

Wiley also publishes its books in a variety of electronic formats. Some content that appears in print may not be available in electronic books.

Designations used by companies to distinguish their products are often claimed as trademarks. All brand names and product names used in this book are trade names, service marks, trademarks or registered trademarks of their respective owners. The publisher is not associated with any product or vendor mentioned in this book. This publication is designed to provide accurate and authoritative information in regard to the subject matter covered. It is sold on the understanding that the publisher is not engaged in rendering professional services. If professional advice or other expert assistance is required, the services of a competent professional should be sought.

Library of Congress Cataloging-in-Publication Data

Liu, Duixian.
Advanced millimeter-wave technologies : antennas, packaging and circuits / Duixian Liu ... [et al.].
p. cm.
Includes bibliographical reference and index.
ISBN 978-0-470-99617-1 (cloth)
1. Millimeter wave devices. 2. Millimeter waves. I. Liu, Duixian.
TK7876.5.A38 2009
621.381–dc22 2008041821

A catalogue record for this book is available from the British Library.

ISBN 9780470996171 (H/B)

Set in 10/12pt Times by Sunrise Setting Ltd, Torquay, UK.

Contents

LIST OF CONTRIBUTORS

Jerry Aguirre
Kyocera America Inc., USA

Makoto Ando
Tokyo Institute of Technology, Japan

Paul S. Andry
Thomas J. Watson Research Center/IBM, USA

Aydin Babakhani
California Institute of Technology, USA

Trevor S. Bird
CSIRO ICT Centre, Sydney, NSW, Australia

Joseph P. Dougherty
The Pennsylvania State University, USA

Karu P. Esselle
Department of Electronics, Macquarie University, Sydney, NSW, Australia

Brian A. Floyd
Thomas J. Watson Research Center/IBM, USA

Brian Gaucher
Thomas J. Watson Research Center/IBM, USA

Janusz Grzyb
Huber & Suhner AG, Switzerland, formerly IBM T. J. Watson Research Center, USA

Ali Hajimiri
California Institute of Technology, USA

M. Himdi
IETR, University of Rennes 1, France

Jiro Hirokawa
Tokyo Institute of Technology, Japan

Nils Hoivik
Vestfold University College, Norway

Tatsuo Itoh
Department of Electrical Engineering, University of California, Los Angeles, USA

Sanggeun Jeon
Korea University, Seoul, Korea

O. Lafond
IETR, University of Rennes 1, France

Anthony Lai
HRL Laboratories, LLC, Malibu, California, USA

Michael T. Lanagan
The Pennsylvania State University, USA

Cheng Jung Lee
Rayspan Corporation, San Diego, California, USA

Duixian Liu
Thomas J. Watson Research Center/IBM, USA

Neville C. Luhmann, Jr
University of California, Davis, USA

Yasunao Katayama
Thomas J. Watson Research Center/IBM, USA

Arun Natarajan
Thomas J. Watson Research Center/IBM, USA

Erik Öjefors
University of Wuppertal, Germany

Hsueh-Yuan Pao
Lawrence Livermore National Laboratory, USA

Jean-Olivier Plouchart
Thomas J. Watson Research Center/IBM, USA

Ullrich Pfeiffer
University of Wuppertal, Germany

Khalid Z. Rajab
The Pennsylvania State University, USA

Ramesh Ramadoss
Auburn University, Norway

Scott K. Reynolds
Thomas J. Watson Research Center/IBM, USA

Anders Rydberg
Uppsala University, Sweden

Barry C. Sanders
Institute for Quantum Information Science, University of Calgary, Alberta, Canada

Zuowei Shen
University of California, Davis, USA

Michelle L. Steen
Thomas J. Watson Research Center/IBM, USA

Cornelia K. Tsang
Thomas J. Watson Research Center/IBM, USA

Alberto Valdes-Garcia
Thomas J. Watson Research Center/IBM, USA

Andrew R. Weily
CSIRO ICT Centre, Sydney, NSW, Australia

Thomas Zwick
Institut für Hochfrequenztechnik und Elektronik (IHE), Universität Karlsruhe (TH), Germany

Preface

This book is intended for a wide range of researchers, engineers, managers and the wireless industry at large who are interested to learn and influence the future direction of wireless. Its focus is on millimeter-wave (mmWave) antennas and packaging, but context and relevance is provided by including a systems perspective to give the reader an understanding of the importance of each element and the hidden depths beyond the seemingly simple topics. The goal of this book is not to showcase problems solved, but to educate everyone in this new and exciting area of research that holds the promise further to invigorate and fuel the wireless industry and pull together the brightest minds to solve some of the toughest technical challenges we have ever faced as a wireless industry. What the reader will find between these covers is the work of a small subset of people who have begun to scratch the surface enough for others to see the few brilliant gems hidden in the depths of granite-hard challenges.

At the time of writing, there were no books available on the subject of mmWave chip, antenna and packaging co-design. In order to fill this void, the authors decided to leverage some of the most recent efforts and pull them together into a coherent story that builds from a bottoms-up approach into useful and interesting systems.

Acknowledgements

The authors would like to acknowledge the many people and organizations that made this work possible, especially seed funding from programs including NASA, under contract NAS3-03070, and DARPA, contracts N66001-02-C-8014 and N66001-05-C-8013. As in every field, there are pioneers who began in this area years ago. There is no clear delineation of by whom and when work in the mmWave field began, but there was a defining demonstration of its potential as early as 1895 with J.C. Bose, and the field has been rich with many contributors since. Likewise, Japan stands out as the first country to promote strongly the use of mmWaves, beginning decades ago. Therefore, it only seems fitting to mention from a very long list a few key people who helped to inspire this work: J.C. Bose [1], Johann-Friedrich Luy [2], Keiichi Ohata [3,4], Peter F.M. Smulders [5], Herbert Zirath [6], Peter Russer [7], Hiroyo Ogowa [8,9], Ted Rappaport [10], Larry Larson [11], John Cressler [12]. Also, fundamental to the success of any new technical area is university involvement. A few of the key researchers and their universities who took up this challenge early on and helped to make great strides, as seen in recent ISSCC and other symposia, journals and conferences, are: Professor Ali Hajimiri of California Institute of Technology; Professors Robert Broderson and Ali Niknejad of UC Berkeley; Professors John Cressler and Joy Laskar of Georgia Institute of Technology; Professor Charles Sodini of the Massachusetts Institute of Technology (MIT); Professors Behzad Razavi, Frank Chang [8] and Tatsuo Itoh of UCLA; Professors Larry Larson, Gabriel Rabiz and Larry Milstein of UCSD; Professor John Long of Delft University; Professors Ken O of the University of Florida; Dr Efstratios Skafidas of Melbourne/NICT Australia; Professors Linda Katehi and Jennifer T. Bernhard of the University of Illinois; Professor John Volakis of Ohio State University; Professor Koichi Ito of Chiba University, Japan; Professor Yue Ping Zhang of Nanyang Technological University, Singapore; and Professor Jri Lee of Taiwan National University.

References

[1] T. Sarka and D. Sengupta, 'An appreciation of J. C. Bose's pioneering work in millimeter and microwaves', *History of Wireless* **9** (1977), pp. 291–310.

[2] J.-F. Luy and B. Adelseck, 'Silicon MMICs for millimeter wave communication links', *1998 IEEE International Conference on Electronics, Circuits and Systems* **1** (1998), pp. 51–4.

[3] K. Ohata, K. Maruhashi, M. Ito, S. Kishimoto, K. Ikuina, T. Hashiguchi, N. Takahashi, and S. Iwanaga, 'Wireless 1.25 Gb/s transceiver module at 60 GHz-band', *2002 IEEE International Solid-State Circuits Conference ISSCC*, 2002. *Digest of Technical Papers,* 1: pp. 298–468, 2002.

[4] O. Keiichi, M. Kenichi, I. Masaharu, and N. Toshio, 'Millimeter-wave broadband transceivers', *NEC Journal of Advanced Technology* **2**(3) (2005), pp. 211–16.

[5] P. Smulders, 'Exploiting the 60 GHz band for local wireless multimedia access: prospects and future directions', *IEEE Communications Magazine* **40** (2002), pp. 140–7.

[6] H. Zirath, C. Fager, M. Garcia, P. Sakalas, L. Landen, and A. Alping, 'Analog MMICs for millimeter-wave applications based on a commercial 0.14-μm pHEMT technology', *IEEE Transactions on Microwave Theory and Techniques* **49**(11) (2001), pp. 2086–92.

[7] P. Russer, 'Si and SiGe Millimeter-Wave Integrated Circuits', *IEEE Transactions on Microwave Theory and Techniques* **46**(5) (1998) pp. 590–603.

[8] H. Ogowa, 'A study on millimeterwave research in NICTi', Japan PROJECT REPORT, http://www.aptsec.org/Program/ICT/WebHRDICT/Batch-5/MillimiterWave.pdf.

[9] Y. Shoji, Y. Hashimoto, and H. Ogawa, 'Fiber-optic broadband signal distribution link based on a millimeter-wave self-heterodyne transmission/optical remote heterodyne detection technique', *IEICE Transactions* **88-C**(7) (2005), pp. 1465–74.

[10] C. R. Anderson, and T. S. Rappaport, 'In-building wideband partition loss measurements at 2.5 and 60 GHz', *IEEE Transactions on Wireless Communications* **3**(3) (2004), pp. 922–8.

[11] L. Larson, 'SiGe HBT BiCMOS technology as an enabler for next generation communications systems', *12th GAAS Symposium*, Amsterdam, 2004.

[12] J. D. Cressler, 'SiGe Research Activities'. http://users.ece.gatech.edu/~cressler/research/Cressler%20Georgia%20Tech%20SiGe%2012-05.pdf.

1

Introduction

Brian Gaucher

There is an unusual confluence of three major disruptive and threshold events taking place that are fundamentally reshaping the wireless industry. First, wireless has become a critical, accepted and necessary part of everyday life, e.g. the number of mobile phone users growing at approximately one billion per year and younger generations and countries skipping the PC in favor of handheld wireless devices [1]. The second threshold event is the now rapidly growing high-definition video, automotive radar and high-resolution imaging markets, which have created a sudden need for extremely broadband gigabits per second (Gbps), highly integrated, low-cost and low-power wireless devices in the millimeter-wave (mmWave) frequency bands, which previously only the military could afford. The third threshold event occurring is silicon technology and tools have been developed with suitable performance characteristics to enable radio design, integration and operation at mmWave frequencies; here we specifically discuss the range 60–194 GHz, although the literature is now showing silicon working at hundreds of gigahertz [2].

The enormous reliance of consumers and enterprises on wireless and the evolving mobile Internet is having a disruptive influence on the telco industry, e.g. wireless subscriptions now outnumber landline phone subscriptions and Apple's iPhone™ or Google's GPhone™ [3,4] are forcing mobile carriers into 'openness', effectively reinventing wireless networks and the way they operate and make money. This reliance and evolution speaks to the convenience factor as well as perceived new utility that this technology is providing consumers and enterprise users.

Likewise in the automotive industry, radar units are now options for high-end vehicles and may become mandatory under Intelligent Highway programs around the world, owing to the potential increased safety they can provide. This means a new market for tens of millions of mmWave systems per year.

The growing high-definition multimedia revolution will also require a significant portion of devices to be wireless. This is creating demand for high-speed bandwidth that goes well

Advanced Millimeter-wave Technologies: Antennas, Packaging and Circuits
Duixian Liu, Ulrich Pfeiffer, Janusz Grzyb and Brian Gaucher © 2009 John Wiley & Sons, Ltd

beyond what wireless systems of today can handle. If one looks at the lowest bandwidth requirement for uncompressed high definition television (HDTV), it is about 1.5 Gbps and with some minor coding to make it more robust to multi-path and fading, it easily tops 2 Gbps as discussed within the IEEE 802.15.3c [5]. Today's conventional WiFi delivers unprecedented performance for both office and home use, but tops out at 54 Mbps, with some proprietary systems going as high as 108 Mbps. There is hope that 802.11n and ultra wide band (UWB) systems will get as high as 480 Mbps. UWB systems that utilize limited spectrum between 3.1 and 10.6 GHz have met with marginal success so far, but also promise up to 960 Mbps over a meter or two. At the time of writing, such systems have met with only marginal success in both data rate and distance. Thus, wireless is exploding in use and rapidly evolving from convenience to need across multiple industries and will continue to grow in this direction. In order to satisfy these future WiFi, HDTV, radar and other system needs of speed, capacity, security and robust performance over distance, completely new mmWave (60–194 GHz) solutions will be required.

1.1 Challenges

There is a growing need to solve the huge technical hurdles to cost-effectively leverage the vast unlicensed bandwidth available at mmWave frequencies and satisfy these growing demands. Wireless HDTV is a good example of this and worthy of further exploration here. This application is demanding dramatically higher data rates on the order of 10–100 times current rates. Indeed, they are increasing much faster than current wireless systems can handle and an alternative solution is needed. Given the data rate, capacity and quality of service (QoS) requirements, this can only reasonably happen in a spectrum location where there is suitable worldwide bandwidth on the order of gigahertz with rules that allow one to close a reasonable link budget. These issues have been the fuel and motivation for looking upward in spectrum. As one climbs the spectrum ladder, the first frequency allocation where all of this has the possibility of working well across the varied application space is the 60 GHz band. At 60 GHz there is 3–7 GHz of worldwide bandwidth available depending upon the country; see Figure 1.1 for a sample of countries. In terms of available bandwidth and allowable rules such as transmit power, lack of incumbent users, simple flexible transmission rules, etc. 60 GHz is a boon but it also comes with significant challenges, e.g. ability to do this in a low-cost, physically suitable and robust manner; likewise, the 77 GHz band is the direction the automotive industry is taking, especially in Europe, for automotive radar systems, also called adaptive cruise control (ACC) and Collision Avoidance Systems. There is also increased development activity in the 94, 120 and 194 GHz bands which are being utilized for homeland security applications such as radar, imaging systems, remote sensing, active denial and many others.

Given this set of events, opportunities and constraints, wireless designers have begun developing mmWave system architectures, circuits, antennas and packages; but as expected, they face enormous challenges of simulation, design, integration, physical realization, packaging and test of complete systems that are literally orders of magnitude more difficult than 2.4 and 5 GHz WiFi systems of today; yet to be successful they have to be nearly the same cost.

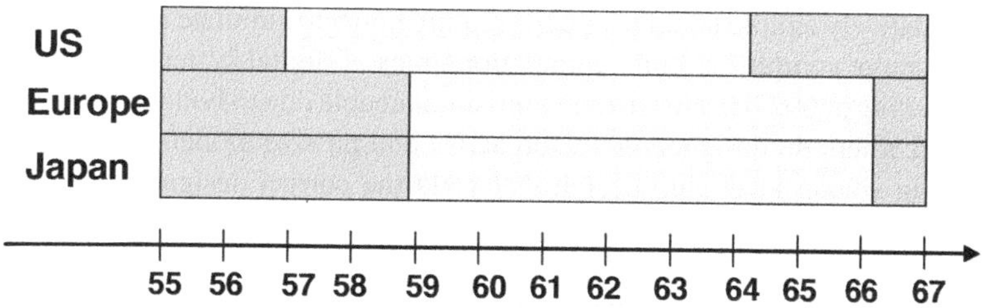

Figure 1.1: The 60 GHz spectrum chart.

If designers can find ways around these challenges, then there are significant benefits that are well worth the effort. The bandwidth and flexible open rules are the most obvious, but there are other less obvious ones also, depending upon if the glass is viewed as half full or half empty. At 60 GHz the wavelength in free space is approximately 5 mm, so circuit designers have the option to use transmission line structures as matching elements and resonant structures, in ways impossible to think about at 2.4 or 5 GHz. Similarly, on-chip filtering becomes possible and on-chip or in-package antennas are now a choice. Traditional microwave board elements such as Lange couplers, 90° hybrids, rat-race structures and many others are now small enough for on-chip consideration. Antenna beam-forming, steering and spatial-power combining become viable system design considerations even in consumer-level solutions, something previously only enjoyed by high-end military systems. Owing to this wavelength consideration, levels of integration go beyond what is achievable at 2.4 and 5 GHz, e.g. including filters and antennas on chip or in the chip package. Indeed, the whole area of packaging is flipped on its head when considering including antennas within the package itself. One's packaging mindset shifts from containing radiofrequency (RF) radiation to intentionally radiating specific frequencies, while attenuating others. This will be one of the themes for this book.

Then there is the glass half empty point of view. At mmWave frequencies, the world of consumer level systems, circuit, antenna and packaging design, is largely unknown. For example, at 2.4 and 5 GHz, a designer takes the dielectric constant of PC boards for granted, something that is known and can be relied upon; but this is not the case at e.g. 60 GHz and higher. At such frequency extremes, each material has to be characterized and relevant data extracted from samples. There are also new metamaterials and devices that might prove invaluable, e.g. electronic band gap (EBG), anisotropic approaches, new polarization techniques, etc., but as yet they remain largely untried and untested at mmWave frequencies. Simple interconnects that work well at 5 GHz may have untenable loss at 60 GHz, where each and every interface outside the chip has to be considered in the link budget. As designers pull the antenna design into the chip or chip package, there are many new and important considerations and design options that need to be taken into account and traded off, which affect the whole system. Then there are the circuit design challenges too. At 60 GHz basic transistors, as good as they have become recently, run out of gas in this frequency range, e.g. gain is considerably lower than 5 GHz, isolation decreases, power generation is much more challenging and losses are much higher [6]. On top of all of this, today's design tools have not been tested in any significant way at 60 GHz and even the smallest variation or error may have a significant impact on the end product's performance. Although time will help, today even relatively simple 60 GHz circuits require hours to simulate and it is not unusual to wait days for more complex circuits. Simulation times of digital systems that include analog front ends running at 60 GHz are currently un-simulateable due to both convergence and time required. And where do designers get valid active and passive models and macros they can trust for use in silicon level chip design that yield the correct designs at the first attempt? Even basic RF switches and switch elements need special consideration for device selection, design, use and simulation.

Thus, the challenges and benefits are many, making this an incredibly rich and deep area of research in the coming years, with streams of patents emerging around all of the above, virtually reinventing the wireless industry. As all of this takes shape, we will begin to benefit as consumers of this technology; we will find 60 GHz will enable 1–10 Gbps wireless solutions in a wide range of products such as next-generation WiFi, wireless HDTV, MP3 players and cell phones. These latter devices will be used as electronic wallets that can nearly instantly download, pay for and store a HD-movie for transport to the home and wirelessly upload it to home theater systems. There will be kiosks that act as displays for the hard drives and systems that reside in your cell phone and so much more.

As we master the design idiosyncrasies of 60 GHz, the wireless industry will move up to 77 GHz to develop cost-effective ACC for cars and Intelligent Highway Systems, where there exist a whole new set of packaging and antenna challenges on top of extreme environmental conditions. As next-generation silicon-based terahertz (THz) imagers evolve, inspectors will more easily be able to detect non-metallic weapons and explosives from a safe distance using THz imaging techniques and high-performance computing. This will make airports and critical entry points safer for everyone. However, to achieve all or any of the above in any cost-effective manner requires designers to solve a myriad of extremely challenging problems.

1.2 Discussion Framework

With a reasonable motivation of why the mmWave frequency is important and useful and the basics of what the challenges are, the next step is to establish a simple consistent framework that can be used throughout the book that addresses the system architecture, antennas, circuits and packaging in a holistic manner. Rather than highlight the multitude of architectural solutions for all wireless architectures and applications, e.g. direct conversion, low-intermediate frequency (IF), heterodyne, super-heterodyne, etc., and performance targets, we choose a single commonly applied approach of the super-heterodyne architecture [7,8]. We use this as our reference point for discussion purposes. It is not the simplest architecture, but is general purpose and contains all the elements of virtually any wireless system one can imagine. Based on that architectural approach, we call out circuits, antennas and packaging options that reference this and which then be addressed in the remainder of the book.

1.3 Circuits

The super-heterodyne architecture is one of the more complex and therefore more encompassing architectures, making it a good choice to frame the larger circuit, antenna and packaging challenges. Although not optimal for all, it could be used for any of the

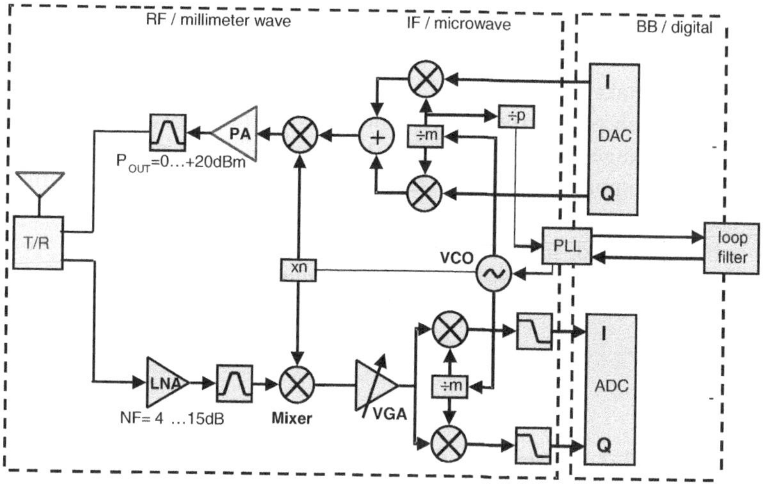

Figure 1.2: Generalized 60 GHz super-heterodyne block diagram.

aforementioned applications; Gbps wireless, radar, imaging etc. A simplified block diagram of a 60 GHz radio architecture is shown in Figure 1.2. It is based on a single-voltage controlled oscillator (VCO) super-heterodyne (multiple mixing stages) time-division duplex (switches from transmit to receive (T/R) rather than simultaneous transmit and receive) design with variable-IF frequency (byproduct of a single local oscillator (LO)). The high-frequency 60 GHz signals connect directly from integrated antennas to the T/R switch and to the low noise amplifier (LNA) input or the power amplifier (PA) output to avoid the need for external packaging and waveguide structures with their associated size, weight and power losses.

1.4 Antenna

For reasons to be discussed in detail throughout the book, mmWave antennas will nearly always be integrated with the chip or chip package. The integrated antenna has two major functions: the first is an efficient radiator or collector and the second as a bandpass filtering function. The natural bandpass filtering provided by the antenna helps both to limit the noise bandwidth prior to the LNA and to provide some image rejection. It is critical to any wireless system, but at mmWave frequencies the antenna becomes even more so because of the potential detrimental effects of interconnection losses, distance to RF electronics, match and proximity effects of nearby materials, receiver noise figure and transmit power. At mmWave frequencies power generation is extremely difficult and 'expensive' DC power wise, making antenna efficiency paramount. Designers cannot afford to waste hard-fought-for RF power only to lose it to 'simple' but lossy interconnects. For consumer level (cost-sensitive) products, performance needs and ease of use, it is a virtual requirement that the antenna be an integral part of the chip package from start to finish. This places new constraints across the whole antenna/package simulation, design and test space.

1.5 RF Electronics

The RF electronics make up the body of any wireless device and consist of a receiver and transmitter where each place critical and sometimes conflicting requirements on both the antenna and the package. For example, loss between the antenna and the RF electronics decreases transmit output power and increases system noise figure directly; thus, each RF subsystem needs to be located close to the antenna. Since there are physical limitations and both cannot be minimized, tradeoffs must be made as to which should be the closest. To establish a common vocabulary for the rest of the book, it is worth reviewing both receive and transmit chains.

1.5.1 Receiver

As shown in Figure 1.2, the signal at the output of the receive antenna is amplified by an LNA with enough gain (>10–15 dB) to establish the system noise figure (<6 dB). The LNA drives an integrated bandpass image filter, which is designed to eliminate received noise power amplified by the LNA at the IF image frequency to minimize the noise. The output of the LNA drives a mixer that translates the mmWave signal to an IF frequency in the range of 9.2 GHz. This IF frequency is chosen to provide easy image rejection at the RF and to support very high data rates. The mixer then drives a variable-gain IF amplifier, which increases the dynamic range of the receiver.

A single VCO operating in the band around 17.6 GHz is multiplied by three to generate the LO for the up and down mixers. The VCO is also divided by two to generate the 8.8 GHz LO signals for the IF quadrature mixers, which translate the received signal to baseband frequency I and Q channels. The baseband I and Q channels are band limited by variable bandwidth low-pass filters. From here the signals are either A/D (Analog to Digital) converted or detected directly depending upon the chosen modulation. The only cost effective, power efficient and performance friendly design approach at mmWave frequencies is to implement fully integrated solutions in silicon; going off chip for any mmWave frequency element could prove disastrous.

1.5.2 Transmitter

The transmit path uses the same variable-IF technique as the receive path to minimize the complexity of the system design. The band limited IQ signals are combined to a real IF signal by the quadrature up mixer. The IF signal is amplified if necessary and translated to the 60 GHz carrier frequency by a second mixer. The 60 GHz signal is increased to the desired transmit power required to close the link budget by the power amplifier (PA). The output of the PA drives an integrated bandpass filter, which suppresses the out-of-band image and carrier feed-through products, as well as broadband noise from the PA. The image filter connects to a tuned antenna designed to pass the minimum amount of spectrum necessary for operation at the desired data rate. As mentioned above, full integration of receiver, transmitter and synthesizer/LO subsystem should be done and thought of as a system. The antenna also needs to be considered closely in the design since single-ended or differential approaches can dramatically impact realizability and performance.

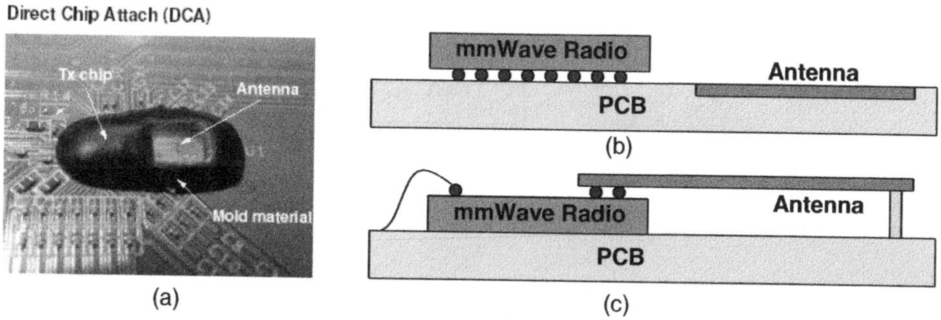

Figure 1.3: DCA with (a) integrated antenna using (b) flip chip or (c) wirebond techniques.

1.6 Packaging

mmWave packaging is an expansive and open set of topics which will be a significant focus for this book. To set the stage, two simple models that have been validated through tests will be shown, but there are a multitude of others such as multi-chip modules (MCM), low-temperature co-fired ceramic (LTCC), silicon microelectromechanical systems (MEMS), microwave chip on flex (MCOF), injection molding, wafer level silicon, etc. The latter ones will be discussed, but the first two will be used to highlight many of the common techniques and to frame the challenges.

Figure 1.3(a) shows a mmWave chip on printed circuit board (PCB) through direct chip attach (DCA). The DCA approach in Figure 1.3(b) highlights one of the most basic and simple approaches. The mmWave radio die is flipped chip attached to a reasonably similar thermal coefficient of expansion (TCE) material board with, e.g., C4 solder or gold stud bumps. In this configuration, an obvious approach for the antenna is to make it a part of the main PCB, as shown in Figure 1.3(b). Here the antenna is part of the main PCB. The main benefits are extremely low cost, ease of assembly and readiness for high-volume manufacturing. The challenges in this approach are matching TCE with acceptable materials, PCB losses in the C4 or stud bump and the Transmission line (T-line) losses to get to the antenna, PCB induced T-line resolution and tolerances, and net radiation efficiency of the antenna.

Figure 1.3(c) shows a slightly more complex approach where the die is simply bonded to a generic PCB and wirebonded in place. The antenna is connected via C4s or stud bumps. The benefits of such a configuration compared to Figure 1.3(b) are less TCE sensitivity, significantly lower losses between the antenna and the die, additional freedom of antenna materials to enhance performance, line resolution and tolerances and much better bandwidth and radiation efficiency through the use of a cavity structure. The challenges are slightly higher antenna and assembly costs.

There are significant technical challenges to make either of these approaches viable. As previously mentioned, basic material properties at mmWave frequencies are unknown, e.g. dielectric constant (Er) and loss tangent. In both approaches shown in Figures 1.3(b) and (c), mold and underfill materials are required to protect the die, bonds and the cavity

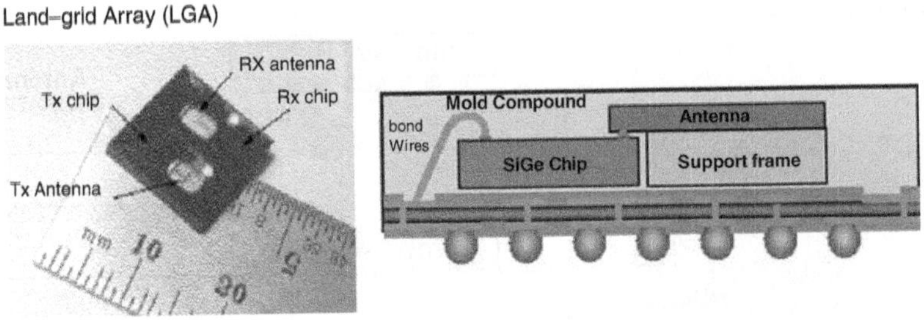

Figure 1.4: Plastic LGA suitable for injection molding, along with side-view schematic.

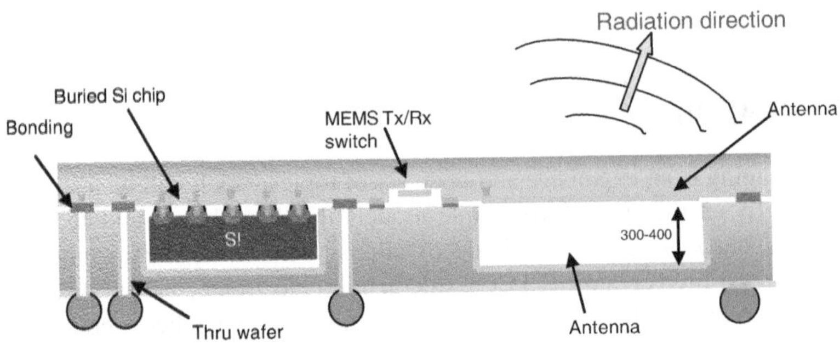

Figure 1.5: LTCC or silicon packaging approach (from reference [9], reproduced by permission of © 2007 IEEE).

in Figure 1.3(c), yet the effect of covering RF electronics at mmWave frequencies with such materials is largely unknown, e.g. isolation and coupling effects, antenna and match detuning. The interconnection losses, coupling and tolerances of mmWave die to PCB or antenna through C4s or gold stud bumps is just beginning to be understood through experiment and analysis.

A hybrid of Figure 1.3(c) and injection molding is illustrated in the land grid array (LGA) approach of Figure 1.4. On the left is an actual plastic LGA mmWave (60 GHz) transceiver shown with antennas exposed for clarity. On the right is a schematic representation of what is inside the LGA. This approach has the highly desirable benefit of standard plastic packaging, deconvolving the mmWave transceiver subsystem from the PCB. This allows a more general tape and reel manufacturing approach. The challenges of such a configuration are similar to that of Figure 1.3(c).

Other approaches that were built and tested, that make good candidates for archetypical packaging, are LTCC and silicon. Figure 1.5 shows a generalized schematic representing both of these options.

Here the RF electronics chip or die is mounted within a substrate material. These substrates offer reasonably low-loss short interconnects to efficient antenna structures. They can also include other passive devices such as decoupling capacitors (caps) resistors, MEMS devices such as transmit/receive (T/R) switches and hermetic packaging. The challenges beyond those previously mentioned include: increased cost, green sheet effects of shrinkage that need to be accounted for in LTCC, the dimensional requirements of antennas versus through wafer vias and wafer-bonding challenges of silicon.

In summary, the above discussion is the framework that will be used throughout the book, but it represents just a simple skeleton of topics that will be covered in much more detail. As the reader can imagine, this work extends far beyond simple packaging, circuits and antennas. As an example, if these packages with integral antennas can be made with outer dimensions less than half a wavelength, then integrated arrays of these could be built that form the basis of phased array systems. This could be taken to an extreme, where an entire wafer of silicon circuits could be wafer bonded to a passive silicon wafer with the antenna directly above the RF electronics and form the entire basis of a phased array system.

Such substrate approaches are also amenable to include other new devices such as EBG devices that could be used in a variety of ways at mmWave frequencies, e.g. filters. New metamaterials could also be incorporated for additional antenna flexibility. Lenses could be added. Each of these technologies is in its infancy at mmWave frequencies. Taken together these topics are a rich and open set of fields that are only now beginning to be explored. Throughout this book the reader will find samples of theory and measurement results for all of them. It is intended to show the beginnings of what is possible for consumer level electronics in the 'greenfield' mmWave bands and to stimulate further investigation.

1.7 Organization and Flow of this Book

The flow of this book follows a bottom-up approach, creating a foundation of building block elements and works up to the system level with examples. Each chapter is reasonably independent and a self-contained entity with its own references where the reader may delve deeper into any of the topics, and it is also an integral element of subsequent chapters.

Chapter 2 begins with an introduction to packaging that establishes the basic and necessary elements needed for mmWave packages. These include those found at lower frequencies, with special emphasis on the unique mmWave material properties and interconnects and how these packaging elements relate to both performance and cost. This leads into a discussion on the importance of system package co-design at mmWave frequencies.

Chapter 3 shifts the discussion to the electrical properties of these packaging elements and materials at mmWave frequencies, specifically dielectric constant and loss tangent. Although basic packaging materials are fairly well characterized into the microwave frequency region, there is a dearth of information above this range. Materials are proposed and their advantages and disadvantages are discussed. In addition to describing a number of common material types and suggesting new ones for various mmWave packaging applications, significant emphasis is also placed upon measurement techniques and material characterization in this extreme frequency range. These include extrapolating microwave techniques such as resonant cavity and transmission line approaches to mmWave frequencies and utilization of time domain spectroscopy.

Chapter 4 builds upon the materials discussion of Chapter 3 to describe a number of key mmWave interconnect types. These interconnects are used between basic devices, circuits, antennas and other structures. The results of this chapter will be built upon and used in subsequent chapters eventually evolving to system level examples. These interconnects described herein, utilize specific materials that complement each interconnect type, while detailing the critical nature of each, at mmWave frequencies. This interconnect discussion includes everything from mechanical issues and tolerances to electrical characteristics and performance. Printed transmission lines (T-lines) on various substrate materials, typical of low and microwave frequencies are presented for use on PCBs and integrated circuits (ICs) such as silicon (Si) or gallium arsenide (GaAs) and extrapolated to mmWave frequencies. Wave guide interconnects are also reviewed and their potential highlighted. With this as a foundation, various packaging approaches are described that make use of these. The design and use of each interconnect type is applied to various packaging approaches and applications and their advantages and disadvantages are highlighted throughout the chapter.

Chapter 5 discusses the use of multilayer technologies to print millimeter antennas. Building upon the packaging, materials and interconnect foundation of Chapters 2–4, we start to create useful circuit level building block elements beginning with the antenna. This discussion begins with critical issues that affect antennas, which are compounded at mmWave frequencies such as metallic and ohmic losses, spurious radiation and surface waves. The characterization of these effects in multilayer materials at mmWave frequencies is described and their impact on various antenna types and resulting performance is highlighted. This is used and applied to a number of antenna design examples from the most basic patch antenna to the more complex antenna arrays. This is then put in context by describing various packaging approaches and how one would create optimum feed networks for each.

Chapter 6 complements the printed antenna chapter with a presentation of planar waveguide-type slot arrays. It begins with single layer slotted waveguide array antennas and works into novel waveguide slot antennas and arrays. Along the way it describes reflection cancellation techniques, method of moment (MoM) analysis using Eigen-mode basis functions. The design and analysis of a number of other important antenna types and feed structures are also presented that are important to mmWave applications and packaging.

Chapter 7 introduces the basics of antenna design from materials selection to feed structures for specific 60 GHz applications and packaging. Flip-chip mounting is suggested as a potential interconnect technique amenable to the challenges of 60 GHz. An air-suspended superstrate antenna design is proposed as a packaging solution for a specific mmWave design that works well in a number of PCB and plastic packaging approaches. The design is detailed along with key characterization techniques and results. Then the design is analyzed for manufacturability and suitability for low-cost packaging through tolerancing analysis using High Frequency Structural Simulator (HFSS™). The HFSS results are then used to determine system performance. With this as a foundation, additional antenna types are investigated such as a 2×4 patch array for high gain, single and dual loop based antennas for circular polarization along with assembly details. Then, additional substrate packages are introduced that hold the potential for higher integration and lower cost such as LTCC and silicon carrier.

Chapter 8 takes the work of low-cost highly integrated antennas of Chapter 7 one step further by introducing the concept of monolithic integrated antennas, which are antennas directly integrated within the IC itself. The promises of these are further cost reductions and smaller size. One can then consider using these to address the needs and costs of commercial and consumer markets. The approach described uses the same substrate as the active radio circuits and adds the antenna, thereby saving the cost of interconnects and significantly altering the packaging goals. Electrically small antennas are defined along with their challenges of performance and issues of substrate modes and efficiency.

Manufacturing issues are discussed and radiator selection criteria are suggested. With this basis, a variety of antennas are investigated including patches, dipoles, loops and inverted-F structures. Their designs and performance are presented and the practical implementations of these into various packages are discussed and how they are measured.

Chapter 9 moves the discussion forward into the very new area of metamaterials for mmWave antenna applications. Simply defined, metamaterials are artificial materials with engineered properties not readily available in nature. In particular, metamaterials with simultaneous negative permittivity and negative permeability, known as left-handed (LH) metamaterials, will be shown to have good application to antenna designs, because of their unique characteristics of negative refractive index, fundamental backward wave, and infinite wave with non-zero group velocity. The chapter begins with three methods to realize LH metamaterials; the resonant method, based on pairing split ring resonators (SRRs) with metal wires, the second is the transmission line (TL) approach based on realizing LH TL unit-cells of series capacitance and shunt inductance. The third method is the evanescent-mode using dielectric resonators inside a cutoff background. Using these methods and different design approaches, several antenna types are investigated, including small resonant backward-wave antennas, fundamental-mode leaky-wave antennas, small circularly polarized antennas, planar and dual-band antennas.

Chapter 10 introduces EBG materials and antennas that leverage the small wavelength at mmWave frequencies. EBG materials will be shown to have properties that are suitable for controlling and manipulating the propagation of electromagnetic waves. EBG materials are created from dielectric and/or metallic structures that are periodic in one or more dimensions. One- and two-dimensional structures will be presented and analyzed and then a discussion ensues of how 'defects' can be added to create devices and waveguides that control the propagation of electromagnetic waves. High impedance ground planes are introduced that reduce surface waves and increase efficiency and are shown to be useful for corrugated horn antennas and other structures. It is then shown how EBG substrates can improve printed antenna efficiency. Low-profile high-gain antennas using EBGs are presented next, followed by high-gain one-, two- and three-dimensional resonator-type antennas and concludes with metamaterial and Woodpile EBG waveguide horn antennas and arrays.

Chapter 11 further builds on the mmWave integration theme through the introduction of RF switches. RF switches are used in many common wireless standards from cell phones to WiFi networks for duplexing and switching between frequency bands and modes. Metrics of reliability, power efficiency and cost are introduced along with key small and large signal specifications and these are used to compare various architectures and implementations and to determine the impact of switch performance on a communication system. Basic models are created that allow the simulation and optimization of key parameters such as on/off resistance and capacitance in relation to insertion loss, isolation and power handling. Inductor matching techniques are introduced to extend and improve performance at mmWave frequencies. Example implementations/devices in 65 nm CMOS (FET) are presented and are compared to other technologies such as GaAs (FETs and Diodes).

Chapter 12 describes the technology behind RF MEMS devices and their application to mmWave antenna systems. The basic principles behind micromachining techniques are introduced (and discussed further in Chapter 18). The concept of electrostatic actuation for switches is covered, followed by a description of two classes of RF MEMS switches, a comparison with solid-state devices, and performance and design considerations for RF

MEMS switches. MEMS switch reliability and power handling challenges are then discussed and various approaches for integration of RF MEMS switches with antennas. This leads to a discussion on the use of MEMS technology for implementation of tunable, steerable and reconfigurable antennas. The chapter is concluded with a look into the future of RF MEMS and NEMS for mmWave systems.

Chapter 13 takes the fundamentals of antenna designs presented thus far and begins a discussion of phased arrays. It begins with an applications discussion highlighting the importance of phased arrays and discusses the basics of beam forming networks and radiating elements. The phased array antenna concept is introduced and used to discuss a design approach that accounts for mutual coupling in the practical design of antenna arrays. The continuous line source antenna is used as a basis from which many antenna concepts are derived. The infinite array methodology is described and used to design a slot coupled patch antenna. Beam-forming networks are discussed in terms of their spatial filtering process through the appropriate use of complex weighting of the elements in the array. The chapter is concluded with a brief discussion of the fabrication of an antenna array designed for W-band operation in a LTCC material technology.

Chapter 14 takes the phased array concepts and importance highlighted in Chapter 13 another step forward as integrated phased arrays, essentially trying to address the cost of commercial and consumer products such as automotive radar, imaging, sensing and other systems through very high levels of integration. The discussion begins with basic principles of phased arrays, their benefits and the challenges of integration, specifically in silicon. Silicon-based antennas are discussed along with fundamental phased array architectures from RF, IF to LO. A candidate fully integrated mmWave silicon-based transceiver is presented with each block detailed and its performance summarized. Then a discussion of alternative methods ensues based upon direct antenna modulation for higher efficiency and security. It is presented in the context of a 60 GHz system with experimental results. The discussion moves into large-scale integrated phased array systems in CMOS technology at 6, 10.35 and 18 GHz along with experimental results.

Chapter 15 extends the application space of mmWave to imaging, which among many other applications, expands our vision by letting us 'see' things under poor visibility conditions. There exist significant military and commercial benefits for such systems from surveillance, precision targeting, navigation, search and rescue, to the commercial realm of aircraft landing aids, airport operations, and highway traffic monitoring in fog. It also suggests that mmWave frequencies are just the beginning and that THz frequencies will become important as well, if they can be achieved in low-cost technologies. Passive mmWave imaging is first presented, which essentially works off black body radiation; that is, passive emission and reflection of mmWave energy. Then active mmWave imaging is presented where a mmWave source is used to illuminate a scene, while the reflected energy is analyzed for content. This is useful for increasing the sensitivity of systems and potentially seeing through building materials. Advantages and disadvantages are presented for each approach. Several examples of commercial systems are presented. The chapter concludes with a survey of the state of the art in imaging technologies.

Chapter 16 provides a high-level overview of mmWave systems tying many of the preceding topics together. It begins with an outlook for low-cost, high-volume mmWave systems and moves into a discussion on the technology, integration and packaging aspects key to making it a commercial success. It suggests what the first high-volume applications might be in the consumer, automotive and military spaces and what the cost drivers are for each. One of the first 60 GHz fully integrated silicon based transceivers is presented and described in detail from design to performance. This chip-set, mounted on a simple PCB with integral antennas, is used to create a high definition streaming video demonstration similar to that envisioned by the IEEE 802.15.3c developing standard. The system design, modulation and demodulation are presented. This chip-set and antenna is then used to highlight how mmWave solutions might achieve low-cost commercial and consumer level packaged solutions through a Chip-on-Board (CoB) example.

Chapter 17 addresses some of the measurement challenges at mmWave frequencies. Up to this point, it was taken for granted that the reader understood the challenges of mmWave equipment, characterization and measurement techniques, but given that mmWave is only now beginning to be pervasive with strong pushes in multiple standards organizations such as the IEEE, ECMA and WirelessHD, the authors felt that this book would not be complete without some emphasis on the special mmWave measurement techniques required. This chapter highlights the fact that virtually any measurement at these frequencies requires great care. Even simple cables, connectors and probes used to contact the device under test (DUT) have non-negligible effects on the measurement result. Thus, special calibration techniques are essential and detailed throughout this section. A brief overview of modern vector error calibration methods are presented followed by special mmWave measurement techniques where the effect of the launching structures are de-embedded based on lumped element models of the launching structure. Two special measurement techniques for highly integrated mmWave systems are described. The first measurement technique allows measurement of the complex impedance and the antenna radiation pattern in an anechoic chamber. Finally, a novel non-destructive IC package characterization method is described that uses a recursive untermination method for non-destructive *in-situ* S-parameter measurements of multi-port packages.

Chapter 18 extends the MEMS work started in Chapter 12, with a more detailed discussion on micromachining and silicon processing. It begins with a review of mmWave packaging technologies and introduces silicon based packaging through examples and applications highlighting advantages and disadvantages along the way. Process options for lithography, machining, metallization, wafer thinning and assembly are presented. The chapter concludes with an example silicon-based mmWave package with integral antenna and its measurement results.

Through this brief introduction, it is hoped that the reader perceives the subject matter of this book as the beginning of a new and exciting wireless world and it is as wide a set of topics as it is rich and deep in novelty, complexity and challenge. Although the surface may only be scratched here, uncovering a few shiny gems, the authors' goal is that the reader is as captured by this field as they are and will become similar pioneers in this new area, contributing to and growing this body of work so that all may reap the benefits of this emerging set of technologies and may launch a new generation of wireless applications and devices.

References

[1] R. Martin, 'Cell phones face extinction as smartphones take over', *InformationWeek*, Tuesday April 1, 6:00 AM ET.

[2] Georgia Tech/IBM announce new chip speed record: Silicon-germanium transistor operates 250 times faster than average cell phone, 500 GHz, June 20, 2006, http://www.gatech.edu/newsroom/release.html?id=1019 [December 29, 2008].

[3] D. Riley, 'GPhone prototype debuts at mobile world congress', *Tech Crunch*, February 11, 2008.

[4] A. Rubin, 'Where's my Gphone?', *Google Blog*, November 11, 2007.

[5] IEEE 802.15 WPAN millimeter wave alternative PHY Task Group 3c (TG3c), http://www.ieee802.org/15/pub/TG3c.html [December 29, 2008].

[6] B Gaucher, B. Floyd, S. Reynolds, U. Pfeiffer, J. Grzyb, A. Joseph, E. Mina, B. Orner, H. Ding, R. Wachnik, and K. Walter, 'Silicon germanium based millimetre-wave ICs for Gbps wireless communications and radar systems', *Semiconductor Science and Technology* **22**(1) (2007), S236–S243.

[7] S. Reynolds, B. Floyd, U. Pfeiffer, T. Beukema, J. Grzyb, C. Haymes, B. Gaucher, and M. Soyuer, 'A silicon 60 GHz receiver and transmitter chipset for broadband communications', *IEEE J. Solid-State Circuits* **41**(12) (2006), 2820–30.

[8] B. Floyd, S. Reynolds, U. Pfeiffer, T. Beukema, J. Grzyb, and C. Haymes, 'A silicon 60GHz receiver and transmitter chipset for broadband communications', *IEEE ISSCC Dig. Tech. Papers*, February 2006, pp. 184–5.

[9] N. Hoivik, *et al.*, 'High-efficiency 60 GHz antenna fabricated using low-cost silicon micromachining techniques', *IEEE International Symposium on Antennas and Propagation*, Honolulu, Hawaii, June 9–15, 2007, pp. 5043–46.

2

Millimeter-wave Packaging

Ullrich Pfeiffer

List of Acronyms

μPGA Micro PGA

ASK Amplitude shift keying

BCA Bare chip attach

BCC Bump chip carrier

BEOL Back-end-of-the-line

BGA Ball grid array

BiCMOS Bipolar CMOS

BIST Built-in self-test

BT Bismaleide-triazine

C4 Controlled collapse chip connection

CLCC Ceramic leaded chip carrier

CMOS Complementary metal–oxide–semiconductor

CNC Computer numerical control

COB Chip on board

Advanced Millimeter-wave Technologies: Antennas, Packaging and Circuits
Duixian Liu, Ulrich Pfeiffer, Janusz Grzyb and Brian Gaucher © 2009 John Wiley & Sons, Ltd

COG Chip on glass

COL Chip on lead

CPGA Ceramic pin grid array

CPW Co-planar wave-guide

CQFP Ceramic quad-flat package

CSOP C-lead SOP

CSP Chip scale package

CTE Coefficient of thermal expansion

DCA Direct chip attach

DFP Dual flat package

DIP Dual inline package

EM Electro magnetic

FBGA Fine ball grid array

FC Flip chip

FC-PGA Flip chip pin grid array

FET Field-effect transistor

FLFP Flat lead frame package

FLGA Fine land grid array

HMIC Heterolithic microwave integrated circuit

IC Integrated circuit

ISM Industrial, scientific and medical

ITRS International Technology Roadmap for Semiconductors

JEDEC Joint Electronic Device Engineering Council

JLCC J-leaded ceramic or metal chip carrier

KGD Known good dies

LCC Leaded chip carrier

LCP Liquid crystal polymers

LGA Land grid array

LNA Low-noise amplifier

LTCC Low-temperature cofired ceramic

MCM Multichip module

MCP Multichip package

MEM Microelectromechanical

MIC Microwave hybrid integrated circuit

MIM Metal insulator metal

MLF Micro lead frame

MLMS Multilithic Microsystems™

MMIC Monolithic microwave integrated circuit

MTBF Mean time between failures

NRE Non-recurring engineering

OFDM Orthogonal frequency-division multiplexing

PA Power amplifier

PBGA Plastic BGA

PCB Printed circuit board

PDIC Planar dielectric IC

PDIP Plastic dual-inline package

PECVD Plasma-enhanced chemical vapour deposition process

PGA Pin grid array

PLCC Plastic leaded chip carrier

PPGA Plastic pin grid array

PSK Phase shift keying

QFJ Quad flat J-lead

QFN Quad flat-pack no-lead

QFP Quad flat package

QUIP Quad in-line package

RF Radio frequency

RMS Root mean square

SCP Single chip packaging

SiGe Silicon germanium

SIMMWIC Silicon monolithic millimeter-wave integrated circuit

SiP System in a package

SMC Surface mounted component

SMD Surface mount device

SMT Surface mount technologies

SO Small outline

SoC System on a chip

SOIC Small outline integrated circuit

SOJ J-leaded small-outline package

VCO Voltage-controlled oscillator

WLAN Wireless local area network

WLP Wafer level package

WPAN Wireless personal area network

2.1 Introduction

Many of the millimeter-wave (mmWave) components which with we are familiar today, such as waveguides, horn antennas, polarizers, dielectric lenses and prisms, were already invented and used a hundred years ago. Hertz [5] demonstrated the generation of microwave frequencies in 1888 [1], some twenty years after James Clerk Maxwell published his equations predicting the existence of electromagnetic radiation propagating at the speed of light in 1865. Signaling experiments at mmWave frequencies were first carried out by J. C. Bose in Calcutta [2,3] and P. Lebedew in Moscow [4].

Surprisingly, Bose demonstrated remote wireless signaling at mmWave frequencies in 1895, two years before Marconi presented his wireless signaling experiment on Salisbury Plain in England in May 1897. While Bose used frequencies as high as 60 GHz, Marconi transmitted at much lower frequencies (328 kHz). The mmWave components, e.g. transmitter, detector, and horn antenna used by J. C. Bose in his experiments are shown in Figure 2.1 [5].

Today one may wonder why mmWave and radio communication systems have not penetrated everyday applications in the same way? Is it not true that we have been unable

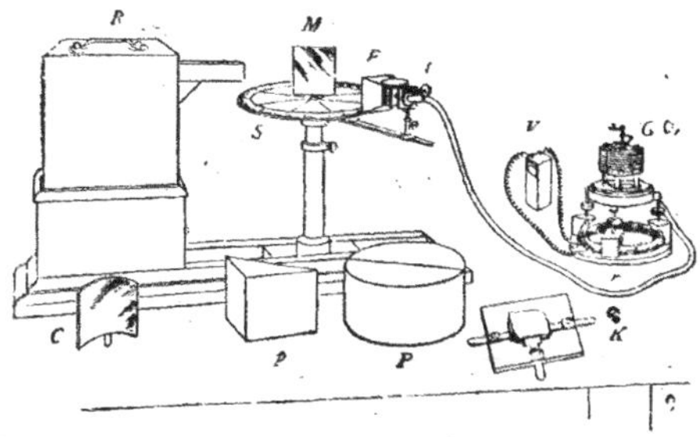

R, radiator; S, spectrometer-circle; M, plane mirror; C, cylindrical mirror; p, totally reflecting prism; P, semi-cylinders; K, crystal-holder; F, collecting funnel attached to the spiral spring receiver; t, tangent screw, by which the receiver is rotated; V, voltaic cell; r, circular rheostat; G, galvanometer.

Figure 2.1: J. C. Bose (1858–1937) and his experimental setup with mmWave components such as horn antennas, polarizers, dielectric lenses and prisms to transmit and detect mmWave radiation at the Royal Institution, London, UK, January 1897 [5], reproduced by permission of © 1997 IEEE.

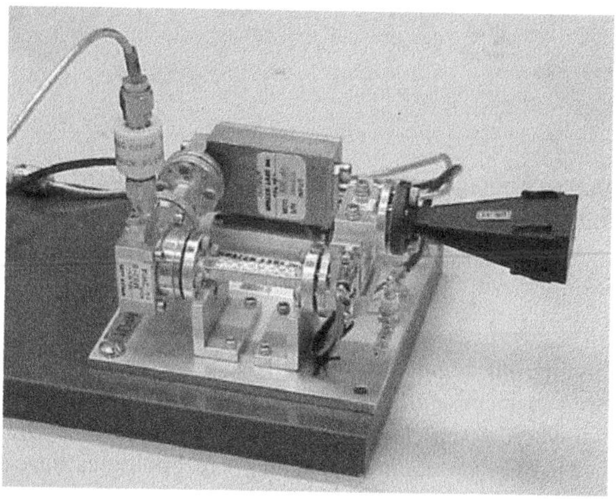

Figure 2.2: A 60 GHz radio receiver using individually packaged devices, each in a solid metal housing (split-block) with waveguide interconnects. Reproduced by permission of © IBM.

to integrate mmWave components in the same way as we have shrunk the size of cell phones and computer chips? Why are many of today's mmWave systems still similar in size to those used a hundred years ago? Figure 2.2, for instance, shows a photo of a bulky 60 GHz radio receiver that has been built from discrete components.

The answer to this question is simple. First, one needs active devices that are fast enough to operate in the mmWave frequency range, and second, one has to be able monolithically to integrate many of them with their interconnects on a single die. The latter point is

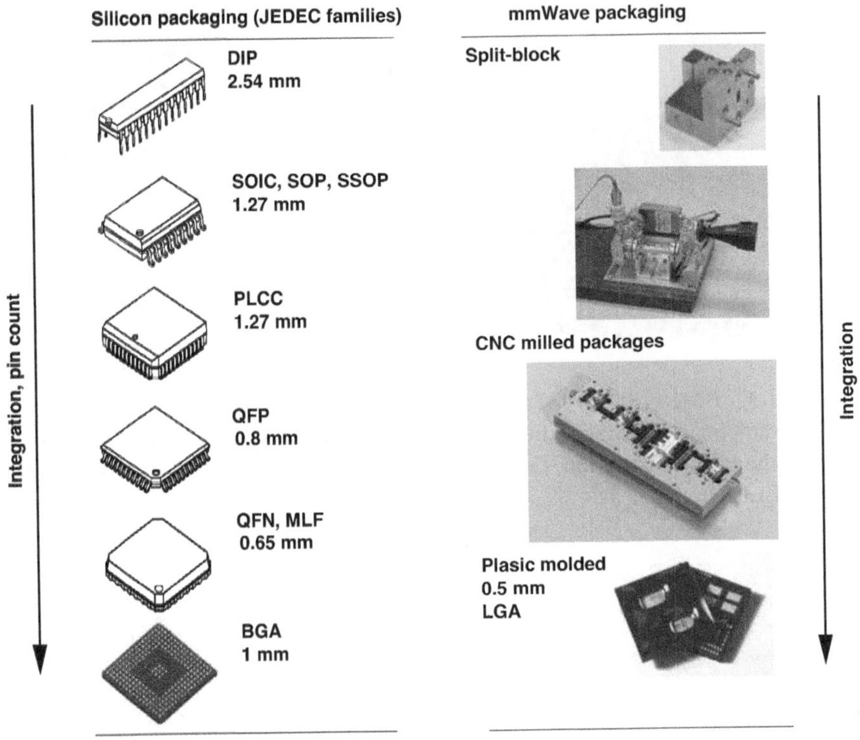

Figure 2.3: Packaging technologies defined by the JEDEC-standard compared with mmWave packaging technologies.

where mmWave circuits fall behind substantially and where silicon technologies at lower frequencies have been so successful. The mmWave system integration was primarily driven on the package-level where complex multichip modules (MCMs) or monolithic microwave integrated circuits (MMICs, pronounced 'mimics') technologies have been employed. In mmWave packaging, the package, however, is part of the mmWave circuit, and the package requirements for mmWave and low-frequency devices are not the same. The problem is that the mmWave circuit performance is strongly influenced by the interaction of the electromagnetic field with all the adjacent conductors and insulators. Higher frequencies often demand low-loss materials with well-characterized dielectric properties. High-precision machining is often needed, paired with high-resolution photo-lithography process technologies and accurate alignment techniques. With the market demand for high data-rate communication, cell phones and other consumer-oriented wireless portable products, the drive towards an mmWave system in a package integration, with application-specific integrated circuits (ASICs), memory devices and mmWave devices in the same package, adds to the breadth and complexity in mmWave packaging technologies. A comparison of low-frequency package body types from the Joint Electronic Device Engineering Council (JEDEC) package standard with mmWave package options is shown in Figure 2.3.

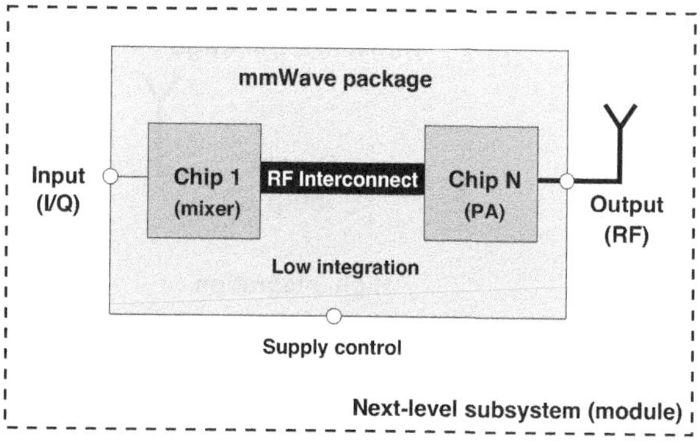

Figure 2.4: Illustration of first-level packaging of mmWave chips. The mmWave circuit performance is strongly influenced by the quality of the interconnects between chips and to the antenna.

This introduction begins with a general definition of mmWave packaging in Section 2.1.1. Future directions, performance trade-offs, and emerging markets are explored in Section 2.1.2. A short review of commonly used packaging technologies is given in Section 2.2 before low-cost packaging options and emerging technologies are addressed in Section 2.3 and Section 2.4, respectively. The discussion includes chip-scale packaging, wafer-level assembly and low-cost plastic molding processes. Specifically, chip-on-board (COB) and no-lead packages such as quad flat-pack no-lead (QFN) packages are presented. Finally, a mmWave package design flow is shown in Section 2.5 that is tailored to fast prototyping and quick turnaround times, making the design of emerging technologies more robust against process variation, material losses and assembly tolerances.

2.1.1 Definition of Packaging

The packaging of electronic components is a multidisciplinary task involving various sciences and technologies. Packaging involves physical, electrical, mechanical, and chemical engineering technologies, which creates a problem when defining packaging. Definitions can be quite complex depending on to whom you are talking. The international technology roadmap for semiconductors (ITRS), for instance, defines packaging and assembly as: *'the final manufacturing process transforming semiconductor devices into functional products for the end user'* [6]. Packaging, therefore, has been around since the early days of electronic products and has become one of the most critical cost and performance factors in today's electronic industry. Packaging technologies are changing rapidly to adopt to emerging markets, innovation and technology, where many new acronyms and abbreviations are being created.

Common practice is to distinguish between different packaging levels. The first-level package is illustrated schematically in Figure 2.4. Such a package may be used in a next-level subsystem, a module or a second-level package. A package may be comprised of one or a number of different chips of similar or different technologies that are to be connected on the package level. Interconnects conduct signals and power through a circuit by metal in the form

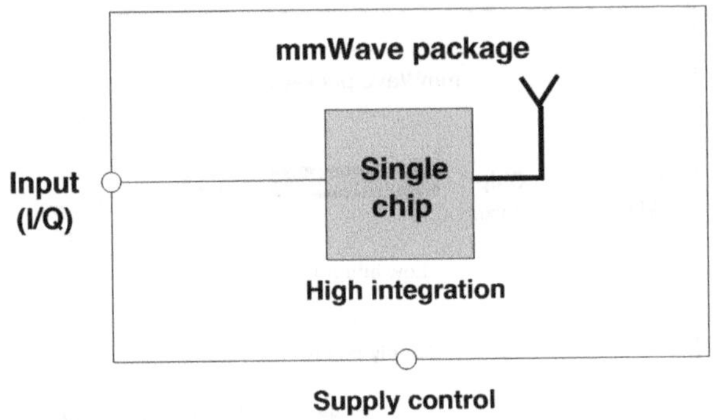

Figure 2.5: Illustration of a single-chip package including an in-package antenna. The electrical performance of the next-level subsystem interconnects can be substantially relaxed.

of wires, contacts, foils, plating, and solders. The chips are enclosed and insulated from others by the package to provide an environmental protection, thermal dissipation, and physical support of the circuit. The package may include passive components, microelectromechanical systems (MEMS) or antennas to complement its radiofrequency (RF) functionality.

The integration of highly complex mmWave systems must be carried out cost effectively, with a high degree of miniaturization and flexibility. Improvements can be made on the chip or package level. The ultimate goal is at least a mmWave system in a package (SiP), or, if feasible, a mmWave system on a chip (SoC). A single-chip package, as indicated in Figure 2.5, confines all high-frequency interconnects within the chip, and hence, high-frequency interconnects do not add to the complexity of the package. Monolithic integrated antennas and antennas integrated inside a package are attractive because they relax the requirements for the next-level subsystem interconnects. This way, only the low-frequency base-band or intermediate frequencies are conducted through package pins while the high-frequency carrier signal is radiated from inside. This opens up new avenues for low-cost mmWave packaging technologies where surface mount technologies (SMTs) may be used to mount packages directly onto the surface of a low-cost printed circuit board (PCB). Clearly, the recent progress of SiGe, BiCMOS and CMOS semiconductor technologies points towards a mmWave SoC and hence could bring Moore's Law into the mmWave arena. The push of MMICs alone proved to be an elusive goal over recent years, given the touchy nature of some III/V semiconductor materials such as GaAs.

At 60 GHz, for instance, the dimensions of low-gain antennas may be on the order of millimeters, which does not add significantly to the overall package size. The small form-factor of the integrated antenna package also allows application to phased array systems to enable digital beam forming or spatial power combining. Figure 2.6 shows a conceptual drawing of a mmWave in-package antenna that is directly bonded to a mmWave chip and fully encapsulated in a low-cost plastic-molded packaging technology, such as in a QFN package. To realize the benefits of the integrated antennas, a novel strategy for the antenna integration is required that allows its performance to be as independent as possible from unknown mold material parameters.

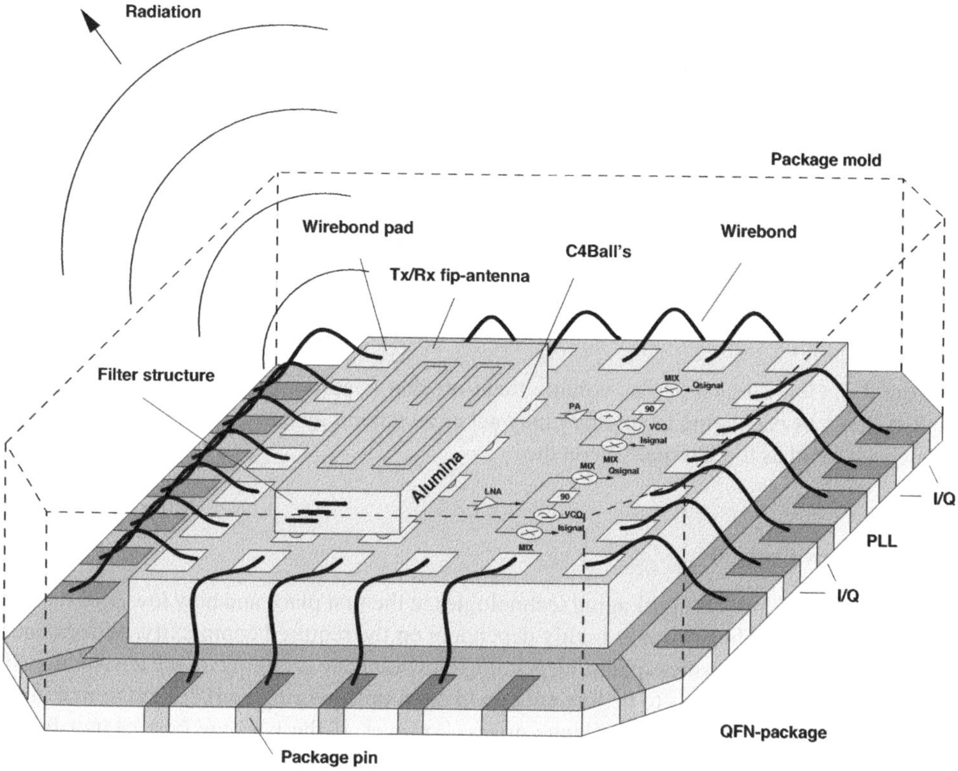

Figure 2.6: Conceptual drawing of a mmWave in-package antenna that is directly bonded to a mmWave chip and fully encapsulated in a low-cost plastic molded packaging technology. Reproduced by permission of © IBM.

2.1.2 Packaging Challenges and Future Directions

Challenges exist in all phases of the assembly and packaging process from the design through manufacturing, test and reliability of electronic products. Packaging of mmWave components is particularly challenging because of the associated complexity both in design and fabrication. Historically, the small wavelength involved often demanded high-precision machining, accurate alignment, or high-resolution photo-lithography. Dielectric and conductor losses have been of concern and demanded high-quality packaging materials. Furthermore, mmWave circuits have usually exhibited low integration levels and required low-loss interconnects between components. As a result, the art of mmWave packaging and assembly has involved high-quality but expensive and bulky waveguides, filters and machined metal housings [7–11].

This section summarizes some of the most important challenges in the assembly and packaging of mmWave components. Some challenges are of a technical nature, others are business-related. Although the focus of this book chapter is on emerging packaging technologies, it is important to understand the selection criteria that are key to growth in the electronic market.

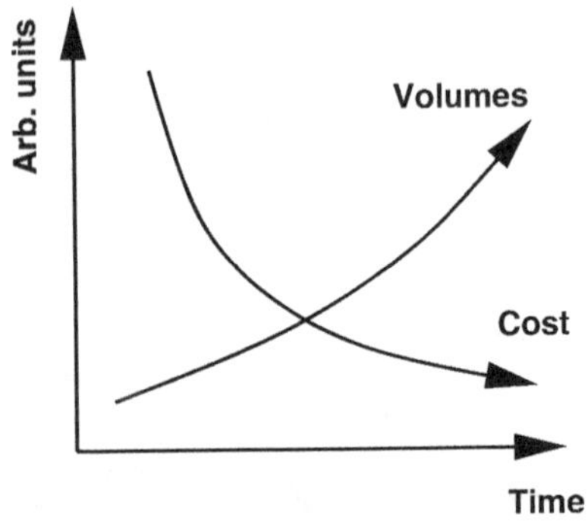

Figure 2.7: General cost versus volume trade-off that is assumed for silicon-based circuits and systems. A packaging technology may act as a technology barrier if it cannot follow the same cost trend as the internal silicon die.

2.1.2.1 Low Cost

Why do we need low-cost packaging technologies in the first place and how low does the cost really have to be? The answer is highly dependent on the required complexity, performance, market volumes and business plans. Depending on whom one is talking to, the answer may be that the packaging cost has to be low enough to open up emerging markets and to maximize profit margins. The real cost challenge arises, however, if the package cost cannot follow the die cost reduction. Figure 2.7 illustrates the general cost versus volume trade-off that is typically assumed for silicon chips. A packaging technology may end up being a technology barrier if it cannot follow this silicon die cost trend.

2.1.2.2 Performance

Package performance challenges are of particular interest at mmWave frequencies. Performance measures may be conductor and/or dielectric losses, frequency limitations, ambient temperature, etc. An endless number of performance trade-offs exist and it is important to understand the most critical ones. See Chapter 4 for more details. Here, the cost–performance trade-off is particularly inspiring and helps an understanding of the potential of emerging markets and its impact on packaging technologies.

Figure 2.8 illustrates the cost–performance trade-off. On the horizontal axis the cost of a packaging technology increases from left to right and on the vertical axis the electrical performance goes up from bottom to top. For each packaging technology one could place a dot in the chart to indicate its relative position. Over time such dots may move, depending on market conditions and volumes. From a business point of view, areas with maximized profit margins are preferred and emerging markets have to be identified.

Most of today's mmWave packaging technologies are likely to be located at the top-right quadrant, where excellent package performance has been achieved, but where the mmWave package is far too expensive and cannot follow the die cost reduction. A simple transition from the top-right to the top-left side has been prevented so far by technology barriers because the existing packaging technologies could not achieve a real breakthrough in cost

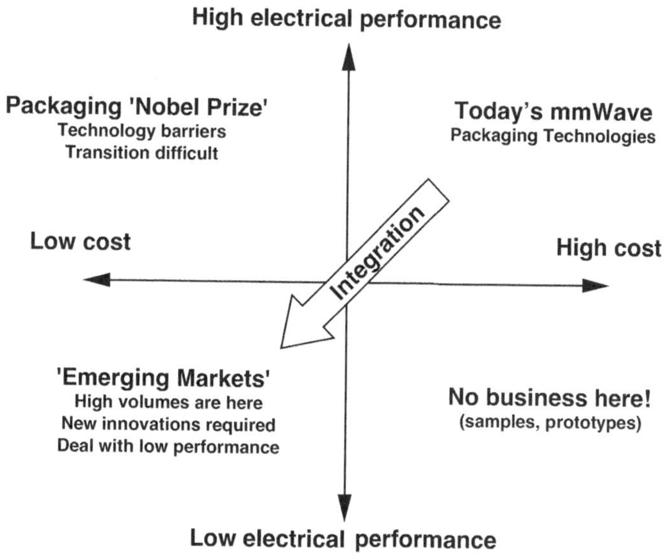

Figure 2.8: Illustration of cost–performance trade-offs in packaging technologies. Integration of mmWave circuits already at the chip level is key for the emergence of future mmWave applications.

reduction. Industry and research institutions are thus trying to come up with the *holy grail in packaging*, where the package does not alter the chip performance and is basically for free. A real breakthrough in this region may not be impossible, but is unlikely to happen anytime soon. At first glance the situation is hopeless, but emerging markets may appear on the lower side of the chart. If mmWave systems could be integrated on a single chip, then the packaging requirements drop substantially, bringing down the cost per package. The overall system cost may be low enough to drive high-volume consumer electronic applications. Driving the integration level of mmWave circuits on the chip level is key for the success of future mmWave applications.

2.1.2.3 Time to Market

Package design cycles may be lengthy. Much depends on the experience of designers and the quality of electromagnetic simulations. The fact that most of today's systems and packages are based on discrete components, with individually tuned connection in between, slows down the time to market for such systems.

2.1.2.4 Reliability

Great emphasis must be placed on system reliability. The reliability of an electronic system, consisting of a group of components, is the probability of operating continuously over a specific period of time with no failure. The reliability is expressed in percent. Component failure rates are strongly dependent on temperature and vibration. Analogue circuits are seen to be much more sensitive to temperature than digital counter parts. Therefore, low-temperature operation is one of the major objectives when designing mmWave electronic systems.

Considering an initial number of components $N_i(0)$ of type i at time 0, one can calculate the number of components $N_i(t, T)$ that have operated failure-free over a time period t at a temperature T, based on an exponential distribution of the form

$$N_i(t, T) = N_i(0) \cdot e^{-\alpha_i(T) \cdot t}$$

where $\alpha_i(T)$ is the temperature-dependent failure rate of component i. The life time of that component τ_i is often referred to as mean time between failures (MTBF)

$$\alpha_i(T) = \frac{1}{\tau_i(T)} = \frac{1}{\mathrm{MTBF}_i}$$

Based on these equations one can calculate the probability of zero failures for an electronic package or module over a given time period. Even if components with good failure rates (e.g. one part per one million hours) are used, the overall failure rate of the package increases rapidly with the number of components inside. Many data communication systems have to follow contingency plans to achieve high link reliability. In the long term, however, reliability of complex systems can only be improved further if most of the components are integrated on the chip level.

2.1.2.5 Test and Known Good Dies

Known good dies (KGDs) are tested unpackaged (bare) chips as they are usually required for packaging. integrated circuit (IC) suppliers often offer several levels of KGD, where each successive level entails a more rigorous test plan. Conventual die tests with wafer probes are slow and probe needles damage the IC pads. Simple chip carriers with pressure contacts have poor contact performance and cannot be considered at mmWave frequencies. As such, future mmWave systems on chip will require designers to invent new ways of chip testing.

Low-cost and high-volume self tests may use on-chip power detection circuits to implement a built-in self-test (BIST) mechanism. A BIST mechanism is widely used in industry to verify the internal functionality of integrated circuits at lower frequencies. A similar mechanism could be implemented at mmWave frequencies to reduce the cost and complexity of external test equipment. The test-cycle duration and the complexity of automated test equipment is particularly high at mmWave frequencies. Packages with integrated antennas, as described in Chapters 7 and 8, add to the problem, since no external RF-ports are available and anechoic chambers are required for accurate radiated power measurements. The solution to this problem may be integrated root mean square (RMS) power detectors along the signal chain. Square-law detectors implemented with diodes, e.g. Schottky diodes or npn/field-effect transistors (FETs), may be used to convert a voltage amplitude into a signal proportional to the RF power [12]. The detector output is typically low-pass filtered to provide an adequate dc output level. On-chip power couplers may use tiny capacitors that are integrated inside an amplifier's output transmission lines (T-lines) as described in reference [13]. Such a power detector may be used to test dies without high-frequency probes and external test equipment. Each output pad simply needs to be terminated with a 50 Ω antenna or probe to provide accurate power readings. No external test equipment would be needed.

A higher circuit-level integration, therefore, is inherently more reliable than any MCM or MMIC technology can be. However, the fact that the antenna is integrated into the package might cause some increased temperature-activated failures owing to higher stress at the antenna interconnect [14].

2.1.2.6 Other Challenges

Other packaging challenges include signal interconnects, which are one of the most critical aspects in the design of mmWave systems and components. See Chapter 4 for a detailed discussion of various aspects of it. This includes the fundamental limits of wire, coax, or waveguide connection. Materials and their properties at mmWave frequencies are covered in great detail in Chapter 3.

2.2 Review of Microwave Packaging Technologies

This section summarizes some commonly used packaging concepts. Many packaging technologies are well established and have been used for a variety of applications in industry. This includes MMIC technologies, metal housings, and MCMs. More details are available elsewhere in references [15–18] or Chapter 4.

2.2.1 MMICs

Strictly speaking, a MMIC is not a packaging technology. In fact, MMIC technologies are an extension of the IC concept for microwave applications [20]. It refers to a combination of interconnected microwave circuit elements monolithically integrated on a single substrate. The package of a MMIC may be a metal housing or a plastic encapsulation, as will be seen later.

While the first IC was invented by Jack Kilby back in July 1958 [21], the first MMIC was published about 20 years later by Ray Pengelly and James Turner in 1976 [19]. Their paper describes a 5 dB X-band amplifier which was implemented monolithically with lumped element matching structures such as capacitors and inductors (see Figure 2.9). No backside ground was available at that time and the FETs were grounded externally in the package.

The packaging of MMICs is primarily an outgrowth of the microwave hybrid integrated circuit (MIC) technology and the discrete device packaging technology [22, 23]. Fabricating MMICs can be high cost and high risk. Volumes are usually small to medium size with reasonably fast fabrication and turnaround times. The design is key to ensure correct fabrication, minimum costs, and first pass design.

Figure 2.10 shows an illustration of on-chip microstrip interconnects in MMIC and silicon IC process technologies. There are two important differences when it comes to packaging of MMIC or silicon dies. First, MMICs use a high-quality III/V semiconductor material, e.g. GaAs, as a dielectric between the signal conductor and the backside ground. The height between the signal and the ground is about 100 μm. The backside ground may be provided by the package or by the MMIC itself. A similar build-up is possible for silicon ICs if a high-resistivity silicon and hence low-loss dielectric is available in the process. This is known

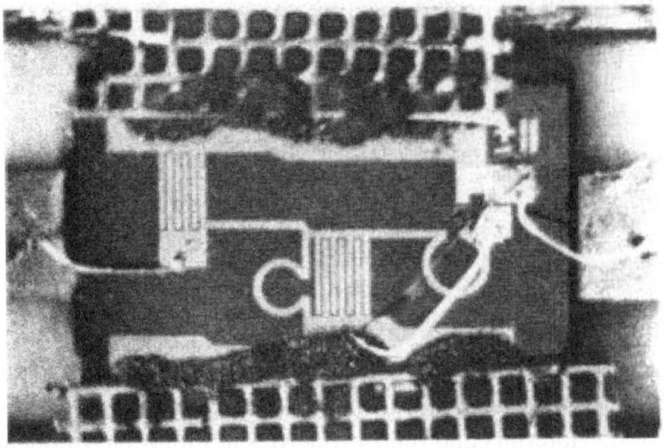

Figure 2.9: First MMIC in 1976 by Ray Pengelly and James Turner [19]. Reproduced by permission of © 1976 IEEE.

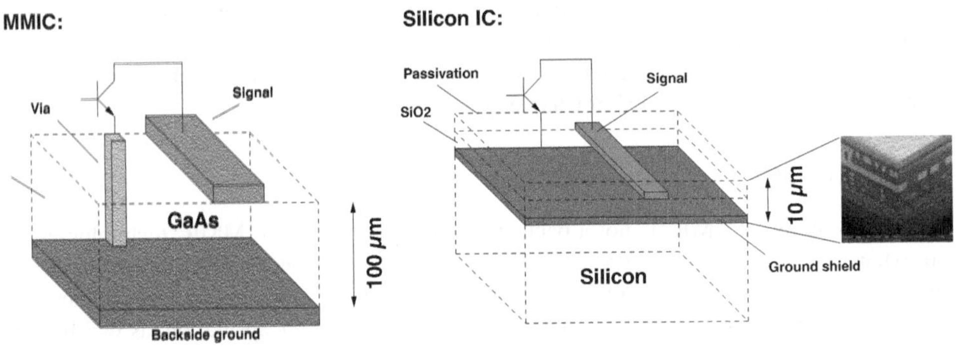

Figure 2.10: Illustration of on-chip microstrip interconnects in MMIC and Silicon IC process technologies. Silicon IC interconnects are confined in a vertical space that is an order of magnitude thinner than in MMIC technologies. Silicon IC surfaces are passivated and interconnect passives are less sensitive to encapsulation materials in contact with the IC surface.

as silicon monolithic millimeter-wave integrated circuits (SIMMWICs) [24]. Unlike these, silicon ICs may also use the metal stack of the so-called back-end-of-the-line (BEOL) for interconnects. In this case, the dielectric material is silicon dioxide (SiO_2) as opposed to the silicon substrate material. The silicon IC interconnects are therefore confined in a vertical space that is an order of magnitude thinner (10 μm) than in MMIC technologies. Second, silicon IC surfaces are passivated with a thick layer of polyimide for reasons of reliability to prevent oxidation of the metal stack. This makes on-chip passives less sensitive to additional IC encapsulation materials. The top surfaces of MMICs, however, are usually not covered by passivation materials and passive devices need to be carefully remodeled to account for the electromagnetic properties of the encapsulation materials.

Figure 2.11: Split-block package housing. Reproduced by permission of © IBM.

2.2.2 CNC Milled Metal Housings

State-of-the-art RF modules require excellent performance and long-term reliability under harsh environmental conditions. Highly integrated packages are often made from solid metals, e.g. aluminum, which are machined by milling machines, operated manually or under computer numerical control (CNC). A split-block package, as shown in Figure 2.11, usually consists of a single device (transistor or MMIC) and a waveguide transition. The package name stems from the fact that the package is fabricated in two halves by the use of a CNC milling machine.

A more complex receiver module is shown in Figure 2.12. The module is used in R&S network analyzers for frequencies up to 24 GHz. The package is made from solid aluminum that is milled to form separate chambers. Wave propagation inside the package is suppressed to provide a good RF isolation between functional units. This is either achieved by a chamber width and height below half the wavelength of the maximum frequency, or, by pieces of absorbing materials which are glued to the package, the lid or the substrates. The surface of the aluminum is gold plated to form a corrosion-resistant electrically conductive layer. The package is protected by a gold-plated aluminum lid for excellent shielding efficiency.

A hybrid circuit technology is used to mount semiconductor components as bare die on a thin-film substrate. The process features bare die wire or ribbon bonding, beam lead, flip-chip, or surface mount device (SMD) mounting. The process supports 4 μm gold conductors, laser trimmable NiCr thin-film resistors, SiN thin-film capacitors, air bridges, and via holes. Lateral fabrication tolerances are within a few micrometers, with substrate thickness such as 5–50 mils (0.127–1.27 mm). Substrates may be fabricated from a variety of materials, such as general purpose Al_2O_3 ceramic with a dielectric constant (k) of 9.9, low loss fused silica for high accuracy filter applications ($k = 3.78$), and aluminum nitride for high thermal conductivity ($k = 8.9$). Transmission lines are realized as microstrip, co-planar waveguide (CPW), or slotline.

Complex modules may consist of a high number of substrates, e.g. 20 or more. Individual substrates are clamped into the package, which simplifies the assembly and disassembly process. The local substrate ground is contacted to the overall package ground via the substrates' gold-plated backside metalization. Isolated silicone rubber pieces, held by

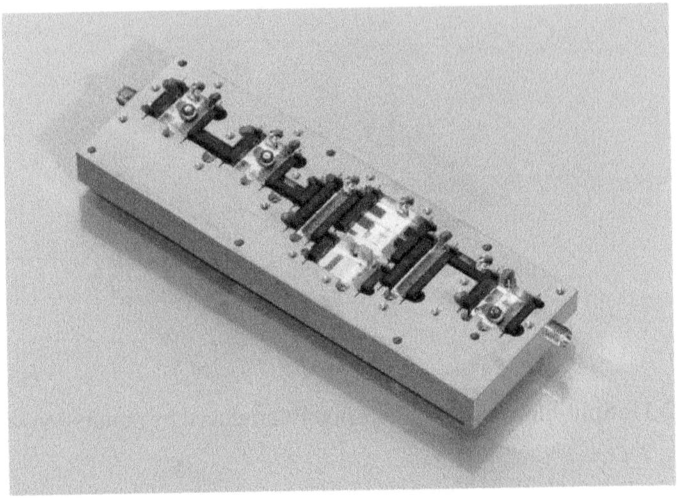

Figure 2.12: CNC-milled mmWave package used in R&S network analyzers for frequencies up to 24 GHz. Reproduced by permission of © 2008 Rohde & Schwarz.

electrically 'stealthy' clickable plastic bars, apply pressure to the substrate where needed, for instance, at the transfer points of a microstrip line.

Substrate interconnects are designed for frequencies up to 40 GHz. Such interconnects are made by gold or silver ribbons, bonded or welded to the substrates. The thermal expansion mismatch between the substrates and the metal housing is accommodated by a sufficient ribbon-bond loop height. Interconnects above 40 GHz need to take the parasitic wire-bond inductance into account to support frequencies up to 70 GHz.

2.2.3 Multi-chip Packages

MCM packaging technology originally evolved out of hybrid micro-circuit electronics. In hybrid circuits many electronic components, made from different semiconductor materials, are mounted on a single layer or multilayer substrate. A MCM packaging technology additionally encapsulates the entire system hermetically. The attraction of a MCM technology, especially for large and complex systems, arises from increasing specialization of semiconductor devices. A system often consists of a larger number of chips, each optimized for a specific task. This makes it desirable to assemble many chips in a single package to reduce system size and external connections. A very high internal pin-count can be implemented inside a MCM, while the MCM package itself has a significantly reduced external pin-count.

Connections between chips are routed internally and only signals such as control, power supply and module input and output signals connect to the outside printed circuit board. With a MCM technology, a ratio of silicon-to-substrate area greater than 30% can be achieved. A MCM technology is classified by its substrate type and the production process used for the multilayer structure. The types are named: MCM-L, MCM-C and MCM-D, where the characters refer to *laminated*, *ceramic* and *deposition*. All three technologies have been successfully applied to mmWave applications over many years. The primary selection criteria are material parameters such as low-loss, good thermal conductivity and precision. See also Chapter 4 for more details.

2.2.3.1 MCM-L

The substrate is based on *laminated* multilayer PCB technology. It is an extension of COB, where the bare dies are mounted directly on a substrate. Multilayer structures are formed by etching patterns in copper foils laminated to both sides of a resin-based organic core. Layers are laminated together with one or more layers of basic resin in between to act as an insulator. Typical insulator materials are polyimide or bismaleide-triazine (BT-epoxy). Core or substrate materials are often epoxy/fiberglass (FR4), aluminum or copper. Interconnections between layers may be formed by 'drilled' vias, which extend all the way through the board, 'blind' vias, which only extend from the surface part-way trough the board, or 'buried' vias, which connect only adjacent inner layers and do not extend to the surface in either direction. The smallest feasible layout structure in a design is referred to as feature size. In MCM-L technology this size can go down to a few tens of a micrometer.

2.2.3.2 MCM-C

The substrate of a MCM-C technology is based on cofired *ceramic* or glass-ceramic materials. This technology evolved from thick-film hybrid technology. The process begins with thin sheets of unfired material which are referred to as green tape. The individual layers are printed with thick-film paste to create the metalization patterns, aligned with the other layers, and are laminated at elevated temperature and pressure. Resistors and capacitors, compatible with the green tape process, may also be fabricated. The feature size depends on the thick film process, which has a typical conductor line width of 200 μm down to 100 μm. low-temperature cofired ceramic (LTCC) MCMs are a good choice for low-loss and good thermal conductivity [25, 26].

2.2.3.3 MCM-D

The interconnects of a MCM-D technology are patterned by *deposition* of dielectrics and conductors on a base substrate of ceramic or silicon. The dielectric materials are used in the liquid state and are applied by spinning. Vias may be opened in the dielectric film by applying photoresistive material and etching, or by using photosensitive materials. The typical process which is used to deposit thin layers of dielectric materials is a plasma-enhanced chemical vapor deposition (PECVD) process to deposit layers of silicon dioxide, for example. The fine-line lithographic technique used in MCM-D technology can produce very high-density interconnections using feature sizes of a few micrometers.

2.3 Low-cost mmWave Packaging

Historically, mmWave circuitry needed to be complemented by high-quality passive devices, which were embedded in the package. Unfortunately, this complexity added additional cost to the final product and prevented a further cost reduction. The recent progress of semiconductor technologies, such as BiCMOS SiGe, CMOS and MMIC technologies, however, has made

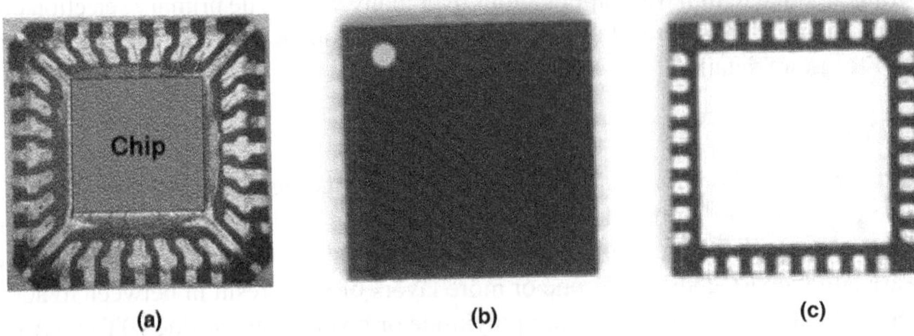

Figure 2.13: QFN package: (a) die paddle before encapsulation of a wire-bonded chip; (b) top view, and (c) bottom view (solder side) of the package. The entire assembly is covered (encapsulated) with a mold compound and a typical 0.5 mm contact pin pitch.

it possible to include many passive device functions on chip and has eliminated them from mmWave packages [27–33]. This opens up new avenues for low-cost packaging technologies.

2.3.1 Low-cost Plastic Molding at mmWaves

The use of conventional low-cost plastic molded chip-scale packaging (CSP) at mmWave frequencies is limited and has only been reported at lower frequencies [34]. This is in part because the typical 1 nH mm^{-1} lead and wire-bond inductances are prohibitive at frequencies above 20 GHz [35] and plastic packaging materials are quite lossy. Moreover, standard mold materials have not generally been characterized at mmWave frequencies [36]. However, the simplest low-cost packaging attempt is to apply low-frequency packaging and assembly procedures to higher frequencies, simply by placing existing MMICs, designed for open-air environment, into encapsulated plastic packages. Although the packaged part may still work quite well, the encapsulant over the MMIC still degrades the performance. Designers must take note of the fact that device parameters such as gain, power, linearity, and stability degrades due to the change in die-operating environment. The challenge is to account accurately for the performance of the packaged die so that the final product will meet the performance target.

For example, a QFN package, as shown in Figure 2.13, is relatively inexpensive and widespread across industry. It was originally designed for low-frequency analog and digital applications. A QFN package has a metal paddle to which the die is glued with conductive epoxy. The die bond pads are wire bonded to metal pins and the entire assembly is covered (encapsulated) with a mold compound. The package does not have internal leads (no-lead) or rewiring capabilities. Multiple dies are usually molded simultaneously on a larger lead-frame. Dicing off the lead-frame singulates the individual packages. The packages are put on tape and reel for placement like any other low-cost SMT reflow solder component. The SMT is a method for constructing electronic circuits in which the surface mounted components (SMCs) are mounted directly onto the surface of PCBs.

Placing MMICs in plastic-molded packages has a considerable effect on the impedance, power, and noise match of the components. In MMIC technologies, transmission lines, inductors, coupled lines, and capacitors are used for matching purposes. Their electric fields extend into the substrate and in the medium above the die. Encapsulants with a higher dielectric constant (ϵ_r) replace the air above the die and perturb the fields. Therefore, encapsulated devices should be accurately simulated and modeled for the correct environment to ensure good circuit performance. As mentioned in Section 2.2.1, silicon dies are passivated with a layer of polyimide, which makes them less sensitive to the mold material. The design and optimization of silicon chips for low-cost plastic molding is not necessarily required.

Today's packaged product offerings include active and passive devices such as amplifiers, mixers, filters, and voltage-controlled oscillators (VCOs) up to 20 GHz, for example. Higher operation frequencies are problematic because of the signal transition from the package to the PCB. The transition includes wire-bonds, package pins, solder and PCB microstrips. The modeling of such transitions is a multiport problem that includes far-end and near-end cross-talk between transitions. The electrical modeling of such interconnects is discussed in detail in Chapter 17. Knowing the transition, the designer can account for the detuning of the circuit by adding additional compensation structures on the chip or on the PCB side. Common practice, however, is to keep the RF wire bonds as short as possible to minimize the associated inductance (typically 1 nH mm^{-1}). This is done by shifting the die location on the die paddle close to the RF pin. This affects the length of other wire-bonds such as supply and control connections, since the die paddle size is fixed for a given number of input/output pins. Caution therefore needs to be taken to ensure sufficient bypassing of the out-of-package capacitors, which have to complement the on-chip bypass capacitors through the wire-bond inductance.

2.3.2 Chip-on-board at mmWaves

Strictly speaking, COB or direct chip attach (DCA) devices are non-packaged devices. The bare die is supplied without a package and is attached, with conductive epoxy glue, for instance, directly to a PCB. These devices require a specific process for assembly, where the chip is wire-bonded to the board first and then protected from mechanical damage and contamination by an epoxy 'glob-top'. In fact, the COB process has better high-frequency performance than QFN packages since the additional pin-layer and solder contacts are omitted from the signal path, allowing a shorter RF connection to the circuit board.

In a COB technology, the designer has the freedom to optimize the PCB footprint of the die within the design rules given by the board manufacture. Figure 2.14 shows an example of such COB design rules. High-quality dielectrics such as glass-fiber reinforced Teflon materials are available. They can be laminated on top of PCB cores to improve the characteristic of microstrip interconnects. Critical, however, is the design of the fanout area around the die, where the space is limited to accommodate enough clearance to adjacent ground and signal conductors.

In a COB technology, the designer is responsible for the correct modeling of the die to circuit board transitions. The greater the flexibility in design, the more variable the interconnect modeling and detailed knowledge of the three-dimensional structure and materials are required to apply electromagnetic simulations successfully. See Chapter 17 for more details about the modeling of such transitions.

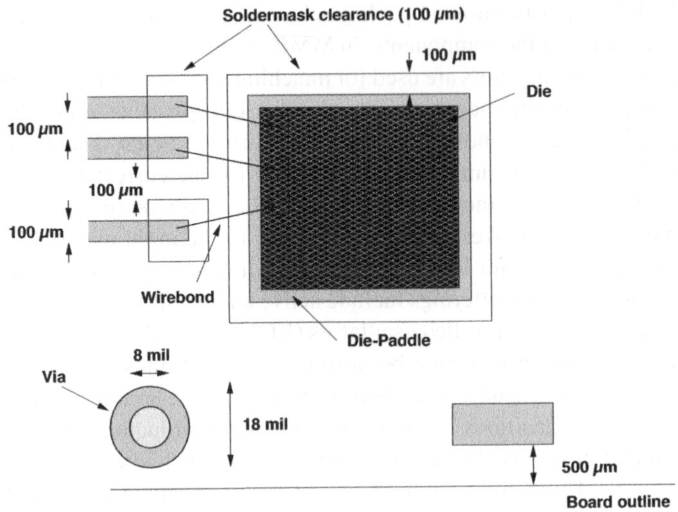

Figure 2.14: PCB design rule example.

2.4 Emerging Packaging Technologies

Packaging technologies are changing rapidly to adopt to emerging markets, innovation and technology. Many new acronyms and abbreviations are being created and details about the fabrication process are often proprietary to a specific vendor. This section reviews only exemplary emerging packaging technologies.

2.4.1 Microcoaxial Wirebonds – Bridgewave

Microcoaxial wirebonds have demonstrated the ability to handle frequencies up to 100 GHz [37]. A coaxial connection with a regular wirebond is formed by the deposition of a dielectric and a final metal coating inside a standard low-cost package. After lithographic patterning, the deposited metal creates the outer conductor coating for the wires as well as the ground connections in the package and on the die. The ground therefore follows along the wirebond and removes the effect of the often unknown ground return path of a package.

2.4.2 Glass Microwave Integrated Circuit (GMIC, HMIC) – TYCO, M/A-COM

A heterolithic microwave integrated circuit (HMIC) uses silicon pedestal embedded in a glass substrate. Conductors can be deposited on top of the glass to form transmission lines or lumped elements, such as spiral inductors, baluns or MIM capacitors. The silicon pedestals may act as conductive vias from the top to the bottom of the die, or there may be diodes formed at the tops of the pedestals. Microstrip-to-waveguide transition is used to launch planar microstrip signals into waveguides or antennas [38, 39].

2.4.3 Epsilon™ Packaging /MLMS™ Devices – Endwave

Epsilon™ Packaging is a subsystem PCB and chip package, all in one. The technology was introduced by Endwave, a manufacturer of transceivers and subsystem modules operating at carrier frequencies in the range 7–100 GHz [40]. It is a multilayer substrate and module package that replaces costly and heavyweight metal mechanics with metalized FR-4 and injection-molded metalized plastics. The Epsilon™ Packaging approach allows COB and SMT components to coexist, thus easing assembly, eliminating troublesome interconnections, and reducing cost. No machined metal parts are required, making it light and mass producible, but still retaining efficient heat extraction. Chips are first attached and wirebonded to a PCB. Next, a metalized plastic lid equipped with a climate-control feature is laminated over the chips for protection. For RF input and output connections up to 100 GHz the lid provides either built-in antennas or waveguide transitions.

The Multilithic Microsystems™ (MLMS™) technology uses flip-chip and electromagnetic coupling methods to minimize expensive semiconductor real estate while eliminating lengthy interconnects and their related variability [41–43]. MLMS™ moves passive circuitry onto an inexpensive proprietary substrate that processes with the ease of silicon, yet works past 100 GHz. Only the discrete active semiconductor devices remain, which are then flip-mounted on top of the MLMS™ substrate. The substrate includes precisely dimensioned heat sinks to keep the chips cool. This technology is extremely rugged, withstanding extreme temperature cycling and mechanical shock. With MLMS™, a complex transmitter or receiver can be placed on a single, compact substrate with no bondwires in the RF path. The performance of circuits fabricated in this technology generally exceeds that of circuits made with MMICs, since the Q of the circuitry is higher and because each discrete device chip can employ whatever materials and process technology (Si, SiGe, GaAs, InP, bipolar, junction transistor (BJT), metal oxide semiconducting field effect transistor (MOSFET), metal semiconductor field effect transistor (MESFET), pseudomorphic high electron mobility transistors (PHEMT), metamorphic high electron mobility transistor (MHEMT), heterojunction bipolar transistor (HBT), heterostructure field-effect transistor (HFET), PIN diode, high-Q Schottky diode, low-barrier diode, high-Q varactor, micro-electro-mechanical systems (MEMS), etc.) are most appropriate for its particular purpose in the circuit.

2.4.4 Plastic Molded MMICs – UMS

Leading MMIC suppliers and foundries are currently extending their product offerings to include lower-cost plastic packaging solutions. Recently, plastic-molded MMICs have been made available that use standard QFN packaging technologies to drive new markets toward large volume production and cost reduction. For instance, Figure 2.15 shows a K-band power amplifier. The amplifier exhibits 19 dB linear gain in the 17–27 GHz frequency band with 28.5 dBm CW output power at 1 dB gain compression and 29.5 dBm in saturation [44]. Such MMIC components are easy to use in standard SMT assembly lines and significantly contribute to the reduction of an overall module cost.

Specific characterization and modeling, however, is required to use standard GaAs pHEMT process technologies. Package co-design is required that uses electrical models of the package interface during the design phase. The SMD plastic molding approach requires provision for stability, since the first external decoupling level is located on the circuit board

Figure 2.15: K-band power amplifier in a plastic molded 28-pin QFN package. Figure courtesy of UMS [44]. Reproduced by permission of © 2007 IEEE.

and is far away from the die. PCBs should also have good RF properties to implement low-loss transmission lines. The packaged MMIC RF performance is only limited by the wire bond length and its associated inductance. The design of compensation structures between the MMIC package leads and the circuit board can handle signals up to 40 GHz.

Maximum power-handling capabilities are of concern and strongly impact the MMIC package size, the PCB design, and the assembly process. The junction temperature of the transistors needs to be carefully investigated and impacts the reliability of a QFN-packaged MMIC. For instance, the thermal resistance between a transistor channel and the die paddle of the package power amplifier shown in Figure 2.15 is on the order of 18.7 C/W. Such SMD plastic package solutions are becoming a standard approach for offering very low-cost and attractive performance.

2.4.5 DCA with Integrated Antenna – IBM

Figure 2.16 shows a cost-effective DCA packaging solution for 60 GHz wireless chipsets capable of multi-Gbps wireless communication [29]. A single-chip transmitter or single-chip receiver is mounted together with 7 dBi cavity-backed folded dipole antennas onto a PCB.

The transmitter achieves a 3 dB beam-width of $\pm 30°$ with a 41 dB total conversion gain in the main radiation direction. The chip and its antenna have been encapsulated using standard mold materials with an interconnect loss of less than 2 dB across the industrial, scientific and medical (ISM) band. The cavity-backed antenna decouples the design from the physical properties of the package such as dimensions and mold materials. The antenna performance shows an extremely low sensitivity to the cavity manufacturing tolerances and uncertainties of the mold material dielectric properties, being a key factor enabling the use of the presented low-cost packaging technology at mmWave frequencies. Envisioned applications of the packaged chipset include one to three or more Gbps directional links using amplitude shift keying (ASK) or phase shift keying (PSK) modulation and 500 Mbps to 1 Gbps omni-directional links using orthogonal frequency-division multiplexing (OFDM)

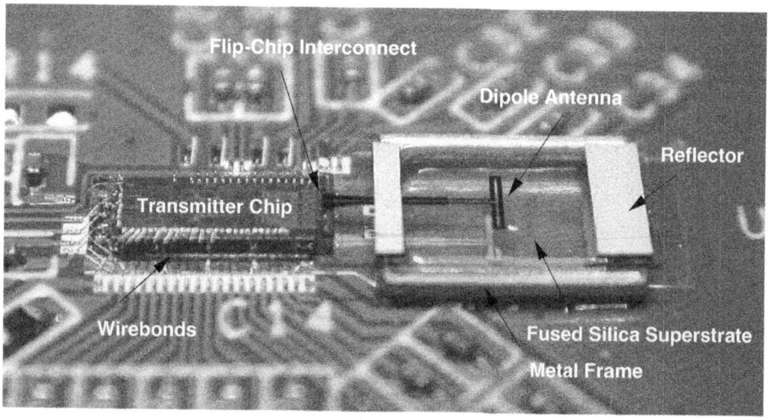

Figure 2.16: Photo of a low-cost DCA microwave packaging technology. The DCA package is shown before encapsulation and protection with a plastic mold material (glob-top). The antenna is flip-chip connected, with the radiation direction being orthogonal to the board surface [29]. Reproduced by permission of © 1976 IEEE.

modulation. The results demonstrate that mmWave chipsets can be packaged today, with a level of integration that is unique for mmWave frequencies systems.

Figure 2.17 shows a conceptual drawing of the package build-up. The antenna is constructed from a fused silica (SiO_2) substrate, which is bonded to a Kovar metal frame using a thermosetting adhesive. A Kovar alloy was chosen because of its superior thermal expansion (3.7 ppm/°C) characteristic, which is matched closely to that of fused silica (0.55 ppm/°C). A chemical etching and photolithographic process was used for frame fabrication to comply with tight tolerances and yet high-volume fabrication. The antenna is flipped to the mmWave circuit chip using a thermal compression flip-chip bonding technique. Simultaneously, the antenna frame is attached to the package base using a silver-filled adhesive. This way, the antenna is suspended in air below the silica superstrate and the metallic base plate of the package acts as a reflecting ground. A uniform ball height of 30 μm could be achieved across the flip-chip ball interconnect. The spacing between the radiating element and the ground influences the bandwidth of the antenna and is defined by the thickness of the Kovar frame. This build-up leads to a direct connection to the antenna through the 30 μm height flip-chip ball interconnect, without the need for vias or electromagnetic coupled substrate interconnects [45,46]. Flip-chip interconnects have shown excellent performance at mmWave frequencies and have been characterized and modeled up to 110 GHz [47].

Figure 2.18 shows a photo of the encapsulated DCA package. The size of the package is 7×11 mm^2. Unlike other suspended antennas [49–52], the Kovar frame serves two purposes. First, it provides the mechanical support for the antenna, and second, it provides a well-controlled electromagnetic (EM) environment, making the antenna performance less sensitive to the surrounding package- and PCB-level dielectric and metal layers. The frame decouples the design of the antenna from the exact physical properties of the package such as dimensions and material parameters – an important design feature simplifying simulation

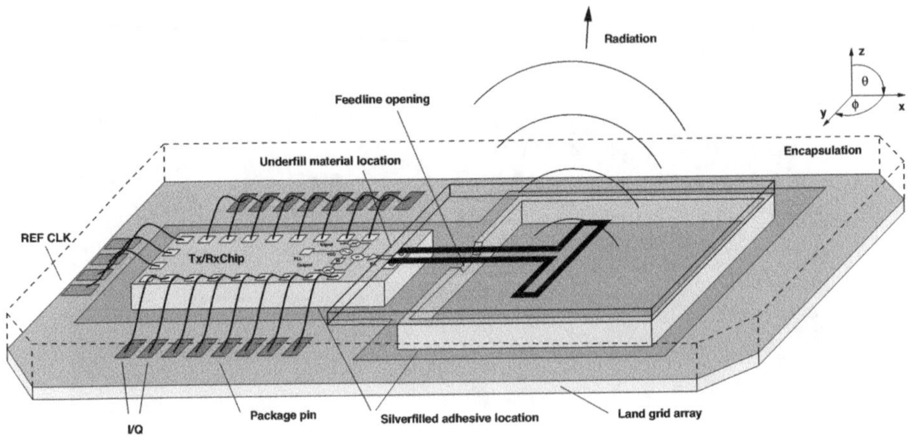

Figure 2.17: Conceptional drawing of the DCA package. The RF signal is radiated directly from inside the package. The package provides a well-controlled EM environment, making the antenna performance less sensitive to the surrounding package- and PCB-level dielectric and metal layers [29]. Reproduced by permission of © 1976 IEEE.

and modeling complexity. The frame avoids any unknown shapes or encapsulants in close proximity to the radiating element. Special care, however, is required for the antenna feedline that interfaces the active circuit to the antenna. A shift in the dielectric constant or loss tangent of any feedline due to encapsulation will detune the antenna impedance match and reduce receiver sensitivity and transmitter efficiency. The feedline therefore passes through a dedicated opening in the frame which was encapsulated prior to the mounting of the antenna. This pre-molding step ensures the hermeticity of the cavity and allows the antenna design process to be virtually independent of the electrical properties of the final package encapsulation material. In this configuration, the folded-dipole antenna achieves 7 dBi gain and ±30°3 dB beamwidth. The simulated interconnect shows a large impedance bandwidth of over 30% defined by a 10 dB return loss and a radiation efficiency above 90% over the entire 59–64 GHz band.

2.4.6 LGA with Integrated Antenna – IBM

Figure 2.19 shows a cost-effective surface mountable land-grid array (LGA) packaging solution for 60 GHz wireless chipsets. A LGA is similar to ball grid array (BGA) packages except that the package does not have spheres on the bottom side of the package. Internal wiring layers may be used for fan-out and routing purposes from the internal bond pads to the bottom side solder pads. The transceiver package includes transmitter and receiver chips with their antennas in a 13 × 13 mm^2 LGA. The area of the LGA is 2.8 times the area of its components (chip plus antenna). The package and antenna build-up provides a well-controlled EM environment, which makes the antenna performance less sensitive to the surrounding package- and PCB-level dielectric and metal structures. This simplifies the simulation and modeling complexity typically involved in mmWave package designs [53].

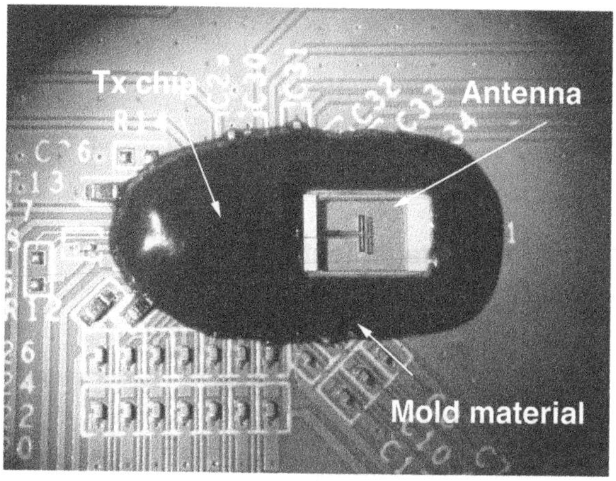

Figure 2.18: DCA mmWave package after encapsulation and environmental protection with a plastic mold material [48]. The antenna is left exposed for illustration purposes and can also be covered with a thin layer of epoxy. Reproduced by permission of © 2006 IEEE.

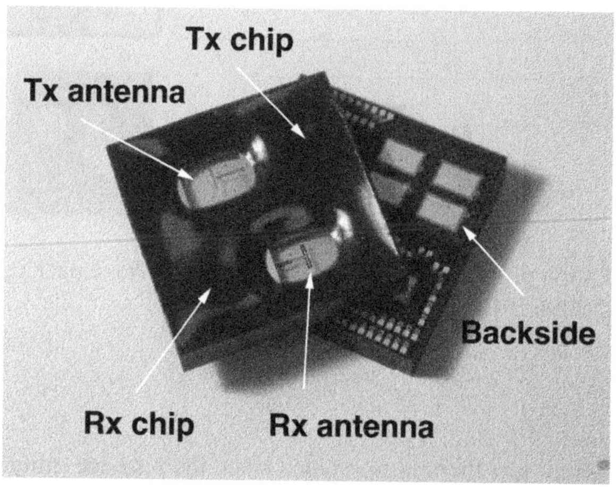

Figure 2.19: Cost-effective surface mountable LGA packaging solution for 60 GHz wireless chipsets. DCA mmWave package after encapsulation [48]. Reproduced by permission of © 2006 IEEE.

In fact, the same antenna design can be used for a variety of low-cost packaging technologies such as DCA, LGA, QFN, etc., without performance degradation.

The packages exhibit a low operating junction temperature and low temperature-activated failure rate. This is achieved due to use of both a low thermal resistance from the junction to the board-level heat-sink and through the coefficient of thermal expansion (CTE) matched materials combined with low-stress molding compound and die-attach adhesives. The DCA

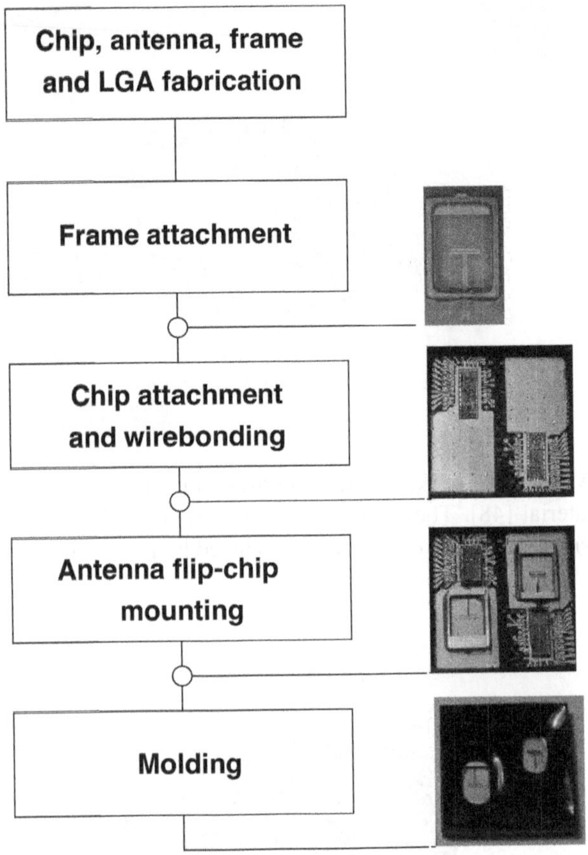

Figure 2.20: Fabrication process flow of the IBM LGA mmWave package [48]. Reproduced by permission of © 2006 IEEE.

package has an inherent low thermal resistance since the package ground connects directly to the board-level ground plane. Although there has been some successful early work in integrating antennas with simple mixers directly in silicon [54], the LGA package represents an individual component with a formal package for environmental protection around it.

Figure 2.20 shows a sketch of the LGA fabrication process [48]. The antenna radiating element is a half-wavelength folded dipole. It is fabricated as a separate component, made of a fused silica substrate which is attached to a Kovar metal frame prior mounting. The folded-dipole antenna achieves 7 dBi gain and ±30°3 dB beam-width. The interconnect shows a very large impedance bandwidth of over 30% defined by a 10 dB return loss and a radiation efficiency of above 90% over the entire ISM band. For more details about the antenna design see Chapter 7.

First, the silicon chip is attached to the LGA die-paddle using a thermosetting adhesive. It is wire-bonded using a standard gold ball-bonding technique. Next, the antenna is connected to the chip using a thermal compression flip-chip mounting technology. Flip-chip interconnects have shown excellent performance at mmWave frequencies and have been characterized and modeled up to 110 GHz [47].

The LGA package uses a thin (250 μm thick) interposer layer with thermal vias that adds little to its thermal resistance. The most critical point in terms of CTE mismatch is the gap between the chip and the Kovar frame that is bridged by the fused silica antenna substrate. The different CTEs between the package ground material, typically copper, and the antenna substrate material can cause a strain on the flip-chip joints. However, this gap is small enough (500 μm) for the stress to be absorbed by an underfill material, dispensed locally between the antenna substrate and the die surface. The length expansion at that point is only of the order of half a micrometer. A low-viscosity mold material was dispensed on the chip and the antenna to provide a final void-free hermetical seal. Deliberately, air is only enclosed within the cavity of the frame. The enclosed air is not of concern since additional stress due to the thermal expansion of air can be absorbed without any delamination, blistering or cracking. Unlike other packaging technologies, the power of the chip is not dissipated within the air-filled cavity, leaving the air closer to the package heat-sink temperature. This greatly relaxes any temperature-activated component failures. Furthermore, glass is used in high-reliability applications and is known for its excellent hermeticity and strength [55]. The air pressure inside the frame increases by 28.8% for a change in ambient temperature from 25 to 85°C. At sea-level this corresponds to a force equivalent to approximately 77 g applied to the area of the fused silica substrate opening (assuming Gay–Lussac's law, constant volume enclosed by the frame). The Kovar metal frame and the fused silica are stiff enough to withstand such forces. Reliability tests over 35 hours (70 temperature cycles) showed failure-free operation, while the chip was operated. The package was cycled from 5°C to 85°C within 30 minutes including 10 minutes soak time at the extrema. After thermal cycling, the package shows no measurable performance degradation in anechoic chamber measurements.

2.4.7 Wafer-level Packaging and Assembly of mmWave Devices

In a conventional packaging process, a wafer, e.g. a silicon wafer, is first singulated or diced into individual chips and then the chips are picked and placed individually in the packaging and assembly process. A wafer-level packaging (WLP) process, however, refers to a packaging technology at wafer level, before the singulation of the wafer. A WLP technology extends the usual wafer fabrication process to include the package assembly, device interconnection, and protection processes. A WLP technology is a true CSP technology, since the resulting package is practically of the same size as the die.

Although WLP technologies have been used for some time for low-frequency analog and digital applications, they have only recently been applied to packaging of mmWave components. With the recent progress of silicon process technologies, signals can be conducted at mmWave frequencies with low loss (see Figure 2.10, Section 2.2.1) and flip-chip interconnects have shown excellent high-frequency performance [47]. Antennas may be integrated in high-resistivity silicon with good efficiency, which makes mmWave WLP an emerging technology [56]. The Non-recurring engineering (NRE) cost of a WLP technology is typically higher, but the costs per packaged part are typically low and become very attractive in high-volume applications. More details about 60 GHz WLP antenna design can

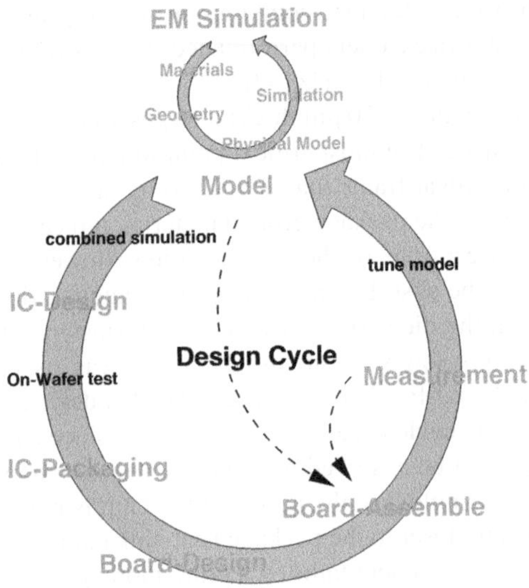

Figure 2.21: Design cycle using IC-packaging and testing with accelerated design cycles using EM simulation.

be found in Chapter 7 and silicon process technologies for WLP are discussed in more detail in Chapter 18.

2.5 Package Codesign at mmWaves

The package and IC chips should be codesigned from the start in order to avoid lengthy and costly redesign cycles. Accurate knowledge of the package in the form of parasitic models are a necessity for high-performance digital and analog integrated circuit design. Typically one of the following two approaches is taken.

First, EM field solvers are frequently applied to extract models usually in the form of RLCG-matrices or S-parameters. However, EM simulations require detailed information of the three-dimensional package and bonding structure, e.g. accurate dimensions, material parameters, and knowledge of the power and ground distribution; information, that might be too complex, proprietary to the package vendor, specific for a single layout only, or otherwise not accessible or not pertinent. Second, measurements in the time or frequency domain are used directly as a package modeling approach. Unlike EM simulations, measurements often require mechanical fixtures and de-embedding techniques to uncover the parts of interest. Moreover, encapsulated internal package ports cannot easily be probed from the outside without partially destroying the sealing [57, 58]. As a result, package measurement setups are often not able to take the real IC package configuration fully into account.

Figure 2.21 shows the IC and package codesign cycle. The cycle starts with a combined simulation of active circuit elements such as transistors together with an initial package model. Knowing the package characteristic, the designer can account for the detuning of the circuit by adding additional compensation structures on the chip or on the PCB side.

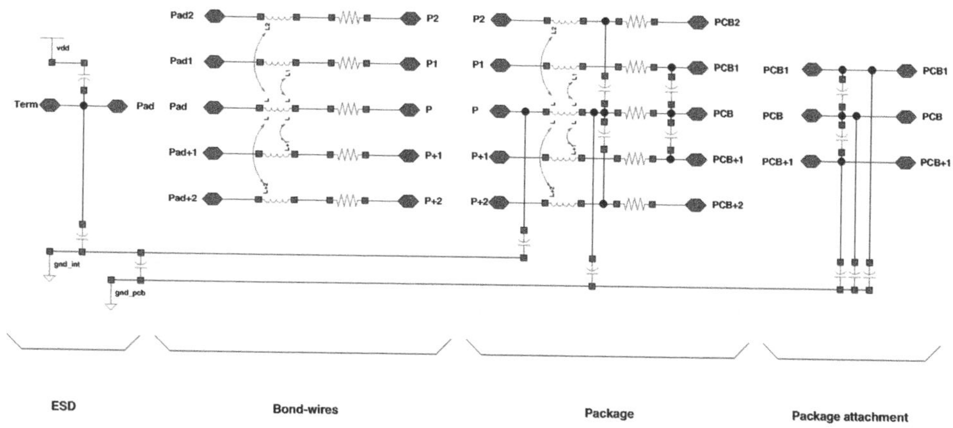

Figure 2.22: Example of a 5-pin (10-port) package model.

The IC may be tested on-wafer after fabrication to verify the accuracy of the circuit design. Next, the chips are packaged while the PCB design starts. Measurement and testing of packaged parts is available only after the PCB assembly. A final verification of the codesign is available only late, at the end of the design cycle.

A first-pass design should be achievable, but due to the complexity of the electromagnetic problem an optimization of the circuit is often required to improve the combined circuit and package performance. Associated circuit and package models need updates that can be obtained from independent EM simulations and measurement cycles as indicated in the figure.

2.5.1 Electromagnetic Modeling of mmWave Packages and Interconnects

The quality of the electrical interconnect between a power amplifier (PA) or a low-noise amplifier (LNA) to an antenna is a critical measure of a packaged chipset performance. It reduces the transmitter efficiency and receiver sensitivity in a radio link, for instance. A well-modeled interconnect is an important design feature, since it enables detailed chip-package cosimulations.

Most simulators can include scattering matrix (S-parameter) data directly in the form of a Touchstone file, for example. The S-parameter data is, however, narrow-band and cannot be used reliably at out-of-band frequencies. A lumped model is often preferred in nonlinear simulations, because it can handle higher-order harmonics. A lumped element circuit model can give more insight into the package parasitic and can be understood by a circuit designer more easily. Therefore, optimization software is often used to fit S-parameter data to a pre-defined package equivalent lumped-element circuit. Figure 2.22 shows an example of a 5 pin lumped package model including the self and mutual inductive coupling of the interconnects.

Although, interconnects and antennas can be highly efficient with little resistance, they may have associated reactance that will change an optimum load impedance seen by a PA and likewise a LNA. A PA, for example, is often designed for maximum power delivery into a 50 Ω load while providing good linearity and high 1 dB Compression point (CP1 dB). A shift

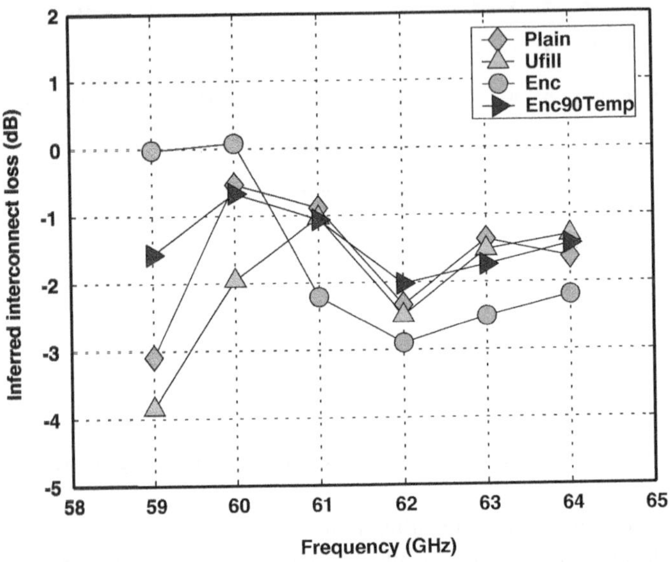

Figure 2.23: Interconnect loss measured at various steps during the assembly. Data was taken after attaching the antenna (Plain), after underfilling the interconnect (Ufill), after molding $\theta = 0, 90°$ (Enc, Enc90), and after thermal cycling tests (Temp) have been completed [29]. Reproduced by permission of © 2006 IEEE.

in an optimum load impedance causes a lower voltage or current compression that will change the gain and compression characteristic of the amplifier respectively. A low insertion- and return-loss are therefore important design features that need to be modeled and co-simulated with the active circuits.

Figure 2.23 shows the measured loss versus frequency for the 60 GHz flip-chip interconnect between a PA and an antenna previously shown in Figure 2.16. Measurements were done after completion of the following processing steps: (i) attaching the antenna, (ii) after underfilling the interconnect, (iii) after molding the antenna compound ($\theta = 0, 90°$), and after thermal cycling tests had been completed. Note, the antenna is designed for encapsulation and therefore shows a higher interconnect loss before molding. The interconnect loss is around 2 dB across the ISM band. This interconnect was codesigned with a two-port lumped equivalent circuit for the flip-chip interconnect in series with a one-port scattering parameter model of the antenna. See Chapter 17, Section 17.6 for more details of a nondestructive measurement technique that has been used to extract the equivalent circuit model of the flip-chip interconnect.

2.5.2 Integrated Antennas

Integrated antennas are discussed in great detail throughout this book; see Chapters 5–8 and 14. This section summarizes some general aspects of integrated antennas as it applies to low-cost package design. The main requirements for the antenna, beyond low-cost and small size, are wide bandwidth (e.g. about 10–15% for 59–64 GHz coverage) and high efficiency [59, 60]. This is usually achieved by choosing a substrate with the lowest possible dielectric constant [59], suspending the antenna in air [60] or adding a superstrate over a planar antenna [50–52].

A packaged chip performance may be measured at various steps during the packaging process to monitor any package-related performance degradation. First, initial on-wafer measurements are required to provide a reference for the conversion gain, noise figure, or maximum available output power that is achievable by a receiver or transmitter chip respectively. Next, such on-wafer results can be verified on the board-level if the input pads to the LNA or output pads from the PA are not obstructed by any objects in close proximity to the probe pads. A final test, however, is required that shows the overall encapsulated performance of the packaged chip including its antenna. Since the antenna cannot be measured *in situ*, the exact interconnect losses are unknown and can only be inferred by accurately calibrated gain measurements in an anechoic chamber.

Acknowledgments

The author would like to thank his former team at the IBM communication technology group (IBM T. J. Watson Research Center, Yorktown Heights, New York, USA) and the IBM SiGe technology group (IBM Burlington, Essex Jct., VT 05452, USA); Pierre Quentin from United Monolithic Semiconductors (UMS), France, and Ralf Juenemann from Rohde & Schwarz, Test and Measurement Division, Munich, Germany, for inclusion of figures, as well as Ed Stoneham from Endwave, San Jose, CA, USA.

References

[1] H. Hertz, *Electric Waves* (London: Macmillan & Co. Ltd). Reprinted by Dover, 1893.

[2] J. Bose, 'On the determination of the wavelength of electric radiation by a diffraction grating' *Proc. Roy. Soc.* **60** (1897), 167–78.

[3] J. Bose, *Collected Physical Papers* (New York: Longmans, Green & Co., 1927).

[4] P. Lebedew, 'Ueber die Dopplbrechung der Strahlen electrischer Kraft', *Ann. Phys. Chem.* **56**(9) (1895), 1–17.

[5] D. Emerson, 'The work of jagadis chandra bose: 100 years of millimeter-wave research', *IEEE Trans. Microw. Theory and Tech.* **45** (1997), 2267–73.

[6] ESIA, JEITA, KSIA, TSIA, and SIA, 'Assembly and packaging', *International Technology Roadmap for Semiconductors*, 2005.

[7] D. Parker, 'Microwave industry outlook – defense applications', *IEEE Trans. Microw. Theory and Tech.* **50** (2002), 1039–41.

[8] J. Schepps and A. Rosen, 'Microwave industry outlook – wireless communications in healthcare', *IEEE Trans. Microw. Theory and Tech.* **50** (2002), 1044–45.

[9] K. Kitazawa, S. Koriyama, H. Minamiue and M. Fujii, '77-GHz-band surface mountable ceramic package', *IEEE Trans. Microw. Theory and Tech.* **48** (2000), 1488–91.

[10] N. Jain, 'Designing commercially viable mm-wave modules', in *IEEE MTT-S Int. Microw. Symp. Dig.*, pp. 565–68, Boston, MA, June 2000.

[11] M. Oppermann, 'Multichip-modules for micro- and millimeter-wave applications – a challenge?', *International Conference on Multi-Chip Modules and High Density Packaging*, pp. 279–84, Denver, CO, April 1998.

[12] G. F. Knoll, *Radiation Detection and Measurement*, 3rd edn (John Wiley & Sons, Ltd/Inc., 2000).

[13] U. Pfeiffer and D. Goren, 'A 20 dBm fully-integrated 60 GHz SiGe power amplifier with automatic level control', *IEEE J. Solid-State Circuits* **42**(7) (2007), 1455–63.

[14] Y. Lin, W. Liu, Y. Guo, and F. Shi, 'Reliability issues of low-cost overmolded flip-chip packages', *IEEE Trans. Adv. Packaging* **28** (2005), 79–88.

[15] S. G. Konsowski and A. R. Helland, *Electronic Packaging of High Speed Circuitry* (McGraw-Hill Professional, 1997).

[16] S. Sacharow and R. Schiffmann, *Microwave Packaging* (Pira International, 1992).

[17] D. L. Monthei, Package Electrical Modeling, Thermal Modeling, and Processing for GaAs Wireless Applications (Dordecht: Kluwer, 2007).

[18] T. S. Laverghetta, Microwave Materials and Fabrication Techniques (Artech House, 1984).

[19] R. Pengelly and J. Turner, 'Monolithic broadband gaas fet amplifiers', *Electron Lett.* **12** (1976), 251–2.

[20] D. McQuiddy, J. Wassel, J. LaGrange, and W. Wisseman, 'Monolithic microwave integrated circuits: An historical perspective', *IEEE Trans. Microw. Theory and Tech.* **32** (1984), 997–1008.

[21] J. Kilby, 'Invention of the integrated circuit' *IEEE Trans. Electron Devices* **23** (1976), 648–54.

[22] T. Midford, J. Wooldridge, and R. Sturdivant, 'The evolution of packages for monolithic microwave and millimeter-wave circuits', *IEEE Trans. Antennas Propag.* **43** (1995), 983–91.

[23] M. Hauhe and J. Wooldridge, 'High density packaging of X-band active array modules' *IEEE Trans. Mfg Technol.* **20** (1997), 279–91.

[24] J. Luy, K. Strohm, J. Buechler, and P. Russer, 'Silicon monolithic millimetre-wave integrated circuits', *IEE Proc. Microw, Antennas and Propagation* **139** (1992), 209–16.

[25] J.-H. Lee, G. DeJean, S. Sarkar, S. Pinel, K. Lim, J. Papapolymerou, J. Laskar, and M. Tentzeris, 'Highly integrated millimeter-wave passive components using 3-D LTCC system-on-package (SOP) technology', *IEEE Trans. Microw. Theory and Tech.* **53** (2005), 2220–29.

[26] J. Heyen, T. von Kerssenbrock, A. Chernyakov, P. Heide, and A. Jacob, 'Novel LTCC/BGA modules for highly integrated millimeter-wave transceivers', *IEEE Trans. Microw. Theory and Tech.* **51** (2003), 2589–96.

[27] S. Reynolds, B. Floyd, U. Pfeiffer, T. Beukema, J. Grzyb, C. Haymes, B. Gaucher, and M. Soyuer, 'Silicon 60-GHz receiver and transmitter chipset for broadband communications', *IEEE J. Solid-State Circuits* **41** (2006), 2820–30.

[28] B. Floyd, S. Reynolds, U. R. Pfeiffer, T. Beukema, J. Grzyb, and C. Haymes, 'A silicon 60 GHz receiver and transmitter chipset for broadband communications', *IEEE Int. Solid-State Circuits Conf.* pp. 184–5, February 2006.

[29] U. Pfeiffer, J. Grzyb, D. Liu, B. Gaucher, T. Beukema, B. Floyd, and S. Reynolds, 'A chip-scale packaging technology for 60-GHz wireless chipsets', *IEEE Trans. Microw. Theory and Tech.* **54** (2006), 3387–97.

[30] Y. Sun, S. Glisic, and F. Herzel, 'A fully differential 60 GHz receiver front-end with integrated PLL in SiGe:C BiCMOS', *European Microw. Conf.*, pp. 198–201, Manchester, UK, September 2006.

[31] A. Natarajan, A. Komijani, X. Guan, A. Babakhani, Y. Wang, and A. Hajimiri, 'A 77 GHz phased array transmitter with local LO-path phase-shifting in silicon', *IEEE Int. Solid-State Circuits Conf.* pp. 182–83, February 2006.

[32] K. Ohata, K. Maruhashi, M. Ito, S. Kishimoto, K. Ikuina, T. Hashiguchi, K. Ikeda, and N. Takahashi, '1.25 Gbps wireless gigabit ethernet link at 60 GHz-band', *Radio Freq. Integr. Circuits Symp.* pp. 509–12, Philadelphia, PA, June 2003.

[33] S. E. Gunnarsson, C. Kärnfelt, H. Zirath, R. Kozhuharov, D. Kuylenstierna, A. Alping, and C. Fager, 'Highly integrated 60 GHz transmitter and receiver MMICs in a GaAs pHEMT technology', *IEEE J. Solid-State Circuits* **40** (2005), 2186–74.

[34] A. Bessemoulin, M. Parisot, and M. Camiade, '1-Watt Ku-band power amplifier MMICs using low-cost quad-flat plastic package', *IEEE MTT-S Int. Microw. Symp. Dig.* 2: 473–6, Fort Worth, TX, June 2004.

[35] J.-Y. Kim, H.-Y. Lee, J.-H. Lee, and D.-P. Chang, 'Wideband characterization of multiple bondwires for millimeter-wave applications', *Asia-Pacific Microwave Conf.* pp. 1265–68, Sydney, Australia, December 2000.

[36] T. Zwick, A. Chandrasekhar, C. Baks, U. R. Pfeiffer, S. Brebels, and B. P. Gaucher, 'Determination of the complex permittivity of packaging materials at millimeter wave frequencies', *IEEE Trans. Microw. Theory and Tech.* **54** (2006), 1001–10.

[37] S. Cahill, E. Sanjuan, and L. Levine, 'Development of 100+ GHz high-frequency microcoax wirebonds', *Proc. of the Int. Symp. on Microelect.* p. 668, San Diego, CA, October 2006.

[38] I. Gresham, N. Jain, T. Budka, A. Alexanian, N. Kinayman, B. Ziegner, S. Brown, and P. Staecker, 'A compact manufacturable 76–77-GHz radar module for commercial ACC applications', *IEEE Trans. Microw. Theory and Tech.* **49**(1) (2001), pp. 44–58.

[39] P. Chinoy, N. Jain, L. Ping, J. Goodrich, and C. Souchuns, 'Manufacture of low-loss microwave circuits using HMIC technology', *IEEE MTT-S Int. Microw. Symp. Dig.* 2: 1137–40, San Diego, CA, 1994.

[40] E. Stoneham, 'Low-cost alternatives for the partitioning and packaging of mm-Wave subsystems', *IEEE MTT-S Int. Microw. Symp. Workshop Presentation on Emerging Packaging Technology and Applications at Millimeter-Wave Frequencies*, Honolulu, Hawaii, June 2007.

[41] E. Stoneham, 'High-precision flip-chip process yields e-band harmonic mixers with potential sub-10-db conversion loss', *Microwave Conf., 2006. 36th European*, pp. 506–9, Manchester, UK, 10–15 September 2006.

[42] E. Stoneham, 'A high-performance 34 to 40 ghz frequency quadrupler fabricated with flip-chip technology', *Microwave Conf., 2006. 36th European*, pp. 1618–20, 10–15, Manchester, UK, September 2006.

[43] D. M. Zeeb, 'A high-performance 14.4 to 19.7 ghz power detector fabricated with flip-chip technology', *Microwave Conf., 2006. 36th European*, pp. 1621–24, Manchester, UK, September 2006.

[44] B. Lefebvre, D. Bouw, J. Lhortolary, C. Chang, S. Tranchant, and M. Camiade, 'A k-band low cost plastic packaged high linearity power amplifier with integrated ESD protection for multi-band telecom applications', *Microwave Symp., 2007. IEEE/MTT-S International*, pp. 825–28, Honolulu, Hawaii, 3–8 June 2007.

[45] G. Strauss and W. Menzel, 'Millimeter-wave monolithic integrated circuit interconnects using electromagnetic field coupling', *IEEE Trans. Components and Packaging Technol.* **19** (1996), 278–82.

[46] L. Zhu and W. Menzel, 'Broad-band microstrip-to-CPW transition via frequency-dependent electromagnetic coupling', *IEEE Trans. Microw. Theory and Tech.* **52** (2004), 1517–21.

[47] U. Pfeiffer and A. Chandrasekhar, 'Characterization of flip-chip interconnects up to millimeter wave frequencies based on a non-destructive in-situ approach', *IEEE Trans. Advd Packaging* **28** (2005), 160–67.

[48] U. Pfeiffer, J. Grzyb, D. Liu, B. Gaucher, T. Beukema, B. Floyd, and S. Reynolds, 'A 60-GHz radio chipset fully-integrated in a low-cost packaging technology', *56th Electronic Components and Technol. Conf. (ECTC)*, pp. 1343–46, San Diego, CA, June 2006.

[49] T. Zwick, D. Liu, B. P. Gaucher, and U. R. Pfeiffer, 'Apparatus and methods for constructing and packaging printed antenna devices', US Patent Application No. 10/881 104, June 2004.

[50] W. Choi, C. Pyo, Y. H. Cho, J. Choi, and J. Chae, 'High gain and broadband microstrip patch antenna using a superstrate layer', *IEEE Antennas and Propag. Symp.* 2: 292–95, June 2003.

[51] N. Alexopoulos and D. Jackson, 'Fundamental superstrate (cover) effects on printed circuit antennas', *IEEE Trans. Antennas Propag.* **32** (1984), 807–16.

[52] C. Gürel and E. Yazgan, 'Bandwidth widening in an annular ring microstrip antenna with superstrate', *IEEE Antennas and Propag. Symp.* pp. 692–95, Newport Beach, CA, June 1995.

[53] D. Griffin and A. Parfitt, 'Electromagnetic design aspects of packages for monolithic microwave integrated circuit-based arrays with integrated antenna elements', *IEEE Trans. Antennas Propag.* **43** (1995), 927–31.

[54] R. Carrillo-Ramirez and R. Jackson, 'A highly integrated millimeter-wave active antenna array using BCB and silicon substrate', *IEEE Trans. Microw. Theory and Tech.* **52** (2004), 1648–53.

[55] R. Traeger, 'Hermeticity of polymeric lid sealants', *Proc. 25th Electronic Components Conf.* p. 361, San Francisco, CA, 1976.

[56] N. Hoivik, D. Liu, C. Jahnes, J. Cotte, C. K. Tsang, C. Patel, U. Pfeiffer, J. Grzyb, J. Knickerbocker, J. Magerlein, and B. Gaucher, 'High-efficiency 60 GHz antenna fabricated using low-cost silicon micromachining techniques', *IEEE Int. Symp. on Antennas and Propag. Conf.* Honolulu, Hawaii, June 2007.

[57] C. Tsai, 'Package inductance characterization at high frequencies', *IEEE Trans. on Components, Packag. and Mfg Technol., Part B* **17** (1994), 175–81.

[58] C. Tsai and W. Yip, 'An experimental technique for full package inductance matrix characterization', *IEEE Trans. on Components, Packag. and Mfg Technol., Part B* **19** (1996), 338–43.

[59] N. Herscovici, 'A wide-band single-layer patch antenna', *IEEE Trans. Antennas Propag.* **46** (1999), 471–74.

[60] M. Faiz and P. Wahid, 'A high efficiency l-band microstrip antenna', *IEEE Int. URSI Conf.* pp. 272–75, Orlando, FL, July 1999.

3

Dielectric Properties at Millimeter-wave and THz Bands

Khalid Z. Rajab, Joseph P. Dougherty and Michael T. Lanagan

3.1 Introduction

Recent years have seen a large amount of interest devoted to the terahertz (THz) region of the electromagnetic spectrum. A lack of sources, especially in the frequency range of 100 GHz to 10 THz, had made investigations into this band difficult. This was despite the promise of applications in the characterization of materials, biomedical imaging, detection of explosives, communications, and a plethora of other fields. The term 'terahertz gap' proved a particularly good description of the gap between the microwave and infrared (IR) bands: the division between solid-state electronics and optical photonics. Lately, this gap has begun to narrow as new, higher-power sources have started to cover a greater range of frequencies, leading to excitement at the range of applications that have begun to open up. To bridge this gap, a better understanding is required of the properties of materials at these frequencies.

The earliest investigations at THz frequencies were reported in the 1960s, not long after the development of the laser. Gebbie *et al.*, in particular, demonstrated an HCN laser operating at 0.337 mm (0.89 THz) [1]. Shortly afterwards, the same group used their setup to characterize the monochromatic refractive indexes and absorption coefficients of various benzene compounds [2], as well as air [3]. In recent years, the development of new THz sources have involved both solid-state electronic devices and optical schemes. Solid-state technologies, extended from the microwave frequencies, have in particular been useful for applications in wireless communication. The two main photonic sources that have been attracting attention are the quantum cascade laser (QCL), as well as traditional optical

Advanced Millimeter-wave Technologies: Antennas, Packaging and Circuits
Duixian Liu, Ulrich Pfeiffer, Janusz Grzyb and Brian Gaucher © 2009 John Wiley & Sons, Ltd

THz generation. The QCL was first demonstrated at far-IR frequencies (70 THz) in 1994 [4], while the first THz band (as we have defined it here) QCL, operating at 4.4 THz, was demonstrated in 2002 [5, 6]. Undoubtedly, the most widely used source for THz materials research and characterization has been time-domain (TD) optical THz generation. The earliest works on this technique were performed at AT&T Bell Laboratories [7] and the IBM Watson Research Center [8]. These two labs also developed, respectively, the electro-optic and photoconductive antenna for THz generation/detection. The electro-optic effect was used to generate Cherenkov radiation, while the technique used currently, namely the free-space electro-optic sampling method, was developed by Wu and Zhang [9]. These works were also the precursors to the technique termed THz time-domain spectroscopy (TDS). This is the most prevalent technique used currently for material characterization at THz frequencies.

3.2 Dielectric Characterization

The dielectric properties of materials are vitally important in the design and application of any particular device at any frequency range. The dielectric behavior of a material is defined by its complex permittivity, $\epsilon = \epsilon' - j\epsilon''$. However, we generally deal separately with the relative permittivity, $\epsilon_r = \epsilon'/\epsilon_0$ (referred to as the dielectric constant κ in the materials community), and the loss tangent, $\tan\delta = \epsilon''/\epsilon'$ (or alternatively quality factor, $Q = 1/\tan\delta$). A variety of characterization techniques exist that would allow one to obtain these dielectric properties, but the availability of so many different techniques can also lead to some confusion; each one of the plethora of material characterization methods has its own advantages and disadvantages, and they are each suitable for use in a particular frequency range, from close to DC up to the infrared frequencies. Above these frequencies, the macroscopic material models begin to fail.

Figure 3.1 shows the electromagnetic spectrum from DC up to the ultraviolet frequencies, with the THz gap in between. At frequencies below the THz gap, we have the electronic regime. Here we can find the radio waves, including the microwaves (300 MHz to 300 GHz). Above 300 GHz, and up to the edge of the visible light spectrum (450 THz to 750 THz), the infrared waves (300 GHz to 450 THz) can be found. The THz gap is, in fact, a part of the near-IR band, and above the gap we have the optical regime. Just as the sources and devices change between these two regimes, the material characterization techniques, along with the response of the materials themselves, change.

3.3 Outside the THz Gap – Material Characterization Techniques

Before we discuss dielectric characterization techniques at THz frequencies, an overview of the available methods outside the THz gap will be provided. This will give the reader the context in which to place the THz techniques, as well as an idea of the frequency ranges that we may cover. As well as splitting the electromagnetic spectrum into electronic and photonic regimes, Figure 3.1 also summarizes a number of the available techniques. These are further discussed in Table 3.1. It is impossible to be truly exhaustive; however, the most widely used methods are included.

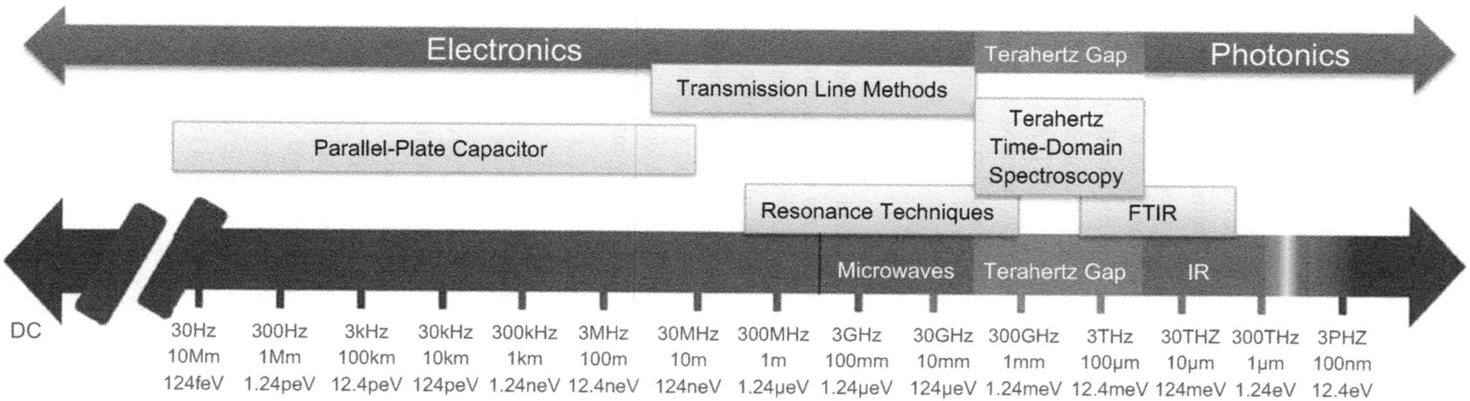

Figure 3.1: Electromagnetic spectrum, from DC up to visible and ultraviolet wavelengths, where macroscopic models of materials begin to fail. The spectrum is split up into the electronic and photonic regimes, with the THz gap in between. Various classes of material characterization techniques are shown, covering different frequency ranges.

Low-frequency techniques, up to around the 10 MHz range, rely on the measurement of quasistatic parameters (i.e. capacitance or impedance). At microwave frequencies, a variety of techniques exist, each of which can loosely be defined as either a resonant or a broadband measurement. The resonant techniques have the ability to measure high-quality factor (high Q), low-loss materials, with small sample sizes, albeit only in a narrow frequency band (or bands). In general, low-loss dielectrics experience very little change in the permittivity over a broad frequency range, so resonant techniques are ideally suited for these materials. They are, however, limited in measuring high-loss (low Q) materials. The broadband techniques, on the other hand, are well suited to measuring high-loss samples, at a large range of frequencies, although they require larger sample dimensions. At the upper end of the frequency scale, reflectivity measurements are used to obtain the broadband dielectric characteristics of a material. The Electromagnetics Division of the National Institute of Standards and Technology have published a series of comprehensive reviews covering a large number of different characterization techniques [10–12]. Afsar *et al.* have also covered a number of techniques [13]. The interested reader is referred to these reviews; however, we will briefly introduce a number of the most important characterization techniques.

3.3.1 Parallel Plate (~DC–30 MHz)

Below 30 MHz, LCR meter or impedance analyzer measurements are used to obtain the permittivity (see, e.g., reference [13]). This involves the measurement of the parallel-plate capacitance of an electroded dielectric. The permittivity is then obtained from the capacitance measurement and the material dimensions. We should also note that capacitor geometries are not limited to parallel plates; interdigital capacitors have been shown to be suitable in certain circumstances, e.g. biosensors for remote complex permittivity determination were demonstrated [14]. This method is more mathematically complicated, however, as complex mapping techniques are needed to reduce the geometry into a solvable problem. This also limits accuracy.

3.3.2 Resonant Cavity (~0.5–50 GHz)

Resonant cavities have the advantage of being high-Q structures. The cavities are usually an easily-analyzed geometry, such as a rectangle or a cylinder, where the resonant modes can be determined as functions of the dimensions and ϵ_r. The general idea is to extract the permittivity, knowing the resonant frequency and the dimensions. The dimensions are measured to within an acceptable level of error (error analysis will be discussed in a later section), while the frequency sweep is usually measured with a vector network analyzer. Subsequently, from the bandwidth of the measured peak, the material's loss tangent or Q may also be extracted. From these techniques we are able to measure extremely high Q (of the order of 10 000), albeit only at a single frequency (or a few frequencies if we use multiple modes).

Open cavity techniques are very widely used, since samples can be conveniently fabricated, and are generally small in size with respect to the wavelength. This is of particular benefit at low frequencies, for example at 1 GHz where the wavelength in free space is 30 cm. The classical geometry for this technique is the dielectric cylinder, as demonstrated by Hakki and Coleman [15], and by Courtney [16]. A more recent method makes use of the

Table 3.1: Summary of material characterization techniques.

Measurement technique	Approximate frequency range	Advantages	Disadvantages	Key references
Parallel plate	DC–30 MHz	Accurate measurements Sub-wavelength samples Simple technique	Uniform field required	[13]
Open cavity (resonant)	0.5–50 GHz	Accurate for low losses High Q Small samples Easily available equipment setup	Not for high losses	[15–17]
Fabry–Perot resonator (resonant)	20–300 GHz	Suitable for thin, low-loss samples High Q High frequency measurements	Limited high-loss accuracy due to metal mirrors	[18, 19]
Split-cylinder cavity (resonant)	10–20 GHz	Accurate for low losses High Q Flexibility with sample area	Inaccurate for high losses Samples must be smooth Difficult to manufacture cavity	[20, 21]
Cavity perturbation (resonant)	1–25 GHz	Accurate Arbitrary sample (with known dimensions)	Approximation errors Not for low losses (rectangular) Difficult to manufacture cavity (cylindrical)	[22]
Ring resonator (resonant)	0.5–40 GHz	Accurate for permittivity Simple calculations Physically intuitive	Difficult to measure loss Loss dominated by conductor	[23, 24]
Coaxial probe (Transmission line)	0.01–25 GHz	Very large frequency range Suitable for liquids and semi-solids	Air gaps cause significant errors Air bubbles (liquids) cause significant errors Thick sample Limited accuracy	[25, 26]

Table 3.1: Continued.

Measurement technique	Approximate frequency range	Advantages	Disadvantages	Key references
Coaxial transmission line	0.5–25 GHz	Broad frequency range Suitable for high loss samples	Difficult to manufacture Inaccurate for low losses	[27–29]
Rectangular waveguide (transmission line)	1–25 GHz	Relatively broadband Suitable for high loss samples	Low-loss inaccuracy Cut-off defines lower frequency Large samples for low frequencies	[27–29]
Planar transmission line	1–25 GHz	Simple technique	Loss dominated by metal Fabrication difficulty	[30]
Free space (transmission line)	1–100 GHz	High frequency measurements Noncontacting Measurement flexibility	Large sample for low frequencies Dielectric lenses required Complicated calibration Flat, smooth samples required	[31]
Terahertz time-domain spectroscopy	0.1–10 THz	Noncontacting Phase coherence Suitable for high/low temperature	Expensive setup Atmospheric absorption bands Smaller spectral range than FTIR measurements	[32–35]
Fourier transform infrared	1–100 THz	Noncontacting Very broadband	Phase incoherent Complicated permittivity fitting procedure Atmospheric absorption bands Limited spectral resolution	[36]

nonradiative dielectric waveguide (NRD), and allows the measurement range to be extended to higher frequencies [17]. As an example of the sample dimensions that would yield a 1 GHz resonance for $\epsilon_r = 10$, the Hakki–Coleman cylinder that would give a TM_{011} resonance here could have a diameter of 10 cm, and a thickness of 8.6 cm. The NRD waveguide LSE_{101} resonance could occur here with dimensions of 7 cm $\times$ 5 cm $\times$ 10 cm. The Fabry–Perot open resonator technique has been shown to be suitable for characterization up to 300 GHz [18,19]. In this case, a coupling aperture in a mirror is used to excite a thin dielectric that is rested on a second planar mirror. The setup has very high quality factors, enabling the measurement of low losses; however, metal losses from the mirrors lead to inaccuracy.

Another resonance technique is the split-cylinder cavity method, demonstrated by Kent [20], and further developed by Janezic and Baker-Jarvis [21]. The fixture here consists of a fixed lower cylindrical cavity, and an upper cavity that is lowered to sandwich the material in-between. The sample is a thin dielectric slab that is larger than the cross-sectional area of the cavity. We may then use the TE_{011} resonant mode of the system to obtain the complex permittivity of the enclosed dielectric.

Other examples of resonant techniques at microwave frequencies include perturbational cavity methods (see, e.g., reference [22]), and ring resonator techniques (see, e.g., references [23] and [24]).

3.3.3 Transmission Line Methods (~0.01–300 GHz)

Broadband characterization methods at microwave frequencies generally consist of various transmission line techniques. The transmission lines may be microstrips, coaxial and rectangular waveguides, or even free space. Roberts and Von Hippel first demonstrated a transmission line technique for measuring permittivities at microwave frequencies, using a waveguide [27]. Nicolson and Ross [28], and Weir [29] later presented their seminal papers on the retrieval of the broadband complex permittivity and permeability of a material, from the inversion of the complex transmission and reflection coefficients. More modern techniques were later extended to the Nicolson–Ross–Weir (NRW) analysis, including an error analysis and the application of robust algorithms to the inversion procedure [37]. A wave propagating along any transmission line will have a phase velocity determined by the structure, so the analysis of the phase of the reflection (S_{11} and S_{22}) or transmission (S_{21} and S_{12}) coefficients will yield the relative permittivity. This, in conjunction with the magnitudes of those coefficients, could allow one to extract $\tan\delta$ as well; and because the techniques do not rely on any particular resonances, they are broadband, allowing characterization over a large frequency range. However, they are by their nature inaccurate when measuring materials with low $\tan\delta$.

Coaxial transmission lines are formed by an outer conductor enclosing a dielectric, with an inner conductor running through the center. They can be used both as an open-ended probe, with S_{11} measurements of a dielectric covering the end of the line, or as a two-port transmission line [25,26]. As a characterization method, they provide results over a broad frequency range, going down to MHz frequencies for the case of the open-ended probe. This technique does, however, have limited ϵ_r accuracy, and as with all transmission line techniques, it is inaccurate when measuring low losses. On the other hand, the two-port coaxial transmission line provides more accurate ϵ_r results. The difficulty with using this technique stems from its awkward geometry. A sample must fit snugly within the conductors.

As a result, this method is most suitable for liquids, powders, and semi-solids (such as biomaterials). Similar techniques take advantage of other structures, such as the rectangular waveguide, in place of the coaxial line. The rectangular waveguide is, however, limited by a lower cut-off frequency of a supported mode. Nevertheless, the geometry is more conducive to measurements of most solid dielectrics.

A microstrip consists of a thin conducting strip on top of one side of a dielectric substrate. The reverse side of the substrate is generally a conducting ground plane. The characterization methods rely on the measurements of the phase of a wave transmitted through the structure, and this phase can be used to extract an effective permittivity, based on the speed of propagation of the wave through the structure. Analytical formulas can then be used to obtain ϵ_r. Similar planar technologies, such as the coplanar waveguides [30], may be used in place of the microstrip.

The free-space measurement technique [31] is a transmission line method in the loosest sense of the term. The transmission line is in fact just the free space in between two antennas (most often, horns). A sample is placed in-between the horns, and the measured S-parameters are used to determine the material characteristics. Although it appears straightforward at first glance, there are in fact experimental difficulties that make it less than trivial to operate. The primary difficulty is with the nature of the experimental setup; theoretically, the material is assumed to have infinite area, while the propagating wave is assumed to be a perfect plane wave. Both of these assumptions are obviously incorrect, but we can try to get as close to the ideal case as possible. A dielectric lens may be used to focus the beam to the center of the sample. Additionally, a sample area of at least a wavelength squared is generally required, which may be prohibitively large at low frequencies. A further difficulty is the calibration of the setup, as there are no connectors. The thru-reflect-line (TRL) calibration method applies well to this technique, however, with a large conducting plate acting as the Reflect standard. Additionally, if the network analyzer in use allows it, TD gating may be used to remove artifacts of the system and improve the calibration. The use of absorbers can also improve measurements. Despite these difficulties, the technique is very useful at millimeter-wave (mmWave) frequencies, up to around 100 GHz. Much above these frequencies, the size of the horn antennas become prohibitively small and traditional microwave frequency sources do not work. This technique is of particular interest to us because it is most similar to, and relies on the same principles as, that of the THz TDS characterization methods that will be described in this chapter.

3.3.4 Fourier Transform Infrared Spectroscopy (~1–100 THz)

At infrared frequencies, Fourier transform infrared (FTIR) spectroscopy can be used to obtain the complex permittivity, or more commonly at these wavelengths, the complex refractive index $\tilde{n} = n - j\kappa$ (where the imaginary part of the refractive index, κ is not to be confused with the materials community's dielectric constant). The absorption coefficient, α, of a material can also be defined as

$$\alpha = \frac{2\omega\kappa}{c} \tag{3.1}$$

A variety of similar techniques exist, as reviewed in reference [36], but a common one makes use of reflectivity measurements. The magnitude of the light reflected off a sample can be measured over a broad frequency range; however, FTIR is insensitive to phase.

Anderman *et al.* [38] showed that the Kramers–Kronig causality relations can be used to obtain a good approximation for the phase. With the measurements of the polar phonon modes, as well as the high-frequency permittivity ϵ_∞, a model dielectric function can be fit to obtain the complex permittivity over the frequency range of interest. This dispersion analysis is not straightforward, however, and numerical estimations must be made to approximate the phase function. Additionally, reflection measurements are sensitive to surface roughness, and hence, the sample surfaces must be smooth. It should also be noted that Fourier transform spectroscopy can be applied at mmWave frequencies, for example with a two-beam polarization interferometer [19], to obtain broadband characterization in the 30–300 GHz range.

3.4 THz TDS (∼0.1–10 THz)

With the increasing interest in the development of applications within the THz gap, material characterization techniques are necessary for frequencies between 100 GHz and 10 THz. These frequencies are not well covered by either the microwave techniques, or by FTIR. THz TDS is one method that can be used to fill this gap. This spectroscopic technique operates with an ultrashort pulse of THz radiation, transmitted through a material. The pulse is probed, either after transmission or reflection. The measured signature can be transformed to the frequency domain. The spectrum subsequently provides amplitude and phase information of the material. This information can be used for a variety of applications, including materials characterization, tomographic imaging (see Chapter 15), and chemical detection.

Before we discuss the dielectric characterization techniques, the experimental setup will be described so as to provide the reader with an overview of the system (see Figure 3.2). The experimental setup can be split into five main operations subsequent to the source producing the laser: beam splitting; signal delay; THz generation; propagation through the sample; and, THz detection. A laser with an appropriately short pulse will generate a beam, which is split so that the majority of its intensity forms the pump beam, used for THz generation. The remaining portion of the beam is used as a probe beam for detection. Once the beam has been split, the pump beam is directed through a delay line. The length of the delay line is swept, so that the entire pulse amplitude is obtained as a function of time. This delayed pump beam is then used for THz generation. The most common method of generating THz waves is by the use of a biased photoconducting gallium arsenide (GaAs) aperture antenna [39,40]. A laser pulse generates carriers in the conduction band of the semiconductor, effectively causing it to turn from an insulator into a conductor. The bias voltage across the antenna accelerates the carriers so that a current is formed, and the rapid switching of the pulses induces the generation of THz radiation, in a manner similar to an antenna working at microwave frequencies. Other generation schemes include the use of crystalline materials. Once the THz waves have been generated, they are free to propagate and interact with the sample. A detector is placed in an appropriate position to measure the signal of interest (for example, transmission or reflection). Photoconducting antennas and electro-optic crystals may also be used for detection. The probe pulse that was split from the pump beam is delayed so that it arrives at the detector at the same time as the THz pulse. For detection with an electro-optic crystal, the polarization is rotated slightly, as the probe pulse travels through the birefringent crystal simultaneously to a point in the THz pulse. This rotation is proportional

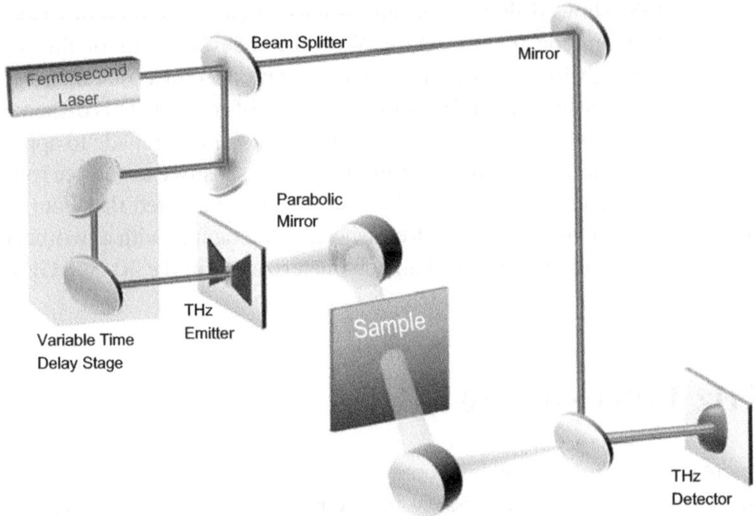

Figure 3.2: Typical setup for THz TDS. The ultrafast laser beam is split into a probe beam and a pump beam. The pump beam is delayed, and then used for THz generation. The THz waves are collimated and focused on the sample, and then again collimated and focused, along with the probe beam, onto a THz detector, so as to obtain the temporal profile of the electric field.

to the amplitude of the THz field, and the direction of rotation is proportional to the sign. Polarizers and photodiodes can be used to determine these polarization changes [9,41,42].

The setup we used is as follows: a diode laser pumped Ti:Sapphire laser emits 60 μs pulses at a center wavelength of 790 nm, operating at 80 MHz. The beam is split so that 90% of the intensity is used as a pump beam for THz generation, with the other 10% used as a probe beam for detection. THz generation was provided by excitation of a biased GaAs photo-conductive aperture antenna, with the THz pulses collimated and focused by a series of off-axis parabolic mirrors through the sample, and onto a 1 mm thick ZnTe crystal.

3.4.1 Transmission

Once the TDS has been measured, the frequency spectrum is obtained with a discrete Fourier transform (DFT). Numerically, this would be done by way of a fast Fourier transform (FFT). A reference spectrum, with no sample present, is first measured. This is best done through a nitrogen-purged atmosphere, so as to minimize water absorption and remove resonances. A second spectrum is then measured with the sample in place.

Figure 3.3 shows the transmitted TDSs through the reference and through a sample specimen. The delay is evident in the pulse traveling through the sample, as light is slowed down due to the material's refractive index. An approximation of the average permittivity of the material over the pulse's measured bandwidth can be determined using this delay, as

$$\epsilon_r = \left(1 + \frac{c}{d}\Delta t\right)^2 \tag{3.2}$$

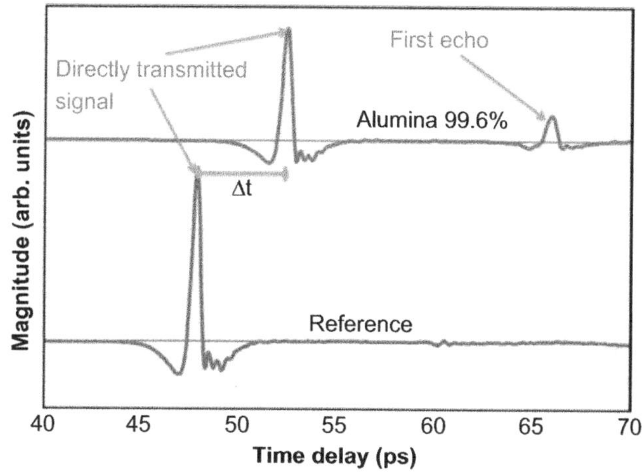

Figure 3.3: Time domain signal.

where c is the speed of light in free space, d is the sample thickness, and Δt, shown in Figure 3.3, is the delay between the peak amplitudes of the reference and sample pulses.

This is, however, just an approximation, and does not give us much more than a crude estimate of the permittivity over the frequency range. To find the frequency-dependent complex permittivity of the sample, an FFT should be performed on the signal. We will now have the complex transmission spectrums of the reference and the sample, the magnitudes and unwrapped phases of which are shown in Figure 3.4. These transmission spectra may be used to investigate the dielectric properties of the sample. We can view the sample's system as a three-medium (air–dielectric–air) Fabry–Perot resonator. The transmission, $t(\omega)$ has been derived by Duvillaret *et al.* [32], from the Fresnel formulas as

$$t(\omega) = \frac{T_{\text{sample}}(\omega)}{T_{\text{ref}}(\omega)} = \tau(\omega) P(\omega, d) \cdot \frac{1}{1 - r(\omega) P^2(\omega, d)} \cdot \exp\left[j \frac{\omega d}{c}\right] \tag{3.3}$$

where $T_{\text{sample}}(\omega)$ and $T_{\text{ref}}(\omega)$ are the complex transmission spectra through the sample and the reference, respectively, and $\omega = 2\pi f$ is the radial frequency. The reflection, transmission, and propagation coefficients are represented by $r(\omega)$, $\tau(\omega)$, and $P(\omega, d)$, respectively, where

$$r(\omega) = \frac{[\tilde{n}(\omega) - 1]^2}{[\tilde{n}(\omega) + 1]^2} \tag{3.4a}$$

$$\tau(\omega) = \frac{4\tilde{n}(\omega)}{[\tilde{n}(\omega) + 1]^2} \tag{3.4b}$$

$$P(\omega, d) = \exp\left[-j \frac{\omega \tilde{n}(\omega) d}{c}\right] \tag{3.4c}$$

For a nonmagnetic material the complex permittivity $\epsilon(\omega)/\epsilon_0 = \sqrt{\tilde{n}(\omega)}$.

Although this solution is exact, in practice (3.3) is not the best method for solving the material characteristics. Temporal echoes, as seen in Figure 3.3, are caused by reflections in

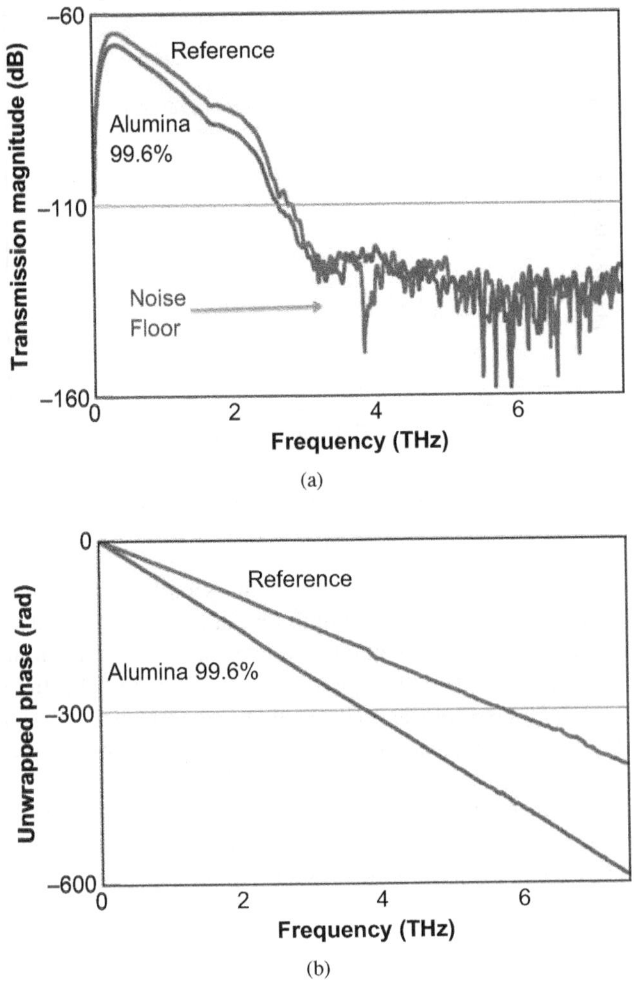

Figure 3.4: Reference and sample transmission spectra: (a) magnitude; (b) unwrapped phase.

the slab that lead to time-delayed transmitted pulses. The equation is derived by assuming that all temporal echoes have been measured, implying an infinite time measurement. In practice a signal of only a few tens of picoseconds is measured, which may not allow sufficient damping of the transmitted peaks before they can be neglected. Furthermore, reflections from the experimental setup will eventually interfere with measurements. Numerical techniques are also necessary to solve for $\tilde{n}$, rendering this technique impractical.

We will now look at a more efficient and accurate technique that also uses the measured transmission spectra. This technique compares the spectra of two different samples of the same material, but with different thicknesses. The trick here is to treat the reference as a material with *zero* thickness. The solution, as will be outlined below, has been derived separately to treat each of the temporal echoes. The directly transmitted signal, which is represented by the 0th echo, is suitable for use when all echoes are neglected. The spectral component of the transmitted field through a sample, $E_t(\omega)$, can be expressed as [33, 43]

$$E_t(\omega) = E_i(\omega)\tau(\omega)P(\omega, d)\sum_{p=0}^{\infty}[r(\omega)P^2(\omega, d)]^p \tag{3.5}$$

where r, τ and P were defined in (3.4) and $E_i(\omega)$ is the incident field. This equation is consistent with (3.3) if we simplify the geometric series with

$$\sum_{p=0}^{N} x^p = \frac{x^{N+1}-1}{x-1} \Rightarrow \sum_{p=0}^{\infty}[r(\omega)P^2(\omega, d)]^p = \frac{1}{1 - r(\omega)P^2(\omega, d)} \tag{3.6}$$

E_t corresponds with the transmission through the sample, T_{sample}. The spectrum T_{ref} is not, however, the incident field, but the field transmitted through the reference. This is just the incident field modified by a phase term, such that $T_{\text{ref}} = E_i \cdot \exp(j\omega d/c)$, and so (3.5) is equivalent to (3.3), with the summation term representing the Fabry–Perot effect. Although the form of (3.3) appears simpler than that of (3.5), we wish to take advantage of the nature of the infinite geometric series, as each successive term will represent a temporal echo. Thus, using (3.5), we can handle a finite number of temporal echoes.

Defining the sample's transmission coefficient for an echo p

$$t(p, \omega) = \frac{T_{\text{sample}}(p, \omega)}{T_{\text{ref}}(p, \omega)} = \tau(\omega)r^p(\omega)P^{2p+1}(\omega, d)\cdot\exp\left[j\frac{\omega d}{c}\right] \tag{3.7}$$

we can find its magnitude and phase as

$$\begin{aligned} |t(p, \omega)| &= |\tau(\omega)|\cdot|r(\omega)|^p\cdot|P(\omega, d)|^{2p+1} \\ &= \frac{4|\tilde{n}|\cdot[(n-1)^2+\kappa^2]^p}{[(n+1)^2+\kappa^2]^{p+1}} \times \exp\left[-\kappa(2p+1)\frac{\omega d}{c}\right] \end{aligned} \tag{3.8a}$$

$$\begin{aligned} \varphi(p, \omega) &\equiv \arg[t(p, \omega)] \\ &= \arg[\tau(\omega)] + p\cdot\arg[r(\omega)] + (2p+1)\cdot\arg[P(\omega, d)] + \frac{\omega d}{c} \\ &= -[(2p+1)n-1]\frac{\omega d}{c} - 2p\cdot\arctan\left[\frac{2\kappa}{|\tilde{n}|^2-1}\right] \\ &\quad + \arctan\left[\frac{\kappa(|\tilde{n}|^2-1)}{|\tilde{n}|^2(n+2)+n}\right] \end{aligned} \tag{3.8b}$$

Note that equation (3) in [33] contains an error in the phase term. φ represents the difference in phase between the T_{sample} and T_{ref}. The phases of these terms should be unwrapped before the difference is taken. Assuming a relatively low loss sample, i.e. $\kappa \ll n$, we can solve for n and α using (3.8a) and (3.8b), as well as (3.1) to obtain

$$n = \frac{1}{(2p+1)}\cdot\left[1 - \frac{c}{\omega d}\varphi(p, \omega)\right] \tag{3.9a}$$

$$\alpha = \frac{-2}{(2p+1)d}\cdot\ln\left[\frac{(n+1)^{2p+2}}{4n(n-1)^{2p}}|t(p, \omega)|\right] \tag{3.9b}$$

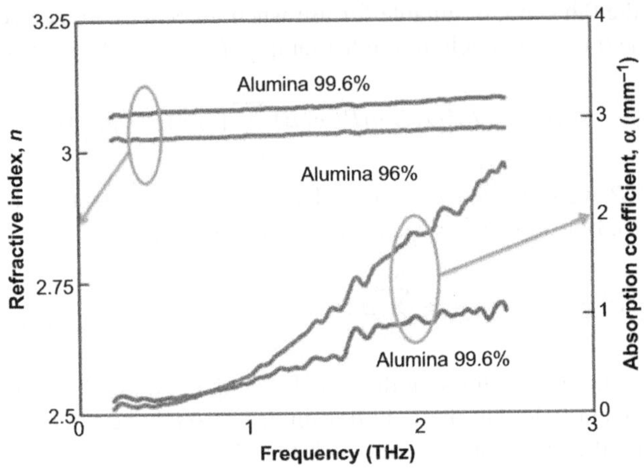

Figure 3.5: Retrieved refractive index, n, and absorption coefficient, α, for two different alumina samples.

The refractive index and absorption coefficient can be found from the measurement of the directly transmitted signal. Substituting $p = 0$ into the equations above, we have

$$n = \left[1 - \frac{c}{\omega d}\varphi(0, \omega)\right] \tag{3.10a}$$

$$\alpha = \frac{-2}{d} \cdot \ln\left[\frac{(n+1)^2}{4n}|t(0, \omega)|\right] \tag{3.10b}$$

Figure 3.5 shows the results of these equations. The above formulas have been modified to determine the permittivities of thin film materials [32, 34].

In Figure 3.6, the real part of the permittivity is plotted for the two different alumina samples. Measurements were performed with THz TDS from around 200 GHz to 2.5 THz. Below these frequencies, a resonant split-cylinder cavity measurement was performed, to obtain the permittivity at 15 GHz. At higher frequencies, FTIR spectroscopy was used to measure the permittivity. There is good correlation between the techniques over the different bands. There is a slight mismatch between the THz TDS and the FTIR measurements of 96% alumina. This can possibly be explained by scattering at higher frequencies, to which FTIR, being a reflection technique, is particularly susceptible. FTIR also relies on complex data fitting and analysis, which may further lead to errors. Nevertheless, it is still important to determine the uncertainty in the THz TDS characterization process.

3.4.2 Error Analysis

The uncertainties in the refractive index and the absorption coefficient, Δn and $\Delta\alpha$ respectively, have been derived in reference [33]. We have modified them to take into account errors due to the sample thickness uncertainty. For a parameter x, the uncertainty may be

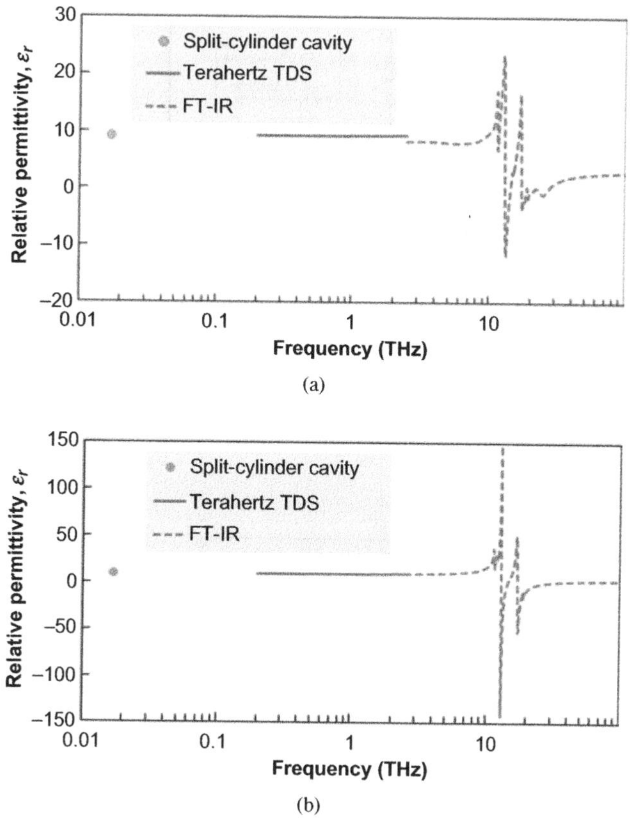

Figure 3.6: Measurements performed using a split-cylinder cavity, THz TDS, and FTIR: (a) permittivity of 96% alumina; (b) permittivity of 99.6% alumina.

expressed as

$$\Delta x = \sqrt{\left(\frac{\partial x}{\partial \varphi}\Delta\varphi\right)^2 + \left(\frac{\partial x}{\partial |t|}\Delta|t|\right)^2 + \left(\frac{\partial x}{\partial d}\Delta d\right)^2} \tag{3.11}$$

The error Δd is determined by the measurement technique in use. $\Delta\varphi$ and $\Delta|t|$ are obtained from the standard deviation of the measured transmission coefficient $t(p, \omega)$. Thus [33]

$$\Delta|t(p, \omega)| = \left[\frac{\langle|t_N(p, \omega)|^2\rangle}{2}\right]^{1/2} \tag{3.12a}$$

$$\Delta\varphi(p, \omega) = \frac{\Delta|t(p, \omega)|}{|t(p, \omega)|} \tag{3.12b}$$

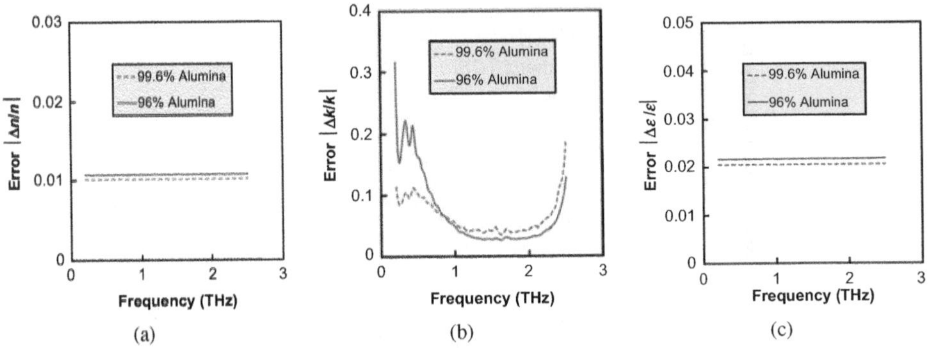

Figure 3.7: Measurement errors: (a) real part of the refractive index, n; (b) imaginary part of the refractive index, κ; (c) real part of the permittivity, ϵ.

$$\Delta n = \frac{c|\varphi|}{\omega d}\sqrt{\left(\frac{\Delta\varphi}{\varphi}\right)^2 + \left(\frac{\Delta d}{d}\right)^2} \tag{3.13a}$$

$$\Delta\alpha = \frac{2}{d}\sqrt{\left(\frac{\Delta|T|}{|T|}\right)^2 + \left(\frac{\Delta n}{n}\right)^2\left[\frac{(n-1)^2}{(n^2-1)}\right]^2} \tag{3.13b}$$

where $\langle\cdot\rangle$ denotes the average value of the enclosed function, and $t_N(p,\omega)$ is the noise in the measured transmission.

In Figure 3.7, the calculated errors in the real and imaginary parts of the refractive index, and in the real part of the permittivity, are plotted with frequency. In common with transmission techniques at microwave frequencies, the accuracy is very good for the real part, but the loss measurements succumb to errors, particularly at higher frequencies.

3.5 Dielectric Properties

3.5.1 Semiconductors

Silicon-based materials allow the packaging of a number of components and devices at low cost, but with high performance. Intrinsic semiconductors will exhibit low losses at mmWave and THz frequencies, and are suitable for antenna and other device applications at these frequencies. High resistivity silicon-based photoconducting switches have been demonstrated, in addition to a number of other devices [45]. Additionally, as was discussed in Section 3.4, photoconducting GaAs has been an important method for generating THz waves.

3.5.2 Ceramic Materials

Applications for low-temperature cofired ceramics (LTCC) and high-temperature cofired ceramics (HTCC) technologies, as well as for ceramic dielectric resonators, are advancing to higher frequencies. The design of ceramic microsystems and electronic packages utilizing these technologies requires the characterization of their dielectric properties into the THz frequency range. Important characteristics required include high permittivity, low loss, and a low-temperature coefficient of resonant frequency τ_f. Broadband investigations have been performed on a number of ceramics using THz TDS, in order to characterize various aspects of the materials [46,47].

3.5.3 Thin Films

An important application of THz TDS has been in the characterization of thin films. These have traditionally been difficult to characterize at microwave frequencies, but modifications to the thick film characterization techniques allow the determination of their properties at THz frequencies [34,43]. Furthermore, the characterization of high-temperature superconducting thin films at THz frequencies has been made possible using this technique. MgB_2, recently discovered to be superconducting at 39 K, has been analyzed using THz TDS, in order to determine its complex, frequency-dependent conductivity $\sigma(\omega)$ (see reference [48]). Surprisingly, the real part of the conductivity exhibited a strong depletion of oscillator strength near 5 meV, which was only around half the value predicted theoretically using the isotropic, weak-coupling scenario. This suggests the existence of complex behavior in this superconductor, opening up a number of avenues for research.

3.5.4 Metamaterials

A number of interesting applications have been proposed for metamaterials, or artificially structured materials, at THz frequencies. Magnetic materials, for example, are particularly rare at THz and optical frequencies. This is due to the fact that the resonant phenomena resulting in magnetic properties occur at much lower frequencies, resulting in very little magnetic response at THz frequencies. Although there are ferromagnetic and antiferromagnetic systems that exhibit magnetic response above the microwave frequency range, their effects are typically weak, and extend over only a narrow band of frequencies. However, artificial materials have been designed and shown to exhibit far stronger magnetic behavior at these frequencies [49,50]. Metamaterials have also been designed in an attempt to address the lack of devices in the THz gap. These include an active THz metamaterial device which effectively acts as a Schottky diode, enabling the modulation of THz transmission by around 50% [51]. The transmission through photonic crystals, formed of periodic arrays of Si_3N_4 spheres, has been measured using THz TDS [52]. This has promise for the development of photonic crystals and all-dielectric metamaterials at THz frequencies. All-dielectric metamaterials have also been designed based on fractal geometries [53]. It is anticipated that all-dielectric metamaterials will prove particularly useful at THz frequencies, as they are not susceptible to the high-conductivity losses of metals. Applications of metamaterials at THz frequencies are covered in more detail in Chapters 9 and 10.

3.5.5 Biomaterials

There are a very large number of biomedical applications that have been proposed at THz frequencies [35,54]. These include the tomographic imaging of biomaterials and tissues to assist with medical procedures or cancer detection, as well as the THz detection of single base-pair differences in femtomolar concentrations of DNA [55], for the purpose of label-free genetic diagnostics. The majority of these techniques are reliant on the determination of a material's parameters and the detection of changes in these parameters resulting from different conditions.

3.5.6 Material Needs

Applications at mmWave and THz frequencies will have a general requirement for low-cost and low-loss materials, as with microwave applications. Unlike at microwave frequencies, however, the trend would be towards lower permittivities, as opposed to the high permittivity resonators and resonant structures in the microwave range. As device applications move towards higher frequencies, the wavelengths, and hence the device sizes will significantly decrease, and so a primary motivation for the use of high permittivity dielectrics – the compression of the waves for device miniaturization – is lost. Furthermore, high permittivity materials may become prohibitively expensive for certain mass manufacturing applications.

Metamaterials will also undoubtedly provide a large range of material properties that can be advantageous to the mmWave and THz engineer. As conductor losses become increasingly prohibitive at these higher frequencies, it may also become advantageous to design all-dielectric metamaterials. In any case, there remains a number of avenues open for research.

References

[1] H. A. Gebbie, N. W. B. Stone, and F. D. Findlay. A stimulated emission source at 0.34 millimetre wave-length. *Nature* **202**(4933) (1964), 685.

[2] H. A. Gebbie, N. W. B. Stone, F. D. Findlay, and E. C. Pyatt. Absorption and refractive index measurements at a wave-length of 0.34 mm. *Nature* **205**(4969) (1965), 377–378.

[3] J. E. Chamberlain, F. D. Findlay, and H. A. Gebbie. Refractive index of air at 0.337-mm wave-length. *Nature* **206**(4987) (1965), 886–887.

[4] J. Faist, F. Capasso, D. L. Sivco, C. Sirtori, A. L. Hutchinson, and A. Y. Cho. Quantum cascade laser. *Science* **264**(4) (1994), 553–556.

[5] R. Koehler, A. Tredicucci, F. Beltram, H. E. Beere, E. H. Linfield, A. G. Davies, D. A. Ritchie, R. C. Iotti, and F. Rossi. Terahertz semiconductor-heterostructure laser. *Nature* **417** (2002), 156–159.

[6] M. Rochat, L. Ajili, H. Willenberg, J. Faist, H. Beere, G. Davies, E. Linfield, and D. Ritchie. Low-threshold terahertz quantum-cascade lasers. *Applied Physics Letters* **81**(8) (2002), 1381–1383.

[7] D. H. Auston, K. P. Cheung, J. A. Valdmanis, and D. A. Kleinman. Cherenkov radiation from femtosecond optical pulses in electro-optic media. *Physical Review Letters* **53**(16) (1984), 1555–1558.

[8] C. Fattinger and D. Grischkowsky. Point source terahertz optics. *Applied Physics Letters* **53**(16) (1988), 1480–1482.

[9] Q. Wu and X.-C. Zhang. Free-space electro-optic sampling of terahertz beams. *Applied Physics Letters* **67**(24) (1995), 3523–3525.

[10] J. Baker-Jarvis, R. G. Geyer, Jr., J. H. Grosvenor, M. D. Janezic, C. A. Jones, B. Riddle, C. M. Weil, and J. Krupka. Dielectric characterization of low-loss materials, a comparison of techniques. *IEEE Transactions on Dielectrics and Electrical Insulation* **5**(4) (1998), 571–577.

[11] J. Baker-Jarvis, M. D. Janezic, B. Riddle, C. L. Holloway, N. G. Paulter, and J. E. Blendell. Dielectric and conductor-loss characterization and measurements of electronic packaging materials. Technical Note 1520, National Institute of Standards and Technology, Boulder, CO, July 2001.

[12] J. Baker-Jarvis, M. D. Janezic, B. F. Riddle, R. T. Johnk, P. Kabos, C. L. Holloway, R. G. Geyer, and C. A. Grosvenor. Measuring the permittivity and permeability of lossy materials: solids, liquids, metals, building materials and negative-index materials. Technical Note 1536, Electromagnetics Division, National Institute of Standards and Technology, Boulder, CO, February 2005.

[13] M. N. Afsar, J. R. Birch, R. N. Clarke, and G. W. Chantry. The measurement of the properties of materials. *Proceedings of the IEEE* **74**(1) (1986), 183–199.

[14] K. G. Ong, J. Wang, R. S. Singh, L. G. Bachas, and C. A. Grimes. Monitoring of bacteria growth using a wireless, remote query resonant-circuit sensor: application to environmental sensing. *Biosensors and Bioelectronics* **16**(4–5) (2001), 305–312.

[15] B. W. Hakki and P. D. Coleman. A dielectric resonator method of measuring inductive capacities in the millimeter range. *IEEE Transactions on Microwave Theory and Techniques* **8**(4) (1960), 402–410.

[16] W. E. Courtney. Analysis and evaluation of a method of measuring the complex permittivity and permeability microwave insulators. *IEEE Transactions on Microwave Theory and Techniques* **18**(8) (1970), 476–485.

[17] K. Z. Rajab, K.-F. Fuh, R. Mittra, and M. Lanagan. Dielectric property measurement using a resonant nonradiative dielectric waveguide structure. *IEEE Microwave and Wireless Components Letters* **15**(2) (2005), 104–106.

[18] R. N. Clarke and C. B. Rosenberg. Fabry–Perot and open resonators at microwave and millimetre wave frequencies, 2–300 GHz. *Journal of Physics E: Scientific Instruments* **15**(1) (1982), 9–24.

[19] M. N. Afsar. Dielectric measurements of millimeter-wave materials. *IEEE Transactions on Microwave Theory and Techniques* **32**(12) (1984), 1598–1609.

[20] G. Kent. Nondestructive permittivity measurement of substrates. *IEEE Transactions on Instrumentation and Measurement* **45**(1) (1996), 102–106.

[21] M. D. Janezic and J. Baker-Jarvis. Full-wave analysis of a split-cylinder resonator for nondestructive permittivity measurements. *IEEE Transactions on Microwave Theory and Techniques* **47**(10) (1999), 2014–2020.

[22] A. Parkash, J. K. Vaid, and A. Mansingh. Measurement of dielectric parameters at microwave frequencies by cavity-perturbation technique. *IEEE Transactions on Microwave Theory and Techniques* **27**(9) (1979), 791–795.

[23] I. Wolff and N. Knoppik. Microstrip ring resonator and dispersion measurement on microstrip lines. *Electronics Letters* **7**(26) (1971), 779–781.

[24] K. Chang. *Microwave ring circuits and antennas*. Wiley Series in Microwave and Optical Engineering (John Wiley & Sons, Ltd/Inc., New York, 1996).

[25] L. L. Li, N. H. Ismail, L. S. Taylor, and C. C. Davis. Flanged coaxial microwave probes for measuring thin moisture layers. *IEEE Transactions on Biomedical Engineering* **39**(1) (1992), 49–57.

[26] J. Baker-Jarvis, M. D. Janezic, P. D. Domich, and R. G. Geyer. Analysis of an open-ended coaxial probe with lift-off for nondestructive testing. *IEEE Transactions on Instrumentation and Measurement* **43**(5) (1994), 711–718.

[27] S. Roberts and A. Von Hippel. A new method for measuring dielectric constant and loss in the range of centimeter waves. *Journal of Applied Physics* **17**(7) (1946), 610–616.

[28] A. M. Nicolson and G. F. Ross. Measurement of the intrinsic properties of materials by time-domain techniques. *IEEE Transactions on Instrumentation and Measurement* **19**(4) (1970), 377–382.

[29] W. B. Weir. Automatic measurement of complex dielectric constant and permeability at microwave frequencies. *Proceedings of the IEEE* **62**(1) (1974), 33–36.

[30] M. Olyphant and J. H. Ball. Strip-line methods for dielectric measurements at microwave frequencies. *IEEE Transactions on Electrical Insulation* **EI-5**(1) (1970), 26–32.

[31] D. K. Ghodgaonkar, V. V. Varadan, and V. K. Varadan. Free-space measurement of complex permittivity and complex permeability of magnetic materials at microwave frequencies. *IEEE Transactions on Instrumentation and Measurement* **39**(2) (1990), 387–394.

[32] L. Duvillaret, F. Garet, and J.-L. Coutaz. A reliable method for extraction of material parameters in terahertz time-domain spectroscopy. *IEEE Journal of Selected Topics in Quantum Electronics* **2**(3) (1996), 739–746.

[33] L. Duvillaret, F. Garet, and J.-L. Coutaz. Influence of noise on the characterization of materials by terahertz time-domain spectroscopy. *Journal of the Optical Society of America B* **17**(3) (2000), 452–461.

[34] Z. Jiang, M. Li, and X.-C. Zhang. Dielectric constant measurement of thin films by differential time-domain spectroscopy. *Applied Physics Letters* (2000), **76**(22), 3221–3223.

[35] B. Ferguson and X.-C. Zhang. Materials for terahertz science and technology. *Nature Materials* **1**(1) (2002), 26–33.

[36] P. R. Griffiths and J. A. De Haseth. *Fourier Transform Infrared Spectrometry*, 2nd edn (Wiley-Interscience, 2007).

[37] J. Baker-Jarvis, E. J. Vanzura, and W. A. Kissick. Improved technique for determining complex permittivity with the transmission/reflection method. *IEEE Transactions on Microwave Theory and Techniques* **38**(8) (1990), 1096–1103.

[38] G. Andermann, A. Caron, and D. A. Dows. Kramers–Kronig dispersion analysis of infrared reflectance bands. *Journal of the Optical Society of America* **55**(10) (1965), 1210.

[39] G. Mourou, C. V. Stancampiano, A. Antonetti, and A. Orszag. Picosecond microwave pulses generated with a subpicosecond laser-driven semiconductor switch. *Applied Physics Letters* **39**(4) (1981), 295–296.

[40] Y. C. Shen, P. C. Upadhya, H. E. Beere, E. H. Linfield, A. G. Davies, I. S. Gregory, C. Baker, W. R. Tribe, and M. J. Evans. Generation and detection of ultrabroadband terahertz radiation using photoconductive emitters and receivers. *Applied Physics Letters* **85**(2) (2004), 164–166.

[41] M. Bass, P. A. Franken, J. F. Ward, and G. Weinreich. Optical rectification. *Physical Review Letters* **9**(11) (1962), 446–448.

[42] Y. Cai, I. Brener, J. Lopata, J. Wynn, L. Pfeiffer, J. B. Stark, Q. Wu, X. C. Zhang, and J. F. Federici. Coherent terahertz radiation detection: Direct comparison between free-space electro-optic sampling and antenna detection. *Applied Physics Letters* **73**(4) (1998), 444–446.

[43] L. Duvillaret, F. Garet, and J.-L. Coutaz. Highly precise determination of optical constants and sample thickness in terahertz time-domain spectroscopy. *Applied Optics* **38**(2) (1999), 409–415.

[44] C. L. Pan, C. F. Hsieh, R. P. Pan, M. Tanaka, F. Miyamaru, M. Tani, and M. Hangyo. Control of enhanced THz transmission through metallic hole arrays using nematic liquid crystal. *Optics Express* **13**(11) (2005), 3921–3930.

[45] R. L. Haupt and J. R. Flemish. Reconfigurable and adaptive antennas using materials with variable conductivity. *Adaptive Hardware Systems (AHS) 2007, Second NASA/ESA Conference*, 5–8 August 2007, pp. 20–26.

[46] S. Kamba, J. Petzelt, E. Buixaderas, D. Haubrich, P. Vanek, P. Kuzel, I. N. Jawahar, M. T. Sebastian, and P. Mohanan. High frequency dielectric properties of $A_5B_4O_{15}$ microwave ceramics. *Journal of Applied Physics* **89**(7) (2001), 3900–3906.

[47] K. Z. Rajab, M. Naftaly, E. H. Linfield, J. C. Nino, D. Arenas, D. Tanner, R. Mittra, and M. Lanagan. Broadband dielectric characterization of aluminum oxide (Al_2O_3). *Journal of Microelectronics and Electronic Packaging* **5**(1) (2008), 2–7.

[48] R. A. Kaindl, M. A. Carnahan, J. Orenstein, D. S. Chemla, H. M. Christen, H. Y. Zhai, M. Paranthaman, and D. H. Lowndes. Far-infrared optical conductivity gap in superconducting MgB_2 films. *Physical Review Letters* **88**(2) (2002), 027003.

[49] T. J. Yen, W. J. Padilla, N. Fang, D. C. Vier, D. R. Smith, J. B. Pendry, D. N. Basov, and X. Zhang. Terahertz magnetic response from artificial materials. *Science* **303**(5663) (2004), 1494–1496.

[50] S. Linden, C. Enkrich, M. Wegener, J. Zhou, T. Koschny, and C. M. Soukoulis. Magnetic response of metamaterials at 100 terahertz. *Science* **306**(5700) (2004), 1351–1353.

[51] H. T. Chen, W. J. Padilla, J. M. O. Zide, A. C. Gossard, A. J. Taylor, and R. D. Averitt. Active terahertz metamaterial devices. *Nature* **444**(7119) (2006), 597–600.

[52] T. Kondo, M. Hangyo, S. Yamaguchi, S. Yano, Y. Segawa, and K. Ohtaka. Transmission characteristics of a two-dimensional photonic crystal array of dielectric spheres using subterahertz time domain spectroscopy. *Physical Review B* **66**(3) (2002), 033111.

[53] E. Semouchkina, Y. Miyamoto, S. Kirihara, G. Semouchkin, and M. Lanagan. Analysis of electromagnetic response of 3-D dielectric fractals of menger sponge type. *IEEE Transactions on Microwave Theory and Techniques* **55**(6 Part 2) (2007), 1305–1313.

[54] P. H. Siegel. Terahertz technology in biology and medicine. *IEEE Transactions on Microwave Theory and Techniques* **52**(10) (2004), 2438–2447.

[55] M. Nagel, P. Haring Bolivar, M. Brucherseifer, H. Kurz, A. Bosserhoff, and R. Büttner. Integrated THz technology for label-free genetic diagnostics. *Applied Physics Letters* **80**(1) (2002), 154–156.

4

Millimeter-wave Interconnects

Janusz Grzyb

Glossary

Scalars

Q	Quality factor
R_q	Surface roughness (RMS)
Z_0	Characteristic impedance
α	Loss attenuation constant
α_c	Conductor loss
β	Propagation constant
δ	Skin depth
ϵ_0	Free space permittivity ($= 8.85419 \cdot 10^{-12}$ F/m)
ϵ_r	Relative dielectric permittivity
$\epsilon_{r_{eff}}$	Effective relative dielectric permittivity
γ	Complex propagation constant
λ	Wavelength
λ_0	free space wavelength
k_0	free space wavenumber
μ_0	Free space permeability ($= 1.25663 \cdot 10^{-6}$ H/m)
μ_r	Relative permeability
σ	Metal conductivity
ω	Angular frequency

Advanced Millimeter-wave Technologies: Antennas, Packaging and Circuits
Duixian Liu, Ulrich Pfeiffer, Janusz Grzyb and Brian Gaucher © 2009 John Wiley & Sons, Ltd

c_0	Light speed ($= 2.99792 \cdot 10^8$ m/s)
$\tan \delta$	Dielectric loss tangent

Abbreviations and Acronyms

3D	3 Dimensional
AFM	Atomic Force Microscope
Au	Gold
BCB	Benzocyclobutene
BGA	Ball Grid Array
CBCPW	Conductor-Backed Coplanar Waveguide
CPW	Coplanar Waveguide
CTE	Coefficient of Thermal Expansion
Cr	Chromium
Cu	Copper
DC	Direct Current
DOE	Design of Experiments
DOP	Degree of Planarisation
DRIE	Deep Reactive Ion Etching
EDP	Ethylene Diamine Pyrocatechol
FDTD	Finite Difference Time Domain
FEM	Finite Element Method
FGCPW	Finite-Ground Coplanar Waveguide
FW-CBCPW	Finite-Width Conductor-Backed Coplanar Waveguide
GCPW	Grounded Coplanar Waveguide
GaAs	Gallium Arsenide
HDD	Hard Disk Drive
HFSS	High Frequency Structure Simulator
IC	Integrated Circuit
LAP	Large Area Processing
LCD	Liquid Crystal Display
LCP	Liquid Crystal Polymer
LNA	Low Noise Amplifier
LTCC	Low Temperature Cofired Ceramics
LTCC-M	Low Temperature Cofired Ceramics on Metal
MCM	Multichip Module
MCM-L	Multichip Module with laminated interconnect
MCM-D	Multichip Module with deposited interconnect
MCM-D/L	Multichip Module with deposited/laminated interconnect

MHDI	Microwave High Density Interconnect
MMIC	Monolithic Microwave Integrated Circuit
Ni	Nickel
PCB/PWB	Printed Circuit Board/Printed Wiring Board
ppm	Parts per million (= 10^{-6})
PTFE	Polytetrafluoroethylene
PVD	Physical Vapour Deposition
RIE	Reactive Ion Etching
RMS	Root-Mean-Square
SEM	Scanning Electron Microscope
SMM	Shielded Membrane Microstrip
SWR	Standing-Wave Ratio
TEM	Transverse Electromagnetic
Ti	Titanium
TOS	Tape on Substrate
TRL	Thru–Reflect–Line
TTL	Transverse Transmission Line
W	Tungsten

4.1 Introduction

This chapter spans different topics related to the millimeter-wave (mmWave) interconnects and interconnect technologies.

To date, there has been no extensive research above 50 GHz undertaken worldwide in the interconnect technology resulting in a comprehensive feasibility study of various system integration concepts for the V and W frequency bands. The reason is that the short wavelengths and increased reflections at discontinuities require a technology with very tight manufacturing tolerances. An overview of a variety of today's available state-of-the-art interconnect technologies that can be considered as candidates for the realization of mmWave systems will be given. Their benefits and limitations will be thoroughly discussed, specifically addressing their potential to realize different transmission media in the V-, W-band and above. Major technology and process requirements influencing the performance of interconnects and integrated passive structures will be presented.

Despite many attractive properties of the printed planar transmission lines, there are numerous potential risks in their use at mmWave frequencies. The main spurious effects include mode coupling or mode conversion, mainly at the discontinuities, which result in the leakage of power into the surface wave, parallel-plate, or other higher order modes. All those phenomena can result in unexpected crosstalk, significant alteration of the guided wavelength, unwanted resonance effects, and excessive loss. This chapter will give a brief explanation of why and when these spurious effects appear and some methods to minimize their parasitic influence.

At lower microwave frequencies, an insertion loss of a straight line section is one of the most important factors determining the performance of interconnect elements, passives, and entire integrated systems. It is, however, no longer that obvious for mmWave frequencies if the contemporary hybrid interconnect technologies with their miniaturization and processing accuracy constraints are considered. When evaluating the performance of such a hybrid mmWave system, a substantially larger number of factors has to be taken into account. Due to the increasingly short wavelengths, the power dissipated along straight transmission lines can be of lesser importance in comparison to the influence of transmission line discontinuities that become electrically large, partly because of various technology limitations. In this regard, some technology build-up development or optimization process in view of the mmWave applications seems to be important. One of its main goals is to achieve a trade-off between two opposite groups of factors. The first includes line loss and antenna performance (if considered to be included), the latter being represented by radiation efficiency and operation bandwidth. The other comprises line dispersion, line width-to-wavelength ratio, intercircuit isolation, grounding path lengths, and major parasitic effects at the transmission line discontinuities, such as reactances, radiation, and excitation of the surface waves or other higher order modes. This trade-off needs very careful judgment considering the performance of different interconnect structures, passives, antennas, and feed networks intended to be integrated. Its major determinants will be analyzed in detail as they should define the proper choice of transmission line topology and the corresponding technology build-up allowing to avoid the complex physical effects deteriorating the performance of a mmWave system.

An overview of the most popular interconnect techniques between the chip and the planar transmission lines will be given, including the matching structures for compensation of the involved discontinuities and other higher order package-related issues.

A discontinuity introduced by the bonding wire can significantly affect the performance of the entire circuit at mmWave frequencies. Nonetheless, wirebonding, well established in consumer electronics, remains a very attractive solution since it is robust and inexpensive. In addition, it has the advantage of being tolerant to the chip thermal expansion, an important requirement for many applications. Low-cost, high-density, and short-transition interconnects supported by the flip-chip technique are considered to be the main advantages of this bonding scheme for millimeter applications. As the wavelength gets smaller, the broadband surface-to-surface electromagnetically coupled transitions without bonding wires have also been arousing an increasing interest in order to avoid the lowpass-type characteristics of galvanic interconnects [1–4].

Although the printed planar lines are of major interest for low-cost mmWave systems, the hollow metal waveguides are required in several cases for high-gain antennas, high-Q oscillators, filters, or device characterization [5, 6]. This chapter will attempt to give a very brief overview of various mmWave integrated waveguide structures, reported in the literature. The well-known planar circuit-to-waveguide transitions, typically involving the use of microstrip and coplanar waveguide (CPW) feeds, will be omitted for space reasons.

4.2 Interconnects at Millimeter-wave Frequencies

For the realization of mmWave modules or systems, an interconnect technology based on printed planar transmission lines or integrated metal waveguides can be considered. Each of the mentioned transmission media types shows a specific set of advantages and constraints, which will be briefly outlined in this chapter.

As the microstrip and CPW lines are the most representative planar lines, their characteristics at mmWave frequencies will be analyzed and compared. The constraints of their usage in hybrid interconnect technologies should be carefully outlined, based on comprehensive understanding of the problems encountered at mmWave frequencies, and analyzed in view of the implications for the design and performance of the structures to be integrated.

At mmWave frequencies, the metal rectangular waveguides can be considered an alternative high-performance transmission medium. Their two main advantages, when compared to the planar transmission lines, are substantially lower current density and fully enclosed nonleaky propagating fields. In a result, they may be the preferred choice for realizing high-Q cavity resonators, low loss coupled-cavity, E- and H-plane filters, or compact antennas [7].

4.2.1 Printed Planar Transmission Lines

The printed planar transmission lines are commonly used as they provide compact and low-weight circuits. However, they are normally inhomogeneous dielectric build-ups. Hence, they do not guide a pure transverse electric magnetic (TEM) mode but lossy dispersive hybrid modes [8, 9]. As a result, some simple relations among transmission line parameters (γ and Z_0), substrate relative dielectric permittivities, and line geometry do not exist.

High-frequency Effects. The dispersive and lossy behavior of the printed lines is related to the influence of mode coupling and energy leakage effects. The general criteria for coupling between modes are that the electromagnetic fields of the two modes should not be orthogonal and that the parasitic mode should have a phase velocity equal to, or lower than the propagating mode [10]. At mmWave frequencies, the substrate modes can show lower velocity than the fundamental modes of operation (e.g. CPW or microstrip mode). The frequency at which both velocities become equal is called the *critical frequency* [10]. Above this frequency, the dominant mode can become leaky. In order to preserve the quasi-TEM behavior of printed lines, the line cross-section should be, in general, small enough compared to the operating wavelength.

Furthermore, the classical concept of characteristic impedance cannot be rigorously employed to the printed lines [8, 9]. As shown by various numerical calculations, different combinations of longitudinal current, voltage, and average propagated power usually lead to different characteristic impedance values at high frequencies at which the hybrid mode phenomena become pronounced [8, 9]. The printed line characteristic impedance should be viewed as a unique property [9] only in the quasi-static sense. The results of different extensive studies, among others performed in references [8, 9, 11], show that the net propagated power, as calculated from the transverse electromagnetic field of the fundamental mode, should be included in any line network quantity formulation. Furthermore, for the strip-based transmission lines, the power–current definition was found to be a better TEM-equivalent than the power–voltage relation [9]. According to the numerical analysis of both possible power-based characteristic impedance formulations performed for microstrip lines, the longitudinal strip current was calculated to show the lower high-frequency dispersion than the corresponding strip center voltage [8, 9] due to the weaker influence of the existing air–dielectric interface on the magnetic fields compared to the electric fields. A survey of numerous measurement results for a variety of microstrip lines, performed by different authors, indicates a good correlation between the measured values and those determined by the power–current relation [9]. Moreover, some other useful arguments supporting the power–current definition for this type of planar transmission line have also been presented. Knorr and Tufekcioglu [12] show that numerical values of the power–current relationship are insensitive to the assumed surface current distribution, whereas those of the power–voltage definition are not [9]. Additionally, the path integral defining a voltage in the power–voltage model for a dispersive frequency range depends on the path between a given pair of endpoints [11]. The loop integral defining a strip current does not show this ambivalence. On the other side, the power–voltage formulation was found to be useful for the slot-based transmission lines, wherein the slot voltage can be successfully applied for the characteristic impedance calculations [9].

Low-frequency Dispersion. The above-mentioned high-frequency dispersion is not the only phenomenon making the propagation characteristics of planar transmission lines a complex function of frequency. Let us analyze the three basic equations routinely used to describe the transmission line parameters in terms of the equivalent distributed circuits model elements

$$\gamma = \alpha + \jmath\beta = \sqrt{(R_{\text{dyn}} + \jmath\omega L_{\text{dyn}})(G_{\text{dyn}} + \jmath\omega C_{\text{dyn}})} \tag{4.1}$$

$$\epsilon_{\text{reff}} = (\beta c_0/\omega)^2 \tag{4.2}$$

$$Z_0 = \sqrt{(R_{\text{dyn}} + \jmath\omega L_{\text{dyn}})/(G_{\text{dyn}} + \jmath\omega C_{\text{dyn}})} \tag{4.3}$$

where β are α are the phase and attenuation constants, respectively, whereas Z_0 is the line characteristic impedance. Using the correspondence between the transmission line parameters, γ and Z_0, and the C_{dyn}, G_{dyn}, L_{dyn}, and R_{dyn} elements of the equivalent distributed circuit model from above, it can be noticed that both the line effective dielectric permittivity, ϵ_{reff}, and the line characteristic impedance are nonlinear, in general, frequency-dependent functions of all the model elements (index 'dyn' stands for frequency dependence). Specifically, they are influenced by the line-loss parameters, represented by G_{dyn} and R_{dyn}, and by the slow-wave effect associated with the line internal inductance L_{int} ($L_{\text{dyn}} = L_{\text{int}} + L_{\text{ext}}$). The latter is caused by noticeable magnetic field penetration inside the conductors and can lead to the considerable low-frequency dispersion if the internal line inductance is comparable to the external line inductance. This phenomenon considerably influences the frequency behavior of thin microstrip lines, wherein the contribution of L_{int} in the total line inductance can be very high [13]. In this case, the measured line effective permittivity and line impedance can be noticeably different from the corresponding values for the lossless lines.

4.2.1.1 Practical Considerations on the Frequency Behavior of Transmission Line Parameters

Line inductance and line resistance. The line inductance, L_{dyn}, and resistance, R_{dyn}, are defined by magneto-quasistatics [13]. Therefore, they are dependent on the variable longitudinal current distribution in nonideal conductors of finite conductivity. This, in turn, effectively results in the frequency dependent characteristics that can be basically subdivided into three different ranges [13, 14]:

- DC range with the homogeneous current density within a metal strip;
- skin effect range with the current flowing only in a shallow area of the conductor surface;
- transition range between the two mentioned above. For the metallization thickness comparable with the skin depth, a simple approximation of the physical behavior in this range is very difficult.

Some attempts to model the frequency dependent behavior of L_{dyn} and R_{dyn} were reported in the literature, specifically for the thin microstrip lines [13], wherein some purely numerical nonlinear functions fitted to the full-wave simulation data were used in the transition frequency range. The formulas assumed the same metallization thickness for both the strip and the ground plane, ideal rectangular strip profiles, and homogeneous conductor materials. Unfortunately, these assumptions are often violated for the hybrid interconnect technologies. In this case, the following additional issues, potentially modifying the behavior of R_{dyn} and L_{dyn}, should be addressed.

1. **Strip profile:** A strip profile can often deviate from the ideal rectangular shape as a result of the nonideal etching process, specifically for a thick metal strip. At mmWave frequencies, the acute angles, being a result of the etching profile, can lead to a significant increase of current density in their vicinity, substantially raising the line attenuation constant and modifying the behavior of R_{dyn} and L_{dyn}.

2. **Surface roughness:** The surface roughness of both metallization and dielectric layers increase the equivalent surface resistance associated with a metal conductor and make it a more complex function of frequency. For the numerous technologies, specifically the laminate-based, the surface roughness can be considerably higher than the corresponding skin depth at mmWave frequencies in typical high-conductivity materials.

3. **Composite metal layer build-up:** A composite metal build-up is usually used. Typically, it consists of the main conductor layer sandwiched between the adhesion or barrier materials, such as Ti, TiW, Cr, or Ni, making the current density distribution within conductors a complex function of frequency. Special care is needed when an intermediate layer of nickel is present in some build-ups because of its magnetic properties screening the field penetration. Analytical derivation of the line inductance can be difficult even for the DC range because of the inhomogeneous current distribution within a composite conductor layer. Furthermore, the measured line DC resistance of a composite metal build-up is an equivalent value taking all conductor conductivities into account. As a result, it cannot be easily used as an input parameter for the approximation of R_{dyn} and L_{dyn} in the skin effect and intermediate frequency ranges.

4. **Mixed metallization thickness:** Different thicknesses or conductor materials applied for both ground and signal metallizations cause some additional differences in the frequency dependence of the field penetration in each of the layers. It results in a modified behavior of R_{dyn} and L_{dyn} when compared to the typically considered metal layers with the consistent thickness and conductivity.

Line capacitance. From quasi-TEM theory, the line capacitance characteristics are covered by electrostatics, thus, being constant with respect to frequency [13]. However, with the frequency increase, some nonTEM contributions become important, leading to some possibly pronounced high-frequency dispersion, as discussed. A broad spectrum of the closed-form analytical equations treating both the quasi-TEM operation regime and the high-frequency dispersion of different transmission lines, in particular microstrip lines, can be found in the literature. They are typically generated by employing multidimensional curve-fitting and are usually valid for only a single-layer substrate. Unfortunately, the multilayer build-ups filled with several layers of different dielectric materials can be commonly found in the modern hybrid interconnect technologies. The two most representative are:

1. a superstrate dielectric configuration created by a passivation layer on top of a printed line;

2. a two-layer dielectric substrate stack-up consisting of the primary dielectric layer and the corresponding adhesion film, being a result of some processing requirements usually for the laminate technologies.

The dispersion characteristics of these multilayer build-ups can be modified to such an extent that it can hardly be incorporated into the existing dispersion models for a single layer substrate. Some empirical attempts to address this issue for the microstrip lines were reported in reference [15], where the concept of a virtual relative permittivity was introduced (see one of the successive paragraphs entitled 'Closed-form analytical models of lossless microstrip lines').

Line conductance. Different loss types exist for the printed lines: dielectric, ohmic, and radiation to the space and surface waves. The latter normally takes place from a number of different discontinuities and will be treated in a more detailed way in the following sections. Ohmic loss, represented by R_{dyn} in the equivalent distributed circuit model given before, can be a very complex function of frequency defined by the multiple process-related parameters, as shown before. A general relation between a dielectric loss, defined by G_{dyn}, and the contributing loss tangents of the corresponding composite dielectrics in a general multilayer substrate build-up takes the form [16, 17]

$$G_{\mathrm{dyn}} = \omega \cdot \left(\epsilon_{\mathrm{r}1} \cdot \frac{\partial C_{\mathrm{dyn}}}{\partial \epsilon_{\mathrm{r}1}} \cdot \tan \delta_1 + \epsilon_{\mathrm{r}2} \cdot \frac{\partial C_{\mathrm{dyn}}}{\partial \epsilon_{\mathrm{r}2}} \cdot \tan \delta_2 + \cdots \right) \tag{4.4}$$

where ϵ_{ri} and $\tan \delta_i$ represent dielectric permittivities and dielectric loss tangents of the constituent linear dielectric layers, respectively. G_{dyn}, in general, also shows a complex frequency dependence, being linearly related to the line partial capacitances filled with the respective dielectric materials.

Summarizing, the transmission characteristics of printed planar lines, specifically in hybrid interconnect technologies, can be very complex frequency-dependent functions, making their analytical formulation very challenging, if not impossible. Their accurate calculation normally demands full-wave simulations with the fields solved inside the metal layers. Even if some problem simplifications could be made in order to arrive at the approximate formulas, the associated uncertainty level may sometimes be high because of difficulties in the prediction of the approximation errors.

4.2.1.2 Microstrip Lines

Microstrip lines are one of the most widely used planar lines. Their transmission properties greatly depend on the substrate parameters, namely its thickness and dielectric permittivity. In this regard, most of them fall into two groups featuring some very distinctive properties. The first are classical microstrip lines normally patterned on an electrically thick substrate carrier and, thus, are dispersive at mmWave frequencies. The other comprises thin microstrip lines operating within the quasi-TEM frequency range.

The microstrip circuits realized using the well-known, traditional, single-layer manufacturing technologies, based on a ceramic or a plastic dielectric substrate with the thickness normally exceeding 0.1 mm, fall into the first group at higher mmWave frequencies. They may experience strong coupling effects to the substrate modes, resulting in highly dispersive and lossy nonquasi-TEM transmission characteristics with poor intercircuit isolation. Another important common drawback associated with these lines is a high line width-to-wavelength ratio, commonly yielding some excessive parasitic reactances, radiation, and high phase shifts at the meander or junction-like transmission line discontinuities. Furthermore, the grounding paths show high short-circuit inductance values due to the usually long grounding vias. For these reasons, microstrip lines have not been the preferred transmission line configuration at mmWave frequencies.

On the other hand, the microstrip interconnect structures found in numerous integrated circuits, typically CMOS or BiCMOS, fall into the other mentioned group. The transmission properties of these lines are primarily dominated by the excessive conductor loss and strong low-frequency dispersion.

With the recent advances in hybrid interconnect technologies, the multilayer substrates became common. They allow the simultaneous use of numerous dielectric materials with different layer thickness varying from a few micrometers to a couple of millimeters, thus, providing unheard-of design flexibility. Moreover, they are capable of delivering a technology setup being optimized in view of the performance of the interconnect structures intended to be realized. For the microstrip lines, it means that the build-ups falling into none of the two mentioned groups are feasible where the disadvantages of both are minimized. This opens the way for the microstrip-based designs to outperform many of the existing CPW-based mmWave solutions. An example of such a microstrip-oriented multilayer technology build-up optimization will be given in Section 4.4.

Classical microstrip lines. A low-frequency dispersion associated with the slow-wave effects caused by the internal strip inductance (see introductory part of this section) can be neglected for classical microstrip lines at mmWave frequencies, in contrast to the high-frequency effects.

Operating frequency limitations. The main high-frequency parasitic phenomena limiting the operating frequency of classical microstrip lines are [18]:

- the lowest order transverse magnetic (TM) surface wave mode, TM_0, that can strongly couple to the quasi-TEM microstrip mode. The *critical frequency* (see the introduction of this section for its definition) for coupling between the TM_0 mode and the quasi-TEM microstrip mode can be calculated from reference [18]

$$f_c = \frac{c \arctan \epsilon_r}{\sqrt{2}\pi h \sqrt{\epsilon_r - 1}} \tag{4.5}$$

which poses the first restriction on a usable substrate thickness, h, without the excessively pronounced overmoding effects.

- the lowest order transverse electric (TE) surface wave mode, TE_1, with a cut-off frequency given by the following relation

$$f = \frac{c}{4h\sqrt{\epsilon_r - 1}} \tag{4.6}$$

h and ϵ_r being the substrate thickness and the substrate relative permittivity, correspondingly.

- the lowest order transverse microstrip resonance mode existing for a sufficiently wide strip. A cut-off frequency for this mode, at which the equivalent strip width (accounting for the side fringing capacitance) can be viewed as a resonant half-wavelength transmission line, is given by reference [18]

$$f_{CT} = \frac{c}{\sqrt{\epsilon_r}(2\,w + 0.8\,h)} \tag{4.7}$$

where h and w are the substrate thickness and the strip width, respectively.

Loss mechanisms. In addition the well-known line dissipative losses, associated with imperfect finite conductivity metals and lossy dielectric substrates, two other parasitic phenomena causing some energy leakage have to be considered for the classical microstrip lines, in particular for the transmission line discontinuities. These are space wave radiation and surface wave propagation. The amount of energy carried out by the surface waves in relation to the radiation loss depends on multiple factors. The major are discontinuity type, operating frequency, dielectric permittivity of the substrate, and its thickness. In general, the surface wave loss is less pronounced at lower frequencies and starts to dominate at very high frequencies. James and Henderson [19] specified that for the substrate thickness, h, satisfying

$$\frac{h}{\lambda_0} \leq q \frac{0.3}{2\pi\sqrt{\epsilon_r}} \tag{4.8}$$

the surface wave loss from an open-end element can be neglected.

An approximate formula for the power radiated from different types of discontinuities was given in reference [20]

$$P_r = 60(k_0 h)^2 F(\epsilon_{reff}) \tag{4.9}$$

where k_0 is the free space wavenumber, and $F(\epsilon_{reff})$ depends on the discontinuity. The open-ends and the right-angle bends for the low-impedance lines on a thick low dielectric constant substrates were identified as the largest contributors to radiation [21]. Commonly used hermetic shielding of the circuits is not always a simple remedy to keep the energy because the high radiation may possibly lead to some considerable intercircuit coupling in the form of numerous package-guided modes [22].

Closed-form analytical models of lossless microstrip lines. There are numerous papers in the literature attempting to describe the behavior of various discontinuities for the classical microstrip lines in terms of the quasi-static models. Unfortunately, it is rather improbable that they are capable of providing any accurate data at mmWave frequencies. Hence, only a short survey of the closed-form analytical equations modeling both the quasi-static and the high-frequency characteristics of a simple lossless microstrip line will be given here, including some multilayer build-ups commonly found in modern hybrid interconnect technologies.

In general, the static line effective permittivity for quasi-TEM lines can be computed using the Schwarz–Christoffel conformal mapping [13]. However, this approach is not that straightforward to formulate for microstrip lines because the dielectric substrate-to-air interface does not coincide with the electric field lines [13]. The variational method treats this boundary in a more general way, and even a multilayer microstrip line can be analyzed using this technique [23–25]. However, it needs a numerical evaluation of some integral and, thus, does not provide an explicit closed-form analytical relation between the line effective permittivity and the line geometry [23, 25]. Furthermore, the characteristic impedance calculated by the variational method shows some deviation from the exact value and should normally be corrected, as shown in reference [23].

Several other approximate solutions for the calculation of the microstrip line effective permittivity are available in the literature [26, 27]. Mostly, the functional approximation formulas of Hammerstad and Jensen [28] are used because they are accurate for a wide range of dielectric constant values and cross-sectional dimensions. A finite metallization thickness of the signal conductor can be taken into account by applying the concept of an equivalent increased conductor width, w_{eq} (see reference [26]). The line characteristic impedance for $\epsilon_{\mathrm{r}} > 1$ is obtained from the effective permittivity concept and the line impedance of the corresponding free space microstrip line. In a result, any explicit closed-form equations for Z_0 are not needed.

In order to expand the validity range of the models to high frequencies, some dispersion formulas should be added. Regarding dispersion of the effective dielectric constant, the following two formulas by Kirschning and Jansen [27,29], and Kobayashi [30] are frequently used. Kirschning and Jansen [27, 29] estimated the accuracy of their dispersion model as better than 0.6% in the following range of values: $0.1 < w/h \leq 100$, $1 < \epsilon_{\mathrm{r}} \leq 20$, and $0 \leq h/\lambda_0 \leq 0.13$; whereas the accuracy of the formula by Kobayashi was specified as better than 0.6% for $0.1 < w/h \leq 10$, $1 < \epsilon_{\mathrm{r}} \leq 128$, and any h/λ_0 (see reference [30]). Both models were originally developed for the microstrip lines with a zero metal strip thickness. The influence of a finite strip conductor thickness can be accounted for by taking advantage of the equivalent increased conductor width, w_{eq}, concept mentioned above [26]. The overall accuracy of both dispersion models depends on the accuracy of the applied quasi-static closed-form equations.

For the reasons given earlier, the high-frequency dispersion for only the power–current definition of microstrip characteristic impedance will be considered here. The most accurate explicit representation of this dispersion was given by Jansen and Kirschning [9]. The method employed for its generation is a multidimensional curve-fitting. The authors of reference [9] estimate the accuracy of their model as better than 1% in the range of $0.1 < w/h \leq 10$, $1 < \epsilon_{\mathrm{r}} \leq 18$, and $0 \leq h/\lambda_0 \leq 0.1$ when compared to a rigorous fully converged numerical hybrid mode solution. The achieved accuracy is partly related to the precision of both static and dynamic values of the line effective permittivity that are the two main constituent model parameters.

A quasi-static effective permittivity of a general lossless multilayer microstrip line with a zero metal thickness can be investigated by two analytical methods. The first, by Svacina [31, 32], and its improved version by Wan and Hoorfar [33], is based on the approximate analysis using the conformal mapping technique. The second, proposed by Verma and Kumar [15] and Verma and Sadr [25], is based on a combination of the transverse transmission line technique (TTL) [24] and the variational method [23, 34]. For a strip of finite thickness, Yamashita [23] proposed a modified variational formula. Nevertheless, the strip thickness, t, is assumed to be thin ($t \ll h$, $t \lessapprox w$). For the thicker conductors, better results can be often achieved by applying the previously mentioned concept of the equivalent conductor width, w_{eq}, directly to the variational or conformal analysis. However, calculation of this corrected line width, requiring the dielectric permittivity of a single-layer substrate configuration to be known [26], presents some difficulty coming from the inhomogeneous dielectric substrate material. Thus, the conversion of a multilayer build-up into an equivalent single layer, represented by a new static equivalent relative permittivity, $\epsilon_{\mathrm{r,eq}}$, is required [34,35].

Dispersion characteristics of the multilayer microstrip lines are generally more complicated than for the single-layer build-ups. The effective relative permittivity of a multilayer microstrip line does not necessarily approach $\epsilon_{\mathrm{r,eq}}$ with a frequency increase, as shown in reference [15], due to the field lines moving towards the dielectric layer of higher permittivity. For some multilayer microstrip line configurations operating at the appropriately high frequencies [15], it can modify the frequency dependence of dispersion characteristics to such an extent that it cannot be modeled by the formulas known for a single-layer microstrip line. To address this issue, Verma and Kumar [15] introduced the concept of a *virtual relative permittivity*. It is a strip width and frequency dependent input parameter of the modified Kirschning and Jansen dispersion model [27, 29], replacing the former relative dielectric permittivity of a single-layer substrate. For a two-layer composite dielectric build-up, a set of empirical equations modeling the virtual relative permittivity as a combination of the quasi-static $\epsilon_{\mathrm{r,eq}}$ and some frequency dependent part was also given.

Thin microstrip lines. Dimensions of the thin microstrip lines are inherently different from their classical counterparts described in the previous paragraph. Specifically, the substrate heights and line widths are of the order of magnitude of the conductor thickness. This leads to some excessive ohmic loss and slow-wave effects, the latter being a result of the nonnegligible internal line inductance, L_{int}, contribution in the total line inductance (see 'Practical considerations on the frequency behavior of transmission line parameters' in the introductory part of this section for the more detailed analysis). As the skin depth can be comparable to the corresponding strip thickness for a very broad frequency range, the line inductance for the thin microstrip lines can be outside of the DC and skin effect ranges [13]. For this reason, a variety of closed-form analytical models available in the literature for the conventional lossless microstrip lines (see the previous paragraph) yields significant approximation errors of the transmission properties for the thin microstrip lines. The most extensive study of the formulas modeling the frequency dependent behavior of thin microstrip lines was reported in reference [13], wherein some purely numerical nonlinear functions fitted to the full-wave simulation data were used for the line inductance in the transition frequency range. For the investigated thin-film microstrip lines ($w = 8-140$ µm, $h = 1.7-50$ µm, $t = 0.8-3.5$ µm; w, h, and t denoting a strip width, a dielectric layer thickness, and a metallization thickness, respectively) on a benzocyclobutene (BCB) dielectric substrate (see Section 4.3.4 for the BCB dielectric properties), the overall reported modeling accuracy was better than 3%, 2%, and 8% for the line characteristic impedance, line propagation constant, and line attenuation, respectively. The largest deviations were observed for the mentioned intermediate frequency region.

The high-frequency dispersion can normally be neglected for the thin microstrip lines at mmWave frequencies.

4.2.1.3 CPW Lines

A CPW has an inherent advantage over the microstrip line in that the short-circuit based elements do not require grounding vias to be used. As the center strip-to-ground spacing is typically substantially lower than that for the classical microstrip lines, the CPW lines show lower high-frequency dispersion. For very low values of this spacing, the low-frequency dispersion, very pronounced for the thin microstrip lines, can also be observed. The CPWs are, in general, considerably less sensitive to the substrate thickness tolerances when compared with the microstrip lines, unless the substrate is very thin. Furthermore, the specified characteristic impedance value can normally be arrived at for different line lateral geometries.

In spite of all these advantages, the CPWs may show some potential dangers at mmWave frequencies, specifically for the hybrid interconnect technologies with a very limited metal pattern resolution. Those risks will be the main scope of the next paragraphs.

An ideal coplanar waveguide consists of a center conducting strip and two semi-infinite side ground planes on an infinitely thick substrate carrier. It can guide two principal modes known as the CPW mode and the coupled slotline mode, the latter normally being considered to be a parasitic effect [36]. As the propagation constant of the CPW mode is lower than for a TEM wave in the substrate, it leads to some energy leakage into the substrate that is strongly dependent on an overall line width (strip with two corresponding slots) [37, 38]. As the ideal coplanar waveguide is impossible to implement, its different practical realizations will be considered here. They will be divided into two separate cases with different boundary conditions on the substrate backside.

CPW lines without conductor-backing. When moving into the mmWave range, mode coupling and energy leakage effects can become an issue. The general criteria for coupling between modes was given at the beginning of this section, where the so-called *critical frequency* was defined [10]. For the coplanar waveguides on an air-suspended dielectric substrate, the critical frequency for the TE and TM surface wave modes can be approximated by reference [10]

$$f_c = \sqrt{\frac{2}{\epsilon_r - 1}}\frac{c}{\pi h}\left(\arctan A + \frac{n\pi}{2}\right) \tag{4.10}$$

where h is the substrate thickness, and A is equal to ϵ_r or 1 for TM modes and TE modes, respectively. Note that the lowest critical frequency for this CPW arrangement is assigned the lowest order TE mode, known as TE_0. For the substrate covered with a ground conductor on one side, all even values of n for TE modes and all odd values of n for TM modes are

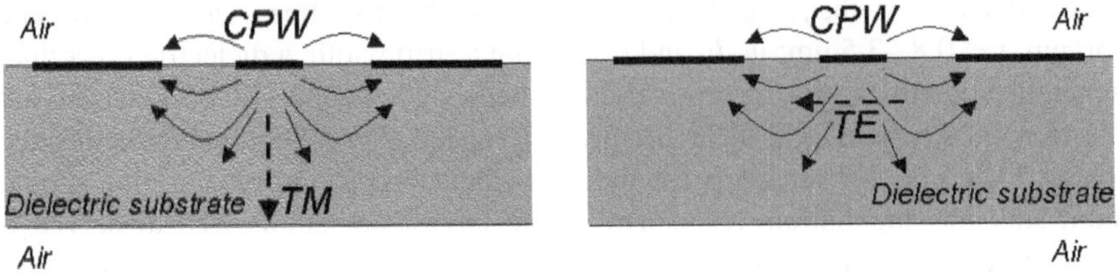

Figure 4.1: Coupling effects (a) between CPW and TM surface wave modes, and (b) between CPW and TE surface wave modes in an air-suspended dielectric substrate. (The solid lines denote the electric field lines of the CPW mode, whereas the dashed lines are associated with the electric field lines of the surface wave modes.)

eliminated. Furthermore, the critical frequencies for all other TE and TM modes are lowered by two because the introduced conducting plane, creating a mirror, effectively doubles the substrate thickness in Equation 4.10. This means that TM_0 is the first mode coupling to the CPW in this case.

A single slot of the coplanar waveguide builds a transverse electric field pattern in the supporting dielectric slab, being, in principle, capable of coupling to the TE modes. However, the complete CPW geometry comprises two complementary slots that cancel each other is electric field, thus minimizing the coupling level. At the same time, a net transverse magnetic field in the substrate away from the line geometry is developed [39], resulting in nonnegligible coupling to the TM modes (see Figure 4.1). Calculation of the field overlap integral for the coplanar waveguide on a dielectric slab predicts greater coupling to the TM_0 mode than to the TE_0 mode [37], correlating very well with the qualitative observations. Therefore, the coplanar waveguide transmission line arrangement ensuring that TE_0 is the mode with the lowest critical frequency is preferred at mmWave frequencies. For this reason, a finite-ground CPW (FGCPW) line with the narrow side ground planes is commonly applied in the mmWave circuits. Accurate prediction of the coupling phenomena for this transmission topology is, however, a more complex task because of different boundary conditions between and beyond the edges of the finite-width ground planes, leading to some interactions between TE_0 and TM_0 modes [40, 41]. To preserve a low leakage rate of the CPW, the cut-off frequencies of the higher order surface wave modes should be shifted above the operating frequency range; This is normally accomplished by substrate thickness reduction. As TE_0 and TM_0 have no cut-off frequency, their parasitic influence should be managed differently. For example, it is imperative to keep the overall line ground–ground separation low in relation to the substrate thickness and to the wavelength in a dielectric substrate for the operating frequency range [37, 39–42].

Conductor-backed CPW lines. Conductor backing of a CPW, common in the wafer-probing environment or in the numerous packaging options, provokes the parallel-plate waveguide modes to propagate [37, 41, 43–45]. The presence of a zero cut-off TEM mode is of special importance because its propagation constant is always higher than for the CPW mode, effectively leading to some power loss and possibly multiple resonance phenomena in the transmission characteristics due to a finite substrate size [42, 46–49]. A large overall line width (center strip with two slots) may strengthen the leakage to this mode, whereas increasing a substrate thickness will reduce it [50]. In order to minimize the field overlap between both modes, the ratio of line width and substrate thickness should be kept below one, as a rule of thumb [51].

Numerous different approaches have been proposed to minimize the leakage rate and the related resonance issues from the conductor-backed coplanar waveguides [52]. The use of vias [51,53], absorbing materials [54], or multiple dielectric layers [55,56] among others was reported. Another practical solution could be the use of the so-called finite-width conductor-backed CPW (FW-CBCPW) with the substantially reduced side ground plane widths. However, a disadvantage of this configuration is that it supports a zero cut-off microstrip-like (MSL) mode, also known as coplanar microstrip (CPM) mode [42,46,47,57] (see Figure 4.2). In general, a straight section of the FW-CBCPW on a single-layer dielectric substrate is not prone to leak power into the MSL mode because the dominant CPW mode always shows a higher phase velocity. At the discontinuities, however, energy can be transferred between the modes, the coupling level being dependent on the details of a particular discontinuity. Moreover, it is important that an overall width of the FW-CBCPW (including the widths of both finite ground planes) is kept below half a wavelength in the dielectric substrate at the highest operating frequency in order to avoid the parasitic resonance-like behavior across the line geometry [58,59]. In reference [42], it was shown that the open-end elements can be very good mode converters, specifically if a very thin substrate is used. In this view, application of the open calibration standards can be detrimental for conductor-backed substrates if some special precautions are not undertaken. Furthermore, this type of conversion can make the mounted CPW-based chips inoperable in some situations. The presence of a backside ground plane has a minimum influence on the propagation characteristics of the CPW if a ratio of the substrate thickness and the line ground–ground spacing is sufficiently high, effectively decoupling the CPW fields from those of the MSL mode [37, 57]. Unfortunately, the thicker substrates increase the risk of launching some surface waves. Hence, an optimum thickness has to be found for the frequency range of interest, taking both coupling mechanisms into consideration.

Connecting the top and bottom ground planes of a FW-CBCPW suppresses the MSL modes, but it can trigger some parasitic waveguide modes. The leakage mechanism for this line configuration was recognized. The authors in references [60, 61] noticed that two waveguide modes starting to propagate at some frequency were responsible for the anomalous behavior of both the propagation constant and the line characteristic impedance for the measured FW-CBCPW, as observed in references [17, 62, 63]. The first waveguide mode is a perturbed TE_{10} (see Figure 4.3), and it can be seen as the counterpart of the MSL mode in a FW-CBCPW with the floating top ground planes. The second one with an electric wall symmetry [61] is similar to the coupled slotline mode in the same line configuration. The coupling mechanism between the dominant CPW mode and the TE_{10}-like waveguide mode is of importance because both show a magnetic wall symmetry and a nonnegligible field overlap. The coupling to the other can be neglected due to its electric wall symmetry. A low ratio of the line ground–ground spacing and the line height can be used to decrease the field overlap between the CPW mode and the TE_{10} waveguide mode. Narrowing the top ground planes, which moves the waveguide cut-off frequencies upwards, may also be applied for the same purposes.

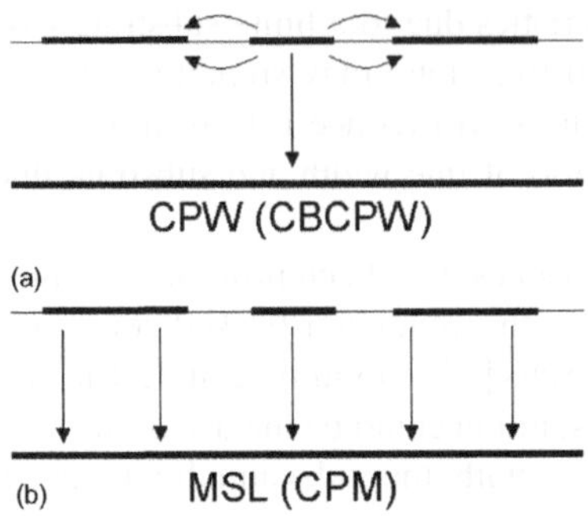

Figure 4.2: Two fundamental modes of an overmoded FW-CBCPW: (a) CPW (CBCPW) and (b) MSL (also known as CPM) [42, 46, 47, 57] (based on the original figure from reference [46], reproduced by permission of © 1993 IEEE). With respect to a standard CPW without conductor-backing, the CPW, or rather CBCPW, mode in a conductor-backed substrate shows the stronger electric field lines passing through the substrate. For very thin substrates, its characteristics resemble more the microstrip-like field distribution than the coplanar waveguide.

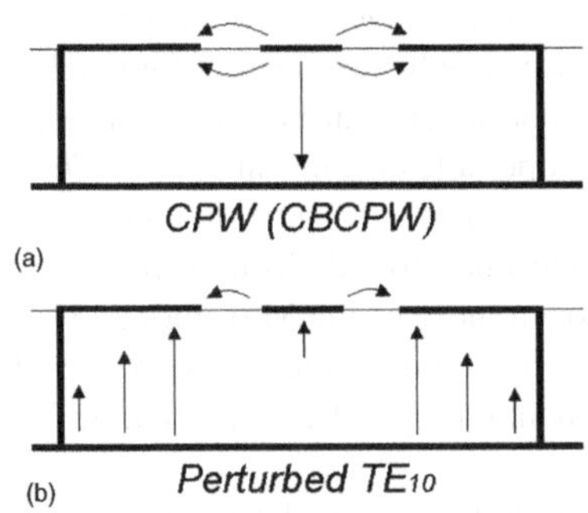

Figure 4.3: (a) The dominant CPW (CBCPW) mode and (b) the perturbed TE_{10} waveguide mode with a magnetic wall symmetry [17,60,61,63] in an overmoded FW-CBCPW with the connected top and bottom ground planes.

CPW lines in hybrid interconnect technologies at mmWave frequencies. Coplanar waveguides are commonly used in the monolithic microwave integrated circuits (MMICs) capable of a very high metal pattern resolution, wherein the lateral line dimensions are usually electrically small even for millimeter waves, and the low inductance short-circuit elements are feasible. The CPW lines, however, may become very limited in use at mmWave frequencies when realized in hybrid interconnect technologies because of some parasitic effects that will be briefly outlined in this paragraph.

The CPW transmission elements can, in general, be patterned on different substrates with a dielectric permittivity varying from low to high. Both options can generate some feasibility problems when moving to higher frequencies. The performance of coplanar waveguide interconnects on the high dielectric constant substrates (typically ceramics) can dramatically deteriorate because of the surface wave excitation and the presence of some other parasitic, substrate boundary condition dependent modes, as discussed previously. Coupling to these parasitic modes, generally line ground–ground spacing dependent, can be considerably higher for the hybrid interconnect technologies than for the MMICs due to the substantial differences in a minimum feasible metal pattern resolution (a minimum slot width of 20 μm even for the very precise MCM-D process is assumed for high yield; see Section 4.3.4 for the detailed technology description). Some additional issues must be considered when choosing the low dielectric constant substrates, such as the very popular microwave laminates. Their inherent advantage is that they commonly allow the processing of large panels in combination with better overmoding characteristics when compared to the high permittivity materials of the same thickness. On the other hand, their use results in wide CPW lines that show a high line width-to-wavelength ratio at mmWave frequencies. In consequence, most of the meander and junction discontinuities may exhibit large reactances and phase shifts, often precluding the realization of numerous distributed passives with the required matching or isolation characteristics. This drawback can be solved partially using a two-layer '*quasi-CPW*' configuration formed by an elevated center conductor above the ground planes on both its sides [17, 64], a technique which became feasible in modern multilayer hybrid technologies with the availability of thin layers. This modified topology enables narrower lines for the fixed characteristic impedance value, but at the cost of sometimes considerably higher line loss.

In order to be more specific, the performance of different interconnect structures and distributed passives at 50–70 GHz, realized in the finite-ground CPW topology on a low permittivity substrate build-up, will be briefly investigated. The FGCPW was chosen to minimize the influence of the previously discussed parasitic mode coupling phenomena (see previous paragraph, 'CPW lines without conductor-backing'). The analysis will be conducted for a three-layer spin-coated BCB ($\epsilon_r = 2.65$) with an overall thickness of 45 μm on a very popular 500 μm thick Rogers 4003 laminate carrier ($\epsilon_r = 3.38$), the Multichip Module with deposited/laminated interconnect (MCM-D/L) technology build-up evaluated within the frame of the LIPS project [17, 65] (see Section 4.3.4 for more detailed description of the BCB dielectric). Any other representative low permittivity dielectric build-up could be used instead. The lines are located on the top metallization layer, and the corresponding tunnels for suppression of the parasitic coupled slotline mode are placed on the next lower layer, 15 μm below.

The lowest *critical frequency* for the chosen dielectric build-up suspended in air is associated with the TE_0 surface wave mode and falls around 125 GHz, according to Equation 4.10. If the conductor-backing is added, the calculated value is 100 GHz and assigned with the lowest order TM_0 mode. Even if the predictions given by the cited equation do not completely apply for a FGCPW, as explained before in the previous paragraph entitled *'CPW lines without conductor-backing'*, a fair estimation of the mode-coupling mechanisms can be derived. A side ground plane width was chosen to be 300 μm. On the basis of numerous investigations performed for different basic FGCPW interconnect elements, the following set of critical disadvantages, substantially limiting their electrical performance above approximately 50 GHz, was found.

1. **Transmission discontinuities.** Due to the low dielectric permittivities of the chosen materials, a 50 Ω line requires a 230 μm wide center strip, assuming a minimum

Table 4.1: Simulated line attenuation and wavelength at 60 GHz for different FGCPW lines on a composite 45 μm/ 500 μm thick BCB/ROGERS 4003 dielectric substrate. Simulations were performed with *HFSS* [67]. A 3 μm thick gold metallization and a BCB loss tangent of 0.002 were assumed (based on the original table from reference [68], reproduced by permission of © 2002 IEEE).

Configuration	Geometry (μm)	Z_0 (Ω)	α (dB mm^{-1}) at 60 (80 GHz)	λ (mm) at 60 GHz
FGCPW	$w/s = 230/20$	50	0.076 (0.092)	3.61
FGCPW	$w/s = 166/52$	70	0.052 (0.063)	3.59
FGCPW	$w/s = 90/90$	100	0.05 (0.06)	3.59

'w' and 's' denote the strip width and the slot width, respectively.

slot width of 20 μm, which results in a line ground–ground spacing approaching one-tenth of the guided wavelength at 60 GHz (see Table 4.1). This, in turn, leads to the electrically large discontinuities, being contrary to their MMIC counterparts. Specifically addressed are the transmission elements requiring the use of bridges or tunnels for the parasitic coupled slotline mode suppression [66], such as bends and junctions. The impact of the excess capacitance introduced by bridge or tunnel for the wide center strips can be more pronounced than the inherent junction parasitics, hence, being difficult to compensate.

2. **Modeling issues.** A large line cross-section also raises the modeling issues of transmission elements, the characteristics of which start to depend on the boundary conditions located on the backside of a finite thickness substrate and on the increasing influence of the substrate modes. A simple equivalent model-based approach, applied in most of the circuit simulators, is able to give only very rough approximations and in many cases is likely to fail completely. As a result, the full-wave electromagnetic simulations are required to predict the performance of numerous transmission line discontinuities, not to mention the substantial effort required for their appropriate compensation. Without the compensation techniques applied, their poor performance will easily lead to the circuit malfunction of larger and more complex functional blocks owing to the cumulative effects.

3. **Computational effort.** The required full-wave electromagnetic simulations are related to a large computational effort. The three-dimensional (3D) finite element method (FEM) simulations even of a single tunnel-based discontinuity can take a few minutes per frequency point on a Sun Ultra 60 with 2 GB RAM, requiring up to 200 MB memory. Typical distributed passives with a higher number of tunnels lead to the substantially larger computational problems, the complete optimization of which can be time consuming. Full-wave simulations of a 60 GHz Wilkinson power divider, performed by means of *HFSS* [67], required 60 min per frequency point on the same Sun workstation.

4. **Distributed passives.** The feasibility of numerous distributed passives based on a combination of the quarter-wavelength-long transmission line sections ($\lambda/4 = 800$–900 μm at 60–70 GHz for the investigated technology build-up), bends, and T-junctions can become questionable due to a very high line width-to-wavelength ratio. The main reason for this is the high phase shifts (40–50° at 60–70 GHz) generated by the compensated bends or junctions that show a very reactive behavior, being a result of their large cross-sections. Therefore, the theoretically calculated quarter-wavelength-long line sections should be reduced even down to 200–300 μm to account for these phase shifts. Such short lengths impact the feasible spacing between the parallel lines that is forced to be in the same range of values. This, in turn, leads to some parasitic coupling between them, affecting the broadband match at the inputs and the isolation between some appropriate branches. A representative example can be the previously mentioned Wilkinson power divider for which the simultaneous demonstration of a 20 dB match at all ports and a 20 dB isolation between the output ports was practically impossible. A similar problem in the design of a Wilkinson divider at 50 GHz on a high resistivity silicon substrate was encountered in reference [69].

5. **High selectivity elements.** High selectivity elements with poor confinement of the electromagnetic fields, e.g. narrowband coupled-line filters, are also practically impossible to realize. The required high separation between the coupled-line sections forces the use of even larger ground–ground spaces and, thus, larger tunnel or bridge elements. This, combined with the very short line sections at the frequency of interest, causes the overall width of the designed structure to be comparable to its total length, making its frequency characteristics practically dominated by the parasitic behavior of the electrically large discontinuities. In reference [70], it was demonstrated that even MMICs, capable of high metal resolution, have reached their limits in the realization of narrowband coplanar coupled-line filters at 60–95 GHz.

6. **Feasible characteristic impedance values.** As a minimum slot width is set to 20 μm, which should be viewed in any case as a reference value for the hybrid interconnect technologies, the range of feasible characteristic impedance values is lower bounded to 50 Ω for a 270 μm line ground–ground spacing (see Table 4.1). For this reason, some elements such as a 3 dB branch line coupler, requiring 35 Ω lines, are unfeasible. The minimum characteristic impedance value could be decreased to approximately 40 Ω for a thin BCB layer on a ceramic substrate and the same minimum slot width. For comparison purposes, a 30–80 Ω impedance range can be achieved for the GaAs or InP substrates, assuming a minimum strip width of 10 μm and a ground–ground spacing of 100 μm.

7. **Intercircuit isolation.** The intercircuit isolation is low due to the large cross-sectional dimensions of transmission lines with the belonging discontinuities when placed on a nonnegligibly thick substrate, which promotes some space wave radiation, and leakage to surface waves or MSL modes (the latter for a conductor-backed substrate). The open-end elements require special attention in this regard.

8. **Grounding paths.** Long grounding paths, resulting from the large line cross-sections, are responsible for a high inductance of the short-circuit elements that may considerably exceed the corresponding values for their microstrip counterparts in the

optimized microstrip-oriented technology (see Section 4.4). For the same reason, the uncompensated loads, even without grounding vias, may perform worse than those realized in the above-mentioned optimized technology setup. A return loss of the uncompensated, 75 μm long, and 60 μm wide 50 Ω NiCr (38 Ω/square) termination embedded in a 270 μm wide 50 Ω FGCPW transmission line was simulated to be inferior to 20 dB above 17 GHz. Its optimized version with a line ground–ground spacing locally narrowed to 100 μm showed a return loss better than 20 dB up to only 30 GHz.

4.2.2 Metal Rectangular Waveguides

Traditional machining techniques for the metal waveguides operating at mmWave frequencies, specifically above 60 GHz, can be complicated and costly. The development of low-cost fabrication techniques, specifically for their integrated counterparts, is of special importance for the millimeter- and submmWave system integration [7]. Essentially, there are three alternative possibilities to create metal rectangular waveguides:

- micromachining, resulting in the air-filled structures;
- metallized plastic injection mold, also yielding the air-filled devices;
- multilayer technologies, leading to the dielectric-filled geometries.

In spite of numerous indisputable advantages of the metal waveguide structures, such as low loss and very high isolation, there are some possible issues, related mainly to their manufacturing process, that should be considered.

The major drawbacks of the micromachined waveguides, compared to the classical quasi-TEM lines, are listed below [71]:

- high manufacturing cost related to a large number of nonstandard processing steps;
- fabrication of the high volume hollow structures makes them more sensitive to the processing defects, possibly leading to some yield issues;
- a one wavelength section of an air-filled waveguide is excessively long when compared to the corresponding quasi-TEM lines on a dielectric substrate; consequently, they are possibly too expensive for the monolithic integration below about 200 GHz.

For frequencies below 100 GHz, still very unpopular metallized plastic injection mold technologies can considerably relax the manufacturing cost related to the micromachining process. Although being limited in the integration capacity when compared to the classical micromachining techniques, numerous truly three-dimensional waveguide parts for mmWave applications were already reported in the literature [4,72–77].

The modern multilayer interconnect technologies are capable of some additional features for creation of the metal waveguide structures when compared to the micromachining [7,71], such as the following.

- The cross-sectional area of a dielectric-filled rectangular waveguide can be decreased by the dielectric permittivity value of a filling material, assuming the fixed f_0/f_c ratio

(f_o and f_c are the operating frequency and the waveguide mode cut-off frequency, respectively). The cost for that is an increased material dependent dielectric loss.

- The dielectric-filled waveguides in combination with a multilayer build-up allow some other innovative forms or geometries to be integrated, such as finlines or dielectric slab waveguides.

The filling materials are, in practice, limited to those showing the lowest loss tangent and possibly constant dielectric permittivity at mmWave frequencies. Many of them, such as spin-coated polyimide and glass dielectrics based on polysiloxan polymers, must be avoided due to the presence of unwanted molecular resonances [71]. Furthermore, they should allow very smooth surfaces to be created for minimum ohmic loss. In this respect, the thin-film BCB and spin-on glasses are a perfect choice, but their achievable layer thickness is too small for practical low-loss waveguide realizations owing to cracking problems in the curing process. Low temperature cofired ceramic (LTCC) and thick-film photoimageable materials can deliver the appropriate thickness, but the dielectric loss tangents associated with most of the available tapes and the imperfect multilayering can become an issue (see Sections 4.3.3 and 4.3.4 for the more detailed analysis of various technology considerations for the LTCC and MCM-D technologies).

Various attempts to develop the integrated metal waveguide structures by each of the above-mentioned processing methods were reported by different authors. A concise summary of the major results will be given in Section 4.3.

In spite of the multiple efforts to fabricate the integrated metal waveguides, a consistent mmWave integration technology with the active devices fully integrated within the waveguide structures does not exist. Today's planar MMICs are incompatible with the metal waveguides and require the use of different transitions, very often with poor performance. A new interesting vision for implementing a completely integrated metal waveguide-based system was reported in references [7, 71]. In this approach, the MMICs are assumed to be realized in a finline, or a slotline topology and appropriately positioned inside the waveguide channel for a maximum coupling between the waveguide and the active devices on the chip side.

4.3 Interconnect Technology Options for Millimeter-wave Applications

There are a number of technologies available that can potentially be used at mmWave frequencies. They will be briefly presented in this Section [17].

4.3.1 Basic Technological Requirements

An extensive list of stringent requirements has to be fulfilled in order for the interconnect technology to be used at mmWave frequencies. The exact set depends on the transmission line topology of choice. For the printed planar lines, it typically includes the following factors.

- **Metal pattern resolution:** Metal pattern resolution and the corresponding etch accuracy. The minimum feature size of the metal pattern with its across-wafer and wafer-to-wafer variations are one of the most important factors defining the capabilities of the technology for mmWave frequencies. The realization of precise and narrowband passive structures, such as bandpass filters or patch antennas, is very demanding or even inaccessible for many of the hybrid interconnect technologies available.

- **Metallization process:** Metallization process applied, in particular, an exact metal layer build-up with all adhesion and barrier materials applied. In view of the high metal pattern dimensional control required, a thick composite layer stack-up should be avoided, especially a copper–nickel–gold composition that is commonly applied on top of a substrate in the hybrid technologies. Its use can lead to potential difficulties in a pattern definition due to the possible undercut issues, and, thus, should be limited to the local areas such as bonding pads, if possible.
- **Strip profile:** Strip profile, very often deviating from the nominal rectangular shape.
- **Surface roughness:** Surface roughness for both metallization and dielectric layers.
- **Thickness control:** Thickness control of dielectric layers. It is normally a parameter of interest for the circuits realized in a microstrip topology and for the multilayer structures. It can show a primary influence on the degree of planarization for the latter.
- **Dielectric substrate properties:** Dielectric properties of the substrate carrier or of the individual layers for a multilayer technology.
- **Passivation layer:** Dielectric properties and thickness control of the solder mask or the passivation layer, if applied. Both options are commonly utilized to protect some conductor materials, such as copper, or resistive layers against the harmful influence of numerous environmental factors, specifically oxidation. Their application can be helpful in avoiding the use of the above-mentioned composite metal build-ups. As long as the spin-coated passivation layers are very well defined in terms of their processing accuracy and their dielectric properties, this is usually not the case for the solder mask materials conventionally applied in the printed circuit board (PCB) industry that are also typically very lossy at mmWave frequencies.
- **Degree of Planarization:** Degree of Planarization (DOP), important for the multilayer technologies.
- **Layer-to-layer alignment:** Layer-to-layer alignment for multilayer technologies. It is essential for the successful realization of numerous multilayer passive interconnect structures, specifically those based on the electromagnetic coupling principle. As far as the modern multilayer multichip module with deposited interconnects (MCM-D) technologies are capable of delivering a very high alignment accuracy in the range of few micrometers, it is impractical to expect it to be better than 25–50 μm for the printed circuit boards.
- **Via geometry:** Minimum via and via chain geometry, defined by its diameter, landing pad size, aspect ratio, and a via-to-via spacing. Electrical characteristics of vias for millimeter waves are one of the major designer concerns. The minimum via diameter is mainly related to the mechanical properties of the insulating materials used and the corresponding processing options available. As it normally increases with the substrate thickness, the high aspect ratio (height /diameter) vias are highly

preferred. The minimum landing pad size is mainly a result of the finite registration capabilities. For the PCB technologies, it can be even two up to three times larger than the corresponding via diameter. The available via-to-via spacing can be critical for the intercircuit isolation and the energy leakage phenomena. For the multilayer technologies, it is also important if the stacked via or only the staircased-like configuration is available.

- **Chip cavity:** Processing of cavities for the chip placement. The usable die attachment methods primarily depend on the transmission line topology for the active circuits intended to be interconnected and the available integration technology options. The microstrip-based MMICs can be embedded into cavities in order to suppress the influence of parasitic effects related to the bonding method applied, normally wirebonding (see Figure 4.12 in Section 4.3.2 for the cavity formation example). A very high accuracy for both the cavity processing and the chip placement is a must for the repeatable chip-to-package transitions at mmWave frequencies.

- **Resistive and capacitive layers:** Availability of additional thin resistive and capacitive layers for the integrated resistors and capacitors, respectively. The functionality and the overall performance of an integrated microwave module can be considerably increased if both layer options are available. The realization of low inductance loads and series resistors with the via-less connections to transmission line elements is key for mmWave passives. Furthermore, the magnetic properties of the applied resistive layers should be carefully verified for the operating frequency range.

The impact of some of the above-listed factors on the performance of basic planar interconnect structures will be studied in the next few paragraphs of this subsection, whereas the influence of some others will be outlined in a more detailed manner separately for each of the later presented interconnect technology options.

For the integrated metal waveguide structures, summarized in Section 4.2.2, a different set of technological requirements can clearly be derived, partially overlapping with the list given above. However, it will not be considered in detail for reasons of space.

4.3.1.1 Line Loss Versus Metallization Process

A microstrip configuration is chosen for the line loss study, being one of the most representative transmission line topologies. Nevertheless, most of the findings can be applied for the other printed planar guiding structures, e.g. CPWs.

The following factors related to the metallization process should be addressed when considering the transmission line loss at mmWave frequencies:

- choice of the main conducting material;
- the presence of some additional adhesion and barrier layers;
- type of metal deposition process and the corresponding pattern generation method;
- metal strip thickness;
- strip width and strip sidewall profile;

- surface roughness of an underlying dielectric.

Copper conductors are often used in hybrid technologies for their high conductivity ($\sigma = 5.88 \times 10^7\ \Omega\,m^{-1}$). However, they need a passivation layer against oxidation and a barrier layer to avoid electromigration [78]. Gold conductors have slightly lower conductivity ($\sigma = 4.55 \times 10^7\ \Omega\,m^{-1}$), but they require no passivation or barrier layers. A choice between different conductor materials in view of the ohmic loss at 60–80 GHz is immaterial for a typical conductor thickness above 1–1.5 μm, provided that a single-layer homogeneous nonmagnetic material is applied. This behavior is a result of the reduced cross-sectional area for the current flow in a conducting layer that is defined by the skin depth, being lower for the higher conductivity materials and higher for the metallizations with lower conductivity. In order to avoid excessive ohmic loss, metal thickness must be in the range of three to five times the skin depth. The excessively thick conductors will only decrease accuracy of the line width definition.

The situation becomes more complex for a composite metal build-up, being usually the case for copper conductors (see also introductory part of Section 4.2.1) that tend to react chemically with polymers or to diffuse into other metals [78]. To prevent such effects, an intermediate layer of nickel is normally applied as barrier, being a magnetic material. Its thickness, in relation to the main conducting layer thickness, should be carefully chosen for the frequency range of interest because it can considerably increase the line ohmic loss. The magnetic permeability of nickel decreases with frequency, but it cannot be neglected even at mmWave frequencies. The other materials such as Ti, TiW or Cr, used as adhesion layers, contribute considerably less to an overall conductor loss increase even at 100 GHz owing to their low layer thickness, typically below 100 nm.

Another important topic is the line width (the strip and slot width for CPW). For mmWave frequencies, the current density in a metal strip of the microstrip is the highest in its lower corners. As a result, ohmic loss is only slightly dependent on the strip width (see also Table 4.6 in Section 4.4). Of course, some line loss increase can be observed for very narrow strips, but the width dependence is considerably weaker than for DC currents. The factors that become important are the strip sidewall profile and the quality of edge definition. Both are dependent on the plating process applied and the surface quality of an underlying dielectric layer.

The wet etching process is preferred for thin metal layers, for two reasons: first, the unavoidable undercut of the metal strips is low for thin features. Secondly, the wafer inhomogeneity of the etching process, depending on the local etchant concentration, is smaller than for the thick conductors. The thickness of electroplated strips is, in turn, strongly related to the local metal pattern density, typically resulting in a lower conductor thickness uniformity within the wafer. Furthermore, the electroplating process yields rougher surfaces than sputtering. In contrast, the etched conductors show rougher sidewalls than those plated within a resist mask. The etched conductors usually show trapezoidal profiles with possibly very acute angles on the bottom surface for the poor process control, whereas the electroplated are more rectangular in profile. The sidewalls of the latter are determined mainly by the resist mask development process.

Both plating options can normally be applied for the mmWave interconnects in MCM-D technologies. The scanning electron microscope (SEM) photos of the copper conductors for the MCM-D technology build-up from Section 4.4 [17, 65] are shown in Figure 4.4. The above-discussed process properties for both plating options can be easily identified.

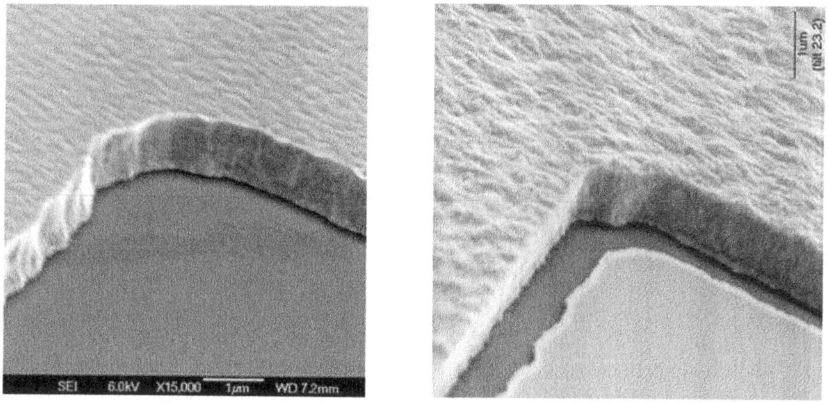

(a) (b)

Figure 4.4: SEM pictures of the (a) sputtered/etched and (b) electroplated copper conductors for the MCM-D technology developed within the framework of EU research project LIPS [17,65,79] (see Section 4.4). Courtesy of Acreo AB, participant of the LIPS consortium [64].

A desirable, almost rectangular edge profile without acute angles can also be observed in both cases. The situation is different with laminated dielectric layers (multichip module with laminated interconnect MCM-L or PCB) that normally need thick copper for planarization of the rough laminate surfaces and for high vertical elongation related to the high coefficient of thermal expansion (CTE) values in the z-direction. Thick metal layers are essentially electroplated due to the faster deposition rates when compared to sputtering. Unfortunately, the conductor surface roughness increases with the metal thickness. A representative example for the achievable sidewall profile and the metallization surface quality of the electroplated copper in this case can be seen in Figure 4.5, wherein the microstrip line cross-section for a 38 μm/25 μm thick *Speedboard C/Biac LCP* composite laminate build-up developed within the EU research project LIPS [17, 65, 79, 80] is shown (see Section 4.3.2 for the detailed technology build-up definition). The plating process is basically composed of the surface cleaning and the sputtering of a 500 nm copper on a 200 nm chromium, followed by a 9.5 μm electrolytic copper [81]. A smooth sidewall profile, rectangular in shape, can be identified. The measured peak-to-peak surface roughness of the metal layer is around 2 μm.

Another important factor influencing the line conductor loss is the metallization surface roughness. From the mechanical point of view, surface roughness promotes adhesion between building layers. Therefore, some special techniques (e.g. plasma etching) are very often used to artificially increase its value. At mmWave frequencies, however, where the skin depth, δ, in a conducting layer is very low ($\delta = 0.22$ μm for copper and 0.25 μm for gold at 80 GHz), it is a critical parameter for ohmic loss minimization. If its root-mean-square (RMS) value, R_q, nears, or exceeds the skin depth, the line ohmic loss (α_c') can even be doubled [82] due to the considerable increase in the current flow path

$$\alpha_c' = \alpha_c \cdot c_{R_q} \tag{4.11}$$

Figure 4.5: Microstrip line cross-section for a 38 μm /25 μm thick *Speedboard C/Biac LCP* composite laminate build-up developed within the EU research project LIPS [65, 80] (see Section 4.3.2 for the detailed technology setup). Dielectric permittivities for both materials are 2.57 and 2.95 at 60 GHz, respectively. The metallization layer consists of a sputtered 500 nm thick copper on a 200 nm thick chromium, followed by a 9.5 μm thick electrolytic copper.

where

$$c_{R_\mathrm{q}} = 1 + \frac{2}{\pi} \cdot \tan^{-1}\left[1.4 \cdot \left(\frac{R_\mathrm{q}}{\delta}\right)^2\right] \tag{4.12}$$

The metallization roughness is determined by the roughness of an underlying dielectric layer and the plating process quality. The spin-coated BCB ($\epsilon_\mathrm{r} = 2.65$) [83], a representative dielectric layer for the modern thin-film process (see Section 4.3.4), shows almost mirror-like roughness (see Figure 4.6). Most laminates have very pronounced surface roughness values compared to the thin-film dielectrics. A new generation of materials, among others LCPs such as *Biac LCP* from *Gore* [80], are capable of supporting considerably smoother surfaces, if, however, still incomparable to the spin-on materials, such as BCB. The measured RMS surface roughness of both *Biac LCP* and the corresponding copper metallization on its top, taken by atomic force microscopy (AFM), are 440 nm and 620 nm, respectively (see Figure 4.7). Such values exceed the skin depth in copper at mmWave frequencies (0.2–0.3 μm at 60–80 GHz), leading to some additional increase in conductor loss.

The frequency-dependent line loss for a set of different microstrip lines (see Figure 4.8) on the *Speedboard C/Biac LCP* dielectric build-up is shown in Figure 4.9. It was de-embedded using a principle of the calibration-comparison method [84, 85], wherein a full TRL, or LRRM, off-wafer calibration and a second-tier on-wafer calibration, based on the measurement technique of two transmission lines [86, 87], was performed [16, 17, 88]. The de-embedded line attenuation is within 0.1–0.12 dB mm^{-1} at 60 GHz for the considered strip width range of 30–485 μm. The similar loss value for a wide range of the strip widths is a result of the previously discussed nonuniform current density distribution at the frequency range of interest.

For comparison purposes, the de-embedded line loss for a 50 Ω microstrip line on a 45 μm thick BCB dielectric substrate from Section 4.4 is shown in Figure 4.10. Two different metallization options are given:

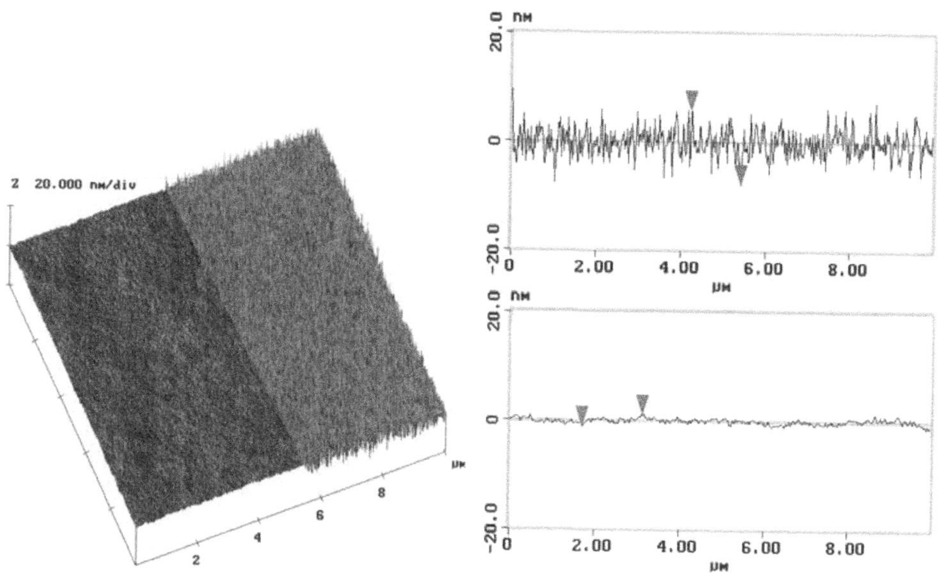

(a) (b)

Figure 4.6: AFM measurements of the BCB surface roughness before and after reactive ion etching (RIE). The two measured profiles are superimposed in (a). The spin-coated BCB is the dielectric of choice for the MCM-D technology from Section 4.4. Courtesy of Acreo AB, participant of the LIPS consortium.

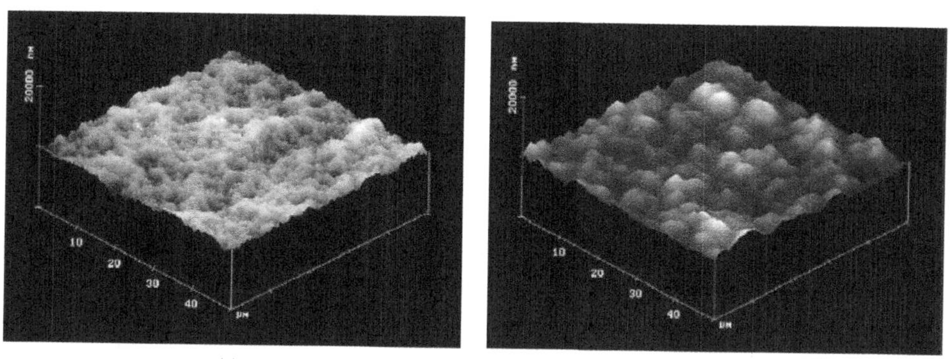

(a) (b)

Figure 4.7: AFM surface roughness measurements of (a) *Biac LCP* ($R_q = 440$ nm), and (b) copper on *Biac LCP* ($R_q = 620$ nm). *Biac LCP* is the dielectric of choice for the MHDI technology developed within the frame of EU research project LIPS [17, 65, 79] (see Section 4.3.2).

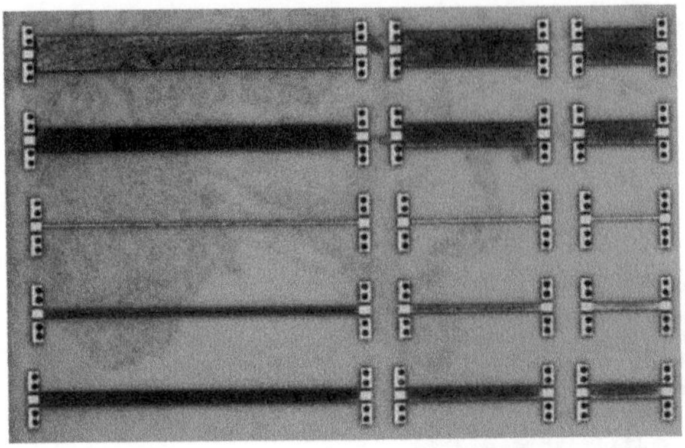

Figure 4.8: A set of 30–485 μm wide microstrip lines on a 38 μm/25 μm thick *Speedboard C/Biac LCP* dielectric build-up. Dielectric permittivities for both materials are 2.57 and 2.95 at 60 GHz, respectively. A 9.5 μm thick copper was used for the metallization layer. The electrodeposited resist equipment was applied to achieve a fine pitch pattern combined with a good line profile definition. (EU research project LIPS [17, 65, 79]; see Section 4.3.2.)

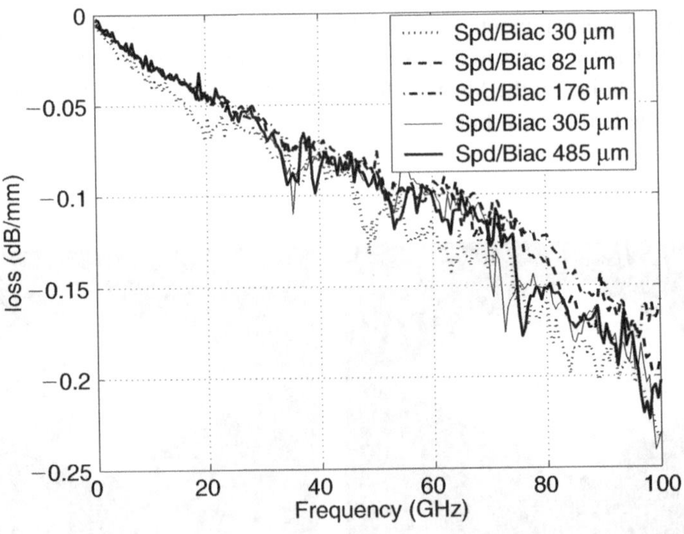

Figure 4.9: De-embedded line loss for a set of 30–485 μm wide microstrip lines on a 38 μm/25 μm thick *Speedboard C/Biac LCP* ('Spd/Biac') composite dielectric substrate; see Section 4.3.2 for the exact technology setup definition. Dielectric permittivities for both materials are 2.57 and 2.95 at 60 GHz, respectively. A copper strip thickness is 9.5 μm. (EU research project LIPS [17, 65, 79].)

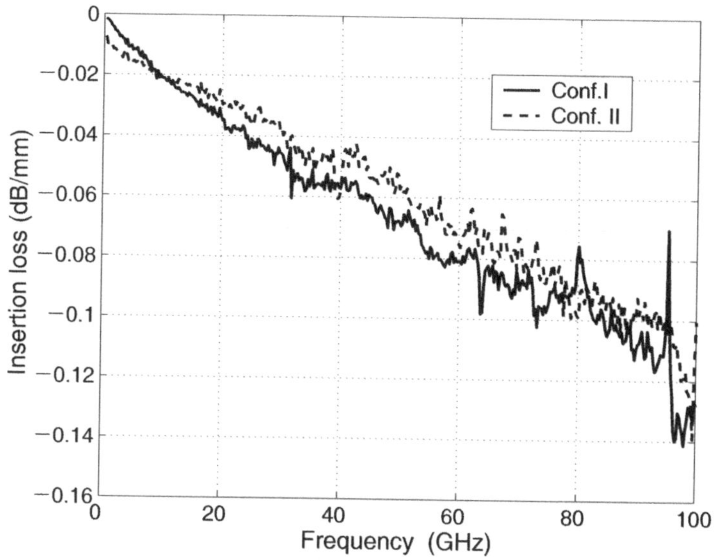

Figure 4.10: De-embedded line loss for a 50 Ω (114 μm wide) microstrip line realized in the BCB-based MCM-D technology from Sections 4.4 and 4.3.4, developed within LIPS consortium [17, 65, 79]; see Section 4.3.4 for the detailed technology description. (Technology configuration I – 45 μm thick BCB, 80 nm/3 μm thick NiCr/Au electroplated strip conductor and 80 nm/3 μm thick TiW/Cu ground metallization; technology configuration II – 45 μm thick BCB, 5 μm thick BCB passivation, 70 nm/1 μm thick Ti/Cu sputtered strip conductor and ground metallization.)

- a 80 nm/3 μm thick NiCr/Au electroplated strip conductor and a 80 nm/3 μm thick TiW/Cu ground metallization;
- a 70 nm/1 μm thick Ti/Cu sputtered metal layer for both the strip and the ground plane; an additional layer of a 5 μm thick BCB passivation was applied to cover the copper conductors.

A very low insertion loss of 0.08 dB mm^{-1} and 0.074 dB mm^{-1} at 60 GHz for both realizations, respectively, can be noted, being very close to the simulated values owing to almost mirror-like BCB surface quality.

The lowest loss (see Figures 4.9 and 4.10) of all the technology build-ups presented in this paragraph corresponds to a 1 μm thick sputtered copper conductor on a 45 μm BCB dielectric from Figure 4.4a. This result strongly emphasizes the meaning of the conducting strip quality for the ohmic loss minimization at mmWave frequencies.

4.3.1.2 Degree of Planarization

The DOP describes the ability of dielectric layers to planarize uneven underground. It can have a major influence on the quality of the related lithography process, especially for modern thin-film multilayer technologies, wherein the thickness of both the dielectric layers and the corresponding conducting strips is very comparable. As it is a parameter directly influencing the local thickness of multilayer geometries, it should be precisely known for a broad spectrum of typical interconnect structures. Unfortunately, it can only be accounted for by extensive measurements, thus emphasizing the importance of a full technology characterization. The planarization is affected by many parameters. The major parameters are:

1. thickness and material properties of the insulating films;
2. number of coatings for deposited dielectric layers;
3. geometry of the underlying patterns;
4. various process properties.

The laminate materials usually show DOP values close to 100%, BCB in the range of 75 to 85% and polyimide around 50% [89]. Technologies supporting a high DOP are highly preferred, as they usually allow a first design attempt to be successful.

4.3.1.3 Technology Performance Characterization and Monitoring

As stated before, the process accuracy is one of the major technology performance defining factors. Hence, it is essential to take advantage of the effective process control methods in order to keep the process performance consistent during manufacturing [90]. Hence, there is a need for the dedicated technology characterization test vehicles (TCTVs) and the corresponding methods capable of measuring the technology parameters responsible for the radiofrequency (RF) performance, including spatial information of their variations [79]. They must be placed at representative locations within the wafer according to the spatial occurrence of parameter variations. An example of such a TCTV, developed for the MCM-D technology setup from Sections 4.3.4 and 4.4, is shown in Figure 4.11. With the information provided by the spatial description, it is possible to optimize the process control flow in the manufacturing [91] or to embed the well-characterized technology variations into the design cycle.

4.3.1.4 Characterization of the Substrate Dielectric Properties

In order accurately to characterize the capabilities of the technologies intended to be applied, especially of those newly developed, a wideband determination of dielectric properties for the applied insulating layers is a must for millimeter waves. The reason is that they are normally not provided by the manufacturers at those frequencies. This section is not intended to give an extensive overview of the characterization techniques but to point out some important issues relating to the determination of dielectric substrate properties from the printed line measurements commonly used in the industry.

Cavity resonators, free-space, open-ended coaxial probe, and transmission/reflection-based approaches are mostly used for the complex dielectric permittivity de-embedding purposes. All these methods normally require a measurement cell made up of a section of coaxial line or rectangular waveguide filled with the bulk sample material to be characterized. However, building the bulk samples for some materials, such as very thin deposited dielectric layers, is impossible. Furthermore, in many situations, an *in situ*

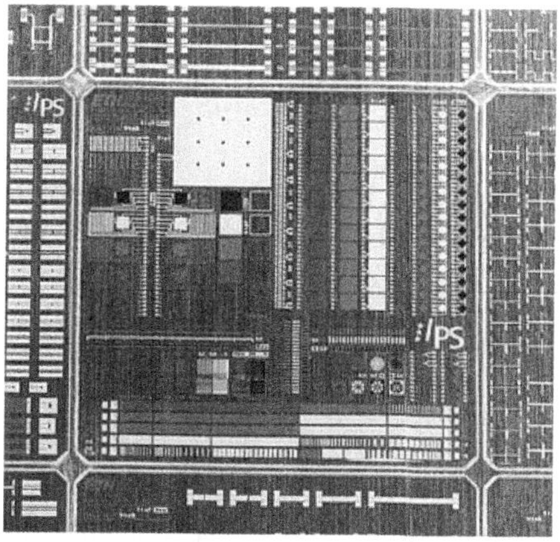

Figure 4.11: A 22 mm × 22 mm large TCTV for the MCM-D technology from Sections 4.3.4 and 4.4. (EU research project LIPS [17,65,79].)

extraction is greatly preferred, meaning that the measurement cells are created following the standard process flow of the technology build-up intended to be used. As most of the interconnect technologies are supposed to work with printed lines, choosing them as a standard measurement cell is a typical approach in the industry. Unfortunately, transmission characteristics of planar transmission lines, specifically in hybrid interconnect technologies, can be very complex frequency-dependent functions at mmWave frequencies, as discussed in Section 4.2.1. Hence, the accurate prediction of material dielectric properties from the printed line measurements can, in practice, be a tedious procedure, requiring multiple physical phenomena, such as high-frequency dispersion, slow-wave effects associated with the field penetration in a metal strip, and different types of loss present on the line, to be addressed. In this regard, special precautions require the use of a classic printed-resonator measurement technique because of some key disadvantages listed below.

1. Narrowband extraction at some frequency points only.

2. In practice, the Q-factor and the resonant frequency de-embedding techniques are mostly formulated assuming an ideal model of the resonator system [92–94]. However, due to some effects associated with the printed lines and the external feed circuitry, the measured responses are distorted. The following phenomena may affect the measured S-parameters of a real resonator [92–94].

    - A basic error is the assumption of dispersionless line characteristics or constant dielectric permittivity around the resonance [94]. The quality factor Q is typically defined by the following relation $Q = \beta/2\alpha$ (β-line propagation constant, α-line attenuation constant) where

$$\beta = (\omega/c)\sqrt{\epsilon_{\text{reff}}} \tag{4.13}$$

    This equation, however, is improper from the theoretical point of view if dispersion is present. Due to the frequency-dependent characteristics of ϵ_{reff} for dispersive lines, the new correct expression for Q should be applied [94]

$$Q = \omega(\partial\beta/\partial\omega)/(2\alpha) \tag{4.14}$$

where β is defined by (4.13). As the quality factor Q of the printed resonators may be low, the measured bandwidth can be relatively large, and the assumption of dispersion-free lines is not necessarily correct.

- The exact frequency-dependent loss and electrical delay introduced by dispersive feed transmission lines should, in principle, be known to account for their influence [93].
- The resonators are gap-coupled for the printed lines, being a source of additional loss from the open-ends and of the frequency detuning caused by some energy stored in a coupling structure. The coupling level is frequency dependent, and only for high-Q systems (loaded $Q > 1000$) may it be valid to assume it constant in the frequency band around the resonance. However, this is not implicitly true for the printed line resonators [93,94].
- Impedance mismatch between the feed structure of a measured resonator and the reference impedance of a measurement system may exist in the incompletely calibrated measurement setup. It is not uncommon that the calibration procedure is performed using only the off-wafer calibration substrates with the CPW line standards (probe-tip measurements assumed). Different substrate dielectric constants, feed geometries, and line topologies for both off- and on-wafer structures cause the feed parasitics to remain influencing the measurements at mmWave frequencies. This mismatch may be responsible for some periodic changes in the magnitudes of both reflection and transmission coefficients, and some phase nonlinearities in reflection coefficients [92,93].
- The printed line resonators can show considerable radiation around the resonant frequencies, especially for thicker substrates and higher frequencies. This should be taken into account in a precise manner in the extraction process of the line Q-factor [95]. Radiation can also cause crosstalk between both coupling structures for transmission mode resonators [93].

Taking the above into account, a different approach based on the transmission/ reflection measurements, taking advantage of the powerful on-wafer TRL calibration technique [96–98], may become more efficient in use, if properly applied [16, 17, 88]. In this technique, the transmission properties of the measured printed line standards are an inherent outcome of the calibration procedure. By applying multiple line standards of different length, the complex line propagation constants can be de-embedded for a vary broad frequency range. Furthermore, due to the self-calibration property of this technique, the influence of feed structures is fully eliminated. The problem of possible radiation from the printed resonators at the measured resonant frequencies is solved, which is of importance for the accuracy of extracted line loss attenuation. The open-end effects associated with the gap-coupling structures are also avoided.

4.3.2 MCM-L

A classic MCM-L or a laminate-based microwave high density interconnect (MHDI) is considered a low-cost technology and can be regarded as a laminated PCB scaled to meet the requirements and dimensions of an MCM. Generally, the laminate layers are clad with a metal foil, resulting in rough metal surfaces at the laminate side. The conductor widths and spaces as well as the via dimensions are typically 100–150 μm. Poor dimensional and electrical stability of the laminate materials with both temperature and humidity, and a high CTE are additional issues. The quality of standard MCM-L integrated passives is relatively poor due to the low quality dielectric materials used.

However, an emerging set of polymer dielectrics in sheet/film or liquid formulation offers a combination of low loss and low CTE [99]. The thinner films (20–60 μm per layer) used in the new organic build-up technologies [100] offer the ability to design and implement the modules at microwave frequencies. A better control of the etching tolerances is achieved by electroless and electrolytic copper plating, and UV lithography. Aggressive feature sizes of 25–50 μm for the line widths and spacings, and a 50 μm microvia technology were shown to be possible for the organic process [99, 101]. A high-quality factor for the passive components within a multilayer organic-based process was reported [99, 100], allowing successful integration of the complete passive RF front-end functional building blocks for lower microwave frequencies. A quality factor of over 160 for a 3.4 nH (0.6×0.6 mm^2) inductor with a self-resonant frequency of 11.5 GHz was achieved in reference [100]. Quality factors greater than 250 at 2.4 GHz were obtained for some embedded capacitors [99].

An attempt to develop a laminate high-density interconnect and packaging technology for mmWave applications was reported in references [17, 65, 102]. For compatibility with low-cost manufacturing, the developed technology favors the processing steps that are performed for a large number of modules at the same time [103–107]. A general substrate build-up of the interconnect unit is shown in Figure 4.12, wherein the active and passive devices are positioned in cavities, allowing the wirebond suppression. A 150 μm thick Microlam 410 resin from Gore [80] laminated on top of a copper substrate, considered a heat sink, is used to create cavities by means of an excimer laser (see Figure 4.13). Connecting vias are defined directly above the chip active surface thanks to some alignment marks left on the substrate. For the chosen composite dielectric layer build-up of a 38 μm/25 μm thick *Speedboard C/Biac LCP*, covering the cavities, reliable drilling with an aspect ratio equal to one could be achieved. The minimum required via pad overlap was found to be about 30 μm. The MMICs are protected by a few micrometers thick BCB layer on top to be compatible with the lamination process. They are glued with a conductive adhesive, and the space around them is filled with an epoxy resin to arrive at the required planarization before the lamination. A photolithographic process is used for a 10 μm thick copper metallization in order to arrive at the appropriate repeatability at mmWave frequencies [81]. The line widths and gap spaces larger than 50 μm are easily reproducible with an in-panel variation better than ±2.5 μm. Even lines as narrow as 30 μm can be properly defined (see Figure 4.8).

Microlam 410 prepreg from Gore [80] is an epoxy resin coated on a microporous expanded polytetrafluoroethylene (PTFE). It shows good electrical properties ($\epsilon_r = 3.4$, $\tan\delta = 0.008$ at 10 GHz), high transition temperature (225°), and excellent isotropic CTE of 19 ppm/°C matched to copper. The excellent surface planarity and uniformity of this material allows fine lines and spacings.

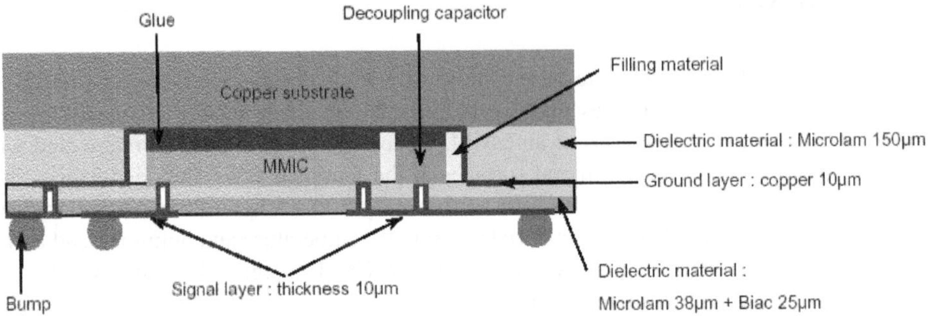

Figure 4.12: A general substrate build-up of the laminate interconnect technology developed within the European research project LIPS [17, 65, 102] (based on the original figure from [102], reproduced by permission of © 2004 IEEE). (EU research project LIPS [17, 65,79].)

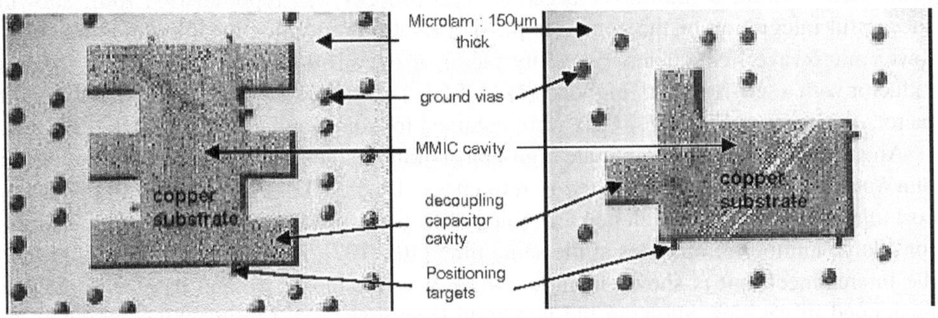

Figure 4.13: Ablation of cavities in a 150 µm thick layer of Microlam 410; see Figure 4.12 for the detailed technology build-up. (Courtesy of Thales Airborne Systems, LIPS consortium participant.)

Biac LCP is a LCP laminate also from Gore [80]. LCP is a thermoplastic polymer considerably cheaper than the Teflon-like dielectrics. It shows an interesting combination of electrical, thermomechanical, and chemical properties. Its mechanical stability is a few times higher than for a polyimide. Its CTE of 16 ppm/°C is close to that of a copper conductor, resulting in low warpage after etch. High moisture and chemical resistance make LCP an ideal candidate for the aggressive operating environments [108]. It absorbs up to a hundred times less water than polyimide films, resulting in stable electrical and mechanical properties [109]. LCPs have not been widely used because of their weak interaction with copper. A substantially roughened copper foil surface can be used in the conventional lamination process in order to improve the bonding strength, being of less interest at mmWave frequencies. Recently, a surface-activated bonding, applying some surface cleaning process, has been developed for LCPs to arrive at a high bonding strength with a simultaneous low surface roughness of 100 nm [109]. Furthermore, the high transition temperature of LCP

(335°) normally requires the use of prepregs as a bonding ply. In references [110, 111], it was reported that LCP provides a nearly constant dielectric permittivity and a low dielectric loss tangent ($\epsilon_r = 3$, $\tan\delta = 0.004$ at 35 GHz) up to 110 GHz. Despite all of these advantages, the potential of LCP for microwave and mmWave circuits has still not been fully explored.

With the recent development of the low- and high-melting-point temperature LCP dielectrics, a multilayer lamination process without the use of adhesive layers became possible [112]. Multiple design examples of various types of package and embedded passive, taking advantage of this new technology, have already been reported in the literature [113–115], but mostly at microwave frequencies.

In spite of the above outlined recent process developments for the laminate technologies, many high-performance interconnect structures or integrated passives, specifically those based on a tight coupling between the line sections, are still out of reach above 50 GHz owing to deficient accuracy control for the small metal feature sizes and for the thickness of insulating layers. Nevertheless, this set of technologies seems to show the potential to realize various low-cost packaging schemes for the single active circuits, or the standard functional mmWave units consisting of several chip bare dies with some optional integrated passives.

4.3.3 LTCC

Recently, LTCC has gained more attention for its claimed low-cost processing [116]. One of the basic disadvantages of this technology, limiting its accuracy, is the shrinkage of ceramic tapes during firing. Usually, the tapes shrink between 12 and 16% in both horizontal dimensions and between 15 and 25% vertically, according to reference [117]. Common shrinkage tolerances are ±0.2% and ±0.5% for both directions, respectively, and are dependent on the particular conductor material pattern on each layer [117]. Thus, fabrication of the multilayer substrates, typically $5'' \times 5''$ large, can result in the aggregate length and width differences between layers as large as 250 μm, possibly leading to some critical alignment issues for the high-density designs incorporating fine lines and spacings, if inadequately addressed [118]. Taking this into account, the three-dimensional capabilities of this technology are extensively exploited to achieve miniaturization. Unfortunately, modeling of truly three-dimensional structures can become complex even at lower microwave frequencies, thus, limiting the use of LTCC in more demanding designs. A practical realization of this three-dimensional integration scheme is at mmWave frequencies, where the influence of high-frequency parasitic phenomena such as mode coupling and energy leakage can be more pronounced. Another basic limitation of the LTCC process is a coarse metal definition. The minimum line widths and spaces for the regular thick-film patterning techniques are normally 75–125 μm with a common tolerance of ±10–20 μm (see references [116, 119–122]). A typical conducting layer thickness is around 10–15 μm [122, 123]. The other two issues relating to LTCC, considerably influencing the mmWave performance, are usually too high loss tangent values in the range of 10^{-2} and a low quality of surface finish [119, 124–130]. Furthermore, the ceramic-based tapes show high dielectric constants, normally 6–10 (see [116, 119]). This, combined with a coarse resolution of the thick-film patterning process, results in a low reproducibility of the mmWave circuits. Recently, many efforts were undertaken to improve the quality of the LTCC process.

To minimize the shrinkage issues, a tape-on-substrate technology (TOS) was introduced [131–133], wherein the sequential laminating and firing of each tape layer on a substrate made of Al_2O_3, BeO, or AiN was proposed. With this approach, some critical shrinkage-related alignment problems can be eliminated. However, TOS is a sequential process requiring an expensive ceramic substrate carrier [117]. As a result, the manufacturing costs are relatively high. Another solution addressing the shrinkage issues was the development of Low Temperature Cofired Ceramic on Metal (LTCC-M) [134] with a specifically defined multilayer ceramic build-up that was attached to a metal core. By suitable combination of the ceramic firing with the core attachment process into one processing step, a more cost-efficient solution than TOS could be achieved. As long as LTCC-M eliminates shrinkage in the substrate plane, the vertical axis is still an issue.

A family of the photopatterned thick-film materials has been developed to address the missing accuracy of a thick-film technology. The *FODEL* material system [124, 135, 136] from Du Pont Inc., including photoimageable gold and silver conductors, yields 50 μm lines and spaces with an overall accuracy of ± 5 μm, and 125 μm large vias. The surface finish of a fired layer deposited on a 96% alumina substrate was estimated to be in the range 0.4–0.5 μm, and a postfired conductor thickness around 10 ± 2 μm. The shrinkage issues also apply to this technology. Another state-of-the-art photoprocessable thick-film material system, *KQ* [132, 133, 137], from Heraeus Inc. consists of a thick-film gold and a novel low dielectric constant ($\epsilon_r = 3.9$), low loss ($\tan\delta = 0.0018$ at 40 GHz) LTCC tape. It is compatible with the above-mentioned TOS processing technique. The lines and spaces as narrow as 25 μm can be well defined for a 5 μm thick conductor. The smallest resolved vias are 75 μm in diameter. The surface finish of a fired layer on a 96% alumina is reported to be about 0.4 μm. A line attenuation of 0.04 dB mm^{-1} and 0.165 dB mm^{-1} at 40 GHz for the 290 μm and 80 μm wide 50 Ω microstrip lines on a 130 μm and a 40 μm thick dielectric tape, respectively, was measured for this material system [138]. For comparison purposes, a line loss of 0.035–0.05 dB mm^{-1} at 40 GHz for a 400 μm wide microstrip line on various 330 μm thick, low-loss, dielectric tapes realized using standard screen printing techniques was presented in reference [119]. While substantially finer lines and spaces are available with the photopatterned thick films, this potentially reduces the claimed cost benefits of LTCC.

4.3.3.1 Application Scenarios

As long as the frequency is low or the system specifications are relaxed, a regular LTCC process may be successfully used. Above 50 GHz, however, its applicability for the practical integration of various high performance, planar, passive elements can be limited. For the microstrip realizations, dielectric tapes should offer very consistent and predictable thickness. As the thin dielectric layers are normally necessary for operation above 50 GHz, it imposes the accuracy limit on the thickness of LTCC tapes that may become inaccessible. Moreover, thin dielectric tapes imply the need for precisely defined narrow strips and spaces that are required for the repeatable performance of many passive structures. Taking a high dielectric permittivity of LTCC tapes into account, the precise CPW patterns with an appropriately narrow ground–ground spacing, necessary for avoiding the excessive coupling and leakage to substrate modes at mmWave frequencies, may also be out of reach.

Nonetheless, LTCC can be interesting for the implementation of some integrated metal waveguide structures above 50 GHz because the typical dimensions for their realization with ceramic tapes are in the range 0.5–1.5 mm, substantially relaxing the requirements on a metal pattern accuracy [139]. Unfortunately, waveguides need a significantly higher amount of real estate when compared to the planar quasi-TEM lines. Additionally, there are no commercially available mmWave active devices and MMICs that are directly compatible with this transmission medium, thus requiring integrated transition design.

An attempt to develop a rectangular waveguide operating at 60–90 GHz (TE_{10} mode of operation), implemented in a thick-film photoimageable TOS technology, was presented in reference [140]. The standard dimensions, 3.1 mm × 1.5 mm, of the hollow rectangular waveguide could be scaled down to 1.22 mm × 0.6 mm for the applied photoimageable dielectric paste ($\epsilon_r = 7$). Unpolished 99.6% alumina was used as a substrate carrier. The solid sidewalls of the rectangular waveguide were replaced by a dense row of 100 μm-wide rectangular vias. Only two prints were applied due to some limitations of the multilayering process, resulting in a waveguide height of 18 μm only, and hence an excessive loss of 0.5 dB mm^{-1} in the E-band (60–90 GHz).

Another, 1.2 mm × 0.6 mm large, dielectric-filled rectangular waveguide realized in LTCC, operating at 75–90 GHz, was reported in reference [139]. Its sidewalls were constructed by the lined via holes and edges of the metallized planes. A low loss ($\tan\delta = 0.0008$ at 60 GHz) glass-ceramic of relative dielectric constant equal to five was used as the filling material. The reported insertion loss was 0.085 dB per wavelength at 83 GHz. Due to the multilayer flexibility of LTCC, the laminated waveguide structures could be wired in three dimensions. The fundamental interconnect structures such as bends, branches, power dividers, and vertical connections between upper and lower layers with the sufficient performance were also presented. A three-dimensional feed network, consisting of the demonstrated waveguide parts, was applied for an antenna array comprising 16 × 16 laminated resonator antenna (LRA) elements [141], resulting in an overall gain of 28.1 dB and a radiation efficiency of 23% at 76.5 GHz.

4.3.4 MCM-D

A thin-film multilayer MCM-D technology is fabricated by a sequential deposition of conductors, normally Cu or Al, and insulating layers, usually polyimide or BCB, on a ceramic, silicon, metal, or laminate substrate carrier [142]. Thin dielectric layers are commonly deposited by the spin-coating process, yielding a very uniform, precisely defined thickness and an almost mirror-like surface finish, the latter being of special importance for the ohmic loss minimization at mmWave frequencies. Vias are developed by laser ablation, reactive ion etching, or wet etching. Thin metallization layers are deposited by sputtering. For the thicker conductors, some additive processing by electroless plating or electroplating may also be done. Curing of dielectrics is processed at considerably lower temperatures compared to LTCC, ranging from 200 °C for BCB to 400 °C for polyimide. The minimum metal pattern feature size is typically around 20 μm but can also be scaled down to a few micrometers, depending on the conductor layer thickness and the exact metallization process flow. The spin-coated dielectrics, in general, show substantially lower dielectric permittivities than the ceramic materials. This yields larger physical dimensions for a given electrical length, leading to easier dimensional control. A high reproducibility of small metal pattern features and a precise dielectric thickness control offered by these technologies make them very interesting candidates for the realization of integrated interconnect structures. The MCM-D technologies are also capable of supporting a common integration of low-cost digital and the high-performance analog electronics. In references [143–147], it was demonstrated that a rather high manufacturing cost associated with the standard process flow can be considerably reduced if a large area processing option can be used.

The ability of the BCB-based MCM-D technology for the integration of various high-performance RF circuits has already been demonstrated for microwave frequencies [142, 148–158]. Some integrated antennas realized using this technology were also presented [159, 160].

The spin-coated BCB shows some important advantages: low dielectric constant (2.64–2.66 up to 110 GHz [16, 17, 88]) with high-temperature stability, low-loss tangent at mmWave frequencies (0.0008–0.002), photosensitivity, and insensitivity to the moisture absorption (self-protection feature). It allows the fine line resolution, precise thickness definition, typically 0.5 μm per layer (see Table 4.2), and shows a very low surface roughness (see Figure 4.6 in Section 4.3.1). The low BCB dielectric permittivity is capable of supporting similar phase velocities for even and odd modes of the coupled-line sections, being thus, advantageous for the design of broadband high directivity elements based on coupled-line topology [158].

As a typical dielectric layer thickness in MCM-D process, including the BCB-based technology, is in the range of a few micrometers, and a standard stack-up includes no more than three layers, the realization of low-loss microstrip lines at mmWave frequencies is impossible. As a result, the coplanar waveguides are usually preferred in the MCM-D integrated microwave modules. In this case, the thin dielectrics can be coated on various substrates with a dielectric permittivity ranging from low to high. Unfortunately, both substrate options can be potentially dangerous for the CPW-based designs at higher mmWave frequencies, as shown in Section 4.2.1.3.

An attempt to develop a mmWave interconnect technology was presented in references [161–164], wherein the composite BCB/Kapton multilayer build-up was applied on top of the alumina substrate. The 50 μm large vias were used to connect the packaged devices that were embedded in recessed cavities formed in a substrate carrier. A microstrip line on an organic BCB/Kapton dielectric substrate was proven to be a viable transmission line configuration at mmWave frequencies. An insertion loss of 0.12 dB mm^{-1} at 110 GHz for a 240 μm wide 50 Ω microstrip line on a three-layer BCB/Kapton dielectric build-up with an overall thickness of 110 μm was measured [161]. Furthermore, it was also demonstrated that the stacked vias can be considered to be a low parasitic vertical interconnection scheme if combined with the proper technology build-up. The high-performance terminations up to 50 GHz and Wilkinson power divider at 60 GHz were also shown to be possible due to the incorporation of an integrated thin-film resistor layer into the main technology process [163, 164]. Recently, the use of spin-coated BCB films has also been reported for the successful embedding of MMICs into the micromachined cavities on a silicon host substrate [165, 166]. The approach from reference [166] utilized gold bumps on the chip pads and on the metallized silicon surface that were fabricated prior to the BCB coating to eliminate the need for regular drilled and plated vias, which was shown to be possible due to the excellent BCB planarity.

Another effort to expand the existing interconnect MCM-D integration technologies up to 60–100 GHz applications was carried out within the European research project LIPS [17, 65, 68, 167]. Within the frame of this project, a modified three-layer, BCB-based MCM-D technology set with a performance-oriented, optimally chosen dielectric thickness was developed (see Section 4.4 for details on finding the optimum technology build-up).

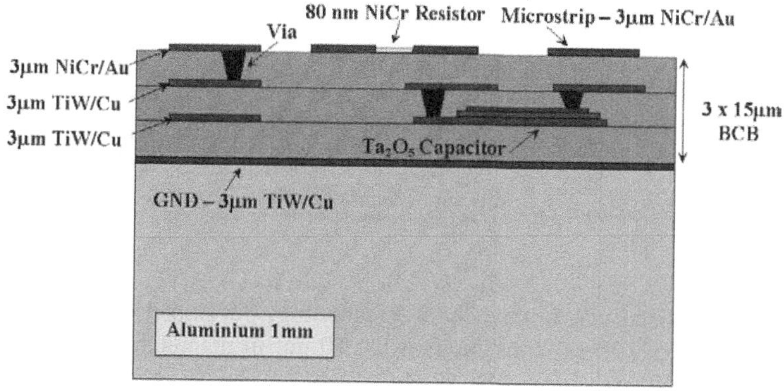

Figure 4.14: A layer build-up of the MCM-D technology (configuration I) developed within the EU research project LIPS [17, 65, 79].

In view of the constraints related to the mmWave usage of CPW structures in the hybrid interconnect technologies (see Section 4.2.1), a microstrip line configuration on an optimized dielectric substrate build-up was selected as the main transmission line topology, similarly to the above-mentioned BCB/Kapton multilayer thin-film process. As a typical layer thickness of the photosensitive BCB dielectric is in the range of a few micrometers for the classical two- and three-layer MCM-D technology, increasing this thickness to its optimum value, defined by the performance of the elements to be integrated, was one of the main technology development goals within the project. It is important to mention that there is a practical limit on the maximum feasible thickness of a single BCB layer and of an overall BCB layer stack. The problem is defined by considerable differences in the CTE for different materials participating in an overall technology build-up, which develops some nonnegligible mechanical tension during different processing steps, typically in the temperature range of 20–250°C.

A final BCB stack-up consisting of three layers, each 15 μm thick, was fixed. Two technology setups were investigated.

1. Configuration I: 3 × 15 μm thick BCB and the additional 5 μm thick local BCB passivation for the exposed NiCr resistors, electroplated NiCr/Au (80 nm/3 μm thick) as the top metallization and electroplated TiW/Cu (80 nm/3 μm thick) for all internal layers (see Figure 4.14).

2. Configuration II: 3 × 15 μm thick BCB and the additional layer of a 5 μm thick BCB passivation, sputtered Ti/Cu (70 nm/1 μm) for all metal layers.

As the microstrip structures are based on the BCB dielectric only, the choice of the main substrate carrier plays no role from the electrical performance point of view because the carrier and the thin-film layers are separated by a buried ground plane (see Figure 4.14). For completeness of the information, the first configuration was coated on aluminum, whereas the other was on a low-resistivity silicon. The smallest allowed via diameter of 60 μm was necessary to result in the appropriate via yield (see Figure 4.15). As a result of the chosen technology setup, the stacked grounding vias for microstrip lines are impossible; only the

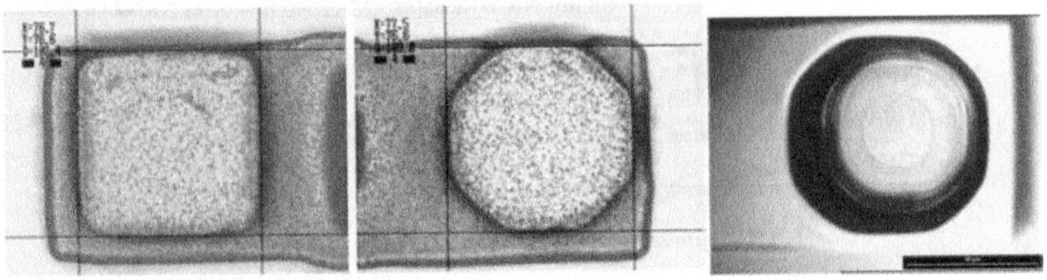

(a) (b)

Figure 4.15: Metallized vias in the MCM-D technology developed within the EU research project LIPS [17, 65, 79]: (a) configuration I – 70 μm in diameter; (b) configuration II – 60 μm in diameter.

staircased- or staggered-like with a minimum via spacing of 15–20 μm are allowed. The straight vertical vias with a higher height/width ratio could be created by laser ablation or anisotropic etching, but with additional cost. A very high DOP [168] above 90% could be achieved owing to the favorable relation between dielectric layer height and metal strip thickness.

The NiCr (38 Ω/square) and Ti (17 Ω/square) layers are deposited on the entire wafer surface to avoid some additional processing steps for the definition of resistors. With this approach, they are formed by only the local conductor patterning. The layer thickness tolerance for both materials results in a typical across-wafer resistance variation of 0.5 Ω/square. The resistive elements located on the substrate top surface support the via-less connections with the main transmission line structures, being of importance at mmWave frequencies. As both NiCr and Ti do not show magnetic properties and can still be considered to be thin at the frequency range of interest (60–100 GHz), their influence on the line loss can be neglected (see Figure 4.10 in Section 4.4).

In order to measure the process variations and to investigate their influence on the performance of various integrated passives, a set of technology characterization test vehicles (TCTVs) was developed [79]. Table 4.2 summarizes the measured basic technology parameters with the corresponding $\pm 3\sigma$ deviations. The parameters of a low profile thin-film build-up from another European research project LAP [143–145], representing a typical MCM-D technology, are included as a reference. Table 4.3 presents the statistical variations of the line characteristic impedance and resonance frequency for some given 50 Ω microstrip line section as a result of the $\pm 3\sigma$ technology parameter deviations from Table 4.2. A more detailed analysis shows that the characteristic impedance value is mainly affected by the dielectric layer thickness variations within the supported technological parameter values. However, the location of a resonance frequency at 77 GHz is highly dependent on the exact dielectric permittivity characterization. About one-third of its variation is due to the permittivity value uncertainty even for the very narrow $\pm 3\sigma$ deviation range of 2.64 ± 0.015 (see Table 4.2), hence, underlining the importance of an accurate in-house material characterization.

Summarizing, the family of MCM-D technologies with their repeatability and accuracy is potentially capable of meeting the high-performance requirements needed for numerous precise interconnect structures and integrated passives at mmWave frequencies, specifically

Table 4.2: Parameter description of the thin-film technologies developed within the frame of two EU research projects: LAP [143–145] and LIPS [17,65,79]. The variations are specified as $\pm 3\sigma$ deviations.

Parameter	Low-profile thin-film	Configuration I	Configuration II
Metal thickness (t) (μm)	4.0 ± 0.5	3.0 ± 0.5	1 ± 0.15
Line width accuracy (w) (μm)	± 2	± 1.5	± 1.0
Metal conductivity (σ)* ($\Omega\,m^{-1}$)	$5.84 \times 10^7 \pm 0.04$	$4.53 \times 10^7 \pm 0.02$	$5.88 \times 10^7 \pm 0.0$
BCB thickness/layer (h) (μm)	6.5 ± 0.5	15.0 ± 0.8	15.0 ± 0.6
BCB permittivity (ϵ_r) *	2.64 ± 0.015	2.64 ± 0.015	2.64 ± 0.015

* Parameter uncertainties represented as variations.

Table 4.3: Variations of the resonance frequency and line characteristic impedance for a 50 Ω microstrip line section as a result of the $\pm 3\sigma$ technology parameter deviations from Table 4.2.

Parameter	Low-profile thin-film	Configuration I	Configuration II
Characteristic impedance (Ω)	$Z_0 = 50 \pm 5.3$	$Z_0 = 50 \pm 2.9$	$Z_0 = 50 \pm 2.2$
Resonance frequency (GHz)	$f_{res} = 77 \pm 0.60$	$f_{res} = 77 \pm 0.34$	$f_{res} = 77 \pm 0.26$

above 50 GHz, definitely outperforming a known set of LTCC and MCM-L technologies in this regard.

4.3.5 Flexible Substrates

There are a number of important advantages related to the flexible interconnect technologies [169].

- Higher potential to absorb stress when compared to the standard thick and rigid substrates, being advantageous for the flip-chip assemblies. For that reason, the flexible substrates are commonly applied where heavy life-time cycles or vibration absorption are in demand.

- Better heat transfer through a thin flex substrate.

- Elimination of the plated through holes and compatibility with the microvia technology, leading to an increased routing density.

- Flexible boards can be bent, being important for various three-dimensional packaging schemes that require stacking or folding. The bending angle is a function of both the layer thickness and the corresponding metal pattern.

- Flexible electronics is a good candidate for the roll-to-roll processing, making it low cost.

- A considerable amount of work is ongoing to develop a number of complementary technologies such as ultra thin (10–20 μm) foldable chips and RF passive components supposed to be embedded within a flex substrate that can lead to the revolutionary packaging and system integration schemes in the near future.

On the other hand, the lack of intrinsic rigidity leads to numerous difficulties in the assembly and handling of the flex substrates, commonly requiring some new tooling.

Most of the flex boards use a thermoplastic, nonreinforced dielectric film, such as polyimide or polyester, as a dielectric layer. Depending on its properties and on the corresponding metallization process flow, the flexible technologies can deliver a metal pattern resolution resembling more a laminated PCB or a thin-film MCM-D. The development of high-resolution flex technologies is driven mainly by consumer applications such as liquid crystal displays (LCDs) and hard disk drive (HDD), and by modern chip-scale packages (CSPs) requiring minimum line/space feature size as small as 20 μm [169, 170].

A flexible circuit technology is popular for microwave applications. Due to the above-mentioned high-density interconnect capabilities, it can also be of special interest for mmWave frequencies. One of the first microwave chip-on-flex technologies was developed by Lockheed Martin [171, 172], wherein the MMIC chips were directly mounted on a multilayer polyimide flexible board, thus eliminating the need for wirebonding. The rigidity of an overall assembly was provided by the plastic mold encapsulating bare active devices. A laser was used to ablate the vias and to pattern the thin-film metallization, resulting in the fine line/space features. A thickness of the applied polyimide layer was chosen to be 25 μm and 50 μm. A C-band Tx/ Rx integrated module delivering a 3.5 W RF power was reported to demonstrate the proposed concept. A similar technology with some additional feedthroughs in the form of vertical copper wires, embedded into the plastic mold, was demonstrated for a 4 GHz microwave BGA package [173]. Recently developed LCPs with their numerous advantages (see Section 4.3.2) were also applied for flexible circuit electronics. A 3.1–10.6 GHz broadside coupled Marchand balun fabricated in a multilayer adhesiveless LCP flex process was reported in reference [112]. The LCP flex has also been successfully applied to form a near-hermetic package for RF MEMS switches at X-band [113].

4.3.6 Silicon Micromachining

Micromachining techniques can be applied to any semiconductor substrate. However, the use of silicon is especially interesting for cost reasons and the capability for direct integration of active devices. Numerous quasi-TEM planar transmission lines can be developed with micromachining, very often showing the superior electrical performance at mmWave frequencies. Basically, there are two techniques for implementing this novel approach. The first takes advantage of a membrane support for transmission line elements [174, 175]. The other utilizes some integrated shielding cavities [174, 176, 177].

4.3.6.1 Membrane-supported Technology

A membrane-supported transmission line may be viewed as an evolution of conventional planar lines, specifically microstrip or CPW lines (Figure 4.16). It supports almost pure and nondispersive TEM wave propagating in a near homogeneous environment. Homogeneity is accomplished by a micrometer thick diaphragm supporting the lines while ground is normally

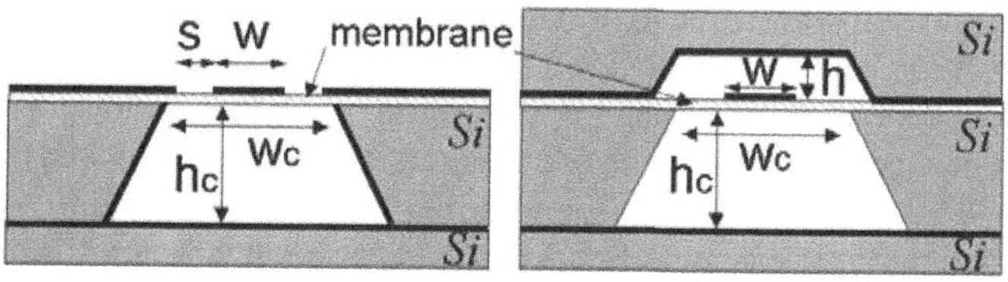

(a) (b)

Figure 4.16: (a) Microshield line, and (b) SMM line (based on the original figures from reference [175], reproduced by permission of © 1996 IEEE).

provided by a metallized micromachined cavity [174, 175]. Two different cavity shapes can be recognized: pyramidal and vertical. The former is fabricated using the anisotropic etchant such as ethylene diamine pyrocatechol (EDP) or potassium hydroxide (KOH) while the latter by means of deep reactive ion etching (DRIE) techniques [178]. The most representative membrane build-up was presented in reference [174]. It is a three-layer $SiO_2/Si_3N_4/SiO_2$ material stack-up with a typical thickness of 7000, 3000, and 4000 Å for each of the consecutive dielectric layers, respectively. Such a thin membrane makes the transmission line transparent to propagating signals up to 3 THz. Membrane dimensions as large as 2 mm × 19 mm were reported [174].

Two basic membrane-based transmission line topologies were proposed: microshield line and shielded membrane microstrip (SMM) [174, 175].

Microshield line. The geometry of a microshield line [174] can be considered as a CPW topology situated above a metallized, air-filled cavity (see Figure 4.16). Its formation requires two wafers: the first with a micromachined cavity, and the other with a metallized ground plane to enclose it from the bottom [174]. The typical cavity heights, h_c, and widths, w_c, for the circuits operated at 10–40 GHz were designed to be 350 μm and 800–2000 μm, respectively.

A seventh-order Chebyshev (0.5 dB passband ripples) stepped impedance lowpass filter with the cut-off frequency at 25 GHz [174] and 85 GHz realized in the membrane-supported microshield line configuration was reported in reference [175]. It showed a passband insertion loss of 0.5 dB and 1 dB for both realizations, respectively. The following line dimensions, w and s (see Figure 4.16 for dimensions definition), were used for the low and high impedance sections: 40/680 μm (265 Ω) and 1320/40 μm (45 Ω) for the first filter, and 20/280 μm (277 Ω) and 540/20 μm (63 Ω) for the second. The corresponding cavity widths were 1800 μm and 800 μm. The filters were fed by a 75 Ω (350/35 μm) and a 92 Ω (220/50 μm) line, respectively.

Shielded membrane microstrip. Thin dielectric membranes also provide the basis for the SMM line configuration [175]. It is essentially a shielded microstrip line suspended in air, resulting in the low high-frequency dispersion and low dielectric loss (see Figure 4.16). The practical SMM lines require a three-wafer assembly [175]. Formation of an upper half micromachined ground plane cavity with a partially removed silicon requires specific attention, as its depth, h, is crucial for the transmission line parameters and, thus, must be precisely controlled. However, due to the temperature and etch area dependencies of the etch rate, this control can become complicated. A ±5 μm variation for the nominal cavity depth of 100 μm was reported in reference [175], resulting in 1% deviation for both the center frequency and the operation bandwidth of a five-section coupled-line 94 GHz bandpass filter with a nominal bandwidth of 4.3%. Some further reduction of depth, h, often preferable at such high frequencies, can possibly make the precise etch control an issue for some narrowband structures.

Three different five-section coupled-line bandpass filters with equal-ripple Chebyshev characteristics operating at 94 GHz were implemented in the SMM configuration [175]. The circuit designs were normalized to 90 Ω, being the characteristic impedance of the feed lines. The heights of both upper and lower shielding cavities were 100 μm and 500 μm, respectively. The following insertion loss was achieved for each of the realizations:

1. center frequency, f_c, of 94.7 GHz, relative bandwidth, B, of 6.1%, and passband insertion loss of 3.6 dB;

2. center frequency, f_c, of 95 GHz, relative bandwidth, B, of 12.5%, and passband insertion loss of 2.2 dB;

3. center frequency, f_c, of 94.9 GHz, relative bandwidth, B, of 17.7%, and passband insertion loss of 1.4 dB.

4.3.6.2 Micromachined Shielded Quasi-planar Circuits

The micromachining may also be applied to develop some completely shielded planar circuits (see Figure 4.17) [174, 176, 177] printed directly on an underlying substrate without the membrane support. The material of choice is typically a polished high resistivity silicon wafer. This build-up requires the use of two silicon wafers, the first with the upper half shielding cavities and the other accommodating planar circuits on the substrate-filled lower half cavities [174]. The two wafers can be attached using the regular adhesion methods or Si-to-Si electrobonding [174, 176, 177]. An additional variable thickness micromachining option [177], feasible with the selective time etching techniques, is able to extend the process flexibility to the design of shielded circuits with a locally reduced substrate thickness (see Figure 4.17). A practical application of this technique for the design of different microstrip stepped impedance lowpass filters with the cut-off frequencies at 20 GHz and 35 GHz was demonstrated in reference [179]. With this approach, an increased impedance ratio between the high and low impedance line sections could be achieved, resulting in the sharper frequency characteristics near cut-off frequency and in a higher stopband attenuation when compared to the conventional planar realizations. Two different techniques were applied [179]. The first aimed at the characteristic impedance decrease for the low impedance line sections by applying a local substrate thinning while in the other, some local micromachining of an air cavity was used to increase the characteristic impedance values for the high impedance line sections.

An insertion loss of 0.075 dB mm^{-1} at 40 GHz for a completely shielded line was reported in reference [176]. The line dimensions, w and s (see Figure 4.17 for the definition of dimensions), were 180 μm and 130 μm, respectively. The upper cavity was 280 μm high (h), whereas the lower was 350 μm (h_s). Both were 800 μm wide (w_c). A 43 Ω microstrip line embedded in a lower shielding cavity, realized in a locally thinned silicon

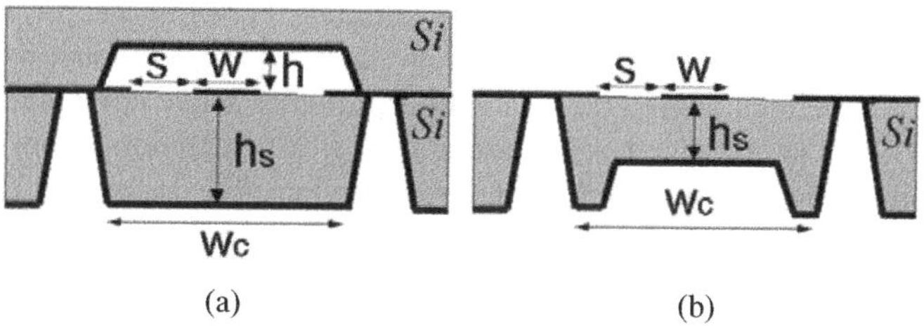

Figure 4.17: (a) Completely shielded planar line, and (b) partially shielded microstrip line on a locally thinned silicon wafer (based on the original figures from references [174] and [177], reproduced by permission of © 1997 IEEE).

wafer ($h_s = 100$ μm), showed an insertion loss of 0.14 dB at 40 GHz [180]. The line and cavity dimensions were: line width, w, of 94 μm, distance to planar ground plane, s, of 89 μm, and cavity width, w_c, of 413 μm.

4.3.6.3 Application Scenarios for the Micromachined Planar Transmission Lines

The evident drawback of the membrane-supported microshield line structures is a high value of the minimal feasible characteristic impedance. As the lateral dimensions of cavities accommodating the line geometries should become smaller with the frequency increase in order to keep the waveguide modes below cut-off, this minimum impedance value can grow substantially. An overall width of a 50 Ω transmission line can exceed 1 mm when realized in the proposed technology, being comparable to the operating wavelength at higher mmWave frequencies. This, in turn, may potentially lead to the nonnegligible performance degradation of meander- and junction-like discontinuities (see Section 4.4.1). For these reasons, the designed circuits should be normalized to the very high reference impedance values. As they will still normally need to be matched to some other low impedance devices, this may call for the use of impedance transforming sections with a high impedance ratio involved or some other matching structures, potentially leading to some operation bandwidth limitations and increased overall loss. The integration of some in-line passive circuits may be considered the most practical application for this approach. The main advantage of this line configuration is the elimination of power loss associated with the surface waves and space wave radiation, and the improvement of an intercircuit isolation due to the inherent shield.

The micromachined shielded structures without the membrane support, briefly summarized in the previous paragraph, may deliver some higher routing functionality as the presence of silicon wafer under the printed line considerably lowers their strip width. Nevertheless, they are no longer nondispersive, and they require a high resistivity silicon to be used for the dielectric loss minimization. Furthermore, the underlying silicon substrate should be substantially thinned to avoid the surface wave excitation at mmWave frequencies. The SMM configuration also appears to support the required routing flexibility at mmWave frequencies if an appropriately hollow cavity, defining the microstrip line height, is machined. With the reported process tolerances of the etch depth to date [175], some difficulties in the circuit realizations at higher mmWave frequencies can arise if a very precise frequency location or a narrow bandwidth is a must.

Micromachining can definitely support the ultimate performance for many mmWave planar transmission lines, but at the expense of state-of-the-art fabrication techniques. A truly monolithic implementation with the high-performance active devices requires multiple semiconductor layers to be formed using the epitaxial growth techniques, making it expensive when compared to the solutions employing some drop-in active devices.

4.3.6.4 Micromachined Waveguides

Several micromachined waveguide structures have been reported in the literature. Only a couple of them are outlined briefly here.

- A 2.5 mm × 1.25 mm large rectangular waveguide operating at 75–110 GHz was presented by McGrath *et al.* [181]. It consists of two half-sections split along its broadside. The vertical walls are formed by KOH etching of the polished silicon wafers, and a Cr/Au (200 Å/3.5 μm) metallization layer is used for the waveguide walls. A very low loss of 0.04–0.06 dB per wavelength across most of the 75–110 GHz band (WR-10) was measured.

- The waveguide designed by Collins *et al.* [182] and Digby *et al.* [183] is 1.84 mm wide and shows a cut-off frequency of 81.5 GHz for the TE_{10} mode. It is fabricated by first metallizing a semiconductor wafer with a Ti/Au layer to form the bottom walls. Next, a photoresist former, defining the internal waveguide dimensions, is prepared on top of this layer by means of photolithographic techniques and is then coated with gold. Finally, an air-filled rectangular waveguide geometry is created by removal of the photoresist with a solvent [182, 183]. The maximum waveguide height was limited to only 100 μm owing to photoresist thickness limitations. A theoretically calculated insertion loss of the designed waveguide geometry was found to vary between 0.227 dB and 0.33 dB per wavelength over the 90–110 GHz frequency range. The measured attenuation was between 0.133 dB and 3.33 dB, being substantially higher than the computed attenuation. This high loss was attributed to some mismatch issues in the measurement setup.

4.3.6.5 Organic Micromachining

An ultra-thick negative photoresist, *SU*-8, introduced about a decade ago, is capable of the processing features of up to 1 mm in height with the large aspect ratios [184, 185]. This property was used by Collins *et al.* [186, 187] to form a substantially higher waveguide geometry than that above detailed and reported in references [182, 183]. The structure was fabricated by consecutive spinning of *SU*-8 onto a semiconductor substrate, exposing, baking, and development until a final air-filled waveguide cross-section of 2.54 mm × 0.7 mm was achieved. It was then sputtered with gold to form the inside walls and was covered with a previously presputtered lid. An insertion loss of 0.2–0.3 dB per wavelength at 85–100 GHz was reported. This relatively high loss was attributed to imperfect join between the waveguide sidewalls and the top lid.

The same photoresist was used in reference [188] as a spacer on a low resistivity silicon wafer to develop low-loss micromachined coplanar waveguide interconnects. An insertion loss of 0.018 dB mm^{-1} at 20 GHz was reported for a 350 μm wide (280 μm and 70 μm for the strip and slot width, respectively) grounded coplanar waveguide printed on top of a 200 μm thick *SU*-8 resist.

4.3.7 Plastic Injection Molding

Modern plastic injection molding can become a viable technology of choice for the realization of various waveguide components and packages at mmWave frequencies [4, 72–77], thus replacing more expensive micromachining processes. Its main advantages are a high-volume production capacity combined with a low component weight. With its excellent geometrical accuracy as low as 5 μm [75], uniformity, and reproducibility [4], it is capable of supporting the most complex design features. High-temperature thermoplastic polymers such as polyphenylene sulfide (PPS) are the main candidates to fulfill a broad range of thermal, mechanical, and metal plating requirements [75]. LCP and polyetherimide (PEI) materials were also reported to be useful [74, 75]. A physical vacuum deposition (PVD) of copper/silver for the plastic metallization was applied in reference [75], whereas a copper/nickel conducting layer was used in reference [74].

A successful high-yield realization of the plastic injection-molded parts requires specific rules to be obeyed in the design process, typically including:

- the use of round corners, replacing the ideal right-angle features for the reliable metallization process;
- a constant wall thickness over the entire three-dimensional geometry [4, 72–75] to avoid possible tensions within the material;
- some mechanical balance of the designed part for form stability [4, 72];
- a small draft angle of the waveguide walls, typically below 1°, for the smooth detachment of a plastic form from the corresponding metallic mold [74].

For the highest manufacturing accuracy of a designed prototype, the molding tool should be optimized with respect to its specific geometrical details and the strain characteristics of the chosen plastic material. Even though some precalculation of dimensional changes for the molded part due to material shrinkage in the cooling process is possible to some extent, its accuracy is insufficient for the first-time-right fabrication of some very challenging mmWave components such as high selectivity, high order filters, and diplexers [72]. Therefore, some preproduction tuning in the form of additional adjustable elements such as posts implemented within the molding tool may be required [4, 72, 75]. Once this process is finished, no additional post-tuning in the production process should be needed owing to a high reproducibility of the plastic injection mold technology.

One of the simplest application scenarios, where the plastic injection mold could replace a more expensive micromachining, is related to the micropackaging concept [189, 190] that was developed to eliminate the unwanted propagation of parasitic modes and related resonance effects inside the electrically large metallic packages by introducing some small, shielded structures to follow the individual circuit paths. A successful realization of this concept by means of a plastic injection mold technology was already reported for both 28 GHz [4, 72] and 60 GHz [75, 76] communication systems.

A high-performance bandpass filter consisting of seven iris-coupled cavity resonators was demonstrated for the Ka-band wireless Internet access applications [74]. The authors in references [4,72] have utilized the plastic injection mold approach to build both LO filter block and diplexer unit for a 28 GHz communication front-end. A 58 GHz high gain (34 dBi), flat panel antenna comprising 256 box horn radiating elements with a waveguide-based feed network was presented in reference [73]. A very extensive use of metallized plastics has been recently reported in references [75,76] for a 60 GHz wireless point-to-point link, wherein a broad range of complex waveguide components, particularly including a specially designed cover part realizing the previously mentioned micropackaging concept, a 38 dBi gain, flat slot antenna array, and a diplexer filter with a Tx/Rx isolation of 80 dB, was demonstrated. A unique contactless, double-ridge waveguide interface was designed for a direct EM coupling between the diplexer input and the corresponding LTCC-integrated subharmonically pumped mixer, thus ensuring that there are no 60 GHz components on the substrate level except for the mixing Schottky diode pairs.

4.4 Performance-oriented Interconnect Technology Optimization

In this section, the development of a general interconnect technology platform for 60–100 GHz applications utilizing a modified multilayer BCB-based MCM-D process with a performance-oriented, optimally chosen dielectric thickness will be analyzed briefly [17,68, 167]. To a detailed description of the ultimate technology setup, developed within the EU research project LIPS [65], see in Section 4.3.4. The numerous simulation and measurement results of different passive structures will also be given, the high performance of which was only possible with the optimized technology build-up.

Due to the numerous issues and risks associated with the CPW interconnect structures realized in the contemporary hybrid interconnect technologies, as outlined in Section 4.2.1, a microstrip line configuration realized on only a BCB-based dielectric substrate was chosen as the main transmission medium. In order to find the optimum dielectric layer thickness, a set of various factors, determining the performance of a broad range of integrated interconnect structures and passives within a general microstrip-oriented interconnect technology, was defined (see also Section 4.2 for the more comprehensive discussion on various parasitic phenomena limiting the performance of mmWave interconnects). The most important of them are gathered in Table 4.4. Note that the most commonly applied microstrip patch antennas are considered to be integrated within the same substrate carrier, preventing excessive loss usually associated with the mmWave transitions connecting front-end modules with external radiators. It is clear that simple patch antennas are unable to meet required performance specifications for some demanding applications, but they are included here for consideration as a capability indicator for a given technology build-up.

The process of finding an optimum BCB thickness for the modified MCM-D technology presented in this section, should be viewed as an example to explain the influence of different high-frequency phenomena on the performance of mmWave interconnects and to indicate the importance of the proper combination of transmission line topology with the corresponding technology build-up for the successful realization of a mmWave system. Any other modern multilayer technology could be used instead for these demonstration purposes.

Table 4.4: Factors determining the performance of a general mmWave, microstrip-oriented interconnect system taken into account in the process of BCB thickness optimization. Parameters marked with (+) increase the system performance with a thicker dielectric substrate, whereas those marked with (−) decrease it.

Parameter	BCB dielectric thickness increase
Line loss	(+)
Microstrip antenna efficiency (patch antenna in particular)	(+)*
Microstrip antenna operation bandwidth (patch antenna in particular)	(+)
Line dispersion	(−)
Line width/wavelength ratio	(−)
Radiation at discontinuities	(−)
Reactances at discontinuities	(−)
Intercircuit isolation	(−)
Shorting ground inductance	(−)
Design simplicity	(−)

*For the substrate thickness exceeding some critical value, the overall radiation efficiency of the microstrip antennas, in particular, patch antennas, decreases because of the parasitic influence of surface waves starting to dominate over the conductor and dielectric dissipation losses; see Equation 4.8 in Section 4.2.1.

4.4.1 Performance-oriented BCB Dielectric Thickness Optimization

At lower microwave frequencies, the insertion loss of a straight transmission line section is a primary factor determining the performance of an overall interconnection system. For this reason, thick dielectric substrates are normally preferred for the circuits realized in microstrip configuration as they imply wider strips and, thus, lower line loss. Due to the short wavelengths at 60–100 GHz, this loss may no longer be a dominant parameter because of the increasing influence of transmission discontinuities that become electrically large for the contemporary hybrid interconnect technologies reaching their miniaturization and processing accuracy limits. Too-thick substrates cause strong coupling to surface waves. This, in turn, results in the highly dispersive and lossy nonquasi-TEM transmission line elements with a poor intercircuit isolation. A high strip width-to-wavelength ratio leads to considerable radiation and high phase shifts at the meander- or junction-like transmission line discontinuities, making the system design very challenging or even impossible.

The main goal of the dielectric substrate (in our case, BCB) thickness optimization is to achieve a trade-off between the microstrip line loss and microstrip antenna performance (defined by radiation efficiency and operation bandwidth) on one side, and the low line dispersion, low strip width-to-wavelength ratio, high intercircuit isolation, short grounding paths, and minimum parasitic effects at the transmission line discontinuities (reactances, space wave radiation and excitation of surface waves) on the other (see Table 4.4). It calls for careful consideration of the performance of different structures intended for integration, typically including some basic interconnect elements, distributed passives, or even antennas and antenna array feed networks.

Table 4.5: Simulated attenuation of a 50 Ω microstrip line versus BCB thickness. Simulations were performed with *HFSS* [67]. A 3 μm thick copper conductor and the BCB loss tangent of 0.002 were assumed in the simulations (from reference [68], reproduced by permission of © 2002 IEEE).

BCB thickness (μm)	50 Ω line width (μm)	α (dB mm^{-1}) at 60 GHz (at 80 GHz)
15	30	0.18 (0.21)
30	70	0.1 (0.115)
45	112	0.067 (0.086)
60	150	0.058 (0.072)
100	240	0.045 (0.055)

A brief analysis of the above-mentioned factors in view of their relation to the BCB dielectric thickness follows.

Table 4.5 gathers the simulated loss of a 50 Ω microstrip line for different values of the BCB thickness. Notice that a 45–60 μm thick layer already supports reasonably low loss that decreases only slightly for the thicker BCB. In particular, a quarter-wavelength-long line section on a 45 μm thick BCB dissipates only 1.25% of the incident power at 100 GHz, 80% of which is associated with ohmic loss. The same line section on a 100 μm thick BCB shows only a slightly better power loss of 0.84%. Table 4.6 shows the simulated line loss and wavelength for different line geometries on a 45 μm thick BCB. Note that the microstrip line loss is only slightly dependent on strip width because the highest current density at 60–100 GHz is located in the corners of a strip cross-section. Thus, the loss for a 50 Ω transmission line can be considered as a representative value. Moreover, the lines show no high-frequency dispersion up to 100 GHz and above, due to the low dielectric constant of BCB and its relatively low layer thickness. A line width-to-wavelength ratio is another very important factor determining the performance of different functional elements. It shows an approximate linear dependence on the BCB thickness for the considered range of dielectric layer thickness values. Changing the BCB thickness from 45 μm to 100 μm results in a considerable increase in the line width-to-wavelength ratio from 5.5% to 12% for a 50 Ω line at 100 GHz. Furthermore, a 45 μm thick 50 Ω microstrip line reduces this ratio by a factor of three when compared with the CPW transmission line configuration on a 45 μm/ 500 μm thick composite BCB/ ROGERS 4003 dielectric substrate from Section 4.2.1 (see Table 4.1).

In order to investigate a direct influence of the BCB thickness on the performance of different distributed passives and antenna array feed networks, the frequency behavior of a variety of transmission discontinuities was simulated. As the detailed investigation results are beyond the scope of this chapter, only some simple examples are presented.

- The first element is an open-ended 50 Ω transmission line that shows a 1% and a 5.5% power loss at 100 GHz on a 45 μm and on a 100 μm thick BCB, respectively. The corresponding open-end length extension values constitute 1.2% and 3.5% of the operating wavelength at 100 GHz. As the radiation at discontinuities is an increasing squared (space waves) or cubed (surface waves) function of frequency, its influence

Table 4.6: Simulated line attenuation and wavelength at 60 GHz for different microstrip lines on a 45 μm thick BCB. Simulations were performed with *HFSS* [67]. A 3 μm thick copper conductor and the BCB loss tangent of 0.002 were assumed in the simulations (based on the original table from [68], reproduced by permission of © 2002 IEEE).

Configuration	Geometry (μm)	Z_0 (Ω)	α (dB mm^{-1}) at 60 GHz (at 80 GHz)	λ (mm) at 60 GHz
Microstrip	$w = 194$	35	0.068 (0.082)	3.28
Microstrip	$w = 112$	50	0.067 (0.083)	3.35
Microstrip	$w = 62$	70	0.072 (0.086)	3.42
Microstrip	$w = 28$	100	0.082 (0.098)	3.5

Parameter 'w' denotes a strip width.

on the mmWave system performance can easily outweigh a simple line loss associated with the typical straight transmission line sections.

- The next structure is a 50–100–100 Ω T-junction with a right-angle geometry. For a 45 μm thick BCB, a return loss better than 30 dB at the 50 Ω port and a 1.2% power loss at 80 GHz is possible to achieve. The same junction on a 100 μm thick BCB shows a return loss inferior to 17 dB and a 7% loss, mainly attributed to radiation. The latter could be reduced to only 4% after some geometry optimization by changing the angle between two output branches.

- The last example is a 50 Ω right-angle meander line consisting of eight chamfered bends and nine line sections (each section 500 μm long). The structure was designed so that the influence of parasitics associated with the bends is cumulative at 80–90 GHz (for the more detailed description of the test configuration, see below). The worst simulated return loss within the considered frequency band of 1–100 GHz is 26 dB (at 95 GHz) and 15 dB (at 85 GHz) for the design on a 45 μm and on a 100 μm thick BCB, respectively. Furthermore, the power lost on the meander and on the equivalent transmission line of the same length differ by only 3% at 100 GHz for the former while for the latter by as much as 28%.

As a typical dielectric layer thickness for the MCM-D technologies is around a few micrometers, and a standard dielectric build-up includes no more than three layers (see Section 4.3.4), the microstrip patch antennas realized using only thin-film dielectrics may show a radiation efficiency in the range of 25% at frequencies even as high as 80 GHz [146, 147, 191]. The use of typically a few hundred micrometers thick substrate carrier supporting the thin deposited layers, commonly applied to enhance the patch antenna performance at lower microwave frequencies, cannot be applied because of the strong surface wave effects. As a consequence, some process enhancement is needed in order to obtain thicker films and, therefore, an increase in radiation efficiency.

Figure 4.18 presents the simulated efficiency of a directly fed rectangular patch antenna at 77 GHz with a width-to-length ratio of 1.5 as a function of a dielectric layer thickness. The efficiency rises slowly for BCB thicker than 45–55 μm (not shown) and achieves its maximum of 81% for a 100 μm thick film. A further thickness increase will result in a lower radiation

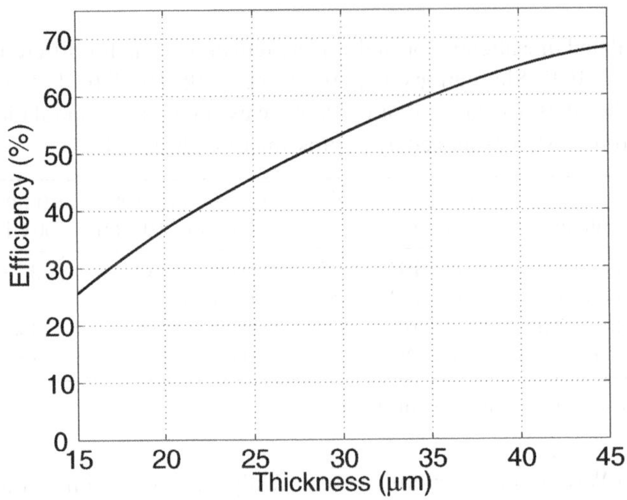

Figure 4.18: Simulated radiation efficiency of a directly fed rectangular patch antenna at 77 GHz with a width-to-length ratio of 1.5 as a function of the BCB thickness. The simulations were performed with *HFSS* [67]. A 3 μm thick copper metallization and the BCB loss tangent of 0.002 were assumed. An additional 5 μm thick BCB passivation on top was used. (Based on the original figure from [68], reproduced by permission of © 2002 IEEE.)

efficiency owing to the more pronounced influence of surface waves (see Equation 4.8 in Section 4.2.1 and Table 4.4). For a 45 μm thick BCB, the following values of the main antenna parameters are achievable for a single linearly polarized patch: a radiation efficiency of 65–68%, a directivity of 7.2–7.5 dB, and an input impedance bandwidth of 1.6 GHz (SWR = 2). The exact values within the given ranges depend mainly on the patch width.

To conclude, it was found out that the performance of a variety of the examined discontinuities, including open-end elements, junctions, and meanders, rapidly deteriorates for BCB thicker than 45–55 μm while the line loss and patch antenna efficiency tend to saturate with a dielectric thickness increase above 55 μm. In a result, the specified BCB thickness range, determined by both groups of the above given contrary factors, can be considered an optimum defining some 'performance plateau' for most of the mmWave interconnect structures intended for integration. This desired overall BCB thickness was the main goal of the technology development process, as described in Section 4.3.4. An ultimate BCB stack-up consisting of three layers, each 15 μm thick, turned out to be achievable, corresponding to the defined optimum range.

The measured and simulated performance of some more important interconnect elements and distributed passives, taking advantage of the optimized technology build-up, are discussed below.

4.4.2 Transmission Line Discontinuities and Distributed Passives

The measured line loss for a 50 Ω microstrip line realized in both developed MCM-D technology options (see Section 4.3.4) was given in Figure 4.10 when considering the basic

Table 4.7: Comparison of the open-ended 50 Ω microstrip lines. The following dielectric substrate build-ups are considered: a 45 μm thick BCB for the developed technology MCM-D setup, a hypothetical 100 μm thick BCB, and a 100 μm thick GaAs.

Open-end configuration	Line extension at 100 GHz; in μm and as a percentage of the operating wavelength	Power loss at 100 GHz; in % of the input power
114 μm wide line on a 45 μm BCB	28 μm (1.35%)	1
260 μm wide line on a 100 μm BCB	68 μm (3.5%)	5.5
70 μm wide line on a 100 μm GaAs	38 μm (3.8%)	2.3

technological requirements for various mmWave technology options. An ultra-low loss of 0.074 dB mm^{-1} at 60 GHz was attributed to a high surface quality of the BCB film and its very low-loss tangent at mmWave frequencies. Some other de-embedded line loss and line effective permittivity values for a broad set of 30–300 μm wide microstrip lines in the technology configuration I are shown in Figure 4.19. Note that the measured loss is weakly dependent on strip width, excluding the very narrow lines ($w < 40$ μm), as stated in Section 4.3.1. All lines show practically no high-frequency dispersion in the measured frequency range of 1–100 GHz, being a result of the low BCB dielectric constant and its appropriately chosen thickness. The loss and dispersion characteristics of the presented microstrip lines are comparable to those realized using more expensive micromachining techniques, wherein a dielectric substrate is removed [174, 192, 193].

4.4.2.1 Open-end Elements

The equivalent length extension values for a wide spectrum of the open-ended 25–100 Ω microstrip lines (28–300 μm line width) are de-embedded to be within 25–31 μm. They show no high-frequency dependence in the entire measured frequency range of 1–100 GHz because of the negligible surface wave effects, once again due to the low BCB dielectric permittivity and its optimized layer thickness. The reported values constitute 1.1–1.6% of the operation wavelength at 100 GHz and can still be considered as a design parameter of secondary importance. The *HFSS* [67] simulated return loss of the considered open-end elements is within 0.035–0.07 dB at 100 GHz, being equivalent to a 0.8–1.5% power loss (1% for a 50 Ω line), and is attributed to the space wave radiation only. This is different from the open circuits on a 100 μm thick GaAs substrate, wherein about 35% of the lost power at 100 GHz is due to the surface waves [194].

The frequency characteristics of the open-ended 50 Ω microstrip lines on a 100 μm thick GaAs (strip width of 70 μm) and on a hypothetical 100 μm thick BCB substrate (strip width of 260 μm) were also simulated by means of *HFSS* and *Ensemble* [67] and compared to the corresponding values for the developed technology build-up (see Table 4.7). Notice that both the length extension and the power loss values for the simulated substrate options are substantially higher than those for the proposed 45 μm thick BCB dielectric build-up.

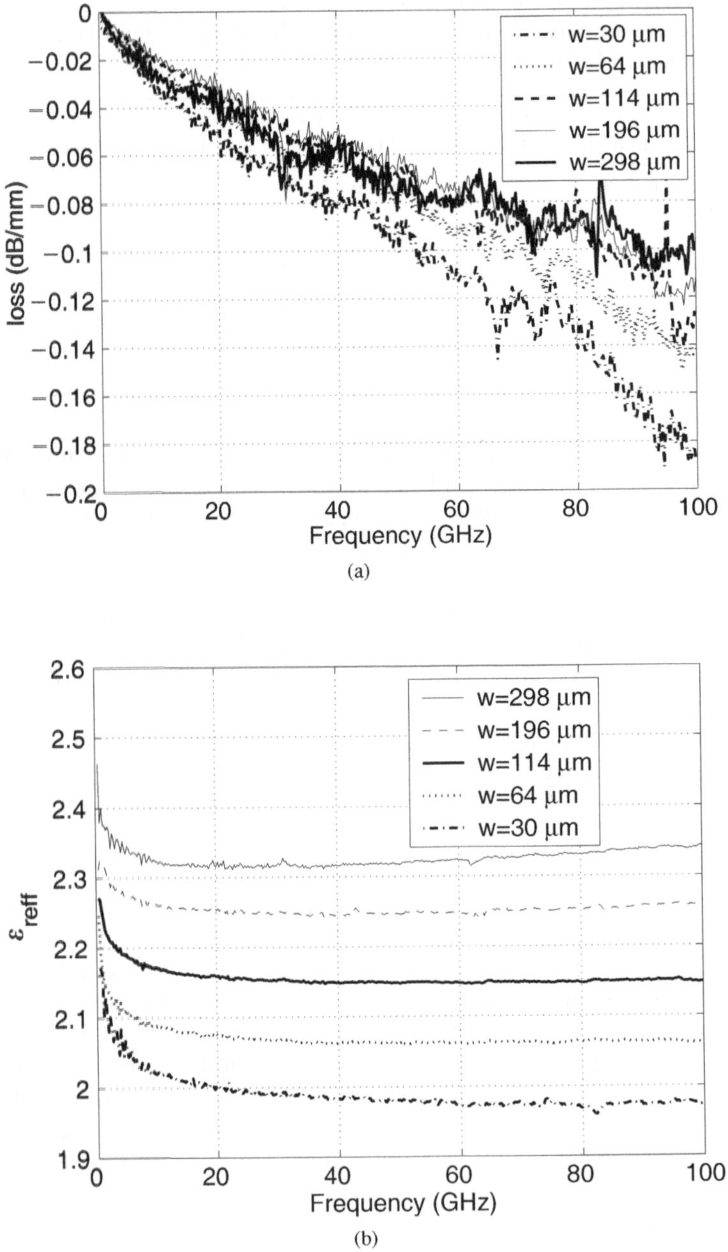

Figure 4.19: De-embedded: (a) line loss and (b) line effective permittivity for a variety of microstrip lines realized in the MCM-D technology configuration I ('w' denotes a microstrip width).

4.4.2.2 Short-circuit Elements

A basic geometry of the transmission line grounding path is the same as that used for the compensated loads, analyzed later (see paragraph *'Compensated loads'*) and shown in Figure 4.24(a).

As only a staircased via geometry was allowed for the three-layer dielectric build-up, the terminating path was divided into two, using three vias per shorting branch, in order to minimize its inductance. A via diameter of 70 μm and a space between vias of 30 μm were used. The de-embedded short-circuit inductance values for a typical set of 100–50 Ω grounded transmission lines cover the range of 5–9 pH at the low frequency end and increase to 8–12 pH at 100 GHz. The corresponding equivalent length extension values are within 10–32 μm and 15–45 μm. They constitute 0.7–2.2% of the operating wavelength at 100 GHz, and their influence on the circuit design is still far from being dominant. Notice that the short-circuit elements corresponding to the high impedance lines can be a better choice in terms of the length extension value than the open circuits (see Table 4.7), even if grounded using quite a complex via arrangement. Frequency dispersion of the considered short-circuit inductances can be attributed mainly to a three-dimensional current distribution at the short-circuit–transmission line interface and to the via inductance, increasing in frequency (towards a quarter-wavelength resonance for a via connected to ground at one end). For comparison purposes, the short-circuit inductance value for a 50 Ω line grounded using a single straight via of the same diameter is only about 25% lower, according to the *HFSS* simulations [67]. Furthermore, a very low return loss of around 0.035 dB (0.8% power loss) at 100 GHz for the entire considered set of grounded lines was predicted by *HFSS*.

4.4.2.3 Via Chains

The layout and the corresponding measured S-parameters of a via chain between two different 50 Ω microstrip line geometries located on the top and intermediate (15 μm below) metal layers are shown in Figure 4.20. The chain consists of three and two transmission line sections on the top and the lower layer, respectively, and four vias connecting them. The measured return loss is better than 26 dB up to 100 GHz, indicating that the excessive shunt capacitance to ground associated with the via transitions is still acceptably low. An insertion loss of the measured chain and of the equivalent 1.34 mm-long line differ by only 0.03 dB at 100 GHz, identifying a very low additional power lost at the vertical transitions.

4.4.3 Bends

The parasitics associated with bends define the frequency behavior of meander elements, commonly used as delay and slow-wave lines or for the phasing of radiating elements in the antenna arrays. A successful design of those structures at mmWave frequencies requires very small reactances and low radiation at the single bend discontinuities. Furthermore, a low electromagnetic coupling between meander line sections is also desirable.

Different bend geometries for a variety of microstrip line widths ranging from 28 up to 200 μm (100–35 Ω) were characterized. In order to minimize the influence of calibration noise at mmWave frequencies on the measured bend characteristics, some number of bends and transmission line sections were cascaded, resulting in the meander line structures. The measured performance of a 50 Ω right angle meander line from Figure 4.21, consisting of

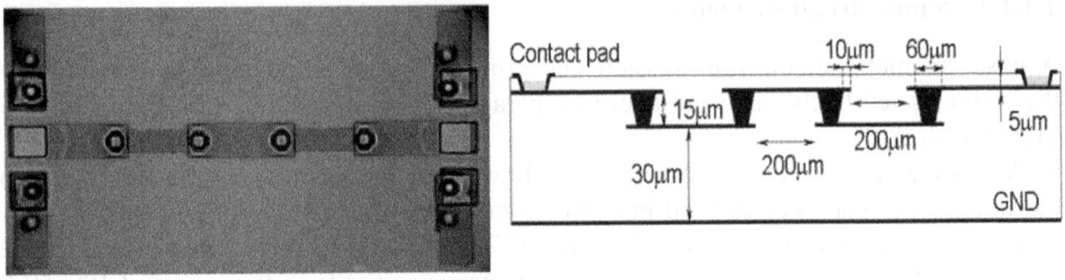

(a)

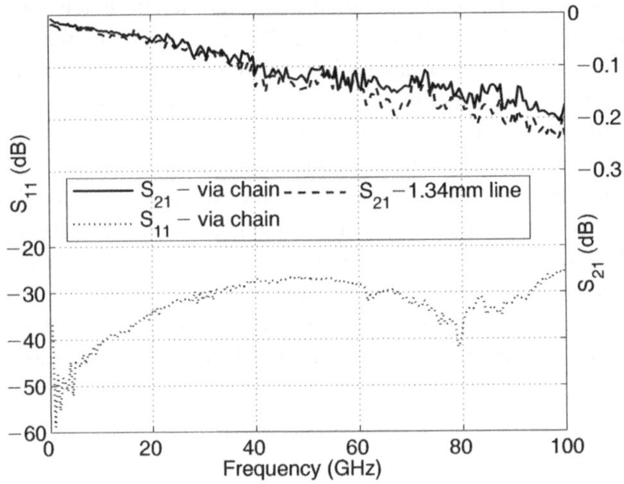

(b)

Figure 4.20: A via chain between the 50 Ω microstrip lines located on the top and intermediate (15 μm below) metal layers of the MCM-D substrate (see Figure 4.14 and Section 4.3.4): (a) layout geometry, (b) measured S-parameters. (Each of the microstrip line sections is 200 μm long. The line widths are 114 μm and 72 μm for the top and the intermediate layer, respectively. The vias are 60 μm in diameter, and the corresponding via landing pad on the lower layer is 80 μm large. 'S_{21}–1.34 mm line' denotes the measured insertion loss of an equivalent 1.34 mm-long line consisting of the 770 μm and 570 μm-long line sections on the top and the intermediate metal layer, respectively.)

eight chamfered bends and nine transmission line sections, is presented in Figure 4.22. A 500 μm separation between the parallel line sections was chosen so that the influence of bend parasitics is cumulative at about 80–90 GHz. For comparison purposes, the *Ensemble* [67] simulated performance of a similar 50 Ω (250 μm wide strip) meander line on a hypothetical 100 μm thick BCB substrate is also included. The mean path lengths, including bends, are 5.15 mm and 5.9 mm for both structures, respectively. The worst measured return loss of the meander line on a 45 μm thick BCB is equal to 26 dB at 95 GHz. Moreover, it is superior to 35 dB up to 75 GHz, being comparable to the measured return loss of an equivalent, 5.15 mm-long, straight transmission line. Both the meander and the equivalent transmission line show practically the same power loss up to 75 GHz. The difference between both increases by only

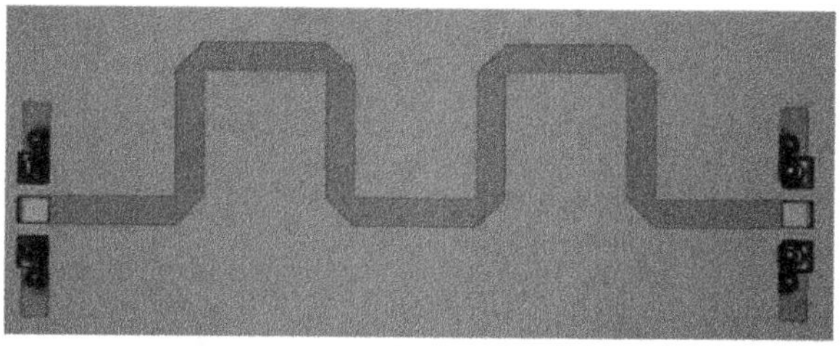

Figure 4.21: Layout of the measured 50 Ω (114 μm wide) meander line on the MCM-D substrate; see Figure 4.14 and Section 4.3.4 (Each of the transmission line sections is 500 μm long, resulting in an overall mean path length of 5.15 mm, including chamfered bends. A 50% meter was used to decrease the excess capacitance at the bends and to lower the radiation loss at higher frequencies.)

3% at 100 GHz. For a 100 μm thick BCB substrate, the transmission through the meander line deteriorates rapidly beyond 60 GHz. The return loss approaches 15 dB at 85 GHz, and 37% of a total input power is lost at 100 GHz while only 10% is dissipated along the equivalent, 5.9 mm-long, straight transmission line section.

4.4.3.1 T- and X-junctions

Due to its asymmetric structure, a simple T-junction can give rise to the considerable phase and amplitude imbalance between its orthogonal branches. A variety of uncompensated T-junctions composed of the line widths ranging from 28 up to 114 μm (100–50 Ω) were characterized. All of them show a very good phase imbalance below 0.5–3° up to 100 GHz. Similarly, the phase deviations of the reflection coefficients from the ideal ±180° stay within 3°. Furthermore, the absolute values of the phase shifts associated with the transmission coefficients are kept below 1–4° at 100 GHz, very close to an ideal value of zero degrees (defining the phase shifts, the reference plane located directly at the crossing point of the three lines is assumed). The highest amplitude imbalance of 0.25 dB and 0.8dB at 100 GHz for the transmitted and reflected signals at the orthogonal ports, respectively, is observed for the largest considered 50–50–50 Ω junction. However, it can be further reduced to 0.06 dB and 0.2 dB at 100 GHz, correspondingly, by simple chamfering of the outer junction edge. The full-wave simulated power loss for the considered uncompensated right-angle geometries is only 1.2–1.6% at 100 GHz.

The X-junctions show a slightly lower radiation loss because none of the port currents terminates at the strip edge. Their amplitude imbalance is also superior due to the higher level of symmetry. For the largest considered 50–50–50–50 Ω arrangement, the amplitude imbalance between transmitted signals at the orthogonal ports is equal to 0.17 dB only at 100 GHz.

4.4.3.2 Compensated Loads

The influence of resistor dimensions on an input match of the uncompensated 50 Ω load is studied in Figure 4.23. The same grounding via arrangement as that shown in Figure 4.24(a) was applied (see also the previous paragraph 'Short-circuit elements'). Notice that the

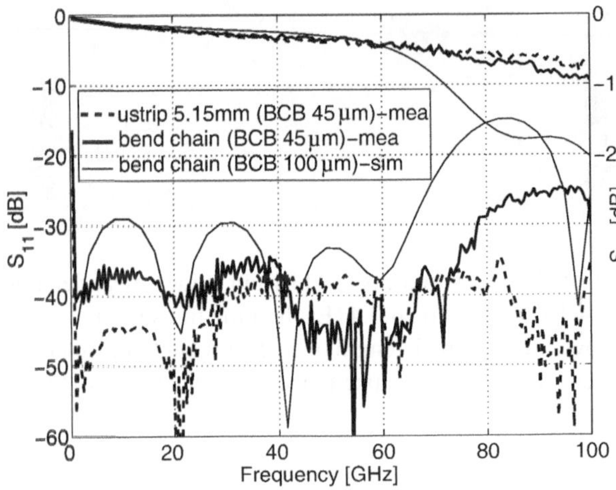

(a)

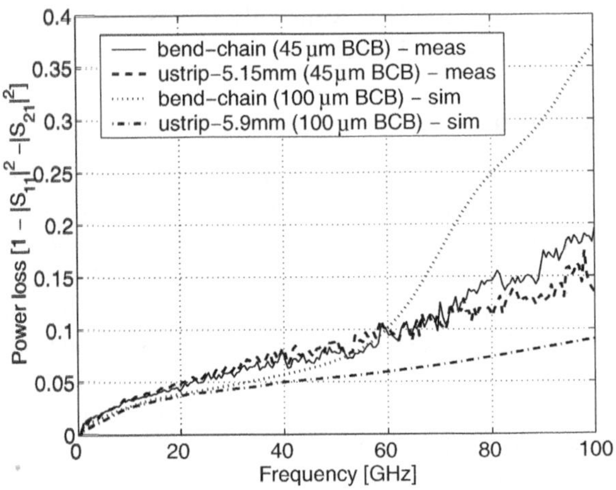

(b)

Figure 4.22: Performance comparison between the 50 Ω meander lines on a 45 µm (see Figure 4.21 for the layout definition) and on a 100 µm thick BCB substrate: (a) return and insertion loss, (b) power lost along the meander lines. ('bend-chain (45 µm BCB)-meas' and 'bend-chain (100 µm BCB)-sim' denote the measured and *Ensemble* [67] simulated characteristics of the meander lines on a 45 µm and on a 100 µm thick BCB, respectively, whereas 'ustrip-5.15 mm (45 µm BCB)-meas' and 'ustrip-5.9 mm (100 µm BCB)-sim', the measured and simulated characteristics of the equivalent transmission line sections of the same length as the corresponding meander structures.)

matching characteristics of the longer and wider load (158/120 µm) are only a little worse (1 dB difference at 80 GHz) than those of its smaller counterpart (79/60 µm) because the same grounding path inductance is transformed to the input through a longer transmission

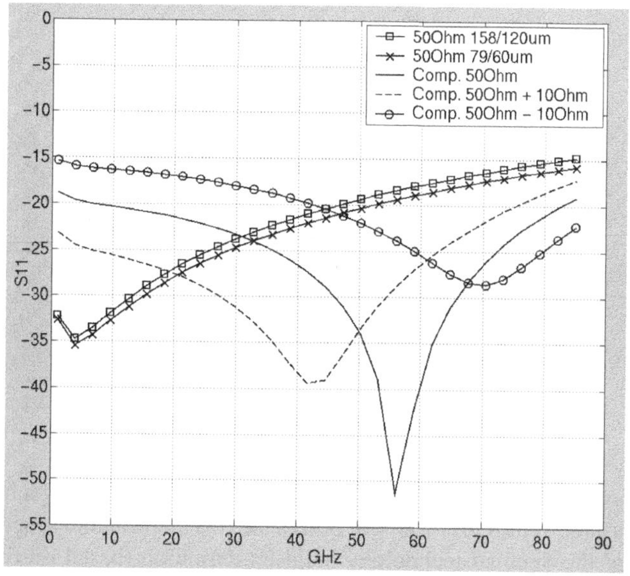

Figure 4.23: *HFSS* [67] simulated return loss for two different uncompensated 50 Ω loads and one compensated geometry from Figure 4.24(a); all realized using the developed MCM-D technology setup (see Figure 4.14 and Section 4.3.4). The influence of a 20% sheet resistance tolerance on the performance of the latter is also shown (from reference [68], reproduced by permission of © 2002 IEEE). ('50 Ohm 158/120 μm' and '50 Ohm 79/60 μm' denote a return loss of the 158/120 μm and 79/60 μm large uncompensated loads, whereas 'Comp. 50 Ohm', 'Comp. 50 Ohm + 10 Ohm' and 'Comp. 50 Ohm − 10 Ohm' an input match of the compensated structure, including a ±20% sheet resistance tolerance.)

line section associated with the wider resistor. Both geometries show a return loss inferior to 20 dB beyond 50 GHz and require some compensation, irrespective of the grounding via configuration (stacked or staircased), as shown in reference [167].

The measured return loss of the 50 Ω and 70 Ω compensated loads used as terminations for the multiport measurements of various distributed passives at 50–60 GHz is shown in Figure 4.24(b). The parasitic influence of a grounding path inductance is compensated by adding two simple open stubs in the front and by appropriate reduction of the resistor length. The 38 Ω and 61 Ω DC resistance values were applied for the 50 Ω and 70 Ω loads, respectively. Due to the sheet resistance tolerances, the measured loads do not show perfect match at 60 GHz, as indicated by the nominal full-wave simulations (see Figure 4.23 for the tolerance analysis results), but still superior to 25 dB. The 70 Ω termination is better matched at low frequencies due to a smaller relative reduction of the resistor length required for compensation purposes.

4.4.3.3 Ultra-wideband Load

In this section, an example of an alternative 50 Ω load, the geometry of which is shown in Figure 4.25(a), will be presented. The proposed topology indicates that the design of an ultra

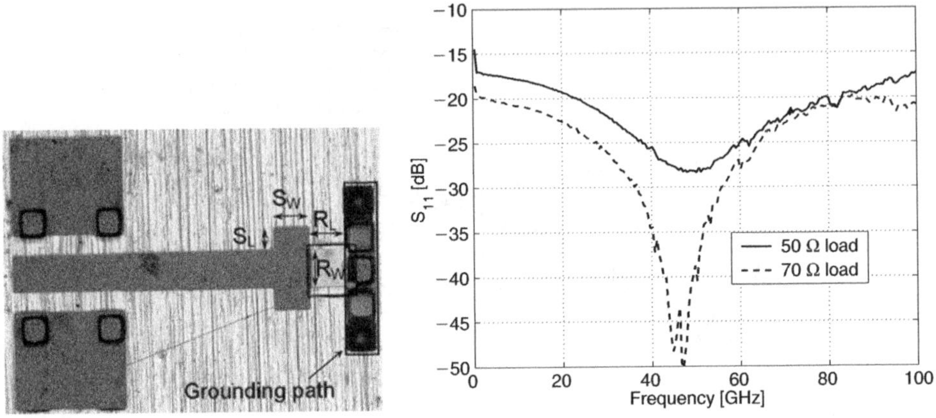

(a) (b)

Figure 4.24: Compensated loads on a 45 μm thick BCB substrate: (a) geometry of the 50 Ω version, (b) measured return loss for both 50 Ω and 70 Ω terminations; see Figure 4.10 and Section 4.3.4 for the detailed technology build-up definition (based on the original figure from [167], reproduced by permission of © 2003 IEEE). (The length, R_L, and the width, R_W, of the intrinsic resistor are 121/120 μm and 100/62 μm for the 50 Ω and 70 Ω loads, respectively, whereas the length, S_L, and the width, S_W, of the front stub are 69/110 μm and 60/61 μm, correspondingly.)

wideband termination at high mmWave frequencies is possible even with dimensionally large resistors and relatively long grounding paths.

A staircased grounding path of the first resistor, $R1$, is the same as that applied for the compensated loads from the previous paragraph. An input impedance associated with only this simple 59 Ω termination is equal to $71 + j30\,\Omega$ and $100 + j53\,\Omega$ at 100 GHz and 160 GHz, respectively, being very inductive at high frequencies. The second resistor, $R2$, together with an open-ended line section located behind creates a tuning structure, and it is connected in parallel with $R1$. Its classical arrangement consisting of an electrically small resistive element and an open stub is able to compensate for the inductive behavior of the main resistor, $R1$, at only a single frequency, as reported in reference [162] for a 60 GHz termination realized in a multilayer BCB/Kapton organic MCM [161–164] (see Section 4.3.4). At lower frequencies, the real part of its input impedance is approximately equal to the DC resistance of $R2$, and the imaginary part is defined by the stub capacitive behavior. However, if the resistive surface of $R2$ constitutes most of an overall length of the entire tuning structure, as shown in Figure 4.25(a), its frequency characteristics can be considerably different. First, the real part of its input impedance can be substantially lower than the DC resistance of $R2$ even at lower frequencies, depending on the percentage coverage of a total length with the resistive material. Next, the avoidance of a quarter wavelength resonance of the tuning structure is possible due to the transmission properties of a lossy long resistor, contrary to the properties of the above-mentioned classical arrangement.

A 250 μm long NiCr resistor, $R2$, applied in the presented concept, shows a DC resistance of 158 Ω, whereas an input impedance of the entire 335 μm long tuning structure is equal

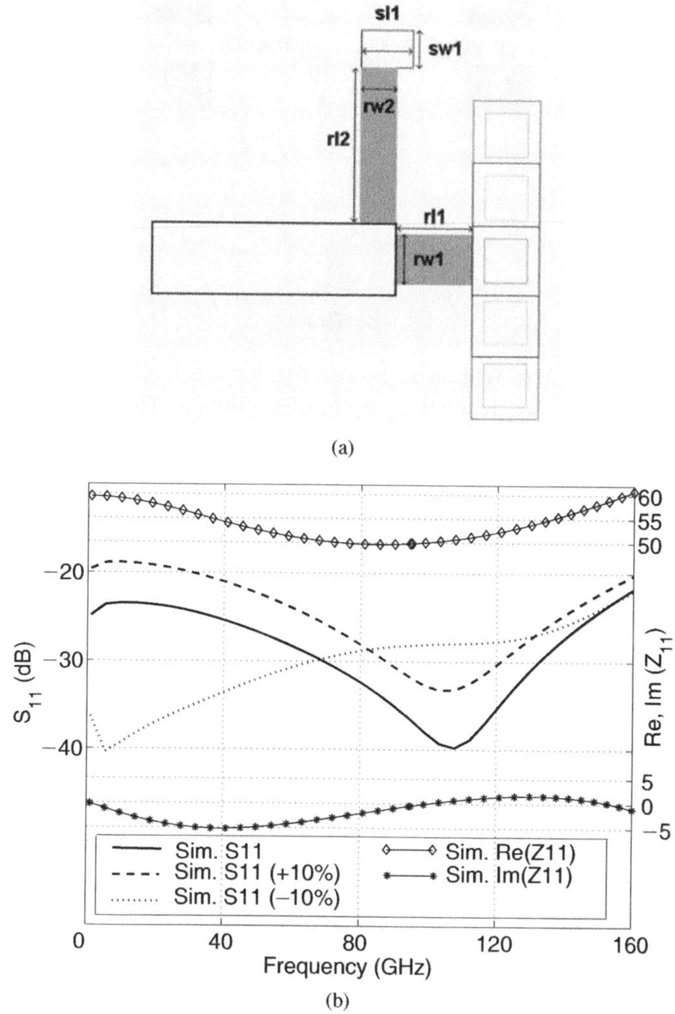

Figure 4.25: An ultra-wideband 50 Ω load on a 45 μm thick BCB dielectric substrate: (a) geometry, (b) simulated input impedance ('Sim. Z11') and return loss ('Sim. S11'), including a 10% sheet resistance tolerance; see Figure 4.14 and Section 4.3.4 for the detailed technology build-up definition. (The length, $rl1$, and the width, $rw1$, of the main resistor are 125 μm and 80 μm, respectively, whereas $rl2$ and $rw2$ of the open-ended resistor are 250 μm and 60 μm, respectively. For the stub, the respective dimensions are 65 μm ($sw1$) and 85 μm ($sl1$).)

to $86 - j6586\,\Omega$ at 1 GHz. Furthermore, an equivalent 335 μm long standard open stub, totally covered with a highly conductive material such as copper, should resonate at around 150 GHz. However, an input impedance of the arrangement implemented in this design concept stays very capacitive at 150 GHz and well above, owing to the fact that 75% of

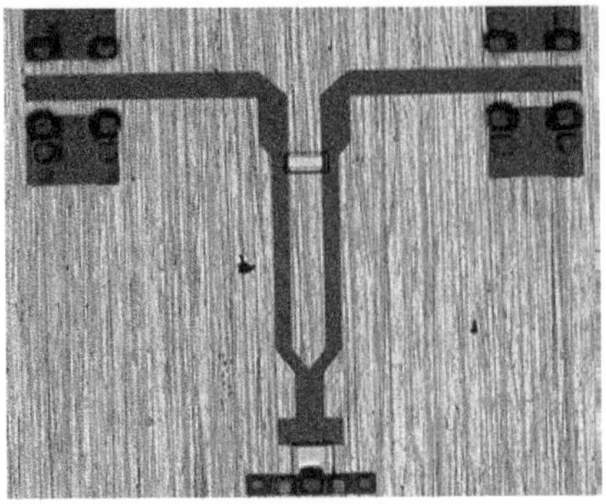

Figure 4.26: Layout of the Wilkinson power divider at 60 GHz; see Figure 4.14 and Section 4.3.4 for the detailed technology build-up definition. The length and width of the series NiCr resistor are 132 μm and 50 μm, respectively. The overall element dimensions are 1150 μm × 360 μm (from reference [167], reproduced by permission of © 2003 IEEE).

its total length constitutes the resistor $R2$. Its exact values at 100 GHz and 160 GHz are $84 - j54\ \Omega$ and $96 - j32\ \Omega$, respectively. When combined with the previously given input impedance values for the uncompensated load, $R1$, a zero input reactance of the entire compensated load at both frequencies is arrived at.

Figure 4.25(b) shows the ultimate *HFSS* [67] simulated return loss and the corresponding input impedance frequency characteristics of the proposed ultra-wideband termination. A constant and frequency independent value of the NiCr sheet resistance was assumed in the simulations, verified by the measurements up to 100 GHz. An input match for the nominal design is superior to 22 dB up to 160 GHz, whereas for a 10% sheet resistance tolerance, it is still better than 20 dB for the major part of the considered band. Very stable input impedance values (50–60 Ω for the real part and $-5 \mp 2\ \Omega$ for the imaginary part) crossing the zero value abscissae axis at 100 GHz and 160 GHz can be seen.

4.4.3.4 Wilkinson Power Divider at 60 GHz

The layout of a Wilkinson power divider operating at 60 GHz and its corresponding measured performance are depicted in Figures 4.26 and 4.27. It shows an ultra-low insertion loss of 0.2 dB (excluding 3 dB power split) at 60 GHz and superior to 0.25 dB for 28–72 GHz, almost equal to the full-wave simulated values. A wideband return loss better than 20 dB within the frequency band of 50–70 GHz and 35–70 GHz for the input and output branches, respectively, and an isolation of 20 dB within 50–70 GHz were obtained. Note that due to the optimized BCB dielectric layer thickness (see Section 4.4.1), it was possible to place both output branches and both parallel quarter wavelength long line sections in a short distance of 132 μm, equal to the length of a series resistor, without losing the high isolation of the divider. As a result, an extensive use of any bend discontinuities, commonly found in the layout of a Wilkinson power divider, could be avoided. From the performance point of view, it was also important to locate a NiCr resistor layer on the top metallization level, enabling via-less connections with the remaining structures.

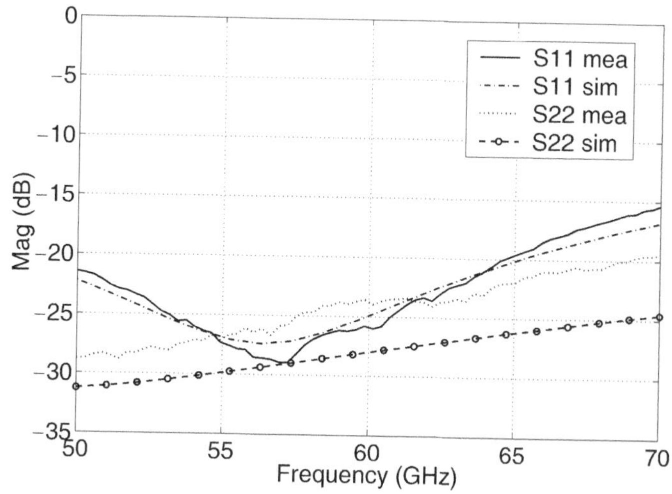

(a)

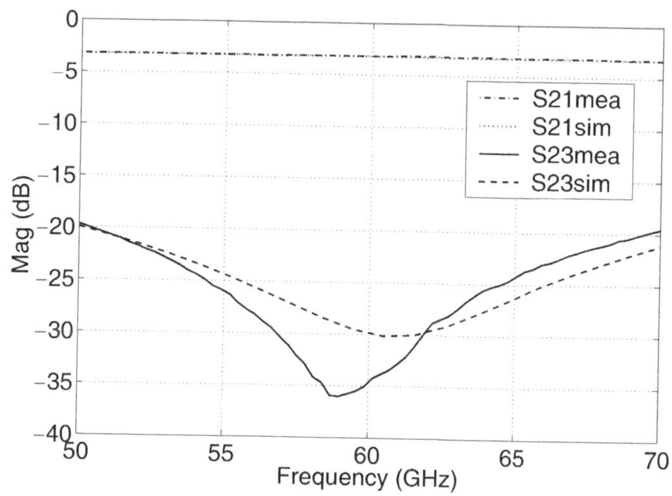

(b)

Figure 4.27: Wilkinson power divider at 60 GHz: (a) return loss at the input and output ports, (b) isolation and transmission. ‘s11 mea’, ‘s11 sim’, ‘s22 mea’ and ‘s22 sim’ denote the measured and simulated (*Ensemble* from Ansoft Corp. [67]) match at the input and output ports, respectively, whereas ‘s21 mea’, ‘s21 sim’, ‘s23 mea’ and ‘s23 sim’ stand for the measured and simulated transmission and isolation between both output branches (from reference [167], reproduced by permission of © 2003 IEEE).

Compared to the performance of a Wilkinson power divider at 60 GHz reported in reference [163], realized in a multilayer BCB/Kapton technology [161–164], a noticeably better isolation (30 dB versus 15 dB) could be achieved due to the superior dielectric layer thickness. Moreover, an insertion loss of the design presented in this paragraph is comparable to that of an ultra-low loss 33 GHz Wilkinson power divider (loss of 0.19 dB) reported in reference [195] and realized in a more expensive micromachined membrane-based microstrip line configuration.

4.4.3.5 Single-section Line Coupler at 78 GHz

Microstrip directional couplers consisting of the parallel coupled lines suffer from poor directivity owing to the inhomogeneous dielectric build-up, resulting in different phase velocities of the propagating even and odd modes. The directivity becomes worse as the coupling is decreased or as the substrate dielectric permittivity is increased [196]. Several methods of improving the directivity for such couplers were proposed, e.g. adding lumped capacitors at each end [197], the use of inductive feedback [198], or applying a dielectric overlay of different permittivity on top of the coupled lines [199]. Figure 4.28(a) shows the geometry of a 13 dB single-section line coupler operating at 78 GHz that was realized in the MCM-D technology configuration II with a 5 μm thick BCB passivation layer (see Section 4.3.4 for the detailed technology definition). The coupler shows a very flat and stable measured coupling level of 12.80 ± 0.5 dB within 60–100 GHz and an insertion loss of 0.35–0.6 dB for the transmitted signal (0.1–0.35 dB when taking the coupled power into consideration). The corresponding input impedance match and isolation within the same frequency range are measured to be better than 30 dB. The *Ensemble* [67] simulated values of the latter are even superior to 40 dB. The difference between both is mainly attributed to the limited calibration and de-embedding accuracy at such high frequencies.

A low permittivity environment created by the BCB dielectric, equalizing the even and odd-mode phase velocities, and excellent high-frequency behavior of the bend discontinuities at all four ports of the coupler result in the performance comparable with the realizations reported for the micromachined shielded membrane-supported microstrip topology [200].

4.5 Chip-to-package Interconnects at Millimeter-wave Frequencies

A brief summary of the most common chip-bonding schemes with special emphasis on their mmWave properties follow.

Over 90% of all available chips today are interconnected using wirebonding. The following factors are mainly responsible for such wide use of this technique:

- a mature, robust, and high yield process compatible with practically all dice types;
- the support of a well-established low-cost infrastructure;
- handling flexibility of different chip and package form factors;
- the capability to accommodate different chip sizes within the same package.

Unfortunately, the wirebonded interconnects show very pronounced lowpass frequency characteristics at mmWave frequencies owing to the electrically long bonding loops, thereby requiring extensive use of frequency compensation techniques in order to achieve an acceptable mmWave performance [2]. Furthermore, the reproducibility of short wire

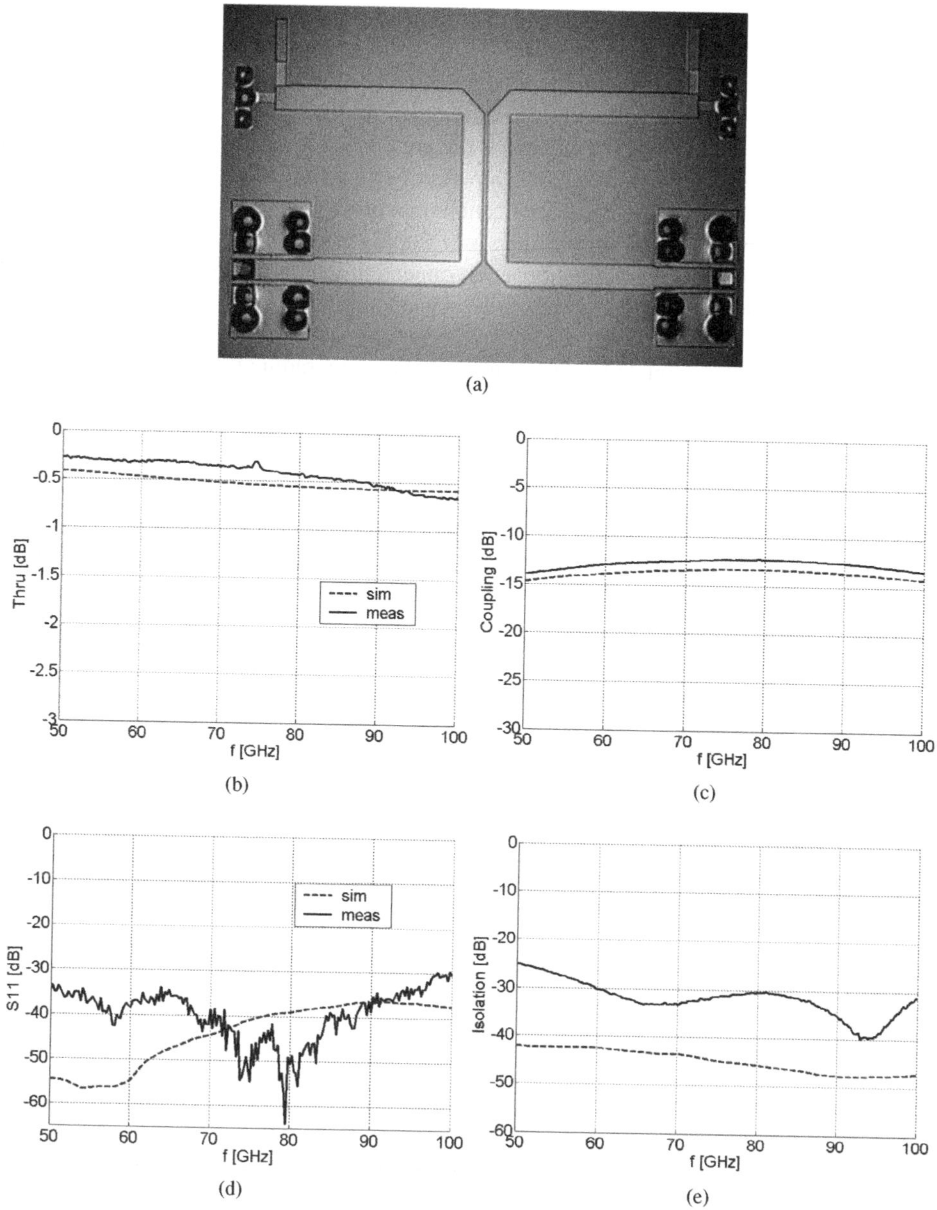

Figure 4.28: Measured and simulated performance of a 13 dB single-section line coupler at 78 GHz: (a) layout, (b) transmission, (c) coupling, (d) return loss, and (e) isolation. Simulations were performed with *Ensemble* from Ansoft Corp. [67]. The coupler is realized in the MCM-D technology configuration II with a 5 μm BCB passivation layer (see Section 4.3.4 and Figure 4.14). The length, width, and separation of the coupled line section are 696 μm, 105 μm, and 20 μm, respectively. ('sim' and 'meas' denote the simulation and measurement results, respectively.)

lengths can be a serious issue, effectively calling for some innovative adaptive bonding schemes [201].

The fact that essentially only chip periphery can be used in the wirebonding process was a driving force in the flip-chip bonding technique in the modern high density packaging industry, where its capability to use the entire die area for interconnection purposes enabled a significantly higher I/O count for a fixed chip size. Furthermore, the electrical performance of today's available high-speed digital applications can profit from the lower switching noise, possibly owing to the higher number of power supply pads and the substantially lower lead inductance. Owing to the very short interconnection length and the readily repeatable assembly process, flip-chip has gained considerable attention in mmWave applications [202–208]. However, close proximity of the chip active surface to the mounting substrate can lead to some parasitic package-related phenomena potentially degrading the overall performance of the entire mmWave module.

In view of the lowpass nature of the classical galvanic interconnects and due to the relatively short wavelengths at mmWave frequencies, especially for high dielectric permittivity materials such as ceramic, silicon, or GaAs, some alternative methods of interconnection based on electromagnetic coupling receive more attention [1–4, 209].

In general, the usable die attach methods in each particular case depend primarily on a combination of the applied MMIC transmission line topology with the available hybrid interconnect technology.

4.5.1 Wirebonding

4.5.1.1 Electrical Performance of the Wirebonds Without Compensation

The electrical performance of wirebonds depends on multiple factors related mainly to the chip mounting scheme and to the transmission line topologies to be connected. Two typical mounting scenarios are:

- the active die placed on top of the substrate carrier;
- the chip located in a substrate recess or side-by-side with another integrated or hybrid circuit.

Most existing MMICs today are realized either in a microstrip or a CPW transmission line configuration. Therefore, different combinations of transmission line types on both the chip and the substrate side can be expected to be bonded together, leading to some differences in the parasitic behavior of the entire assembly.

Chip on top of the substrate. Bonding the chips located on top of the substrate, although low-cost in terms of assembly, involves the use of rather long wires and can be expected to result in severe performance degradation at mmWave frequencies, independently of the transmission line of choice. Interconnecting microstrip-based MMICs with the corresponding microstrip elements on the mounting substrate additionally requires some vias to be applied for the galvanic contact between two separate ground planes.

An extensive full-wave characterization of a single, ball-bonding, 25 μm-diameter wire grounding a 100 μm thick GaAs chip mounted on top of some substrate carrier was reported in reference [210]. A minimum required total wire length was obtained on the statistical basis and was estimated to be 480 μm. The frequency behavior of such a wire can be viewed as a usable guideline for all other chips involving the use of a similar mounting scheme. It was shown that radiation dominates other types of loss above 70 GHz. Furthermore, the wire inductance can be considerably higher than its static value owing to the radiation effects, possibly leading to some crosstalk and package resonance issues [210]. Indications were that the radiation phenomenon was proportional to the height of a vertical wire section connected with the bonding pad owing to the missing direct image currents. Changing a distance between the bonding wire crest and the chip bonding pad from 70 μm up to 210 μm resulted in a tenfold and a hundredfold increase of the wire inductance and radiation resistance at 100 GHz, respectively, owing to the proximity of a wire self-resonance frequency [210]. Moreover, a parallel resonance between the bonding wire and the corresponding 70 μm × 70 μm large bonding pad came into play around 100 GHz, indicating the importance of a pad size for the performance of mmWave circuits [210].

Applying thin ribbons instead of wires for the galvanic connection of thin microstrip chips with the corresponding microstrip or CPW lines on top of the mounting substrate was proven to be feasible for the lower mmWave frequencies [4, 211, 212] because the lower stiffness of a thin ribbon cross-section allows easier bending and, thus, a shorter and more compact looping process. A return loss of a single ribbonbond transition between the microstrip line on a top 127 μm thick alumina substrate and the coplanar line on a 635 μm thick alumina mounting slab was measured to be superior to 15 dB up to 40 GHz [211,212]. A comparable performance was reported for a similarly arranged transition (one substrate on top of the other), but between two microstrip lines patterned on the separate 127 μm thick ceramic substrates [4].

To conclude, the wirebond flatness was identified as a critical parameter directly influencing the operation bandwidth of mmWave circuits. Therefore, different mounting arrangements facilitating a side-by-side bonding between the substrates of a similar height (e.g. chip in a recess) are usually preferred as they are capable of keeping the wirebond flat and of reducing the connection length.

Chip side-by-side bonding. A return loss of a typical 25 μm-diameter wire loop connecting two 50 Ω microstrip lines placed on a 254 μm thick alumina substrate in a distance of 470 μm was measured to be around 6 dB at 35 GHz [201, 213, 214], indicating rather poor matching even at the lower mmWave frequencies. The height of a wire crest over the substrate carrier was 100 μm, being a typical value. Adding an additional second wire in its optimum position allowed a 20% decrease of the equivalent total interconnect inductance and, thus, a 5 dB improvement in return loss, as reported in references [213, 214]. The optimum position was shown to correspond to the spacing between the wires approximately equal to the microstrip width, which minimized the mutual inductance between both wires.

In reference [215], two open-ended 50 Ω microstrip lines on a 100 μm thick alumina substrate placed face-to-face at a distance of 400 μm were wirebonded, resulting in a very poor return loss of 2.5 dB at 80 GHz. The wire height over the dielectric substrate was assumed to be 100 μm in the center of the wire loop. The average characteristic impedance and line effective permittivity of such a connecting wire, modeled as an equivalent uniform transmission line section, were calculated to be 157 Ω and 1.63, respectively.

For the coplanar wirebond interconnects, three wires have to be used to connect the center strips and both side ground planes. The average equivalent characteristic impedance and line effective permittivity for a simplified uniform raised coplanar waveguide model of this interconnection scheme was reported in reference [216] for a typical range of ground-wire-to-ground-wire spacings and as a function of the wirebond distance to the underlying dielectric substrate. It was shown that beyond 100 µm, the latter is no longer an important parameter and, thus, the line effective permittivity approaches 1 for most of the wire length. Then, the corresponding characteristic impedance was estimated to be around 160–190 Ω for a ground-wire-to-ground-wire spacing of 120–160 µm. Some measured insertion loss values were given for this type of wirebonding scheme connecting two coplanar GaAs chips. They were around 0.9, 4 and 6 dB at 80 GHz for a 50, 410 and 700 µm long wire, respectively. According to the presented simplified modeling approach, it was indicated that the maximum wire length of around 100 µm is allowed for a 10 dB return loss at 94 GHz [216]. The same matching could be achieved with an almost 200 µm long and 17 × 50 µm wide ribbon bond. Furthermore, ribbons offered a significant insertion loss reduction, especially for the longer connections. An insertion loss of only 0.5 dB and a return loss superior to 20 dB at 75 GHz were reported in reference [207] for a 200 µm long, 50 Ω coplanar side-by-side connection between a GaAs chip and an alumina substrate, both of the same height.

4.5.1.2 Compensation of the Wirebond Parasitics

In view of some representative performance examples for various wirebond interconnection schemes presented in the previous paragraph, it should be obvious that some further reduction of the bonding discontinuity at mmWave frequencies is, in practice, unfeasible without applying some compensation techniques.

According to the basic filter theory, the wirebond interconnect can be treated as a lowpass filter. Hence, increasing the filter order for some predefined cut-off frequency, a considerably larger center inductor value is required, directly implying the use of longer bonding wires without affecting the interconnect operation bandwidth [217]. A five-stage lowpass filter prototype is commonly applied for the wire compensation purposes, requiring the combination of a low and a high impedance line section on either side of the bonding loop [201, 217]. In reference [201], it was pointed out that even a 40 dB return loss for the nominal design is feasible if the total electrical airbridge length is shorter than Z_0/Z_b, where Z_0 and Z_b are the reference impedance for the bonding pads and the average equivalent airbridge characteristic impedance, respectively. Such a low value of return loss requires, however, the very tight assembly and manufacturing tolerances that are, in practice, impossible to keep. Errors in the location of bonding pads, mechanical tolerances of the bonded chips or substrates, and limited control of the wire loop shape lead to variations of the bonded interconnection length that are nonnegligible at mmWave frequencies. Moreover, the typically achievable air gap length between the chips placed side-by-side is in the range of 100 µm [215].

A microstrip wirebond interconnect between two 127 µm thick glass chips involving the use of the above presented compensation scheme, was reported in reference [217]. Two parallel 432 µm long, 25 µm-diameter ball bonds connecting two microstrip line sections showed an average wire loop length tolerance of ±50 µm, resulting in the measured return and insertion loss superior to 12 dB and 0.3 dB, accordingly, from DC up to 80 GHz, being considerably below the nominal values. In order to arrive at the improved wirebond performance, taking some assembly tolerances into account, an innovative bonding technique was presented in reference [201], wherein the mutual inductance variation between two flat parallel wires as a function of the spacing between them was applied to compensate for a variable separation between the bonding pads. The latter was monitored with a camera and used as an input parameter for some algorithm calculating the appropriate bond position and the loop height in a real time mode [201]. A return loss superior to 20 dB up to 100 GHz was found to be possible for the interconnects between two 5 mil-thick *GaAs* chips and a ±35 μm chip placement accuracy.

It should be emphasized that the compensation networks discussed above are symmetric, thus requiring some additional area on the chip side. It can be an important limitation for the commercially available MMICs with the very restricted access to the contact pads for some possible modifications. Furthermore, the output impedance at the pads with the applied compensation network is normally different from the common 50 Ω reference and optimized for the specific assembly scheme. For this reason, a simplified asymmetric matching topology on only one interconnection side is very often applied, even at the cost of about a 30% decrease in operation bandwidth, as given in reference [216]. Its practical implementation was reported in reference [209] for the connection of a 125 μm thick bare die placed in the topside air cavity formed in a four-layer Ferro LTCC substrate. A low insertion loss of 0.25 dB up to 40 GHz for a single transition was indicated while a return loss of 10 dB and 20 dB was measured at 40 GHz and 28 GHz, respectively.

4.5.1.3 Modeling of the Wirebonds

At lower frequencies, the wirebonds are often modeled as lumped inductors, which can be to some extent justified for the very short interconnection lengths applied. For millimeter waves, transmission line properties of the wires can become very pronounced, making their analytical equivalent circuit modeling rather challenging owing to the three-dimensional field complexity involved in the computations. Those models, however, can be invaluable in giving some parametric insight into the frequency behavior of some common bonding configurations. They may also be very helpful for fast approximate computation of the compensation networks outlined earlier. However, they cannot be simply generalized over a wide spectrum of all possible assembly and packaging schemes. For this reason, a final full three-dimensional simulation seems to be a must for the accurate prediction of their performance at very high frequencies.

Different attempts to develop simple but accurate analytical models of the bonding wire interconnects for some particular mounting scenarios were reported in the literature. A quasi-static model of both single and double bonding wire microstrip interconnects in the form of four cascaded uniform transmission line sections, accounting for the inherent bonding nonuniformity, was presented in references [213, 218]. A side-by-side bonding scheme with an air gap between two separate substrates was assumed. The equivalent transmission line parameters for each section were calculated at the midpoint by applying a conformal mapping analysis from reference [215]. The bonding regions were also included in the model. From the systematic comparison with the full-wave finite difference time domain (FDTD) simulation results within a frequency range 5–50 GHz, it was found that the proposed model is very precise for flat bonds, but it loses its accuracy as the wire height increases. No loss mechanisms were included in this modeling approach.

The three cascaded sections of a raised CPW transmission line configuration were identified in reference [216] as a sufficient modeling approach at mmWave frequencies for a flat side-by-side coplanar bonding scheme. Two of them accounted for the short wire lengths in close proximity to the chip-bonding pads. Although the model was shown to be accurate for some bonding examples, its parameters were extracted from the full-wave *HFSS* [67] simulations, and no analytical formulas for their calculation were given.

4.5.2 Flip-chip Bonding

4.5.2.1 Millimeter-wave Characteristics of the Flip-chip Interconnects

A flip-chip bonding technique is normally applied in combination with a CPW transmission line configuration on both sides of the connection because it offers a very short electrical length of transition and, thus, a wide application potential at mmWave frequencies. From this point of view, the intrinsic parasitics associated with such a single bump transition are considerably smaller when compared with the wirebonding. However, the upside-down chip mounting onto an underlying substrate involves some other parasitic phenomena that can lead to significant performance deterioration of an overall packaged mmWave system if some design precautions are not carefully considered. These phenomena are as follows.

- Chip detuning effects, being a result of the close proximity of a mounting substrate carrier.
- Parasitic coupling to the substrate modes that are able to propagate in a motherboard.
- Underfill materials used for an enhanced assembly reliability, possibly leading to some additional severe detuning effects and some excessive loss for various passive structures on the chip level. An underfill process with a low-viscosity, stress-relieving polymer is usually required owing to the thermal mismatch between the active die, typically silicon, and the mounting substrate. Of special interest are the laminated organic boards with normally high coefficients of thermal expansion, the fatigue life of which can be reduced even by a factor of 100 compared to the ceramic materials if no underfill epoxy is applied.

In order to avoid the influence of some of the above-listed parasitic effects, some modified flip-chip interconnect configurations, involving the use of a microstrip line transmission topology on only the substrate level or on both sides of the transition, were proposed. Application of the microstrip lines on an appropriately thin motherboard substrate is often capable of delivering a superior mmWave performance as compared with CPWs (see Sections 4.2.1 and 4.4) and allows some parasitic substrate mode coupling effects related to the conventional CPW-to-CPW flip-chip bonding to be mitigated [219–222]. Furthermore, connecting the face-up mounted microstrip chips with the corresponding microstrip lines on a substrate carrier additionally eliminates the chip detuning issues [223, 224]. For all the advantages introduced by both modified flip-chip interconnections, the inherent high performance of a single bump transition is no longer that evident because a rather complex use of additional hot and grounding vias is needed. For this reason, the design of some compensation networks is necessary to improve their performance at mmWave frequencies.

In summary, in spite of its inherently low parasitics associated with the bump transition, the flip chip due to its electromagnetic field proximity between the mounted chip and the corresponding substrate carrier can be a source of serious performance degradation of the entire mmWave module if the package- and assembly-related issues are neglected at the design stage.

Intrinsic properties of a bump transition with and without compensation. Primarily, the mmWave characteristics of a CPW-to-CPW flip-chip interconnection will be considered here, being a dominant mounting scenario. A set of the major geometrical parameters affecting the performance of a single transition includes [157, 220, 221, 225]:

- bump diameter;
- bump height;
- bump pad length;
- distance between the chip edge and the corresponding chip pad, defining a dielectric overlap between the mounting substrate and the chip die;
- pitch of the transition area determined by the separation between both ground bumps;
- width of the signal bump pad, and the spacing between signal and ground pads.

An extensive study of the influence of the above-mentioned parameters on the performance of a typical flip-chip assembly connecting the 50 Ω CPW lines on a GaAs chip (line ground–ground spacing of 54 μm) and a ceramic substrate carrier (line ground–ground spacing of 120 μm) was given in references [220, 221, 226]. The bump height was fixed at 22 μm, whereas the bump diameter and the bump pad length varied from 25 to 35 μm and 30 to 60 μm, respectively. It was found that the frequency behavior of an overall transition can be described by an equivalent shunt capacitance up to the highest considered frequency of 70 GHz, indicating the importance of an open-end capacitance associated with dielectric loading of the bumping pads. On the other hand, the contribution of inductive effects assigned to the vertical part of the transition was minor. Changing the bump height within a range 10–25 μm led to only slight return loss variations above 40 GHz, whereas increasing the bump pad length from 30 μm up to 50 μm caused a significant return loss deterioration from 33 dB to 22 dB [220].

Taking the above into account, the bump diameter and the corresponding pad size are critical parameters directly influencing the open-end effects and should be kept low. A short distance between the bumping pad and the chip edge, below 50 μm, can also play a nonnegligible role in defining the open-end capacitance value owing to its direct influence on the fringing electrical fields [220, 221, 226]. However, for typical chip dicing tolerances, it is almost unlikely to keep this distance smaller, and its influence on the transition properties can, in practice, be neglected [221].

Some further improvement of the transition performance can be achieved by increasing the pitch of ground bumps and the spacing between signal and ground pads, wherein a longer ground current path is effectively applied to compensate for the capacitive behavior of the bonding pads [157, 221, 225]. This approach was used in reference [157] to show that even 100 μm large and 35 μm high bumps connecting the 50 Ω CPW lines on a GaAs chip and on a composite BCB/borosilicate glass MCM-D substrate were capable of delivering a return loss superior to 21 dB up to 75 GHz if an optimized ground bump pitch was applied.

The design of experiments (DOE) analysis performed in reference [225] for a broad value range of various geometrical parameters influencing the flip-chip transition performance showed that the inductive effects are no longer completely negligible if the ratio of bump height to bump pad size is large enough. This can be used for some additional performance optimization, as shown in reference [157], wherein an increase of the bumping height from 35 μm up to 100 μm resulted in improved mmWave matching characteristics (21 dB versus 25 dB at 75 GHz) for the flip chip interconnection between the above-mentioned 50 Ω CPW lines on a GaAs chip and on a mixed BCB/borosilicate glass MCM-D substrate.

Several other methods can be successfully applied to compensate for the parasitic behavior of the flip-chip transitions involving the use of larger bumps. In general, the compensation techniques presented in Section 4.5.1 for the wirebond connections, based on a five-stage lowpass filter synthesis, will also work in this case. However, taking the significantly lower parasitics of the bump assembly into account, a simple high impedance line section or at most a two-stage high/low impedance line arrangement on the motherboard side is normally sufficient to achieve reflection coefficients as good as 20 dB up to 80–100 GHz [157, 221, 227]. An alternative approach can be to stagger the signal and ground bumps on the chip level [228], virtually decreasing the influence of open-end effects associated with the signal bump pads. Although effective, it requires an additional chip area, which is typically prohibited. The potential of both compensation techniques was verified for the previously mentioned flip-chip interconnection between the 50 Ω CPW lines on a GaAs chip and a ceramic substrate carrier [220, 221, 226]. A bump diameter and a bump pad area were fixed to 35 μm and 60×60 μm^2, respectively, while the separation between both ground bumps was 200 μm. For the stagger arrangement, the signal and ground pads were placed 145 μm apart. A return loss superior to 26 dB up to 70 GHz was simulated for both compensation networks, being a significant improvement when compared to 19 dB for the initial design. The simulations were verified by measurements, additionally indicating a reproducible ultra-low loss below 0.2 dB up to 90 GHz for a single transition.

A compensated flip-chip transition connecting a coplanar GaAs chip with the corresponding thin-film microstrip line on a 20 μm thick BCB substrate was presented in reference [219], wherein a simple matching network in the form of two cascaded high and low impedance line sections on the motherboard side was optimized for 38 GHz applications. The 80 μm-diameter bumps as high as 100 μm were applied for reasons of cost fabrication. A return loss superior to 20 dB up to 55 GHz for a single interconnect was demonstrated.

Some different modified flip-chip interconnection between a face-up mounted microstrip chip and a 50 Ω microstrip line on a 250 μm thick LTCC substrate was designed for mmWave applications below 40 GHz [223]. The presented transition basically consists of two parts. In the first, the chip microstrip line is connected to a high impedance coplanar line section by means of a hot-via configuration through the chip substrate followed by a bump assembly. Then, the CPW transmission line configuration is converted into the microstrip line, which additionally includes some short, low impedance line section for fine-tuning purposes [223]. The full-wave simulations of a complete single transition indicated a return loss better than 30 dB up to 40 GHz, whereas its uncompensated counterpart showed a return loss inferior to 20 GHz above 18 GHz.

Circuit modeling of the flip-chip interconnects. The equivalent circuit model of a flip-chip transition between CPW lines can consist of only a shunt capacitance in its simplest form, as shown in reference [221]. For the higher accuracy, a more extended π-version including both shunt capacitances and series inductance, can be applied [157, 228]. Some frequency-dependent conductive elements were also added in reference [228] to catch the radiation and surface wave dependent substrate losses; however, no physical insight was provided. Another lumped element T-model featuring some scalability in terms of the major physical transition dimensions was given in reference [229]. A linear regression was applied to arrive at the equivalent element parameter values extracted from numerous electromagnetic simulations of the variable transition geometry.

In spite of all the advantages of the equivalent circuit models, the ultimate accurate prediction of the flip-chip interconnections require the use of full-wave three-dimensional simulations. Furthermore, taking a large diversity of flip-chip packaging scenarios, there are no relevant existing ready-to-use models for many of them, and the mentioned electromagnetic simulations are indispensable.

Chip detuning. It is obvious that the electrical parameters of various circuits on a flip-chip mounted die, specifically transmission line structures, can be significantly influenced by the proximity effects to both dielectric and metal surfaces on a motherboard. Specifically, the presence of the latter underneath the chip surface leads to the severe detuning and should normally be avoided [216,220,226]. As shown in references [220,226], the line characteristic impedance and the line propagation constant for a 50 μm wide 50 Ω CPW line on a GaAs chip mounted face down onto the nonmetallized alumina substrate change by less than 1.5% for a 20 μm high bump, whereas an equally spaced metallized top surface of the motherboard leads to deviations as high as 7% and 5%, respectively. As a ground-to-ground spacing is of primary importance for the CPW field confinement, increasing its value will raise the detuning effects significantly. To minimize their influence, a distance between both side ground planes should be kept below the bump height value if the mounting substrate is metallized on its top, as indicated in references [206, 207, 216], whereas for a nonmetallized surface, it can be easily increased by three times [221].

An interesting study on the influence of a bumping height on the performance of a coplanar GaAs Lange coupler flip-chip mounted on an alumina substrate was presented in references [230, 231], wherein the perturbation of transmission line parameters for both even and odd modes due to the mounting substrate proximity had to be investigated. It was found that the bump heights as small as 30 μm are satisfactory to keep the odd mode parameters practically unchanged, whereas modification of the even mode properties is still nonnegligible for bumps as high as 100 μm. The main reason for that is a different field confinement for both modes. The transmission parameters of the former are primarily determined by the slot width while those of the latter are strongly related to the ground–ground spacing of the coupler that has to accommodate four strips and three slots of the finger elements, being comparatively large. It was shown that the coupler amplitude balance and its directivity were the most sensitive parameters, the deviation of which was nonmarginal even for the chip-to-motherboard spacing in the range 50–100 μm.

The flipped microstrip-based MMICs with a common substrate thickness in the range of 100 μm require considerably higher bumps to be applied when compared with the CPW structures owing to a lower field confinement at the substrate-to-air boundary. For a 50 Ω line (82 μm wide strip) on a 100 μm thick GaAs substrate, a 50 μm high bump at least is needed for a propagation constant deviation below 1.5% while the line characteristic impedance variation is less pronounced [230, 231].

An interesting concept to avoid the parasitic proximity effects between chip and substrate carrier in the flip-chip face-down assembly was presented in reference [208]. A novel multi-layer MMIC technology platform based on the thin-film inverted microstrip lines enabled a ground metal layer on the top chip surface to be placed, thereby providing an electromagnetic shield between MMIC and motherboard. A flip-chip mounted 76 GHz receiver amplifier was proven to deliver almost the same performance as its unpackaged counterpart.

The influence of underfill. It was stated in the introductory part of this section that the underfill process may be required for reasons of reliability. As expected, it can potentially lead to severe detuning effects of the on-chip circuits and some additional excess loss due to a relatively high loss tangent of the commonly applied underfill epoxies. Therefore, its presence should be taken into account during the design cycle. Choosing a low loss and low dielectric permittivity underfill material can be important in minimizing its parasitic influence.

As indicated in reference [232], the 50 Ω CPW lines on a GaAs chip, flip-chip mounted on an alumina board by means of 75 μm high and 150 μm large bumps, have shown a 23% increase in the line effective permittivity at mmWave frequencies when underfilled with U300 epoxy from Epoxy Tech. Inc. ($\epsilon_r = 4.1$ and $\tan\delta = 0.009$ at 100 kHz). Another study on the influence of underfill on the transmission characteristics of passive structures and on the properties of RF transistors was given in reference [233]. A variation of the line characteristic impedance and the line propagation constant as high as 11% was specified for the CPW lines on alumina mounted face down onto another ceramic motherboard when applying an epoxy material with a dielectric permittivity of 3.2 to fill a 40 μm large gap between both substrates. Furthermore, the pad capacitances in the equivalent circuit model of the characterized underfilled metal semiconductor field effect transistor (MESFET) device were found to be enlarged by 50%, leading to the transistor transit frequency decrease from 31 GHz down to 25 GHz. The operating frequency range of a flip-chip mounted three-stage 60 GHz LNA from reference [234] was found to shift by as much as 15% when underfilled, whereas the corresponding in-band gain was reduced from 24 dB to 19 dB.

Substrate mode coupling at the flipped-chip–motherboard interface. Previously, it was shown that a single flip-chip interconnection can deliver the excellent performance at mmWave frequencies by means of rather simple compensation techniques.

However, a global behavior of the entire packaged module comprising some flip-chip mounted devices can be of concern owing to the possible parasitic modes propagation in a motherboard [204, 222]. Coupling to these modes is expected to happen at the chip inputs and outputs owing to the discontinuity effects introduced by bumping transitions [204, 222]. The number and the corresponding field pattern of the parasitic modes depend on many factors, such as the transmission line topology on both the chip and the motherboard level, the presence of conductor-backing, the number and location of signal and ground vias, and the size of both the mounting substrate carrier and the flipped chip die [204, 222]. The parasitic phenomena induced by the propagating substrate modes can lead potentially to poor isolation especially for high-gain circuits, serious resonance issues, and complete system malfunction in the worst case. The global motherboard interconnect system designed in a coplanar style can be potentially hazardous in combination with the flip-chip assembly owing to the conducting backplane, typically existing in the packaging environment, that supports the parallel-plate waveguide modes with no cut-off frequency [235] (see also Section 4.2.1 for more detailed information on the parasitic mode coupling effects). Locally, the presence of such modes under the chip surface is nearly independent of the planar transmission line type applied (CPWs and microstrip lines are mainly treated) on both the chip and substrate carrier level because two parallel, partially overlapping ground planes normally exist for the considered transmission line arrangements. Only the properties of those modes will differ, depending on the particular combination of the CPW and microstrip line topology on both sides of a bumping transition.

A comprehensive study of the mode conversion issues at a single flip-chip transition connecting two 50 Ω CPW lines, one of them on a GaAs chip and the other on the corresponding ceramic substrate, was given in references [220, 221]. The simulated coupling to the parallel-plate waveguide mode in the forward direction was around −17 to −20 dB, being comparable to the reflection coefficient values for the CPW mode. Furthermore, the coupling level was shown to be slightly dependent on the bumping height. Only very short bumps in the range of 10 μm could potentially lead to the very pronounced coupling effects in excess of −5 dB as the basic electromagnetic criteria for critical coupling between modes were fulfilled (see Section 4.2.1). The conversion level into a backward propagating parallel-plate waveguide mode was simulated to be below −30 dB up to 60 GHz. The measurements performed for some test chips flipped onto the conductor-backed substrate carrier of a finite size indicated that the local mode conversion levels for a single flip-chip transition as small as −20 dB can lead to some severe resonance-like phenomena with the resonant peaks reaching almost 0 dB at multiple frequencies, depending on the board size and the relevant boundary conditions along its surface [220, 221].

The performance of both microstrip and coplanar MMICs mounted face down onto a substrate carrier with the microstrip feeds was studied in reference [222], wherein a two-path circuit model was developed to investigate the isolation issues for the considered flip-chip mounting schemes. The chip lateral dimensions were electrically large, resulting in some resonant effects. Due to the selected transmission line topology on the board level, the use of grounding vias through the substrate was necessary. Furthermore, multiple grounding vias were applied to shift the resonant peaks to higher frequencies, thus improving the isolation at lower frequencies. The lower resonant frequency reported for the flipped microstrip chips, being a result of the higher effective dielectric permittivity of an equivalent, parallel-plate waveguide structure created between the chip and substrate ground planes, was anticipated to be the main reason for a 5–10 dB isolation loss when compared to the corresponding CPW chips of the same size.

4.5.3 Alternative Chip Interconnection Methods

4.5.3.1 Electromagnetic Coupling

As the wavelength at mmWave frequencies becomes shorter, an alternative chip-bonding technique involving the principle of electromagnetic coupling, typically based on the overlapping quarter-wavelength-long line sections, was shown to be possible [1–4,209]. The same concept can be successfully applied for the hermetically sealed package feedthroughs [2, 4] and some other interconnection structures on the substrate carrier level realized by means of the modern multilayer interconnect technologies [3, 236, 237].

Depending on the transmission line topology on both sides of the transition, a number of possible coupling arrangement can be applied, a good survey of which can be found in references [1–3].

A basic idea of a surface-to-surface transition between CPW lines was given in reference [238], wherein the coupling was realized by applying a quarter-wavelength-long coupled line section consisting of a three-strip finite-ground CPW (FGCPW) on each side of the separating dielectric substrate. On the basis of the coupled line theory, it was shown that with a substrate thickness increase, the transmission line parameters (characteristic impedance and effective dielectric permittivity) for even and odd modes come closer to each other, resulting in a lower coupling level. In order to reduce the lateral size of the coupling region, very thin substrates should be utilized. The overall length and width of a 50 Ω transition at 35 GHz for a 100 μm thick GaAs chip mounted on top of a 250 μm thick quartz substrate, was calculated to be 790 μm and 1400 μm, respectively, being electrically large [238]. Another transition of this type was reported in reference [236] for 75–110 GHz applications, wherein two FGCPWs on both sides of a 100 μm thick silicon substrate were connected. The coupling region was 280 μm long and 900 μm wide (including the width of both side ground planes). A return loss superior to 15 dB within 80–110 GHz for a back-to-back arrangement was measured, whereas an insertion loss for a single transition was estimated to be better than 0.6 dB over the entire measured frequency band of 75–110 GHz.

The approach similar to that above-mentioned was proposed to connect microstrip lines with CPWs [239], whereby the ground plane of the microstrip chip had to be locally patterned in the coupling region. In this case, a general asymmetric coupled line theory for inhomogeneous media [240] was applied for design purposes owing to the missing up-down symmetry of the considered build-up. A 35 GHz transition between the CPW lines on a 127 μm thick alumina and the corresponding microstrip lines on a 100 μm thick GaAs chip was computed to be 880 μm long and 1 mm wide, the latter being a nonnegligible fraction of a wavelength. An additional matching structure on the side of a mounting substrate was proposed in references [1–3] for some bandwidth enhancement purposes of the basic transition topology from above. A return loss of such an improved transition between a 150 μm thick microstrip chip and the CPW feed on a 381 μm thick alumina substrate in a back-to-back configuration was measured to be better than 10 dB within 29–40 GHz. Another broadband, electromagnetically coupled vertical microstrip-to-coplanar waveguide interconnect structure was presented in reference [241], wherein a 10 dB defined return loss in the frequency band 3.2–11.2 GHz was reported. The wideband frequency characteristics were achieved owing to a significant increase in the width of both coupled strips, leading to the very pronounced frequency dependent behavior of the coupling region. The coupled-strip length was indicated to be as high as half a wavelength at the highest operating frequency [242], thus being different from the classical approach involving the use of quarter-wavelength-long line sections.

A principle of direct electromagnetic coupling (*direct* means no slot in the ground plane) was also reported in the literature [1, 2] to connect on-chip and on-substrate microstrip lines. The use of additional via holes on the motherboard level was required in this case for electrical connection between two separate ground planes of both transmission lines. A back-to-back transition prototype consisting of two stacked RT Duroid substrates, 635 μm and 1.27 mm thick, was measured to deliver a 13 dB defined return loss within 7.5–13.5 GHz [2].

The electromagnetic coupling was also shown to be applicable to the on-wafer testing of microstrip chips with the CPW probes [4, 237, 243–246]. The virtual short-circuit elements realized by means of the quarter-wavelength-long structures printed on both sides of the microstrip input and output signal lines were used to replace grounding via holes within the measurement pad configuration. Both rectangular and radial stubs were demonstrated to provide the efficient RF ground connections, as shown in references [236, 237, 243, 244]. A return loss of a single transition of this type designed for the microstrip chip on a 254 μm thick alumina substrate was measured to be superior to 10 dB within 7–33 GHz [244]. Its W-band version was also reported in reference [236] for a 100 μm thick high resistivity silicon substrate. Two transitions measured back-to-back have shown a return loss better than 15 dB for 75–96 GHz. Another way of avoiding grounding via holes when converting a FGCPW to a microstrip line, was proposed in reference [246]. Its operation principle is based on a quarter-wavelength-long three-conductor coupled microstrip line section that can also be viewed simply as a finite-width conductor-backed CPW. The W-band measurements of this transition type on a 120 μm thick high resistivity silicon substrate in a back-to-back configuration indicated a return loss superior to 17 dB from 83 GHz up to 100 GHz. An insertion loss was de-embedded to be around 0.4 dB, including an 860 μm long line section in-between.

It is important to note that the conductor-backing issues for coplanar waveguides should be carefully considered when designing electromagnetically coupled CPW transitions owing to the potential risk of conversion to some undesired parasitic microstrip-like modes, specifically at the open-end discontinuities [237, 245] (see also Section 4.2.1).

4.5.3.2 Via-in-pad for Connecting the Cavity-embedded Chips

The use of short vias through a thin dielectric overlay, normally in the form of laminated or spun layers, applied on top of the active circuits embedded in the recessed cavities in a substrate carrier was also proven to deliver a repeatable high-performance interconnection option at mmWave frequencies [17, 65, 102, 103, 161–164, 166] if combined with the properly chosen technology stack-up. From the point of view of the MMIC performance perturbation, relative dielectric permittivities of the covering dielectric materials should be as close to one as possible. A more detailed description of some of the technology options supporting this type of chip-bonding scheme was given in Sections 4.3.2 and 4.3.4. Especially interesting seems to be the approach from reference [166] (see Section 4.3.4), wherein the preformed gold bumps on the chip pads were used as connecting vias through a 36 μm thick BCB layer, allowing some of the processing steps associated with a regular via formation to be avoided.

Acknowledgments

The author would like to acknowledge all the project partners of the European Union LIPS consortium (project reference: IST-2000-30128) for their cooperation, invaluable effort in the technology development process and great contribution to the results presented herein. Very special thanks go to P. Bodö, J. Haglund, D. Chouvaev, P. Blomqvist, M. Danestig from Acreo AB (Sweden), J. Galière, J. L. Valard, E. Estèbe from Thales Airborne Systems (former Thales Research and Technology, France), L. Hellequin, I. Favier, F. Bernand from Thales Microwave SA (France), H. Nilsson, M. Lindgren from Kitron Development AB (Sweden),

S. Grosmaire, P. Guyon from Cimulec SA (France), and D. Cottet, M. Klemm, I. Ruiz from Swiss Federal Institute of Technology (Switzerland).

The author would also like to acknowledge the fruitful teamwork with his former communication technology group at IBM T. J. Watson Research Center, Yorktown Heights, New York, USA. Very special thanks go to Duixian Liu for encouraging the author to write this chapter.

References

[1] G. Strauss and W. Menzel, 'Millimeter-wave MMIC Interconnects using electromagnetic field coupling', in *3rd IEEE Top. Electrical Performance of Electronic Packaging* (Monterey, CA), pp. 142–144, November 1994.

[2] G. Strauss and W. S. Menzel, 'Millimeter-wave monolithic integrated circuit interconnects using electromagnetic field coupling', *IEEE Trans. Comp. Packag. Technol.* **19** (1996), pp. 278–282.

[3] G. Strauss and W. Menzel, 'A novel concept for mm-wave MMIC interconnects and packaging', in *IEEE MTT-S Int. Dig.* (San Diego, CA), pp. 1141–1144, May 1994.

[4] W. Menzel and J. Kassner, 'Novel techniques for packaging and interconnects in mm-wave communication front-ends', in *IEE Seminar* (London, UK), pp. 1/1–1/18, 2000.

[5] L. Hyvönen and A. Hujanen, 'A compact MMIC-compatible microstrip to waveguide transition', in *IEEE MTT-S Int. Dig.* (San Francisco, CA), pp. 875–878, 1996.

[6] W. Grabherr, B. Huder, and W. Menzel, 'Microstrip to waveguide transition compatible with MM-wave integrated circuits', *IEEE Trans. Microwave Theory Tech.* **42**(9) (1994), pp. 1842–1843.

[7] S. Lucyszyn, S. R. P. Silva, I. D. Robertson, R. J. Collier, A. K. Jastrzebski, I. G. Thayne, and S. P. Beaumont, 'Terahertz multi-chip module (T-MCM) technology for the 21st century ?' in *IEE Colloquium on Multi-chip Modules and RFICs* (London, UK), pp. 6/1–6/8, May 1998.

[8] R. H. Jansen and N. H. L. Koster, 'New aspects concerning the definition of microstrip characteristic impedance as a function of frequency', in *IEEE MTT-S Int. Dig.*, pp. 305–307, 1982.

[9] R. H. Jansen and M. Kirschning, 'Arguments and accurate model for the power-current formulation of microstrip characteristic impedance', *AEU*, vol. Band 37, pp. 108–112, 1983.

[10] R. Collier, 'Coupling between coplanar waveguides and substrate modes', in *29th European Microwave Conf.*, pp. 11–25, December 1989.

[11] R. B. Marks and D. F. Williams, 'A general waveguide circuit theory', *J. Res. Nat. Inst. Stand. Technol.* **97** (1992), pp. 533–562.

[12] J. B. Knorr and A. Tufekcioglu, 'Spectral-domain calculation of microstrip characteristic impedance', *IEEE Trans. Microwave Theory Tech.* **23** (1975), pp. 725–728.

[13] F. Schnieder and W. Heinrich, 'Model of thin-film microstrip line for circuit design', *IEEE Trans. Microwave Theory Tech.* **49** (2001), pp. 104–110.

[14] W. Heinrich, 'Full-wave analysis of conductor losses on MMIC transmission lines', *IEEE Trans. Microwave Theory Tech.* **38** (1990), pp. 1468–1472.

[15] A. K. Verma and R. Kumar, 'New empirical unified dispersion model for shielded- suspended-, and composite substrate microstrip line for microwave and mm-wave applications', *IEEE Trans. Microwave Theory Tech.* **46** (1998), pp. 1187–1192.

[16] J. Grzyb, I. Ruiz, and G. Tröster, 'Extraction of material complex permittivities for composite substrate MCM-L technologies up to 100 GHz', in *Proc. 7th IEEE Workshop on Signal Propagation on Interconnects (SPI)* (Siena, Italy), pp. 143–146, May 2003.

[17] J. Grzyb, *MCM Integration Technologies for 60–80 GHz Applications*. PhD thesis, Swiss Federal Institute of Technology, Zurich, Switzerland, 2004.

[18] T. C. Edwards, *Foundations for Microstrip Circuit Design* (John Wiley & Sons, Ltd/Inc., 1981).

[19] J. R. James and A. Henderson, 'High-frequency behaviour of microstrip open-circuit terminations', *IEE J. Microwaves, Opt. and Acoust.* **3** (1979), pp. 205–218.

[20] L. Lewin, 'Radiation from discontinuities in strip-line', *Proc. Inst. Elec. Engr* **107C** (1960), pp. 163–170.

[21] E. Belohouberk and E. Denlinger, 'Loss considerations for microstrip resonators', *IEEE Trans. Microwave Theory Tech.* **23** (1975), pp. 522–526.

[22] A. A. Oliner, 'Package effects cause by leaky modes at higher frequencies in microwave integrated circuits', in *29th EuMC* (Munich), pp. 122–125, 1999.

[23] E. Yamashita, 'Variational method for the analysis of microstrip like transmission lines', *IEEE Trans. Microwave Theory Tech.* **16** (1998), pp. 529–535.

[24] R. Crampagne, M. Ahmadpanah, and J. L. Guiraud, 'A simple method for determining the Green's function for a large class of MIC lines having multilayered dielectric structures', *IEEE Trans. Microwave Theory Tech.* **26** (1978), pp. 82–87.

[25] A. K. Verma and G. H. Sadr, 'Unified dispersion model for multilayer microstrip line', *IEEE Trans. Microwave Theory Tech.* **40** (1992), pp. 1587–1591.

[26] R. K. Hoffmann, *Integrierte Mikrowellenschaltungen* (Berlin: Springer, 1983).

[27] K. C. Gupta, R. Garg, and I. Bahl, *Microstrip Lines and Slot Lines*, 2nd edn (Norwood, MA: Artech House, 1996).

[28] E. Hammerstad and O. Jensen, 'Accurate models for microstrip computer-aided design', in *IEEE MTT-S Int. Dig.*, pp. 407–409, 1980.

[29] M. Kirschning and R. H. Jansen, 'Accurate model for effective dielectric constant of microstrip with validity up to millimetre-wave frequencies', *Electron. Letters* **18** (1982), pp. 272–273.

[30] M. Kobayashi, 'A dispersion formula satisfying recent requirements in microstrip CAD', *IEEE Trans. Microwave Theory Tech.* **36** (1988), pp. 1246–1250.

[31] J. Svacina, 'Analysis of multilayer microstrip lines by a conformal mapping method', *IEEE Trans. Microwave Theory Tech.* **40** (1992), pp. 769–772.

[32] J. Svacina, 'A simple quasi-static determination of basic parameters of multilayer microstrip and coplanar waveguide', *IEEE Microwave Guided Wave Lett.* **2** (1990), pp. 385–387.

[33] C. Wan and A. Hoorfar, 'Improved design equations for multilayer microstrip lines', *IEEE Microwave Guided Wave Lett.* **10** (2000), pp. 223–224.

[34] A. K. Verma, A. Bhupal, Z. Rostamy, and G. P. Srivastava, 'Analysis of rectangular patch antenna with dielectric cover', *IEICE Trans. (Japan)* **E-74** (1991), pp. 1270–1276.

[35] K. F. Lee and J. S. Dehele, 'Circular disc microstrip antenna with an airgap', *IEEE Trans. Antennas Propagat.* **32** (1984), pp. 880–884.

[36] C. P. Wen, 'Coplanar waveguide: a surface strip transmission line suitable for nonreciprocal gyrometric device applications', *IEEE Trans. Microwave Theory Tech.* **17** (1969), pp. 1087–1090.

[37] M. Riaziat, R. Majidi-Ahy, and I.-J. Feng, 'Propagation modes and dispersion characteristics of coplanar waveguides', *IEEE Trans. Microwave Theory Tech.* **38** (1990), pp. 245–250.

[38] D. B. Rutledge, D. P. Neikirk, and D. P. Kassiligam, 'Integrated-circuit antennas', in *Infrared and Millimeter Waves 10* (K. J. Button, ed.) (New York: Academic, 1983) pp. 1–90.

[39] E. M. Godshalk, 'Surface wave phenomenon in wafer probing environments', in *40th ARFTG Conf. Dig.* (Orlando, FL), pp. 10–19, December 1992.

[40] H. Shigesawa, M. Tsuji, and A. Oliner, 'A new mode-coupling effect on coplanar waveguides of finite width', in *IEEE MTT-S Int. Dig.* (Dallas, TX), pp. 1063–1066, May 1990.

[41] A. O. M. Tsuji, H. Shigesawa, 'New Interesting leakage behaviour on coplanar waveguides of finite and infinite widths', *IEEE Trans. Microwave Theory Tech.* **39** (1991), pp. 2130–2137.

[42] R. W. Jackson, 'Mode conversion at discontinuities in finite-width conductor-backed coplanar waveguide', *IEEE Trans. Microwave Theory Tech.* **37** (1989), pp. 1582–1589.

[43] J. Ke, I. Tsai, and C. Chen, 'Dispersion and leakage characteristics of coplanar waveguides', *IEEE Trans. Microwave Theory Tech.* **40** (1992), pp. 1970–1973.

[44] R. Collier, D. Edgar, J. Yip, A. Jastrzebski, D. Li, K. Elgaid, H. McLelland, S. Ferguson, F. Williamson, A. Ross, F. Doherty, and I. Thayne, 'A study of high frequency performance of coplanar waveguide as a function of substrate thickness', in *30th Eur. Microwave Conf.* (Paris, France), pp. 354–356, 2000.

[45] H. Shigesawa, M. Tsuji, and A. Oliner, 'Conductor-backed slot line and coplanar waveguide: Dangers and full-wave analyses', in *IEEE MTT-S Int. Dig.* (New York), pp. 199–202, 1998.

[46] C.-C. Tien, C.-K. C. Tzuang, S. Peng, and C.-C. Chang, 'Transmission characteristics of finite-width conductor-backed coplanar waveguide', *IEEE Trans. Microwave Theory Tech.*, **41** (1993), pp. 1616–1623.

[47] W.-T. Lo, C.-K. C. Tzuang, S.-T. Peng, C.-C. Tien, C.-C. Chang, and J.-W. Huang, 'Resonant phenomena in conductor-backed coplanar waveguides (CBCPW's)', *IEEE Trans. Microwave Theory Tech.* **41** (1993), pp. 2099–2107.

[48] S.-P. Liu and C.-K. Tzuang, 'Scattering analyses of asymmetric conductor-backed CPW open-end discontinuity problem', *IEEE Microwave Guided Wave Lett.* **7** (1997), pp. 130–132.

[49] W. H. Haydl, 'Resonance phenomena and power loss in conductor-backed coplanar structures', *IEEE Microwave Guided Wave Lett.* **20**, pp. 514–516.

[50] W. McKinzie and N. G. Alexopoulos, 'Leakage losses for the dominant mode of the conductor-backed coplanar waveguide', *IEEE Microwave Guided Wave Lett.* **2** (1992), pp. 65–66.

[51] W. H. Haydl, 'On the use of vias in conductor-backed coplanar circuits', *IEEE Trans. Microwave Theory Tech.* **50** (2002), pp. 1571–1577.

[52] N. K. Das, 'Methods of suppression or avoidance of parallel-plate power leakage from conductor-backed transmission lines', *IEEE Trans. Microwave Theory Tech.* **44** (1996), pp. 169–181.

[53] M. Yu, R. Vahldieck, and J. Huang, 'Comparing coax launcher and wafer probe excitation for 10 mil conductor backed CPW with via holes and airbridges', in *IEEE MTT-S Int. Dig.* (Atlanta, GA), pp. 705–708, 1993.

[54] K. Jones, 'Suppression of spurious propagation modes in microwave wafer probes', *Microwave J.* **32**(11) (1990), pp. 173–174.

[55] R. W. Jackson, 'Consideration in the use of coplanar waveguide for millimeter-wave integrated circuits', *IEEE Trans. Microwave Theory Tech.* **34** (1986), pp. 1450–1456.

[56] M. Magerko and K. C. L. Fan, 'Configuration considerations for multi-layered packaged conductor-backed coplanar waveguide MICs', in *IEEE MTT-S Int. Dig.* (San Diego, CA), pp. 1697–1700, 1994.

[57] D. Pavlidis and H. Hartnagel, 'The design and performance of three-line microstrip couplers', *IEEE Trans. Microwave Theory Tech.* **24** (1976), pp. 631–640.

[58] G. Ponchak, S. Robertson, F. Brauchler, J. East, and L. Katehi, 'Finite width coplanar waveguide for microwave and millimeter wave integrated circuits', in *ISHM Digest* (USA), pp. 517–521, 1996.

[59] J. Papapolymerou, F. Brauchler, J. East, and L. P. B. Katehi, 'W-band finite ground coplanar monolithic multipliers', *IEEE Trans. Microwave Theory Tech.* **47** (1999), pp. 614–619.

[60] K. Wu and R. Vahldieck, 'Rigorous analysis of the characteristic impedance in conductor-backed miniature coplanar waveguides considering multiple layers of lossy and finite thickness metal', in *IEEE MTT-S Int. Dig.* (Albuquerque, NM), pp. 987–990, 1992.

[61] K. Wu, R. Vahldieck, J. Fikart, and H. Minkus, 'The influence of finite conductor thickness and conductivity on fundamental and higher-order modes in miniature hybrid MIC's (MHMIC's) and MMIC's', *IEEE Trans. Microwave Theory Tech.* **41** (1993), pp. 421–430.

[62] Y. Shih, 'Broadband characterization of conductor-backed coplanar waveguide using accurate on-wafer measurement techniques', *Microwave J.* **34** (1991), pp. 95–103.

[63] J. Grzyb and G. Tröster, 'Finite-Width CBCPW characteristic impedance de-embedding with probe-tips', *IEEE Microwave and Wireless Components Lett.* **14** (2004), pp. 213–215.

[64] J. Grzyb, I. Ruiz, M. Klemm, and G. Tröster, 'Quasi-CPW lines on MCM-D for the realization of via-less transmission elements and distributed passives', in *11th IEEE Top. Meet. Electtrical Performance of Electronic Packaging* (USA), pp. 51–54, 2002.

[65] LIPS, *Low Cost Interconnect, Packaging and Subsystem Integration Technologies for Millimetre-wave Applications*. Available at http://cordis.europa.eu/fetch?ACTION=D&CALLER=PROJ_IST&RCN=61018 [December 30, 2008].

[66] E. A. Soliman, P. Pieters, and E. Beyne, 'Suppression of the parasitic modes in CPW discontinuities using MCM-D technology – application to a novel 3-dB power divider', *IEEE Trans. Microwave Theory Tech.* **46** (1998), pp. 2426–2430.

[67] 'Ansoft Corp.' http://www.ansoft.com/.

[68] J. Grzyb, I. Ruiz, M. Klemm, and G. Tröster, 'MCM-D technology for realization of low cost system-on-package concept at 60-80 GHz', in *11th IEEE Top. Meet. Electrical Performance of Electronic Packaging* (USA), pp. 55–58, 2002.

[69] T. M. Weller, R. M. Henderson, K. J. Herric, S. V. Robertson, R. T. Kihm, and L. P. B. Katehi, 'Three-dimensional high-frequency distribution networks – Part I: optimisation of CPW discontinuities', *IEEE Trans. Microwave Theory Tech.* **48** (2000), pp. 1635–1642.

[70] E. Rius, G. Prigent, H. Happy, G. Dambrine, S. Boret, and A. Cappy, 'Wide- and Narrow-band bandpass coplanar filters in the W-frequency band', *IEEE Trans. Microwave Theory Tech.* **51** (2003), pp. 784–791.

[71] S. Lucyszyn, 'The future of on-chip terahertz metal-pipe rectangular waveguides implemented using micromachined and multilayer technologies', *IEE Colloquium on Terahertz Technology and its Applications* (London, UK) pp. 10/1–10/10, April 1997.

[72] W. Menzel, J. Kassner, and U. Goebel, 'Innovative packaging and fabrication concept for a 28 GHz and communication front-end', *IEICE Trans. on Electronics*, vol. E82-C, no. 11, pp. 2021–2028, 1999.

[73] R. Dolp, W. Mayer, and W. Grabherr, 'Industrialization of a 58 GHz high gain flat panel antenna design using injection molding technique', in *29th EuMC* vol. 3 (Munich, Germany), pp. 13–15, 1999.

[74] H. Asao, N. Yoneda, M. Mukuda, K. Yamasaki, O. Kamohara, Y. Yoshino, and K. Henmi, 'Metal-plated plastic waveguide filter using injection molding process', in *IEEE MTT-S Int. Dig.*, pp. 941–944, 2003.

[75] M. Wagner, D. Stanelli, P. Nuechter, and U. Goebel, 'Compact 60 GHz diplexer in metallized plastic technology for gigabit wireless links', in *34th EuMC*, vol. 2, pp. 1009–1012, October 2004.

[76] P. Serbe, V. Schwarz, and U. Goebel, '*Sensity*™ Link 60 - A wireless point-to-point transparent ethernet bridge', in *8th European Conference on Fixed Wireless Networks and Technologies* (France) 2007.

[77] U. Goebel, F. Rehme, and D. Schmidt, 'New microwave packaging technologies for high-rel and low-cost applications', in *28th EuMC* (Amsterdam), 1998.

[78] R. R. Tummala, E. J. Rymaszewski, and A. G. Klopfenstein, *Microelectronics Packaging Handbook* 2nd edn (New York: Chapman and Hall, 1997).

[79] D. Cottet, *Characterisation of High Density Substrates for use at Millimetre-wave Frequencies.* PhD thesis, Swiss Federal Institute of technology, September 2003.

[80] 'Gore electronics products', http://www.goreelectronics.com/ [December 30, 2008].

[81] T. Bergstresser, R. Hilburn, H. Kaplan, and R. Le, 'Adhesiveless copper on polyimide substrate with Nickel-Chromium tiecoat', in *Proc. of IPC Printed Circuits Expo.* (Long Beach, CA), 23–27 March 2003.

[82] K. E. G. Pitt, C. E. Free, D. Li, and P. Barnwell, 'The influence of material properties on the design of MCM and microstrip structures', *J. Mater. Sci.: Mater. in Electron* **10** (1999), pp. 519–524.

[83] The Dow Chemical Company, 'CYCLOTENE advanced electronics resins', http://www.dow.com/cyclotene/ [December 30, 2008].

[84] D. Williams, U. Arzt, and H. Grabinski, 'Accurate characteristic impedance measurement on silicon', in *IEEE MTT-S Int. Dig.* (Baltimore, MD), pp. 1917–1920, June 1998.

[85] D. Williams, U. Arzt, and H. Grabinski, 'Characteristic impedance measurement error on lossy substrates', *IEEE Microwave Guided Wave Lett.* **11** (2001), pp. 299–301.

[86] D. Pozar, *Microwave Engineering*, 2nd edn (John Wiley & Sons Ltd/Inc., 1998).

[87] G. Carchon, B. Nauwelaers, W. D. Raedt, D. Schreurs, and S. Vandenberghe, 'Characterising differences between measurement and calibration wafer in probe-tip calibrations', *Electron. Lett.* **35** (1999), pp. 1087–1088.

[88] J. Grzyb, I. Ruiz, and G. Tröster, 'An investigation of the material and process parameters for thin-film MCM-D and MCM-L technologies up to 100 GHz', in *Proc. 53rd Electron. Components & Technol. Conf. (ECTC2003)* (New Orleans, LA), pp. 478–486, May 2003.

[89] P. Chiniwalla, R. Manepalli, K. Farnsworth, M. Boatman, B. Dusch, P. Kohl, and S. A. Bidstrup-Allen, 'Multilayer planarization of polymer dielectrics', *Trans. Adv. Packag.* **24** (2001), pp. 41–53.

[90] A. H. Landzberg, K. H. Brown, and J. K. Ho, 'Process and tool monitoring', in *Microelectronics Manufacturing Diagnostics Handbook*, 6, 1st edn (Van Nostrand Reinhold, 1993).

[91] E. D. Perfecto, K. S. Desai, and G. McAfee, 'MCM-D/C Yield improvements through effective diagnostics', *Int. J. Microcircuits and Electron. Packag.* **22**(4) (1999), pp. 411–417.

[92] D. Kajfez, 'Linear fractional curve fitting for measurements of high Q factors', *IEEE Trans. Microwave Theory Tech.* **42** (1994), pp. 1149–1153.

[93] K. Leong and J. Mazierska, 'Precise measurements of the Q factor of dielectric resonators in the transmission mode accounting for noise, crosstalk, delay of uncalibrated lines, coupling loss, and coupling reactance', *IEEE Trans. Microwave Theory Tech.* **50** (2002), pp. 2115–2127.

[94] S. Toncich and R. E. Colli, 'Data reduction method for Q measurements of strip-line resonators', *IEEE Trans. Microwave Theory Tech.* **40** (1992), pp. 1833–1836.

[95] A. E. Fathy, V. A. Pendrick, B. D. Geller, S. M. Perlow, E. S. Tormey, A. Prabhu, and S. Tani, 'An innovative semianalytical technique for ceramic evaluation at microwave frequencies', *IEEE Trans. Microwave Theory Tech.* **50** (2002), pp. 2247–2252.

[96] G. F. Engen and C. A. Hoer, 'Thru-Reflect-Line: an improved technique for calibrating the dual six-port automatic network analyser', *IEEE Trans. Microwave Theory Tech.* **27** (1979), pp. 987–993.

[97] C. A. Hoer and G. F. Engen, 'Calibrating a dual six-port or four-port analyser for measuring two-ports with any connectors', in *IEEE MTT-S Int. Dig.* (Canada), pp. 665–668, 1986.

[98] R. R. Pantoja, M. J. Howes, J. R. Richardson, and R. D. Pollard, 'Improved calibration and measurement of the scattering parameters of microwave integrated circuits', *IEEE Trans. Microwave Theory Tech.* **37** (1986), pp. 1675–1680.

[99] V. Sundaram, F. Liu, S. Dalmia, J. Hobbs, E. Matoglu, T. N. M. Davis, J. Laskar, M. Swaminathan, G. E. White, and R. R. Tummala, 'Digital and RF integration in system-on-a-package (SoP)', in *51st Electron. Components & Technol. Conf.* (San Diego, CA), pp. 646–650, 2002.

[100] M. F. Davis, A. Sutono, S.-W. Yoon, S. Mandal, N. Bushyager, C.-H. Lee, K. Lim, S. Pinel, M. M. A. Obatoyinbo, S. Chakraborty, J. L. E. M. Tentzeris, T. Nonaka, and R. R. Tummala, 'Integrated RF architectures in fully-organic SOP technology', *IEEE Trans. Adv. Packag.* **25** (2002), pp. 136–142.

[101] S. Dalmia, J. M. Hobbs, V. Sundaram, M. Swaminathan, S. H. Lee, F. Ayazi, G. White, and S. Bhattacharya, 'Design and optimisation of high-Q RF passives on SOP-based organic substrates', in *Proc. of ECTC Conf.*, pp. 495–503, 2002.

[102] J. Galiere and E. E. J. L. Vallard, 'Millimetre-wave MMIC packaging compatible with surface-mount technology (SMT)', in *12th GAAS Symposium*, (Amsterdam), pp. 591–593, 2004.

[103] B. Reig, C. Marcoux, and E. Estebe, 'Collective wiring technology for microwave modules', *Microwave Enging* November (2002), pp. 29–33.

[104] F. Deborgies, T. Lemoine, R. May, and M. Oudart, 'Low cost highly integrated S-band receiver', in *29th EuMC* (Munich, Germany), vol. MF-WED3-1, 1999.

[105] T. Barbier, Y. Mancuso, F. Mazel, B. Reig, and P. Monfraix, '3D wideband package technologies for smart antennas', in *Wireless Design Conf.* (London), 2002.

[106] E. Saint-Etienne, 'Silicon packaging and RF solder-free interconnect for X-band SAR T/R module', in *Gallium Arsenide and Related III-V Compounds Applic. Symp., GAAS*, 2001.

[107] T. Lemoine, M. Oudart, D. Rimbert, I. Favier, and Y. Mancuso, '3D T/R modules for integrated active antennas', in *APMC Proc.* (Jokohama), December 1998.

[108] J. W. Balde, 'Packaging at the turning point – Technology and management issues', in *Proc. IMAPS France*, pp. 22–24, May/June 2001.

[109] T. Suga, A. Takahashi, M. Howlander, K. Saijo, and S. Oosawa, 'A lamination technique of LCP/Cu for electronic packaging', in *IEEE Polytronic Conf.*, pp. 177–182, 2002.

[110] G. Zou, H. Gronqvist, J. P. Starski, and J. Liu, 'Characterization of liquid crystal polymer for high frequency system-in-a-package applications', *IEEE Trans. Adv. Packag.* **25** (2002), pp. 503–508.

[111] D. Thomson, P. Kirby, and M. M. T. J. Papapolymerou, 'W-band characterization of finite ground coplanar transmission lines on liquid crystal polymer (LCP) substrates', in *ECTC Conf. Proc.* (New Orleans, LA), pp. 1652–1655, May 2003.

[112] A. C. Chen, M. J. Chen, and A.-V. Pham, 'Design and fabrication of ultra-wideband baluns embedded in multilayer liquid crystal polymer flex', *Trans. Adv. Packag.* **30** (2007), pp. 533–539.

[113] M. J. Chen, A.-V. H. Pham, N. A. Evers, C. Kapusta, J. Iannotti, W. Kornrumpf, J. J. Maciel, and N. Karabudak, 'Design and development of a package using LCP for RF/Microwave MEMS switches', *IEEE Trans. Microwave Theory Tech.* **54** (2006), pp. 4009–4015.

[114] K. Aihara and A.-V. Pham, 'Development of thin-film liquid crystal polymer surface mount packages for Ka-band applications.', in *IEEE MTT-S Int. Dig.* (San Francisco, CA), pp. 956–959, June 2006.

[115] A. C. Chen, A. Pham, and R. E. L. III, 'A 6-18 GHz push-pull power amplifier with wideband even-order distortion cancellation in LCP module.', in *IEEE MTT-S Int. Dig.* (Honolulu, HI), pp. 1079–1082, June 2007.

[116] L. J. Golonka, 'Technology and applications of low temperature cofired ceramic (LTCC) based sensors and microsystems', *Bull. Polish Acad. Sci.* **54**(2) (2006), pp. 221–231.

[117] S. Q. Scrantom and J. C. Lawson, 'LTCC Technology: Where we are and where we're going – II', in *IEEE MTT-S Int. Dig. Technol. for Wireless Applic.*, pp. 193–200, February 1999.

[118] M. Richtarsic and J. Thornton, 'Characterization and optimization of LTCC for high density large area MCM's', in *7th Int. Conf. on Multichip Modules and High Density Packag.* (Denver, CO), pp. 92–97, 1998.

[119] R. Kulke, W. Simon, C. Guenner, G. Moellenbeck, D. Koether, and Rittweger, 'RF-benchmark up to 40 GHz for various LTCC low loss tapes', in *Proc. of 39th IMAPS-Nordic Conf.* (Stockholm), 2002.

[120] P. Barnwell and M. P. O'Neill, 'Microwave MCM-C utilising low loss LTCC and photo-patterning process', in *Int. Conf. on High Density Packag. and MCMs* (Denver, CO), pp. 237–241, 1999.

[121] J. Page, D. I. Amey, R. Draudt, M. Inman, and S. J. Horowitz, 'MCM-C cost comparison: Fodel® vs. Diffusion Patterning™ vs. Green Tape™', in *6th Int. Conf. on Multichip Modules* (Denver, CO), pp. 53–58, April 1997.

[122] B. Dziurdzia, S. Nowak, W. Gregorczyk, and M. Ciez, 'Photo patterned thick-film resonators', in *14th Int. Conf. on Microwaves, Radar and Wireless Commun.*, vol. 2, pp. 467–471, 2002.

[123] S. Consolazio, K. Nguyen, D. Biscan, K. Vu, and A. F. A. Ramos, 'Low Temperature Cofired Ceramic (LTCC) for wireless applications', in *IEEE MTT-S Int. Dig. Technologies for Wireless Applic.* (Vancouver, BC), pp. 201–205, February 1999.

[124] D. Amey and S. J. Horowitz, 'Microwave material characterization', in *Proc. Int. Symp. Microelectron.* (Minneapolis, MN), pp. 494–499, 1996.

[125] D. I. Amey and S. J. Horowitz, 'Microwave characterization of packaging materials', in *3rd Int. Symp. on Advanced Packaging Materials* (Braselton, GA), pp. 158–161, March 1997.

[126] D. I. Amey and J. P. Curilla, 'Microwave properties of ceramic materials', in *41th Electron. Comp. Technol. Conf. (ECTC)* (Atlanta, GA), pp. 267–272, May 1991.

[127] D. Amey, P. Barnwell, R. Brown, F. Gaughan, S. Horowitz, A. London, R. Novak, D. Slutz, and D. Wicox, 'Ceramic technology for integrated packaging for wireless', in *IEEE Radio Frequency Integrated Circuit Symp.* (Anaheim, CA), pp. 63–66, 1999.

[128] R. L. Brown, P. W. Polinski, and A. S. Shaikh, 'Manufacturing of microwave modules using low-temperature cofired ceramics', in *IEEE MTT-S Int. Dig.* (San Diego, CA), pp. 1727–1730, 1994.

[129] D. I. Amey, S. J. Horowitz, and R. L. Keusseyan, 'High frequency electrical characterization of electronic packaging materials: environmental and process considerations', in *Int. Symp. on Advanced Packaging Materials* (Braselton, GA), pp. 123–128, 1998.

[130] D. I. Amey and S. J. Horowitz, 'Materials performance at frequencies up to 20 GHz', in *IEMT/IMC Proc.*, pp. 331–336, 1997.

[131] W. A. Vitriol, C. J. Sabo, R. L. Brown, and R. G. Pond, 'Development of a new tape dielectric technology for thick film multilayer applications', in *ISHM Proc.* (Dallas, TX), 1986.

[132] P. G. Barnwell and L. Wood, 'A novel thick-film on ceramic MCM technology offering MCM-D performance', in 6th *Int. Conf. on MCMs* (Denver, CO), pp. 48–52, April 1997.

[133] P. Barnwell, C. E. Free, and C. S. Aitchison, 'A novel thick-film on ceramic microwave technology', in *APMC Proc.* (Yokohama, Japan), pp. 495–498, December 1998.

[134] B. Geller, B. Thaler, A. Fathy, M. J. Liberatore, H. D. Chen, G. Ayers, V. Pendrick, and Y. Narayan, 'LTTC-M: An enabling technology for high performance multilayer RF systems', in *IEEE MTT-S Int. Dig.* (Anaheim, CA), pp. 189–192, June 1999.

[135] B. Dziurdzia and S. Nowak, 'FODEL photoimageable technology in microwave applications', in *12th Int. Conf. Microwave and Radar, Mikon'98* (Krakow, Poland) , pp. 445–450, May 1998.

[136] C. E. Free, P. G. Barnwell, and C. S. Aitchison, 'A new thick-film technology for mm-wave MCM and microwave devices', in *Int. Symp. Microelectron. Proc.* (San Diego, CA), pp. 370–373, November 1998.

[137] C. Free and Z. Tian, 'A new LTCC fabrication technology for planar millimeter-wave circuits', in *APMC Proc.* (Singapore), vol. 3, pp. 962–965, November 1999.

[138] P. Barnwell and J. Wood, 'Fabrication of low cost microwave circuits and structures using an advanced thick film technology', in *IEMT/IMC Symp. Proc.* (Tokyo, Japan), pp. 327–332, April 1998.

[139] H. Uchimura, T. Takenoshita, and M. Fujii, 'Development of a laminated waveguide', *IEEE Trans. Microwave Theory Tech.* **46** (1998), pp. 2438–2443.

[140] M. S. Aftanasar, P. R. Young, I. D. Robertson, and S. Lucyszyn, 'Fabrication of dielectric-filled rectangular waveguide using thick-film processing', 6th *IEEE High Frequency Postgraduate Student Colloquium* (Cardiff, UK), pp. 82–87, 2001.

[141] H. Uchimura and T. Takenoshita, 'A ceramic planar 77 GHz Antenna Array', in *IEEE MTT-S Int. Dig.* (Anaheim, CA), pp. 453–456, 1999.

[142] S. Donnay, P. Pieters, K. Vaesen, W. Diels, P. Wambacq, W. D. Raedt, E. Beyne, M. Engels, and I. Bolsens, 'Chip-package codesign of a low-power 5-GHz RF front end', *Proc. IEEE*, vol. 88, pp. 1583–1596, October 2000.

[143] LAP, *Low Cost Large Area Panel Processing of MCM-D Substrates and Packages*. Available at http://cordis.europa.eu/data/MSS_PROJ_FP4_SE/ACTIONeqDndSESSIONeq29799200595 ndDOCeq10ndTBLeqEN_PROJ.htm.

[144] M. Scheffler *et al.*,'LAP: Low cost Large Area Panel processing of MCM-D substrates and packages – achievements and results', in *Proc. Int. Electron. Packag. Tech. Conf.* (Hawaii), July 2001.

[145] D. Cottet, J. Grzyb, and G. Tröster, 'The impact of low cost high density substrates on RF structure integration', in *IEEE/IMAPS Int. Workshop on Chip-Package Codesign (CPD2000)* (Zurich, Switzerland), pp. 88–92, March 2000.

[146] J. Grzyb and G. Tröster, 'MM-wave microstrip patch and slot antennas on low cost large area panel MCM-D substrates – A feasibility and performance study', *IEEE Trans. Adv. Packag.* **25** (2002), pp. 397–408.

[147] J. Grzyb, D. Cottet, and G. Tröster, 'MM-wave microstrip and novel slot antennas on low cost large area panel MCM-D substrates – feasibility and performance study', in *Proc. IEEE 51st Electron. Components & Technol. Conf. (ECTC2001)* (Lake Buena Vista, FL), May 2001.

[148] G. Carchon, K. Vaesen, S. Brebels, W. D. Raedt, E. Beyne, and B. Nauwelaers, 'Multilayer thin-film MCM-D for the integration of high-performance RF and microwave circuits', *IEEE Trans. Comp. Packag. Technol.* **24** (2001), pp. 510–519.

[149] J. Ryckaert, S. Brebels, B. Come, W. Diels, D. Hauspie, S. Stoukatch, K. Vaesen, W. D. Raedt, and S. Donnay, 'Single-package 5GHz WLAN RF module with embedded patch antenna and 20dBm power amplifier', in *IEEE MTT-S Int. Dig.* (USA), pp. 1037–1040, 2003.

[150] G. Carchon, W. D. Raedt, and B. Nauwelaers, 'Integration of CPW quadrature couplers in multilayer thin-film MCM-D', *IEEE Trans. Microwave Theory Tech.* **49** (2001), pp. 1770–1776.

[151] P. Pieters, S. Brebels, W. D. Raedt, and E. Beyne, 'Broadband coplanar couplers in multilayer thin film MCM-D technology', in *IEEE MTT-S Int. Dig.* (Anaheim, CA), pp. 1717–1720, 1999.

[152] P. Pieters, S. Brebels, W. D. Raedt, and E. Beyne, 'Distributed microwave MCM-D circuits for X- and K-Band applications', in *Int. Conf. High Density Packaging and MCM-s* (Denver, CO), pp. 444–449, 1999.

[153] P. Pieters, K. Vaesen, G. Carchon, S. Brebels, W. D. Raedt, and E. Beyne, 'Integration of passive components in thin film multilayer MCM-D technology for wireless front-end applications', in *IEEE APMC Dig.* (Australia), pp. 221–224, 2000.

[154] J. Grzyb, D. Cottet, and G. Tröster, 'Integrated passive elements on low cost MCM-D substrates', in *Proc. High Density Interconnect Conf.* (Santa Clara, CA), pp. 256–261, April 2001.

[155] D. Cottet, J. Grzyb, M. Scheffler, and G. Tröster, 'Experimental analysis of design options for spiral inductors integrated on low cost MCM-D substrates', in *IEEE 51st Electron. Components & Technol. Conf. (ECTC 2001)* (Lake Buena Vista, FL), pp. 824–830, May 2001.

[156] G. Carchon, W. D. Raedt, and B. Nauwelaers, 'Accurate measurement and characterization of reciprocal 3-ports, application to CPW T-junction in thin-film multi-layer MCM-D', in *IEEE APMC* (Sydney, NSW, Australia), pp. 453–456, December 2000.

[157] G. Carchon, *Measurement, Modelling and Design of Monolithic and Thin-Film Microwave Integrated Circuits*. PhD thesis, IMEC, 2001.

[158] A. C. Chen, A.-V. Pham, and R. E. L. III, 'Development of low-loss broad-band planar baluns using multilayered organic thin films', *IEEE Trans. Microwave Theory Tech.* **53** (2005), pp. 3648–3654.

[159] U. Dahlgren, J. Svedin, H. Johansson, O. Hagel, H. Zitath, and C. Karlsson, 'An integrated millimeter wave BCB patch antenna HEMT receiver', in *IEEE MTT-S Int. Dig.* (San Diego, CA) pp. 661–664, 1994.

[160] E. Soliman, S. Brebels, and E. Beyne, 'Circularly polarized aperture antenna fed by CPW and built in MCM-D technology', *Electron. Lett.* **35**(4) (1999), pp. 250–251.

[161] A. H. Pham, J. Laskar, V. B. Krishnamurthy, H. S. Cole, and T. Sitnik-Nieters, 'Ultra low loss millimeter wave multichip module interconnects', *IEEE Trans. Comp., Hybrids, Manuf. Technol.* **21** (1998), pp. 302–308.

[162] S. Manohar, Z. Jin, A. Pham, V. Krishnamurthy, W. M. D. Bates, B. Schmanski, R. Saia, and L. Sprinceanu, 'Development of microwave/millimeter wave integral passives for multi-layer organic MCMs', in *IEEE MTT-S Int. Dig.* (Boston, MA), pp. 1879–1882, 2000.

[163] A. Pham, R. Ramachandran, J. Laskar, V. Krishnamurthy, D. Bates, W. Marcinkiewicz, B. Schmanski, P. Piacente, and L. Sprinceanu, 'Development of a millimeter-wave system-on-a-package utilizing MCM integration', *IEEE Trans. Microwave Theory Tech.* **49** (2001), pp. 1747–1749.

[164] A. H. Pham, V. B. Krishnamurthy, D. Bates, W. Marcinkiewicz, R. S. B. Schmanski, and L. Sprinceanu, 'Development of integral passive components for multilayer organic MCMs at millimeter wave frequencies', *Trans. Adv. Packag.* **25** (2002), pp. 98–101.

[165] P. Abele, J. Konle, D. Behammer, E. Sonmez, K.-B. Schad, A. Trasser, and H. Schumacher, 'Wafer level integration of a 24 GHz and 34 GHz differential SiGe-MMIC oscillator with a loop antenna on a BCB membrane', in *IEEE MTT-S Int. Dig.* (USA), pp. 1033–1036, 2003.

[166] R. Carrillo-Ramirez and R. Jackson, 'A technique for interconnecting millimeter wave integrated circuits using BCB and bump bonds', *IEEE Microwave Wireless Comp. Lett.* **13** (2003), pp. 196–198.

[167] J. Grzyb, M. Klemm, and G. Tröster, 'MCM-D/L technology for realization of low cost system-on-package concept at 60-80 GHz', in *EuMC Proc.* (Munich, Germany), pp. 963–966, 2003.

[168] P. Chiniwalla, R. Manepalli, K. Farnsworth, M. Boatman, B. Dusch, P. Kohl, and S. Bidstrup-Allen, 'Multilayer planarisation of polymer dielectrics', *IEEE Trans. Adv. Packaging* **24** (2001), pp. 41–53.

[169] B. Chou, S. Beilin, H. Jiang, D. Kudzuma, M. Lee, M. McCormack, T. Massingill, M. Peters, J. Roman, Y. Takahashi, and V. Wang, 'Multilayer high density flex interconnect', in *Int. Conf. Multichip Modules High Density Packag.* (Denver, CO), pp. 270–278, 1999.

[170] J. Geissinger, F. Keller, S. Trevino, and R. Kamei, 'Tape based CSP package supports fine pitch wirebonding', in *Electronics Manufacturing Technology (IEMT) Symp., 2002* (San Jose, CA), pp. 41–45, 2002.

[171] P. D. Cooper, P. A. Piacente, and R. J. Street, 'Multichip-on-flex plastic encapsulated MHDI- low cost substrateless manufacturing for microwave and millimeterwave modules', in *IEEE MTT-S Int. Dig.* (San Francisco, CA), pp. 219–222, 1996.

[172] M. McNulty, J. Schnell, and D. Nixon, 'Microwave multichip modules using low cost microwave chip on flex packaging technology', in *Int. Conf. Multichip Modules High Density Packag.* (Denver, CO), pp. 262–267, 1998.

[173] A.-V. H. Pham, A. Sutono, J. Laskar, V. Krishnamurthy, D. Lester, E. Balch, and J. Rose, 'Development of microwave multilayer plastic-based multichip modules', *Trans. Adv. Packag.* **24** (2001), pp. 37–40.

[174] L. Katehi, G. Rebeiz, T. M. Weller, R. F. Drayton, H.-J. Cheng, and J.F.Whitaker, 'Micromachined circuits for millimeter- and sub-millimeter-wave applications', *IEEE Antennas Propagat. Mag.* **35** (1993), pp. 9–17.

[175] S. V. Robertson, L. B. Katehi, and G. M. Rebeiz, 'Micromachined W-Band filters', *IEEE Trans. Microwave Theory Tech.* **44** (1996), pp. 598–606.

[176] R. Drayton and L. Katehi, 'Development of self-packaged high frequency circuits using micromachining techniques', *IEEE Trans. Microwave Theory Tech.* **43** (1995), pp. 2073–2080.

[177] R. Drayton, R. M. Henderson, and L. P. B. Katehi, 'High frequency circuit components on micromachined variable thickness substrates', *Electron. Letters* **33** (1997), pp. 303–304.

[178] E. H. Klaasen, K. Petersen, J. M. Noworolski, J. Logan, N. I. Maluf, J. Brown, W. Storn, and G. T. A. Kovacs, 'Silicon fusion bonding and deep reactive ion etching: A new technology for microstructures', in *8th Int. Conf. on Solid-State Sensors and Actuators*, vol. 1 (Stockholm), pp. 556–559, June 1995.

[179] R. Drayton, S. P. Pacheco, J. Wang, J.-G. Yook, and L. Katehi, 'Micromachined filters on synthesized substrates', *IEEE Trans. Microwave Theory Tech.* **49** (2001), pp. 308–314.

[180] S. R. Banerjee and R. F. Drayton, 'Micromachined low dispersion interconnects for optoelectronic array packaging', in *IEEE MTT-S Int. Dig.* (Seattle, WA), pp. 1565–1568, 2002.

[181] W. R. McGrath, C. Walker, M. Yap, and Y.-C. Tai, 'Silicon micromachined waveguide for millimeter-wave and submillimeter-wave frequencies', *IEEE Microwave Guided Wave Lett.* **3** (1993), pp. 61–63.

[182] C. L. Collins, J. W. Digby, R. D. Pollard, R. E. Miles, and G. M. Parkhurst, 'W-band measurements of 100 μm height micro-machined air-filled rectangular waveguides', in *IEEE MTT-S Int. Dig.* (Denver, CO), pp. 1439–1442, 1997.

[183] J. W. Digby, G. M. Parkhurst, C. E. McIntosh, G. M. Parkhust, and B. M. Towlson, 'Fabrication and characterization of micromachined rectangular waveguide components for use at millimeter-wave and terahertz frequencies', *IEEE Trans. Microwave Theory Tech.* **48** (2000), pp. 1293–1302.

[184] P. D. Curtis, S. Iezekiel, R. E. Miles, and C. R. Pescod, 'SU-8 as a material for integrated all-optical microwave filters', *Microwave Enging* April (2001), pp. 51–53.

[185] J. M. Shaw, J. D. Gelorme, N. C. LaBianca, W. W. Conley, and S. H. Holmes, 'Negative photoresists for optical lithography.', *IBM J. Res.* **41** (1997), pp. 81–94.

[186] C. L. Collins, R. E. Miles, R. D. Pollard, D. P. Steenson, and J. W. Digby, 'Technique for micromachining millimetre-wave rectangular waveguide', *Electron. Letters* **34** (1998), pp. 996–997.

[187] C. L. Collins, R. E. Miles, J. W. Digby, and G. M. Parkhurst, 'A new micro-machined millimeter-wave and terahertz snap-together rectangular waveguide technology', *IEEE Microwave Guided Wave Lett.* **9** (1999), pp. 63–65.

[188] D. P. Newlin, A.-V. H. Pham, and J. E. Harriss, 'Development of low loss organic-micromachined interconnects on silicon at microwave frequencies', *IEEE Trans. Components, Packag. Technol.* **25** (2002), pp. 506–510.

[189] R. M. Henderson and L. P. B. Katehi, 'Silicon-based micromachined packages for high-frequency applications', *IEEE Trans. Microwave Theory Tech.* **47** (1999), pp. 1563–1569.

[190] L. P. B. Katehi, J. F. Harvey, and E. Brown, 'MEMS and Si micromachined circuits for high-frequency applications', *IEEE Trans. Microwave Theory Tech.* **50** (2002), pp. 858–866.

[191] J. Grzyb, D. Cottet, and G. Tröster, 'MM-wave integrated antennas on low cost MCM-D substrates', in *9th IEEE Top. Meet. Elect. Performance of Electron. Packag.* (Scottsdale, AZ), pp. 269–272, 2000.

[192] I. Jeong, S.-H. Shin, J. Go, J. Lee, C. Nam, and Y. K. D. Kim, 'High-performance air-gap transmission lines and inductors for millimeter-wave applications', *IEEE Trans. Microwave Theory Tech.* **50** (2002), pp. 2850–2855.

[193] J. Haley and R. Drayton, 'Characterization of high-density micromachined interconnects', in *IEEE MTT-S Int. Dig.* (Phoenix, AZ), pp. 1911–1914, 2001.

[194] R. W. Jackson and D. M. Pozar, 'Full-wave analysis of microstrip open-end and gap discontinuities', *IEEE Trans. Microwave Theory Tech.* **33** (1985), pp. 1036–1042.

[195] T. M. Weller, L. Katehi, M. Herman, and P. Wamhof, 'Membrane Technology (MIST-T) applied to microstrip: A 33GHz Wilkinson power divider', in *IEEE MTT-S Int. Dig.* (San Diego, CA), pp. 911–914, 1994.

[196] S. L. March, 'Phase velocity compensation in parallel-coupled microstrip', in *IEEE MTT-S Int. Dig.* (USA), pp. 410–412, 1982.

[197] W. Dydyk, 'Accurate design of microstrip directional couplers with capacitive compensation', in *IEEE MTT-S Int. Dig.* (Dallas, TX), pp. 581–584, 1990.

[198] J.-L. Chen, S.-F. Chang, and C.-T. Wu, 'A high-directivity microstrip directional coupler with feedback compensation', in *IEEE MTT-S Int. Dig.* (USA), pp. 101–104, 2002.

[199] J. L. Klein and K. Chang, 'Optimum dielectric overlay thickness for equal even- and odd-mode phase velocities in coupled microstrip circuits', *Electron. Lett.* **26** (1990), pp. 274–276.

[200] S. V. Robertson, A. R. Brown, L. P. B. Katehi, and G. M. Rebeiz, 'A 10-60 GHz micromachined directional coupler', *IEEE Trans. Microwave Theory Tech.* **46** (1998), pp. 1845–1849.

[201] U. Goebel, 'Dc to 100 GHz chip-to-chip interconnects with reduced tolerance sensitivity by adaptive wirebonding', in *3rd IEEE Top. Meet. Electrical Performance of Electronic Packaging* (Monterey, CA), pp. 182–185, November 1994.

[202] H. Richter, P. Koelzer, and A. Baumgärtner, 'Flip Chip attach and RF-evaluation of GaAs-assemblies', in *MCM and VLSI Packag. Tech and Mfg Technol. Workshop*, pp. 2_8–2_10, 1994.

[203] H. Sakai, Y. Ota, K. Inoue, T. Yoshida, K. Takahashi, S. Fujita, and M. Sagawa, 'A novel milimeter-wave IC on Si substrate using flip-chip bonding technology', in *IEEE MTT-S Int. Dig.* (San Diego, CA), pp. 1763–1766, May 1994.

[204] T. Shimura, Y. Kawasaki, Y. Ohashi, K. Shirakawa, T. Hirose, S. Aoki, H. Someta, K. Makiyama, and S. Yokokawa, '76 GHz flip-chip MMICs for automotive radars', in *IEEE RFIC Symp. Dig.* (Baltimore, MD), pp. 25–28, June 1998.

[205] H. Sakai, Y. Ota, K. Inoue, M. . Yanagighara, T. Matsuno, M. Tanabe, T. Yoshida, Y. Ikeda, S. Fujita, K. Takahashi, and M. Sagawa, 'A millimeter-wave flip-chip IC using micro-bump bonding technology', in *Proc. 43rd IEEE ISSCC* (San Francisco, CA), pp. 408–409, 482, February 1996.

[206] T. Hirose, K. Makiyama, K. Ono, T. M. Shimura, S. Aoki, Y. Ohashi, S. Yokokawa, and Y. Watanabe, 'A flip-chip MMIC design with coplanar waveguide transmission line in the W-band', *IEEE Trans. Microwave Theory Tech.* **46** (1998), pp. 2276–2282.

[207] G. Baumann, H. Richter, A. Baumgärtner, D. Ferling, R. Heilig, F. Hollmann, H. Müller, H. Nechansky, and M. Schlechtweg, '51 GHz frontend with flip chip and wire bond

interconnections from GaAs MMICs to a planar patch antenna', in *IEEE MTT-S Int. Dig.* (Orlando, FL), pp. 1639–1642, 1995.

[208] S. Masuda, T. Hirose, S. Yokokawa, M. Nishi, S. Iijima, K. Ono, and Y. Watanabe, 'A new flip-chip MMIC technology with multi-layer transmission line structure for low-cost W-band transceivers', in *23rd IEEE GaAs IC Symp. Dig.* (Baltimore, MD), pp. 118–123, 2001.

[209] W. Simon, R. Kulke, A. Wien, M. Rittweger, I. Wolff, A. Girard, and J.-P. Bertinet, 'Interconnects and transitions in multilayer LTCC multichip modules for 24 GHz ISM-band applications', in *IEEE MTT-S Int. Dig.* (Boston, MA), pp. 1047–1050, 2000.

[210] H.-Y. Lee, 'Wideband characterization of a typical bonding wire for microwave and millimeter-wave integrated circuits', *IEEE Trans. Microwave Theory Tech.* **43** (1995), pp. 63–68.

[211] W. Menzel, 'Packaging and interconnects for milimeter wave circuits: a review', in *Annales de Télécommunications*, vol. 52 (France), pp. 145–154, 1997.

[212] M. Rittweger, 'System technologies for micro- and millimeter wave circuits', tech. rep., 2nd Report of the research program of the German Ministry of Research, Contract No. 13 MV 0218. (in German).

[213] F. Alimenti, P. Mezzanotte, L. Roselli, and R. Sorrentino, 'Modeling and characterization of the bonding-wire interconnection', *IEEE Trans. Microwave Theory Tech.* **49** (2001), pp. 142–150.

[214] F. Alimenti, P. Mezzanotte, L. Roselli, and R. Sorrentino, 'Multi-wire microstrip interconnections: a systematic analysis for the extraction of an equivalent circuit', in *IEEE MTT-S Int. Dig.* (Baltimore, MD), pp. 1929–1932, 1998.

[215] F. Alimenti, U. Goebel, and R. Sorrentino, 'Quasi static analysis of microstrip bondwire interconnects', in *IEEE MTT-S Int. Dig.* (Orlando, FL), pp. 679–682, 1995.

[216] T. Krems, W. Haydl, H. Massler, and J. Rüdiger, 'Millimeter-wave performance of chip interconnections using wire bonding and flip chip', in *IEEE MTT-S Int. Dig.* (San Francisco, CA), pp. 247–250, 1996.

[217] T. P. Budka, 'Wide-bandwidth millimeter-wave bond-wire interconnects', *IEEE Trans. Microwave Theory Tech.* **49** (2001), pp. 715–718.

[218] F. Alimenti, P. Mezzanotte, L. Roselli, and R. Sorrentino, 'An equivalent circuit for the double bonding wire interconnect', in *IEEE MTT-S Int. Dig.* (Anaheim, CA), pp. 633–636, 1999.

[219] N.-H. Huynh, W. Heinrich, K. Hirche, W. Scholz, M. Warth, and W. Ehrlinger, 'Optimized flip-chip interconnect for 38 GHz thin-film microstrip multichip modules', in *IEEE MTT-S Int. Dig.* (Boston, MA), pp. 69–72, 2000.

[220] W. Heinrich, A. Jentzsch, and G. Baumann, 'Millimeter-wave characteristics of flip-chip interconnects for multichip modules', *IEEE Trans. Microwave Theory Tech.* **46** (1998), pp. 2264–2268.

[221] A. Jentzsch and W. Heinrich, 'Theory and measurements of flip-chip interconnects for frequencies up to 100 GHz', *IEEE Trans. Microwave Theory Tech.* **49** (2001), pp. 871–878.

[222] R. W. Jackson and R. Ito, 'Modeling millimeter-wave IC behavior for flipped-chip mounting schemes', *IEEE Trans. Microwave Theory Tech.* **45** (1997), pp. 1919–1925.

[223] F. J. Schmückle, A. Jentzsch, W. Heinrich, J. Butz, and M. Spinnler, 'LTCC as MCM substrate: Design of strip-line structures and flip-chip interconnects', in *IEEE MTT-S Int. Dig.* (Phoenix, AZ), pp. 1903–1906, 2001.

[224] T. E. Kazior, H. N. Atkins, A. Fatemi, Y. Chen, F. Y. Colomb, and J. P. Wendler, 'DBIT- direct backside interconnect technology: a manufacturable, bond wire free interconnect technology for microwave and millimeter wave MMICs', in *IEEE MTT-S Int. Dig.* (Denver, CO), vol. 2, pp. 724–727, 1997.

[225] D. Staiculescu, J. Laskar, and E. M. Tenzeris, 'Design rule development for microwave flip-chip applications', *IEEE Trans. Microwave Theory Tech.* **48** (2000), pp. 1476–1481.

[226] W. Heinrich, A. Jentzsch, and G. Baumann, 'Millimeterwave characteristics of flip-chip interconnects for multi-chip modules', in *IEEE MTT-S Int. Dig.* (Baltimore, MD), pp. 1083–1086, 1998.

[227] N. Iwasaki, F. Ishitsuka, and K. Kato, 'High performance flip-chip technique for wide-band modules', in *5th IEEE Top. Meet. Electrical Performance of Electronic Packaging* (Napa, CA), pp. 207–209, October 1996.

[228] H. H. M. Ghouz and E.-B. El-Sharawy, 'An accurate equivalent circuit model of flip chip and via interconnects', *IEEE Trans. Microwave Theory Tech.* **44** (1996), pp. 2543–2553.

[229] F. M. D. Paola, V. Aliberti, B. Rajaei, N. Rinaldi, and J. N. Burghartz, 'A scalable physical model for coplanar waveguide transition in flip-chip applications', in *Proc. 24th Int. Conf. Microelectron.*, vol. 1 (Miel, Serbia and Montenegro), pp. 345–348, May 2004.

[230] G. Carchon, S. Brebels, P. Pieters, K. Vaesen, S. Vanderberghe, D. Schreurs, W. de Raedt, and B. Nauwelaers, 'The influence of bumping height on the performance of flip-chipped microwave circuits, application to coplanar Lange couplers', in *ISRAMMT 99* (Malaga, Spain), pp. 755–758, 1999.

[231] G. Carchon and B. Nauwelaers, 'The influence of bumping height on the performance of flip-chipped CPW Lange couplers', *Microwave and Optical Technol. Lett.* **29** (2001), pp. 263–267.

[232] Z. Feng, W. Zhang, B. Su, K. C. Gupta, and Y. C. Lee, 'RF and mechanical characterization of flip-chip interconnects in CPW circuits with underfill', in *IEEE MTT-S Int. Dig.* (Baltimore, MD), vol. 3, pp. 1823–1826, June 1998.

[233] G. Baumann, E. Müller, f. Buchali, D. Ferling, H. Richter, and W. Heinrich, 'Evaluation of glob top and underfill encapsulated active and passive structures for milimeter wave applications', in *Proc. 27th EuMC* (Jerusalem, Israel), pp. 26–31, September 1997.

[234] H. Kusamitsu, Y. Morishita, K. Maruhashi, M. Ito, and K. Ohata, 'The flip-chip bump interconnection for millimeter-wave GaAs MMIC', in *Int. Conf. Multichip Modules High Density Packag.* (Denver, CO), pp. 47–52, April 1998.

[235] K. Beilenhoff and W. Heinrich, 'Excitation of the parasitic parallel-plate mode at coplanar discontinuities', in *IEEE MTT-S Int. Dig.* (Denver, CO), vol. 3, pp. 1789–1792, 1997.

[236] J.-P. Raskin, G. Gauthier, L. P. Katehi, and G. M. Rebeiz, 'W-band single-layer vertical transitions', *IEEE Trans. Microwave Theory Tech.* **48** (2000), pp. 161–164.

[237] J.-P. Raskin, G. Gauthier, L. P. Katehi, and G. M. Rebeiz, 'Mode conversion at GCPW-to-microstrip-line transitions', *IEEE Trans. Microwave Theory Tech.* **48** (2000), pp. 158–161.

[238] R. W. Jackson and D. W. Matolak, 'Surface-to-surface transition via electromagnetic coupling of coplanar waveguides', *IEEE Trans. Microwave Theory Tech.* **35** (1987), pp. 1027–1032.

[239] J. J. Burke and R. W. Jackson, 'Surface-to-surface transition via electromagnetic coupling of microstrip and coplanar waveguide', *IEEE Trans. Microwave Theory Tech.* **37** (1989), pp. 519–525.

[240] V. Tripathi, 'Asymmetric coupled transmission lines in an inhomogeneous medium', *IEEE Trans. Microwave Theory Tech.* **23** (1975), pp. 734–739.

[241] L. Zhu and W. Menzel, 'Broad-band microstrip-to-CPW transition via frequency-dependent electromagnetic coupling', *IEEE Trans. Microwave Theory Tech.* **52** (2004), pp. 1517–1522.

[242] L. Zhu and K. Wu, 'Ultrabroad-band vertical transition for multi-layer integrated circuits', *IEEE Microwave Guided Wave Lett.* **9** (1999), pp. 453–455.

[243] G. Strauss, P. Ehret, and W. Menzel, 'On-wafer measurements of microstrip-based MMICs without via holes', in *IEEE MTT-S Int. Dig.* (San Francisco, CA), pp. 1399–1402, 1996.

[244] J. Kassner and W. Menzel, 'In-circuit testing of complex circuits using on-wafer probing and electromagnetic coupled ground interconnects', in *IEEE MTT-S Int. Dig.*, pp. 1863–1866, 2000.

[245] G. Zheng, J. Papapolymerou, and M. M. Tenzeris, 'Wideband coplanar waveguide RF probe pad to microstrip transitions without via holes', *IEEE Microwave Wireless Comp. Lett.* **13** (2003), pp. 514–516.

[246] G. P. Gauthier, L. P. Katehi, and G. M. Rebeiz, 'W-band finite ground coplanar waveguide (FGCPW) to microstrip line transition', in *IEEE MTT-S Int. Dig.* (Baltimore, MD), pp. 107–109, 1998.

5

Printed Millimeter Antennas – Multilayer Technologies

O. Lafond and M. Himdi

5.1 Introduction and Considerations for Millimeter-wave Printed Antennas

5.1.1 Introduction

Printed antenna technologies are worthwhile owing to the antennas' low profile and weight and to a good capability to become interconnected with active devices (amplifiers, mixers or others). Moreover, if a low-cost substrate is used, these technologies can be competitive for industrial and commercial development. However, some problems can appear with printed antenna technologies, especially in the millimeter-wave (mmWave) range:

- metallic loss;
- dielectric loss;
- spurious radiation due to feeding lines;
- loss due to surface wave modes.

Indeed, metallic loss (in neper per meter) increases with frequency. Metallic loss of one microstrip line can be calculated using the following formula:

$$\alpha_c = \frac{\pi \mu_0 f}{Z_0 W} \delta_s \tag{5.1}$$

where δ_s is the skin depth, Z_0 the characteristic impedance and W the line width.

Advanced Millimeter-wave Technologies: Antennas, Packaging and Circuits
Duixian Liu, Ulrich Pfeiffer, Janusz Grzyb and Brian Gaucher © 2009 John Wiley & Sons, Ltd

The dielectric loss increases with frequency and loss tangent can be higher if we compare with lower frequencies. Moreover, surface waves can easily appear owing to the ratio between substrate thickness and operating wavelength. All losses have to be quantified at mmWave range, because substrate manufacturers commonly supply loss tangent and dielectric constant data only up to 10 GHz but not in the mmWave range. Therefore, it becomes necessary to characterize substrates up to 100 GHz for a satisfactory estimation of the resonant frequency and antenna efficiency. To characterize substrates, a number of techniques can be employed.

- Coaxial probe [1]: This is an open-ended coaxial transmission line probe with a cutoff section. The material is measured by setting the probe on the flat face of a sample or by immersing it into a liquid. The reflected signal can be measured and related to ϵ_r. This technique gives good results up to 20 GHz, but not for mmWaves.

- Transmission line techniques [2, 3]: The dielectric sample is placed inside an enclosed transmission line offering a rectangular waveguide or a coaxial line. With this technique, complex ϵ_r and μ_r can be computed from the measurement of the reflected signal (S11) and transmitted signal (S21) after the calibration procedure. These techniques usually yield good results but it becomes difficult to make dielectric-shaped mmWave samples in coaxial or waveguide lines.

- Open resonant cavity with high Q [4, 5]: This method provides high Q because it has two large concave mirrors facing each other with a dielectric sample placed in the center. With this method, it becomes possible to assess the real part and the imaginary part of the complex permittivity. Open resonators are very convenient for the measurement of low-loss materials because higher Q values can be made at millimeter-wavelength range compared to a closed resonant cavity.

- Free space technique [6, 7]: A layer of the dielectric under test of a given thickness is placed between two antennas facing each other. Horn/lens antennas are often used to provide a plane wave. These antennas enable diffraction effects on the dielectric sample to be avoided. Calibration before measuring permits multiple reflections between two horns thru the surface of the sample to be removed. By this technique, ϵ_r and $\tan\delta$ can be obtained from the reflection and transmission coefficients. It is suitable for the mmWave range because the size of antennas is reduced and thin substrate samples can be used. Even if high-Q cavity and free-space technique give good results for mmWaves, it is complex to implement them.

- It is possible, therefore, to have a good estimation of the dielectric constant and loss tangent while using printed resonant circuits (ring or open stubs). This last technique is helpful in characterizing substrates to be used for printed antennas as it enables global losses to be estimated in lines. With the resonant ring [8] (see Figure 5.1), the first step consists in optimizing the printed ring to obtain resonant frequencies in the considered bandwidth. Two microstrip lines are coupled with this ring ($R = n\lambda_g/2\pi$). There is a gap between the lines and the ring, that is equivalent to a capacitor which enables energy coupling. The S21 parameter and resonant frequency are measured. These values reach ϵ_{eff}. After this measurement, the design is analyzed using electromagnetic software to determine the dielectric constant. For a printed stub [9], the technique is similar but in this case, a parallel open stub is printed ($L = (2n + 1)\lambda_g/4$) (see

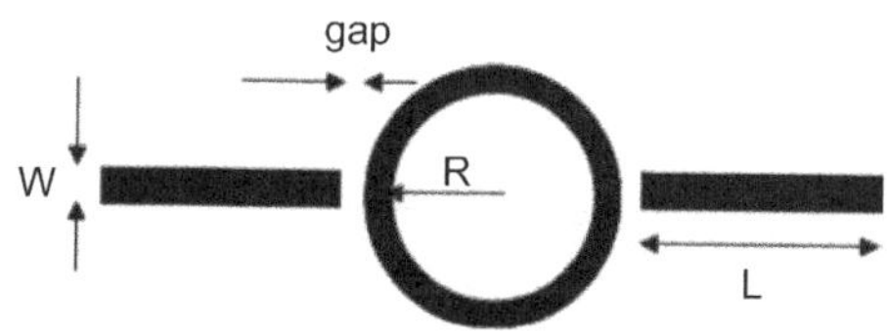

Figure 5.1: Resonant ring.

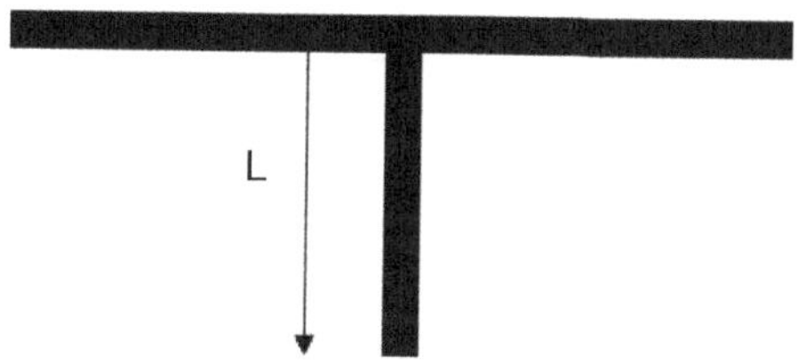

Figure 5.2: Open stub.

Figure 5.2). The measured S21 parameter presents resonances that will permit the dielectric constant to be calculated by comparison with the electromagnetic software result. Additionally, two microstrip lines of different lengths are manufactured to estimate loss tangent. The total loss is measured. Subsequently a new simulation is carried out to extract tan δ.

Surface mode wave loss appears more commonly in the mmWave spectrum because of the ratio between substrate thickness and operating wavelength. Indeed, some authors [10, 11] have reported on this problem and showed some patch antenna efficiency results versus dielectric substrate. The efficiency is therefore given by the following formula:

$$\eta = 1 - \sqrt{\epsilon_r - 1}\frac{h}{\lambda}\left[3.4 - \frac{370}{\epsilon_r}\left(\frac{h}{\lambda}\right)^2\right] \tag{5.2}$$

The same authors provide an efficiency comparison for antennas printed on substrates with $\epsilon_r = 2.55$ and $\epsilon_r = 12$ respectively. It is specified that the efficiency is lower with high permittivity dielectric substrates. Moreover, in this case, the efficiency due to surface mode waves decreases quickly with substrate thickness.

In conclusion, it seems primordial to use low thickness and low permittivity substrate to design millimeter printed antennas. On the other hand, a feeding line network for antenna arrays can be printed on high dielectric constant substrates because they are more suitable for active component interconnection.

Finally, spurious radiation due to the feeding line can have a dramatic influence on radiating patterns, especially for printed antenna arrays. A number of results will be detailed below to pinpoint these discontinuity radiation problems. One solution to reduce unwanted radiation is to design multilayer printed antennas in order to separate the radiating layer from the feeding line layer.

Table 5.1: PTFE substrate characterization at 100 GHz with free space and high-Q Fabry–Perot techniques.

Data at 10 GHz	Free space	High-Q Fabry–Perot
$\epsilon_r = 6.2$ $\tan\delta = 0.002$	$\epsilon_r = 7.4$ $\tan\delta = 0.002$	$\epsilon_r = 7.5$ $\tan\delta = 0.002$

Table 5.2: Quartz substrate characterization at 100 GHz with free space and high-Q Fabry–Perot techniques.

Data at 10 GHz for fused quartz	Free space	High-Q Fabry–Perot
$\epsilon_r = 3.8$ $\tan\delta = 0.0001$	$\epsilon_r = 3.8$ $\tan\delta = 0.0002$	$\epsilon_r = 3.8$ $\tan\delta = 0.00025$

5.1.2 Results for Substrate Characterization Using Free Space and High-Q Techniques

Characterization results for this first part concern PTFE Ceramic substrate (RT duroid 6006). The characteristics at 10 GHz (given by the manufacturer) are:

$$\epsilon_r = 6.2$$

$$\tan\delta = 0.002$$

This substrate has been characterized as between 75 and 110 GHz using the free-space technique and high-Q Fabry–Perot cavity [12]. Comparative results are presented in the following Table 5.1. There is quite a difference between the data and measures concerning the dielectric constant but the loss tangent is stable up to 110 GHz. In this case, shift in the dielectric constant between 10 and 100 GHz is very strong. Therefore, if we rely on the manufacturer's data to make printed antennas at 100 GHz, the shift of resonant frequency between calculation and measure will be dramatic.

Fused quartz and foam substrates (ϵ_r very close to air) are often used in the millimeter-wavelength range owing to their low loss tangent. These two dielectrics have been characterized [12] within the same bandwidth and give the following results detailed in Tables 5.2 and 5.3 respectively. The new measured data show that fused quartz is low loss and very stable up to very high frequencies, which is why quartz is quite a good substrate for making high-efficiency antennas. Concerning Divinycell foam substrates, values prove that they can be employed to make millimeter antennas because loss tangent is low and the dielectric constant close to 1, meaning that high-directivity antennas with good efficiency can be obtained.

5.1.3 Results of Substrate Characterization Using Printed Resonant Circuits

In all of the following cases, resonator stubs and microstrip lines have been printed so as to find out the dielectric constant and loss tangent for several kinds of commercial soft substrate: polymetyl–pentene (TPX), RT duroid 5880, RT duroid 6006, aln or 35 NQ substrate.

Table 5.3: Divinycell foam substrate characterization at 100 GHz with free space and high-Q Fabry–Perot techniques.

Data at 10 GHz for foam substrate	Free space	High-Q Fabry–Perot
$\epsilon_r = 1.14$ $\tan\delta = 0.002$	$\epsilon_r = 1.12$ $\tan\delta = 0.0028$	$\epsilon_r = 1.26$ $\tan\delta = 0.0043$

Table 5.4: Characterization of TPX, RT duroid 6006 substrates in the mmWave range.

TPX ($h = 0.127$ mm)	Duroid 6006 ($h = 0.254$ mm)
$\epsilon_r = 2.14$, $\tan\delta = 0.001$ at 10 GHz	$\epsilon_r = 6.2$, $\tan\delta = 0.002$ at 10 GHz
$\epsilon_r = 2.17$, $\tan\delta = 0.0035$ at 60 GHz	$\epsilon_r = 7$ at 45 GHz

The RT duroid 5880 ($h = 0.127$ mm, $\epsilon_r = 2.2$ and $\tan\delta = 0.0009$ at 10 GHz) was considered first. A parallel open stub was optimized to bring the resonant frequency (S21 parameter) to around 60 GHz. Two microstrip lines were used to estimate insertion losses. The objective here is to obtain a fair appraisal of the dielectric constant and loss tangent to study printed antennas for high bite rate indoor communication systems. Measurement and simulation are performed to deduce these parameters. The comparison between the S21 magnitude measurement (dB) and simulation [13] in terms of resonant frequency enables the dielectric constant value to be obtained (Figure 5.3). A good agreement was found for $\epsilon_r = 2.24$.

The second step concerns the measurement of two different length microstrip lines, one to calibrate the vector analyzer and the other to estimate the total loss (54 dB m^{-1} in this case at 60 GHz). Next, electromagnetic software is used to calculate the S21 magnitude of this second line. This result is then compared with the measurement and the loss tangent is extracted. Measurement and simulation results concerning S21 magnitude are in good agreement for $\tan\delta = 0.004$.

At 60 GHz, the dielectric constant is close to the 10 GHz manufacturers, data (2.24 compared with 2.2). Conversely, the loss tangent is higher at 60 GHz (0.004 compared with 0.0009), which accounts for a drop in the printed antennas' efficiency. The value of the total loss at 60 GHz for a microstrip line printed on this substrate is confirmed by BAE system company in reference [14] where a loss of 50 dB m^{-1} was recorded.

Using the same approach (printed resonator stub and microstrip lines), other results are presented in Tables 5.4 and 5.5 for TPX, RT duroid 6006, aln (ϵ_r close to alumina) and 35 NQ substrates [12]. The first one was used to design multilayer printed antennas in the 40 and 60 GHz band and the last two were characterized to manufacture stacked patch, multilayer printed arrays for multimedia communication in the 50 GHz band. Manufacturers data and estimated values in the mmWave bandwidth are compared and S21 magnitude parameters are given to show the resonant frequencies (Figures 5.3–5.7).

The last substrate (35 NQ) was not used for printed patches but for parasitic printed patches in order to protect printed antennas against spurious space radiation. The loss tangent

Table 5.5: Characterization of aln and 35 NQ substrates in the mmWave range.

aln ($h = 0.127$ mm)	35 NQ ($h = 0.127$ mm)
$\epsilon_r = 9$, $\tan\delta = 0.004$ at 1 MHz	$\epsilon_r = 3.5$, $\tan\delta = 0.0009$ at 1 MHz
$\epsilon_r = 8.5$, $\tan\delta = 0.0065$ at 50 GHz	$\epsilon_r = 3.4$, $\tan\delta = 0.0175$ at 50 GHz

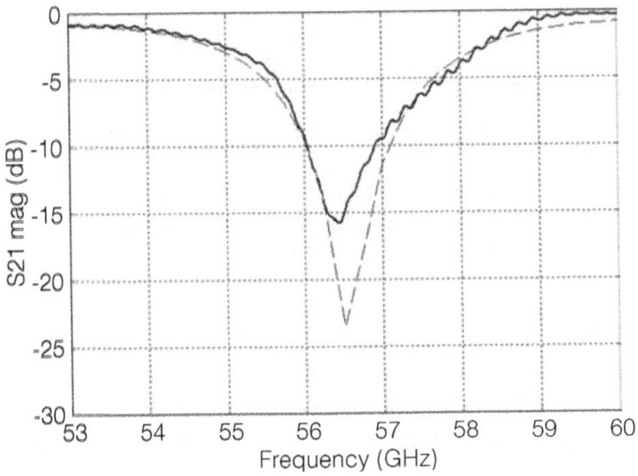

Figure 5.3: Magnitude of transmission coefficient (in decibels) at 60 GHz for RT duroid 5880 substrate.

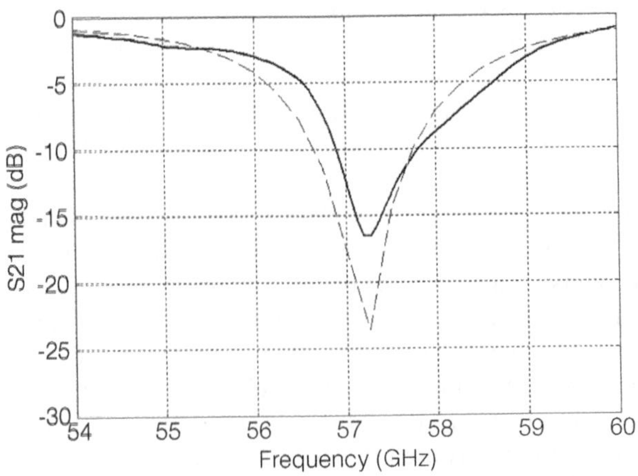

Figure 5.4: Magnitude of transmission coefficient (in decibels) at 60 GHz for TPX substrate.

for this substrate is unsatisfactory ($\tan\delta = 0.016$) in the 50 GHz band and results concerning total loss for a microstrip line are shown in Figure 5.8 (S21 magnitude).

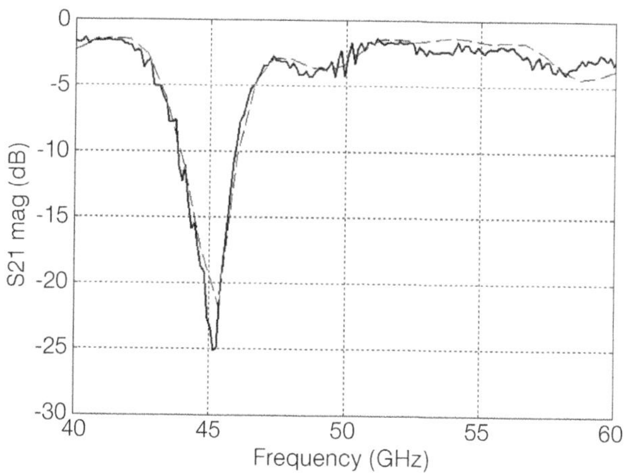

Figure 5.5: Magnitude of transmission coefficient (in decibels) at 45 GHz for duroid 6006 substrate.

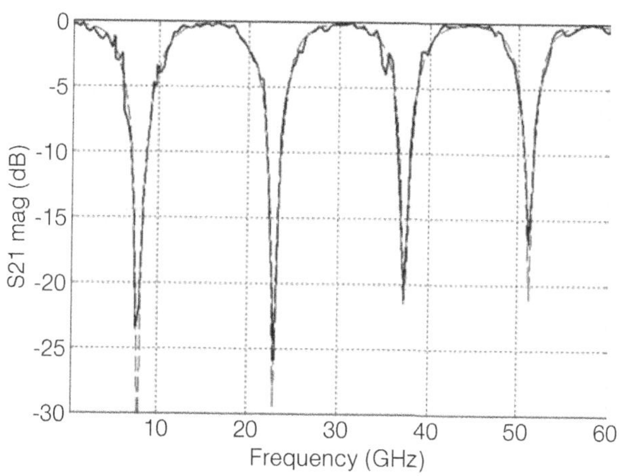

Figure 5.6: Magnitude of transmission coefficient (in decibels) at 50 GHz for aln substrate.

The characterization of substrates in the mmWave range makes it possible to point out the electrical parameters (ϵ_r and $\tan\delta$), thereby enabling antennas' efficiency to be appraised with greater accuracy. Indeed, the efficiency of millimeter printed antenna is often low owing to dielectric and metallic losses, especially for large arrays or multibeam antennas containing a very long feeding line network. In the following section, several examples of antenna arrays are given to exemplify the efficiency and impact of the substrate choice.

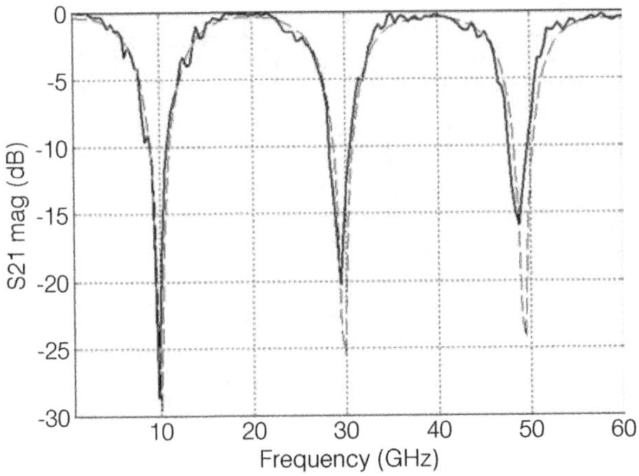

Figure 5.7: Magnitude of transmission coefficient (in decibels) for 35 NQ substrate.

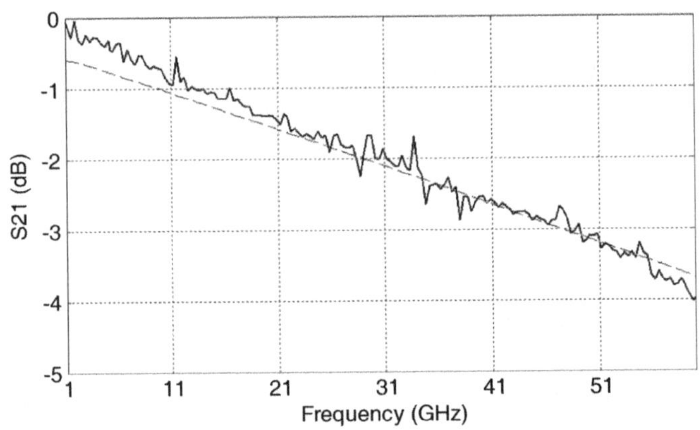

Figure 5.8: Total insertion loss in a 20 mm-long microstrip line printed on a 35 NQ substrate.

5.1.4 Substrate Choice: Impact on Antenna Efficiency

As was pointed out in the preceding section, the choice of a substrate thickness, dielectric constant and loss tangent becomes essential when it comes to mmWaves in order to avoid losses and also to enhance efficiency. The following results clarify this.

The first example is explained by reference [15] and deals with a four-patch single-layer array in the 38 GHz band printed on alumina substrate ($\epsilon_r = 9.8$). The design layout is presented in Figure 5.9. With this design, efficiency is rather poor (25%), due mainly to the high dielectric constant. To confirm this, a very similar array in the 60 GHz band (see Figure 5.10) was printed on fused quartz substrate using thin film technology [16]. For this

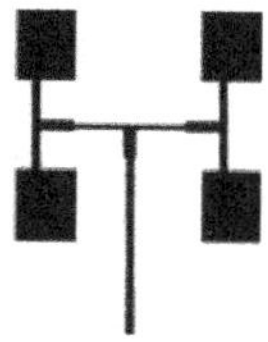

Figure 5.9: Four-patch single-layer array at 38 GHz printed on alumina substrate.

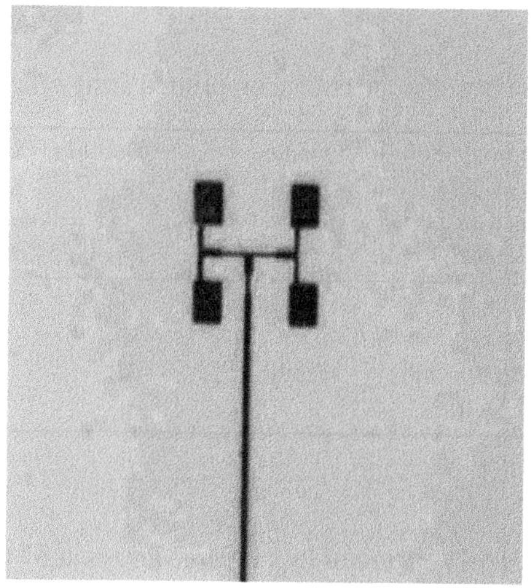

Figure 5.10: Four-patch single-layer array at 60 GHz printed on fused quartz substrate.

second array, the efficiency is significantly better (70%) at 60 GHz owing to a lower dielectric constant (3.8) and low loss tangent for the fused quartz substrate.

The third example, still in the 60 GHz band, concerns a four-patch array with circular polarization [9]. The polarization quality is increased through the principle of sequential rotation [17]. This array is shown in Figure 5.11. It is printed on RT duroid 5880 ($\epsilon_r = 2.23$, $h = 0.127$ mm and $\tan\delta = 0.004$). The measured efficiency is 60%. In this case, the antenna gain decreases, mainly owing to loss tangent value but also because of the considerable line length in the feeding network.

Table 5.6 summarizes the efficiency of several identical four-patch arrays printed on different substrates (quartz, RT duroid 5880, RT duroid 6006, alumina), and details the main loss contributors for the different cases: dielectric, metallic, surface wave.

In larger arrays, loss effects are even more obvious since the feeding line network is composed of longer lines to excite all patches. For mmWaves, the efficiency can become quite reduced.

Let us consider a 256 patch single-layer array [8] printed on a RT duroid 5880 substrate ($\epsilon_r = 2.2$, $h = 0.254$ mm and $\tan\delta = 0.002$). This design with a parallel feeding line network has been studied at around 37 GHz. The measured gain equals 28 dB at 37 GHz compared

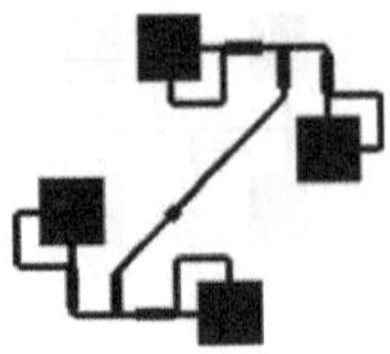

Figure 5.11: Circularly polarized array at 60 GHz with four patches printed on RT duroid 5880.

Table 5.6: Impact of substrate choice on printed array efficiency at 60 GHz.

	Duroid 5880 $\epsilon_r = 2.24$ $\tan\delta = 0.004$ $h = 0.127$ mm	Quartz $\epsilon_r = 3.8$ $\tan\delta = 0.0003$ $h = 0.137$ mm	Duroid 6006 $\epsilon_r = 7$ $\tan\delta = 0.002$ $h = 0.254$ mm	Alumina $\epsilon_r = 9.8$ $\tan\delta = 0.0005$ $h = 0.127$ mm
Gain (dB)	12	11	9	8
Efficiency (%)	60	70	50	25
Loss origin	Metallic and dielectric	Metallic	All	Surface waves

Table 5.7: Nature of loss in large arrays at 37 GHz.

	Theoretical losses (dB)
Theoretical directivity	32
Radiating element losses	1.05
Dielectric losses	0.38
Metallic losses	0.7
Radiating losses	1.2
Surface wave losses	0.2
Connector losses	0.2
Dielectric losses	0.38
Theoretical gain	28.2
Measured gain	28

with 32 dBi for directivity. This gain corresponds to a 40% efficiency for an array of this type. From this result, it is essential to understand the nature of the main loss such as that detailed in Table 5.7. In this particular case, surface wave loss is minimal because of the low thickness and low dielectric constant of the RT duroid 5880 substrate. Radiating element efficiency and metallic loss are the main contributors.

Various antenna arrays and their respective efficiency depicted in this part definitely show that in the field of mmWaves, a low dielectric constant, low thickness and loss tangent susbtrates have to be chosen to optimize large arrays' efficiency.

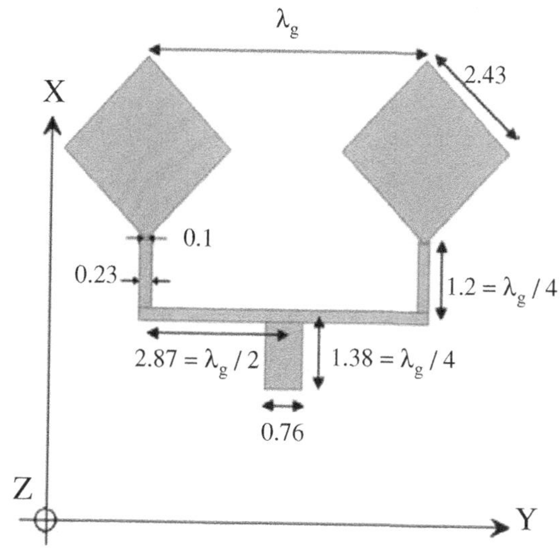

Figure 5.12: Design with T-junction between patches.

5.1.5 Feeding Line Influence on Radiating Patterns

Another point to explore for millimeter antenna arrays concerns the influence of spurious radiation due to the feeding lines [18] on radiating patterns. To illustrate this phenomenon, two similar printed antenna arrays composed of two patches are featured (Figures 5.12 and 5.13). The only difference between designs lies in the T-junction, which is added in the first case only. Both arrays are fed by a 50 Ω coaxial line and the H-plane radiating patterns (Figure 5.14) clearly show the T-junction influence. As shown in this figure, the side lobe level is lower (−22 dB/−15 dB) for the array with the T-junction. This result proves that discontinuities can have an advantageous influence on radiating patterns. A second example will show that the radiating effects due to the feeding lines and to discontinuities can be turned into an advantage when designing antenna arrays. Let us consider a meander antenna (Figures 5.15 and 5.16). To make a meander array, a four-corner/cell unit has to be duplicated. Looking at the orientation of the magnetic current, it appears that each cell radiates energy. The design finally results in a horizontal electric field polarization, meaning that a high directivity antenna can be optimized. A prototype was made at 40 GHz [8] on duroid 5880 substrate ($h = 0.254$ mm) and showed very convincing results in terms of radiating patterns (Figure 5.17). The cross-polarization level is acceptable (−15 dB).

However, in most cases, lines and steps have a spurious influence on radiating patterns. The following instance of a single-layer 256-patch array in the 40 GHz band [8] clearly illustrates their negative effects. This first design is based on a serial feeding line network (Figure 5.18). The measured H-plane radiating pattern at 40 GHz is presented in Figure 5.19. The result brings out a high cross-polarization level that is due to the microstrip step and discontinuities radiation in the feeding line network. A second example is a 256-patch array at the same frequency (40 GHz) but with a parallel feeding line network (Figure 5.20). The measured H-plane radiating pattern is featured in Figure 5.21 and shows a high sidelobe

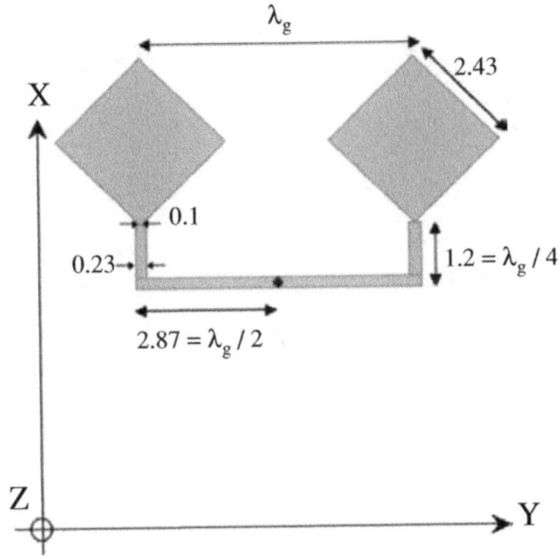

Figure 5.13: Design with no T-junction between patches.

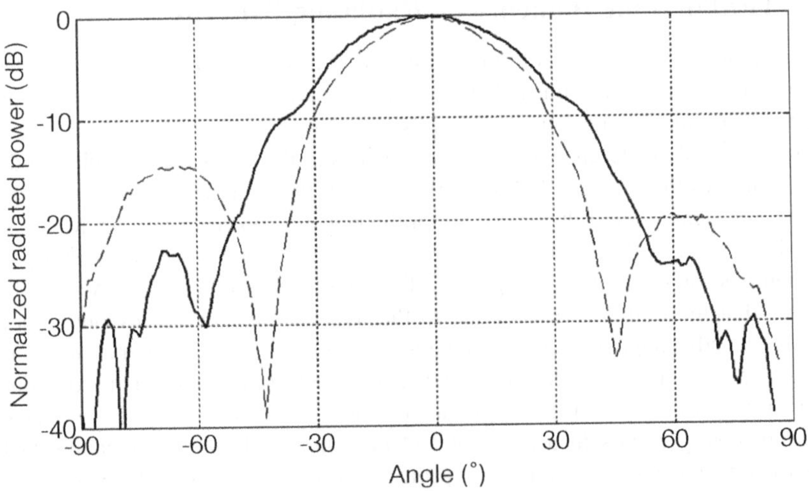

Figure 5.14: Measured radiated fields in the H-plane at 38.6 GHz for the meander antenna. Solid and dotted curves denote the design with and without T-junction respectively.

and cross-polarization levels for a 45° incidence angle. Finally, a third example with a mixed architecture (Figure 5.22) was tested. The result is presented in Figure 5.23. For this last array, spurious radiation due to discontinuities has a disadvantageous impact on cross-polarization.

One solution to this major drawback, is to use multilayer technologies [9] in order to separate the radiating element layer from the feeding line network layer. A very similar

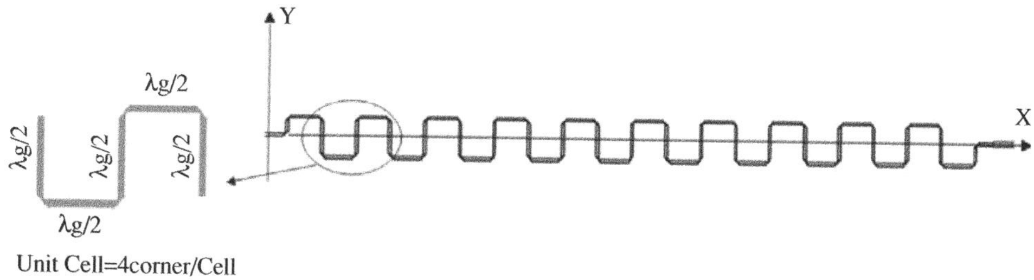

Figure 5.15: Meander antenna design.

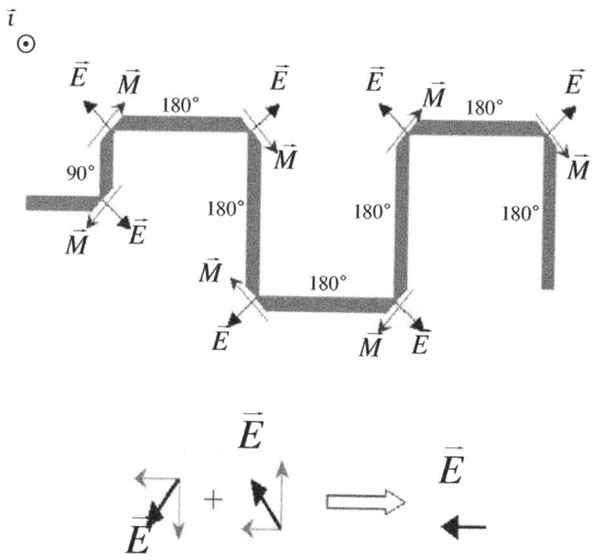

Figure 5.16: Magnetic current orientation for meander antenna design.

array composed of 256 patches at 60 GHz is shown in Figure 5.24 for the radiating layer and in Figure 5.25 for the feeding line network. However, in this case, an aperture coupled patch antenna has been chosen as the radiating element. The measured E-plane radiation at 60 GHz is plotted in Figure 5.26. If we compare it with the result presented earlier, it is clear that the cross-polarization level (−22 dB/−10 dB) has been considerably enhanced. The multilayer technology seems to be a good solution for mmWaves, with the capacity to avoid spurious radiation and to improve sidelobe and cross-polarization levels. Moreover, the multilayer technology allows active devices (monolithic microwave integrated circuits (MMICs), microelectromechanical systems (MEMS)) to be placed on a different layer and to keep them separate from the radiating elements. Accordingly, the next section will deal with multilayer technologies based on low dielectric constant materials intended to optimize radiating patterns and efficiency. More detail will be given on transitions and on how to

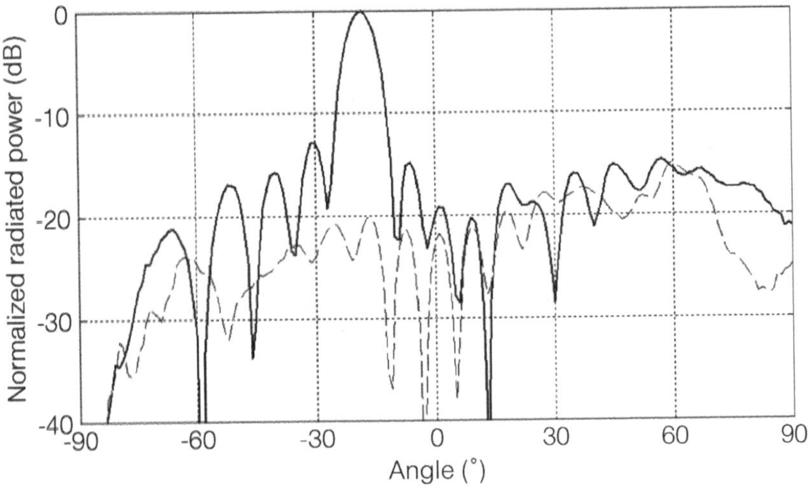

Figure 5.17: Radiating patterns at 40 GHz for the meander antenna.

Figure 5.18: A 256-patch single array at 40 GHz; design with serial feeding line network.

couple energy between layers. Finally, a variety of antennas will be presented with the objective of showing how to obtain directivity or particular radiating patterns (sectorial, cosecant, multibeams).

5.2 Multilayer Interconnection Technology

5.2.1 Introduction

The use of multilayer technologies to design printed antenna arrays involves careful consideration of transitions and interconnections between different layers. Taking, for instance, a two-layer microstrip circuit, with glass Teflon substrates for both layers, then

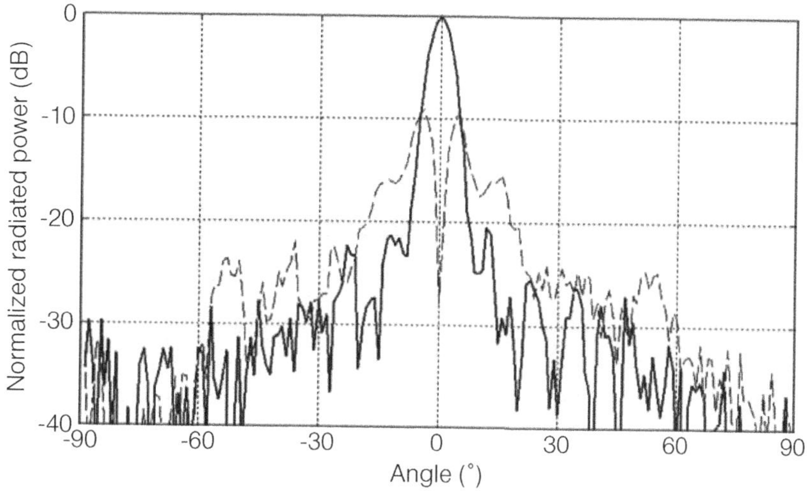

Figure 5.19: H-plane co-polar (solid curve) and cross-polar (dotted curve) radiating patterns.

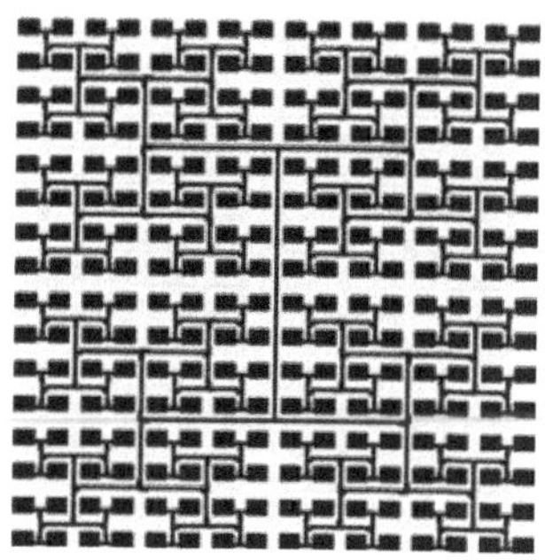

Figure 5.20: A 256-patch single array at 40 GHz; design with parallel feeding line network.

it will be essential to take into account the transition technique to connect lines or to couple energy between the lines.

This coupling mechanism can be achieved using different devices, depending on whether it is applied between microstrip lines or between microstrip lines and antennas. Different options can be defined as:

- coaxial transition between lines (see Figure 5.27);
- aperture coupled patch antenna to couple energy between line and antenna (see Figure 5.28);
- slot transition between lines with one or two outputs (see Figure 5.29).

As pointed out in the preceding section, the substrate thickness must be low compared to the wavelength in order to achieve proper efficiency for mmWave printed antennas;

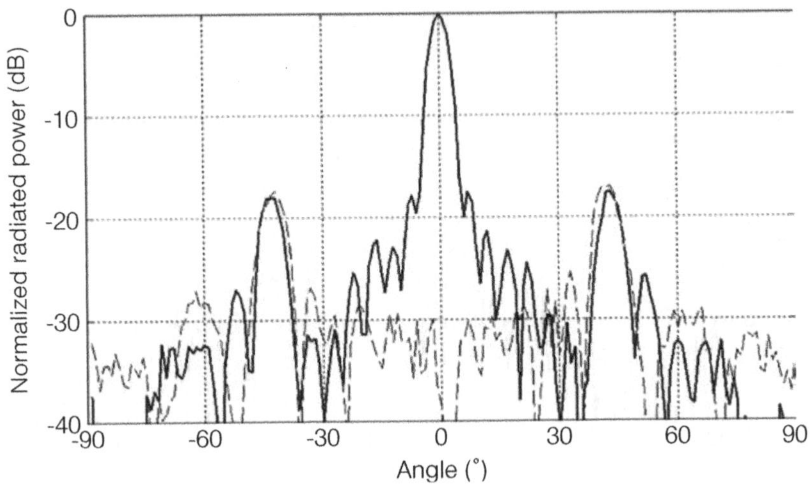

Figure 5.21: H-plane co-polar (solid curve) and cross-polar (dotted curve) radiating patterns.

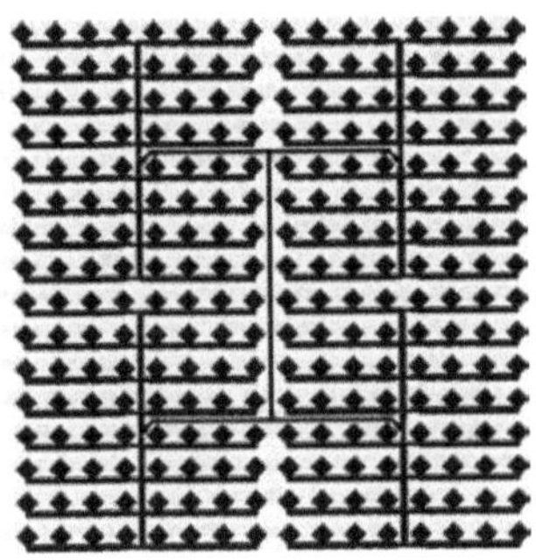

Figure 5.22: A 256-patch single array at 40 GHz; design with mixed feeding line network.

for instance, thickness often equals 0.1 or 0.2 mm between 30 and 100 GHz. Therefore, the printed circuits are very thin and flexible. Then, it becomes useful to apply a thick ground plane to rigidify the antenna. Additionally, this thick metallic support allows the active components to be placed more easily.

For slot transition and aperture coupled patch antenna, the coupling slot will be engraved in the support. It will then be essential to take into account the fact that a thick ground plane has been added. Hence, it will be necessary to increase the slot length or to change the slot shape to permit the coupling between layers. At low frequencies (several gigahertz), a ground plane is often, in fact, very thin ($\leq 0.001\lambda_0$). It is then possible to consider this slot as infinitely thin. On the contrary, at mmWaves, this parameter becomes more influential. For instance, if we consider only a 0.2 mm-thick slot, it corresponds to $0.05\lambda_0$ at 77 GHz. The impact could be strong on the S11 parameter and on the coupling between lines or between line and patch [19, 20].

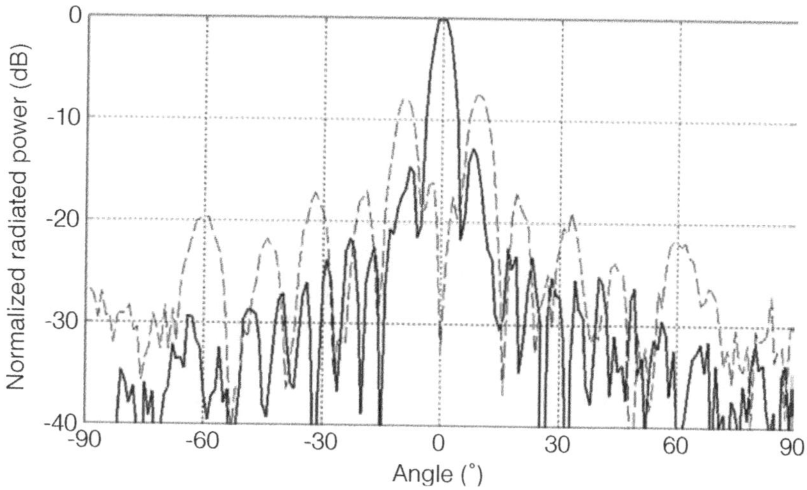

Figure 5.23: H-plane co-polar (solid curve) and cross-polar (dotted curve) radiating patterns.

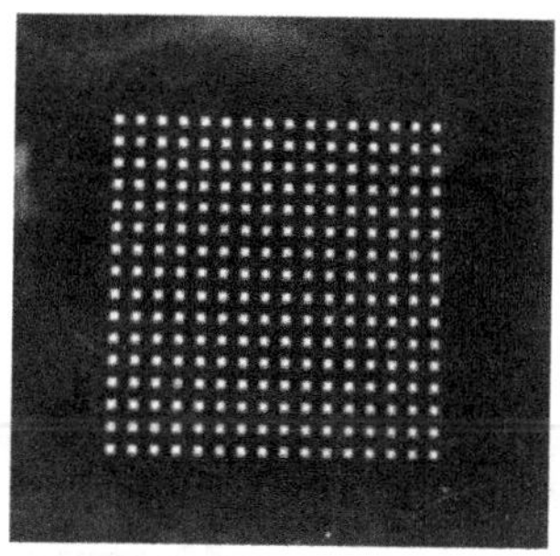

Figure 5.24: Antenna array composed of 256 aperture coupled patches at 60 GHz. Radiating layer shown.

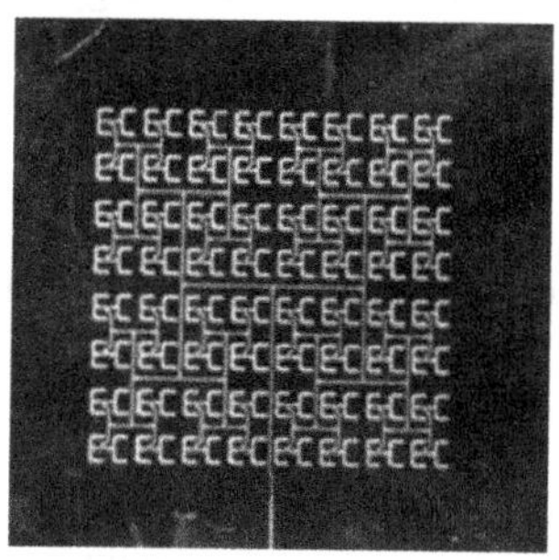

Figure 5.25: Antenna array composed of 256 aperture coupled patches at 60 GHz. Feeding lines network layer shown.

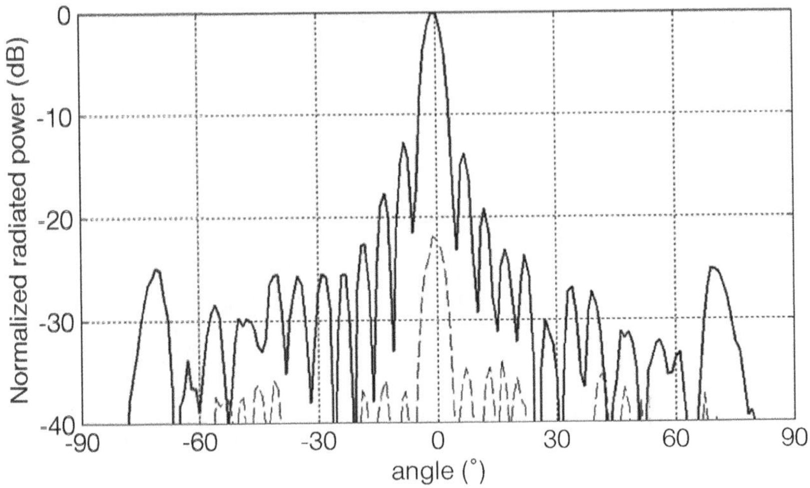

Figure 5.26: Antenna array composed of 256 aperture coupled patches at 60 GHz. E-plane co-polar (solid curve) and cross-polar (dotted curve) radiating pattern measurements.

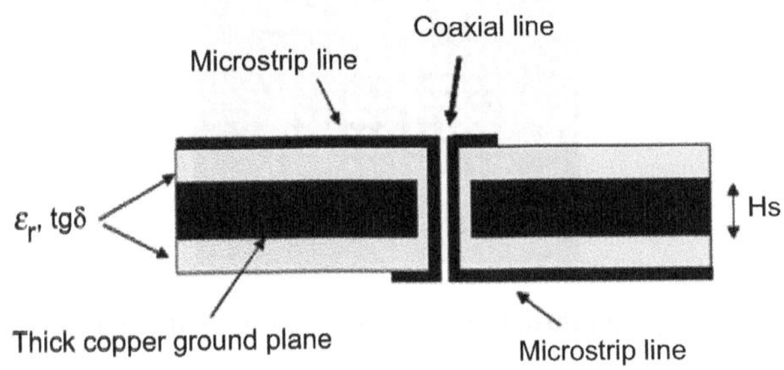

Figure 5.27: Transition between layers to couple energy; coaxial transition between lines.

5.2.2 Multilayer Technologies on Soft Substrate with Thick Ground Plane

In this section, multilayer technologies will be detailed for coaxial transition, microstrip line–slot–microstrip line transition or aperture coupled microstrip patch in the mmWave range. In most cases, soft substrates (polytetrafluoroethene (PTFE), RT duroid) are used. We show that the developed technologies are also compatible with active circuits and antennas. A number of examples will be given between 24 and 77 GHz to prove the capability of these technologies to achieve low loss circuits and antennas.

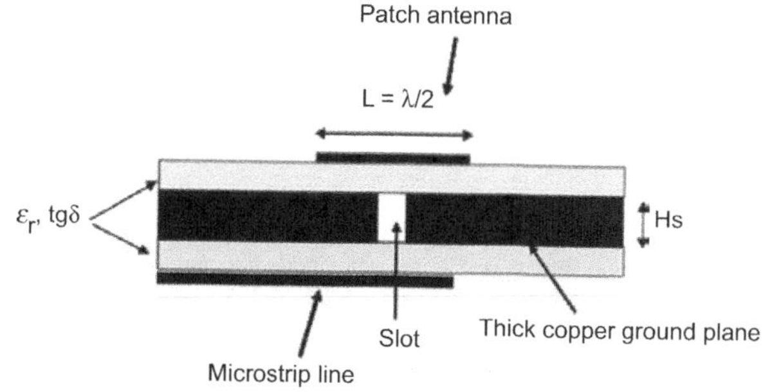

Figure 5.28: Transition between layers to couple energy; aperture coupled patch antenna.

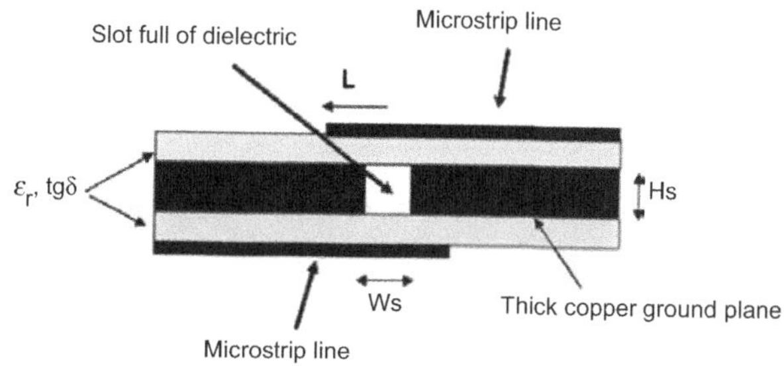

Figure 5.29: Transition between layers to couple energy: slot transition between lines.

5.2.2.1 Technological Processes

Two technological processes are presented here: one associating polypropylene or TPX substrate with a thick metallic support, the other allowing commercial substrates to be joined with the same thick support. The objective was to select the best material to optimize antenna efficiency and cost (commercial application).

In the first case, the considered substrates being noncommercial, they offer no ready-made thickness calibration. Therefore, several sheets of substrates (TPX, for instance) are first stacked and pressed on a thick support (Figure 5.30 and Figure 5.31) at a high temperature (170°C) to achieve the desired thickness (0.127 or 0.254 mm for mmWave designs). Next, a copper sheet is added via oxide deposition for grappling. This technology has been developed in collaboration with IETR, France Telecom and the company AVI&PESCHARD [21, 22]. The operation can be repeated on the other side of the thick support in order to obtain multilayer circuits. In the case of engraved slots in the ground plane, a dielectric will be used to fill the slots.

This second process is very similar to that previously described. The main difference lies in the substrate. Commercial substrates are used (RT duroid 5880, RT duroid 3003, RT duroid 6006 or others) and are joined to the thick ground plane by means of glue or solder. First, the initial copper film is removed from the substrate, next the two substrate layers are stacked

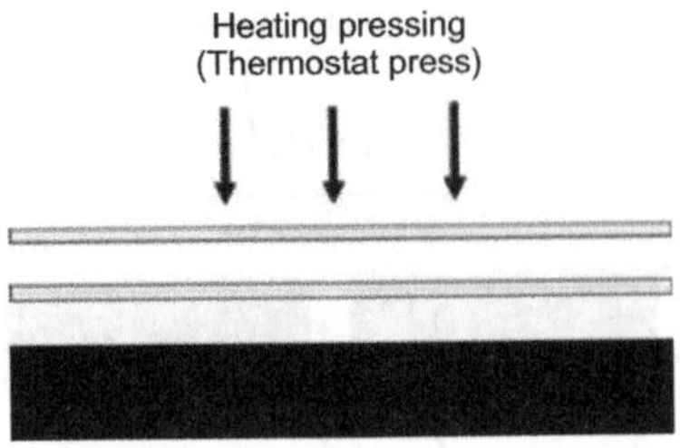

Figure 5.30: Process based on polypropylene or TPX; First step.

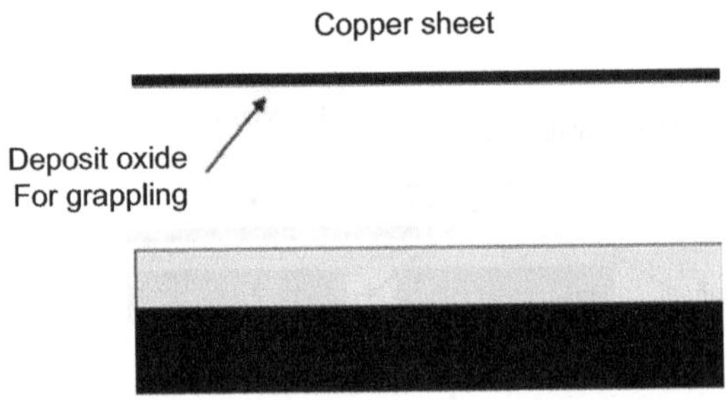

Figure 5.31: Process based on polypropylene or TPX; second step.

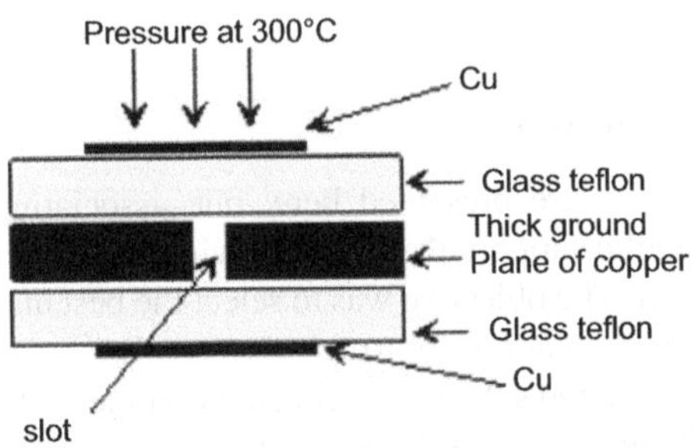

Figure 5.32: Process based on commercial substrates and a thick ground plane.

and pressed at high temperature (Figure 5.32). This process requires oxidation of the metallic support and the use of microwave glue to achieve a good bonding.

5.2.2.2 Two-layer Microstrip-to-microstrip Transition using a Coaxial Line

Two microstrip lines are connected via a coaxial line which is drilled in the thick ground plane. Millimeter transitions have been studied in reference [23] up to 40 GHz to feed

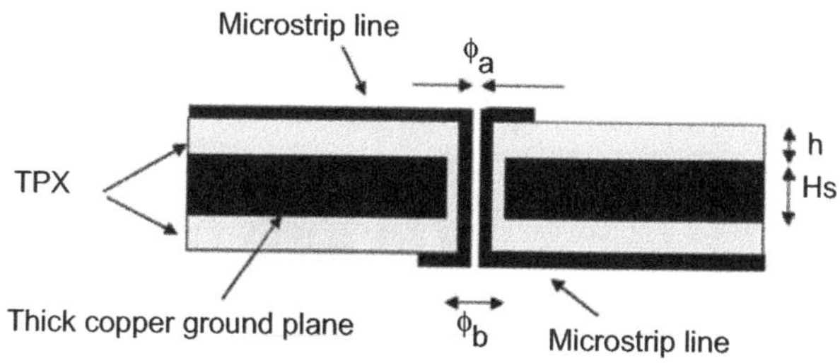

Figure 5.33: Two-layer microstrip-to-microstrip transition using a coaxial line; cross-section view.

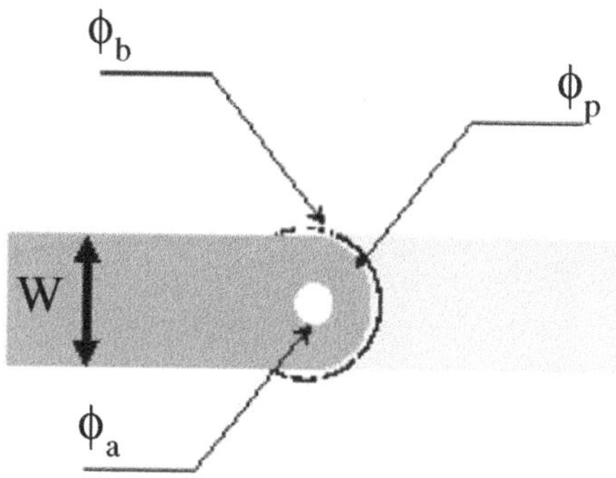

Figure 5.34: Two-layer microstrip-to-microstrip transition using a coaxial line; top view.

multilayer printed arrays. A cross-section view and a top view of this device are shown in Figures 5.33 and 5.34 respectively.

Technological constraints limit some of the parameters:

- inner diameter of coaxial line $\phi_a \geq 0.2$ mm
- outer diameter of coaxial line $\phi_b \geq 2$Hs where Hs is the metallic support thickness.

One transition has been simulated (using HFSS) on TPX substrate ($\epsilon_r = 2.15$ and $h = 0.254$ mm) between 20 and 50 GHz. The parameters are: $\phi_a = 0.3$ mm, $\phi_b = 1$ mm, $W = 0.78$ mm (50 Ω line), Hs = 0.5 mm. Figure 5.35 shows a cross-section photography featuring the metallized hole and the thick copper plane.

Simulation and measurement results are compared in Figures 5.36 and 5.37 for magnitude of S21 and S11 respectively . This transition provides very good results in terms of insertion loss (S21 $\leq$ 0.5 dB up to 40 GHz).

The objective is to feed printed arrays with this transition. An example is given in Figure 5.38 for the radiating layer and in Figure 5.39 for the active layer. The 16-patch antenna array is divided into two subarrays composed of eight patches, each one being associated with an MMIC amplifier at 40 GHz. A tapering amplitude is applied to reduce the side lobe level.

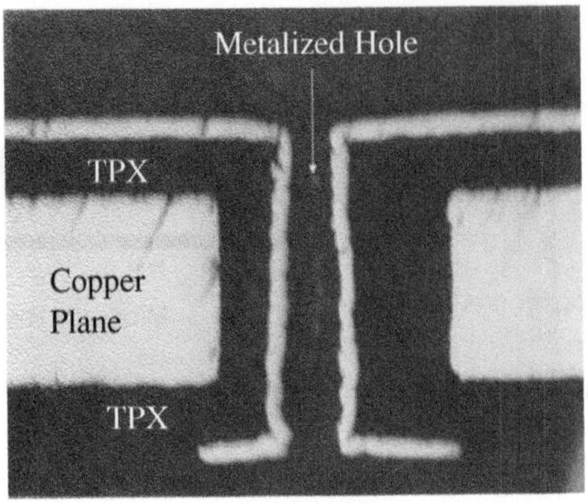

Figure 5.35: Two-layer microstrip-to-microstrip transition using a coaxial line; cutting plane of prototype.

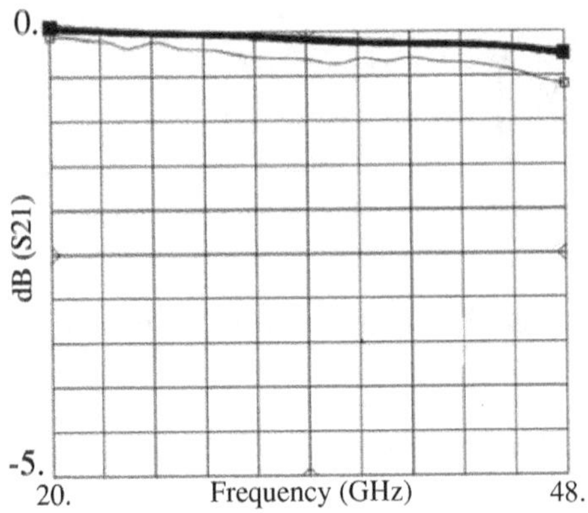

Figure 5.36: Two-layer microstrip-to-microstrip transition using a coaxial line. Comparison between simulation (thick curve) and measurement (fine curve) for S21 magnitude parameter.

For the two-layer microstrip-to-microstrip transition using a coaxial line, if the inner diameter of the line is too large, the transition will behave as a radiating aperture. Then, the higher the frequency, the smaller ϕ_a and ϕ_b have to be kept. Therefore, this kind of transition cannot give acceptable results for very high frequencies ($\geq$60 GHz). Hence, for higher frequencies, it is advised to use an electromagnetic coupling transition via-engraved slot in the ground plane.

5.2.2.3 Two-layer Microstrip-to-microstrip Transition using a Coupling Slot

This transition allows energy (without contact) to couple between top and bottom layers (Figures 5.40 and 5.41) [24]. In most cases, one layer corresponds to the radiating element level and the other one to the feeding line network with the active components layer.

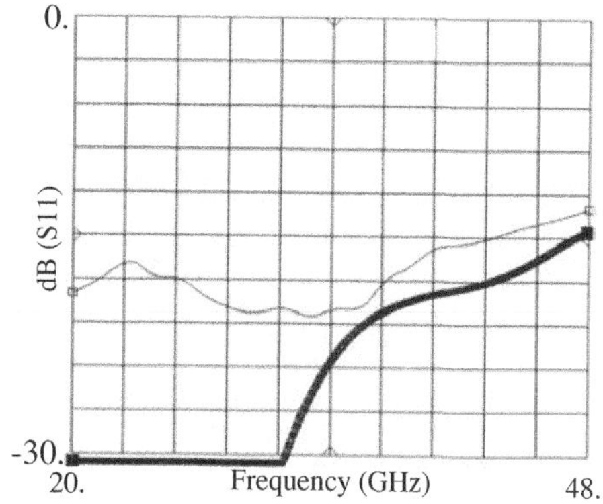

Figure 5.37: Two-layer microstrip-to-microstrip transition using a coaxial line. Comparison between simulation (thick curve) and measurement (fine curve) for S11 magnitude parameter.

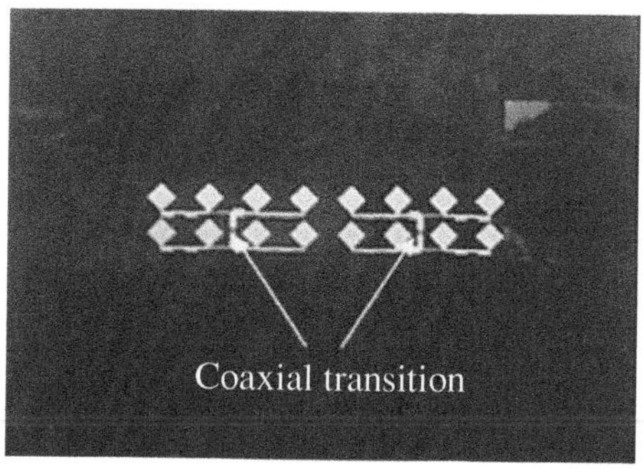

Figure 5.38: Active 16-patch array; radiating layer.

At mmWaves, the ground plane thickness cannot be considered electrically thin. Therefore, the results for S11 and S21 are directly linked to ground plane thickness, and are essential to consider for an accurate assessment of the electromagnetic coupling. A photograph featuring the transition for TPX substrate used for both layers is presented in Figure 5.42. Note that the slot is filled with dielectric. While this device offers only one input, it is possible for it to fit into one or two outputs so as to feed another array or subarrays. Moreover, because of technological considerations, the slot width (Ws) must be larger than half of the ground plane thickness (Hs). Many examples could be given for transitions for frequencies ranging from 24 to 60 GHz. As shown below, an example is detailed to indicate how to achieve good matching and low loss transition at 30–50 GHz. In this case, the ground plane thickness equals 0.5 mm and the substrate used is TPX for both layers ($h = 0.25$ mm and $\epsilon_r = 2.1$). The results with regard to the magnitude parameters for S11 and S21 are outstanding T–T Figures 5.43 and 5.44, where measurements are compared with three-dimensional software simulations (HFSS).

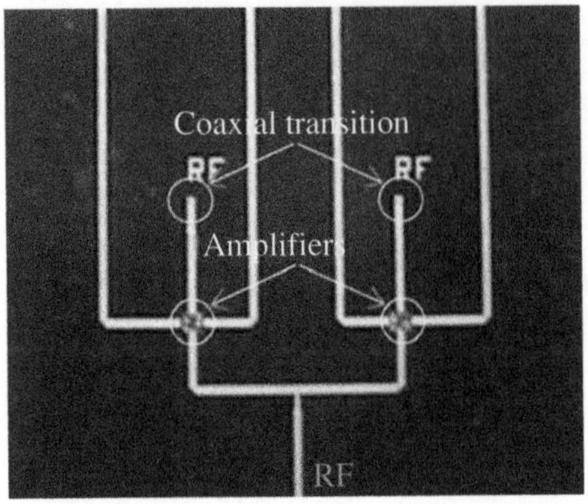

Figure 5.39: Active 16-patch array; active layer and lines.

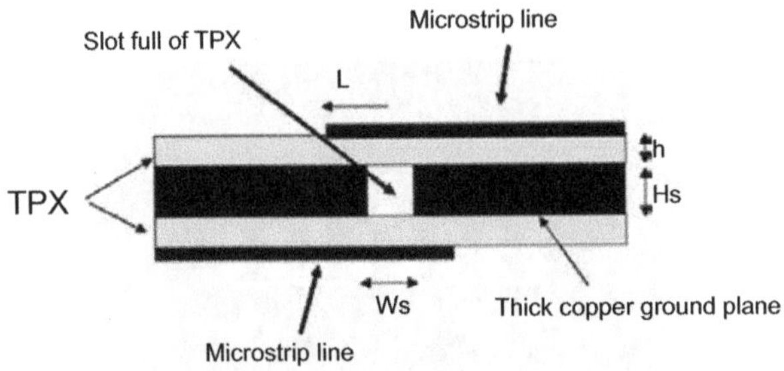

Figure 5.40: Two-layer microstrip-to-microstrip transition using a coupling slot; cross-section, of view.

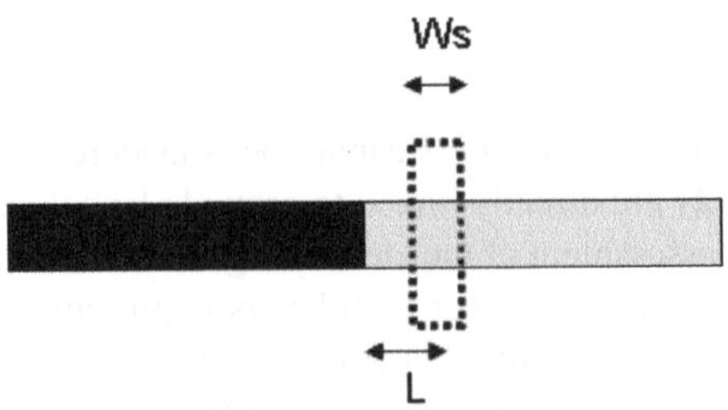

Figure 5.41: Two-layer microstrip-to-microstrip transition using a coupling slot; top view.

At 40 GHz, for example, loss is close to 0.4 dB and good matching can be achieved for a large bandwidth (20 GHz) around this frequency. Some other good results can also be obtained for higher frequencies also (60 GHz). This two-layer transition with coupling slot was used to feed an eight-patch antenna array divided into two four-patch subarrays

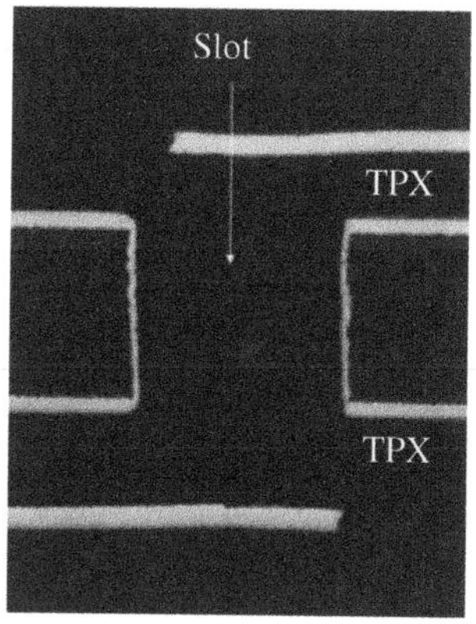

Figure 5.42: Two-layer microstrip-to-microstrip transition using a coupling slot; cutting plane of the prototype.

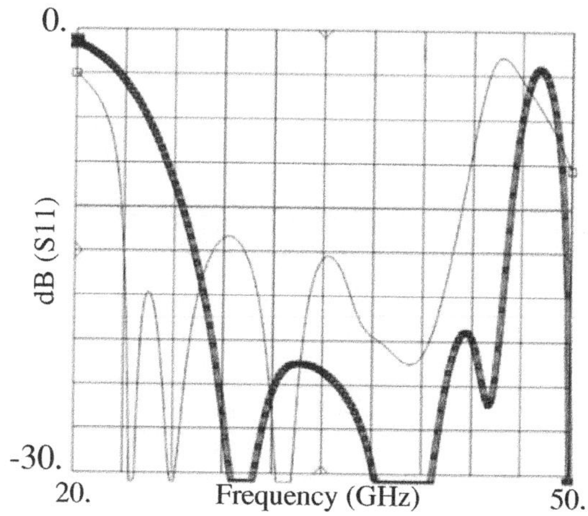

Figure 5.43: Two-layer microstrip-to-microstrip transition using a coupling slot. Comparison between simulation (thick curve) and measurement (fine curve) for S11 magnitude parameter.

(Figure 5.45). The radiating pattern result for the array fed by this transition was compared with the result for a single-layer array (Figure 5.46). A substantial improvement of the side lobe level could be observed for the array fed via the transition.

A very similar transition was made with a 0.2 mm-thick metallic support and TPX substrate ($h = 0.127$ mm and $\epsilon_r = 2.17$). Complete results are reported in terms of magnitude of reflection (S11) and transmission coefficient (S21). Moreover, insertion phase shift is carried out for this transition. The latter result is significant as it shows the S21 phase parameter versus frequency. Very low transition loss level is achieved (0.4 dB) between 52 and 62 GHz (Figure 5.47). Three-dimensional software is essential to design this transition

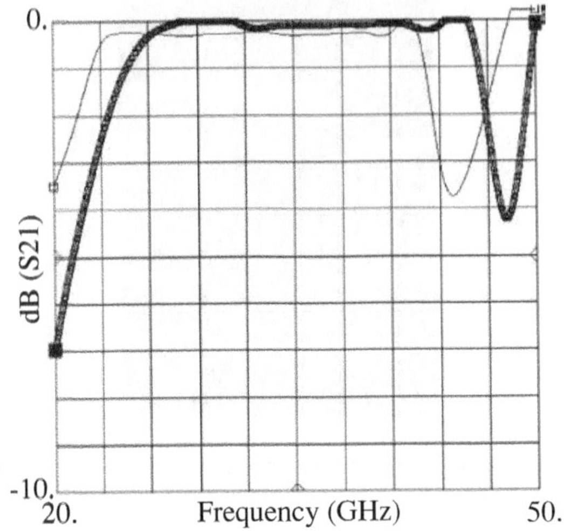

Figure 5.44: Two-layer microstrip-to-microstrip transition using a coupling slot. Comparison between simulation (thick curve) and measurement (fine curve) for S21 magnitude parameter.

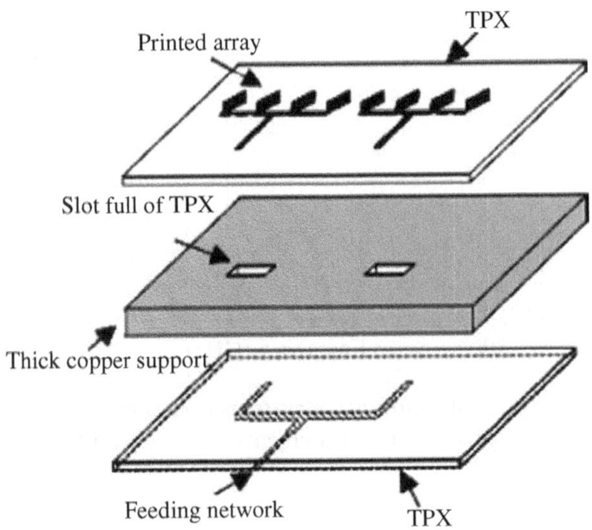

Figure 5.45: Eight-patch antenna array fed by a two-layer microstrip-to-microstrip transition using a coupling slot; design.

adequately when it comes to mmWaves and when a thick ground plane is added. Indeed, if this transition is used in Butler matrix, for instance, where the aim is to design multibeam antennas, it is vital to know the insertion phase of this device with the greatest accuracy (Figure 5.48).

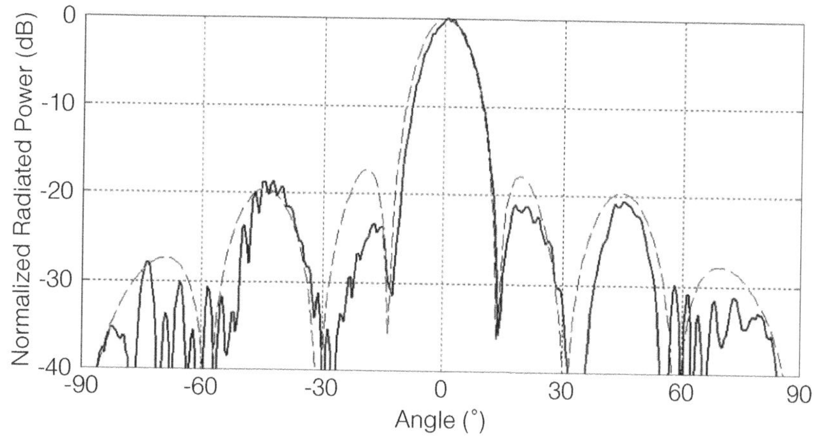

Figure 5.46: Radiating pattern at 40 GHz for a single-layer array (solid curve) and for a double-layer array (dotted curve).

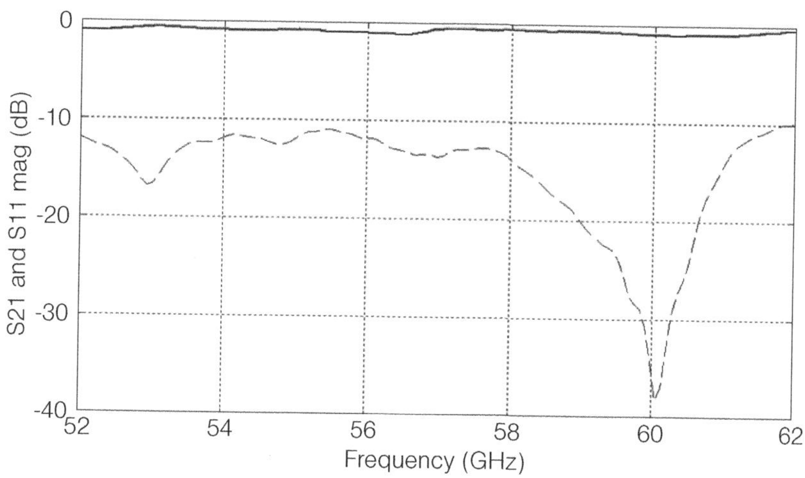

Figure 5.47: Two-layer microstrip-to-microstrip transition in the 52–62 GHz band using a coupling slot; measure of S11 and S21 magnitude parameter.

Even if the ground plane thickness is limited, it is essential to take this into account. However, in some particular cases, this thickness could be higher. In such cases, rectangular slots cannot achieve good coupling between layers. Then, the slot shape has to be changed to optimize the transition. Such an example is presented [25] where the thickness is 2 mm ($0.33\lambda_0$ at 50 GHz). An H shape thick slot has been optimized to achieve good power transfer (Figure 5.49). This shape allows both the slot size to be reduced and the parasitic radiation power to be decreased. The two microstrip lines are printed on Glass Teflon substrate (Rogers duroid 5880) offering the same thickness (0.127 mm). The slot is engraved in a thick

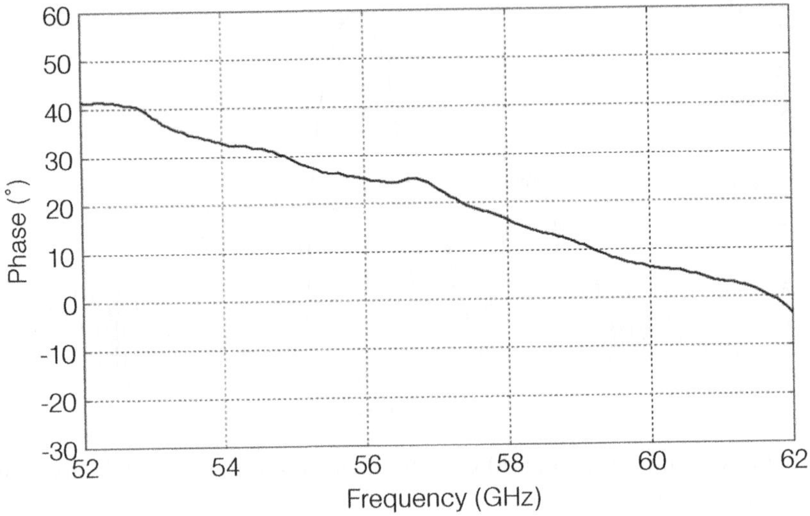

Figure 5.48: Two-layer microstrip-to-microstrip transition in the 52–62 GHz band using a coupling slot; measure of S21 phase parameter.

ground plane using the electro erosion technique. This transition has been optimized with the finite integration method. Dielectric loss (due to tan δ) and conductor loss (due to σ) were not considered in the simulations in order to point out the insertion loss of the thick slot transition only. Some good simulation results (Figure 5.50) were recorded. Between 45 and 55 GHz, the calculated return loss (S11) is better than -20 dB and insertion loss (S21) is lower than 0.5 dB. After the simulation of this transition, measurements were performed on an XF 8510C network analyzer and with the help of a V-band Anritsu Wiltron test fixture. The measurement results of the S parameters are represented in Figure 5.51. In the considered bandwidth, the insertion loss (S21) is close to 0.8 dB. For the same frequencies, the magnitude of reflection coefficient (S11) is better than -17 dB. We can see that the measured loss is close to the simulation result (0.5 dB).

This transition has been extended by two outputs to feed a 16-patch array divided into two subarrays (Figures 5.52 and 5.53). The array is composed of 16 stacked patches to increase both the gain and the bandwidth in the Q band (47.2 to 50.2 GHz) [25]. This antenna has been designed for FAFR in spatial communication. The two-output transition is shown in Figure 5.53. It allows the two subarrays to be excited with a 180° phase shift and to cancel cross-polarization. Simulation results (CST Microwave Studio) concerning the transition are presented in Figure 5.54 for magnitude parameters and in Figure 5.55 for phase shift between outputs.

This transition was used to feed a 16-patch antenna array that had to be matched between 47.2 and 50.2 GHz and that was required to present a stable directivity in this bandwidth. Actually, the two outputs of the transition are opposite in phase and feed two eight-patch antenna arrays in order to reduce the cross-polarization of the antenna as a whole. To comply with the specifications, the authors optimized a multilayer antenna array with stacked patches. These parasitic patches will be printed on 35 NQ substrate. This substrate will allow the

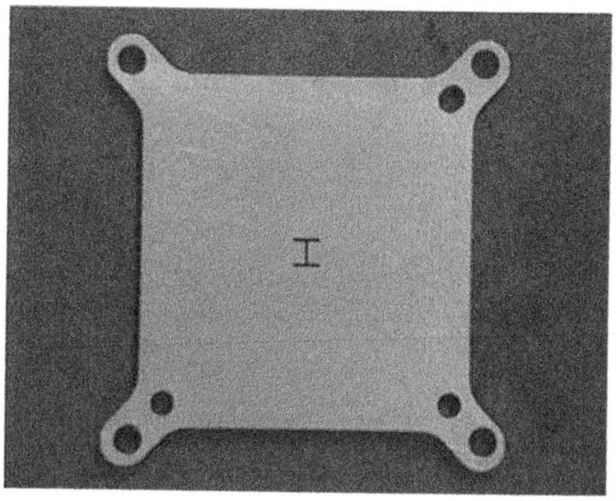

Figure 5.49: Design of transition with the H-shape slot in the Q band.

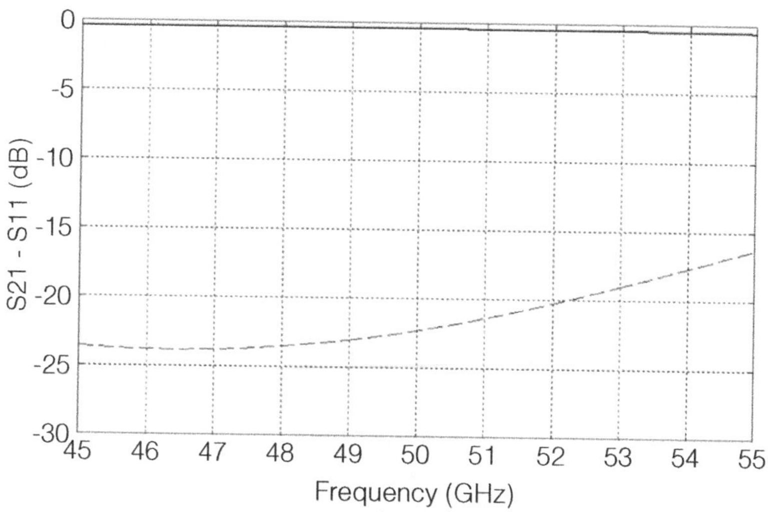

Figure 5.50: Transition with the H-shape slot in the Q band; calculation for magnitude of S11 (dotted curve) and S21 (solid curve).

antenna array to be protected against spatial radiation. An air gap ($h = 0.38$ mm) was provided between patches and parasitic patches to optimize bandwidth and power gain. All patches were tapered to decrease the sidelobe level. This antenna was simulated with CST Microwave Studio. The S11 parameter is still better than -10 dB for the operating bandwidth. The magnitude of S11 is presented in Figure 5.56, as measured. The antenna is matched ($S11 \leq -15$ dB) in the considered bandwidth (47.2–50.2 GHz). The theoretical directivity calculated by CST is close to 19.7 dBi. The measured power gain is plotted in Figure 5.57. Appraising the antenna's efficiency is clearly essential, at this point. It involves estimating the aperture efficiency with Equation (5.3) where D_a stands for the printed antenna array's directivity and for the antenna's surface (18 mm $\times$ 18 mm) respectively. The directivity of

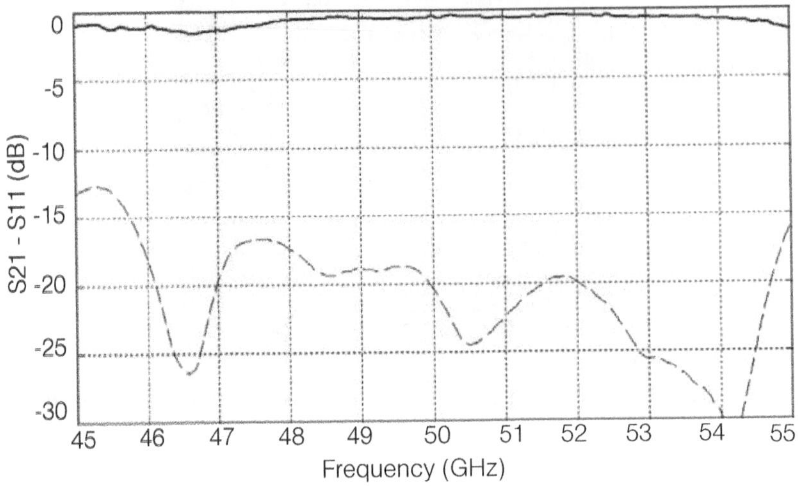

Figure 5.51: Transition with the H-shape slot in the Q band: measurement for magnitude of S11 (dotted curve) and S21 (solid curve).

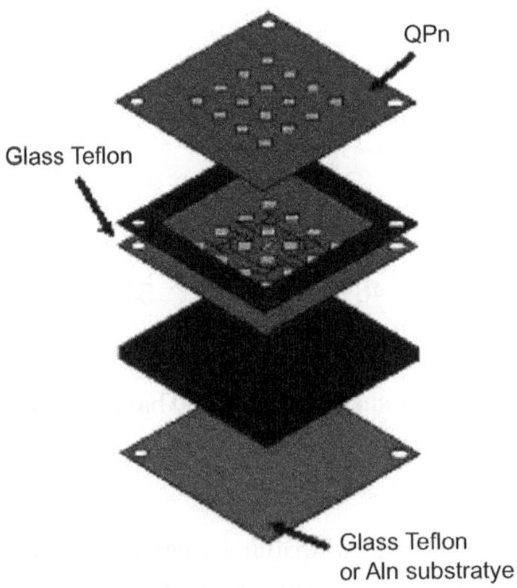

Figure 5.52: Multilayer stacked patch antenna array in the Q band; global design.

this antenna array is close to 20.3 dBi. Next, the antenna efficiency is calculated by means of Equation (5.4), resulting in 51% at 47.2 GHz, 47% at 48.7 GHz, and 42% at 50.2 GHz. The efficiency is close to 50% which was the expected value. Efficiency decreases at 50.2 GHz

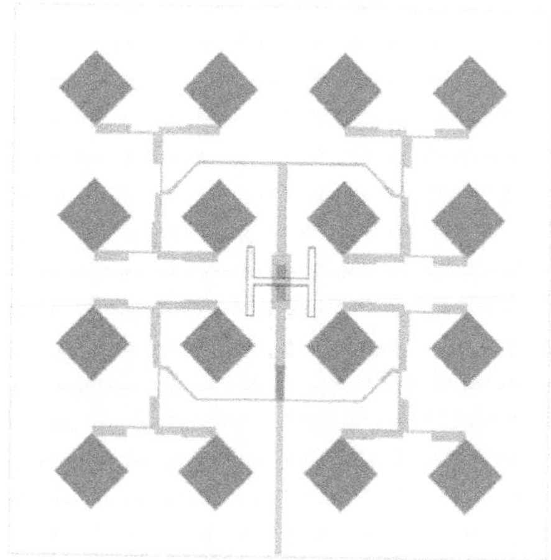

Figure 5.53: Multilayer stacked patch antenna array in the Q band; transition with two outputs to excite antenna array.

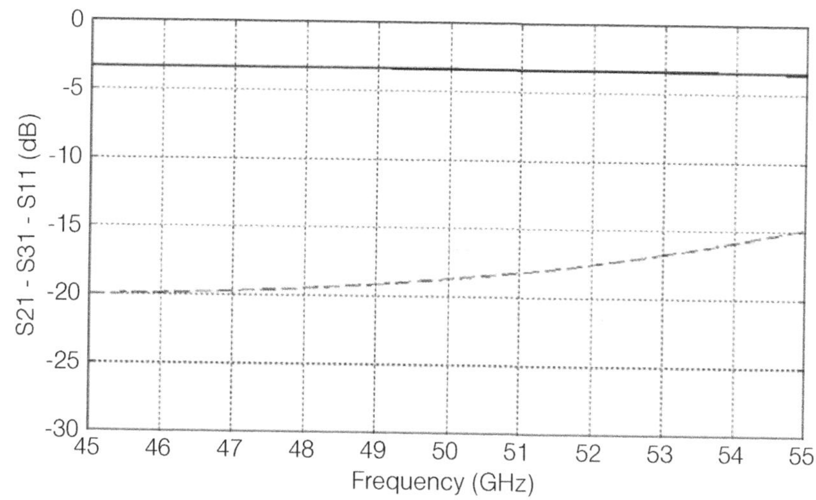

Figure 5.54: Two-layer transition with two outputs using the H-shape slot; magnitude of S11 and S21 (CST microwave simulation).

due to aperture efficiency (84%).

$$D = 10\log\left(\frac{4\pi S}{\lambda^2}\right),\ \eta_s = \frac{D_a}{D} \tag{5.3}$$

$$\eta_t = \frac{G_a}{D} \tag{5.4}$$

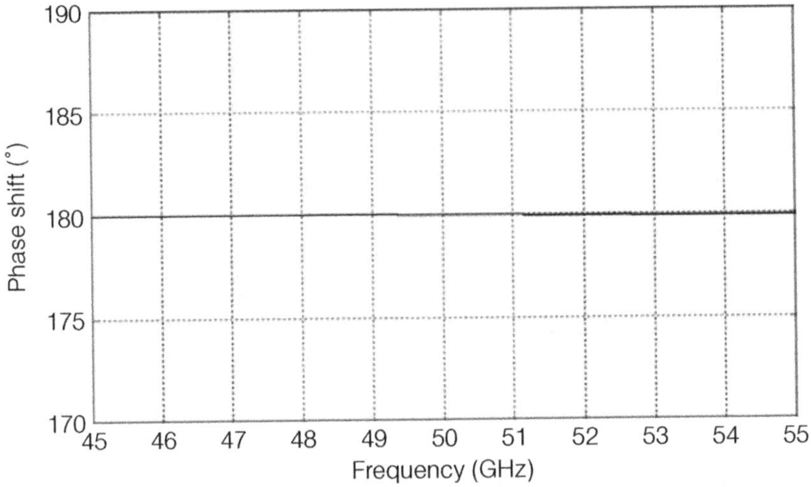

Figure 5.55: Two-layer transition with two outputs using the H-shape slot; phase shift between the two outputs (CST microwave simulation).

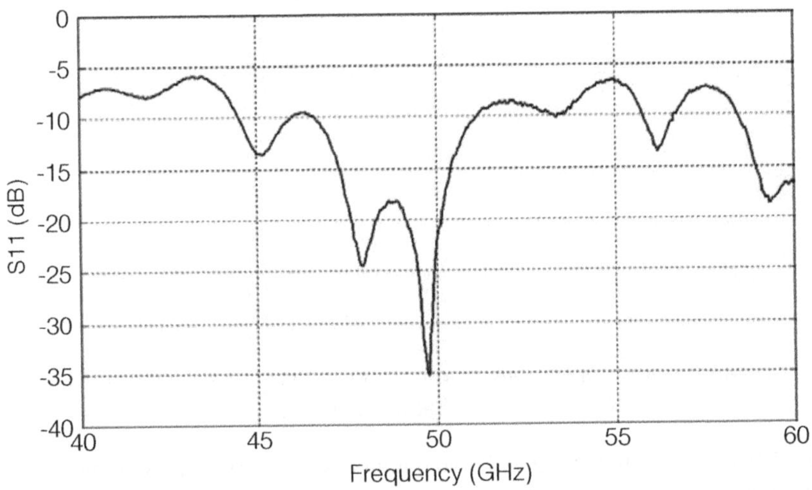

Figure 5.56: Multilayer stacked patch antenna array in the Q band; measured magnitude of S11 parameter.

Concerning radiating patterns at 48.7 GHz for the two principal planes (H and E), the measured results are plotted in Figures 5.58 and 5.59. The side lobe level is lower than −20 dB. The cross polarization level is good (−30 dB) for all angles. In the E-plane, the measured side lobe level is better than −14 dB. The radiating pattern assymmetry can be

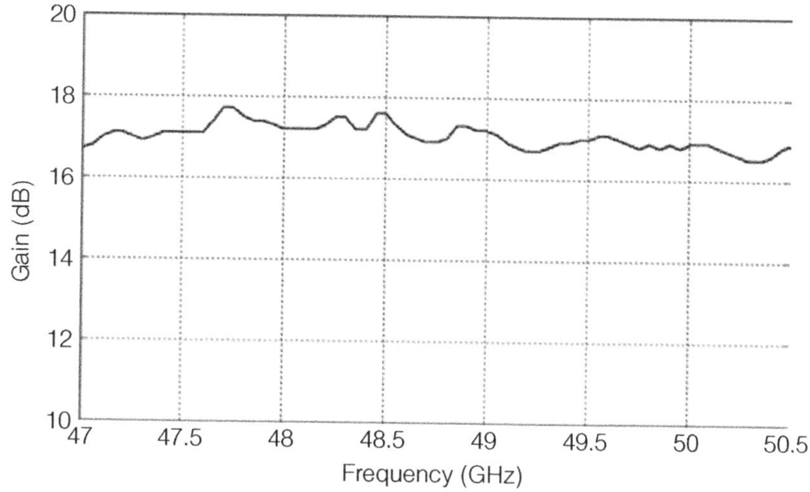

Figure 5.57: Multilayer stacked patch antenna array in the Q band; measured gain (dB).

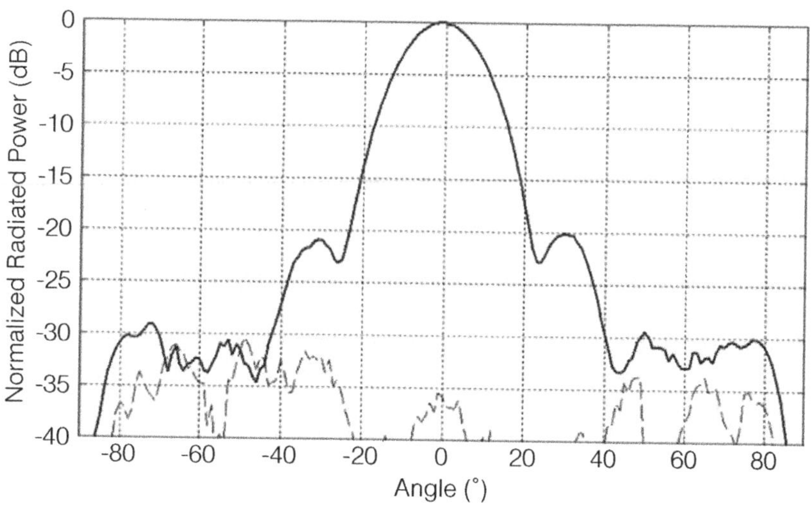

Figure 5.58: Multilayer stacked patch antenna array in the Q band; radiating patterns at 48.7 GHz in the H-plane with co-polarization (solid curve) and cross-polarization (dotted curve).

due to diffraction on the V-connector and metallic support. Very promising results have been obtained with this antenna prototype in the Q band.

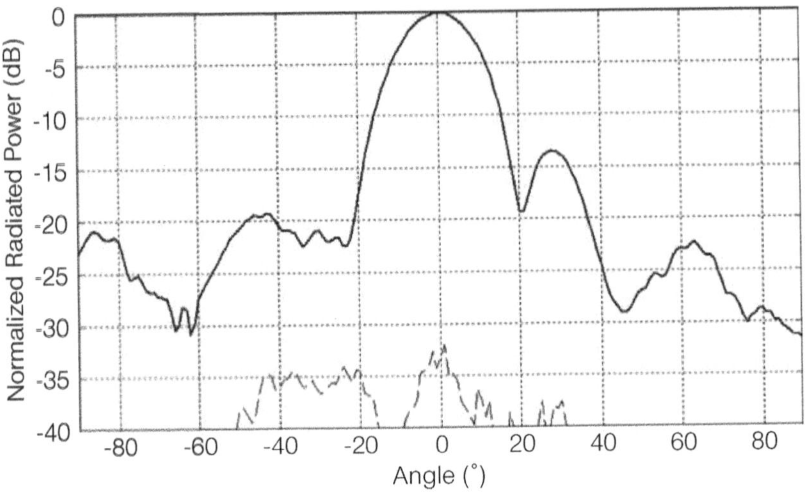

Figure 5.59: Multilayer stacked patch antenna array in the Q band; radiating patterns at 48.7 GHz in the E-plane with co-polarization (solid curve) and cross-polarization (dotted curve).

5.2.2.4 Aperture Coupled Patch Antenna

The aim is not to couple energy between lines via a slot but to couple energy between the microstrip line and the printed patch antenna. The design is well known (Figure 5.60) and has been studied by several authors [26,27]. This antenna is based on multilayer technology with a feeding line printed on the bottom substrate and a patch on the top substrate. The patch is fed using a slot engraved in the thick ground plane. At low frequency, the ground plane thickness does not affect the input impedance and the resonance frequency, but for mmWaves, this parameter has to be taken into account to assess the input impedance. The authors in references [19] and [27] have demonstrated the impact of a thick ground plane. The cavity method method has been adapted to calculate this antenna. Indeed, the input impedance (real) at the resonance frequency decreases very quickly when the ground plane thickness increases (Figure 5.61). For example, for a double-layer antenna at 60 GHz, the copper metallization of substrate corresponds to a slot thickness which equals 35 μm (0.007 λ_0). In this case, the input impedance will be divided by two (30 Ω/65 Ω) as compared with the theoretical case where the slot is considered as infinitely thin. To prove the point, an aperture coupled antenna was studied in the 60 GHz band with a duroid 5880 substrate ($h = 0.127$ mm) for both layers. The two substrate plates were soldered and the overall slot thickness was 35 μm. First, the patch and slot dimensions were optimized (cavity method) without taking this thickness into account. Then, a comparison between calculations and measurements was plotted in Figure 5.62. It shows a considerable discrepancy in the input impedance. Next, the cavity method was used again, accounting this time for the thickness factor. This second analytical calculation reveals considerable improvement. It is plotted in Figure 5.63. A very good agreement is achieved in this case, proving that the impact of slot thickness can be very strong for mmWaves.

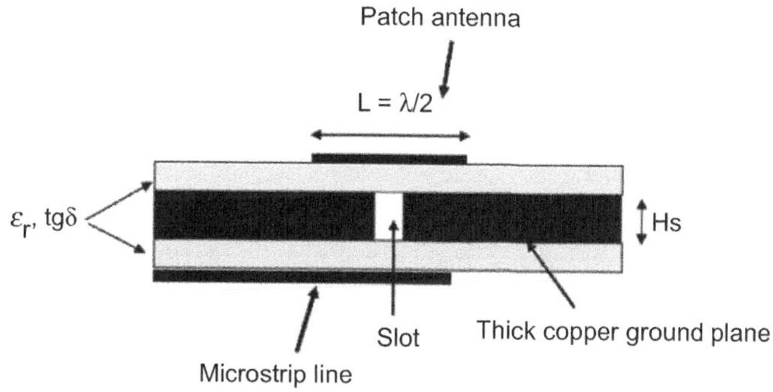

Figure 5.60: Thick aperture coupled patch antenna.

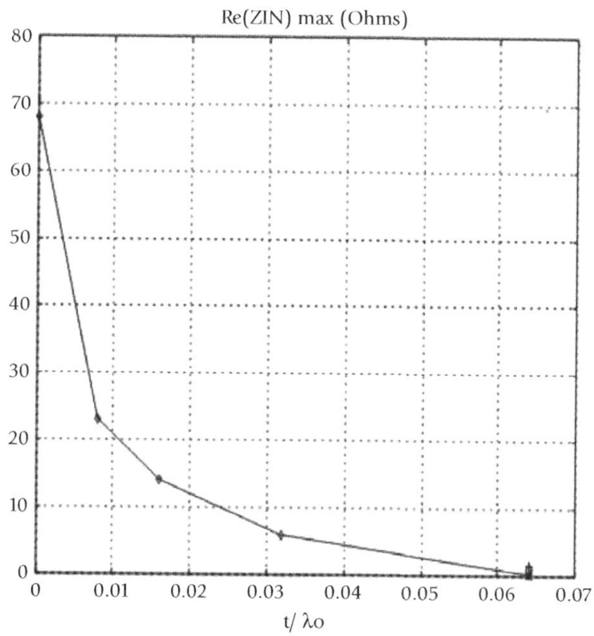

Figure 5.61: Thick aperture coupled patch antenna. Input impedance decrease at resonance frequency versus slot thickness.

Another example of an aperture coupled patch antenna at 55 GHz is given below. A TPX substrate ($h = 0.127$ mm) is used for both layers and a 0.2 mm-thick ground plane with a coupling slot was added between layers. The various physical parameters of this antenna were:

- square patch: 1.467 mm;
- slot: 0.92 mm × 0.25 mm;
- slot thickness: $t = 0.2$ mm.

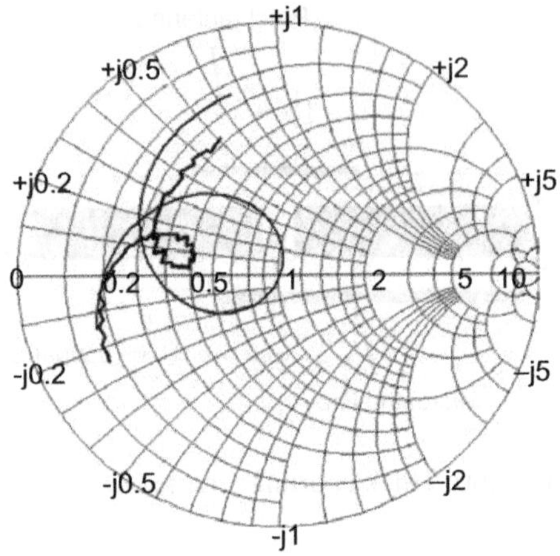

Figure 5.62: Thick aperture coupled patch antenna results in the 57–62 GHz band; input impedance result for calculation (infinitely thin slot) and measurement ($t = 35$ μm).

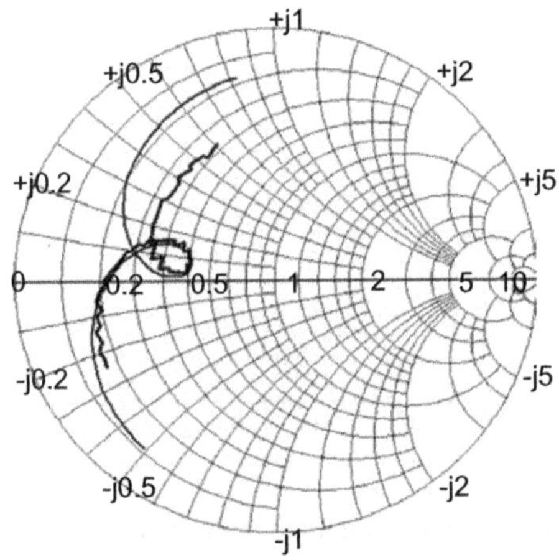

Figure 5.63: Thick aperture coupled patch antenna results in the 57–62 GHz band; input impedance result for calculation ($t = 35$ μm) and measurement ($t = 35$ μm).

The H-plane radiation pattern at 54.5 GHz is plotted in Figure 5.64. Good agreement is shown concerning co-polarization between simulations and measurements. The measured gain is 5.2 dB, corresponding to a 70% efficiency.

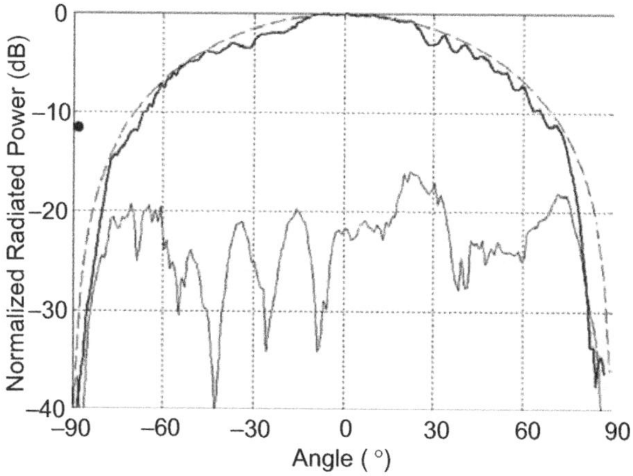

Figure 5.64: Aperture coupled patch antenna with thick ground plane on a TPX substrate; radiating pattern at 54.5 GHz in H-plane.

5.3 Multilayer Antenna Array with Shaped Beam

In this section, many examples of antenna arrays based on multilayer technology will be studied. The aim is to obtain shaped radiating patterns:

- directive pattern in one or two planes;
- sectorial or cosecant pattern;
- highly directive gain;
- multibeam antennas.

Most of the antenna arrays have been simulated using the Moment Method [13]. Only radiating patterns can be obtained because it does not allow for thick metallic ground planes with this software. Then, each radiating multilayer element has been optimized by means of the cavity method or three-dimensional software (CST Microwave Studio) so as to take the thick slot parameter into account and accurately calculate the input impedance. Concerning the experimental result, the input impedance of antenna was measured using a network analyzer (HP 8510C). As for radiating patterns, the IETR millimeter anechoic chamber was used up to 110 GHz.

5.3.1 Directive Pattern with Passive Linear Array

The first example concerns a multilayer linear antenna array composed of six aperture coupled patches in the 60 GHz band [13]. See Figure 5.65 for the feeding line network with amplitude tapering. This array has been designed on a TPX substrate ($h = 0.127$ mm and $\epsilon_r = 2.17$ at 60 GHz). This substrate with a low dielectric constant has been chosen to optimize efficiency. A thick copper ground plane ($t = 0.2$ mm) was added to rigidify

Figure 5.65: Multilayer linear antenna array with six aperture coupled patches; design of feeding line network.

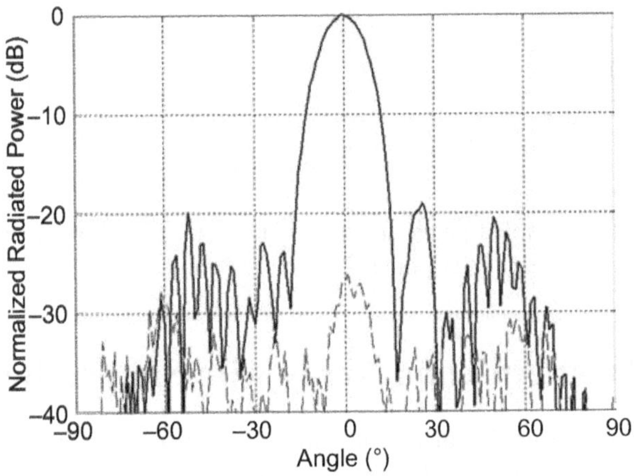

Figure 5.66: Multilayer linear antenna array with six aperture coupled patches; measured co-polarization (solid curve) and cross-polarization (dotted curve) radiating patterns at 59.5 GHz.

the device. For all linear arrays, a 30×30 mm^2 square ground plane was used. Amplitude tapering was applied to reduce the sidelobe level (−20 dB). The measured radiating pattern at 59.5 GHz is presented in Figure 5.66. The sidelobe level is close to prediction (−20 dB) and the level of cross-polarization is satisfactory (better than −27 dB) at the same frequency. The beam width equals 15° for this plane. The antenna array is matched at 59 GHz (Figure 5.67). The measured gain is presented in Figure 5.68. The highest gain (10.5 dB) is obtained at 59.2 GHz corresponding to a 55% efficiency. Multilayer technology makes it possible to separate the feeding line network from the radiating element and to avoid spurious radiation due to lines. Conversely, however, this technology can cause back side radiation because of slots and lines. This is why it is necessary to measure this back side radiation level, plotted in Figure 5.69. The result is satisfactory because it is lower than −17 dB. An equivalent antenna array has been made on a glass Teflon substrate (RT duroid 5880), giving very similar results in terms of radiating pattern and efficiency.

In a 60 GHz indoor high-bit rate communication system, it can be advisable to use circular polarization to reduce multipath propagation [28]. For this reason, a circularly polarized aperture coupled patch was studied. The design is based on a nearly square-patch offering two cross slots in the ground plane (Figure 5.70) [29]. The simulated axial ratio for this radiating element has been plotted in Figure 5.71, showing that good circular polarization

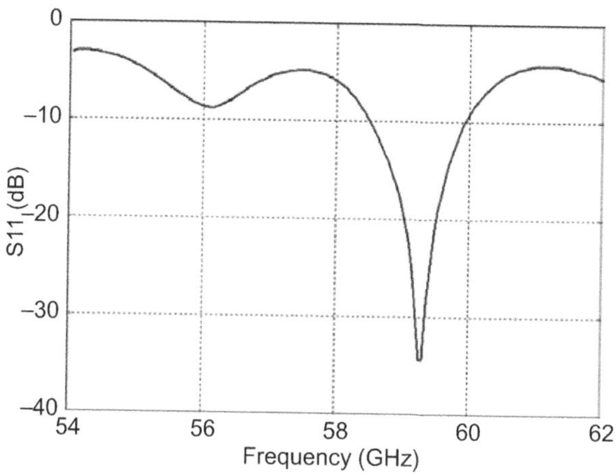

Figure 5.67: Multilayer linear antenna array with six aperture coupled patches; measured return loss (in decibels) in the 54–62 GHz band.

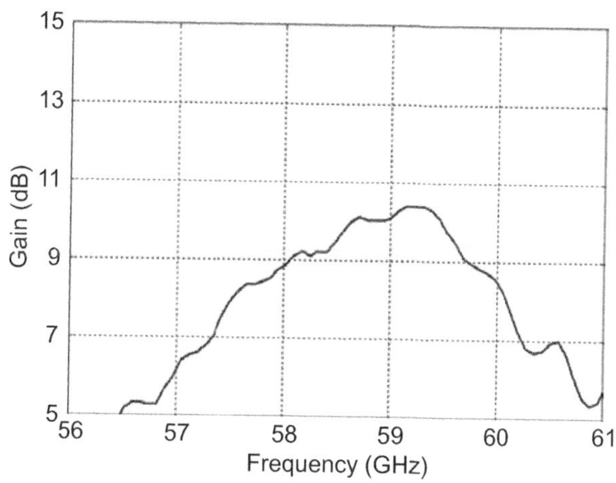

Figure 5.68: Multilayer linear antenna array with six aperture coupled patches; measured gain (in decibels) in the 54–62 GHz band.

can be obtained at 59.5 GHz. This aperture coupled patch is, then, used to design a six-patch linear array in order to obtain a directive pattern in the H-plane. First, a measured axial ratio is shown in Figure 5.72 for frequencies between 58 and 62 GHz. A good circular polarization becomes visible at around 59.5 GHz. Next, the H-plane radiation pattern is plotted in Figures 5.73 and 5.74 for two frequencies: 59.5 and 59.8 GHz. For measurement purposes, a circularly polarized emitter antenna was used to determine left and right polarization levels. Good polarization rejection is obtained, confirming-good axial ratio result.

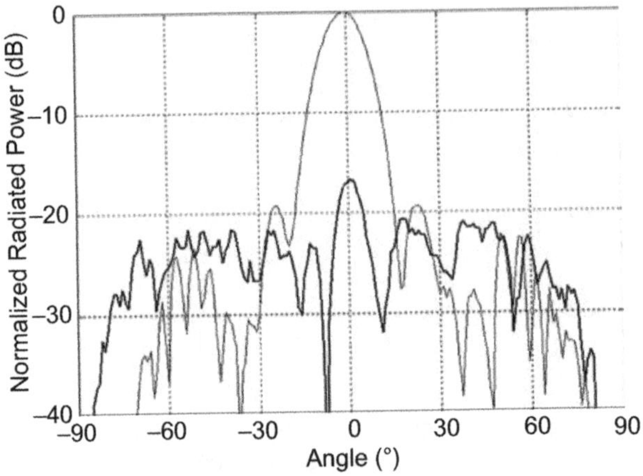

Figure 5.69: Multilayer linear antenna array with six aperture coupled patches; measured back side radiation (thick curve) and front side radiation (fine curve).

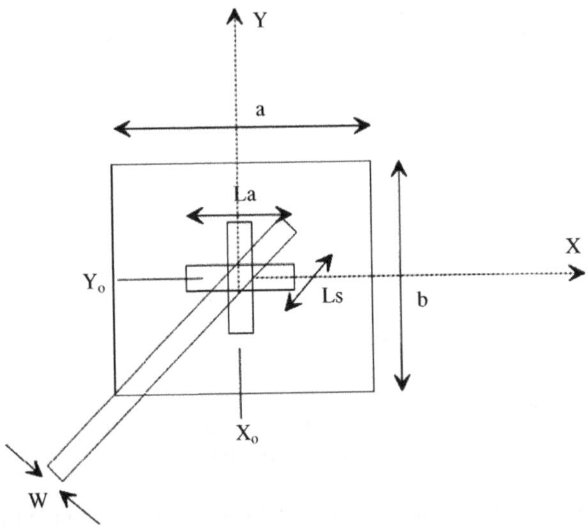

Figure 5.70: Layout of the circularly polarized aperture coupled patch antenna on a TPX substrate with a 0.2 mm thick ground plane.

5.3.2 Sector Beam with Linear Array

For base station antennas in indoor communication systems (Figure 5.75), it could be worthwhile to use a sector beam antenna array [30]. Then, two six-patch sector antenna arrays at 57.5 GHz are presented, hereafter using the previously mentioned technologies. One is based on a TPX substrate and the second on a RT duroid 5880. In both cases, the same

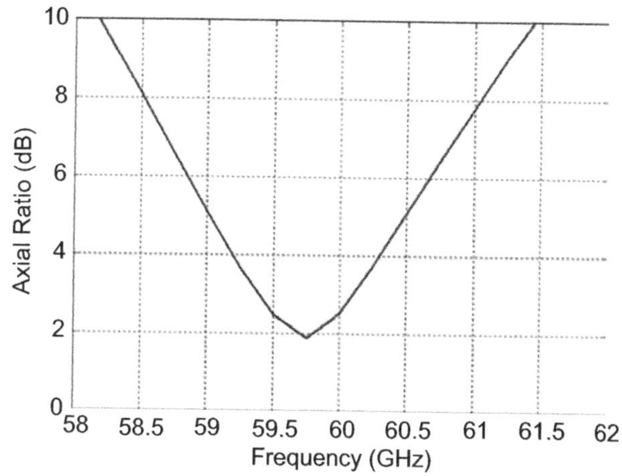

Figure 5.71: Circularly polarized aperture coupled patch antenna on a TPX substrate with a 0.2 mm thick ground plane; simulated axial ratio.

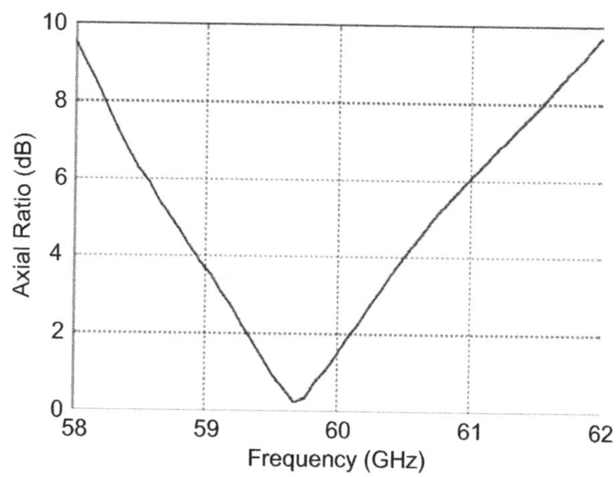

Figure 5.72: Measured axial ratio for a six-patch antenna array.

0.2 mm thick ground plane has been added between the layers. A feeding line network layer of one sector antenna is presented in Figure 5.76. The radiating elements are tapered in amplitude and phase to obtain the expected sector beam coverage. For each patch, the normalized amplitude and phase are

$$(0.15, 0°), (0.282, 180°), (1, 0°), (1, 0°), (0.282, 180°), (0.15, 0°)$$

Once all these requirements have been met, the measured radiating patterns at 57.5 GHz can be compared for both the RT duroid 5880 (Figure 5.77) and TPX (Figure 5.78) designs. The results for each antenna array are summarized in Table 5.8. The results are very similar

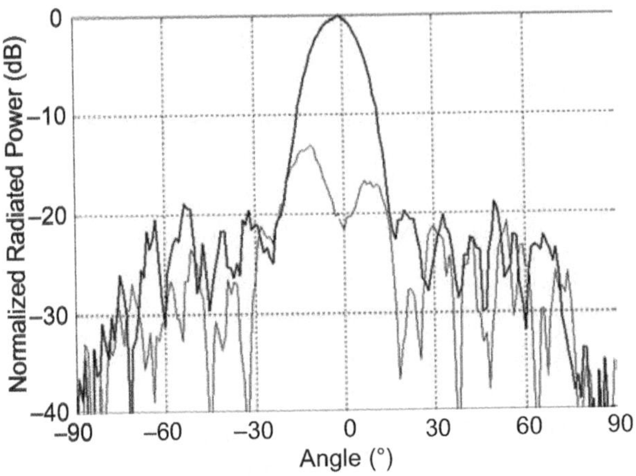

Figure 5.73: Measured co-polarization (solid curve) and cross-polarization (fine curve) H-plane radiation patterns at 59.5 GHz.

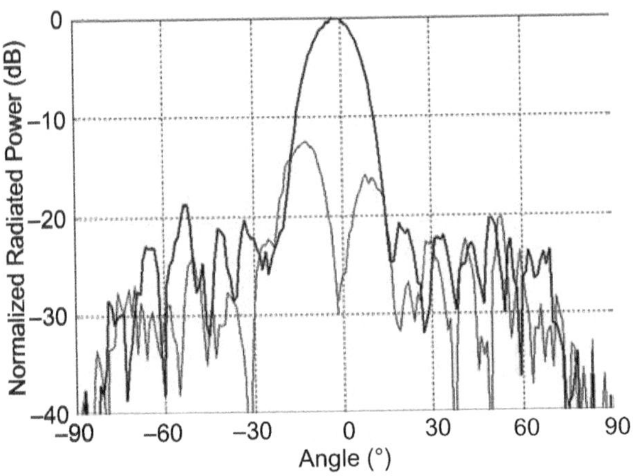

Figure 5.74: Measured co-polarization (solid curve) and cross-polarization (fine curve) H-plane radiation patterns at 59.8 GHz.

for these two sector beam antenna arrays and match the theoretical results with regard to the radiating patterns. All results confirm that both technologies can be used at millimetre-wave to fabricate aperture coupled microstrip antenna arrays.

Clearly, with indoor communication systems, this particular kind of sector beam array can be made quite suitable to obtain circular polarization. In this case, the source element will be chosen as depicted in the previous section. The results with TPX are acceptable. They offer an axial ratio equal to 1 dB around 59.5 GHz (Figure 5.79). The radiating pattern in

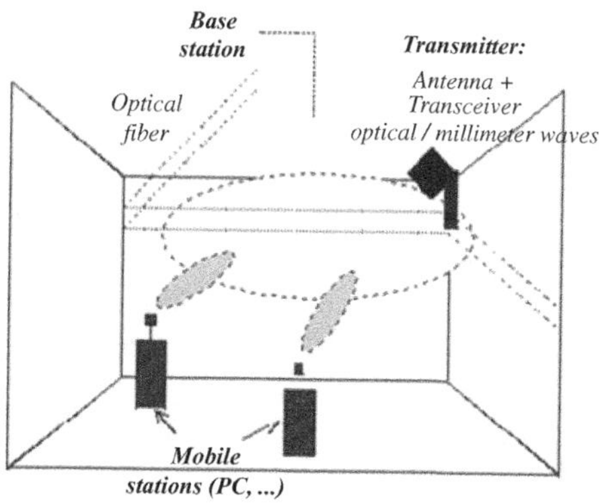

Figure 5.75: Antennas configuration for an indoor communication system with one transmitter and several mobile stations.

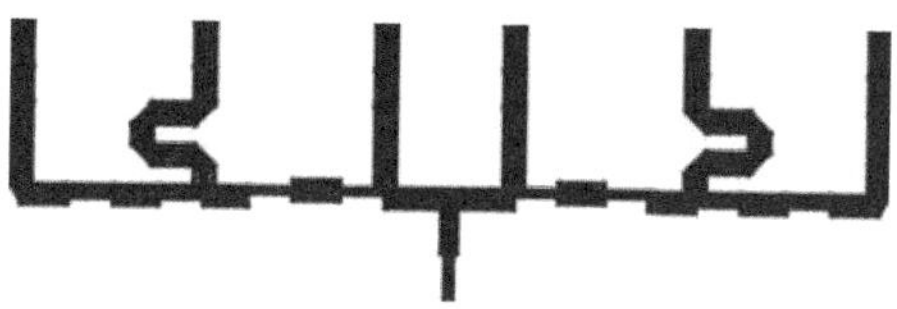

Figure 5.76: Feeding line network of linear sector beam antenna array; amplitude and phase tapering is applied.

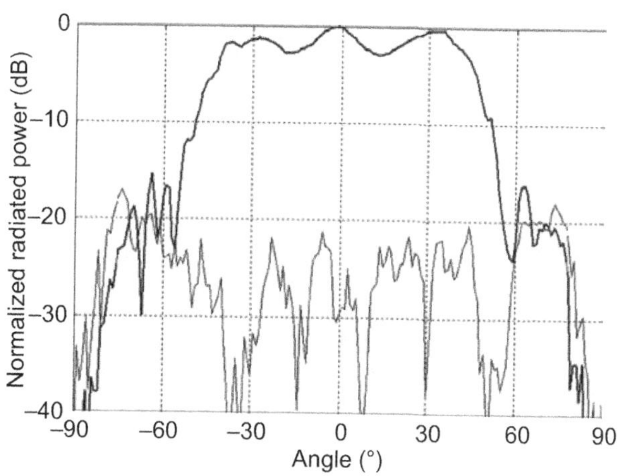

Figure 5.77: Measured co-polarization (thick curve) and cross-polarization (fine curve) radiating patterns for the sector beam array on a RT duroid 5880 substrate.

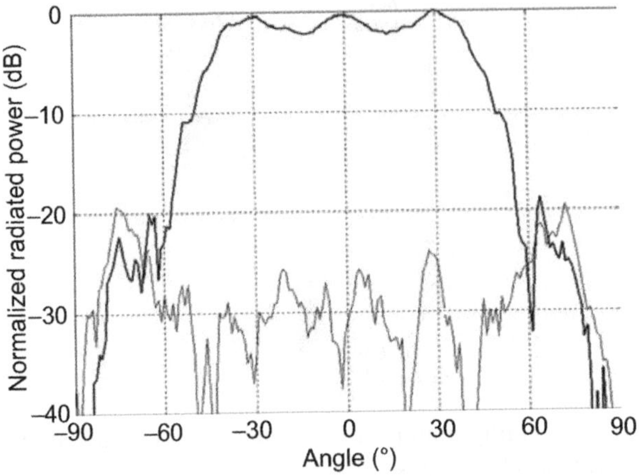

Figure 5.78: Measured co-polarization (thick curve) and cross-polarization (fine curve) radiating patterns for the sector beam array on a TPX substrate.

Table 5.8: Comparison of radiating patterns for both TPX and RT duroid 5880 sector antenna designs.

Substrates	Beamwidth	Side lobe level (dB)	Cross-polar level (dB)	Efficiency (%)
RT duroid 5880	80	≤ -16	≤ -19	55
TPX	84	≤ -20	≤ -20	58

the H-plane at the same frequency offers a good ripple as well as a satisfactory polarization rejection (Figure 5.80).

5.3.3 Cosecant Beam with Linear Array

In indoor communication systems, it is also possible to select a cosecant beam antenna for the base station. It allows several mobile stations to be illuminated which are not equally distant from the emitter (Figure 5.81). The aim of the cosecant beam is to even up the radiating pattern and to compensate path loss due to spacing between the emitter and the stations.

To obtain this kind of radiating pattern, patches require accurate phase and amplitude tapering. For instance, in a six-patch linear array, the computed radiating pattern, plotted in Figure 5.82, can be obtained with the following tapering values (Table 5.9) and with a distance between patches equal to $0.5\lambda_0$ at 60 GHz:

Based on the tapering values from the table, an aperture coupled six-patch antenna array with linear polarization was achieved at 60 GHz band. A RT duroid 5880 substrate ($h = 0.127$ mm and $\epsilon_r = 2.23$ at 60 GHz) was used and a thick ground plane was added. The measured radiating pattern at 59 GHz in the H-plane is plotted in Figure 5.83. There is a discrepancy between the computed and measured data, which can be due to a difference in the length and width of the thick slots, which may induce tapering.

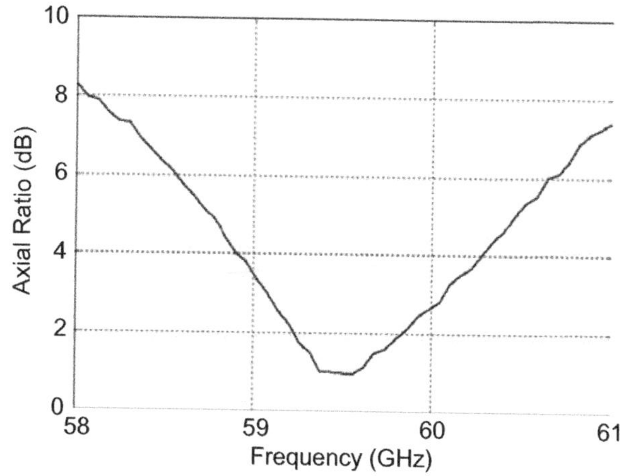

Figure 5.79: Multilayer sector beam antenna array with circular polarization; measured axial ratio versus frequency.

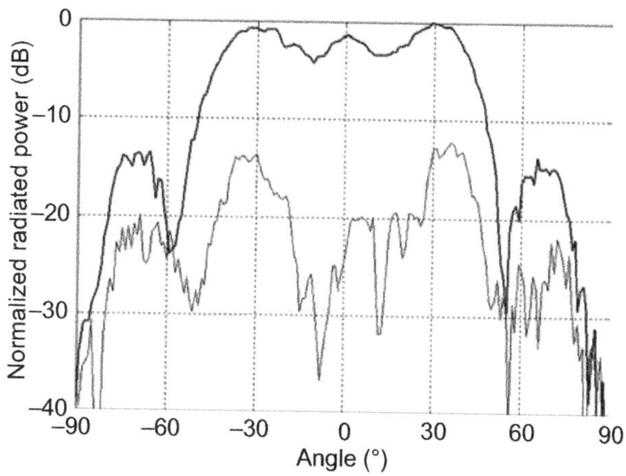

Figure 5.80: Multilayer sector beam antenna array with circular polarization – radiating co-polarization (thick curve) and cross-polarization (thin curve) patterns.

Table 5.9: Antenna array tapering values.

Patch number	Tapering (linear amplitude/phase)
1 (on left)	0.7/–50°
2	0.1/20°
3	1.0/–60°
4	1.0/0°
5	0.53/10°
6 (on right)	0.4/70°

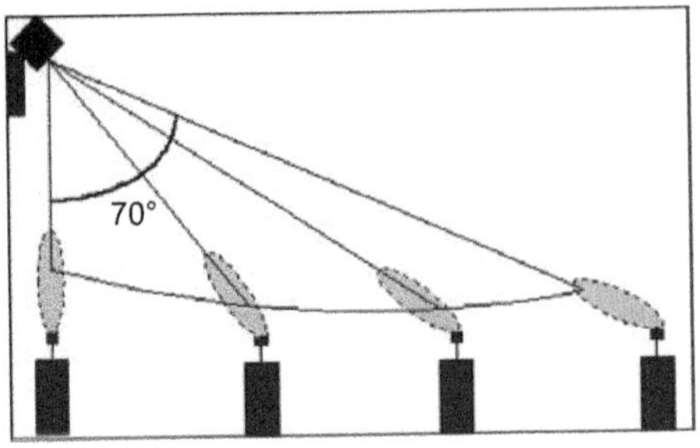

Figure 5.81: Antenna configurations for an indoor communication system; cosecant beam antenna for the base station.

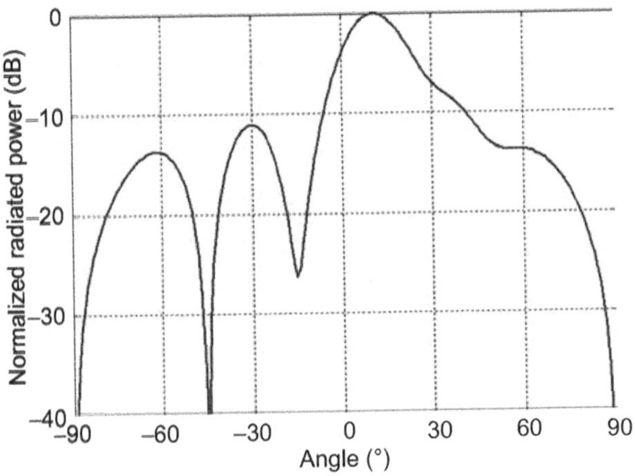

Figure 5.82: Multilayer cosecant beam antenna array; computed radiating pattern at 60 GHz in the H-plane.

5.3.4 Highly Directive Antennas

In order to obtain a pencil beam, a two-dimensional linearly polarized antenna array with 36 patches (6 × 6) has been designed. Figures 5.84 and 5.85 show both sides of this array which has been manufactured on a TPX substrate with a 0.2 mm thick ground plane. The radiating elements are identically tapered in amplitude on both E and H principal planes so as to reduce the side lobe level. The radiating co- and cross-polarization patterns at 59 GHz are plotted in Figure 5.86 for the E-plane and in Figure 5.87 for the H-plane. The E and H-plane beamwidths are 15° and 16° respectively. The cross-polarization level is lower than −27 dB in both planes. The side lobe level is around −22 dB in the E-plane and −20 dB in the H-plane. The efficiency of this array reaches an approximate 40% at 59 GHz.

Some specific applications require very highly directivity antennas (e.g. up to 32 dBi or more). In this case, it is conceivable to use microstrip technology even though the measured gain will be lower than directivity owing to dielectric and metallic loss. It seems worthwhile

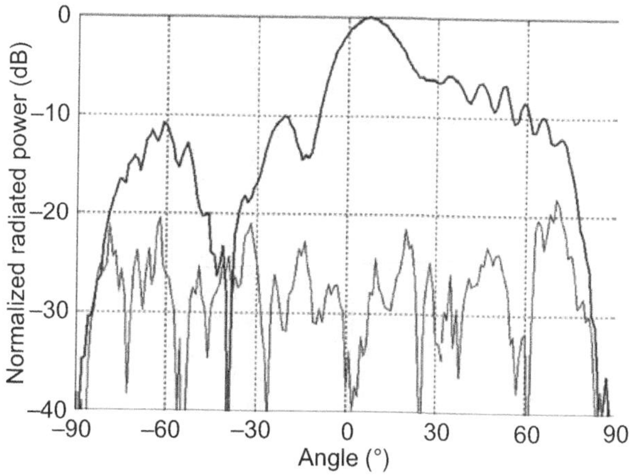

Figure 5.83: Multilayer cosecant beam antenna array – measured co-polarization (thick curve) and cross-polarization (thin curve) radiating patterns at 59 GHz in the H-plane.

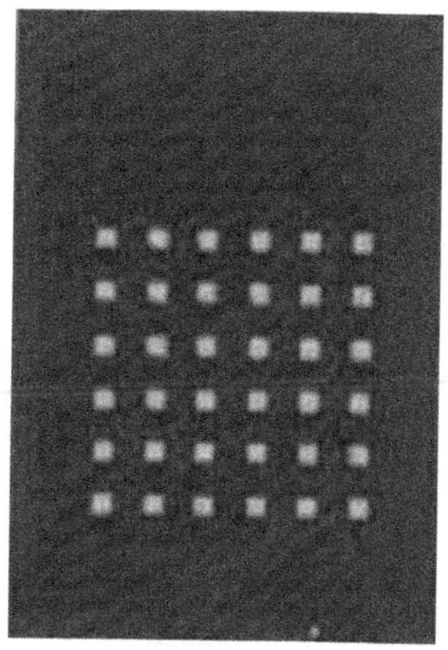

Figure 5.84: Two-dimensional antenna array; radiating array side with a $0.72\lambda_0$ spacing between patches.

nevertheless to design such a high directivity multilayer antenna based on an aperture-coupled patch antenna array. An example is shown in Figures 5.88 and 5.89 featuring a 256-patch 60 GHz antenna array (16 × 16). It was patterned on a TPX substrate with a 0.2 mm-thick ground plane. In this case, a $100 \times 100\ mm^2$ square ground plane was used. When large arrays are optimized, it is true that a thick ground plane becomes quite useful to rigidify the structure and when such arrays are designed without a thick support, prototypes can be too flexible. The two layers are presented so as to visualize both the radiating array and the feeding line network. No tapering has been applied to increase directivity and gain, so the theoretical side lobe level is −13 dB and the distance between patches has been optimized

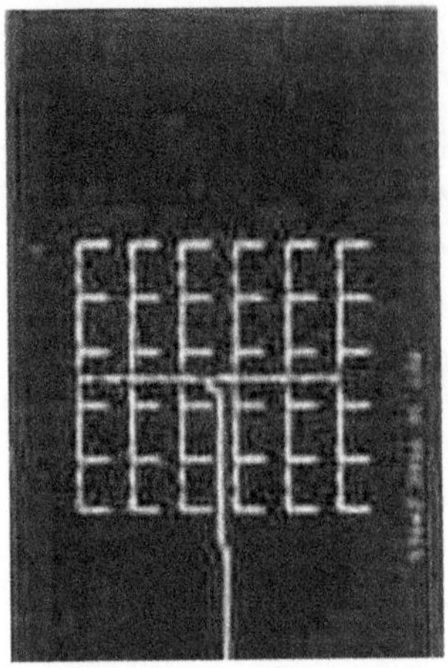

Figure 5.85: Two-dimensional antenna array; feeding line network amplitude taper in both planes.

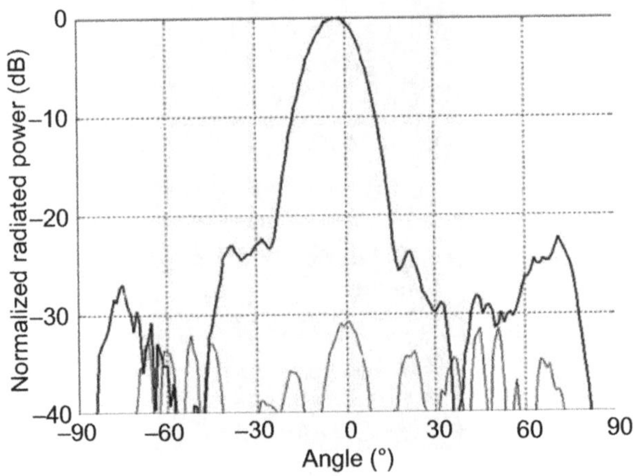

Figure 5.86: Two-dimensional antenna array; measured co-polarization (thick curve) and cross-polarization (thin curve) radiating patterns in the E-plane.

as $0.74\lambda_0$. A parallel feeding line network has been chosen to increase the bandwidth if compared to a serial feed. For such an antenna array, a typical radiating pattern in the H-plane has been plotted in Figure 5.90. The typical beamwidth is equal to 4.5° and the side lobe level is -13 dB. In that case, the theoretical directivity is close to 32 dBi. The first radiating pattern measurement results at 60.5 GHz have been plotted in Figure 5.91 for the H-plane and in Figure 5.92 for the E-plane.

Radiating characteristics are detailed in Table 5.10. Two frequencies are considered separately. Results prove the stability of all three radiating patterns. The last result concerns

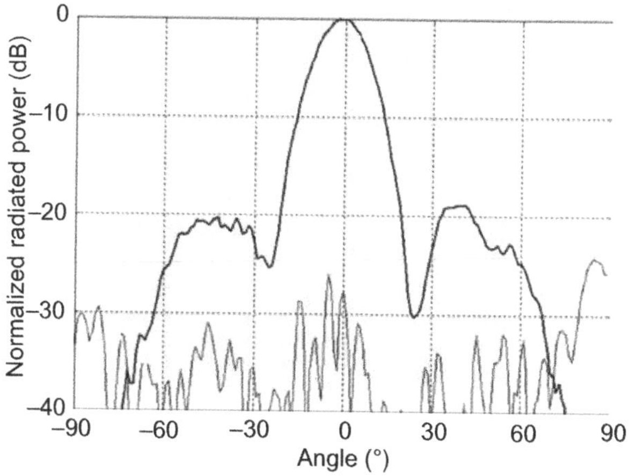

Figure 5.87: Two-dimensional antenna array; measured co-polarization (thick curve) and cross-polarization (thin curve) radiating patterns in the H-plane.

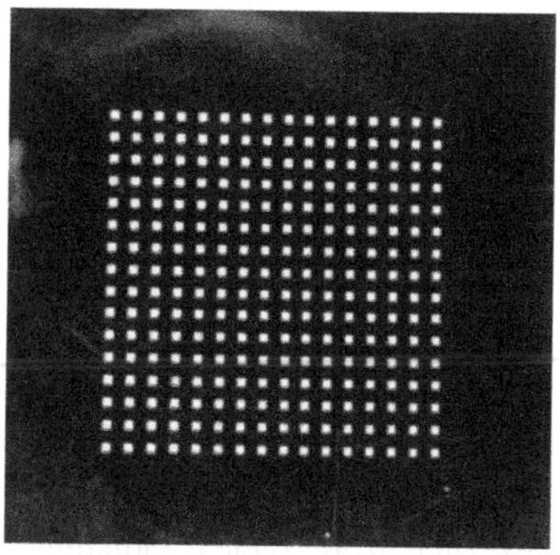

Figure 5.88: A multilayer 256-patch array; radiating layer.

the measured gain which is important to consider. Typically, at mmWaves, insertion loss due to copper conductivity and loss tangent are considerable, thus reducing antenna efficiency to a low level, especially for large arrays. In this last case, the measured gain was initially 25 dB. However, a very long additional line (50 mm) had to be connected to the antenna input for measurement purposes, involving a line loss of 2.4 dB. Taking this loss into account, the true measured gain is close to 27.5 dB. Therefore, this gain level corresponds to a 30% efficiency. This value can be accounted for in the following way.

- Loss in the feeding line network: 4.8 dB.
- Radiating element loss: 1.5 dB.

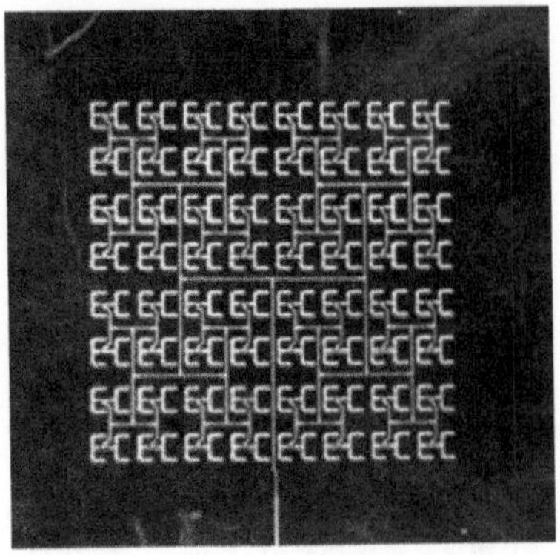

Figure 5.89: A multilayer 256-patch array; parallel feeding line network layer.

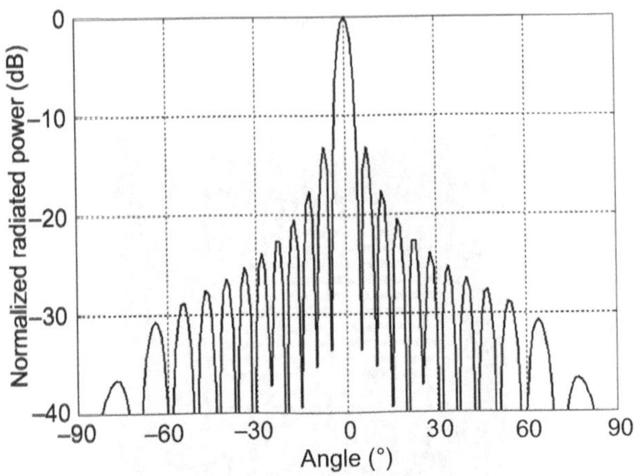

Figure 5.90: A multilayer 256-patch array; typical radiating pattern in the H-plane for the 256-patch array.

Table 5.10: Radiating characteristics for a 256-patch array.

	58.5 GHz E-plane	58.5 GHz H-plane	60.5 GHz E-plane	60.5 GHz H-plane
Beamwidth	4.3°	4.5°	4.3°	4.4°
Side lobe level (dB)	−13.5	−12	−12	−13
Cross-polarization level (dB)	≤ −20	≤ −23	≤ −21	≤ −24

Finally, let us examine the back side parasitic radiation due to slots and lines. The measured back side radiating pattern is plotted in Figure 5.93. A front-to-back radiation ratio of 13 dB is obtained.

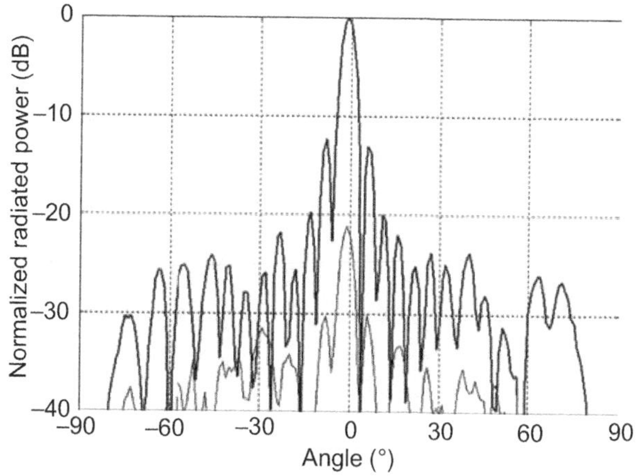

Figure 5.91: A multilayer 256-patch array; measured co-polarization (thick curve) and cross-polarization (thin curve) radiating patterns at 60.5 GHz in the H-plane.

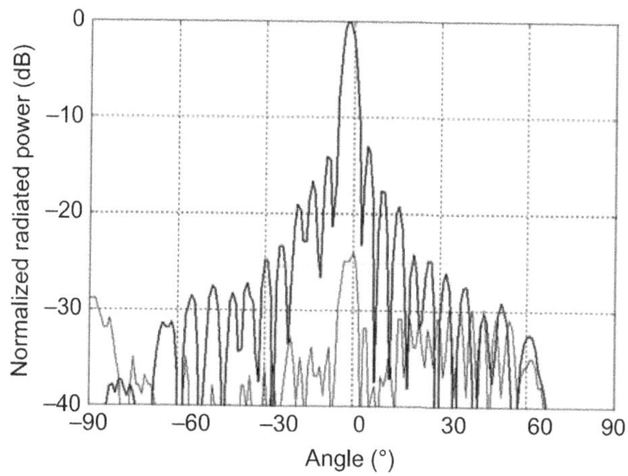

Figure 5.92: A multilayer 256-patch array; measured co-polarization (thick line) and cross-polarization (thin line) radiating patterns at 60.5 GHz in the E-plane.

Another example of high directivity antenna at 77 GHz can be given and has been studied for automotive radar application (automotive cruise control). The design is presented in reference [31]. The objective was to develop a highly directive antenna for a conventional monopulse system. In this case, two receiving beams are formed: a sum beam (Σ) and a difference beam (Δ). The actual antenna array is joined by a 180° hybrid coupler to allow both sum and difference inputs.To keep the feed line as simple as possible, a linear series array has been chosen. The constraint of straight lines between elements leads to a one-guided wavelength spacing (or a 0.75 free-space wavelength). In order to have the same half-beam width for sigma (Σ) and delta (Δ) inputs, the two half arrays were located

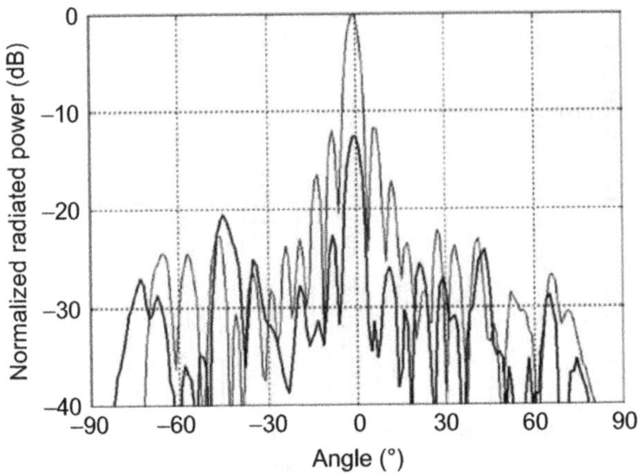

Figure 5.93: A multilayer 256-patch array; back side radiation (thick curve) and front side radiation (thin curve).

closer together, whereby the synthesis method can perform the same amplitude excitations, taking into account the nonidentical spacing in the H-plane of each patch. The optimization technique is based on the simplex method [32], which uses the Dantzig algorithm [33] and yields symmetrical or nonsymmetrical pattern synthesis.

The objective of the Automotive Cruise Control radar antenna is to obtain:

1. the half-beam width $= 6°$ for both Δ and Σ input in the H-plane;
2. side lobe level ≤ -20 dB;
3. the separation angle between two maxima for Σ input 10°.

The results are:

1. number of elements: 16 in the H-plane and 20 in the E-plane;
2. distance between patches for each subarray $0.75\lambda_0$;
3. distance between subarrays $1.2\lambda_0$.

The used substrate is TPX ($\epsilon_r = 2.2$, $h = 127$ μm). The corner-fed patch dimensions are 1.25×1.25 mm^2. The corner-fed patch is realized on one layer of dielectric and provides high input impedance which is well suited for the series' array. The manufactured array is shown in Figure 5.94. A one-guided wavelength (microstrip line) was used between two patches. The impedance transformers are necessary to obtain the right amplitude excitations. A two-step quarter-wavelength transformer can be used for each cell. If necessary, four quarter-wave transformers can be inserted when the spacing between patches is one wavelength. The final input impedance matching is obtained with two quarter-wavelength transformers. Very good performances were obtained for this array. Results for sum and difference radiating patterns are plotted in Figures 5.95 and 5.96.

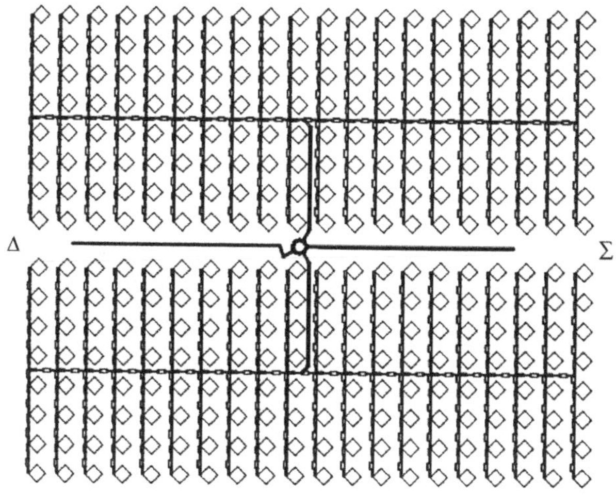

Figure 5.94: Layout of 320-elements planar array at 77 GHz for automotive cruise control radar antenna.

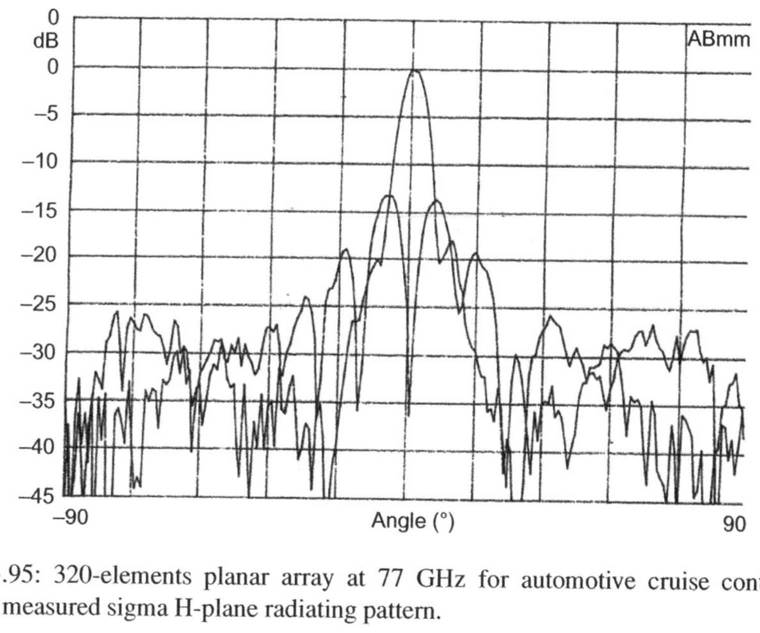

Figure 5.95: 320-elements planar array at 77 GHz for automotive cruise control radar antenna; measured sigma H-plane radiating pattern.

5.3.5 Multibeam Antenna

In this section, several designs of multibeam antenna will be presented: first, Butler-matrix-fed antenna arrays for indoor communication systems in the 60 GHz band; second, a couple

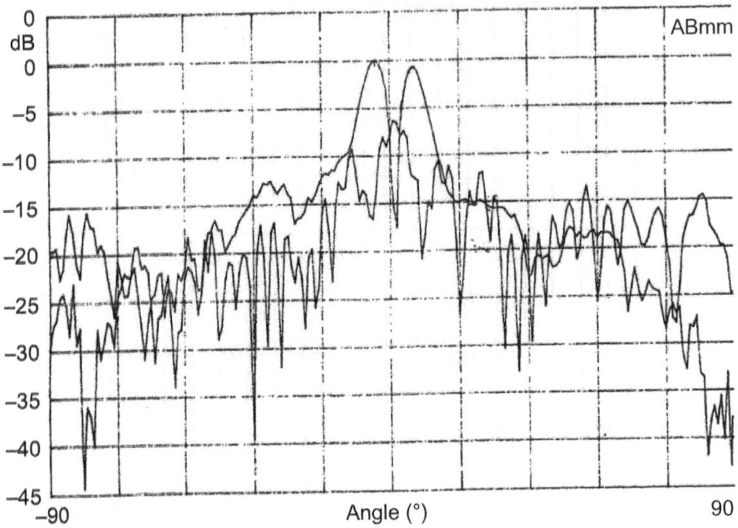

Figure 5.96: 320-elements planar array at 77 GHz for automotive cruise control radar antenna; measured delta H-plane radiating pattern.

of two-beam printed arrays in the 24 GHz band making use of the Doppler effect and designed for vehicle speed measurement.

5.3.5.1 Multibeam Antenna Based on the Butler Matrix

Quite a few applications such as, for example, automotive radar systems, call for multibeam antennas. There are many different ways to achieve this kind of radiating pattern:

- constrained lenses [34]: these are often designed using waveguide technologies so as to reduce loss at mmWaves [35, 36]; however, constrained lenses can also be made through the use of printed technologies [37];
- homogeneous dielectric lenses associated with different sources [38];
- inhomogeneous dielectric lenses associated with different sources [39];
- folded reflect-array antennas [40];
- Butler-matrix and antenna arrays [41].

The Butler matrix makes it possible to turn an antenna array into a multibeam system. In addition, the Butler-matrix system and microstrip technologies make a good match. The Butler matrix is made of hybrids (3 dB, 90°), phase shifters, and elements which allow the microstrip lines to be crossed in order to feed different patches with the appropriate phases. In fact, at low frequencies, some bonding wires are used to cross the microstrip lines. At mmWaves, they generate spurious radiation. Two other solutions will then be presented to avoid the inconvenience of bonding wire crossing:

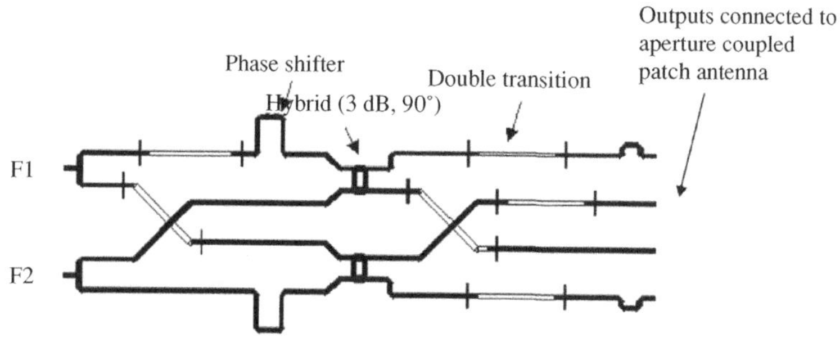

Figure 5.97: Multibeam antenna based on multilayer Butler-matrix and aperture coupled patches.

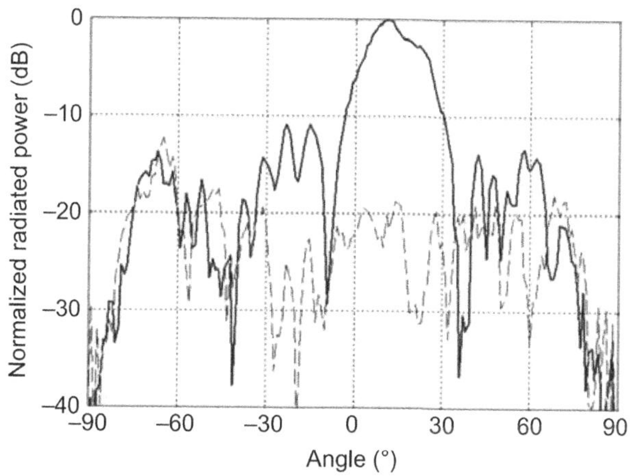

Figure 5.98: Radiating co-polarization (solid curve) and cross-polarization (dotted curve) patterns for configuration F1.

- by two layer microstrip line-to-microstrip line transition;.
- by double hybrid (3 dB, 90°).

In the first case, a two-layer microstrip line-to-microstrip line transition using a coupling slot can be used [42]. A thick ground plane has been added to rigidify the structure. A first Butler-matrix design with only two beams is described below. In this case, the phase shift is set to 45° so that two tilt angles can be generated (±15°) depending of the feeding input port (F1 or F2) (Figure 5.97). Four aperture coupled patches were added at the outputs and the design was optimized at 60 GHz for indoor communication applications. RT duroid 5880 substrate was chosen ($h = 0.127$ mm and $\epsilon_r = 2.23$). The distance between patches is equal to $0.7\lambda_0$.

Measurement results of radiating patterns at 60 GHz are plotted in Figures 5.98 and 5.99 for F1 and F2 configurations respectively. Measurement and theoretical results agree with

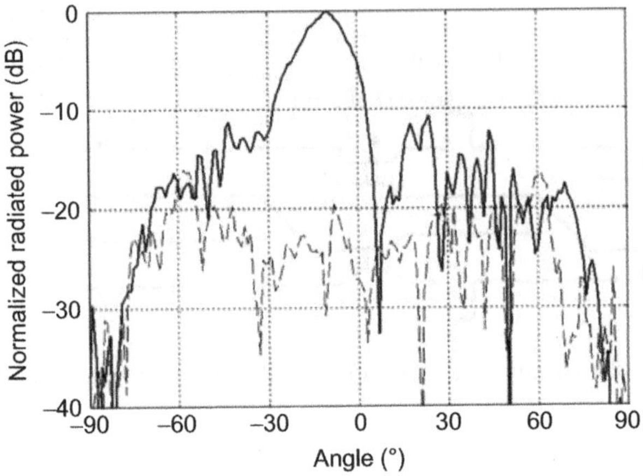

Figure 5.99: Radiating co-polarization (solid curve) and cross-polarization (dotted curve) patterns for configuration F2.

acceptable accuracy and the side lobe level matches predictions. Nevertheless, even though radiating patterns are satisfactory, the gain and efficiency are not. The feeding line network is not optimum in terms of line length and number of two-layer transitions. In this case, a poor efficiency (25%) is obtained. To achieve better efficiency, a new four-beam design comprising a double hybrid (3 dB, 90°) can be optimized in order to avoid crossing points within the Butler matrix. This matrix is shown in Figure 5.100. Theoretical results of the double hybrid are plotted in Figure 5.101 for magnitude of S parameters. Very low level insertion loss is obtained (0.8 dB). The four configurations under scrutiny (F1–F4) reveal tilt angles of ±15° and ±45°. The distance between patches is $0.5\lambda_0$ so as to keep correct side lobe level, whatever the tilt angle. The actual matrix and antenna array are presented in Figure 5.102 wherein two distinct layers are shown. In addition, the magnitude of S11 has been measured for each configuration (F1–F4) in order to see if all configurations are properly matched. The magnitude of S11 is better than −10 dB (Figure 5.103) whatever the configuration. The measured radiating patterns are plotted in Figure 5.104 for all configurations (F1–F4). Good results can be shown in terms of tilt angle and side lobe level. These results prove that achieving good multibeam antenna performances is possible even at mmWaves. With regard to efficiency, the result is better than for the first Butler-matrix design. It is actually close to 40% due to low insertion loss in the cross-over and to the feeding network optimization.

5.3.5.2 Two-beam Antenna for Speed Measurement at 24 GHz

A printed array has been manufactured on a polypropylene substrate backed by a rigid metallic support (multilayer technology). The objective was to experiment with a two-beam antenna array to measure the speed of a vehicle. Speed measurement can be extracted by the doppler principle, provided that tilt beams are emitted below the vehicle (Figure 5.105). The optimized printed array is presented in Figure 5.106. It is composed of 16 × 6 patches. The measured tilt angle are close to specification (±40°) and are plotted in Figure 5.107 along with the radiating patterns.

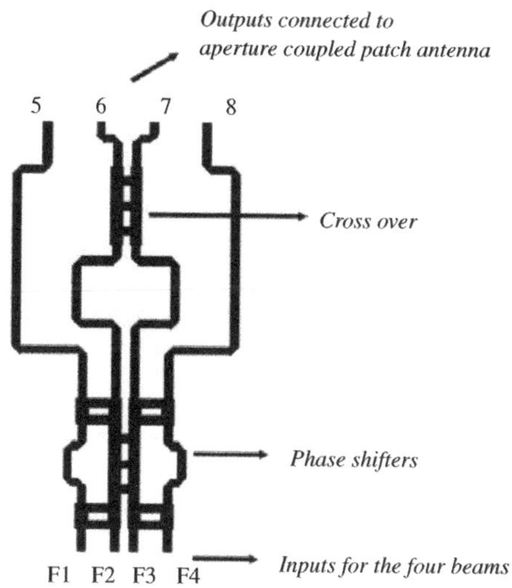

Figure 5.100: Layout of the four-beam Butler matrix with double hybrid to cross lines.

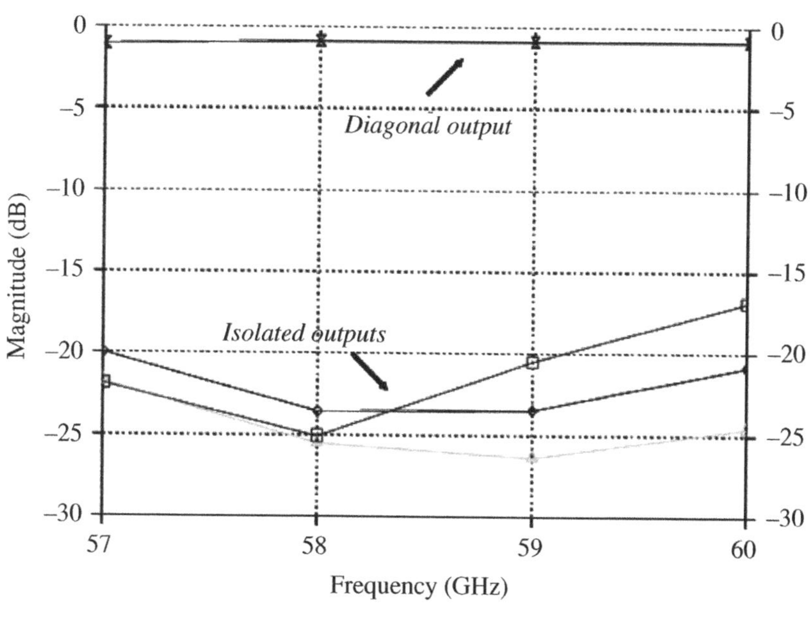

Figure 5.101: Four-beam Butler matrix with double hybrid to cross lines. A magnitude of S parameters results in a 60 GHz band.

5.4 Measurement Disturbances: Connector and Diffraction Problems for Printed Antennas

Measuring the performances of mmWave antennas can be troublesome and will generate a variety of difficulties such as:

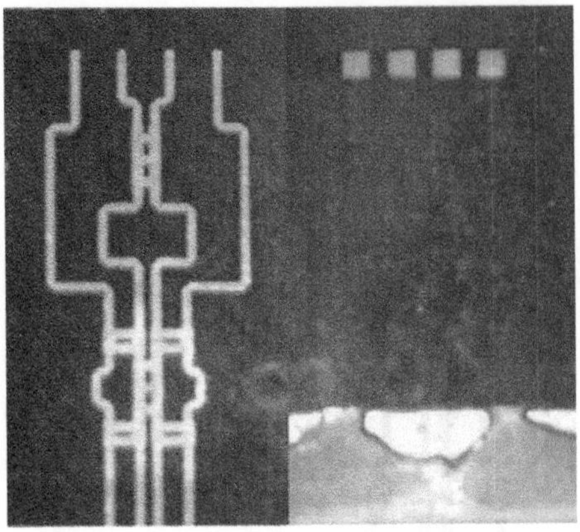

Figure 5.102: Four-beam Butler matrix with double hybrid to cross lines. Feeding line network and radiating layers.

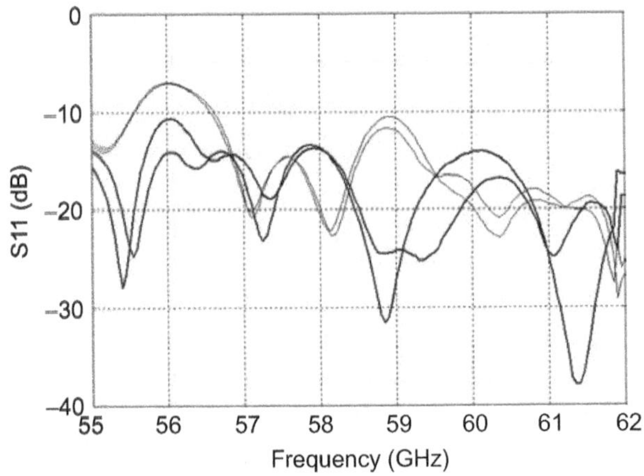

Figure 5.103: Four-beam Butler matrix with double hybrid to cross lines. Return loss results (dB) for all configurations.

- problems with measuring the S11 parameter because of bonding wires and connectors;
- problems with the assessment of radiating patterns owing to ground plane and connector diffraction effects.

To address these questions, several examples will be examined for a radiating single source (a patch) or an antenna array at 38 and 60 GHz.

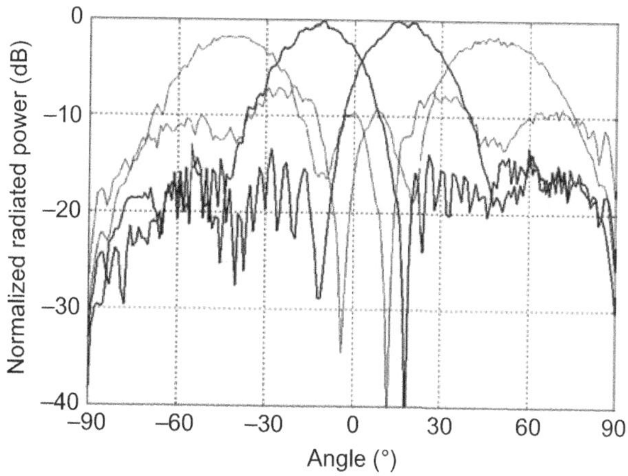

Figure 5.104: Four-beam Butler matrix with double hybrid to cross lines. Radiating patterns at 60 GHz for all configurations.

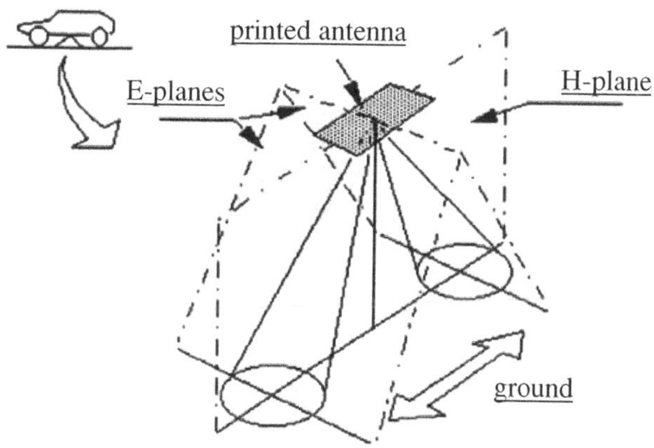

Figure 5.105: Doppler speed measurement with two-beam antenna.

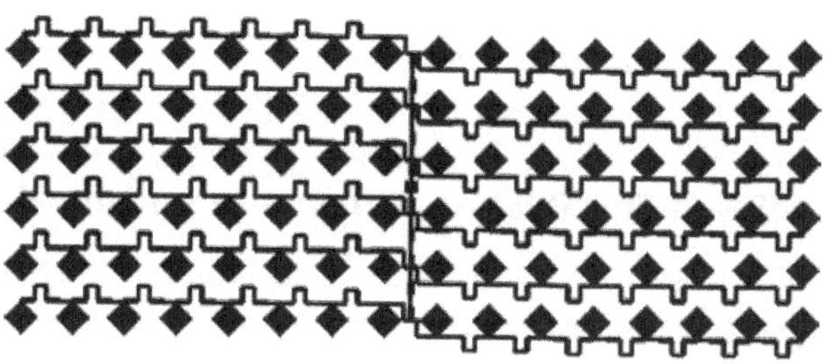

Figure 5.106: Printed antenna array at 24 GHz with serial feeding line network to obtain two-beam radiating patterns.

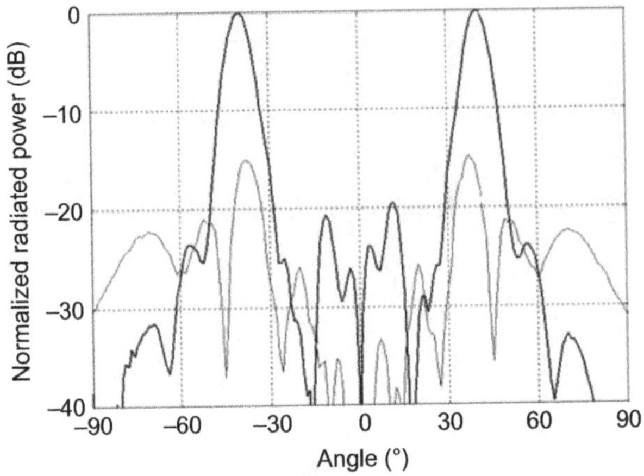

Figure 5.107: Printed antenna array at 24 GHz with serial feeding line network. Measured radiating patterns at 24 GHz.

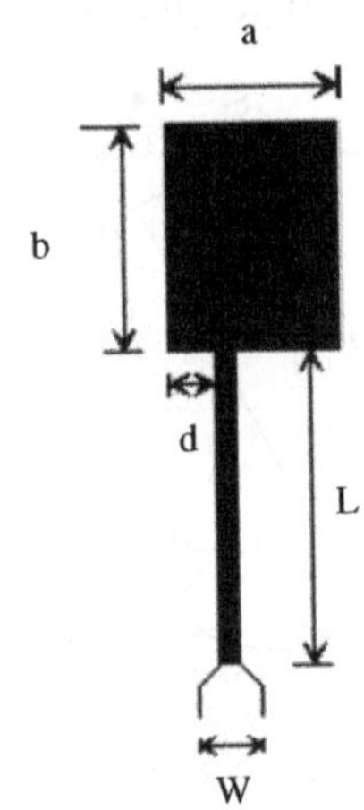

Figure 5.108: Printed patch at 60 GHz on a INP substrate; $a = 0.656$ mm, $b = 0.850$ mm, $d = 0.18$ mm, $w = 0.08$ mm, $L = 0.975$ mm.

5.4.1 Impact of Bonding Wire on Antenna Input Impedance

First, this section will consider how the connecting circuitry may affect S11 measurements. The impact of the bonding wire on the input impedance will be brought to light.

The example concerns a patch antenna in the 60 GHz band printed on a INP substrate ($\epsilon_r = 12.8$, $h = 95$ μm) (Figure 5.108) [43]. This antenna was then connected to a 50 Ω microstrip line printed on an alumina substrate. A bonding wire was added between the two printed circuits to ensure proper connection. The influence of the wire length was simulated

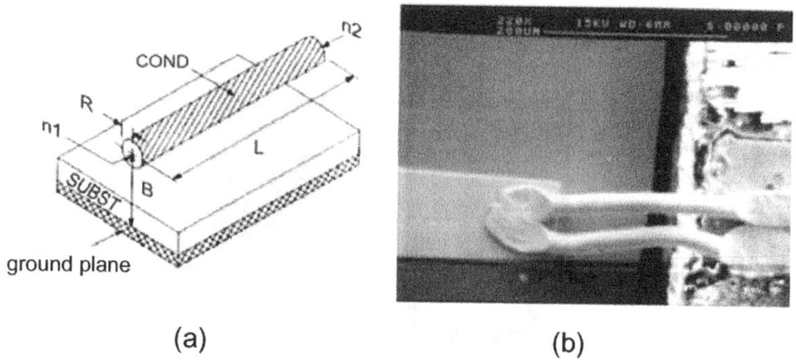

Figure 5.109: Impact of bonding wire length: (a) simulation with ADS; and (b) photography of double bonding wire.

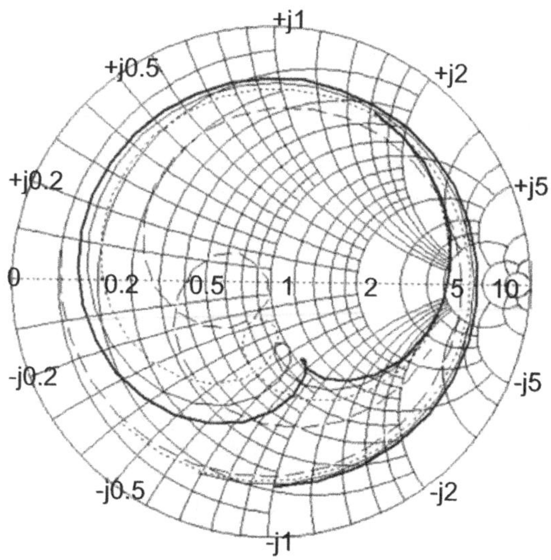

Figure 5.110: Input impedance for the 57–62 GHz band. The following configurations are given: without wire (dash curve), $L = 100$ μm (dot curve), $L = 150$ μm (solid curve), $L = 200$ μm (thick curve).

by means of a circuit design simulation software (ADS) as featured in Figure 5.109. The results of input impedance have been plotted in Figure 5.110 for different lengths of wire (between 0 and 200 μm). They prove that the input impedance in the mmWave range is dependent on the length of the wire.

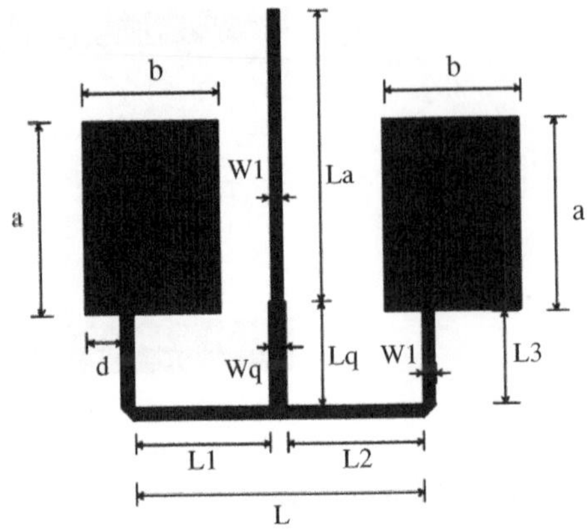

Figure 5.111: Passive antenna array at 38 GHz printed on alumina substrate. The major antenna dimensions are: $a = 1680$ µm, $b = 1202$ µm, $d = 322$ µm, $L = 2518$ µm, $L1 = 1186$ µm, $L2 = 1186$ µm, $Lq = 897$ µm, $L3 = 800$ µm, $La = 10150$ µm, $Wq = 146$ µm; $Wl = 115$ µm.

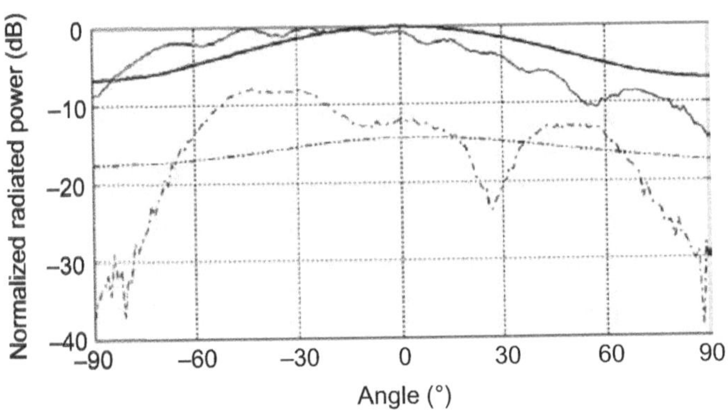

Figure 5.112: E-plane radiating patterns at 38 GHz for the passive antenna array printed on an alumina substrate.

5.4.2 Impact of Diffraction Effects on the Ground Plane and on the Connecting Circuitry

To illustrate the diffraction influence on the radiating ground plane, a two-patch antenna array at 38 GHz has been studied. First, a passive prototype was printed on an alumina substrate ($\epsilon_r = 9.9$ and $h = 0.127$ mm). Dimensions of patches and lines are given in Figure 5.111. Because of the high dielectric constant of alumina, the radiating patterns were very large, particularly in the E-plane. Theoretical and measured radiating patterns were compared for the E-plane (Figure 5.112) and H-plane (Figure 5.113) respectively. It appears that diffraction on the ground plane and on the connector device (K connector and fixture) caused some

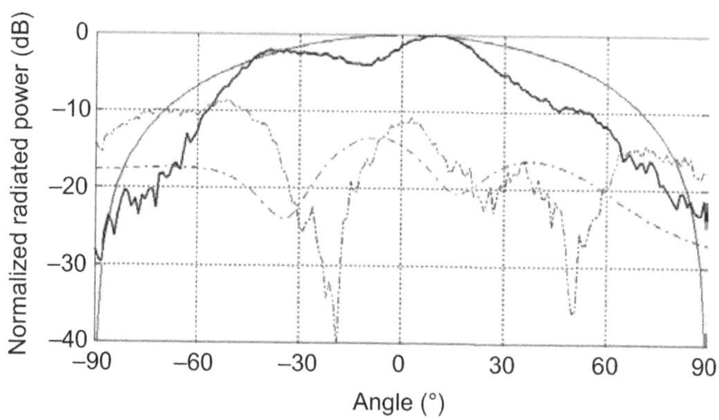

Figure 5.113: H-plane radiating patterns at 38 GHz for the passive antenna array printed on an alumina substrate.

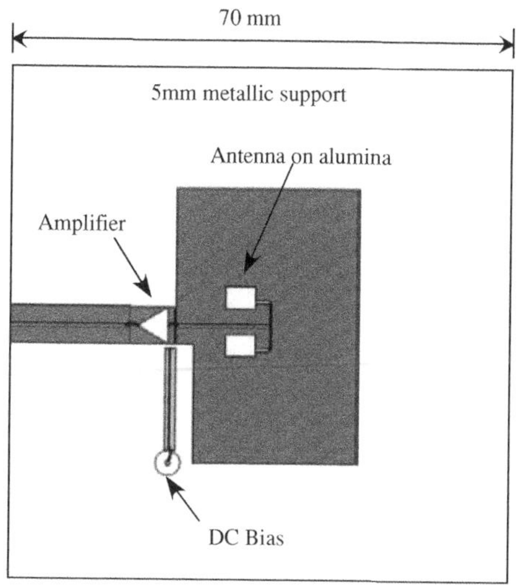

Figure 5.114: Active antenna array at 38 GHz printed on an alumina substrate; global design with amplifier.

asymmetry in the measurement results of the E-plane. As for the H-plane, some diffraction-linked disturbances could also be observed in the form of ripples. The measured gain is very poor; barely 2 dB. Later, an active array with an amplifier was studied so as to improve radiating patterns and gain level. Such a prototype is depicted in Figure 5.114. The radiating pattern of this active antenna is plotted in Figure 5.115 for the E-plane and in Figure 5.116 for the H-plane, and compared with the theoretical results of the passive array. The experimental results proved that the E-plane radiating pattern is symmetric. The cross-polarization level is lower for this active device, particularly in the E-plane (−15 dB compared with −8 dB for

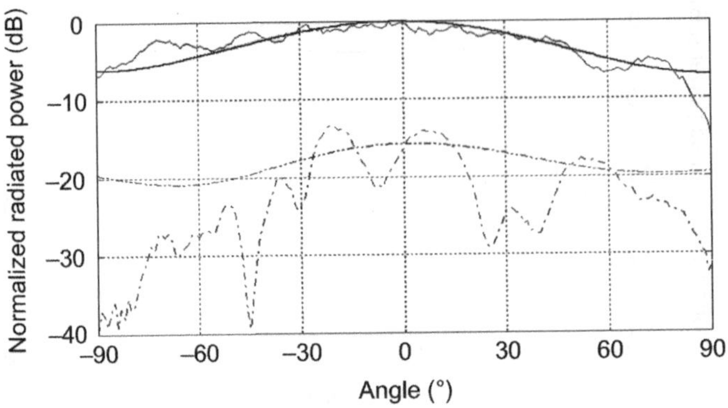

Figure 5.115: E-plane radiating patterns at 38 GHz for the active antenna array printed on an alumina substrate.

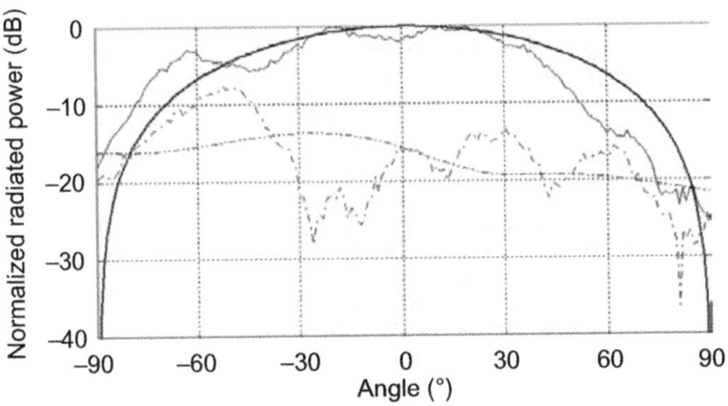

Figure 5.116: H-plane radiating patterns at 38 GHz for the active antenna array printed on an alumina substrate.

the passive antenna). These results are very telling and prove the advantage of active antennas (over passive arrays) when the active circuit is very close to the antenna.

In this final example, some interesting results are presented, concerning active printed arrays and the purity of their radiating patterns. First, let us consider a V-band multilayer printed antenna array, the objective being to optimize the integrated V-band active feed for satellite antennas. A passive array has already been described in Figure 5.52. Figure 5.117 features the passive prototype and Figure 5.118 pictures the active array. In either case, multilayer technology has been employed, using the H-shape slot coupling between both the feeding line layer and the radiating array layer. The objective here was to show the impact of diffraction effects on the V-connector frame which was used to measure the passive prototype. The radiating patterns of both the passive and active antenna arrays have been plotted and compared. The H and E-planes radiating patterns at 48.7 GHz for the passive antenna are shown in Figures 5.119 and 5.120. It appears that while side lobe and cross-polarization

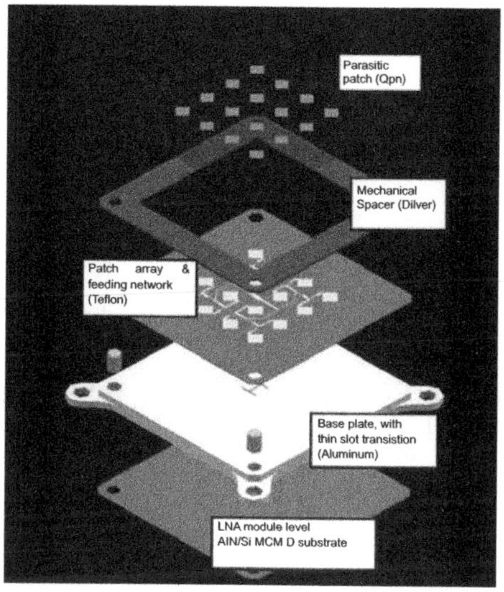

Figure 5.117: Printed multilayer antenna array in the Q band with stacked patches; global design.

Figure 5.118: Printed multilayer antenna array in the Q band with stacked patches; photo of the active prototype.

levels are quite satisfactory in the H-plane, there still remains a serious problem caused by diffraction in the E-plane and a very significant increase of the side lobe level for $\theta = 80^\circ$ tilt angle shows. The problem clearly finds its origin in diffraction on the mechanical support of the V-connector, needed to feed the antenna array during measurements.

Finally, an active antenna was made [44]. It combined a passive array and some active devices (MMIC amplifier, filter and MEMS switch). In this case, the V-connector was included in a global antenna packaging, allowing diffraction effects to be reduced.

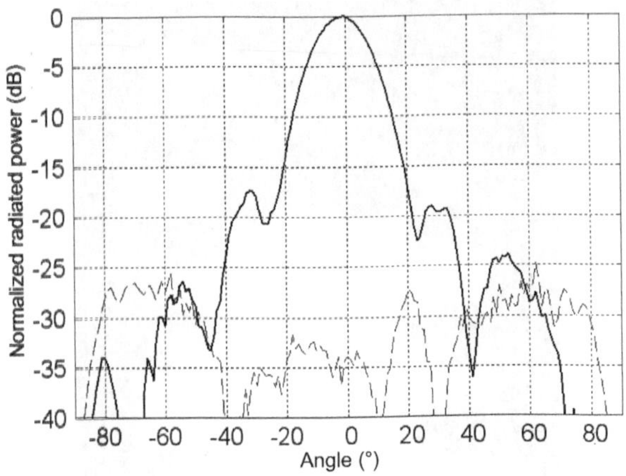

Figure 5.119: Printed multilayer antenna array in the Q band with stacked patches. H-plane co-polarization (solid curve) and cross-polarization (dotted curve) radiating patterns at 48.7 GHz for the passive prototype.

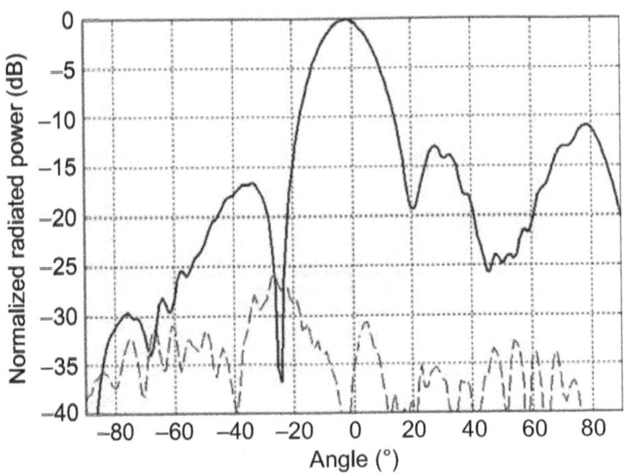

Figure 5.120: Printed multilayer antenna array in the Q band with stacked patches. E-plane co-polarization (solid curve) and cross-polarization (dotted curve) radiating patterns at 48.7 GHz for the passive prototype.

The radiating patterns have been plotted in Figures 5.121 and 5.122 for the E and H-planes and compared with the theoretical results (CST Microwave Studio) [45]. Very good agreement could be observed between simulation and measurement. In fact, diffraction effects, previously reported for the E-plane in the case of a passive antenna array, could no longer be seen with this new active array.

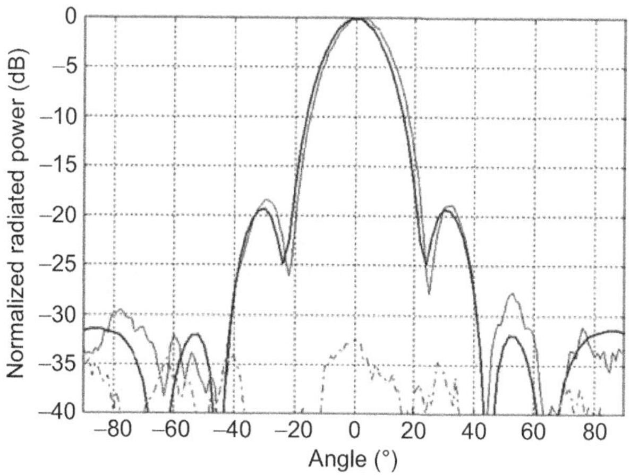

Figure 5.121: Printed multilayer antenna array in Q band with stacked patches. H-plane co-polarization (solid curve) and cross-polarization (dotted curve) radiating patterns at 48.7 GHz for active prototype.

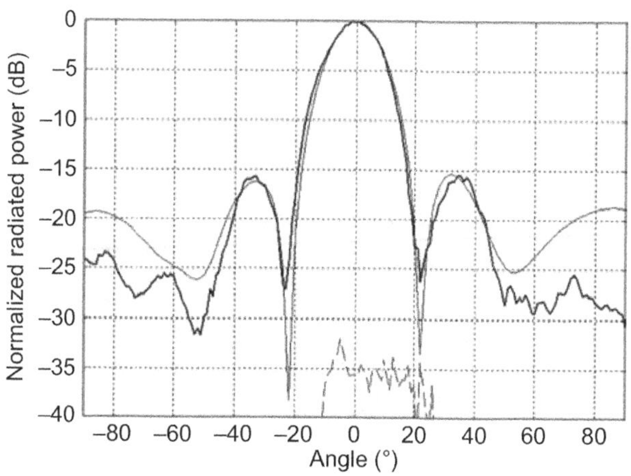

Figure 5.122: Printed multilayer antenna array in Q band with stacked patches. E-plane co-polarization (solid curve) and cross-polarization (dotted curve) radiating patterns at 48.7 GHz for active prototype.

5.5 Conclusion

In this chapter, we have a described the problems when printed antennas are designed in the mmWave frequency range. For example, efficiency can become poor due to dielectric and metallic loss. The characterization of substrates becomes essential in this frequency range in order to obtain a good estimation of efficiency. Moreover, the radiation effects of the feeding line network can include very disadvantageous effects in terms of side lobe or cross-polarization levels.

The advantages to multilayer technology can be shown to reduce spurious radiating effects. Moreover, the multilayer technologies are well suited to the active antenna concept, because they allow the active layer to be separated from the radiating layer. Different multilayer transitions via the coaxial line or slots are described to couple energy between lines and radiating elements.

Different examples of multilayer printed array have been demonstrated to obtain directive, sectorial or cosecant patterns. We have also shown that it is possible to design high gain printed antennas' up to 60 GHz. Multibeam antennas based on the Butler matrix have also been optimized with the objective of obtaining several different beams for an indoor communications system or a vehicle radar.

Acknowledgments

The authors take this opportunity to thank Dr Y. Cailloce (Alcatel Alenia Space) and Dr M. El Haj Sleimen for their contributions to this research and for the work on passive and active millimeter printed antenna arrays in the IETR laboratory from 1993 to 1999.

References

[1] HP Application 1217-1. 'Basics of measuring the dielectric properties of materials'.

[2] A. M. Nicolson and G. F. Ross, 'Measurement of the intrinsic properties of materials by time-domain technique', *IEEE Trans. Instrum. Measmt* **19**(4) (1970), pp. 377–382.

[3] W. B. Weir, 'Automatic Measurement of complex dielectric constant and permeability at Microwaves frequencies', *Proc. IEEE* **62**(1) (1974), pp. 33–36.

[4] A. L. Cullen and P. K. Yu, 'The accurate measurement of permittivity by means of an open resonator', *Proc. Roy. Soc. A.* **325** (1971), pp. 493–509.

[5] M. N. Afsar, 'Dielectric Measurement of Millimeter-Wave Materials', *IEEE Trans. Microw. Theory and Techq.* **32** (1984), pp. 1598–1609.

[6] S. Trabelsi and S. O. Nelson, 'Free space measurement of dielectric properties of cereal grain and oilseed at microwaves frequencies' (Bristol: Institute of Physics Publishing, 2003).

[7] D. K Ghodgaonkar and V. V. Varadan, 'A free space method for measurement of dielectric constants and loss tangents at microwave frequencies', *IEEE Trans. Instrum. and Measmnt* **37**(3) (1989), pp. 789–793.

[8] M. El Haj Sleimen, 'Etude de réseaux d'antennes imprimées en millimétrique', PhD Thesis, University of Rennes 1, 1999.

[9] O. Lafond, 'Conception et technologies d'antennes imprimées multicouches à 60GHz', PhD Thesis, University of Rennes 1, 2000.

[10] D. M. Pozar, 'Rigorous closed form expressions for the surface wave loss of printed antennas', *Electron. Lett.* **26**(13) (1990), pp. 954–956.

[11] A. Van De Capelle 'Microstrip antennas and arrays', Int. Res. Report, K. U. Leuven, October 1988.

[12] O. Lafond and M. Himdi, 'Substrates characterisation (ϵ_r, tan δ) up to millimeter-wavelength', in *Proc. ANTEM 2004*, pp. 20–23, Ottawa (Canada), July 2004.

[13] Ansoft Designer V2.0, Ansoft Corporation.

[14] ACE Deliverable 2.1 – D3, 'Report on facilities assessment', Document No FP6 – IST 508009/2.1 – D3, Chap. 4, p. 15.

[15] Y. Cailloce, 'Antennes actives et réseaux d'antennes en millimétrique', PhD Thesis, University of Rennes 1, 1997.

[16] R. Sauleau, 'Etude de résonateurs de Perot Fabry et d'antennes imprimées en ondes millimétriques. Conception d'antennes à faisceau gaussien', PhD Thesis, University of Rennes 1, 1999.

[17] T. Teshirogi, M. Tanaka and W. Chujo, 'Wideband circularly polarized array antenna with sequential rotations and phase shift of Elements', in *Proc. ISAP'85*, Tokyo, Japan, pp. 117–120.

[18] M. El Haj Sleimen, M. Himdi and J. P. Daniel, 'Dicontinuity effects in printed arrays', *Microw. and Opt. Technol. Lett.* **21**(3) (1999), pp. 226–229.

[19] M. Himdi, O. Lafond, S. Laignier and J. P. Daniel, 'Extension of cavity method to analyse aperture coupled microstrip patch antenna with thick ground plane', *Electron. Lett.* **34**(16) (1998), pp. 1534–1536. 1998.

[20] O. Lafond, M. Himdi and J. P. Daniel, 'Aperture coupled microstrip patch antenna with thick ground plane in millimeter-waves', *Electron. Lett.* **35**(17) (1999), pp. 1394–1396.

[21] European patent, EP0149394A3, 'Process for making a panel made of polypropylene as basis component, which has several metallic layers, and panel made according to this process'.

[22] French Patent no 2753724, 'Procédé de traitement de surface d'un objet en Polyméthyl-Pentène en vue de sa métallisation', March 1998.

[23] M. El Haj Sleimen, M. Himdi, J. P. Daniel, N. Haese and P. A. Rolland, 'Millimeter-wave array fed through thick slots filled with dielectric', *Microw. and Opt. Technol. Lett.* **22**(1) (1999), pp. 51–53.

[24] O. Lafond, M. Himdi, J. P. Daniel and N. Haese-Rolland, 'Microstrip / thick-slot / microstrip transitions in millimeter-waves', *Microw. and Opt. Technol. Lett.* **34**(2) (2002), pp. 100–103.

[25] O. Lafond, M. Himdi, O. Vendier and Y. Cailloce, 'Thick-slot transition and antenna arrays in the *Q* band', *Microw. and Opt. Technol. Lett.* **44**(1) (2005), pp. 24–29.

[26] D. M. Pozar, 'Microstrip antenna aperture coupled to a microstrip line', *Electron. Lett.* **21**(2) (1985), pp. 49–50.

[27] M. Himdi, 'Analyse et synthèse d'antennes imprimées alimentées par fentes, application aux réseaux', PhD Thesis, University of Rennes 1, 1990.

[28] C. Loyez, N. Haese, O. Lafond, P. Lefevre, G. Lewandowski and P. A. Rolland, 'Indoor propagation channel considerations in 60 GHz high data rate communications', *Wireless 2000*, Paris, October 2000.

[29] O. Lafond, M. Himdi and J. P. Daniel, 'Cavity method to analyse aperture coupled patch antenna with a tilted feeding line and circularly polarisation. Extension to millimeter-waves (60 GHz)', Cost 260: Smart antennas, Aveiro, Portugal, 2–5 November 1999.

[30] O. Lafond, M. Himdi and J. P. Daniel, 'Thick slot coupled printed antennas arrays for a 60 GHz indoor communication system', *Microw. Opt. Technol. Lett.* **25**(2) (2001), pp. 105–108.

[31] M. Himdi, 'Conception et technologies des antennes imprimées', HDR, University of Rennes 1, January 2000.

[32] M. Boguais, J. P. Daniel and C. Terret, 'Deux méthodes de synthèse de réseaux d'antennes, application aux antennes imprimées', In *Proc. JINA*, Nice, France, pp. 310–311, 1986.

[33] G. B. Dantzig and W. Orchard-Hays, 'The product form for the inverse in the simplex method', *Mathematical Tables and other Aids to Computation*, April 1954, pp. 64–67.

[34] W. Rotman and R. F. Turner, 'Wide angle microwave lens for line source application', *IEEE Trans.* **11** (1963), pp. 623–632.

[35] E. Rausch, A. Peterson and W. Wiebach, 'A low-cost high performance electronically scanned MMW antenna', *Microw. J.* **40**(1) (1997), pp. 20–32.

[36] H. H. Fuchs and D. Nubler, 'Design of Rotman lens beam-stearing of 94 GHz antenna array', *Electron. Lett.* **35**(11) (1999), pp. 854–855.

[37] European Project IST-2000-28276, 'MIPA: MEMS based integrated phased array antennas', Alcatel Space, BOSCH GMBH, IMEC, Coventor, CNES, IETR.

[38] X. Wu, G. V. Eleftheriades and T. E. Van Deventer-Perkins, 'Design and characterization of single- and multiple-beam mm-wave circularly polarized substrate lens antennas for wireless communications', *IEEE Trans. Microw. Theory and Techq.* **4**(3) (2001), pp. 431–441.

[39] US patent 5 781 163, 'Low profile hemispherical lens antenna array on a ground plane'.

[40] W. Menzel and R. Leberer, 'Folded reflectarray antennas for shaped beam applications', in *Proc. of EuCAP2006*, Nice, France.

[41] J. Butler and R. Lowe, 'Beam forming matrix simplifies design of electrically scanned antennas', *Electron. De.* **9** (1961), pp. 170–173.

[42] O. Lafond and M. Himdi, 'Multibeam antenna in millimeter-waves', in *Proc. of the 32nd European Microw. Conf. (EuMC)*, Milan, September 2002.

[43] Y. Cailloce, 'Antennes actives et réseaux d'antennes en millimétrique', PhD Thesis, University of Rennes 1, March 1997.

[44] O. Vendier *et al.*, 'Use of RF MEMS and micromachined parts to realize highly integrated V band active feed for satellite antennas', *Europ. Microw. Week*, Manchester (UK), 10–15th September 2006.

[45] L. Le Coq, G. Godi, B. Fuchs, O. Lafond, R. Sauleau and M. Himdi, 'Far-field millimetric band antenna test facility: positioning procedure using phase measurements', in *Proc. of Europ. Conf. on Antennas and Propagation*, Nice, France 2006.

6

Planar Waveguide-type Slot Arrays

Jiro Hirokawa and Makoto Ando

6.1 Introduction

A planar waveguide-type slot array antenna is an attractive candidate for millimeter applications requiring high gain because the waveguide has small transmission loss and the slot manufacturing is cost effective. The loss in the feeding line determines the antenna efficiency, while the choice of radiating elements as well as waveguide structure affects the manufacturing cost. The waveguide-type slot array keeps high efficiency in the high gain range. However, waveguide-type slot arrays have not been used commercially with the exception of a few military applications. The key problem has been the high manufacturing cost of the waveguides. We have developed several types of single-layer waveguide array that realize high efficiency and mass producibility. This chapter presents the development of our planar waveguide-type slot antennas.

Section 6.2 discusses the equivalent length of a round-ended straight slot because slots in waveguide arrays usually have semicircular ends (round-ended slots) [1,2] and finite wall thickness in milling or etching, which are not negligible especially in the millimeter-wave (mmWave) band. A new rectangular slot approximation of a round-ended slot is derived as a linear combination of conventional equal-area and equal-perimeter approximation. The new approximation is accurate to the order of 0.25% for a wide variety of parameters such as slot width-to-length ratio, wall thickness and dielectric constant of the filling material inside the waveguide. This is particularly useful since a rectangular slot allows us to use conventional methods of moment (MoM) [3].

Section 6.3 presents the alternating-phase fed single-layer slotted waveguide arrays [4,5] with chokes. The unique advantage of the alternating-phase fed array is that the electrically

Advanced Millimeter-wave Technologies: Antennas, Packaging and Circuits
Duixian Liu, Ulrich Pfeiffer, Janusz Grzyb and Brian Gaucher © 2009 John Wiley & Sons, Ltd

tight contact between the slot plate and the narrow walls of the waveguides on the base plate is not required in principle because adjacent waveguides are excited in alternating phase and because electric currents on the narrow walls do not flow across the contacts [5–7]. Furthermore, a choke structure is introduced along the periphery of the aperture in order to eliminate the energy leakages at the periphery, where the above condition does not hold [8]. Therefore, the alternating-phase fed array can be mass-produced at low cost just by fixing the slotted plate tacked on the feed structure with screws at the periphery.

Section 6.4 deals with the center feed single layer slotted waveguide array [9]. By feeding the array not from its end, but from the center of the waveguide array, the long line effect is halved and bandwidth enhancement is expected. The symmetrical structure of the center feed would also contribute to the main beam staying in the boresight. One difficulty of the center feed array is the blockage occupied with the feed waveguide in the center of the array aperture which results in reduction of efficiency as well as the high sidelobes [10]; sidelobe suppression by controlling slot excitation is also conducted by using genetic algorithm.

Section 6.5 shows a simple, compact and low-loss Butler matrix using a single layer and hollow waveguide at 22 GHz. The full single-layer configuration is realized by short slot couplers for both hybrids and crossing junctions. The propagation loss in air is remarkably high in the mmWave band. Base station antennas for mobile communications using high frequency require higher gain. The combination of a beam-switching circuit and a slotted waveguide array is good for this purpose, and could help to simplify adaptive signal processing. The integration of a beam-switching circuit and a slotted waveguide array in a body is also discussed. This integration reduces the connection loss between them.

Section 6.6 presents the application of the radial line slot antenna (RLSA) [11] to higher frequency up to 60 GHz. It also presents the design and the characteristics of a conical beam RLSA for 60 GHz band wireless local area network (LAN) systems. RLSA has a very simple structure fed by a point at the center of a radial line.

Section 6.7 covers the post-wall waveguide-fed parallel plate slot array antennas [12]. The post-wall waveguide is an array of metalized via-holes with a narrow spacing in a grounded dielectric substrate. The antenna can be produced easily at low cost by conventional printed circuit board (PCB) fabrication techniques such as via-holing, metal plating and etching. We investigate the dependence of the antenna efficiency on the size (or the gain).

Section 6.8 presents transformers to connect a 60 GHz cost-effective module and a post-wall waveguide array antenna. We focus upon a transition between a coaxial line and a post-wall waveguide. The structure of the transformers is well known; however, confirmation of practical realization in the mmWave band is required. The discussion on the fabrication tolerance is an important issue in the mmWave band.

6.2 Equivalent Length of a Round-ended Straight Slot

6.2.1 Waveguide with a Round-ended Slot

Figure 6.1(a) shows a waveguide with a round-ended straight slot. A round-ended, wide straight slot which has semi-circular ends is cut in the broad wall (thickness t) of an infinitely long rectangular waveguide (width a, height b). The waveguide is filled with dielectric material (dielectric constant ε_r), while the slot and the external regions are hollow. An infinite ground plane is embedded so as to regard the external region as a half-free space. The slot

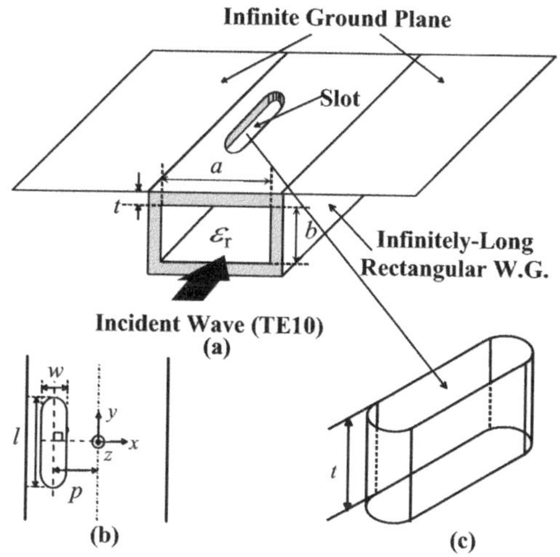

Figure 6.1: Analysis model and antenna parameters: (a) waveguide with a round-ended slot; (b) slot configuration; (c) round-ended slot-shaped waveguide (from reference [13], reproduced by permission of © 2004 IEICE).

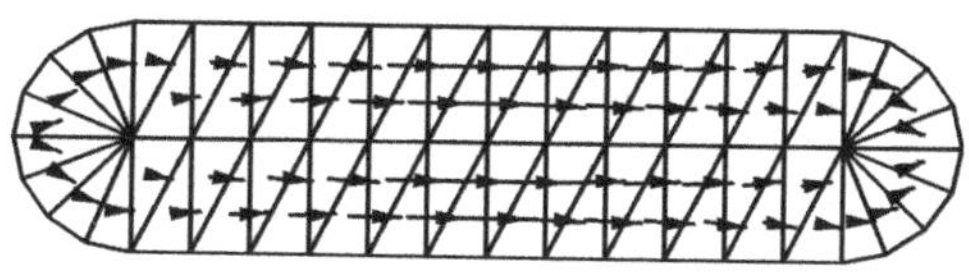

Figure 6.2: FEM meshes and magnetic current basis function for the round-ended slot (from reference [13], reproduced by permission of © 2004 IEICE).

configuration is shown in Figure 6.1(b). The slot length, width and offset from the waveguide axis are l, w and p, respectively. Figure 6.1(c) shows a round-ended slot-shaped waveguide which is a waveguide with the cross section of the round-ended slot.

6.2.2 Comparison Between Calculation and Measurement

The MoM/finite-element method (FEM) analysis is conducted for the model shown in Figure 6.1 [13, 14]. The magnetic current basis function on the aperture of a round-ended slot ($w/l = 0.25$) and the FEM meshes are shown in Figure 6.2. The division number in the semicircular section is eight, and the inner rectangular section is divided into two along the width and 12 along the length. The frequency characteristics of S_{21} for different slot width-to-length ratios w/l of 1/16, 1/8 and 1/4 ($l = 40$ mm) are shown in Figure 6.3 for the model whose antenna parameters are listed in Table 6.1. The phase reference plane of S_{21} is under the slot center. Even for an extremely wide slot of $w/l = 1/4$, the calculated results coincide well with the measured ones. For a fixed l, the bandwidth becomes wider and the resonant frequency becomes higher as w becomes larger.

Table 6.1: Antenna parameters 1.

a (mm)	b (mm)	t (mm)	l (mm)	w (mm)	p (mm)	ε_r
58.1	29.1	1.6	40	10	15	1

Table 6.2: Antenna parameters 2.

a (mm)	b (mm)	t (mm)	l (mm)	w (mm)	p (mm)	ε_r
19.05	9.525	1.27	10	2.5	5	2.08

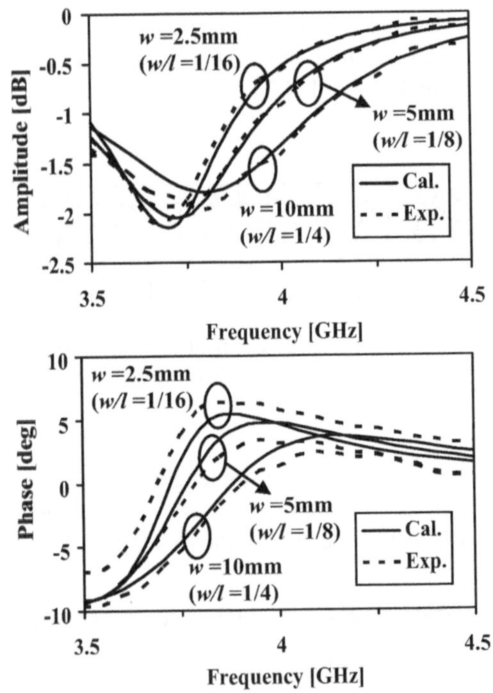

Figure 6.3: Frequency characteristic of S_{21} of the model with antenna parameters listed in Table 6.1 with variation of w/l ($l = 40$ mm) (from reference [13], reproduced by permission of © 2004 IEICE).

The frequency characteristics of S_{21} for a thick waveguide wall ($t = 11.6$ mm, $0.155\lambda_0$ where λ_0 is a free space wavelength) are also shown in Figure 6.4 for the model listed in Table 6.1. The calculated result coincides well with the measured one. For comparison,

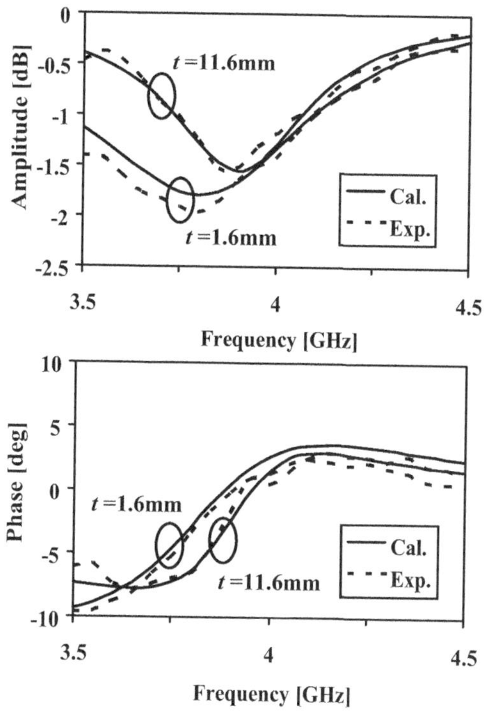

Figure 6.4: Frequency characteristic of S_{21} of the model with antenna parameters listed in Table 6.1 with variation of t (from reference [13], reproduced by permission of © 2004 IEICE).

the result for $t = 1.6$ mm is also shown in Figure 6.4. For frequencies below the resonance, the amplitude of S_{21} of a thick waveguide wall is larger than that of a thin waveguide wall because the wall thickness region, which is regarded as a waveguide, is cutoff. The effect of the wall thickness is accurately taken into account in the analysis.

6.2.3 Equal-area and Equal-perimeter Rectangular Slots for a Round-ended One

As mentioned above, rectangular slots with relatively large width can be analyzed accurately by conventional MoM [3]. Therefore, it is worth deriving the equivalent rectangular slot which is electrically equivalent to the round-ended slot. First, two types of slot are considered. Equal-area and equal-perimeter rectangular slot are shown in Figure 6.5(b) and (c), respectively. The equivalent lengths for these slots are expressed as follows.

$$
\begin{aligned}
l_a &= l - \left(1 - \frac{\pi}{4}\right) w \\
l_p &= l - \left(2 - \frac{\pi}{2}\right) w
\end{aligned}
\tag{6.1}
$$

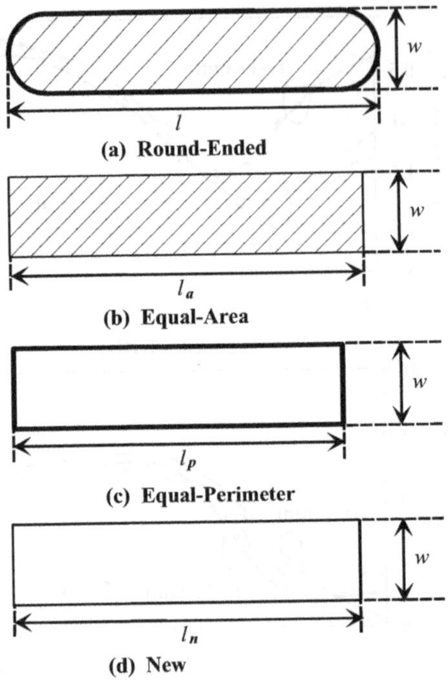

Figure 6.5: Equivalent rectangular slots for a round-ended slot (from reference [13], reproduced by permission of © 2004 IEICE).

where l_a and l_p are lengths of the equal-area and equal-perimeter slots, respectively. Slot lengths l_a and l_p are defined so as to make the area of Figure 6.5(b) and perimeter of Figure 6.5(c) equal to those of Figure 6.5(a), respectively.

The equivalence of two models is investigated by applying MoM/FEM. The resonant frequencies of two conventional rectangular slots, equal-area and equal-perimeter, are compared with that of a round-ended slot. The differences between the resonant frequencies of two rectangular slots (f_{reso}) and that of a round-ended slot (f^r_{reso}), all of which are calculated by MoM/FEM, are plotted in Figure 6.6 as a function of slot width-to-length ratio w/l for the model with antenna parameters listed in Table 6.1. They are normalized by f^r_{reso}. As the slot width increases, the resonant frequency of the equal-perimeter slot increases over that of the round-ended slot, while that of the equal-area slot decreases. Equal-area slots always provide a better approximation than the round-ended ones. However, it is also noted that for w/l larger than 0.2, the error becomes notable and about 0.5%.

Figure 6.7 shows the differences of resonant frequencies of the equal-area and equal-perimeter slots from that of the round-ended one as a function of dielectric constant ε_r in the waveguide. For each dielectric constant, the slots are redesigned with constant offset of 4.5 mm and constant w/l of 1/4 so that the resonant frequency is in the operating frequency range of the 12 GHz standard waveguide filled with dielectric material in Table 6.2. Slot lengths l are listed in Table 6.3 for each examined dielectric constant ($\varepsilon_r = 1, 2, 5, 8, 10$). For wide variation of the inner dielectric constant from 1 to 10, the equal-area slot is a reasonable

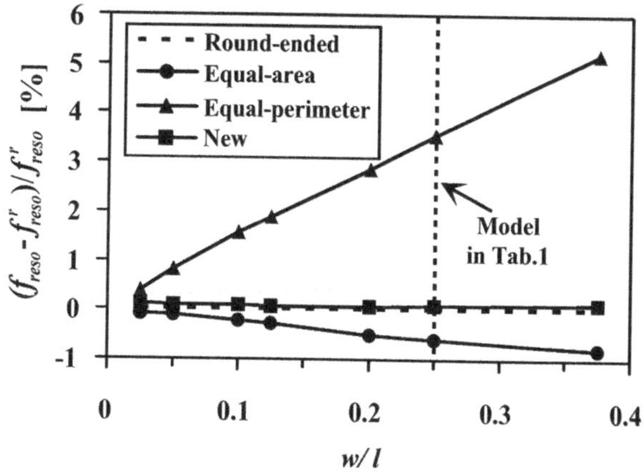

Figure 6.6: Differences of resonant frequencies as a function of w/l (only w is varied in Table 6.1, $p = 15$ mm) (from reference [13], reproduced by permission of © 2004 IEICE).

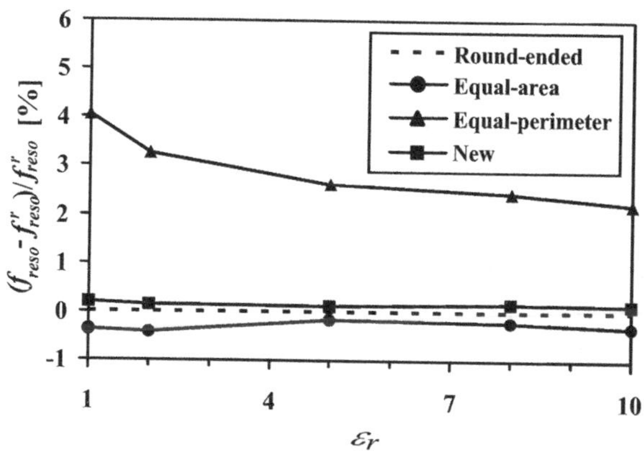

Figure 6.7: Differences of resonant frequencies as a function of ε_r ($w/l = 1/4$, $p = 4.5$ mm, a, b, t are listed in Table 6.2 and l is listed in Table 6.3) (from reference [13], reproduced by permission of © 2004 IEICE).

Table 6.3: Slot parameters.

ε_r	1	3	5	8	10
l (mm)	12	16	16	20	20

approximation for the round-ended slot. The difference in resonant frequencies of the equal-area slot is less than 0.5% for $w/l = 1/4$.

6.2.4 New Definition of an Equivalent Rectangular Slot

Here, a more accurate equivalent rectangular slot approximation, referred to as new slot, is proposed for practical application. The following two findings are utilized: (1) as illustrated in the above-mentioned figures, the approximation accuracy is determined mainly by slot width-to-length ratio w/l; (2) it is observed from Figure 6.6 that the errors of both equal-area and equal-perimeter slots are almost linearly dependent upon w/l. These suggest that the linear combination of Equation (6.1) gives a new rectangular slot approximation for the round-ended slot, the width of which is the same as the round-ended slot:

$$l_n = \frac{6l_a + l_p}{7} = l - \frac{2(4-\pi)}{7}w \tag{6.2}$$

The resonant frequencies calculated by MoM/FEM analysis for this new rectangular slot approximation are included in Figure 6.6 by notation 'new'. The error in terms of MoM/FEM is less than 0.25% for all the combinations of a, b, p, t, w/l and ε_r. Almost perfect equivalence to the round-ended slot is observed. Good results are also obtained for dielectric filled waveguides in Figure 6.7.

6.3 Alternating-phase Fed Single-layer Slotted Waveguide Array and its Sidelobe Suppression

6.3.1 Alternating-phase Fed Arrays

Figure 6.8 presents the structure of the alternating-phase fed single-layer slotted waveguide array, which consists of two parts: a slot plate as the broad wall and a corrugated base plate which accommodates radiating and feed waveguide in the same layer. The frame is used to reinforce screw mounting of the slot plate over the base plate. The feed waveguide consists of T-junctions and works as a multiple-direction power divider to the radiating waveguides. The spacing between adjacent coupling windows is a half guide wavelength to excite the adjacent radiating waveguides, which are 180° out of phase. The width of the feed waveguide is set so that the wavelength is twice that of the broad wall width of the radiating waveguides. Divided waves into the radiating waveguides are controlled by width and offset from the coupling windows in the T-junctions. The radiation pattern in the E-plane is associated with power-dividing distribution in the series of T-junctions in the feed waveguide into the radiating waveguides.

The radiating waveguides have a resonant shunt slot array on the broad wall [15]. Longitudinal slots with staggered offsets are arrayed at intervals of approximately a half of the guided wavelength, so that the slots are excited in-phase. Thus, all the slots over the aperture are excited in-phase. The radiation pattern in the H-plane of the alternating-phase fed array reflects the excitation distribution of the slot array and length and offset of each slot in the array are varied to synthesize the aperture distribution. The coupling strength of each slot is controlled mainly by the slot offset, while the slot length is set to be a resonant slot. The slot spacing along the waveguide axis is constant, as is usually the case with a resonant shunt slot array, to realize uniform phase distribution [15].

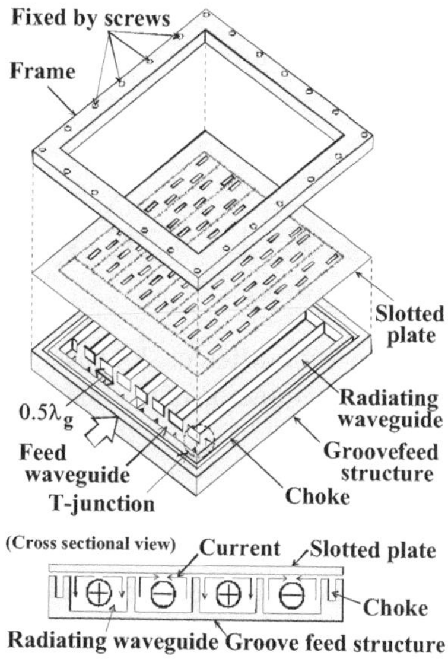

Figure 6.8: Configuration of an alternating-phase fed single-layer slotted waveguide array (from reference [16], reproduced by permission of © 2005 IEICE).

6.3.2 Array Design

6.3.2.1 Slot Design

For accurate slot design, it is important to take into account, the mutual couplings between slots in the external region since the slots on adjacent waveguides are arranged very closely in the alternating-phase fed array. The authors have established an approximate electro magnetic (EM) analysis model for slot coupling in the environment of uniform aperture illumination, where a combination of perfectly electrical conducting (PEC) walls and periodic boundary walls is placed in the external region to simulate the mutual couplings approximately [8]. The relationship among the slot offset, length and the coupling strength is analyzed by the MoMs for the model [17, 18] and is reserved for later use as a design chart. In the aperture distribution, the couplings are first assigned for each slot by an equivalent circuit of the array [19, 20]. Then offset and length of each slot required for the coupling are determined by using the design chart obtained in EM analysis. Finally, the slot offsets and lengths of the array are fine-tuned by taking the effects of finite size of the array into account. The detail of design of the slot array for uniform aperture illumination is described in the literature [8]. This design technique is extended to the aperture synthesis of Taylor distribution as well by simply changing the assignment of coupling strength of each slot in the equivalent circuit of the array. The details of the slot design for Taylor distribution is beyond the scope of this section.

An important design of the alternating-phase fed array is to suppress grating lobes in the diagonal planes, because the periodicity of the staggered slot arrangement in the diagonal directions becomes approximately one wavelength in the free space. This corresponds to the critical condition of whether or not the grating lobes are visible. The gain and the aperture

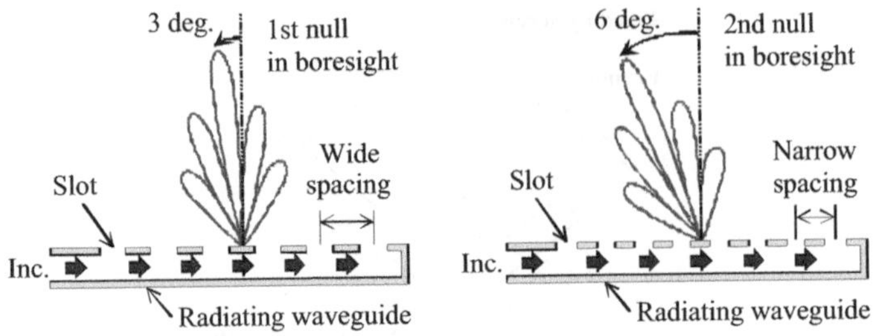

Figure 6.9: Backward beam tilting with different slot spacing for grating lobe suppression (from reference [16], reproduced by permission of © 2005 IEICE).

efficiency are degraded considerably by the grating lobes although their level seems not to be so high. Therefore, it is important to reduce the diagonal spacing. This is accomplished by introducing backward beam tilting, in which the main beam is slightly tilted toward the feed point from the boresight of the array, as shown in Figure 6.9. The beam-tilting technique [21] is often used for suppressing the reflection from the slots at the feed point. The slot spacing s is perturbed from the half of the guide wavelength according to the tilting angle through the following condition [22],

$$l_s = \frac{\lambda_g}{2} + \frac{m}{N}\lambda_g \tag{6.3}$$

where the tilting angle is specified by m, that is, the mth null of the radiation pattern of the array falls on the boresight, as shown in Figure 6.9; λ_g is the guide wavelength and N is the number of slots in the array. The negative number of m corresponds to the backward beam tilting and results in smaller spacing than a half of the guided wavelength. The backward beam tilting reduces the slot periodicity in the diagonal directions and then the grating lobes, appearing in the diagonal plane, are suppressed. The gain of the alternating-phase fed arrays with different beam tilting angles ($m = -1$ and $m = -2$) is compared.

6.3.2.2 Design of the Waveguide T-junctions

Here, the design of a unit waveguide T-junction as well as the feed waveguide for E-plane pattern synthesis is briefly summarized [23]. Figure 6.10 presents the analysis model for a unit T-junction [24, 25]. The structure is two-dimensional because it is uniform in a perpendicular direction. The incident wave propagating in the feed waveguide couples to the coupling window and is divided into the radiation waveguide. Amplitude and phase of the divided wave are controlled by width w and offset d of the coupling window, respectively. The reflected wave from the window is canceled out by an inductive wall installed in the feed waveguide. The position of the inductive wall (p and q) is optimized so that the reflection is minimized. The reflection from each T-junction is so negligible that traveling wave operation in the feed waveguide is expected.

Design of the T-junctions for Taylor distribution with -25 dB sidelobe level and $\bar{n} = 5$ is described [23]. Taylor distribution is widely used for high gain and low sidelobe antennas. The Taylor line source for the desired sidelobe level is synthesized by the Woodward–Lawson method [26]. In this case, 8 dB amplitude taper with constant phase distribution is desired

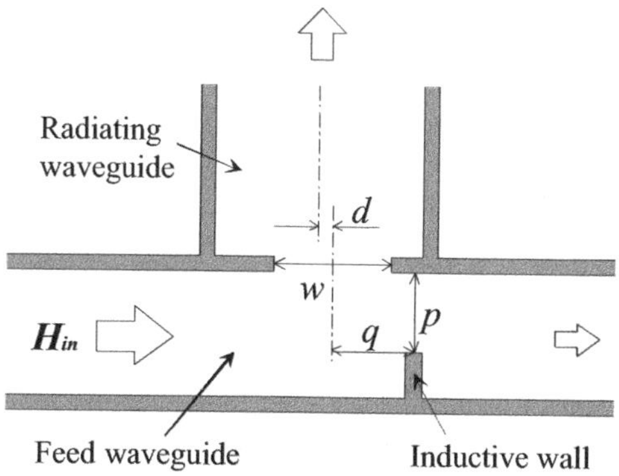

Figure 6.10: Analysis model for a unit T-junction (from reference [16], reproduced by permission of © 2005 IEICE).

from the center to the termination of the feed waveguide. The frequency is set to be 76.5 GHz. The widths of the broad wall of the feed waveguide and the radiating one are 2.46 mm and 2.53 mm, respectively. Height and thickness of the narrow walls separating the waveguides are 1.47 mm and 0.78 mm, respectively. Figure 6.11 (a) presents the specified coupling strength of each T-junction for the Taylor distribution, where 24 radiating waveguides are assumed. In this figure, the junction is numbered from the termination to the feed point and the results of 12 waveguides are presented, since the T-junctions are arrayed symmetrically with respect to the center. The coupling strength is gradually increased from the feed to the termination. Figure 6.11(b) presents variation of the window width w and the offset d of the T-junctions. The window width w varies corresponding to the coupling strength in Figure 6.11(a) while the offset d is chosen so that the phase of the divided waves in the adjacent waveguides may be 180° out of phase.

6.3.3 Measurements

6.3.3.1 Test Antennas

The test antennas consist of 24 radiating waveguides with a slotted area of 80 mm in the feed waveguide direction by 84 mm in the radiating waveguide direction [27]. The number of slots in a radiation waveguide depends upon the angle of backward beam tilting; 29 slots for 3° beam tilting ($m = -1$), while 31 slots for 6° beam tilting ($m = -2$). Common dimensions and parameters of the test antennas are summarized in Table 6.4.

Figure 6.12 presents a photograph of the test antenna. The material of the slotted plate and the groove feed structure is aluminum. The slotted plate with a thickness of 0.1 mm is processed by etching. The groove feed structure with a depth of 1.47 mm is manufactured by machining process. The manufacturing accuracy is specified within ±0.01 mm. Both parts are just tacked and screwed at the periphery of the antenna. A metal frame with a relatively large thickness is put on the slotted plate to realize close contact between the slotted plate and the feed structure around the choke, as shown in Figure 6.8. A WR-15 standard waveguide is connected to the back of the feed structure.

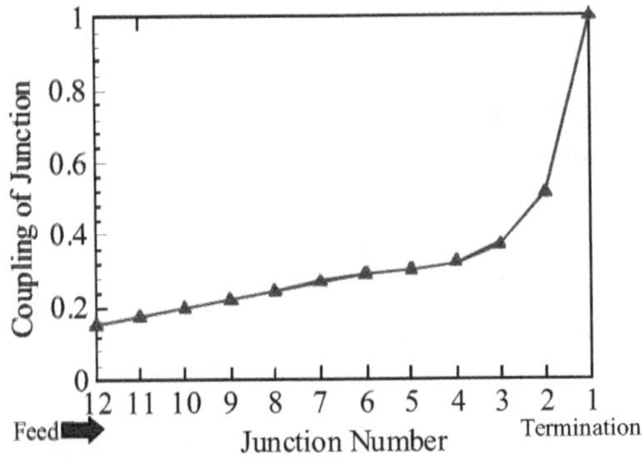

(a)

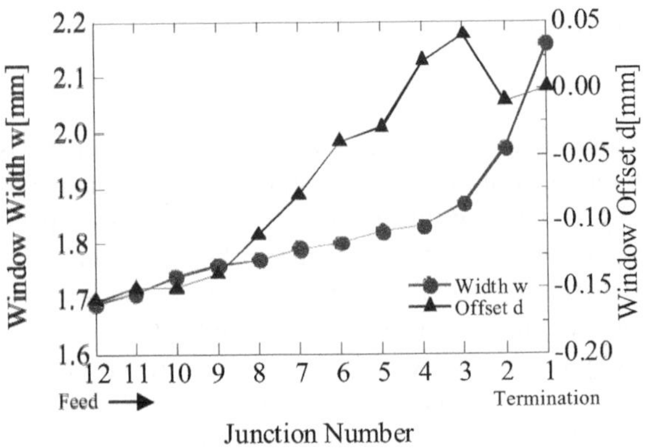

(b)

Figure 6.11: Design parameters of T-junctions for Taylor distribution with −25 dB sidelobes (a) coupling distribution; (b) window width w and window offset d (from reference [16], reproduced by permission of © 2005 IEICE).

Table 6.4: Parameters of the test antennas.

Design frequency	76.5 GHz	
Number of waveguides	24	
Width of radiating waveguides	2.53 mm	
Width of feed waveguide	2.46 mm	
Height of waveguides	1.47 mm	
Thickness of narrow walls	0.78 mm	
Thickness of a slotted plate	0.1 mm	
Direction of beam tilting	Backward	
Beam tilting angle	3°	6°
Number of slots	29	31
Slot spacing	2.88	2.69 mm
Aperture area	80 × 84 mm	

Figure 6.12: Photograph of the test antenna (from reference [16], reproduced by permission of © 2005 IEICE).

Gain and radiation patterns of the test antennas are measured in an anechoic chamber where a standard gain horn of 25 dBi is used as a reference antenna. Aperture field distributions are obtained by near field measurement. A truncated waveguide (WR-12) is used as a probe and is scanned at the height of 16 mm ($4\lambda_0$) over the aperture. The scanned area is 120 mm by 120 mm and the scanning step is 1.5 mm. The measured data is transformed from the scanning plane to the antenna aperture.

6.3.3.2 Performance of the Array with Uniform Aperture Distribution

Influences of loss due to the material of the antenna and the grating lobes due to the staggered slot arrangement on the gain and efficiency are discussed experimentally. Figure 6.13 presents frequency dependence of the measured gain and the corresponding efficiency of the test antennas designed for uniform aperture illumination. This figure also includes results of an antenna of stainless steel with the same design as that of three-degree beam tilting ($m = -1$) [23]. Compared with the antennas of aluminum and stainless steel with three-degree beam tilting, 34.2 dBi gain with 48% efficiency is obtained by the aluminum antenna while 32.6 dBi gain with 33% efficiency by the stainless steel antenna. The efficiency is increased by 15% by replacing the material to aluminum. The estimated losses of aluminum and stainless steel waveguide with the conductivities of 3.8×10^7 and 1.4×10^6 S m^{-1} are -0.24 and -1.26 dB, respectively, where 60 mm is used for the effective length of the waveguides in the antenna. The excessive loss of about 1 dB reasonably accounts for the measured antenna efficiency degradation of 15%.

The highest gain of 34.8 dBi with 57% efficiency is obtained by the aluminum antenna with six-degree beam tilting. This efficiency is comparable to the in-phase fed waveguide array processed by brazing contact and is two times higher than conventional typical planar arrays in the range of high gain over 30 dBi and high frequencies. The efficiency is enhanced by around 10% in comparison with the antenna of the same material with three degree beam tilting. Figure 6.13 compares the calculated gain and efficiency of the antennas with the different beam tilting angles of 3° and 6°, where the waveguide loss is ignored.

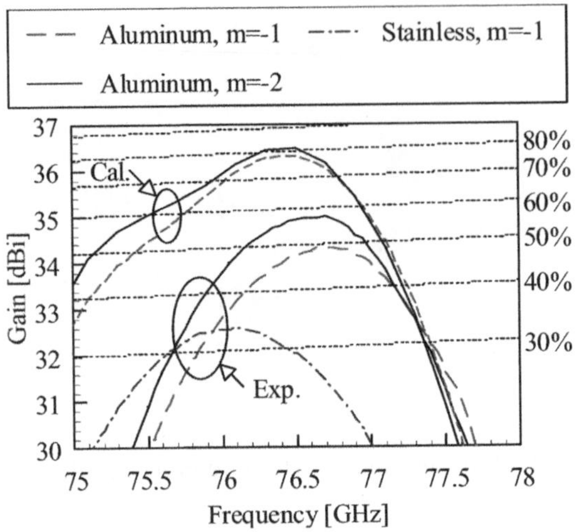

Figure 6.13: Measured gain and efficiency of the test antennas for uniform aperture distribution with 3° beam tilting ($m = -1$) and 6° beam tilting ($m = -2$) (from reference [16], reproduced by permission of © 2005 IEICE).

The discrepancy seems to be very small. It means that it is difficult to evaluate the small grating lobes accurately in the calculation, where the slots are approximated to small magnetic dipoles [28]. The influence of the grating lobes due to the staggered slots on gain and efficiency are compared quantitatively by the measurements.

The effects of misalignment of the slot plate in the single layer slotted waveguide arrays were discussed in reference [29]. Figure 6.14 presents predicted efficiency degradation due to the alignment error of the slotted plate in the X-direction. Taking 5% loss of the waveguides due to the conductivity of aluminum into account, the excessive loss of about 20% in measurement may be due partly to this alignment error of 0.06 mm or 1.5% of a wavelength at 76 GHz. Fine alignment accuracy is required at 76 GHz, as is usually the case with other mmWave antennas, although the need for electrical contact is relaxed.

The phase distribution is slightly tapered along the feed waveguide (X-axis). The operating frequency of the T-junctions in the feed waveguide is shifted to 76.75 GHz. This is because the slotted plate, just tacked and fixed by screws, slightly affects the operation of the T-junctions, while its effect is not taken into account in the design. This should be solved in the future.

The first sidelobe level is -13 dB in the radiation patterns of the uniform antenna with six-beam tilting in the H-plane at 76.5 GHz and in the E-plane at 76.75 GHz. The measured patterns and the calculated ones are in good agreement.

6.3.3.3 Performance of the Array with Taylor Distribution

Figures 6.15 present the radiation patterns of the Taylor antennas. The H- and E-plane patterns are measured at 76.5 and 76.75 GHz, respectively. Good agreement between the measured patterns and the calculated ones is observed in the figures. The measured sidelobe

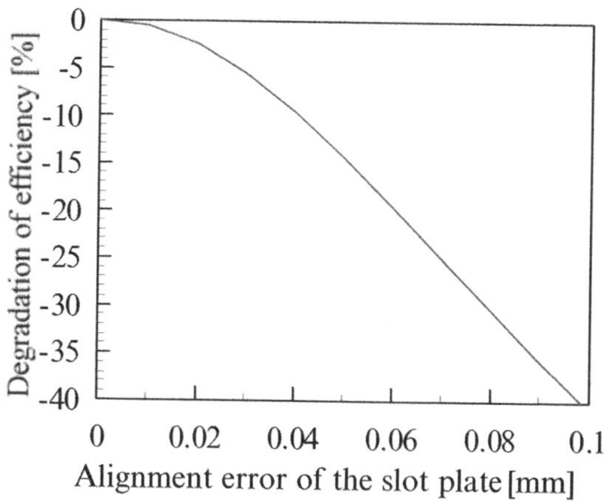

Figure 6.14: Degradation of efficiency due to alignment error of the slotted plate (from reference [16], reproduced by permission of © 2005 IEICE).

level is −20 dB on the H-plane and −23 dB in the E-plane of the antenna with 6° beam tilting. Even for the calculated H-plane patterns, the sidelobe level exceeds the specified one of −25 dB by a few decibels. This is because coupling strengths of slots near both ends of the waveguides are not evaluated accurately, where the analysis model for slot coupling uses the periodic and PEC conditions to simulate the external mutual couplings.

The measured reduction of gain and efficiency due to the tapered amplitude distribution is within 1 dB and 10%, respectively, in comparison with the uniform antennas. The difference in the efficiency due to the different beam tilting angles is 10% similarly, as was explained previously. Satisfactory performance of the sidelobe suppression as well as the high efficiency is confirmed by the test antennas for Taylor distribution.

6.4 Center Feed Single Layer Slotted Waveguide Array

6.4.1 Structure of a Center Feed Array

The overall configuration of the center-feed single-layer slotted waveguide array is explained in Figure 6.16. The feed waveguide, which works as a multiple-way power divider consisting of a cascade of waveguide cross-junctions, is installed at the center of the array. Each cross-junction has four inductive posts to control the amplitude and phase of the divided power [9]. An antenna input aperture is cut on the bottom at the center of the feed waveguide. The radiation waveguides are placed with a spacing of half the guide wavelength in the feed waveguide for excitation in the alternating phase. The radiating shunt slots are longitudinally spaced by a half the guide wavelength while they are transversally offset on both sides of the waveguide axis. Each radiating element consists of a slot and an inductive wall which cancels out the reflection from the slot; the radiation waveguide is free of cumulative reflection at the input even if the beam is directed at the boresight. EM design of a slot and an inductive wall is performed by MoM with uniform line current approximation on the inductive wall

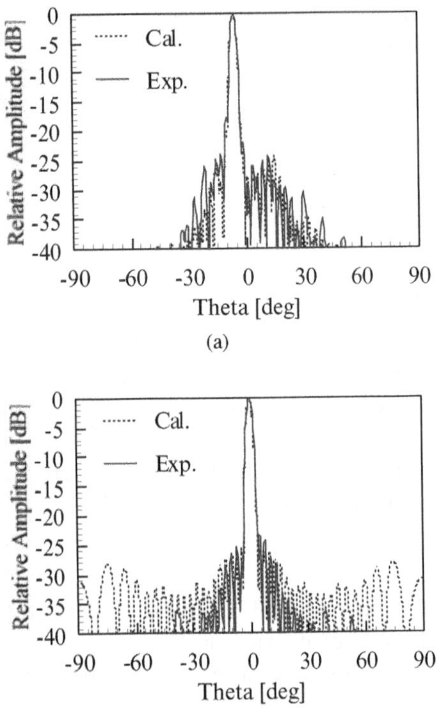

Figure 6.15: Radiation patterns of the test antennas with Taylor distribution (a) H-plane; (b) E-plane (from reference [16], reproduced by permission of © 2005 IEICE).

surface. A choke is accommodated in the periphery of the antenna, which prevents the leakage through the gap between the slot plate and the groove structure at the periphery of the array [8]. It dispenses the electrically perfect contact and the screws are used simply for assembly. The structure of center feed arrays has a higher degree of symmetry than the conventional end fed ones. One difficulty is the blocked area in the center of the aperture occupied with the cross junctions; the typical blockage is about 2.5 free space wavelength in width. The growing sidelobes as well as the efficiency decrease should be taken into account in the design.

6.4.2 Suppression of Sidelobes due to Aperture Blockage by Center Feed Waveguide

A center fed waveguide array typically has a blockage region at the center of the aperture with a width of 2.5 free space wavelengths. This give rises to high sidelobe levels. Figure 6.17 shows the prediction of antenna efficiency and the sidelobe levels as functions of array size. A square aperture ($2N \times 2N$) and magnetic dipole source model is used in the calculation. The sidelobe of a center fed waveguide array mainly depends on the ratio of array size and blockage region. It is required to increase the number of slots for some applications requiring

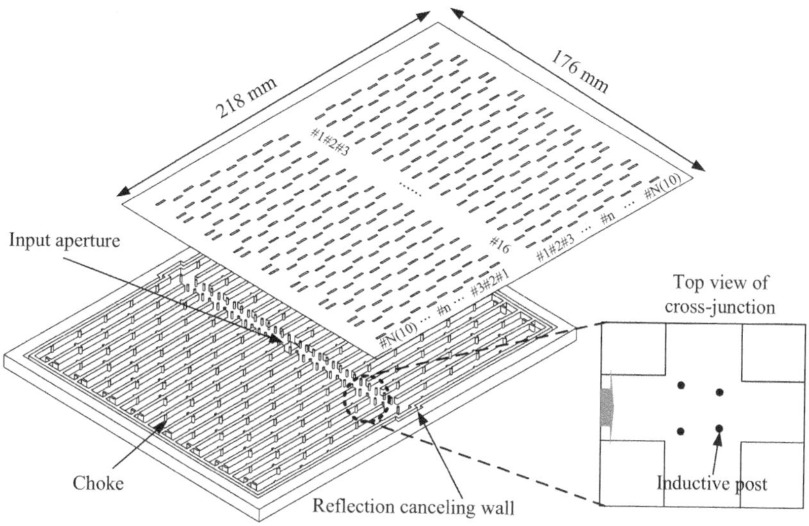

Figure 6.16: Center feed single-layer slotted waveguide array (cross-junction) (from reference [30], reproduced by permission of © 2006 IEEE).

lower sidelobe. The efficiency degradation as well as the sidelobe level growth is associated with the blocking area ratio. For example, the aperture 160 mm × 200 mm (16 × 20) in a 26 GHz band with the initial first sidelobe level of −13 dB for a 20-slot design of uniform illumination increases up to −9.4 dB with a 1.5 guided wavelength blockage as shown in Figure 6.17. Here, the first sidelobe level is suppressed by using a genetic algorithm [31]. In this step, only the excitation amplitudes are optimized for the half side array owing to symmetry of the structure. The genetic algorithm optimized a 10-element array with 6-bit amplitude accuracy. The maximum relative-sidelobe level is used for the cost function [32, 33]. Figure 6.18 shows the resulting radiation pattern after genetic algorithm optimization where the optimized excitation amplitude is given in Figure 6.19. The radiation pattern has a reduced sidelobe level of −16 dB. Figure 6.20 shows the slot and inductive wall parameters to realize desired excitation amplitude and reflection canceling.

6.4.3 Experimental Results

6.4.3.1 Reflection

A test antenna of the center fed waveguide arrays is designed and fabricated at 26 GHz band as shown in Figure 6.21. The antenna consists of two sets of radiation waveguides, one on each side of the feed waveguide. Each set consists of 16 radiation waveguides, and each radiation waveguide has 10 slots. The aperture area is around 160 mm × 180 mm. The overall reflection at the input port is as high as −10 dB at the design frequency of 25.3 GHz. They are in remarkable agreement although the structure is very large and complicated. This implies that the fabrication of parts, the slot plate, the base plate and the posts in cross junctions, as well as the contact by the choke with simple screws is accurate and works well. From the simulation, the high reflection comes mainly from the nonoptimized length of feeding

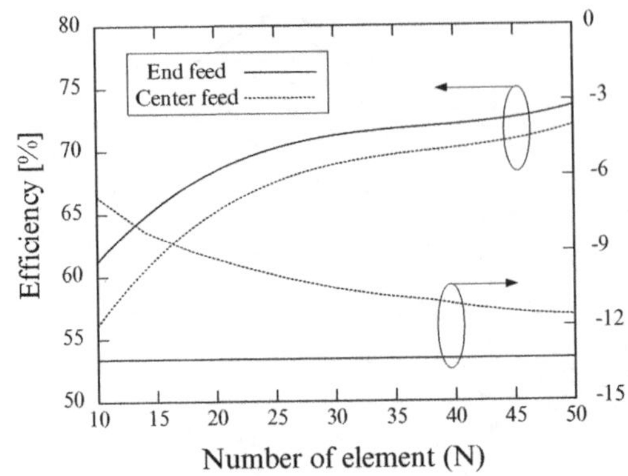

Figure 6.17: Efficiency and sidelobe level as function of antenna size (from reference [30], reproduced by permission of © 2006 IEEE).

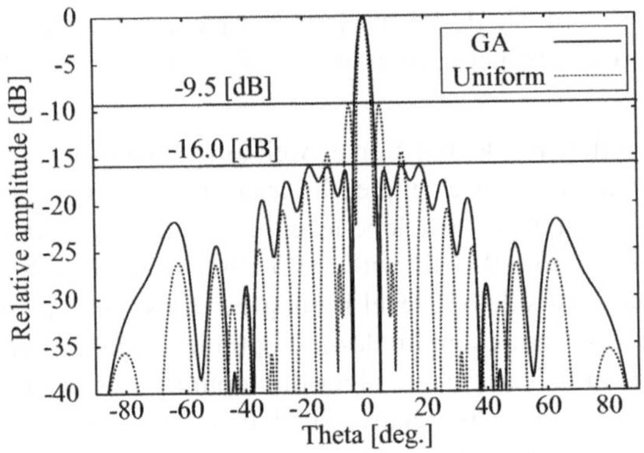

Figure 6.18: Suppressed sidelobe level by using genetic algorithm (from reference [30], reproduced by permission of © 2006 IEEE).

aperture at the antenna input. If the length of feeding aperture was changed from 5.9 mm to 5.0 mm, the reflection at 25.5 GHz would be minimized as is predicted by HFSS simulation. The inductive posts will also be replaced by walls to simplify fabrication in the near future.

6.4.3.2 Near Field Distribution

Figure 6.22 shows the two-dimensional amplitude distribution at 25.5 GHz. The feed waveguide with cascaded cross junctions is placed at the center along the vertical axis. As is expected, weak amplitude is observed at the center of the arrays because there is no slot above on the feed waveguide. Uniform amplitude and phase distributions with small ripples are observed at 25.5 GHz. These confirm the desired operation of the cross-junction fed waveguide. The amplitude has the blocking at the center and is tapered down toward the

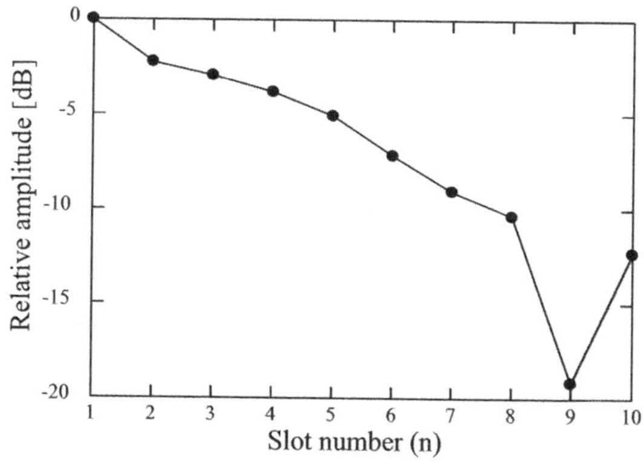

Figure 6.19: Slot excitation determined by genetic algorithm (from reference [30], reproduced by permission of © 2006 IEEE).

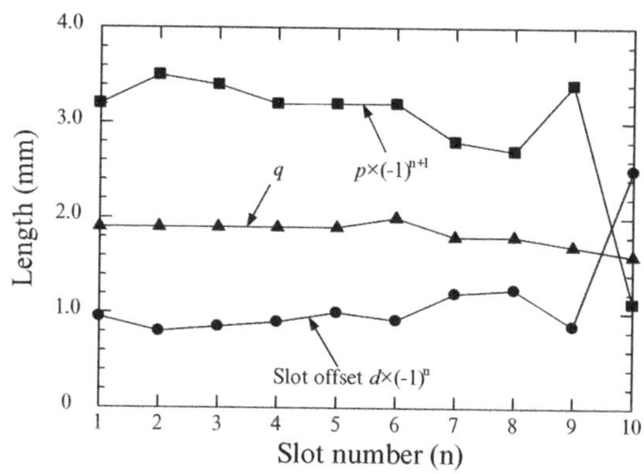

Figure 6.20: Slot parameters for desired slot coupling (from reference [30], reproduced by permission of © 2006 IEEE).

aperture end following the design as shown in Figure 6.19 for sidelobe suppression, while the uniform phase is observed.

6.4.3.3 Radiation Pattern

Figure 6.23 shows the measured H-plane pattern for −16 dB sidelobe suppression design. This pattern is associated with the excitation of the radiating slot array. The maximum sidelobe of measured result is −14.7 dB at the design frequency. It is slightly higher than the calculated one of −16 dB, but reasonable agreement was obtained. The measured radiation pattern in the E-plane coincides with that for uniform illumination with the sidelobe −13 dB. As an important advantage of the center fed array confirmed in these figures, the main beam direction in the E- and H-planes stays in the boresight across a wide frequency range and is not squinted with frequency.

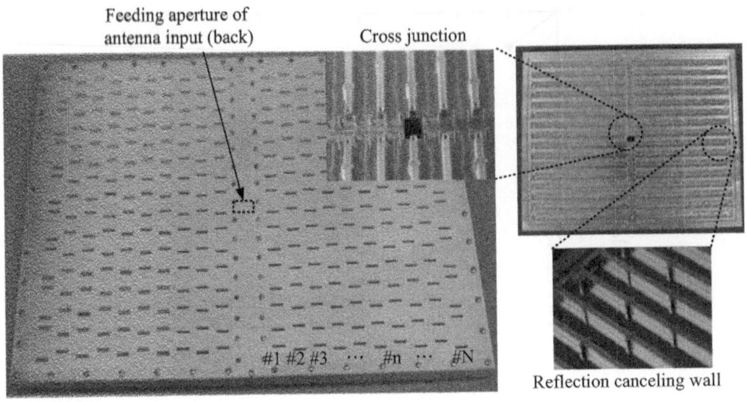

Figure 6.21: Fabricated model antenna (from reference [30], reproduced by permission of © 2006 IEEE).

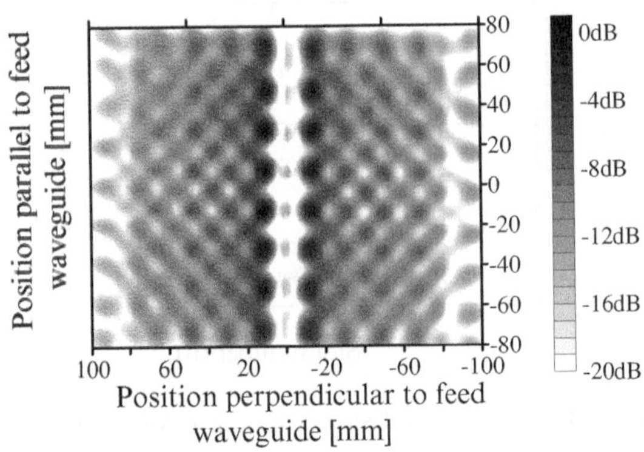

Figure 6.22: Aperture distribution of the aperture field at 25.5 GHz (from reference [30], reproduced by permission of © 2006 IEEE).

6.4.3.4 Gain and Efficiency

Figure 6.24 shows the frequency characteristics of the gain and efficiency of the antenna. The measured gain is compared with the end fed waveguide array for commercial use of fixed wireless access (FWA) system. The maximum gain is 30.5 dBi, which corresponds to 46% efficiency for the aperture size. This is a bit lower than the end fed waveguide arrays. However, the enhancement of frequency bandwidth is observed as expected. The degradation of the gain and efficiency comes from the blockage region and the aperture tapering for sidelobe suppression.

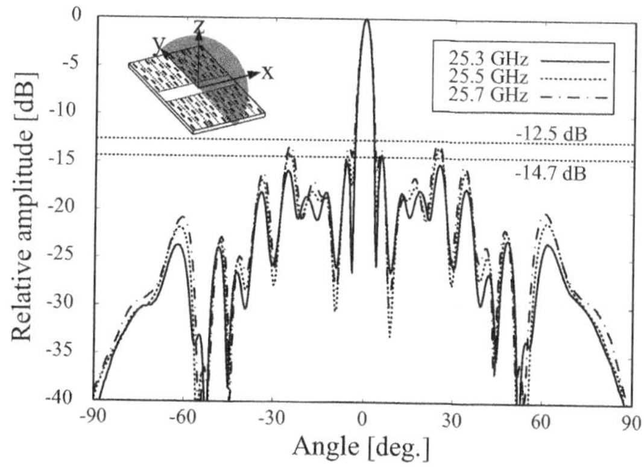

Figure 6.23: Radiation pattern in the *yz*-plane (H-plane) (from reference [30], reproduced by permission of © 2006 IEEE).

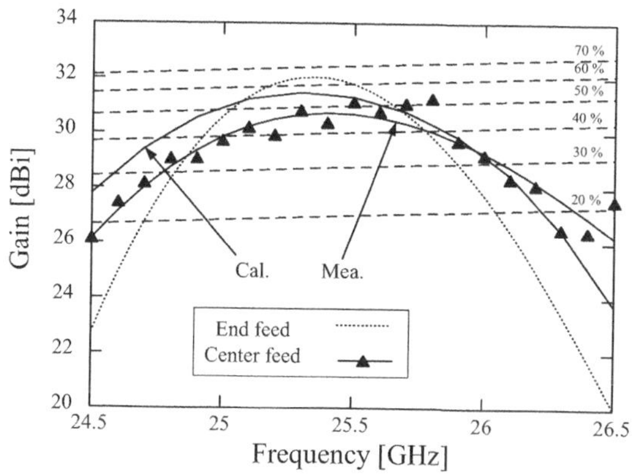

Figure 6.24: Gain and efficiency of the antenna (from reference [30], reproduced by permission of © 2006 IEEE).

6.4.4 Polarization Isolation between two Center-feed Single-layer Waveguide Arrays Arranged Side-by-Side

We demonstrate the feasibility of a challenging system where frequency is fully reused by the use of polarization isolation only [34, 35]. A FWA system with this concept is presented in Figure 6.25. Two center-fed single-layer slotted waveguide arrays with orthogonal polarization are used in exactly the same frequency band for transmission and reception. In order to completely reuse the frequency two times [36, 37], approximately 100 dB of transmission-reception isolation is required. A preliminary scenario is to realize this isolation by the combination of an antenna isolation of 50 dB and a cross-polarization compensating algorithm circuit of 50 dB. The latter dispenses with the diplexer, and the use of microwave integrated circuits realizes the miniaturization and economization of equipment. The arrays

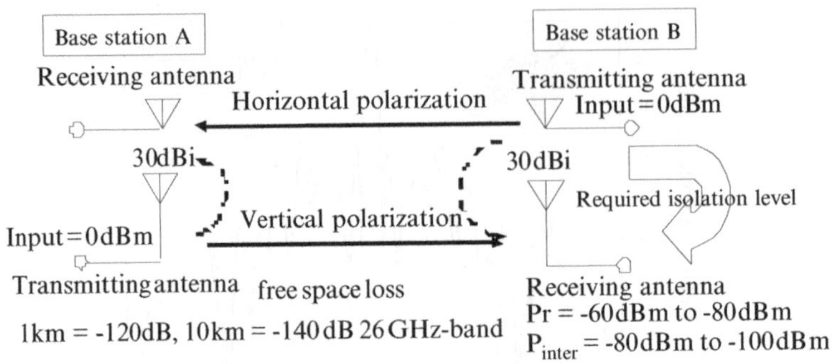

Figure 6.25: Dual polarization wireless system for two-times frequency reuse.

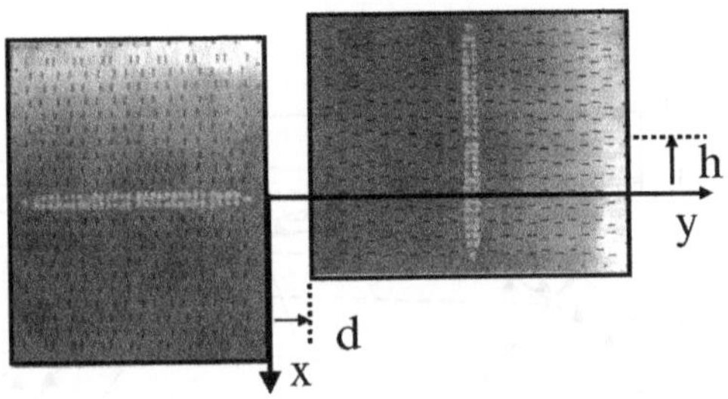

Figure 6.26: Isolation between two trial manufactured antennas arranged with distance d and position h.

are arranged side-by-side in the same plane: one is for transmitting and the other is for receiving in the FWA system.

The effects of arrangement are discussed in more detail. We prepare two center-fed single-layer waveguide arrays which have the same structure. We measure the isolation for different values of distance d $(= 0, 1, 2, 3\lambda)$ and position h $(= 0, 1, 2, 3\lambda)$ as shown in Figure 6.26. In Figure 6.27, the measured isolation results are summarized as functions of d and h. The results indicate a serious degradation of isolation due to increasing h, but an improvement in isolation due to increasing d. In order to confirm these general results qualitatively, we conducted a series of simulations. Figure 6.28 shows the full size arrays used in the simulation; isolation is evaluated for variety of distances d $(= 0, 1, 2, 3\lambda)$ and positions h $(= 0, 1, 2, 3\lambda)$.

Figure 6.29 shows the results of isolation between two arrays at 25.3 GHz. If the distance d is increased, the isolation improves, but if the position h is increased, isolation degrades. Almost the same tendency as in Figure 6.27 is observed. This phenomenon can be summarized as follows. The residual cross-polarization coupling between two arrays is effectively canceled out at the receiving antenna output, resulting in remarkably high isolation, due to the symmetrical structure and arrangement of the paired arrays.

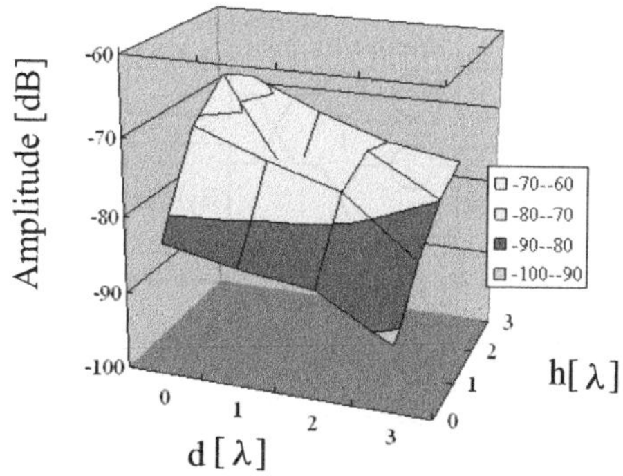

Figure 6.27: Position dependence of isolation at 25.3 GHz (measured).

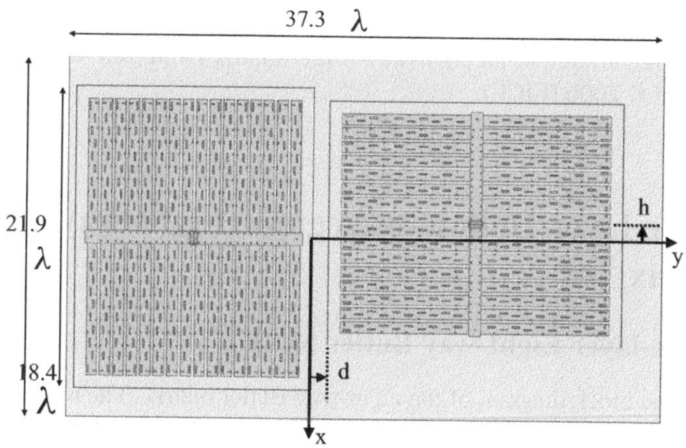

Figure 6.28: Full array model for simulation of position dependence of isolation.

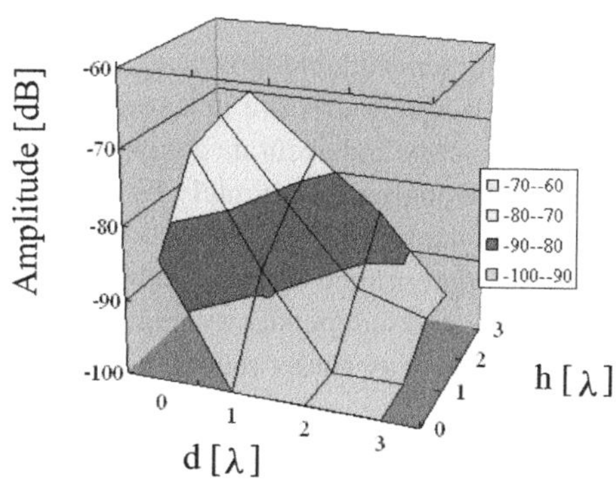

Figure 6.29: Position dependence of isolation between two arrays at 25.3 GHz (calculated).

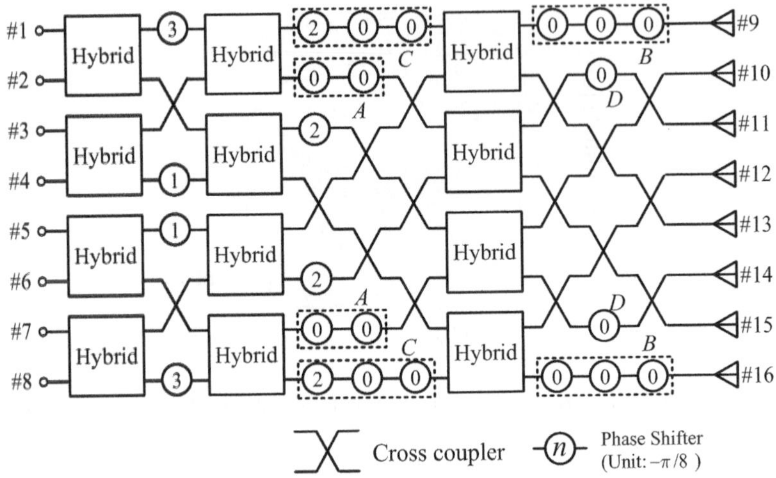

Figure 6.30: Configuration of an eight-way Butler matrix (from reference [39], reproduced by permission of © 2006 IEICE).

6.5 Single-layer Hollow-waveguide Eight-way Butler Matrix

6.5.1 Single-layer Eight-way Butler Matrix

Figure 6.30 is the block diagram of the eight-way Butler matrix. The numbers in circles are phase shifters and they correspond to the amount of the phase shift in units of $-\pi/8$. The Butler matrix consists of quadrature hybrid couplers and fixed phase shifters. The Butler matrix with four or more input ports generally needs three-dimensional crossings where two transmission lines intersect; this is a well-known difficulty for realizing a planar structure. A completely planar crossing structure of the waveguide, called a 'cross-coupler' is realized here using a short-slot coupler [38].

The size-reduced short-slot coupler is adapted to waveguide hybrids and waveguide cross-couplers. All components of the Butler matrix are in a single layer which contributes to not only low-cost fabrication, but also easy and accurate analysis. The structure is uniform along its height. Each component can be analyzed accurately by two-dimensional MoM including only TE_{n0} modes [40]. The components are arranged according to Figure 6.30. If the distance between adjacent components is determined by the condition that higher-mode coupling is negligibly small, they can be designed independently. Figure 6.31 shows the top view of the single-layer hollow-waveguide eight-way Butler matrix.

6.5.2 Design of the Couplers

The couplers are designed at 22.0 GHz. The input and output waveguides are WR42 standard waveguides (10.668 mm ×4.318 mm). The waveguide width a is 10.668 mm and the wall

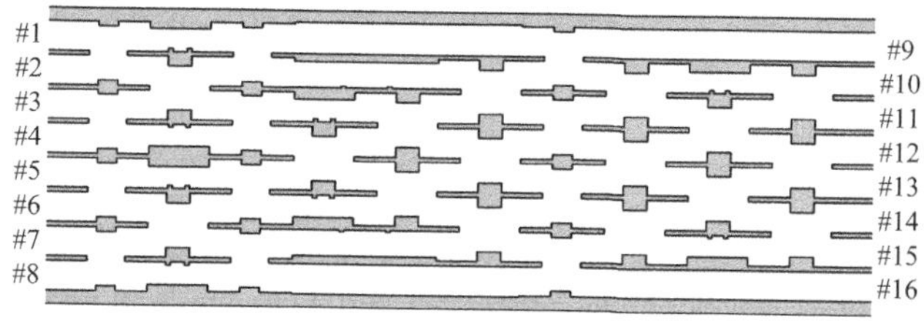

Figure 6.31: Top view of a planar eight-way Butler matrix (from reference [39], reproduced by permission of © 2006 IEICE).

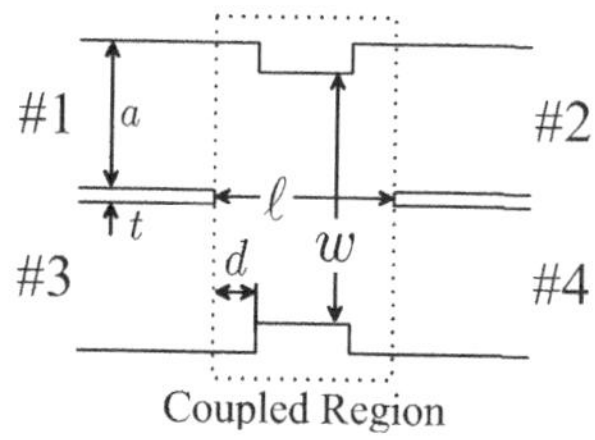

Figure 6.32: Structure of the short-slot directional coupler (from reference [39], reproduced by permission of © 2006 IEICE).

thickness t is 1.6 mm. The calculated transmission loss of the waveguide is 0.003 dB cm^{-1} at 22.0 GHz, where the conductivity of silver is assumed to be $\sigma = 6.1 \times 10^7$ S m^{-1}.

The couplers are designed by the following steps.

1. The initial parameters are determined by the equations for the phase constants of the two propagating modes in the coupled regions.

2. The parameters are modified by the MoM analysis.

3. The previous step is repeated by changing the offset d in Figure 6.32 until the reflection characteristics are sufficiently improved.

6.5.2.1 Initial Design

Figure 6.32 shows the proposed structure of the short-slot coupler. A step structure is introduced in the coupled region instead of reflection-suppression posts in the conventional structure. The proper ratio of the phase constants between the two propagating modes for both the desired coupling ratio and the reflection cancellation can be derived as follows [41]:

$$\begin{cases} \dfrac{\beta_2}{\beta_1} = \dfrac{2n}{2n+1} = \dfrac{2}{3}, \dfrac{4}{5}, \dots & \text{(for hybrid)} \\ \dfrac{\beta_2}{\beta_1} = \dfrac{2n-1}{2n+1} = \dfrac{1}{3}, \dfrac{3}{5}, \dots & \text{(for cross-coupler)} \end{cases} \tag{6.4}$$

where β_1 and β_2 are the phase constants of the TE_{10} and TE_{20} modes in the coupled region. The realizable ratio of the phase constant is limited in the actual fabrication. $\beta_2/\beta_1 = 2/3$ ($n = 1$) for the hybrid and $\beta_2/\beta_1 = 3/5$ ($n = 2$) for the cross-coupler, which corresponds to $\ell = 3\lambda_{g1}/4$ and $5\lambda_{g1}/4$, respectively (where λ_{g1} is the guide wavelength of the TE_{10} mode).

The phase constants of the two propagating modes in the coupled region are selectively designed by the width w of the coupled region. The phase constant decreases rapidly around each cutoff. The cutoff wavelengths of the TE_{10} and TE_{20} mode in the coupled region are $\lambda_{c10} = 2w$ and $\lambda_{c20} = w$, respectively. The difference in the phase constant $\beta_1 - \beta_2$ can be enlarged so that the coupled length becomes small. The width w is chosen so that the ratio of the phase constants is satisfied. The coupled length ℓ is also determined by the equation described in the previous paragraph. The above-mentioned ratio of the phase constants for the hybrid is obtained with $w = 17.24$ mm, where the phase constants are $\beta_1 = 0.424$ rad mm $^{-1}$ and $\beta_2 = 0.282$ rad mm $^{-1}$ and the coupled length ℓ is 11.13 mm. The ratio for the cross-coupler is obtained with $w = 16.25$ mm where the phase constants are $\beta_1 = 0.419$ rad mm $^{-1}$ and $\beta_2 = 0.251$ rad mm $^{-1}$ and $\ell = 18.76$ mm.

6.5.2.2 Modification by MoM Full-wave Analysis

The structure of the coupler is uniform along its height. It is analyzed by the MoM which includes only TE_{n0} modes uniform along the height [40]. The offset d in Figure 6.32 is set to zero in the beginning of the MoM modification using the initial parameters. The reflection of the TE_{20} mode is neglected in the design using Equation (6.4) [41]. Since the ℓ of the cross-coupler corresponds to $5\lambda_{g1}/4 = 3\lambda_{g2}/4$ (λ_{g2} is the guide wavelength of the TE_{20} mode), the reflections of the TE_{10} and TE_{20} modes at the both ends of the coupled region cancel out. Therefore, the total reflection of the cross-coupler with $d = 0$ mm is suppressed over a narrow bandwidth around the design frequency. On the other hand, ℓ of the hybrid corresponds to $3\lambda_{g1}/4 = \lambda_{g2}/2$ and the reflections of the TE_{20} mode add. The reflections of the TE_{10} mode at the both ends cancel out. The sufficient reflection suppression is not obtained for the hybrid with $d = 0$ mm. This step is repeated by changing the parameter d until the reflection characteristics are sufficiently improved to -40 dB or less.

The design parameters of the couplers are summarized in Table 6.5. In the hybrid, the coupled length ℓ is 14.07 mm, which is 62% of the value in the conventional structure using reflection-suppressing posts. In the cross-coupler, ℓ is 22.19 mm, which corresponds to 49%. The length of the conventional structure simply estimated from the phase-constant difference without full-wave analysis.

Figure 6.34 shows the calculated frequency characteristics of the designed couplers. In the hybrid, the reflection $|S_{11}|^2$ and the isolation $|S_{31}|^2$ are suppressed to less than -30 dB over 21–23 GHz. The phase difference between $\angle S_{21}$ and $\angle S_{41}$ is $90° \pm 0.02°$ over a 30% bandwidth. In the cross-coupler, the reflection $|S_{11}|^2$ and the isolation $|S_{21}|^2$, $|S_{31}|^2$ are suppressed less than -20 dB in a 6.0% bandwidth and -15 dB in a 10.4% bandwidth, respectively. The transmission phase $\angle S_{41}$ of the cross-coupler with a 22.19 mm length is equal to that of 28.0 mm length straight waveguide. Therefore, the cross-coupler has a phase delay comparable to a straight waveguide with 22.19 mm length. The bandwidth of the cross-coupler is narrower than that of the hybrid because the coupled length is longer.

If the stepped structure in Figure 6.32 is not installed in the coupled region ($w = 2a + t = 22.936$ mm), an undesired TE_{30} mode can propagate (in the coupled region). The stepped

Table 6.5: Design parameters of the couplers.

(a) Initial parameters	Hybrid	Cross-coupler
Coupled length ℓ	11.13 mm	18.76 mm
Width w	17.24 mm	16.25 mm
(b) Modified parameters with $d = 0$ mm	**Hybrid**	**Cross-coupler**
Coupled length ℓ	12.33 mm	19.90 mm
Width w	16.70 mm	16.15 mm
(c) Modified parameters with d	**Hybrid**	**Cross-coupler**
Coupled length ℓ	14.07 mm	22.19 mm
Width w	19.45 mm	15.99 mm
Offset d	3.45 mm	6.80 mm

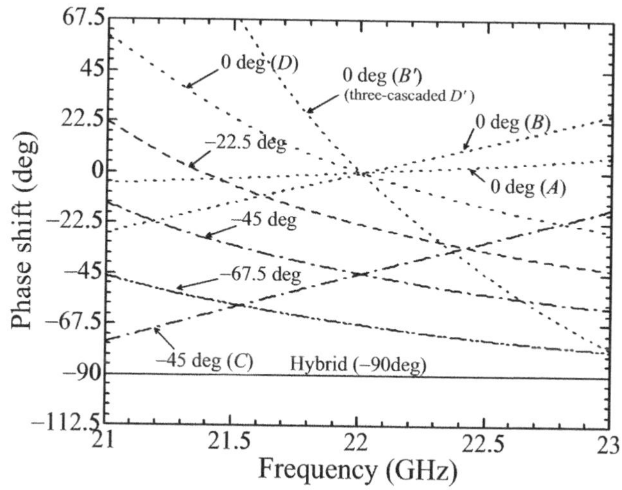

Figure 6.33: Characteristics of designed phase shifters (22.0 GHz) (from reference [39], reproduced by permission of © 2006 IEICE).

structure also helps to suppress the TE_{30} mode. In the conventional design, the waveguide width a is usually chosen under the condition $2a + t < \lambda_{c30} = 1.5\lambda$, where λ_{c30} is the cutoff wavelength of the TE_{30} mode.

6.5.3 Design of Phase Shifters for the Eight-way Butler Matrix

In the eight-way Butler matrix, the waveguide phase shifters require phase shifts of 0°, −22.5°, −45° and −67.5° at the proper positions shown in Figure 6.30. The phase shift is defined as the difference in transmission phase between the cross-coupler and a phase shifter. The length of the phase shifter is equal to the coupled length ℓ of the cross-coupler

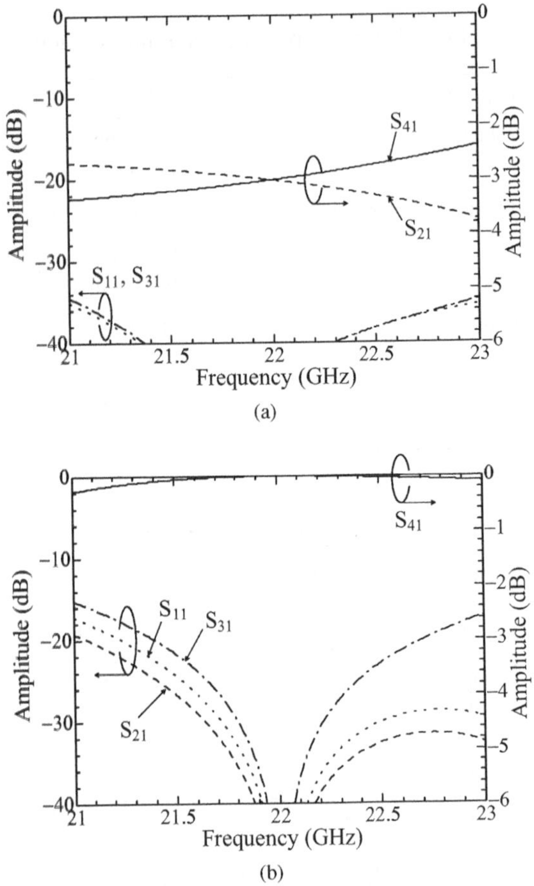

Figure 6.34: Characteristics of designed couplers (22.0 GHz): (a) hybrid; (b) cross-coupler (from reference [39], reproduced by permission of © 2006 IEICE).

($\ell = 22.19$ mm). The reference planes of the phase are placed at the ends of the coupled region. The phase shifters are designed by the MoM [40]. The input–output waveguide width a is 10.668 mm. The phase shifts are realized by changing the phase constant of the TE_{10} mode as is shown in Figure 6.35. The reflection is suppressed by the positions of two inductive walls.

A narrower or wider waveguide width a_s of the phase shifter gives positive (progression) or negative (delay) phase, respectively. Required phase shifts in the Butler matrix are 'negative'. In addition, the cross-coupler has phase delay compared with an equal-length straight waveguide as is described in the previous section. Since the wider-width phase shifters with phase delay are difficult to be arrayed due to space limitation in the Butler matrix, the narrower-width phase shifters with phase progression are adopted in the Butler matrix. The required phase delay is equivalently realized by a progressive phase shifter adding 360° progression. It causes frequency dependence of the phase shift. The required

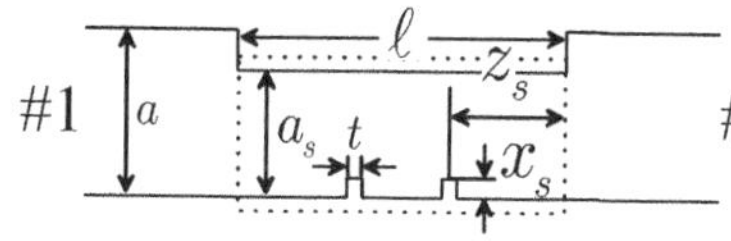

Figure 6.35: Structure of the phase shifter (from reference [39], reproduced by permission of © 2006 IEICE).

Table 6.6: Design parameters of the phase shifters.

	0°(*D*)	−22.5°	−45°	−67.5°
a_s	7.74 mm	7.90 mm	8.05 mm	8.25 mm
ℓ	22.19 mm	22.19 mm	22.19 mm	22.19 mm
x_s	0.96 mm	1.04 mm	1.07 mm	1.15 mm
z_s	8.17 mm	7.98 mm	7.86 mm	7.68 mm
	0°(*A*)	0°(*B*)	−45°(*C*)	
a_s	9.45 mm	10.64 mm	11.08 mm	
ℓ	52.78 mm	83.37 mm	83.37 mm	
x_s	0.60 mm	—	—	
z_s	17.96 mm	—	—	

phase shift is obtained only at the design frequency. The design parameters of the phase shifters are summarized in Table 6.6. Figure 6.33 shows the frequency characteristics of the phase shift. The frequency variation is larger in a phase shifter with a smaller amount of phase shift.

The phase shifters indicated by the dotted lines in Figure 6.30 (and referred to as the cascaded phase shifters) are replaced by one phase shifter in the fabrication (phase shifters *A*, *B* and *C*). The lengths of the united phase shifters are twice or three times that of the coupled length of the cross-coupler. The definition of the phase shift for the united phase shifter is the phase difference from the transmission phase of the cascaded cross-coupler. The reflection of the cascaded phase shifter is suppressed individually. The united phase shifter has the required phase shift equivalent to the cascaded phase shifter and the reflection suppression. The design parameters and the frequency characteristics are added to Table 6.6 and Figure 6.33. The phase shifters *B* and *C* do not require the reflection-suppressing walls because the waveguide width is almost equal to the input waveguide width. The phase shifter B' in Figure 6.33 is the cascade of three phase shifters *D*. The frequency variation of the phase shift *B* is much better than that of B'. The frequency tendency of the phase shifters *A*, *B* and *C* is opposite to that of the others.

6.5.4 Characteristics of the Butler Matrix

A signal input at different input ports in a Butler matrix produces a different phase taper among the output ports [42]. Inputs from the eight ports give eight switching beams in an

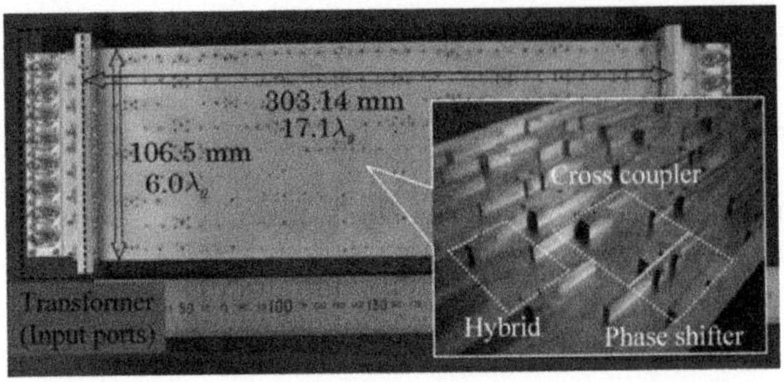

Figure 6.36: Photograph of the fabricated Butler matrix (from reference [39], reproduced by permission of © 2006 IEICE).

eight-way Butler matrix. The ideal dividing characteristics are $1/8 = -9.03$ dB in amplitude and a linear phase taper dependent on the input ports.

Each component is designed individually, and is then arranged according to Figure 6.30. The distance between adjacent components is determined to cause no higher-mode coupling. The higher modes can be calculated by MoM analysis as well as the dominant mode. The maximum amplitude of the TE_{20} mode at the end of the coupled region of the couplers is -12.5 dB normalized by an input TE_{10} mode. The distance between the couplers is required to be 8.4 mm in order to attenuate TE_{20} modes below -40 dB. This attenuated level is set to be smaller by about 20 dB than the obtained level of reflection and isolation around -20 dB in the couplers. The level is enough not to affect the operation of an adjacent coupler.

Figure 6.31 shows the top view of the single-layer hollow-waveguide eight-way Butler matrix. The structure is uniform along its height. The reflection-suppressing posts are not required in the couplers and their additional losses can be reduced. The total size of the matrix is 303.14 mm × 106.544 mm which corresponds to $17.1\lambda_g \times 6.0\lambda_g$ at 22 GHz. The total length of the matrix consists of $3\ell_H + 7\ell_C + 9\ell_S + 2\ell_I = 3 \times 14.07\text{ mm} + 7 \times 22.19\text{ mm} + 9 \times 8.4\text{ mm} + 2 \times 15\text{ mm} = 303.14$ mm, where ℓ_H is the length of the hybrid, ℓ_C is the length of the cross-coupler, ℓ_S is the distance between adjacent couplers, ℓ_I is the input and output ports. The matrix is compact and simple.

Figure 6.36 shows the photograph of the fabricated Butler matrix. It is divided into a bottom structure and a top plate in the fabrication. The bottom structure is milled from a solid brass plate. The top plate is brass of 0.5 mm thickness. They are plated with copper and silver in order to reduce the conductor loss. Then they are combined by soldering.

Coaxial-waveguide transformers are also designed and fabricated. They are connected to the Butler matrix with the waveguide flanges. The scattering parameters are measured with waveguide calibration. The reference planes are located at the waveguide flange and the measured scattering parameters do not include the characteristics of the transformers. On the other hand, isolations include the characteristics of the transformer, since they are measured by coaxial calibration. The reflection of the transformer is about -15 dB and it cannot affect the measured results very much in the coaxial calibration measurement. A coaxial load is connected at each dummy port through the waveguide transformer. Since the load is for a lower frequency band, its reflection is larger than -15 dB in the measured frequency band around 22 GHz.

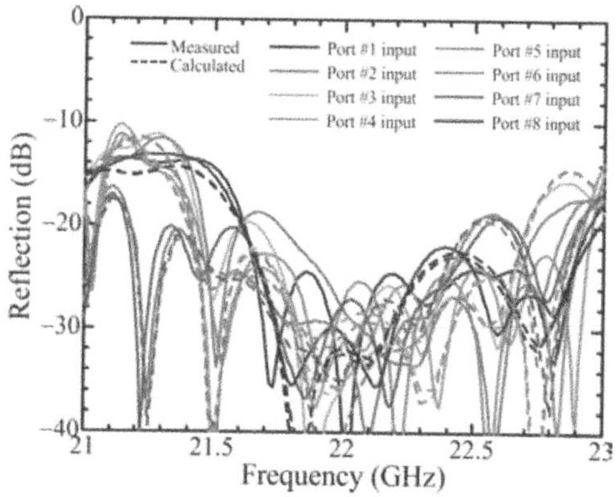

(a)

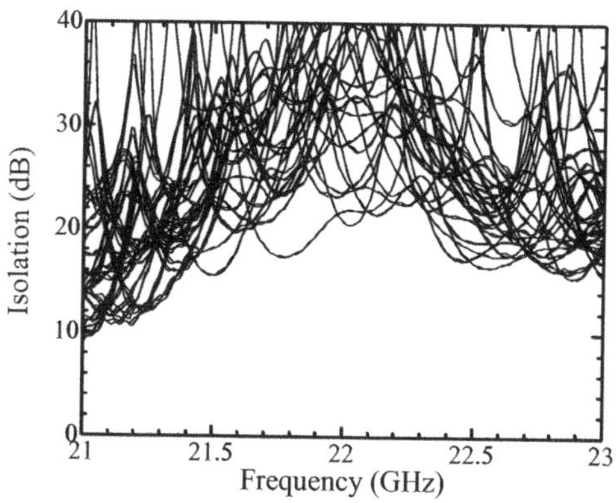

(b)

Figure 6.37: Reflection and isolation of the Butler matrix: (a) reflection; (b) isolation (from reference [39], reproduced by permission of © 2006 IEICE).

6.5.4.1 Dividing Characteristics

The calculated and measured frequency characteristics of the Butler matrix are shown in Figures 6.37 to 6.39. The average amplitude and phase difference are defined as the averages of the values at the output ports at the design frequency. The characteristics of the full structure of the eight-way Butler matrix are calculated by Ansoft HFSS with conductor loss ($\sigma = 6.1 \times 10^7$ S m^{-1}, silver).

Figure 6.37 shows the frequency characteristics of the reflections to the input ports and the isolations among the input ports of the Butler matrix. The calculations and measurements are in very good agreement. The reflection is less than -20 dB in a 5% bandwidth. The isolations are more than 15 dB over a 15% bandwidth. If the multiple ports are excited with a complex weight at a time, various radiation patterns are realized; the input impedance or the reflection should be discussed based upon Figure 6.37.

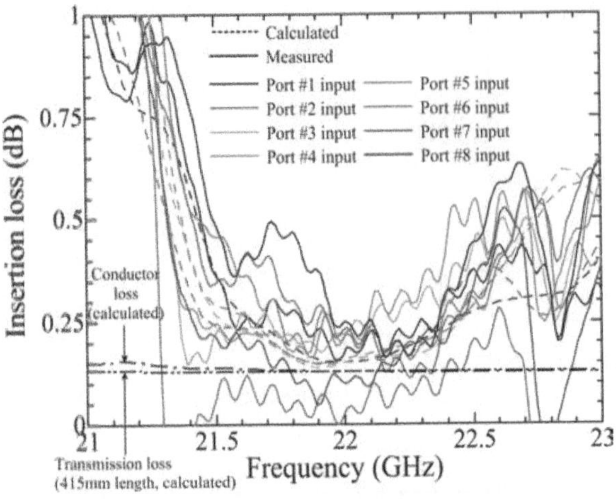

Figure 6.38: Insertion loss of the Butler matrix (from reference [39], reproduced by permission of © 2006 IEICE).

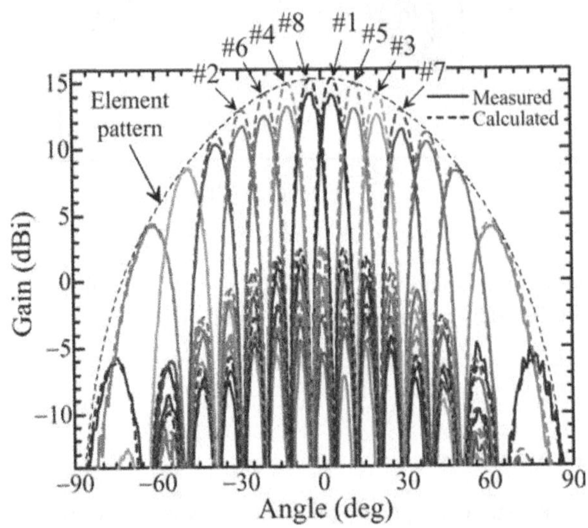

Figure 6.39: Radiation pattern from output ports (from reference [39], reproduced by permission of © 2006 IEICE).

The measured amplitude error for each port is less than 1.7 dB. The amplitude error for input port #*n* is defined as the difference between the maximum and minimum amplitudes among all the output ports when port #*n* is excited. Although the measured error is larger than the calculated one, it does not affect radiation patterns well compared with phase characteristics. The transformer and the Butler matrix are connected with a flange in the measurement. The measured isolation at some output port could include an error of a leakage from adjacent output ports through the flange because the contact of the flange could be imperfect and the wall thickness of 0.1 wavelengths between the adjacent output ports could be too small. In addition, the loads connected to the adjacent dummy ports are imperfectly matched. They increase the measured amplitude error. The measured phase error for each port is less than 1.9°. The bandwidth of the phase characteristics of the Butler matrix is narrow due to the frequency dependence of the unit characteristics of the phase shifters. The phase error for input port #n is defined as the difference between the ideal and the average of the phase difference between adjacent output ports.

Figure 6.38 shows the insertion loss of the Butler matrix. The insertion loss for input port #n is estimated by the ratio of the total output power ($= \sum_{m=9}^{16} |S_{m,n}|^2$ for $n = 1, \ldots, 8$) over the input power (= 1). Figure 6.38 shows the frequency characteristics of the insertion loss of the Butler matrix. At the design frequency, the insertion losses for each port are less than 0.16 dB in the calculation and around 0.25 dB in the measurement. The insertion loss of a standard coaxial-waveguide transformer is 0.3–0.7 dB. The insertion loss of the Butler matrix is very low. The insertion loss calculated from the measured divided power with error could be smaller than zero. The calculated value is almost equal to the sum of the losses of the components estimated individually: $3 \times Loss_{\text{Hybrid}} + 7 \times Loss_{\text{Cross-coupler}} + 9 \times Loss_{\text{Spasing}} + 2 \times Loss_{\text{Input}} = 3 \times 0.017\ \text{dB} + 7 \times 0.020\ \text{dB} + 9 \times 0.003\ \text{dB} + 2 \times 0.005\ \text{dB} = 0.23\ \text{dB}$.

The insertion loss consists of conductor loss and reflection loss. The conductor loss corresponds to the power absorbed inside the Butler matrix, and is estimated from the total power ($= \sum_{m=1}^{16} |S_{m,n}|^2$ for $n = 1, \ldots, 8$) to all the input and output ports. The difference between the conductor loss and the insertion loss corresponds to the reflection loss. The conductor loss is almost equal to the transmission loss of a straight waveguide of 415 mm-length, which corresponds to the average length of eight paths between the input and output ports. Since the reflection and isolation are sufficiently suppressed around the design frequency, the conductor loss is dominant. Therefore, the insertion loss of the Butler matrix is determined by the transmission loss of the waveguide and it shows the following two facts. The low measured loss of 0.25 dB can be realized by only the waveguide with low transmission loss; that is, a hollow waveguide with high-conductivity walls, where the transmission loss is 0.003 dB cm^{-1}. The coupler length dominates the average path length, and the half-sized coupler contributes not only the size-reduction of the overall Butler matrix, but also loss reduction.

6.5.4.2 Radiation Pattern

Figure 6.39 shows the patterns radiating directly from the output ports of the Butler matrix. It shows the calculated patterns for the directivity. The measured patterns show the gain by the comparison method with a standard gain horn. A large ground plate of 810 mm × 500 mm is embedded around the waveguide aperture at the output ports. The calculated patterns are derived from the ideal division of the Butler matrix at the output ports and an element pattern. The element pattern is simply calculated by assuming only the electric field of the TE_{10} mode on the waveguide aperture. It becomes a cosine-like distribution. The element spacing is $a + t = 12.268$ mm and corresponds to $0.90\lambda_0$ (the wavelength in free space), where a is the waveguide width and t is the wall thickness.

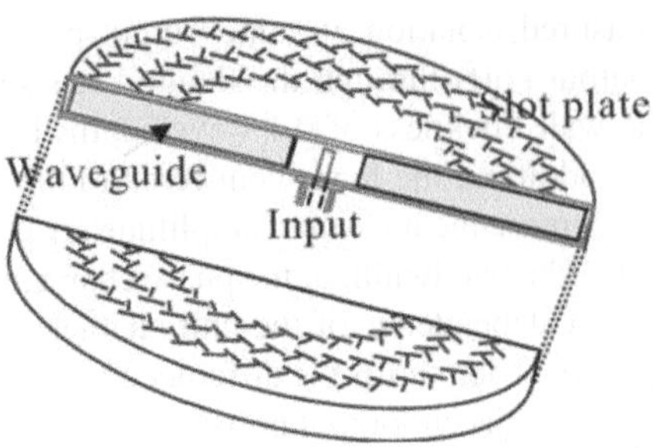

Figure 6.40: Structure of radial line slot antennas (from reference [43], © 2008 IET).

The gain reduction of 1–2 dB in the measurement is observed in comparison with the calculated directivity. This comes from a large reflection at the output port in the radiation, which affects the full angle range. The losses of the Butler matrix and transformer are also affected slightly. The discrepancy between measurement and calculation becomes small over the angle range, 45–90°, because the element pattern estimated from the measurement pattern is 1–2 dB higher than the calculated one over this range, which includes an error of the aperture field distribution without neglecting the reflection.

The main beam direction is switched for each input port. The main beam directions, the null directions and the sidelobe levels in the experiment are in good agreement with the calculation. High grating lobe levels due to the large element spacing of $0.90\lambda_0$ are observed. Although the grating lobe suppression condition is $a + t < 0.5\lambda_0$, the waveguide width is below the cutoff of the TE_{10} mode in that case. The grating lobes cannot be avoided in the H-plane arranged hollow-waveguide structure.

6.6 Radial Line Slot Antennas

6.6.1 High Gain Radial Line Slot Antennas with a Boresight Beam

6.6.1.1 Basic Structure and Loss in Radial Line Slot Antennas

Figure 6.40(a) shows the structure of a single layer RLSA [44]. Two conductor disks compose a radial waveguide and a dielectric material is inserted in the waveguide to keep its height constant. The slot pairs for circular-polarization are arrayed spirally on the top plate to realize a boresight beam. The power is fed by the coaxial feed at the center of the radial waveguide. It is transferred into a radially outward travelling wave, which radiates from slots gradually.

The loss in the radial line is estimated first. Transmission loss in the parallel plate waveguide is the sum of the conductor loss and the dielectric loss. Figure 6.41 shows the loss per cm versus the height of the waveguide for various values of the permittivity of the dielectric. Typical values of conductivity for copper and the permittivity for ceramics and polytetrafluoroethylene (PTFE) are considered. The dielectric loss dominates the total loss if the height is large, while the conductivity does if the height is small. Therefore, the combination of lower tan δ (or ε_r) and larger waveguide height generally contributes to low loss characteristics of the transmission line. RLSAs in the 12 GHz band have realized the

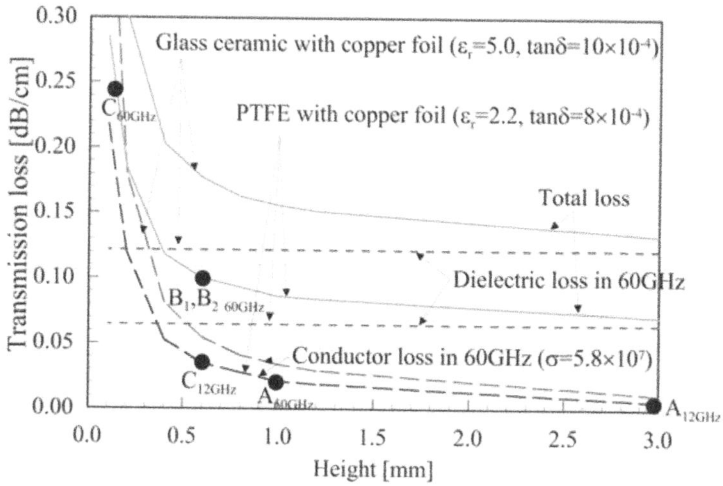

Figure 6.41: Transmission loss per cm versus the height of the waveguide (from reference [43], © 2008 IET).

excellent antenna efficiency of more than 85% by adopting low loss foam dielectric ($\varepsilon_r \approx 1$, $\tan\,\delta \approx 0$) and large waveguide height ($h \approx 0.15 \sim 0.3\lambda_g$) as is indicated by 12 GHz in Figure 6.41. It would be small even in the 60 GHz band as is denoted by $A_{60\ \mathrm{GHz}}$. On the other hand, in the case of other planar structures such as the microstrip and triplate line, the height of the line should be limited to about $0.03\lambda_g$ to prevent additional loss due to unwanted radiation from lines. Therefore, the overall transmission loss grows with the frequency much faster than the radial line, as is shown in Figure 6.41 by $C_{12\ \mathrm{GHz}}$ and $C_{60\ \mathrm{GHz}}$ in 12 GHz and 60 GHz, respectively.

We fabricate two types of RLSAs for the 60 GHz band that have different structures. The first (Type A) consists of three separate parts to be stacked up [45,46]. The transmission loss remains small as is estimated by A_{60} GHz in Figure 6.41. However, the misalignment of the two disks must be carefully minimized, as the frequency becomes higher. Figure 6.42 shows the gain reduction of a typical RLSA with 30–35 dBi in gain as a function of the alignment error of the top plate; in the 60 GHz band, the alignment error of 0.3 mm in Type A causes a gain reduction of about 0.5 dB (10% in efficiency). As an alternative structure to cope with this problem, the second (Type B) utilizes the existing copper–PTFE–copper substrate, which assures structural accuracy as well as productivity. Larger permittivity of PTFE ($\varepsilon_r \approx 2.2$) results in the loss four times larger than Type A as is suggested by $B_{60\ \mathrm{GHz}}$ in Figure 6.41; but it is about 1/3 of microstrip line ($C_{60\ \mathrm{GHz}}$) and is much smaller. The transmission loss in RLSA using the substrate is less than about 0.5 dB for the diameter of 100 mm, for example.

6.6.1.2 Conventional Type RLSA with Foaming Polyethylene (Type A)

The conventional RLSAs consists of three separate parts: a slot plate, the foam polyethylene with negligible tan δ with the thickness of 1mm and a bottom plate. These three parts are fixed to each other by the spray adhesive. The cylindrical edge at the aperture periphery of the

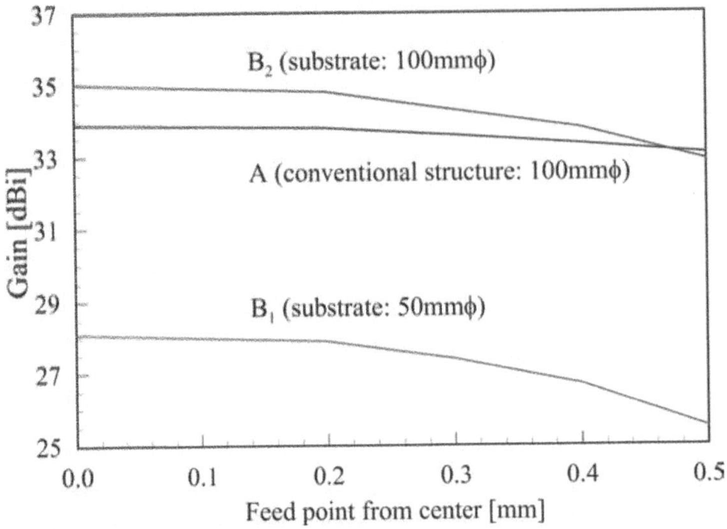

Figure 6.42: Gain reduction as a function of the alignment error of the top plate (from reference [43], © 2008 IET).

Table 6.7: Design parameters.

	A	B1	B2
Antenna diameter	100 mm	50 mm	100 mm
Number of rounds	9.2	5.75	13.5
Permittivity	1.08	2.2	2.2
Waveguide height	1.0 mm	0.635 mm	0.635 mm
Design frequency	60 GHz	60 GHz	60 GHz
Number of slots	1290	442	1904

antenna is opened. Each part can be fabricated easily. The important technique in fabricating this structure is to maintain the accuracy for the alignment of the two disks.

An antenna with a diameter of 100 mm is fabricated based upon the theoretical design procedure for uniform illumination [44]. The design parameters are shown in Table 6.7. The slot length gradually increases with the distance from the center. The associated residual loss power is 20%.

6.6.1.3 RLSA using the Substrate (Type B)

RLSAs with a new structure are also fabricated. They use a substrate consisting of three layers: upper foil, dielectric (PTFE) and bottom metal plate. Therefore, they are suitable for mass production. The cylindrical periphery of the antenna is shorted by plating. The PTFE ($\varepsilon_r \approx 2.2$, $\tan\delta \approx 8 \times 10^{-4}$) with the thickness of 0.6 mm is used as a dielectric material.

Slots are etched on the top plate. The dielectric loss is larger but the alignment error of slots can be negligible in comparison with Type A.

Two antennas with different diameter (B1; 50 mmϕ and B2; 100 mmϕ) are fabricated. The design parameters are also shown in Table 6.7. The termination loss is roughly calculated by 13% and 8% for Types B1 and B2, respectively. From Figure 6.41, the transmission loss of the substrate for Type B is about 0.1 dB cm^{-1} and is a bit larger than that of A; the predicted loss for type B with 50 mmϕ to 100 mmϕ is less than 0.5 dB and is about one-third that for the triplate and microstrip. In test antennas, the bottom plate of the substrate is thick (6 mm) to fix the coaxial feeder by screws.

6.6.1.4 Measurement

The measured reflection of the three antennas is less than −15 dB around 60 GHz, which is acceptable.

Far-field radiation patterns are obtained by Fourier transform of the near-field measurement data [47, 48]. The beam shapes are not perfect but reasonably symmetrical. The measured sidelobe level of −10 dB is a bit higher than that for uniform illumination (−13 dB). The cross-polar level is also a bit high suggesting the possible alignment error of the slot plate in this type. The associated axial ratio is about 1 dB. The symmetry of the beam shape of Type B2 is considerably degraded. The possible reason for it is the alignment error of the center feed point, which would excite asymmetric field in the radial waveguide. The sidelobe levels of antennas B1 and B2 are −15 and −11 dB, respectively; the deviation from the ideal one of −13 dB is almost compatible with that of Type A. The cross-polar level of B1 and B2 are −13 and −18 dB, respectively, with the associated axial ratio of 4 dB and 2 dB, respectively.

Figures 6.43(a) and (b) show the measured and the predicted gain for Types A and B, respectively. A gain of 33 dBi and the efficiency of about 50% is measured for Type A in Figure 6.43 (a), while the peak gain of Types B1 and B2 in Figure 6.43(b) are 27.3 dBi and 33.4 dBi, respectively. The remarkable efficiency of 54% and 55% is realized for B1 and B2, respectively. The gain is lower than the prediction by about 1 dB for types A and B1 and 2 dB for Type B2. The possible reason for the gain reduction in these test models seems to be the positioning error of the slot plate for Type A and that of the feed point for Type B. From the pattern symmetry and the gain reduction in Figure 6.43(b), the positioning error is roughly estimated to be about 0.5 mm for Type B2. For all this incompleteness, the efficiency of more than 50% realized here is about twice as high as other types of planar array in this range of gain [29]. These results confirm the high potential of RLSAs in the application to mmWave frequency. At the same time, the potential of substrate RLSAs is also demonstrated to be compatible with the conventional type RLSA.

6.6.2 Small Aperture Conical Beam Radial Line Slot Antennas

6.6.2.1 Structure of Concentric Array RLSA (CA-RLSA)

Figure 6.44 shows the structure of concentric-array (CA-)RLSA. Two conductor disks compose a radial waveguide and a dielectric material is filled in the waveguide to create a slow wave structure. The slot pairs are arrayed on the top disk, each of which is a unit radiator of circular polarization. A CA-RLSA has many rounds of slot arrays arranged concentrically,

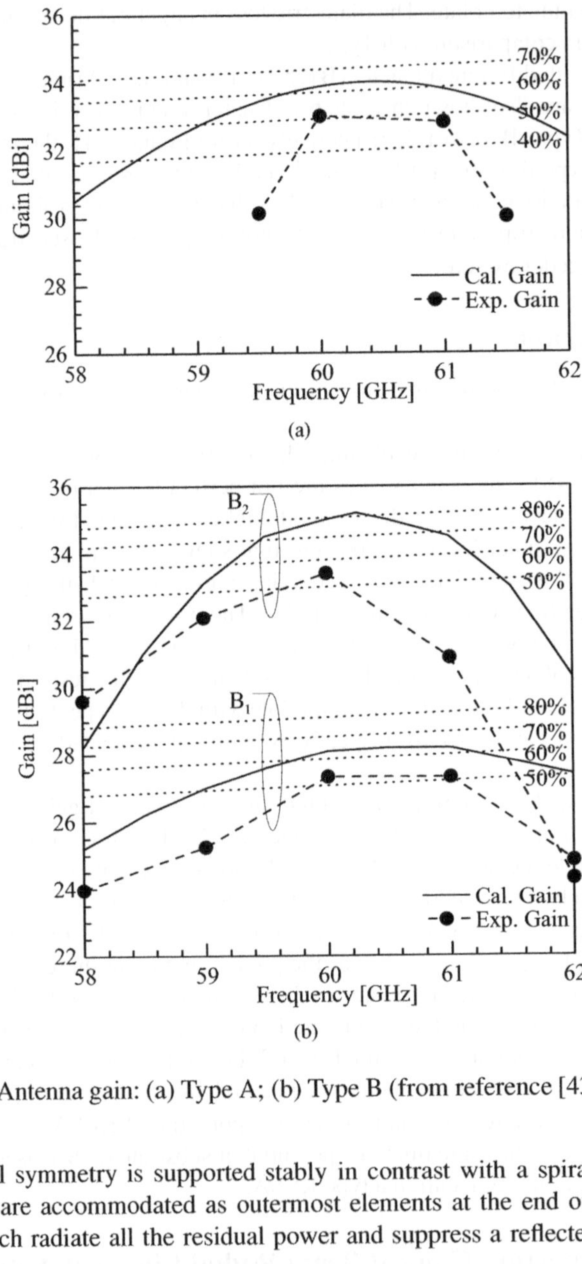

Figure 6.43: Antenna gain: (a) Type A; (b) Type B (from reference [43], © 2008 IET).

where rotational symmetry is supported stably in contrast with a spirally arrayed RLSA. Matching slots are accommodated as outermost elements at the end of the shorted radial waveguide, which radiate all the residual power and suppress a reflected inward travelling wave.

Power is fed at the center of the radial waveguide. In CA-RLSAs, two types of feed are selectively adopted depending upon types of antenna beam. One is a cavity resonator exciting a rotating mode in order to produce a boresight pencil beam [50]. The other is a coaxial feed exciting a rotationally symmetric mode in order to produce a conical beam [51], where the structural symmetry results in the radiation pattern with its null in the boresight. The slot design as well as the antenna diameter determine the shape of the conical beam.

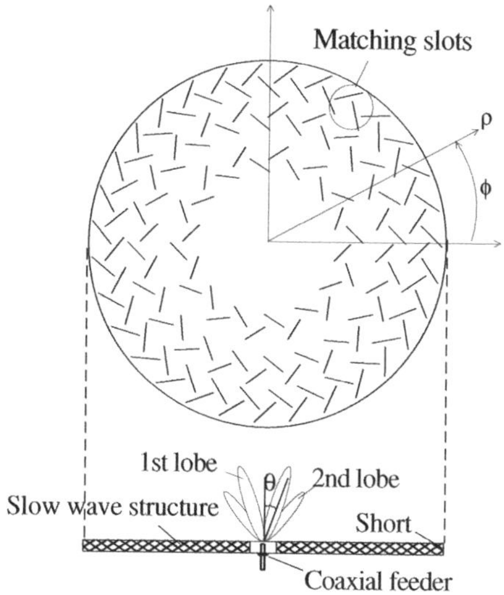

Figure 6.44: Structure of a CA-RLSA (from reference [49], reproduced by permission of © 1999 IEICE).

A coaxial feed is adopted and conical beam antennas are designed for the base station antenna in wireless local area network LAN systems.

6.6.2.2 Beam Shape Control of Conical Beam RLSAs

The important parameters, which determine the slot excitation, are the slot length and the radial position. MoM analysis, taking the mutual coupling effects into account, reveals that the length and position of the slots are the leading parameters for controlling the amplitude and phase of the slot excitation, respectively; larger spacing between the pairs corresponds to progressive phase delay towards the end of the aperture while longer slot length corresponds to larger amplitude of slot excitation.

The behavior of the shape of the conical beam is assessed as a function of the slot excitation distribution for a two-round CA-RLSA. The basic structural parameters from existing substrate are shown in Table 6.8. MoM analysis is adopted to estimate the slot excitation for given slot parameters. Figure 6.45 illustrates the example of the conical beam radiation pattern as well as their parameters, where the slot excitation distribution is uniform. The direction of peak radiation as well as the levels of lobes are focused upon. In two-round CA-RLSA, only the parameters of slot pairs in the inner (1st) rounds are varied, since the slot pairs in the outermost (2nd) round are the matching slots designed a priori. The slot length and the radial position of pair #2 are controlled while pair #3 is designed previously.

Figure 6.46 shows the beam direction and the directivity of the first and the second lobes as a function of the slot position of #2 or radiation phase difference between the inner and the outer rounds, where the radiation amplitude is set to be equal. From the figure, as the radiation phase difference becomes larger, the directivity of the first lobe becomes smaller

Table 6.8: Analysis parameters.

Number of rounds	2
Diameter	18.0 mm
Design frequency	60.0 GHz
Dielectric constant	2.2
Waveguide height	0.635 mm
Slot pair spacing in ϕ-direction	$0.5\lambda_0$

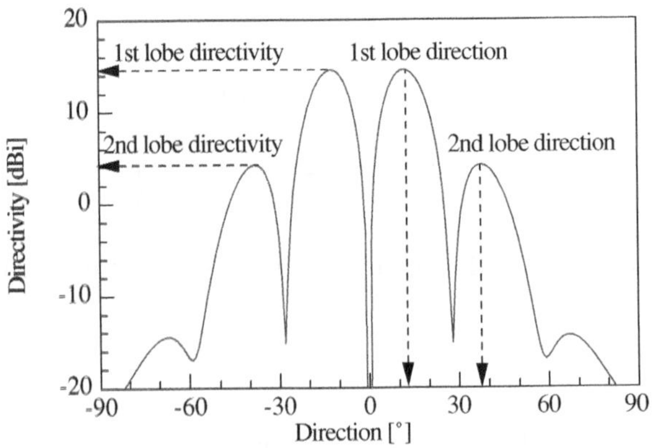

Figure 6.45: Example of conical radiation pattern (slot excitation distribution is uniform) (from reference [49], reproduced by permission of © 1999 IEICE).

and the second lobe becomes larger. Beam directions of the first and the second lobe remain almost stable as around 10° and 30°, respectively.

The dependence of beam shapes and the directivity upon the slot length or the amplitude difference between the inner and the outer rounds is also discussed. We have confirmed that the beam direction and the directivity of both the first and the second lobe remain almost unchanged for the change in amplitude difference, if the radiation phase is set to be equal. Therefore, the phase difference is set to be 180° for wider control of the pattern. Figure 6.47 shows the beam direction and the directivity. A positive value of the radiation amplitude difference indicates that the length of the inner slots is longer and its excitation is stronger. As the radiation amplitude of inner slots becomes larger, the first lobe is suppressed, although beam directions of the first and the second lobes do not change substantially. A 30° beam is obtained by enhancing the second lobe and suppressing the first lobe at 10°, where the relative amplitude and the phase of the inner round is set to be about +9 dB and 180°, respectively. These results suggest the limited controllability of the conical beam RLSA.

6.6.2.3 Fabrication of Model Antennas

Higher accuracy is required for dimensions of the slots and the feed circuit in manufacturing mmWave antennas. In the conventional RLSA, the radial waveguide is constructed by

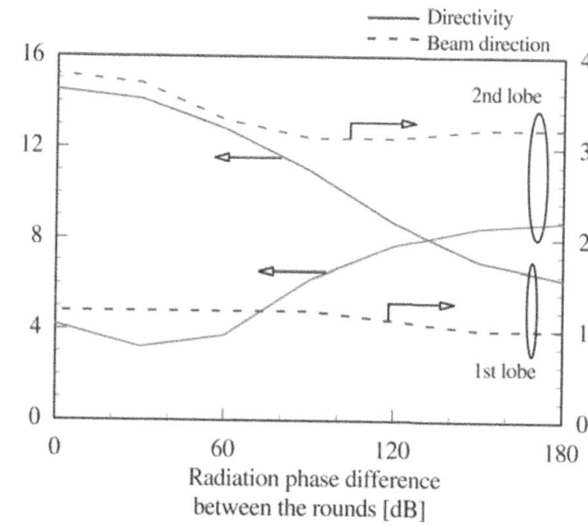

Figure 6.46: Beam direction and the directivity (radiation amplitude of both rounds are equal) (from reference [49], reproduced by permission of © 1999 IEICE).

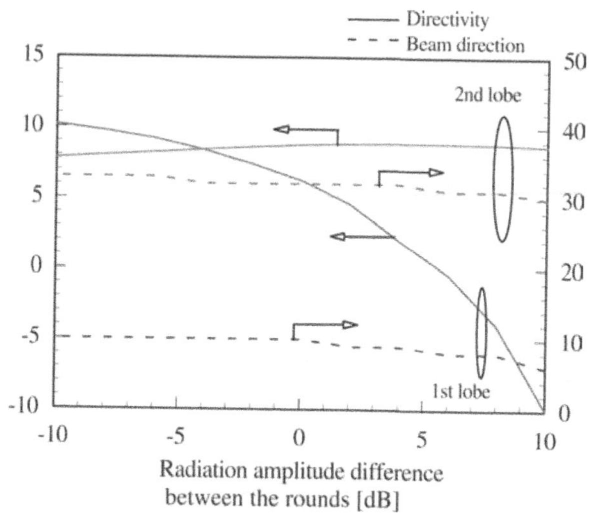

Figure 6.47: Beam direction and the directivity (radiation phase difference between two rounds is 180°) (from reference [49], reproduced by permission of © 1999 IEICE).

stacking up three separate sheets: a slot plate, a foam dielectric sheet with relatively low permittivity and a bottom plate. However, accuracy of the waveguide height and the alignment of the slot plate are insufficient for mmWave. For accurate mass-production in the future, a PTFE substrate that consists of three layers, a copper foil, a dielectric material

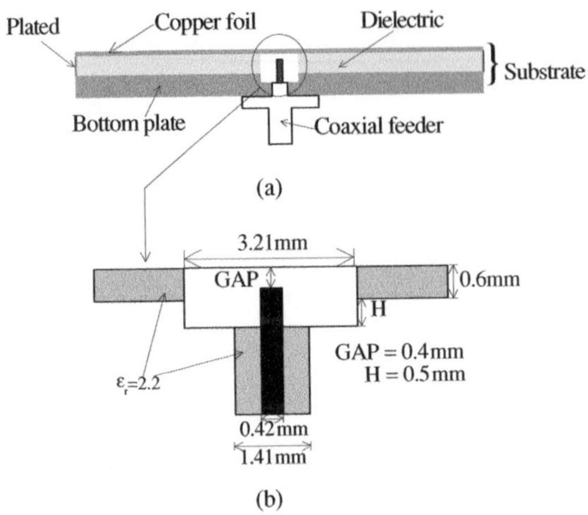

Figure 6.48: Configuration of RLSA using PTFE substrate (a) waveguide; (b) coaxial feeder (from reference [49], reproduced by permission of © 1999 IEICE).

($\epsilon_r \approx 2.2$) and a bottom metal plate, is newly adopted for mmWave conical beam RLSAs, as shown in Figure 6.48(a). The periphery of the antenna is shorted by plating. Slots are cut on the copper foil by etching. A coaxial feeder with step structure, shown in Figure 6.48(b), is attached at the center of the substrate [52].

Typical manufacturing errors of the slot length and the position in the etching are less than 50 μm, which is acceptable in terms of antenna performance. The alignment error of the feeding point is of the order of 0.1 mm, which corresponds to about 0.3 dB gain degradation. The critical parameter for the voltage standing wave ratio (VSWR) is the GAP in Figure 6.48(b). The tolerance for errors of GAP for reflection below −25 dB is $0.05\lambda_0$ for the step structure ($H \neq 0$ mm), which is five times larger than that for nonstep structure ($H = 0$ mm). The manufacturing error of the model antenna cannot be identified.

The two- and three-round antennas are fabricated. The parameters are shown in Table 6.9. Antenna A (three-round) is designed to realize uniform amplitude and phase, and the main beam direction is relatively small. Antenna B (two-round) is designed to realize the main beam direction of 30°.

6.6.2.4 Experimental Results

The reflection of the model antennas is minimum at 59.6 GHz. The −15 dB bandwidth is about 500 MHz and is relatively narrow. The reflection is caused mainly by a coaxial feed with the alignment error.

A radiation pattern is calculated by Fourier transforming the above near-field data [47]. The principal polarization (right-hand circular polarization) pattern of antenna A at 60 GHz is shown in Figure 6.49. In the antenna A, the measured main lobe direction and the shape are well predicted by the calculation. The 3 dB down beam width is 3.4°. The angle of the second lobe is near to the predicted one; however, the amplitude of the second lobe is smaller. Measured cross polarization is about −12 dB, which is 8 dB larger than the predicted one.

Table 6.9: Test antenna parameters.

	A	B
Diameter	25.0 mm	18.0 mm
Number of rounds	3	2
Number of slots	134	60
Main beam direction	9°	30°
Dielectric constant	2.2	
Waveguide height	0.635 mm	
Slot pair spacing in ϕ-direction	$0.5\lambda_0$	

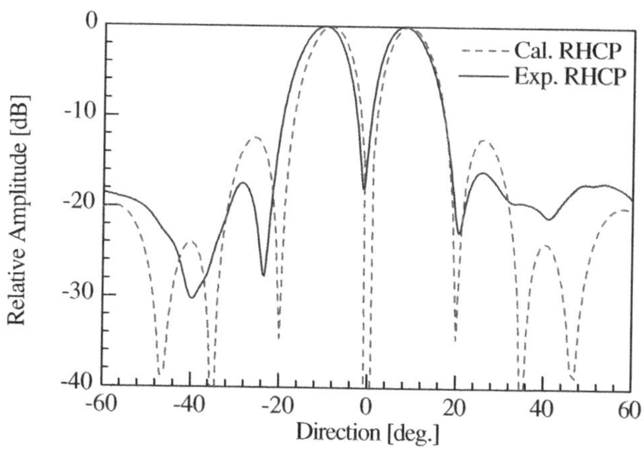

Figure 6.49: Co-polarized (RHCP) radiation pattern of antenna A (three-round) (from reference [49], reproduced by permission of © 1999 IEICE).

The radiation pattern of principal polarization of antenna B is also shown in Figure 6.50. The angle of peak radiation is about 25°, which is 5° smaller than the predicted one. The measured pattern is missing from the first lobe and deviates from the predicted one. One reason for the asymmetric measured patterns of both antennas is that an infinite conductor plane is assumed as the slot aperture in the calculation, while the measurement is conducted with a small slot plate. Other reasons may be manufacturing errors, such as the alignment error of the feed, and incomplete shorted wall at the aperture periphery which disturbs the symmetry of the inner field.

The measured and the predicted gain in the peak direction are shown in Figure 6.51. Because of the incomplete rotational symmetry of the field distribution, the gain varies in the ϕ-direction. Dots on the measured result indicate the average value, while error bars indicate the gain variation in the ϕ-direction. In antennas A and B, the gain variation is about 1.5 dB and 2.5 dB, respectively. The average gain of antenna A and B at 60 GHz is 14.8 dBi and 5.7 dBi, respectively. The measured gain is about 2.5 dB smaller than the predicted gain.

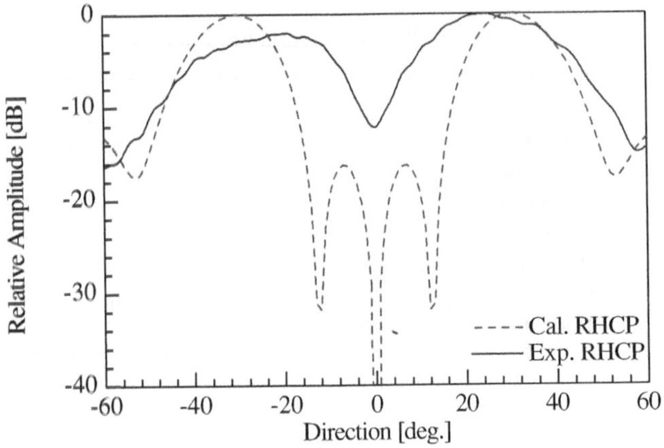

Figure 6.50: Co-polarized (RHCP) radiation pattern of antenna B (two-round) (from reference [49], reproduced by permission of © 1999 IEICE).

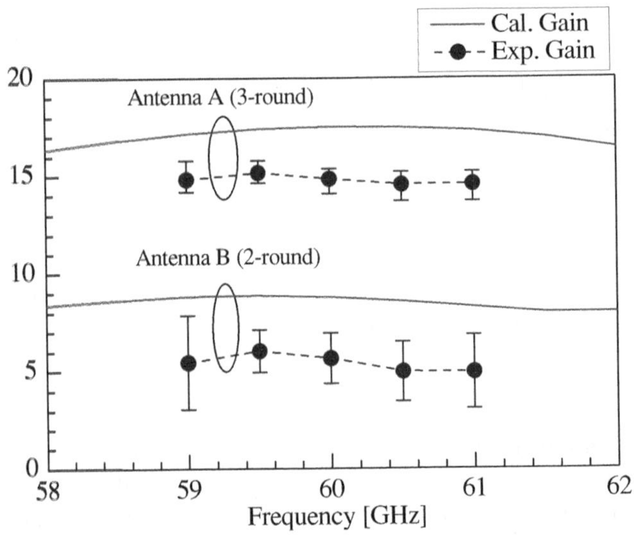

Figure 6.51: Measured and predicted gain (from reference [49], reproduced by permission of © 1999 IEICE).

6.7 Post-wall Waveguide-fed Parallel Plate Slot Arrays

6.7.1 Transmission Loss in Post Waveguide

Loss factors of the post-wall waveguide are investigated by using the Ansoft HFSS simulator under various loss conditions at 76.5 GHz. We can consider three types of loss: the dielectric

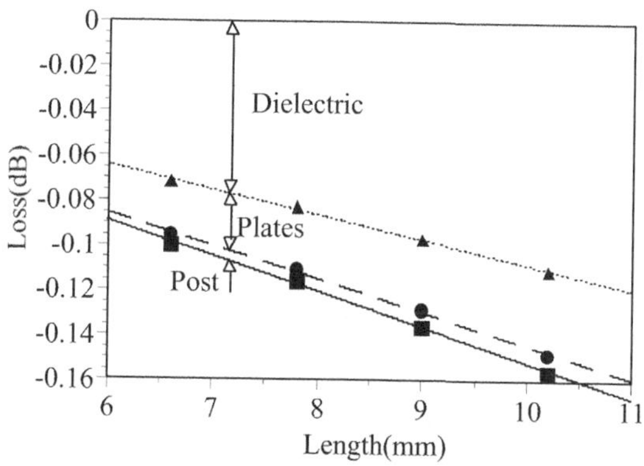

Figure 6.52: Simulated loss per millimeter under various loss conditions of straight post-wall waveguide at 76.5 GHz: squares, results for including all the losses; circles, results for including the loss of the dielectric and the plate; triangles, results for including only the dielectric loss.

loss (d), the conductor losses of the parallel plates (p) and the post walls (w). We confirm that the leakage from the post spacing is negligible in the simulation. The conductivity in both the parallel plates and the post walls are assumed to be 5.76×10^7 S m^{-1} for pure copper. The waveguide dimension and the dielectric parameters are the same as those in the model antennas. Figure 6.52 shows the losses as functions of the waveguide length under three loss conditions. Each waveguide in the simulation has a constant length (1.5 mm $\times$ 2) of metal walls at the both ends to define ports and a variable length of the post walls between them. The external region of the post-wall waveguides is terminated by absorbing boundaries at a proper distance from the post walls. The squares in this figure are the results for including all the losses ($d + p + w$). The circles are for including the loss ($d + p$) of the dielectric and the plate. The triangulars are for including only the dielectric loss (d). The gradients correspond to the loss per millimeter. By taking the difference in the results under the three conditions, these types of loss can be evaluated separately as shown in Figure 6.52. In the post-wall waveguide, the dielectric loss is 0.0110 dB mm^{-1}, the parallel plate loss is 0.0037 dB mm^{-1} and the post wall loss is 0.0010 dB mm^{-1}. The dielectric loss is much larger than the conductor losses. The total loss is 0.0157 dB mm^{-1}, which is about two-thirds the measured loss. The values of the conductivity and the dielectric loss tangent used in the simulations are not accurate in the mmWave band. They should be measured accurately in this band. The dielectric loss is similarly estimated to be 0.0087 dB mm^{-1} and the parallel plate loss is 0.0030 dB cm^{-1} in parallel plates.

6.7.2 Structure

Figure 6.53 shows the structure of the post-wall waveguide-fed parallel-plate slot array [4, 54]. A post-wall feed waveguide is located at the edge of a parallel plate waveguide. Coupling windows are placed with a spacing of the guide wavelength in the feed waveguide to excite in phase. The guide wavelength is longer than the wavelength in the parallel plates so that the undesired wave can propagate obliquely like grating lobes. Two additional windows are

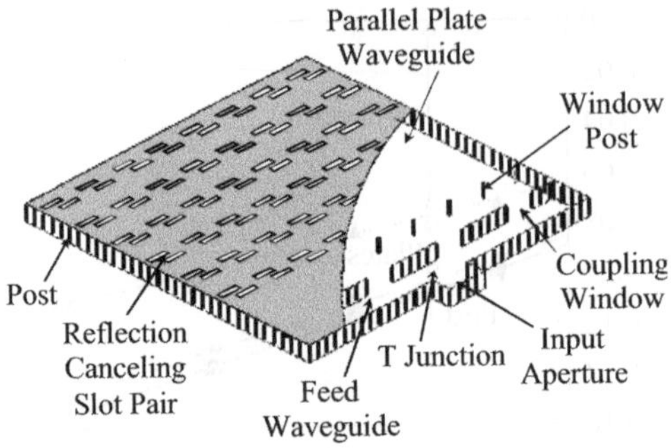

Figure 6.53: Post-wall waveguide fed parallel plate slot array (from reference [53], reproduced by permission of © 2000 IEICE).

made in the face of each coupling window to suppress the undesired wave [55]. A post is put in front of each coupling window in the feed waveguide to suppress the reflected wave due to the window coupling [54]. An input aperture is cut on the bottom of the parallel plates at the center of the feed waveguide, together with a T-junction and a 90° E-bend. All the radiating slots are paired with a spacing of about a quarter of wavelength in the parallel plates to achieve traveling wave propagation [11, 56, 57]. The slot pairs are arrayed with one-wavelength spacing to excite in phase and to radiate a main beam in the boresight.

6.7.3 Antenna Efficiency as a Function of the Size

Model antennas are fabricated for experiments in the 61.25 GHz band. Uniform distributions are designed each with the aperture sizes of $20 \times 20\ \text{mm}^2$, $30 \times 30\ \text{mm}^2$, $40 \times 40\ \text{mm}^2$ and $80 \times 80\ \text{mm}^2$. A fiberglass reinforced dielectric substrate of PTFE is used with rolled copper on the both surfaces. The height of the substrate is 1.2 mm and the loss is $\tan \delta = 0.00085$ at 10 GHz. Metal-surface via-holes with 0.5 mm diameter are arrayed with the spacing of 1.0 mm. The slot array elements consisting of a set of slots with 0.2 mm width are etched on the upper plate to estimate the inner field in the oversized waveguide.

Antenna gains and efficiencies are confirmed experimentally in the 61.25 GHz bands. Measured results are presented for only the antennas designed for uniform distribution. Figure 6.54 shows the frequency dependencies of the gain for various aperture area of the antenna. As the aperture size becomes larger, the peak increases. The bandwidth decreases due to the long line effect of the series arrays of the coupling windows and the slots. The peaks of the measured gain at 61.5 GHz are 20.9 dBi with 58.6% efficiency in $20 \times 20\ \text{mm}^2$, 24.6 dBi with 60.6% in $30 \times 30\ \text{mm}^2$, 26.7 dBi with 56.9% in $40 \times 40\ \text{mm}^2$ and 32.7 dBi with 55.6% in $80 \times 80\ \text{mm}^2$. The thin lines in Figure 6.54 are the antenna gains eliminating the reflection loss. These results reveal that the post-wall antenna has the potential to improve the efficiency up to 60% after suppressing the losses associated with the reflection. Figure 6.55 summarizes the efficiency as a function of the gain. The range of the measured efficiency is from 55.6% in $80 \times 80\ \text{mm}^2$ to 60.6% in $30 \times 30\ \text{mm}^2$ for various sizes of antenna. The antenna efficiencies of the 61.25 GHz model are improved by around 10% as compared with those of the 76.5 GHz model. The measured efficiency of a $20 \times 20\ \text{mm}^2$ size

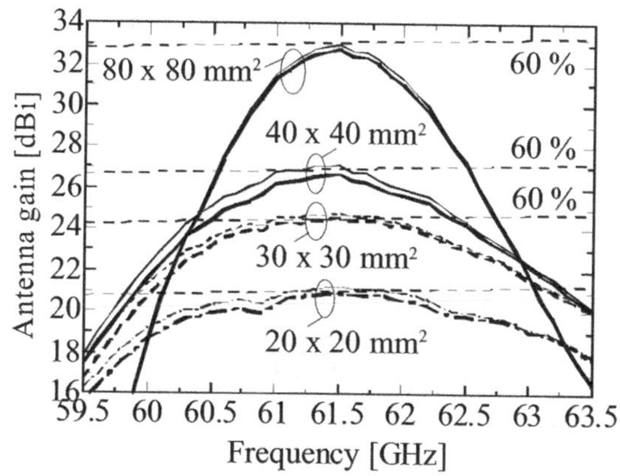

Figure 6.54: Frequency dependence of gain for various aperture area of antenna.

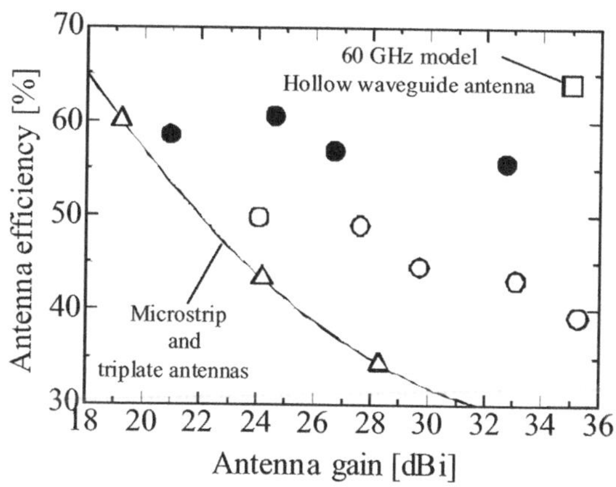

Figure 6.55: Antenna efficiency as a function of gain: black dots, 61.25 GHz model post-wall antennas; white dots, 76.50 GHz model post-wall antennas.

antenna is 20.9 dBi with 58.6% and approaches the level of microstrip and triplate antennas. A 80 × 80 mm^2 size antenna gives 32.7 dBi with 55.6% at 61.5 GHz and can be comparable to the efficiency of hollow waveguide antenna [58] after suppressing the reflection loss.

6.7.4 Sidelobe Suppression and 45° Linear Polarization

A −12 dB amplitude taper of Taylor distribution is adopted in both directions of the 76.5 GHz antenna [59]. The aperture size is 52 × 49 mm^2. The height of PTFE substrate is 0.762 mm and the post diameter is 0.3 for this frequency band. Figure 6.56 shows the amplitude distribution at 77.0 GHz. The strongest field is located around the center of the aperture. The amplitude taper is about −8 dB in both directions, parallel and perpendicularly to the feed. The radiation patterns show successful sidelobe suppression in both planes. The sidelobe

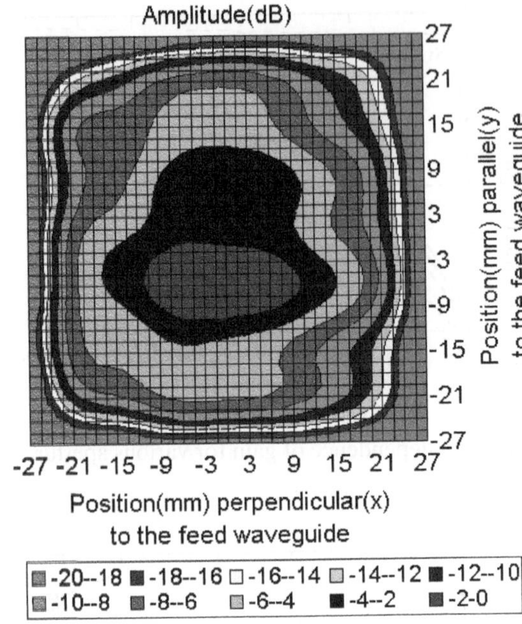

Figure 6.56: Near-field amplitude distribution of the sidelobe-suppressed antenna (from reference [53], reproduced by permission of © 2000 IEEE).

level is −18.1 dB in the E-plane and −18.0 dB in the H-plane. The 3 dB beamwidth is 4.8° in the E-plane and 4.6° in the H-plane.

Figure 6.57 shows the 45° tilted linearly-polarized antenna [60]. A 45° linear polarization has been proposed in mmWave car radar application since radiation with the orthogonal polarization from cars coming from the opposite direction does not affect the radar operation. The slot configuration consists of one radiating slot tilted for a desired polarization and two reflection-canceling slots without any radiation in the boresight. They are separated by a quarter of a wavelength. The amplitude becomes weak around the left side near the feed waveguide in the near field distribution. This may be caused by the oblique propagation of the scattering wave of the tilted radiating slots in the parallel plates. The cross polarization level is below −28 dB in the boresight in the radiation pattern. The two parallel slots for the reflection cancellation in the slot trios are excited out of phase as expected.

6.8 Coaxial-line to Post-wall Waveguide Transformers

6.8.1 Transformer Using a Quasi-coaxial Structure and a Post-wall Waveguide

Figure 6.58 shows a bird's-eye view of the integration with a RF device and a post-wall planar antenna. The packaged RF circuit with the microstrip line interface is mounted on the upper side of print-circuit board in which the post-wall planar antenna is formed. Figure 6.59

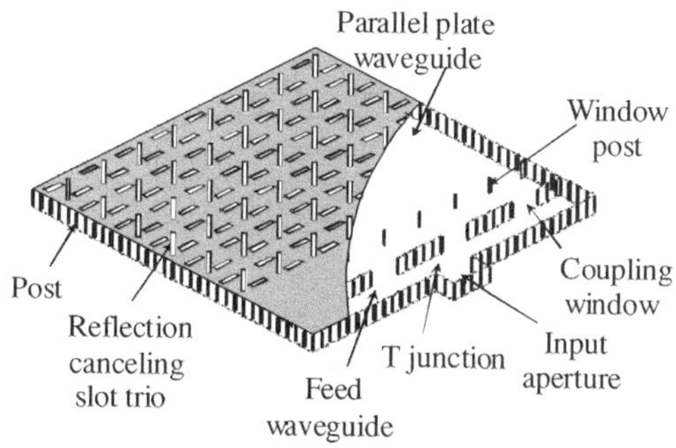

Figure 6.57: 45° tilted linearly-polarized antenna (from reference [60], © 2008 IET).

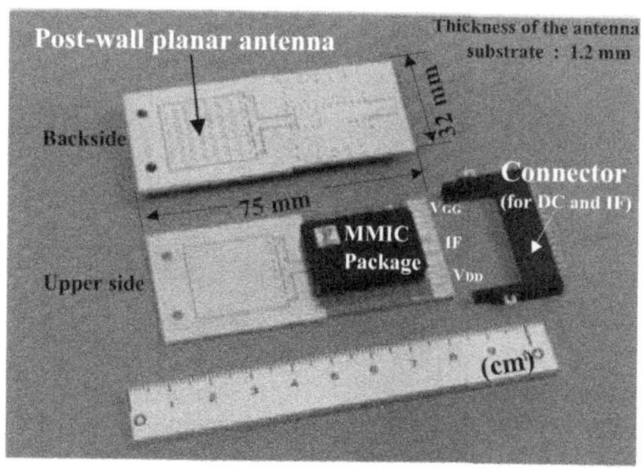

Figure 6.58: A cost-effective 60 GHz module with a post-wall planar antenna (from reference [61], reproduced by permission of © 2006 IEICE).

shows a proposed transformer between a RF device and a planar slot-array antenna using a post-wall waveguide. Figure 6.59(a) presents the sectional view of the transformer. The RF devices are packaged to keep their reliability and has the microstrip line as the interface. The package size is $26 \times 20 \times 4.5$ mm^3. The output power from the RF device is fed through the connection marked in a dotted elliptic circle in the figure. The slot antenna followed by the post-wall waveguide is formed in a PCB. One side length of the antenna is designed to be around 50 mm and the antenna can offer a gain of 25 dBi at 60 GHz. The length of the post-wall waveguide is approximately 10 mm.

Figure 6.59(b) specifies the connection structure between a microstrip line and a post-wall waveguide via a quasi-coaxial structure. Some techniques to connect a microstrip line and a waveguide were reported in references [62–64]. An integrated low temperature co-fired ceramic (LTCC) laminated waveguide-to-microstrip line [62] and a coaxial-to-microstrip transition [63] can be candidates, but need a multilayer substrate and are a bit more complicated from the manufacturing and design points of view. A transition between a microstrip line and a waveguide fabricated on a single layer dielectric substrate [64] yields promising characteristics in terms of reflection and transmission in mmWave bands.

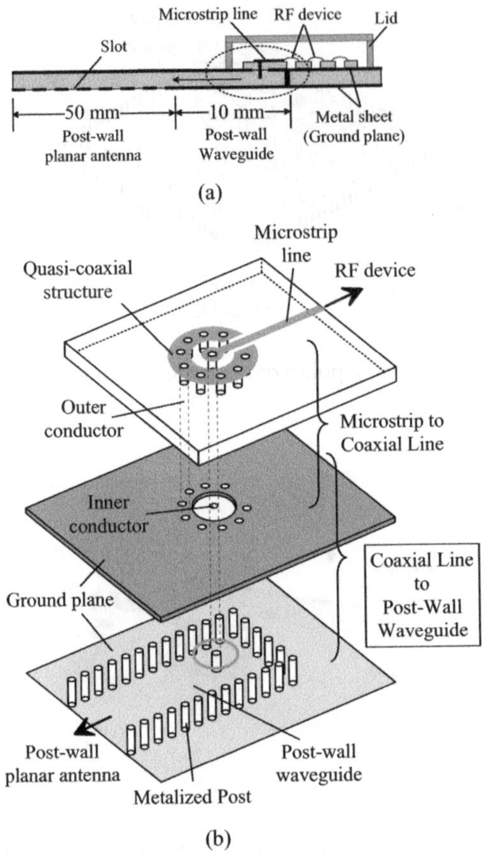

Figure 6.59: A proposed cost-effective transformer between a RF device and a planar slot-array antenna using a post-wall waveguide: (a) the sectional view of the transformer; (b) the precise connection structure between a microstrip line and a post-wall waveguide (from reference [61], reproduced by permission of © 2006 IEICE).

The microstrip line of this structure, however, is connected perpendicularly to a rectangular waveguide. The transformer is not suitable when a microstrip line-based RF circuit is placed in parallel on the backside of a waveguide planar antenna as indicated in Figure 6.59. A compact and cost-effective transformer can be suitable for the integration with a 60-GHz module and a planar slot-array antenna.

The proposed structure consists of two key technologies. One is the quasi-coaxial structure, which is composed of several metalized posts located coaxially around an inner conductor. These posts serve as the outer conductor of the coaxial structure. We have already developed the transformer between a quasi-coaxial structure and a microstrip line in the previous work [65]. The through-loss was measured to be 1.2 dB at 60 GHz, where the receptacle in the V-connector was used for the measurement in place of the post-wall waveguide.

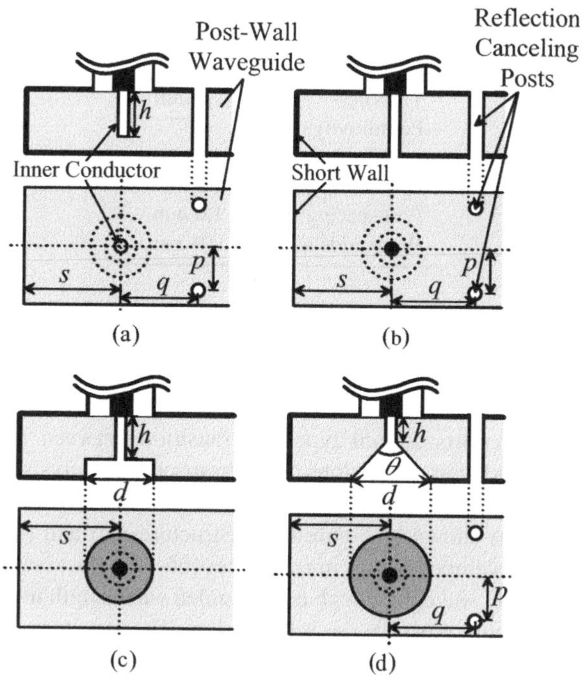

Figure 6.60: Transformers between coaxial line and post-wall waveguide in PTFE substrate: (a) open-ended; (b) short ended; (c) short-stepped; (d) short-taper-stepped (post-walls are replaced with conducting walls at the equivalent position in the analysis) (from reference [61], reproduced by permission of © 2006 IEICE).

Another part of the transformer is a connection between a coaxial line and a post-wall waveguide. The post-wall waveguide and following planar antenna are simply and cost-effectively fabricated by densely arranging metalized posts in the same PCB.

We focus upon a transition between a coaxial line and a post-wall waveguide in order to realize the overall transformer between a microstrip line and a post-wall waveguide, in conjunction with the other transition between a quasi-coaxial structure and a microstrip line in the previous work [65]. From the point of view of mass-production in mmWave bands, the overall characteristics of the transformer greatly depend on the shape and the dimension of the inner conductor.

The PTFE substrate is chosen as the PCB. The substrate parameters are listed up in Table 6.10. A PTFE is low-loss material and a large antenna of gain up to 35 dBi can be realized in the substrate. Manufacturing such as suspended via-holes and its metallization is difficult because the substrate is fiberglass-reinforced and mechanically strong. Various types of transition between a coaxial line and a post-wall waveguide in the PTFE substrate are proposed and discussed in Chapter 3. The manufacturing reliability, the frequency characteristics of the reflection and the transmission loss are investigated and verified experimentally in the 60 GHz band.

Table 6.10: Parameters of PTFE substrate.

Thickness	1.2 mm
Permittivity	2.17
tan δ	0.00085
Post diameter	0.5 mm
Post spacing	1.0 mm
Waveguide width	3.08 mm

6.8.2 Transformer between a Coaxial Line and a Post-wall Waveguide in PTFE Substrate

The literature [66–68] reports several types of transition between a coaxial line and a rectangular waveguide and their discussions concentrate on the analysis and the design of the transformers. We investigate four kinds of coaxial-line to post-wall waveguide transformers in the PTFE substrate, as shown in Figure 6.60. Structures (c) and (d) are proposed here while (a) and (b) were developed before in references [66–68]. Post-walls of waveguides are replaced with a metal-wall waveguide with equal guided wavelength in the design [12].

Feed structures in Figure 6.60 are described below. The structure (a) is an open-ended transformer [66, 68]. The inner conductor is suspended in the middle of a dielectric substrate. The input impedance is controlled over a wide range by changing the insertion length h in a substrate and the position of the short wall s. This structure is not, however, suitable for a PTFE substrate since the reflection characteristics are sensitive for the insertion length h while the control of the length by metallization of a blind alley is very difficult.

Figure 6.60(b) presents the conventional short-ended transformer [67]. The inner conductor penetrates a PTFE substrate and then metallization of the inner conductor is easy. However, the inner conductor excessively perturbs the electric field in the waveguide and the input impedance is much larger than 50 Ω. Reflection suppression over a broad bandwidth would be difficult although the additional posts may be used to suppress the reflection but only in a narrow band.

A short-stepped transformer is shown in Figure 6.60(c) [69]. The inner conductor in a PTFE substrate has a stepped structure at the end. The stepped structure lowers the input impedance and wide band matching to a coaxial line would be expected. Precise manufacturing and metallization of the inner conductor are possible in a PTFE substrate since the inner conductor penetrates the substrate. Figure 6.60(d) shows an alternative realization of Figure 6.60(c) [69]. The short-taper-stepped transformer is installed at the end of the inner conductor. It is expected that the metalizing-liquid flows into the end of the inner conductor more easily.

6.8.2.1 Design

The transformers are designed at 60 GHz to suppress the reflection over a wide frequency range for given dimensions of post-wall waveguides. In the design, a post-wall waveguide is replaced with a metal-wall waveguide to have equal guided wavelength [12]. Table 6.10 indicates the parameters of the PTFE substrate. The diameter of the inner conductor is chosen to be 0.3 mm. An FEM-based electromagnetic field simulator 'Ansoft HFSS' is utilized for the design.

Initial parameters are introduced for each structure, as described below. A cylindrical metallic post that extends into the waveguide is a reactive element in waveguide matching. In a short-ended structure, when h is equal to the waveguide height, the axial current induced by the dominant TE_{10} mode is constant along the post surface so that the inductance is too large. To weaken the EM coupling of the inner field in the waveguide, the short-ended post would be placed for the matching at the position of around $0.5\lambda_g$ (λ_g is guide wavelength in the waveguide) from the shorting wall.

On the other hand, the open-ended structure equivalently works as a series circuit of inductance and capacitance. The metallic post itself has inductance and the capacitance is generated between the open-end of the post and the facing waveguide wall. The input impedance can be controlled by changing the length of the inner conductor h and series resonance is realized when h is equal to about $0.25\lambda_\varepsilon (= \lambda_0/\sqrt{\varepsilon_r})$. The distance between the inner conductor and the shorting wall would be approximately $0.25\lambda_g$ in order to obtain the strong coupling with the inner field of the waveguide. The taper-stepped structures in (c) and (d) are used for impedance reduction so that the position of the shorting wall s would be around 0.25 λ_g as well. Additional reflection canceling posts, which fully penetrate the waveguide, would be installed to assist the suppression of the reflection if required. These posts are located at around $0.25\lambda_g$ and $0.50\lambda_g$ from the inner conductor in the open-ended and short-ended structures, respectively.

Based upon above initial parameters, the insertion length h (in (a), (c) and (d)), the angle of gradient θ (in (d)), the step width d (in (c)), the short position s and the reflection canceling posts position p, q (in (a), (b) and (d)) are determined by iteration in a few turns. Let the shorting posts position s be measured from the center of the inner conductor to that of the shorting posts.

Figure 6.61 shows the calculated frequency characteristics of the reflection for various angles of the inner conductor, where the transformers are designed to minimize the reflection. The structures of $\theta = 90°$, $120°$ and $180°$ give bandwidth wider than 10% with respect to the reflection below -15 dB. Even without reflection canceling posts, a wide bandwidth for reflection suppression is obtained in case of $\theta = 180°$. The taper-stepped structures decrease the input impedance and match to a coaxial line over a wide frequency bandwidth. On the other hand, when the declining angle θ is $0°$ and $60°$, the bandwidths become narrow, 2.0% and 5.9%, respectively. The size of the taper-stepped structure is small and the input impedance of the structure does not decrease, so that the suppression of the reflection is difficult in the sufficient broad bandwidth.

Table 6.11 summarizes the parameters of structure (a)–(d) after fine optimization. The structures of $\theta = 180°$ and $120°$ are applied as short-stepped (c) and short-taper-stepped structures (d), respectively. The parameter p is around 0.35 λ_ε irrespective of the structure.

Figure 6.62 summarizes the calculated frequency characteristics of the reflection for the transitions between a coaxial line and a post-wall waveguide. The reflection of structure (a) has a very wide frequency range of between 55 GHz and 65 GHz with the reflection below -20 dB. In the short-ended structure (b), the bandwidth less than -15 dB is not wide, 2.0 %, as expected. On the contrary, the short-stepped (c) and the short-taper-stepped structures (d) give wide bandwidths of 16.7% and 12.8%, respectively, where the reflections are below

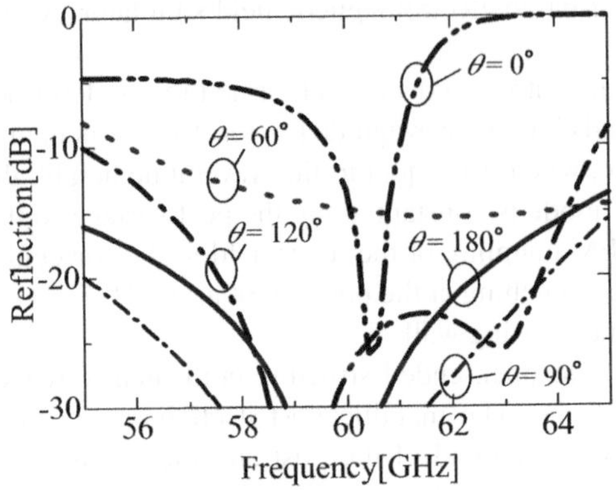

Figure 6.61: Calculated results of reflection characteristics for various angles of inner conductors in PTFE substrate (from reference [61], reproduced by permission of © 2006 IEICE).

Table 6.11: Determined parameters of transitions.

	s mm (λ_g)	h mm (λ_ε)	p mm (λ_ε)	q mm (λ_g)
Open-ended	2.89 (0.70)	0.90 (0.27)	1.20 (0.35)	1.50 (0.37)
Short-ended	1.99 (0.49)	—	1.20 (0.35)	1.80 (0.44)
Short-stepped ($d = 1.4$ mm)	1.25 (0.31)	0.70 (0.21)	—	—
Short-taper-stepped ($d = 2.8$ mm $\theta = 120°$)	3.39 (0.83)	0.39 (0.11)	1.20 (0.35)	3.40 (0.84)

−15 dB. The structure (c) and (d) are comparable to the open-ended structure (a) in terms of the bandwidth of the reflection.

6.8.2.2 Fabrication

Figure 6.63 includes cross-sectional photos of the fabricated structures (a)–(c). In the structure (a), the metalizing does not reach into the end of the inserted inner conductor. The metalizing-liquid tends to be reluctant to creep into the narrow gap. This tendency causes a significant difference between analysis and measurement. On the other hand, the surface of the inner conductor in (b) is metalized smoothly. The stepped structure in (c) is grooved by using a particular T-type drill. The surface and the metallization are almost smooth while the roughness of the metallization around the discontinuity of the stepped structure is observed. Figure 6.63(d) shows a photo from above. The metalizing has been successfully done over

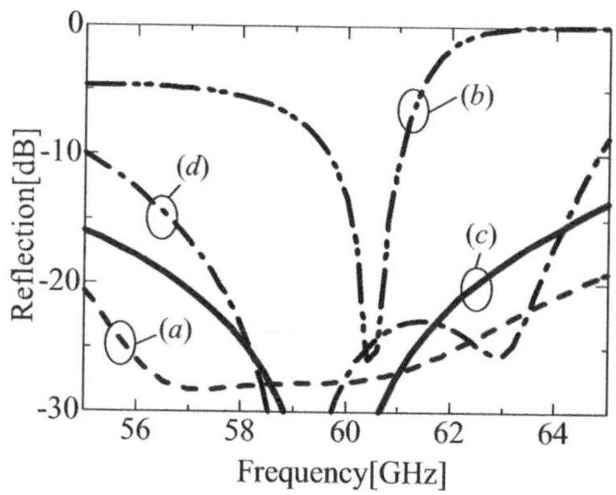

Figure 6.62: Calculated results of reflection characteristics of transitions between a coaxial line and a post-wall waveguide in PTFE substrate (from reference [61], reproduced by permission of © 2006 IEICE).

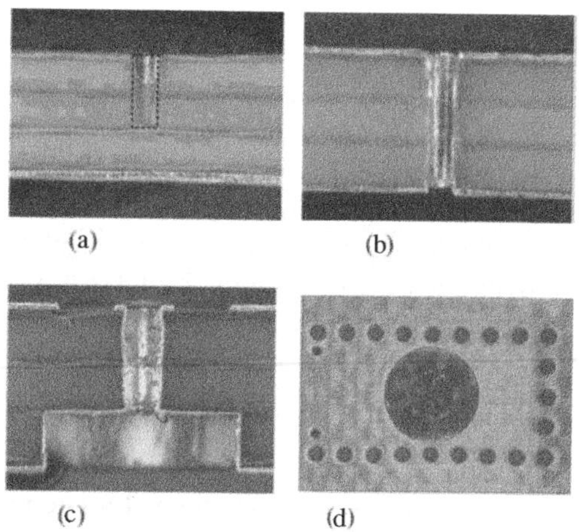

Figure 6.63: Photos of fabricated transitions between a coaxial line and a post-wall waveguide in PTFE substrate: (a) open-ended; (b) short-ended; (c) short-stepped; (d) short-taper-stepped (from reference [61], reproduced by permission of © 2006 IEICE).

the whole inner conductor. The taper structure is smooth and is well suited for metalization, which would reduce the chances of fabrication error.

6.8.2.3 Measured Frequency Characteristics of the Reflection

In the following measurements, the receptacle in the V-connector is installed in place of the post-wall waveguide. Reflection coefficients of the transitions themselves are extracted by using the time-gate function of the vector network analyzer. Figure 6.64 shows the measured frequency characteristics of the reflection for the structure (a)–(d). For the structure (a),

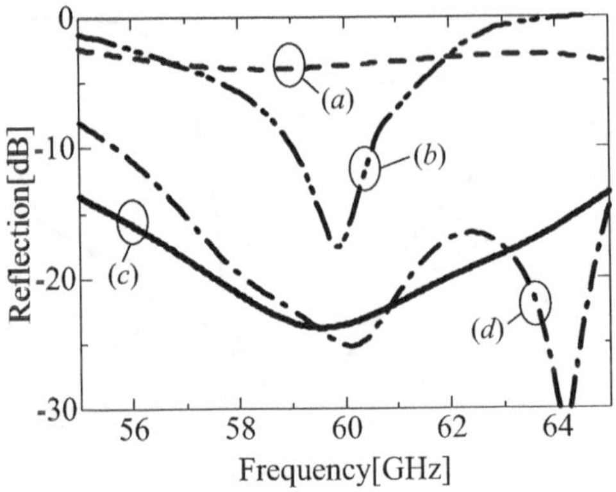

Figure 6.64: Experimental results of reflection characteristics of transitions between a coaxial line and a post-wall waveguide in PTFE substrate (from reference [61], reproduced by permission of © 2006 IEICE).

a serious increase of reflection is observed in the measurement; the reflection is larger than approximately -3 dB, although the broadband suppression of the reflection is predicted in calculation. The short-ended structure (b) has a very narrow bandwidth, 1.1%, for a reflection less than -15 dB. The frequency range below -15 dB of the structure (c) is as wide as 55.6–64.4 GHz (14.7%) while the calculation matches well with the measured bandwidth. The structure (d) also provides a wide bandwidth, 7.9 GHz (13.2%). The discrepancy in (b)–(d) between analysis and measurement is acceptable so that accurate manufacturing is confirmed. The structure (c) and (d) fulfill the required bandwidth of 7.0 GHz for a reflection less than -15 dB and can be candidates for mmWave band wireless systems.

6.8.2.4 Transmission Characteristics of Short-stepped Structure (c)

We fabricated various lengths of straight post-wall waveguide. Each post-wall waveguide is terminated by a few posts at both ends and the transformers with the short-stepped structure (c) are installed as the ports for back-to-back measurement of the transmission coefficients. Receptacles are installed on the structure (c) at both ports in the measurements as shown in Figure 6.65. From the measured transmission characteristics as a function of the waveguide length, the insertion loss of the transformer and the transmission loss per centimeter are identified. Figure 6.66 shows the frequency dependence of the measured transmission loss of each waveguide, the loss per centimeter and the insertion loss for the structure (c). The thin lines show the transmission loss after eliminating the reflection loss of the input aperture in order to compensate for the reflection loss in each waveguide. The transmission loss increases as the waveguide becomes longer. The loss of the post-wall waveguide is around 0.13 dB cm^{-1} and the insertion loss of the transformer only is about 0.13 dB at 60.0 GHz. Assuming 1 cm as the typical size of the connection between the transformer and antenna, we can estimate the total connector loss of about 0.26 dB for the structure (c).

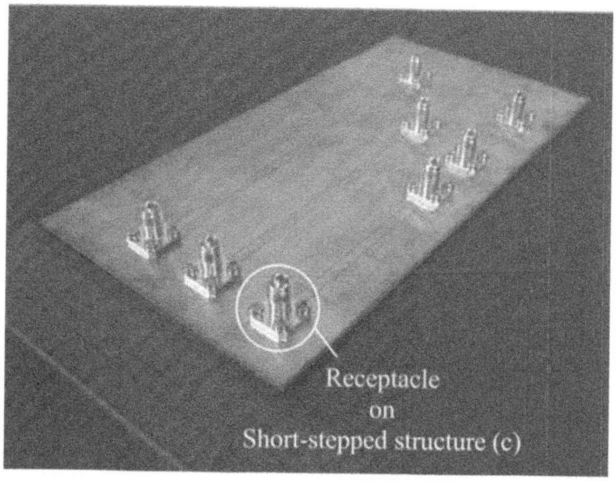

Figure 6.65: Experimental model for transmission characteristics of short-stepped structure (c) in PTFE substrate (from reference [61], reproduced by permission of © 2006 IEICE).

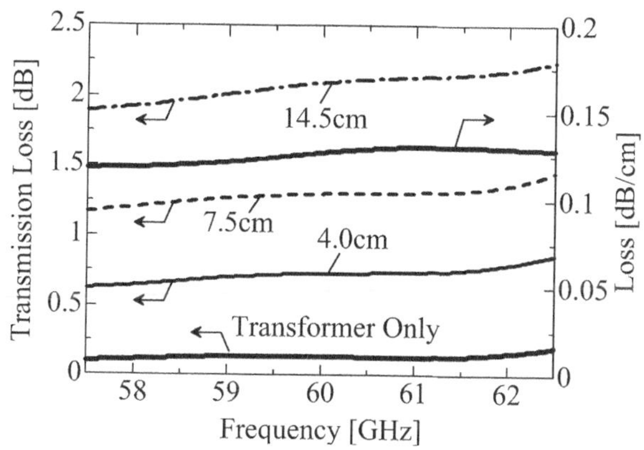

Figure 6.66: Measured transmission characteristics of short-stepped structure (c) in PTFE substrate (from reference [61], reproduced by permission of © 2006 IEICE).

6.8.2.5 Overall Reflection and Loss of a Short-stepped Structure (c) with a Prototype High Gain Antenna

Overall characteristics of the transformer including a large high gain antenna in the PTFE substrate is demonstrated. We manufacture a post-wall waveguide planar antenna [12,53,59] fed by the short-step structure (c) as shown in Figure 6.67. A post-wall fed waveguide in the antenna has several coupling windows to excite a TEM wave in a parallel plate waveguide, where slot pairs are arrayed and designed to obtain a uniform aperture distribution. The design of this feed waveguide is not mature and is still narrow band [53, 59]. Figure 6.68 shows the frequency dependence of the overall reflection at the input port and the gain of

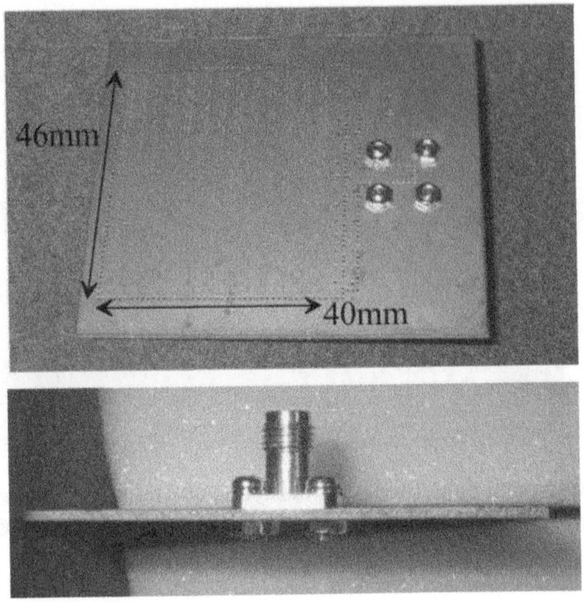

Figure 6.67: Coaxial line fed post-wall waveguide planar slot array antenna (from reference [61], reproduced by permission of © 2006 IEICE).

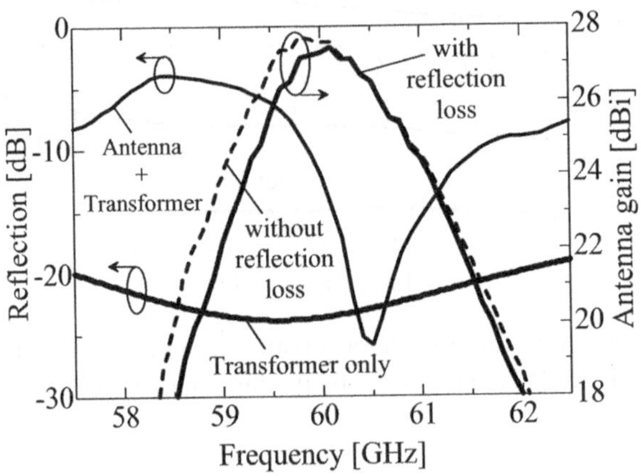

Figure 6.68: Reflection and gain characteristics of the antenna fed by the short-stepped structure (c) (from reference [61], reproduced by permission of © 2006 IEICE).

the antenna. The reflection is around −12.2 dB (6.0% loss) at 60.0 GHz. The peak of the measured antenna gain is 27.3 dBi with 58.2% efficiency at 60.0 GHz for the aperture size $40 \times 46\ \text{mm}^2$. The frequency range in which the gain is larger than 25 dBi is 59.4–60.9 GHz (2.5%). The gain loss due to the reflection is about 0.3 dB (100% − 6% = 94%) around 60 GHz, but is much larger otherwise. In order to identify the gain loss due to these, the thin dotted line in Figure 6.68 indicates the antenna gain without the reflection loss. This result reveals that the antenna with the low loss transformer would have the potential for excellent efficiency up to 60%, and 3.0% bandwidth lager than 25 dBi. The reflection characteristics of the short-stepped transformer only is also included in Figure 6.68. This specific array of the antenna is too narrowband to fully utilize the wideband characteristics of the transformer discussed here.

References

[1] H. Y. Yee, 'Impedance of a narrow longitudinal slots in a slotted waveguide array', *IEEE Trans. Antennas Propagation* **22** (1974), pp. 589–592.

[2] A. J. Sangster and A. H. I. McCormick, 'Moment method applied to round-ended slots', *IEE Proc. Pt. H* **134**(3) (1987), pp. 310–314.

[3] R. W. Lyon and A. J. Sangster, 'Efficient moment method analysis of radiating slots in a thick-walled rectangular waveguide', *IEE Proc. Pt. H* **128**(4) (1981), pp. 197–205.

[4] N. Goto, 'A planar waveguide slot antenna of single layer structure', *IEICE Tech. Rept*, AP88–39, July 1988.

[5] N. Goto, 'A waveguide-fed printed antenna', *IEICE Tech. Rept*, AP89–3, April 1989.

[6] R. Hirose, M. Ando and N. Goto, 'A design of a multiple-way power divider for a single layered slotted waveguide array', *IEICE Tech. Rept*, AP90–130, February 1991.

[7] K. Sakakibara, Y. Kimura, A. Akiyama, J. Hirokawa, M. Ando and N. Goto, 'Alternating phase-fed waveguide slot arrays with a single-layer multiple-way power divider', *IEE Proc. Microw. Antennas Propagation* **144**(6) (1997), pp. 425–430.

[8] Y. Kimura, T. Hirano, J. Hirokawa and M. Ando, 'Alternating-phase fed single-layer slotted waveguide arrays with chokes dispensing with narrow wall contacts', *IEE Proc. Microw. Antennas Propagation* **148**(5) (2001), pp. 295–301.

[9] S.-H. Park, J. Hirokawa and M. Ando, 'A planar cross-junction power divider for the center feed in single-layer slotted waveguide arrays', *IEICE Trans. Commun.* **85**(11) (2002), pp. 2476–2481.

[10] A. C. Ludwig, 'Low sidelobe aperture distribution for blocked and unblocked circular apertures', *IEEE Trans. Antennas Propagation* **30**(5) (1982), pp. 933–946.

[11] N. Goto and M. Yamamoto 'Circularly polarised radial line slot antennas', *IEICE Tech. Rept*, AP80–57, August 1980.

[12] J. Hirokawa and M. Ando, 'Single-layer feed waveguide consisting of posts for plane TEM wave excitation in parallel plates', *IEEE Trans. Antennas Propagation* **46**(5) (1998), pp. 625–630.

[13] M. Zhang, T. Hirano, J. Hirokawa and M. Ando, 'Analysis of a waveguide with a round-ended wide straight slot by the method of moments using numerical-eigenmode basis functions', *IEICE Trans. Commun.* **87**(8) (2004), pp. 2319–2326.

[14] T. Hirano, J. Hirokawa and M. Ando, 'Method of moments analysis of a waveguide crossed slot by using the eigenmode basis functions derived by the edge-based finite-element method', *IEE Proc. Microw. Antennas Propagation* **147**(5) (2000), pp. 349–353.

[15] R. C. Johnson and H. Jasik (ed.), 'Antenna Engineering Handbook' (New York: McGraw-Hill, 1984), ch. 9.

[16] Y. Kimura, M. Takahashi, J. Hirokawa, M. Ando and M. Haneishi, 'An alternating-phase fed single-layer slotted waveguide array in 76 GHz band and its sidelobe suppression', *IEICE Trans. Electron.* **88**(10) (2005), pp. 1952–1960.

[17] J. Hirokawa, K. Sakurai, M. Ando and N. Goto, 'Matching-slot pair for a circularly polarized slotted waveguide array', *IEE Proc. Microw. Antennas Propagation, Pt. H* **137**(6) (1990), pp. 367–371.

[18] H. Seki, 'An alternative representation of electromagnetic fields in a rectangular waveguide with an aperture in its walls', *IEICE Natl. Conv.* **1**(16) (1984), p. 16.

[19] Y. Tsunoda and N. Goto, 'Nonuniformly spaced slot array antenna with low sidelobe pattern', *IEE Proc., Pt. H* **133**(2) (1986), pp. 155–158.

[20] Y. Miura, T. Shirosaki, T. Taniguchi, Y. Kazama, Y. Kimura, J. Hirokawa and M. Ando, 'A low-cost and very small wireless terminal integrated on the back of a flat panel array for 26 GHz band fixed wireless access systems', *IEEE Topical Conf. Wireless Commun. Tech.*, s21p08, Honolulu, Hawaii, October 2003.

[21] R. E. Collin and F. J. Zucker, 'Antenna Theory' (New York: McGraw-Hill, 1969) pt. 1, sec. 14.8.

[22] K. Sakakibara, J. Hirokawa, M. Ando and N. Goto, 'Periodical boundary condition for evaluation of external mutual couplings in a slotted waveguide array', *IEICE Trans. Commun.* **79**(8) (1996), pp. 1156–1764.

[23] Y. Kimura, M. Takahashi, J. Hirokawa, M. Ando and M. Haneishi, '76 GHz alternating-phase fed single-layer slotted waveguide arrays with suppressed sidelobes in the E-plane', *IEEE AP-S Intl. Symp.* **3** (2003), pp. 1042–1045.

[24] J. Hirokawa, M. Ando and N. Goto, 'Waveguide π-junction with an inductive post', *IEICE Trans. Electron.* **75**(3) (1992), pp. 348–350.

[25] T. Takahashi, J. Hirokawa, M. Ando and N. Goto, 'A single-layer power divider for a slotted waveguide array using π-junctions with an inductive wall', *IEICE Trans. Commun.* **79**(1) (1996), pp. 57–62.

[26] W. L. Stutzman and G. A. Thiele, 'Antenna Theory and Design' (John Wiley & Sons Ltd/Inc. 1981), ch. 10.

[27] Y. Kimura, M. Takahashi and M. Haneishi, 'Sidelobe suppression in the H-plane and experimental investigation of alternating-phase fed single-layer slotted waveguide arrays in millimeter-wave band', *IEICE Commun. Conf.*, B-1-102, September 2004.

[28] T. Nagatsuka, M. Ando and N. Goto, 'A calculation of the directive gain of a radial line slot antenna', *IEICE Natl. Conv.*, 747, March 1985.

[29] K. Sakakibara, J. Hirokawa, M. Ando and N. Goto, 'Single-layer slotted waveguide arrays for millimeter wave applications', *IEICE Trans. Commun.* **79**(12) (1996), pp. 1765–1772.

[30] S.-H. Park, Y. Tsunemitsu, J. Hirokawa and M. Ando, 'Center feed single layer slotted waveguide array', *IEEE Trans. Antennas Propagation* **54**(5) (2006), pp. 1474–1480.

[31] R. L. Haupt, 'An introduction to genetic algorithm for electromagnetics', *IEEE Antennas Propagation Mag.* **37**(2) (1995), pp. 7–15.

[32] R. L. Haupt, 'Thinned arrays using genetic algorithm', *IEEE Trans. Antennas Propagation* **42**(7) (1994), pp. 993–999.

[33] M. Shimizu, 'Determining the excitation coefficient of an array using genetic algorithm', *IEEE AP-S Int. Symp.* **1** (1994), pp. 530–533.

[34] Y. Tsunemitsu, Y. Miura, Y. Kazama, S.-H. Park, J. Hirokawa and M. Ando, 'Polarization isolation between two high-gain slotted waveguide arrays arranged side-by-side', *IEICE Commun. Conv.* B-1-210, September 2003.

[35] Y. Tsunemitsu, J. Hirokawa, M. Ando, Y. Miura, Y. Kazama and N. Goto, 'Polarization isolation characteristics between two center-feed single-layer waveguide arrays arranged side-by-side', *ACES J.* **21**(3) (2006), pp. 240–247.

[36] Y. Tsunemitsu, Y. Miura, Y. Kazama, S.-H. Park, J. Hirokawa, M. Ando and N. Goto, 'Polarization isolation between center feed waveguide arrays arranged side-by-side', *IEICE General Conv.*, B-1-172, March 2004.

[37] Y. Tsunemitsu, Y. Miura, Y. Kazama, S.-H. Park, J. Hirokawa, M. Ando and N. Goto, 'Polarization isolation between two center-feed single-layer waveguide arrays arranged side-by-side', *IEEE AP-S Int. Symp.* **3** (2004), pp. 2380–2383.

[38] H. J. Riblet, 'The short-slot hybrid junction', *Proc. IRE* **40** (1952), pp. 180–184.

[39] S. Yamamoto, J. Hirokawa and M. Ando, 'A single-layer hollow-waveguide eight-way Butler matrix', *IEICE Trans. Electron.* **89**(7) (2006), pp. 1080–1088.

[40] Y. Leviatan, P. G. Li, A. T. Adams and J. Perini, 'Single-post inductive obstacle in rectangular waveguide', *IEEE Trans. Microw. Theory Tech.* **31**(10) (1983), pp. 806–812.

[41] S. Yamamoto, J. Hirokawa and M. Ando, 'A half-sized post-wall short-slot directional coupler with hollow rectangular holes in a dielectric substrate', *IEICE Trans. Electron.* **88-C**(7) (2005), pp. 1387–1394.

[42] J. Butler and R. Lowe, 'Beam-forming matrix simplifies design of electronically scanned antennas' *Electron. Des.* **9**(8) (1961), pp. 170–173.

[43] A. Akiyama, T. Yamamoto, J. Hirokawa, M. Ando, E. Takeda and Y. Arai, 'High gain radial line slot antennas for millimeter wave applications', *IEE Proc. Microwaves, Antennas Propagation* **147**(2) (2000), pp. 134–138.

[44] M. Takahashi, J. Takada, M. Ando and N. Goto, 'Characteristics of small-aperture single-layered radial line slot antennas', *IEE Proc., Pt. H* **139**(1) (1992), pp. 79–83.

[45] M. Ando, J. Hirokawa, T. Yamamoto, A. Akiyama, Y. Kimura and N. Goto, 'Novel single-layer waveguides for high-efficiency millimeter-wave arrays' *IEEE Trans. Microw. Theory Tech.* **46**(6) (1998), pp. 792–799.

[46] A. Akiyama, T. Yamamoto, J. Hirokawa and M. Ando, 'Low cost and mass-producible planar antennas for millimeter wave LAN systems', *Proc. 3rd Workshop on Personal Wireless Commun.* 1998, pp. 121–128.

[47] D. T. Paris, W. M. Leach Jr and E. B. Joy, 'Basic theory of probe compensated near-field measurement', *IEEE Trans. Antennas Propagation* **29**(3) (1978), pp. 373–379.

[48] A. Akiyama, K. Sakurai, J. Hirokawa and M. Ando, 'Near-field measurement system for millimeter wave', *IEICE Natl. Conv.*, B-1-127, September 1997.

[49] A. Akiyama, J. Hirokawa, M. Ando, E. Takeda and Y. Arai, '60 GHz band small aperture conical beam radial line slot antennas', *IEICE Trans. on Electron.* **82**(7) (1999), pp. 1229–1235.

[50] S. Hosono, J. Hirokawa, M. Ando, N. Goto and H. Arai, 'A rotating mode radial line slot antenna fed by a cavity resonator', *IEEE AP-S Int. Symp.* **3** (1994), pp. 2200–2203.

[51] J. Takada, A. Tanisho, K. Ito and M. Ando, 'Circularly polarized conical beam radial line slot antenna', *Electron. Lett.* **30**(21) (1994), pp. 1729–1730.

[52] M. Kitamura, T. Yamamoto, J. Hirokawa and M. Ando, 'Design of a coaxial-to-radial line adapter with a step structure', *IEICE Natl. Conv.*, B-1-183, March 1997.

[53] J. Hirokawa and M. Ando, 'Efficiency of 76 GHz post-wall waveguide-fed parallel plate slot arrays', *IEEE Trans. Antennas Propagation* **48**(11) (2000), pp. 1742–1745.

[54] J. Hirokawa, M. Ando and N. Goto, 'A single-layer multiple-way power-divider for a planar slotted waveguide array', *IEICE Trans. Commun.* **75**(8) (1992), pp. 781–787.

[55] N.Goto, 'A technique of grating lobe suppression and an application to planar waveguide slot arrays for dual frequency use', *IEICE Tech. Rept.*, AP87-10, May 1987.

[56] J. Hirokawa, M. Ando and N. Goto, 'Waveguide-fed parallel plate slot array antenna', *IEEE Trans. Antennas Propagation* **40**(2) (1992), pp. 218–223.

[57] M. Ando, K. Sakurai, N. Goto, K. Arimura and Y. Ito, 'A radial line slot antenna for 12 GHz satellite TV reception', *IEEE Trans. Antennas Propagation* **33**(12) (1985), pp. 1347–1353.

[58] M. Ando, J. Hirokawa, T. Yamamoto, A. Akiyama, Y. Kimura and N. Goto, 'Novel single-layer waveguides for high-efficiency millimeter-wave arrays', *IEEE Trans. Microwave Theory Tech.* **46**(6) (1998), pp. 792–799.

[59] J. Hirokawa and M. Ando, 'Sidelobe suppression in 76-GHz post-wall waveguide-fed parallel-plate slot arrays', *IEEE Trans. Antennas Propagation* **48**(11) (2000), pp. 1727–1732.

[60] J. Hirokawa and M. Ando, '45-degree linearly-polarized post-wall waveguide-fed parallel plate slot arrays', *IEE Proc. Microwaves, Antennas Propagation* **147**(6) (2000) pp. 515–519.

[61] T. Kai, Y. Katou, J. Hirokawa, M. Ando, H. Nakano and Y. Hirachi, 'A coaxial line to post-wall waveguide transition for a cost-effective transformer between a RF-device and a planar slot-array antenna in 60-GHz band', *IEICE Trans. Commun.* **89**(5) (2006), pp. 1646–1653.

[62] Y. Huang, K.-L. Wu and M. Ehlert, 'An integrated LTCC laminated waveguide-to-microstrip line T-junction', *IEEE Microw. Wireless Components Lett.* **13**(8) (2003), pp. 338–339.

[63] S. A. Wartenberg and Q. H. Liu, 'A coaxial-to-microstrip transition for multilayer substrates', *IEEE Trans. Microw. Theory Tech.* **52**(2) (2004), pp. 584–588.

[64] H. Iizuka, T. Watanabe, K. Sato and K. Nishikawa, 'Millimeter-wave microstrip line to waveguide transition fabricated on a single layer dielectric substrate', *IEICE Trans. Commun.* **85**(6) (2002), pp. 1169–1177.

[65] U. Azuma, J. Hirokawa, M. Ando, H. Nakano and Y. Hirachi, 'Low loss connection between millimeter-wave RF planar circuits and radial waveguide through coaxial structure', *IEICE Tech. Rept.*, AP2002-121, January 2003.

[66] A. G. Williamson, 'Coaxially fed hollow probe in a rectangular waveguide', *IEE Proc. Pt. H* **132**(5) (1985), pp. 273–285.

[67] A. G. Williamson, 'Analysis and modeling of a coaxial-line/rectangular-waveguide junction', *IEE Proc., Pt. H* **129**(5) (1982), pp. 262–270.

[68] R. B. Keam and A. G. Williamson, 'Broadband design of coaxial line/rectangular waveguide probe transition', *IEE Proc. Microw. Antennas Propagation* **141**(1) (1994), pp. 53–58.

[69] Y. Kato, J. Hirokawa, M. Ando, H. Nakano and Y. Hirachi, 'Coaxial-line feed for millimeter-wave post-wall waveguides', *IEICE Tech. Rept*, AP2003-259, January 2004.

7

Antenna Design for 60 GHz Packaging Applications

Duixian Liu

7.1 Introduction

The increasing capabilities of high-speed silicon germanium (SiGe) technology have made millimeter-wave (mmWave) frequencies attractive for low-cost applications. An overview of the capabilities can be found in reference [1]. Promising applications could be high data rate wireless personal-area networks (WPANs) at 60 GHz [2], automotive radars at 76–77 GHz or 78–81 GHz [3], and imaging at 94 GHz [4, 5]. These applications subsequently require small and low-profile packaged high-performance systems available at moderate to low cost. Such mmWave systems require not only radiofrequency integrated circuits (RFICs), but also a wide range of high-quality passive components, antennas, switches and other devices, which must be densely packaged with high-performance interconnects. It is well known that electronic device packages play a key role in the function of any semiconductor product. Electronic packages conduct signals through a circuit by metal in the form of wires, contacts, foils, plating, and solders. The package further insulates circuits from others and provides environmental protection and physical support of the circuit. Packaging of mmWave components is particularly challenging because of the associated complexity both in the electrical and physical design and fabrication. The small wavelength at mmWave frequencies often demands high-precision machining, accurate alignment, or high-resolution photolithography. A mmWave radio architecture with a baseband modulated interface can relax the frequency requirements, i.e. packages of highly integrated I/Q radios, for example, do not have to conduct frequencies higher than the baseband if the antenna can be integrated together with the chip in a single package [6]. However, integrating an antenna into a package is not an easy task. It is worth noting that there are many papers

Advanced Millimeter-wave Technologies: Antennas, Packaging and Circuits
Duixian Liu, Ulrich Pfeiffer, Janusz Grzyb and Brian Gaucher © 2009 John Wiley & Sons, Ltd

discussing integrating the antennas into the radiofrequency (RF) chips [7–15]. However, directly integrated antennas show very poor performance with antenna gain ranging from −15 to −7 dBi [7, 8]. Micromachined antennas or antennas using high resistivity silicon have gains ranging from −4 to 5 dBi [9–15], but impedance bandwidth is still too narrow for practical applications. These antennas are also not cost-effective due to substantial increase in chip area.

Integrating an antenna into a package is not trivial. The following important factors have to be considered [16].

7.1.1 Material Selection

For mmWave integrated antenna design, two kinds of material are involved. The first one is for the antenna substrate or superstrate, and the second one for packaging. Most antenna substrate material properties are specified at very low frequencies only. Some well-known RF substrates have only minimum material specifications, e.g. at 10 GHz, well below the currently emerging 60 GHz frequency band. Two high-performance substrate materials, Teflon and fused silica, have been evaluated for 60 GHz applications [17]. Teflon has been used for waveguide slot arrays [18], while fused silica has been used for integrated antennas [6, 19–21]. Due to its excellent thermal stability with a 2.5 ppm/°C coefficient of thermal expansion (CTE), fused silica is more suitable for packaging applications. The dielectric constant of fused silica is 3.8 with very low loss tangent at 60 GHz band. Silicon (Si) with high resistivity ($\gg 1$ Ω-cm) can also be used as a substrate material, having an 11.8 dielectric constant. Experiments indicate that silicon substrate with resistivity equal to or better than 1 kΩ-cm yields good antenna efficiency [22].

Figures 7.1 and 7.2 show the measured dielectric properties for a 250 μm-thick non-crystalline amorphous fused silica sample [17]. A very stable dielectric constant of 3.8 and a loss tangent of below 0.0015 up to 90 GHz were observed. Fused silica is already well established for metal deposition processing which enables smaller tolerances than the normal etching process used in standard printed circuit board (PCB) manufacturing. Figure 7.1 also shows the measured dielectric constant for 1 and 10 kΩ-cm silicon while Figure 7.2 also shows the measured loss tangent of the 1 kΩ-cm silicon. Figure 7.2 indicates that the loss tangent for the 1 kΩ-cm silicon is about 0.002 in the 60 GHz band adequate for most applications. These two materials are used in the antenna designs to be discussed later.

Packaging (mold) materials are related to RF chip encapsulation. Encapsulation of RF chips is required to protect the circuit from damage resulting from handling of the package during assembly or the environment (e.g. water absorption, dust, etc.). The encapsulation includes underfill (for flip-chip devices) and glob top. The underfill strengths the flip-chip attachment between the chip and its carrier and acts as a mechanical constraint between the low CTE die and higher CTE substrate. The difference in CTE can result in mechanical strain and failure of the solder interconnects during thermal cycling. The encapsulants are composite materials typically composed of epoxy filled with SiO_2 or SiN. The encapsulation is generally detrimental to RF device performance, especially to the antenna in a package. The encapsulant materials are non-uniform and degrade performance because the electric field generated by the RF circuit in the encapsulant is non-uniform resulting in lossy behavior. Typical packaging materials are very lossy at 60 GHz [17]. There exist some less lossy mold materials, but they are difficult to use and are impractical for high-volume manufacturing.

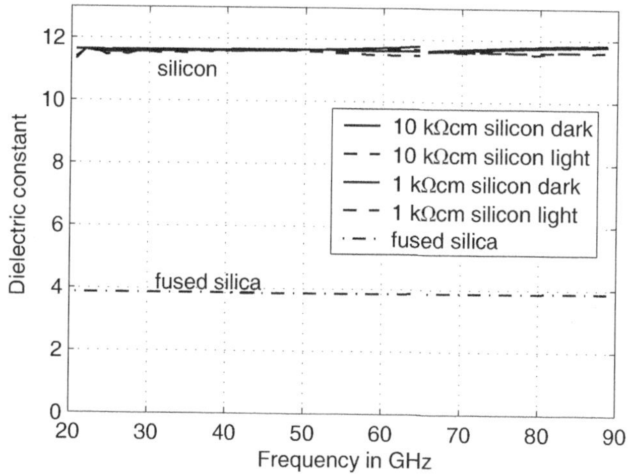

Figure 7.1: Extracted dielectric constant ε_r of both high-resistivity silicon wafers and fused silica over frequency (from reference [17], reproduced by permission of © 2006 IEEE).

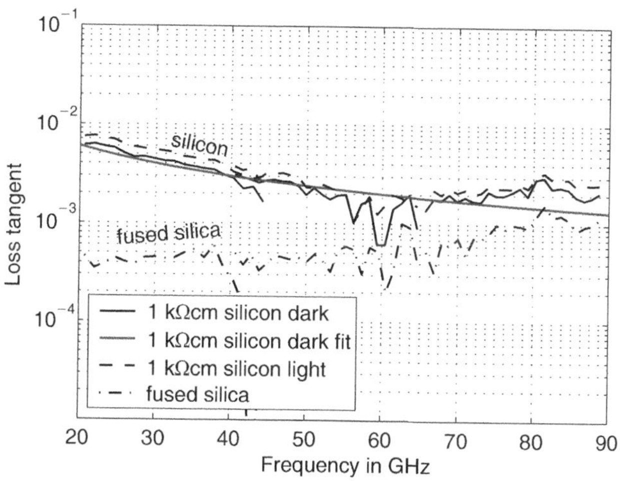

Figure 7.2: Extracted dielectric loss tangent of the 1 kΩ-cm silicon wafer and fused silica over frequency (from reference [17], reproduced by permission of © 2006 IEEE).

7.1.2 Antenna Feed Line

Microstrip lines are widely used at microwave and the lower end of mmWave bands but their application as PCB-based antenna feed lines at 60 GHz is very limited due to dispersion and other issues. Figure 7.3 shows the problems of 50 Ω microstrip lines designed with fused silica ($\epsilon_r = 3.8$). It is clear that the line can be used if the substrate thickness is thinner than 100 μm. However, it is difficult to make microstrip lines on 100 μm or less substrate owing to

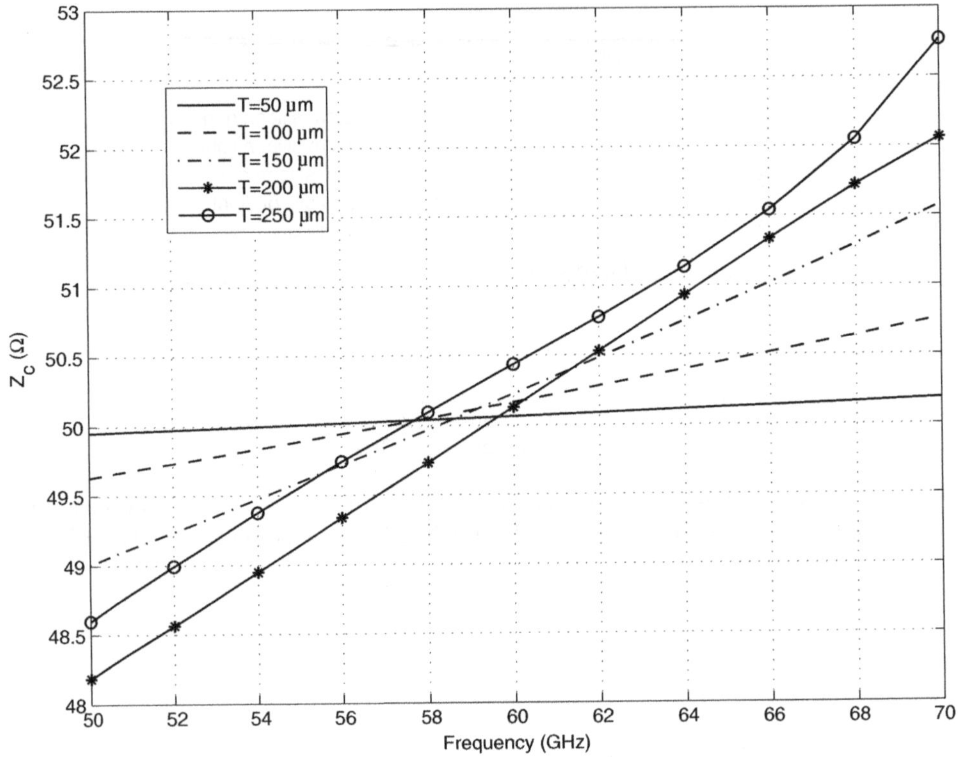

Figure 7.3: Microstrip line characteristic impedance as a function of frequency at different substrate thicknesses for fused silica.

both mechanical weakness and lack of such substrates, particularly for single substrate layer designs. For a realistic 250 μm substrate thickness, the width for a 50 Ω line will be 690 μm, which is about one quarter wavelength (about 680 μm) at 60 GHz, too wide for a practical microstrip line. At this thickness, even simulation tools based on the method of moments will complain that it is too thick to simulate reliably.

Therefore, coplanar strips (CPSs) or coplanar waveguides (CPWs) are usually used. If fused silica is used as the antenna substrate, the characteristic impedance of CPS lines has to be greater than 70 Ω or the CPW line greater than 50 Ω. Otherwise the separation between the metal strips will be less than 20 μm which is the minimum separation required by most thin-film manufacturing processes. If CPS or CPW is used as an antenna feed line, the antenna can be connected to an RFIC with a flip-chip attachment without using vias, which are required if a microstrip feed line is used.

7.1.3 Flip-chip Mount

Antenna attachment is not easy at 60 GHz or higher frequencies. For lower frequencies, wirebond techniques can be easily used without significant impedance mismatch and insertion loss. However, wirebonds at 60 GHz can easily produce 0.5 dB insertion loss.

Table 7.1: Measured S_{11} of wirebond at different lengths (unit: μm). Reproduced by permission of © 2007 IEEE.

CPW – CPW		CPW – GSG Pads on Si	
Transition#/separation/length	S11 (dB)	Transition#/separation/length	S11 (dB)
1 / 58 / 200	−13	1 / 68 / 184.8	−9.1
2 / 58 / 100	−19	2 / 68 / 110	−9.1
3 / 58 / 222	−15	3 / 68 / 405	−10.5
4 / 58 / 70	−22	4 / 68 / 237	−6

From reference [16], © 2007 IEEE, reprinted with permission.

Impedance mismatch is another issue. Figure 7.4 shows wire bond from CPW to CPW both on fused silica and from CPW on fused silica to ground–signal–ground (GSG) pads on Si connection respectively. The 50 Ω CPW was made on fused silica substrate of 300 μm-thick with the center conductor 150 μm wide and the gap to ground 20 μm wide. Four transitions were made for each case. Table 7.1 shows the measured S_{11} for each transition. For the CPW to CPW in the fused silica case, the separation between the two CPW lines is 58 μm, but 68 μm for the CPW on fused silica to GSG pads in the silicon case. Since wirebonds introduce extra inductance, the match will deteriorate if the wire bond is too long. Therefore, flip-chip mounting is preferred. Compared to wirebond, flip-chip attachment has the following advantages [23].

- Eliminating package bond wires reduces the required board area, package height and weight.
- Flip-chip attachment offers the highest speed electrical performance of any assembly method due to the reduced interconnection inductance.
- Flip-chip provides the greatest input/output connection flexibility. Wire bond connections are limited to the perimeter of the die, driving die sizes up as the number of connections increases. Flip-chip connections can use the whole area of the die, accommodating many more connections on a smaller die. Area connections also allow three-dimensional stacking of die and other components.
- Flip-chip is mechanically the most rugged interconnection method when completed with an adhesive 'underfill'.
- Flip-chip can be the lowest cost interconnection for high-volume automated production, with costs below US $0.01 per connection.

7.1.4 Electromagnetic Interference Issues

Planar antennas typically excite strong surface waves. These surface wave not only reduce antenna efficiency but also make antenna performance dependent on the package environment. Furthermore they also affect active circuits nearby. Therefore some mechanisms, such

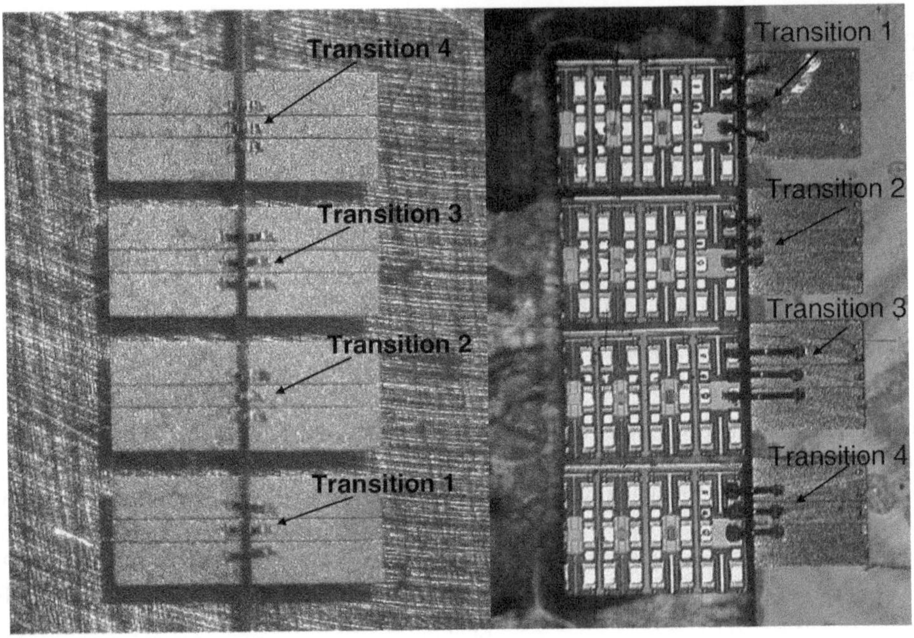

Figure 7.4: Wirebond connection CPW to CPW, both on fused silica (left) and CPW on fused silica to GSG on silicon (right) (from reference [16], reproduced by permission of © 2007 IEEE).

as metal rings and cavities, have to be designed to suppress the surface waves. Figure 7.5 shows an Ansoft HFSS model for this study. In this study, a folded dipole is printed on the bottom side of a 150 μm-thick silicon substrate with a 11.9 dielectric constant and 1 kΩ-cm resistivity. The back cavity has a $2198 \times 2198 \times 399\ \mu m^3$ size. The cavity was implemented in a $2600 \times 2600 \times 600\ \mu m^3$ size silicon block with 1 Ω-cm resistivity, so the silicon cavity wall thickness is 200 μm. The internal wall of the cavity is plated with 1 μm-thick copper. The rings were assumed to be conducting perfectly with zero thickness during the simulations.

Figures 7.6 and 7.7 show the effects on input impedance of no metal ring, with ring 1 only, with ring 2 only, and with both rings. It is clear that the rings definitely affect the antenna impedance. Ring 2 has the most dominate effect. The ring 1 effect is small due to the fact that the copper-plated cavity wall touches ring 1, so most surface waves have already been suppressed by the metal walls. For the low-cost package introduced in references [6, 19, 20], two metal bars were used owing to the radiation nature of folded dipoles. Experiments indicate that the bars effectively eliminate the surface waves, ensuring good electrical performance which correlates quite well with simulations. In other types of antenna, the ring structure is preferred.

7.1.5 Packaging Effects

As discussed in Section 7.1.1, packaging materials are typically very lossy, especially at the 60 GHz band, compared with the antenna substrate. In principle the antenna should not be

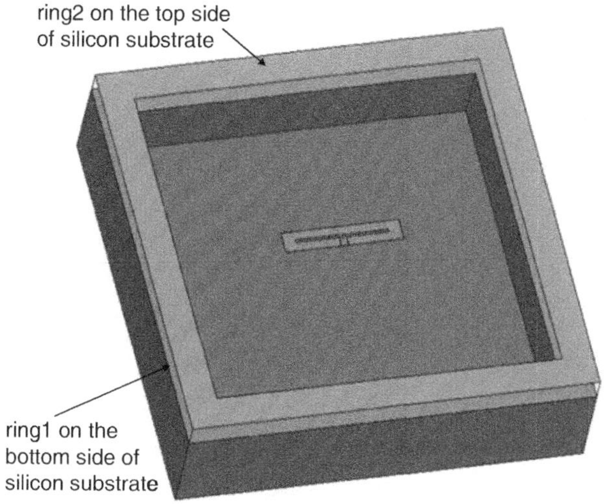

Figure 7.5: Simulation model for ring effects (from reference [16], reproduced by permission of © 2007 IEEE).

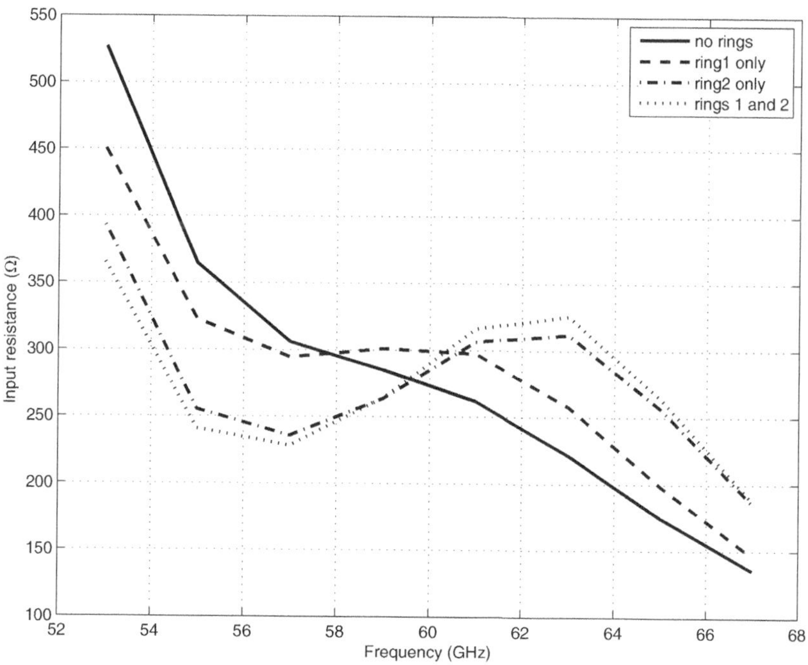

Figure 7.6: Simulated antenna input reactance (from reference [16], reproduced by permission of © 2007 IEEE).

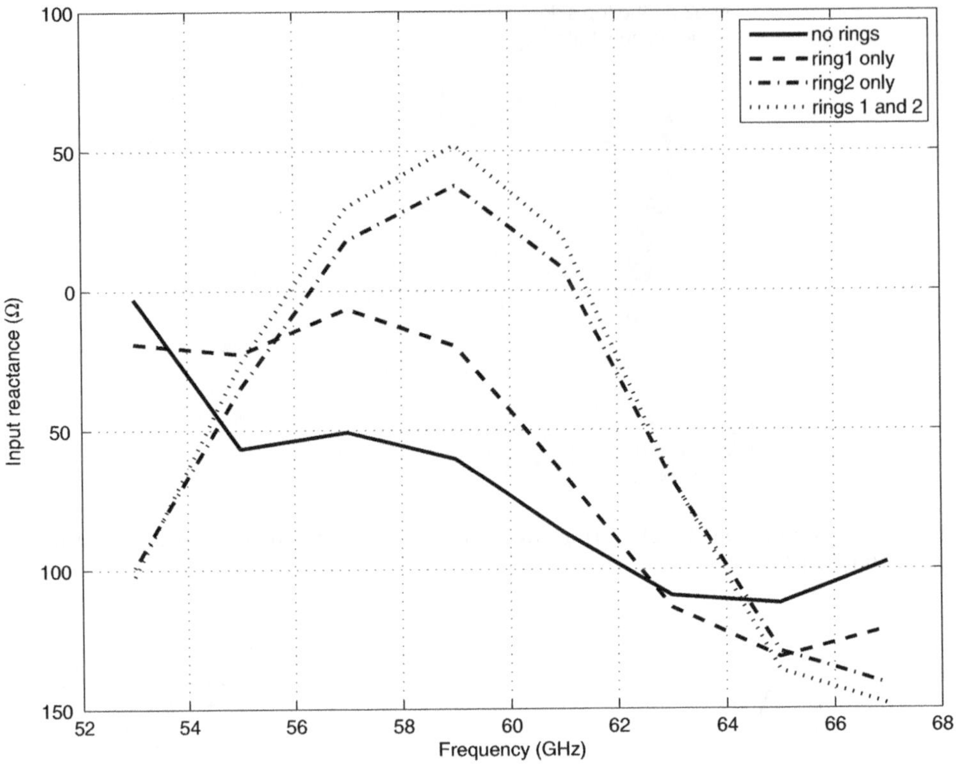

Figure 7.7: Simulated antenna input resistance (from reference [16], reproduced by permission of © 2007 IEEE).

covered by any packaging materials, but this is impractical for a real manufacturable solution; but if the antenna system is designed properly, taking into account appropriate surrounding materials (see the next section), the antenna efficiency degradation can be made low if, e.g., the cover thickness over the antenna radiation portion is minimized and accounted for in the design.

Package environment affects antenna impedance, bandwidth, radiation and efficiency. These effects can be minimized by utilizing a metal cavity around the antenna and designing a wide bandwidth antenna.

7.1.6 Antenna Design

The goal is to find a cost-effective approach for an off-chip antenna design and construction that would show the potential for its integration with a 60 GHz SiGe chipset, described in [1, 24], in a low-cost plastic chip-scale packaging technology [6, 19, 20] or silicon carrier based wafer-scale packaging technology [22]. Obtaining the maximum bandwidth-gain product with a given design constraint is the basic challenge for the design of printed planar antennas [25, 26]. An approach that combines a fused-silica superstrate and an air

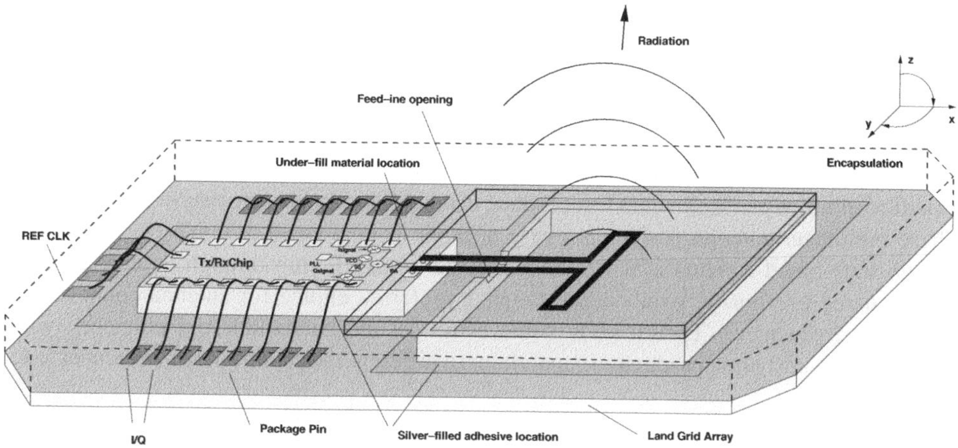

Figure 7.8: Conceptual drawing for the build-up of an LGA package (from reference [6], reproduced by permission of © 2007 IEEE).

substrate suitable for flip-chip attachment and packaging is presented in [19, 20], where the antenna is printed on the bottom side of the substrate and suspended in air over the reflecting ground. This solution results in a wide bandwidth of more than 30% and an antenna efficiency of over 90%. Figure 7.8 shows the conceptual drawing of the proposed land-grid-array (LGA) package structure with an integrated antenna [6]. Except for the 60 GHz RF signal, standard wirebonding is used. With the antenna integrated in the package, all 60 GHz signals are confined within the package while all other low-frequency signals are connected to the package external leads. The mold material in the antenna feed-line opening, as well as the package encapsulation material, are omitted for better clarity. The following sections discuss the details of this antenna design and packaging concept.

7.2 Air-suspended Superstrate Antenna

The packaged antennas in this chapter are based on air-suspended superstrate antenna concepts; therefore it is important to have a good understanding of the air-suspended superstrate antenna.

Depending on the country, the 60 GHz band spans from 57 to 66 GHz with a near world-wide overlap from 59 to 64 GHz. The major challenges for a practical packaged antenna design is to achieve a wide antenna bandwidth of 10% or better while maintaining a high-efficiency antenna and being manufacturable in today's low-cost mass production processes. In PCB antennas such as patch antennas, bandwidth and efficiency can be traded against each other but it is very difficult to improve both parameters simultaneously. As shown in references [25] and [26], the bandwidth-efficiency 'product' can be optimized by using the lowest dielectric constant substrate available. A patch suspended in air was investigated in reference [25]. However, such designs lack mechanical stability and can not be realized with the dimensions for mmWave applications. An L-band microstrip antenna using a very thin substrate and foam (dielectric constant of nearly 1) between the substrate and ground plane achieves about 3% bandwidth with 84% efficiency [26]. The same idea is behind the 802 GHz tapered slotline antenna on a 1.7 μm thin SiO_2/Si_3N_4 dielectric membrane presented in reference [27]. In reference [28] the dielectric constant of the silicon substrate was minimized by drilling closely spaced holes in the substrate. An overview of other mmWave and terahertz antennas can be found in references [29] and [30]. Of course, stacked aperture-coupled patch antennas can produce bandwidths of up to one octave [31]. However, these multilayer antenna designs are very difficult to implement at mmWave frequencies. A substrate-based antenna is typically connected to an RF chip through wirebonding; but as discussed previously, wirebonding is not a preferred choice for antenna connection at mmWave frequencies. Therefore, to achieve the desired bandwidth and efficiency while enabling low-cost mass production capabilities, a different concept from the references listed above is required.

Placing a superstrate with high-dielectric constant over a PCB antenna can increase the antenna gain drastically as shown in references [32–34]. The detailed investigation on the fundamental effects of a superstrate or cover on a printed circuit antenna is presented in reference [35], and it demonstrates that efficiencies approaching 100% can be achieved by selecting the right material thicknesses and properties to minimize surface waves [36]. Other methods to achieve 100% efficiency also exist [37] and [38] but their applications are very limited and difficult to implement in practical applications, especially at mmWave frequencies. In references [39–41], it was found that a dielectric cover on a PCB antenna decreases the resonant frequency as expected, but also increases the half-power bandwidth of the antenna. Here, a new antenna design is proposed which combines the findings of the above-cited publications with a special requirement that the antenna is suitable for integration into a mmWave chip package.

Figure 7.9 shows a side view sketch of the proposed antenna design stackup [19]. An antenna structure on a substrate of thickness T is flipped onto an integrated circuit (IC) or carrier with solder balls ('interconnect') as shown in Figure 7.9. The ground of the IC package base also acts as the reflecting ground plane for the antenna with a spacing H. This results in a superstrate configuration with air between the antenna and the ground plane and a higher dielectric material above the antenna structure. Wider bandwidth than standard PCB antennas can be achieved with this stackup while maintaining high antenna efficiency. For mechanical support, a spacer with solder balls might be placed on the other end of the antenna superstrate, as shown in Figure 7.9. In a fully integrated transceiver IC [1, 24], all input–output (IO) signals except the ones to and from the antennas are baseband signals. Therefore and also because the IC is only partly covered by the antenna superstrate, all other IO signal pads of the IC can be connected to the package leads via ordinary bond wires as in standard IC packages. The whole structure can be covered with a mmWave transparent lid or by using a low loss and low dielectric mold material. A dielectric property investigation of mold materials as well as antenna substrate materials can be found in reference [17].

One problem with the proposed antenna stackup is that microstrip feed lines cannot be used. A microstrip line with the package base as ground is not realizable, since the spacing H is too large for a good microstrip transmission line. In principle, an inverted microstrip line [42] with the ground plane on the top of the superstrate would be possible, but first this would limit the antenna choices to PCB antennas with a radiating aperture in the upper ground plane. Second, to achieve reasonably narrow microstrip lines for the mmWave frequencies,

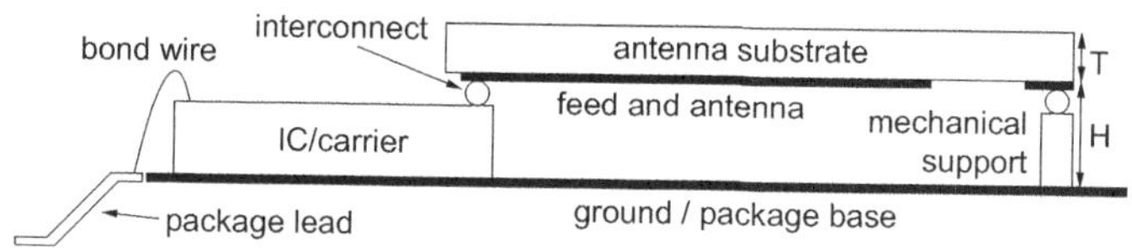

Figure 7.9: Concept of antenna and IC or carrier configuration (from reference [19], reproduced by permission of © 2006 IEEE).

very thin superstrates would be needed that are not realizable in most materials. Third, at the interconnect, the ground signal has to be connected to the IC via closely spaced solder balls. This would require vias through the superstrate layer with very tight spacing, which is impossible for most materials that are suitable for mmWave applications (e.g. fused silica, alumina). Thus, the most viable transmission line structures to connect the antenna with the flip-chip solder balls are either finite ground coplanar waveguides (FGCPW) with three solder balls in a GSG configuration or CPS lines with two solder balls in a ground–signal (GS) or signal–signal differential configuration. Using the balun first presented in reference [43], a very broadband and low-loss transition between FGCPW and CPS transmission lines can be achieved that consumes very little space.

7.2.1 Air-suspended Superstrate Antenna Designs

As discussed in Section 7.1.1, fused silica is an excellent material in both electrical and thermal properties. Furthermore, gold antenna and feed-line structures on fused silica can be manufactured with tight tolerances of around 10 μm as required here. Therefore, most of the antenna designs in this chapter are based on fused silica. The folded dipole was chosen due to its small dimensions and the fact that it perfectly fits to the FGCPW or CPS feed lines.

Figures 7.10 and 7.11 show a single folded dipole and a dual folded dipole design for the 60 GHz band respectively. The single folded dipole design is targeted for the transmitter that requires a 100 Ω differential feed line [24], while the dual folded dipole is targeted for the receiver that requires a 50 Ω CPW feed line [24]. The main goal for the designs was a good antenna impedance match (better than 10 dB return loss) between 59 and 64 GHz and a high efficiency in the frequency range.

Why is the folded dipole chosen here instead of the regular dipole? The input impedance of a dipole on fused silica superstrate would be around 40 Ω. However, the lowest manufacturable transmission line characteristic impedance for CPS lines on fused silica (based on state-of-the-art manufacturing capabilities) is around 100 Ω. Therefore, the ordinary dipole would be difficult to match to a CPS feed line. To overcome this problem a folded dipole was used instead, since its input impedance is about four times of that of a normal dipole [44, Ch. 5.2]. Background on folded dipoles, which have been in use as wire antennas as well as PCB antennas for a long time, can be found in references [45–48]. In reference [49], a dual dipole antenna with a CPS feed on a thin dielectric membrane was proposed for 246 GHz. A dual folded dipole for 5.2 GHz or 5.8 GHz but with a microstrip feed was presented in reference [50]. In reference [51], a double-folded slot antenna was chosen, but its input impedance was only around 20 Ω at the antenna resonant frequency of

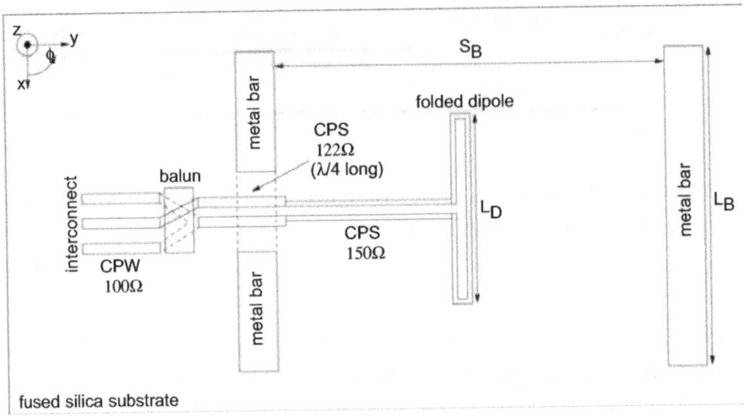

Figure 7.10: A layout sketch of a folded dipole with metal bars for radiation pattern improvement (from reference [19], reproduced by permission of © 2006 IEEE).

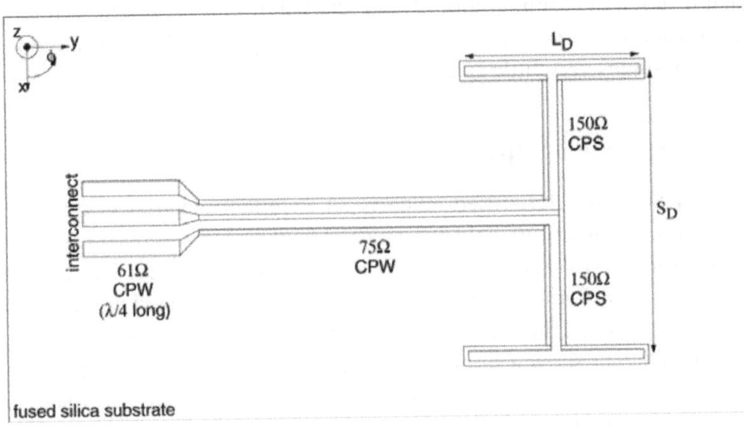

Figure 7.11: A layout sketch of a dual folded dipole array (from reference [19], reproduced by permission of © 2006 IEEE).

around 94 GHz. Another important feature of the folded dipole antenna is its high efficiency, as will be seen later, owing to its high input impedance.

Extensive simulations indicate that the optimal height H is about 500 μm, which perfectly fits the typical thickness of packaged SiGe ICs plus solder ball heights. The superstrate thickness $T = 300$ μm is a compromise, since thinner fused silica samples would become difficult to manufacture (material becomes brittle) and a thicker superstrate leads to reduced efficiency due to more energy staying in the dielectric layer as parasitic surface waves. This air-superstrate combination is also consistent with the configuration that yields good efficiency results presented in reference [35].

The antenna in Figure 7.10 is a half-wavelength folded dipole of length $L_D = 1.4$ mm on fused silica. It has a 150 Ω input impedance and is fed by a 150 Ω CPS transmission line followed by a quarter-wavelength-long 122 Ω CPS line to transform the input impedance to 100 Ω. At this point the antenna could be connected to a 100 Ω IC terminal via two solder balls of about 150 μm pitch, which perfectly fits to the IC design presented in references [1, 24]. However, to measure the antenna performance with a GSG probe, a broadband balun [43] and a 100 Ω FGCPW line are added to the design. The FGCPW line has a 150 μm pitch, which again fits to the IC design in references [1,24]. A folded dipole has an omnidirectional radiation pattern in the plane perpendicular to the folded dipole. Since any radiation parallel to the substrate is very undesirable, additional metal bars have been placed parallel to the folded dipole to minimize the radiation parallel to the superstrate as shown in Figure 7.10. The length of the bars is $L_B = 3$ mm and their spacing is $S_B = 3.6$ mm, but antenna performance is not very sensitive to these dimensions. The whole fused silica sample has a 12.5×3.4 mm^2 size. Most of the area is required to allow fixturing in the measurement setup.

The dual folded dipole configuration, as shown in Figure 7.11, has two important features. First, it provides another mechanism beside the metal bars shown in Figure 7.10 to minimize the radiation parallel to the superstrate. Second, it naturally forms FGCPW lines required in many mmWave RFIC designs. Two of the same folded dipoles as before are combined via 150 Ω CPS lines into a 75 Ω FGCPW line. The dipoles are spaced at half a free space wavelength $S_D = 2.5$ mm. At the end of the 75 Ω FGCPW line, a quarter-wavelength-long 61 Ω FGCPW line transforms the input impedance to 50 Ω at the interconnect point. The pitch of the FGCPW line at the interconnect is again 150 μm, which perfectly fits to the IC design presented in references [1,24]. The second design was realized on samples of the same size as for the first design (12.5×3.4 mm^2).

With the use of a quarter wavelength transformer and the balun as presented above, both designs can be made with either a FGCPW interconnect with an impedance between 50 Ω and 150 Ω or a CPS interconnect with an impedance between 100 Ω and 200 Ω as the IC requires.

7.2.2 Air-suspended Superstrate Antenna Evaluation

Since a flip-chip attachment is required, connectors are not necessary in the antenna design. This eliminates the tremendous effort required to de-embed the connector effects on antenna performance. This makes the antenna characterization much easier at mmWave frequencies. A probe-based measurement setup proposed in reference [52] enables the simultaneous measurements of the input impedance right at the end of the coplanar feed line and the far field radiation patterns. Thus the antenna impedance/matching can be measured exactly at the flip-chip attachment point to the RF IC. Due to the measurement setup using a vector network analyzer, waveguides, and a waveguide rotational joint, the measurement frequency range is limited from 50 to 65 GHz and only radiation patterns with a 180° angular range can be obtained without setup change. Figure 7.12 shows photos of the single folded dipole antenna mounted in the sample holder and connected by the coplanar probe. A thin-film metal patterning process on fused silica was used to make the antenna samples. As can be seen on the left side of Figure 7.12, the antenna has to be mounted upside-down during the measurement. This will allow probing from the top, and the antenna ground plane can be provided by placing a copper ground ‘bridge’ over the antenna as shown in the right side of

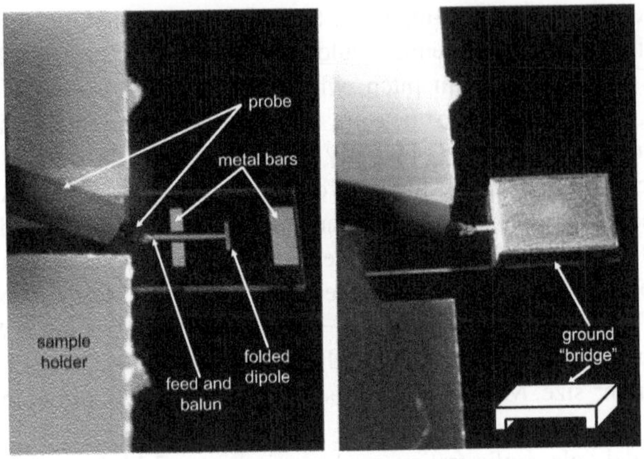

Figure 7.12: Photo of the single folded dipole in measurement setup with and without ground 'bridge' (from reference [19], reproduced by permission of © 2006 IEEE).

Figure 7.12. This measurement arrangement naturally reduces the probe assembly effort on antenna performance.

Ansoft's HFSS was used to design these antennas. The sample holder and the probe assembly were not considered in the simulations. In fact, the sample holder and the probe assembly have negligible effects on antenna performance, as discussed above. Although the maximum measured frequency is limited to 65 GHz, simulations were performed up to 70 GHz to show antenna performance tendencies.

Figure 7.13 shows the input match (S_{11} in decibels) of the single folded dipole for $Z_0 = 100\ \Omega$. Very good agreement between the measured and simulated results with the ground 'bridge' can be observed. A good match was achieved from 58 to 67 GHz which even leaves some tolerance margin against the target bandwidth. To evaluate the influence of the ground dimensions the simulation was repeated for the infinite ground plane. From this simulation result, it can be seen that even with a larger ground plane the bandwidth can be expected to stay the same.

The measured and simulated antenna gain is shown in Figure 7.14. Within the frequency band of interest, 7–9 dBi gain was achieved. The simulations show that the antenna has an efficiency better than 90%. Figures 7.15 and 7.16 show the measured and simulated co-polarization patterns of the antenna in the planes at $\phi = 0°$ and $\phi = 90°$ (see Figure 7.10 for coordinate system definition) respectively. Again the measured and simulated results are in good agreement. This provides confidence that the simulated antenna efficiency is close to the actual one. Note that the simulated radiation pattern shown in Figure 7.15 is asymmetric. This is caused by an imperfect 180° phase difference feed to the folded dipole due to the balun.

Beside a good agreement between measurements and simulations, the radiation patterns show that the energy radiated parallel to the superstrate was successfully limited by the bars shown in Figure 7.10.

The input match of the dual folded dipole array (see Figure 7.11) for $Z_0 = 50\ \Omega$ is shown in Figure 7.17. Again a good antenna match was achieved in the frequency range of 58 GHz

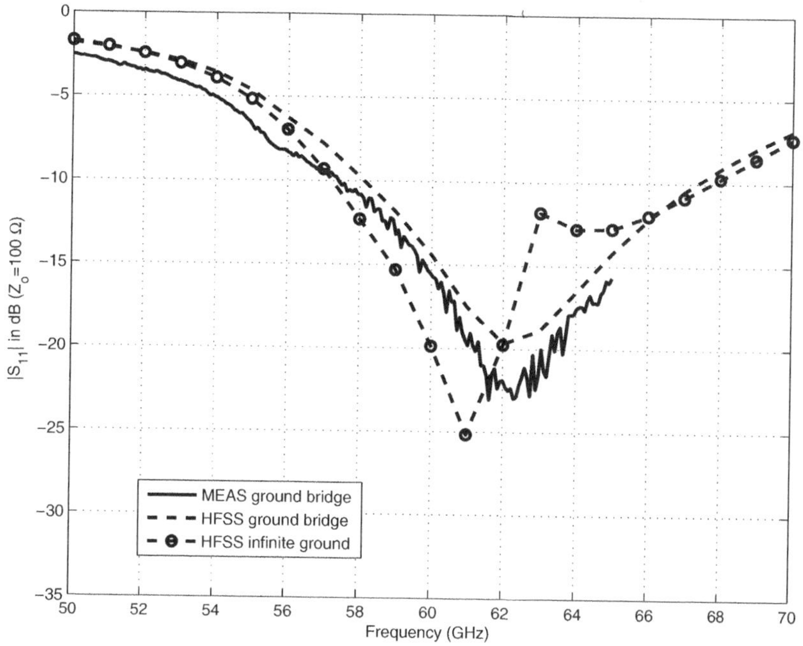

Figure 7.13: Measured and simulated antenna input match of the single element folded dipole (from reference [19], reproduced by permission of © 2006 IEEE).

to above 66 GHz, exceeding the target bandwidth for both cases: with the ground 'bridge' as well as for the infinite ground. The dual dipole array also achieves an antenna gain at $\theta = 0°$ of 7–9 dBi inside the target frequency range. The fact that the gain of the dual dipole array is not higher than for the single dipole proves the effectiveness of the bars used to limit the horizontal radiation of the single folded dipole. The antenna efficiency was simulated to be at least 85% in the target frequency range as shown in Figure 7.18.

The co-polarization radiation patterns in the $\phi = 0°$ and $\phi = 90°$ planes (see Figure 7.11 for coordinate system definition) are given in Figures 7.19 and 7.20 respectively. Again the antenna shows a nearly hemispherical radiation pattern as would be desirable for applications such as WPANs.

In both design cases, an impedance bandwidth better than 13% and a gain of about 8 dBi can be achieved. The efficiency is difficult to measure at 60 GHz band, but is expected to be close to 90% simulated efficiency owing to good agreements between simulation and measurement results.

7.3 Packaged Antennas

The antenna described above is supposed to be flip-chip mounted to a front-end SiGe chip, and requires an additional spacer with solder balls attached on the other side of the substrate as a mechanical support. This may result in an unstable mechanical design. Furthermore,

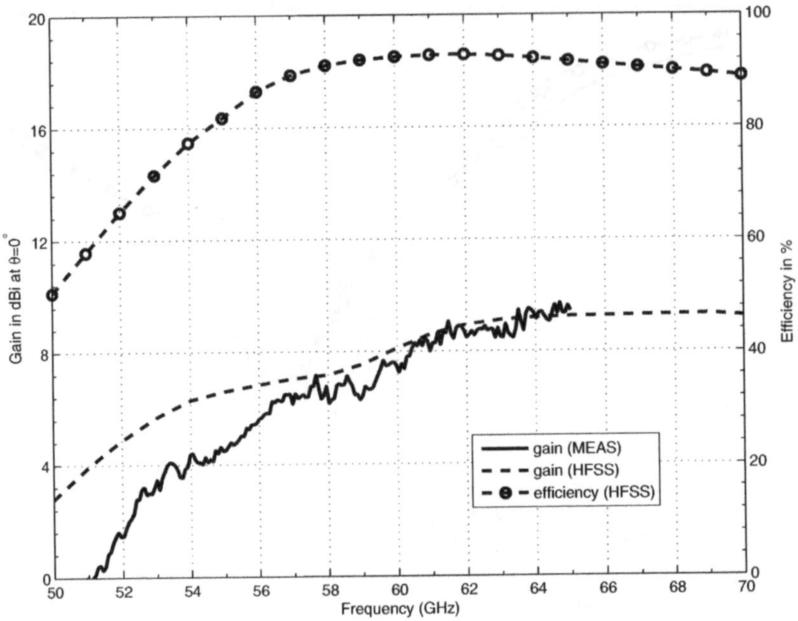

Figure 7.14: Measured and simulated gain and efficiency of the single element folded dipole with ground 'bridge' (from reference [19], reproduced by permission of © 2006 IEEE).

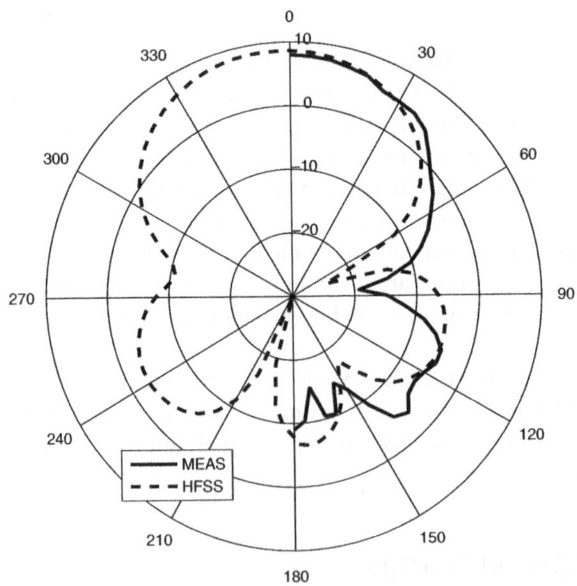

Figure 7.15: Measured and simulated co-polarization pattern of the single element folded dipole in the $\phi = 0°$-plane (from reference [19], reproduced by permission of © 2006 IEEE).

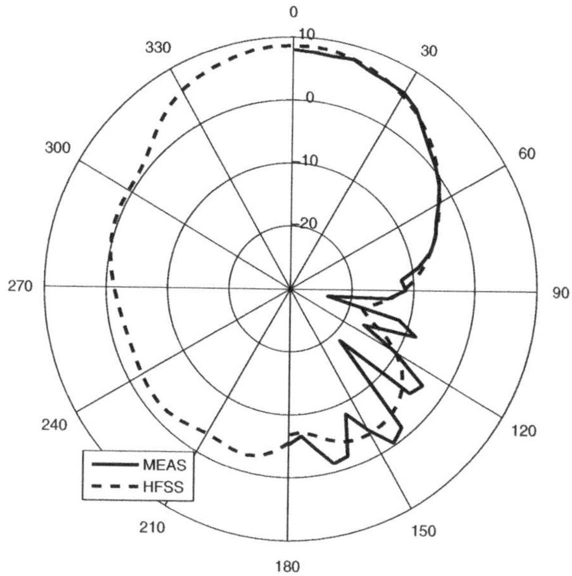

Figure 7.16: Measured and simulated co-polarization pattern of the single element folded dipole in the $\phi = 90°$-plane (from reference [19], reproduced by permission of © 2006 IEEE).

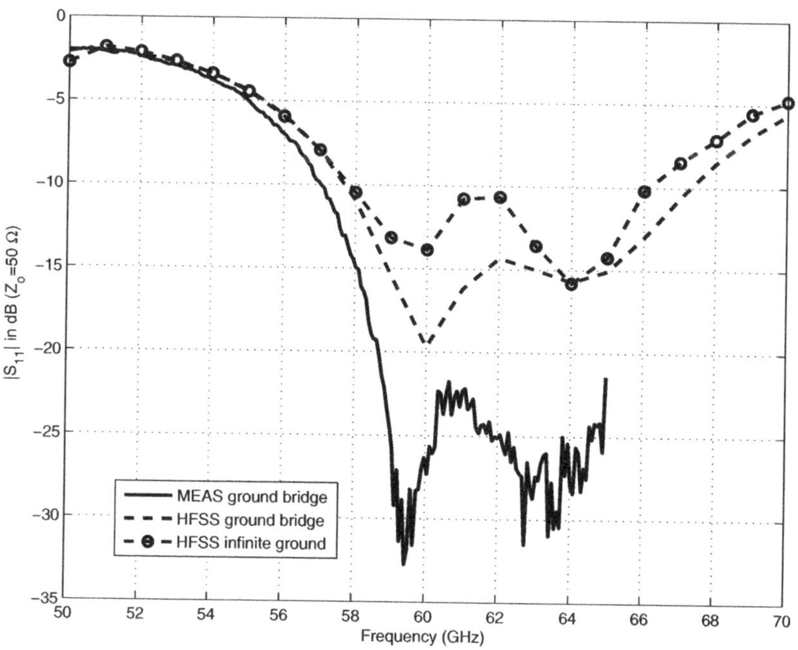

Figure 7.17: Measured and simulated antenna input match of the dual folded dipole array (from reference [19], reproduced by permission of © 2006 IEEE).

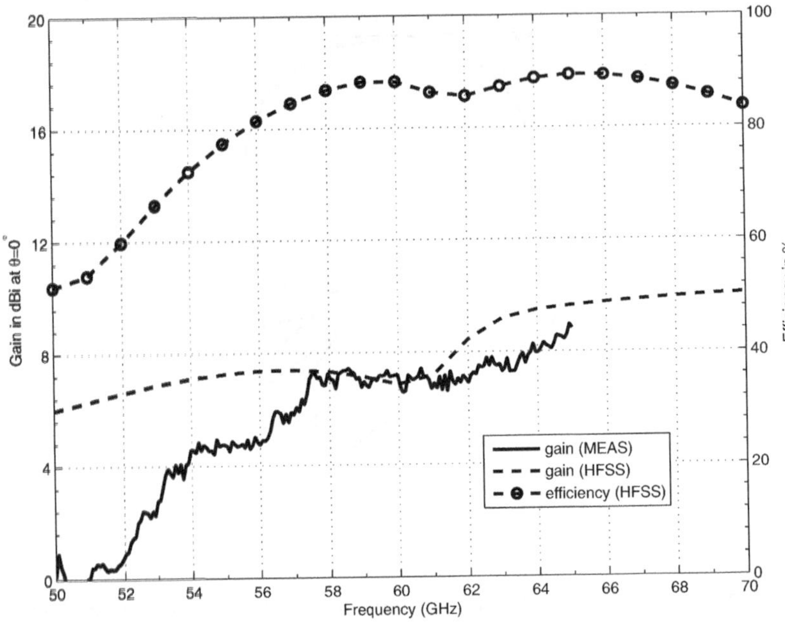

Figure 7.18: Measured and simulated gain and efficiency of the dual folded dipole array (from reference [19], reproduced by permission of © 2006 IEEE).

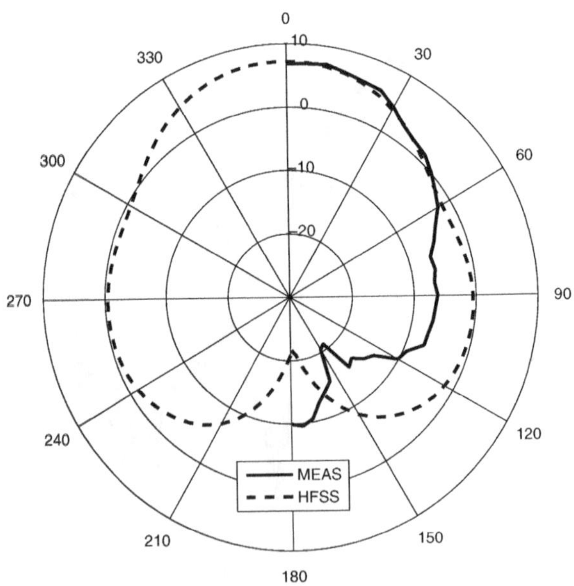

Figure 7.19: Measured and simulated vertical co-polarization pattern of the dual folded dipole array in the $\phi = 0°$-plane (from reference [19], reproduced by permission of © 2006 IEEE).

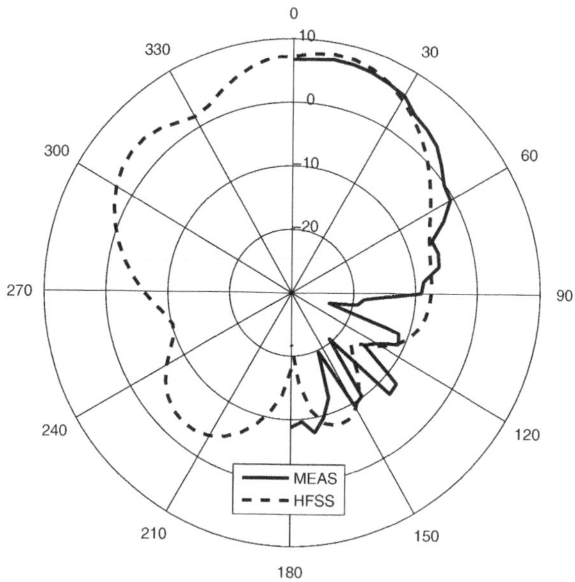

Figure 7.20: Measured and simulated vertical co-polarization patterns of the dual folded dipole array in the $\phi = 90°$-plane (from reference [19], reproduced by permission of © 2006 IEEE).

the antenna assumes air between the superstrate and the ground plane. Thus, taking into account that standard mold materials are lossy at mmWave frequencies with a dielectric constant typically between 3 and 5, it may have limited use within a low-cost plastic package technology.

However, the antenna design concept described above can be extended so that the antenna could keep its performance when encapsulated with a lossy mold material. It is achieved by applying a metal frame (see Figures 7.21 and 7.22) surrounding the antenna substrate [20]. This metal frame in the form of a cavity provides not only the mechanical support for the antenna substrate but also a natural barrier to prohibit the mold material from flowing into the space between the antenna superstrate and the ground, which is the key in preserving the antenna bandwidth and efficiency. Metal walls of the cavity provide additional very important features. Their proximity substantially influences the antenna input match and its radiation pattern. This phenomenon was found to be very useful for the antenna bandwidth enhancement purposes. The shape and size of the applied cavity and the folded dipole structure were optimized according to the antenna input impedance bandwidth. Also, two additional metal bars on the top of the fused silica substrate were added (see Figures 7.21 and 7.22) to minimize the surface wave effects that cause radiation in the antenna plane, although the detailed analysis of those effects are beyond the scope of this chapter.

For the packaged antenna, a standard 250 μm-thick substrate was used as a compromise between the mechanical stability and the surface wave effects and to reduce the manufacturing cost. The distance between the folded dipole and the ground plane is still 500 μm. The internal cavity size is 3×4.2 mm^2. The cavity is formed with a Covar metal frame.

Figure 7.21: Packaged antenna in the measurement setup (from reference [20], reproduced by permission of © 2006 IEEE).

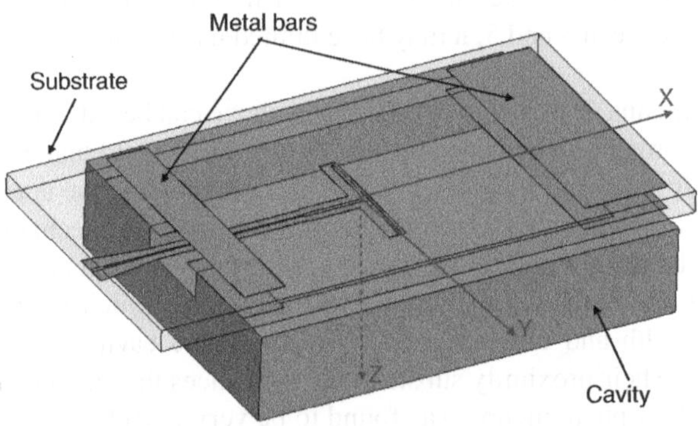

Figure 7.22: A three-dimensional view of the antenna geometry (from reference [20], reproduced by permission of © 2006 IEEE).

A Covar alloy was chosen because of its superior thermal expansion (3.7 ppm/°C) characteristics, which matches closely to that of fused silica (0.55 ppm/°C). A chemical etching and photolithography process was used for frame fabrication to comply with tight tolerances and yet high-volume fabrication. The 1.52 mm-long folded dipole is fed by a quarter-wavelength transformer line followed by a uniform 100 Ω CPS line within the cavity and then by a 100 Ω tapered CPS line. Some portion (0.7 mm long) of the antenna feed line passes the frame through the opening in the wall for measurement and chip attachment purposes.

The antenna, with its surrounding cavity, was measured in an anechoic chamber using the probe-tip based measurement setup [52]. For that purpose, it was mounted in a special sample holder and connected with a coplanar probe (see Figure 7.21). Only the front part of the build-up, including the most important feed line, could be encapsulated for test purposes because of the mounting limitations in the chamber. Epoxy mold was applied to the feed line section between the probing pads and just outside of the wall frame opening. The mold material has a 4.4 dielectric constant and a 0.02 loss tangent measured at 60 GHz [17].

The simulated 10 dB input impedance bandwidth exceeds 30% as shown in Figure 7.23. Because of the frequency limitations of the measurement setup, it could not be experimentally confirmed beyond 65 GHz, but very good agreement between the measurement and simulation results can be seen within the measured band of 50–65 GHz. The simulated antenna efficiency is above 90% within the entire 50–75 GHz frequency range. The measured and simulated antenna gains shown in Figure 7.24 are also in good agreement, which allows one to believe that the simulated antenna efficiency is very accurate. Both vertical co-polarization radiation patterns for $\phi = 0°$ and 90° are presented in Figures 7.25 and 7.26, respectively. Very good correlation between the measured and simulated patterns can be noticed here as well.

7.3.1 Cavity Size Effects on Antenna Performances

Since the cavity places a very important constraint for the packaged antenna, it is very important to study its influence on the antenna performance.

Figures 7.27 and 7.28 show the simulated antenna cavity depth and size effects on antenna match and gain respectively. Figures 7.27(a) and 7.27(b) show the influence of the cavity depth and size variations on the antenna impedance match. Notice that even with ±20% and ±33% changes in the cavity depth and size, respectively, most of the input impedance bandwidth is still preserved. The antenna gain deviation (see Figures 7.28(a) and 7.28(b)) due to the same variations is below 0.5–1 dB for most of the portion of the considered 50–75 GHz band.

The proximity of the cavity metal walls can be used in the input impedance bandwidth enhancement process. Moreover, the metal cavity provides a well-controlled electromagnetic environment, making the antenna performance insensitive to a large extent to the surrounding package and PCB-level dielectric and metal structures. The cavity decouples the design of the antenna from the exact physical properties of the package such as dimension and material parameters, an important design feature simplifying simulation and modeling complexity. It is important to note that the antenna performance shows an extremely low sensitivity to the cavity manufacturing and attachment tolerances, which are of special importance at mmWave frequencies.

Figures 7.24 and 7.28 show that there exists a gain dip around 60 GHz frequency. This gain dip is caused by the small cavity size as studied in reference [53]. For some given cavity depth and the folded dipole dimensions, a wide impedance bandwidth can be achieved with a smaller cavity size, while smooth (no gain dip) and large antenna gain can be achieved with larger cavity sizes to some extent [53]. Therefore, a tradeoff is necessary between antenna

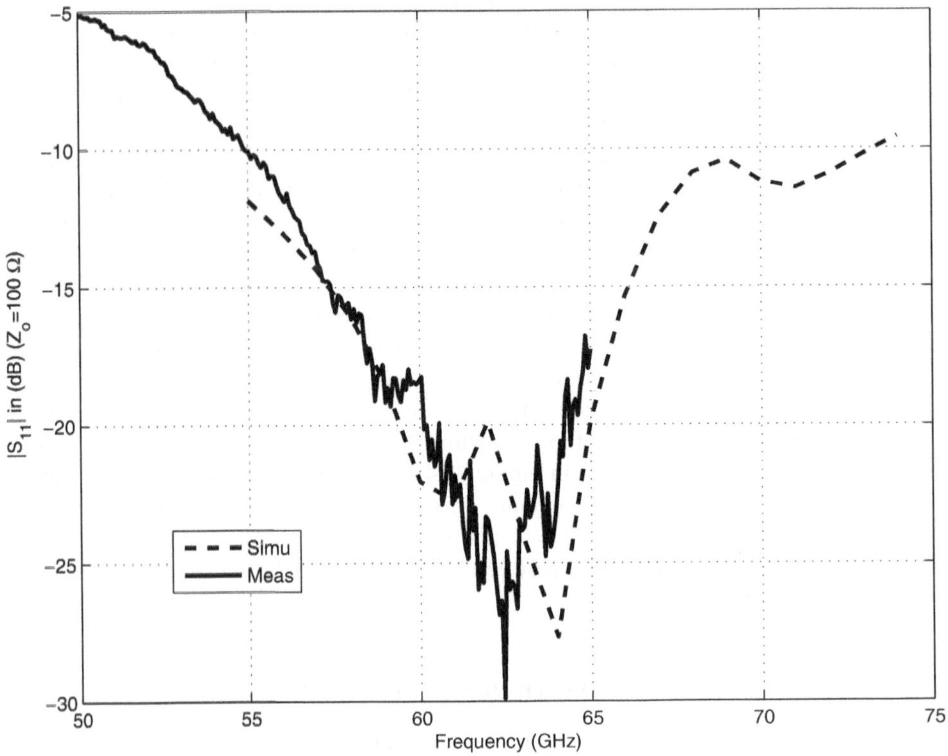

Figure 7.23: Measured and simulated S_{11} of the packaged antenna (from reference [20], reproduced by permission of © 2006 IEEE).

gain and impedance bandwidth. However, for package applications, a small cavity size is extremely important, assuming that antenna gain values are reasonably constant in the band of interest.

7.3.2 Packaging Effects on Antenna Performance

In addition to the cavity depth and size influence on antenna performance, the package molding materials and construction also play an important role in the package. The molding (or cover) not only protects the chip and the antenna structure, but also influences the antenna design and its performance, so the molding effects on antenna performance have to be studied in detail as well.

A three-dimensional HFSS simulation model, which includes the antenna and a silicon chip within an envisioned LGA package placed on top of the PCB, is shown in Figure 7.29 [54]. Referring to Figure 7.29, a folded dipole antenna is printed on the bottom side of a 250 μm-thick fused silica 'superstrate'. A 500 μm-thick 'metal frame' is attached to the superstrate bottom and the metallized package base 'package GND', creating a 500 μm deep and 2.8 × 4 mm² size air-filled cavity. The LGA package uses a 250 μm-thick FR4

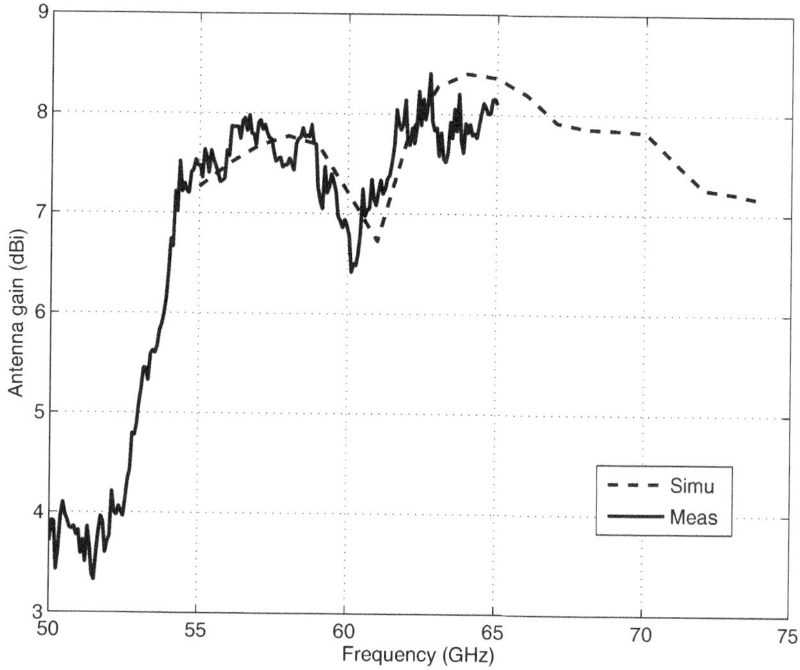

Figure 7.24: Measured and simulated gain of the packaged antenna (from reference [20], reproduced by permission of © 2006 IEEE).

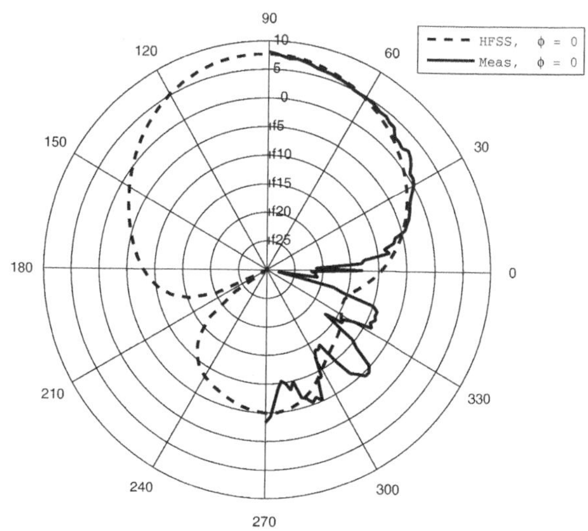

Figure 7.25: Measured and simulated copolarization patterns for $\phi = 0°$ (0° for X-direction and 270° for Z-direction) (from reference [20], reproduced by permission of © 2006 IEEE).

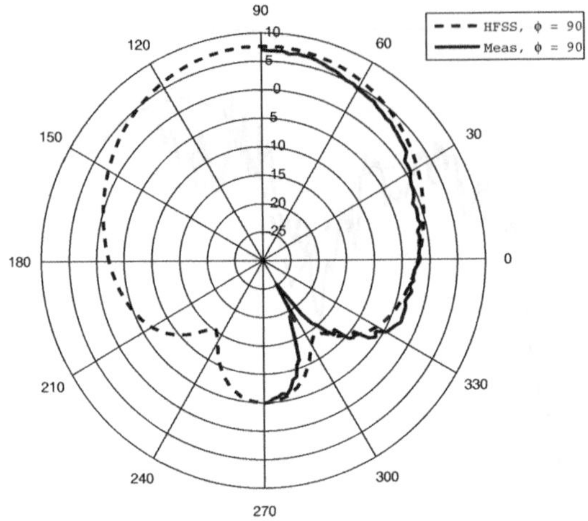

Figure 7.26: Measured and simulated copolarization patterns for $\phi = 90°$ (0° for Y-direction and 270° for Z-direction) (from reference [20], reproduced by permission of © 2006 IEEE).

epoxy base layer with thermal vias that are able to connect 'package GND' on its top with the corresponding GND on the board level 'PCB GND'. The antenna was designed to be encapsulated ('encapsulant' with a mold material of $\epsilon_r = 2.5$). A 'radiation window' above the antenna superstrate is kept open in the encapsulation process for the maximum radiation efficiency. The 1.462 mm-long folded dipole is fed by a 150 Ω quarter wavelength transformer followed by a uniform 100 Ω CPS transmission line within the cavity in series with another 0.83 mm-long CPS line, the lateral dimensions of which were adjusted to account for the influence of encapsulation material. Metal bars on the top and bottom of the fused silica superstrate were placed to minimize radiation in the antenna plane.

Figures 7.30–7.33 show the influence of the dielectric constant and loss tangent uncertainties of the encapsulant on the antenna S_{11}, gain and radiation efficiency, respectively. The antenna is assumed to be detached from the board ('Package base' and 'PCB' were removed from the three-dimensional model). A loss tangent of 0.02 was assumed in the simulations in Figures 7.30 and 7.31. Note that even very high ϵ_r variation from 1.5 to 4 merely impacts the input match and gain. Moreover, the molded antenna using even a very lossy material with a 0.1 loss tangent still shows an efficiency better than 83% (Figure 7.32). Figure 7.33 presents some radiation efficiency values for the radiation window completely covered by the encapsulant. That option could simplify the molding process. Even a lossy cover with a 0.04 loss tangent as thick as 200 μm results in an efficiency higher than 85%. Two other results are also given for the encapsulant with $\epsilon_r = 3.5$. It is clear that higher permittivity leads to some loss increase for the same loss tangent. Covering the antenna radiation window will cause less than 5% efficiency reduction as shown in Figures 7.32 and 7.33.

The sensitivity investigations of the antenna impedance match and gain to a variable environment are shown in Figures 7.34 and 7.35. A few typical mounting scenarios are considered: antenna on a 250 μm or a 1000 μm-thick FR4 base without any metal plane in its proximity (CONF1 and CONF2 in Figures 7.34 and 7.35), the mentioned CONF1 over

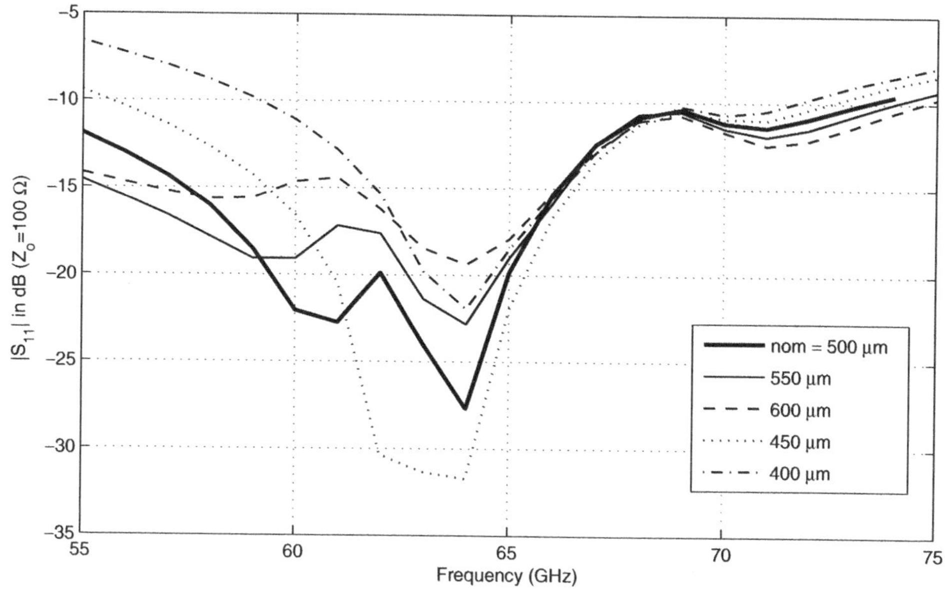

(a)

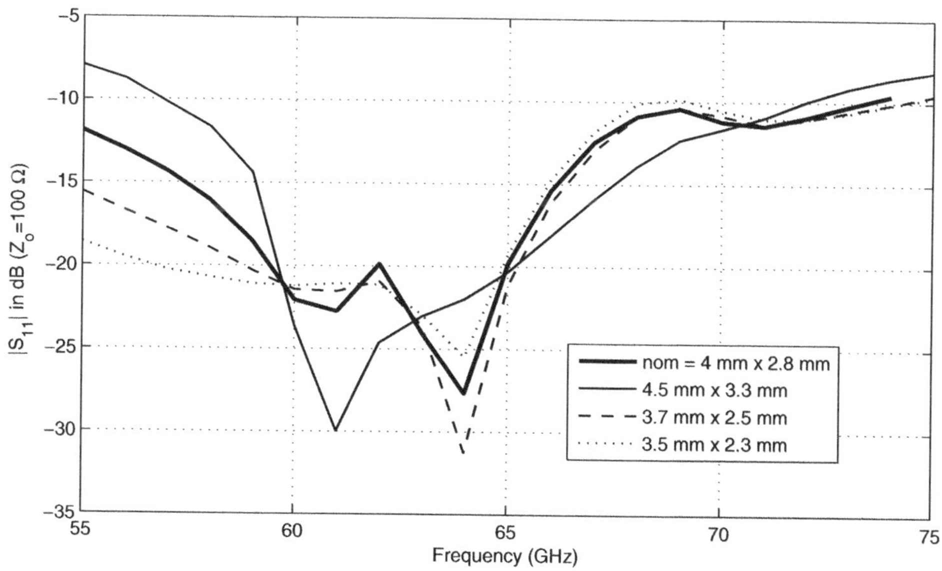

(b)

Figure 7.27: Cavity effects on antenna S_{11}: (a) depth; and (b) size (from reference [20], reproduced by permission of © 2006 IEEE).

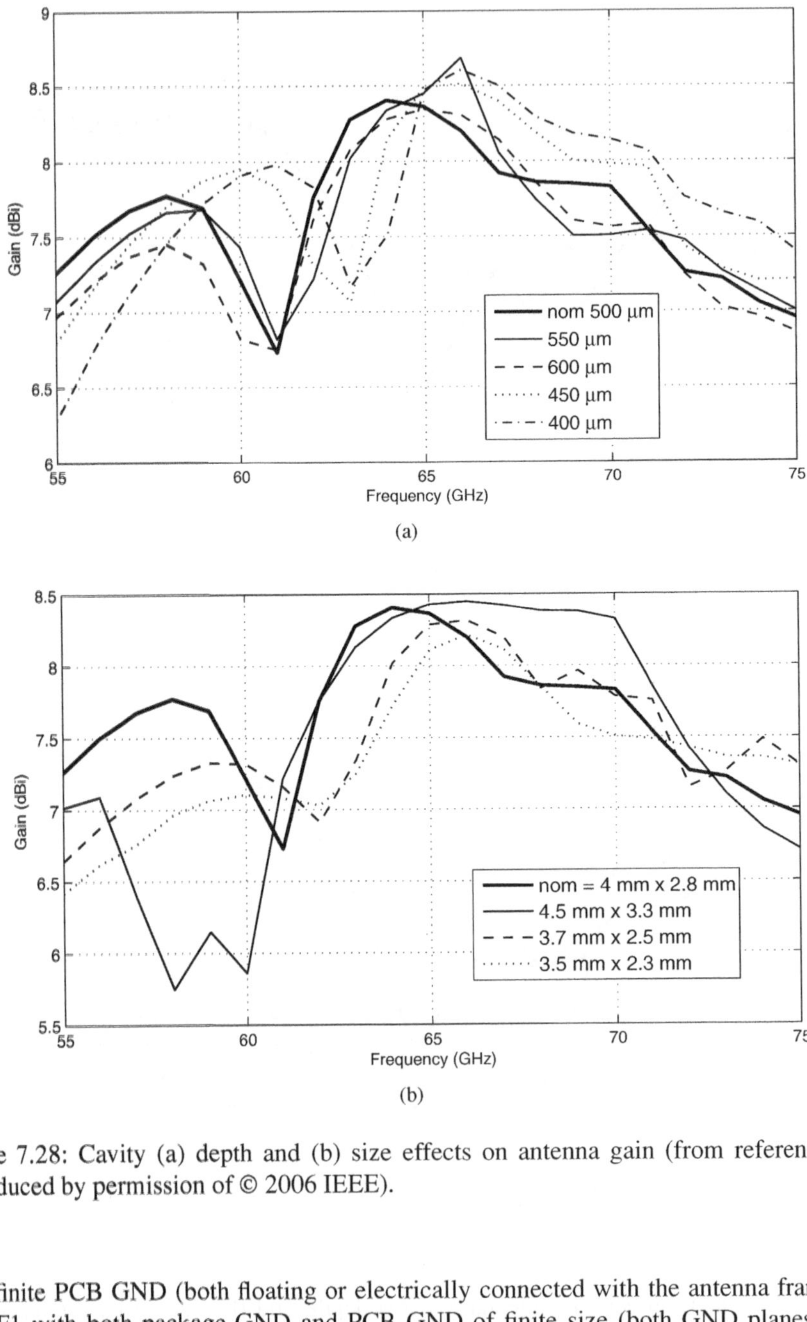

Figure 7.28: Cavity (a) depth and (b) size effects on antenna gain (from reference [20], reproduced by permission of © 2006 IEEE).

an infinite PCB GND (both floating or electrically connected with the antenna frame) and CONF1 with both package GND and PCB GND of finite size (both GND planes extend over the frame by 1 mm on each side) connected with the frame. The chip and part of the encapsulant surrounding the antenna frame were removed from the three-dimensional HFSS model for all those simulations. Only the front side of the model with the antenna feed line was covered. It can be seen that the antenna impedance match is extremely stable.

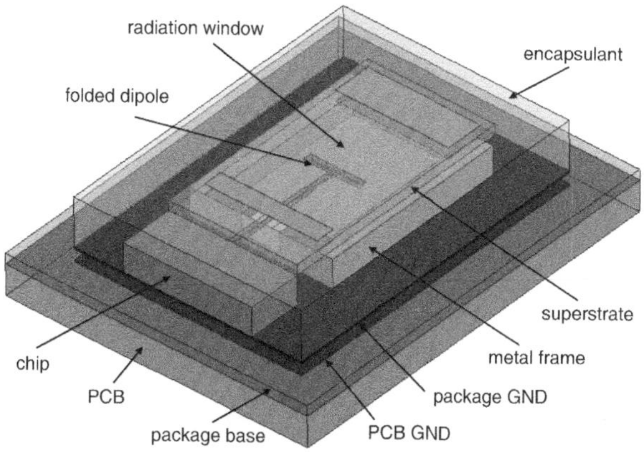

Figure 7.29: HFSS three-dimensional view of the antenna package (from reference [54], reproduced by permission of © 2007 IEEE).

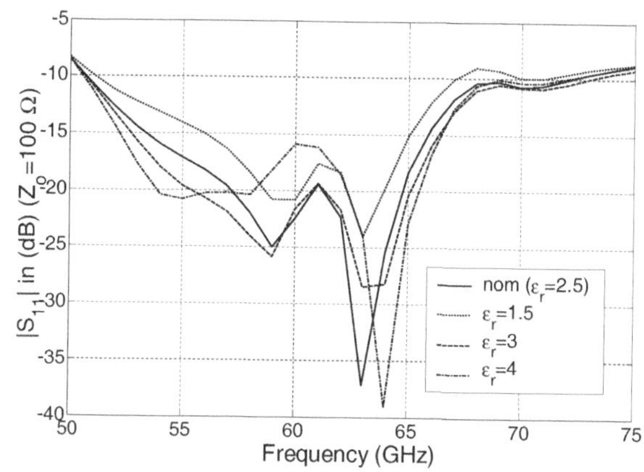

Figure 7.30: Simulated antenna $|S_{11}|$ for different ϵ_r of the mold (from reference [54], reproduced by permission of © 2007 IEEE).

The simulated gain is also consistent for all the considered cases with some reasonably low variation. The influence of the mold height and width surrounding the frame on the impedance match and gain is analyzed in Figures 7.36 and 7.37. The above-mentioned CONF1 with package GND and PCB GND of finite size (a real packaging environment) is considered to be a reference for the simulations. The results for two different mold heights, H, of 0.75 mm and 1 mm, and two different mold widths, W, of 1 mm and 2 mm, extending beyond the frame, are presented. Note that the overall height of the frame with the fused silica substrate is 0.75 mm. The same encapsulation schemes were applied with and without the presence of the silicon chip. Notice that the antenna impedance match variation is minor for all cases, opposite to the gain curves corresponding to the models without the chip proximity.

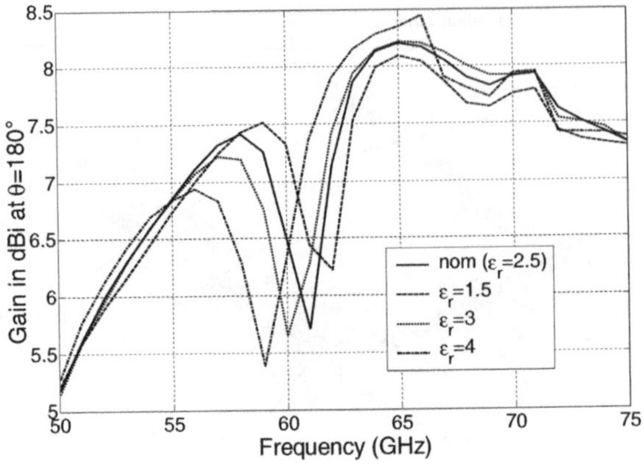

Figure 7.31: Simulated antenna gain for different ϵ_r of the mold (from reference [54], reproduced by permission of © 2007 IEEE).

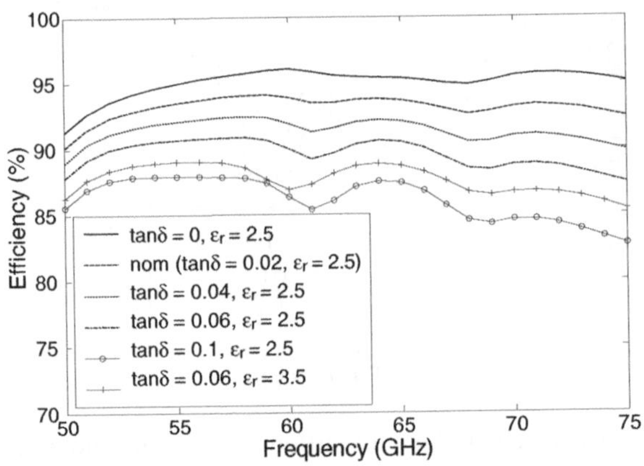

Figure 7.32: Simulated antenna radiation efficiency for different tan δ of the mold (from reference [54], reproduced by permission of © 2007 IEEE).

The silicon chip plays a very important role in stabilizing the gain over the broad frequency range, 50–75 GHz. Moreover, the gain characteristics with the chip present is flatter than for the reference configuration.

From these extensive simulations, it is clear that the performance of this packaged antenna is insensitive to the variable package and board level surrounding environments, which is very important at mmWave frequencies.

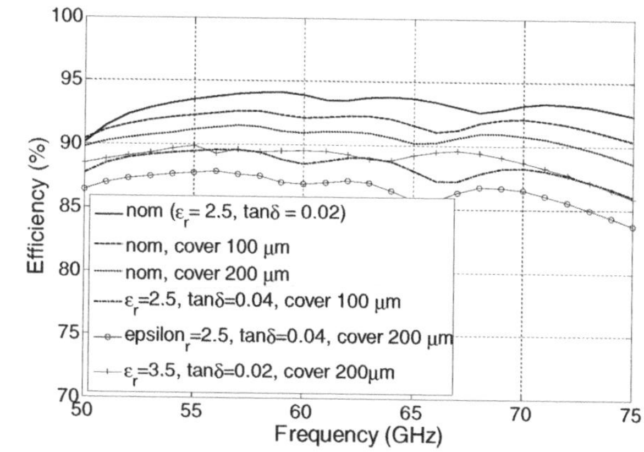

Figure 7.33: Simulated antenna radiation efficiency for different ϵ_r of the mold including cover (from reference [54], reproduced by permission of © 2007 IEEE).

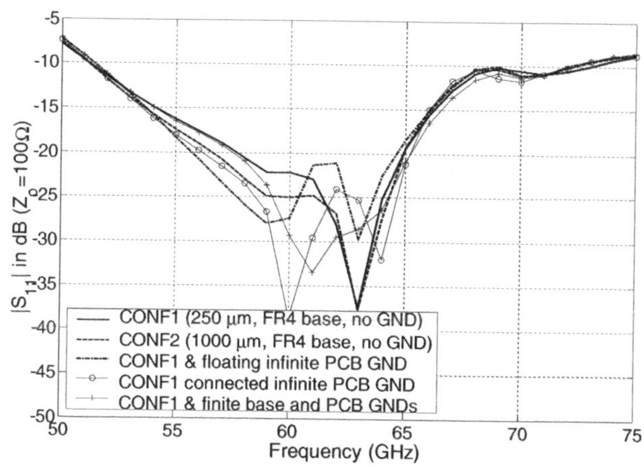

Figure 7.34: Simulated antenna $|S_{11}|$ for different mounting scenarios (from reference [54], reproduced by permission of © 2007 IEEE).

7.3.3 Antenna in System Performance

Whatever the antenna performance is, in freestanding, the antenna has to perform well when it is used in a real wireless system. The packaged chip and antenna performance was measured at various steps during the packaging process to monitor any package-related performance degradation. First, initial on-wafer measurements of the radio chip are required to provide a reference for the conversion gain, noise figure or maximum available output power that is achievable by the receiver or transmitter chip respectively (see reference [6] for details). Next, such on-wafer results can be verified at board-level if the input pads to the low-noise amplifier of the receiver or the output pads to the power amplifier of the transmitter are not obstructed

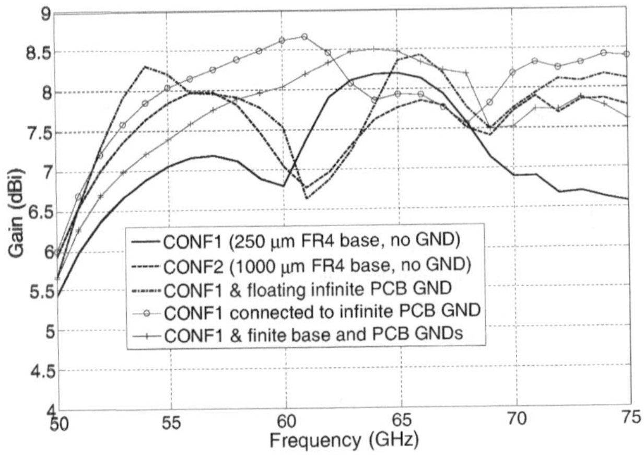

Figure 7.35: Simulated antenna gain for different mounting scenarios (from reference [54], reproduced by permission of © 2007 IEEE).

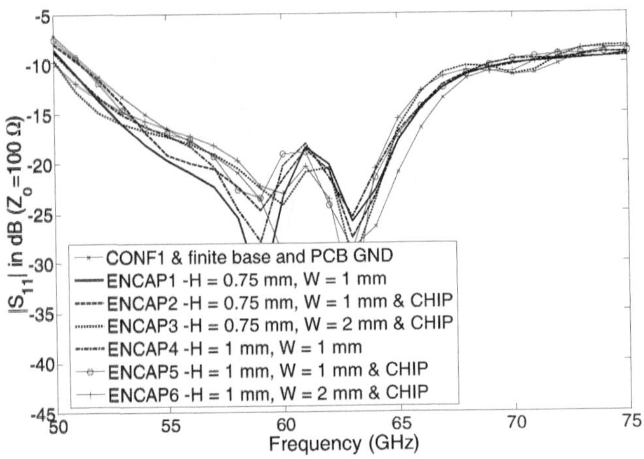

Figure 7.36: Simulated antenna $|S_{11}|$ for different mold height (H) and width (W) (from reference [54], reproduced by permission of © 2007 IEEE).

by any objects in close proximity to the probe pads. A final test, however, is required that shows the overall encapsulated (molded) performance of the packaged chip, including its antenna. Since the antenna cannot be measured *in situ*, the exact interconnect loss is unknown and can only be inferred by accurately calibrated gain measurements in an anechoic chamber. To do this, a similar measurement setup such as that described in reference [52] was used here to provide accurate gain calibration and radiation pattern measurements. Unlike passive antennas, however, the active radio circuit includes a frequency conversion. A network analyzer-based gain calibration and measurement technique therefore cannot be used [6].

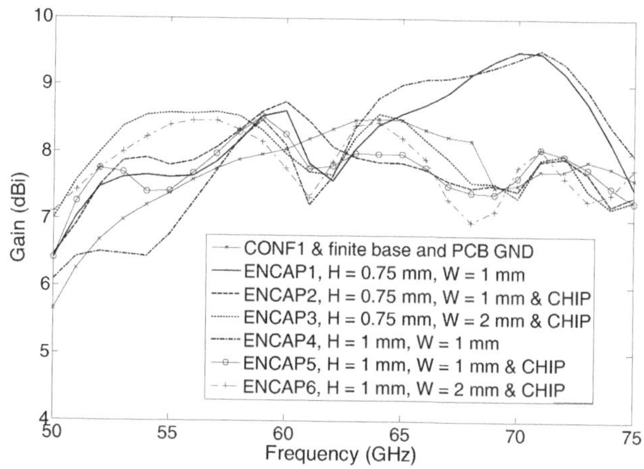

Figure 7.37: Simulated antenna gain for different mold height (H) and width (W) (from reference [54], reproduced by permission of © 2007 IEEE).

The following data was taken at 60 GHz with a 10 dB back-off from the 1 dB compression point of the transmitter. The transmitter on-wafer conversion gain at that point was 35 dB. Figure 7.38 shows the vertical radiation pattern along $\theta = 0°$ to $+120°$ for 60 GHz before and after molding with a standard low-cost mold material in $\phi = 0°$ direction (see Figure 7.8 for the coordinating system definition). The $\phi = 90°$ direction of the molded package is assumed to be symmetrical around θ and is shown in Figure 7.39. The cross-polarization is 20 dB down (measured from $-45°$ to $45°$ only). The overall gain in the main radiation direction is about 41 dB with a 3 dB beamwidth of $\pm 30°$. The pattern is smooth, implying little package and board-level interference. The molded package shows a slightly higher disturbance at $\pm 30°$ owing to interference with, and refraction at, the package boundary.

7.4 A Patch Array

As discussed in the previous sections, the cavity-backed folded dipole superstrate antenna shows the potential for cost-effective integration with the 60 GHz SiGe chipset [1, 24] in a plastic LGA package [6]. A metal frame, used to create an air-filled cavity below the flipped antenna substrate, is crucial for that approach. The metal wall proximity was used in order to enhance the input impedance bandwidth [20]. Furthermore, the frame also provides a barrier for the mold flow into the cavity, a key in preserving the antenna performance. Moreover, it shows very low sensitivity to the frame manufacturing tolerances. The antenna presented in reference [20] has about 7 dBi gain, which is adequate for wireless USB-type applications with a data rate up to 3 Gbps up to 1.5 m range. However, for point-to-point indoor applications up to 10 m range, antenna gains as high as 15 dBi, depending on applications, are required. In this case, array antennas are required.

For frequencies up to the low end of the mmWave band, microstrip lines can be realized easily on PCB with substrate as thin as 300 μm. For these applications, microstrip lines

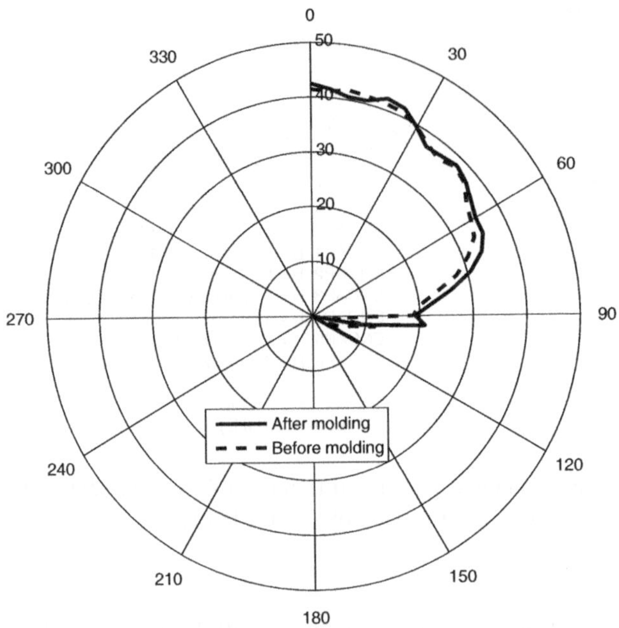

Figure 7.38: Antenna in system radiation pattern at 60 GHz and $\phi = 0°$ before and after molding.

are well behaved and have acceptable dispersion, the line characteristic impedance is nearly constant in the band of interest, and the fringe field is weak, meaning the coupling between feed lines and between feed lines and the radiating elements is weak. However, for frequencies in the 60 GHz band and above, the arrays cannot be easily implemented on substrate as thin as 100 um or thinner owing to mechanical reliability and the manufacturing process. As a result, the substrate is usually too thick for microstrip lines operating at mmWave and higher frequencies. Due to the thick substrate issues, microstrip line impedance changes significantly with frequencies. Also, the microstrip line fringe field is very strong owing to the small ratio of the microstrip line width to the substrate thickness. In fact, the microstrip lines cannot be treated as simple feed lines anymore, but should be treated as radiating elements as well. Therefore, it is almost impossible to design patch arrays on PCB, even of a small size array. Besides microstrip line issues at 60 GHz band, there is also an interest in packaging the array antenna with the 60 GHz radios designed in references [1, 24] in a similar packaging style proposed in reference [20]. One important feature in reference [20] is that the antenna is connected to the 60 GHz radio in the flip-chip attachment. This antenna-to-RF chip connection provides minimum insertion loss and impedance mismatch.

In the proposed feed network [55], CPS and CPW lines are used in the feed structure. Since CPS and CPW lines have a reliable return current path, the line characteristic impedance is almost constant in the band of interest regardless of the substrate thickness. Also, the field primarily stays between the signal line and the ground line/lines, so the fringe field is weak.

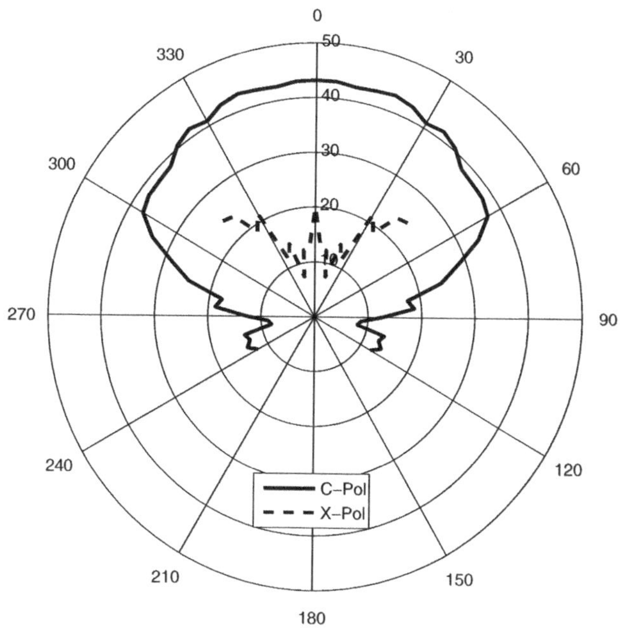

Figure 7.39: Antenna in system radiation pattern at 60 GHz and $\phi = 90°$ after molding for co- and cross polarization.

As a result, the coupling between feed lines and between feed lines and the radiating elements is weak. This makes the feed structure and radiating elements design much easier. Figure 7.40 shows a two-dimensional view of an eight-element patch array using the proposed feed structure.

Referring to Figure 7.40, each patch element has a short 'microstrip' line. Two patches are connected to a differential CPS line. Only the input impedance of the patch and microstrip line combination at the CPS connection point is of concern. This impedance should be about half of the CPS line impedance. A CPW feed is formed by two CPS line connections. For array applications, a transition from CPW to CPS is necessary. A bridge based on the same ideas as the balun in reference [43], very easy to implement with thin-film technologies, is required for this transition and much easier to realize than vias at mmWave frequencies and above. Note that a three-way CPW to CPW connection is difficult, since extra air-bridges are required.

Unlike traditional patch arrays, each patch pair is fed on the opposite sides of the two patches. This is possible since the CPS is a differential feed line that naturally provides 180° phase difference to the two patches.

The patches and feed-line structures are implemented on the bottom side of a 300 μm-thick fused silica of 3.8 dielectric constant. The ground plane of the patches is separated 200 μm from the bottom side of the fused silica. The cavity below the fused silica is 200 μm deep with a 8.6×11.6 mm^2 size. The cavity wall is 500 μm-thick and has a 1.1 mm wide cutout for passing the feed line. The patches are 1.68 mm long and 1.6 mm wide. The patches

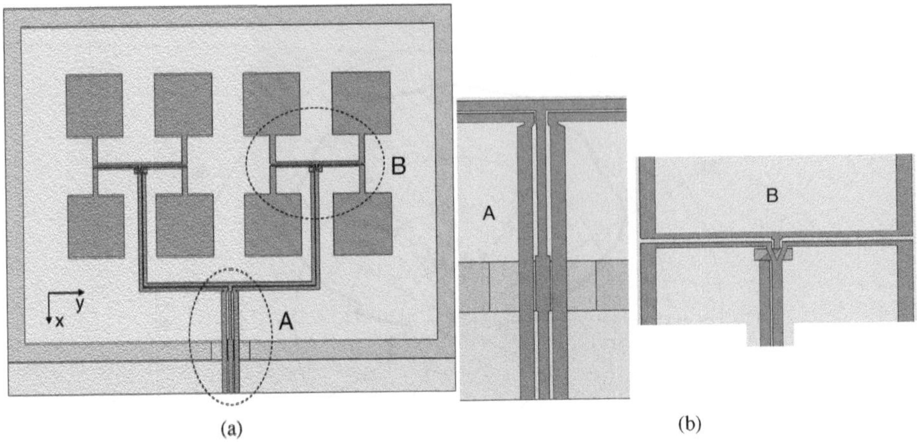

Figure 7.40: The 2 × 4 patch array structure: (a) two-dimensional view; (b) details of the feed structure.

are separated 3.28 mm in the x-direction and 2.5 mm in the y-direction. The antenna has a $10.49 \times 12.6\ \text{mm}^2$ overall size. The patch and feed-line structures use 2 μm-thick gold.

To be used with the 60 GHz SiGe chipset [1, 24] in the same way as discussed in reference [20], epoxy mold (with $\epsilon_r = 4$ and 0.02 loss tangent measured at 60 GHz) is used at the gap between the metal frame and the SiGe chip during the simulation.

Figure 7.41 shows the simulated antenna impedance match. It is clear that the antenna is well matched to 50 Ω and with adequate bandwidth for 60 GHz applications. Note that the antenna was designed slightly above the 60 GHz band with an anticipation that encapsulation will lower the frequency slightly. Figure 7.42 shows the simulated antenna gain. The maximum antenna gain is 14.8 dBi at 61 GHz, and the 3 dB bandwidth is more than 9 GHz. Figures 7.43 and 7.44 show the simulated antenna radiation patterns on the $\phi = 0°$ plane (E-plane) and the $\phi = 90°$ plane (H-plane) of the antenna, respectively.

The array feed-line structures are suitable for mmWave applications where microstrip lines do not function very well. Since CPS and CWP lines are used, the feed-line loss is reduced tremendously so that large arrays are possible for mmWave applications. Although the corporate-fed patch array is presented here, the feed-line structures can be used for slot arrays or a combination of series and corporate-fed patch arrays. Since air-bridges are used in the feed-line structures, thin-film technologies are the preferred manufacturing choice for these arrays.

7.5 Circularly Polarized Antenna

For applications such as wireless USB, the operating distance is limited to about a meter. A single antenna with about 7 dBi gain at 60 GHz will provide the necessary antenna gains. However, these devices typically operate in indoor environments. As a result, multi-path fading due to multiple reflections and diffractions deteriorates the radio system performance.

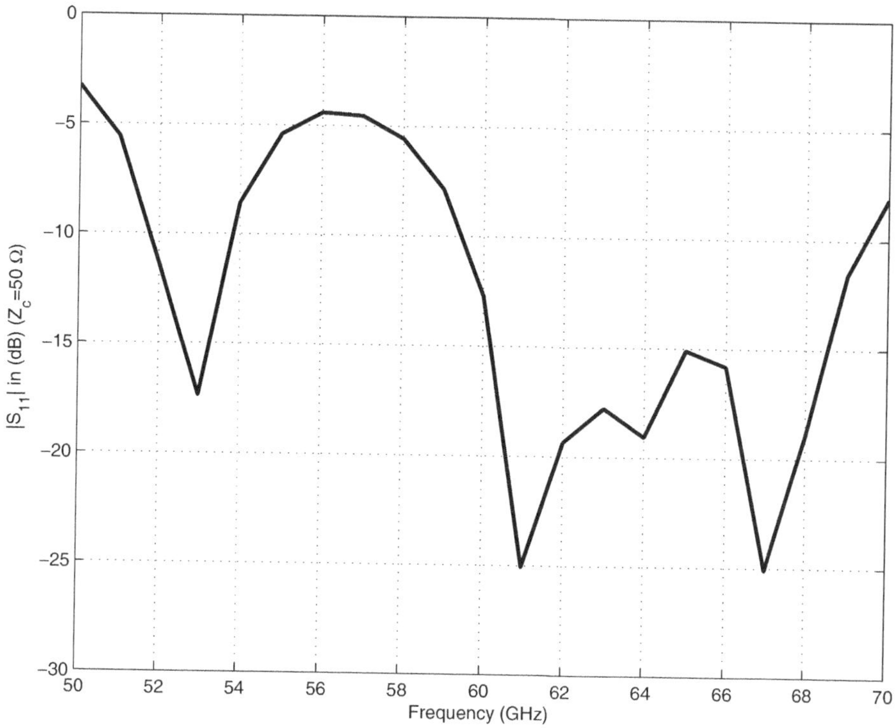

Figure 7.41: The simulated $|S_{11}|$ of the eight-element patch array.

Numerous studies indicate that circularly polarized (CP) antennas can be used to mitigate the multipath effect on the system performance.

Many CP antenna designs exist. Microstrip patch-related CP antennas either do not have the required impedance and axial ratio (AR) bandwidth or the antenna feed networks, typically microstrip lines, are not suitable for mmWave applications or are difficult for flip-mounting. As discussed previously, wire bonds between semiconductor chips and antennas are usually avoided to minimize insertion loss and inductance for impedance matching. Many cavity-backed planar CP slot or printed antennas, including spirals, require the distance between the antenna element and the background plane to be a quarter wavelength (about 1250 μm at 60 GHz) or larger. This distance is too large for mmWave packaging applications (typically less than 500 μm). Most of these antennas require microstrip feed lines, another reason that these antennas are not suitable for mmWave packaging. Furthermore, many of these antenna designs require a feed network that is perpendicular to the antenna-radiating elements. This means that vias are required for the feed networks. However, vias should be avoided primarily owing to the manufacturing process. After applying antenna performance, manufacturing process, interconnection, and packaging requirements, one can eliminate almost all the available CP antenna designs in the open literature.

Wire or metal strip CP antenna structures in Figure 7.45 can be considered as the origin of the proposed CP antennas [56]. Figure 7.45(a) shows an open square loop CP antenna.

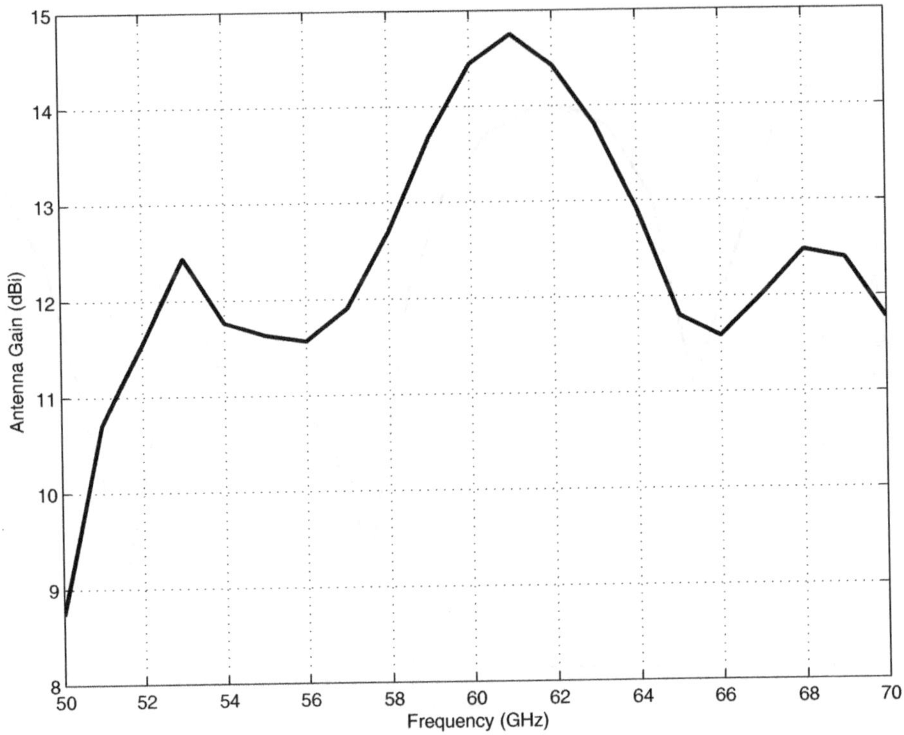

Figure 7.42: The simulated gain of the eight-element patch array.

Note that the general shape for the antenna structure in Figure 7.45(a) is an open rhombic loop. For an extreme case, an open circular loop also works. To improve the antenna bandwidth, a parasitic open loop can be used. Figure 7.45(b) shows a perpendicularly-fed double open rhombic loop CP antenna. If the bandwidth requirement is not critical, the two parasitic loops can be removed. The antenna structures in Figure 7.45(b) have good performance if the background plane is at least a quarter wavelength away from the antenna structure. Again this distance separation is too large and not realistic in real applications. Also, it is very hard to have a coplanar feed structure. Therefore, they are not suitable for mmWave package applications.

Thus, in package applications, the antenna structures in Figure 7.45(b) do not have a wide enough bandwidth. One important feature of these designs is that the left and right structures resonate at the same frequency. Another problem is how to modify the structures so that CPW feed lines can be used. It transpires that these antenna designs can be modified easily to have a CPW feed line and to have a wide bandwidth by letting the left and right loops resonate at two different frequencies as shown in Figure 7.46 [57]. As mentioned above, the exact shape of the loops is not critical.

Figure 7.47 shows a sketch of the proposed planar CP antenna structure. The antenna also has two open loops. Loop1 is formed from the right branch and the right ground of the CPW

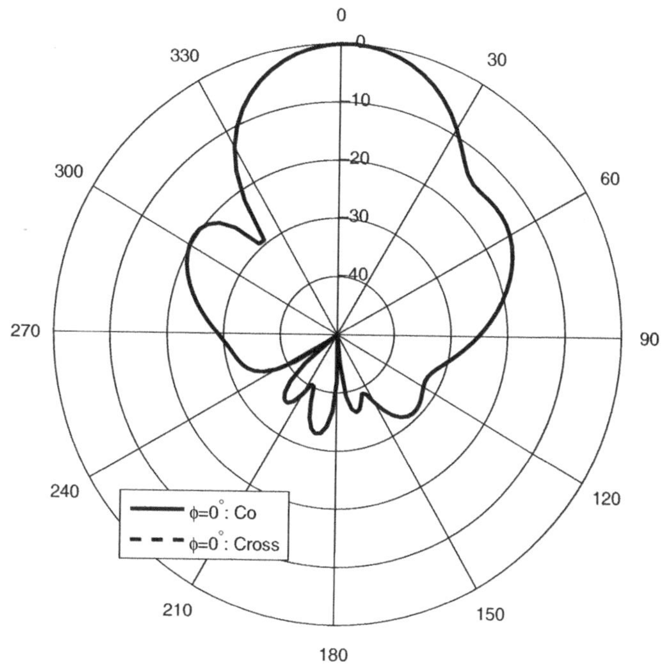

Figure 7.43: The simulated radiation patterns for $\phi = 0°$ of the eight-element patch array at 61 GHz.

feed line, while Loop2 is formed from the left branch and the left ground of the CPW feed line. These two loops form the CP antenna. Each loop produces a narrow band CP wave. The operating frequencies of these two CP waves are offset, so the combination of these two CP waves produces a relatively wide band CP wave. The antenna impedance match is realized by changing the CPW parameters, possibly with a quarter wavelength transformer. The proposed antenna structure has three major advantages over the structures in Figure 7.45(b). First, the spacing between the antenna structure and the background plane can be smaller than one-eighth of a wavelength, instead of at least one-quarter of a wavelength, owing to a relatively wider bandwidth. Second, one can use a coplanar feed line. Third, the antenna structure is more compact than those in Figure 7.45(b). Furthermore, a metal cavity and a metal ring for suppressing substrate surface waves can be easily applied. If a much wider bandwidth is required, two parasitic loops can also be used.

To demonstrate the proposed CP antenna concept, an initial design is shown in Figure 7.48. The fused silica in this case is 300 µm-thick, the cavity is 500 µm deep with a $3849 \times 3890\ \mu m^2$ size. The metal wall is 500 µm-thick and has a $1100 \times 300\ \mu m^2$ opening for passing the CPW feed line.

Figure 7.49 shows the simulated S_{11} of the antenna. The 10 dB return loss bandwidth is about 10% even without optimization. Note that for simplicity, the metal strip width is uniform; however, non-uniform width can be used to improve the bandwidth. The 3 dB axial ratio in the main direction is also about 10%, as shown in Figure 7.50, but shifted

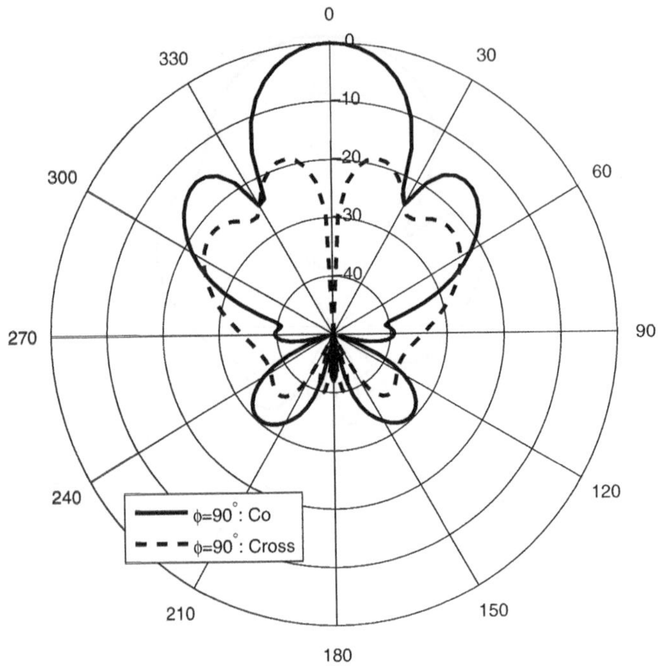

Figure 7.44: The simulated radiation patterns for $\phi = 90°$ of the eight-element patch array at 61 GHz.

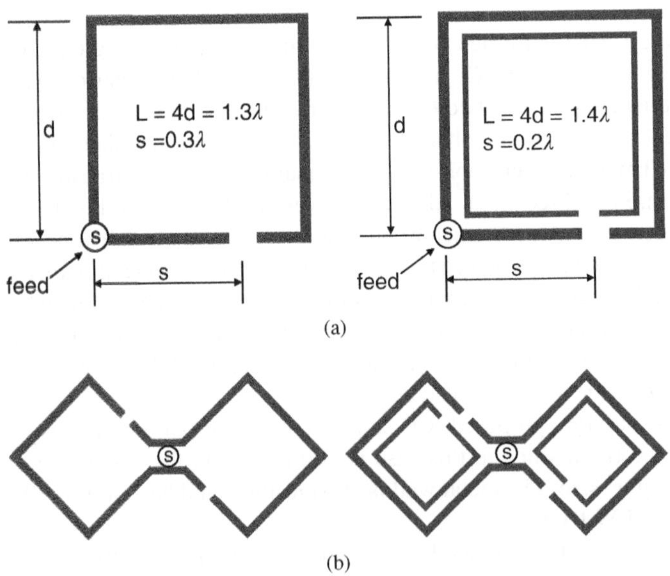

Figure 7.45: CP antenna designs with/without coupled loops: (a) single loop; (b) dual loops.

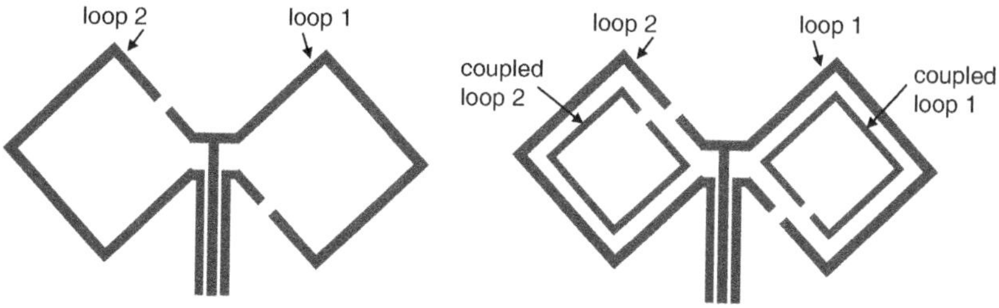

Figure 7.46: CP antenna designs with CPW feed line.

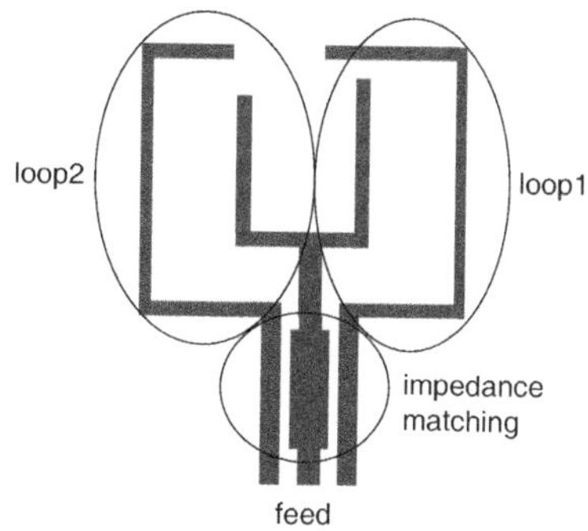

Figure 7.47: CP antenna designs with CPW feed line and impedance transformer.

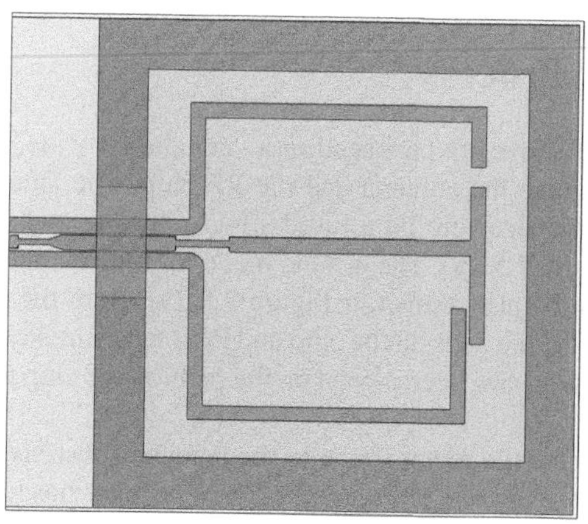

Figure 7.48: An initial CP antenna design.

by about 2 GHz toward lower frequencies. Figure 7.51 indicates that the antenna is right-hand polarized. The left-handed polarization can be obtained by left-right mirroring of the antenna structure along the center line of the center conductor in the CPW feed line.

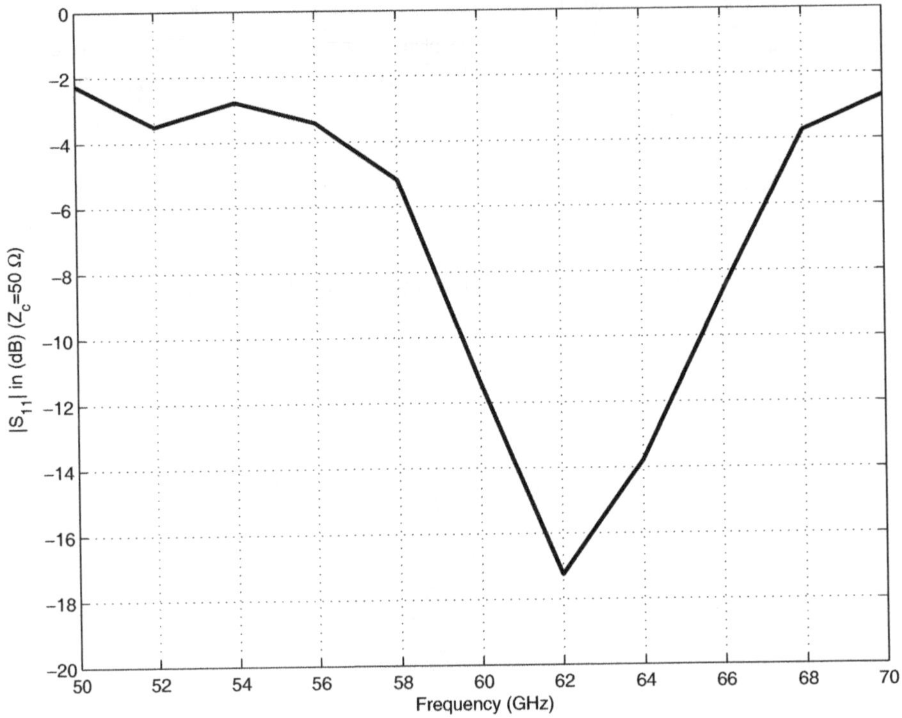

Figure 7.49: Simulated S_{11} of the initial CP antenna design.

7.6 Assembly Process

All antennas discussed above are packaged in a very similar way. This section will explain steps necessary to package the antenna and the RF chip, The antenna itself has two main components: the antenna structure on a fused silica substrate and a metal frame (or cavity walls) as shown in Figure 7.52(a). The ground plane for the antenna is provided by the PCB or package base ('board') also shown in Figure 7.52(a). Here the silicon chip has already been attached to the PCB, the pads on the chip and PCB have already been connected through bondwires, and gold studs have been placed on the pads on the chip to which the antenna will be connected.

The first step is to glue the metal frame to the antenna substrate with a conductive paste such as the silver epoxy as shown in Figure 7.52(b). The paste has to be dried before moving to the next step. To reduce the drying time, this subassembly can be placed in a suitable oven. The second step is to attach the subassembly to the chip through the flip-chip process as shown in Figure 7.52(c). Alignment is very critical in this step. After the flip-chip process, a thin glue is used to underfill the gap between the bottom of the antenna substrate and the top of the chip. This underfill is necessary mechanically to strengthen the connection between the antenna and the chip and to prevent the overflow of the liquid encapsulant in the next step. The metal frame is also glued to the ground plane on the PCB with the conductive paste.

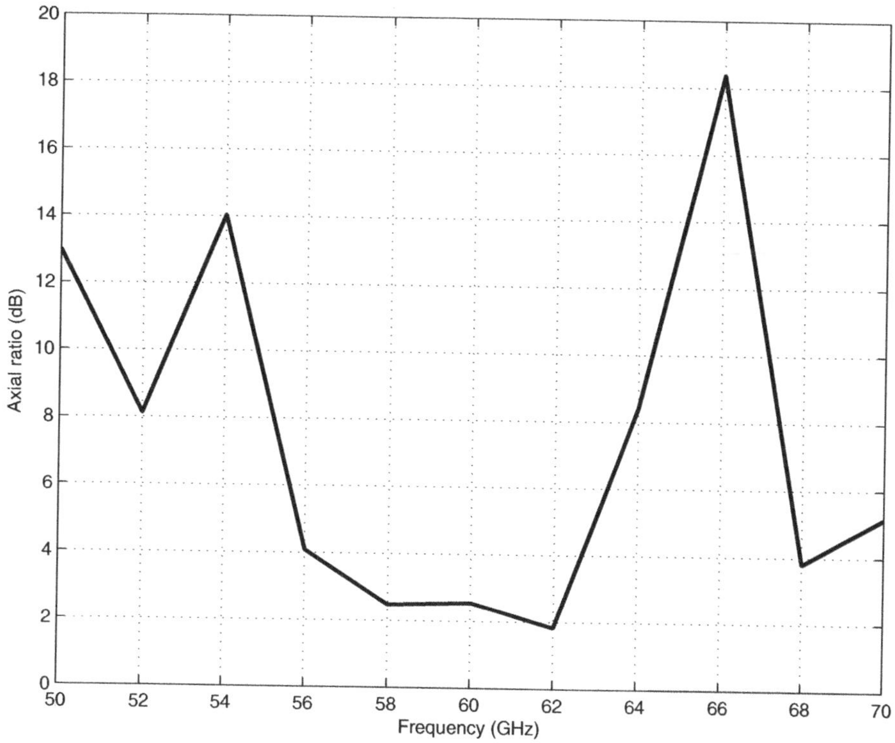

Figure 7.50: Simulated axial ratio of the initial CP antenna design.

The metal frame and the ground plane form the cavity for the antenna design. The final step is to fill the gap between the cavity and the chip (referring to the left figure of Figure 7.52(d)) with a low-loss encapsulant. This has to be done carefully to prevent the encapsulant from flowing into the opening on the cavity wall. The encapsulant is also applied on the whole antenna and chip assembly as shown in the right figure of Figure 7.52(d). A window for antenna radiation is preferred, otherwise about 5% radiation efficiency reduction will occur as discussed previously. The encapsulated package is then to be dried in an oven. The final package will resemble Figure 7.52(e).

7.7 Advanced Packaging Application

The above packaging process is very simple and flexible in principle. However, it is not suitable for mass production. Furthermore, since several manual steps are involved and it is time consuming, it can be a very expensive packaging process. This section will introduce two possible packaging processes suitable for mass production with a relatively low cost.

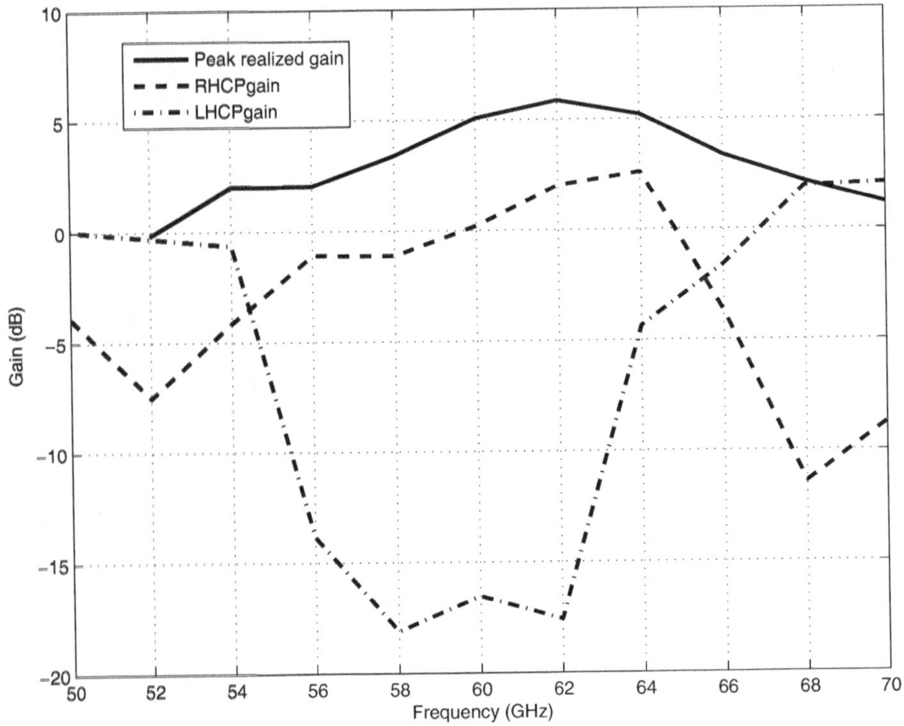

Figure 7.51: Simulated gains of the initial CP antenna design.

7.7.1 LTCC-based Packages

Low temperature co-fired ceramic (LTCC) multilayer technology has the capability to integrate passive components such as filters and antennas with RFICs in a single, cost-effective package. The active RFIC circuits can be integrated into the package either with wire bonding or flip-chip technology. Major benefits of the LTCC technology are low dielectric and conductor loss, good thermal conductivity, stability, hermeticity and low cost owing to the mass production possibility. The main challenge of the LTCC technology at the mmWave region has been the fabrication accuracy/tolerance of narrow conductors or spacing. The LTCC tape shrinking problems during sintering are also a big problem, but can be accounted for during the design process. Conductor widths and spacings of 50 μm can be realized reliably with the latest LTCC technology. For mmWave antenna designs using LTCC technology, the higher dielectric constant, typically greater than 5, is always a concern, since a high dielectric constant limits an antenna's bandwidth and efficiency often, owing to strong surface waves. However, the effective dielectric constant can be reduced by using cavities, even embedded cavities.

LTCC technology has been exploited for mmWave antenna designs [58–60] and antenna in package (AiP) applications [61, 62]. Figure 7.53 shows the exploded view of the AiP

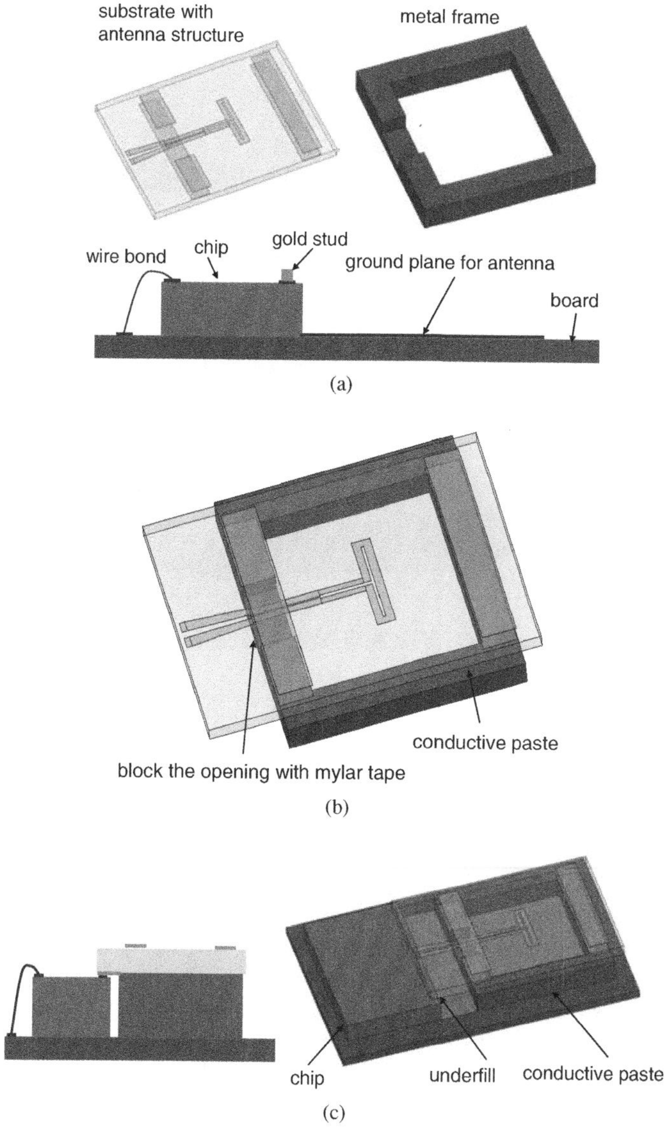

Figure 7.52: Assembly process: (a) three major components; (b) substrate and frame subassembly; (c) flip-chip attachment; (d) glob-top; (e) final package.

proposed in reference [62] for the 60 GHz radios [1,24]. For showing the metal traces clearly, the package is placed upside down.

The signal traces for the chip connection in the package are not shown purposely to highlight the antenna design. The AiP ceramic material is LTCC ferro-A6 with a relative dielectric constant and loss tangent of 5.9 and 0.002, respectively. The AiP metallic materials are gold and silver. There are four cofired laminated ceramic layers for the package. The first

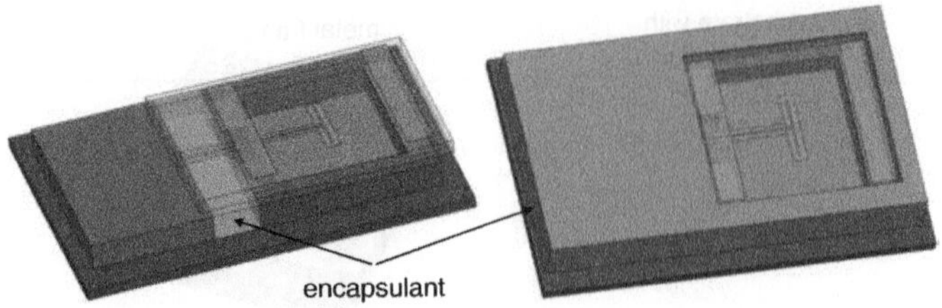

(d)

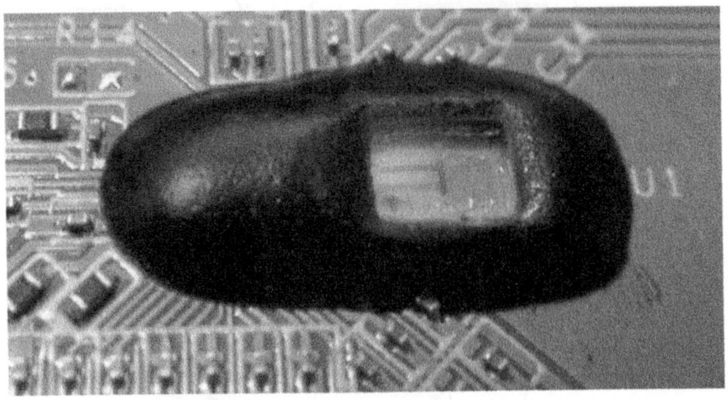

(e)

Figure 7.52: Continued.

ceramic layer (actually the package top) is 0.385 mm thick, the second ceramic layer is 0.285 mm thick with a 3.8×2 mm^2 opening, the third layer is 0.21 mm thick with a 5×3.2 mm^2 opening, and the fourth layer is 0.385 mm thick with a 5×3.8 mm^2 opening. These openings form a stepped cavity that can house the highly integrated 60 GHz radio die. There are also four metallic layers for the package. The first buried layer provides the metallization for the package ground plane and antenna guard ring, the second buried layer the metallization for the slot antenna and signal traces, the third buried layer the metallization for the signal traces, and the fourth exposed layer (actually the bottom of the package) the metallization for the antenna ground plane and solder ball pads. The size of the whole AiP is $12.5 \times 8 \times 1.265$ mm^3.

Figure 7.54 shows the details of the slot antenna. It consists of a triangular slot radiator, a guard-ring director, a ground-plane reflector, and a fence of vias. A well-controlled electromagnetic environment is created for the antenna part of the AiP, which makes the antenna performance less sensitive to the radio chip, PCB dielectric and metallic structures – an important design feature for system integration. The triangular slot with $l = 2.064$ mm (about one guided wavelength λ_g) is inductively fed with a 50 Ω CPW line. The CPW pitch is 250 μm. The guard ring is $w = 0.925$ mm (about $\lambda_g/2$) wide. The gaps between the guard ring and the slot radiator are $s = 0.442$ mm and $g = 0.8156$ mm. The fence of vias shorts the outer metal edge of the slot radiator to the reflector. A quasi cavity-backed slot antenna is thus realized. The ceramic material under the slot radiator is modulated with air holes to

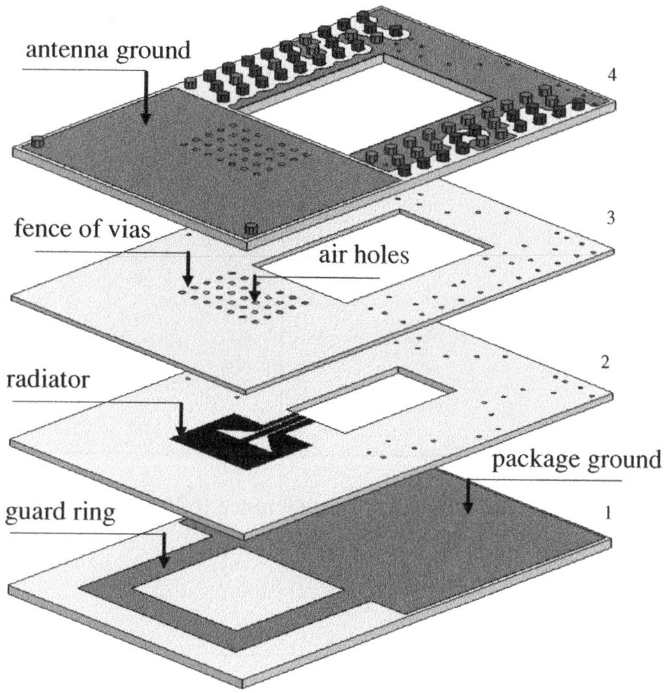

Figure 7.53: Explored view of the AiP (from reference [62], reproduced by permission of © 2007 IEEE).

reduce surface waves owing to a lower effective dielectric constant. Figure 7.55 shows the simulated-with-HFSS and measured performance of the AiP where it is seen that they are in good agreement. The simulated and measured return loss values are better than 7 dB from 59 to 65 GHz, indicating an acceptable matching to a 50 Ω source at these frequencies. It should be mentioned that the simulated and measured impedance values from 59 to 65 GHz exhibit capacitive reactance as shown in Figure 7.56. The matching to the 50 Ω output of the radio chip will be improved if the inductive reactance from the bondwires is properly exploited and a quarter wavelength transformer is used. The simulated and measured E- and H-plane radiation patterns at 61.5 GHz are shown in Figures 7.57 and 7.58 respectively. It is clear that the H-plane patterns are similar to but the E-plane patterns are different from those of a conventional cavity backed slot antenna. A shaped-beam co-polarization pattern can be seen in the E-plane with the main beam in the directions from 45° to 60°. The shaped-beam pattern in the E-plane is mainly caused by the package ground. The measured and calculated peak gain values for the slot AiP in the main beam direction are 11 and 9.5 dBi at 61.5 GHz, respectively with a simulated efficiency of 94%. The gain difference between the simulated and the measured is due to the fact that the chip cavity portion was not included in simulation to reduce the computation time. The chip cavity portion has less effect on the antenna impedance calculation.

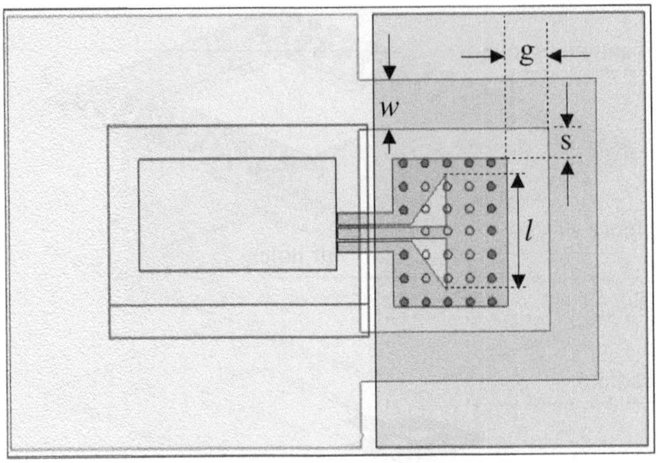

Figure 7.54: Details of the slot antenna (from reference [62], reproduced by permission of © 2007 IEEE).

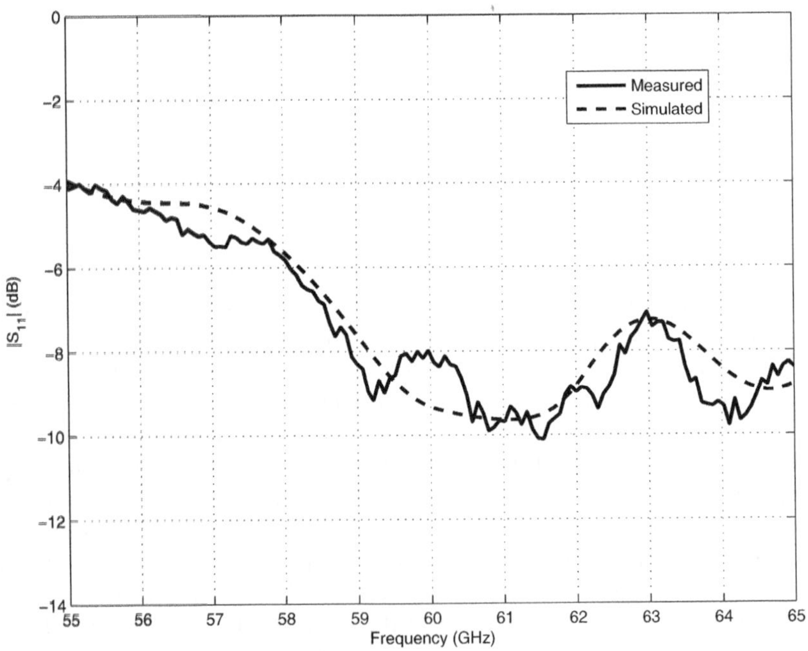

Figure 7.55: Measured and simulated S_{11} of the antenna in LTCC package (from reference [62], reproduced by permission of © 2007 IEEE).

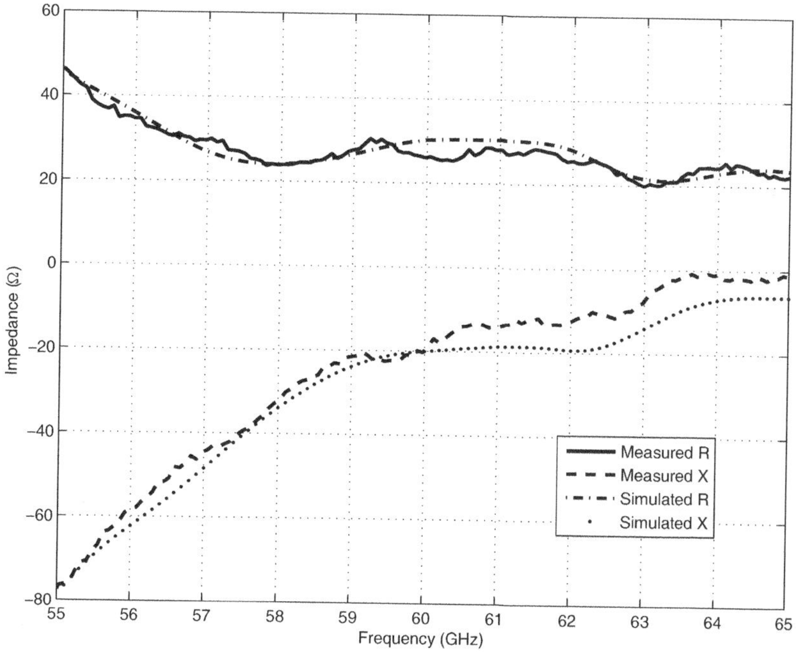

Figure 7.56: Measured and simulated impedance of the antenna in LTCC package (from reference [62], reproduced by permission of © 2007 IEEE).

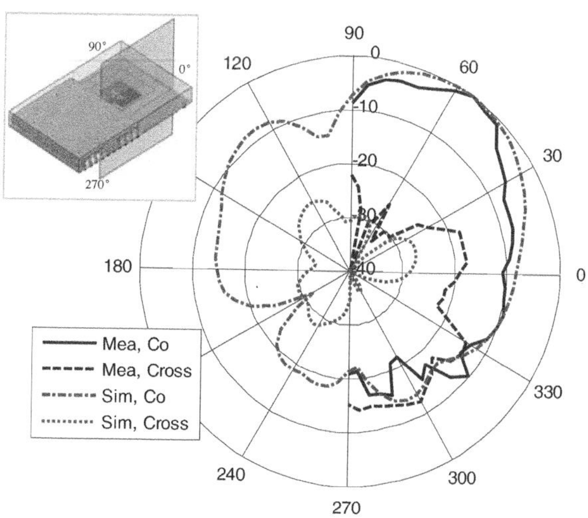

Figure 7.57: Measured and simulated radiation patterns: E-plane (or front) (from reference [62], reproduced by permission of © 2007 IEEE).

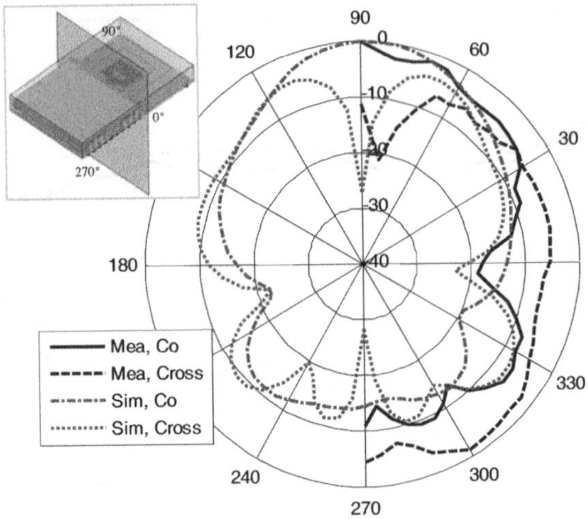

Figure 7.58: Measured and simulated radiation patterns: H-plane (or side) (from reference [62], reproduced by permission of © 2007 IEEE).

7.7.2 Silicon-based Packages

Comparing to the LTCC, molding, or advanced polymers such as liquid crystal polymers-based packages, silicon-based packages can be much more functional owing to increased integration levels. A wide range of high-quality passive components, antennas, switches, and other devices with high-performance interconnects can be densely packaged with RFICs using Si technologies at low cost. It is well known that electronic device packaging plays a key role in the function and cost of any product. Performance at mmWave frequencies is often limited by the minimum possible spacing of discrete components on the package and by the quality of the interconnect technology available. Furthermore, the packaging often limits the miniaturization of the system and may degrade reliability, predominantly when the thermal expansion coefficients of different components vary widely.

A miniature, multi-functional Si-based packaging technology which can reduce the size and cost and increase the performance of a wide range of mmWave systems is proposed [22]. High-density capacitors, low-temperature coefficient resistors, high-Q inductors, low-loss transmission lines, filters and antennas can all be built within the Si package using standard semiconductor fabrication methods with very high manufacturing precision compared to conventional packaging technologies. Fine pitch metal bumps can be used to attach RFICs and other components to the package, while gold bonding ring can be used to provide hermetic sealing where required [63]. Vias through the Si package eliminate inductive bond wires and minimize parasitics at mmWave frequencies. By integrating both antenna and an RFIC in one single package, all high frequency signals are confined within the package and only baseband signals are connected to the external package [6]. Figure 7.59 illustrates the concept of this Si-based package containing a protected RFIC chip, through wafer vias, a microelectromechanical system (MEMS) antenna switch, and an integrated antenna all combined into one low-profile unit.

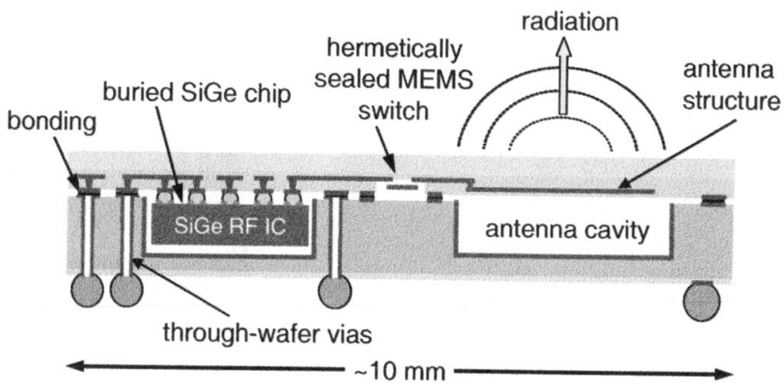

Figure 7.59: Schematic of an all Si-based low-cost package for 60 GHz transceiver (from reference [22], reproduced by permission of © 2007 IEEE).

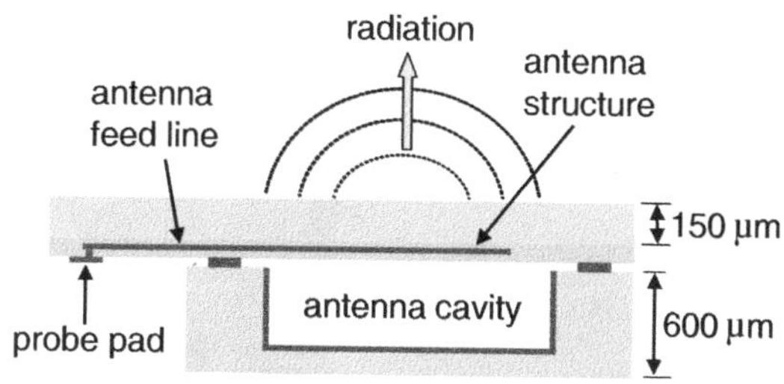

Figure 7.60: Schematic of the cavity backed antenna structure.

However, before the whole Si-based package can be designed, one has to evaluate the feasibility and to demonstrate the validity of an all Si-based mmWave package. Therefore, a 60 GHz antenna with the desired performance, bandwidth, and efficiency was designed in the high dielectric ($\epsilon_r = 11.9$) Si environment [22]. The antenna system consists of two pieces of Si joined together as illustrated in Figure 7.60, where the top part contains the antenna fabricated on high-resistivity (1000 Ω-cm) Si wafers and the bottom part the metallized cavity. The top Si also contains the two probe pads for antenna evaluation.

The antenna is fabricated from a 1.2 μm-thick Cu film patterned with a damascene process. In order to follow conventional interconnect processing, a 70 nm-thick SiN layer together with a 1.0 μm-thick SiO_2 layer separates the Cu from the Si substrate. To maximize antenna performance, the Si wafers were thinned after processing from 725 μm to 150 μm using a back-side grinding process. A modified folded dipole antenna (or dipole with a T-match) is used to improve the antenna bandwidth and efficiency. The antenna is connected to the 100 Ω coplanar strip line through a quarter wavelength transformer. A coplanar open loop is used to suppress the surface waves in the silicon substrate.

The bottom part is fabricated using a doped Si wafer in which a $2200 \times 2200 \times 400\ \mu m^3$ cavity was etched using deep Si reactive ion etching. The cavity was metallized with a TaN/Ta adhesion layer, a sputtered Cu seed layer, and an 8 μm-thick electroplated Cu layer. The unwanted metal surrounding the cavity was removed using a chemical-mechanical

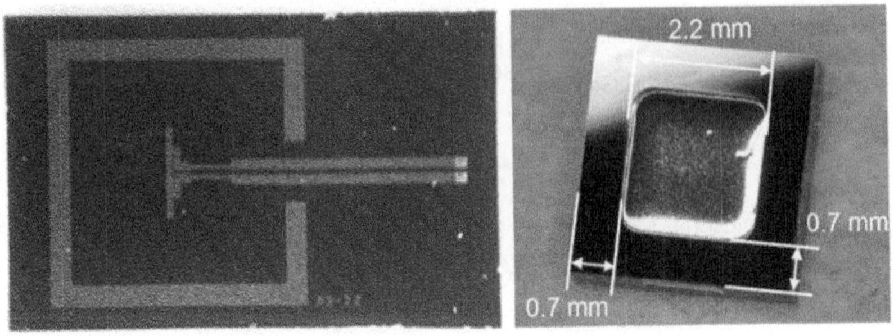

(a) (b)

Figure 7.61: Photograph (a) of antenna, and (b) cavity (from reference [22], reproduced by permission of © 2007 IEEE).

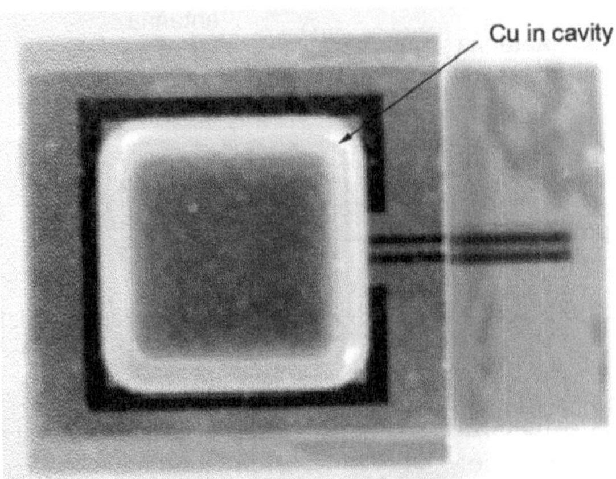

Figure 7.62: IR photograph of assembled antenna. The Cu layer inside the cavity is clearly identified as a white ring inside the darker Cu features on the antenna substrate (from reference [22], reproduced by permission of © 2007 IEEE).

polishing (CMP) process. All processing steps applied are the same as those used in conventional Cu-interconnect technology, and therefore these mmWave packages/antennas can be manufactured at low cost in existing semiconductor fabs.

Following dicing of both the antenna and the cavity wafer (see Figure 7.61), the assembly of the individual antennas was done by hand. A small amount of adhesive was used to keep the two Si pieces together, and care was taken to avoid any adhesive covering the antenna features. Simulations show that the resonant frequency of the antenna is very sensitive to any misalignment between the cavity and the ring around the antenna. Optical inspection of the assembled antennas using an IR camera indicated alignment accuracy around 50–100 μm (see Figure 7.62), which is acceptable for a first-pass demonstration. To fully characterize the Si-based antennas, all devices were mounted on a specially designed sample holder made from

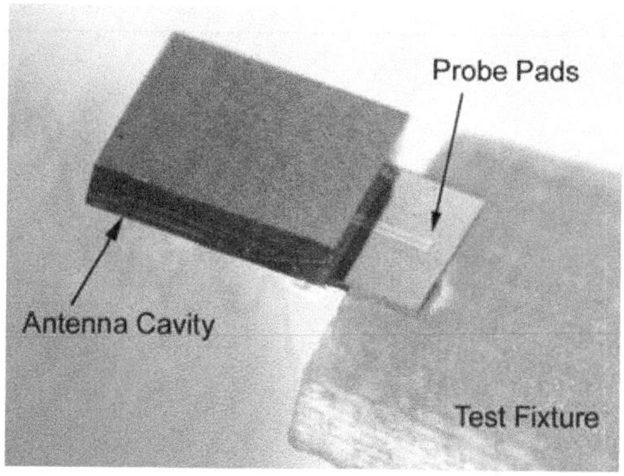

Figure 7.63: Photograph of assembled antenna mounted on FR4 test fixture. Probe access to antenna is obtained by mounting the device upside down (from reference [22], reproduced by permission of © 2007 IEEE).

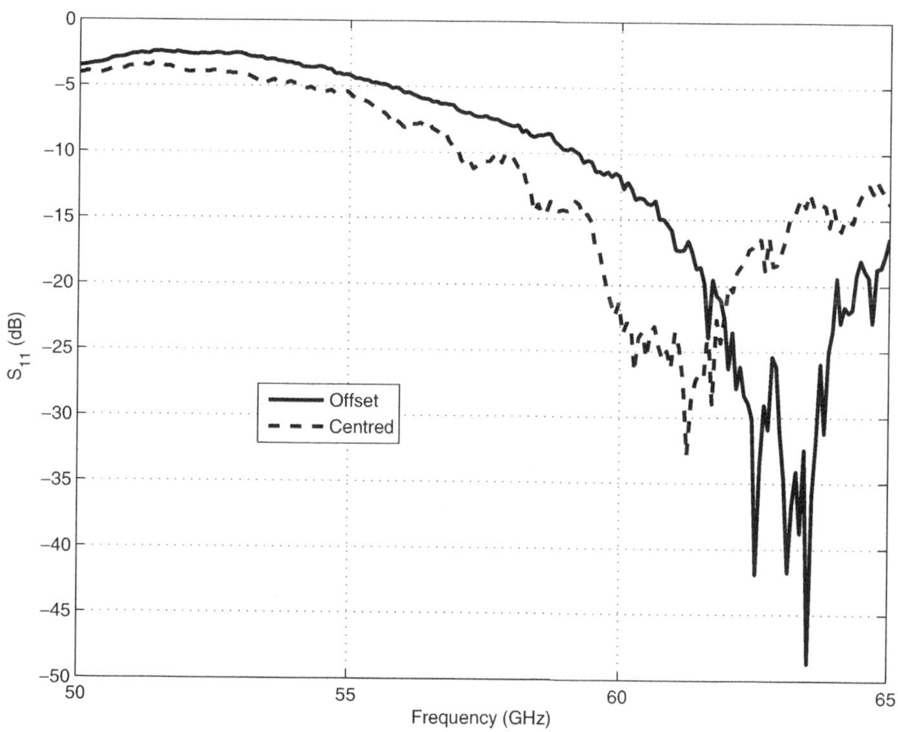

Figure 7.64: Measured antenna input match of two identical antenna structures using high-resistivity Si ($Z_0 = 100\ \Omega$) (from reference [22], reproduced by permission of © 2007 IEEE).

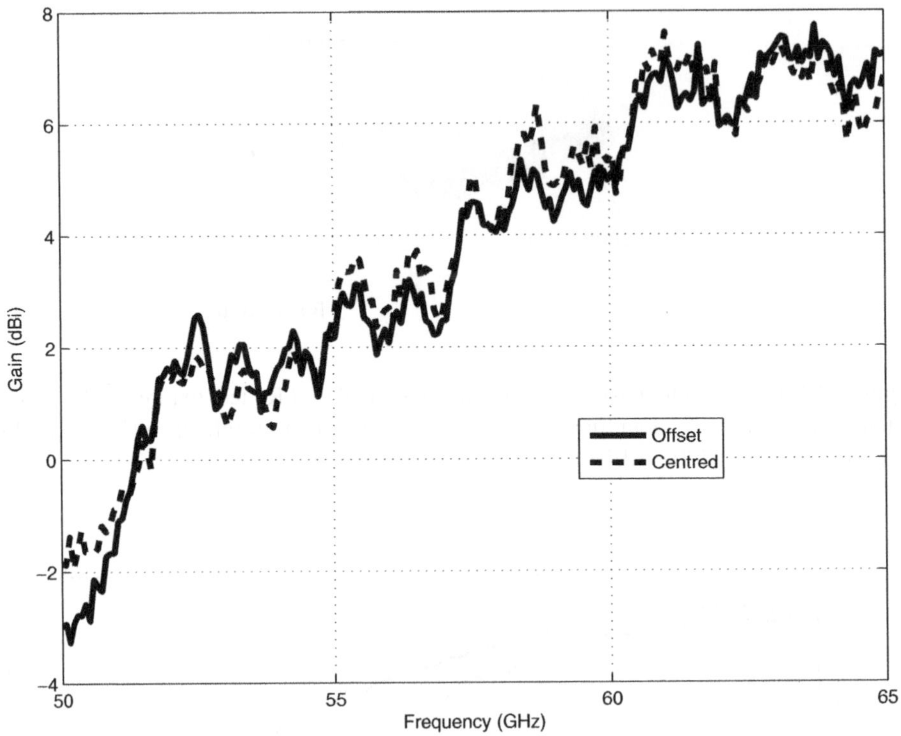

Figure 7.65: Measured gain versus frequency for the Si-based antennas fabricated on high-resistivity silicon substrates (from reference [22], reproduced by permission of © 2007 IEEE).

FR4 and connected using a coplanar probe, as shown in Figure 7.63. To facilitate suitable access to the probe pads, the antennas are mounted upside down with radiation direction downwards. The cavity backed antenna was measured in an anechoic chamber as a standalone part using a probe-based measurement system described in reference [52], with the maximum measured frequency limited to 65 GHz due to the measurement system setup.

The antenna match was measured for two identical antenna geometries fabricated using high-resistivity Si substrates. The measured antenna match for $Z_0 = 100\ \Omega$ (required by the power amplifier [6]) is shown in Figure 7.64, where the difference in resonant frequency between antenna A and B can be explained from misalignment between the cavity and antenna structure. From optical inspection of antenna A it was determined that the misalignment was around 100 μm towards the feed line. The measured return loss is better than 10 dB for 59–65 GHz for both devices, with a bandwidth greater than 10%. Figure 7.65 shows the measured antenna gain for the two antennas versus frequency. The gain is between 6 and 8 dBi from 60 to 65 GHz, which is comparable to previously demonstrated high-efficiency packaged antennas fabricated using low-loss fused silica [20]; so the antenna efficiency should be better than 85% within the band of interest. Note that when radiating through a thin layer of Si, the efficiency of the antenna was not adversely degraded by

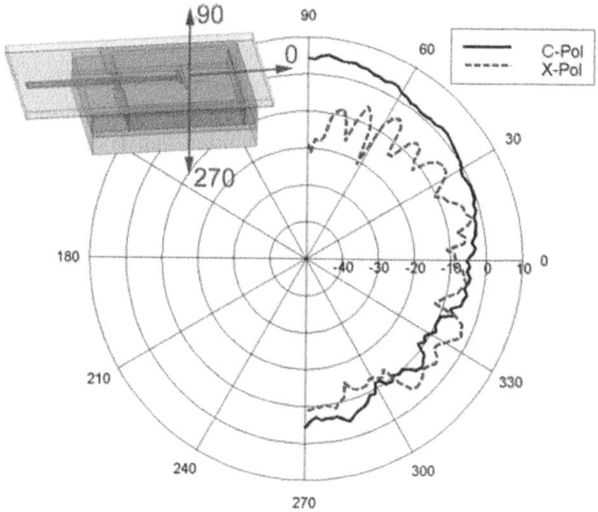

Figure 7.66: Measured 'front' radiation pattern at 61.5 GHz (from reference [22], reproduced by permission of © 2007 IEEE).

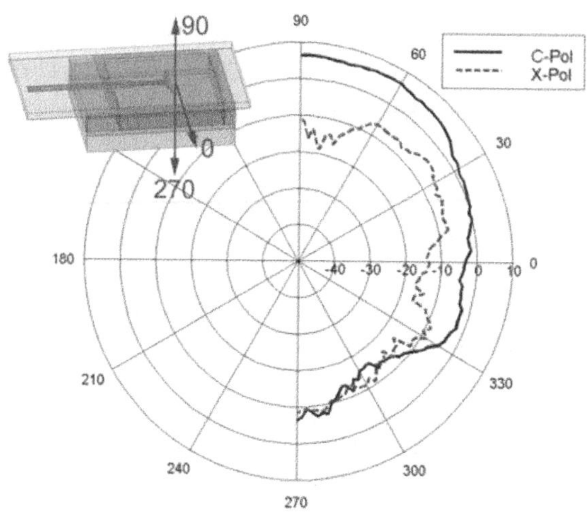

Figure 7.67: Measured 'side' radiation pattern at 61.5 GHz (from reference [22], reproduced by permission of © 2007 IEEE).

any misalignment between the antenna and cavity since the antenna match is good. The radiation patterns for $\phi = 0°$ and $90°$ (front and side) measured at 61.5 GHz are presented in Figures 7.66 and 7.67, respectively. The patterns are similar to those of small cavity backed dipole antennas. The effect of the cavity on the radiation pattern can also be observed.

Acknowledgments

The author would like to thank DARPA (contracts N66001-02-C-8014 and N66001-05-C-8013) and NASA (contract NAS3-03070) for partial funding and support. He would also like to thank M. Soyuer, M. Oprysko, S. Gowda, B. Floyd, and B. Gaucher for the mmWave project leadership and management support. Many people participated in the mmWave antenna projects, particularly T. Zwick, who did the air-suspended superstrate antenna study and designed the mmWave chamber, and in fact laid the foundation for the 60 GHz antenna projects, J. Grzyb, who designed the packaged antenna and studied the antenna integration issues, U. Pfeiffer, who did the antenna in system evaluation, N. Hoivik, who led the antenna in silicon projects, R. John, who did all the antenna assembly, and C. Baks, who implemented the mmWave chamber and made all the testing fixtures. At the time of work all of them were with the IBM T. J. Watson Research Center. Y. P. Zhang and his team from NTU of Singapore contributed to the LTCC antenna in package effort.

References

[1] S. K. Reynolds, B. A. Floyd, U. R. Pfeiffer, T. J. Beukema, T. Zwick, J. Grzyb, D. Liu and P. B. Gaucher, 'Progress toward a low-cost millimeter-wave silicon radio', in *Proc. IEEE Custom Integrated Circuits Conf. (CICC)*, pp. 563–570, San Jose, CA, September 2005.

[2] P. Smulders, 'Exploiting the 60 GHz band for local wireless multimedia access: prospects and future directions', *IEEE Commun. Mag.* **40**(1) (2002), pp. 140–147.

[3] W. J. Fleming, 'Overview of automotive sensors', *IEEE Sensors J.* **1**(4) (2001), pp. 296–308.

[4] T. Hirose, M. Sato, T. Oki, H. Sato, K. Sawaya and K. Mizuno, 'Development of a 94-GHz passive millimeter-wave imaging sensor', *IEIC Tech. Rep.* **106**(403) (2006), pp. 35–40.

[5] D. Notel, J. Huck, H. Essen, H. H. Fuchs, F. Kloppel and P. Pergande, 'A demonstrator approach for the technology of a passive millimeter wave imaging system and the related image processing', *Infrared and Millimeter Waves* and *13th Int. Conf. on Terahertz Electron.* vol. 1, pp. 319–320, Williamsburg, VA, 2005.

[6] U. R. Pfeiffer, J. Grzyb, D. Liu, B. Gaucher, T. Beukema, B. A. Floyd and S. K. Reynolds, 'A chip-scale packaging technology for 60-GHz wireless chipsets', *IEEE Trans. Microwave Theory Tech.* **54**(8) (2006), pp. 3387–3397.

[7] K. Kihong, H. Yoon and K. K. O, 'On-chip wireless interconnection with integrated antennas', *IEEE Int. Electron Devices Meeting* (2000) pp. 485–488, San Francisco, CA.

[8] Y. P. Zhang, M. Sun and L. H. Guo, 'On-chip antennas for 60-GHz radios in silicon technology', *IEEE Trans. Electron Devices* **52** (2005), pp. 1664–1668.

[9] E. Öjefors, E. Sönmez, S. Chartier, P. Lindberg, A. Rydberg and H. Schumacher, 'Monolithic integration of an antenna with a 24 GHz image-rejection receiver in SiGe HBT technology', in *Proc. 35th Eur. Microwave Conf.*, Paris, France, October 2005.

[10] I. Papapolymerou, R. F. Drayton and L. P. Katehi, 'Micromachined patch antennas', *IEEE Trans. Antennas Propag.* **46**(2) (1998), pp. 275–283.

[11] M. Zheng, Q. Chen, P. Hall and V. Fusco, 'Broadband microstrip patch antenna on micromachined silicon substrates', *Electron. Lett.* **34**(1) (1998), pp. 3–4.

[12] D. Neculoiu *et al.*,'The design of membrane-supported millimeter-wave antennas', in *Proc. Int. Semicond. Conf.*, CAS, vol. 1, pp. 65–68, Romania, 2003.

[13] P. Caudrillier, A. Takacs, O. Pascal, H. Aubert, P. Pons and R. Plana, 'Compact circularly polarized radiating element for Ka-band satellite communications', in *Proc. IEEE Antennas Propag. Soc. Int. Symp.* vol. 1, pp. 18–21, San Antonio, TX, 2002.

[14] D. Neculoiu, P. Pons, M. Saadaoui, L. Bary, D. Vasilache, K. Grenier, D. Dubuc, A. Muller and R. Plana, 'Membrane supported Yagi–Uda antennae for millimetre-wave applications', *IEE Proc. Microwaves, Antennas and Propag.*, vol. 151, no. 4, pp. 311–314, August 2004.

[15] Q. Chen, V. Fusco, M. Zheng and P. Hall, 'Micromachined silicon antennas', in *Proc. Int. Conf. on Microwave and Millimeter Wave Technol. (ICMMT)*, pp. 289–292, Beijing, China, August 1998.

[16] D. Liu and B. Gaucher, 'Design considerations for millimeter wave antennas within a chip package', in *Proc. IEEE Int. Workshop on Anti-counterfeiting, Security, Identification*, pp. 13–17, Xiamen, China, April 16–18, 2007.

[17] T. Zwick, A. Chandrasekhar, C. Baks, U. R. Pfeiffer, S. Brebels and B. P. Gaucher, 'Determination of the complex permittivity of packaging materials at millimeter-wave frequencies', *IEEE Trans. Microwave Theory Tech.* **54**(3) (2006), pp. 1001–1010.

[18] K. Hashimoto, T. Kai, J. Hirokawa and M. Ando, 'A center-feed post-wall waveguide parallel plate slot array', in *Proc. IEEE Antennas Propag. Soc. Int. Symp. and UNSC/URSI and AMEREM Meetings*, pp. 3043–3046, Albuquerque, New Mexico, July 9–14, 2006.

[19] T. Zwick, D. Liu and B. Gaucher, 'Broadband planar superstrate antenna for integrated mmWave transceivers', *IEEE Trans. Antennas Propag.* **54**(10) (2006), pp. 2790–2796.

[20] J. Grzyb, D. Liu, U. Pfeiffer and B. Gaucher, 'Wideband cavity-backed folded dipole superstrate antenna for 60 GHz applications', in *Proc. IEEE Antennas Propag. Soc. Int. Symp. and UNSC/URSI and AMEREM Meetings* pp. 3939–3942, Albuquerque, New Mexico, July 9–14, 2006.

[21] B. Gaucher, T. Beukema, S. Reynolds, B. Floyd, T. Zwick, U. Pfeiffer, D. Liu and J. Cressler, 'MMW-transceivers using SiGe HBT technology', in *Proc. IEEE Topical Meeting on Silicon Monolithic Integrated Circuits in RF Systems*, pp. 81–84, Atlanta, GA, September 2004.

[22] N. Hoivik, D. Liu, C. V. Jahnes, J. M. Cotte, C. Tsang, C. Patel, U. Pfeiffer, J. Grzyb, J. Knickerbocker, J. H. Magerlein and B. Gaucher, 'High-efficiency 60 GHZ antenna fabricated using low-cost silicon micromaching techniques', in *Proc. IEEE Antennas Propag. Soc. Int. Symp.*, pp. 5043–5046, Honolulu, Hawaii, June 10–15, 2007.

[23] G. A. Riley, 'Introduction to flip chip: what, why, how', http://www.flipchips.com/RileyWP01.pdf, December 15, 2008.

[24] B. A. Floyd, S. K. Reynolds, U. R. R. Pfeiffer, T. Zwick, T. Beukema and B. Gaucher, 'SiGe bipolar transceiver circuits operating at 60 GHz', *IEEE J. Solid-State Circuits* **40**(1) (2005), pp. 156–167.

[25] N. Herscovici, 'A wide-band single-layer patch antenna', *IEEE Trans. Antennas Propag.* **46**(4) (1998), pp. 471–474.

[26] M. M. Faiz and P. F. Wahid, 'A high efficiency L-band microstrip antenna', in *Proc. IEEE Antennas Propag. Soc. Int. Symp.*, pp. 272–275, Orlando, FL, July 1999.

[27] P. R. Acharya, H. Ekström, S. S. Gearhart, S. Jacobsson, J. F. Johansson, E. L. Kollberg and G. M. Rebeiz, 'Tapered slotline antennas at 802 GHz', *IEEE Trans. Microw. Theory Tech.* **41**(10) (1993), pp. 1715–1719.

[28] G. P. Gauthier, J. P. Raskin, L. P. B. Katehi and G. M. Rebeiz, 'A 94-GHz aperture-coupled micromachined microstrip antenna', *IEEE Trans. Antennas Propag.* **47**(12) (1999), pp. 1761–1766.

[29] G. M. Rebeiz, 'Millimeter-wave and terahertz integrated circuit antennas', *Proc. IEEE* **80**(11) (1992), pp. 1748–1770.

[30] T. S. Bird, J. S. Kot, N. Kikolic, G. L. James, S. J. Barker, F. Cooray and D. G. Bateman, 'Millimetre-wave antenna and propagation studies for indoor wireless LANs', in *Proc. IEEE Antennas Propag. Soc. Int. Symp.*, pp. 336–339, Seattle, WA, June 1994.

[31] S. D. Targonski, R. B. Waterhouse and D. M. Pozar, 'Design of wide-band aperture-stacked patch microstrip antennas', *IEEE Trans. Antennas Propag.* **46** (1998), pp. 1245–1251.

[32] D. R. Jackson and N. G. Alexópoulos, 'Gain enhancement methods for printed circuit antennas', *IEEE Trans. Antennas Propag.* **33**(9) (1985), pp. 976–987.

[33] L. Bernard, R. Loison, R. Gillard and T. Lucidarme, 'High directivity multiple superstrate antennas with improved bandwidth', in *Proc. IEEE Antennas Propag. Soc. Int. Symp.*, vol. 2, pp. 522–525, San Antonio, TX, June 2002.

[34] W. Choi, C. Pyo, Y. H. Cho, J. Choi and J. Chae, 'High gain and broadband microstrip patch antenna using a superstrate layer', in *Proc. IEEE Antennas Propag. Soc. Int. Symp.*, vol. 2, pp. 292–295, Columbus, OH, June 2003.

[35] N. G. Alexópoulos and D. R. Jackson, 'Fundamental superstrate (cover) effects on printed circuit antennas', *IEEE Trans. Antennas Propag.* **32**(8) (1984), pp. 807–816.

[36] N. G. Alexópoulos and D. R. Jackson, 'Fundamental superstrate effects on printed circuit antenna efficiency', in *Proc. IEEE MTT-S Int. Microw. Symp. Digest*, vol. 84, pp. 475–476, San Francisco, CA, May 1984.

[37] D. R. Jackson, J. T. Williams, A. K. Bhattacharyya, R. Smith, S. J. Buchheit and S. A. Long, 'Microstrip patch designs that do not excite surface waves', *IEEE Trans. Antennas Propag.* **41** (1993), pp. 1026–1037.

[38] M. A. Hickey, M. Qiu and G. V. Eleftheriades, 'A reduced surface-wave twin arc-slot antenna for millimeter-wave applications', *IEEE Microwave and Wireless Components Lett.* **11**(11) (2001), pp. 459–461.

[39] K. M. Luk, W. Y. Tam and C. L. Yip, 'Analysis of circular microstrip antennas with superstrate', *IEE Proc.* **136**(3) (1989), pp. 261–262.

[40] K.-L.Wong, Y.-T. Cheng and J.-S. Row, 'Resonance in a superstrate loaded cylindrical-rectangular microstrip structure', *IEEE Trans. Microw. Theory Tech.* **41**(5) (1993), pp. 814–819.

[41] C. S. Gürel and E. Yazgan, 'Bandwidth widening in an annular ring microstrip antenna with superstrate', in *Proc. IEEE Antennas Propag. Soc. Int. Symp.*, pp. 692–695, Newport Beach, CA, June 1995.

[42] S. Yamaguchi, Y. Imai, T. Shibata, T. Otsuji, M. Hirano and E. Sano, 'An inverted microstrip line IC structure for ultra high-speed applications', *IEEE MTT-S Int. Microwave Symp. Digest*, Vol. 3, pp. 1643–1646, Orlando, FL, May 1995.

[43] S. Kim, S. Jeong, Y.-T. Lee, D.-H. Kim, J.-S. Kim, K.-S. Seo and S. Nam, 'Ultra-wideband (from DC to 110 GHz) CPW to CPS transition', *IEE Electron. Lett.* **38**(13) (2002), pp. 622–623.

[44] W. L. Stutzman and G. A. Thiele, *Antenna Theory and Design* (New York: Wiley & Sons Ltd/Inc., 1998).

[45] J. C. W. Harrison and R. W. P. King, 'Folded dipoles and loops', *IRE Trans. Antennas Propag.* **9**(2) (1961), pp. 171–187.

[46] W. T. Whistler, 'The microstrip dual folded dipole as a phased array element', in *Proc. IEEE Antennas Propag. Soc. Int. Symp.*, vol. 25, pp. 1024–1027, Blacksburg, VA, June 1987.

[47] D. A. McNamara and L. Botha, 'On the functioning of folded dipole antennas on conducting masts', *IEEE Trans. Veh. Technol.* **42**(4) (1993), pp. 393–398.

[48] M. C. van Beurden, A. B. Smolders, M. E. J. Jeuken, G. H. C. van Werkhoven and E. W. Kolk, 'Analysis of wide-band infinite phased arrays of printed folded dipoles embedded in metallic boxes', *IEEE Trans. Antennas Propag.* **50**(9) (2002), pp. 1266–1273.

[49] D. F. Filipović, W. Y. Ali-Ahmad and G. M. Rebeiz, 'Millimeter-wave double-dipole antennas for high-gain integrated reflector illumination', *IEEE Trans. Microw. Theory Tech.* **40**(5) (1992), pp. 962–967.

[50] F.-R. Hsiao and K.-L.Wong, 'Omnidirectional planar folded dipole antenna', *IEEE Trans. Antennas Propagat.* **52**(7) (2004), pp. 1898–1902.

[51] G. P. Gauthier, S. Raman and G. M. Rebeiz, 'A 90-100 GHz doublefolded slot antenna', *IEEE Trans. Antennas Propag.* **47**(6) (1999), pp. 1120–1122.

[52] T. Zwick, C. Baks, U. R. Pfeiffer, D. Liu and B. P. Gaucher, 'Probe based MMW antenna measurement setup', in *Proc. IEEE Antennas Propag. Soc. Int. Symp.*, vol. 1, pp. 747–750, Monterey, CA, June 2004.

[53] D. Liu, 'Cavity size effects on the performance of superstrate folded dipole antennas for 60 GHz applications', in *Proc. IEEE Int. Workshop on Antenna Technol.*, pp. 71–74, Chiba, Japan, March 4–6, 2008.

[54] J. Grzyb, D. Liu and B. Gaucher, 'Packaging effects of a broadband 60 GHz cavity-backed folded dipole superstrate antenna', in *Proc. IEEE Antennas Propag. Soc. Int. Symp.*, pp. 4365–4368, Honolulu, Hawaii, June 10–15, 2007.

[55] D. Liu, B. Gaucher and R. Sirdeshmukh, 'Antenna array feed line structures for millimeter wave applications', *US Patent Application Number 11/772464*, July 2, 2007.

[56] R. Li, G. DeJean, J. Laskar and M. M. Tentzeris, 'Investigation of circularly polarized loop antennas with a parasitic element for bandwidth enhancement', *IEEE Trans. Antennas Propag.* **54**(12) (2005), pp. 3930–3939.

[57] D. Liu, 'Planar circular polarized antennas', *US Patent Application Number 12/031282*, February 14, 2008.

[58] A. Lamminen, J. Säily and A. Vimpari, 'Design and processing of 60 GHz antennas on low temperature co-fired ceramic (LTCC) substrates', in *Proc. 4th ESA Workshop on Millimetre-Wave Technol. and Applic.*, pp. 43–48, Espoo, Finland, February 15–17, 2006.

[59] A. Panther, A. Petos, M. G. Stubbs and K. Kautio, 'A wideband array of stacked patch antennas using embedded air cavities in LTCC', *IEEE Microw. Wireless Compon. Lett.* **15**(12) (2005), pp. 916–918.

[60] Y. Huang, K.-L. Wu, D.-G. Fang and M. Ehlert, 'An integrated LTCC millimeter-wave planar array antenna with low-loss feeding network', *IEEE Trans. Antennas Propag.* **53**(3) (2005), pp. 1232–1234.

[61] T. Seki, N. Honma, K. Nishikawa and K. Tsunekawa, 'A 60-GHz multilayer parasitic microstrip array antenna on LTCC substrate for system-onpackage', *IEEE Microw. Wireless Compon. Lett.* **15**(5) (2005), pp. 339–341.

[62] Y. P. Zhang, M. Sun, K. M. Chua, L. L. Wai, D. Liu and B. Gaucher, 'Antenna-in-package in LTCC for 60 GHz radio', in *Proc. IEEE Int. Workshop on Antenna Technol.*, Cambridge, UK, March 21–23, 2007.

[63] B. Min and G. M. Rebeiz, 'A low-loss silicon-on-silicon DC–110-GHz resonance-free package', *IEEE Trans. Microwave Theory Tech.* **54**(2) (2006), pp. 710–716.

8

Monolithic Integrated Antennas

Erik Öjefors and Anders Rydberg

8.1 Introduction

It is expected that millimeter-wave (mmWave) radio front-ends will become important in a number of low-cost consumer applications in the future. The license exempt 24.05–24.25 GHz as well as the 59–64 GHz industrial, scientific and medical (ISM) frequency bands are of particular interest for short-range communication while the 77 GHz and 94 GHz bands are primarily targeted by radar applications. However, the high cost of the radio front-end components has so far limited the use of mmWave technology to professional applications such as directional radio links or long distance radars. In these applications the design objective is usually system performance. Hence, low-loss packaging techniques are commonly used together with waveguide technology and highly directive antennas such as parabolic dishes.

A different set of requirements applies for the design of the upcoming low-cost consumer products employing mmWave technology. Most of the foreseen communication applications will be of the short-range type, such as indoor computer networks or cable-replacing radio links in audio-visual entertainment systems. Although not performing as well as their III-V counterparts, mmWave circuits manufactured in silicon and silicon germanium (SiGe) semiconductor technologies are increasingly used, since they offer lower manufacturing costs as well as better integration with the supporting low frequency and digital circuits. In the targeted applications, low-directivity antennas such as simple dipole or patch radiators are often sufficient, since low-antenna gain can be accommodated in the link budgets. Simple, low-directivity radiators also help to avoid the pointing losses which would occur in a mobile communication scenario if no beam-steering is used.

Packaging and assembly of the mmWave radio front-end is currently a significant cost of the total product. Hence, monolithic integration of all high-frequency components on a single semiconductor chip has the potential of lowering the price of transceiver units to a level acceptable for mass market applications. Monolithic integrated front-ends would require little

Advanced Millimeter-wave Technologies: Antennas, Packaging and Circuits
Duixian Liu, Ulrich Pfeiffer, Janusz Grzyb and Brian Gaucher © 2009 John Wiley & Sons, Ltd

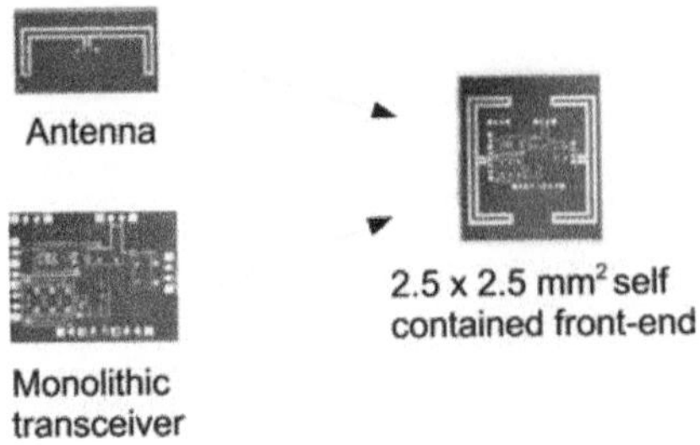

Figure 8.1: Suggested monolithic integration of a mmWave communication or radar front-end with an on-chip antenna.

high frequency expertise by the system integrator as no mmWave off-chip interconnects are needed. An example of such an on-chip system is shown Figure 8.1, where a single chip 24 GHz short range communication front-end is depicted.

Despite their potential cost and integration benefits, on-chip antennas have so far not seen widespread use. One reason is that antennas are one of the most challenging RF-components to integrate on-chip. Key antenna properties, such as size and bandwidth, are governed by the fundamental electromagnetic requirements of radiating elements and thus do not obey the same scaling properties as circuits implemented in rapidly improving semiconductor technology. To fulfill the cost-driven requirements of circuit and chip-size miniaturization, there is in many cases a need to consider the use of compact radiator types, at least at the lower mmWave bands such as the K and Ka bands (<40 GHz). However, the performance compromises dictated by such miniaturization must be carefully balanced against application requirements on antenna bandwidth, efficiency and radiation pattern. Circuit integration issues, such as cross-talk between the radio transceiver and the on-chip antenna, also need to be addressed.

Adding to the difficulties of on-chip antenna integration, the semiconductor substrate is in many cases a poor dielectric for the antenna. In silicon processes the substrate can be very lossy owing to the low bulk resistivity of the wafers normally used for the manufacturing of the active devices. Loss minimization, such as the use of modified substrates and micromachining techniques, is thus important for the implementation of high efficiency on-chip radiators.

In this chapter, practical low directivity mmWave radiators suitable for monolithic integration with active devices in commercial semiconductor processes are reviewed. Silicon is the semiconductor material primarily discussed, but several results are also valid for III-V process technologies such as gallium arsenide (GaAs) and indium phosphide (InP). Emphasis is placed on the selection and design of compact radiators on radiators which can be closely integrated with the on-chip electronics.

8.2 Monolithic Antenna Integration Challenges

8.2.1 Antenna Size

Due to the high cost of a processed semiconductor wafer there is a desire to minimize the size of the individual chips. Although many antenna designs using silicon as a substrate

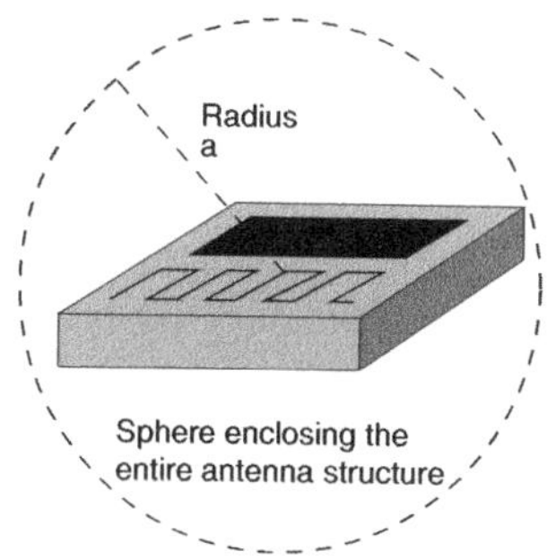

Figure 8.2: A sphere enclosing the entire antenna, including any (un)intentional parasitic radiators on-chip, can be used to classify the electrical size of the on-chip antenna.

material are available in the literature, most of them are primarily suited to cases where the silicon wafer is used as a passive carrier or multi-chip module (MCM) substrate owing to their relatively large area requirements. The die area available for an on-chip radiator will be dependent on the size and complexity of the active circuit as well as the packaging and interconnects used. The required space for a full-size radiator, such as a half-wave dipole, is not always available and a reduced size radiator, or electrically small antenna, might thus be needed.

The electrical size of an antenna is defined by enclosing the entire antenna structure in a sphere with radius a as illustrated in Figure 8.2. The sphere should be large enough to include any local groundplanes and other parasitic radiating structures which also have to be considered part of the antenna. An electrically small antenna is commonly defined as an antenna which fits within the so-called radiansphere , which has the radius

$$a = \frac{1}{k} = \frac{\lambda_0}{2\pi} \tag{8.1}$$

where k is the wavenumber and λ_0 is the free space wavelength at the frequency of interest. At 24 GHz, the radius of the radiansphere is $a = 1.9$ mm, thus indicating that on-chip antennas for typical die sizes of $< 10\ \text{mm}^2$ have to be classified as electrically small at the lower end of the mmWave frequency range.

The properties of small antennas have been extensively studied by Chu [1] and Wheeler [2] and relations between the antenna bandwidth and efficiency with respect to the radiator size have been derived. These results were re-examined by Mclean [3] to improve the accuracy for electrically larger radiators arriving at the expression

$$Q = \frac{1}{ka} + \frac{1}{k^3a^3} \tag{8.2}$$

for the obtainable minimum quality factor Q. Although Equation (8.2) sets a bound on the minimum quality factor, and hence the maximum bandwidth, it is important to remember that the equation only takes into account radiation losses. As electrically small antennas, especially if implemented in a lossy semiconductor environment, often suffer from poor efficiency the usable bandwidth can be larger. Hence, an electrically small radiator with an unusually large bandwidth will often show a poor efficiency.

Electrically small radiators with no reflectors present tend to have a directivity close to the 1.8 dBi value observed for a short dipole antenna. Superdirectivity can theoretically be used to realize antennas with high directivity in a small aperture but robustness and loss issues make such approaches less practical for monolithic mmWave applications. Hence, a die size which is small compared to the wavelength will yield an on-chip antenna with an almost isotropic radiation pattern. If a more directive beam is needed, beam shaping can be more effectively accomplished through the use of reflectors, lenses or parasitic radiators in the package than by a more elaborate on-chip radiator design.

8.2.2 Substrate Modes

The high dielectric constant of the semiconductor wafer provides some advantages as well as challenges in monolithic antenna design. On one hand, it allows the size of the antenna to be reduced as the dimensions of a resonant radiator are inversely proportional to the square root of the effective dielectric constant of the surrounding environment. On the other hand, the semiconductor makes the antenna susceptible to substrate modes. Conditions and design guidelines for implementation of antennas on thick dielectric substrates were considered for imaging antennas [4] and a general set of rules for mmWave antenna implementation have been derived [5]. In the case of a grounded substrate, such as conductor-backed dipole or patch antenna, the cutoff frequency f_c for substrate modes with number $n = 0, 1, 2 \ldots$ can be calculated as

$$f_c = \frac{nc}{4d\sqrt{\varepsilon_r - 1}} \tag{8.3}$$

where c is the speed of light, d the thickness of the substrate and ε_r the dielectric constant of the material. Since the cutoff frequency with $n = 0$, which corresponds to the TM_0 mode, is zero, this mode will always propagate, although the power coupled into the mode is small if the substrate is thin compared to the free-space wavelength λ_0. Unless any special substrate mode reduction techniques are used, it is generally advisable to avoid the excitation of any higher order modes by selecting a thin enough substrate to ensure that the cutoff frequency f_c for mode $n = 1$ is above the frequency of operation.

Although the problem of power coupling from the radiator into substrate modes is a serious one, the implications for monolithic antennas on diced substrates might be less severe than those indicated by the initial analysis. Most guidelines and analytical formulas are based on the assumption of an infinite dielectric substrate. Particularly at the lower millimeter-wave frequencies, such as the 24 GHz band, the size of the diced chip is often comparable to or smaller then the wavelength of the substrate guided waves. Hence, power coupled into the substrate is re-radiated at the chip edges with little attenuation or phase shift compared to the waves radiated from the on-chip antenna and might contribute to the main lobe of the antenna. In these cases careful modeling using a three-dimensional electromagnetic simulation package is usually needed to predict the substrate coupling and antenna performance.

8.2.3 Antenna Efficiency

Monolithic integrated antennas have traditionally been implemented in III-V monolithic microwave integrated circuit (MMIC) processes as the necessary device performance for high microwave and mmWave applications were only available using such semiconductor materials. As these processes use a semi-insulating GaAs or InP substrate, dielectric losses were not a major efficiency issue. In most present commercial complementary metal–oxide–semiconductor (CMOS), bipolar CMOS (BiCMOS) and SiGe heterojunction bipolar transistor (HBT) processes the only substrate available is low resistivity silicon. High frequency integrated circuits are predominantly designed using lumped passive components, such as spiral inductors, or microstrip transmission lines with both the signal conductor and the ground-plane implemented in the interconnect metallization layers present in the process back-end. Hence, there has been little demand for an increase in the substrate resistivity.

Unless loss minimization techniques, such as selective micromachining, can be used, low antenna efficiency is usually obtained in silicon technologies owing to the large substrate losses. Realized mmWave antennas on such substrates [6] typically have an antenna efficiency below 10%. However, an increased use of silicon-on-isolator (SOI) and silicon-on-sapphire (SOS) substrate in future CMOS process can also be expected, which merits an investigation of the possibility of monolithic integration of antennas with circuits manufactured on high resistivity wafers.

8.3 Manufacturing Techniques for Enhanced Antenna Performance

To address the issue of high dielectric losses a number of strategies can be used to improve the on-chip antenna efficiency. Three main methods can be identified as shown in Figure 8.3: The use of higher bulk substrate resistivity, addition of low loss layers such as benzocyclobutene (BCB) or polyimide, or selective removal of the substrate in critical areas through bulk micromachining. Traditional III-V manufacturing technologies such as GaAs and InP processes make use of a semi-insulating substrate, and the conductive substrate losses are hence low. In silicon, the typically substrate resistance used in a CMOS or BiCMOS process is usually lower than 50 Ω cm. High resistivity silicon wafers can, however, be used in some bipolar processes [7] although the high-temperature manufacturing steps encountered in the processing of the active devices can cause a the bulk resistivity of the wafer to drop. Proton implantation [8] of a selected substrate areas has been used to increase the silicon substrate resistivity. Special equipment which is not commonly available at a silicon processing plant is, however, needed. A problem with the high-resistive substrate approach to monolithic antenna integration is that it does not solve the problem of excitation of substrate modes.

Another method to reduce the high dielectric losses of the silicon substrate is to increase the distance between the radiator and the bulk silicon through the use to a thick organic layer such as polyimide or BCB. These materials have low dielectric constants and very low losses but are difficult to apply in thicknesses more than 20–30 μm. In combination with a thick metallization low-loss transmission line, structures have successfully been implemented on low resistivity silicon wafers [9, 10]. However, in the case of antennas, the application of additional dielectrics only brings a limited increase in antenna efficiency unless the substrate is shielded by a groundplane as, for instance, in the case of patch antenna.

Silicon bulk micromachining offers a way largely to eliminate substrate losses by etching away the substrate locally under critical areas of an antenna structure. An advantage of the

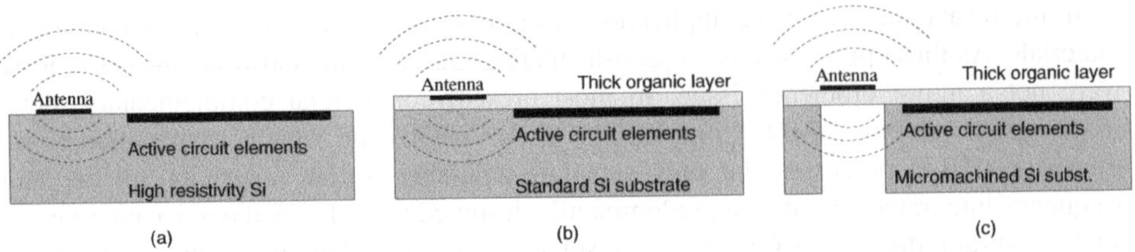

Figure 8.3: Cross-section of substrate with (a) high resistivity silicon, (b) organic polymer coating and (c) micromachined substrate.

bulk micromachining technique it that it reduces the problem of substrate mode excitation in addition to lowering the dielectric losses. Classically, bulk micromachining in silicon has been performed by potassium hydroxide (KOH) wet etching [11], yielding slanted silicon walls around the etched away area. Hence, the method is well suited for forming large membranes to support patch antennas [12]. However, large membranes also mean that most of the semiconductor substrate is removed, thus restricting the size of the active circuit to a few devices such as the diodes used in a membrane antenna mixer [13].

To be able to integrate bulk micromachined radiators with active electronics in a space-efficient way, the use of a modern dry etching technique based on deep reactive ion etching (DRIE) [14, 15] concept is needed. The advantage of the DRIE process is that very high-aspect ratios can be obtained, thus allowing integration of bulk micromachined membranes with arbitrary shape in close vicinity of the active devices. The main disadvantage of the DRIE process is the comparatively high manufacturing cost caused by the long processing times needed to etch through a wafer and release the membranes. However, ongoing development of the etching technique will help to reduce this cost disadvantage.

8.4 Selection and Design of the On-chip Radiator

The choice of a suitable antenna type for integration on the same chip as the driving electronics is to a large extent determined by the available space, circuit requirements (differential versus single ended feed) as well as the general layout of the chip and packaging. In contrast to a typical microstrip III-V MMIC, no substrate backside metallization or substrate vias are usually available in a silicon active device process. Hence, circuit ground mostly consists of local groundplanes defined by the circuit designer in the interconnect metallization layers. This fact makes the implementation of a conventional patch antenna with a microstrip feed difficult, even if the dielectric losses of the substrate can be reduced through micromachining or the use of a high resistivity substrate.

In many cases a differential circuit topology is favored over a single-ended owing to its increased voltage swing, lesser dependence on a well-defined circuit ground and the inclusion of differential circuit topologies as, for example, the widely used double-balanced Gilbert cell mixer. Hence, the use of an antenna with a balanced feed, such as a dipole or full wave loop, is advantageous as the need for a balun is avoided. Where a single-ended antenna is preferred, as in an architecture based on coplanar waveguide (CPW) transmission lines, an unbalanced slot antenna or a monopole type of antenna can be used, provided that the circuit layout can supply the necessary ground plane.

In this section the main types of compact low directivity broadside radiators suitable for monolithic integration in semiconductor processes will be described. Although most of the design examples will be presented for the lower mmWave frequencies in the 20–40 GHz K-band, the antennas can be scaled to higher frequencies as all critical dimensions are relative to the wavelength. However, it has to be considered that substrate waves and other effects might appear with increasing frequency unless the thicknesses of the dielectric layers are also scaled accordingly.

8.4.1 Patch Antennas

The patch antenna is a popular radiator type in planar integration technologies and consists of a half wavelength microstrip resonator backed by a dielectric substrate, as illustrated in Figure 8.4(a)–(d). Due to the use of conductor-backing, the back-lobe of the antenna is eliminated, thus increasing the directivity with 3 dB compared to a half-wavelength dipole in free space. The obtained broadside radiation pattern has a typical directivity ranging from 5 to 7 dBi, depending on the geometry of the patch element as well as the properties of the substrate. The required length L of the patch in order to obtain the desired half-wavelength resonance of the patch can be calculated [16] as

$$L = \frac{1}{2 f_{\mathrm{r}} \sqrt{\varepsilon_{\mathrm{eff}}} \sqrt{\mu_0 \varepsilon_0}} - 2\Delta L \tag{8.4}$$

where f_{r} is the desired resonance frequency, $\varepsilon_{\mathrm{eff}}$ is the effective dielectric constant for a microstrip transmission line of the same width W as the patch and ΔL is the extension of the electrical length of the patch caused by the fringing electric fields present at the radiating edges.

The width W of the patch mainly determines the conductance at the radiating edges and hence the input of the impedance. Together with the feed-point position the width is used to control the input impedance of the antenna. Although the patch radiator is normally fed by an unbalanced microstrip line connected to one of the radiating edges, the antenna can also be differentially excited through the use of two feed lines connected either to each radiating edge or symmetrically along the side of the patch.

8.4.1.1 High Resistivity Patch Substrate

In the III-V MMIC technologies GaAs and InP the patch antenna is the most common choice for an on-chip radiator since a backside metallization and a low-loss substrate is already present. Normally, the substrate is thinned down to a thickness of approximately 100 μm, thus reducing the problem of substrate wave excitation with mmWave antennas. Hence, the patch radiator can be implemented in the microstrip interconnect layer using the backside metallization as a groundplane as illustrated in Figure 8.4(a).

The patch element can usually be fed directly from the edge, since microstrip transmission lines utilizing the same groundplane as the radiator are normally available in a III-V-process. An example is the V-band antenna integrated self-oscillating mixer [17] implemented in a commercial 0.15 μm pseudomorphic high electron mobility transistor (pHEMT) process. A

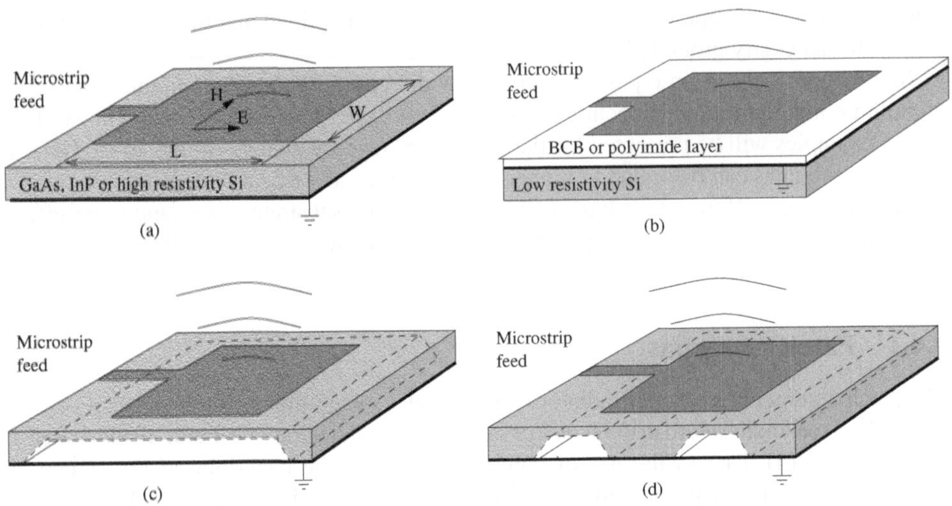

Figure 8.4: Monolithic patch integration through the use of (a) high resistivity semiconductors, (b) thick spin-on polymers, (c) bulk micromachining, or (d) selective micromachining.

patch radiator with dimensions $L = 680$ μm and $W = 900$ μm, which was implemented on the 100 μm thick substrate, yielded a resonance frequency 58.46 GHz. A differential feed technique was used with two microstrip feed lines connected through insets to the radiating edges of the patch.

Since the substrate resistivity is low ($< 50\ \Omega$ cm) in most commercially available silicon technologies, it is not feasible due to loss reasons to implement a patch antenna using the bulk semiconductor substrate as the antenna dielectric. In addition, the standard wafers are thicker than the III-V counterparts, thus creating problems with substrate modes, especially at higher mmWave frequencies, unless the substrate is thinned. K-band patch antennas suitable for monolithic integration have, however, been successfully realized in experimental silicon processes where a high resistivity substrate, backside metallization and microstrip transmission lines have been used [18].

8.4.1.2 Patch Manufactured in an Above-IC process

One approach to patch antenna integration in silicon processes with low substrate resistivity is the use of an above-integrated circuit (IC) process, as shown in Figure 8.4(b). A groundplane is placed between the patch radiator and the bulk silicon material and the antenna dielectric is implemented by the use of the available oxide layers in the process back-end or by adding thick polymer materials to the processed wafer through post-processing methods. As the antenna can be placed on top of other components, this building technique allows extremely compact integration of the radiator with the driving electronics. The main drawback is the limited achievable bandwidth caused by the restrictions in the maximum thickness of the back-end oxides or the spin-on organic dielectric. Typically, it is difficult to obtain a dielectric

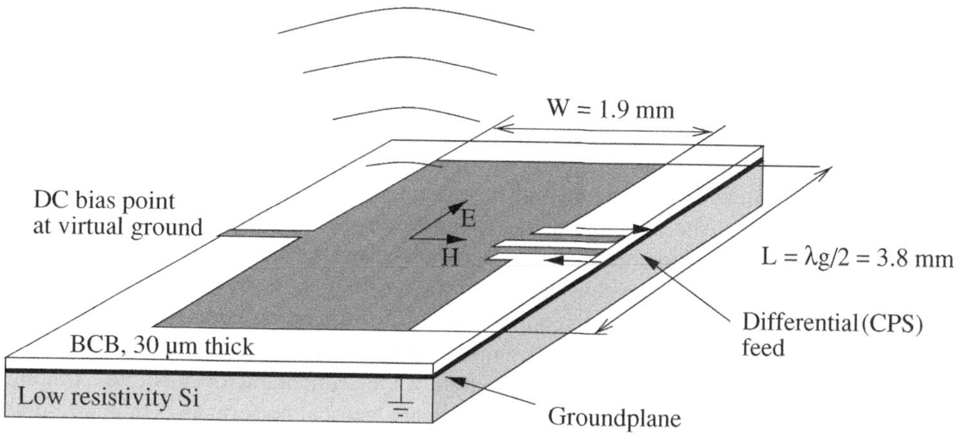

Figure 8.5: Differential 24 GHz patch antenna manufactured using an polymer above-IC process [19].

thickness of more than 30 μm through the use of conventional spin-on polymer materials. As a thick dielectric is required to obtain a wide bandwidth, this integration method is primarily suited for narrow-band applications or frequencies above 100 GHz.

An example of a patch antenna implemented using a thick polymer layer on silicon is shown in Figure 8.5. The patch is isolated from the low resistivity substrate by a ground plane and 30 μm thick BCB layer. Although the reported implementation [19] was placed on a separate silicon carrier, this integration technique can be used as an above IC technology as well. The required patch length can be calculated using the transmission line model of the patch antenna and the microstrip parameters for the polymer substrate. With the used technology a patch length of 3.2 mm was required to obtain resonance at the electrical length $\lambda_g/2$ at 24 GHz. A patch width of 1.9 mm was selected in order to suppress higher order modes on the patch radiator. To avoid the need for baluns and through BCB vias, a differential type of feed along the long, non-radiating, side of the patch was used to connect the radiator directly to a differential circuit. The correct distance between the two feedpoints to obtain a $2 \times 50\ \Omega$ input impedance was determined by electromagnetic (EM) simulation to 270 μm. At the opposite end of the patch with respect to the feedpoint a DC line for biasing of the connected circuit is connected to the virtual ground-point, which appears at the symmetry line of the patch.

The use of a thin dielectric in the implemented radiator limited the measured bandwidth at 24 GHz to 0.5 GHz. The gain of the passive radiator was not reported, but based on measurements of the radiated power with an integrated oscillator, 2 dBi gain can be estimated. However, since the impedance match between the oscillator and the antenna was poor (reported return loss −2.5 dB) this estimate is conservative and the intrinsic antenna gain, compensated for return loss, is likely to be higher.

8.4.1.3 Bulk Micromachined Patch Substrate

Bulk micromachining of silicon wafers equipped with patch radiators, illustrated in Figure 8.4(c), has been used in the K-band [20] as well as at 94 GHz [21] in order to reduce the dielectric and substrate wave propagation losses normally associated with the semiconductor material. A silicon membrane, formed by stopping the backside bulk etching before the entire thickness of the substrate has been removed, was used to support the patch over the groundplane. Alternatively, the patch can be suspended on a thin membrane created by a polymer or silicon oxide/nitride layer deposited before the removal of the bulk silicon.

Since the effective dielectric constant ε_{eff} of the resulting dielectric is reduced, this post-processing technique increases the required size of a resonant radiator. No active circuits can be placed within the aperture as all semiconductor material is removed in the vicinity of the radiator, thus leading to a poor utilization of the semiconductor substrate. Although membrane-suspended patch antennas yield excellent electrical performance, their size often makes them too large to be integrated on the same semiconductor chip as active circuits. However, this type of bulk micromachined patch antennas can find use in a three-dimensional integration wafer level packaging concept where different substrates are used for the antenna and the active electronics.

A modified version [22] of the bulk micromachined patch antenna is illustrated in Figure 8.4(d). In this type of implementation the silicon is only etched away in the high electric field region at the radiating edges of the patch, thus leaving a large part of the silicon die available for co-integration of circuits with the radiator. The reported patch radiator was designed for an operating frequency of 13.9 GHz and had the dimensions $L = 4$ mm, $W = 5.9$ mm. The patch was equipped with 0.7 mm-wide trenches at the radiating edges of the patch which increased the bandwidth to 720 MHz compared to 515 MHz for a similar radiator without a micromachined substrate. Also, the measured gain was improved by 1.2 dBi when trenches were introduced in patch substrate. However, since the substrate is exposed to the fields from the patch in this implementation, high resistivity silicon (10 kΩ cm) still has be used to avoid excessive dielectric losses. Hence, patch antennas of this type cannot be easily integrated in many silicon processes.

8.4.2 Dipole and Slot Antenna

The half-wavelength dipole and slot antennas, depicted in Figure 8.6 belong to a class of fundamental radiator types which have seen several on-chip implementations. The radiator consists of either a metallic strip or a slot in a groundplane of $L_{\text{d}} = L_{\text{s}} = \lambda_{\text{g}}$/two-length. Both the dipole and the slot radiators occupy a relatively small area on-chip, but the electrical aperture of these antennas are larger than the size of the printed pattern. The dipole is sensitive to metallic and dielectric objects in its vicinity and an area of ground clearance where no other interconnect metallization is present is required. Similarly, the slot antenna needs a sufficiently large groundplane around the rectangular opening to ensure an undisturbed current distribution.

In contrast to the patch radiators presented in the previous section, these radiators tend to have a backlobe which, in the presence of the substrate dielectric, can be stronger than the front lobe. However, a directed beam, radiating from the top of the chip, is desired in many applications. Such a beam can be achieved by placing a reflector or cavity behind the on-chip antenna. Ideally, the reflector should be placed at a distance of $\lambda_{\text{g}}/4$ where λ_{g} is

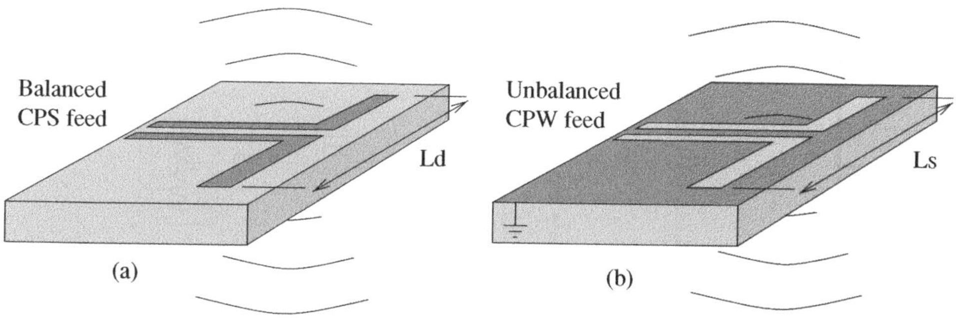

Figure 8.6: On-chip half wavelength: (a) dipole and (b) slot antenna.

the wavelength of the medium. In an on-chip antenna application this condition is difficult to fulfill as the chip is usually too thin to support such a spacing. The problem of substrate waves excitation in the high dielectric constant semiconductor die also restricts the maximum spacing of the reflector to the antenna, with a dielectric thickness of $d < \lambda_g/10$ typically recommended. Hence, the radiator antenna will in many cases sit close to the reflector, which in turn reduces the radiation resistance at resonance. Co-design and simulation of the antenna with its beamforming reflectors and cavities is necessary in this case.

Although the basic dipole or slot radiators are low-directivity antennas, the gain can be improved through the use of parasitic elements, provided that space is available on-chip. Yagi-Uda arrays can be used in order to create a beam with the main lobe along the surface of the chip [23, 24].

8.4.2.1 Compact Dipole Radiators

There is often a need, especially at the lower mmWave frequencies, to reduce the length of the dipole radiator in order to fit the available space on the semiconductor die. Different length reduction techniques based on the creation of a slow wave structure such as the zig-zag and meander dipole have been proposed in the literature [25].

As illustrated by Figure 8.7, the dipole is often placed at the edge of the diced chip, which is also advantageous from an efficiency point of view as it minimizes the coupling to the substrate. The presence of the silicon substrate as well as different loading techniques created by folding the sections of the radiator can be used to fit the radiator to the available length of the silicon die. The radiation resistance of the antenna will, however, drop by the loading techniques, since the radiation from the parallel sections of the dipole will cancel in the far-field. A reduction of radiation resistance yields a lower input impedance at resonance as well as a lower bandwidth. The efficiency of the dipole usually also drops, as the radiation resistance is decreased while the resistance due to losses remain almost constant. To maximize the radiation resistance for a given maximum radiator length, the use of optimum meander shapes such as the meander bow-tie radiators [26] can be considered. The low impedance of the shortened dipole can be matched to circuits and transmission lines using a distributed matching network such as a standard T-match [16].

Compact half-wave dipole radiators have seen applications in wireless interconnects and clock distribution. A 2 mm-long zig-zag dipole manufactured in a 0.18 μm CMOS process

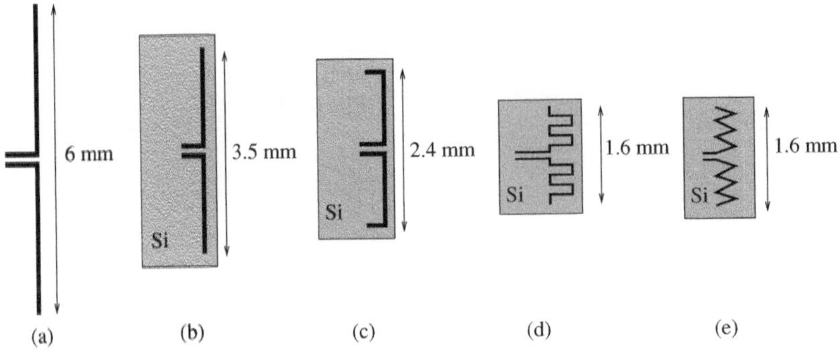

Figure 8.7: Example of the possible length reduction of a 24 GHz half wavelength dipole from the (a) free space length to (b) loaded by the substrate, and optional length reduction techniques: (c) top-loaded; (d) meander dipole; (e) zig-zag dipole.

with a 15–25 Ω cm silicon substrate was used for 15 GHz clock distribution [6]. Up to −53 dB transmission gain was obtained between two on-chip transmit and receive antennas with 6.7 mm spacing. Other monolithic antennas for chip-to-chip interconnection have been manufactured with a proton implantation technique [27] where the resistivity of a standard 10 Ω cm silicon substrate was increased to 0.1 MΩ cm. The measured transmission gain between two 2 mm-long on-chip dipoles with 10 mm spacing was −36.5 dB at 18 GHz, thus indicating that the substrate resistivity is an important factor for the antenna gain.

An example of a reduced-length dipole suitable for on-chip integration [28] is shown in Figure 8.8. Two 24 GHz meander dipoles were suspended on a 3.3×0.8 mm^2 large BCB membrane in order to minimize the coupling to the 20 Ω cm silicon substrate. The radiator length was reduced from approximately 6 mm to 3 mm by the use of a meandered line. As seen in Figure 8.9, the length reduction causes the antenna impedance to drop to less than 20 Ω at resonance, thus yielding a poor return loss in a 50 Ω system. However, the low input impedance might be useful for direct connection to certain radio front-end building blocks such as power amplifiers.

8.4.2.2 Folded Dipole Antenna

Half-wavelength dipole antennas implemented in an IC environment will usually have a significantly lower radiation resistance compared to a dipole in free space. The reduced radiation resistance can be caused by the presence of the substrate as well as metallic objects, such as conductor backing, in the vicinity of the radiator. Currents induced in such nearby bodies will cancel part of the far-field radiation and reduce the input impedance at resonance. Dipole-length reduction techniques such as the use of bending and top loading of the radiator will also reduce the radiation resistance, as illustrated in the previous section.

Although antennas can be matched to the driving electronics using a suitable network, the impedance of a dipole radiator itself can be increased by folding of the radiator. The closed

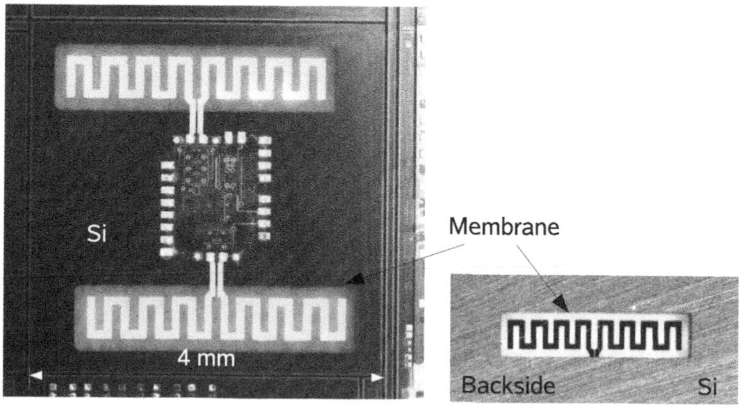

Figure 8.8: Micrograph of membrane-supported 24 GHz meander antennas based on the design in reference [28], here integrated with a SiGe transceiver.

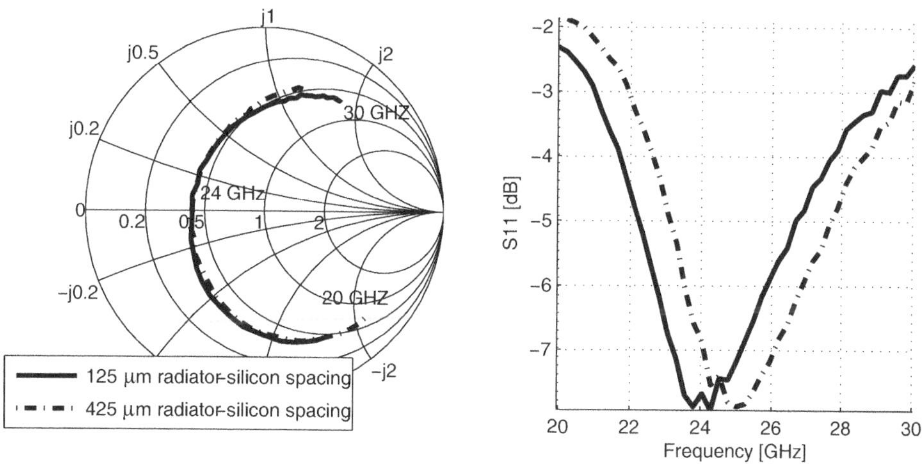

Figure 8.9: Measured impedance and return loss of the antenna in Figure 8.8 with 125 μm (solid curve) and 425 μm (dot-dash curve) spacing between the meander radiator and the edge of the membrane.

form equation [29] for the input impedance of the folded dipole, can be determined as

$$Z_{\text{in}} = \frac{2(1+a)^2 Z_{\text{d}} Z_x}{(1+a)^2 Z_{\text{d}} + 2Z_x} \tag{8.5}$$

where a is the ratio between the currents on the driven and parasitic branch, Z_{d} is the input impedance of a non-folded dipole with the same shape as the folded radiator while Z_x represents the impedance of the quarter wave shorted stub formed by the two CPS transmission lines from the feed-point to each end of the radiator. Since Z_x goes to infinity

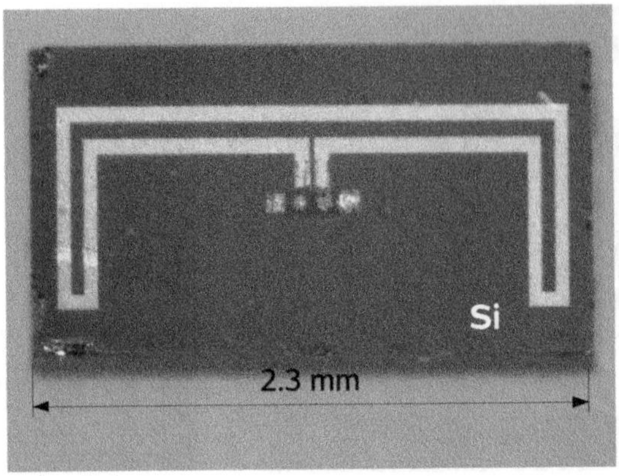

Figure 8.10: Folded top-loaded 24 GHz dipole manufactured in a commercial SiGe process, presented in reference [30].

at resonance, the dipole impedance at resonance is multiplied by the factor $(1+a)^2$. Often, equal width is used for both branches of the folded radiator, thus yielding a current division ratio a of unity and an impedance step-up factor of four. However, by changing the widths of the individual branches of the radiator the current ratio a and thereby also the real part of the impedance Z_{in} at resonance can be tuned to the value required for the active mmWave front-end circuits.

An example of an on-chip folded dipole manufactured in a commercial SiGe process is depicted in Figure 8.10. Silicon wafers with a bulk resistivity of 1500 Ω cm, made available as a process compatible option by the manufacturer, were used to improve the antenna efficiency. In order to reduce the overall length of the radiator to 2.1 mm, the dipole branches are bent back along the edges of the chip, thus effectively creating a top load of the dipole. The disadvantage of this top-loading technique is a reduced radiation resistance as well as an increase in the cross-polarization level of the antenna.

By the use of folding, the impedance of the radiator is increased from a value of less than 20 Ω to close to 50 Ω, which yields a low return loss in a 50 Ω system as shown in Figure 8.11. The high level of return loss at frequencies far from the radiating resonance at 24 GHz indicates that excessive substrate losses are not present.

The measured radiation patterns for the top-loaded antenna (Figure 8.12) show typical dipole characteristics with minimums in the E-plane along the axis of the main radiator section. An increase in the cross-polarization is caused by the top-loading sections. The maximum gain of the folded dipole is -2 dBi. Main factors contributing to the gain reductions are losses in the nearby $p+$ layers of the substrate as well as reduced wafer resistivity during processing.

8.4.2.3 Slot Antennas

Through the field equivalence principle, the slot antenna can be considered as a complementary antenna to the dipole which in a planar implementation is obtained by replacing

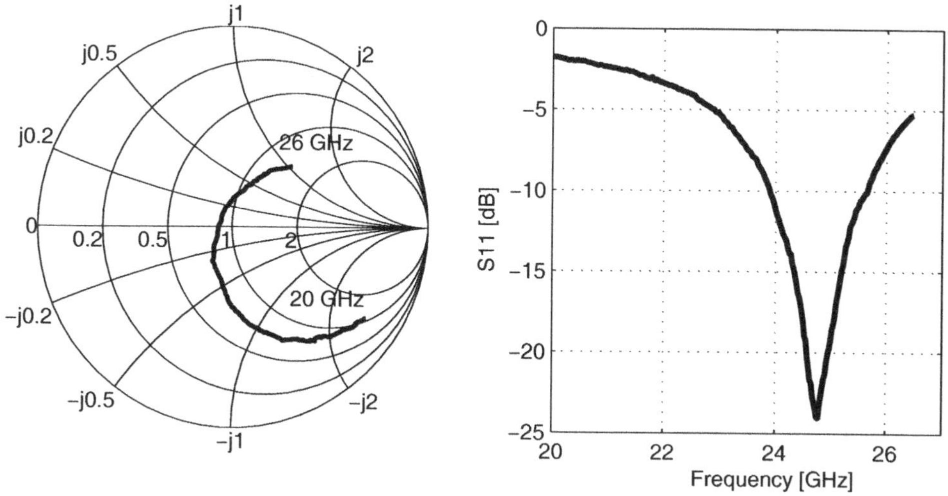

Figure 8.11: Measured return loss in a differential 50 Ω system for the folded dipole antenna presented in Figure 8.10.

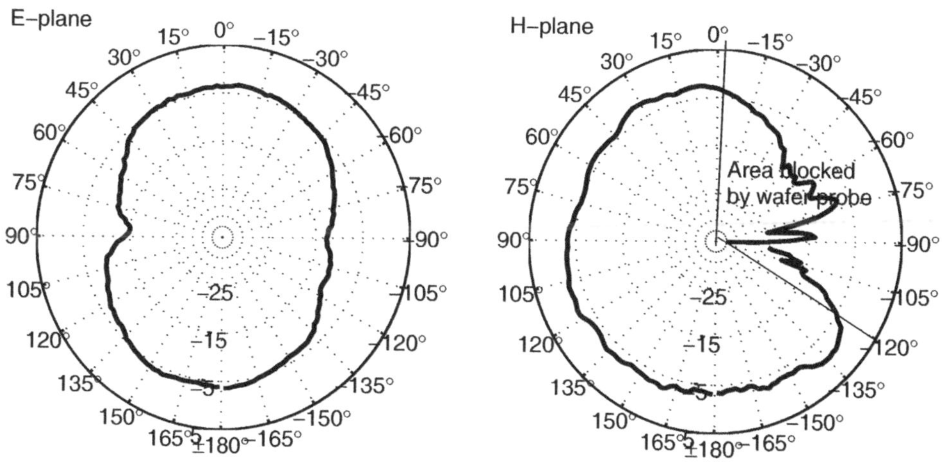

Figure 8.12: Measured E- and H-plane gain of the folded dipole in Figure 8.10.

the metalized surface in the dipole implementation with of similar shape in a groundplane. Many properties of the slot antenna, such as input impedance at resonance, can be derived directly from results for a similar shaped dipole radiator. However, it is important to realize that the field equivalence principle can be directly applied only if the slot antenna is implemented in an infinite ground plane. With on-chip radiators this is rarely the case, since the groundplane is normally severely limited by the die size. As the circuit ground planes also have resonant lengths determined by their size and shape, it is necessary to consider the

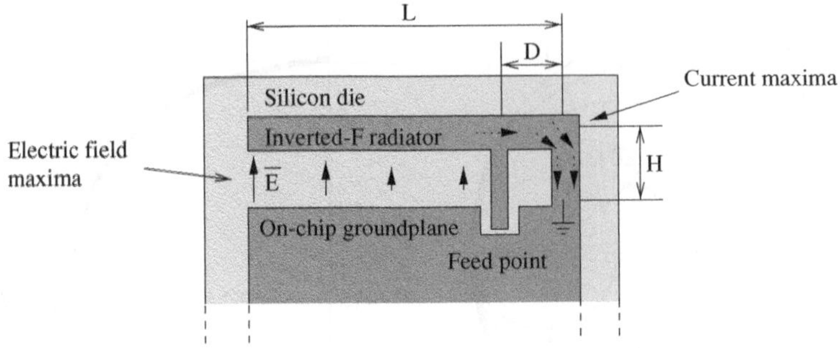

Figure 8.13: Principle of the IFA. Current distribution indicated by dashed line arrows, electric field by solid line arrows.

interaction with the slot radiator which is most efficiently done using a three-dimensional EM simulation package.

Rectangular slot antennas can be folded in the same way as dipole antennas in order to adjust the input impedance and improve the return loss. An example is the folded slot antenna manufactured on a CMOS grade silicon substrate using polyimide as an interface layer [31]. The rectangular slot with length $L_S = 2.2$ mm was manufactured on a 1 Ω cm silicon substrate which had been coated with a 20 μm thick polyimide layer to reduce the losses. By the use of folding the high-input impedance of a half-wavelength slot radiator was reduced to facilitate the match to a 50 Ω system. The measured return loss at the resonance frequency 30.5 GHz was −14.6 dBi. Neither gain nor efficiency figures were reported but distortion of the radiation pattern indicated the presence of a strong substrate mode.

8.4.3 Inverted-F Antenna

In the cases where the allowable area for an on-chip radiator is severely limited, the inverted-F antenna (IFA) may be used. The IFA, which is essentially a modified monopole antenna, was originally invented as a low-profile missile antenna [32] although it has found wide use as a compact radiator for mobile communication [33].

As illustrated in Figure 8.13, the important design parameters of the inverted-F when implemented on a planar substrates are the distance from groundplane H, the length of the top load L and the position D of the feed point relative to the ground connection of the radiator. As in the case of the inverted-L antenna [34], a large distance H from the ground is desirable is order to maximize the radiation resistance and bandwidth of the radiator, while the additional length L is selected to obtain resonance at the operating frequency. In an on-chip antenna application, where the circuit metallization serves as antenna ground, H is restricted by the area the antenna is allowed to occupy. An advantage of the inverted F is the ease with which the antenna can be matched to a wide range of impedances by selection of the feed-point distance D from the short-circuit.

Due to the fact that an L-shaped monopole is used as the radiating element, the polarization purity of the IFA is relatively low. In Figure 8.13 the H long short-circuit sections of the radiator will contribute to a vertical polarization while the L long top-loading section will give rise to a horizontal component. As the quarter-wave IFA is a monopole type of

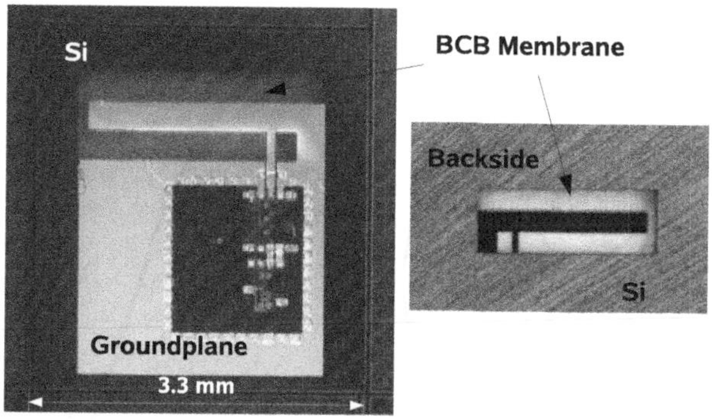

Figure 8.14: The 24 GHz on-chip IFA with membrane-suspended radiator described in reference [35], here with an integrated oscillator in SiGe technology.

antenna, the size of its groundplane is also critical. When implemented on-chip together with active devices, in-circuit ground-planes typically have to serve as the antenna ground. Although the groundplanes of at least $\lambda_g/4$ size should be possible to implement on-chip at millimeter wavelength, the currents on these relatively small groundplanes will contribute to the radiation of the antenna. Hence, in most cases the entire chip has to be considered as part of the radiator, and electromagnetical simulations, including the diced chip size and as well as the outline of the metallizations present on the die, are necessary to model the antenna.

Monolithic mmWave IFAs have been demonstrated using silicon integrated circuit compatible post-processing techniques. A 60 GHz on-chip IFA with dimension $H = D =$ 100 μm, and a conductor width of 10 μm was implemented in a CMOS compatible process [24] using a substrate with 10 Ω cm resistivity. A range of different lengths L from 0.45 mm for the fundamental quarter wave mode up to 2 mm was evaluated in order to excite higher order modes on the radiator. To minimize substrate losses in the presence of the CMOS grade substrate, a 20 μm-thick silicon oxide back-end was deposited on top of the 6 μm oxide already present on the substrate. Despite these loss-minimization techniques, reported gain of the antenna was, however, only −19 dBi and the simulated efficiency was 3.5%. Less than −10 dB return loss was obtained over a 55–67.5 GHz frequency range.

The gain and efficiency of IFAs implemented on low-resistivity silicon can, as in the case of the other radiator types, be improved by selective micromachining of the bulk substrate. A 24 GHz micromachined IFA [35] using the silicon IC-compatible BCB membrane post-processing method described in this chapter is shown in Figure 8.14. The antenna has been manufactured on a 3.8×3.8 mm^2 large, 11–15 Ω cm silicon substrate where the chip size was selected to simulate the size of a complex integrated circuit. The inverted-F radiator is suspended on a 2.4 mm^2 large BCB membrane to reduce the losses. As shown in Figure 8.15 a return loss of −16 dB is obtained at 24 GHz and the −10 dB bandwidth is 2 GHz. An inductive impedance behavior at frequencies away from resonance is observed, since the radiator arm creates a shunt inductance to ground close to the feed point. The size of the impedance locus can be tuned by the selection of a suitable feed-point D to improve the impedance match to a 50 Ω system.

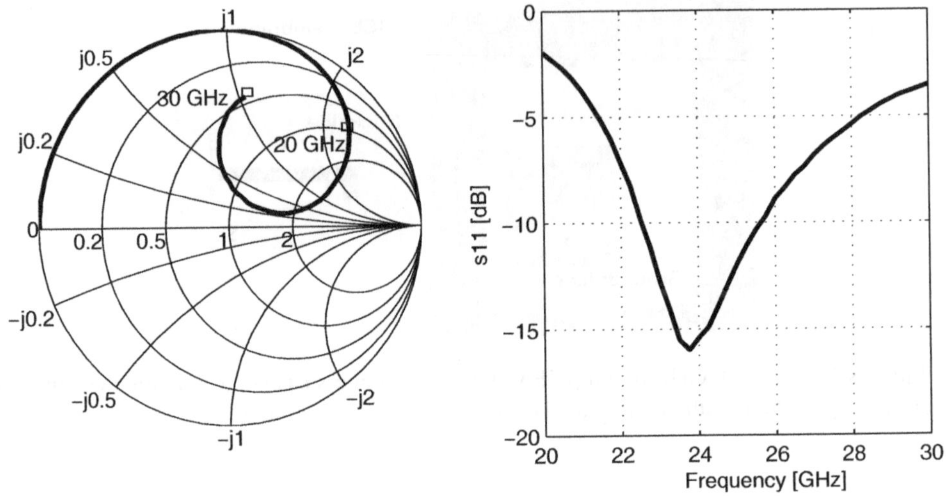

Figure 8.15: Measured impedance and return loss of the IFA presented in Figure 8.14.

The radiation pattern, shown in Figure 8.16 with the principal polarization axis indicated in Figure 8.13 resembles a conventional monopole antenna, with zeros in the E-plane. The measured gain is −0.7 dBi, which, compared with a simulated directivity of 2.1 dBi, indicates better than 50% efficiency. The relatively low gain of this inverted-F antenna can be partly explained by the low directivity. However, another source of gain degradation is substrate losses in the vicinity of the ground-plane. As a small ground-plane representative of typical circuit sizes is used as a counter-poise for the radiator, significant currents will flow outside the membrane area where they are subject to the losses of the low resistivity substrate.

8.4.4 Loop Antennas

The rectangular, often close to square, shape of semiconductor chips make the loop radiator a suitable choice for an on-chip antenna when the free space wavelength at the operating frequency is large compared with the die size. In such cases the loop can be implemented at the perimeter of the chip while leaving the silicon area in the center of the chip free for implementation of the active electronics. The feasibility of such compact integration of loop antennas with circuits has been demonstrated in conventional PCB technology [36].

It is important to distinguish between the electrically small loop antenna primarily used in low-frequency applications such as 13.56 MHz radiofrequency identification (RFID) applications and the full wavelength loop. In the former, the current can be assumed to be constant and the radiation resistance is very small, which yields extremely small bandwidth and poor efficiency. However, for millimeter wavelength applications, there is usually no need for such extreme miniaturization. Hence, only the full wave loop will be treated here. An advantage of the full wave rectangular loop geometry is that circular polarization, polarization diversity as well as dual band [37] operation can be obtained by suitable loading of the loop or by introduction of short circuits and multiple feed points.

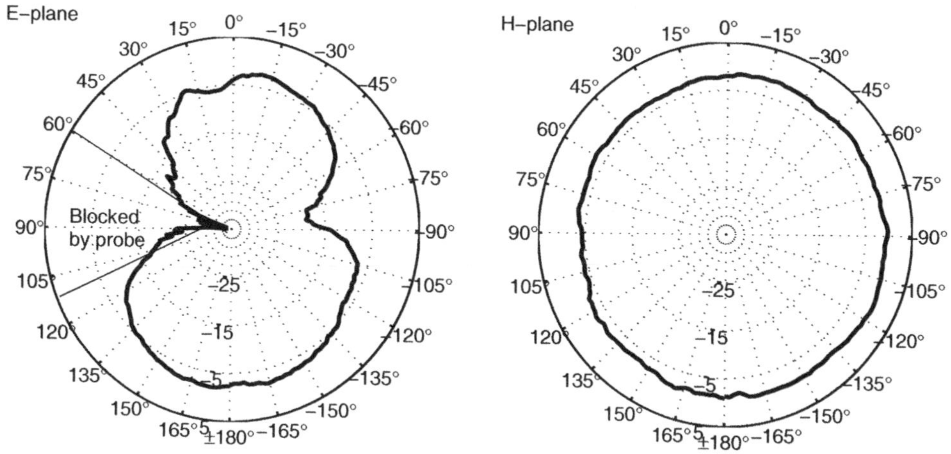

Figure 8.16: Measured E- and H-plane gain of the IFA in Figure 8.14.

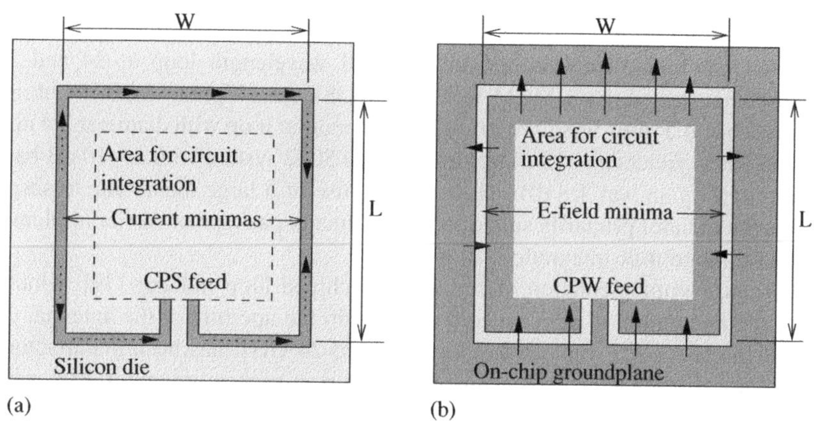

Figure 8.17: Compact integration of full wavelength (a) wire loop and (b) slot loop radiators with circuits on a silicon die. The electrical current in the wire radiator is indicated by dashed line arrows while the electric field in the vicinity of the slot radiator is shown by solid line arrows.

The full wavelength loop, which can be implemented either as a wire or slot loop as shown in Figure 8.17, has a side length of $\lambda_g/4$ for a square geometry where λ_g is dependent on the substrate type. While λ_g can be as small as $\lambda_0/2$ for a radiator implemented directly on a silicon, the effective wavelength is close to its free space value in the case of a micromachined substrate. The directivity of the square full wavelength loop is 3.3 dBi at resonance with no reflectors present and without consideration of substrate effects.

Figure 8.18: Micromachined 24 GHz full-wavelength loop antenna similar to the passive radiator in reference [39]; here with an integrated SiGe receiver circuit. A guard ring metallization connected to the silicon using substrate contacts as well as to the zero potential symmetry point of the loop radiator is present in the design.

8.4.4.1 Wire Loop Antenna

A suspended membrane implementation of a full wavelength loop at 24 and 36 GHz have been demonstrated in an MCM technology [38]. An elliptical BCB membrane with dimensions $5.6 \times 4.5\ \text{mm}^2$ was used to suspend an annular loop with diameter 2.4 mm in air. The loop antennas yielded a -15 dB return loss in a 50 Ω system with a -10 dB bandwidth and the estimated gain was 1.8 dBi. However, the use of a large membrane to suspend the loop makes the radiator primarily suited for an antenna in package or carrier implementation rather than for monolithic integration.

An alternative implementation of the micromachined loop radiator [39], which leaves spaces for the integration of active electronics within the aperture of the antenna, is shown in Figure 8.18. The $3 \times 3\ \text{mm}^2$ large loop, tuned to 24 GHz, has been manufactured on a 20 μm-thick BCB layer which has been added as a post-processing option to a wafer from the semiconductor foundry. To minimize the substrate losses, the silicon is etched away in 360 μm-wide trenches centered around the metallic conductor in the loop, thus leaving the radiator suspended on a BCB membrane. Silicon bridges for mechanical support are left between the center piece and the surrounding silicon at the voltage minimums at the feed point as well as the opposite end of the loop.

The impedance and return loss is plotted in Figure 8.19 for loops with as well as without circuit metallization present in the loop. The useful series resonance is obtained at 24 GHz, which is also the frequency where the electrical length of the loop corresponds to a full wavelength, since the effective dielectric constant of the suspended membrane ε is close to one. The loop where circuit metallization, including the guard ring connected to substrate contacts, is present has a lower measured impedance at resonance as currents induced in this guard-ring cause a reduction in the loop radiation resistance.

The measured radiation pattern of the wire loop (Figure 8.20) displays a compressed H-plane compared to a normal half wavelength dipole radiator. The reason is the array action

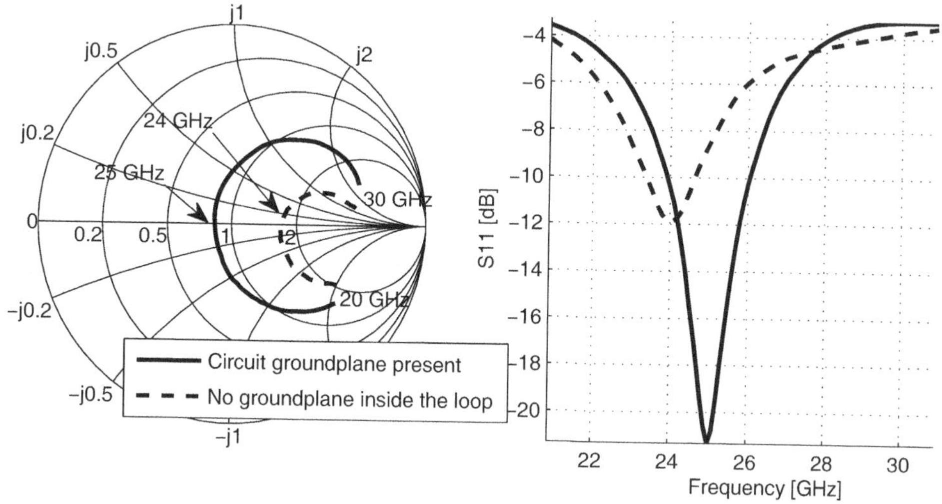

Figure 8.19: Measured impedance and return loss of the loop antenna in Figure 8.18 measured with (solid line) and without (dashed line) a circuit metallization and guard ring in the center of the loop.

created by the current elements at the opposite sides of the loop. With no circuit metallization in the loop, 1 dBi gain is measured, compared to a calculated directivity of 3.3 dBi. Substrate losses in the silicon remaining in the vicinity of the loop is the main loss mechanism and a higher efficiency can be obtained by increasing the width of the micromachined trenches. Larger membranes do, however, reduce the available space for circuits. For a similar radiator with the circuit metallization shown in Figure 8.18 added, the measured gain dropped to 0.5 dBi.

8.4.4.2 Slot Loop Antenna

As any metallization present inside the parameter of the loop tends to reduce the radiation resistance, and hence decrease the antenna bandwidth and efficiency, the complementary slot version of the loop can be considered. Since the slot loop requires that the majority of the surface of the chip is covered by a ground plane, it is well suited for integration of CPW circuits within the aperture of the radiator.

Early examples of slot loop antennas in silicon technology include a microshield implementation [40] where the entire slot loop is suspended on a membrane. A more practical approach to on-chip integration of a full wavelength loop is the square slot loop antenna manufactured using a selectively bulk micromachined substrate [41]. In this case trenches were etched in the high electric field regions in the vicinity of the slot-loop. However, high resistivity silicon was used as the substrate and the main purpose of the micromachining step was to increase the size of the manufactured loop rather than to minimize conductivity losses in the substrate.

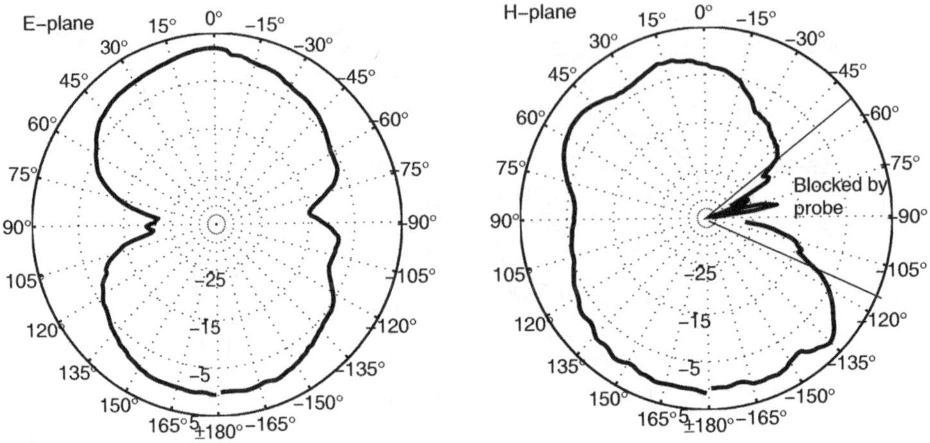

Figure 8.20: Wireloop (Figure 8.18) E- and H-plane gain (in decibels), 0° corresponds to top of the chip. Part of the H-plane is shadowed in the measurement setup.

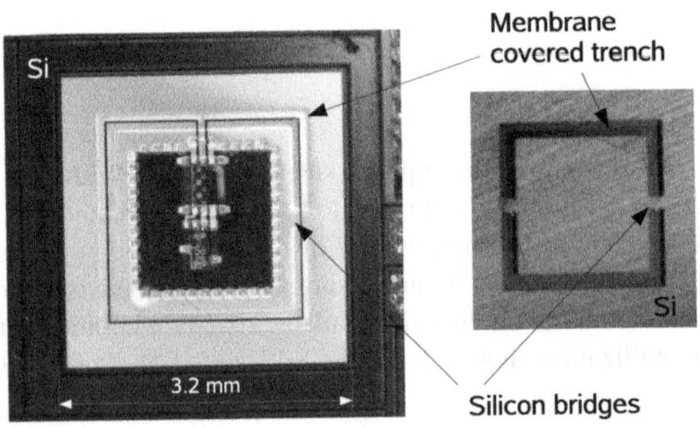

Figure 8.21: Micromachined K-band slotloop antenna with integrated oscillator based on passive radiator described in reference [39].

A K-band slot-loop manufactured using the same process as for the wire loop antenna is depicted in Figure 8.21. The size of the slot loop is 2.2×2.2 mm^2 and the size of the groundplane, as well as the diced chip used in the measurements, is 3.2×3.2 mm^2. Below the slot of the antenna, 200 µm-wide trenches are etched through the use of DRIE, leaving the slot suspended on the BCB membrane. As in the case of the wire loop antenna, silicon bridges are left across the membrane to support the center section of the silicon. These bridges are placed in the voltage minimums in the loop found along the two non-fed edges.

In the impedance plot (Figure 8.22) it is clearly seen that the second radiating resonance of the loop is a parallel resonance. Due to the use of a relatively small ground plane, the groundplane mode resonance, coincides with the slot resonance, thus increasing the bandwidth of the antenna. The ground plane resonance together with the substrate loading

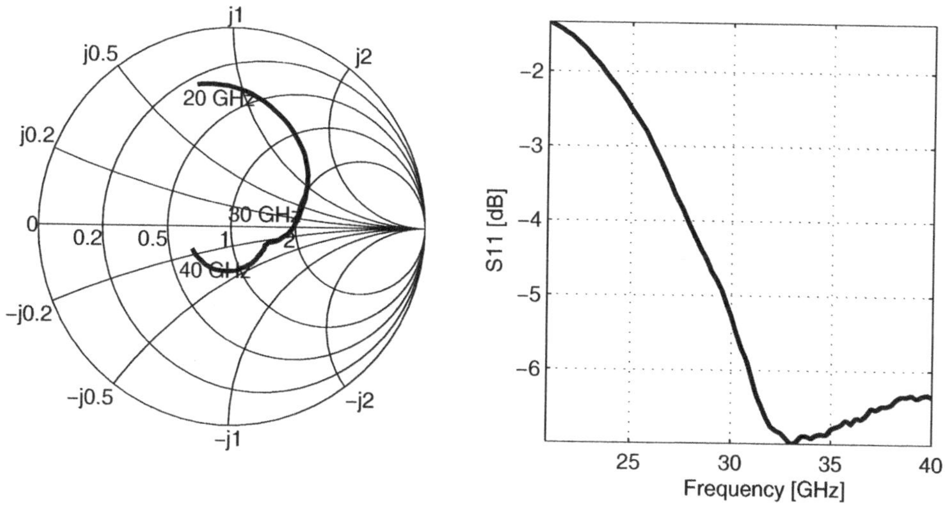

Figure 8.22: Slot loop antenna (Figure 8.21) measured impedance and return loss.

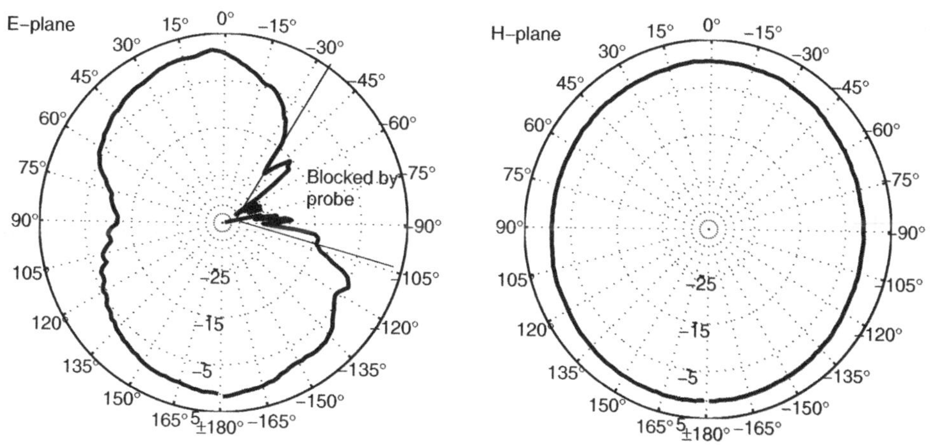

Figure 8.23: Slot-loop (Figure 8.21) measured E- and H- plane gain (in dB), 0° corresponds to top of the chip.

also decreases the input impedance of the loop from the free space value. Hence, the loop can relatively easily be matched to a standard 50 Ω system.

The radiation pattern of the slot-loop (Figure 8.23) has the same characteristics as the wireloop, where the main difference is that the E-plane instead of the H-plane is compressed. The peak gain of the loop is 1.5 dBi. Compared with the theoretical directivity of 3.3 dBi for a square loop, it is clear that relatively good antenna efficiency can be obtained with the use of micromachined trenches in selected areas of the loop.

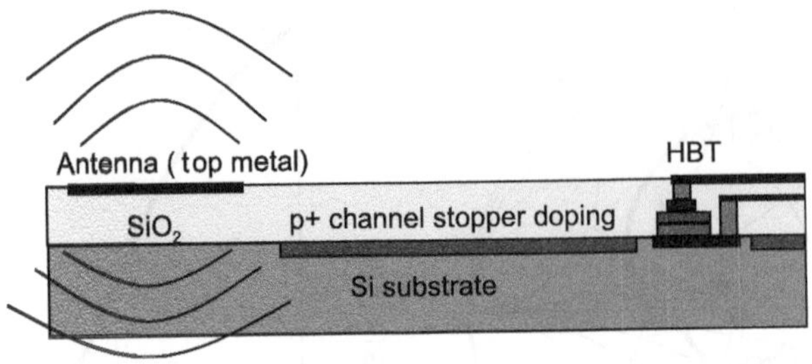

Figure 8.24: Antenna integration in a typical Si/SiGe process substrate stack. Channel stopper dopings are removed underneath the antenna.

8.5 Circuit Integration

A typical antenna integration scenario for an on-chip radiator in a silicon technology is illustrated in Figure 8.24. To minimize the coupling to the substrate, the radiator is usually implemented in the top metal layer offered by the semiconductor process. Another reason to implement the antenna in the top metal layer is that this interconnect layer is often dedicated to low-loss passives and thus implemented using a thicker metallization, which leads to lower ohmic losses in the radiator. Depending on the back-end options offered by the manufacturer, the thickness of the oxide/nitride stack can range from a few micrometers up to more than 10 µm. Since the thickness of the oxide is not large enough to implement a reflector-backed dipole or the ground plane for a patch at mmWave frequencies, all metallization layers below the radiator must be removed to prevent the radiator from being electromagnetically short-circuited.

In a silicon bipolar or BiCMOS technology the surface of the substrate is usually heavily doped, typically using a $p+$ channel stopper doping for a Si/SiGe NPN device process, to suppress parasitic devices from forming in the bulk substrate. As in the case of spiral inductors, this doping layer must be blocked out in the vicinity of the antenna to prevent efficiency loss due to currents induced in this heavily doped layer.

8.5.1 Cross-talk

A concern with monolithic integrated antennas is the risk of cross-talk between the on-chip radiator and the passive elements in the mmWave active front-end. Zero-IF receiver and transmitter architectures are particularly sensitive to cross-talk between the antenna port and the local oscillator (LO). In the receiver case, LO leakage which couples to the antenna can cause DC-offset problems, an increased noise floor and blocking of the front-end in extreme cases while transmitters can suffer from instability and increased phase noise.

Spiral inductors, commonly used as reactances at lower mmWave frequencies due to space considerations, are particularly susceptible to cross-talk due to their large physical size. Although few studies of antenna-to-inductor cross-talk are available, coupling between adjacent spiral inductors has been studied [42]. An effective way of reducing such cross-talk is to surround each of the inductors with a ground shield, as shown in Figure 8.25. EM simulations of the cross-talk between a spiral inductor and an on-chip dipole, spaced 500 µm

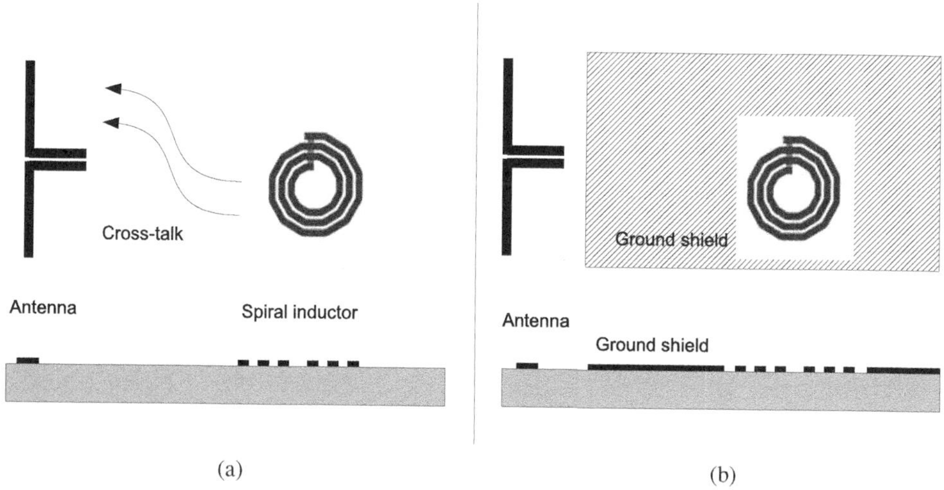

Figure 8.25: Improvement of antenna to spiral inductor isolation through the use of ground shields: (a) without ground shield; (b) with ground shield.

apart [30] indicate an improvement in isolation from 50 dB without the ground shield to 72 dB with shields. To maximize the efficiency of the shielding, substrate contacts should be placed around the perimeter of the shield to tie the substrate to the same potential as the shield. The disadvantage of introducing a ground shield is the reduction of the inductance and quality factor caused by currents in the shield.

The power distribution network can also be a source of cross-talk with the antenna as long lines spanning the circuit can become resonant at the operating frequency of circuit. Efficient decoupling of the powered subcircuits, such as a combination of inductor-capacitor (LC)- and resistor-capacitor (RC)-networks [43], is thus important to prevent signal leakage from the circuit into the antenna.

The problem of cross-talk issues can be minimized by the choice of a suitable receiver or transmitter architecture. A comparison between a conventional zero-IF receiver front-end and a subharmonic downconversion chain developed for on-chip integration with a radiating element [44] is presented in Figure 8.26. Subharmonic downconversion is used together with a fully differential circuit architecture. By employing a differential local oscillator running at half of the received frequency, very little leakage to the on-chip antenna is obtained, since mainly odd LO harmonics are present at the oscillator output in addition to the subharmonic LO signal.

8.5.2 Monolithic Integrated Antenna Examples

An example of a top-loaded half-wavelength dipole antenna with an integrated oscillator was demonstrated at 10 GHz [45] using a 0.25 μm Si/SiGe BiCMOS process. The antenna was integrated together with the oscillator on a 4.48×2.7 mm large silicon chip. From measurements of the radiated power, an antenna gain of -7.5 dBi was estimated.

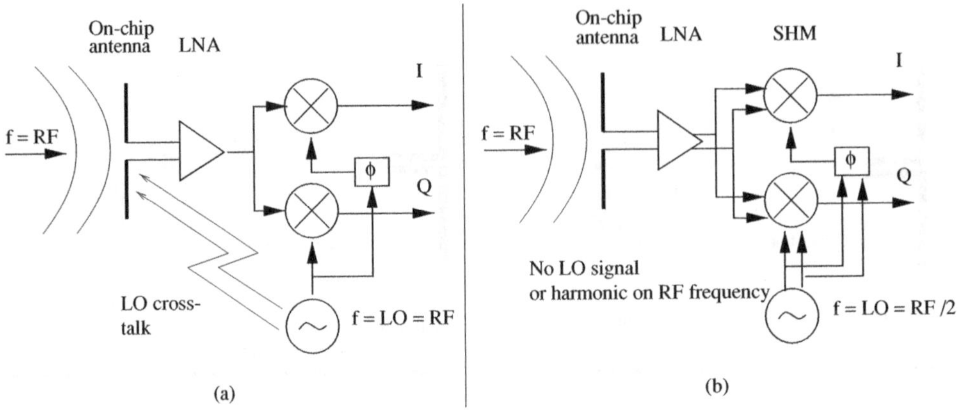

Figure 8.26: Local oscillator to antenna cross-talk reduction through the use of a differential subharmonic mixer (SHM) front-end architecture (b) (differential subharmonic mixer arch) instead of a conventional fundamental downconversion (a) (zero IF receiver architecture).

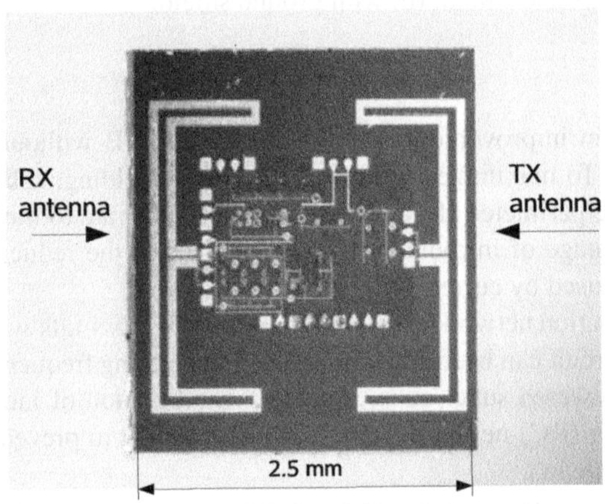

Figure 8.27: Monolithic integrated transceiver for the 24 GHz ISM band in a SiGe technology with on-chip receive (left) and transmit (right) antennas [30].

The folded dipole antenna described in Sec. 8.4.2 has been integrated with the 24 GHz monolithic receiver manufactured in a 80 GHz SiGe technology [30] shown in Figure 8.27. The receive (left) antenna is directly connected to the differential low noise amplifier (LNA) of the receiver. An additional (right) antenna is connected to a switchable local oscillator buffer to allow the chip to be used as a simple transmitter. The received antenna is tuned to provide the impedance of $70 + j40\ \Omega$ which corresponds to Γ_{opt} of the LNA. The measured system noise figure was 8.8 dB, which is in close agreement with the measured −2 dBi gain of the antenna and 6.6 dB noise figure of a separately characterized receiver circuit. Hence, it can be concluded that substrate coupled noise from the on-chip circuits to the antenna does not cause a major noise performance degradation.

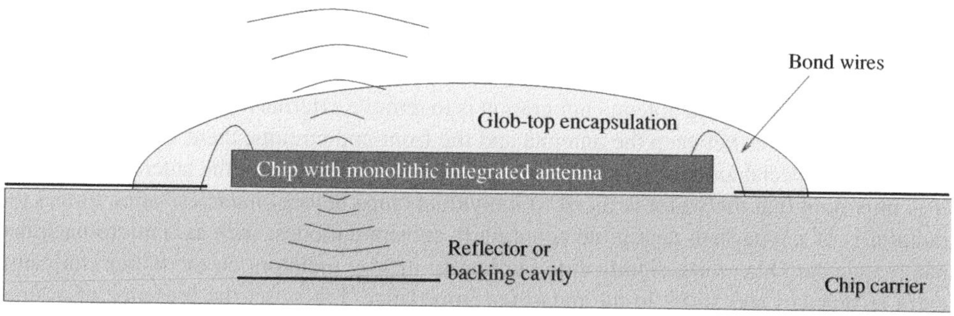

Figure 8.28: Glob-top encapsulation of integrated circuits with on-chip antennas.

One of the most advanced examples of monolithic antenna integration to this date is the 77 GHz phased array [46] implemented in a SiGe process, described in more detail in Chapter 14.

8.6 Packaging of Integrated Circuits with On-chip Antennas

One of the main strengths of monolithic integrated antennas is that no mmWave interconnects are needed in the package if the up- and down-conversion chains as well as the local oscillator circuits are implemented on-chip. However, the presence of an on-chip radiator implies that the package has to be transparent to the radiation. Additionally, the package can be used to enhance the gain of the antenna by shaping the lobe or removal of the back lobe through the use of a reflector. Although few examples of packaged monolithic antennas are available in the literature, some in-package integrated systems such as chip scale packaging methods developed for 60 GHz transceivers [47] share several of the requirements of monolithic antenna packaging and can thus be studied.

For a broadside chip radiating toward the top of the package, a transparent window or encapsulation is needed to cover the chip. A cost-effective solution which also minimizes the problems of package resonances is the use of an organic material to form a glob top over the wirebonded die. Measurements of several commercial glob-top materials show acceptable performance in the millimeter bands [48].

In Figure 8.28 an example of glob-top encapsulation of a circuit with monolithic integrated antenna is shown. The chip with the integrated radiator can be connected to the package either using flip chip or through wirebonding. Particularly in the flip chip case it is important that the carrier substrate is designed to provide suitable electrical boundary conditions for the on-chip radiator. Depending on the ratio between the size of the glob-top packaging and the wavelength, the packaging polymer can also serve as a lens [49] in order to increase the gain of the on-chip antenna.

8.7 Monolithic Antenna Measurement Techniques

As the purpose of on-chip antenna integration is to remove external high-frequency interconnects and interfaces between the antenna and the front-end circuits, there is often a need to manufacture special antenna test structures in order to be able to test the antenna separately. It is important that the wafer is diced to individual chips before characterization unless the radiator is of a type with negligible coupling to substrate modes, such as a micromachined patch antenna. Otherwise, significant coupling to nearby radiators on the wafer can cause extra resonances and shifts in the measured impedance. Likewise, the presence of a large substrate will cause detuning if the radiator has been designed for a specific chip size.

The impedance and return loss of a passive on-chip radiator is commonly measured using a wafer probe station. However, since neither the wafer probe nor a standard probe station has been designed for antenna measurements, precautions have to be taken in order to avoid excessive measurement errors. If the antenna has been designed for free space operation or for the use of a specially designed reflector or cavity, care has to be taken not to place the antenna under test (AUT) directly on the metal chuck of the probe station. A combination of a transparent foam material with low dielectric constant on top of a microwave absorber can be used to prevent reflections [50]. The foam material is needed to bring the absorber out of the near-field region of the probe. Reflections from other nearby metallic objects such as the probe holder should also be minimized by the use of absorbers.

Balanced antennas such as dipole and wire loop antennas should ideally be measured using either a differential wafer probe or through the use of an on-chip balun. Sometimes satisfactory measurements can be made using a conventional ground–signal–ground (GSG) or ground–signal (GS) probe where one of the antenna connections are connected to the ground. However, there is a risk that the current will flow on the shield of the probe, thus effectively making the probe a part of the antenna and altering the measured impedance.

Radiation pattern measurements of monolithic antennas are often difficult to perform since the AUT normally needs to be fed using a wafer probe. Either a mobile wafer probe setup can be mounted in an ordinary antenna measurement setup or a setup where the reference horn antenna sweeps around the AUT and wafer probe can be used [51]. Since the wafer probe and probe holder are normally much larger than the AUT, part of the radiation pattern can usually not be measured, owing to shadowing.

The most relevant test of an integrated circuit with an on-chip antenna is a system test carried out in transmit or receive mode. Such tests are needed to verify the cross-talk levels and stability of the complete system that cannot be tested with the front-end circuit and antenna separately. In the case of a transmitter or a transceiver in transmit mode, the radiated power can easily be measured. This measurement will not only verify the antenna gain and power amplifier output but also the impedance matching between these circuit blocks. Receiver chains with on-chip antennas are most effectively characterized [30] by measuring the ratio G/T of the complete system where G is the conversion gain and T the total system noise temperature. By illuminating the device under test (DUT) with a known field strength, both G and T can be calculated from spectrum analyzer measurements of the received signal. Measurements of the conversion gain G alone are not sufficient, as a monolithic integrated antenna can pick up substrate noise from the active parts of the circuit. Zero-IF receiver implementations should additionally be characterized for DC offset that can be caused by local oscillator leakage to the integrated antenna.

8.8 Summary

Monolithic integration of antennas is still in its infancy and current mmWave technology has so far been using off-chip antenna technology. However, as the ever-increasing amount of integration will make single-chip mmWave circuits common in the future, removing the last off-chip high frequency connection will also remove the last obstacle in achieving a truly self-contained mmWave system-on-chip. The use of higher operating frequencies will be a key driving factor for monolithic antennas as the size of the radiators reduce while off-chip interconnects become harder to realize.

The most commonly anticipated problem of monolithic antenna integration is their perceived need for an excessive area on chip. Through several examples in this chapter it has been illustrated that compact monolithic integration of low directivity radiators with circuits is possible in the low mmWave frequency range and above. Since the performance of all the presented antenna types is quite similar, the suitable antenna type for a specific application will primarily be governed by factors such as the required feed type, presence of on-chip ground-planes as well as the size and geometry of the semiconductor die. When the application calls for a higher gain antenna, the package can in many cases be used to increase the directivity of the radiator.

One of the remaining problems for widespread adoption of monolithic antenna technology is the poor efficiency of the implemented radiators, particularly in commercial silicon device technologies. However, the increased availability of thick spin-coated polymer layers, thick oxide layers in the semiconductor process backend as well as selective bulk micromachining of the semiconductor substrate can help to reduce the substrate losses to acceptable levels.

Acknowledgments

The authors would like to acknowledge Dr Katia Grenier and Professor Robert Plana at CNRS-LAAS in Toulouse for their help with the manufacturing of several of the micromachined antennas presented as integration examples in this chapter. Many of the described Si/SiGe circuit designs have been developed by the University of Ulm and the authors would like to thank Professor Hermann Schumacher, Dr Ertugrul Sönmez and Dr Peter Abele for their support.

References

[1] L. J. Chu, 'Physical limitations on omni-directional antennas', *J. Appl. Phys.* **19** (1948), pp. 1163–1175.

[2] H. A. Wheeler, 'Small antennas', *IEEE Trans. Antennas Propag.* **AP-23** (1975), pp. 462–469.

[3] J. S. McLean, 'A re-examination of the fundamental limits on the radiation of electrically small antennas', *IEEE Trans. Antennas Propag.* **44** (1996), pp. 672–676.

[4] D. B. Rutledge and M. S. Muha, 'Imaging antenna arrays', *IEEE Trans. Antennas Propag.* **30** (1982), pp. 535–540.

[5] D. M. Pozar, 'Considerations for millimeter wave printed antennas', *IEEE Trans. Antennas Propag.* **31** (1983), pp. 740–747.

[6] B. Floyd, C. M. Hung and K. K. O, 'Intra-chip wireless interconnect for clock distribution implemented with integrated antennas, receivers, and transmitters', *IEEE J. Solid-State Circuits* **37** (2002), pp. 543–552.

[7] A. Schuppen, J. Berntgen, P. Maier, M. Tortschanoff, W. Kraus and M. Averweg, 'An 80 GHz SiGe production technology', *III-Vs Review* **14**(6) (2001), pp. 42–46.

[8] K. T. Chan, A. Chin, Y. B. Chen, Y.-D. Lin, T. S. Duh and W. J. Lin, 'Integrated antennas on Si, proton-implanted Si, and silicon-on-quartz', in *IEEE IEDM Tech. Dig.* pp. 40.6.1–40.6.4, Washington, DC, 2001.

[9] X. Huo, K. J. Chen and P. C. Chan, 'Silicon-based high-Q inductors incorporating electroplated copper and low-K BCB dielectric', *IEEE Electron Device Lett.* **23**(9) (2002), pp. 520–522.

[10] K. Grenier *et al.*, 'MEMS above IC technology applied to a compact RF module', in *Proc. SPIE – Int. Soc. Opt. Enging*, vol. 5836, (Seville, Spain), pp. 504–515, 2005.

[11] K. E. Bean, 'Anisotropic etching of silicon', *IEEE Trans. Electron Devices* **25** (1978), pp. 1185–1193.

[12] G. P. Gauthier, A. Courtay and G. M. Rebeiz, 'Microstrip antennas on synthesized low dielectric-constant substrates', *IEEE Trans. Antennas Propag.* **45**(8) (1997), pp. 1310–1314.

[13] D. Neculoiu *et al.*, 'The design of membrane-supported millimeter-wave antennas', in *Intl. Semiconductor Conf.* vol. 1, pp. 65–68, Sinaia, Romania, 2003.

[14] E. H. Klaassen, K. Petersen, J. M. Noworolski, J. Logan, N. I. Maluf, J. Brown, C. Storment, W. McCulley and G. T. A. Kovacs, 'Silicon fusion bonding and deep reactive ion etching; a new technology for microstructures', in *Dig. Tech. Papers Transducers'95/Eurosensors IX* (Stockholm, Sweden), pp. 556–559, 1995.

[15] F. Larmer and P. Schilp, 'Method of anisotropically etching silicon', German Patent No. DE 4,241,045, 1994.

[16] C. A. Balanis, *Antenna Theory, Analysis and Design*, 2nd edn (John Wiley & Sons Ltd/Inc., 1997).

[17] W. Choi, C. Cheon and Y. Kwon, 'A V-band self oscillating mixer active integrated antenna using a push-pull patch antenna', in *IEEE Intl. Microwave Symp. Dig.*, pp. 630–633, San Francisco, CA, June 2006.

[18] M. M. Kaleja, A. Grbl, F. X. Sinnesbichler, G. Olbrich, K. Strohm, J.-F. Luy and E. Biebl, 'Si/SiGe HBT active integrated antenna on high resistivity silicon substrate', in *IEEE Intl. Microwave Symp. Dig.* (Boston, MA), pp. 1899–1902, June 2000.

[19] P. Abele, E. Öjefors, K.-B. K.-B. Schad, E. Sonmez, A. Trasser, J. Konle and H. Schumacher, 'Wafer level integration of a 24 GHz differential SiGe-MMIC oscillator with a patch antenna using BCB as a dielectric layer', in *33rd European Microwave Conf.* (Munich, Germany), pp. 293–296, October 2003.

[20] I. Papapolymerou, R. F. Drayton and L. P. Katehi, 'Micromachined patch antennas', *IEEE Trans. Microwave Theory Technol.* **46** (1998), pp. 295–306.

[21] G. Gauthier, J. Raskin, L. Katehi and G. Rebeiz, 'A 94-GHz aperture-coupled micromachined microstrip antenna', *IEEE Trans. Antennas Propag.* **47** (1999), pp. 1761–1766.

[22] Q. Chen, V. Fusco, M. Zheng and P. Hall, 'Micromachined silicon antennas', in *Proc. Int. Conf. on Microwave and Millimeter Wave Technol.* (Beijing), pp. 289–292, 1998.

[23] D. Neculoiu, P. Pons, M. Saadaoui, L. Bary, D. Vasilache, K. Grenier, D. Dubuc, A. Muller and R. Plana, 'Membrane supported yagi-uda antennae for millimetre-wave applications', *IEE Proc. Microwaves, Antennas and Propag.* **151** (2004), pp. 311–314.

[24] Y. Zhang, M. Sun and L. Guo, 'On-chip antenna for 60 GHz radios in silicon technology', *IEEE Trans. Antennas Propag.* **52** (2005), pp. 1664–1668.

[25] H. Nakano, H. Tagami, A. Yoshizawa and J. Yamauchi, 'Shortening ratios of modified dipole antennas', *IEEE Trans. Antennas Propag.* **32** (1984), pp. 385–386.

[26] M. Ali and S. S. Stuchly, 'A meander-line bow-tie antenna', in *IEEE AP-S Int. Symp. Dig.*, vol. 3, pp. 1566–1569, Baltimore, MD, 1996.

[27] A. B. M. H. Rashid, S. Watanabe and T. Kikkawa, 'High transmission gain integrated antenna on extremely high resistivity Si for ULSI wireless interconnect', *IEEE Electron Device Lett.* **23** (2002), pp. 731–733.

[28] E. Öjefors, F. Bouchriha, K. Grenier, A. Rydberg, and R. Plana, 'Compact micromachined dipole antenna for 24 GHz differential SiGe integrated circuits', in *European Microwave Conf.*, vol. 2, pp. 1080–1084, Amsterdam, The Netherlands, 2004.

[29] R. Lampe, 'Design formulas for an asymmetric coplanar strip folded dipole', *IEEE Trans. Antennas Propag.* **33**(9) (1985), pp. 1028–1031.

[30] E. Öjefors, E. Sönmez, S. Chartier, C. Schick, P. Lindberg, A. Rydberg and H. Schumacher, 'Monolithic integration of a folded dipole antenna with a 24-GHz receiver. in SiGe HBT technology', *IEEE Trans. Microwave. and Technol.* **55** (2007), pp. 1467–1475.

[31] J. Papapolymerou, C. Schwartzlow and G. Ponchak, 'A folded-slot antenna on low resistivity Si substrate with a polyimide interface layer for wireless circuits', in *Topical Meeting on Silicon Monolithic Integrated Circuits in RF Systems.* Ann Arbor, MI, pp. 215–218, 2001.

[32] R. King, C. Harrison and D. Denton, 'Transmission-line missile antennas', *IRE Trans. on Antennas and Propag.* **8** (1960), pp. 88–90.

[33] K. Fujimoto and J. R. James, *Mobile Antenna Systems Handbook*, 2nd edn (Artech House, 2001).

[34] Z. Chen, 'Note on impedance characteristics of L-shaped wire monopole antenna', *Microwave and Opt. Technol. Lett.* **26** (2000), pp. 22–23.

[35] E. Öjefors, K. Grenier, L. Mazenq, F. Bouchriha, A. Rydberg and R. Plana, 'Micromachined inverted-F antenna for integration on low resistivity silicon substrates', *IEEE Microwave Wireless Components Lett.* **15** (2005), pp. 627–629.

[36] G. Forma and J. Laheurte, 'Compact oscillating slot loop antenna with conductor backing', *Electron. Lett.* **32** (1997), pp. 1633–1635.

[37] D. Llorens, P. Otero and C. Camancho-Pealosa, 'Dual-band, single CPW port, planar slot antenna', *IEEE Trans. Antennas Propag.* **51** (2003), pp. 137–139.

[38] P. Abele, J. Konle, D. Behammer, E. Sonmez, K.-B. Schad, A. Trasser and H. Schumacher, 'Wafer level integration of a 24 GHz and 34 GHz differential SiGe-MMIC oscillator with a loop antenna on a BCB membrane', in *IEEE Intl. Microwave Symp. Dig.*, vol. 2, pp. 1033–1036, Philadelphia, PA, 2003.

[39] E. Öjefors, H. Kratz, K. Grenier, A. Rydberg and R. Plana, 'Micromachined loop antennas on low resistivity silicon substrates', *IEEE Trans. Antennas Propag.* **54** (2006), pp. 3593–3601.

[40] L. Rexberg, N. Dib and L. Katehi, 'A microshield line loop antenna for sub-mm wavelength applications', in *IEEE AP-S Intl. Symp. Dig.*, vol. 4, pp. 1890–1893, Chicago, IL, 1992.

[41] Q. Chen, V. Fusco, M. Zheng and P. Hall, 'Silicon active slot loop antenna with micromachined trenches', in *IEE Nat. Conf. Antennas Propag., Conf. Publ.* no. 461, pp. 253–255, York, UK, 1999.

[42] J. H. Mikkelsen, O. K. Jensen and T. Larsen, 'Measurement and modeling of coupling effects of CMOS on-chip co-planar inductors', in *Topical Meeting on Silicon Monolithic Integrated Circuits in RF Systems Proc.*, pp. 37–40, Atlanta, GA, September 2004.

[43] E. Sönmez, S. Chartier, A. Trasser and H. Schumacher, 'Isolation issues in multifunctional Si/SiGe ICs at 24 GHz', in *Proc. of Int. Microwave Conf. (IMS)*, (Long Beach, CA), June 2005.

[44] P. Lindberg, E. Öjefors, E. Sönmez and A. Rydberg, 'A SiGe HBT 24 GHz sub-harmonic direct-conversion IQ-demodulator', in *Silicon Monolithic Integrated Circuits in RF Systems*, Atlanta, GA, September 2004.

[45] F. Touati and M. Pons, 'On-chip integration of dipole antenna and VCO using standard BiCMOS technology for 10 GHz applications', in *Conf. on Eur. Solid-State Circuits. ESSCIRC'03.*, pp. 493–496, Estoril, Portugal, 2003.

[46] A. Babakhani, X. Guan, A. Komijani, A. Natarajan and A. Hajimiri, 'A 77 GHz phased array transceiver with on-chip dipole antennas in silicon: Receiver and antennas', *IEEE J. Solid-State Circuits* **41** (2006), pp. 2795–2806.

[47] U. R. Pfeiffer, J. Grzyb, D. Liu, B. Gaucher, T. Beukema, B. A. Floyd and S. K. Reynolds, 'A chip-scale packaging technology for 60 GHz wireless chipsets', *IEEE Trans. Microwave Theory Technol* **54** (2006), pp. 3387–3397.

[48] L. Li, B. Cook and M. Veatch, 'Measurement of RF properties of glob top and undercapsulant materials', in *Proc. Elect. Perf. Electron. Packag.* **1** (2001), pp. 121–124.

[49] L. Bergstedt and H. Zirath, 'US patent 6424318: Method and arrangement pertaining to microwave lenses', July 2002.

[50] L. Roy, M. Li, S. Labonte and N. Simons, 'Measurement techniques for integrated-circuit slot antennas', *IEEE Trans. Instrumn and Measmt* **46** (1997), pp. 1000–1004.

[51] T. Zwick, C. Baks, U. Pfeiffer, D. Liu and B. P. Gaucher, 'Probe based MMW antenna measurement setup', in *IEEE AP-S Int. Symp. Dig*, vol. 1, pp. 747–750, Monterey, CA, 2004.

9

Metamaterials for Antenna Applications

Anthony Lai, Cheng Jung Lee and Tatsuo Itoh

9.1 Introduction

Metamaterials are artificial materials with engineered properties not readily available in nature. There has been a tremendous amount of research into metamaterials over the last several years to expand the range of electromagnetic applications and properties [1–3]. In particular, metamaterials with simultaneous negative permittivity and negative permeability, known as left-handed (LH) metamaterials [4], have been applied to antenna applications. The unique characteristics of LH metamaterials such as supporting a negative refractive index, fundamental backward wave, and infinite wave with non-zero group velocity have made them attractive for antenna applications. Small resonant backward-wave antennas and fundamental-mode leaky wave antennas capable of continuous scanning from backfire to endfire are just some of the novel antennas realized by using LH metamaterials not possible with conventional materials alone. Needless to say, LH metamaterials are an exciting research area with potential impact across multiple engineering problems.

Currently, there are at least three methodsto realize LH metamaterials. The first method, known as the resonant approach, is based on pairing split ring resonators (SRRs) with metal wires to provide an effective negative permeability and negative permittivity, respectively [5, 6]. The second method, known as the transmission line (TL) approach, is based on realizing a LH TL unit cell, which consists of a series capacitance and shunt inductance. Since parasitic right-hand (RH) effects are unavoidable, a composite right/left-handed (CRLH) TL is a practical realization of a LH TL [7–9]. The third method, known as the evanescent-mode approach, achieves negative permeability and negative permittivity with dielectric resonators placed inside a cutoff background [10].

Advanced Millimeter-wave Technologies: Antennas, Packaging and Circuits
Duixian Liu, Ulrich Pfeiffer, Janusz Grzyb and Brian Gaucher © 2009 John Wiley & Sons, Ltd

In this chapter, CRLH TLs and evanescent-mode LH metamaterials are discussed with regard to antenna applications. The broadband, low-loss, and planar nature of CRLH TLs are used to realize novel resonant antennas and antenna feeding networks. Evanescent-mode LH metamaterials are used to realize dominant-mode leaky wave antennas capable of backward, broadside, and forward scanning with potential loss reduction and high-power handling in comparison to CRLH TLs. Although the majority of the presented examples are for microwave applications, the same concepts can be applied to millimeter-wave (mmWave) applications as discussed in Section 9.4. In particular, LH metamaterials have been experimentally realized from 650 MHz [11] all the way up to visible frequencies [12–14].

9.2 Left-handed Metamaterials: Transmission Line Approach

The TL approach towards LH metamaterials is based on realizing a TL that supports a fundamental backward wave. A backward wave has a phase velocity which is in the opposite direction of its group velocity. As a result, the refractive index of a structure supporting a backward wave is negative. A LH TL, whose unit cell can be represented by a series capacitance (C_{L}) and shunt inductance (L_{L}), can support a fundamental backward wave [15]. However, the LH TL is not physically possible because the group velocity of the supported wave exceeds the speed of light in vacuum as the operational frequency increases [2]. In contrast, a CRLH TL can support a fundamental backward wave and is physically possible. The CRLH TL is a practical realization of a LH TL which includes parasitic RH effects that naturally occur with any physical LH TL implementation.

The equivalent unit cell circuit model of a CRLH TL is shown in Figure 9.1(a) and consists of a series LH capacitance (C_{L}), series RH inductance (L_{R}), shunt RH capacitance (C_{R}), and shunt LH inductance (L_{L}). The dispersion diagram (ω versus β) of the CRLH TL unit cell is shown in Figure 9.1(b). This diagram shows that the CRLH TL has a LH ($v_g v_p < 0$) region at low frequencies and a RH ($v_g v_p > 0$) region at high frequencies. In the LH region, a backward wave is supported and phase advance occurs. While in the RH region, a forward wave is supported and phase delay occurs. A stop-band region can be observed when the series resonance given by

$$\omega_{\mathrm{se}} = \frac{1}{\sqrt{C_{\mathrm{L}} L_{\mathrm{R}}}} \tag{9.1}$$

and shunt resonance given by

$$\omega_{\mathrm{sh}} = \frac{1}{\sqrt{C_{\mathrm{R}} L_{\mathrm{L}}}} \tag{9.2}$$

are not equal; this is known as the unbalanced case. This stop-band can be eliminated when $\omega_{\mathrm{se}} = \omega_{\mathrm{sh}}$, which is known as the balanced case. An infinite wavelength with a non-zero group velocity can be supported by a balanced CRLH TL, while for the unbalanced case, a stop-band occurs which does not allow for guided wave propagation. For either the balanced or unbalanced case, the CRLH TL can be used as an infinite wavelength resonator as discussed in Section 9.2.1.

The CRLH TL unit cell shown in Figure 9.1(a) is an ideal circuit model and has no physical size associated with it. To physically realize a CRLH TL unit cell, either

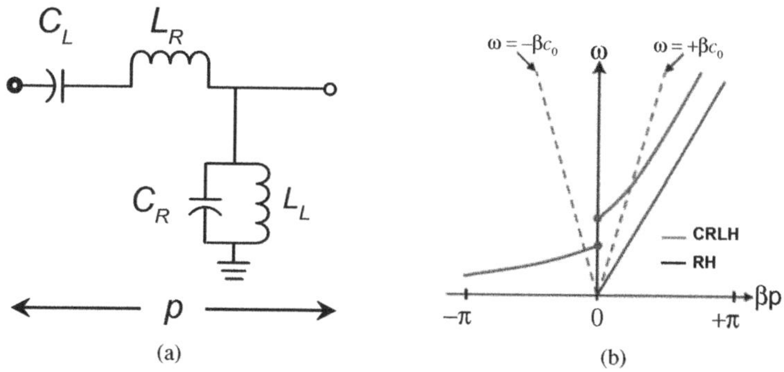

Figure 9.1: CRLH TL unit cell: (a) equivalent circuit model of period p; (b) general dispersion diagram showing unbalanced case, $\omega_{se} \neq \omega_{sh}$; ω_{se} and ω_{sh} can be interchanged.

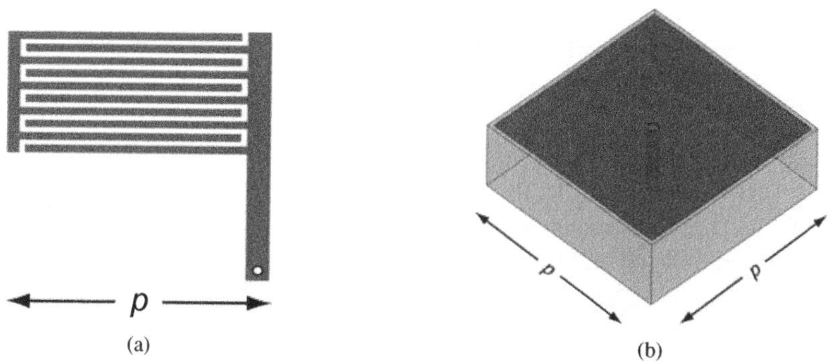

Figure 9.2: Distributed implementation of CRLH TL unit cell shown in Figure 9.1(a) on microstrip technology: (a) interdigital capacitor unit cell; series capacitance from interdigital capacitor and shunt inductance from shorted stub; (b) Sievenpiper mushroom structure; series capacitance from edge coupling from adjacent unit cells and shunt inductance from shorting post [16].

lumped components, distributed technology, or a combination of the two can be used. In general, distributed implementation is preferred over lumped component implementation for fabrication simplicity and for use in radiative applications. Due to its planar nature and popularity, microstrip technology is commonly used to realize CRLH TLs for antenna applications. Two common implementations of CRLH TL unit cells for antenna applications are shown in Figure 9.2.

9.2.1 Composite Right/Left-handed Resonator Theory

The CRLH TL's unique dispersion characteristic can be exploited to realize novel resonators not possible with conventional RH TLs. As a result, novel resonant antennas can also be realized by using CRLH TLs. To understand the unique resonance conditions of the CRLH

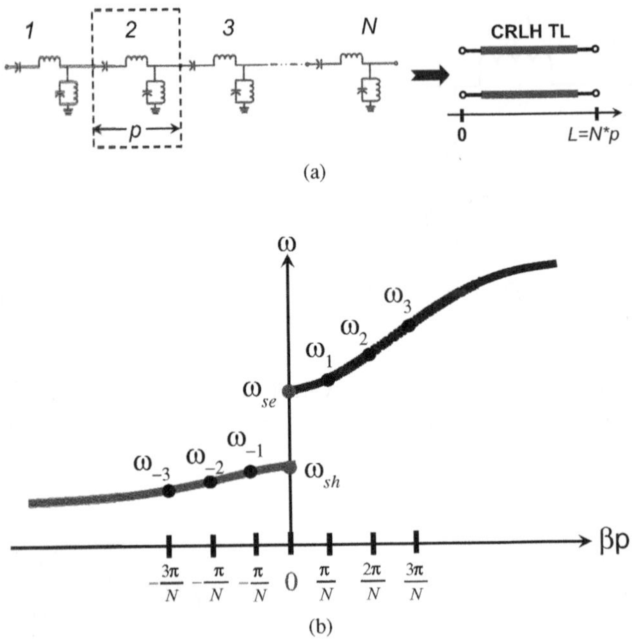

Figure 9.3: CRLH resonator. (a) CRLH resonator consisting of N unit cells; (b) dispersion diagram of CRLH resonator showing resonance modes; ω_{se} and ω_{sh} can be reversed. From reference [18], reproduced by permission of © 2005 IEEE.

metamaterial TL, a review of resonator theory is presented in this section. A conventional RH open- or short-circuited TL of physical length, L, can be used to realize a resonator with resonance condition described by

$$\beta_n = \frac{n\pi}{L} \tag{9.3}$$

where β_n is the phase constant of resonance mode n [17]. For the conventional TL resonator, n has to be a non-zero, positive integer. In the case of a CRLH TL resonator, n can be a positive integer, negative integer, or even zero. Figure 9.3(a) shows a CRLH TL of length, $L = Np$ with its resonance modes shown in Figure 9.3(b).

A CRLH TL of length, $L = Np$ can support a total of $2N - 1$ resonances; $N - 1$ negative resonances, $N - 1$ positive resonances, and a zeroth-order ($n = 0$) resonance. The negative resonances have the same field distribution as the positive resonances, but at lower frequencies. By using the $n = -1$ resonance of the CRLH TL resonator, a half-wavelength field distribution can be obtained at a much lower frequency than a conventional TL resonator of comparable length and substrate makeup. In Section 9.2.2, $n = -1$ resonant antennas with size reductions up to 90% in comparison to conventional half-wavelength resonant antennas are presented. In the case of the $n = 0$ resonance, an electrical length of zero is obtained. Therefore, the $n = 0$ resonance is *independent* of the CRLH TL resonator's physical length. In Section 9.2.3 and 9.2.4, infinite wavelength resonant antennas and series feed networks with arbitrary physical lengths are discussed, respectively. For an unbalanced CRLH TL, there exist two possible $n=0$ resonances at ω_{se} and ω_{sh}. The $n=0$ mode for the unbalanced case is determined by the boundary condition applied to the input and output of the CRLH TL resonator. In the case of short-circuit boundary conditions, the shunt components of the CRLH unit cell are eliminated. Therefore, ω_{se} determines the $n=0$ resonance for short-circuit boundary conditions. The input impedance for short-circuit boundary conditions is given by

$$Z_{\text{in,s.c.}} = -jZ_o \tan(\beta L) \overset{\beta\to 0}{\approx} -jZ_o\beta L = NZ \tag{9.4}$$

In the case of open-circuit boundary conditions, the series components of the CRLH unit cell are eliminated. Therefore, ω_{sh} determines the $n=0$ resonance for open-circuit boundary conditions. The input impedance for the case of open-circuit boundary conditions is given by

$$Z_{\text{in,o.c.}} = -jZ_o \cot(\beta L) \overset{\beta\to 0}{\approx} -jZ_o/\beta L = 1/NY \tag{9.5}$$

In addition, the dispersive nature of the CRLH TL means that its resonance conditions are not harmonics of each other unlike a RH TL resonator. In general, the negative and zero resonance modes of the CRLH TL are of interest for antenna applications

9.2.2 Small Resonant CRLH TL Antennas

In this section, the $n=-1$ resonance mode of the CRLH TL is used to realize small resonant metamaterial antennas. A half-wavelength field distribution is supported under the $n=-1$ resonance mode whose operational frequency can be arbitrary because of the CRLH TL's non-linear dispersion as shown in Figure 9.1(b). Since the dispersion relation of the CRLH TL is determined by the unit cell's equivalent circuit parameters (i.e. C_L, C_R, L_L, and L_R), the operational frequency of the $n=-1$ resonance mode can be engineered by properly designing the unit cell [19].

To design a small CRLH TL resonant antenna, an initial unit cell is first analyzed and its dispersion diagram is plotted. The initial dispersion diagram of a CRLH TL unit cell with $\omega_{sh} < \omega_{se}$ is plotted in Figure 9.4 and is represented by a solid line. The other three dispersion curves represent the dispersion relation when L_L, C_L and both L_L and C_L are increased and the other parameters are unchanged. When L_L is increased, ω_{sh} and the LH cutoff frequency determined by

$$\omega_{\text{LH,cutoff}} \approx \frac{1}{2\sqrt{C_L L_L}} \tag{9.6}$$

will be decreased. When C_L is increased so that the product $C_L L_R$ is greater than $C_R L_L$, ω_{se} becomes lower than ω_{sh} and the LH cutoff frequency will also be decreased. Similarly, if both L_L and C_L are increased, the dispersion diagram will be shifted to an even lower frequency band. For example, if $N=4$ for a given CRLH TL, the $n=-1$ mode will occur at $\beta p/\pi = 0.25$ which equates to a 3.0 GHz initial operation frequency. This operational frequency will be reduced from 3.0 GHz to 1.2 GHz if the series capacitance and shunt inductance are increased as shown in Figure 9.4. Consequently, if the physical size of the unit cell can remain small and the value of L_L and C_L can be simultaneously increased, a small antenna can be realized.

A prototype small resonant antenna based on the $n=-1$ resonance mode is shown in Figure 9.5, which consists of two substrate layers and three metal layers [20]. The antenna

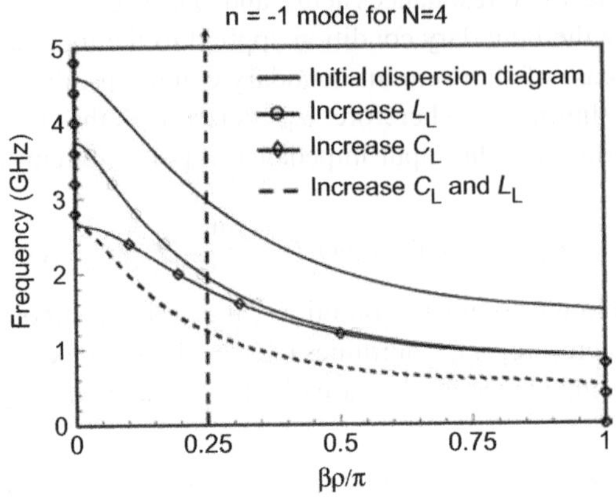

Figure 9.4: Comparison of the different dispersion relation based on the different value of equivalent circuit parameters. From reference [20], reproduced by permission of © 2006 IEEE.

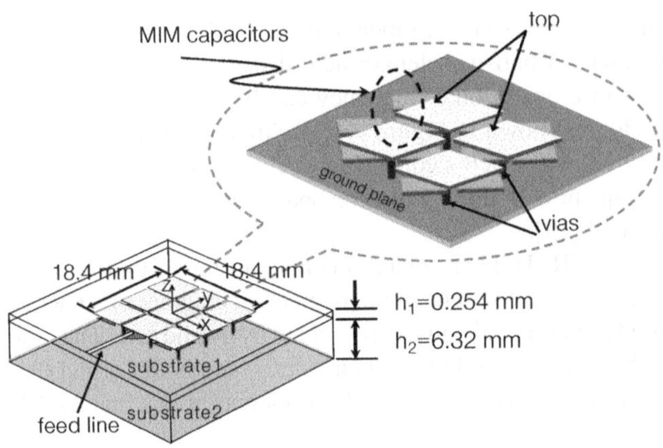

Figure 9.5: Configuration of the proposed small antenna prototype. From reference [20], reproduced by permission of © 2006 IEEE.

consists of 3 × 3 CRLH TL unit cells based on the Sievenpiper mushroom structure. Each unit cell includes a 6 × 6 mm^2 square patch on top, a metallic via connected to the ground, and four metal–insulator–metal (MIM) capacitors (2.7 × 2.7 mm^2) linked to the adjacent cell in both x and y directions. MIM capacitors were fabricated on a thin substrate ($h_1 =$ 0.254 mm) with a permittivity of 10.2 (substrate 1) in order to maximize series capacitive coupling. The shunt inductance was implemented by using vias with radius of 0.12 mm and height of 6.584 mm. Substrate 2 consists of a thick substrate ($h_2 = 6.32$ mm) with a permittivity of 2.2.

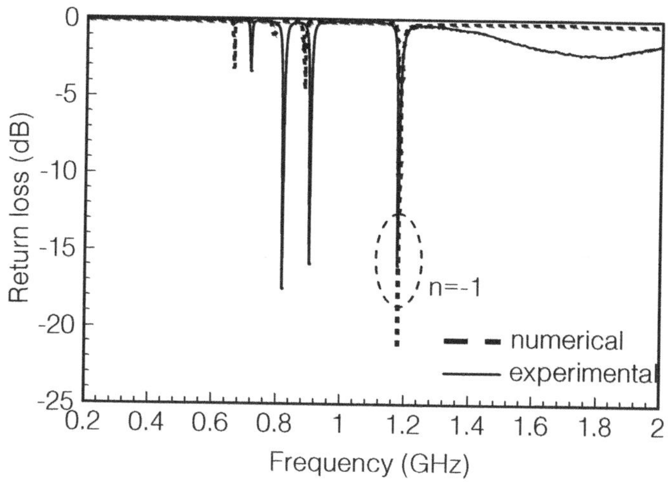

Figure 9.6: Numerical and experimental return loss of small antenna. From reference [20], reproduced by permission of © 2006 IEEE.

The resonant length is in the y-direction, which consists of three unit cells. Each unit cell provides a phase advance of $\pi/3$ radians at 1.17 GHz. Therefore, a half-wavelength resonance occurs at 1.17 GHz. A 3×3 unit-cell configuration is chosen to provide adequate gain and efficiency for practical implementation. Edge feeding (0.1 mm gap) is used to match the antenna to a 50 Ω system impedance. The numerical and experimental return loss of the antenna is illustrated in Figure 9.6. An experimental return loss of −16.0 dB is obtained at 1.17 GHz. The antenna measures $18.4 \times 18.4 \times 6.574$ mm^3 or $1/14\lambda_0 \times 1/14\lambda_0 \times 1/39\lambda_0$ in terms of free-space wavelength. The experimental E- and H-plane radiation patterns of the antenna at 1.17 GHz are shown in Figure 9.7. A maximum gain of 0.6 dBi at broadside with a radiation efficiency of 26% were measured.

9.2.2.1 Small Circular-polarized Antenna

To demonstrate the ability to engineer the CRLH TL unit cell for different operational frequencies, the small antenna of Figure 9.5 is modified to operate at 2.4 GHz. In addition, the small antenna of Figure 9.5 is modified for circular-polarization. The proposed small 2.4 GHz circular-polarized antenna is shown in Figure 9.8(a). The CRLH TL unit cell is a scaled-down version of the unit cell used in Figure 9.5. Two orthogonal edge feeds and a 90° hybrid coupler are used to excite circularly polarized modes in two orthogonal directions. The numerical electric-field distribution of the antenna under single edge excitation is shown in Figure 9.8(b); the minimum and maximum field only occurs at the center and edge of the antenna, respectively, indicating a half-wavelength distribution. The fabricated antenna measures $12.4 \times 12.4 \times 3.414$ mm^3 and is $1/10\lambda_0 \times 1/10\lambda_0 \times 1/36\lambda_0$ in terms of free-space wavelength. Figure 9.8(c) shows the 90% footprint size reduction achieved with the metamaterial antenna in comparison to a conventional circular-polarized patch antenna fabricated on substrate 2.

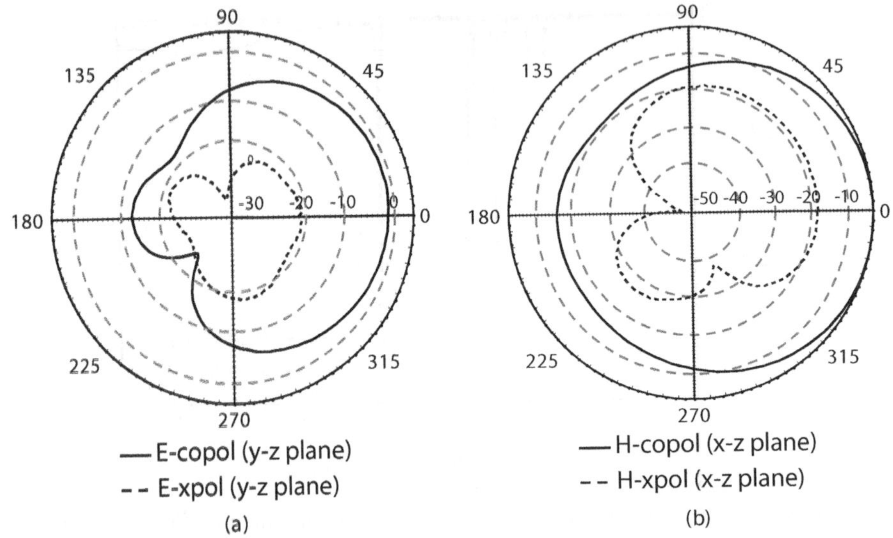

Figure 9.7: Normalized experimental radiation patterns for antenna shown in Figure 9.5. (a) E-plane (y–z plane). (b) H-plane (x–z plane). From reference [20], reproduced by permission of © 2006 IEEE.

The experimental return and insertion loss of the circular-polarized antenna are shown in Figure 9.9. At 2.416 GHz a return loss of −31 dB and −17 dB are obtained at the two input ports. In addition, the insolation loss between the two input ports is less than −30 dB, which indicates that the two input ports are not coupled.

The radiation pattern of the antenna is measured and shown in Figure 9.10; a hybrid coupler is used to provide the 90° phase difference between the antenna's input ports for circular polarization. The maximum antenna gain is 2.17 dBi at 2.416 GHz and the cross polarization is approximately 23 dB. In addition, the minimum axial ratio of 1.2 dB is achieved at broadside and a 3 dB axial ratio beamwidth of 116° is obtained.

9.2.2.2 Compact Dual-band Antenna

A two-dimensional isotropic CRLH TL unit cell based on the Sievenpiper high-impedance structurewas used to realize the antennas of Figure 9.5 and 9.8(a). If an anisotropic CRLH TL unit cell is used, the dispersion relations for orthogonal directions can be different. The resonant frequencies in the orthogonal directions are independent of each other and can be arbitrarily chosen by engineering the unit cell. In this section, an example of a compact dual-band antenna based on an anisotropic CRLH TL unit cell is discussed. The model of the proposed antenna consisting of 2 × 2 unit cells is shown in Figure 9.11(a) along with its cross-sectional views in the x–z and y–z planes in Figure 9.11(b) and 9.11(c), respectively. The substrates are the same as the ones used for realizing the small antenna of Figure 9.5. The anisotropic nature of the CRLH TL unit cell is achieved by making it different in orthogonal directions. In this example, for a wave traveling in the x-direction, the LH series capacitance is due only to the edge coupling between the unit cells. While, for a wave

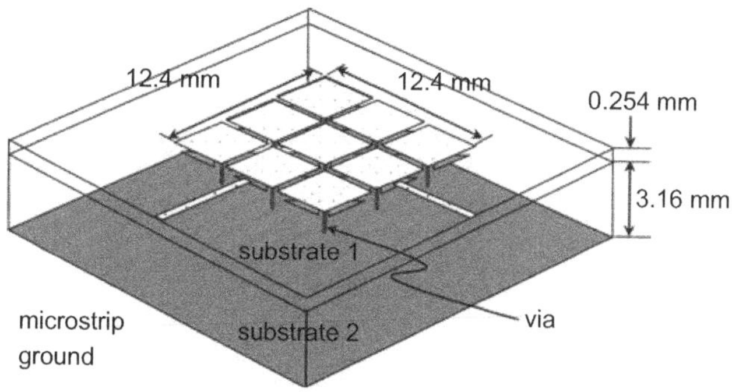

(a)

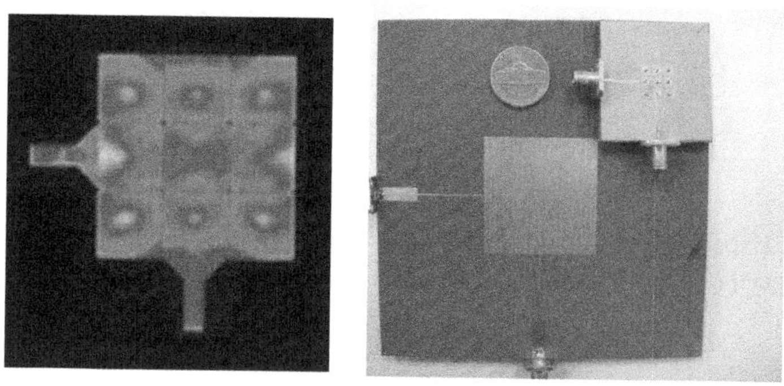

(b) (c)

Figure 9.8: Circularly polarized antenna based on CRLH TL. (a) Configuration of the circularly polarized antenna. (b) Field distribution on the proposed circularly polarized antenna. (c) Comparison of CRLH TL circularly polarized antenna and conventional circularly polarized patch antenna. From reference [20], reproduced by permission of © 2006 IEEE.

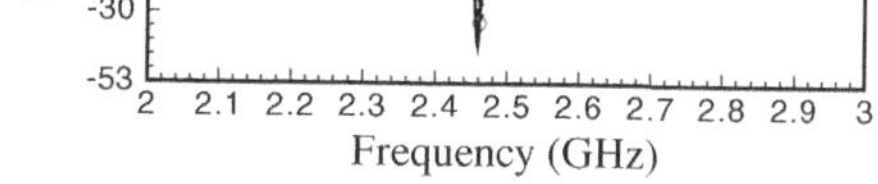

Figure 9.9: Experimental S-parameters of the antenna structure shown in Figure 9.8(a). From reference [20], reproduced by permission of © 2006 IEEE.

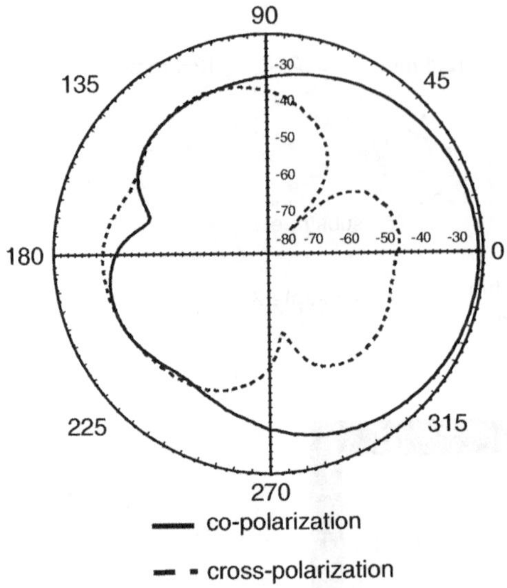

Figure 9.10: Experimental radiation patterns of circular-polarized antenna of Figure 9.8(a). From reference [20], reproduced by permission of © 2006 IEEE.

traveling in the y-direction, the LH series capacitance is due to edge coupling and also from the MIM capacitance. As a result, the dispersion curve is at a lower frequency band in the y-direction. The dispersion diagram along the x-direction and y-direction of the CRLH TL unit cell is shown in Figure 9.12. The $n = -1$ resonance mode is used to achieve broadside radiation at two frequencies; 2.4 GHz in the x-direction and 1.9 GHz in the y-direction.

By using Ansoft HFSS, a commercial finite-element-method (FEM) solver, the position of a single feed line was optimized to excite two $n = -1$ modes in the x- and y-directions. The feed line is placed 1.05 mm from the center of the antenna with an edge coupling gap of 0.1 mm. The experimental return loss of the dual-band antenna is shown in Figure 9.13; a return loss of -18.4 dB and -6.8 dB at 1.96 GHz and 2.37 GHz, respectively. The normalized radiation patterns of the antenna at 1.96 GHz and 2.37 GHz are shown in Figure 9.14 and 9.15, respectively.

The experimental antenna gains in the broadside direction are -3 dBi and -2.3 dBi at 1.96 GHz and 2.37 GHz, respectively. An antenna radiation efficiency of 25.4% and 28.9% are achieved at 1.96 GHz and 2.37GHz, respectively. The width, length, and height of the dual-band antenna are $1/17\lambda_0$, $1/17\lambda_0$, and $1/19\lambda_0$, respectively in terms of free-space wavelength at 2.37 GHz. This indicates a 96% area reduction compared to a conventional patch antenna built on substrate 2.

9.2.3 Infinite Wavelength Resonant Antennas

In this section, the $n = 0$ resonance mode supported by a CRLH TL resonator is used to realize infinite wavelength resonant antennas with monopolar radiation patterns [22]. As discussed in Section 9.2.1, the $n = 0$ resonance is independent of the resonator's physical

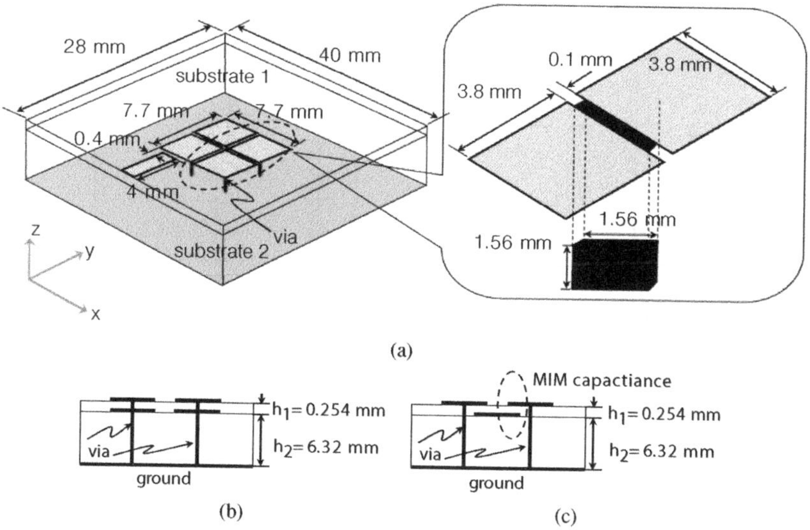

Figure 9.11: Dual-band antenna based on anisotropic CRLH TL unit cell; (a) configuration of the dual-band antenna; (b) x–z plane view of the antenna (vias are in the front and back of the middle layer patches); (c) y–z plane view of the antenna. From reference [21], reproduced by permission of © 2006 IEEE.

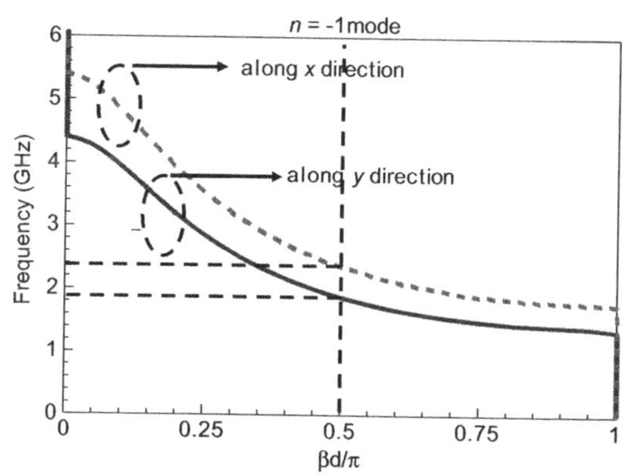

Figure 9.12: Dispersion diagram along the x and y directions. From reference [21], reproduced by permission of © 2006 IEEE.

size. As a result, the infinite resonant antenna can be made arbitrarily small or large. In the case of open-circuit boundary conditions, the resonant frequency is determined by the shunt resonance of the unit cell. To demonstrate the size independence, infinite wavelength antennas consisting of two, four, and six CRLH TL unit cells are presented. The CRLH TL unit cell used to realize these three antennas are shown in Figure 9.16(a) and the general

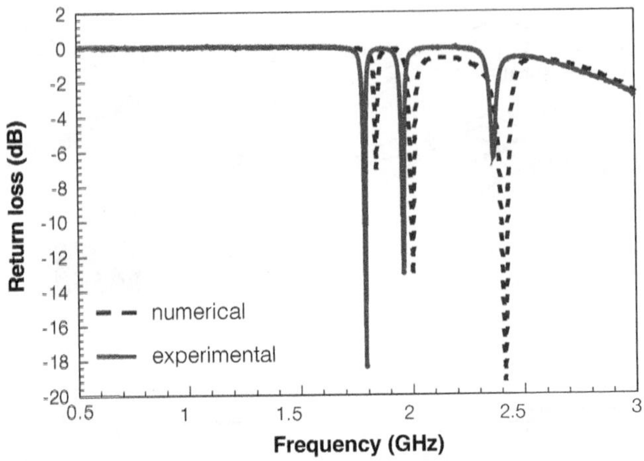

Figure 9.13: Numerical and experimental return loss of the proposed compact dual-band antenna. From reference [21], reproduced by permission of © 2006 IEEE.

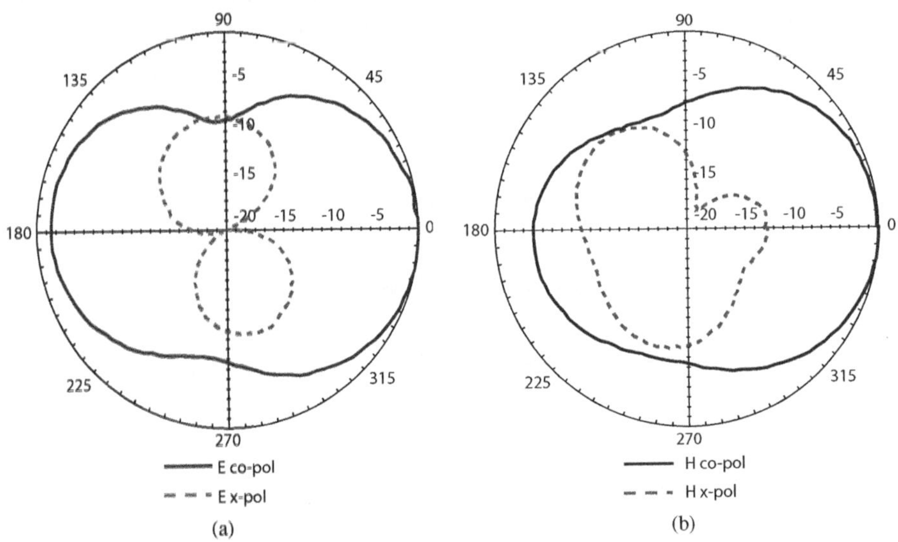

Figure 9.14: Experimental radiation pattern at 1.96 GHz: (a) E-plane (x–z plane); (b) H-plane (y–z plane). From reference [21], reproduced by permission of © 2006 IEEE.

model of the realized antennas are shown in Figure 9.16(b). A Rogers RT/duroid 5880 substrate with permittivity of 2.2 and height of 1.57 mm is used to implement the antennas. The shunt resonance of the unit cell is 3.51 GHz.

The CRLH TL unit cell of Figure 9.16(a) was chosen to implement the infinite wavelength antennas because at $n = 0$, the electric field is a maximum along the antenna's perimeter to

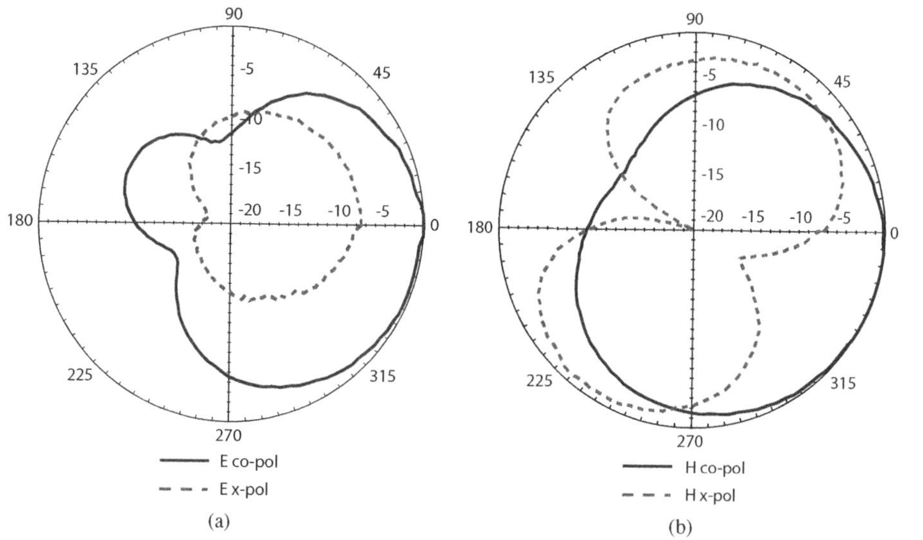

Figure 9.15: Experimental radiation pattern at 2.37 GHz: (a) E-plane (y–z plane); (b) H-plane (x–z plane). From reference [21], reproduced by permission of © 2006 IEEE.

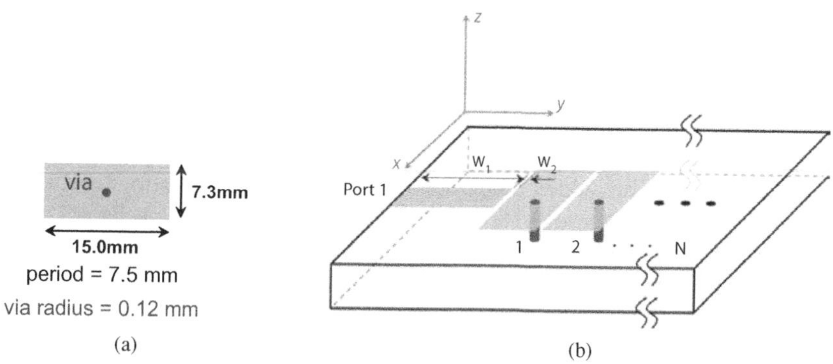

Figure 9.16: Infinite wavelength resonant antenna with monopolar radiation: (a) unit cell top view; (b) model of antenna consisting of N unit cells. From reference [22], reproduced by permission of © 2007 IEEE.

produce a monopolar radiation pattern. The numerical input impedance ($Z_{\text{in}} = R + jX\ \Omega$) for the three infinite wavelength antennas ($N = 2$, 4, and 6) are shown in Figure 9.17. The resonant frequency of the antenna is defined as the frequency where the resistance reaches a maximum, independent of the value of reactance [23]. Figure 9.17 shows that the input impedance follows the trend predicted by (9.5) as N is increased.

The input impedance for the two unit cell CRLH TL antennas is quite high for quarter-wavelength matching. Therefore, proximity coupling is used to match the antenna to a 50 Ω

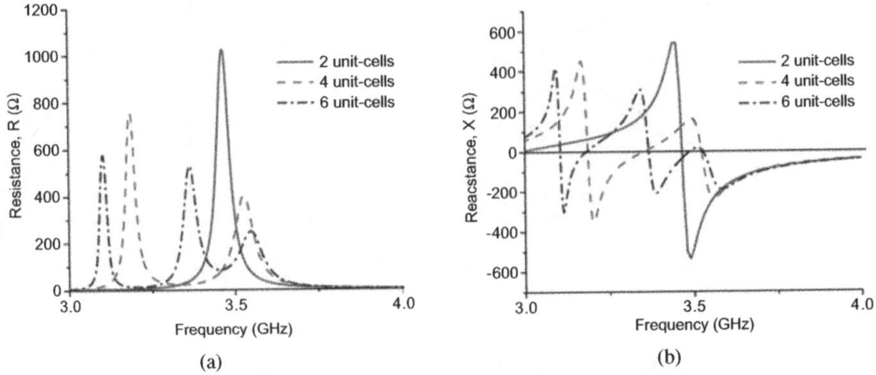

Figure 9.17: CRLH antenna input impedance: (a) real part (R); (b) imaginary part (X). From reference [22], reproduced by permission of © 2007 IEEE.

line as shown in Figure 9.16(b) with $w_1 = 15.0$ mm and $w_2 = 0.2$ mm. For the CRLH TL-based antenna an experimental return loss of -12.34 dB is obtained at $f_0 = 3.38$ GHz. The electrical size of the antennas is $\lambda_o/6 \times \lambda_o/6 \times \lambda_o/57$ at f_0. The numerical and experimental radiation patterns of the two unit cell CRLH TL infinite wavelength antenna are shown in Figure 9.18 and reveal the expected monopolar radiation pattern. A maximum gain of 0.87 dBi with a 70% radiation efficiency is experimentally obtained. In addition, the x–y plane radiation pattern and cross-polarization (normalized relative to co-polarization) of the CRLH TL antenna are shown in Figure 9.18(c) and Figure 9.18(d), respectively. Figure 9.18(c) illustrates the omni-directional coverage in the x–y plane provided by the monopolar antenna, while Figure 9.18(d) shows that the cross-polarization is less than the co-polarization.

Additional unit cells are added along the y-direction to create the four and six unit cell antennas as depicted in Figure 9.16(b). A single-section quarter-wavelength transformer is used to match each four and six unit cell antenna to a 50 Ω line. The experimental infinite wavelength frequency, return loss, gain, and radiation efficiency of the four unit cell antenna is 3.52 GHz, -17.33 dB, 4.50 dBi and 88%, respectively. The experimental infinite wavelength frequency, return loss, gain, and radiation efficiency of the six unit cell antenna is 3.55 GHz, -11.17 dB, 5.17 dBi, and 91%, respectively. The electrical size of the four unit cell antenna is $\lambda_o/6 \times \lambda_o/3 \times \lambda_o/53$ at f_0 and the electrical size of the six unit cell antenna is $\lambda_o/6 \times \lambda_o/2 \times \lambda_o/53$ at f_0. Although the antennas become physically larger, the infinite wavelength frequency remains approximately constant. In addition, gain increases as the antenna becomes physically larger. The radiation patterns of the four and six unit cell CRLH antennas are similar to those of the two unit cell CRLH antenna and therefore are not shown.

9.2.3.1 Dual-mode Antenna

In Section 9.2.2.2, a dual-band CRLH TL antenna was discussed; for the two operational frequencies, the radiation patterns of the antenna were similar. It may be desirable to have a compact antenna with radiation pattern selectivity at two different frequencies for certain

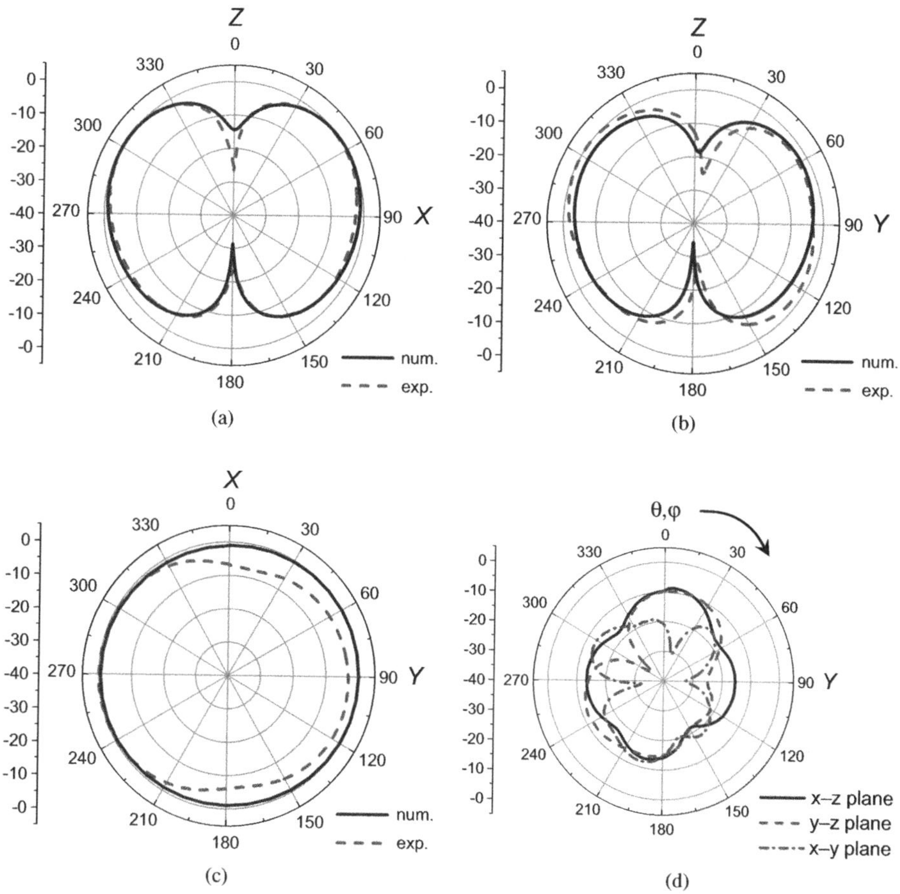

Figure 9.18: Two unit cell CRLH TL antenna radiation patterns. (a) $\varphi = 0°$ (x–z plane); (b) $\varphi = 90°$ (y–z plane); (c) $\theta = 90°$ (x–y plane); (d) experimental cross-polarizations normalized to co-polarizations. From reference [22], reproduced by permission of © 2007 IEEE.

wireless applications. Such a dual-mode antenna can be achieved by using two different resonance modes of a CRLH TL antenna. In reference [18], the $n = 0$ and $n = -1$ resonance modes were used to realize a monopolar pattern and a broadside pattern, respectively, at two different frequencies.

The proposed dual-mode antenna is based on Figure 9.16(b) and consists of three CRLH TL unit cells each measuring 4.8×15 mm^2 with a period of 5.0 mm. The radius of the shorting post is 0.12 mm. A return loss of -10.21 dB and -9.2 dB are experimentally obtained at $f_0 = 4.00$ GHz and at $f_{-1} = 3.57$ GHz, respectively. The antenna size is $\lambda_o/5 \times \lambda_o/5 \times \lambda_o/50$ at f_0 and the antenna size is $\lambda_o/5.7 \times \lambda_o/5.7 \times \lambda_o/54$ at f_{-1}. The experimental radiation patterns at f_0 and at f_{-1} are shown in Figure 9.19(a) and Figure 9.19(b), respectively. A maximum gain of 2.3 dBi with an efficiency of 75% was

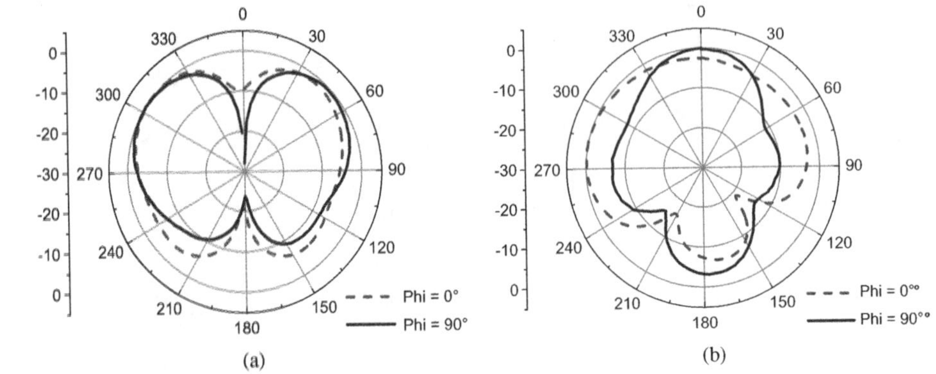

Figure 9.19: Experimental radiation patterns of dual-mode antenna. (a) $n = 0$ mode showing monopolar radiation; (b) $n = -1$ mode showing broadside pattern. From reference [18], reproduced by permission of © 2005 IEEE.

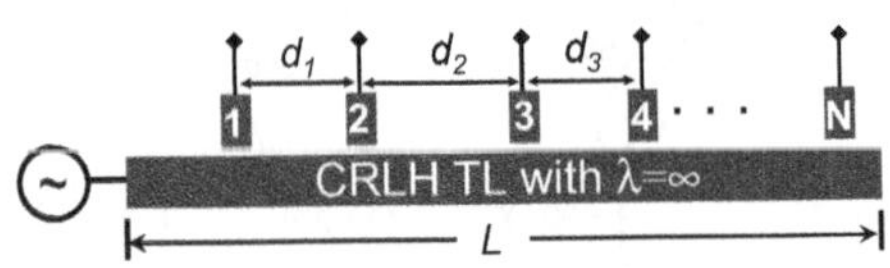

Figure 9.20: Novel feed network for equally excited, arbitrary spaced antenna array. From reference [25], reproduced by permission of © 2005 IEEE.

achieved at f_0, while a maximum gain of -2.5 dBi with an efficiency of 25% was achieved at f_{-1}. The two operational frequencies are not harmonics of each other owing to the dispersive nature of the CRLH TL unit cell. Since the unit cell's dispersion relation is determined by its unit cell, the operational frequencies of the antenna can be controlled by modifying the unit cell and/or the number of unit cells.

9.2.4 *N*-port Infinite Wavelength Series Feed Network

Under $n = 0$ resonance, the magnitude and phase along the CRLH TL resonator are all equal. This phenomenon can be exploited to realize an N-port series divider with arbitrary length, N output ports, and spacing of output ports [24]. As a result, such a divider can be used for sparse antenna arrays that require equal magnitude and phase feeding [25]. Figure 9.20 shows the concept of the infinite wavelength divider. In addition, this divider can be used to address the size and design complexity of conventional power dividers [26] for equal magnitude and phase division.

To demonstrate the novel feed network, the interdigital CRLH TL unit cell was used; the dimensions of the unit cell are shown in Figure 9.21. The substrate used is Rogers duroid/RT 5880 with dielectric constant of 2.2 and height of 1.57 mm. The CRLH TL unit cell was designed to have a shunt resonance of 2.31 GHz and is unbalanced. To implement the feed

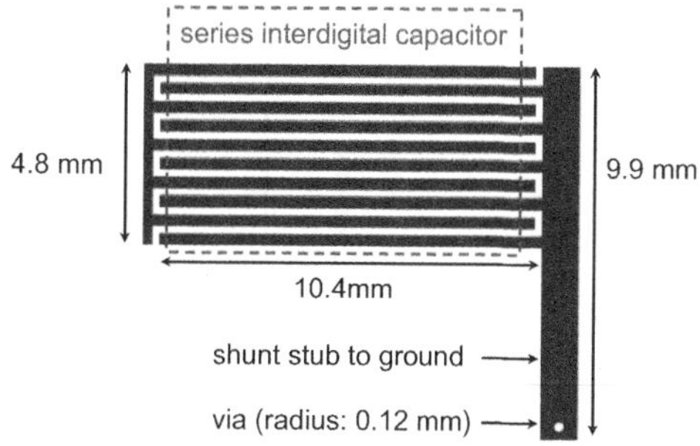

Figure 9.21: CRLH TL unit cell; width of fingers are 0.3 mm, width of stub is 1.0 mm, and gaps are all 0.2 mm. From reference [24], reproduced by permission of © 2005 IEEE.

network, thirteen unit cells of Figure 9.21 were cascaded. The completed six-port series-fed array is shown in Figure 9.22. Quasi-Yagi antennas were placed at the five output ports to demonstrate equally excited, arbitrary sparse array application of this feed network. To enforce the required open-circuit boundary condition, the CRLH TL is open-ended. Port 1 is used as the feed port and the other five ports are used for output. A 1.5 pF capacitor is used to maximize the transferred power from the feed port. In addition, the output ports are connected to the CRLH TL via 0.2 pF capacitors in order to increase the output coupling and not to affect the infinite wavelength frequency. The quasi-Yagi antennas were designed to have a broadband response ($|S_{11}| < -10$ dB) from 2.07 GHz to 2.80 GHz.

With the quasi-Yagi antennas connected, the infinite wavelength frequency is shifted to $f = 2.38$ GHz since each antenna's impedance is not exactly 50 Ω. The measured radiation pattern of the series-fed antenna array at $f = 2.38$ GHz is compared with the theoretical pattern of an equally excited antenna array with the same spacing between elements in Figure 9.22(b). In addition, the measured radiation pattern of the antenna array at $f =$ 2.63 GHz is also plotted in Figure 9.22(b) to show that broadside radiation does not occur once the elements are out-of-phase with each other. Since this series feed network is operating in the fast-wave region, radiation leakage can occur. An infinite wavelength, traveling wave-based series feed network was demonstrated in reference [27] where the radiation leakage is avoided by using a metal shield.

9.3 Left-handed Metamaterials: Evanescent-mode Approach

In both the resonant and TL approaches towards LH metamaterials, metallic structures are required to provide both negative permeability and permittivity. As a result, the design of the unit cell for these approaches has to avoid concentration of high current regions to reduce metallic loss. In addition, metallic structures are not good conductors at terahertz and higher frequencies [10]. To reduce metallic loss and to address LH metamaterials for higher frequencies, the evanescent-mode approach towards LH metamaterials was proposed in reference [10]. In the evanescent-mode approach, dielectric resonators (DRs) under

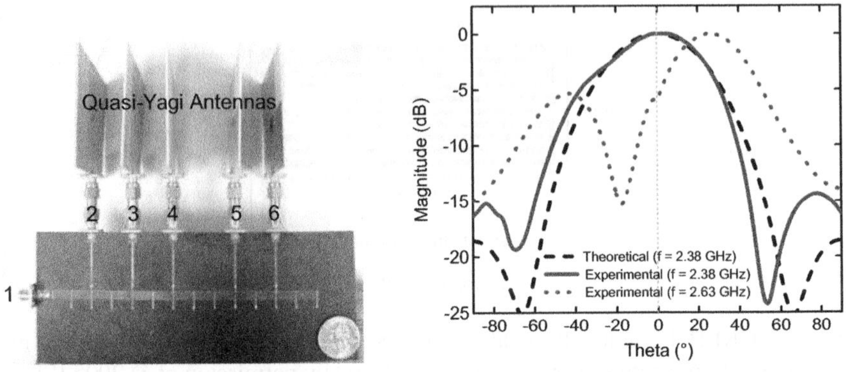

(a) (b)

Figure 9.22: Sparse antenna array: (a) top view; spacing between 2 and 3: 0.184λ, spacing between 3 and 4: 0.184λ, spacing between 4 and 5: 0.276λ, spacing between 5 and 6: 0.184λ, where $f = 2.38$ GHz; (b) theoretical and experimental radiation patterns. From reference [25], reproduced by permission of © 2005 IEEE.

resonance providing negative permeability are placed inside a cutoff waveguide which provides negative permittivity.

A one-dimensional LH evanescent mode metamaterial consisting of five unit cells is shown in Figure 9.23(a) along with its unit cell dispersion diagram shown in Figure 9.23(b). The unit cell has a period, $p = 6.0$ mm with a parallel plate waveguide consisting of Rogers duroid/RT 5880 with $\varepsilon_{\mathrm{BG}} = 2.2$ and $d = 5.0$ mm and a DR with $\varepsilon_{\mathrm{DR}} = 38$, $a = 2.55$ mm $h = 2.03$ mm, where $\varepsilon_{\mathrm{BG}}$ and $\varepsilon_{\mathrm{DR}}$ are the substrate and DR permittivities, respectively. The TE_1 cutoff of the parallel plate waveguide is used to provide the required negative permittivity, while the DR's $\mathrm{TE}_{01\delta}$ resonance provides the required negative permeability. The TE_n modes supported by the parallel plate waveguide without DRs is determined by

$$f_{c,n} = \frac{nc}{2d\sqrt{\varepsilon_{\mathrm{BG}}}} \tag{9.7}$$

where c is the speed of light in vacuum. The effective permittivity of the parallel plate waveguide, $\varepsilon_{\mathrm{eff}}$, under TE_n mode operation is given by

$$\varepsilon_{\mathrm{eff}} = \varepsilon_{\mathrm{BG}}\left[1 - \left(\frac{f_{c,n}}{f}\right)^2\right] \tag{9.8}$$

which shows that when f is less than f_c, a negative effective permittivity is achieved [29]. Using (9.7), $f_{c,1}$ is 20.2 GHz and the DR's $\mathrm{TE}_{01\delta}$ resonance is numerically determined to be 10.5 GHz. Figure 9.23(b) shows that a fundamental backward wave is supported below the TE_1 parallel plate waveguide mode.

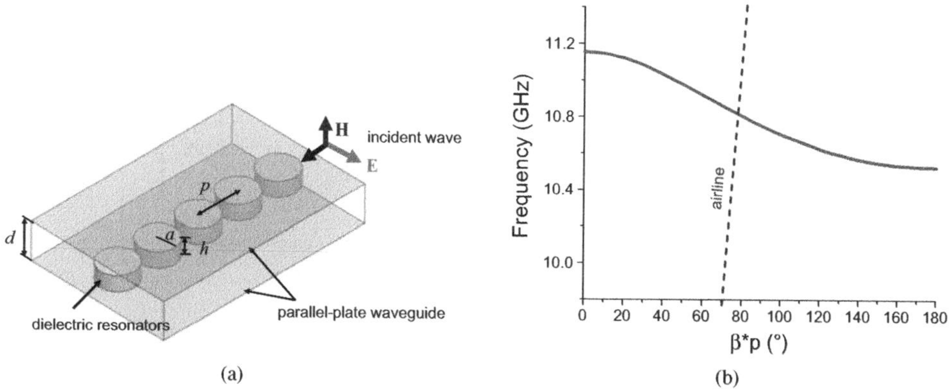

Figure 9.23: One-dimensional LH TL based on evanescent mode approach: (a) TL realization based on five unit cells; (b) dispersion diagram of unit cell showing support of a fundamental backward wave. From reference [28], reproduced by permission of © 2006 IEEE.

9.3.1 Leaky Wave Antennas Based on Evanescent-mode LH Metamaterials

An unique feature of LH metamaterials is that part of its fundamental mode lies within the fast-wave region, where β is less than the free space wavenumber, k_o. As a result, LH metamaterials can be used to realize leaky wave antennas capable of backfire ($\theta = -90°$) to endfire ($\theta = +90°$) scanning [30]. In this section, leaky wave antennas based on evanescent-mode LH metamaterials are presented.

9.3.1.1 Backward Scanning Leaky Wave Antenna

The backward wave of the one-dimensional LH metamaterial shown in Figure 9.23(a) lies in the fast-wave region from 10.8 GHz to 11.2 GHz. By modifying the unit cell so that the field concentration is near an open aperture, a backward scanning leaky wave antenna was realized in reference [28]. The one-dimensional leaky wave antenna is shown in Figure 9.24.

Rectangular waveguides with a cross-sectional area of $5 \times 5\ \text{mm}^2$ are used to feed and terminate the LH metamaterial structure. The rectangular waveguides are designed to operate under the TE_{10} mode; WR-90 adapters were used to excite the input and output rectangular waveguides. Since a backward wave is supported, backward scanning can be realized and the scan angle, θ is determined by

$$\theta(\omega) = \arcsin\left(\frac{\beta(\omega)}{k_o}\right) \tag{9.9}$$

The experimental radiation patterns of the leaky wave antenna are shown for $f = 10.7$ GHz, $f = 10.9$ GHz, and $f = 11.0$ GHz, in Figure 9.25(a), 9.25(b) and 9.25(c), respectively. From the radiation patterns, it can be confirmed that backward scanning with $\theta = -57°$, $\theta = -44°$, and $\theta = -39°$ at 10.7 GHz, 10.9 GHz, and 11.0 GHz is obtained, respectively.

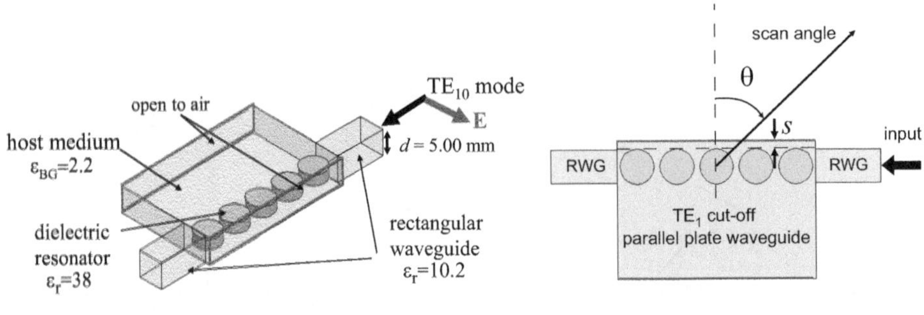

(a) (b)

Figure 9.24: One-dimensional LH leaky wave antenna based on evanescent-mode approach: (a) perspective view; (b) top view; $s = 1.0$ mm. From reference [28], reproduced by permission of © 2006 IEEE.

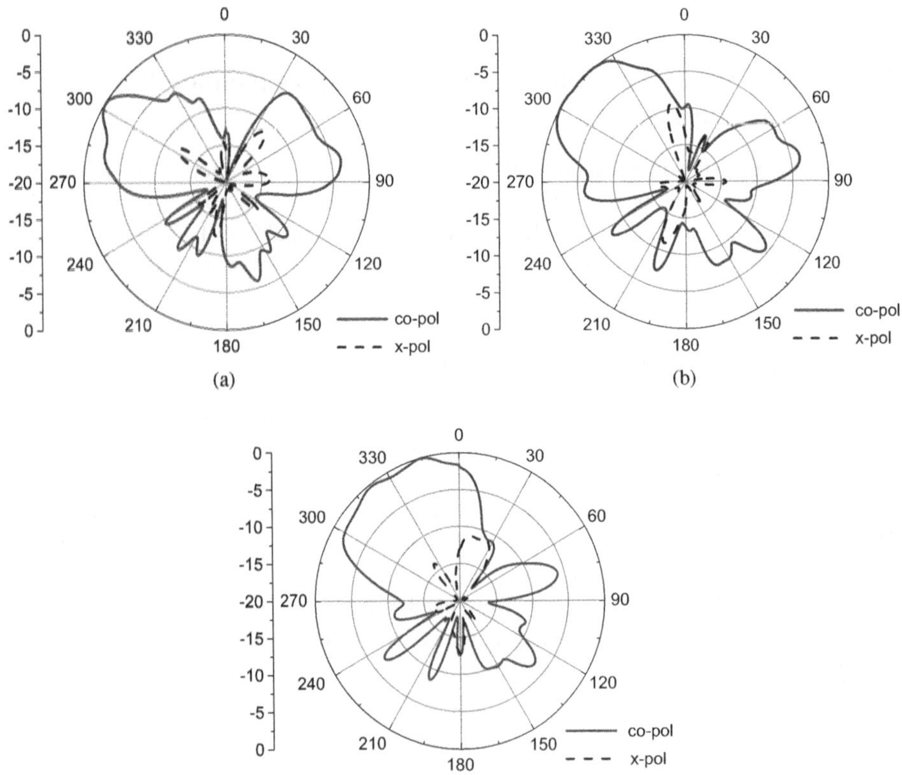

Figure 9.25: Experimental radiation patterns of Figure 9.24; cross-polarization is less than 10 dB: (a) $f = 10.7$ GHz; (b) $f = 10.9$ GHz; (c) $f = 11.0$ GHz. From reference [28], reproduced by permission of © 2006 IEEE.

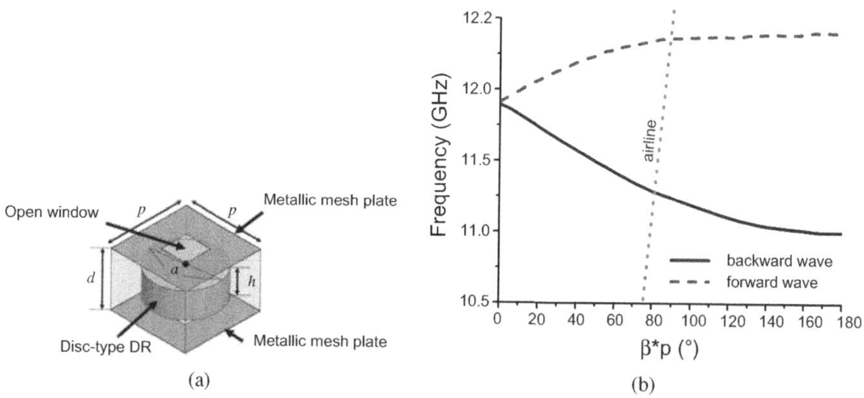

Figure 9.26: Balanced CRLH unit cell based on evanescent-mode approach: (a) unit cell details; (b) dispersion diagram. From reference [31], reproduced by permission of © 2007 IEEE.

9.3.1.2 Balanced CRLH Leaky Wave Antenna

In Section 9.3.1, only the backward wave supported by an evanescent-mode LH unit cell was used to accomplish backward scanning. However, in order to achieve continuous backfire-to-endfire scanning including broadside scanning, an evanescent-mode balanced CRLH unit cell is required. The evanescent-mode LH unit cell of Figure 9.26(a) is used to demonstrate this concept. Open aperture windows are placed on both the upper and lower parallel plate metallization. By adjusting the metal plate distance, d, a balanced CRLH condition can be achieved as shown in Figure 9.26(b). The optimized unit cell has a period, $p = 6.0$ mm with a parallel plate waveguide consisting of Rogers duroid/RT 5880 with $\varepsilon_{BG} = 2.2$ and $d = 4.25$ mm, aperture window area of 2×2 mm^2, and a DR with $\varepsilon_{DR} = 38$, $a = 5.10$ mm $h = 2.03$ mm. Figure 9.26(b) shows that backward scanning can be achieved around, 11.25 GHz $< f <$ 11.9 GHz, forward scanning around 11.9 GHz $< f <$ 12.3 GHz, and broadside scanning at 11.9 GHz.

Five unit cells of Figure 9.26(a) were cascaded to form a one-dimensional balanced CRLH TL and rectangular waveguides ($\varepsilon_r = 10.2$) of cross-sectional area 5×6 mm^2 were used for the input and output as shown in Figure 9.27. The experimental radiation patterns at $f = 11.7$ GHz, $f = 11.82$ GHz and $f = 12.1$ GHz are shown in Figure 9.28(a), 9.28(b) and 9.28(c), respectively. From the radiation patterns, it can be confirmed that backward scanning with $\theta = -30°$, broadside scanning with $\theta = 0°$, and forward scanning with $\theta = +30°$ at 11.7 GHz, 11.82 GHz and 12.0 GHz is obtained, respectively.

9.4 mmWave Metamaterial Antenna Applications

The metamaterial concepts used to realize the microwave resonant and leaky wave antennas of sections 9.2.2 to 9.3.1.2 can be directly applied to mmWave frequencies. Currently, there is relatively little research and publications on metamaterial antennas at mmWave frequencies, unlike at microwave frequencies. This disparity is mainly due to the commercial need

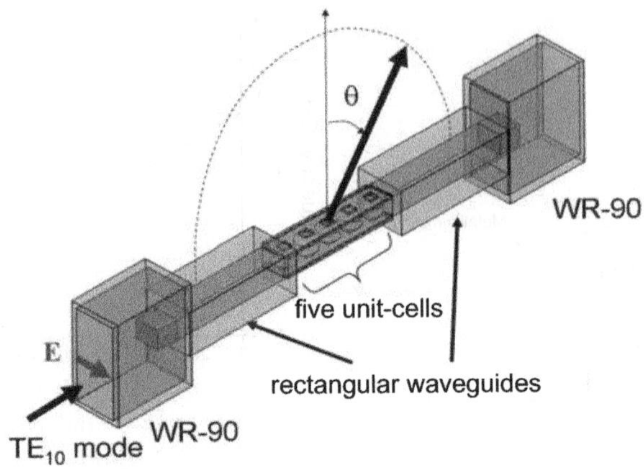

Figure 9.27: Balanced CRLH leaky wave antenna based on evanescent-mode approach. From reference [31], reproduced by permission of © 2007 IEEE.

of small microwave frequency antennas for cellular and wireless networking applications. In addition, fabrication and measurement are more cost-effective and readily available at microwave frequencies for demonstrating metamaterial devices, especially for university-based research. However, with the increasing popularity of mmWaves for multimedia and automotive applications, research into mmWave metamaterial antennas is growing. To illustrate the application of mmWave LH metamaterials, a 94 GHz CRLH TL for a series antenna feed [32] and a W-band CRLH TL leaky wave antenna [33–35] are discussed in this section. In particular, the main differences in terms of fabrication and functionality between the microwave frequency counterpart of these two examples are pointed out.

9.4.1 94 GHz CRLH TL Feed Network

To eliminate the insertion beam shift with frequency due to the insertion phase-shift of series-fed microstrip patch antenna arrays, a 94 GHz 0° phase-shift metamaterial TL was proposed in reference [32]. For this 0° TL, a 94 GHz CRLH TL was numerically implemented by using a scaled down unit cell similar to Figure 9.21 built on a substrate with dielectric constant of 11.7 and height of 0.10 mm. Four of these CRLH unit cells were cascaded to form a CRLH TL with a phase advance of +145.5° at 94 GHz. The 0° phase-shift metamaterial TL is implemented by connecting the CRLH TL with a 0.472 mm section of microstrip TL, which has a −145.5° phase delay at 94 GHz. The 0° phase-shift metamaterial TL was used to feed two slot-fed microstrip patch antennas spaced 1 mm away. This metamaterial TL can be tailored to provide a relative 0° phase shift and have its physical length smaller than a one wavelength RH TL.

A similar 0° metamaterial series feed network was experimentally demonstrated in reference [36] for 1.9 GHz. However, the phase advance line of reference [36] used lumped components to realize the required series capacitance and shunt inductance. Lumped components are not practical for mmWave applications due to their relatively large size and their low self-resonance frequency. As a result, distributed technology should be used to

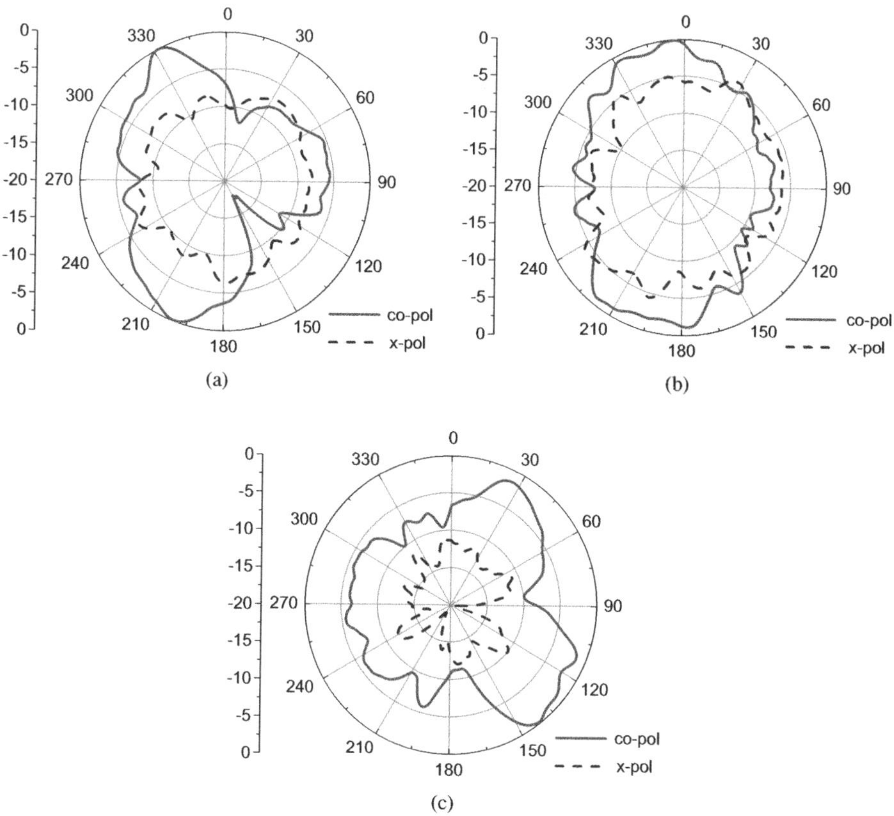

Figure 9.28: Experimental radiation patterns of Figure 9.27: (a) $f = 11.7$ GHz; (b) $f = 11.82$ GHz; (c) $f = 12.1$ GHz. From reference [31], reproduced by permission of © 2007 IEEE.

implement metamaterials at mmWave frequencies. In comparison to the infinite wavelength series feed network of Section 9.2.4, the metamaterial TLs of references [32] and [36] are operated in the slow-wave region to avoid any radiation leakage and based on a traveling wave.

9.4.2 W-band CRLH TL Leaky Wave Antenna

To address the requirement of beam scanning antennas for mmWave automotive radar systems used in adaptive cruise control (ACC) systems, Toyota Central Research has proposed and realized CRLH leaky wave antennas based on the TL approach towards LH metamaterials [33–35]. Evanescent-mode leaky wave antennas are not currently practical for automotive radar applications, since planar structures are preferred for their space and fabrication cost effectiveness.

Figure 9.29 shows the proposed W-band CRLH leaky wave antenna for continuous backward and forward beam scanning. The antenna is designed on substrate with permittivity

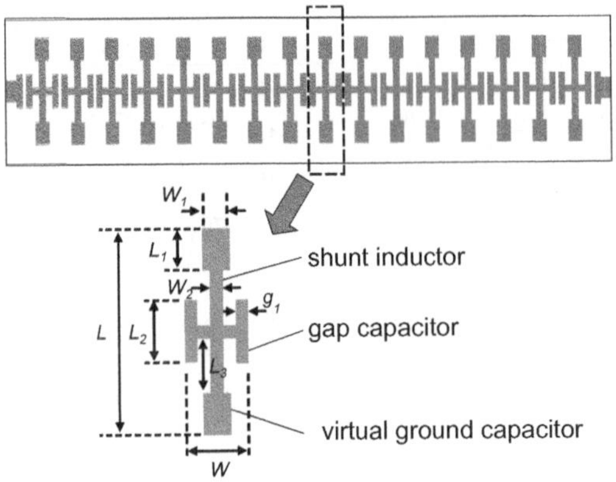

Figure 9.29: W-band CRLH leaky wave antenna consisting of sixteen unit cells with period 0.6 mm; $L = 1.62$ mm, $W = 0.5$ mm, $g_1 = 0.1$ mm, $L_1 = 0.3$ mm, $L_2 = 0.9$ mm, $L_3 =$ 0.3 mm, $W_1 = 0.2$ mm, and $W_2 = 0.1$ mm. From reference [33], reproduced by permission of © 2005 IEEE.

of 2.2 and thickness of 0.127 mm. Sixteen unit cells with period of 0.6 mm were cascaded to form the antenna. The gap capacitor between unit cells achieves the required LH capacitance and the shunt inductor with virtual ground capacitor [37] achieves the required LH inductance for realizing a CRLH TL. The authors chose to use a via-free CRLH TL unit cell for accurate mmWave circuit fabrication; interdigital capacitors were not used in order to reduce conductor losses from narrow traces. In addition, a symmetric CRLH TL unit cell was used to implement the CRLH leaky wave antenna in order to decrease cross-polarization in comparison to the asymmetric CRLH TL unit cell leaky wave antenna demonstrated in reference [30]. The dimensions of the unit cell were chosen to achieve a balanced CRLH TL with $\beta = 0$ at 77 GHz. By using the LH fast-wave region of the CRLH TL unit cell's dispersion curve, backward scanning is achieved and forward scanning is achieved by operating in the RH fast-wave region as discussed in Section 9.3.1.2. The structure was simulated in Ansoft HFSS and a backward scanning range of $-30°$ to $0°$ was achieved by varying the operational frequency from 75.1 GHz to 77.5 GHz, while a forward scanning range of $0°$ to $+30°$ was achieved from 77.5 GHz to 82.0 GHz. A maximum gain of 10.0 dBi was obtained throughout the scanning range [33].

Since automotive radar systems have a narrow frequency bandwidth, the leaky wave antenna of Figure 9.29 is not practical if the whole scanning range is required. As a result, the leaky wave antenna of Figure 9.29 was modified in reference [34] for mechanical tunability of the CRLH TL's dispersion curve. A movable Teflon substrate was placed at various heights above the CRLH leaky wave antenna. By varying the height of the Teflon substrate from 0.025 mm to 0.1 mm, an experimental scanning angle range of $-17°$ to $+24°$ at a fixed frequency of 76 GHz was achieved. A peak gain of 10.0 dBi was maintained throughout the scanning range. A mechanically tunable evanescent-mode leaky wave antenna was also experimentally demonstrated in reference [38] at 11 GHz.

Although the mechanically tunable CRLH leaky wave antenna solves the problem of frequency dependence, it is still not very practical because of the required mechanical parts.

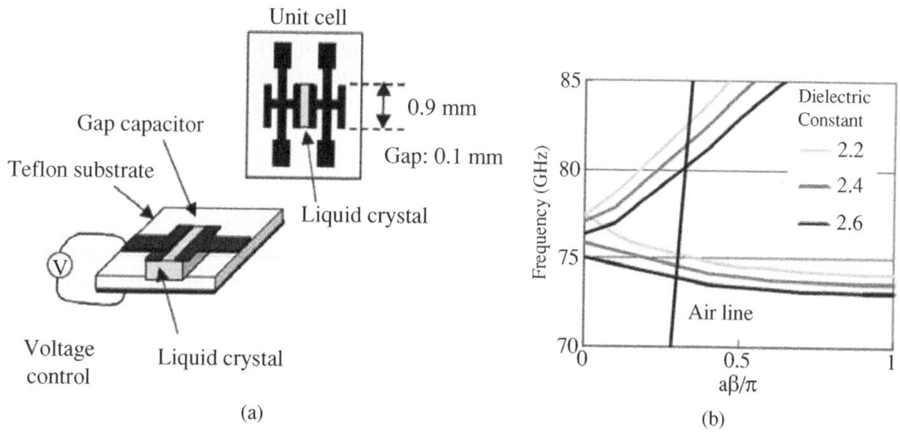

Figure 9.30: Proposed idea for electronically controlled 77 GHz CRLH leaky wave antenna: (a) unit cell with liquid crystal implementation; (b) numerical dispersion diagram for varied dielectric constants. From reference [35], reproduced by permission of © 2006 IEEE.

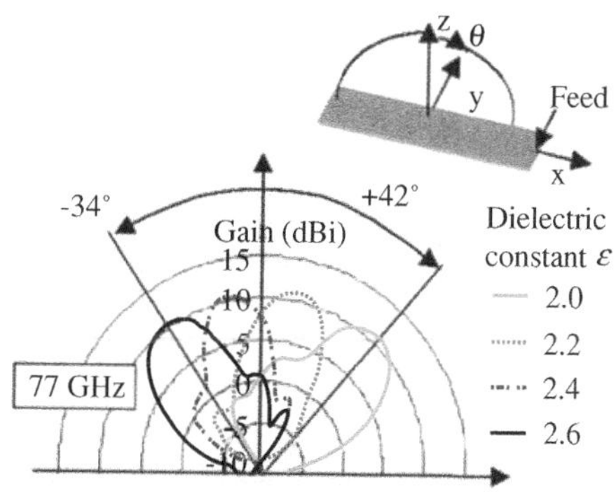

Figure 9.31: Numerical radiation patterns of CRLH leaky wave antenna when the dielectric constant of the liquid crystal changes. From reference [35], reproduced by permission of © 2006 IEEE.

An electronically scanned CRLH leaky wave antenna for automotive radar system implementation was proposed in reference [35]. The proposed antenna is based on embedding liquid crystal between the capacitive gaps of the CRLH TL unit cell as shown in Figure 9.30(a). The first electronically scanned CRLH leaky wave antenna was demonstrated in reference [39] at microwave frequencies by embedding varactor diodes into the unit cell; by varying the bias voltage to the diodes, the dispersion of the unit cell can be electronically controlled. However, varactor diodes are too lossy at mmWave frequencies [35]. Liquid crystal has low loss in the mmWave band and has a low voltage requirement with a dielectric constant that can be varied between 2.0 to 2.6 [35]. The variation of the unit cell's dispersion diagram as a function of the liquid crystal's dielectric constant is shown in Figure 9.30(b).

Numerical simulations of the proposed leaky wave antenna show that a scanning range of $-34°$ to $+42°$ at 77 GHz can be obtained by varying the liquid crystal's dielectric constant between 2.0 and 2.6. The numerical radiation patterns of the antenna are illustrated in Figure 9.31; a 10.0 dBi gain is obtained throughout the scan range.

9.5 Conclusions

Antenna applications of LH metamaterials realized by the TL and evanescent-mode approach have been presented. In particular, small backward wave resonant antennas, infinite wavelength antenna feeding networks, and leaky wave antennas exploiting the unique characteristics of LH metamaterials were discussed. These examples show the practical aspects of LH metamaterials for microwave and mmWave antenna applications.

References

[1] R. W. Ziolkowski and N. Engheta, 'Metamaterial special issue introduction", *IEEE Trans. Antennas Propag.* **51** (2003), pp. 2546–2549.

[2] C. Caloz and T. Itoh, *Electromagnetic Metamaterials: Transmission Line Theory and Microwave Applications* (New York: John Wiley & Sons Ltd/Inc., 2005).

[3] D. Schurig, J. J. Mock, B. J. Justice, S. A. Cummer, J. B. Pendry, A. F. Starr and D. R. Smith, 'Metamaterial electromagnetic cloak at microwave frequencies', *Science* **314** (2006), pp. 977–980.

[4] V. G. Veselago, 'The electrodynamics of substances with simultaneously negative values of ε and μ', *Soviet Physics Uspekhi* **10** (1968), pp. 509–514.

[5] J. B. Pendry, A. J. Holden, D. J. Robbins and W. J. Stewart, 'Magnetism from conductors and enhanced nonlinear phenomena', *IEEE Trans. Microwave Theory Tech.* **47** (1999), pp. 2075–2084.

[6] R. A. Shelby, D. R. Smith and S. Schultz, 'Experimental verification of a negative index of refraction', *Science* **292** (2001), pp. 77–79.

[7] C. Caloz, H. Okabe, T. Iwai and T. Itoh, 'Transmission line approach of left-handed (lh) materials', *IEEE AP-S/URSI Int. Symp. Dig.*, p. 39, June 2002.

[8] G. V. Eleftheriades, O. Siddiqui and A. K. Iyer, 'Transmission line models for negative refractive index media and associated implementations without excess resonators', *IEEE Microwave Wireless Compon. Lett.* **13** (2003), pp. 51–53.

[9] A. A. Oliner, 'A periodic-structure negative-refractive index medium without resonant elements', *IEEE AP-S/URSI Int. Symp. Dig.*, p. 41, June 2002.

[10] T. Ueda, A. Lai and T. Itoh, 'Negative refraction in a cut-off parallel-plate waveguide loaded with two-dimensional lattice of dielectric resonators', *Eur. Microwave Conf.*, September 2006.

[11] H. Iizuka and P. S. Hall, 'Orthogonally polarised dipole antenna using left handed transmission lines', *Eur. Microwave Conf.*, September 2006.

[12] N. I. Landy, H. Chen, J. F. O'Hara, J. M. O. Zide, A. C. Gossard, C. Highstrete, M. Lee, A. J. Taylor, R. D. Averitt and W. J. Padilla, 'Metamaterials for novel terahertz and millimeter wave devices', *Int. Symp. on Signals, Syst. and Electron.*, July 2007.

[13] N. Wongkasem, A. Akyurtlu, J. Li, A. Tibolt, Z. Kang and W. D. Goodhue, 'Novel broadband terahertz negative refractive index metamaterials: analysis and experiment', *Prog. in Electromagn. Res.*, vol. PIER 64, pp. 205–218, 2007.

[14] G. Dolling, M. Wegener, C. M. Soukoulis and S. Linden, 'Negative index metamaterial at 780 nm wavelength', *Opt. Lett.* **32** (2007), pp. 53–55.

[15] S. Ramo, J. R. Whinnery and T. Van Duzer, *Fields and Waves in Communication Electronics* 2nd edn (New York: John Wiley & Sons Ltd/Inc., 1984).

[16] D. Sievenpiper, L. Zhang, F. J. Boras, N. G. Alexopulos and E. Yablonovitch, 'High-impedance electromagnetic surfaces with a forbidden frequency band', *IEEE Trans. Microwave Theory Tech.* **47** (1999), pp. 2059–2074.

[17] A. Sanada, C. Caloz and T. Itoh, 'Zeroth order resonance in composite right/left-handed transmission line resonators', *Asia-Pacific Microwave Conference*, vol. 5, pp. 1588–1592, November 2003.

[18] A. Lai, K. Leong and T. Itoh, 'Dual-mode compact microstrip antenna based on fundamental backward wave', *Asia-Pacific Microwave Conf.*, December 2005.

[19] C.-J. Lee, K. M. K. H. Leong and T. Itoh, 'Design of resonant small antenna using composite right/left-handed transmission line', *IEEE AP-S/URSI Int. Symp.*, June 2005.

[20] C.-J. Lee, K. M. K. H. Leong and T. Itoh, 'Composite right/left-handed transmission line based compact resonant antennas for RF module integration', *IEEE Trans. Antennas Propag.* **54** (2006), pp. 2283–2291.

[21] C.-J. Lee, K. M. K. H. Leong and T. Itoh, 'Compact dual-band antenna using anisotropic metamaterial', *Eur. Microwave Conf.*, September 2006.

[22] A. Lai, K. M. K. H. Leong and T. Itoh, 'Infinite wavelength resonant antennas with monopolar radiation pattern based on periodic structures', *IEEE Trans. Antennas Propag.* **55** (2007), pp. 868–876.

[23] E. Chang, S. A. Long and W. F. Richards, 'An experimental investigation of electrically thick rectangular microstrip antennas', *IEEE AP-S/URSI Int. Symp.*, vol. 34, pp. 767–772, June 1986.

[24] A. Lai, K. M. K. H. Leong and T. Itoh, 'A novel n-port series divider using infinite wavelength phenomena', *IEEE-MTT Int. Symp.*, June 2005.

[25] A. Lai, K. M. K. H. Leong, and T. Itoh, 'Novel series divider for antenna arrays with arbitrary element spacing based on the composite right/left-handed transmission line', *Eur. Microwave Conf.*, October 2005.

[26] W. L. Stutzman and G. A. Thiele, *Antenna Theory and Design* 2nd edn (New York: John Wiley & Sons Ltd/Inc, 1998).

[27] N. V. Nguyen and C. Caloz, 'Tunable arbitrary n-port CRLH infinite wavelength series power divider', *Electronics Lett.* **43** (2007), pp. 1292–1293.

[28] T. Ueda, A. Lai, N. Michishita and T. Itoh, 'Leaky-wave radiation from left-handed transmission lines composed of a cut-off parallel-plate waveguide loaded with dielectric resonators', *Asia Pacific Microwave Conf.*, December 2006.

[29] D. M. Pozar, *Microwave Engineering* 3rd edn (New York: John Wiley & Sons Ltd/Inc., 2005).

[30] L. Liu, C. Caloz and T. Itoh, 'Dominant mode (dm) leaky-wave antenna with backfire-to-endfire scanning capability', *Electron. Lett.* **38** (2002), pp. 1414–1416.

[31] T. Ueda, N. Michishita and T. Itoh, 'Composite right/left handed metamaterial structures composed of dielectric resonators and parallel mesh plates', *IEEE-MTT Int. Symp.*, June 2007.

[32] Z. Qi, Z. Zhongxiang, X. Shanjia and D. Wenwu, 'Millimeter wave microstrip array design with crlh-tl as feeding line', *IEEE AP-S/URSI Int. Symp. Dig.*, vol. 3, pp. 3413–3416, June 2004.

[33] S. Matsuzawa, K. Sato, A. Sanada, H. Kubo and S. Aso, 'Left-handed leaky wave antenna for millimeter-wave applications', *IEEE Int. Workshop on Antenna Technol.: Small Antennas and Novel Metamat.*, pp. 183–186, March 2005.

[34] S. Matsuzawa, K. Sato, Y. Inoe and T. Nomura, 'Steerable composite right/left-handed leaky wave antenna for automotive radar applications', *Eur. Microwave Conf.*, September 2006.

[35] K. Sato, S. Matsuzawa, Y. Inoe and T. Nomura, 'Electronically scanned left-handed leaky wave antenna for millimeter-wave automotive applications', *IEEE Int. Workshop on Antenna Technol.: Small Antennas and Novel Metamat.*, pp. 420–423, March 2006.

[36] M. A. Antoniades and G. V. Eleftheriades, 'A broadband series power divider using zero-degree metamaterial phase-shifting lines', *IEEE Microwave Wireless Compon. Lett.* **15** (2005), pp. 808–810.

[37] A. Sanada, M. Kimura, I. Awai, H. Kubo, C. Caloz and T. Itoh, 'A planar zeroth-order resonator antenna using left-handed transmission line', *Eur. Microwave Conf.*, October 2004.

[38] N. Michishita, A. Lai, T. Ueda and T. Itoh, 'Tunable dielectric resonator-based left-handed leaky wave antenna', *Int. Symp. on Antennas and Propag.*, August 2007.

[39] S. Lim, C. Caloz and T. Itoh, 'Metamaterial-based electronically controlled transmission line structure as a novel leaky-wave antenna with tunable radiation angle and beamwidth', *IEEE Trans. Microwave Theory Tech.* **52** (2004), pp. 2678–2690.

10

EBG Materials and Antennas

Andrew R. Weily, Trevor S. Bird, Karu P. Esselle and Barry C. Sanders

10.1 Introduction

Millimeter-wave (mmWave) and sub-mmWave frequency bands bridge the region between microwaves and infrared/optics. For this reason many technologies used in mmWave antennas are borrowed from the fields of both microwaves and optics. One such technology from the field of optics, that has received considerable attention in recent years, is that of electromagnetic bandgap (EBG) materials .

EBG materials have properties that are suitable for a variety of mmWave antenna applications. Their unique properties have created new possibilities for controlling and manipulating the propagation of electromagnetic waves. EBG materials are created from dielectric and/or metallic structures that are periodic in one or more dimensions [1]. Within the EBG material there is a range of frequencies where propagating modes can be fully suppressed in one or more dimensions. This range of frequencies is known as the EBG. It is the electromagnetic analogue of solid state bandgaps of semiconductors that are due to the periodic potential of a crystal lattice of atoms or molecules. By breaking the periodicity in these EBG materials over one or more periods (known as creating a defect), many useful components can be created such as antennas, waveguides, resonators, 90° waveguide bends, power dividers and filters. Further novel properties of some types of EBG material are the ability to produce an effective negative index of refraction, or to produce self-collimation.

EBG materials provide significant advantages for suppressing and directing radiation when used in antennas. Many novel antennas based on EBG technology have been described in the literature and are already finding useful applications. The operating mechanism of these devices can be grouped broadly into the following five classes: (i) EBG substrates and high impedance ground planes that reduce surface waves and increase radiation

Advanced Millimeter-wave Technologies: Antennas, Packaging and Circuits
Duixian Liu, Ulrich Pfeiffer, Janusz Grzyb and Brian Gaucher © 2009 John Wiley & Sons, Ltd

efficiency, or lower the profile of the design; (ii) reflector antennas made from all-dielectric EBG materials or high impedance ground planes; (iii) resonator antennas that create high directivity over a narrow bandwidth; (iv) sources embedded in metamaterials that have high directivity due to a dielectric constant less than one, or the limited angular propagation allowed within the material; and (v) directive EBG horn antennas and arrays that allow efficient coupling from an EBG waveguide to free space.

Section 10.2 of this chapter gives a brief overview of EBG materials, components and high impedance ground planes. Section 10.3 describes printed antennas on substrates made from EBG materials. Section 10.4 reviews high-gain EBG antennas composed of one-dimensional, two-dimensional and three-dimensional EBG materials as well as partially reflecting surface (PRS) and metamaterial based designs. Section 10.5 describes EBG horn antennas, waveguides and arrays based on the woodpile three-dimensional EBG material. Section 10.6 describes other miscellaneous EBG antennas and components. Many examples of EBG antennas that operate at mmWave frequencies have been included. However, since EBG antennas are a fairly recent innovation, devices that operate at microwave frequencies and show excellent potential to be scaled to mmWave frequencies are also included. In some cases modifications are discussed that may be required when scaling devices from microwave to mmWave frequencies. A good review of EBG devices operating at microwave frequencies can be found in reference [2].

10.2 EBG Materials and Components

10.2.1 One-dimensional, Two-dimensional and Three-dimensional EBG Materials

This section gives a brief overview of EBG materials. Although this topic is a relatively new one, there are several books dedicated entirely to it, and readers requiring greater depth should refer to these texts [1,3–7]. Some useful introductory articles are also available [8,9]. The study of electromagnetic propagation in periodic dielectric materials dates back to 1887 when Lord Rayleigh (John William Strutt) investigated the reflective properties of crystalline minerals [10]. These minerals were in fact one-dimensional periodic structures, and exhibited high reflection to visible light when viewed from certain angles due to the periodic nature of the crystal. One-dimensional and two-dimensional periodic structures have been used in microwave and antenna engineering for many years and are used in filters, travelling-wave tubes, frequency selective and corrugated surfaces, and array antennas. The stopband in microwave devices (such as travelling-wave tubes made from waveguides with periodic loading elements) was limited to one-dimensional and higher dimensional stopbands were not sought.

Recent research in the field of EBG materials was inspired by the seminal papers of Yablonovitch [11] and John [12], who both predicted the existence of an EBG over a three-dimensional space. Each researcher proposed the concept for a different purpose: Yablonovitch was seeking to inhibit spontaneous emission to improve the efficiency of semiconductor lasers, while John was seeking a dielectric lattice that would serve as a photonic analogue to electron localization (i.e. photon localization) that occurs in amorphous semiconductors. It is the concept of a two-dimensional or three-dimensional periodic structure made from materials with a substantial difference in dielectric constant that sets the work of Yablonovitch and John apart from earlier work by Lord Rayleigh and that in microwave and antenna engineering. However, the important prior work in this field, which is often classified as belonging under the umbrella of EBG materials, is important prior knowledge.

Unfortunately, due to the different disciplines all working in the field of EBG materials, various terminology has arisen to describe identical concepts. In general, literature from the physics and optics community use the names photonic crystals and photonic bandgap (PBG). Literature from the microwave and antenna engineering community tends to use the nomenclature EBG materials and EBG, to emphasize that the concepts are not limited to optical wavelengths where photonic behaviour is evident but works equally well over the entire electromagnetic spectrum. In this chapter the latter terminology is used.

As defined previously, EBG materials comprise periodic dielectric and/or metallic media. For frequencies in the bandgap, the periodicity of the material is proportional to the wavelength scale (typically a half wavelength). EBG materials may be classified as one-dimensional, two-dimensional or three-dimensional, according to the periodicity of the lattice as shown in Figure 10.1. The ratio of the dielectric constants (or relative permittivity) of the two different materials is known as the 'dielectric contrast', while the spatial period of the lattice (or repeat distance) is known as the 'lattice constant' and denoted by the symbol a. For the classical one-dimensional EBG material (also known as the Bragg mirror or Bragg multilayer stack) shown in Figure 10.1(a), the thickness of the dielectric sheets should be one quarter of the guided wavelength for high reflection. A wave normally incident on the Bragg stack will undergo reflections from the multiple interfaces, and for frequencies within the bandgap of the material these reflected waves will add together in phase to create a mirror, or highly reflective surface. In Bragg stacks, a dielectric contrast of greater than one is enough to guarantee the existence of a bandgap in a single direction. For small dielectric contrasts the EBG will be very narrow, while a large dielectric contrast will give a wide EBG. Bragg stacks have been used in optics for many years to create dielectric mirrors, anti-reflective coatings and Fabry–Perot filters [13]. For two-dimensional and three-dimensional materials, such as those shown in Figures 10.1(b) and (c), simply creating a periodic dielectric structure does not guarantee the presence of a bandgap. As the dimensions of the periodicity increase, so does the complexity of the material and its computational analysis.

A periodic structure can be created by translating a 'unit cell' in one, two or three dimensions. This translation length is the lattice constant, and the translation vector is called the 'primitive lattice vector'. For the Bragg stack the primitive lattice vector $\mathbf{a} = a\hat{\mathbf{x}}$ and the complete set of lattice vectors $\mathbf{R} = n\mathbf{a}$, where n is an integer. Using this terminology, which is borrowed from solid state physics, the periodic dielectric permittivity function for position vector $\mathbf{r}$ may be defined as $\epsilon(\mathbf{r}) = \epsilon(\mathbf{r} + \mathbf{R})$. To analyse the electromagnetic propagation through an infinite periodic structure, it is only necessary to analyse the field in the unit cell, since all other fields in the lattice will be related to those of the unit cell. Unit cells for a Bragg stack, two-dimensional array of dielectric rods, and the woodpile EBG material are presented in Figure 10.2. The various periodic structures in Figure 10.1 are each created from the unit cells shown in Figure 10.2 by translating the unit cells in one, two and three dimensions using the appropriate lattice vectors $\mathbf{R}$.

The band diagram is of fundamental importance to the field of EBG materials, as it defines the region over which propagation is forbidden and hence a material's electromagnetic properties. It is based on the solutions to Maxwell's equations cast in eigenmode form as

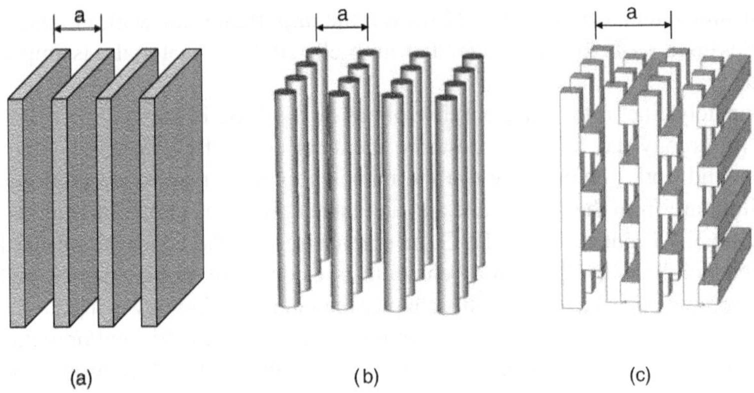

Figure 10.1: Configuration of three different types of EBG material composed of periodic dielectric materials in: (a) one dimension; (b) two dimensions; and (c) three dimensions.

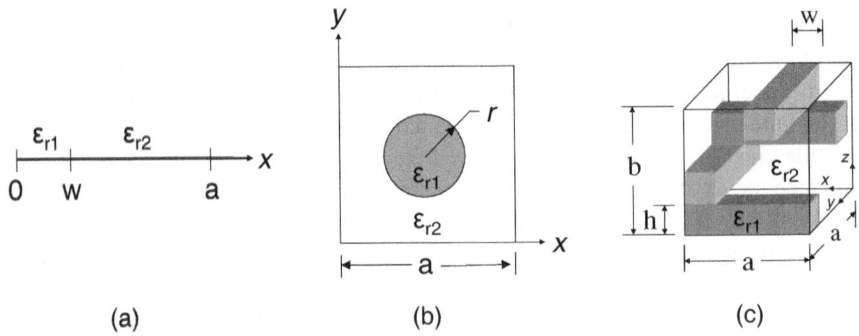

Figure 10.2: Unit cell for the EBG materials periodic in: (a) one dimension; (b) two dimensions; and (c) three dimensions.

follows [1]

$$\nabla \times \frac{1}{\epsilon(\mathbf{r})} \nabla \times \mathbf{H}(\mathbf{r}) = \left(\frac{\omega}{c}\right)^2 \mathbf{H}(\mathbf{r}) \tag{10.1}$$

From Floquet's theorem (also known as Bloch's theorem) solutions to the fields in a periodic structure can be expressed as:

$$\mathbf{H}(\mathbf{r}) = e^{j\mathbf{k}\cdot\mathbf{r}} \mathbf{H}_n(\mathbf{r}) \tag{10.2}$$

where $\mathbf{H}_n(\mathbf{r})$ is itself a periodic function that satisfies

$$\mathbf{H}_n(r) = \mathbf{H}_n(\mathbf{r} + \mathbf{R}) \tag{10.3}$$

The solution of Equation (10.1) using Floquet's theorem yields discrete eigenvalues $\omega_n(\mathbf{k})$ for each wavevector, where the subscript n is known as the band index. These discrete eigenvalues form continuous functions with variation of wavevector **k**: the continuous functions are known as bands. To obtain the band diagram of the EBG material $\omega_n(\mathbf{k})$ is plotted versus **k**, in a similar manner to plotting the dispersion relation of a waveguide. The variation of $\omega_n(\mathbf{k})$ versus **k** is known as the band structure and the electromagnetic properties of the material can be inferred from this plot.

Another property of the Floquet modes is that they are periodic functions of wavevector **k**. Considering the simplest case scenario of the Bragg stack periodic in the x-direction, then $\omega(k_x) = \omega(k_x + 2n\pi/a)$, where n is an integer. This result is important because it restricts the number of wavevectors that need to be analysed to completely characterize the periodic structure. The restricted region of wavevectors is known as the first Brillouin zone, and for one-dimensional periodic structures (with lattice constant a) it is defined as $-\pi/a \leq k_x \leq \pi/a$. All solutions obtained for wavevectors outside of the first Brillouin zone will be equal to a solution of a wavevector within the zone. To generalize this result to two dimensions and three dimensions a reciprocal lattice vector, **G**, is defined as follows

$$\mathbf{G} \cdot \mathbf{R} = 2n\pi \tag{10.4}$$

where n is an integer.

Then, for eigenvalues in two dimensions and three dimensions

$$\omega_n(\mathbf{k}) = \omega_n(\mathbf{k} + \mathbf{G}) \tag{10.5}$$

The first Brillouin zone of a one-dimensional EBG material is a line. For a two-dimensional EBG material it is a surface, and for a three-dimensional EBG material the first Brillouin zone is a volume.

Figure 10.3(a) shows the first Brillouin zone for a two-dimensional EBG material with a square lattice. As previously explained, this zone removes the redundancy in wavevectors due to the periodicity of the Floquet modes. However, the first Brillouin zone also contains redundant wavevectors due to symmetries in the periodic structure, such as mirror planes. To remove these redundancies the irreducible Brillouin zone is defined. In Figure 10.3(a) this is the triangle with corners labelled Γ, X, M and these corners correspond to regions of high symmetry. The irreducible Brillouin zone contains every wavenumber required completely to characterise the EBG material, since it removes redundancy in both periodicity and symmetry.

The analysis of two-dimensional EBG materials, like most two-dimensional electromagnetic problems, can be separated into two different polarizations: transverse electric (TE) and transverse magnetic (TM). TE polarization has three components (E_x, E_y and H_z) with the electric field parallel to the plane of periodicity (xy-plane) and the magnetic field perpendicular (z-directed). TM polarization has three components (H_x, H_y and E_z) with the magnetic field parallel to the plane of periodicity (xy-plane) and the electric field perpendicular (z-directed). For a complete bandgap to occur in a two-dimensional EBG material there must be no propagation for both TE and TM polarizations. A band diagram for a square lattice of dielectric rods in air is shown in Figure 10.3(b). The dielectric constant of the rods is $\epsilon_r = 9.8$ (corresponding to alumina) and the radius of the rods normalized to the lattice constant is $r/a = 0.237$. The boundary of the irreducible Brillouin zone of Figure 10.3(a) (marked by the symmetry points Γ, X and M) forms the horizontal axis of the band diagram. The bandgap for this material is shaded in the figure and extends from

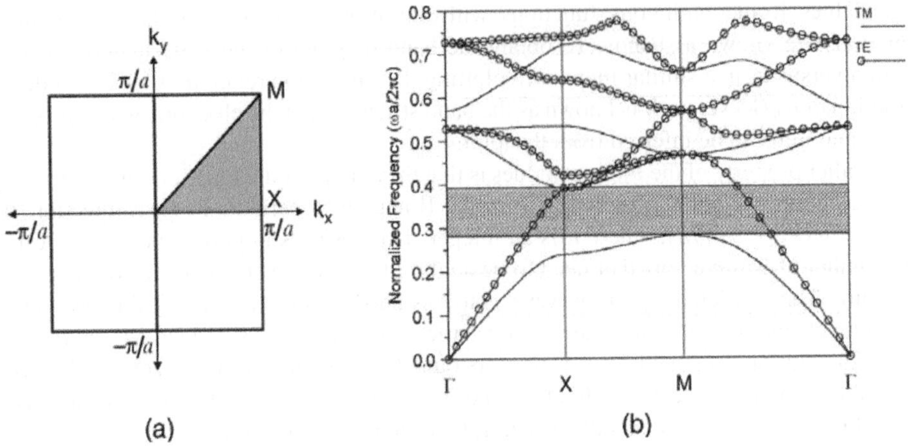

Figure 10.3: (a) First Brillouin zone for the two-dimensional square lattice. The irreducible Brillouin zone is the shaded triangle with corners labelled Γ, X, M. (b) Band diagram for the square lattice of cylindrical dielectric rods in air.

$\omega a/2\pi c = 0.283$ to 0.393 with a gap width to mid-gap ratio of 32.5%. This bandgap exists for TM polarizations only and so is incomplete. Four bands have been calculated for each polarization (TE and TM) and only the first TM bandgap has been shaded.

The band diagram of Figure 10.3(b) has been calculated using RSOFT's BandSOLVE commercial software, which implements the plane wave expansion method [14]. There are also various freely available software packages from the Internet that can be used to calculate band diagrams. One software that is particularly useful for Unix based operating systems is the MIT Photonic Bands package [15]. There are extensive online tutorials for the package and the technique used to implement it is also fully described in the literature [16]. Band diagrams can also be computed using the more traditional high-frequency electromagnetic software packages based on the finite element method and finite integration technique, such as HFSS [17] and CST Microwave Studio [18]. In these packages band diagrams may be obtained by combining eigenmode solvers with periodic boundary conditions [19], but the calculation of band diagrams tends to be less efficient than those based on the plane wave expansion method.

Although materials with a complete three-dimensional bandgap were predicted in 1987, it was not until 1990 that theorists were able to identify realistic material candidates for realizing such band gaps. Ho, Chan and Soukoulis showed using the plane-wave expansion method that dielectric spheres arranged to form a diamond lattice possess a complete bandgap, and that the minimum dielectric contrast for a complete bandgap to exist is 4:1 [20]. They showed that a complete bandgap existed for dielectric spheres in an air background and also for air spheres in a dielectric background. Shortly after this discovery, the first measurements verifying a three-dimensional bandgap were reported for a dielectric structure at microwave frequencies. This structure, now known universally as 'Yablonovite' in honour of its inventor Eli Yablonovitch, is constructed by drilling three sets of holes into the dielectric at angles 35.26° from the vertical axis [21]. The fabrication method for Yablonovite is much

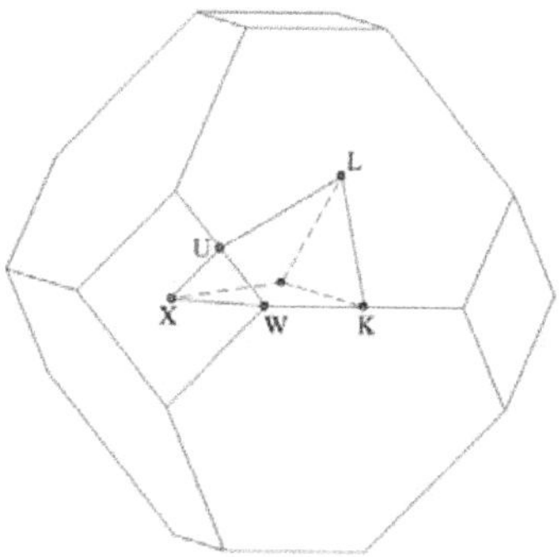

(a)

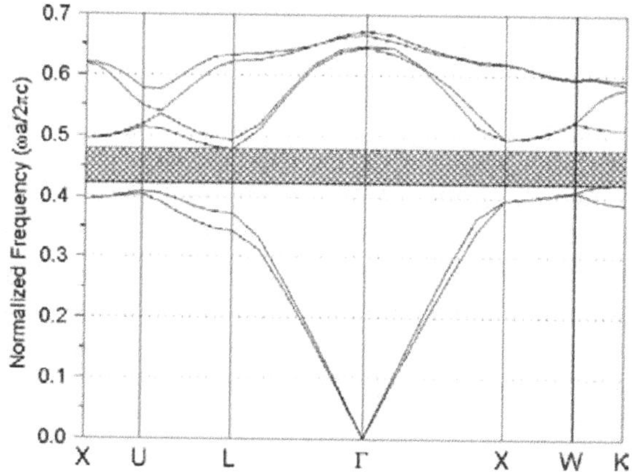

(b)

Figure 10.4: (a) First Brillouin zone for the fct lattice. (b) Band diagram for the woodpile EBG material. The shaded region corresponds to the bandgap of the material, since there are no modes present. (From reference [91], reproduced by permission of © 2005 IEEE.)

simpler than that for a diamond lattice of spheres. Experimental results showed that the ratio of hole diameter to lattice constant $d/a = 0.469$ gave the widest bandgap, and using a material with a dielectric constant of 13 gave a gap width to mid-gap ratio of ~20%.

Another three-dimensional EBG material known as the woodpile or layer-by-layer period structure is depicted in Figure 10.1(c), and its unit cell is given in Figure 10.2(c). It is periodic in all three dimensions and contains a complete bandgap [22, 23]. Referring to Figure 10.2(c), this material has a lattice constant of a, rod width w, rod height h, and unit cell height of b. Consecutive layers are orthogonal to each other, parallel rods spaced two layers apart are offset by $a/2$, and there is a four-layer stacking sequence. The lattice structure of the woodpile is face-centred tetragonal (fct). If $b = \sqrt{2}a$, then the lattice structure becomes face centred cubic (fcc). Figure 10.4(a) shows the first Brillouin zone for the fct lattice. The irreducible Brillouin zone is the polyhedron with corners labelled Γ, X, W, K, L, U. As in the case for two-dimensional materials, the irreducible Brillouin zone contains every wavenumber required to completely characterize the EBG material. The irreducible Brillouin zone boundary forms the horizontal axis of the band diagram in the Figure 10.4(b), and the symmetry points (Γ, X, W, K, L, U) are marked to show their relative positions. The band diagram was calculated once again using RSOFT's BandSOLVE software. Six bands are plotted for an alumina woodpile ($\epsilon_r = 8.4$, $w = h = 0.286a$, and $b = 1.144a$) in Figure 10.4(b) that shows the bandgap extending in normalized frequency from $\omega a/2\pi c = 0.423$ to 0.479, with a gap width to mid-gap ratio of 12.4%.

The woodpile EBG material has been implemented at mmWave frequencies using several different techniques and materials. One of the earliest methods stacks layers of high resistivity (110) silicon wafers on top of one another in a layer-by-layer fashion that repeats every four layers. Each silicon wafer is micromachined using an etchant of aqueous potassium hydroxide (KOH) to create an array of parallel rods. The anisotropic etching properties of the KOH enable fast micromachining of the (110) plane of the silicon wafer, and leave the {111} plane virtually unaffected. This method produced a woodpile with $w = 340$ μm, $h = 390$ μm, and $a = 1275$ μm that had a bandgap centred around 94 GHz [24]. By using a double-etch technique (i.e. etching different patterns on the two sides of the wafer) it is possible to create two layers of parallel rods in a single silicon wafer and hence double the operating frequency of the EBG material, compared to the single-etch process. The angle between the rods etched on the top and bottom surfaces is 70.5° to conform to the different crystal planes of the silicon. By over-etching the silicon it is also possible to provide a tuning mechanism for the bandgap. This technique produces woodpiles with bandgap centres that range from 340 GHz to 375 GHz [25]. Another method of manufacturing woodpiles that is independent of the material used is with a dicing saw typically used for dicing semiconductor wafers into smaller circuits. The dicing saw cuts trenches in the top and bottom of the wafer material to form two layers of parallel rods. The array of trenches on the top is at 90° to those on the bottom. To obtain a four-layer stacking sequence, alternate wafers of rods need to be offset by $0.5a$ from those in the wafer below. Mechanically machined woodpiles have been implemented in: (1) high resistivity silicon with bandgaps centred at 265 GHz [26] and 500 GHz [27]; (2) ceramic Zr/Sn titanate ($\epsilon_r \approx 37$) with a bandgap centred at 500 GHz [28]. Three-dimensional EBG materials have also been fabricated from the ceramic Zr/Sn titanate by a combination of dicing to create channels in the dielectric and laser ablation to create holes for operation at 500 GHz [28]. Other techniques that have been used to fabricate woodpiles are: rapid prototyping of alumina [29, 30]; and deep reactive ion etching (RIE) of high resistivity silicon [31].

10.2.2 EBG Waveguides and Components

When part of the periodic structure is removed (or additional dielectric material added) to create defects, electromagnetic waves can be guided in the defect as the surrounding EBG material form boundary conditions on a closed structure. The guided wavelength depends on the gap width and periodicity of the EBG. Three-dimensional EBG materials such as the woodpile have the advantage of a complete EBG i.e. that propagation may be suppressed in all three dimensions. In such a material it is possible to create a waveguide in which the electromagnetic confinement depends solely on the EBG of the material for frequencies in the bandgap, as opposed to index guiding in dielectric waveguides and optical fibres, or the use of metallic walls such as those used in rectangular and circular metallic waveguides. For example, if a rod is removed from the woodpile structure of Figure 10.1(c), waves can be guided for frequencies in the bandgap of the material [32, 33]. By combining such structures, a range of useful mmWave components are formed. Examples of a waveguide and 90° waveguide bend are shown in Figure 10.5, while a power divider and phase-shifter are presented in Figure 10.6. In the power divider of Figure 10.6(a), the signal at the input port P1 is split into two equal signals at the output ports P2 and P3. Woodpile waveguides, 90° bends and power dividers have been implemented experimentally at microwave frequencies by two different groups [34, 35]. Best results are obtained when an efficient rectangular-waveguide to EBG-waveguide transition is used, such as that described in reference [35].

The dispersion relation for EBG waveguides is usually calculated using a 'supercell' analysis, where the unit cell is expanded by several lattice constants in the dimensions transverse to the direction of propagation. A supercell for the woodpile EBG material is shown in Figure 10.7(a) and has been expanded by seven lattice constants in the x- and

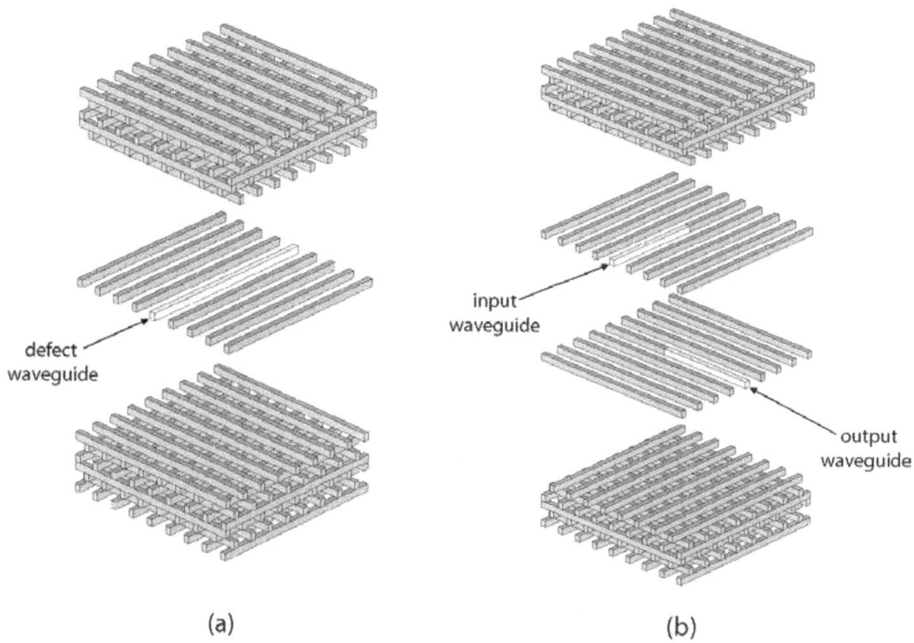

Figure 10.5: Exploded view of woodpile EBG components: (a) waveguide; and (b) 90° waveguide bend. The grey rods represent a high dielectric constant material such as alumina or silicon and the white rods represent air.

y-directions, and one lattice constant in the z-direction. It can be seen in the figure that a single dielectric rod has been removed from the centre of the supercell structure to create the waveguide. In the analysis of this supercell, periodic boundary conditions are used for all six boundaries.

Figure 10.7(b) shows the dispersion relation for the woodpile waveguide calculated using the finite difference time domain (FDTD) method with periodic boundary conditions [36]. For this calculation the sine-cosine method was used to implement the periodic boundary conditions in FDTD [37]. The horizontal axis is the normalized wave vector, the left vertical axis plots frequency (for $a = 10.6$ mm) and the right vertical axis plots normalized frequency. The woodpile material is made from alumina ($\epsilon_r = 9.3$, $\tan\delta = 0.0002$) with $w = 0.25a$, $h = 0.3a$ and $b = 1.2a$ that has an EBG that extends from $a/\lambda = 0.414$ to 0.477. This figure shows that the single mode non-dispersive operating range is from $a/\lambda = 0.428$ to 0.456 (i.e. about 6.4%). Above $a/\lambda = 0.456$ there are two modes present (hence it is multi-mode), while below $a/\lambda = 0.428$ the slope of the dispersion relation changes rapidly, leading to variations in group velocity and high levels of dispersion in the waveguide. A significant increase in the single-mode operating range of the waveguide can be obtained through the use of a different type of defect that adds a strip of dielectric directly above and below the missing rod. Using this technique in a woodpile made from GaAs material, it is possible to achieve a single-mode bandwidth of 15.1% [38].

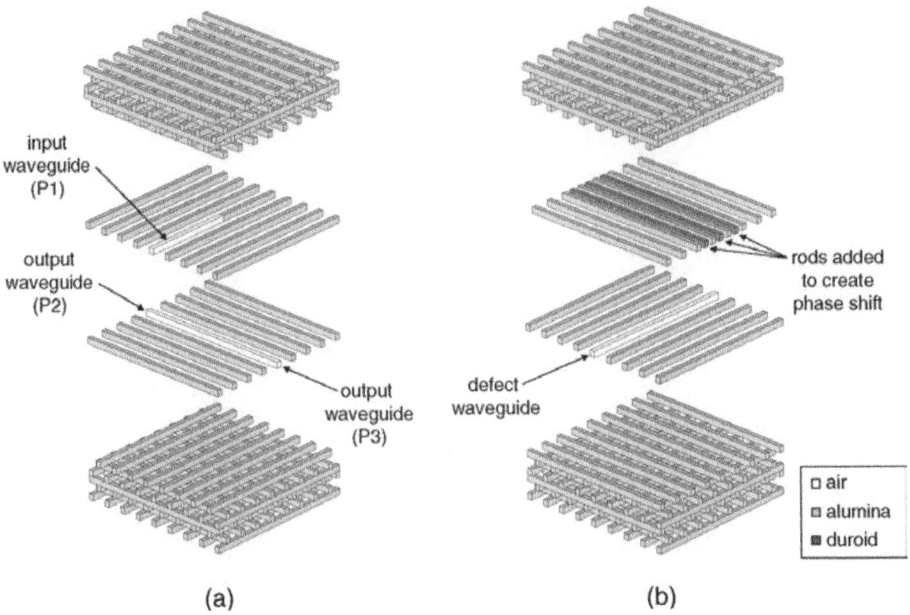

Figure 10.6: Exploded view of woodpile EBG components: (a) power divider; and (b) phase shifter. The light grey rods represent a high dielectric constant material such as alumina or silicon, the white rods represent air and the dark grey rods represent a low dielectric constant material (such as duroid or Teflon).

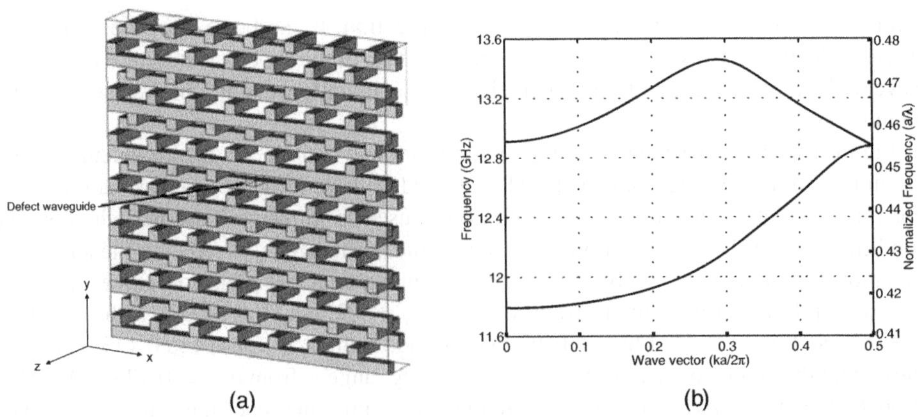

Figure 10.7: Woodpile EBG waveguide: (a) $7 \times 7 \times 1$ supercell used for analyzing waveguide dispersion relation; and (b) dispersion relation. (From reference [105], reproduced by permission of © 2006 IEEE.)

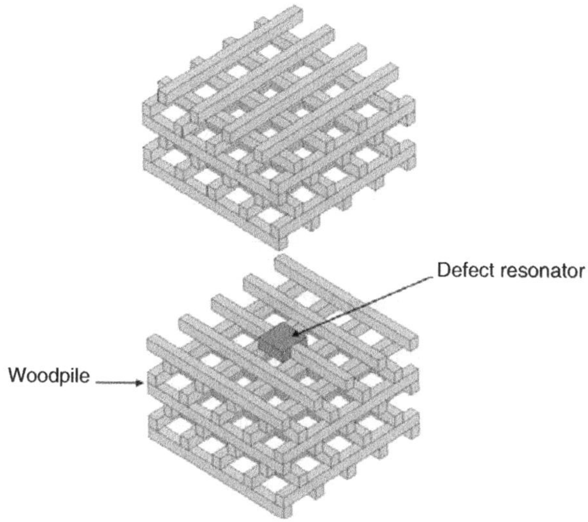

Figure 10.8: Exploded view of a woodpile EBG defect resonator, created by adding dielectric material to the woodpile.

If a portion of a dielectric rod is removed from the woodpile EBG material, then a defect resonator or point defect will be created. For frequencies in the bandgap, this structure is equivalent to a resonant cavity where the electromagnetic confinement is created by the EBG material. Defect resonators can also be created by adding dielectric material to a uniform lattice as shown in Figure 10.8. Defect resonators have been demonstrated at microwaves by adding and removing dielectric material from an alumina woodpile lattice [39], and at mmWave frequencies by adding dielectric to a silicon woodpile [40]. Defect resonators can also be analysed using the supercell method, and for the case of the woodpile the unit cell is expanded in all three dimensions (instead of only two dimensions for waveguide analysis). A study of the mode behaviour of a woodpile defect resonator has shown that the aspect ratio of the dielectric inclusion determines the number of modes present and the spacing of the modes [41]. The number of layers of dielectric rods above and below the defect also strongly influences the quality factor of the defect resonator.

The next logical extension of a defect resonator is an EBG filter. By coupling defect resonators together, a multipole filter can be created. An example is a theoretical channel drop filter that uses a two-dimensional EBG material of dielectric rods in air [42]. A two-pole Tchebyshev filter has been realized at 45 GHz using silicon micromachined EBG resonators [43]. The filter used a two-dimensional rectangular lattice of air holes in silicon for horizontal confinement of the EBG resonators, and metal plates for vertical confinement. The filter achieved a 1.4% equi-ripple bandwidth with an insertion loss of 2.3 dB. A channel drop filter has been implemented at microwave frequencies using an air defect resonator in the woodpile EBG material [44].

A phase shifter based on the woodpile EBG waveguide is shown in Figure 10.6(b). By sequentially adding low dielectric constant rods into the voids of the woodpile directly above the defect waveguide, a phase delay is created without significantly deteriorating the transmission performance. This component has been demonstrated at microwave frequencies [45]. At 12.4 GHz a woodpile made from alumina with a single RT duroid rod placed above the defect waveguide gave a 10° phase delay. Larger phase shifts were obtained by adding more duroid rods. Results were demonstrated for a mechanical phase shifter, but electrical control could be achieved at mmWave frequencies through the use of a woodpile defect waveguide partially infiltrated with a nematic liquid crystal, where the dielectric constant of the liquid crystal is voltage controlled.

10.2.3 High Impedance Ground Planes

High impedance surfaces formed by closely spaced quarter-wavelength deep corrugations find many applications, such as in the walls of corrugated horn antennas. The concept of a high impedance ground plane based on a planar metallic structure was first proposed in 1999 [46,47], and has the advantages of compactness compared to conventional corrugated surfaces as well as compatibility with printed circuit manufacturing techniques. An example of a high-impedance ground plane based on a square lattice is shown in Figure 10.9(a). It is often described as an array of metallic 'mushrooms' or 'thumbtacks' electrically attached to a metallic ground plane using vias. Both the height of the high-impedance ground plane and the lattice constant between metallic elements is much less than a quarter wavelength, so the surface can be represented by an effective medium. The surface impedance of the effective medium is modelled as a parallel LC circuit, where C represents the capacitance between the edges of the metallic patches, and L represents the inductance of the vias and ground plane between vias. The surface impedance of the effective medium may then be described by [47]

$$Z_s = \frac{j\omega L}{1 - \omega^2 LC} \tag{10.6}$$

with a resonant frequency of

$$\omega_0 = \frac{1}{\sqrt{LC}} \tag{10.7}$$

when $\omega < \omega_0$ TM surface waves are supported, while for $\omega > \omega_0$ TE surface waves are supported. However, when $\omega \approx \omega_0$ the surface impedance is significantly greater than the intrinsic impedance (377 Ω), and no bound surface waves (TE or TM) can propagate. Hence there is a bandgap for bound surface waves at resonance. The existence of the bandgap can be confirmed from a band diagram in a similar fashion to the volumetric EBG materials described previously. A band diagram for the material is shown in Figure 10.9(b). It is calculated from a unit cell using the finite element method. The first band is a TM wave, the second band is TE, and the bandgap is shaded in the figure. The thin dotted lines represent the light line ($f = ck/2\pi$).

The lower band edge for bound surface waves is defined by the top of the first band, while the upper band edge is defined by the crossing of the second band and the light line. The TE waves of the second band that are on the left of the light line in the ΓX panel (and to the right of the light line in the MΓ panel) are leaky waves, and so are not bound to the surface of the high-impedance ground plane.

Another interesting property of high-impedance ground planes is the reflection phase. At the resonant frequency of the surface, the tangential magnetic fields will be close to zero, thus

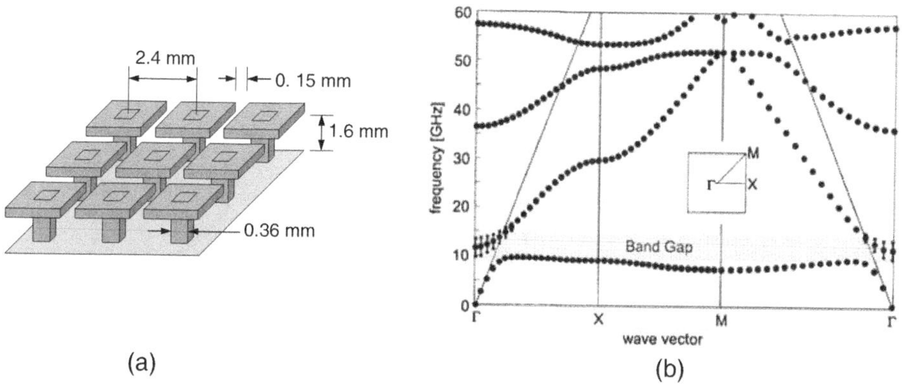

Figure 10.9: High impedance ground plane based on a square lattice: (a) configuration with dielectric substrate removed; (b) surface wave band diagram. (From reference [47], reproduced by permission of © 1999 IEEE.)

forming an artificial magnetic conductor (AMC) as opposed to an electric conductor where tangential electric fields are zero. The reflection phase for normal incident waves will be zero, in contrast to a perfect metallic conductor which has a reflection phase of π. This concept is illustrated for a high impedance ground plane with a triangular lattice in Figure 10.10 [47]. The bandwidth of an AMC is usually defined as the region where the reflection phase is between $\pi/2$ and $-\pi/2$. For many high-impedance surfaces, this region also corresponds to the surface wave bandgap, as shown in Figure 10.10(b). AMCs have also been realized using printed structures such as the uniplanar compact EBG (UC-EBG), which does not use vias [48]. Polarization dependent AMCs and compact slotted designs have also been presented [49].

The ability of high-impedance ground planes to suppress surface waves and to produce a 0° reflection phase makes them very useful for antenna design. The suppression of surface waves allows for: (1) an improvement in efficiency in antennas that require a metallic ground plane; (2) improved radiation patterns; and (3) improved front-to-back ratio since there is less radiation from the edge of the ground plane. A 0° reflection phase allows for design of a lower profile antenna than would be possible with a metallic ground plane. A horizontal wire radiator can be placed almost directly on top of an AMC and still radiate efficiently, since the currents on the surface add in phase with the currents of the wire radiator. The reduction of the antenna profile is important at microwave frequencies, but at mmWave frequencies antennas tend to be inherently low-profile due to their much shorter wavelengths. Therefore for mmWave antenna applications, the surface wave suppression is the more attractive feature of the high-impedance ground plane.

To illustrate the surface wave suppression, an example is provided from the literature [47]. Radiation patterns for a vertical monopole antennawith a metallic ground plane are compared with the same monopole on a high impedance ground plane. The operating frequency is 35 GHz, the monopole is 3 mm long, and the ground plane is 50 mm × 50 mm. The high-impedance ground plane has a triangular lattice, with a substrate height of 1 mm, a lattice constant of 1.4 mm, which gives a surface wave bandgap at 35 GHz. A schematic drawing

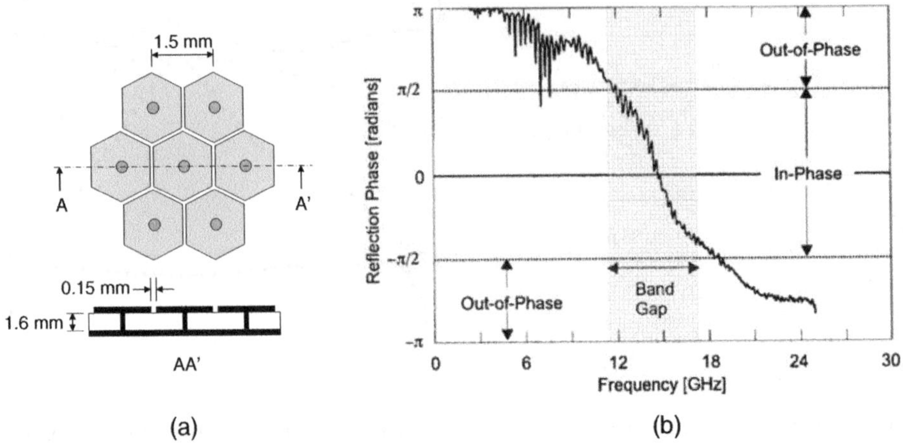

Figure 10.10: High-impedance ground plane based on a triangular lattice: (a) plan and cross-sectional views; (b) measured reflection phase for normal incidence. (From reference [47], reproduced by permission of © 1999 IEEE.)

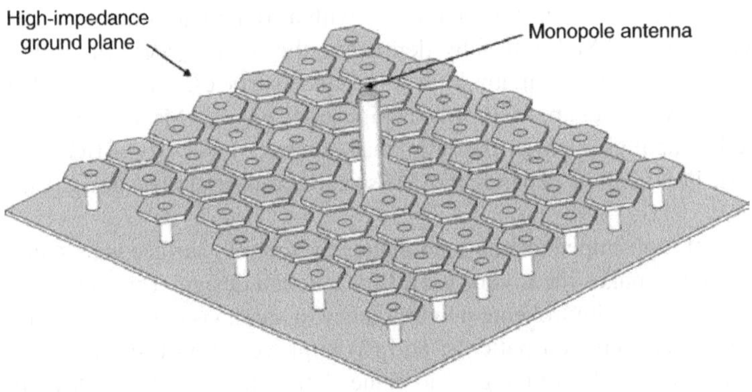

Figure 10.11: Schematic drawing of a monopole antenna on a triangular lattice high-impedance ground plane (with substrate removed to show the vias).

of the monopole antenna on the high impedance surface is shown in Figure 10.11. The measured radiation pattern for the monopole on a metallic ground plane is presented in Figure 10.12(a) and shows significant ripple, rear radiation, and radiation in the azimuth direction (90° and 270°) due to the currents induced on the metal. In contrast, the patterns for the monopole on the high impedance ground plane given in Figure 10.12(b) are smooth, and show reduced radiation towards the rear and azimuth directions as well as useful power in the forward direction.

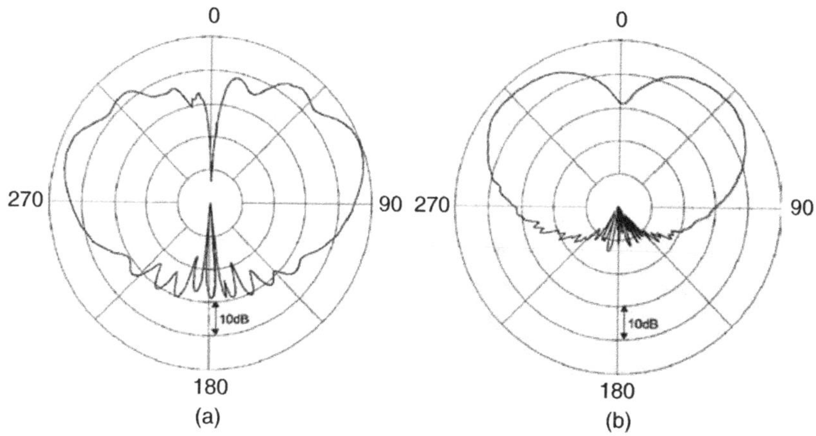

Figure 10.12: Measured radiation patterns at 35 GHz for a monopole antenna on: (a) a metallic ground plane; (b) a high-impedance ground plane. (From reference [47], reproduced by permission of © 1999 IEEE.)

10.3 Printed Antennas on EBG Substrates

It is well known that printed antennas do not generally, operate very well on high dielectric constant substrates. A printed dipole placed on a substrate with relative permittivity ϵ_r will radiate $(\epsilon_r)^{1.5}$ times more power into the substrate than into air [50]. The concept of using an EBG material as a substrate to improve the efficiency of printed antennas mounted on semiconductor substrates was first proposed in 1993 by Brown *et al.* [50]. Their implementation used a bow-tie antenna placed on the so-called 'Yablonovite' EBG material [21] containing a bandgap from around 13 to 16 GHz. The bow-tie antenna was chosen because of its non-resonant properties. For frequencies within the bandgap of the EBG material, it was demonstrated that the antenna radiated into the air, not the substrate. Hence its efficiency was substantially improved by using the EBG material as a substrate.

Dipole antennas have also been implemented on woodpile substrates, where the performance of the antenna has a strong dependence on the relative position of the dipole with respect to the woodpile surface [51, 52]. Printed slot antennas placed on the surface of woodpile substrates are also able to give a 2–3 dB increase in radiated power [53]. Edge-fed microstrip patch antennas have shown a gain improvement of 1.6 dB when placed on a UC-EBG substrate [54]. Aperture coupled microstrip patches on thick high dielectric constant substrates show a gain improvement of 3 dB when placed on UC-EBG substrates [55]. Similar improvements in gain and front-to-back ratio has been obtained for patch antennas on EBG materials formed from a two-dimensional square lattice of air holes in high dielectric constant substrates [56]. Printed spiral antennas located on EBG substrates made from a triangular lattice of air holes obtained improvements in bandwidth, gain, front-to-back ratio, and have a lower profile than classical designs [57]. Curl antennas on high-impedance ground planes are able to produce circular polarization, and the antenna height is greatly reduced compared to implementations on conducting ground planes [58]. Planar EBG structures such

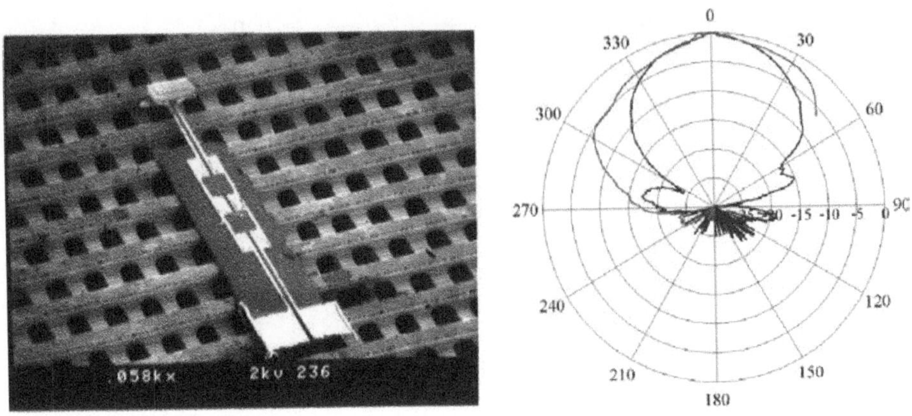

(a) (b)

Figure 10.13: (a) Dipole antenna placed upon a woodpile EBG material operating at 500 GHz. (b) Measured E- and H-plane radiation patterns for dipole antenna. (From reference [2], reproduced by permission of © 2003 IEEE.)

as the high-impedance ground plane and UC-EBG substrate have also been shown to be useful for phased array applications, where they significantly lower the mutual coupling between array elements due to surface waves [59, 60].

A successful implementation in the sub-mmWave (or terahertz) region of a printed dipole antenna mounted on a woodpile EBG material is shown in Figure 10.13(a). The intended application of this antenna is terahertz imaging, where the dipole antenna forms a single pixel of a much larger array [61]. The image to be captured is projected onto the array by a lens or reflector system. Terahertz arrays are usually implemented from waveguide-based technology, but the cost of this approach restricts the number of pixels that can be built, which in turn limits the resolution of the imaging system. Use of printed antennas would reduce the cost considerably, but the coupling between pixels is typically higher than that of waveguides. Hence, the woodpile EBG material is used to reduce coupling between pixels. Referring to Figure 10.13(a), a simple dipole 300 μm in length is fed by a balanced transmission line fabricated on a quartz substrate for operation at 500 GHz. Note that the dipole is parallel to the dielectric rods on the surface of the woodpile, and placed halfway between two adjacent rods. The figure also shows a low-pass filter and two bonding pads for external connection, and a Schottky diode bonded between the two arms of the dipole. The woodpile EBG material is fabricated from high resistivity silicon and exhibits a complete bandgap from 480 GHz to 540 GHz. Figure 10.13(b) shows the measured radiation patterns for the antenna at 500 GHz. The directivity of the dipole on the woodpile material is 11 dBi, whereas without the woodpile it is 2.1 dBi. From the patterns it is evident the radiation is low in the directions $\phi = 90°$ and 270°, hence the surface wave modes and inter-element coupling is greatly reduced.

Improvements in the performance of the dipole antenna located on a woodpile substrate can be obtained by changing the location of the two dielectric rods closest to dipole (on the surface of the woodpile) [62]. Displacement of these rods to create a wider air gap under the dipole leads to: (1) more symmetric radiation patterns with a small improvement in directivity; and (2) increased impedance of the dipole, which makes it easier to match to transmission lines such as the coplanar stripline that generally have a high impedance when implemented at mmWave frequencies. Similar improvements can also be obtained by displacing only a portion of the dielectric rod closest to the dipole, instead of the entire rod.

10.4 High Gain PRS, EBG and Metamaterial Antennas

10.4.1 High Gain PRS and Fabry–Perot Antennas

In high-gain applications requiring the lowest possible profile, resonator antennas created using a partially reflecting surface (PRS) present an attractive option. The advantages of PRS antennas are similar to those of EBG resonator antennas in that they provide high directivity and a relatively simple configuration. The typical implementation of these antennas is a PRS material placed approximately half a wavelength above a ground plane containing some type of source antenna [63–65]. If ϕ_1 is the phase angle of the reflection coefficient of the PRS $(-\pi < \phi_1 < \pi)$ the centre frequency is given by

$$f_o = \frac{c}{4\pi d}[\phi_1 + (2m - 1)\pi] \qquad m = 1, 2, \ldots \tag{10.8}$$

where d is the separation between the PRS material and ground plane, and m denotes the number of reflections.

PRS materials are usually realized by a frequency selective surface (FSS), which is a planar periodic surface. The dimensions of the FSS material are varied to obtain a given reflection coefficient, which in turn determines the directivity of the PRS antenna. The profile of the PRS antenna can be significantly reduced through the combination of a PRS material with an AMC or a metamaterial ground plane [66, 67]. Dual band PRS antennas have also been demonstrated at microwave frequencies [68–70]. If the PRS material and ground plane are shaped into two concentric circles (with the ground plane forming the inner circle) and the source antenna placed on the curved ground plane, it is possible to create an antenna useful for base-station applications [71]. Such an antenna has a broad beam in the azimuth plane with a narrow beam width in the elevation plane. The PRS material can also be combined with a metallic polarizer to generate circular polarization [72]. Circular polarization has been demonstrated using a ring-shaped FSS as the PRS and a circularly polarized patch antenna as the source [73].

Another implementation that shares many similarities with PRS antennas is the Fabry–Perot resonator antenna [74, 75]. This antenna has the characteristic feature of locating the source antenna outside of the resonant cavity, in contrast to the PRS antennas previously mentioned, where the source antenna is located within the cavity. A planar antenna or array couples energy to the Fabry–Perot cavity in the near-field region. The upper PRS material is either curved or requires a flat non-uniform planar grid to obtain a Gaussian beam. The PRS materials consist of an inductive metal mesh, where the width of the metal strips and their lattice constant determine the reflection coefficient of the material and the Q factor of the cavity. Two implementations of Fabry–Perot resonator antennas are shown in Figure 10.14. The separation between the PRS materials is half-a-guide wavelength, while the length of the coupling region is slightly less than half-a-guide wavelength and depends on the parameters

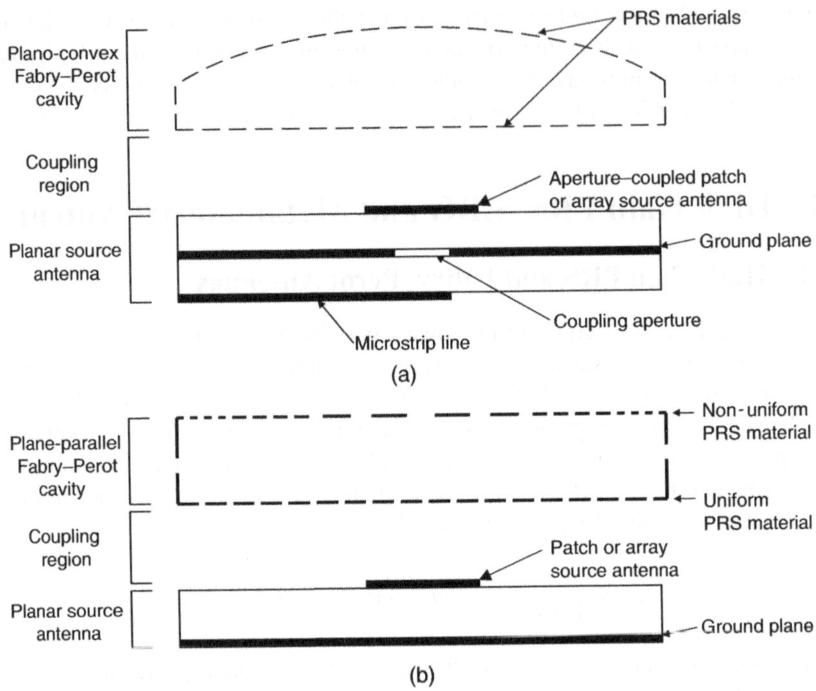

Figure 10.14: Configuration of two Fabry–Perot cavity antennas: (a) plano-convex; (b) plane-parallel with non-uniform upper PRS material. (From reference [74], reproduced by permission of © 2003 IEEE.)

of the lower PRS material and the source antenna. Efficiency of the antenna depends upon both the insertion loss of the Fabry–Perot cavity and the losses of the printed source.

Fabry–Perot antennas have been successfully implemented at V-band using fused quartz substrates to realize both the source antenna and a plane-parallel cavity [74]. Measured and computed radiation patterns for a Fabry–Perot antenna with a microstrip patch antenna source are shown in Figure 10.15. Radiation patterns of the patch antenna without the Fabry–Perot cavity are also shown in this figure. At 57.6 GHz the antenna achieves a gain of 17.3 dBi, theoretical directivity of 21.2 dBi, efficiency of 41% and side lobe levels of −22 dB and −25 dB in the E- and H-planes respectively. A plano-convex Fabry–Perot cavity antenna has been implemented at 57.7 GHz. Using a 2×2 aperture-coupled microstrip array as the source, this Fabry–Perot antenna achieves a measured gain of 15 dBi, theoretical directivity of 18.5 dBi, efficiency of 45%, and the side lobes are below −27 dB [75]. Plane parallel resonators have the advantage of being more cost effective to manufacture than plano-convex resonators due to the stringent tolerances required at mmWave frequencies.

10.4.2 High-gain One-dimensional EBG Resonator Antennas

High-gain antennas made with one-dimensional EBG materials have the advantage of being relatively simple in structure. A typical configuration of such an antenna is shown in

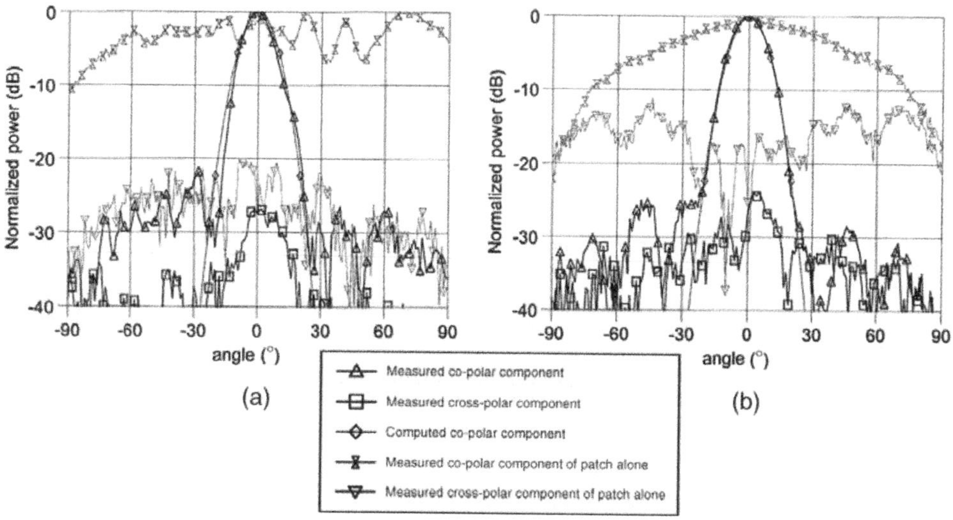

Figure 10.15: Measured radiation patterns at 57.6 GHz of a plane-parallel Fabry–Perot cavity antenna excited by a patch antenna: (a) E-plane; and (b) H-plane. (From reference [74], reproduced by permission of © 2003 IEEE.)

Figure 10.16. It consists of a ground plane, a source antenna located on the ground plane, and a one-dimensional EBG material located approximately half-a-wavelength above the ground plane. The one-dimensional EBG material and ground plane form a high-Q resonant cavity, which modifies the radiation characteristics of the source antenna. Typically, these antennas will have high-directivity, low side lobes and a narrow radiation bandwidth. Depending upon the design, the antenna can radiate a broadside pencil beam or a conical beam scanned to a particular angle [76]. For highest directivity, the dielectric contrast between the dielectric layers of the one-dimensional EBG material should be large. Increasing the number of dielectric layers also increases the directivity at the expense of the radiation bandwidth. There is a direct relationship between the Q-factor of the resonant cavity and the directivity of the antenna.

The effect of the EBG material on the source antenna can be interpreted in several ways. The EBG/cavity combination can be viewed as a spatial filter that provides angle-dependent transmission from the resonator to free space. The greater the variation in transmission with the angle, the greater the directivity of the antenna. Alternatively, the directive radiation properties may be interpreted as leaky wave propagation through the multilayer structure [76].

Research on one-dimensional EBG resonator antennas is essentially an extension of the concept of gain enhancement of printed antennas using single [77] and multiple superstrates [78]. Resonator antennas using one-dimensional EBG materials in combination with probe-fed patch antennas have achieved directivities of 20 dBi at 5 GHz using two layers of alumina [79], 27.3 dBi at 20 GHz using three alumina layers [80] and 24.4 dBi at 12 GHz using three slabs of FR4 material [81]. By using a perturbing dielectric slab, it has been shown that a dual frequency one-dimensional EBG antenna can be created with accurate control

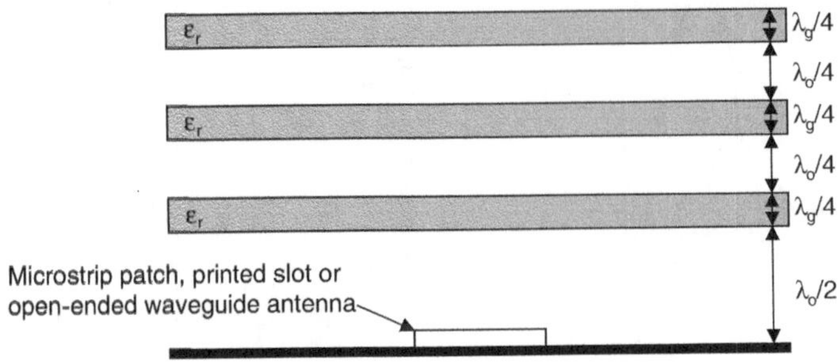

Figure 10.16: Configuration of a high-gain one-dimensional EBG antenna.

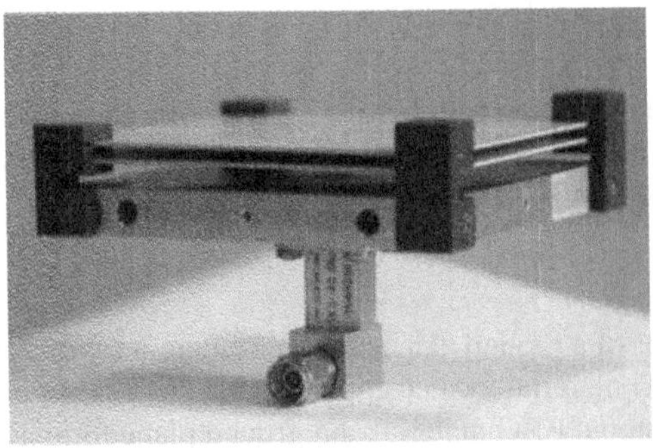

Figure 10.17: Configuration of a mmWave one-dimensional EBG resonator antenna. (From reference [85], reproduced by permission of © 2006 John Wiley & Sons, Inc.)

of the design frequencies [82]. Unfortunately, the radiation bandwidth of EBG resonator antennas is typically quite low, which limits their usefulness. Directivity can be traded for radiation bandwidth and vice versa by adjusting the quality factor of the resonator, but this technique may not satisfy more challenging design specifications. It has been shown that use of an array of microstrip patches instead of a single source increases the radiation bandwidth of the one-dimensional EBG antenna [83]. A combination of dual resonators with an array of slots can increase the operating range even further, giving a gain bandwidth of 13.2% and maximum gain of 22.7 dBi at 12 GHz [84].

A one-dimensional EBG resonator antenna that operates at 38.9 GHz has been reported in the literature [85]. A photograph of this antenna is shown in Figure 10.17. The one-dimensional EBG material consists of two slabs of Rogers TMM10 ($\epsilon_r = 9.2$) dielectric that are 0.635 mm thick, and the resonator is fed by a single rectangular slot in the ground plane connected to a WR-28 waveguide. The measured gain is 19 dBi at 38.9 GHz with a radiation bandwidth of 1.3%. The measured and computed radiation patterns are shown in Figure 10.18, while gain versus frequency is shown in Figure 10.19 for one, two and no dielectric slabs.

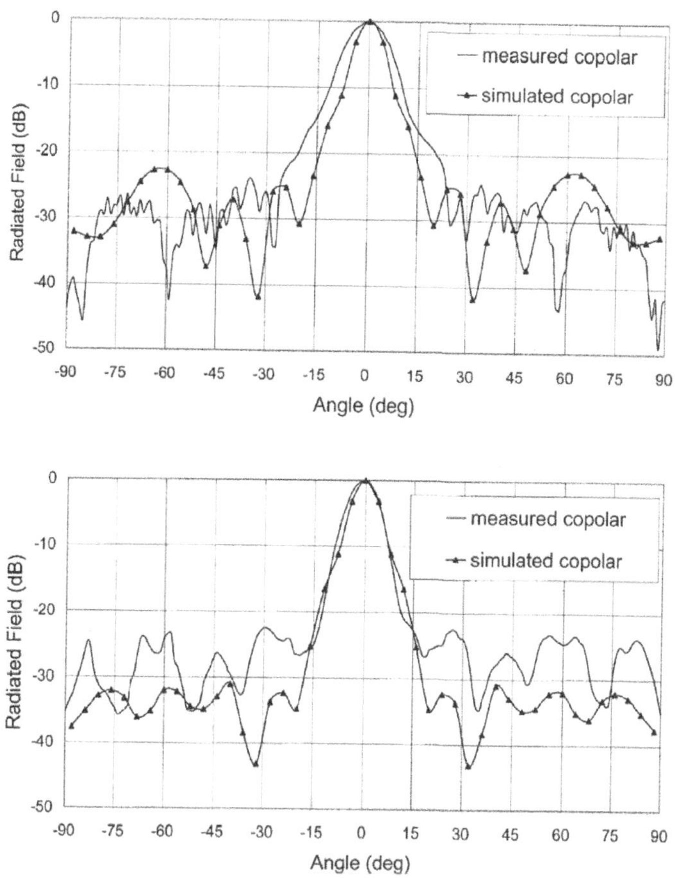

Figure 10.18: Measured and computed E- and H-plane radiation patterns for the one-dimensional EBG resonator antenna at 38.9 GHz. (From reference [85], reproduced by permission of © 2006 John Wiley & Sons, Inc.)

A circularly polarized one-dimensional EBG resonator antenna has been implemented at 11.9 GHz using a feed element of four linearly polarized slots sequentially rotated in phase and space. Each slot was fed by a stripline, and an external feed network of two 90° hybrids and a single 180° hybrid was used to create the sequential phase shifts of 0°, 90°, 180° and 270°. Peak measured gain for the antenna was 19.6 dBic at 11.9 GHz, gain bandwidth was 1.3% and the measured axial ratio was less than 1 dB from 11.6 GHz to 12.2 GHz [86].

10.4.3 High-gain Two-dimensional EBG Resonator Antennas

Instead of using the one-dimensional EBG material as a superstrate, as shown in Figure 10.17, it is also possible to use a two-dimensional EBG material such a periodic array of dielectric rods [87–89]. A typical microwave implementation is shown in Figure 10.20. The two-dimensional EBG material is composed of a square lattice of alumina dielectric rods in air,

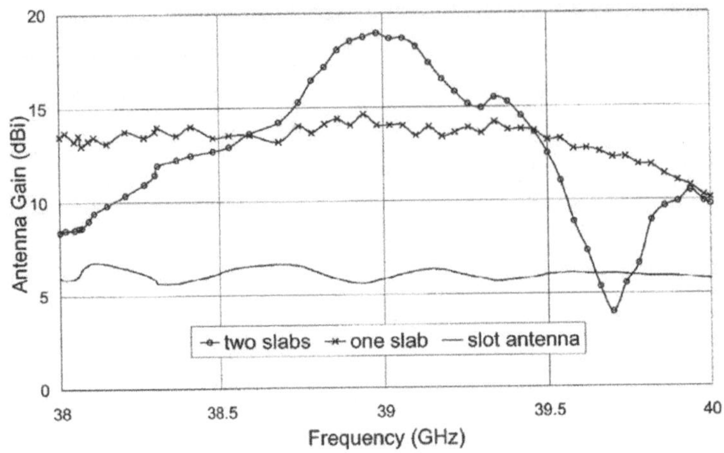

Figure 10.19: Measured gain for the one-dimensional EBG resonator antenna for: two TMM 10 slabs; one TMM 10 slab and the slot antenna without any EBG material. (From reference [85], reproduced by permission of © 2006 John Wiley & Sons, Inc.)

with a lattice constant of a = 22.36 mm, and rod radius of $r = 5.3$ mm ($r/a = 0.237$). The band diagram for this EBG material is shown in Figure 10.3 and the bandgap for TM polarizations extends from $\omega a/2\pi c = 0.283$ to 0.393. To obtain the bandgap frequencies for the lattice constant of $a = 22.36$ mm, the normalized frequency is multiplied by c/a to obtain $f = 3.80$ GHz to 5.27 GHz. The measured gain reported for this antenna is 19 dBi at 4.75 GHz (computed gain 20.5 dBi), and the two-dimensional EBG superstrate material increases the theoretical gain of the microstrip patch antenna from 7 dBi to 20.5 dBi. The antenna exhibits directive radiation patterns, and very low side lobes as shown in the comparison of measurement with theory in Figure 10.21. Some relevant design parameters of this antenna are the placement of the first layer of dielectric rods approximately $\lambda/2$ above the metallic ground plane and the aperture size is 4λ by 4λ in area. A further possible modification of this antenna is to incorporate a second two-dimensional EBG material into the substrate of the patch antenna, located on the ground plane [89]. This second EBG material suppresses the surface waves generated by the microstrip patch antenna and gives a further increase in the antenna directivity. Resonator antennas based on two-dimensional EBG materials have yet to be demonstrated at mmWave frequencies, but it should be reasonably straightforward to scale the design. The probe-fed microstrip patch antenna shown in Figure 10.20 is less suitable for mmWave operation, and could be replaced by an edge-, aperture or proximity coupled microstrip patch antenna. Other alternatives are the waveguide-fed slot of Figure 10.17, or a printed slot antenna fed by either a microstrip or coplanar waveguide.

10.4.4 High-gain Three-dimensional EBG Resonator Antennas

A natural extension of the one-dimensional and two-dimensional EBG resonator antennas is to use a three-dimensional EBG material as the antenna superstrate. Such an antenna using a

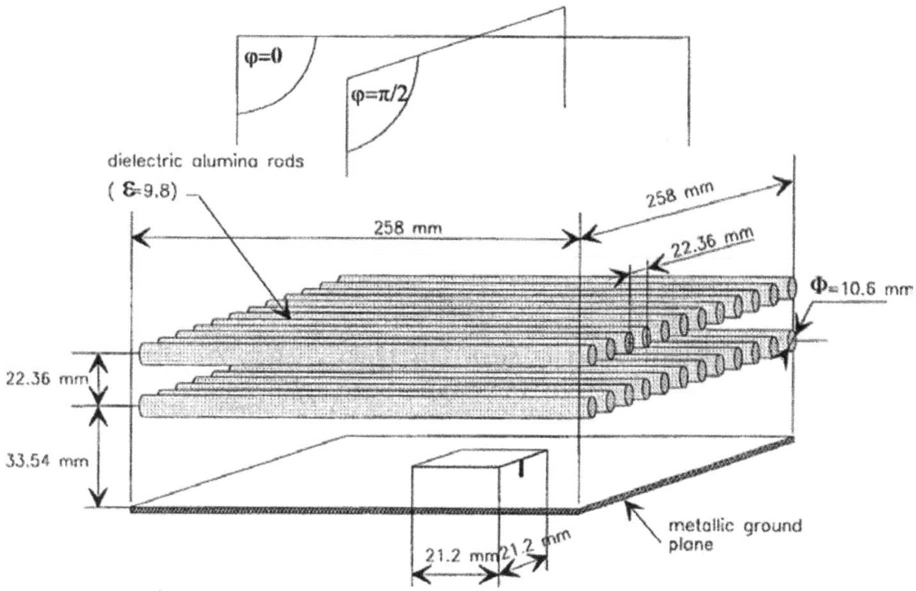

Figure 10.20: Configuration of a two-dimensional EBG resonator antenna. (From reference [87], reproduced by permission of © 2002 IEEE.)

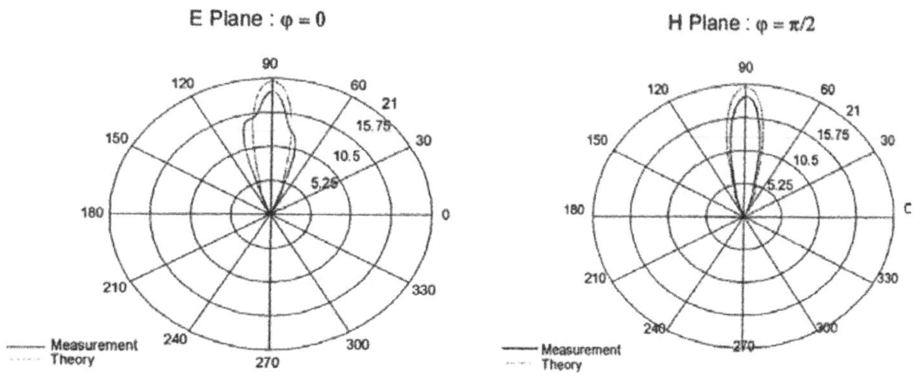

Figure 10.21: Measured and computed radiation patterns at 4.75 GHz. (From reference [87], reproduced by permission of © 2002 IEEE.)

woodpile EBG material is presented in Figure 10.22, with one quarter of the EBG material removed to show the slot antenna more clearly. The first reported three-dimensional EBG resonator antenna, realized at microwave frequencies, used a Fabry–Perot cavity built using woodpile materials [90]. When the thicknesses of the cavity walls were equal, a symmetric radiation pattern was obtained with two directive beams pointing in opposite directions. Making one wall thicker than the other resulted in a single highly directive beam.

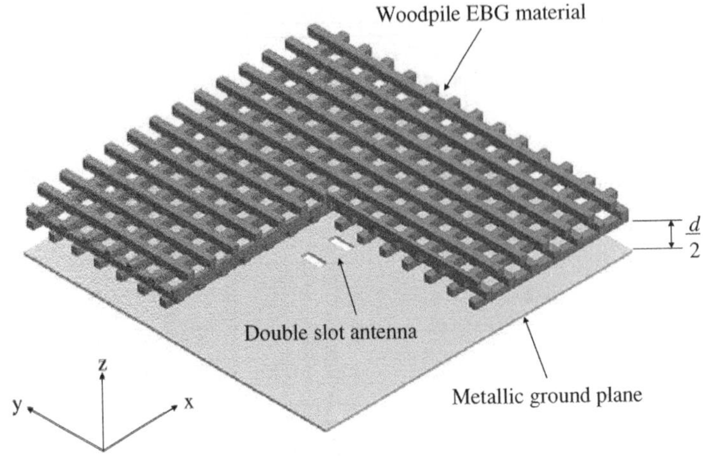

Figure 10.22: Cut-away drawing of a three-dimensional EBG resonator antenna. (From reference [91], reproduced by permission of © 2005 IEEE.)

To reduce the profile of the antenna, a metallic ground plane can be used as depicted in Figure 10.22 [91]. This antenna uses a double slot antenna fed by a square waveguide from beneath the ground plane to obtain a measured gain of 19.8 dBi at 12.565 GHz, with measured half power beamwidths of 13.2° and 13.3° in the E-plane and H-plane, respectively. The woodpile material is made from alumina with a dielectric constant of 8.4. The band diagram for this EBG material is shown in Figure 10.4 and extends in normalized frequency from $\omega a/2\pi c = 0.423$ to 0.479. The design is easily implemented at mmWave frequencies by scaling its dimensions, which is examined through a simulation. Referring to Figure 10.4, scaling the parameters of the design from [91] by a factor of 1/3.0522 changes the operating frequency from 12.565 GHz to 38 GHz. Hence the parameters of the woodpile become $a = 3.67$ mm, $w = h = 1.05$ mm, which gives a bandgap of 34.58 GHz to 39.16 GHz. The woodpile is placed 3.8 mm above the ground plane, slot dimensions are 1.31 mm × 3.145, slot separation 4.456 mm, and the square waveguide inner dimensions 6.29 mm × 6.29 mm. A plot of the radiation patterns obtained using CST Microwave Studio is given in Figure 10.23. The plot shows highly directional radiation in both principal planes, and the realized gain is 22.2 dBi.

To increase the operating frequency of this antenna even further, accurate fabrication of the woodpile material becomes an issue. This problem has been addressed by using a rapid prototyping technique called ceramic extrusion free-forming to manufacture the woodpile. A W-band (75–110 GHz) woodpile EBG resonator antenna has been constructed using this technique and its performance measured [92]. The woodpile material used in the design has lattice parameters $a = 1.67$ mm, $w = h = 0.41$ mm, and was made from high-purity alumina with a dielectric constant of 9.6. Radiation patterns were measured for a woodpile with lateral dimensions of 30 × 30 mm^2 placed on a 120 × 120 mm^2 ground plane. The measured half-power beamwidth (HPBW) of the three-dimensional EBG resonator antenna at 93.3 GHz is 14° and 10° in the E-plane and H-plane, respectively. This gives the antenna a directivity of approximately 23.6 dBi.

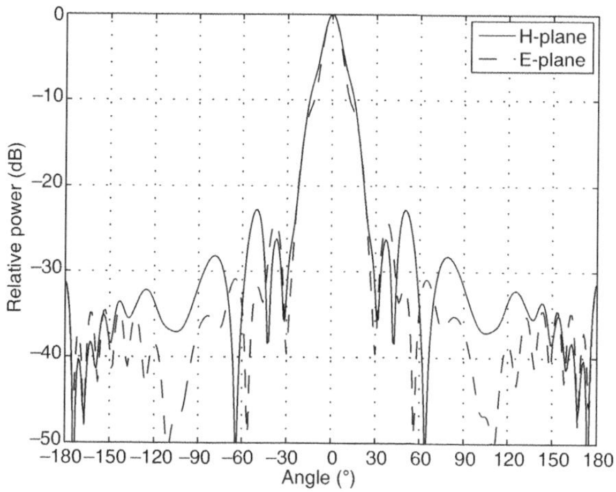

Figure 10.23: Computed radiation patterns at 38.2 GHz of a three-dimensional EBG resonator antenna.

10.4.5 High-gain Metamaterial Antennas

Another method of obtaining a directive antenna is to place a source within a metamaterial that has a refractive index between zero and one. This concept was first demonstrated theoretically in 1971 using an artificial dielectric material [93]. A ground plane is usually placed on one side of the metamaterial and the source is often a dipole or patch antenna (or a line source in theoretical analyses). The simplest implementation of a suitable metamaterial is an array of parallel metallic wires, which behave like a plasma below its cut-off frequency (the plasma frequency) determined by the wire spacing and radius. A two-dimensional planar grid of metallic wires, compatible with printed circuit techniques, has also been used to implement the metamaterial as shown in Figure 10.24 [94]. Below the plasma frequency (ω_{p}) the effective permittivity of the metamaterial is negative, while above the plasma frequency the effective permittivity is positive. The following equation for effective permittivity describes this plasma-like behaviour:

$$\epsilon_{\mathrm{eff}}(\omega) = 1 - \frac{\omega_{\mathrm{p}}^2}{\omega^2} \qquad (10.9)$$

Directive metamaterial antennas exploit the frequency just above ω_{p}, where $0 < \epsilon_{\mathrm{eff}} < 1$. A simple explanation of the behaviour of the antenna can be obtained using Snell's law and ray optics, since at the metamaterial/air interface the refracted rays from the source will be normal to the interface, thus yielding high directivity. This simple analysis is valid for an infinitely thick metamaterial, but it has been shown that when a ground plane is placed under the metamaterial the directivity is due to the excitation of a leaky wave with low attenuation [95].

Directive metamaterial antennas have been implemented at 14.65 GHz using planar copper grid structures spaced with foam, and fed by a monopole antenna, and obtained a maximum directivity of 25.7 dBi [94]. A measured gain of 23.4 dBi was obtained using metamaterial realized with an array of parallel copper wires fed by a quarter-wavelength monopole at an operating frequency of 10.25 GHz [96]. Another implementation used a ring

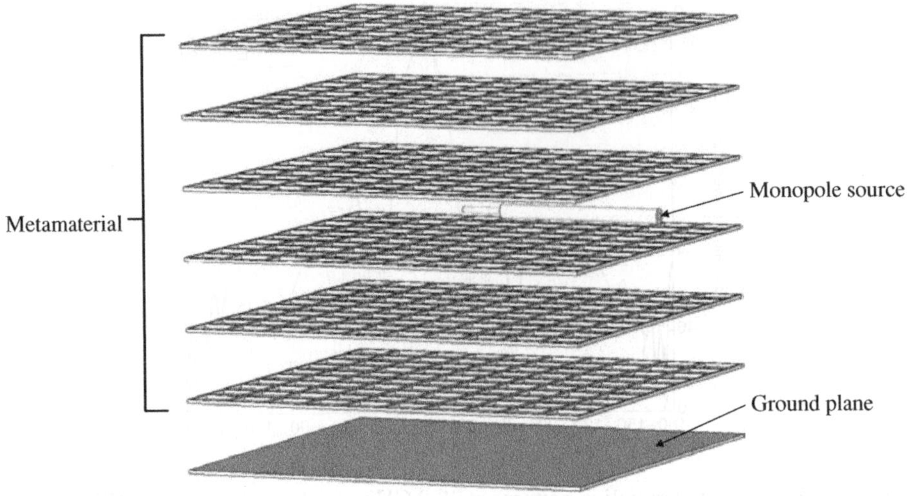

Figure 10.24: Perspective view of a directive metamaterial antenna.

aperture lattice and patch antenna feed to obtain a peak gain of 17.2 dBi at 2.55 GHz [97]. Although these examples operate at microwave frequencies, directive metamaterial antennas show good potential for implementation at mmWave frequencies as well, particularly when printed circuit techniques are used. However, antenna efficiencies are likely to be lower at millimeter wavelengths owing to the increased losses of the metals.

Directive radiation can also be obtained by placing a source in an EBG material, and operating at the frequency corresponding to the band edge. At the band edge of a correctly designed EBG material, the dispersion properties limit propagation to a small angular region which enhances the directivity. This is a completely different operating mechanism to the high-gain EBG resonator antennas also presented in this chapter. Antennas with high directivity operating at EBG material band edge have been demonstrated in an expanded simple cubic EBG material [98] and a two-dimensional square lattice of dielectric rods [99]. The operating bandwidth of these antennas is quite narrow and the radiation patterns have a strong dependence on the electrical size of the EBG material.

10.5 Woodpile EBG Waveguides, Horn Antennas and Arrays

10.5.1 Woodpile EBG Sectoral Horn Antennas

A useful extension of the defect waveguide described in Section 10.2.2 is the woodpile EBG horn antenna. This horn enables efficient coupling from the defect waveguide to free space and vice versa. In addition, the parameters of the horn may be varied to meet certain directivity requirements. One of the unique features of such an antenna is that the walls are made entirely of a periodic dielectric material, as opposed to traditional horn antennas where the walls are made of metal. EBG horn antennas have been proposed by several researchers [100–105].

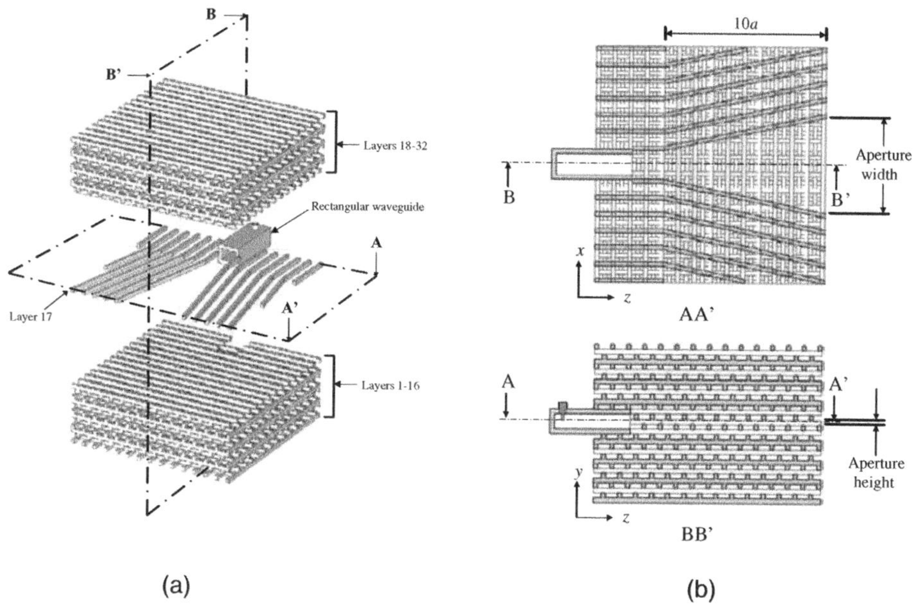

Figure 10.25: Diagrams of the woodpile EBG sectoral horn antenna: (a) exploded perspective view; and (b) cross-sections taken through planes AA′ and BB′. (From reference [105], reproduced by permission of © 2006 IEEE.)

Like the defect waveguide, the electromagnetic confinement of the EBG horn antenna is a result of the bandgap of the material. The simplest implementation of the antenna is a sectoral horn, where the walls of the waveguide are flared in a single layer of the woodpile corresponding to the H-plane of the device. An exploded view of a woodpile EBG sectoral horn antenna is shown in Figure 10.25(a). The vertical and horizontal cross-section views are shown in Figure 10.25(b) to clarify further the structure of the antenna. The section AA′ corresponds to the H-plane of the antenna, and section BB′ corresponds to the E-plane. The radiation pattern is more directive in the H-plane since it contains the taper, while the E-plane pattern is less directive due to its narrower aperture (as is typical in sectoral horns). In the implementation shown in Figure 10.25 a metallic rectangular waveguide has been used to couple energy to the defect EBG waveguide. The internal dimensions of the rectangular waveguide and depth of protrusion into the EBG material are crucial for efficient operation. Inner rectangular waveguide dimensions of $1.311a$ by $0.9a$ and depth of protrusion of $2.25a$ give good coupling for alumina and silicon woodpiles. With a well-designed transition the sectoral horn will operate over almost the same single mode non-dispersive operating range as the woodpile waveguide.

The theoretical performance of a woodpile EBG sectoral horn antenna made from high resistivity silicon ($\epsilon_r = 11.7$, $\tan\delta = 0.0015$) is now examined. The woodpile material has the parameters of $w = 0.25a$, $h = 0.25a$ and $b = 1.0a$ with a complete EBG that extends from $\omega a/2\pi c = 0.405$ to 0.487 in normalized frequency, as calculated by the plane-wave expansion method [14]. The configuration of the antenna is shown in Figure 10.25 and

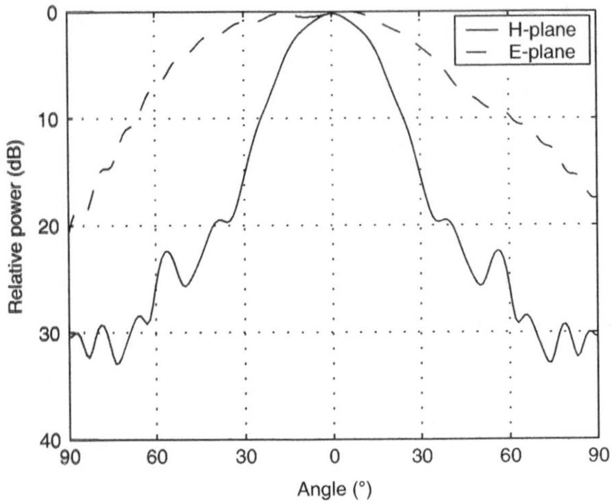

Figure 10.26: Computed H- and E-plane radiation patterns at $\omega a/2\pi c = 0.449$ for a woodpile EBG sectoral horn antenna made from silicon. (From reference [105], reproduced by permission of © 2006 IEEE.)

consists of 32 layers of silicon rods arranged to form a woodpile. The lengths of the rods are $14.25a$ and a single rod is removed from the centre of layer 17 to form the defect waveguide. At a distance of $4.25a$ from the edge of the lattice, the rods on layer 17 bend away from the centre of the device to create a sectoral horn antenna with a flare angle of 22.6° in the H-plane. The width of the radiating aperture in terms of lattice constants is $5.75a$ and the height is $0.25a$. Since the rods flare only in a single layer, the surrounding woodpile material maintains its original bandgap. Keeping the flare angle small gives a smooth transition from the defect waveguide to the flared section of the horn. The computed radiation patterns for the antenna at $\omega a/2\pi c = 0.449$ are shown in Figure 10.26. This plot shows the radiation characteristics of the antenna in the two principal planes. Notice the slight beam squint of the radiation pattern in the E-plane due to the asymmetry of the lattice in this plane (with respect to the centre of the antenna) and diffraction effects from the ends of the lattice material. The computed HPBW at $\omega a/2\pi c = 0.449$ is 74.4° in the E-plane and 27.2° in the H-plane, and the gain is 11.1 dBi.

Experimental results were obtained for a woodpile EBG sectoral horn antenna made from alumina rods at Ku-band [105]. The measured impedance bandwidth extended from 12.4 GHz to 13.0 GHz (a bandwidth of 4.7%) and measured radiation patterns gave very good agreement with FDTD simulations. The H-plane HPBW of the prototype was stable across the impedance bandwidth, but the E-plane HPBW changed from 61.8° at 12.4 GHz to 38.7° at 13.0 GHz. As a result of the E-plane patterns narrowing with frequency, the measured gain varied from 11.4 dBi at 12.4 GHz to 14.1 dBi at 13.0 GHz for the alumina prototype.

10.5.2 Woodpile EBG Array Antennas

The woodpile EBG sectoral horn antenna is a useful element for linear array antennas. The periodic stacking sequence of the woodpile EBG material can be exploited when creating

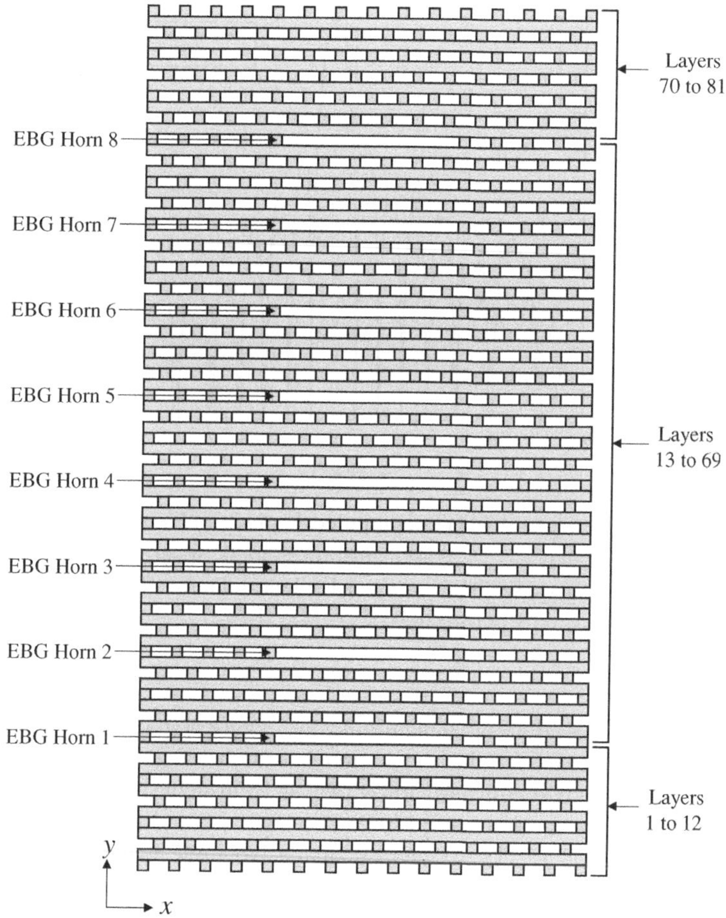

Figure 10.27: Eight-element linear array of woodpile EBG sectoral horn antennas. (From reference [105], reproduced by permission of © 2006 IEEE.)

such a linear array. Increasing the number of array elements allows the directivity of the antenna to be significantly increased, to suit a given application. The aperture configuration of an eight-element linear array made from high resistivity silicon sectoral horns is shown in Figure 10.27. The layer-by-layer stacking sequence of the woodpile material, which repeats every four layers, creates restrictions on the separation of the array elements. For elements to be placed directly above one another (in the y-direction or E-plane in Figure 10.27) the element separation is limited to multiples of $4h$ (i.e. separation $= 4nh$, where $n = 1, 2, 3 \ldots$). Hence the minimum separation is $4h$, which is important as it limits the scan range in phased array implementations.

The array of Figure 10.27 is made from 81 layers of high resistivity silicon rods, and the element separation is $8h$. At the normalized frequency of $\omega a/2\pi c = 0.449$, this array spacing is equivalent to 0.9λ, hence the antenna is suitable for high-directivity fixed-beam

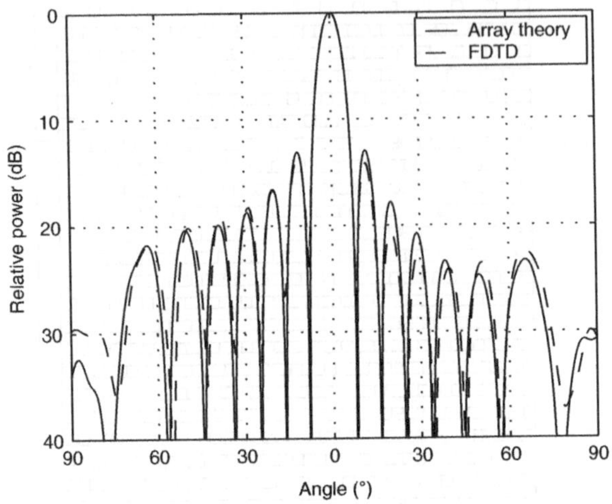

Figure 10.28: Array theory and FDTD computed E-plane radiation patterns for the eight-element linear array at $\omega a/2\pi c = 0.449$ with an element separation of $8h$. The patterns obtained from array theory use the computed E-plane patterns from Figure 10.26. (From reference [105], reproduced by permission of © 2006 IEEE.)

applications and not phased arrays which typically require a spacing of 0.5λ. The mutual coupling between elements for this separation has been calculated to be -31.5 dB at $\omega a/2\pi c = 0.449$, which is significantly lower than an equivalent sized metallic sectoral horn in a ground plane at the same spacing [105]. The eight sectoral horn elements are placed at layers 13, 21, 29, 37, 45, 53, 61 and 69 of the woodpile material and each antenna is identical to that shown in Figure 10.25, with identical rectangular-waveguide to EBG-waveguide transitions and woodpile parameters.

Radiation patterns of the linear array have been calculated using both the FDTD method and array theory. For the FDTD calculation the entire array was modelled with all eight ports excited simultaneously with the same amplitude and phase. The resulting E-plane radiation pattern at a normalized frequency of $\omega a/2\pi c = 0.449$ is compared with simple array theory in Figure 10.28. The radiation pattern calculated from array theory used the radiation pattern of Figure 10.26 multiplied by the array factor, and gives good agreement with FDTD results. The computed radiation pattern shows that the side lobe levels are below -13.6 dB, the E-plane HPBW is 7.0°, and the gain is 22.2 dBi. The H-plane radiation pattern is very similar to Figure 10.26, and the H-plane HPBW is 28.2° (close to the value of the isolated horn). It is envisaged that an appropriate corporate power divider network could also be implemented from the woodpile EBG material, thus forming a single compact high-gain EBG device that could be fabricated at mmWaves by rapid prototyping or RIE techniques.

The linear array can also be used in scanned beam applications. To obtain a wide scan range the elements should be placed closer together than in the previous fixed beam example. A theoretical scanned beam array made from 16 woodpile EBG sectoral horn elements has been examined in reference [105], where the element separation was $4h$. At an operating frequency of $\omega a/2\pi c = 0.442$ this spacing corresponds to 0.44λ, and the theoretical scan range is ±90°. However, the closer spacing also increases the mutual coupling between EBG horn elements. The theoretical analysis showed that the performance of the phased array was significantly improved by adding four extra parasitic elements at either end of the linear array, where each parasitic element was terminated with a matched load. Without the parasitic elements an image beam appears at wide scan angles due to the close spacing of the array elements, the consequent higher mutual coupling and partial reflection of the beam from the woodpile at the edge of the array. Use of four parasitic elements was found to be the best compromise between performance and increased size of the array, and a wide theoretical scan range was demonstrated.

10.6 Miscellaneous EBG Antennas and Components

This section briefly discusses other EBG antennas that operate on different principles to those already described. Several authors have examined using the reflecting properties of EBG materials within the bandgap to realize reflector antennas. A parabolic reflector antenna made from seven dielectric layers ($\epsilon_r = 2.38$) that operates from 38 to 43 GHz was reported that gave similar performance to an equivalent-sized metallic parabolic reflector for frequencies in the bandgap [79, 106]. Outside of the bandgap, the reflection of the EBG structure is much lower than a metallic reflector, hence it may be useful for stealth applications. Planar reflectors made from two-dimensional and three-dimensional EBG materials were combined with dipole antennas to obtain similar radiation performance to planar metallic reflectors [107, 108]. The two-dimensional EBG material consisted of a square lattice of dielectric rods ($\epsilon_r = 2.38$), while the three-dimensional EBG materials were a body-centred cubic structure and the woodpile. Tunable impedance surfaces have also been used in reflector antenna configurations, where tuning the reflection phase of the surface enabled two-dimensional beam steering over a range of ±40° for both polarizations, with a bandwidth of around 8% [109]. The tunable surface was implemented by loading a high-impedance ground plane with varactor diodes, that tuned the resonant frequency of the LC resonators, and hence their reflection phase.

A mmWave subharmonic mixer that combines an EBG material, coplanar stripline (CPS) and microstrip circuits, and rectangular waveguide technology is shown in Figure 10.29. The RF frequency of the mixer is 250 GHz, the LO frequency is 115–135 GHz, and the IF spans 2.5–3.5 GHz [110]. The EBG material is used to improve the radiation performance of the dipole antenna that receives the RF signal. The Schottky diode mixing elements and RF filter are implemented in CPS, and the LO signal is supplied by a WR-8 rectangular waveguide. Future mixer configurations could implement the LO waveguide using EBG defect waveguide technology, which may improve the economy of manufacture of the device. The main application of the mixer is in imaging systems.

10.7 Summary

This chapter has reviewed EBG materials and some antenna structures formed from them that are suitable for operation at mmWave frequencies. Various concepts associated with EBG materials such as irreducible Brillouin zones, band structure, band diagrams, complete bandgaps and minimum dielectric contrasts have been described, and examples of materials

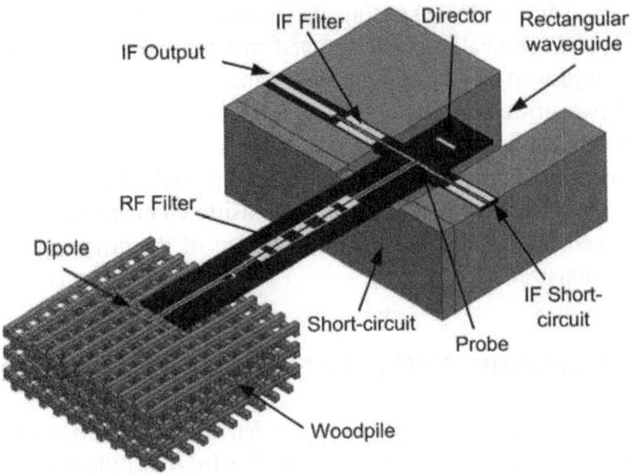

Figure 10.29: Subharmonic mixer that uses woodpile EBG technology. Only the lower half of the rectangular waveguide is shown to reveal internal features (From reference [110], reproduced by permission of © 2007 IEEE.)

given in one- to three-dimensions. High-impedance ground planes have been examined, and some of their properties such as surface wave bandgaps and reflection phase have been described. Several approaches to implementing mmWave antennas using EBG materials have been presented, with an emphasis on devices that achieve high directivity.

Planar antennas placed on EBG substrates have improved efficiency, increased directivity, and less surface wave coupling. Resonator antennas are able to achieve directivities typically in excess of 20 dBi, by using cavities created from one-dimensional, two-dimensional or three-dimensional EBG materials. Partially reflecting surfaces are attractive options for implementing high-gain resonator antennas when a low profile is required. Metamaterial antennas are able to obtain high directivity by operating near the plasma frequency, where the effective dielectric constant of the metamaterial is positive but less than one. Woodpile EBG sectoral horn antennas, that rely completely on the bandgap of the material for electromagnetic confinement, were described and it was shown how they can be used to form linear arrays to increase directivity. Such arrays may become attractive at mmWave frequencies if they are fabricated using rapid prototyping methods. There is also potential to integrate many of the various components described in this chapter such as horn antennas, waveguides, bends, power dividers, filters and phase shifters into a single compact sub-system for future millimeter and sub-mmWave applications.

References

[1] J. D. Joannopoulos, S. G. Johnson, J. N. Winn and R. D. Meade, *Photonic Crystals: Molding the Flow of Light,* 2nd edn (Princeton, NJ: Princeton University Press, 2008) http://ab-initio.mit.edu/book.

[2] P. de Maagt, R. Gonzalo, Y. C. Vardaxoglou and J. M. Baracco, 'Electromagnetic bandgap antennas and components for microwave and (sub)millimeter wave applications', *IEEE Trans. Antennas Propagat.* **51**(10) (2003), pp. 2667–2677.

[3] K. Sakoda, *Optical Properties of Photonic Crystals* (Springer, Berlin, 2001).

[4] S. G. Johnson and J. D. Joannopoulos, *Photonic Crystals: The Road from Theory to Practice* (Kluwer, Boston, MA, 2002).

[5] K. Busch, S. Lölkes, R. B. Wehrspohn and H. Föll, *Photonic Crystals: Advances in Design, Fabrication, and Characterization* (Wiley-VCH, Weinheim, Germany, 2004).

[6] J.-M. Lourtioz, H. Benisty, V. Berger, J.-M. Gerard, D. Maystre and A. Tchelnokov, *Photonic Crystals: Toward Nanoscale Photonic Structures* (Springer, Berlin, 2005).

[7] N. Engheta and R. W. Ziolkowski, *Metamaterials: Physics and Engineering Explorations* (Wiley-IEEE Press, 2006).

[8] E. Yablonovitch, 'Photonic crystals: semiconductors of light', *Scien. Am.* **285**(6) (2001), pp. 47–55.

[9] S. G. Johnson and J. D. Joannopoulos, 'Introduction to photonic crystals: Bloch's theorem, band diagrams, and gaps (but no defects)', 3 Feb. 2003 (http://ab-initio.mit.edu/photons/tutorial/photonic-intro.pdf), December 18, 2008.

[10] Lord Rayleigh, 'On the maintenance of vibrations by forces of double frequency, and on the propagation of waves through a medium endowed with a periodic structure', *Phil. Mag.* **24**(147) (1887), pp. 145–159.

[11] E. Yablonovitch, 'Inhibited spontaneous emission in solid state physics and electronics', *Phys. Rev. Lett.* **58**(20) (1987), pp. 2059–2062.

[12] S. John, 'Strong localization of photons in certain disordered lattices', *Phys. Rev. Lett.* **58**(23) (1987), pp. 2486–2489.

[13] A. Macleod, *Thin-Film Optical Filters* 2nd edn (Bristol, Adam Hilger Ltd, 1986).

[14] BandSOLVE 1.0 User Manual (RSOFT Design Group Inc., Ossining, NY, 2002).

[15] S. G. Johnson and J. D. Joannopoulos, MIT Photonic Bands software (http://ab-initio.mit.edu/mpb), December 18, 2008.

[16] S. G. Johnson and J. D. Joannopoulos, 'Block-iterative frequency-domain methods for Maxwell's equations in a planewave basis', *Optics Express* **8**(3) (2001), pp. 173–190.

[17] *High Frequency Structure Simulator 10 User Manual*, Ansoft Corporation, Pittsburgh PA, 2005 (www.ansoft.com), December 18, 2008.

[18] *CST Microwave Studio 2008 User Manual,* CST Darmstadt Germany (www.cst.com).

[19] R. T. Remski, 'Analysis of photonic bandgap surfaces using ansoft HFSS', *Microwave J.* **43**(9)(2000), pp. 190–198.

[20] K. M. Ho, C. T. Chan and C. M. Soukoulis, 'Existence of a photonic gap in periodic dielectric structures', *Phys. Rev. Lett.* **65**(25) (1990), pp. 3152–3155.

[21] E. Yablonovitch, T. J. Gmitter and K. M. Leung, 'Photonic band structure: the face-centred-cubic case employing nonspherical atoms', *Phys. Rev. Lett.* **67**(17) (1991), pp. 2295–2298.

[22] K. M. Ho, C. T. Chan, C. M. Soukoulis, R. Biswas and M. M. Sigalas, 'Photonic band gaps in three dimensions: new layer-by-layer periodic structures', *Solid State Commun.* **89**(5) (1994), pp. 413–416.

[23] H. S. Sozuer and J. P. Dowling, 'Photonic band calculations for woodpile structure', *J. Mod. Opt.* **41**(2) (1994), pp. 231–239.

[24] E. Ozbay, E. Michel, G. Tuttle, R. Biswas, M. Sigalas and K. M. Ho, 'Micromachined millimeter-wave photonic band-gap crystals', *Appl. Phys. Lett.* **64**(16) (1994), pp. 2059–2061.

[25] E. Ozbay, E. Michel, G. Tuttle, R. Biswas, K. M. Ho, J. Bostak and D. M. Bloom, 'Double-etch geometry for millimeter-wave photonic band-gap crystals', *Appl. Phys. Lett.* **65**(13) (1994), pp. 1617–1619.

[26] A. Chelnokov, S. Rowson, J.-M. Lourtioz, L. Duvillaret and J.-L. Coutaz, 'Terahertz characterization of mechanically machined 3D photonic crystal', *Electron. Lett.* **33**(23) (1997), pp. 1981–1983.

[27] R. Gonzalo, B. Martinez, C. Mann, H. Pellemans, P. H. Bolivar and P. de Maagt, 'A low cost fabrication technique for symmetrical and asymmetrical layer-by-layer photonic crystals at submillimeter wave frequencies', *IEEE Trans. Microwave Theory Tech.* **50**(10) (2002), pp. 2384–2393.

[28] B. Martinez, I. Ederra, R. Gonzalo, B. Alderman, L. Azcona, P. G. Huggard, B. de Hon, A. Hussain, S. R. Andrews, L. Marchand and P. de Maagt, 'Manufacturing tolerance analysis, fabrication, and characterization of 3-D submillimeter-wave electromagnetic-bandgap crystals', *IEEE Trans. Microwave Theory Tech.* **55**(4) (2007), pp. 672–681.

[29] W. J. Chappell, C. Reilly, J. Halloran and L. P. B. Katehi, 'Ceramic synthetic substrates using solid freeform fabrication', *IEEE Trans. Microwave Theory Tech.* **51**(3) (2003), pp. 752–760.

[30] Y. Lee, X. Lu, Y. Hao, S. Yang, R. Ubic, J. R. G. Evans and C. G. Parini, 'Rapid prototyping of ceramic millimetre wave metamaterials: Simulations and experiments', *Microwave Opt. Technol. Lett.* **49**(9) (2007), pp. 2090–2093.

[31] C. Mann *et al.*, 'Microfabrication of 3D terahertz circuitry', in *IEEE MTT-S Int. Symp. Dig.* (2003), pp. 739–742.

[32] A. Chutinan and S. Noda, 'Highly confined waveguides and waveguide bends in three–dimensional photonic crystal', *Appl. Phys. Lett.* **75**(24) (1999), pp. 3739–3741.

[33] A. Chutinan and S. Noda, 'Design for waveguides in three-dimensional photonic crystals', *Jpn. J. Appl. Phys.* **39**(4B) (2000), pp. 2353–2356.

[34] M. Bayindir, E. Ozbay, B. Temelkuran, M. M. Sigalas, C. M. Soukoulis, R. Biswas and K. M. Ho, 'Guiding, bending and splitting of electromagnetic waves in highly confined photonic crystals', *Phys. Rev.* B **63** (2001), 081107.

[35] A. R. Weily, K. P. Esselle, T. S. Bird and B. C. Sanders, 'Experimental woodpile EBG waveguides, bends and power dividers at microwave frequencies', *Electronics Lett.* **42**(1) (2006), pp. 32–33.

[36] A. Taflove and S. C. Hagness, *Computational Electrodynamics: The Finite-Difference Time-Domain Method*, 2nd edn (Boston, MA, Artech House, 2000).

[37] P. Harms, R. Mittra and W. Ko, 'Implementation of the periodic boundary condition in the finite-difference time-domain algorithm for FSS structures', *IEEE Trans. Antennas Propagat.* **42**(9) (1994), pp. 1317–1324.

[38] S. Kawashima, M. Okano, M. Imada and S. Noda, 'Design of compound-defect waveguides in three-dimensional photonic crystals', *Opt. Express* **14**(13) (2006), pp. 6303–6307.

[39] E. Ozbay, G. Tuttle, M. Sigalas, C. M. Soukoulis and K. M. Ho, 'Defect structures in a layer-by-layer photonic band-gap crystal', *Phys. Rev.* B **51**(20) (1995), pp. 13961–13965.

[40] A. Chelnokov, S. Rowson, J.-M. Lourtioz, L. Duvillaret and J. L. Coutaz, 'Light controllable defect modes in three-dimensional photonic crystal', *Electronics Lett.* **34**(20) (1998), pp. 1966–1967.

[41] M. Okano, A. Chutinan and S. Noda, 'Analysis and design of single-defect cavities in a three-dimensional photonic crystal', *Phys. Rev.* B **66**(16) (2002), 165211.

[42] S. Fan, P. R. Villeneuve, J. D. Joannopoulos and M. A. Haus, 'Channel drop tunneling through localized states', *Phys. Rev. Lett.* **80**(5) (1998), pp. 960–963.

[43] T. Euler and J. Papapolymerou, 'Silicon micromachined EBG resonator and two-pole filter with improved performance characteristics', *IEEE Microwave and Wireless Components Letters* **13**(9) (2003), pp. 373–375.

[44] P. Kohli, C. Christensen, J. Muehlmeier, R. Biswas, G. Tuttle and K. M. Ho, 'Add-drop filters in three-dimensional layer-by-layer photonic crystals using waveguides and resonant cavities', *Appl. Phys. Lett.* **89**(23) (2006), 231103.

[45] A. R. Weily, K. P. Esselle, T. S. Bird and B. C. Sanders, 'Woodpile EBG phase shifter', *Electronics Lett.* **42**(25) (2006), pp. 1463–1464.

[46] D. Sievenpiper, 'High-impedance electromagnetic surfaces', Ph.D. dissertation, Dept. Elect. Eng., Univ. California at Los Angeles, Los Angeles, CA, 1999.

[47] D. Sievenpiper, L. Zhang, R. F. J. Broas, N. G. Alexópoulos and E. Yablonovitch, 'High-impedance electromagnetic surfaces with a forbidden frequency band', *IEEE Trans. Microwave Theory Tech.* **47**(11) (1999), pp. 2059–2074.

[48] K. Ma, K. Hirose, F. Yang, Y. Qian and T. Itoh, 'Realization of magnetic conducting surface using novel photonic bandgap structure', *Electronics Lett.* **34** (1998), pp. 2041–2042.

[49] F. Yang and Y. Rahmat-Samii, 'Reflection phase characterizations of the EBG ground plane for low profile wire antenna applications', *IEEE Trans. Antennas Propagat.* **51**(10) (2003), pp. 2691–2703.

[50] E. R. Brown, C. D. Parker and E. Yablonovitch, 'Radiation properties of a planar antenna on a photonic-crystal substrate', *J. Opt. Soc. Am. B.* **10** (1993), pp. 404–407.

[51] S. D. Cheng, R. Biswas, E. Ozbay, S. McCalmont, G. Tuttle and K. M. Ho, 'Optimized dipole antennas on photonic band-gap crystals', *Appl. Phys. Lett.* **67**(23) (1995), pp. 3399–3401.

[52] M. M. Sigalas, R. Biswas, Q. Li, D. Crouch, W. Leung, R. Jacobs-Woodbury, B. Lough, S. Nielsen, S. McCalmont, G. Tuttle and K. M. Ho, 'Dipole antennas on photonic band-gap crystals – Experiment and simulation', *Microwave Opt. Technol. Lett.* **15**(3) (1997), pp. 153–158.

[53] W. Y. Leung, R. Biswas, S. D. Cheng, M. M. Sigalas, J. S. McCalmont, G. Tuttle and K. M. Ho, 'Slot antennas on photonic band gap crystals', *IEEE Trans. Antennas Propagat.* **45**(8) (1997), pp. 1569–1570.

[54] Y. Qian, R. Coccioli, D. Sievenpiper, V. Radisic, E. Yablonovitch and T. Itoh, 'A microstrip patch antenna using novel photonic band-gap structures', *Microwave J.* **42** (1999), pp. 66–76.

[55] R. Coccioli, F. R Yang, K. P. Ma and T. Itoh, 'Aperture-coupled patch antenna on UC-PBG substrate', *IEEE Trans. Microwave Theory Tech.* **47**(11) (1999), pp. 2123–2130.

[56] R. Gonzalo, P. de Maagt and M. Sorrolla, 'Enhanced patch-antenna performance by suppressing surface waves using photonic-bandgap substrates', *IEEE Trans. Microwave Theory Tech.* **47**(11) (1999), pp. 2131–2138.

[57] T. H. Liu, W. X. Zhang, M. Zhang and K. F. Tsang, 'Low profile spiral antenna with PBG substrate', *Electron. Lett.* **36**(9) (2000), pp. 779–780.

[58] F. Yang and Y. Rahmat-Samii, 'A low-profile circularly polarized curl antenna over an electromagnetic bandgap (EBG) surface', *Microwave Opt. Technol. Lett.* **31**(4) (2001), pp. 264–267.

[59] F. Yang and Y. Rahmat-Samii, 'Microstrip antennas integrated with electromagnetic band-gap (EBG) structures: a low mutual coupling design for array applications', *IEEE Trans. Antennas Propagat.* **51**(10) (2003), pp. 2936–2946.

[60] Z. Iluz, R. Shavit and R. Bauer, 'Microstrip antenna phased array with electromagnetic bandgap substrate', *IEEE Trans. Antennas Propagat.* **52**(6) (2004), pp. 1446–1453.

[61] R. Gonzalo, I. Ederra, C. Mann and P. de Maagt, 'Radiation properties of terahertz dipole antenna mounted on photonic crystal', *Electron. Lett.* **37**(10) (2001), pp. 613–614.

[62] I. Ederra, R. Gonzalo, B. Martinez, L. Azcona, B. Alderman, P. Huggard, B. P. de Hon, M. van Beurden, L. Marchand and P. de Maagt, 'Modifications of the woodpile structure for the improvement of its performance as substrate for dipole antennas', *IET Microwaves, Antennas Propagat.* **1**(1) (2007), pp. 226–233.

[63] G. V. Trentini, 'Partially reflecting sheet arrays', *IEEE Trans. Antennas Propagat.* **4**(4) (1956), pp. 666–671.

[64] A. P. Feresidis and J. C. Vardaxoglou, 'High gain planar antenna using optimized partially reflective surfaces', *IEE Proc. Microwaves Antennas Propagat.* **148**(6) (2001), pp. 345–350.

[65] Y. J. Lee, J. Yeo, R. Mittra, Y. Lee and W. S. Park, 'Design of high directivity electromagnetic bandgap (EBG) resonator antenna using a frequency selective surface (FSS) superstrate', *Microwave Opt. Technol. Lett.* **43**(6) (2004), pp. 462–467.

[66] A. P. Feresidis, G. Goussetis, S. Wang and J. C. Vardaxoglou, 'Artificial magnetic conductor surfaces and their application to low-profile high-gain planar antennas', *IEEE Trans. Antennas Propagat.* **53**(1) (2005), pp. 209–215.

[67] S. Wang, A. P. Feresidis, G. Goussetis and J. C. Vardaxoglou, 'High-gain subwavelength resonant cavity antenna based on metamaterial ground planes', *IEE Proc. Microwaves Antennas Propagat.* **153**(1) (2006), pp. 1–6.

[68] D. H. Lee, Y. J. Lee, J. Yeo, R. Mittra and W. S. Park, 'Design of a novel thin film frequency selective surface superstrate for dual-band directivity enhancement', *IET Microwaves, Antennas Propagat.* **1**(1) (2007), pp. 248–254.

[69] E. Rodes, M. Diblanc, E. Arnaud, T. Monédière and B. Jecko, 'Dual-band EBG resonator antenna using a single-layer FSS', *IEEE Antennas and Wireless Propagation Lett.* **6** (2007), pp. 368–371.

[70] A. Pirhadi, M. Hakkak, F. Keshmiri and R. K. Baee, 'Design of compact dual band high directive electromagnetic bandgap (EBG) resonator antenna using artificial magnetic conductor', *IEEE Trans. Antennas Propagat.* **55**(6) (2007), pp. 1682–1690.

[71] G. K. Palikaras, A. P. Feresidis and J. C. Vardaxoglou, 'Cylindrical electromagnetic bandgap structures for directive base station antennas', *IEEE Antennas and Wireless Propagation Lett.* **3** (2004), pp. 87–89.

[72] M. Diblanc, E. Rodes, E. Arnaud, M. Thèvenot, T. Monédière and B. Jecko, 'Circularly polarized metallic EBG antenna', *IEEE Microw. Wireless Compon. Letters* **15**(10) (2005), pp. 638–640.

[73] D. H. Lee, Y. J. Lee, J. Yeo, R. Mittra and W. S. Park, 'Directivity enhancement of circular polarized patch antenna using ring-shaped frequency selective surface superstrate', *Microwave Opt. Technol. Lett.* **49**(1) (2007), pp. 199–201.

[74] R. Sauleau, P. Coquet, T. Matsui and J.-P. Daniel, 'A new concept of focusing antennas using plane-parallel Fabry–Perot cavities with nonuniform mirrors', *IEEE Trans. Antennas Propagat.* **51**(11) (2003), pp. 3171–3175.

[75] R. Sauleau, P. Coquet and T. Matsui, 'Low-profile directive quasi-planar antennas based on millimeter wave Fabry–Perot cavities', *IEE Proc. Microwaves Antennas Propagat.* **150**(4) (2003), pp. 274–278.

[76] D. R. Jackson, A. A. Oliner and A. Ip, 'Leaky-wave propagation and radiation for a narrow-beam multiple-layer dielectric structure', *IEEE Trans. Antennas Propagat.* **41** (1993), pp. 344–348.

[77] D. R. Jackson and N. G. Alexopoulos, 'Gain enhancement methods for printed circuit antennas', *IEEE Trans. Antennas Propagat.* **33** (1985), pp. 976–987.

[78] H. Y. Yang and N. G. Alexopoulos, 'Gain enhancement methods for printed circuit antennas through multiple superstrates', *IEEE Trans. Antennas Propagat.* **35** (1987), pp. 860–863.

[79] M. Thèvenot, C. Cheype, A. Reineix and B. Jecko, 'Directive photonic-bandgap antennas', *IEEE Trans. Microw. Theory Tech.* **47** (1999), pp. 2115–2122.

[80] C. Serier, C. Cheype, R. Chantalat, M. Thèvenot, T. Monédière, A. Reineix and B. Jecko, 'one-dimensional photonic bandgap resonator antenna', *Microwave Opt. Technol. Lett.* **29**(5) (2001), pp. 312–315.

[81] A. R. Weily, K. P. Esselle, B. C. Sanders and T. S. Bird, 'High-gain 1D EBG resonator antenna', *Microwave Opt. Technol. Lett.* **47**(2) (2005), pp. 107–114.

[82] L. Leger, R. Granger, M. Thèvenot, T. Monédière and B. Jecko, 'Multifrequency dielectric EBG antenna', *Microwave Opt. Technol. Lett.* **40**(5) (2004), pp. 420–423.

[83] L. Leger, T. Monédière and B. Jecko, 'Enhancement of gain and radiation bandwidth for a planar 1-D EBG antenna', *IEEE Microw. Wireless Compon. Lett.* **15**(9) (2005), pp. 573–575.

[84] A. R. Weily, K. P. Esselle, T. S. Bird and B. C. Sanders, 'Dual resonator one-dimensional EBG antenna with slot array feed for improved radiation bandwidth', *IET Microwaves, Antennas Propagat.* **1**(1) (2007), pp. 198–203.

[85] R. Alkhatib, E. Marzolf and M. Drissi, 'Waveguide-fed directive antennas based on focusing system for millimeter-wave applications', *Microwave Opt. Technol. Lett.* **48**(8) (2006), pp. 1592–1594.

[86] A. R. Weily, K. P. Esselle, T. S. Bird and B. C. Sanders, 'High gain circularly polarized 1-D EBG resonator antenna', *Electronics Lett.* **42**(18) (2006), pp. 1012–1013.

[87] C. Cheype, C. Serier, M. Thèvenot, T. Monédière, A. Reineix and B. Jecko, 'An electromagnetic bandgap resonator antenna', *IEEE Trans. Antennas Propagat.* **50**(9) (2002), pp. 1285–1290.

[88] Y. J. Lee, J. Yeo, R. Mittra and W. S. Park, 'Application of electromagnetic bandgap (EBG) superstrates with controllable defects for a class of patch antennas as spatial angular filters', *IEEE Trans. Antennas Propagat.* **53**(1) (2005), pp. 224–235.

[89] M. Qiu and S. He, 'High directivity patch antenna with both photonic bandgap substrate and photonic bandgap cover', *Microw. Opt. Technol. Lett.* **30**(1) (2001), pp. 41–44.

[90] R. Biswas, E. Ozbay, B. Temelkuran, M. Bayindir, M. M. Sigalas and K.-M. Ho, 'Exceptionally directional sources with photonic-bandgap crystals', *J. Opt. Soc. Am. B* **18**(11) (2001), pp. 1684–1689.

[91] A. R. Weily, L. Horvath, K. P. Esselle, B. C. Sanders and T. S. Bird, 'A planar resonator antenna based on a woodpile EBG material', *IEEE Trans. Antennas Propagat.* **53**(1) (2005), pp. 216–223.

[92] Y. Lee, X. Lu, Y. Hao, S. Yang, R. Ubic, J. R. G. Evans and C. G. Parini, 'Directive millimetre-wave antenna based on freeformed woodpile EBG structure', *Electronics Lett.* **43**(4) (2007), pp. 195–196.

[93] K. C. Gupta, 'Narrow-beam antennas using an artificial dielectric medium with permittivity less than unity', *Electron. Lett.* **7** (1971), pp. 16–18.

[94] S. Enoch, G. Tayeb, P. Sabouroux, N. Guérin and P. Vincent, 'A metamaterial for directive emission', *Phys. Rev. Lett.* **89**(21) (2002), pp. 213902.

[95] G. Lovat, P. Burghignoli, F. Capolino, D. R. Jackson and D. R. Wilton, 'Analysis of directive radiation from a line source in a metamaterial slab with low permittivity', *IEEE Trans. Antennas Propagat.* **54**(3) (2006), pp. 1017–1030.

[96] D. Bonefačić, S. Hrabar and D. Kvakan, 'Experimental investigation of radiation properties of an antenna embedded in low permittivity thin-wire-based metamaterial', *Microwave Opt. Technol. Lett.* **48**(12) (2006), pp. 2581–2586.

[97] Z. B. Weng, N. B. Wang, Y. C. Jiao and F. S. Zhang, 'A directive patch antenna with metamaterial structure', *Microwave Opt. Technol. Lett.* **49**(2) (2007), pp. 456–459.

[98] S. Enoch, B. Gralak and G. Tayeb, 'Enhanced emission with angular confinement from photonic crystals', *Appl. Phys. Lett.* **81**(9) (2002), pp. 1588–1590.

[99] I. Bulu, H. Caglayan and E. Ozbay, 'Highly directive radiation from sources embedded inside photonic crystals', *Appl. Phys. Lett.* **83**(16) (2003), pp. 3263–3265.

[100] R. L. Moore, M. P. Kesler, J. G. Maloney and B. L. Shirley, US Patent 5,689,275, 1997.

[101] J.-M. Lourtioz, H. Benisty, V. Berger, J.-M. Gerard, D. Maystre and A. Tchelnokov, 'Photonic crystal THz imaging system', in *Photonic Crystals: Toward Nanoscale Photonic Structures* (Springer, Berlin, 2005) pp. 385–386.

[102] A. R. Weily, K. P. Esselle and B. C. Sanders, 'Photonic crystal horn and array antennas', *Phys. Rev. E* **68**(1) (2003), 16609.

[103] C. Chen, S. Shi, D. W. Prather and A. Sharkawy, 'Beam steering with photonic crystal horn radiators', *Opt. Engg.* **43**(1) (2004), pp. 174–180.

[104] A. R. Weily, K. P. Esselle and B. C. Sanders, 'Layer-by-layer photonic crystal horn antenna', *Phys. Rev. E* **70**(3) (2004), p. 37602.

[105] A. R. Weily, T. S. Bird, K. P. Esselle and B. C. Sanders, 'Linear array of woodpile EBG sectoral horn antennas', *IEEE Trans. Antennas Propagat.* **54**(8) (2006), pp. 2263–2274.

[106] M. Thèvenot, A. Reineix and B. Jecko, 'A dielectric photonic parabolic reflector', *Microwave Opt. Technol. Lett.* **21**(6) (1999), pp. 411–414.

[107] M. P. Kesler, J. G. Maloney and B. L. Shirley, 'Antenna design with the use of photonic band-gap materials as all-dielectric planar reflectors', *Microwave Opt. Technol. Lett.* **11**(4) (1996), pp. 169–174.

[108] G. S. Smith, M. P. Kesler and J. G. Maloney, 'Dipole antennas used with all-dielectric, woodpile photonic band-gap reflectors: gain, field patterns, and input impedance', *Microwave Opt. Technol. Lett.* **21**(3) (1999), pp. 191–196.

[109] D. F. Sievenpiper, J. H. Schaffner, H. J. Song, R. Y. Loo and G. Tangonan, 'Two-dimensional beam steering using an electrically tunable impedance surface', *IEEE Trans. Antennas Propagat.* **51**(10) (2003), pp. 2713–2722.

[110] I. Ederra, L. Azcona, B. E. J. Alderman, A. Laisné, R. Gonzalo, C. M. Mann, D. N. Matheson and P. de Maagt, 'A 250 GHz subharmonic mixer design using EBG technology', *IEEE Trans. Antennas Propagat.* **55**(11) (2007), pp. 2974–2982.

11

Millimeter-wave Electronic Switches

Jean-Olivier Plouchart

11.1 Introduction

RF and millimeter-wave (mmWave) switches are used in most of the wireless applications for duplexing and switching between frequency bands and mode [1, 2]. In the context of the system integration of multiple wireless standards, and since only one of the multiple radios is on at a given time, RF and mmWave switches are a critical function in keeping the system reliability, power efficiency and cost down [1]. In modern multi-transceiver systems, the switch functions are pervasive, and have onerous power-handling requirements. In fact, the switch function often limits the system efficiency and therefore mobile system operating time. Owing to these stringent requirements, electronic switches have been traditionally implemented in compound semi-conductor technologies [3]. Recent advances in silicon technologies have allowed designers to explore the design of high-frequency switches in complementary metal–oxide–semiconductor (CMOS) technology [4]. Also, the application of new switches in nanometer silicon technologies is emerging using digitally assisted analog circuits to compensate process variability. In this chapter we will first review switch applications in wireless communication systems. Next, we will define the relevant switch specifications, and show the impact of switch performance on a communication system. In the fifth section, we will use a small-signal approach to design a mmWave switch, and apply it in the sixth section to the design of a mmWave switch in 65 nm CMOS technology. In the sixth section we will also address some of the large signal design techniques. In the final section we will review some examples of switch implementations in a variety of technologies and provide an overview of the design challenges for solid-state technologies.

Advanced Millimeter-wave Technologies: Antennas, Packaging and Circuits
Duixian Liu, Ulrich Pfeiffer, Janusz Grzyb and Brian Gaucher © 2009 John Wiley & Sons, Ltd

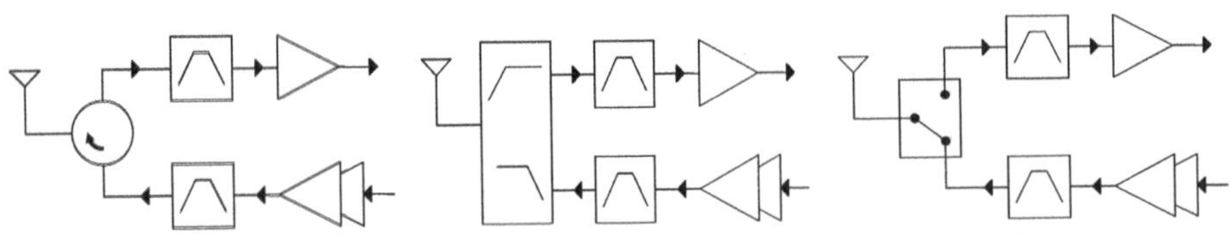

Figure 11.1: Front-end duplexing options: (a) circulator; (b) filter-bank or diplexer; (c) switching.

Table 11.1: Front-end duplexing option comparison.

	Circulator	Diplexer	Switch
Advantages	Load-pulling protection	Linearity Allows full-duplex	Highly scalable
Disadvantages	Bulky, heavy due to ferrite Narrow band	High-Q resonator required Losses from multiple filters	Linearity Bias and control signals

11.2 Switch Applications in mmWave Wireless Communication Systems

The switch function is usually taken for granted, but it can be extremely challenging to design, especially at mmWave frequency. As shown in Figure 11.1 in a classical transceiver architecture, where the antenna is shared between the transmitting and receiving sections, several options are possible. The first one is the use of a circulator relying on ferrite technology. The advantage of the circulator is the load-pulling protection, but it is usually bulky, heavy and narrow band and is therefore difficult to use for mobile applications.

The second option is the use of a diplexer, which allows full-duplex communication, as required, for example, for code division multiple access (CDMA) application. The other advantage of the diplexer is its linearity, due to the use of passive filters. Despite its intrinsic linearity advantage, it is worth noting that finite isolation between the transmitter and receiver implies higher linearity requirements for the receiver. The disadvantages are that a high-Q resonator is required and the losses can be as high as those of other solutions. The third option is the use of an electronic switch. This can be used in the half-duplex system to share the transmitting and receiving antenna. The advantage of the electronic switch is that it is highly scalable. However, it requires some biasing circuits and control signals. Table 11.1 summarizes the trade-off between the different options.

As shown in Figure 11.2, multiple switches can be used in a multi-band wireless communication TDMA system such as GSM and DCS1800. Depending on the transmitting or receiving band used, different banks of filters, either the INB1/OUTB1 or INB2/OUTB2 transmit and receive section, are selected. Another application, shown in Figure 11.3, is in

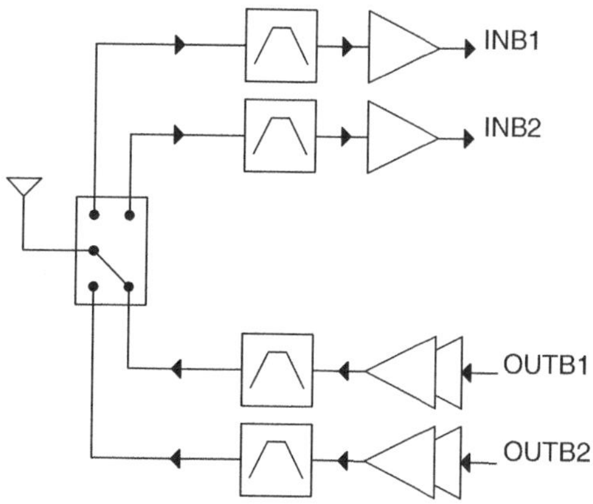

Figure 11.2: Multi-band front-end switch applications.

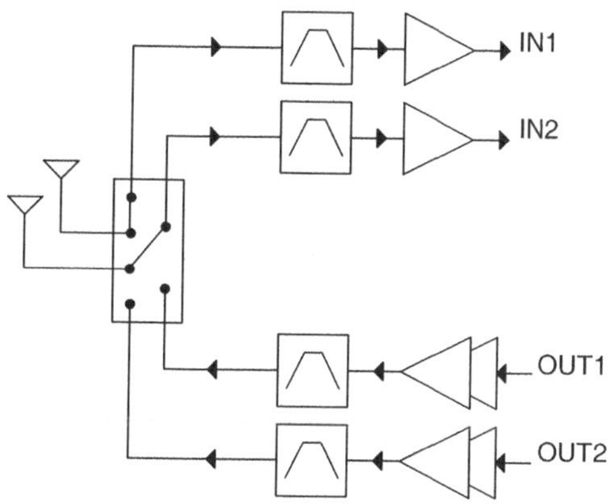

Figure 11.3: 2 × 2 multiple input multiple output front-end switch applications.

the multiple input multiple output (MIMO) receiver using spatial diversity. Spatial diversity improves the signal-to-noise ratio in the presence of multiple transmission paths, where some frequencies tend to be attenuated at one antenna but not at the other. Therefore MIMO can improve the transmission bit rate. For a 2 × 2 MIMO transceiver, a switch with six high-frequency ports is required.

Electronic switches are critical not only in the front-end but also to enable new circuit functions. An example is shown in Figure 11.4, where switches are used to switch delay line and generate programmable phase shift in order to enable phase array systems. The specification for this type of switch can be very different from a switch used in a front-end transceiver. For this type of circuit switch high-voltage handling as compared to a front-end switch is usually not required.

Another emerging application of the high-frequency switch is in CMOS or bipolar CMOS (BiCMOS) circuits where the switch function is used inside to compensate the manufacturing

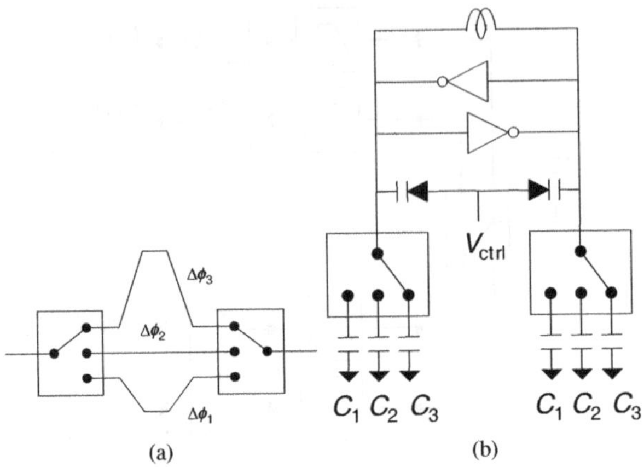

Figure 11.4: Example of switch in circuit applications: (a) delay line or phase shift switch; (b) capacitor switch in voltage-controlled oscillator.

variability of RF and mmWave functions. An example of such a circuit is the use of a binary weighted capacitor bank in a voltage-controlled oscillator (Figure 11.4). This programmable capacitor bank can be used to compensate process, supply and temperature variations. With the use of low-cost, high-density digital CMOS circuits, analog-to-digital and digital-to-analog, a complex calibration algorithm can be implemented to assist the analog and high-speed circuits in adapting to process and environmental variability.

As shown in Figures 11.2, 11.3 and 11.4, switches are used in a variety of configurations. The simplest switch is the single pole single throw (SPST). The SPST switch is a simple on–off switch. A single pole double throw (SPDT) switch can be built by using two SPST switches that are alternatively connected to a single pole. The SPDT switch is used, for example, to connect an antenna to either the receiver or the transmitter. The concept can be expanded to an arbitrary number of α throws and β poles in order to design a αPβT switch. In MIMO applications αPβT switches are used to connect multiple antennas to multiple transmitters and receivers.

11.3 Switch Specifications

The most important specifications are provided in Table 11.2. Small-signal performances are measured with a vector network analyzer. For the SPST series switch, the insertion loss (IL) is defined as the measured scattering coefficient of transmission $-S_{21}$ or $-S_{12}$ in dB when the switch is turned on. The SPST switch can also be used in parallel, and in that case the IL is defined as the measured scattering coefficient of transmission $-S_{21}$ or $-S_{12}$ in dB when the switch is turned off. The series and parallel SPST switch return loss (RL) are defined as the measured scattering coefficient of reflection $-S_{11}$ and $-S_{22}$ in dB when the switches are turned on for the series configuration and turned off for the parallel configuration. The series and parallel SPST switches isolation (I) are defined as the measured scattering coefficient of transmission $-S_{21}$ or $-S_{12}$ in dB when the switch is turned off for the series configuration

Table 11.2: Key switch specifications.

Small signal	Large signal	Control
Insertion loss	Power handling	Actuation voltage
Return loss	Distortion, 1 dB	On–off, off–on
Isolation	Compression	Switching time
Bandwidth	Intermodulation, IP3	

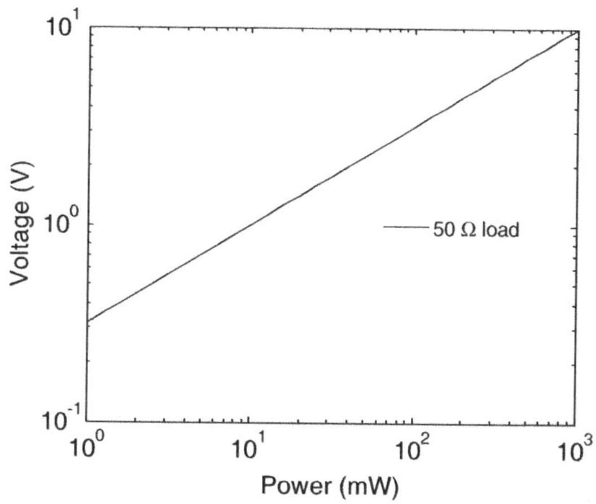

Figure 11.5: Voltage on 50 Ω load versus power.

and turned on for the parallel configuration. In the next sections, as commonly used in most publications, we will plot the scattering coefficients without inverting their signs. Therefore the insertion, isolation and return loss will appear as negative numbers in the plots. The bandwidth is defined as the frequency range where the IL and I is better than the system target specification. The switch frequency bandwidth can also be defined when the IL degrades by an arbitrary amount. These definitions can be expanded easily to αPβT switches.

Large signal performances are measured with a synthesizer exhibiting broad output power control, and with a high-dynamic range spectrum analyzer. This measurement can be extremely challenging, especially at mmWave frequency. One of the most important switch specifications is how much power it can handle without degrading the switch reliability. In a 50 Ω load environment, as shown in Figure 11.5, the switch will have to sustain voltage as high as 10 V for a 1 W power and 1 V for a 10 mW power delivered on a 50 Ω load-impedance.

In real applications the antenna can be close to some metal plate and the impedance seen by the switch is other than 50 Ω. The switch will have to sustain higher voltage. The specification is given by providing the voltage standing wave ratio (VSWR), which can be as high as 1 to 15 for RF applications. However, for mmWave applications the VSWR requirement is not as stringent since the wavelength is of the order of a few millimeters and

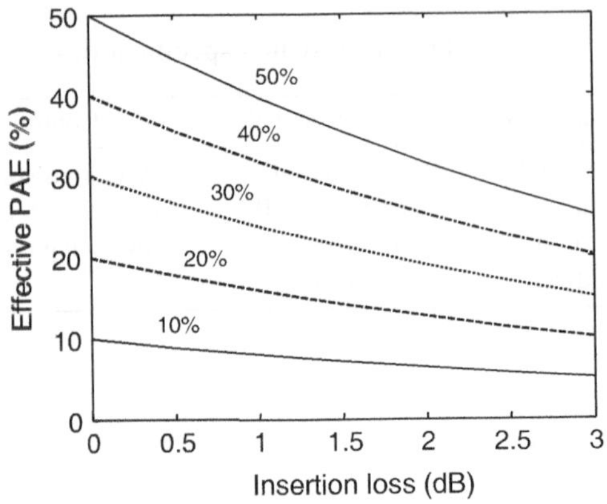

Figure 11.6: Insertion loss impact on effective power added efficiency for power amplifier with PAE of 10 to 50%.

a metal plate would be for the worst scenario at several wavelengths away from the antenna where the transmitted power would already be significantly reduced. Therefore the reflected power would not be able to induce a significant VSWR. Other large signal parameters, resulting from the non-linear switch property, are of how much distortion and intermodulation the switch generates. The distortion is measured by increasing the input carrier power up to the point where the IL is compressed by 1 dB. For multi-carrier communication systems low intermodulation is required in order to avoid jamming the adjacent channels. The intermodulation is measured by injecting two carrier frequencies f_1 and f_2 at the switch input and by measuring the intermodulation powers at $(2f_1 - f_2)$ and $(2f_2 - f_1)$. When the f_1 and f_2 carrier power is increased, the intermodulation powers at $(2f_1 - f_2)$ and $(2f_2 - f_1)$ increase three times faster than the carrier power f_1 and f_2. The extrapolated point where the intermodulation and carrier power intersect is defined as the third-order intercept point or IP3. The other important switch specifications relate to the switch control such as actuation voltage and the time required for the switch to change state from on to off and off to on.

11.4 Impact of Switch Performance on Communication System

One of the most important factors for mobile communication systems using batteries is the impact of the switch insertion loss on the power amplifier power added efficiency (PAE). As shown in Figure 11.6, an RF amplifier with a typical 50% PAE will have an effective PAE of 40% for a 1 dB insertion loss, and this therefore reduces significantly the transmit time.

For mmWave applications, typical PAE reported for power amplifiers is of the order of 10 to 20%. In that case the impact of a 1 dB IL mmWave switch is only a PAE reduction by 1 to 3% (Figure 11.6). Therefore, at mmWave frequency more losses can be tolerated in the switch, without impacting significantly the system efficiency. This is valid as long as mmWave power amplifier efficiency remains low. One can argue that any field effect transistor (FET) technology improving the solid-state mmWave amplifier's efficiency is also

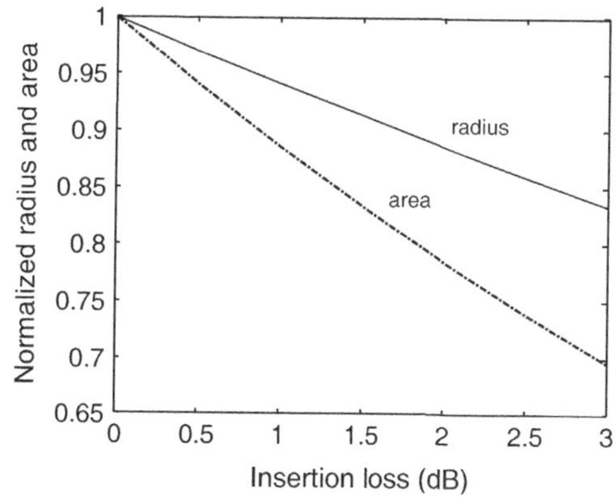

Figure 11.7: Insertion loss impact on receiver sensitivity.

likely to improve the electronic switch's IL. Therefore FET technologies for switch and power amplification are tightly coupled. However, it is worth noting that any IL degrades by the same amount as the transmitted output power, and therefore the link budget. This could be an unacceptable penalty at the mmWave frequency, since it is difficult to generate a large amount of power at such a frequency. On the receiver side IL reduces the receiver sensitivity and therefore the radius and area that can be covered by the communication system. As shown in Figure 11.7 a 1 dB insertion loss reduces the maximum radius and area covered by the communication system by more than 5 and 10% respectively. This can be an issue because at mmWave frequency it is difficult to generate power and the link budget can therefore be tight.

11.5 Small-signal mmWave Switch Design

11.5.1 Series SPST Switch First-order Model

A first-order model of a switch can be a resistor to model the on state and a capacitor to model the off state. This simple model can be used to model diode and FET switches. For a series switch, the IL decreases as the on-resistor decreases. The insertion loss in dB for a series switch can be computed simply from the switch r_{on} in a Z_0 environment by

$$IL_{\text{series}} = 20\log(1 + r_{\text{on}}/2Z_0)\ [\text{dB}] \tag{11.1}$$

With this first-order model, the IL does not depend on frequency. The IL increases at 0.085 dB per Ω off switch on-resistor in a 50 Ω system. For a given on-resistor the insertion loss can be decreased by increasing the reference impedance Z_0. This might be possible for on-chip signals, or if the antenna is integrated on-chip. That can be a very interesting knob to increase mmWave system performance.

For a series switch, the I increases as the off-capacitor decreases. The I in dB for a series switch can be computed simply from the switch c_{off} in a Z_0 environment by

$$I_{\text{series}} = 10\log[1 + (4\pi F c_{\text{off}} Z_0)^{-2}]\ [\text{dB}] \tag{11.2}$$

The I degrades or decreases as the frequency F increases.

11.5.2 Shunt SPST Switch First-order Model

The insertion loss in dB for a shunt switch can be computed simply from the switch c_{off} in a Z_0 environment by

$$IL_{\text{shunt}} = 10\log[1 + (\pi F c_{\text{off}} Z_0)^2]\ [\text{dB}] \tag{11.3}$$

The shunt IL degrades or increases as the frequency F and off-capacitor increase.

For a shunt switch, the I increases as the on-resistor decreases. The I in dB for a series switch can be computed simply from the switch r_{on} in a Z_0 environment by

$$I_{\text{shunt}} = 20\log[1 + Z_0/2r_{\text{on}}]\ [\text{dB}] \tag{11.4}$$

For a given on-resistor the I can be increased by increasing the reference impedance Z_0. The I does not depend on frequency with this first-order model.

11.5.3 Series–shunt SPST Switch First-order Model

The insertion loss in dB for a series–shunt switch can be computed from the switch r_{on} and c_{off} in a Z_0 environment by

$$IL_{\text{series–shunt}} = 10\log\left[\left(1 + \frac{r_{\text{on}}}{2Z_0}\right)^2 + (\pi F c_{\text{off}})^2 (Z_0 + r_{\text{on}})^2\right]\ [\text{dB}] \tag{11.5}$$

The isolation in dB for a series–shunt switch can be computed from the switch r_{on} and c_{off} in a Z_0 environment by

$$I_{\text{series–shunt}} = 10\log\left[\left(1 + \frac{Z_0}{2r_{\text{on}}}\right)^2 + \left(\frac{1}{4\pi F c_{\text{off}} Z_0}\right)^2 \left(1 + \frac{Z_0}{r_{\text{on}}}\right)^2\right]\ [\text{dB}] \tag{11.6}$$

Both the $IL_{\text{series–shunt}}$ and the $I_{\text{series–shunt}}$ degrade as frequency increases. However, as we will see in the following sections, a better trade-off between isolation and insertion-loss can be achieved.

11.5.4 Switch Figure-of-merit

From the previous section we can derive a first-order switch figure-of-merit (FOM) based on r_{on} and c_{off}. The switch FOM is a time constant given by

$$FOM_{\text{switch}} = r_{\text{on}} c_{\text{off}}\ [\text{s}] \tag{11.7}$$

The best solid-state technology will provide the lowest FOM switch time constant. Also, r_{on} is inversely proportional to the total switch size, and proportional either to the lithography for planar devices, or the semi-conductor layer thicknesses for vertical devices. Therefore the designer can scale r_{on} to target a specified IL. However, to the first order the FOM_{switch} is technology, and lithography or epitaxy dependent only. Therefore, for a given technology, the circuit designer can only trade I for IL.

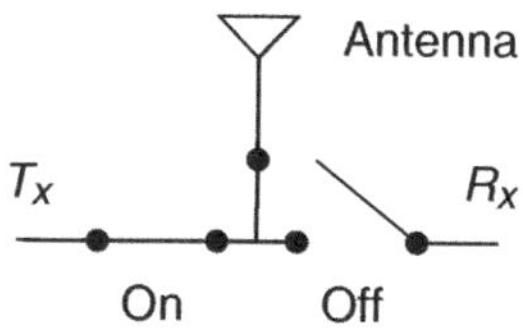

Figure 11.8: SPDT switch using two switches in the series.

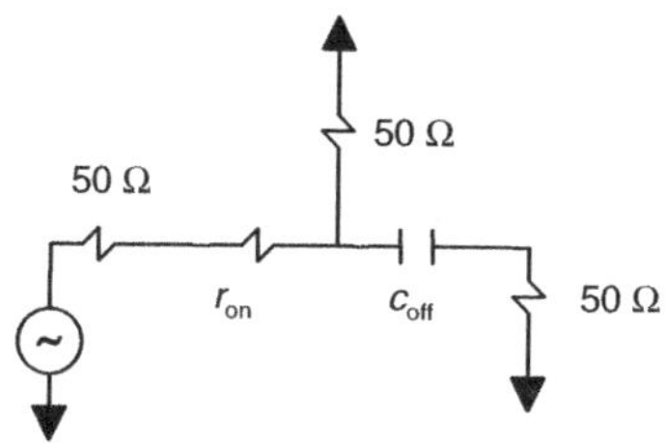

Figure 11.9: First-order electrical SPDT switch model driven in a 50 Ω environment.

11.5.5 SPDT with Series Switches

Figure 11.8 shows one of the simplest SPDT switches that can be designed using only two switches in the series. Figure 11.9 shows the equivalent electrical circuit using a first-order model driven in a 50 Ω environment.

From Equation (11.1), for an arbitrary target IL of 0.34 dB, the series switch r_{on} is 4 Ω. Figure 11.10 shows the insertion loss versus frequency for a 4 Ω on-resistor, for different switch technology time-constant FOMs ranging from 25 to 200 fs. As the frequency is decreased, all the technologies converge to a 0.34 dB IL. As the frequency increases the off-capacitor impedance decreases, the isolation from the second port decreases and therefore the IL increases. If we can tolerate an IL as high as 1 dB, technologies of less than 100 and 50 fs can be used up to 50 and 98 GHz respectively.

In order to achieve isolation higher than 10 dB, 100 and 50 fs technologies can be used up to 50 and 98 GHz respectively (Figure 11.11). A 25 fs technology would be ideal for a broadband switch design up to 140 GHz, but it is a great challenge for solid-state technologies. A 25 fs technology FOM is equivalent to a 6.4 THz cut-off frequency FOM, which to the best of the author's knowledge has not yet been reported for any electronic switch. It is also clear that a 200 fs technology is not suitable for the design of a broadband SPST switch, since, for example, there is only a 3.2 dB difference between the I and IL at 60 GHz.

11.5.6 SPDT with Series and Shunt Switches

One way to improve isolation without degrading significantly the IL is by adding a switch in parallel for each switch in the series, as shown in Figures 11.12 and 11.13. Although both switches are implemented with the same technology as with $r_{on} \times c_{off}$ FOM, for the parallel

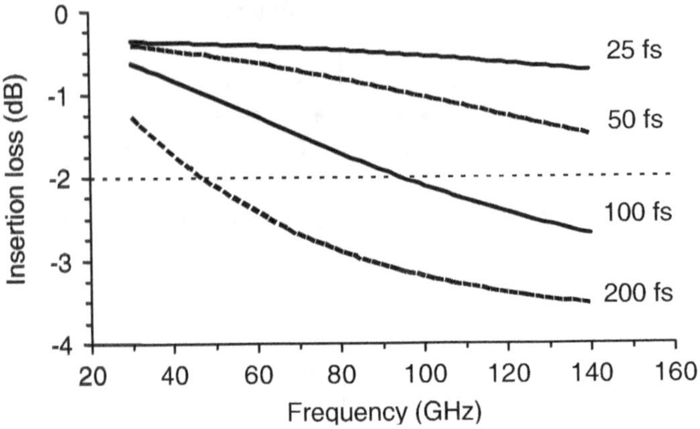

Figure 11.10: Insertion loss for simple SPDT switch with $r_{on} = 4\ \Omega$.

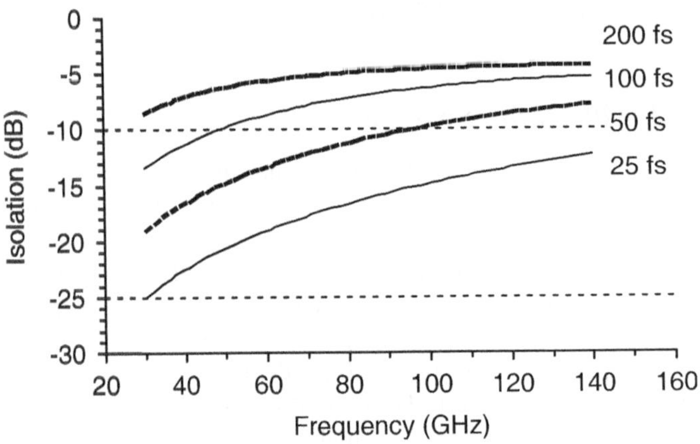

Figure 11.11: Isolation for simple SPDT switch with $r_{on} = 4\ \Omega$.

switch the on-resistor is chosen as r_{shunt} and can be different from the series on-resistor r_{series}. This can be accomplished with solid-state technologies by scaling the switch width and/or length.

We choose first an arbitrary 12 Ω shunt resistor. We will next show how to optimize the shunt switch resistor. As shown in Figure 11.14 the IL does not degrade significantly for the 25–100 fs technologies. However the isolation is improved by more than 13 dB at 60 GHz between the simple series and series–shunt SPDT switches for the 25–100 fs technologies (Figure 11.15).

Another important specification not yet verified here is the RL. The RL is simulated in Figure 11.16 for a 4 and 12 Ω series and shunt switch respectively. A 12 dB RL cannot be achieved with a 200 fs technology. However, it can be achieved with 50 and 100 fs

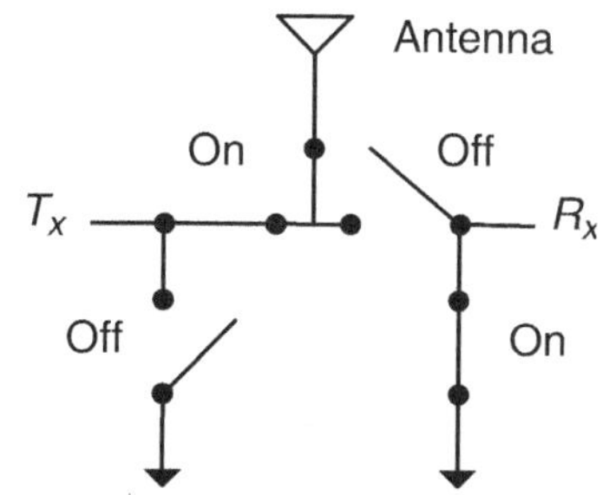

Figure 11.12: SPDT switch using four switches in series and parallel.

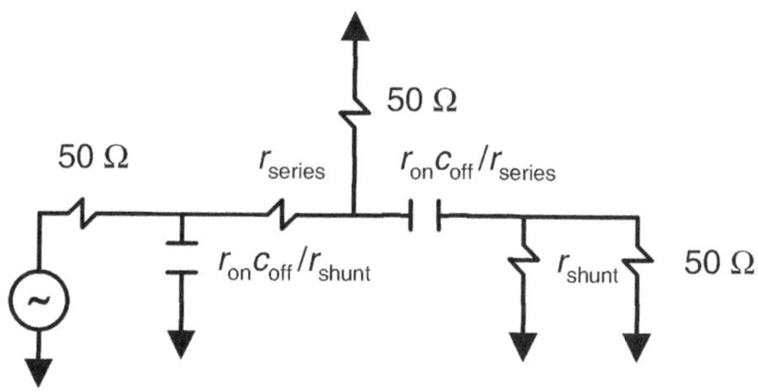

Figure 11.13: First-order electrical SPDT switch model driven in a 50 Ω environment using series–shunt switches.

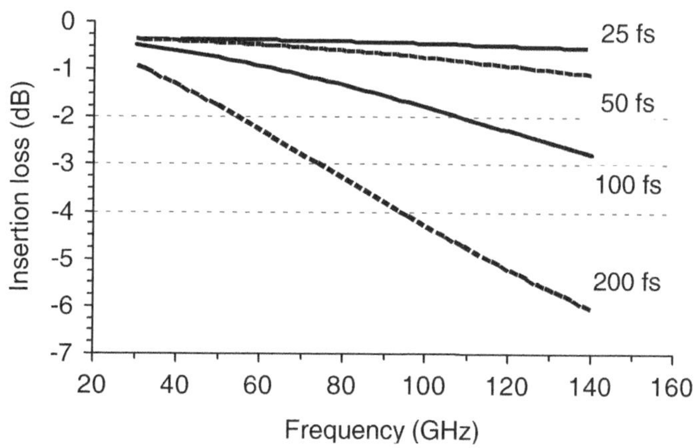

Figure 11.14: Insertion loss for series–shunt SPDT switch with $r_{series} = 4\ \Omega$, $r_{shunt} = 12\ \Omega$.

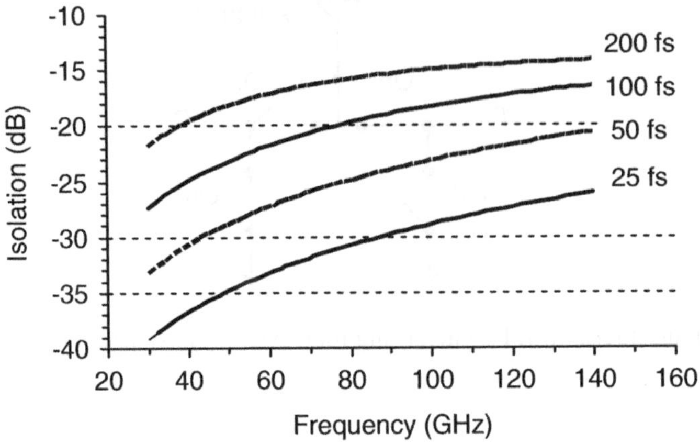

Figure 11.15: Isolation for series–shunt SPDT switch with $r_{series} = 4\ \Omega$, $r_{shunt} = 12\ \Omega$.

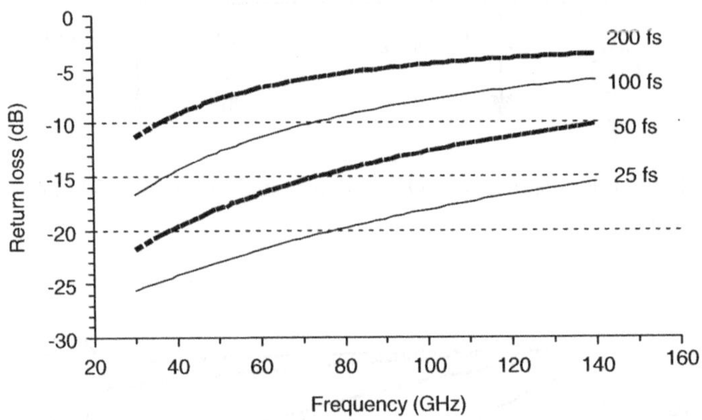

Figure 11.16: Return loss for series–shunt SPDT switch with $r_{series} = 4\ \Omega$, $r_{shunt} = 12\ \Omega$.

technologies up to a maximum frequency of 109 and 55 GHz respectively. With a 25 fs technology we could achieve a RL better than 15 dB up to 140 GHz.

11.5.7 SPDT with Series and Shunt Switches and Matching Inductor

Poor return loss results in mismatch loss and therefore degrades IL. One way to improve the return loss is to create a lumped-element transmission line by using series inductors as shown in Figure 11.17. If r_{series} and r_{shunt} impedances are much smaller than c_{series} and c_{shunt} impedances, then c_{series} and c_{shunt} are effectively in parallel. We have a T section of a lumped-element transmission line with l_{series} in parallel with l_{series} and c_{shunt} plus c_{series}.

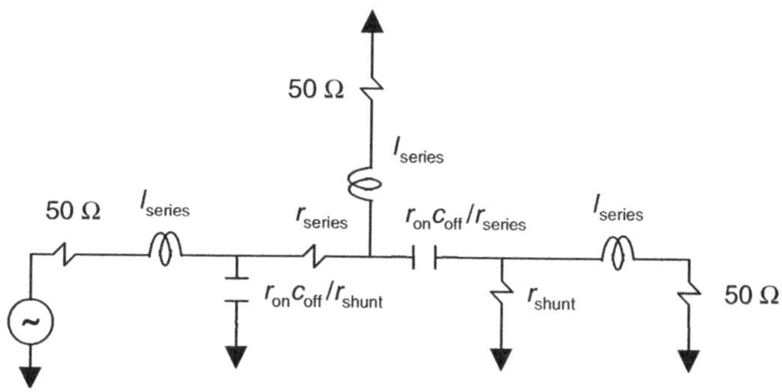

Figure 11.17: SPDT with inductor using two switches in series and parallel driven in a 50 Ω environment.

We can therefore calculate the l_{series} as

$$l_{series} = Z_0^2(c_{series} + c_{shunt})\ [H] \tag{11.8}$$

Up to now we have chosen an arbitrary r_{shunt} of 12 Ω; however, the shunt resistor can be optimized to achieve minimum IL. If the shunt resistor is too large, the I in the off section is degraded, and therefore the IL is degraded. If the shunt resistor is too small, the shunt capacitor is too large and therefore the IL is degraded. To simplify the optimization procedure, we chose to fix the frequency at 60 GHz and the series resistor at 4 Ω. As calculated from equation (11.1), a 4 Ω series resistor leads to 0.34 dB attenuation in a 50 Ω environment. As shown in Figure 11.18, a minimum IL is achieved at 60 GHz, depending on the technology FOM, for different optimum values of r_{shunt}. The optimum r_{shunt} value decreases as the technology FOM is decreased. This is because of the $r_{on} \times c_{off}/r_{shunt}$ dependency of the shunt capacitor. For 50, 100 and 150 fs technologies a minimum 60 GHz IL of 0.23, 0.3 and 0.85 dB respectively is achieved for a shunt resistor of 1.2, 3.3 and 7.2 Ω respectively. At 60 GHz the return loss is also dependent on the shunt resistor, as shown in Figure 11.19. For the 50 fs technology a RL of 33 dB is achieved for a r_{shunt} of 1.3 Ω. The sharp transition of less than half an ohm around the minimum could be a challenge for the modeling and manufacturing of such a high-performance switch. For technology FOM beyond 100 fs the transition is more relaxed and therefore the required manufacturing control is not as tight. For the 150 fs technology a 24.7 dB minimum RL is achieved for a r_{shunt} of 8.9 Ω. At the minimum IL for the 150 fs technology, achieved with a r_{shunt} of 7.2 Ω, the RL is 19.7 dB. Figure 11.19 shows also that it is difficult to achieve a RL better than 15 dB at 60 GHz with a 200 fs technology.

As shown in Figure 11.20, the 60 GHz isolation degrades as r_{shunt} is increased. The isolation is better than 28.4, 22.28, 18.8 and 17.7 dB for the 50, 100, 150 and 200 fs technologies respectively. At the minimum IL for the 150 fs technology, achieved with a r_{shunt} of 7.2 Ω, the isolation is 21.3 dB.

Owing to the model flexibility, we can now fix the r_{series} and r_{shunt} at 4 Ω and 7.2 Ω respectively, and verify the switch performance as a function of technology FOM. A sweet spot is achieved for a 138 fs technology, as shown in Figure 11.21. At 60 GHz, for a 138 fs technology, values of 0.7, 29.3 and 21.9 dB for IL, RL and I are achieved respectively.

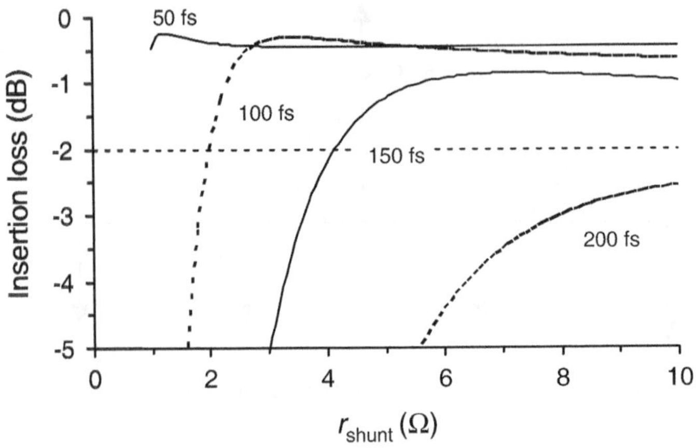

Figure 11.18: Insertion loss at 60 GHz for series–shunt SPDT switch with inductor for $r_{series} = 4\ \Omega$ as function of r_{shunt}.

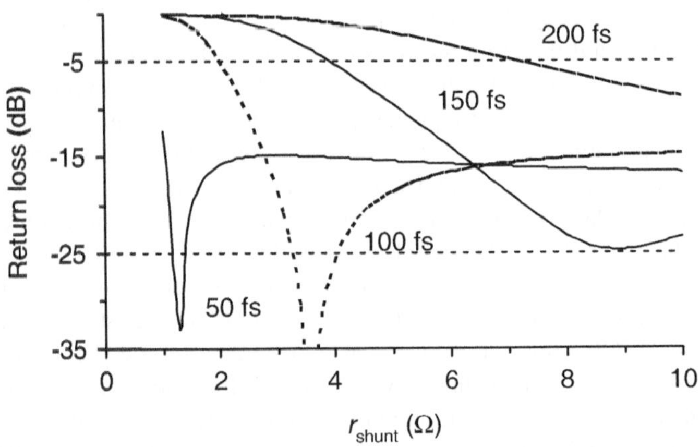

Figure 11.19: Return loss at 60 GHz for a series–shunt SPDT switch with inductor for $r_{series} = 4\ \Omega$ as a function of r_{shunt}.

For a SPDT switch with inductor with r_{series} and r_{shunt} at 4 Ω and 7.2 Ω respectively we can compute the IL versus frequency. As shown in Figure 11.22, for the 50 fs technology an IL of less than 0.56 dB is achieved up to 140 GHz. For the 100, 150 and 200 fs technologies, an IL of less than 1 dB is achieved up to 93.8, 63 and 45 GHz respectively. For the SPDT switch without inductor, an IL of less than 1 dB is achieved up to 129, 65.2 and 32.2 GHz for the 50, 100 and 200 fs technologies respectively. Therefore, by using the inductor matching technique, we can increase the switch bandwidth by 40 and 44% for the 200 and 100 fs technologies respectively.

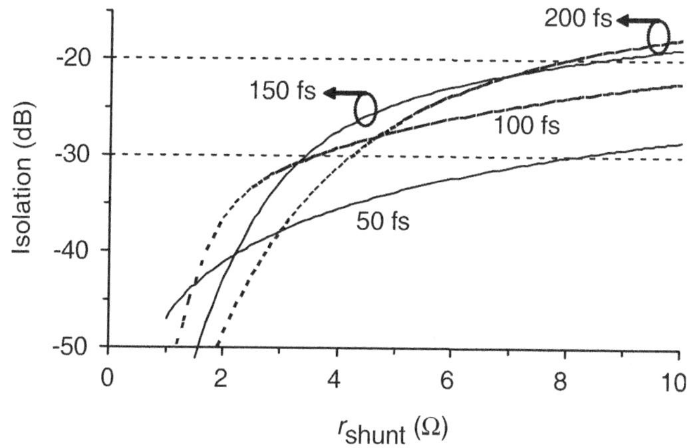

Figure 11.20: Isolation at 60 GHz for a series–shunt SPDT switch with inductor for $r_{series} = 4\ \Omega$ as a function of r_{shunt}.

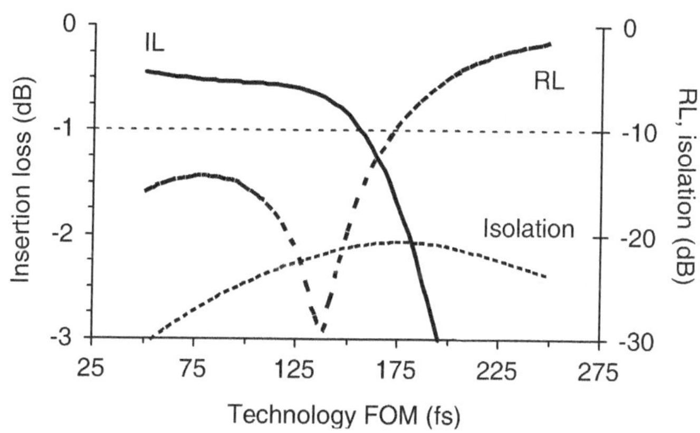

Figure 11.21: Insertion, return loss and isolation at 60 GHz for a series–shunt SPDT switch with inductor for $r_{series} = 4\ \Omega$, $r_{shunt} = 7.2\ \Omega$ as a function of technology FOM.

The I shown in Figure 11.23 is better than 23.3 dB up to 140 GHz for the 50 fs technology. For the 100, 150 and 200 fs technologies, the I is better than 20.7 dB up to 140 GHz. The I decreases with frequency and reaches a minimum, beyond which the I increases again, owing to the rapid increase of the IL. By using the inductor matching technique, the 60 GHz isolation is 30.7, 24.6, 21.2 and 21.2 dB for the 50, 100, 150 and 200 fs technologies respectively. This is a 3.5, 3 and 4.1 dB isolation improvement for the 50, 100 and 200 fs technologies over the SPDT switch without inductor respectively.

The most important improvement achieved by using the inductor-matching technique is for the RL. As shown in Figure 11.24, a better than 14.5 dB RL is achieved up to 47.6, 63, 93.8 and 140 GHz for the 200, 150, 100 and 50 fs technologies respectively. For the SPDT

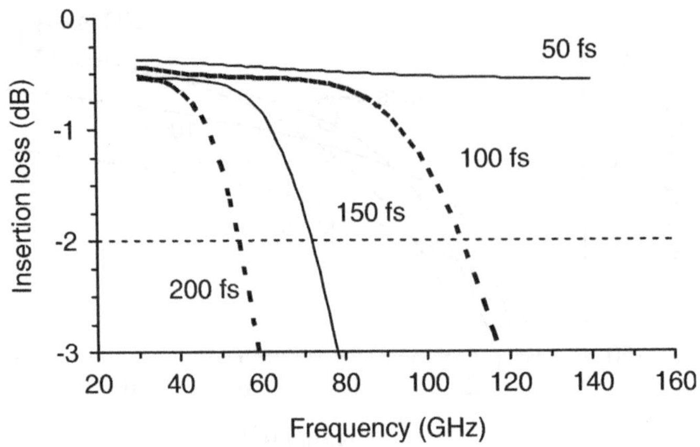

Figure 11.22: Insertion loss for a series–shunt SPDT switch with inductor for $r_{series} = 4\ \Omega$, $r_{shunt} = 7.2\ \Omega$.

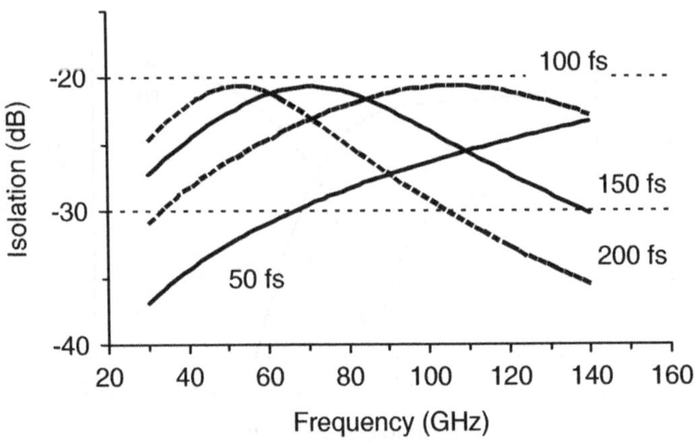

Figure 11.23: I for a series–shunt SPDT switch with inductor for $r_{series} = 4\ \Omega$, $r_{shunt} = 7.2\ \Omega$.

switch without inductor (Figure 11.16), a better than 14.5 dB RL is achieved up to 38.8, and 78.4 GHz for the 100 and 50 fs technologies, respectively. This is a bandwidth increase of more than 78% for the SPDT switch with inductor matching as compared with the SPDT switch without inductor matching.

Another advantage of using the series inductor is that the package inductor lead can be part of the matching inductor. This is an efficient way of using the package parasitic inductor as the switch-matching network. However, several challenges remain for mmWave switch packages. For mmWave application, as we have seen, low $r_{on} \times c_{off}$ is required and therefore the matching inductor is small. This leads to the use of advanced packages exhibiting a low parasitic inductor. The second challenge is the ground parasitic inductor, which degrades the switch performance. Once again low lead parasitic packages are required, or design

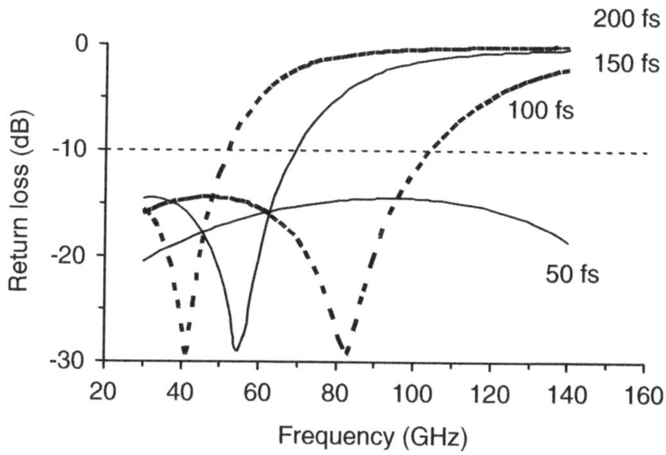

Figure 11.24: Return loss for a series–shunt SPDT switch with inductor for $r_{\text{series}} = 4\ \Omega$, $r_{\text{shunt}} = 7.2\ \Omega$.

techniques such as the use of a coplanar waveguide (CPW) or differential architecture can in certain cases be used.

11.6 Solid-state Switch Implementation

11.6.1 PIN Diode Switch

The p-type, insulator, n-type semiconductor (PIN) diode is a two-terminal device widely used for RF and mmWave switch applications [5], owing to its excellent switch FOM. The PIN diode is based on a high-resistivity intrinsic semi-conductor sandwiched between highly doped p-type and n-type regions, as shown in Figure 11.25. The PIN diode IV DC characteristics are shown in Figure 11.26. When the PIN diode is reverse biased or the V_{PN} is negative, there are no charges in the intrinsic region I. The PIN diode is off, and a first-order equivalent model is a capacitor $c_{\text{off-pin}}$. The $c_{\text{off-pin}}$ capacitor is proportional to the diode area and the dielectric constant, and inversely proportional to the intrinsic semiconductor thickness T_i (11.9). Therefore, the I can be increased by increasing the intrinsic layer thickness T_i, and/or reducing the area A

$$c_{\text{off-pin}} = \frac{\epsilon_0 \epsilon_i A}{T_i}\ [\text{F}] \tag{11.9}$$

where ϵ_0 is the vacuum permittivity, and ϵ_i is the intrinsic region I relative dielectric constant.

When the PIN diode is forward biased or the V_{PN} is positive, charges are injected in the intrinsic region I. Charges stay in the intrinsic I region for an average carrier lifetime τ_c. This results in an average charge Q in I, which lowers the I region resistivity. The PIN diode is in the on state, and a first-order equivalent model is a resistor $r_{\text{on-pin}}$ (Figure 11.26). The PIN

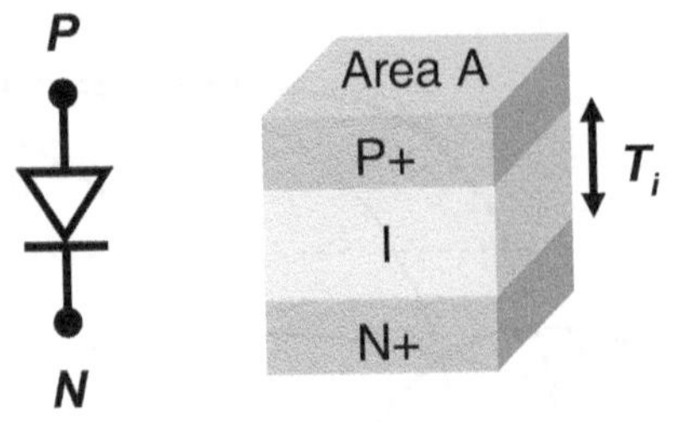

Figure 11.25: PIN diode symbol and three-dimensional material cross-section.

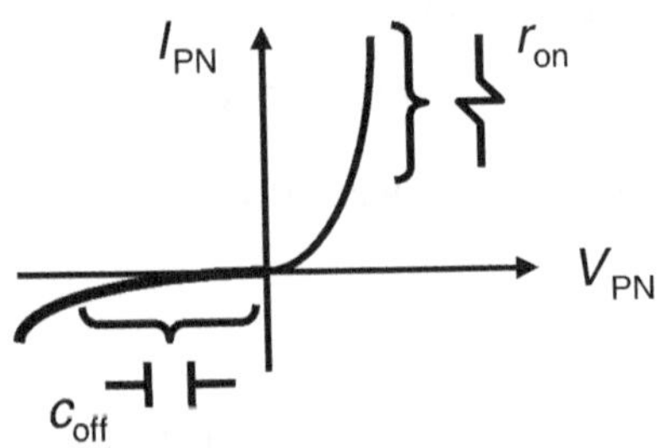

Figure 11.26: PIN diode I–V characteristics and first-order equivalent circuit.

diode on resistor can be calculated using the following equation [5]

$$r_{\text{on-pin}} = \frac{T_i^2}{(\mu_n + \mu_p)Q} \ [\Omega] \tag{11.10}$$

where T_i is the intrinsic I region thickness, μ_n is the electron mobility, μ_p is the hole mobility and $Q = I_{\text{PN}}\tau_c$ is the I region charge. The insertion loss can be decreased by decreasing the intrinsic layer thickness T_i, and/or increasing the carrier mobility, lifetime and bias current. The first-order $r_{\text{on-pin}}$ equation does not take into account the semi-conductor contact resistance P+ and N+. The semi-conductor contact resistance can be a significant factor limiting how low the on resistor can be. As we can see, one of the drawbacks of PIN diode switches is that a forward DC bias current is required to turn the switch on.

The PIN diode FOM is $r_{\text{on-pin}} \times c_{\text{off-pin}}$; therefore, we have

$$FOM_{\text{pin}} = \frac{\epsilon_0 \epsilon_i A T_i}{(\mu_n + \mu_p) I_{\text{PN}} \tau_c} \ [\text{s}] \tag{11.11}$$

The FOM_{pin} can be decreased or improved by decreasing the area A, the I region thickness, and/or increasing the carriers mobility, lifetime and bias current. Several important limitations must be considered; first, decreasing the area can limit how much power the diode can handle; second, higher mobility materials such as compound semi-conductors can have higher contact resistance than lower mobility material such as silicon, thus negating the material mobility improvement advantage. In narrow band applications such as wireless

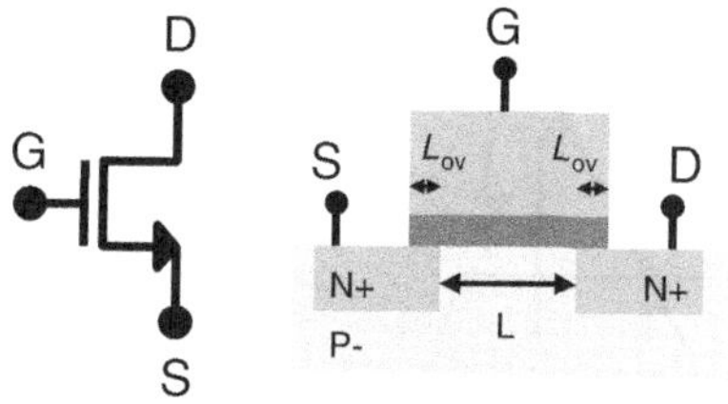

Figure 11.27: FET symbol and simplified cross-section.

applications, quarter wavelength transmission lines can be used to transform a short in an open and complex design switch structure [5].

11.6.2 NFET Switch

The FET device is a three-terminal device that can also be used as a switch for RF and mmWave applications. The FET can be implemented in a variety of semi-conductor materials. For RF and mmWave applications, the most prevalent technologies are the NFET for silicon, and the metal epitaxial semiconductor FET (MESFET) and high electron mobility transistor (HEMT) for compound semi-conductors. A simplified cross-section of an NFET is shown in Figure 11.27. By applying to the NFET a positive voltage between gate (G) and source (S) to a voltage higher than the threshold voltage V_{th}, an inversion layer can be formed between source (S) and drain (D) along the gate dielectric [6]. When the drain-to-source voltage is less than the overdrive voltage ($V_{\text{GS}} - V_{\text{th}}$), the NFET is in triode mode and the drain-to-source triode current can be modeled by reference [6]

$$I_{\text{DS}} = \mu_n C_{\text{diel}} \frac{W}{L} \left[(V_{\text{GS}} - V_{\text{th}}) V_{\text{DS}} - \frac{V_{\text{DS}}^2}{2} \right] \text{[A]} \tag{11.12}$$

where

$$C_{\text{diel}} = \frac{\epsilon_0 \epsilon_{\text{diel}}}{T_{\text{diel}}} \ [\text{F/M}^2] \tag{11.13}$$

C_{diel} is the gate capacitor per unit area for a gate dielectric thickness T_{diel} and dielectric constant ϵ_{diel}. This equation does not take into account depletion effects from the Si substrate and gate. Since the gate thickness scaling T_{diel} has reached a few atomic layers, advanced CMOS technologies do not use only silicon dioxide (SiO_2) with a 3.9 relative dielectric constant, but a complex compound of SiO_2 and silicon nitride (Si_3N_4) with a 7.5 relative dielectric constant, or more recently hafnium oxide (HfO_2) with a 25 relative dielectric constant. W is the total NFET width, L is the channel length, and μ_n is the electron mobility in the channel.

From the drain source current in triode and for $V_{\text{DS}} \ll 2(V_{\text{GS}} - V_{\text{th}})$, we can derive the NFET on resistor (Figure 11.28)

$$r_{\text{on-nfet}} = \frac{L}{\mu_n C_{\text{diel}} W (V_{\text{GS}} - V_{\text{th}})} \ [\Omega] \tag{11.14}$$

The IL can be reduced by reducing L or improving the lithography, and/or increasing the carrier mobility, the gate capacitor, the total width of the NFET and the overdrive voltage.

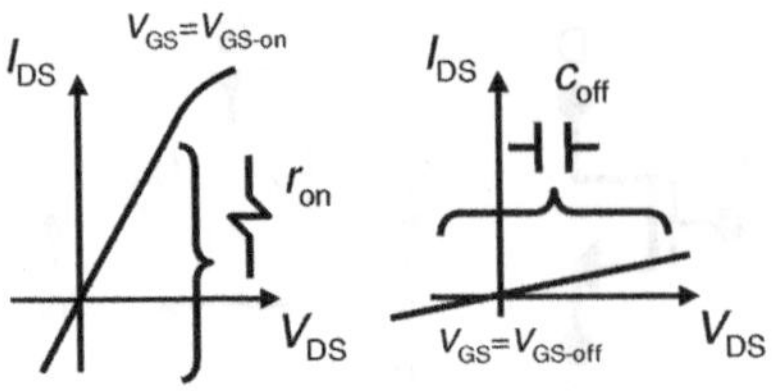

Figure 11.28: On and off FET I–V characteristics and first-order equivalent circuit.

By applying a voltage between gate (G) and source (S) lower than the threshold voltage V_{th}, the channel resistance between drain and source becomes very high, and has a much higher impedance than the fringe capacitor between gate and source and gate and drain. The NFET switch is in off mode. As shown in Figure 11.27, there is an overlap between the gate and the source and drain N+ diffusions. The overlap distance is L_{ov}. For a symmetrical NFET, the off capacitor is (Figure 11.28)

$$C_{\text{off-nfet}} = \frac{2\epsilon_0 \epsilon_{\text{diel}} W L_{\text{ov}}}{T_{\text{diel}}} \text{ [F]} \tag{11.15}$$

The NFET FOM is $r_{\text{on-nfet}} \times c_{\text{off-nfet}}$; therefore, we have

$$FOM_{\text{nfet}} = \frac{2LL_{\text{ov}}}{\mu_n (V_{\text{GS}} - V_{\text{th}})} \text{ [s]} \tag{11.16}$$

The FOM_{nfet} can be decreased or improved by decreasing L, L_{ov}, and/or increasing the carrier mobility and the overdrive voltage. Therefore, the best switch FET technologies will exhibit, if we can neglect source and drain contact resistance, high mobility carriers and advanced lithography. Several approaches are possible to improve carrier mobility. The first is to use high-mobility materials such as GaAs, InP, or GaN. The second is to use a heterojunction such as GaAs/AlGaAs to design a quantum well separating carriers from their doping, thus exhibiting high-mobility carriers. The third is to use stress engineering to increase electron or hole mobility, as, for example, by using SiGe in source and drain diffusion to stress the PFET channel.

11.6.3 Small-signal 65 nm CMOS mmWave Switch Design

We can now evaluate the performance of a 65 nm bulk CMOS technology for the implementation of a mmWave switch. A Berkeley short-channel insulated-gate field-effect transistor (BSIM) model is used to simulate the switch performance in a 50 Ω environment (Figure 11.29). A large resistor is inserted in series with the FET gate to allow voltage division of the drain voltage. This is important for large signal performance optimization. In order to achieve a target switch resistor of 5 Ω a 56 μm large FET is required. The simulated technology FOM is 182 fs, and as shown in Figure 11.30, an IL of less than 1 dB is achieved up to 46 GHz. However, the RL is less than 12.7 dB, starting from 30 GHz. The isolation is better than 18.1 dB. Therefore, with this switch architecture, the predicted performance level might not be sufficient for applications beyond 30 GHz.

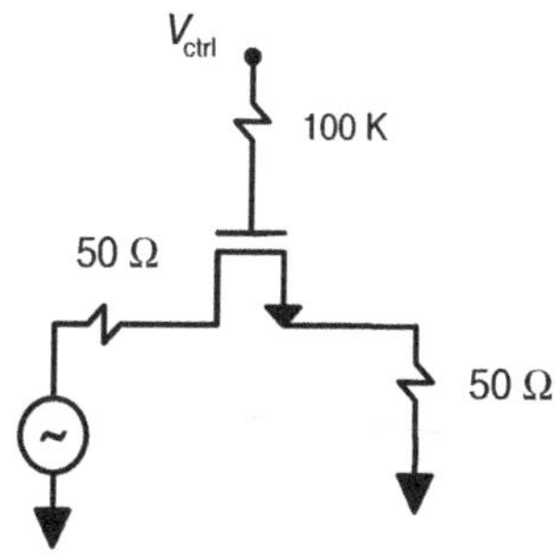

Figure 11.29: NFET switch driven in a 50 Ω environment.

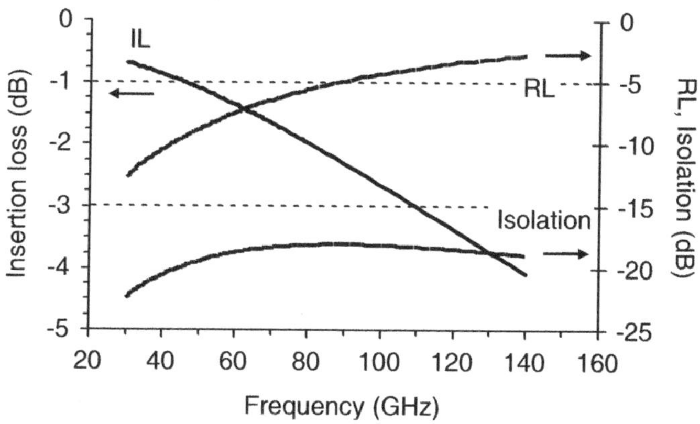

Figure 11.30: Insertion loss for a 65 nm CMOS NFET switch with $r_{series} = 5\ \Omega$, $C_{off} = 36$ fF, $T_{tech} = 182$ fs, $W_{total} = 56$ μm.

One way to improve the switch performance is to add series and parallel inductors, as shown in Figure 11.31. The 520 pH and 3.5 nH inductors are used to resonate the FET parasitic capacitors, in order to improve IL and RL. As shown in Figure 11.32, an IL, a RL and an isolation of 0.18, 24, and 43 dB is simulated at 40 GHz respectively. One drawback of this circuit switch topology is that it has a bandpass characteristic. For example, for a more than 15 dB target RL, the operating frequency is from 38.9 to 42.1 GHz ($F_{max}/F_{min} = 8.2\%$), and the IL is less than 0.31 dB. This approach can be used to build more complex switches such as series–shunt SPDT switches to improve I, for example.

11.6.4 Large-signal 65 nm CMOS mmWave Switch Design

The 40 GHz FET with inductor switch was simulated at different levels of input power using periodic steady-state non-linear simulation. As shown in Figure 11.33, the simulated 1 dB compression point is achieved for an input power of 3.6 dBm. This limitation in power handling is due to the limitation of the CMOS bulk FET device voltage breakdown.

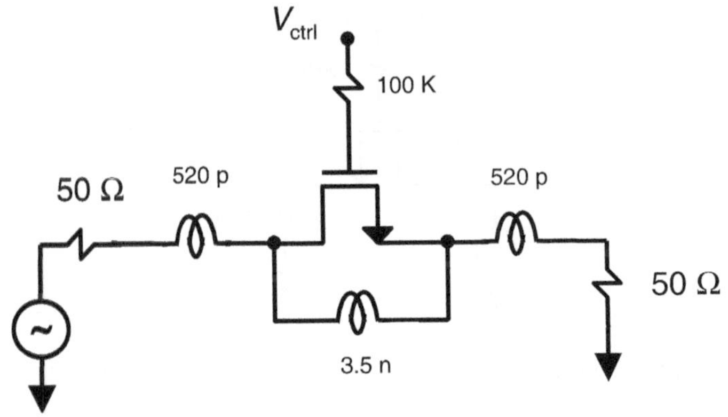

Figure 11.31: NFET switch with inductors driven in a 50 Ω environment.

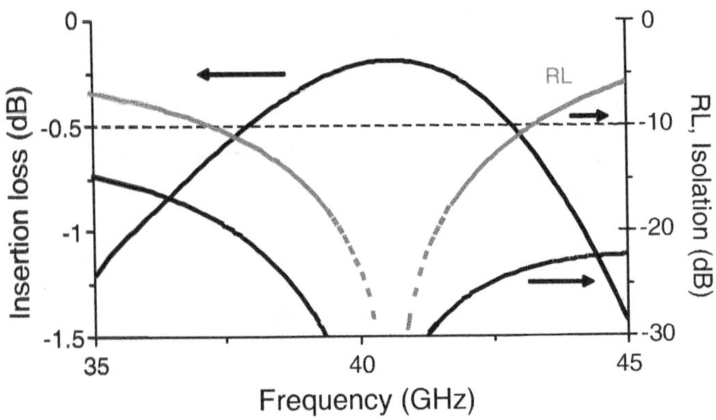

Figure 11.32: Insertion loss for 65 nm CMOS NFET switch with $r_{\text{series}} = 5\ \Omega$, $C_{\text{off}} = 36$ fF, $T_{\text{tech}} = 182$ fs, $W_{\text{total}} = 56$ μm.

Another source of linearity limitation is the large non-linear variable capacitances from the source and drain diffusions.

At large voltages the switch starts to turn on, therefore FET devices can be stacked as shown in Figure 11.34 to divide by the total number of stacked devices the voltage signal across the drain and source FETs. This can be achieved for CMOS technologies on semi-insulated substrate only, such as, for example, sapphire or high-resistivity silicon-on-isolator (SOI) substrate. For CMOS bulk technologies, the voltage handling is limited by the silicon P–N junction breakdown.

Another challenge is with the BSIM model. The BSIM model, as shown in Figure 11.35, exhibits a discontinuity in the on resistor at 0 V across the drain and source. Another way to see the BSIM discontinuity issue at $V_{\text{DS}} = 0$ is by verifying the FET model Gummel symmetry by applying a positive voltage V_x on the drain and a negative voltage $-V_x$ on

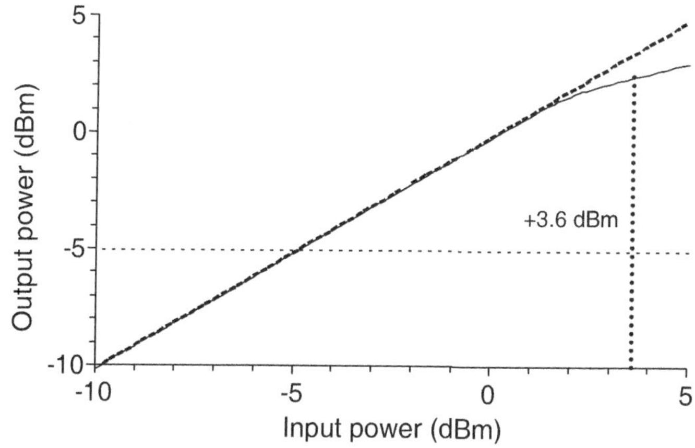

Figure 11.33: 1 dB compression point at 40 GHz for 65 nm CMOS NFET switch with $r_{series} = 5\ \Omega$, $C_{off} = 36$ fF, $T_{tech} = 182$ fs, $W_{total} = 56\ \mu$m.

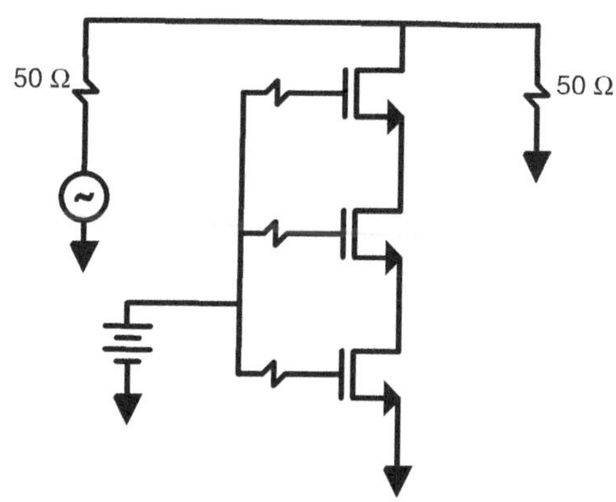

Figure 11.34: Increasing power handling by stacking FET switches.

the source [7]. It can be shown that due to the device symmetry the drain current is an odd function of V_x. Consequently, all odd order derivatives $\partial^{2n+1} I_{DS}/\partial^{2n+1} V_x$ should be continuous at $V_x = 0$, and all even derivatives $\partial^{2n} I_{DS}/\partial^{2n} V_x$ should exist and be equal to zero at $V_x = 0$ [7]. For BSIM3 and 4 the second derivative of the drain current $\partial^2 I_{DS}/\partial^2 V_x$ is not defined at $V_{DS} = 0$ because the drain current first derivative has a kink at $V_{DS} = 0$ [7]. All non-linear simulations must therefore be verified in hardware.

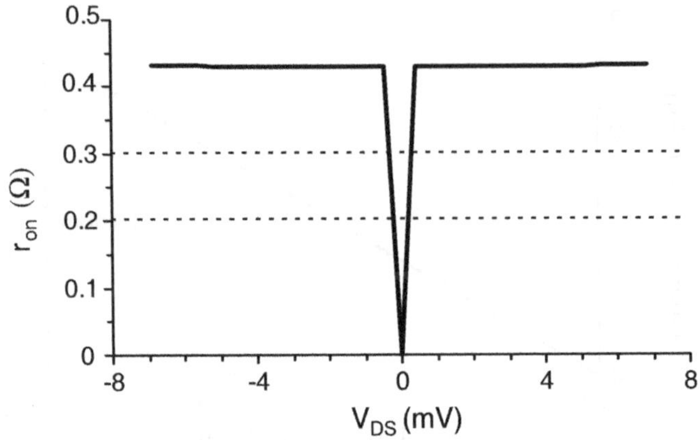

Figure 11.35: BSIM r_{on} discontinuity at $V_{DS} = 0$ V.

11.7 Comparison of Electronic Switch Implementations

11.7.1 Performance Comparison of PIN Diode Switches

Table 11.3 shows a comparison of published PIN diode performances. Some of the best reported PIN diodes use compound semi-conductors such as InGaAs/InP. The InGaAs/InP technology FOM can be as low as 30 fs which is equivalent to an outstanding 5.3 THz cut-off frequency F_c [8], and is more than two times better than for GaAs. One of the best reported technology FOM for silicon PIN diode is 40 fs, which is 33% higher than for the best compound technology. Figures 11.36 and 11.37 show the reported PIN diode IL and I values versus frequency. A shunt–shunt SPST switch exhibits at 20 GHz a 1.2 dB IL and a 28 dB I and is integrated in a small area of 0.71 mm by 1 mm [9]. A CPW SPST GaAs PIN diode switch exhibits measured 0.6 dB IL and 20 dB I from 55 to 75 GHz [10]. The pin diode is in a CPW configuration to reduce ground inductance, and low loss and high isolation are achieved by proper selection of the diode radius, semiconductor properties and air-bridge dimensions [10]. The silicon PIN diode SPDT switch exhibits 2.5 dB IL and 25 dB I at 83 GHz [11]. A high I of 40 dB and an IL of 1.8 dB are achieved at 94 GHz by a SPST switch using two shunt InGaAs pin diodes [8]. The chip dimension is 0.8 mm by 1.7 mm. Owing to lower series resistance because of the higher electron mobility and lower contact resistance as compared with GaAs, InGaAs PIN diodes exhibit improved isolation and lower power dissipation. Another advantage of InGaAs is its lower energy bandgap as compared to GaAs, leading to a turn on voltage 350 mV lower ($V_{onInGaAs} = 0.4$ V, $V_{onGaAs} = 0.75$ V), leading to reduced DC power consumption. For example, a DC biasing on-current as low as 0.3 mA per diode is reported in reference [8].

11.7.2 Performance Comparison of CMOS Switches

Figures 11.38 and 11.39 show some of the best reported CMOS switch IL and I versus frequency. Despite all the drawbacks of CMOS technology for high-frequency switch

Table 11.3: PIN diode technology and performance comparison.

Measured performance	InGaAs/InP [9]	Si [11]	InGaAs/InP [8]	GaAs [10]
R_{on} (Ω)	1.3	4	2.5	2.2
C_{off} (fF)	24	10	12	30
Technology FOM (fs)	31.2	40	30	66
Technology F_c (THz)	5.1	4	5.3	2.4

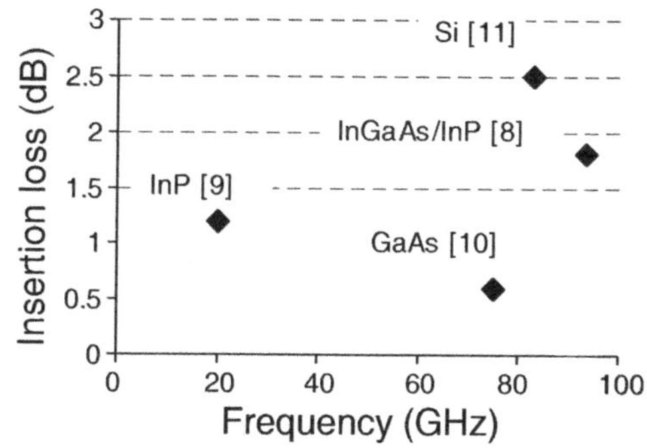

Figure 11.36: Reported PIN diode IL versus frequency for different semi-conductor technologies and designs.

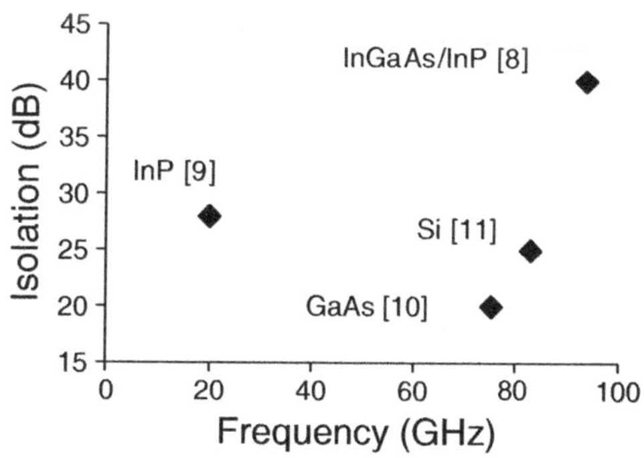

Figure 11.37: Reported PIN diode I versus frequency for different semi-conductor technologies and designs.

applications, recent publications demonstrate great advances in the level of performance for mmWave CMOS switches. A triple-well 130 nm CMOS switch with a measured 2 dB bandwidth of 28 GHz was reported in reference [12]. The transmit/receive switch is based on an improved transistor layout with asymmetric drain and source region, which reduces the drain and source feed-through for body-floated RF switches, and achieves an IL of

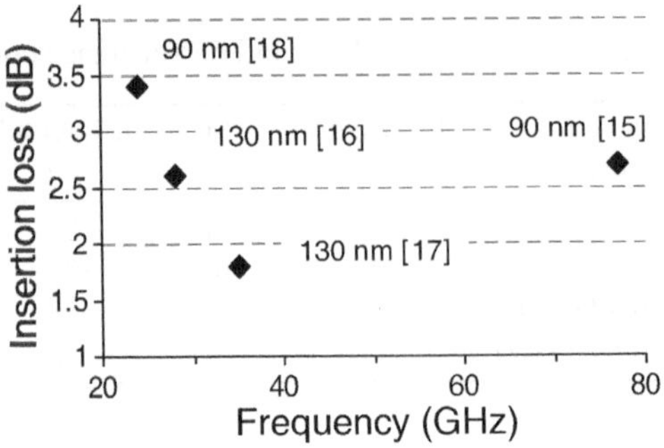

Figure 11.38: Reported NFET IL versus frequency for different technology nodes and designs.

2 dB at 28 GHz. This CMOS switch does not use any inductor and the reported area is only 0.015 mm^2. The isolation is 15 dB at 28 GHz, and the 1 dB compression point is 26.5 dBm [12]. In reference [13] the author took advantage of the triple-well 130 nm technology and some layout techniques to design a switch with a high-substrate resistor. The author found that the shunt switch and the series switch with a high-substrate resistance network have a lower insertion loss than in a standard design. Using this new design, a SPST switch exhibits a measured IL of 1.8 dB at 35 GHz, an I of 32 dB I, a RL of 20 dB a 1 dB compression better than 22 dBm and an IIP3 of 31 dBm [13]. Despite the use of inductors, the switch core dimension is only 115 μm ×240 μm [13]. One of the issues with CMOS switches is that junction diodes limit the signal voltage swing; however in reference [14] a record 28.7 dBm 1 dB compression is reported at 24 GHz. The SPDT switch designed in a 90 nm CMOS technology exhibits a measured 3.4 dB IL, a 22 dB isolation and an RL better than 10 dB. The transmit and receive sections use different switch topologies to minimize the power leakage in from the transmit to the receive section, therefore improving linearity. Some other circuit design techniques such as the traveling-wave concept can also be used to increase the switch frequency application. In reference [15], a 50–94 GHz SPDT switch is reported using a 90 nm CMOS technology. The four shunt switches are distributed along a transmission line, and the drain diffusion parasitic capacitors are embedded into the artificial line to increase the frequency performance. A SPDT switch exhibits 2.7 dB IL for 29 dB isolation and 20 dB RL at 77 GHz. The reported R_{on} is 22 Ω, the C_{off} is 12 fF, and the technology FOM is 264 fs [15].

11.7.3 Performance Comparison of III-V Switches

Compound HEMT technologies are suitable for high-frequency applications owing to the high mobility, low substrate parasitic and high-resistivity substrate. Figures 11.40 and 11.41 show some of the best reported III-V HEMT switch IL and I values versus frequency. In reference [16], the authors propose to use impedance-transformation networks to compensate the drain-source capacitance effect for the off state at high frequencies. The mmWave MMIC SPDT switch was implemented in a 0.15 μm InGaAs/AlGaAs/GaAs. The pHEMT switch exhibits a measured 4 dB IL and 30 dB I from 53 to 61 GHz [16]. The chip size is 2 mm^2,

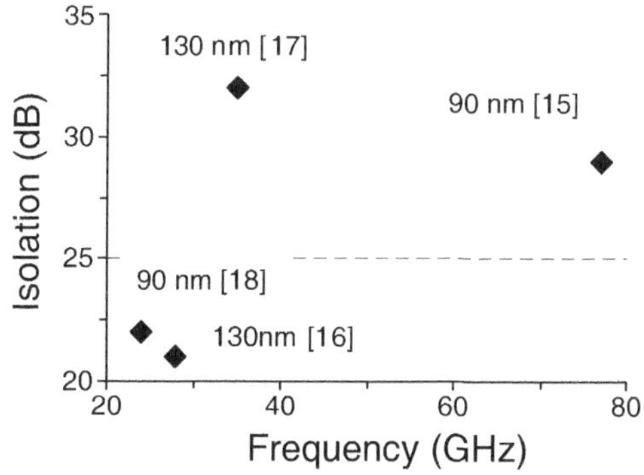

Figure 11.39: Reported NFET I versus frequency for different technology nodes and designs.

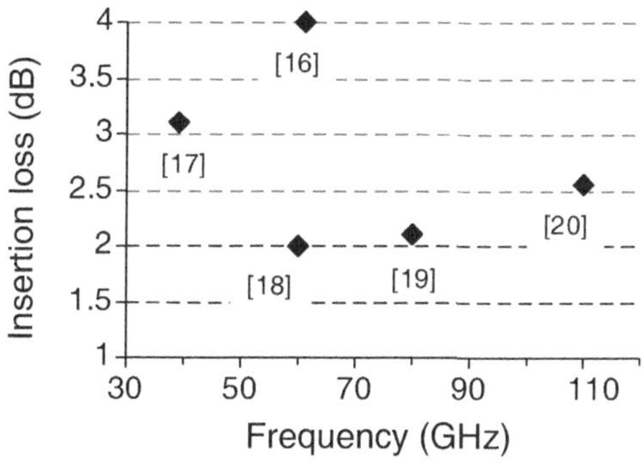

Figure 11.40: Reported III-V IL versus frequency for different switch designs.

and the technology FOM is 240 fs [16]. In references [17] and [18], the authors propose to integrate the SPDT switch with the filter, thus integrating two functions in one, and reducing the overall IL of the front-end. At 39 GHz, a 3.1 dB IL, a 30 dB I, a 15 dB RL and a 15 dBm 1 dB compression are measured [17]. At 60 GHz, a 2 dB IL, a 32 dB I, a better than 10 dB RL and a 15.3 dBm 1 dB compression are measured [18]. Both switches were designed in a 0.15µm InGaAs/AlGaAs/GaAs technology. An ultra-wide bandwidth switch can be designed by using distributed switches along a transmission line. In references [19] and [20], the authors propose the use of a very wide HEMT of 400 µm to design a traveling-wave switch. At 80 GHz, a 2.1 dB IL and a 25.5 dB I are measured [20]. At 110 GHz, a 2.55 dB IL, a 22 dB I, a 10 dB RL and a better than 26.5 dBm 1 dB compression are measured [19].

11.7.4 Performance Comparison of mmWave Switches

Table 11.4 shows a comparison of some of the best mmWave switches reported in the literature. Owing to the advances in lithography, modern CMOS technologies can exhibit

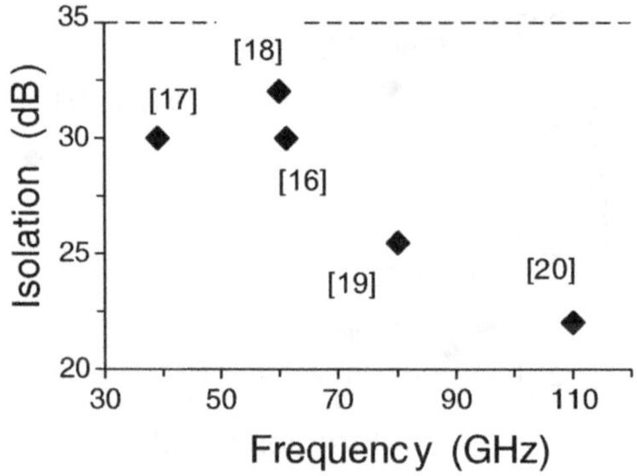

Figure 11.41: Reported III-V I versus frequency for different switch designs.

Table 11.4: Fundamental semi-conductor properties.

Measured performance	90 nm CMOS bulk FET [15]	GaAs HEMT [18]	InGaAs/InP PIN diode [8]
Frequency (GHz)	77	60	94
IL (dB)	2.7	2	1.8
Isolation (dB)	29	32	40
Technology FOM	264 fs	324 fs	30 fs

a better FOM than some compound technologies for FET switches. Despite a 23% higher FOM, the IL and isolation are better than for CMOS. This could be attributed to the advantage of a high-resistivity substrate used in compound technology. However, the best IL and I are achieved by PIN diode as shown in Table 11.4. The advantage of the PIN diode is its simplicity as compared to the FET, and the FOM can be 6 to 11 times better than for FET switches.

Figures 11.42 and 11.43 show, for each of the PIN diode, the CMOS NFET, and the compound HEMT, some of the three best-reported switch IL and I values versus frequency. For the IL and I, the PIN diode performance advantage is clearly visible. However, for IL and I, CMOS switches near the compound technologies' level of performance. As reported in recent publications [12, 15], CMOS designers are taking advantage of triple-well technology effectively to float the substrate terminal and improve IL, I and linearity. CMOS designers have also been very creative in coming up with new topologies to enhance switch performance, from lumped to distributed, and new biasing schemes. At RF frequency, some excellent levels of performance have been reported by CMOS technology on insulated substrate, such as CMOS on sapphire [1], and SOI-CMOS on high-resistivity substrate [22]. These emerging CMOS on high-resistivity substrate will certainly be very suitable technologies for future mmWave applications.

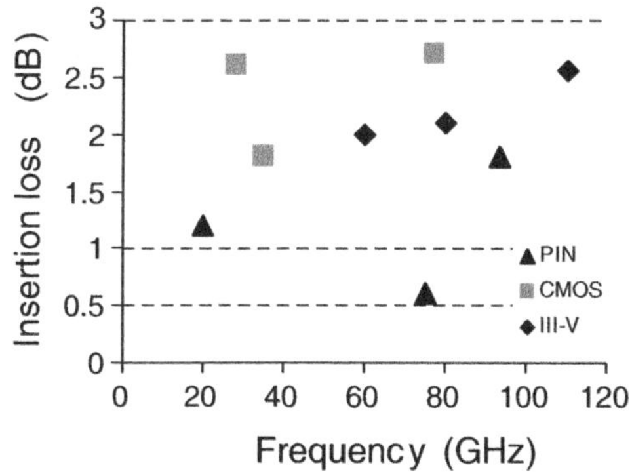

Figure 11.42: IL versus frequency technology and switch design comparison.

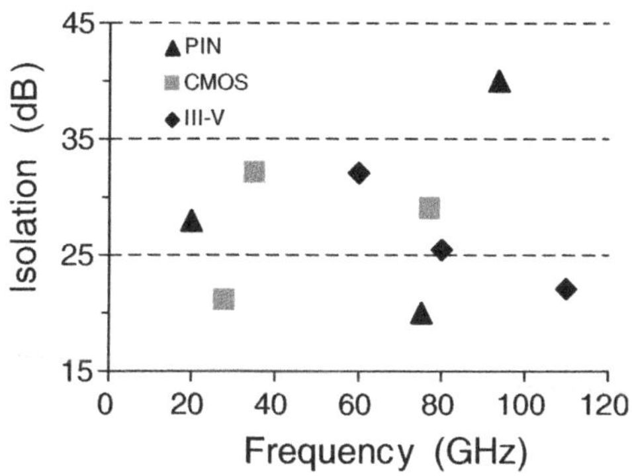

Figure 11.43: I versus frequency technology and switch design comparison.

11.7.5 Power Handling for Different Semi-conductor Technologies

For solid-state technologies, the most important properties are derived from the semi-conductor band structure. The theoretical semi-conductor maximum power handling is proportional to the power of four of the bandgap energy E_g between the conduction and valence bands, the square of the saturation velocity V_s, and inversely proportional to the frequency F ($P_{max} \propto E_g^4 V_s^2 / F^2$, [21]). As shown in Table 11.5, GaN compound semi-conductors can offer intrinsic power handling as much as 340 times higher than silicon [21]. One of the drawbacks of compound semi-conductors is that contact resistance is much higher than for silicon. This higher contact resistance can decrease the higher mobility advantage of compound semi-conductors.

Table 11.5: Fundamental semi-conductor properties.

Property	Si	GaAs	GaN
E_g (eV)	1.1	1.4	3.4
V_s (10^7 cm/s)	0.7	0.8	2.5
Relative maximum power handling	1	3.4	340

Table 11.6: Overview of solid-state switch technology challenges.

PIN diode	SOS CMOS	GaAs	GaN
High DC current implementation area	Low-voltage FET	IM3 Integration challenges with control	Very negative pinch-off Integration challenges with control

11.7.6 Solid-state Switch Technology Challenges

As shown in Table 11.6, the challenges are for the PIN diode, the high DC bias current and the implementation area. For the silicon on sapphire, the low-voltage FET is a challenge in handling large voltage. For the GaAs technology, the third-order intermodulation and integration of control circuits are a drawback. GaN has the same issue as GaAs with the integration of circuit control, and GaN requires a very negative pinch off voltage, which can be difficult to generate.

Even though solid-state switches are very successfully integrated in current systems, some intense research work is on-going to replace or complement solid-state switches with MEMS switches.

Acknowledgments

The author would like to thank M. Soyuer, S. Gowda, B. Floyd and B. Gaucher for the mmWave projects leadership and management support.

References

[1] Dylan Kelly, 'CMOS RF switch design and applications', *IEEE Compound Semiconductor IC Symposium, RF and High Speed CMOS Short Course*, San Antonio, TX, November 2006.

[2] Y. Mimino, K. Nakamura, Y. Hasegawa, Y. Aoki, S. Kuroda and T. Tokumitsu, 'A 60 GHz millimeter-wave MMIC chipset for broadband wireless access system front-end', *IEEE MTT Symposium*, Seattle, WA, 2002.

[3] V. Ziegler, C. Gassler, C. Wolk, F.-J. Berlec, R. Deufel, M. Berg, J. Dickmann, H. Schumacher, E. Alekseev and D. Pavlidis, 'InP-based and metamorphic devices for multifunctional MMICs in mm-wave communication systems', *International Conference on Indium Phosphide and Related Materials*, pp. 341–344, Williamsburg, VA, May 2000.

[4] Q. Li and Y. P. Zhang, 'CMOS T/R switch Design: Towards ultra-wideband and higher frequency', *IEEE Journal of Solid-State Circuits* **42**(3) (2007), pp. 563–570.

[5] Application Note 1002, 'Design with PIN diodes', SKYWORKS http://www.skyworksinc.com.

[6] B. Razavi, *Design of Analog CMOS Integrated Circuits* 1st edn (McGraw-Hill, 2000), pp. 17–18.

[7] K. Joardar, K. K. Gullapalli, C. C. McAndrew, M. E. Burnham and A. Wild, 'An improved mosfet model for circuit simulation', *IEEE Transactions on Electron Devices* **45**(1) (1998), pp. 134–148.

[8] E. Alekseev, D. Pavlidis and D. Cui, 'InGaAs PIN diodes for high-isolation W-band monolithic integrated switching applications', *IEEE Cornell Conference on Advanced Concepts in High Speed Semiconductor Devices and Circuits*, pp. 332–340, Ithaca, NY, 4–6 August 1997.

[9] J. G. Yang, H. Eom, S. Choi and K. Yang, '2–38 GHz broadband compact InGaAs PIN switches using a 3-D MMIC technology', *IEEE International Conference on Indium Phosphide and Related Materials*, pp. 542–545, Matsue, Japan, May 2007.

[10] T. Buber, N. Kinayman, Y.-H. Yun and J. Brogle, 'Low-loss high-isolation 60–80 GHz GaAs SPST PIN switch', *IEEE MTT Symposium*, Philadelphia, PA, 2003.

[11] A. Klaassen and J.-M. Dieudonne, '77 GHz monolithic MMIC Schottky- and PIN-diode switches based on GaAs MESFET and silicon SIMMWIC technology', *IEEE MTT Symposium*, Orlando, FL, 1995.

[12] Q. Li, Y. P. Zhang, K. S. Yeo and W. M. Lim, '16.6- and 28-GHz Fully Integrated CMOS RF Switches With Improved Body Floating', *IEEE Transactions on Microwave Theory and Techniques*, **56**(2) (2008), pp. 339–345.

[13] B.-W. Min and G. M. Rebeiz, 'Ka-band low-loss and high-isolation switch design in 0.13-μm CMOS', *IEEE Transactions on Microwave Theory and Techniques* **56**(6) (2008), pp. 1364–1371.

[14] P. Park, D. H. Shin, J. J. Pekarik, M. Rodwell and C. Patrick Yue, 'A high-linearity, LC-tuned, 24-GHz T/R switch in 90-nm CMOS', *IEEE Radio Frequency Integrated Circuits Symposium*, Atlanta, GA, June 2008.

[15] S.-F. Chao, H. Wang, C.-Y. Su and J. G. J. Chern, 'A 50 to 94-GHz CMOS SPDT switch using traveling-wave concept', *IEEE Microwave and Wireless Components Letters* **17**(2) (2007), pp. 130–132.

[16] K.-Y. Lin, Y.-J. Wang, D.-C. Niu and H. Wang, 'Millimeter-wave MMIC single-pole-double-throw passive HEMT switches using impedance-transformation networks', *IEEE Microwave and Wireless Components Letters* **17**(5) (2003), pp. 1076–1085.

[17] S.-F. Chao, C.-C. Kuo, Z.-M. Tsai, K.-Y. Lin and H. Wang, '40-GHz MMIC SPDT and multiple-port bandpass filter-integrated switches', *IEEE Transactions on Microwave Theory and Techniques* **55**(12) (2007), pp. 2691–2699.

[18] Z.-M. Tsai, Y.-S. Jiang, J. Lee, K.-Y. Lin and H. Wang, 'Analysis and design of bandpass single-pole double-throw FET filter-integrated switches', *IEEE Transactions on Microwave Theory and Techniques* **55**(8) (2007), pp. 1601–1610.

[19] H. Mizutani and Y. Takayama, 'DC-110-GHz MMIC traveling-wave switch', *IEEE Transactions on Microwave Theory and Techniques* **48**(5) (2000), pp. 840–845.

[20] H. Mizutani, N. Iwata, Y. Takayama and K. Honjo, 'Design considerations for traveling-wave single-pole multithrow MMIC switch using fully distributed FET', *IEEE Transactions on Microwave Theory and Techniques* **55**(4) (2007), pp. 664–671.

[21] M. Poulton, 'Design techniques for high linearity, high efficiency GaN PA architectures," *IEEE Compound Semiconductor IC Symposium*, 'GaN Circuits and Applications' Short Course, San Antonio, TX, November 2006.

[22] C. L. Chen, C. K. Chen, P. W. Wyatt, J. M. Knecht, D.-R. Yost, P. M. Gouker, P. D. Healey and C. L. Keast, 'Fully depleted SOI RF switch with dynamic biasing', *2007 IEEE Radio Frequency Integrated Circuits Symposium*, pp. 175–178, Honolulu, HI, June 2007.

12

MEMS Devices for Antenna Applications

Nils Hoivik and Ramesh Ramadoss

12.1 Introduction

As the demand for more sophisticated and versatile wireless and telecommunication systems continues to grow, it is easy to forget that quite often the source of the limiting factors for the development of new cutting-edge wireless technology are the simplest electronic elements such as switches, inductors, and capacitors. These devices are crucial elements for the operation of any wireless system, and there is a continued demand for both smaller and lighter systems with increased performance and functionality [1]. These requirements place a considerable demand on the circuit power dissipation, design, and circuit compatibility, which consequently increase the design complexity, manufacturing cost and overall weight of current components and devices. Based on this need, radiofrequency micro-electro-mechanical systems, commonly termed as RF MEMS, are currently being developed for wireless applications. RF MEMS enables the design of new products with enhanced functionality, low insertion loss, high isolation and reduced power consumption. By moving into the design of three-dimensional RF MEMS components, the possibility to remove the performance-degrading elements, such as parasitic capacitances and inductances that plague discrete off-chip RF devices becomes a reality.

Recently, RF MEMS devices have opened up a wide range of technological approaches to enhance reconfigurable antenna systems. Furthermore, this technology might even enable new antenna concepts previously not practically usable or possible to fabricate with traditional technologies [2]. Particularly in terms of reduction in losses and power consumption, MEMS switches have demonstrated excellent performance compared to their conventional counterparts.

Advanced Millimeter-wave Technologies: Antennas, Packaging and Circuits
Duixian Liu, Ulrich Pfeiffer, Janusz Grzyb and Brian Gaucher © 2009 John Wiley & Sons, Ltd

Over the past decade, there have been many different approaches to actuate RF MEMS switches. Magnetic, thermal, electrostatic and piezoelectric actuation methods have all been demonstrated with great success. Nevertheless, electrostatic actuation is probably the most commonly used approach owing to ease of implementation. However, piezoelectric actuation has recently gained much research interest owing to the promise of significantly lower actuation voltage compared to electrostatic devices (3–5 V versus 35–90 V, respectively), and the ability to control the switch actuation in both directions using the polarity of the applied voltage. Magnetic and thermal actuation schemes have been set aside in the last few years owing to challenges in integration. Magnetic actuation requires a permanent magnet to be present in the package, and thermal actuation exhibits relatively slow response times and consumes a great deal of power owing to the high current requirement compared to electrostatic switches.

This chapter describes the technology behind RF MEMS devices and their application to millimeter-wave (mmWave) antenna systems. The first section covers the basic principles behind micromachining techniques. The concept of electrostatic actuation for switches is covered in Section 12.3. Section 12.4 describes the two classes of RF MEMS switch, comparison of RF MEMS switches to solid-state devices, and performance and design considerations for RF MEMS switches. MEMS switch reliability and power handling challenges are discussed in Section 12.5. Section 12.6 covers various approaches for the integration of RF MEMS switches with antennas. Section 12.7 describes the use of MEMS technology for implementation of tunable and steerable reconfigurable antennas. Lastly, Section 12.9 looks into the future use and outlook of RF MEMS and nano-electro-mechanical systems (NEMS) for mmWave systems.

12.2 Micromachining Techniques

MEMS devices are generally fabricated using techniques and tools used in and adapted from semiconductor device fabrication. Typical sizes for most MEMS devices range from micrometers to several millimeters, depending on the application. The application of MEMS in RF technology can generally be grouped into two categories: active, in other words moving devices which involve mechanical motion (e.g. switches and varactors) and static components such as micromachined transmission lines and suspended T-line resonators. The extensive possibilities of micromachining combined with mechanical movement is the key feature of RF MEMS. By designing these devices to be mechanically actuated by means of electrical control signals, a whole new concept in increased functionality can be generated. Similar to electronically tuned solid-state devices, a MEMS device exhibits a wide range of desirable values. This tunabiliy can be used for adjustment of another RF device or component.

Micromachining is commonly divided into two categories: *surface micromachining* and *bulk micromachining*. Surface micromachining is typically used for switches, varactors, inductors, etc., whereas bulk micromachining is used for fabrication of T-lines, waveguides, resonators, etc. These processes are not independent, and are often combined to create RF MEMS devices or full systems. As stated previously, these processes are all established based on conventional semiconductor manufacturing methods, where photolithography forms the basis to define the micro-scale features. The versatility of the process allows for a wide range

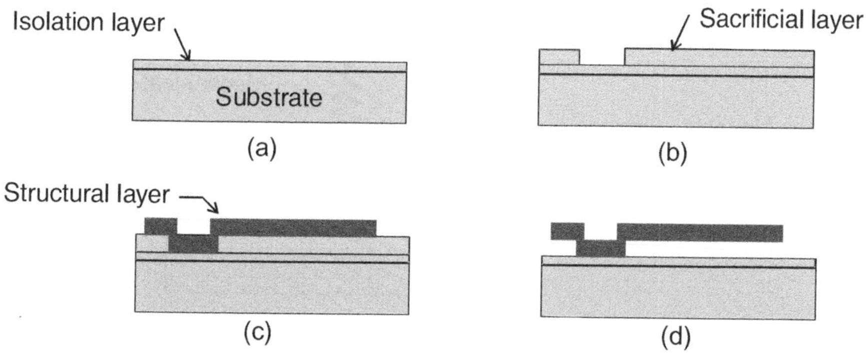

Figure 12.1: Illustration of surface micromachining with a sacrificial layer used to create a free-standing MEMS structure. Compared to conventional semiconductor manufacturing where all layers are utilized in the final product, the sacrificial layer temporarily supports the structural layer during fabrication and is removed at the last stage of processing. (a) Deposit isolation layer; (b) deposit and pattern sacrificial layer; (c) deposit and pattern structural layer; (d) remove sacrificial layer.

of materials to be patterned and defined, making it therefore the defining process for both surface and bulk micromachined devices.

Surface micromachining is based upon modern integrated circuit (IC) fabrication principles and involves growing, or adding thin films on top of a Si wafer or any other suitable substrate. These films are then selectively patterned by photolithography and regions are removed either by wet or dry chemical processes [3]. Deposited materials are typically glass, silicon, dielectrics or metals films. Compared to conventional IC fabrication, MEMS processes use fewer layers (1–10 versus 100) and comprise of free-standing structures.

The free-standing structure created in any MEMS process is obtained by inclusion of a so-called sacrificial (or release) layer as shown in Figure 12.1. This layer defines a gap between moving and non-moving parts, and is typically removed at the very end of the fabrication of a MEMS device. The purpose of the sacrificial layer (soluble or removable) is temporarily to support the structural layers during subsequent fabrication steps. Commonly used sacrificial layers include metals (Au, Ni, Al, etc.), dielectrics (SiO_2 and Si_3N_4) and polymers (photoresist, polymethyl) and polyimides [3, 4].

Bulk micromachining is usually referred to as a subtractive process, in which devices and structures are fabricated by etching into the substrate. Silicon is the most commonly used substrate material; however, etching on both Quartz and Sapphire has been demonstrated for RF MEMS components such as suspended transmission lines and resonators. Possibly one of the greatest benefits of a bulk process is that MEMS devices can be created from single crystalline Si (i.e. micromachined resonators), which makes it feasible to directly integrate MEMS devices on complementary metal–oxide–semiconductor (CMOS) substrates.

For etching into Si substrates, two major types of bulk etching are employed: *isotropic* and *anisotropic*. In isotropic etching, the etch rate is uniform in all directions, whereas in anisotropic etching the etch rate depends on the crystallographic orientation of the substrate. Bulk micromachining based on wet chemical etching is used widely for fabrication of RF

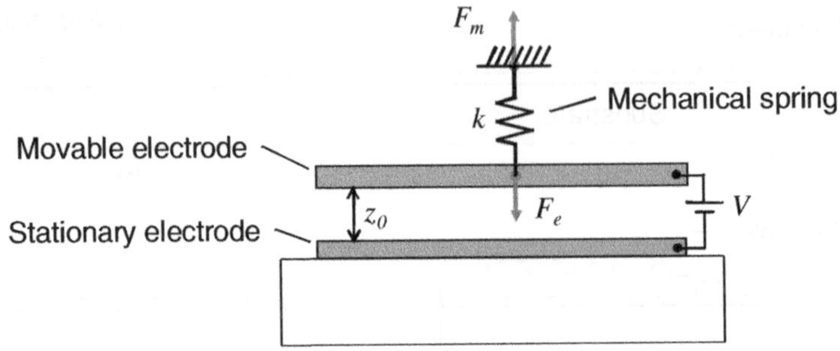

Figure 12.2: One-dimensional model of an electrostatically actuated switch with electrodes separated by a distance z_0. The mechanical restoring force is a function of the spring constant k, and the electrostatic force is controlled by the applied voltage V.

MEMS components. The etchants used are primarily based on selectivity to etch masks, metals and other exposed materials or by crystal orientation. Selectivity can also be obtained by doping the Si substrate where heavily doped regions can be made to etch more slowly [4].

Recently, a well-established dry etching process named deep reactive ion etching (D-RIE) has evolved into the de-facto processing approach to create very high-aspect-ratio structures in silicon. This process allows for near vertical sidewalls, which is very useful for defining RF MEMS devices with definite vertical dimensions.

For further details on micromachining and silicon processing, see chapter 18. For further information on MEMS fabrication, the reader is directed to excellent reference books [3] and [4].

12.3 MEMS Switches – Principle of Operation

In order to obtain mechanical movement of the structural layer, RF MEMS switches utilize electrostatic, magnetic, electrothermal or piezoelectric actuation principles [5]. The actuator converts electrical energy into mechanical movement which opens or closes the switch. Probably the most widely adapted actuation principle used in RF MEMS is electrostatic actuation, owing to the relatively simple fabrication process and low power consumption. As of today, the commercially available RF MEMS switches are electrostatic. Therefore, this chapter will focus on electrostatic actuation.

For electrostatically actuated switches, the most common approach is to use a parallel plate actuator where a movable top plate (structural layer) is brought into contact with a lower (fixed) layer by means of an electrostatic force. The electrostatic force acts as an attractive force between the two plates, and the elasticity of the material in the structural layer acts as a mechanical force acting in the opposite direction. Figure 12.2 illustrates the forces in action on an electrostatically actuated switch. The mechanical force (F_m) is controlled by the spring constant, k, and the electrostatic force (F_e) is controlled by the applied voltage, V. Since the mechanical motion is relatively small compared to the structural dimensions, the mechanical restoring force is linear with deformation as long as the elastic limit of the material is not exceeded.

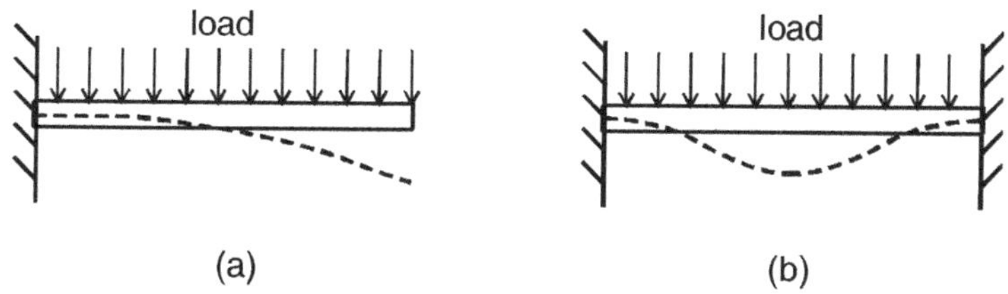

Figure 12.3: Illustration of (a) cantilever (clamped-free) and (b) clamped–clamped beam with a uniform load and the corresponding deformation profile.

12.3.1 Mechanical Spring Constant

The mechanical restoring force can be modeled as a linear spring with spring constant k (N/m), which is the ratio of the applied force to deformation or motion of the movable structure. The spring constant can vary greatly for a given MEMS switch, depending on the design and how it is configured (cantilever, or clamped–clamped). A cantilever beam (as shown in Figure 12.3(a)) is clamped at one end and free at the other end. A bridge, or clamped–clamped beam (as shown in Figure 12.3(b)) is fixed at both ends. The spring constant for these two configurations under a uniformly distributed load (from the electrostatic force) can be expressed as [6]

$$k_{\mathrm{cf}} = \frac{2Ewt^3}{3l^3} \tag{12.1}$$

$$k_{\mathrm{cc}} = \frac{32Ewt^3}{l^3} \tag{12.2}$$

where E is the Young's modulus of the beam material, and w, t and l are the width, thickness and length of the beam, respectively. Depending on the beam width, the biaxial modulus $E' = E/(1 - \nu)$ (ν = Poisson's ratio) is used for beams with $w > 5t$ [7]. It is noteworthy that the spring constant is highly dependent on the thickness and length of the beam, and for a clamped–clamped beam the spring constant k is 48 times higher than for a cantilever beam.

By taking into consideration any residual tensile biaxial stress in the material used to fabricate the beam, the spring constant of the clamped–clamped beam (k'_{cc}) will increase, and can be expressed as [8]

$$k'_{\mathrm{cc}} = \frac{32Ewt^3}{l^3} + \frac{8\sigma_0(1 - \nu)wt}{l} \tag{12.3}$$

where σ_0 is the biaxial residual tensile stress. For a cantilever switch, the residual stress does not affect the overall spring constant of the beam. However, any stress gradients in the material will lead to an undesired deformation of the suspended beam, causing it to curl down, or away from the substrate. These strain gradients usually arise from varying conditions such as deposition rate and temperature during deposition of the films used in the fabrication of the MEMS device. Such residual stress-induced deformations can easily render switches useless as they might end up in a permanently closed state. Thus, any stress or strain gradients in the structural layer is extremely important for the operation of an RF MEMS device, and should

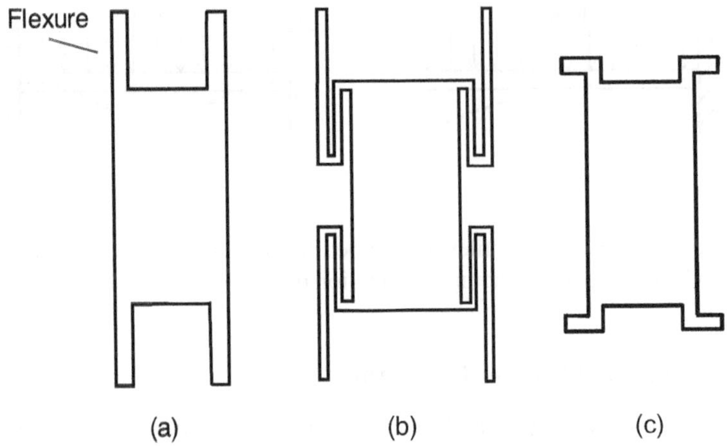

Figure 12.4: Illustration of (a) thin flexures; (b) 'crab-leg' flexures; (c) double flexures, used to control the spring constant for a clamped–clamped beam configuration.

be kept to a minimum value. However, film stresses can also be used as a design feature, and several switch designs cleverly use the built-in film stress to control the shape or deformation of the switch, as will be described later in Section 12.4.2.

The spring constant of the actuator structure can also be controlled by means of compliant flexures. A number of various designs have been demonstrated for RF MEMS switches, of which three different compliant flexures utilized for clamped–clamped configuration are illustrated in Figure 12.4. Calculation of the spring constant for the more elaborate flexures are not as straightforward as the previous examples, and are covered in detail in references [8] and [6].

As stated earlier, the most common electrostatic switch design is the parallel plate actuator. The two conductive plates, or electrodes, are separated by an initial distance z_0, where the top electrode is allowed to move towards the lower stationary electrode (Figure 12.2). As a voltage V is applied between the two electrodes, the electrostatic force generated between the two plates displaces the top electrode by Δz ($\Delta z = z_0 - z$) and thus gives rise to a mechanical restoring force

$$F_{\mathrm{m}} = k\Delta z \tag{12.4}$$

where k is the spring constant of the switch and Δz is the displacement of the top electrode.

12.3.2 Electrostatic Force

In a parallel plate capacitor, the electrostatic energy stored for a given voltage can be expressed as

$$E_{\mathrm{e}} = \frac{1}{2}C(z)V^2 \tag{12.5}$$

with

$$C(z) = \frac{\epsilon_0 \epsilon_{\mathrm{r}} A}{z} \tag{12.6}$$

where $C(z)$ is the parallel plate capacitance, ignoring fringing effects, and ϵ_r is the relative dielectric constant of the medium between the two plates (for dry gases, such as air and N_2 the value of ϵ_r is very close to 1), ϵ_0 the permittivity of free space and A is the area of the actuation electrode.

The electrostatic force (F_e) between the two electrodes (at a constant voltage) is equal to the rate of change of stored electrical energy with displacement and is given by

$$F_e = \frac{\partial E_e}{\partial z} \tag{12.7}$$

Substituting Equation (12.5) into (12.7) yields

$$F_e = \frac{1}{2}\frac{\epsilon_r \epsilon_0 A V^2}{z^2} \tag{12.8}$$

It is worth mentioning that the electrostatic force is independent of polarity, meaning, as long as there is a potential difference (V) between the plates, that there will be an attractive electrostatic force between the two plates.

12.3.3 Pull-in and Release Voltage

The mechanical restoring force (F_m) acts in a direction opposite to the deflection (as illustrated in Figure 12.2), and the total force F_{total} acting on the movable plate is the sum of the electrostatic and mechanical forces:

$$F_{total} = F_e + F_m = \frac{1}{2}\frac{\epsilon_r \epsilon_0 A V^2}{z^2} - k\Delta z \tag{12.9}$$

The movement of the top plate with applied voltage is governed by the relationship between the electrostatic and mechanical restoring forces. Since the mechanical restoring force is a linear function with displacement and the electrostatic force varies with the square of the voltage, there exists a stable equilibrium as long as the displacement (Δz) does not exceed a critical value. This value can be found by solving Equation (12.9) for the voltage (Equation (12.10)) and then solving for z_{cr} by setting the derivative with respect to beam height (z) equal to zero. It can be shown that a stable solution exists only for displacements $0 > z > 2z_0/3$, where the critical position is $z_{cr} = 2z_0/3$,

$$V(z) = \sqrt{\frac{2kz^2(z_0 - z)}{\epsilon_r \epsilon_0 A}} \tag{12.10}$$

If the voltage is increased beyond the critical value, the electrostatic force will overcome the mechanical restoring force, and the top electrode will collapse down and land on the lower electrode, as shown in Figure 12.5. The corresponding applied voltage is referred to as the pull-in voltage (V_{pi}). This voltage is found by using the critical gap distance z_{cr} in Equation (12.10), and is given as

$$V_{pi} = V\left(\frac{2z_0}{3}\right) = \sqrt{\frac{8kz_0^3}{27A\epsilon_r\epsilon_0}} \tag{12.11}$$

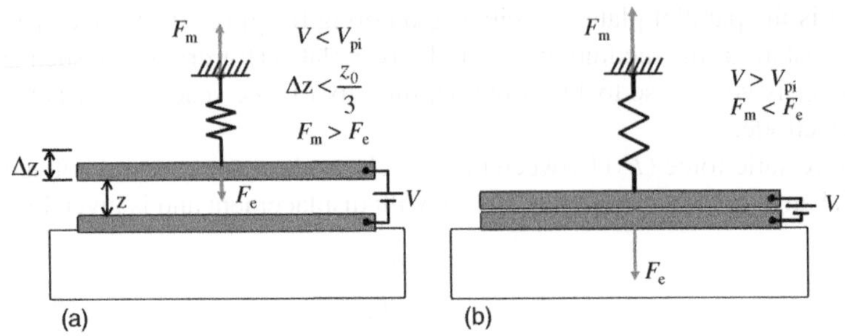

Figure 12.5: Illustration of the two regions of operation in a parallel-plate electrostatic actuator: (a) for an applied voltage less than the pull-in voltage the mechanical restoring force is greater than the electrostatic force, and the displacement, Δz, is less than one-third of the initial gap (z_0). The location of the movable electrode is controllable in this stable operating region. (b) when the voltage is equal to, or above the pull-in voltage, the electrostatic force overcomes the mechanical restoring force and the top plate snaps down onto the fixed electrode.

where k is the spring constant of the switch, z_0 is the gap separating the two electrodes and A is the electrode area. Note that ϵ_r will be equal to 1 if a medium such as a dry gas, air or N_2 is present between the two electrodes.

For compliant switches, where the two actuation electrodes make contact upon pull-in, a dielectric layer is inserted between the two electrodes to prevent a direct electrical short. Thus, the medium between the two electrodes consists of a gas plus a dielectric material, which can be modeled as two capacitors in series. The dielectric layer is usually relatively thin (100–300 nm) and commonly used materials are SiO_2 or Si_3N_4. Adding a dielectric layer changes the pull-in voltage owing to change in the capacitance of the parallel-plate actuator, and the critical distance can be shown to be

$$z'_{cr} = \frac{2z'_0}{3} - \frac{t_d}{3\epsilon_d}, \quad z'_0 = z_0 - t_d \tag{12.12}$$

where t_d and ϵ_d refers to the thickness and relative dielectric constant, respectively, of the added dielectric layer.

This further changes the expression for the pull-in voltage to

$$V'_{pi} = \sqrt{\frac{8k[z'_0 + (t_d/\epsilon_d)]^3}{27A\epsilon_0}} \tag{12.13}$$

Note that if the electrode separation (z'_0) is significantly greater than the thickness of the dielectric layer, V_{pi} effectively reduces to Equation (12.11).

It is clearly evident from Equations (12.1) and (12.3), and the pull-in voltage expression in Equation (12.11) that the mechanical boundary conditions, beam thickness and intrinsic stress of the structural material play a significant role in the operation of an electrostatic switch. Since both the spring constant and the electrode area are linearly dependent on the plate width w, the pull-in voltage is independent of the beam width. This can be of great importance for capacitive switches which will be discussed later.

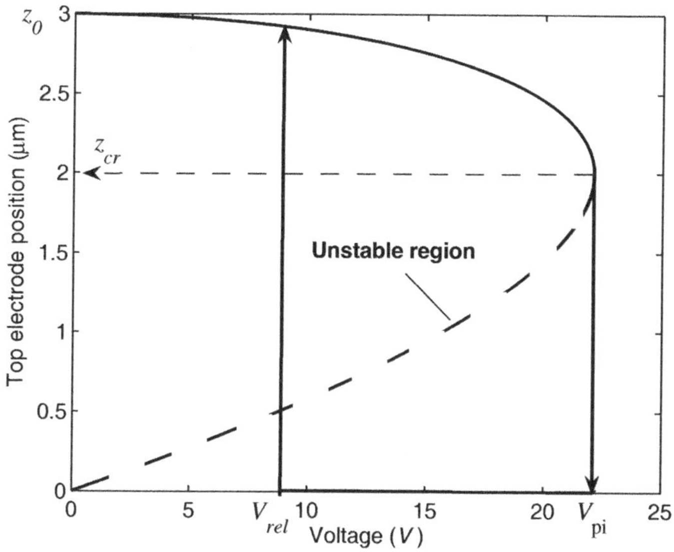

Figure 12.6: Illustration of switch hysteresis as a function of applied voltage. The unstable region refers to the solution of Equation (12.10) past the critical position z_{cr}. After pull-in of the top electrode, the release voltage V_{rel} will have to be lowered below the value for V_{pi}.

Another very important parameter for any RF MEMS switch is the *release voltage* at which the switch opens, or releases from the lower electrode. Since the capacitance in the switch is very high after pull-in (due to the small gap separating the two electrodes), the electrostatic force is subsequently very high. Thus, in order to open the switch, the voltage needs to be lowered well below the initial pull-in voltage level for the mechanical restoring force to overcome the electrostatic force. This causes the parallel plate actuator to exhibit a hysteresis characteristic for V_{pi} and V_{rel} as shown in Figure 12.6.

The release voltage, also referred to as the hold-down voltage, can be estimated using Equation (12.8) by taking into consideration the thickness of the dielectric layer and using the mechanical restoring force from Equation (12.4) for F and solving for voltage

$$V_{rel} = \sqrt{\frac{2k(z_0 - t_d)[t_d + (t_d/\epsilon_d)]^2}{\epsilon_d \epsilon_0 A}} \tag{12.14}$$

For practical purposes, this expression is best served as an estimate for the release voltage as the actual value is usually higher than the estimated value, since the down-state capacitance is very sensitive to air gaps, which significantly reduces the hold-down electrostatic force.

12.4 Contact and Capacitive MEMS Switches

RF MEMS switches have typically been classified into two types: *contact* and *capacitive* switches. The contact switches are sometimes referred to as ohmic switches, since they make and break a resistive contact and operate from DC and up to several GHz, whereas the capacitive switches typically operate at frequencies above 7–8 GHz and up to 100 GHz (or beyond). This section will present some of the principles behind these switches, together with some lumped element models for series and shunt configured devices. Additionally, a few examples of the better known and demonstrated switches will be presented, and the main differences between various switch designs will be pointed out. For additional examples of switches, in addition to a more comprehensive and detailed analysis of RF MEMS devices see references [8], [9] and [10].

12.4.1 Ohmic Contact MEMS Switches – Series Configuration

Ohmic, or DC contact switches operate on the principle of making and breaking a physical contact in the RF transmission line. In many ways, these switches are very similar to old-fashioned relays. For many switches, the typical gap separating the two contact points ranges from between 1 and 3 μm. Upon actuation, the top electrode brings the contact pairs together, shorting the open ends. Since a physical contact is made, these switches can operate at low frequencies – even at DC. A separate actuation electrode is commonly used for applying the bias voltage in order to reduce any interference of the switch structure with the transmission line.

Figure 12.7(a) illustrates an inline series switch where the movable structure acts as a continuation of the transmission line. Depending on the design of the switch, the impedance of the cantilever can play an important role. Ideally, an in-line cantilever switch is designed such that it has a continuous conductive part from its anchor base to the contact point. This is important for a series configured device since the entire switch will pass RF current along its length.

Alternatively, the switch can be configured with a shorting bar using two contact pairs as shown in Figure 12.7(b). This will facilitate in isolating the RF section of the switch from the actuation electrodes and simplifies biasing of the actuation electrodes. The drawback will be an increase in the overall resistance (with two contacts in series) and thus will lead to an increase in the insertion loss. Furthermore, the impedance of the shorting bar can play a significant role, and should not be overlooked [8].

For series MEMS contact devices, the two main design parameters are the up-state capacitance C_t, which affects the isolation, and the down-state resistance R_c, which dominates the insertion loss. These two parameters can be expressed as

$$C_t = C_p + C_{ca} = C_p + \frac{\epsilon_0 A_c}{z_c} \tag{12.15}$$

$$R_c = \frac{\rho_c}{A_c} \tag{12.16}$$

where A_c corresponds to the area of the contact pair, C_p is the parasitic capacitance between the switch and T-line, C_{ca} is the contact area capacitance and z_c is the up-state gap separation between the contacts and ρ_c is the surface resistivity of the metal used in the contact. It is important to note that these two equations are best used as a guide for determining the switch performance. It should be pointed out that the switch isolation performance is affected by fringing effects and parasitic capacitances, and the on-state resistance is greatly affected by contact force. The measured contact resistance in a MEMS switch is usually significantly

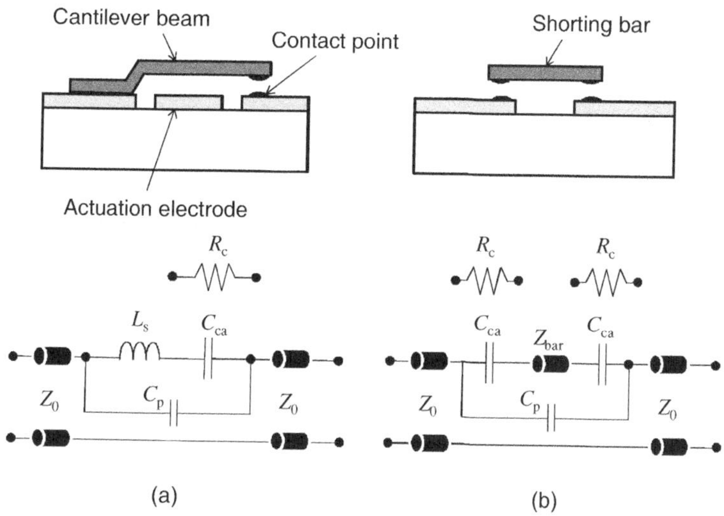

Figure 12.7: Illustration of contact switches, (a) an in-line cantilever contact switch, (b) a two-contact switch with a shorting bar. For the cantilever switch, the actuation voltage is applied between the actuation electrode and the beam itself, whereas for the shorting bar switch the actuation electrodes are not shown.

larger than the theoretical value, since physical contact is made at only a small number of 'contact points' which translates to a smaller contact surface area. The physics behind contact mechanisms in low-force MEMS switches is a topic of significant research efforts [11, 12], and some design features to overcome this challenge will be briefly discussed in Section 12.5.

The ideal series contact switch will result in an open circuit on the T-line in the unactuated state and a short circuit in the down state obtained by applying a bias voltage. The ideal switch exhibits an infinite isolation in the unactuated up-state and zero insertion loss in the down-state. Typical demonstrated values for ohmic MEMS series switches are

- isolation: −50 to −60 dB at 1 GHz;
- isolation: −20 to −30 dB at 20 GHz;
- insertion loss: −0.1 to −0.3 dB at 0.1 to 30 GHz.

Referring to Figure 12.7(a), the impedance of a series-configured cantilever switch in the off-state is given by

$$Z_s = j\omega L_s + \frac{1}{j\omega C_t} \tag{12.17}$$

where C_t is the sum of the parasitic capacitance C_p and the contact area capacitance C_{ca}, and L_s is the inductance of the cantilever beam. For an ideal switch, in which the switch armature is a perfect continuation of the transmission line, the return loss (S_{11}), isolation and insertion loss (S_{12}) for a series load on a transmission line is given by reference [13]

$$S_{11} = S_{22} = \frac{Z_s}{2Z_0 + Z_s} \quad (12.18)$$

$$S_{12} = S_{21} = \frac{2Z_0}{2Z_0 + Z_s} \quad (12.19)$$

Inserting the switch impedance (Equation (12.17)), the return loss and isolation for the switch in the up-state is given by

$$S_{11}[\text{dB}] = 10\log\left[\frac{(1-\omega^2 C_t L_s)^2}{(1-\omega^2 C_t L_s)^2 + (2\omega C_t Z_0)^2}\right] \quad (12.20)$$

and

$$S_{12}[\text{dB}] = 10\log\left[\frac{(2\omega C_t Z_0)^2}{(1-\omega^2 C_t L_s)^2 + (2\omega C_t Z_0)^2}\right] \quad (12.21)$$

Figure 12.8 illustrates the effect of the up-state capacitance on the series switch performance where C_{ca} is varied between 1, 5 and 10 fF. In the up state, the inductance (L_s) has a limited effect on the return loss and isolation, since the impedance is dominated mainly by the up-state capacitance. However, as mentioned previously, for a shorting bar switch, neglecting the high-impedance section of the bar with two contact pairs may lead to erroneous values in the simulation. This is because the high-impedance of the shorting bar section bounded by two high-reactance capacitors on both sides improves the isolation in the up-state [8]. Therefore, such a switch is best modeled using T-line models.

In the down-state, the impedance of the contact switch will be dictated by the contact resistance (replacing the contact-area capacitance in Figure 12.7) and the switch inductance and is given by

$$Z_s = j\omega L_s + R_c \quad (12.22)$$

which yields the following expressions for the return and insertion loss

$$S_{11}[\text{dB}] = 10\log\left[\frac{R_c^2 + (\omega L_s)^2}{(2Z_0 + R_c)^2 + (\omega L_s)^2}\right] \quad (12.23)$$

and

$$S_{12}[\text{dB}] = 10\log\left[\frac{(2Z_0)^2}{(2Z_0 + R_c)^2 + (\omega L_s)^2}\right] \quad (12.24)$$

Compared to when the switch is in the up-state, the down-state inductance significantly affects the return loss of a MEMS series switch. Particularly for antenna systems, where matching is essential, the return loss must be kept to a minimum. For MEMS switches, the inductance L_s can vary between 30–80 pH for 100 µm-long switches [8], and Figure 12.9 shows simulated return loss results for a series switch with an on-state contact resistance of 1 Ω and a varying inductance L_s. Figure 12.10 shows a plot of the estimated insertion loss of an ohmic switch in the on-state for various contact resistance values of 1–3 Ω. It should be pointed out that the insertion loss is dominated by the contact resistance and is nearly independent of frequency. An inductance of 80 pH for the switch armature is included in the simulation.

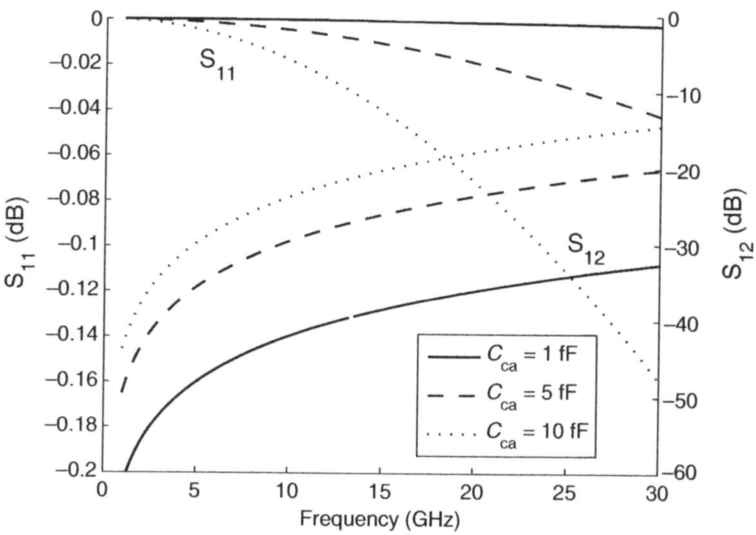

Figure 12.8: Simulated isolation and return loss for a one-contact series configured ohmic switch in the up state. The capacitance of the contact area dominates both the isolation and the return loss. The switch inductance has very little effect on the up-state performance.

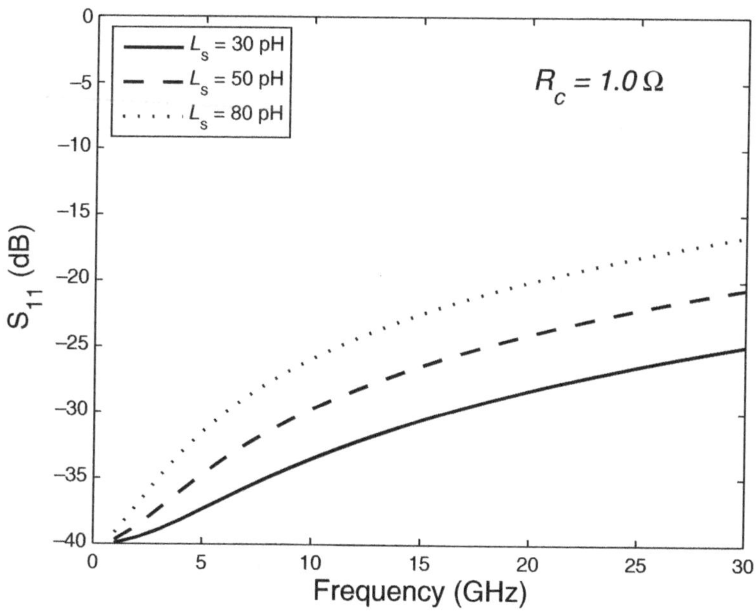

Figure 12.9: Simulated return loss results for a series ohmic switch in down-state with a contact resistance of 1.0 Ω, and with a varying inductance value of of 30, 50 and 80 pH.

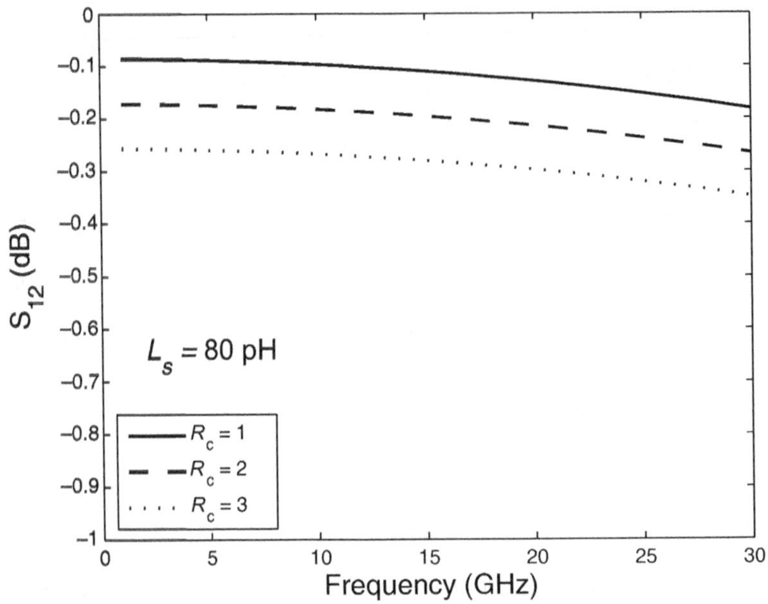

Figure 12.10: Simulated insertion loss results for a series ohmic switch in the down-position. The insertion loss is dominated by the contact resistance and nearly independent of frequency. An inductance of 80 pH for the switch armature is included.

Lastly, if one choses to neglect the contribution from the inductance of the cantilever beam, or the shorting bar, to the switch impedance ($\omega L_s \ll \omega C_t$) the expressions for the isolation (Equation (12.21)) and insertion loss (Equation (12.24)) of the switch are simplified to

$$(Up)IS = 10\log\left(\frac{(2\omega C_t Z_0)^2}{1+(2\omega C_t Z_0)^2}\right) \tag{12.25}$$

$$(Down)IL = 20\log\left(\frac{2Z_0}{R_c + 2Z_0}\right) \tag{12.26}$$

which serve as good estimates of switch performance for known values of up-state capacitance and on-state contact resistance.

12.4.1.1 Contact Switch Examples

One of the more promising contact RF MEMS devices is manufactured by Radant MEMS Inc. This switch consists of a cantilever structure with a pull-down electrode at the very end of two flexures. The cantilever is relatively sturdy, from a MEMS point of view, with a thickness of 7–9 μm electroplated Au making it very stiff and robust, as well as capable of handling relatively high currents. This yields a high spring constant, which again requires a high actuation voltage (60–70 V), but the stiffness of the beam prevents the upper electrode from touching the lower pull-down electrode when actuated. This design feature is particularly useful, since the switch does not need a dielectric material for protecting the lower electrode,

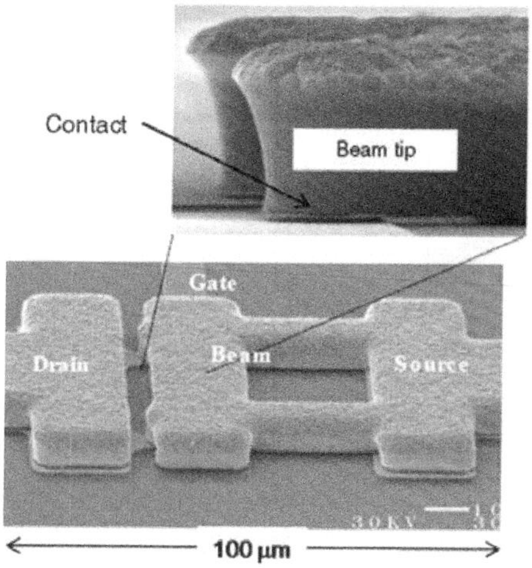

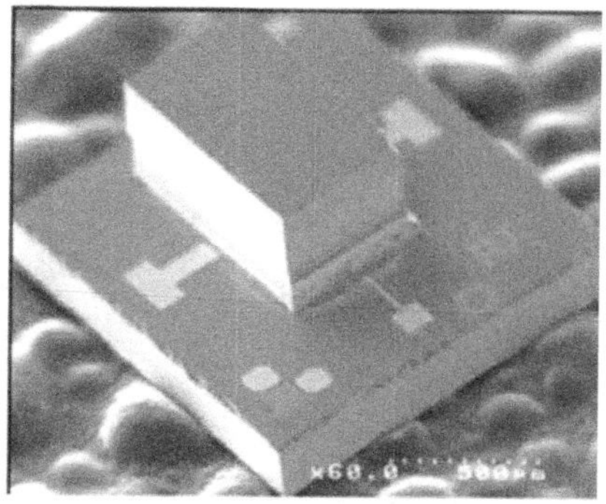

Figure 12.11: Photograph of the Radant MEMS ohmic RF switch. Top left picture shows the tip of the switch where one of the contacts can be seen underneath the beam. Lower left picture shows the overall switch structure where the gate is the actuation electrode. The packaged product is shown in the right picture where a protective lid covers the MEMS switch [14] (reproduced by permission of © 2003 SPIE).

which can have a significant effect on the lifetime of the switch (discussed further in Section 12.4.3). Two contact pairs in parallel are used to lower the overall resistance.

A switch utilizing a shorting bar, to open and short the transmission line, was employed by Hughes Research Laboratory (HRL) and is shown in Figure 12.12. The shorting bar is located at the end of the cantilever beam. In order to create a flat beam, the cantilever is made of three layers (SiN/Au/SiN). This balances the intrinsic stresses and the silicon nitride layer isolates the shorting bar/RF line and the actuation electrode reducing coupling between the two. Lastly, the silicon nitride layer also prevents any short between the actuation electrodes. In contrast to the Radant switch (Figure 12.11) which uses a very stiff beam, the HRL switch consists of two beam sections with different spring constants, one that ensures good contact across the gap when actuated and the second near the anchor point, which provides the required restoring force [15]. This is obtained by varying the length and cross-section of the beam structure between the actuation electrode and contact section. In contrast to the the Radant switch, which does not make contact between the actuation electrodes, the HRL switch requires a dielectric (SiN) to prevent an electrical short.

12.4.2 Broadband Capacitive MEMS Switches – Shunt Configuration

A schematic of a shunt configured MEMS capacitive switch is shown in Figure 12.13. Similarly to the previously described contact switches, the movable electrode is separated from the coplanar waveguide (CPW) center conductor by a gap z_0'. The switch is actuated by a DC voltage applied on the transmission line (S) and the top electrode which is grounded (G). Upon actuation, the switch bridge lands on the dielectric layer present on the microwave

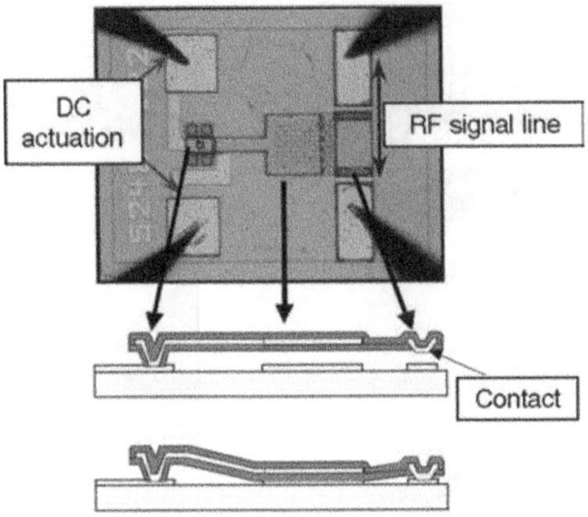

Figure 12.12: Photograph and illustration of HRL contact switch fabricated on a GaAs substrate. The RF portion of the switch has a series gap in the microstrip transmission line that can be short-circuited by a bar of gold deposited at the end of a cantilever beam [15] (reproduced by permission of © 2000 IEEE).

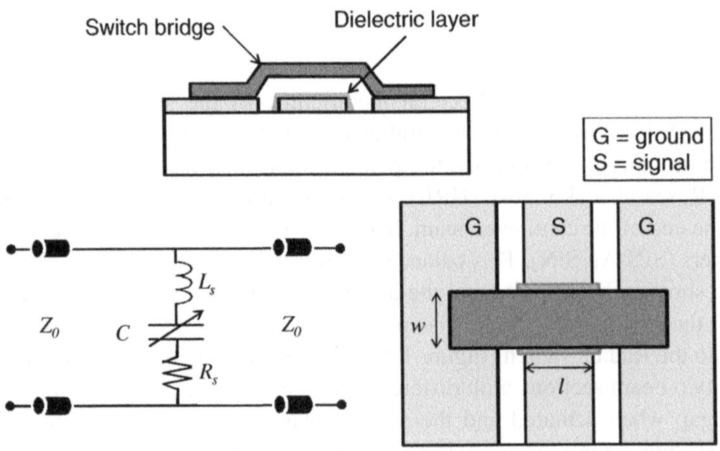

Figure 12.13: Top and side view illustration of a typical capacitive MEMS switch implemented on a CPW transmission line.

line, greatly increasing the capacitance to ground. This causes the switch to act as a short circuit (shunt) at microwave frequencies, resulting in a reflective switch [8].

Compared to a series-configured switch, a capacitive switch placed in shunt between the T-line and ground either leaves the T-line undisturbed or connects it to ground when actuated. The switch does not present any resistive losses in the on-state aside from the intrinsic losses of the T-line itself. This fact often favors shunt switches over series switches, especially at higher frequencies. Therefore, the ideal shunt switch has zero insertion loss in the up-state (no bias-voltage) and infinite isolation when actuated. Typical values for a shunt MEMS switch are:

- isolation: −20 dB at 10–50 GHz;
- insertion loss: 0.04 to 0.1 dB at 5–100 GHz.

The main design parameter for a capacitive switch is the capacitance ratio. When the switch is actuated in the up-state, the capacitance is given by

$$C_{\mathrm{u}} = \frac{1}{(t_{\mathrm{d}}/\epsilon_{\mathrm{d}} A) + (z_0'/\epsilon_0 A)} \tag{12.27}$$

where ϵ_{d} and t_{d} are the relative dielectric constant and thickness of the dielectric layer, respectively and $A = wl$. For a typical MEMS switch, the up-state capacitance is in the range of tens of femtofarads. Upon actuation, when the top bridge is in contact with the lower electrode, the capacitance is given by

$$C_{\mathrm{d}} = \frac{\epsilon_{\mathrm{d}} A}{t_{\mathrm{d}}} \tag{12.28}$$

and the capacitance ratio can be expressed as

$$\frac{C_{\mathrm{d}}}{C_{\mathrm{u}}} = \frac{\epsilon_{\mathrm{d}} z_0'}{t_{\mathrm{d}}} \tag{12.29}$$

Due to the surface roughness of the bridge and the dielectric surface, the down-state capacitance can be 20% less than the theoretical parallel-plate value. Thus, a common approach is to fabricate the switch bridge over a smooth section of the T-line in order to obtain a high capacitance ratio.

The low insertion loss is probably the greatest advantage of a shunt-configured capacitive switch. Referring to Figure 12.13, the impedance of the shunt capacitive switch is given by

$$Z_{\mathrm{s}} = R_{\mathrm{s}} + j\omega L_{\mathrm{s}} + \frac{1}{j\omega C} \tag{12.30}$$

where C refers to the capacitance value in either up (C_{u}) or down-state (C_{d}). For an ideal shunt-configured switch, the S-parameters are given as [13]

$$S_{11} = S_{22} = \frac{-Z_0}{2Z_{\mathrm{s}} + Z_0} \tag{12.31}$$

$$S_{12} = S_{21} = \frac{2Z_{\mathrm{s}}}{2Z_{\mathrm{s}} + Z_0} \tag{12.32}$$

In the up-state, compared to the capacitance (C_{u}) of the switch, the inductance value (L_{s}) and resistance (R_{s}) of the switch do not contribute significantly to the return or insertion loss. Thus, Equation (12.30) is simplified and the return loss can be expressed as

$$S_{11}[\mathrm{dB}] = 10 \log\left[\frac{(\omega C_{\mathrm{u}} Z_0)^2}{4 + (\omega C_{\mathrm{u}} Z_0)^2}\right] \tag{12.33}$$

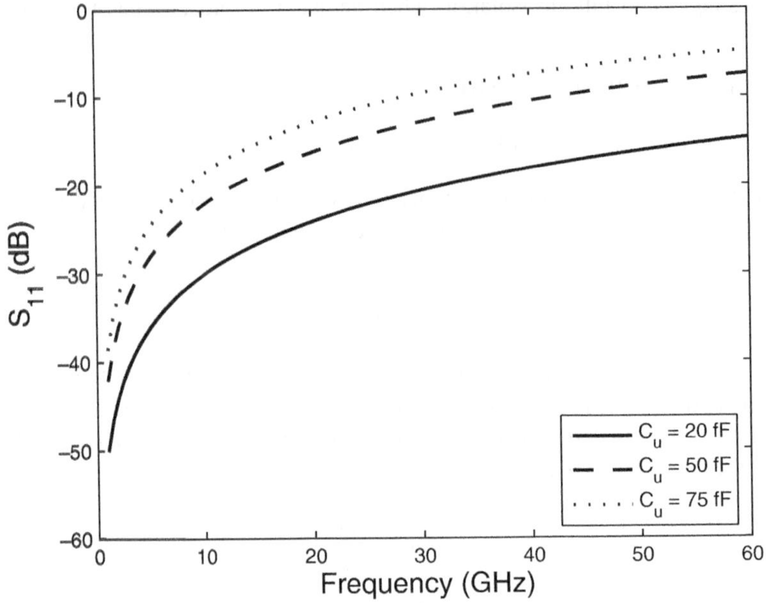

Figure 12.14: Simulated return loss results of a capacitive shunt switch in the up-state. The contribution of the inductance and resistance to the return loss is negligible.

whereas the insertion loss of the T-line is best solved using computer-aided simulation tools.

Figure 12.14 shows the simulated return loss for a capacitive shunt switch where the up-state capacitance is varied between typical values of 20 and 75 fF.

When the switch is actuated (in the down-state), the return loss and isolation of the switch can be shown to be

$$S_{11}[\mathrm{dB}] = 10\log\left[\frac{(\omega C_{\mathrm{u}} Z_0)^2}{(2+2R_{\mathrm{s}})^2 + (2\omega L_{\mathrm{s}} + \omega C u Z_0)^2}\right] \tag{12.34}$$

$$S_{12}[\mathrm{dB}] = 10\log\left[\frac{(2-2\omega^2 L_{\mathrm{s}} C_{\mathrm{d}})^2 + (2\omega C_{\mathrm{d}} R_{\mathrm{s}})^2}{(2-2\omega^2 L_{\mathrm{s}} C_{\mathrm{d}})^2 + (2\omega C_{\mathrm{d}} R_{\mathrm{s}} + \omega C_{\mathrm{d}} Z_0)^2}\right] \tag{12.35}$$

As mentioned earlier, it can be challenging to obtain a high down-state capacitance value when the switch is actuated and is in contact with the dielectric on the T-line. However, the inductance and capacitance of the bridge will exhibit an LC resonance at a frequency f_0 ($f_0 = 1/\sqrt{2\pi(L_{\mathrm{s}} C_{\mathrm{d}})}$) at which the isolation can be enhanced. Figure 12.15 illustrates inductive tuning of the isolation by varying the switch inductance (L_{s}) from 5 to 15 pH for a given down-state capacitance, $C_{\mathrm{d}} = 1.8$ pF and resistance $R_{\mathrm{s}} = 0.1\ \Omega$. The downside to inductively tuned shunt switches is the reduction in bandwidth around the the resonant frequency. This is especially critical at lower frequencies (10–25 GHz) [16].

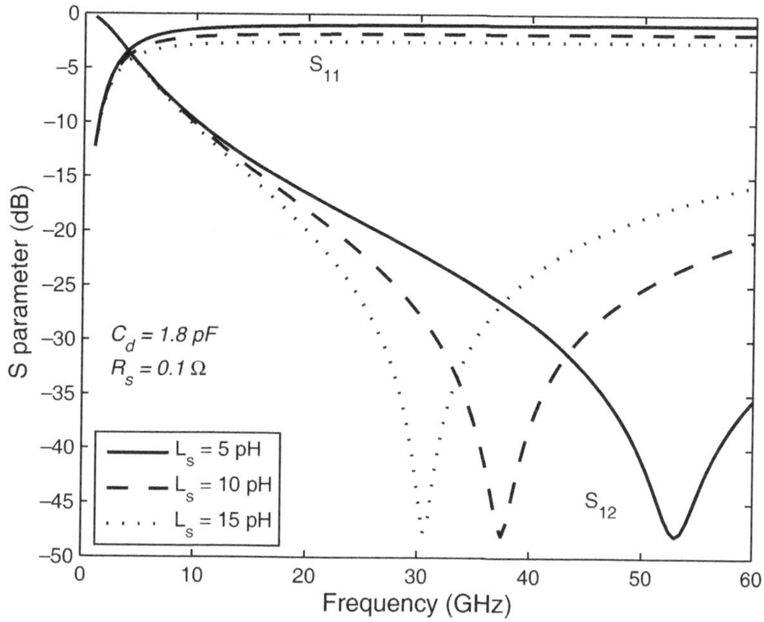

Figure 12.15: Simulated performance of a capacitive shunt switch. The inductance and capacitance of the switch controls the location of the resonant frequency, and enhances the isolation. The return loss is hardly affected by the change in inductance. The downside to inductive tuning of the switch isolation is a narrower bandwidth.

12.4.2.1 Capacitive Switch Examples

One of the most mature capacitive switches is currently under development at MEMtronics, which is based on earlier work done at Raytheon. The shunt switch shown in Figure 12.16 can be fabricated on any type of substrate (high-resistivity silicon, glass or sapphire). A thin layer of dielectric (SiN – 2 kA thick) insulates a 0.5 μm-thick W (tungsten) electrode on the bottom. Using W as the material for the electrode ensures a very smooth surface in order to increase the down-state capacitance. The membrane, or bridge, is formed from a 0.5 μm-thick Al alloy [17]. The small holes visible in the membrane are etch holes which facilitate accelerated access to the sacrificial layer during release.

In contrast to the shunt device, a series-configured cantilever capacitive switch has been developed with great success by MIT Lincoln Laboratory. This particular switch, shown in Figure 12.17, utilizes built-in stresses to create a device with a very low up-state capacitance, and thus a very high capacitance ratio. The switch footprint is very small (approximately 100 × 60 μm) which makes series-shunt and other configurations quite attainable. The switches are fabricated on Si substrate with SiO_2/TaN/SiO_2 deposited using low-pressure chemical vapor deposition (LPCVD) and Al deposited and patterned to form the T-lines (0.5 μm-thick). The sacrificial layer is polyimide, which is spun on and patterned to create ridges in the structural layer. The cantilever switch is created from three layers (plasma-enhanced chemical vapor deposition (PECVD) SiO_2 (2 kA) compressive, Sputtered Al (5 kA)

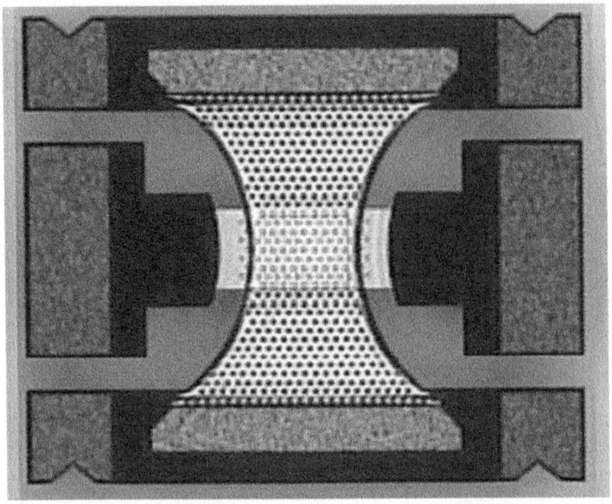

Figure 12.16: Photograph of the MEMtronics capacitive shunt RF switch. The membrane is fabricated from an Al alloy. Tungsten is used as material for the electrode underneath the switch to ensure a smooth surface and high capacitance ratio [18] (reproduced by permission of © 2005 IEEE).

Table 12.1: Performance comparison of RF MEMS switch examples.

Parameter	Radant	HRL*	MEMtronics	MIT-LL
Switch type	Ohmic series	Ohmic (bar)	Cap shunt	Cap series
Voltage (V)	100	30–40	35	40
$R_C(\Omega)$	0.8	< 1	—	—
Capacitance ratio	—	—	20:1	>170:1
Isolation (dB)	20@10 GHz	40@5 GHz	15@10 GHz	20@20 GHz
	14@20 GHz	25@40 GHz	30@35 GHz	14@40 GHz
Return loss	>20@20 GHz	–	>17@35 GHz	>26@20 GHz
Insertion loss	0.5@12 GHz	0.18@5 GHz	<0.2@35 GHz	0.1@20 GHz
	0.7@25 GHz	0.2@15 GHz		0.2@40 GHz

*Device not packaged. Adapted from references [14, 15, 19–21].

and PECVD SiO_2 (2 kA) - tensile) which combined cause the switch to curl away from the substrate.

The typical performance of these RF MEMS switch examples is summarized in Table 12.1. It is noteworthy that all performance parameters, except for the HRL switch, refer to packaged devices. Particularly for discretely packaged devices, such as the Radant and MIT-LL switch, the package affects switch performance at mmWave frequencies, and requires careful design to reflect the inherent broadband performance of these MEMS switches.

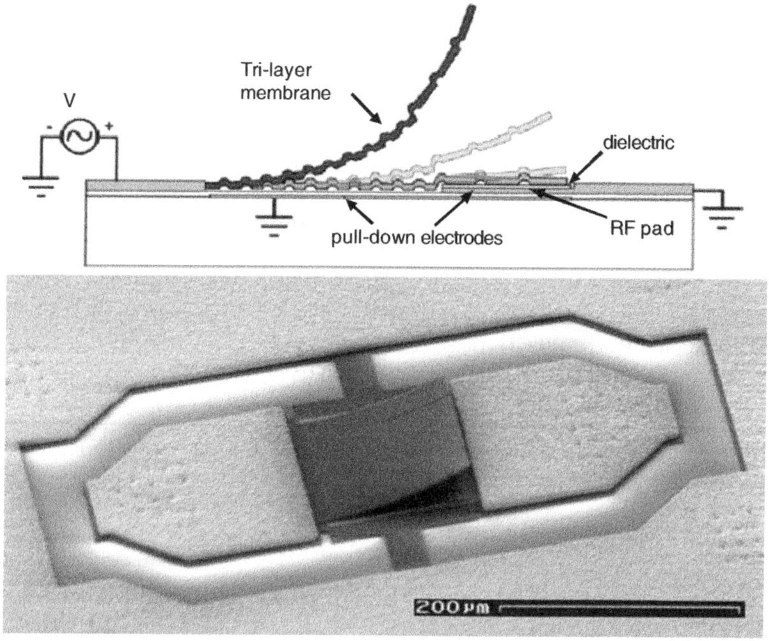

Figure 12.17: Illustration and photograph of MIT Lincoln Laboratory capacitive series configured switch. The illustration shows how the switch membrane curls away from the T-line yielding a low-up state capacitance, and hence a high capacitance ratio (reprinted with permission of MIT Lincoln Laboratory, Lexington, Massachusetts).

12.4.3 Switch Performance and Design Considerations

The previous sections presented a brief introduction to the various aspects of RF MEMS switches. Probably, the most exciting aspect of this technology is the multi-disciplinary nature of the MEMS switches; but within this aspect also lies the challenges facing these devices as both mechanical and electrical considerations have to be taken into account. While each switch has a different circuit model, in essence the fundamental desired property of a switch is a low on-state resistance and low off-state capacitance, and most switches can be quite well modeled as a simple LCR circuit. Table 12.2 summarizes some of the major differences between packaged RF MEMS switches with PIN diode and field-effect transistor (FET) switches.

It could be misleading to use the product of R_{on} and C_{off} as a figure of merit (FOM) for RF MEMS switches. With inherently low capacitance and resistance values, RF MEMS switches exhibit unreasonably high FOM compared to solid-state devices and thus the conventional FOM definition does not offer a true basis for comparison. A better performance comparison of RF MEMS switches can be made by defining an FOM in terms of the insertion loss and isolation in the operating frequency band. Therefore, the FOM may be expressed as

$$FOM = BW\left(\frac{IS_{\text{min}}}{IL_{\text{max}}}\right) \tag{12.36}$$

Table 12.2: Performance comparison of *packaged* MEMS switches to PIN and FET devices.

Parameter	RF MEMS	PIN	FET
Voltage (V)	25–100	±5	3–5
Current (mA)	0	3–20	0–0.05
Switching time	1–40 μs	1–100 ns	1–10 ns
Power handling (W) (ohmic, series)	10* 0.1**	<10	<10
Power handling (W) (capacitive, shunt)	$0.01^{\dagger}$		
Isolation (dB) (1–20 GHz)	>30	>35	>35
Isolation (dB) (20–40 GHz)	>15	>35	
Loss (dB) (1–20 GHz)	0.05–0.1	0.9	2–3
Loss (dB) (20–40 GHz)	0.01–0.3	<2	2–3

*Cold switched >13 B cycles. **>400 B cycles. †Hot switched >100 B cycles.
Contents of table adapted from reference [8] and TriQuint Semiconductor Inc.

Table 12.3: Figure of merit (FOM) comparison of *packaged* RF MEMS switches to solid-state devices.

Switch	Bandwidth (GHz)	IL_{max} (dB)	IS_{min} (dB)	*FOM*
Radant (ohmic)	DC–6	0.35	25	428
Teravicta (ohmic)	DC–6	0.25	33	792
Skyworks UWB PHEMT	0.5-6	0.8	17	117
Radant (ohmic)	18–27	0.43	13	272
MEMtronics (capacitive)	18–27	0.125	24	1728
MIT-LL (capacitive)	18–27	0.15	16	960
QuinStar coaxial PIN	18–27	1.5	28	168

where BW refers to the bandwidth in GHz in the frequency band of interest, e.g. X, Ku or K-band, and IL_{max} is the maximum insertion loss and IS_{min} is the minimum isolation in dB within the chosen band. This relationship offers a direct comparison among various RF MEMS devices when evaluating the performance. Naturally, there are also other parameters to be considered, such as power handling and return loss when evaluating the switch performance. Especially, the wide bandwidth of MEMS switches sets them apart from conventional devices. Second, at high frequencies, the very low insertion loss of capacitive shunt switches significantly increases the FOM of MEMS switches. Table 12.3 illustrates a comparison among switches operating in the S, C and K-bands.

Mechanical considerations: For micromachined structures, intrinsic material stresses are unavoidable and must be considered in any MEMS device. For instance, a clamped–clamped device with compressive stress will buckle and warp uncontrollably. If the device is highly tensile, the actuation voltage will greatly increase. Whereas a single-layer cantilever beam is not sensitive to intrinsic stress, it is very sensitive to any stress gradient within the beam. Any significant stress gradient will cause a cantilever beam either to curl up, or down depending on the distribution of the stress gradient. However, a cantilever beam is more robust against thermal excursions than a clamped–clamped structure as it is not confined by the boundary conditions to expand and contract.

The layer stack comprising the structural layer will also need to be considered. Often a thin adhesion or barrier layer of a refractory metal (e.g. W, TaN or TiW) is deposited prior to the deposition of the structural layer. These layers are usually very high-stress materials and can cause undesired deformations in the switch.

Since the actuation voltage is independent of the beam width, increasing the beam width does not reduce the actuation voltage. However, for devices with separate actuation electrodes it is possible to reduce the actuation voltage by increasing the beam width. It should be pointed out that wider beams with internal stresses will be susceptible to bifurcation, meaning that they will deform in both X and Y directions. This can often be seen on membrane switches, which under tension curl up along the edges.

Lastly, ensuring design for manufacturability is always highly desirable as variations in manufacturing may greatly affect the micromachined device. A design that is robust to process variation such as layer thicknesses and stress levels is key to a successful implementation.

Electrical considerations: Compared to a solid-state device, the greatest benefit of a MEMS switch is the high linearity. For instance, a series switch in the up-state is basically very close to a perfect open. The mechanical resonant frequency of MEMS devices are usually in the low MHz, which means that they do not respond to any imposed RF signal in the GHz range as operation is above the natural resonance. However, the major setback for MEMS devices is their slow response time, especially when compared to solid-state devices. Electrostatic MEMS switches typically have switching times around 1–40 μs. This is significantly longer than solid-state devices, often eliminating the possibility to use MEMS as TR switches; however, many switch applications such as tunable filters, switched and tunable antennas operate at switching speeds suitable for MEMS devices.

For electrostatic switches, the release voltage is different from the pull-in voltage, particularly for devices with a dielectric between the actuation electrodes. This leads to a hysteresis in the actuation characteristics, which also affects the switching time. Invariably, the switching on-time will be different from the off-time, although on-time can be controlled by the voltage applied [8].

The other major challenge to MEMS devices is the relatively high actuation voltage (35–100 V being typical). Driving a switch barely above the actuation voltage is typically not recommended as the switching time is reduced with increasing voltage, and a high enough actuation voltage ensures a more reliable down-state value (resistance for ohmic switches or capacitance for capacitive devices). For years, researchers have attempted to lower the actuation voltage of electrostatic switches, either by reducing the nominal gap or by modifying the spring constant. For a electrostatically actuated switch, a smaller gap can be obtained at the expense of a lower capacitance ratio. Also, designing a low spring constant switch will lead to a very low restoring force potentially making the switch unreliable.

Ohmic switches are very sensitive to variation in the on-state resistance. Switches with one contact pair will have a higher contact force than switches utilizing multiple contact pairs. However, by placing multiple contact pairs in parallel the on-state contact resistance can be reduced, and the switch is able to handle more power since the current is divided among the contacts (as in the contact pair used in the Radant switch, Figure 12.11).

Table 12.4: RF MEMS switches, key design parameters.

	Off		On			
	S_{11}	S_{12}	S_{11}	S_{12}	Frequency range	Design challenge
Ohmic					DC–40 GHz	Contact resistance
Series	C_u	C_u	L_s	R_c		R_c
Shunt	L_s	$L_s(R_c)$	C_u	T-line		R_c
Capacitive					10–100 GHz	Capacitance ratio
Series	C_u	C_u	L_s	C_d		C_d – insertion loss
Shunt	C_d	$C_d L_s$	C_u	T-line		C_d – isolation

As discussed earlier in Section 12.4.2, isolation of a shunt capacitive switch can be increased at lower frequencies by increasing the shunt inductance of the switch. However, there is a limit to how much inductance can be designed into the switch without reducing the spring constant to a very low value. If needed, one can add additional inductive lines extending from the switch anchor point to the ground-plane.

Some key design parameters for the various types and configurations of switch are presented in Table 12.4.

12.4.4 MEMS Varactors

The MEMS varactor, or variable capacitor, has been widely demonstrated in RF circuit applications due to their tuning functionality. In MEMS varactors, the capacitance is varied by adjusting the physical dimensions of the structure, either by reducing the gap or changing the overlapping area of the two capacitor electrodes. MEMS varactors have been designed for analog or digital capacitance change. In analog designs, the capacitance value can be changed continuously with applied voltage, whereas in digital designs the capacitance value is changed in a step-wise fashion.

Electrostatically actuated MEMS varactors operate in a similar fashion to electrostatic switches, where an equilibrium between the restoring spring force and the electrostatic forces will be reached at two-thirds of the initial gap distance. Thus, for the two-plate varactor the maximum tuning range is about 50%. The tuning range can be extended to 100% by introducing multiple plates, or gaps, and separate control and signal electrodes. In this case, the air gap between the control electrode and the suspended electrode can be made larger than the gap between the signal electrode and the suspended electrode. The pull-in due to the control voltage does not occur before the suspended electrode touches the signal electrode. In this configuration, the capacitance due to the dielectric layer defines the theoretical maximum tuning range [22].

12.5 MEMS Reliability and Power Handling

With the promise of MEMS switches as enabling key elements in future high-performance RF systems, the lack of commercially available devices to date surely raises the question of their

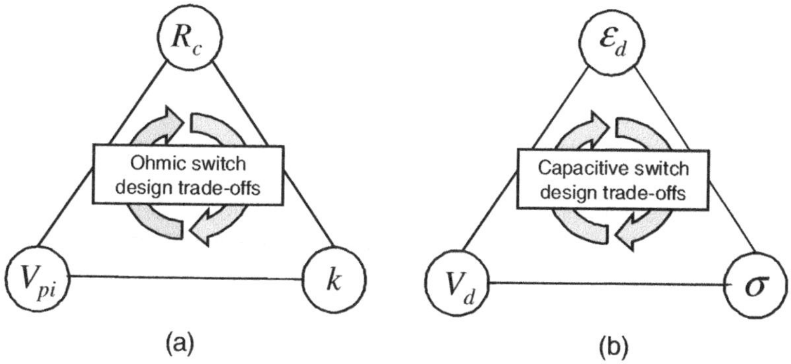

Figure 12.18: (a) Design trade-offs often encountered when designing an ohmic switch. With contact resistance as the critical parameter, both actuation voltage and the restoring force (spring constant, k) must be evaluated. (b) For a capacitive MEMS switch, the three main design parameters which affect switch performance and reliability are the dielectric constant (ϵ_{d}), breakdown voltage (V_{d}) and charge accumulation (σ) in the dielectric material. Whereas a high dielectric constant may be desired to obtain a high capacitance ratio, a large charge accumulation will cause a MEMS capacitive switch to fail in the down-state position.

viability. The reality is that for several years, reliability and packaging issues have prevented the insertion of MEMS switches in military and commercial systems. When compared to solid-state devices, these two challenges are not separate issues, but are highly interdependent and must be treated concurrently. As a consequence of the mechanical nature of MEMS, a cavity needs to be provided to house the switch and thus the packaging requirements are different for a movable device compared to a solid-state device. As has been revealed through research activities over the past several years, this cavity needs to be fully hermetic with a controlled ambient environment. Furthermore, the reliability of a MEMS switch highly depends on the environment in which it is packaged and operated [23, 24].

12.5.1 Reliability and Failure Modes

For ohmic contact MEMS switches, the main reliability issues, such as failure due to stiction, contact welding and contact resistance degradation, have been observed to be the key failure modes. Stiction is the unintentional adhesion of the movable and fixed parts in MEMS devices caused by surface adhesion forces. Failure due to stiction is frequently encountered in electrostatically actuated contact type MEMS relays; however, it is also the main failure mode in capacitive contact switches. Typically, MEMS switches are designed for operation at low actuation voltages, necessitating the design of movable micromechanical parts with relatively low restoring spring forces. Permanent failure due to stiction occurs when the restoring spring force of the movable part is lower than the adhesion forces generated at the contact surfaces. To circumvent this challenge, one successful approach is to design for a high contact force device that overcomes the stiction failures by careful management of the switch contact and return force using high actuation voltage [24]. Figure 12.18(a) illustrates the design trade-offs usually encountered when designing ohmic switches.

Another common reliability issue in ohmic contact MEMS switches is the increase of the contact resistance over many actuation cycles. Specifically, the resistance of the MEMS switches gradually increases with actuation cycles which leads to an unacceptably high insertion loss after several million actuation cycles. Early switches used pure Au as the contact material due to its high conductivity and resistance to surface oxidation. However, the soft nature of Au is undesirable for contact applications. More recent switches use harder metals, such as Ruthenium or Rhodium, or an Au-based alloy with significantly improved mechanical properties [25]. However, with harder contact materials it could be challenging to implement low voltage switches with low contact force [23]. The exact composition of the contact material in today's commercially available switches is carefully protected by the respective companies. Lastly, contact welding occurs when very high current density at the contacts actually causes the contact material to melt and fuse together. This is typically a failure mechanism associated with hot switching (where the RF current is present during the switch transition from off to on) or at very high power levels.

The primary reliability issue and failure mode for capacitive MEMS switches is stiction due to charge accumulation in the dielectric layer, keeping the switch in a permanently closed state. Second, down-state capacitance degradation with increasing actuation cycles and excess charge accumulation invariably causes intermittent performance of the switches, and needs to be kept under control [24].

Over the last few years a significant amount of work has gone into understanding the charging mechanism, and particularly the discharging mechanism within the dielectric of MEMS capacitive switches. For the researchers, it quickly became apparent that the quality of the dielectric material was of great importance for switch reliability, and in particular, trapped charges or defects within the dielectric arising from deposition of the layer needed to be avoided. Both SiN and SiO_2 are commonly used dielectric materials. SiN especially is attractive due to a higher dielectric constant; however, breakdown voltage and charge-injection threshold voltage favors the use of SiO_2, resulting in less charging when used in a MEMS capacitive switch. As illustrated in Figure 12.18(b) the design trade-offs for a capacitive switch have to balance the desire to obtain a high-performance device (typically from a high-capacitance ratio) with charge-injection characteristics and breakdown voltage. In particular, one main difference between a MIM (metal–insulator–metal) capacitor and a MEMS switch is the free surface on the top surface of the dielectric. Furthermore, the metal in the movable structure is often different from the metal used below the dielectric. This is further complicated by realizing that a time-varying junction in the interface between the free-surface and the movable top-metal electrode exists as the switch is cycled, and with the contacting surfaces changing over the lifetime of the switch.

How charge is transferred to, located within and discharged from the dielectric layer is a topic of great research activity, and obtaining this knowledge has significantly helped to increase the lifetime of RF MEMS switches [26–28]. One of the more recent efforts to increase switch reliability has been in terms of reducing the amount of dielectric material present in capacitive switches, at the expense of capacitance ratio. This approach has led to significant improvements in lifetimes to be demonstrated. In addition, design of switches with lower actuation and hold-down voltages, and the use of innovative dielectric materials are among the many other strategies employed further to increase the lifetime of capacitive switches.

12.5.2 Power Handling

The power handling of an RF MEMS switch depends on factors such as contact type (metal or capacitive), RF configuration (series or shunt), switch state (up or down) and actuation mechanism. In this section, power handling of electrostatically actuated RF MEMS switches is discussed.

12.5.2.1 Maximum Power Handling in the Up-state

Maximum power handling in the up-state is limited by switch self-actuation (i.e. RF signal-induced actuation which occurs when the Root Mean Square (RMS) voltage of the incident RF power equals that of the actuation voltage of the switch). Let the amplitudes of the transmitted wave and the reflected wave be V^+ and V^-, respectively. The power transmitted P_t through the switch is then given by

$$P_t = \frac{|V^+|^2}{2Z_0} \tag{12.37}$$

Power transmitted can also be expressed in terms of the input power as

$$P_t = |S_{21}|^2 P_{in} \tag{12.38}$$

Equating Equations (12.37) and (12.38), we obtain

$$|V^+| = |S_{21}|\sqrt{2Z_0 P_{in}} \tag{12.39}$$

using $V^+_{rms} = V^+/\sqrt{2}$ in Equation (12.39) yields

$$|V^+_{rms}| = |S_{21}|\sqrt{Z_0 P_{in}} \tag{12.40}$$

Therefore, with $|V^+_{rms}| \geq V_{pi}$ the maximum input power $P_{max,in}$ can be found from Equation (12.40) as

$$P_{max,in} = \frac{V_{pi}^2}{Z_0}\frac{1}{|S_{21}|^2} \tag{12.41}$$

For a shunt switch with a capacitance C_{sh} across a transmission line of characteristic impedance Z_0, the insertion loss (IL) (due to reflection) can be obtained as

$$IL = \frac{1}{|S_{21}|^2} = 1 + \frac{\omega^2 C_{sh}^2 Z_0^2}{4} \tag{12.42}$$

where $\omega = 2\pi f$ and f is the frequency of the RF signal.

Now, substituting for the IL given in Equation (12.42), the maximum power for a capacitive shunt switch can be written as

$$P_{max,in\text{-}shunt} = \frac{V_{pi}^2}{Z_0}\left(1 + \frac{\omega^2 C_{sh}^2 Z_0^2}{4}\right) \tag{12.43}$$

Similarly, the maximum power handling for a series switch, with up-state capacitance C_{se}, is expressed as

$$P_{max,in\text{-}series} = \frac{V_{pi}^2}{Z_0}\left(1 + \frac{1}{4\omega^2 C_{se}^2 Z_0^2}\right) \tag{12.44}$$

For ohmic contact-type switches, the up-state capacitance is very small (fF) and therefore the maximum power handling in the up-state can be expected to be higher than for capacitive contact switches. Furthermore, ohmic switches fabricated with separate actuation electrodes are rarely susceptible to self-actuation as the contact area is very small.

12.5.2.2 Maximum Power Handling in the Down-state

Maximum power handling in the down-state largely depends on the type of contact (ohmic or capacitive). In capacitive contact RF MEMS switches, when the total effective voltage (DC voltage and RMS RF voltage) across the dielectric layer is above the breakdown voltage the dielectric will suddenly begin to conduct current. This phenomenon is known as dielectric breakdown and usually results in permanent damage of dielectric material in solid dielectrics. In air, dielectric breakdown occurs at an electric field strength of 3×10^6 V/m.

For a dielectric layer of thickness t_d and dielectric breakdown field strength of E_{max}, the breakdown voltage is given by

$$V_{b,max} = t_d E_{max} \tag{12.45}$$

In the switch down-position, the total electric field strength must be less than the dielectric breakdown voltage $V_{b,max}$. Thus, the power-handling capability in the down-state is limited by the breakdown of the dielectric layer in a capacitive-type MEMS switch. However, depending on the switch geometry and design, the current in the switch in the down position might limit the amount of power it can handle rather than the breakdown voltage of the dielectric. The peak current in a shunt switch in the down-position can be expressed as [29]

$$I_{shunt} = \frac{2\omega C_{sh}\sqrt{2P_{in}Z_0}}{\sqrt{4 + (\omega C_{sh} Z_0)^2}} \tag{12.46}$$

and for a series switch,

$$I_{series} = \frac{2\omega C_{se}\sqrt{2P_{in}Z_0}}{\sqrt{1 + (2\omega C_{se} Z_0)^2}} \tag{12.47}$$

where C_{sh} and C_{se} are the down-state capacitance of the shunt and series switch, correspondingly. It is noteworthy that the current in a shunt switch is significantly larger than for the series switch, as shown in Figure 12.19.

For an ohmic contact switch, the power-handling capability in the down state is limited by the maximum current which can pass through the small metal contacts without causing any thermal dissipation induced failure such as microwelding.

A high-power MEMS switch was recently developed by Radant MEMS, with power handling in the range of 1 to 4 W. Radant is currently developing second-generation MEMS switches capable of handling up to 20 W of power. Another example is the fully packaged MIT-LL capacitive switch, which is capable of handling power greater than 2 W for hot switching and 7 W for cold switching.

12.5.2.3 Switch Lifetime

There is no doubt that recent improvements in reliability and packaging make MEMS switches attractive for reconfigurable antenna applications. With the lifetime requirement for military applications in the range of 100 to 500 billion cycles, governmental research

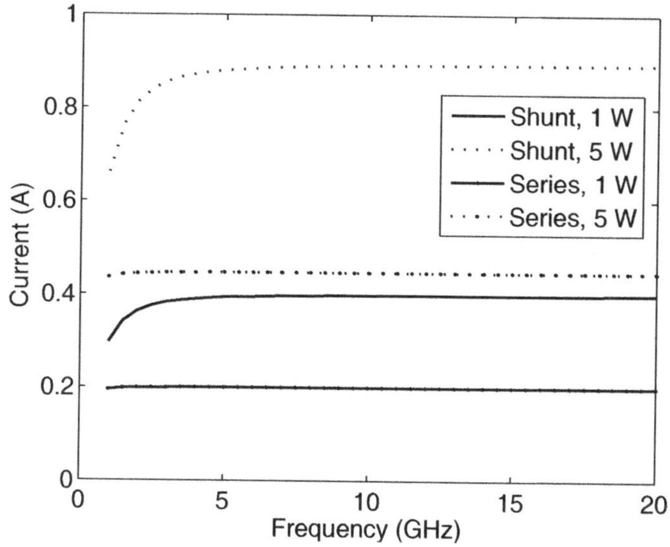

Figure 12.19: Peak currents for both a series and shunt capacitive switch in the down-state. For a given power and frequency, the maximum current in a shunt switch is nearly double that of a series switch.

programs sponsored by DARPA during the past five years, such as the RF MEMS Improvement Program (RF-MIP) and the Harsh Environment Robust Micromechanical Technology (HERMIT) program, have facilitated a steady progress in improving both reliability and packaging approaches of RF MEMS switches. The outcome of these programs prove that MEMS switches can have high reliability, while maintaining excellent RF performance. However, switch requirements in terms of lifetime depends on the type of application. Many successful MEMS switches have been demonstrated with lifetime in excess of 100 million cycles, and commercially available switches from TeraVicta and Radant MEMS have become a reality. Radant MEMS demonstrated 900 billion cycles of operation for their ohmic contact switches, whereas MEMtronics Corporation demonstrated over 100 billion cycles of switch lifetime for capacitive MEMS switches, and MIT-LL have surpassed 290 billion cycles on their series capacitive switch. TeraVicta switches have been tested up to 100 million cycles over the 0° to 70°C operating range of the device.

Reliability and packaging issues are commonly resolved by hermetic sealing of MEMS switches after completion of the micromachining process. Several MEMS switches use a form of wafer-level, or 0-level packaging, in which the switch is hermetically capped directly on the wafer very early in the fabrication cycle. This greatly reduces the fabrication cost, and the package can be made to have very little impact on the RF performance.

An example of a switch fabricated with 0-level packaging is shown in Figure 12.20 where the MEMtronics switch is sealed off at the wafer level using a spin-on polymer [19], creating a cage to protect the switch. The benefit of such an approach is that the cage has virtually no added parasitic and the feed-through provides very low insertion loss up to 110 GHz.

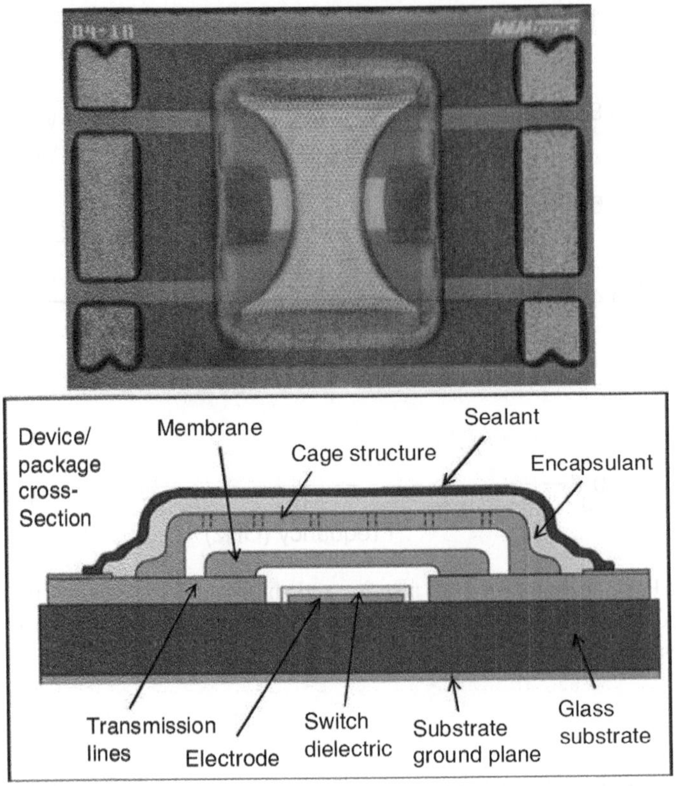

Figure 12.20: Photograph and illustration of 0-level packaging of the MEMtronics RF MEMS switch. The MEMS switch is protected from the environment using a low-dielectric constant package. The wafer-level encapsulation provides a very low-loss package operating up to and above 50 GHz. A cross-section of the micro-scale encapsulated package illustrates the cage, encapsulant, and the sealant layer housing the MEMS switch inside the cavity [19].

In contrast to the low-profile spin-on package used by MEMtronics, Figure 12.21 illustrates the MIT-LL switch which is packaged at wafer level using thermo-compression bonding to a micromachined cavity. A 90 μm-deep metal-coated cavity forms the ground plane and allows free movement of the switch. The use of tungsten thru-wafer vias with Au metal pattern and stud bumps yield very low loss thru-wafer transitions (0.025 dB insertion loss), and allow for the packaged device to be flip-chip integrated [21].

12.6 Integration of MEMS Switches with Antennas

RF MEMS devices can be integrated with antennas to enable frequency as well as pattern reconfiguration. Two different integration approaches are (1) hybrid integration, and (2) monolithic integration. In this section, these two integration approaches are discussed

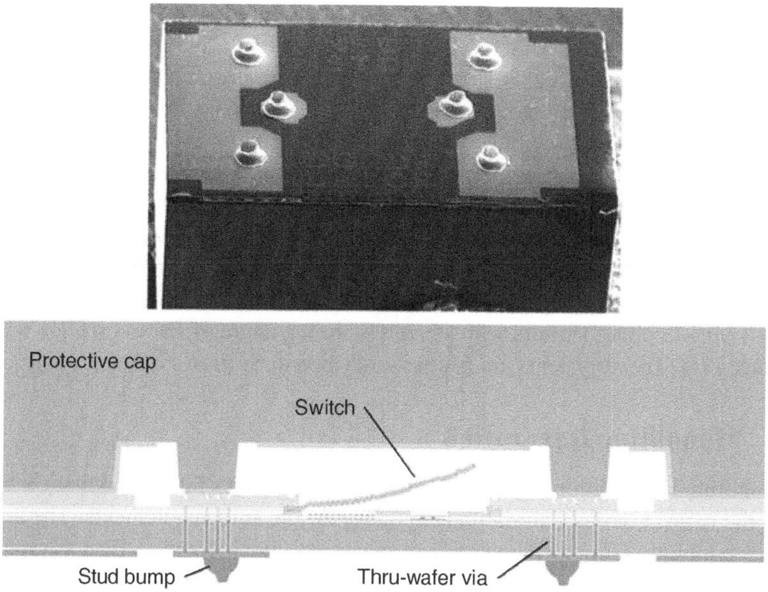

Figure 12.21: Photograph [21] and illustration of packaged MIT-LL capacitive switch. The package is created from thermo-compression bonding of an SOI wafer to the device wafer, and includes thru-wafer vias as transitions. Au stud bumps allow the packaged switch to be integrated using flip-chip assembly methods (reproduced by permission of © 2008 IEEE, reprinted with permission of MIT Lincoln Laboratory, Lexington, Massachusetts).

with examples. Advantages, disadvantages and integration issues associated with these two integration approaches are also highlighted.

12.6.1 Hybrid Integration

Discrete RF MEMS devices are ideal devices to integrate with antennas owing to their low losses, large isolation values and very low power consumption. Second, the biasing network to the switches can be made using high resistive lines (10–120 kΩ), which will not interfere with the antenna performance [8]. With large antenna arrays typically fabricated on composite substrates or low-dielectric constant materials, of which most are incompatible with MEMS fabrication, hybrid integration approach will be required. This is not only due to the material limitations, but also to the fact that the dimensions of antenna arrays are often far greater than a typical MEMS substrate. In this approach, the MEMS devices are fabricated on a separate substrate, singulated, packaged and tested before being transferred onto the antenna substrate by using integration techniques such as wafer bonding, flip-chip bonding, wire-bonding, etc.

Some examples of discretely integrated switches have been demonstrated in reference [30], where an X-band electronically steerable antenna (ESA) containing 25 000 ohmic MEMS switches from Radant MEMS was designed, fabricated and successfully tested. The ESA operated over a 1 GHz bandwidth at X-band with a beam steering angle of up to 60°. Employment of MEMS switches in-lieu of PIN switching diodes (or other semiconductor-based switches) results in improved antenna performance (lower loss and increased bandwidth) as well as a significant reduction in prime power consumption, which is critical for the intended aerostat platform.

In reference [15] the previously described ohmic switch from HRL was discretely integrated with a microwave dipole to tune the resonant length of the dipole for operation at X-band and Ku-band. The printed circuit dipole was fed by a slot-coupled microstrip line in a multi-layer alumina substrate. When the switches were closed, the dipole resonant frequency was 11 GHz, and when the switches were open, effectively shortening the dipole length, the resonant frequency increased to 18 GHz. Last, a MEMS-based ESA using 3-bit distributed MEMS phase shifters with a beam scanning angle of 18° at 60 GHz was reported in reference [31]. These systems are further described in Section 12.8.

12.6.2 Monolithic Integration

In the monolithic integration approach, antennas are typically fabricated first on a substrate, and then the MEMS devices are fabricated on top of the antenna elements. The antenna elements are commonly fabricated on laminate substrates with low dielectric constant by printed circuit board (PCB) processing techniques. Challenges such as process incompatibilities, complexities and cost issues are involved in the integration of the surface micromachined RF MEMS devices with the antenna elements fabricated on PCB substrates. Therefore, it would be attractive to develop a low-cost, printed-circuit compatible, RF MEMS technology for monolithic realization of low-cost MEMS-based ESAs on PCBs.

PCB process-compatible RF MEMS switches [32–35], phase shifters [36, 37], and tunable/reconfigurable antennas [38–41] have been developed by various research groups. The advantages of PCB-compatible MEMS technology include low cost, compatibility with organic laminate substrates, ease of integration with surface mount components and suitability for batch fabrication in large panels and high-volume manufacturing. In references [39] and [40], RF MEMS switches compatible with PCBs have been monolithically integrated with antennas for reconfigurable applications. In reference [41], 3-bit MEMS loaded-line phase shifters fabricated using printed circuit processing techniques have been monolithically integrated with a two-element slot antenna array for demonstration of a MEMS-based ESA.

12.6.3 Integration Issues

Several factors such as packaging, materials and process incompatibilities, and biasing requirements need to be considered when integrating MEMS devices with antennas.

12.6.3.1 Device Packaging

In hybrid integration, MEMS devices can be packaged separately using the best approach for the lowest cost for that particular switch. This allows the packaged MEMS devices to be treated as conventional discrete devices which can be integrated using appropriate integration and packaging techniques where they are needed on the antenna structure. In the monolithic integration approach, after fabrication of the entire assembly (MEMS devices and antennas) the MEMS devices can either be packaged by wafer-capping techniques or individually protected by depositing a protective packaging film at the device locations.

Hybrid integration also offers the ability to assemble known-good-devices with the antenna structure, thus increasing yield whereas the monolithically integrated devices will suffer from a cumulative yield (antenna features + MEMS devices). However, compared to monolithically integrated devices, the added package of a hybrid device might add unwanted return and insertion loss, which has to be included in the antenna design.

12.6.3.2 Materials and Process Compatibilities

In the hybrid integration, the materials used for fabrication of MEMS devices and antennas are separate. Since the MEMS devices and the antennas are fabricated on two separate wafers, the individual processes need not be compatible but the process used for integration must be compatible. For instance, for flip-chip integration the possible interconnect pitch of the MEMS devices needs to be matched on the antenna substrate. This can be a challenging design task, especially for large-scale antenna arrays where the substrate may be fabricated using a fabrication approach with low tolerances and larger design features, i.e. PCB fabrication.

In the monolithic integration approach, the maximum process temperature and thermal coefficient of expansion (TCE) needs to be considered during various stages of fabrication. Differential expansion due to TCE mismatch during process steps would lead to residual stresses, which could significantly change the performance of the MEMS devices. Referring to earlier described examples, MEMS devices are sensitive to imposed stress as it may, for instance, change the initial gap height (electrode separation) or quite possibly render the devices in a permanent closed, or open, position.

12.6.3.3 Biasing Network Design

As mentioned earlier, MEMS devices typically require higher voltage than conventional semiconductor devices and thus would require high voltage DC sources in the antenna assembly. In spite of the high bias voltage requirements, the power consumption for electrostatic switches is low when compared to their semiconductor counterparts.

MEMS devices integrated with RF circuits and antennas can be biased using (1) resistive bias lines, or (2) reactive bias-Tees. In the first case, a resistive bias line is present between the DC power supply and the MEMS devices, as shown in Figure 12.22. The resistive bias line allows a DC current flow from the DC power supply to the MEMS device. The resistance of the bias lines should be designed to provide a high impedance path ($>10\ \text{k}\Omega$ for 0.1λ) from the MEMS device to the DC power supply, and thus preventing the leakage of RF signal into the supply. Resistive bias lines can be fabricated by depositing resistive materials such as NiCr, tantalum, titanium oxide, etc.

Two important factors that need to be considered in a resistive bias network are power dissipation due to ohmic losses, and the corresponding slow switching time due to RC time constant (i.e., the time required for a capacitor, C, to charge to 63% of the full charge through a series resistor, R) for electrostatic type actuators. In the reactive bias-Tee network design, a bias-Tee (shown in Figure 12.22(b)) consisting of a capacitor between the RF input and the antenna, and an inductor between the DC power supply and the antenna is used. The capacitor provides a low-impedance path (between the RF input and the antenna) for the RF signal and a high-impedance path for the DC current (between the DC power supply and the RF input). Thus, the inductor presents a low-impedance path (between the DC power supply

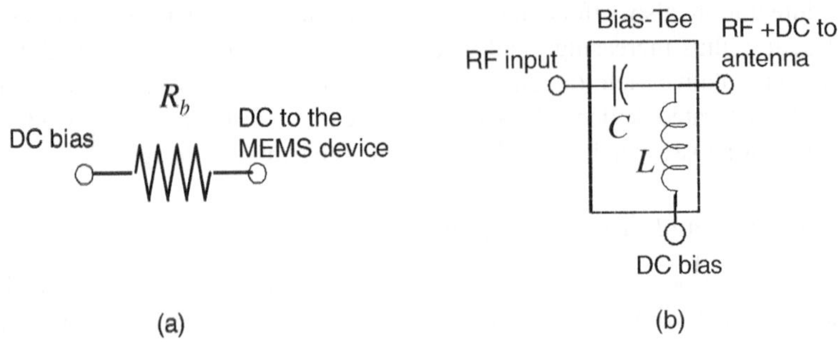

Figure 12.22: Two different schemes used for biasing MEMS devices (a) resistive bias network, and (b) reactive bias-Tee.

and the antenna) for the DC signal, and a high-impedance path (between the RF input and the DC power supply) for the RF signal. Thus, the bias-Tee delivers RF signal and DC to the antenna and isolates the RF input and DC bias terminals.

Since the capacitor between the RF input and the antenna limits low-frequency operation, a bias-Tee is typically used for high-frequency applications. Also, the maximum DC bias voltage applied must be less than the breakdown voltage of the capacitor. A downside to using bias-Tees is the associated reactances that can interact with the antenna and introduce undesirable resonances in the frequency range of interest. In summary, resistive bias lines can be used for broadband applications, but they are lossy. Reactive bias-Tees are low loss but are not suitable for low-frequency applications. Particularly, the routing of bias lines/networks in a system such as an ESA can be complex, and thus, the bias network layout must be optimized so as to avoid interference with the RF performance of the antenna.

12.7 MEMS for Reconfigurable Antennas

With the advent of RF MEMS switches, there has been a considerable interest in the development of multi-band, reconfigurable antennas. Multi-mode, multi-band, and multi-standard mobile devices create the need to simplify RF front ends. Instead of having multiple antennas switched into multiple transceiver chains covering different frequency ranges, a single multi-band tunable antenna assembly would provide size advantages and minimize product packaging complexity. Multi-band antenna technology can be fully utilized for seamless mobility of next-generation wireless devices. Furthermore, multi-band antennas can be created to cover combinations of existing law enforcement bands, Homeland Security applications, cellular bands, wireless local area network (WLAN) bands, Bluetooth, 3G applications, ultra wide band (UWB), for both voice, video, and high-data-rate wireless communications with tunable performance, high RF power handling, and low power consumption.

Two important characteristics parameters of an antenna are: (1) the operating frequency, and (2) the radiation pattern. RF MEMS devices can be integrated with antennas for altering the operating frequency and/or the radiation pattern. These two topics are discussed in the following sections.

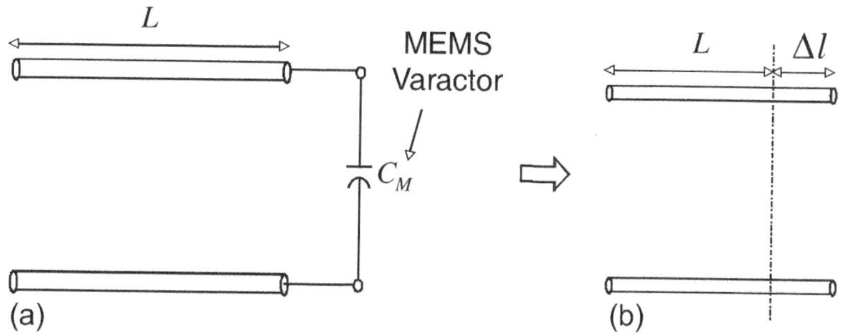

Figure 12.23: Transmission line model for a tunable printed antenna (a) an open-ended transmission line resonator section loaded at the end by a MEMS varactor, and (b) the capacitive load of the MEMS varactor modeled as an equivalent transmission line section of length Δl.

12.7.1 MEMS-based Frequency Reconfigurable Antenna

RF MEMS varactors can be integrated with an antenna to continuously tune the operating frequency of an antenna in a narrow frequency range. RF MEMS switches can also be used to reconfigure the radiating structure or aperture of the antenna and thus enable discrete switching between two or more operating frequencies.

12.7.1.1 Frequency Tuning using MEMS Varactors

In continuous frequency tunable antennas, RF MEMS devices (typically, varactors) are used to continuously tune the operating frequency of the antenna. For example, the length of a microstrip patch antenna can be adjusted to obtain continuous tuning of the operation frequency. Typically, antenna is matched for operation at a fixed frequency. At adjacent frequencies, the antenna provides high return loss due to impedance mismatch. A broadband matching network could be employed to provide a reasonable return loss in the desired operating frequency range.

Printed antennas based on microstrip, slotline or Coplanar lines can be modeled using transmission line theory. Let us consider the case where a printed antenna is modeled as an open-ended transmission line section as shown in Figure 12.23. The characteristic impedance of the line is Z_o and the propagation constant is $\gamma = \alpha + j\beta$. The effective length of the transmission line L_{eff} can be expressed as $L_{\text{eff}} = L + \Delta l$ where Δl is the equivalent line length extension that accounts for the capacitive loading due to the MEMS varactor. For discussion purposes, the discontinuity reactances (such as the open-end capacitance) are ignored in our this model.

The resonant frequency of the antenna can be expressed as a function of the effective length L_{eff} as

$$f_r = \frac{c}{2L_{\text{eff}}\sqrt{\epsilon_{\text{eff}}}} \tag{12.48}$$

where c is the velocity of electromagnetic waves in free space and ϵ_{eff} is the effective dielectric constant of the transmission line configuration used in the antenna. In the absence of the MEMS varactor, the resonant frequency of the antenna is given by

$$f_0 = \frac{c}{2L\sqrt{\epsilon_{\mathrm{eff}}}} \tag{12.49}$$

Assuming L_{eff} remains constant, the ratio of the antenna resonant frequencies in the down- and up-states of the MEMS varactor can be obtained from Equation (12.49) as

$$\frac{f_{\mathrm{d}}}{f_{\mathrm{u}}} = \frac{L + S + \Delta l_{\mathrm{u}}}{L + S + \Delta l_{\mathrm{d}}} \tag{12.50}$$

where Δl_{u} and Δl_{d} are equivalent transmission line sections corresponding to the capacitance of the MEMS varactor in the up and down-positions, respectively.

The frequency ratio (FR) is defined as the ratio of the change in the antenna resonant frequency to the up-state resonant frequency

$$FR = \frac{f_{\mathrm{u}} - f_{\mathrm{d}}}{f_{\mathrm{u}}} \tag{12.51}$$

The advantage of this configuration is that a single antenna element can be tuned to operate in a narrow frequency band. However, electrostatic actuation-based RF MEMS devices exhibit pull-in instability above the pull-down voltage. As a result, frequency tuning can be achieved only in a narrow frequency range below the pull-down voltage. Further, the frequency will remain constant until the applied voltage is decreased to the release voltage and thus the frequency tuning is not typically controllable during the release process.

12.7.1.2 Frequency Switching using MEMS Switches

In this case, RF MEMS devices (typically, switches) are used discretely to switch between various frequencies. This type of antenna can be realized by reconfiguring the physical structure of the antenna using switches. A reconfigurable antenna would require reconfigurable matching network to obtain good impedance matching at various operating frequencies.

Let us consider the case where a frequency switchable antenna is formed by cascading a primary antenna of length L_1 and a secondary antenna of length L_2. A transmission line model of this frequency switchable antenna is shown in Figure 12.24, where a MEMS switch is mounted in series configuration between the primary and secondary antennas. When the MEMS switch in series configuration is OFF (see Figure 12.24(a)), the antenna resonates at a frequency corresponding to the length L_1 of the primary antenna, and the resonant frequency of the antenna is given by

$$f_1 = \frac{c}{2L_1\sqrt{\epsilon_{\mathrm{eff}}}} \tag{12.52}$$

When the MEMS switch in series configuration is ON (see Figure 12.24(b)), the antenna resonates at a frequency corresponding to the combined length $L_1 + L_2$ of the primary and secondary antennas. In this case, the resonant frequency of the antenna is given by

$$f_2 = \frac{c}{2(L_1 + L_2)\sqrt{\epsilon_{\mathrm{eff}}}} \tag{12.53}$$

Naturally, a frequency switchable antenna can also be designed by mounting MEMS switches in shunt configuration between the primary and secondary antennas.

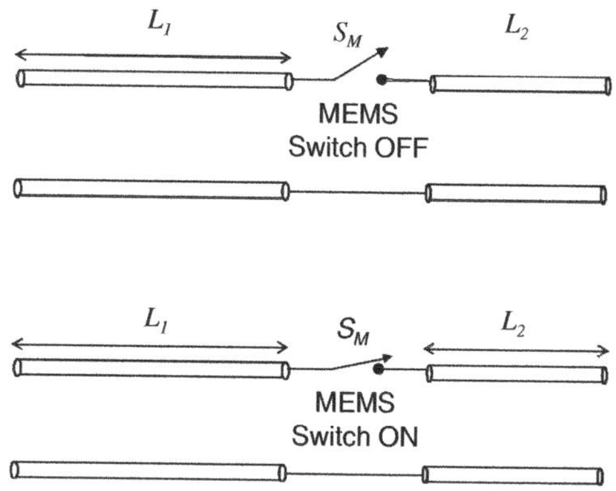

Figure 12.24: Transmission line model for a frequency switchable antenna with MEMS switch in series configuration (a) MEMS series switch OFF, and (b) MEMS series switch ON.

The ratio of the antenna resonant frequencies in the OFF and ON states of the MEMS switch is given by

$$\frac{f_1}{f_2} = 1 + \frac{L_2}{L_1} \tag{12.54}$$

The antenna frequency can be switched between two different frequency bands by choosing an appropriate ratio of lengths between the primary and the secondary antennas. Issues such as electromagnetic coupling and impedance matching need to be considered in the design. A reconfigurable impedance matching network can be used to obtain a good return loss at both operating frequencies.

12.7.2 Example Configurations

In this section, schematic configurations of MEMS-enabled tunable antenna elements such as dipole, slot, microstrip and coplanar patch antennas, as shown in Figures 12.25 and 12.26, are discussed. The configurations can be modified for continuous frequency tuning and discrete frequency switching applications.

12.7.2.1 Microstrip Antennas

A two-dimensional array of RF MEMS cantilever switches were used to demonstrate a two-frequency reconfigurable patch antenna [42]. The conceptual diagram of the array is shown in Figure 12.27(a). A specific metal patch is selected by appropriately selecting the desired row and column actuator control lines in the array. The high-frequency antenna is formed by a solid metal patch (shown as the solid region in Figure 12.27(b)) surrounded by the open-state metallization (shown as the shaded region in the figure). A good agreement in comparison of input impedance between the mock-up switch circuit and the solid antenna (with no open-state switches) was reported. The low frequency antenna is obtained by actuating all the switches (in the shaded region) to form a larger metal patch. The patch demonstrated the capability of changing the radiation frequency from 10 GHz to 20 GHz.

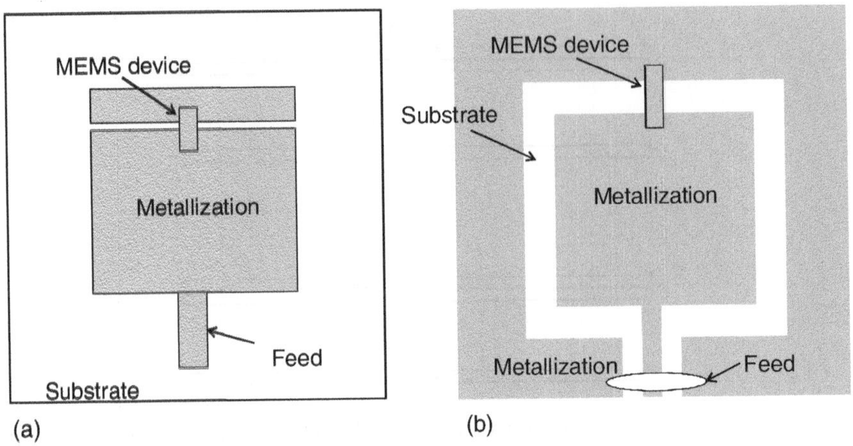

Figure 12.25: Schematic configuration of MEMS enabled reconfigurable antennas: (a) microstrip patch antenna; (b) coplanar patch antenna.

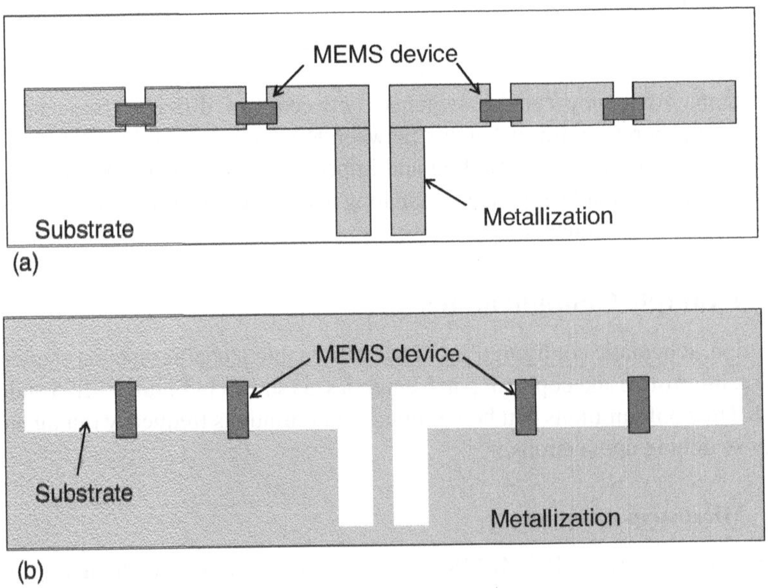

Figure 12.26: Schematic layout of a reconfigurable, (a) dipole antenna, and (b) slot antenna, integrated with MEMS devices.

A two-frequency reconfigurable patch antenna using MEMS capacitive switches was demonstrated in reference [43], where a schematic of the reconfigurable MEMS patch antenna is shown in Figure 12.28. In actuation up position, the patch antenna operates at a nominal frequency determined by the patch dimension and loading due to the actuator up

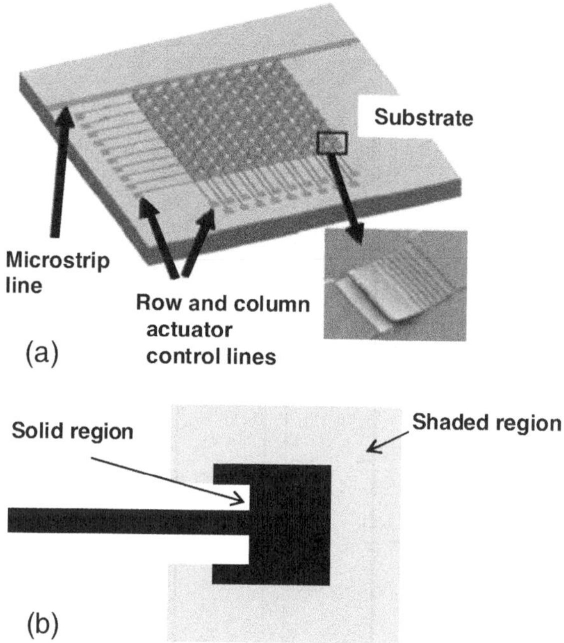

Figure 12.27: Reconfigurable microstrip patch antenna: (a) two-dimensional switch array conceptual diagram, and (b) high-frequency patch antenna (solid region) surrounded by open-state metallization (shaded region). The low-frequency antenna is obtained by actuating all the switches in the shaded region to form a larger metal patch [42] (reproduced by permission of © 2000 IEEE).

state capacitance in shunt with metal strip capacitance. When the actuators (#1 and #2) are actuated, the actuator capacitance increases and this increase in capacitance tunes the patch to a lower operating frequency. The operating frequency of the patch is switchable from 25 GHz to 24.6 GHz when both the actuators are in the OFF (unactuated) and ON (actuated) states, respectively. The obtained frequency shift of 400 MHz is about 1.6% with reference to 25 GHz.

12.7.2.2 Coplanar Patch Antenna

The coplanar patch antenna (CPA) is based on the concept of a half-wavelength, open-ended CPW resonator. The schematic layout of a CPA integrated with a MEMS varactor is shown in Figure 12.25(b). In reference [44], a tunable rectangular CPA designed using a MEMS varactor was reported. The MEMS varactor is monolithically integrated with the antenna on duroid substrate using printed circuit processing techniques. Specifically, the MEMS varactor located at one of the radiating edges capacitively loads the CPA. The resonant frequency of the antenna is tuned electrostatically by applying a DC bias voltage between the MEMS varactor and the actuation pad on the antenna. The deflection of the varactor membrane decreases the air gap, thereby increasing the loading capacitance. The increase in the loading

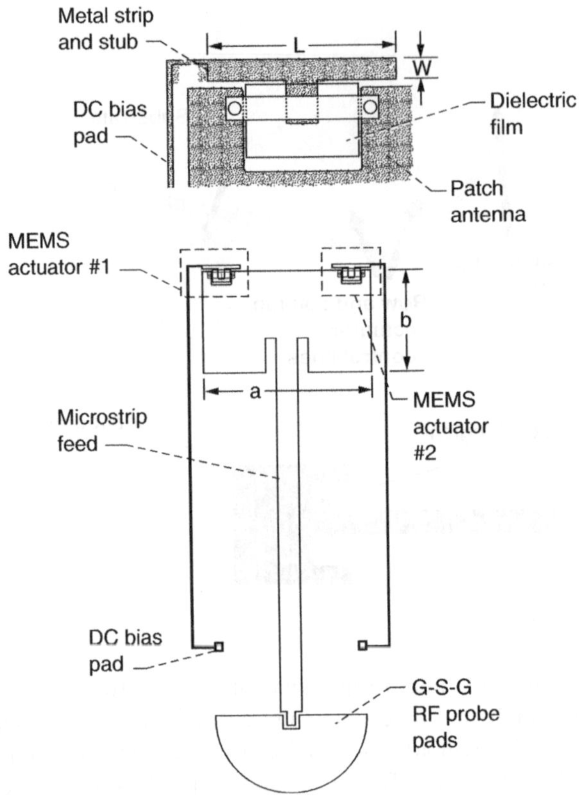

Figure 12.28: Reconfigurable microstrip patch antenna using two independent MEMS actuators to tune the patch to a lower operating frequency. $L = 580$ μm, $W = 50$ μm, $a = 2600$ μm and $b = 1500$ μm [43] (reproduced by permission of © 2001 IEEE).

capacitance results in a downward shift in the resonant frequency of the CPA. The CPA is center-fed at the second radiating edge using a 50 Ω CPW feed line. The CPA operates in the frequency range from 5.185 to 5.545 GHz corresponding to the down and up states of the varactor. The tunable frequency range is about 360 MHz and the return loss is better than 40 dB in the entire tuning range. In this tuning range, the required DC voltage is in the range 0–116 V.

12.7.3 Frequency Tuning by Changing the Effective Dielectric Constant

A typical piston-type frequency tunable microstrip antenna is shown in Figure 12.29. The antenna consists of a microstrip radiating segment (rectangular or circular) supported by two (or) four suspension springs and is capable of movement in the direction normal (i.e. vertical) to the patch. These antennas do not utilize switches to control the resonant frequency; however, the actuation mechanism is similar to electrostatic MEMS switches, and is thus

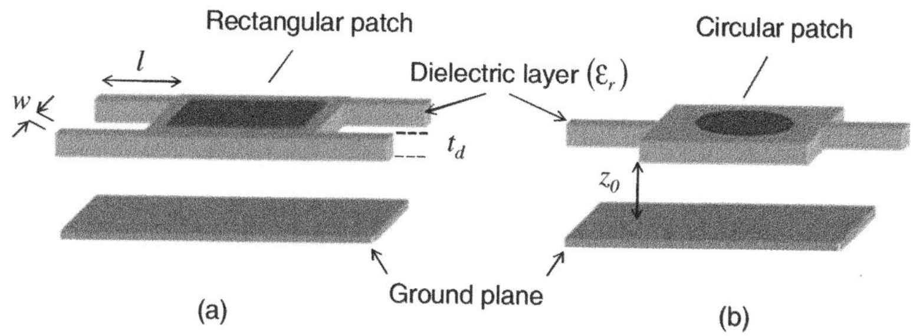

Figure 12.29: Schematic diagram of a frequency tunable microstrip antenna: (a) a rectangular microstrip patch suspended by four suspension springs, and (b) a circular microstrip patch suspended by two suspension springs.

included in this chapter. Arrays of piston-type antenna elements can be employed in reflective array applications.

The operating frequency of the antenna is electrostatically tuned by applying a DC bias voltage between the patch and the ground plane. The movable suspended patch along with the dielectric layer deflects downward toward the fixed ground plane due to electrostatic force of attraction caused by the applied DC bias voltage. This deflection decreases the air gap (z_0) thereby increasing the effective permittivity (ϵ_{eff}) of the antenna. The decrease in the air gap increases the effective permittivity of the antenna resulting in a downward shift in the resonant frequency.

It is well known that the thickness of the dielectric layer of a microstrip patch antenna is a major factor in determining the antenna resonant frequency. Because the dielectric layer of the tunable antenna is a combination of a dielectric layer (t_{d}) and a variable air gap (z_0), the effective permittivity is given by (reference [45])

$$\epsilon_{\mathrm{eff}} = \frac{\epsilon_{\mathrm{r}}(t_{\mathrm{d}} + z_0)}{t_{\mathrm{d}} + \epsilon_{\mathrm{r}} z_0} \tag{12.55}$$

where ϵ_{r} is the dielectric constant of the structural, or dielectric, layer with thickness t_{d} and z_0 is the air gap height between the dielectric layer and the ground plane.

Looking closer at how the antenna is tuned requires investigation into the effective length and radius of the antenna patches.

Rectangular microstrip patch: the effective length of the rectangular patch is given by (reference [45])

$$L_{\mathrm{eff}} = \frac{c}{2 f_{\mathrm{r}} \sqrt{\epsilon_{\mathrm{eff}}}} \tag{12.56}$$

If the fringing capacitance is ignored, the effective permittivity given in Equation (12.55) can be used. The formula for the effective permittivity of a suspended microstrip available in reference [45] can be used to account for the fringing capacitance.

Now, the resonant frequency for the TM_{10} mode is given by (reference [45])

$$f_{10} = \frac{c}{2L\sqrt{\epsilon_{\mathrm{eff}}}} \tag{12.57}$$

Circular microstrip patch: the effective radius of the patch antenna accounting for the fringing capacitance is given by (reference [45])

$$a_{\mathrm{eff}} = a\sqrt{1 + \frac{2t_{\mathrm{d}}}{\pi a \epsilon_{\mathrm{eff}}}\left(\ln\frac{\pi a}{2t_{\mathrm{d}}} + 1.7726\right)} \tag{12.58}$$

where a is the radius of the circular patch of the antenna (shown as the dark circular area in Figure 12.29).

For the fundamental TM_{11} mode, the resonant frequency of the circular disk microstrip patch is expressed as (reference [45])

$$f_{11} = \frac{x_{11}c}{2\pi a_{\mathrm{eff}}\sqrt{\epsilon_{\mathrm{eff}}}} \tag{12.59}$$

where $x_{11} = 1.84118$ from reference [45].

Thus, from Equation (12.55) it can be observed that for $z_0 > 0$ the effective dielectric constant $\epsilon_{\mathrm{eff}} < \epsilon_{\mathrm{r}}$, and when $z_0 = 0$ the effective dielectric constant $\epsilon_{\mathrm{eff}} = \epsilon_{\mathrm{r}}$. Therefore, it can be observed from Equation (12.55) that a decreases in the air gap (z_0) would increase the effective permittivity (ϵ_{eff}) of the antenna. Furthermore, from both Equations (12.57) and (12.59) it can be seen that an increase in the effective permittivity of the antenna would decrease the resonant frequency of the two antenna types. Thus, it is possible to tune the resonant frequency of the suspended microstrip patch antenna by changing the air gap z_0.

The air gap z_0 is controlled electrostatically by applying a DC bias voltage between the patch and the ground plane, in a similar fashion to how MEMS switches are actuated, as described in Section 12.3.3. The voltage required to lower the frequency of the tunable patch antenna can be estimated from Equation (12.11).

Example demonstrations: in the early 1980s, mechanically tunable microstrip patch antennas were reported by Lee *et al.* [46,47]. A 6 mm-diameter circular patch antenna tunable from 16.91 GHz at 0 V to 16.64 GHz at 165 V, a tuning range of 270 MHz, was reported in reference [38]. The schematic diagram of the antenna is shown in Figure 12.29(b). The antenna consists of three layers: a 30 mil RT/duroid 6002 ($\epsilon_{\mathrm{r}} = 2.94$, $\tan\delta = 0.001$) substrate with $\frac{1}{4}$ ounce copper cladding (about 9 μm), a 2 mil-thick flexible Kapton E polyimide film ($\epsilon_{\mathrm{r}} = 3.1$) with 100 A nickel-chrome under 3 μm copper cladding, and a 2 mil-thick adhesive bonding film which serves as the spacer. The copper cladding of the substrate serves as the ground plane of the microstrip patch. The microstrip patch antenna is patterned on the top side of the Kapton polyimide film, which is suspended above the fixed ground plane by four mechanical flexures in the Kapton film. The Spacer film introduces an air gap between the Kapton film and the ground plane. Resistive bias lines are defined in the flexures of the Kapton polyimide film to supply the DC bias voltage to the circular patch. The resistive bias lines prevent leakage of the RF signal into the DC supply. The microstrip patch is excited by aperture coupling using a CPW-fed slot in the ground plane. In reference [48], a $6 \times 6\ \mathrm{mm}^2$ microstrip patch antenna with resonant frequency switchable from 18.34 GHz at 0 V bias to 17.95 GHz at 268 V bias is discussed. The return loss is 40.6 dB and 20.1 dB at 18.34 and 17.95 GHz, respectively. The bandwidth of the antenna is about 0.9 GHz for return loss better than 10 dB in both states. In reference [49], implementation of a tunable reflect array using electrostatically tuned microstrip patch antenna elements has been proposed.

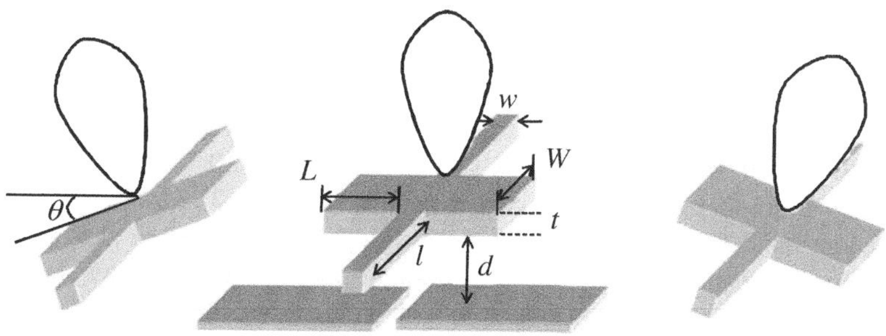

Figure 12.30: Mechanical beam-scanning characterisitics of a torsional-type micromechanical microstrip antenna element (a) actuated to the left side, (b) unactuated, and (c) actuated to the right side.

12.8 MEMS-enabled Antenna Beam Scanning

Two methods used for antenna beam scanning are (1) mechanical beam scanning, and (2) electronic beam scanning. In the former case, beam scanning is accomplished by *mechanical motion* of the antenna structure enabled by MEMS actuation mechanisms such as electrostatic, thermal or piezoelectric. In the latter case, beam scanning is accomplished by adjusting the phase of the signal fed to the antenna elements using MEMS phase shifters.

12.8.1 Mechanical Beam Steering

Mechanical beam-scanning antennas can be designed using micromechanical torsional antenna elements similar to tilt mirrors used in Texas Instruments's digital micromirror device (DMD). The antenna element consists of a flat microstrip segment supported by two torsional springs. The angle of radiation of the antenna can be changed by angular or torsional rotation of the radiating segment suspended by torsional springs as shown in Figure 12.30. Torsional rotation can be accomplished using electrostatic, electromagnetic or electrothermal actuation mechanisms. An array of torsional antenna elements can be used to develop a phased-array antenna system.

For a given torsional antenna element the actuation voltage is, as previously described, dependent on the mechanical spring constant. The spring constant of the torsional beam shown in Figure 12.30 is given by reference [50]

$$k_{\Theta} = 2\frac{Gwt^3}{3l}\left(1 - \frac{192}{\pi^5}\frac{t}{w}\tanh\frac{\pi w}{2t}\right) \quad (12.60)$$

where w, l and t are the width, length and thickness of the torsional beams, and G is the shear modulus of the torsional beam material.

For electrostatic actuation, the applied voltage can be expressed as a function of the tilt angle [50]

$$V = \sqrt{\frac{2k_{\Theta}\Theta^3}{\epsilon_0 W\{(L\sin\Theta/(d - L\sin\Theta)) + \ln[1 - (L/d)\sin\Theta]\}}} \quad (12.61)$$

where W and $2L + w$ are the width and length of the antenna, d is the gap height of the antenna above the ground plane, and Θ is the tilt angle of the antenna.

Example demonstrations: an electrostatic torsional type microstrip antenna patterned on a micromachined fused quartz structure was reported in reference [51]. A mechanical beam steering up to $\pm 10^{\circ}$ was obtained for a voltage lower than 80 V for a 2×1 mm^2 plate suspended by 2 mm long and 10 μm wide torsional springs. For 2×2 mm^2 plates suspended by 1 mm long and 30 μm wide torsional springs, 200 V was required to obtain 10° of deflection. In electrostatic actuation, the gap height between the radiating element and the ground plane must be small for low actuation voltage. On the other hand, a small gap height sets a limit on the tilt angle.

In reference [52], magnetic actuation was employed to develop a two-dimensional mechanical beam-steering antenna. The antenna was fabricated on a single high-resistivity silicon substrate and uses a polymer-based hinge structure. Nickel, a magnetic material, was located on the backside of the plate, which was actuated using an off-the-shelf solenoid coil. A maximum beam-scanning angle of up to 40° was obtained.

12.8.2 Electronic Beam Scanning Using MEMS Phase Shifters

Electronically scanned, phased array antennas consist of multiple stationary antenna elements that are fed coherently and use variable phase or time-delay control at each element to scan the radiated beam in space. ESAs employ switches/phase shifters to alter the phase of individual radiating elements across an antenna aperture and thus enable the radiated beam to steer without any mechanical motion of the antenna system. ESAs offer many advantages over conventional, mechanically scanned gimbaled arrays. The advantages include fast scanning rate, higher range, capability to track multiple targets, low probability of interception, ability to function as a radio/jammer, simultaneous air and ground modes, and synthetic aperture radar. ESAs have broad applicability for both commercial and military applications. Currently, ESAs are utilized in advanced military radars, cellular base stations, some satellite communications applications and other communication systems. Future uses include automotive anti-collision radar, smart navigation systems, and improved wireless communications. Multiple beams can be radiated by a single ESA to enable detection, tracking, and communication with multiple objects simultaneously. ESAs have been used for ground, ship and airborne radars, and in several satellite systems. However, their complexity and cost has limited their wider use in space [51–54]. Phase shifters are critical elements in an ESA that enable the steering of an antenna beam without any mechanical motion of the antenna system. Phase shifters typically represent a significant amount of the cost of producing an ESA and could cost nearly half of the entire ESA [55]. ESAs have traditionally used monolithic microwave integrated circuit (MMIC) phase shifters, which have two key weaknesses. They can contribute substantially to the cost of fabrication and tend to introduce a relatively high insertion loss, reducing the ESA's effective isotropic radiated power (EIRP).

The phase shifts provided by the phase shifter units enable desired steering of the antenna beam. The ESA beam steering angles for various phase shifts can be calculated using the formula

$$\Theta_0 = \cos^{-1}\left(-\frac{\phi}{\beta d}\right) \tag{12.62}$$

Table 12.5: Beam steering angle of the antenna array for various phase shift.

Phase shift (ϕ)	Beam angle (ψ)
0°	0°
15°	3°
30°	6°
45°	10°
60°	13°
75°	16°
90°	20°

where Θ_0 is the direction of antenna beam, ϕ is the phase difference between the signals fed to the two antenna elements, and d is the spacing between the two antenna elements. The beam steering angle is given by

$$\psi = 90° - \Theta_0 \tag{12.63}$$

The simulated beam steering angles of a two-element antenna array for various phase shifts are shown in Table 12.5. It can be observed that the beam angle changes from 0° to 20° as the phase shift of the signal fed to one antenna element is varied from 0° to 90° with respect to the other antenna element in the array. The simulated antenna patterns for various phase shifts are shown in Figure 12.31. It should be pointed out that an increase or decrease in the antenna gain would cause a corresponding increase or decrease in the peak radiated power in various states.

Over the past decade, RF MEMS switches and phase shifters that exhibit excellent characteristics including low insertion loss and low DC power consumption compared to those demonstrates by their GaAs-based MMIC counterparts. With the advent of novel RF MEMS switches and phase shifters, there has been a considerable interest in the development of MEMS-based ESAs [30,31,56]. Two example demonstrations are discussed in this section.

The first example utilized the aforementioned ohmic switch by Radant MEMS Inc where an X-band MEMS-based ESA was reported [30, 56]. The 0.4 m^2 MEMS ESA containing 25 000 MEMS switches shown in Figure 12.32 was successfully demonstrated. The MEMS ESA integrated with a commercial AN/APG-67 airborne multimode radar system successfully detected both airborne and ground-moving targets. The antenna scans electronically in azimuth ±60° with a 1 GHz bandwidth at X-band with 4 bits of phase control. The MEMS ESA is suitable for the development of large apertures (>8 m^2) for applications such as airborne moving target indicators (AMTIs) and ground moving target indicators (GMTIs).

In references [31, 57], a V-band MEMS based has been demonstrated by integrating a VCO, power amplifier, MEMS phase shifters, Wilkinson power divider and microstrip antenna array as shown in Figure 12.33. The MEMS ESA is composed of three parts – a V-band MMIC chipset, a beam-forming network and an antenna array. The V-band MMIC chip set consists of a voltage controlled oscillator (VCO) and a power amplifier. The beam-forming network consists of a Wilkinson power divider and 3-bit distributed MEMS phase shifters using metal-air-metal varactors. The 3-bit phase shifters consist of three 1-bit 22.5°, 45°, and 90° phase shifters. The Wilkinson power divider uses an asymmetric coplanar

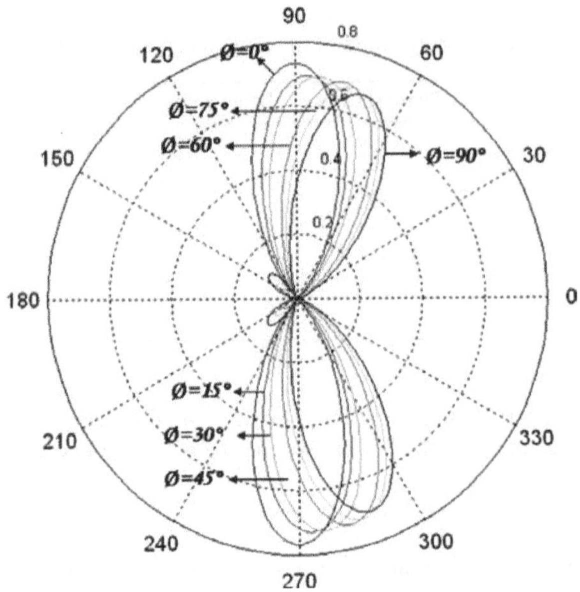

Figure 12.31: Simulated beam steering characteristics of a two-element antenna array.

Figure 12.32: One of the early large-scale successful demonstrations of MEMS in an ESA. The X-band ESA featured 25 000 radant MEMS switches with an aperture area of 0.4 m^2 [56] (reproduced by permission of © 2007 IEEE).

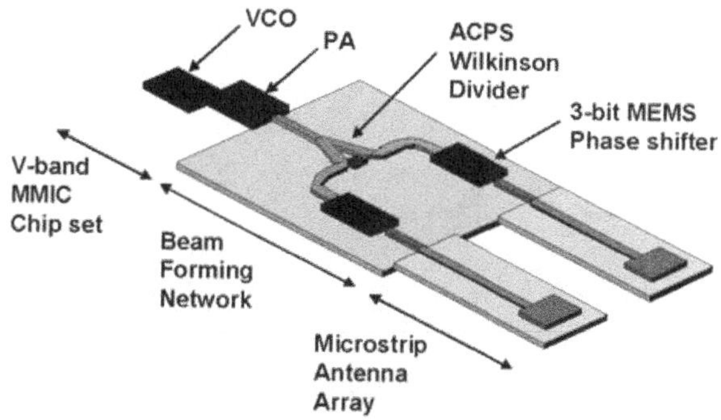

Figure 12.33: Schematic of a V-band MEMS ESA using 3-bit distributed MEMS phase shifters [57] (reproduced by permission of © 2006 IEEE).

stripline (ACPS) transmission line for compact layout. The antenna array consists of a two-element microstrip patch designed at 60 GHz. The overall size of the MEMS ESA is $2.2 \times 0.9 \times 0.4$ cm^3. The ESA provides a beam scanning angle of 18° at 60 GHz.

12.8.3 MEMS-enabled Antenna Pattern Reconfiguration

RF MEMS also enables implementation of antenna pattern reconfiguration such as beam shaping and beam switching. In this section, two example demonstrations are presented.

MEMS actuators have been used to steer the beam and change the shape of the beam of a Vee antenna [58]. In Figure 12.34, the concept and photograph of a MEMS reconfigurable Vee-antenna is shown. The structure contains fixed and moveable rotating hinges activated by scratch drive actuators. One end of the antenna arms is held by moveable rotating hinges locked on the substrate. When the arms of the Vee antenna are moved by actuators (through forward or backward movement), the movable rotation hinges translate the lateral movement of the actuators to circular movement of the antenna arms. The antenna beam can be steered by moving the antenna arms in the same direction with a fixed Vee-angle. The antenna radiation beam shape can be adjusted by changing the Vee-angle. The fabrication process includes deposition and etching of a dielectric material, metal and silicon. The main structure was fabricated using the multi-user MEMS processes (MUMPs) poly-silicon process (now offered through MEMSCAP). Main beam shifts of 30° and 48° were obtained for the 30° and 45° arm rotations (in the same direction) with 75 Vee-angle as no steering case. The beam-shaping capability was demonstrated by rotating the arms in the opposite direction to Vee-angles of 90° and 120°.

In reference [39], a RF MEMS switched diversity antenna monolithically fabricated using the PCB compatible process is reported. A photograph of the antenna configuration is shown in Figure 12.35. The configuration consists of two modified quarter-dime antenna elements (A1 and A2) and two RF MEMS switches (S1 and S2). In order to obtain orthogonal radiation patterns by operating either of the antennas elements, they are oriented 90° to each other. High-impedance quarter-wavelength sections are used for applying the DC bias to the RF MEMS switches. A second set of quarter-wavelength DC block sections are used to isolate

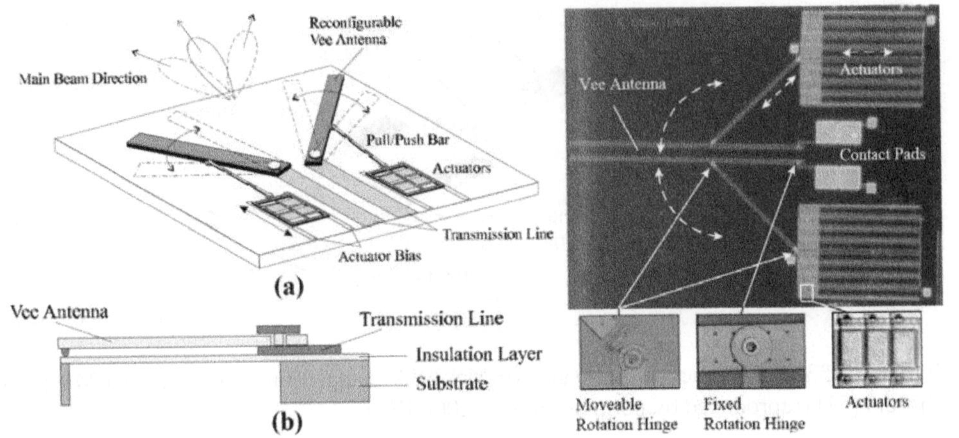

Figure 12.34: MEMS-based reconfigurable Vee antenna: (a) schematic view, (b) cross section, and (c) photograph of the mechanical components [58] (reproduced by permission of © 1999 IEEE).

the DC bias from the RF signal. These sections are located between the output ports of the T-junction and the RF MEMS switches. These sections transform the RF short at the RF MEMS switch location to an RF open at the output port of the T-junction. Thus, the RF signal sees an open at the output port of the T-junction when the RF MEMS switch connected to that port is actuated. Therefore, the signal is directed to the antenna element connected to the other output port. The operating frequency of the antenna is 6.15 GHz. Individual operation of antenna elements (i.e. S1 in the up state, S2 in the down state, or vice versa) provides 180° sector coverage by switching the beam between the antenna elements. Thus, this antenna configuration is capable of providing spatial and angular diversity.

12.8.4 MEMS-enabled Reflect Array Antennas

Reflect array antennas are useful for long-distance communications. There has been considerable interest in the development of printed reflect arrays. The printed reflecting arrays can be conformally mounted on to an existing supporting structure with relatively small incremental mass and volume. The advantages of printed reflect arrays include low-profile, large aperture size and low-insertion loss. Printed reflect arrays have been designed using identical microstrip patch elements integrated with phase-delay lines, variable-size printed dipoles, variable size microstrip patches and variable size circular rings [59].

A printed reflect array antenna (shown in Figure 12.36) consists of two basic elements: a feed antenna and a flat or curved reflecting surface. The reflecting surface consists of many printed antenna elements with no power division network. The feed antenna located above the surface illuminates these antenna elements. The antenna elements are designed to re-radiate the incident field with appropriate phase shifts required for formation of a planar phase front. The main beam can be tilted or scanned by integrating a phase shifter into each antenna element or by slightly changing the resonant frequency of the antenna elements to introduce phase shift due to off-resonance operation. MEMS-based phase shifters and tunable antenna

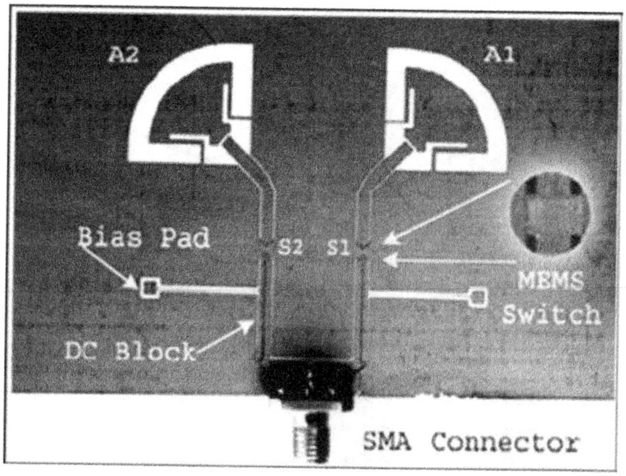

Figure 12.35: Photograph of the RF MEMS switched diversity antenna. The system consists of two antenna elements and two RF MEMS switches located on CPW feed lines that route the signal to the two antennas [39] (reproduced by permission of © 2003 IEEE).

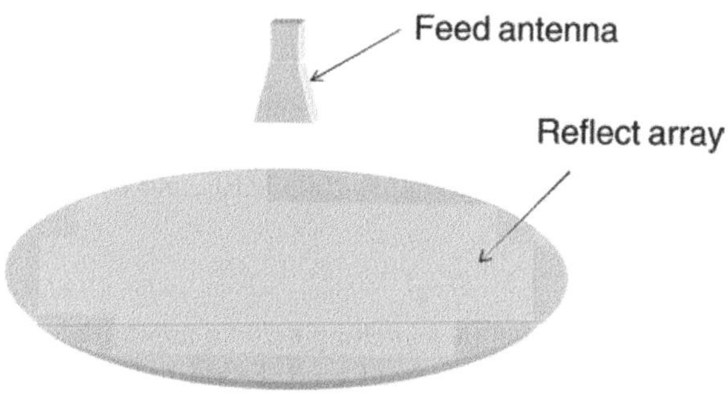

Figure 12.36: Schematic of a printed reflect array antenna consisting of two elements: a feed antenna and a flat, or curved, reflecting surface designed to re-radiate the incident field.

elements can be integrated with reflect arrays to enable electronic beam scanning, frequency and polarization reconfiguration.

The goal of the DARPA RECAP program (1999–2003) was to develop such reconfigurable reflect array systems [60]. During the early development stages, several challenges faced by the RF MEMS technology hampered the development of such reflect array systems. Recent progress in the reliability and packaging makes RF MEMS even more appealing for the development of reconfigurable reflect arrays. MEMS devices along with DC bias lines and associated control networks can be integrated on the backside of the reflect array (e.g. microstrip configuration) which is a great advantage from a packaging standpoint.

In reference [61], a reconfigurable reflect array, shown in Figure 12.37(a), using printed dipole antennas and MEMS switches, was reported. Each unit cell shown in Figure 12.37(b) consists of a set of six switchable dipoles and 12 MEMS switches, shown in Figure 12.37(c).

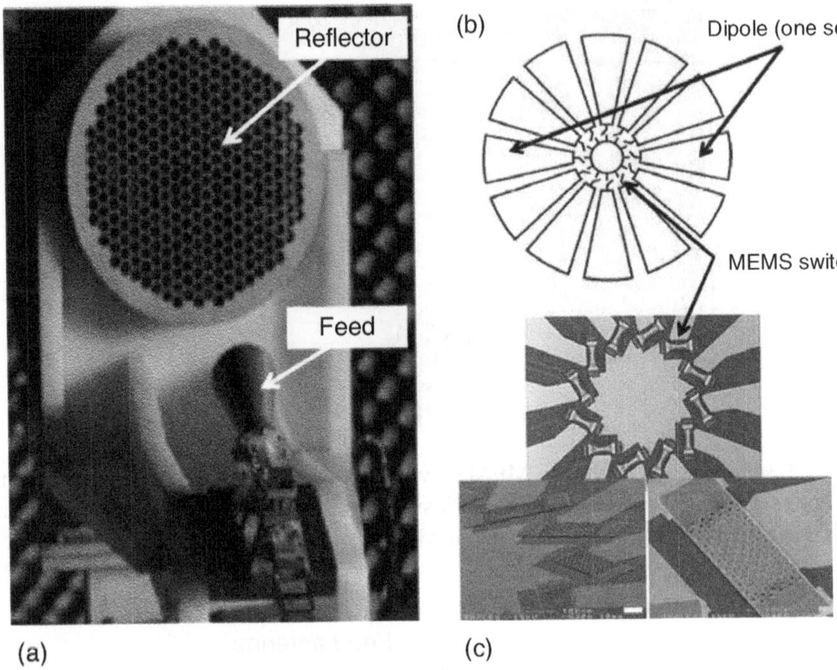

Figure 12.37: A MEMS-based reconfigurable reflect-array using printed dipoles: (a) photograph of the (b) schematic of a unit cell and (c) photographs of the individual MEMS switches [61] (reproduced by permission of © 2003 IEEE, and © Thales).

Only one set of dipoles is activated by switching the two appropriate MEMS switches. The dipoles operating at 5.7 GHz are fabricated on a Teflon-based substrate and the MEMS switches (12 switches per chip) are fabricated separately on a glass substrate. The MEMS chip is bonded onto the antenna substrate by hybrid integration. In reference [49], prototype reflect arrays using fixed height patches have been developed to demonstrate the feasibility of using variable height MEMS-based patch antennas.

12.9 Future Applications/Outlook

In this chapter, principles behind MEMS switches, performance and reliability issues and the use of these devices in various antenna applications have been discussed. Integration of MEMS devices with antennas enable features such as frequency tuning/switching, beam scanning and pattern reconfiguration. The characteristic dimensions of the MEMS devices are in the range of 100 nm to 1000 μm. In recent years, there has been significant interest in the exploration of novel devices with dimensions in the range 1–100 nm. It is believed that over the next few decades, nanotechnology is expected to bring tremendous advances in novel materials, devices and systems, and will surely have a large impact on antenna technologies as well.

Nanotechnology is anticipated to produce revolutionary solutions for the increasingly demanding requirements of future computational, sensing, storage and communication systems. Specifically, there is an increasing research activity in the use of carbon nanotube (CNT) for antenna applications [62]. A CNT is a cylinder made from a graphene sheet, i.e. mono-atomic layer of graphite, typically having radius values of a few nanometers or less, and lengths up to centimeters. CNTs can be either metallic or semiconducting, and either single walled (SWNT) or multi-walled (MWNT). Since their discovery in 1991, CNTs have been investigated for a wide variety of electronic applications such as transistors, wires/transmission lines/interconnects and field emission devices. CNTs have immense potential to realize future nano-scale electronic systems such as computers and wireless communication systems.

With the most recent DARPA program in Science and Technology Fundamentals of N/MEMS the focus of the program is to investigate fundamental key areas to continue the advancement of N/MEMS technology. A deeper understanding of the interfacial physics, and the ability to control, and engineer, the physical interfaces at the nanoscale will enable new and inherently reliable designs of RF N/MEMS devices.

Acknowledgments

The authors would like to thank Dr Chuck Goldsmith for insightful discussions regarding switch reliability and the schematic illustration of design trade-offs, Associate Professor Ulrik Hanke at Vestfold University College for discussions on switch power handling, and lastly, Professor Oddvar Søråsen at the University of Oslo, Norway for reviewing the manuscript.

References

[1] J. J. Yao, 'RF MEMS from a device perspective', *J. Micromech. Microengng.* **10** (2000), pp. R9–R38.

[2] L. Dussopt and G. M. Rebeiz, 'High-Q millimeter-wave MEMS varactors: Extended tuning range and discrete-position designs', *IEEE MTT-S Int. Microwave Symp. Dig.* **3** (2002), pp. 1205–1208.

[3] M. J. Madou, *Fundamentals of Microfabrication* 1st edn (CRC Press, 1998).

[4] G. T. Kovacs, *Micromachined Transducers Sourcebook* (McGraw-Hill, 1998).

[5] H. J. D. L. Santos, G. Fischer, H. A. Tilmans and J. T. van Beek, 'RF MEMS for ubiquitous wireless connectivity: Part 1 – fabrication', *IEEE Microwave Mag.*, pp. 36–49, 2004.

[6] W. C. Young, *Roark's Formulas for Stress and Strain* 6th edn (McGraw Hill, 1989).

[7] S. D. Senturia, *Microsystem Design* (Norwell, MA, Kluwer Academic Publishers, 2001).

[8] G. M. Rebeiz, *RF MEMS Theory, Design, and Technology* (John Wiley & Sons Ltd/Inc., 2003).

[9] H. J. D. L. Santos, *RF MEMS Circuit Design for Wireless Communications* (Artech House Publishers, 2002).

[10] I. Bahl and P. Bhartia, *Microwave Solid State Circuit Design* (John Wiley & Sons Ltd/Inc., 2003).

[11] L. Kogut and K. Komvopoulosa, 'Electrical contact resistance theory for conductive rough surfaces', *J. Appl. Phys.* **94**(5) (2003), pp. 3153–3162.

[12] N. McGruer, G. Adams, L. Chen, Z. Guo and Y. Du, 'Mechanical, thermal and material influence on ohmic-contact-type MEMS switch operation', *19th Int. Conf. on Micro Electro Mech. Syst., 2006*, pp. 230–233, Istanbul, Turkey, 22–26 January, 2006.

[13] D. M. Pozar, *Microwave Engineering* 2nd edn (John Wiley & Sons Ltd/Inc., 1998).

[14] J. Maciel, S. Majumder, R. Morrison and J. Lampen, 'Lifetime characteristics of ohmic MEMS switches', *Proc. SPIE, Reliability, Testing, and Characterization of MEMS/MOEMS III*, vol. 5343, pp. 9–14, 2003.

[15] J. H. Schaffner, R. Y. Loo, C. Quan, R. C. Allison, B. M. Pierce, S. W. Livingston, A. E. Schmitz, T.-Y. Hsu, D. F. Sievenpiper, F. A. Dolezal and L. T. Gregory, 'Microwave components with MEMS switches', *Eur. Microwave Conf., 2000. 30th Paris, France*, pp. 1–4, October 2000.

[16] J. B. Muldavin and G. M. Rebeiz, 'High isolation CPW MEMS shunt switches part 2: Design', *IEEE Trans. on Microwave Theory and Tech.* **48**(6) (2000), pp. 1053–1056.

[17] Z. Yao, S. Chen, S. Eshelman, D. Denniston and C. Goldsmith, 'Micromachined low-loss microwave switches', *J. Microelectromech. Syst.* **8**(2) (1998), pp. 129–134.

[18] C. L. Goldsmith and D. I. Forhand, 'Temperature variation of actuation voltage in capacitive MEMS switches', *IEEE Microwave Wireless Comp. Lett.* **15**(10) (2005), pp. 718–720.

[19] D. I. Forehand and C. L. Goldsmith, 'Zero-level packaging for RF MEMS switches', *2005 Govt Microcircuit Applic. and Critical Tech. Conf.*, pp. 320–323, Las Vegas, NV, April 2005.

[20] C. Goldsmith, A. Malczewski, Z. Yao, S. Chen, J. Ehmke and D. Hinzel, 'RF MEMS variable capacitors for tunable filters', *Int. J. RF and Microwave Computer-Aided Engng* **9** (1999), pp. 362–374.

[21] J. Muldavin and C.O. Bozler and S. Rabe and P. W. Wyatt and C. L. Keast, 'Wafer-scale packaged RF microelectromechanical switches', *IEEE Trans. Microwave Theory and Tech.* **56**(2) (2008), pp. 522–529.

[22] H. Nieminen, V. Ermolov, K. Nybergh, S. Silanto and T. Ryhanen, 'Microelectromechanical capacitors for RF applications', *J. Semiconductor Technol. and Sci.* **12** (2002), pp. 177–186.

[23] Q. Ma, Q. Tran, T.-K. A. Chou, J. Heck, H. Bar, R. Kant and V. Rao, 'Metal contact reliability of RF MEMS switches', *Proc. SPIE, Reliability, Packaging, Testing, and Characterization of MEMS/MOEMS VI*, vol. 6463, January 2007.

[24] C. Goldsmith, J. Maciel and J. McKillop, 'Demonstrating reliability', *IEEE Microwave Mag.* **8**(6) (2007), pp. 56–60.

[25] H. Lee, R. A. Coutu, S. Mall and K. D. Leedy, 'Characterization of metal and metal alloy films as contact materials in MEMS switches', *J. Micromech. Microengng* **16** (2006), pp. 557–563.

[26] X. Yuan, J. Hwang, D. Forehand and C. Goldsmith, 'Modeling and characterization of dielectric-charging effects in RF MEMS capacitive switches', *Microwave Symp. Dig., 2005 IEEE MTT-S International* Long Beach, CA.

[27] S. Mellé *et al.* 'Failure predictive model of capacitive RF MEMS', *Microelectron. Reliability* **45** (2005), pp. 1770–1775.

[28] Z. Peng, X. Yuan, J. C. M. Hwang, D. I. Forehand and C. L. Goldsmith, 'Superposition model for dielectric charging of RF MEMS capacitive switches under bipolar control-voltage waveforms', *IEEE Trans. Microwave Theory and Techn.* **55**(12) (2007), pp. 2911–2918.

[29] U. Hanke, A. M. Bøifot and G. U. Jensen, 'Assessment of RF performance for various capacitive switches', *SINTEF Rapport No. 90-NO050044*, 2005.

[30] J. Maciel, 'Large scale employment of MEMS switches in a passive electronically steerable antenna', *RF MEMS for Antenna Applications Workshop, 2005 Int. IEEE Antennas and Propag. Symp., Washington, DC*, July 2005.

[31] Y. Kwon, 'Reconfigurable RF and mm-wave wave front-ends using MEMS technology', *RF MEMS for Antenna Applications Workshop, 2005 Int. IEEE Antennas and Propagation Symp., Washington, DC*, July 2005.

[32] R. Ramadoss, S. Lee, Y. C. Lee, V. M. Bright and K. C. Gupta, 'RF MEMS capacitive switches fabricated using printed circuit processing techniques', *IEEE/ASME J. Microelectromech. Syst.* (2006), **15**(6), pp. 1595–1604.

[33] H. P. Chang, J. Y. Qian, B. A. Cetiner, F. D. Flaviis, M. Bachman and G. P. Li, 'RF MEMS switches fabricated on microwave laminate printed circuit boards', *IEEE Electron Device Lett.* **24**(4) (2003), pp. 227–229.

[34] R. Lempkowski, K. Lian, M. Eliacin and P. Kulkarni, 'A PWB-based MEMS Switched filter bank using lumped element embedded passives', *Proc. of the 31st Annual Conf. IEEE Ind. Electron. Soc.* pp. 2331–2334, Raleigh, NC, November 2005.

[35] G. Wang, D. Thompson, E. Tentzeris and J. Papapolymerou, 'Low cost RF MEMS switches using LCP substrate', *34th Eur. Microwave Conf.*, vol. 3, pp. 1441–1444, Amsterdam, The Netherlands, October 2004.

[36] R. Ramadoss, A.Sundaram and L. Feldner, 'RF MEMS phase shifters based on PCB MEMS technology', *IEE Electron. Lett.* **41**(11) (2005), pp. 654–656.

[37] N. Kingsley and J. Papapolymerou, 'Organic "Wafer-Scale" packaged miniature 4-bit RF MEMS phase shifter', *IEEE Trans. Microwave Theory and Techn.* **54**(3) (2006), pp. 1229–1236.

[38] R. Jackson and R. Ramadoss, 'A MEMS-based electrostatically tunable circular microstrip patch antenna', *J. Micromech. Microengng* **17**(1) (2007), pp. 1–8.

[39] B. A. Cetiner, J. Qian, H. Chang, M. Bachman, G. Li and F. D. Flaviis, 'Monolithic integration of RF MEMS switches with a diversity antenna on PCB substrate', *IEEE Trans. Microwave Theory and Tech.* **51**(1) (2003), pp. 332–335.

[40] B. Cetiner, H. Jafarkhani, J.-Y. Qian, H. J. Yoo, A. Grau and F. D. Flaviis, 'Multifunctional reconfigurable mems integrated antennas for adaptive MIMO systems', *IEEE Commun. Mag.* **42**(12) (2004), pp. 62–70.

[41] A. Sundaram, M. Maddela, R. Ramadoss and L. M. Feldner, 'Electronically steerable antenna array using PCB-based MEMS phase shifters', *Proc. 31st Annual Conf. IEEE Ind. Electron. Soc.*, pp. 2335–2339, Raleigh, NC, November 2005.

[42] C. Bozler, R. Drangmeister, M. Duffy, S. Gouker, J. Knecht, L. Kushner, R. Parr, S. Rabe and L. Travis, 'MEMS microswitch arrays for reconfigurable distributed microwave components', *IEEE Antennas and Propagation Symp. Dig.*, pp. 587–591, Salt Lake City, UT, 2000.

[43] R. N. Simons, D. Chun and L. P. B. Katehi, 'Microelectromechanical systems (MEMS) actuators for antenna reconfigurability', *IEEE MTT-S Int. Microwave Symp. Dig.*, pp. 215–218, Phoenix, AZ, 2001.

[44] M. Maddela and R. Ramadoss, 'MEMS-based tunable coplanar patch antenna fabricated using PCB processing techniques', *J. Micromech. Microengng* **17**(4) (2007), pp. 812–819.

[45] R. Garg, P. Bhartia, I. Bahl and A. Ittipiboon, *Microstrip Antenna Design Handbook* (Artech House, 2000).

[46] R. K. F. Lee, K. Y. Ho and J. S. Dahele, 'Circular-disk microstrip antenna with an air gap', *IEEE Trans. Antennas and Propagation* **32**(8) (1984), pp. 880–884.

[47] J. S. Dahele and K. F. Lee, 'Theory and experiment on microstrip antennas with airgaps', *IEE Proc.*, vol. 132, pt. H, no. 7, December 1985.

[48] R. V. Goteti, R. Jackson and R. Ramadoss, 'MEMS based electrostatically tunable microstrip patch antenna fabricated using printed circuit processing techniques', *IEEE Antennas and Wireless Propagation Lett.* **5** (2006), pp. 228–230.

[49] J. P. Gianvittorio and Y. Rahmat-Samii, 'Reconfigurable patch antennas for steerable reflectarray applications', *IEEE Trans. Antennas and Propagation* **54**(5) (2006), pp. 1388–1392.

[50] H. Toshiyoshi and H. Fujita, 'Electrostatic micro torsion mirrors for an optical switch matrix', *IEEE J. Microelectromech. Syst.* **5**(4) (1996), pp. 231–237.

[51] D. Chauvel, N. Haese, P.-A. Rolland, D. Collard and H. Fujita, 'A micromachined microwave antenna integrated with its electrostatic spatial scanning', *Proc. IEEE Int. MEMS Workshop, Nagoya, Japan*, pp. 84–89, January 1997.

[52] C.-W. Baek, S. Song, J.-H. Park, S. Lee, J.-M. Kim, W. Choi, C. Cheon, Y.-K. Kim and Y. Kwon, 'A v-band micromachined 2-d beam-steering antenna driven by magnetic force with polymer-based hinges', *IEEE Trans. Microwave Theory and Techn.* **51**(1) (2003), pp. 325–331.

[53] R. J. Mailloux, *Phased Array Antenna Handbook*, 1st edn (Artech House, 1994).

[54] J. C. Rock, 'Phased arrays using dual-wafer fabrication /high-integration processes', *Proc. IEEE Aerospace Conf.*, vol. 2, pp. 914–921, March 6–13, Big Sky, MT, 2004.

[55] T. Watson, 'Affordable phase shifters', *RF Des. Mag.*, October 1, 2003.

[56] J. J. Maciel, J. F. Slocum, J. K. Smith and J. Turtle, 'MEMS electronically steerable antennas for fire control radars', *Proc. 2007 IEEE Radar Conf.*, pp. 677–682, Boston, MA, April 2007.

[57] S. Lee, J.-M. Kim, J.-M. Kim, Y.-K. Kim, C. Cheon and Y. Kwon, 'V-band single-platform beam steering transmitters using micromachining technology', *IEEE MTT-S Int. Microwave Symp. Dig.*, pp. 148–151, San Francisco, CA, June 2006.

[58] J.-C.Chiao, Y. Fu, D. Choudhury and L.-Y.Lin, 'MEMS millimeterwave components', *IEEE MTT-S Int. Microwave Symp. Dig.*, pp. 463–466, Anaheim, CA, 1999.

[59] J. Huang and R. J. Pogorzelski, 'A ka-band microstrip reflectarray with elements having variable rotation angles', *IEEE Trans. Antennas and Propagation* **46**(5) (1998), pp. 650–656.

[60] J. K. Smith, 'Special session – reconfigurable aperture antennas', *IEEE Int. Antennas and Propagation Symp. Dig.*, Salt Lake City, UT, July 2000.

[61] H. Legay, B. Pinte, M. Charrier, A. Ziaei, E. Girard, and R. Gillard, 'A steerable reflectarray antenna with MEMS controls', *IEEE Int. Symp. on Phased Array Syst. Technol.*, pp. 494–499, Boston, MA, October 2003.

[62] P. J. Burke, *Nanotubes and Nanowires* (World Scientific Publishing Company, 2007).

13

Phased Array

Hsueh-Yuan Pao and Jerry Aguirre

13.1 Phased Array Essentials

13.1.1 Introduction

After many years of development, antenna technology has evolved into two major categories: mechanically scanned reflectors and lenses, and electronically scanned arrays. The new challenges in modern millimetre-wave (mmWave) radar and communications systems demand that antennas are capable of scanning narrow beams, have low sidelobes, and potentially have special shaped patterns in order to mitigate noise and clutter. New challenges in the growing mmWave short-range wireless networking market similarly require beam-scanning and beam-forming capabilities from the antenna systems. Given the wide bandwidths available for high data transmission with mmWave frequencies, the accompanying high-speed data transmission performance requirements are narrowing interest down to the 60 GHz band of the mmWave spectrum. Under these requirements, phased array antennas are both important and practical. Phased array antennas produce directive beams by means of arraying radiation elements that are suitably spaced and controlled with appropriate relative amplitudes and phases. The variable phase, or time-delay, at each individual radiation element makes it possible to scan the beams to a desired angle in space while the control of the amplitude of each element shapes the radiation pattern.

The development of a phased array consists of two major parts: the beam-forming network and the radiating element design. The primary goal of this chapter is to introduce the phased array antenna concept and to discuss a practical design approach that accounts for the mutual coupling that exists between radiating elements in any non-ideal mmWave phased array antenna. We begin the discussion of phased arrays with analysis of the continuous line source antenna that forms the basis for many antenna concepts. In the subsequent section, the continuous line source antenna is used to derive the radiated E-field far-field pattern as

Advanced Millimeter-wave Technologies: Antennas, Packaging and Circuits
Duixian Liu, Ulrich Pfeiffer, Janusz Grzyb and Brian Gaucher © 2009 John Wiley & Sons, Ltd

the product of an array factor and an element factor. In Section 13.2, we briefly discuss an approach for designing an antenna element for a phased array that takes into account the effects of mutual coupling. This approach, based on the infinite array methodology, is described and numerically illustrated for a slot couple patch antenna through the use of a commercially available full-wave electromagnetic solver. We demonstrate and quantify the impact on array performance that results from using the infinite array approach on a uniformly illuminated 8×8 64-element array. The design approach of simulating an entire array, otherwise known as the 'brute force' method, is briefly discussed. In section 13.3 we discuss the beam-forming network in terms of the spatial filtering process through the appropriate use of complex weightings of the elements in the array. In particular, we compare and discuss the use of a cosine weighting and Dolph–Chebyshev weightings and the array factors leading to different sidelobe levels. We conclude with a brief discussion of the fabrication of an antenna array designed for W-band operation in a low temperature co-fired ceramic (LTCC) material technology. We provide salient points in the full-wave design process and describe the manufacturing steps taken to achieve a successful design.

13.1.2 Continuous Line Source Antenna

The wire antenna is the oldest, simplest and most basic configuration of the antenna. Stratton and Chu first solved the radiation field of a conducting cylinder rigorously in 1941 [1]. Many antenna concepts evolved from this fundamental form. In this section, we begin the understanding of the phased array antenna by first reviewing the basic characteristics of the continuous line source antenna.

We start with what is conventionally known as the endfire case. In Figure 13.1 a continuous line source antenna having a length l is located symmetrically at the origin of a spherical coordinate system and oriented along the z axis, as shown. The current density associated with this wire is

$$\boldsymbol{J}(\hat{\boldsymbol{r}}') = \delta(x')\delta(y')I(z')\hat{z} \tag{13.1}$$

where the prime coordinate system is associated with the source, and $I(z')$ is the current distribution along the wire.

The vector potential for this continuous line electric current source is given by

$$\begin{aligned}\boldsymbol{A}(x, y, z) &= \frac{\mu}{4\pi}\frac{e^{-ikr}}{r}\hat{z}\iiint_{v'} \delta(x')\delta(y')I(z')e^{ik\hat{\boldsymbol{r}}\cdot(x'\boldsymbol{x}+y'\hat{\boldsymbol{y}}+z'\hat{z})}\,\mathrm{d}v' \\ &= \frac{\mu}{4\pi}\frac{e^{-ikr}}{r}\hat{z}\int_{-l/2}^{l/2} I(z')e^{ikz'\hat{\boldsymbol{r}}\cdot\hat{z}}\,\mathrm{d}z' \\ &= \frac{\mu}{4\pi}\frac{e^{-ikr}}{r}\hat{z}\int_{-l/2}^{l/2} I(z')e^{ikz'\cos\theta}\,\mathrm{d}z' \end{aligned} \tag{13.2}$$

In the far field of an antenna, only the transverse components of the electromagnetic fields are present. These transverse components can be derived from

$$\hat{z} - (\hat{z}\cdot\hat{\boldsymbol{r}})\hat{\boldsymbol{r}} = \cos\theta\hat{\boldsymbol{r}} - \sin\theta\hat{\boldsymbol{\theta}} - \cos\theta\hat{\boldsymbol{r}} = -\sin\theta\hat{\boldsymbol{\theta}} \tag{13.3}$$

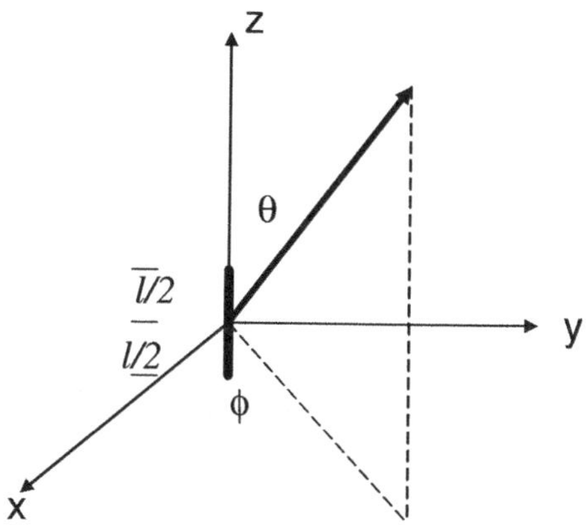

Figure 13.1: A length l wire at the spherical coordinate (endfire case).

where $\hat{\boldsymbol{r}}$ and $\hat{\boldsymbol{\theta}}$ are the radial unit vector and elevation unit vector, respectively, in a spherical coordinate system. Since the field in terms of the vector potential $\boldsymbol{A}$ is given by

$$\boldsymbol{E}(x, y, z) = -i\omega \boldsymbol{A}(x, y, z) \tag{13.4}$$

and from radiation condition is given by

$$\boldsymbol{H}(x, y, z) = \frac{\hat{\boldsymbol{r}} \times \boldsymbol{E}(x, y, z)}{\eta} \tag{13.5}$$

we can write

$$\boldsymbol{E} = \frac{i\omega\mu}{4\pi} \frac{e^{-ikr}}{r} \sin\theta \int_{-l/2}^{l/2} I(z')e^{i(k\cos\theta)z'}\,\mathrm{d}z'\hat{\boldsymbol{\theta}} \tag{13.6}$$

$$\boldsymbol{H} = \frac{E_\theta}{\eta}\hat{\boldsymbol{\phi}} \tag{13.7}$$

where $\hat{\boldsymbol{\phi}}$ is the azimuth unit vector in the spherical coordinate system. In Equation (13.6), $\sin\theta$ in front of the integral is called the taper factor. This factor represents the angular field distribution of the element. In this small current source antenna, the taper factor is the same as a unit length infinitesimal dipole at the origin and depends on the current type, either electric or magnetic, and the direction of current flow. The integral in Equation (13.6) is called the space factor. This factor represents the angular field distribution owing to the current distribution along the wire.

The time-average power density S_r propagating in the r direction is in general given by

$$\boldsymbol{S} = \frac{1}{2}\Re[\boldsymbol{E} \times \boldsymbol{H}^*] = \frac{1}{2}\Re(E_\theta H_\phi^*)\hat{\boldsymbol{r}} = \frac{1}{2\eta}|E_\theta|^2\hat{\boldsymbol{r}} = \frac{(\omega\mu)^2\sin^2\theta}{2\eta(4\pi r)^2}|g(\theta)|^2\hat{\boldsymbol{r}} \tag{13.8}$$

where $\Re(F)$ denotes the real part of F, $*$ denotes the complex conjugate, and

$$g(\theta) = \int_{-l/2}^{l/2} I(z')e^{i(k\cos\theta)z'}\,\mathrm{d}z' \tag{13.9}$$

The radiation pattern of this wire antenna is

$$U(\theta, \phi) = \frac{S_r}{\max_\theta(S_r)} = \frac{\sin^2\theta|g(\theta)|^2}{\max_\theta(\sin^2\theta|g(\theta)|^2)} \tag{13.10}$$

and

$$\sqrt{U(\theta, \phi)} = \sqrt{\frac{S_r}{\max_\theta(S_r)}} = \frac{\sin\theta|g(\theta)|}{\max_\theta(\sin\theta|g(\theta)|)} \tag{13.11}$$

It is clear that the radiation pattern depends only on $\sin\theta$ and $g(\theta)$.

Now let us consider a half-wavelength dipole. In this case we define the wire length as $l = \lambda/2$. The current distribution is described by

$$I(z) = I_0 \cos kz, \quad -\frac{l}{2} \leq z \leq \frac{l}{2} \quad \text{or} \quad -\frac{\lambda}{4} \leq z \leq \frac{\lambda}{4} \tag{13.12}$$

where I_0 is a constant that represents the electric current amplitude.

The electrical far field due the half-wavelength dipole is

$$\begin{aligned} E_\theta &= i\omega\mu \sin\theta I_0 \frac{e^{-ikr}}{4\pi r} \int_{-\lambda/4}^{\lambda/4} \cos kz' e^{ik\cos\theta z'}\,\mathrm{d}z' \\ &= \frac{i\eta I_0}{2\pi} \frac{e^{-ikr}}{r} \frac{\cos((\pi/2)\cos\theta)}{\sin\theta} \end{aligned} \tag{13.13}$$

and the magnetic field is

$$H_\phi = \frac{E_\theta}{\eta} \tag{13.14}$$

We can therefore write the antenna pattern of the half-wavelength dipole as

$$V(\theta, \phi) = \sqrt{U(\theta, \phi)} = \left|\frac{\cos((\pi/2)\cos\theta)}{\sin\theta}\right| \tag{13.15}$$

Now, we consider the broadside case, which is the case when the observer is located in a line perpendicular to the line source. First we transform the spherical coordinate system to the antenna coordinate system by a change of variables

$$\psi = -\theta + \frac{\pi}{2} \tag{13.16}$$

where the following relations are easily derived

$$\cos\psi = \cos\left(\frac{\pi}{2} - \theta\right) = \sin\theta \tag{13.17}$$

and

$$\sin\psi = \cos\theta \tag{13.18}$$

Figure 13.2 shows a continuous line source antenna of length l in the transformed antenna coordinate system.

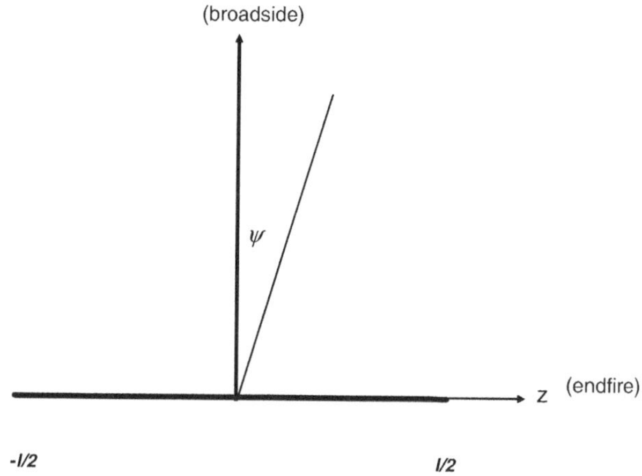

Figure 13.2: A length l wire in an antenna coordinate system (broadside case).

If we combine (13.11) and (13.17), we have the continuous line source antenna pattern for the broadside case of the new antenna coordinate system

$$V(\psi) = \frac{\cos\psi \left| \int_{-l/2}^{l/2} I(z') e^{ikz' \sin\psi} \, \mathrm{d}z' \right|}{\max_\theta [\cos\psi \left| \int_{-l/2}^{l/2} I(z') e^{ikz' \sin\psi} \, \mathrm{d}z' \right|]} \tag{13.19}$$

Now let

$$x = \frac{2z'}{l} \quad \longrightarrow \quad \mathrm{d}x = \frac{2}{l} \, \mathrm{d}z' \tag{13.20}$$

$$u = \frac{\pi l}{\lambda} \cos\theta = \frac{\pi l}{\lambda} \sin\psi \tag{13.21}$$

then from (13.9) we have

$$g(u) = \int_{-1}^{1} I\left(\frac{l}{2}x\right) e^{iux} \frac{l}{2} \, \mathrm{d}x \tag{13.22}$$

We can then define the following

$$\frac{2}{l} g(\theta) = \hat{g}(u) = \int_{-1}^{1} \hat{I}(x) e^{iux} \, \mathrm{d}x \tag{13.23}$$

Equation (13.23) is a Fourier transform. It states that the space factor $g(\theta)$ is the spectrum of the continuous line current distribution. From (13.21) we have

$$\sin\psi = \frac{\lambda u}{\pi l} \tag{13.24}$$

Then the taper factor is easily shown to be

$$T(u) = \cos\psi = \sqrt{1 - \left(\frac{\lambda u}{\pi l}\right)^2} \tag{13.25}$$

and similarly, the antenna pattern is shown to be

$$\hat{V}(u) = \frac{T(u)|\int_{-1}^{1} \hat{I}(x)e^{iux}\,\mathrm{d}x|}{\max_\theta[T(u)|\int_{-1}^{1} \hat{I}(x)e^{iux}\,\mathrm{d}x|]} \tag{13.26}$$

In particular, for a uniform current distribution, we have

$$\hat{I}(x) = I_0 \tag{13.27}$$

and therefore, we can write the space factor as

$$\hat{g}(u) = \int_{-1}^{1} I_0 e^{iux}\,\mathrm{d}x = 2I_0\frac{\sin u}{u} \tag{13.28}$$

The antenna pattern is

$$\hat{V}(u) = \left|\frac{\sin u}{u}\right| = \left|\frac{\sin(\pi l/\lambda)\sin\psi}{(\pi l/\lambda)\sin\psi}\right| \tag{13.29}$$

13.1.3 From Continuous Line Source Antenna to Phased Array Antenna

In the previous section we briefly introduced the continuous line source antenna. In this section we develop the phased array antenna concept by starting with an example of a three-element array as derived from the continuous line source antenna.

This continuous line source antenna, with a length $l = 2$ in the normalized antenna coordinate system, is shown in Figure 13.3. The E-field generated from this antenna is given by

$$E_\theta = i\omega\mu\cos\varphi\frac{e^{-ikr}}{4\pi r}\int_{-1}^{1} I(x)e^{iux}\,\mathrm{d}x \tag{13.30}$$

where

$$u = \frac{2\pi}{\lambda}\sin\varphi \tag{13.31}$$

We subdivide this line source into three equal-length pieces as shown in Figure 13.4. The x-coordinates for the ends of each piece are l, $-l + 2/3$, $-l + 6/3$, respectively. The x-coordinates for the ends of each piece are -1, $-1 + 2/3$, $-1 + 4/3$, and $-1 + 6/3$, respectively. Thus, the E-field of a three-piece continuous line source antenna can be expressed as

$$\begin{aligned} E_\theta &= i\omega\mu\cos\varphi\frac{e^{-ikr}}{4\pi r}\left\{\int_{-1}^{-1+2/3} I(x)e^{iux}\,\mathrm{d}x + \int_{-1+2/3}^{-1+4/3} I(x)e^{iux}\,\mathrm{d}x\right. \\ &\qquad \left. + \int_{-1+4/3}^{-1+6/3} I(x)e^{iux}\,\mathrm{d}x\right\} \\ &= i\omega\mu\cos\varphi\frac{e^{-ikr}}{4\pi r}\sum_{n=0}^{2}\int_{-1+2n/3}^{-1+2(n+1)/3} I(x)e^{iux}\,\mathrm{d}x \end{aligned} \tag{13.32}$$

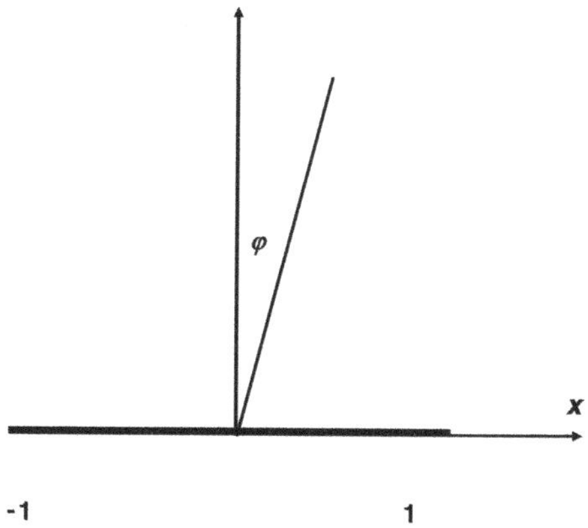

Figure 13.3: A normalized continuous line source antenna (length $2l$) in the antenna coordinate system.

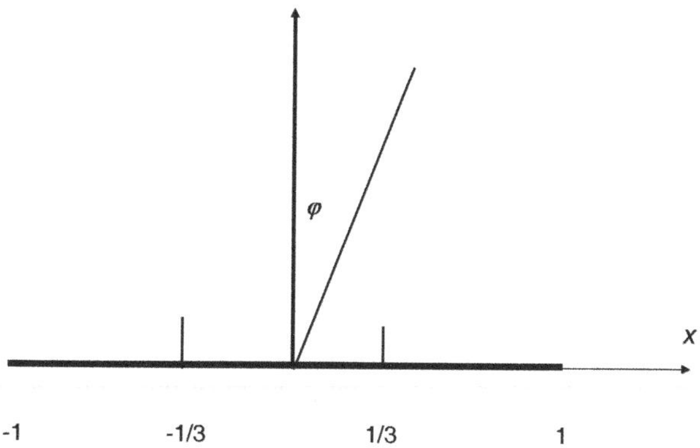

Figure 13.4: A normalized continuous line source antenna divided into three pieces in the antenna coordinate system.

The current distribution is defined as

$$I(x) = \begin{cases} I_0(x) = \hat{I}_0(x)a_0 & -1 < x < -\dfrac{1}{3} \\ I_1(x) = \hat{I}_0\left(x - \dfrac{2}{3}\right)a_1 & -\dfrac{1}{3} < x < \dfrac{1}{3} \\ I_2(x) = \hat{I}_0\left(x - \dfrac{4}{3}\right)a_2 & \dfrac{1}{3} < x < 1 \end{cases} \tag{13.33}$$

Thus, the E-field of a three-piece wise continuous line source antenna can be expressed as

$$E_\theta = i\omega\mu \cos\varphi \frac{e^{-ikr}}{4\pi r} \sum_{n=0}^{2} a_n \int_{-1+2n/3}^{-1+2(n+1)/3} \hat{I}_0\left(x - \frac{2n}{3}\right)e^{iux}\,\mathrm{d}x \tag{13.34}$$

where a_n is usually a complex coefficient so that the amplitude and phase of the element is adjustable for pattern control. Examination of (13.34) shows that it is not difficult to extrapolate the E-field for the general N-piece continuous line source antenna as

$$E_\theta = i\omega\mu \cos\varphi \frac{e^{-ikr}}{4\pi r} \sum_{n=0}^{N-1} a_n \int_{-1+2n/N}^{-1+2(n+1)/N} \hat{I}_0\left(x - \frac{2n}{N}\right) e^{iux}\, \mathrm{d}x \tag{13.35}$$

With the following variable transformation

$$x' = x - \frac{2n}{N} \tag{13.36}$$

the E-field of an N-piecewise continuous line source antenna can be written as

$$\begin{aligned} E_\theta &= i\omega\mu \cos\varphi \frac{e^{-ikr}}{4\pi r} \sum_{n=0}^{N-1} a_n \int_{-1}^{-1+2/N} \hat{I}_0(x') e^{iu(x'+2n/N)}\, \mathrm{d}x' \\ &= i\omega\mu \frac{e^{-ikr}}{4\pi r} \sum_{n=0}^{N-1} a_n e^{iu(2n/N)} \cos\varphi \int_{-1}^{-1+2/N} \hat{I}_0(x') e^{iux'}\, \mathrm{d}x' \end{aligned} \tag{13.37}$$

We define the array factor $A.F.$ as

$$A.F. = \sum_{n=0}^{N-1} a_n e^{iu(2n/N)} \tag{13.38}$$

and the element factor $E.F.$ as

$$E.F. = \cos\varphi \int_{-1}^{-1+2/N} \hat{I}_0(x') e^{iux'}\, \mathrm{d}x' \tag{13.39}$$

Therefore, the E-field can be expressed as the product of the array factor and the element factor [2]

$$E_\theta = i\omega\mu \cos\varphi \frac{e^{-ikr}}{4\pi r} \cdot (A.F.) \cdot (E.F.) \tag{13.40}$$

By comparing (13.40) with (13.6) we see that the element factor in (13.40) is the product of the taper factor, which was the original taper factor $\sin\theta$, and the current distribution of the line source. This derivation accounts for the mutual coupling between the line source segments. This accounting of the mutual coupling is a major challenge in the design of the phased array antenna radiating elements. We will detail aspects of radiating element design in Section 13.2. The array factor demonstrates the nature of how a beam is formed. In Section 13.3 we will describe different beam-forming networks and illustrate that the development of the phased array consists of two parts: the beam-forming networks, and the element design.

Proceeding with our discussion of the characteristics of a phased array, let us examine the discrete form of a linear array of length $l = Nd$, where N is the number of elements and d is the inter-element spacing. This broadside array is shown in Figure 13.5.

We rewrite (13.21) for this discrete array as follows

$$u = \frac{\pi l}{\lambda} \sin\varphi = \frac{\pi N d}{\lambda} \sin\varphi \tag{13.41}$$

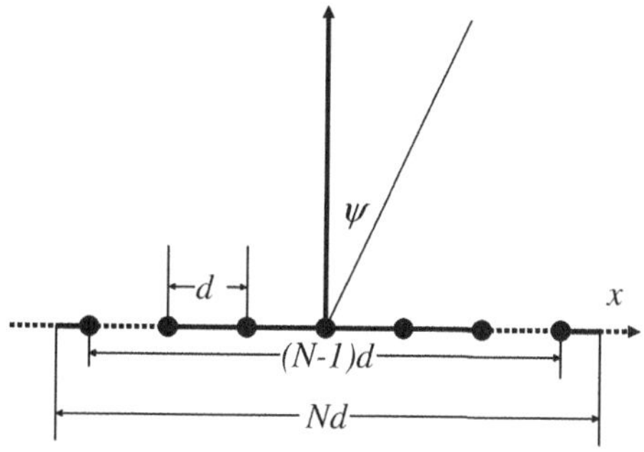

Figure 13.5: A discrete N element linear broadside array with uniform spacing.

By substituting (13.41) into (13.38), the phase of the array factor is

$$u\frac{2n}{N} = \frac{kNd}{2}\frac{2n}{N}\sin\varphi = nkd\sin\varphi \tag{13.42}$$

Let

$$v = kd\sin\varphi \tag{13.43}$$

the array factor then becomes

$$A.F. = \sum_{n=0}^{N-1} a_n e^{inkd\sin\varphi} = \sum_{n=0}^{N-1} a_n e^{inv} \tag{13.44}$$

It is sometimes convenient to take the geometrical center of the array as the phase reference. In this case Equation (13.44) becomes

$$A.F. = \sum_{n=-(N-1)/2}^{(N-1)/2} a_n e^{inv} \tag{13.45}$$

As an example, consider the case where we have a uniformly excited N-element array, where $a_n = 1$ for all a_n. The array factor $A.F.$ is then written as

$$A.F. = \sum_{n=0}^{N-1} e^{inv} \tag{13.46}$$

If we make a variable change as indicated below

$$t = e^{iv} \tag{13.47}$$

we recognize that the array factor is a geometric series that has a well-known sum. The array factor can then be written as

$$\begin{aligned} A.F. &= \sum_{n=0}^{N-1} t^n = \frac{1-t^N}{1-t} = \frac{1-e^{iNv}}{1-e^{iv}} = \frac{e^{i(Nv/N)}(e^{-i(Nv/2)} - e^{i(Nv/2)})}{e^{i(v/2)}(e^{-i(v/2)} - e^{i(v/2)})} \\ &= e^{i(N-1)(v/2)}\frac{\sin(Nv/2)}{\sin(v/2)} \end{aligned} \tag{13.48}$$

The magnitude of the array factor is

$$|A.F.| = \left|\frac{\sin(Nv/2)}{\sin(v/2)}\right| \tag{13.49}$$

and the maximum of the array factor is

$$|A.F.|_{\max} = N \tag{13.50}$$

The normalized pattern of this antenna is

$$\hat{V}(\theta) = \frac{|A.F.|}{|A.F.|_{\max}} = \left|\frac{\sin(Nv/2)}{N\sin(v/2)}\right| = \left|\frac{\sin((Nkd\sin\varphi)/2)}{N\sin((kd\sin\varphi)/2)}\right| \tag{13.51}$$

Since $Nd = l$, we have the following

$$\begin{aligned}\hat{V}(\theta) &= \left|\frac{\sin((kl\sin\varphi)/2)}{N\sin((kd\sin\varphi)/2)}\right| \approx \left|\frac{\sin((kl\sin\varphi)/2)}{((kNd\sin\varphi)/2)}\right| = \left|\frac{\sin((kl\sin\varphi)/2)}{((kl\sin\varphi)/2)}\right| \\ &= \left|\frac{\sin(\pi l/\lambda)\sin\varphi}{(\pi l/\lambda)\sin\varphi}\right|\end{aligned} \tag{13.52}$$

where we assume $kl\sin\varphi \ll 1$.

If we compare (13.52) and (13.29), we find that they are the same since $\psi = \varphi$. This indicates that the pattern of the array factor of a uniformly excited array is identical to that of a uniform continuous line source antenna.

The first null in the pattern described by (13.49) is found by

$$\frac{Nv}{2} = \pi \tag{13.53}$$

which yields

$$v_{\text{null}} = \frac{2\pi}{N} \tag{13.54}$$

(13.54) tells us that the first sidelobe must be beyond v_{null}, and that it is at

$$\frac{Nv}{2} = \frac{3\pi}{2} \tag{13.55}$$

which gives the sidelobe location at

$$v_{\text{sidelobe}} = \frac{3\pi}{N} \tag{13.56}$$

We can conclude that the results indicating the sidelobe locations for the continuous line source antenna can be used to determine the first sidelobe locations of a phased array antenna for cases where the element number N is greater than 10. The reason for this number is that if v is less than 1

$$\frac{3\pi}{N} < 1 \tag{13.57}$$

then $N > 3\pi$. Since N must be an integer, we take $N > 10$ as the resulting condition.

It is clear that the pattern of the array factor (13.52) is a sinc function; therefore, the 3 dB beamwidth of a sinc function can be written as

$$\frac{Nv}{2} = 1.39 \tag{13.58}$$

which gives

$$v_{3\text{ dB}} = \frac{2.78}{N} \tag{13.59}$$

Based on this, we can conclude that the results for the 3 dB beamwidth of the continuous line source antenna can be used to determine the 3 dB beamwidth of the phased array antenna as long as the element number N is greater than 3. The reason for selecting 3 arises from v being less than 1, since

$$\frac{2.78}{N} < 1 \tag{13.60}$$

then $N > 2.78$. N must be an integer, so we assign $N > 3$ the condition stated.

Let us consider a specific case for the coefficient a_n. For the uniform phase case

$$a_n = A_n e^{-inkd \sin \phi} \tag{13.61}$$

where A_n is a real coefficient. The array factor can then be written as

$$A.F. = \sum_{n=0}^{N-1} A_n e^{inkd(\sin \varphi - \sin \phi)} = \sum_{n=0}^{N-1} A_n e^{inkd(v - v_0)} \tag{13.62}$$

where $v_0 = \sin \phi$.

Now let us consider the methods for scanning the beam of a phased array antenna. This is no more than finding the maximum of the array factor. Since it is fundamentally the case that the maximum occurs in the direction where the wave fronts from each array element arrive at the desired position in a constructive manner, we can in a straightforward manner derive the necessary in-phase conditions of the array elements. In the far-zone of the antenna, the maximum occurs for the following condition

$$kd(\sin \varphi - \sin \phi) = 2m\pi \tag{13.63}$$

This condition can be rewritten for the integer value solution as

$$\frac{d}{\lambda}(\sin \varphi - \sin \phi) = m \tag{13.64}$$

where m is the integer, $m = 0, 1, 2, \ldots$. It is obvious that since $\sin \varphi = \sin \phi$ the first maximum takes place at $m = 0$. This suggests that any desired scan angle for the beam can be created by introducing the appropriate additional phase shift to the elements in the array.

In the design of phased array antennas it is important to understand that more than one maximum radiation lobe might exist in the antenna pattern of a phased array antenna, especially in the case when the array antenna is scanned off boresight. The maximum lobe in the direction of intended radiation is referred to as the main beam and any other existing maximum lobes are referred to as grating lobes. In the array factor pattern, these grating lobes have an intensity equal to that of the main beam. The existence and angular location of a grating lobe can be found by examining the antenna array factor equation (13.44). From (13.44) it can be shown that the maxima of the array factor take place whenever the condition of (13.63) is met. We can rewrite this condition in terms of the wavenumber, k, and the inter-element spacing, d, as follows

$$kd \sin \phi_m = \frac{2\pi}{\lambda} d \sin \phi_m = 2m\pi \qquad (13.65)$$

where π_m is the angular maximum lobe location, and m is the integer, $m = 0, 1, 2, \ldots$.

By simply rearranging this equation, we show that

$$\sin \phi_m = \frac{m}{d/\lambda} \qquad (13.66)$$

and therefore can see that the angular location of the grating lobe is dependent on the ratio of array inter-element spacing and the operating frequency of the phased array. Let us illustrate the grating lobe concept by letting $d/\lambda = 2$, which gives

$$\sin \phi_m = \frac{m}{2} \qquad (13.67)$$

The value range for sine function is $-1 < \sin \phi_m < 1$ and thus $m = 0, 1, 2$. At $m = 0$, the maximum radiation takes place at $\phi_m = 0$. At $m = 1$, we have $\sin \phi_m = 1/2$ and the maxima take place at $\phi_m = \pm\pi/6$. At $m = 2$, we have $\sin \phi_m = 1$ and the maxima take place at $m = \pm\pi/2$.

From Figure 13.5 we note that the range of ϕ is

$$-\frac{\pi}{2} \leq \phi \leq \frac{\pi}{2} \qquad (13.68)$$

We call this range the visible region. Thus for $d/\lambda = 2$ there are a total of five maximum lobes in the visible region. As another example, let us allow $d/\lambda = 1$. This element spacing leads to

$$\sin \phi_m = m \qquad (13.69)$$

At $m = 0$, we have the maximum taking place at $\phi_m = 0$ as before. At $m = 1$, we have $\sin \phi_m = 1$, and the maximum takes place at $\phi = \pm\pi/2$. In the visible region, there are three maximum lobes. To have only one maximum in the visible region, we must have $d/\lambda < 1$ to prevent grating lobes. Usually, in practical designs, we take the criteria $d/\lambda < 0.5$ for scanned array and $d/\lambda < 0.7$ for the fixed main-lobe at the boresight array.

13.2 Antenna Element Design for Phased Arrays

In this section we discuss the approach to the element design for a phased array antenna. The element design is arguably the most critical part to phased array antenna performance. To select a proper radiating element for the phased array antenna, a number of conditions must be known: the application, the volume constraints, and the environment in which the phased array antenna will operate. For example, if the array is for an airborne radar application, the requirement will very likely be to have the phased array conformal to the body of the platform. This can be a very stringent requirement on the available volume for the phased array. On the other hand, if the phased array is to operate in a ground-based system, the opportunity to define the form factor will likely have more flexibility. In the extreme environment of space applications, the specifications will be constrained by reliability and weight concerns. Although not expounded upon in the scope of this work, cost can be a major factor in the design of a phased array. If cost is the main driver, then the material of choice is likely to be an organic substrate. If minimizing loss is a main performance driver, then the element may be a horn, open-ended waveguide, or other air-filled type of antenna. If the requirement is to miniaturize the array and minimize loss, then the ceramic material technology becomes a viable technology of choice.

Whatever the main driver in the design of the phased array antenna, the approach to designing an antenna that performs as intended must consider mutual coupling. Mutual coupling is the energy that is incident on antenna radiators from adjacent radiators in the antenna array. This coupled energy from each element adds at the termination of each element and is interpreted as a reflection from that radiating element. This incident energy can be present from internal feed networks to the radiators, or externally to the radiators, and can dramatically affect the overall performance of the phased array as the antenna is scanned to different directions. The coupling between elements depends on many factors: element spacing, element phase, and the type of radiator chosen for the array.

Figure 13.6 shows a few likely candidate radiators. Figure 13.6(a) is a small pyramidal shaped horn. There is a variety of horn shapes and sizes used for affecting the gain, radiation pattern and impedance. These horn antennas can be arrayed and fed with an appropriate network for scanning capability. The cavity-backed antenna in Figure 13.6(b) is a common phased array radiator. In many cases, the cavity is dielectrically loaded to reduce the dimensions, or provide a mechanical barrier to the exterior environment. Another element shown in Figure 13.6(c) is a slot in a waveguide. This radiator is fabricated directly on a rectangular waveguide by machining slot openings at various locations and orientations on the broad or narrow wall of the waveguide. The coupling of the energy between the waveguide and the slot antenna depends on the location of the slot relative to the centerline of the broad wall dimension, or the slot dimensions and angle relative to the centerline of the narrow wall dimension. Another common array element is the slot-coupled microstrip patch antenna shown in Figure 13.6(d). We will take a closer examination of this particular radiator in a subsequent discussion on the use of the infinite array design methodology.

Once the appropriate element has been selected, the design of the phased array can begin using a number of methods. In the following sections we will outline and describe the design of a finite array based on the analysis and use of infinite array method for designing a phased array using a full-wave electromagnetic solver. For the purpose of the discussion we will use HFSS, a high-frequency structure simulator, which is a well-known finite-element solver used throughout the electromagnetic design community. We forego rigorous mathematical formulation and instead approach the design by using the fundamentals of antenna theory to guide the appropriate use of HFSS. With the advent of powerful full-wave solvers such as HFSS, the path toward designing a phased array has become more direct with less risk of design error, increasing of course a higher probability of success after fabrication. A key note in the use of HFSS is that material and process must be known precisely in order to expect close correlation between simulations and any measurements. To design a phased array using a full-wave solver such as HFSS, the main focus should be on the fundamentals of antenna array theory.

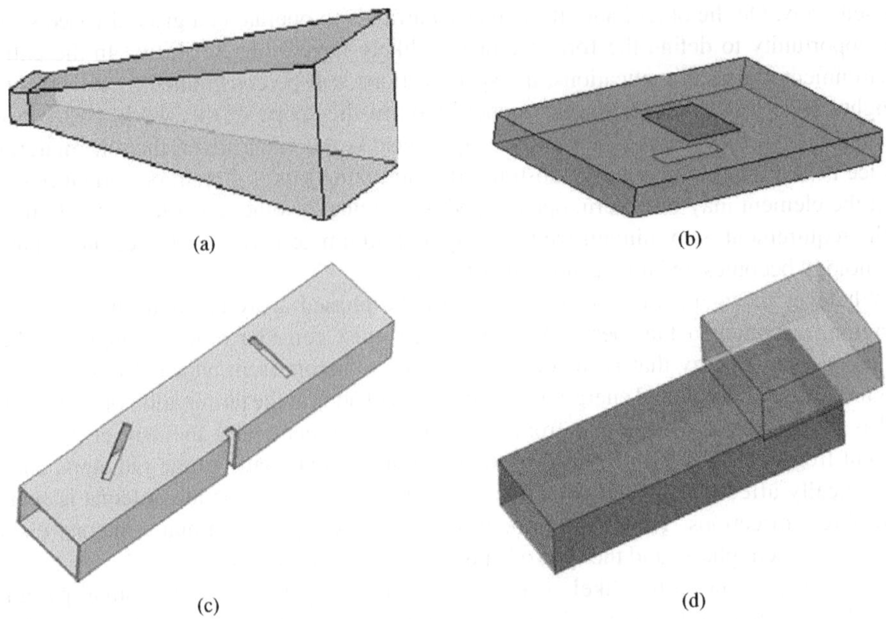

Figure 13.6: A few likely candidate radiators: (a) small horn; (b) slot-coupled patch; (c) slot in the waveguide; (d) waveguide-fed cavity.

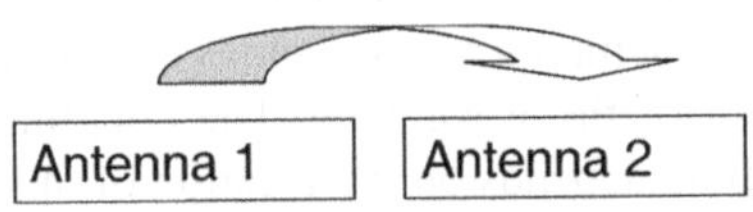

Figure 13.7: Two antennas in close physical proximity.

13.2.1 Mutual Coupling

To design an array with desired performance, it is absolutely necessary to account for the effects of mutual coupling. Antenna array theory is normally presented in the context of having ideal radiating elements in the array, whose radiation patterns combine in space by superposition. However, it is very risky to expect a phased array to work optimally, or at all without accounting for the mutual coupling that exists between antenna elements when they are placed next to each other. Therefore, although the overall radiation pattern of a phased array is still the superposition of every element pattern, the element pattern of each element is strongly affected by mutual coupling and is therefore different for each element in the array, including the ideal element radiating in isolation of all elements. The coupled energy between the elements also has a major impact on the input impedance of each element and subsequently the element pattern. Mutual coupling is illustrated for two elements, as shown in Figure 13.7.

The voltage at the terminals of each antenna can be described by self-impedances, Z_{nn}, and mutual impedances, Z_{mn}. Mathematically, the relationship between the two impedances can be described by the following equations

$$\left.\begin{aligned} E_1 &= i_1 Z_{11} + i_2 Z_{12} \\ E_2 &= i_1 Z_{12} + i_2 Z_{22} \end{aligned}\right\} \tag{13.70}$$

If the second antenna is terminated by a load, Z_2, it is easy to show that the input impedance, Z_1, of antenna one depends on the load impedance Z_2 at antenna 2. This is written explicitly as

$$Z_{\text{in}1} = Z_{11} - \frac{Z_{12}^2}{Z_{22} + Z_2} \tag{13.71}$$

Using these network concepts illustrates the effects on input impedance arising from the close proximity of radiating antenna elements.

We can also describe the effects of mutual coupling on the input impedance as a function of scan angle in a large array by considering the active impedance of the array. The active impedance of an antenna array is defined simply as the input impedance seen by the generator of each element in the array when all other antenna elements are active. This active impedance will result in a measured or simulated return loss resulting from the sum of all coupled energy to each element. Following Hannan [3] we can consider that all coupling terms for the array are computed and available. The reflected wave coefficient, R, for each element is therefore the sum of the products of the coupling coefficients times a phase term associated with the neighboring elements. This is done for each element in the array

$$R(\alpha, \beta) = \sum_{m=-\infty}^{\infty} \sum_{n=-\infty}^{\infty} C_{mn} e^{j(m\alpha + n\beta)} \tag{13.72}$$

where m and n are the column and row location of the element in the array, respectively, C_{mn} is the complex coupling coefficient and α and β are the phases associated with each generator connected to the elements. Clearly, the reflection coefficient is a function of scan angle. Rather than delve into the mathematical considerations forming the foundation for phased array theory, we next consider the design process using a full-wave solver that can compute the active impedance and hence the return loss for each element in an array as a function of scan angle.

13.2.2 Large Array Design Methodology

The practical design of a radiating element in a large phased array antenna can be a very complicated process, since an accurate model of the element performance in a phased array antenna of many elements is difficult to obtain. We attribute this difficulty to the fact that the behavior of each element is greatly affected by the presence of the mutual coupling between surrounding elements in the array. Due to this mutual coupling effect, the input impedance of a radiating element as it resides in a phased array is different from its input impedance if this same antenna is isolated. However, the interactions of the more distant elements usually become negligible for practical approximations. This fact suggests why Wheeler assumed an array of infinite dimensions as an approximation to a finite array of a large number of elements more than a half century ago [4]. In his pioneering paper, Wheeler, for the first time, presented the idea of a 'waveguide simulator'. The technique of using a 'waveguide simulator' for planar array characterization has now been in use for more than 40 years [5].

It is well-known that the radiating ***E*** field, the radiating ***H*** field, and the propagation direction are perpendicular to each other in free space [6]; therefore, the radiating field is a transverse electromagnetic field (TEM). This free space condition must be satisfied as we use the infinite array approach to solve for the radiating fields. The radiating fields generated by the radiating elements in the infinite array must be TEM mode only.

We know that the plane wave is the simplest solution for the fields radiating from elements in the infinite array. Furthermore, the plane wave is closely approximated in the vicinity of a finite planar array. This approximation allows us to describe the active impedances of all radiating elements except those very close to the edge. Based on the authors' extensive phased array antenna design experience, it has been found that only a couple of rows of radiating elements near the array edge of a large array might have the active impedance substantially different from that of an element simulated in an infinite array. In the next section, we will show that the aperture taper is necessary to obtain the decent antenna pattern for phased array antenna beam-forming; therefore, the energy radiating outward toward space, or the information received from space by these edge elements is much less than the energy of those elements in the central part of the array. The mismatched active impedance of the edge elements therefore causes negligible effects on the antenna pattern. If the specification mandates that all radiating elements must be perfectly matched, a row of loaded elements could be added in the edge of the array as the dummy elements.

The methodology used to solve an infinite planar phased array involves solving a hypothetical case of the ordinary transmission line that guides a plane wave. An antenna in the array radiates into such a transmission line, generating the plane wave in a finite confined space. All antennas in the array radiate in the same manner so that a contiguous plane wave is formed in space.

A transmission line of special interest is the rectangular waveguide. Figure 13.8 illustrates the concept: an antenna residing in an infinite phased array radiates into a hypothetical rectangular waveguide, which permits the plane wave traveling along the guide. This hypothetical rectangular waveguide is also called a 'waveguide simulator'. This is the simplest practical example, where a plane wave is able to propagate in a confined channel space. The contiguous plane wave could be realized if the hypothetical rectangular waveguides for all antennas are identical and adjacent to each other.

Figure 13.9 illustrates two types of rectangular waveguide that are able to transmit the plane wave (TEM wave).

Figure 13.9(a) shows the simplest example of a propagating plane wave along a transmission line comprised of a pair of parallel electrical conducting plates separated by a dielectric rectangular cross section. The plane wave could be realized if the separation b is much less than width a.

It is well known that the conventional rectangular waveguide does not support the TEM mode [6]. To allow plane wave propagation we define a hypothetical rectangular waveguide in Figure 13.9(b) for theoretical purposes. The top and bottom surfaces (a dimension) consist of the perfect electric conductor (PEC) on which the tangential electrical field vanish, but the tangential magnetic fields survive [7]. The side surfaces (b dimension) consist of the perfect magnetic conductor (PMC) on which the tangential magnetic field vanish, but the tangential

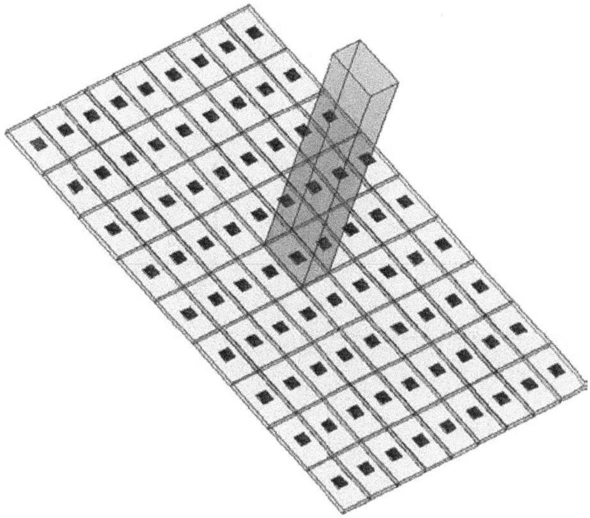

Figure 13.8: An antenna residing on an infinite phased array radiates into a hypothetical rectangular waveguide which permits the plane wave traveling along the guide.

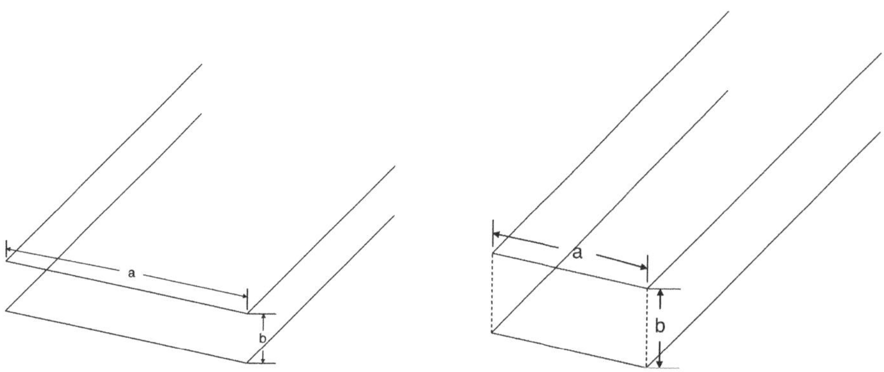

(a) (b)

Figure 13.9: The waveguides support the TEM mode: (a) parallel plate waveguide; (b) hypothetical rectangular waveguide with PEC imposed on the a dimension and PMC imposed on the b dimension.

electrical fields survive [7]. We present PEC by a solid line and PMC by a dashed line in Figure 13.9(b). From image theory, we know that the electrical fields perpendicular to the top and bottom plates extend to infinity as if the guide wave structure becomes the parallel plate waveguide illustrated in Figure 13.9(a). This structure, therefore, permits a plane wave to travel along the waveguide. There is no cutoff frequency in the structure; all frequencies travel along the waveguide at the same velocity. If an antenna with vertical polarization were

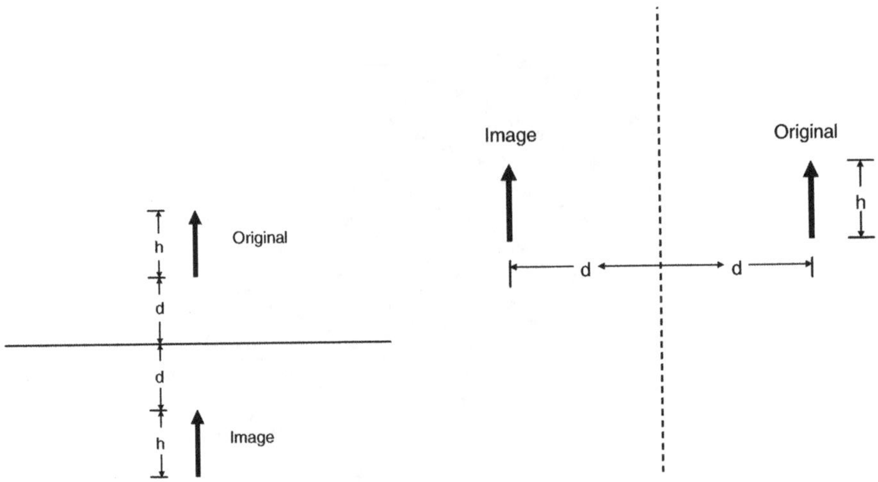

(a) (b)

Figure 13.10: Applications of the image theorem for an antenna adjacent PEC or PMC: (a) an antenna above PEC with its image; (b) an antenna next to PME and its image.

to be placed in the end plane of this hypothetical rectangular waveguide, the power radiated and the only mode propagating along the line would be a plane wave (TEM) if both the transverse dimensions were less than one wavelength.

The PMC defined in the hypothetic rectangular waveguide is not physically realizable. These boundary conditions are used as an approximation for placing an array of identical antennas in the positions of the images of the given antenna as imaged by the hypothetical boundaries.

Figure 13.10(a) illustrates an antenna of height h located at height d above a PEC. This antenna plus an 'image' antenna of height h located at height $-d$ below PEC, radiating in free space, produces a zero tangential $\boldsymbol{E}$ field over the plane bisecting the line joining the two antennas. According to the uniqueness theorem [7] the solution to this problem is also the solution of the configuration in Figure 13.10(a). Therefore, we can remove the PEC and use the image antenna to solve the total fields above the PEC.

In Figure 13.10(b) an antenna of height h located at a distance d on the right side of a PMC is shown. This antenna plus an 'image' antenna of height h located at $-d$ on the left side of PMC radiate in free space. The tangential $\boldsymbol{E}$ fields at both sides of the plane bisecting the line joining the two antennas are the same. According to the uniqueness theorem, the solution to this problem is also the solution of the configuration in Figure 13.10(b), since the tangential $\mathbf{E}$ fields crossing the PMC are continuous [7]. If we remove the PMC and replace it with the image antenna, the solution for the fields is identical to the original problem shown in Figure 13.10(b).

Applying the image theorem, the images on the four boundaries of the hypothetical rectangular waveguide shown in Figure 13.9(b) form a planar array which has infinite width and height. Figure 13.11 depicts this array.

It is clear that:

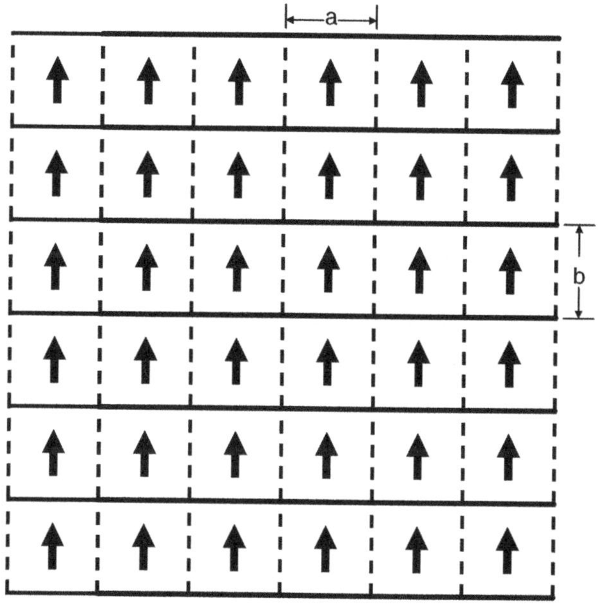

Figure 13.11: The infinite array antenna constructed by images of the hypothetical rectangular waveguide.

- this array radiates a contiguous plane wave in space, since each radiating element contributes to the plane wave in a confined finite space (hypothetical rectangular waveguide);
- the dimensions of the hypothetical rectangular waveguide a and b determine the separations of the adjacent elements;
- the separations of the adjacent antenna elements cannot be too large, since waveguide modes other than the TEM mode might be excited;
- all radiating elements have identical radiation characteristics; we only need to find the radiation behaviors of a single antenna that radiate into the hypothetical rectangular waveguide;
- the values of the self-impedances z_{nn} are equal for all radiating elements while the values of the mutual impedances z_{mn} depend only on the spacing between the elements and on their orientation, and not upon their specific location in the array [8].

In Figure 13.11, we only show the vertically polarized antenna case. If we want to solve the case for the horizontally polarized antennas, we need only simply rotate the hypothetical rectangular waveguide by 90°. With both radiation characteristics of vertical and horizontal polarizations, we are able to solve for the antenna radiation characteristics for any polarization.

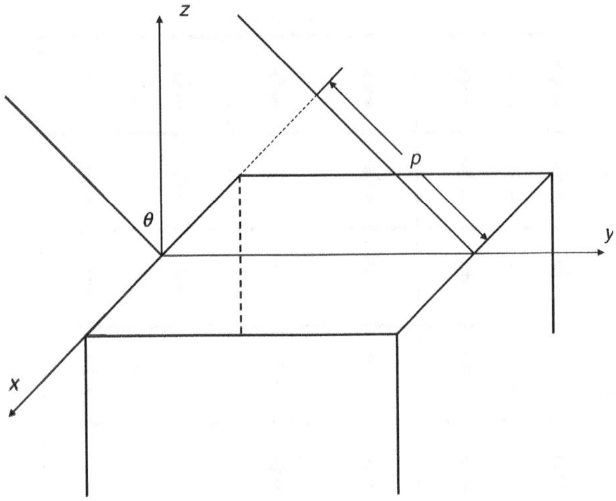

Figure 13.12: The hypothetical rectangular waveguide under a plane wave obliquely incident.

When the array scans, the antennas in the array maintain equal amplitudes, but experience a constant phase shift. For an antenna that radiates into the hypothetical rectangular waveguide, we can visualize that a plane wave is obliquely incident rather than normally incident, as shown in Figure 13.12. In this scenario, the side walls in Figure 13.12 are identical to each other except that the right wall has a phase shift kp compared with the left wall, where $p = a \sin \theta$ is a physical path length difference, a is the dimension of the waveguide, and θ is the scan angle defined in Figure 13.12. If we enforce this constant phase difference in the opposite walls, we are able to obtain the correct radiation characteristics as the array scan. Edelberg and Oliner described this procedure in detail [8, 9], the interesting reader could check these references.

Thus far we have introduced the background for the element design for a large planar phased array by using the infinite array methodology. We have presented only the simplest scenario: the rectangular layout. In the same manner, the radiation solution for many other array layouts are available. The 'waveguide simulator' analysis approach is suitable for other, more complex, configurations such as the triangular array presented by Eisenhart and Park [10].

Now let us use Ansoft HFSS to demonstrate how to design an antenna element by accounting and compensating for the effects of the mutual coupling that exists between elements in a large phased array. Consider the infinite array shown in Figure 13.8.

The radiating elements, in this case a slot-coupled patch antenna, extend in both the x and y directions without end. A unit cell is drawn around a typical antenna element, as shown. The unit cell, which is the hypothetical rectangular waveguide, extends indefinitely representing a pseudo-waveguide structure whose walls are field boundary conditions such that the fields outside the unit cell can be completely ignored. All mutual coupling effects from neighboring elements are taken into account through this approach. It may seem counterintuitive that this infinite array can provide useful information; however, because of its infinite nature it is more

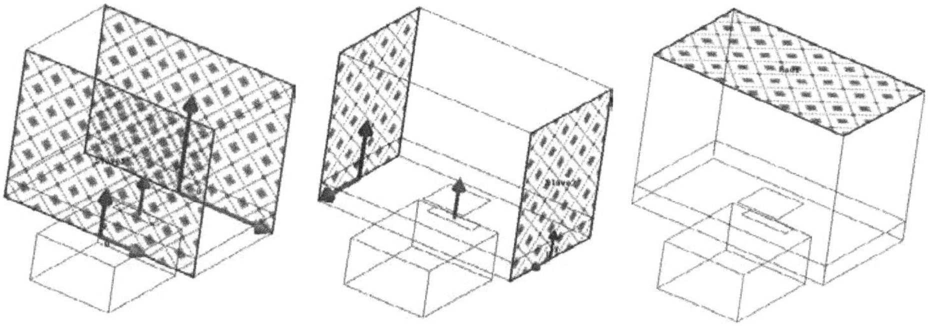

(a) (b) (c)

Figure 13.13: Boundary condition setup of HFSS: (a) master and slave boundary conditions on E-plane walls; (b) master and slave boundary conditions on H-plane walls; (c) radiation boundary conditions on top wall of unit cell.

amenable to have analytical as well as numerical solutions than direct analysis of a very large array. Thus, the design of a very large array can be straightforward if it behaves approximately as an infinite array. Of course, approximating a very large array with an infinite array analysis will introduce errors in the final analysis of the finite array; however, these can be deemed small depending on the actual array size. In the finite array, the edge elements will have different characteristics than the center elements, which more closely approximate an element in an infinite array environment. In the next section, we will discuss the implementation of the infinite array approach to designing a finite array.

In the infinite array environment, it is fairly straightforward to compute the input impedance of a single element radiating in the presence of other identical elements using full-wave solvers such as HFSS. Adjacent elements are modeled as all having the same amplitude, differing only in phase from one another. The single element in the array is identified as a unit cell and the simulation holds that all elements in the infinite array are radiating. Because the field difference between one cell and the next is only different by a phase constant, the full-wave solver is able to take advantage of this field behavior by applying a periodic boundary condition at the four sides of the rectangular box defined by the element spacing. The boundary conditions are applied as master and slave boundary conditions in HFSS. These boundary conditions enable the planes of periodicity to match surface E-field components to within a phase difference associated with the distance separating the walls. Each field point on the slave boundary is forced to match the corresponding point on the master boundary. The top surface of the unit cell is modeled as a boundary condition capable of terminating the problem space in such a fashion as to simulate a boundless space. This type of termination is an absorbing boundary condition. These boundary conditions are illustrated in Figure 13.13.

The port S-parameters for each element can be obtained for the single element defined in such a problem space. Since every other element in the infinite array is radiating, it is appropriate to define the element port as having an active S-parameter. The active S-parameter as we previously discussed is the S-parameter for the radiating element while all other elements are simultaneously radiating. All energy traveling toward the port of each element is interpreted as reflected energy for determining the drive point impedance of the

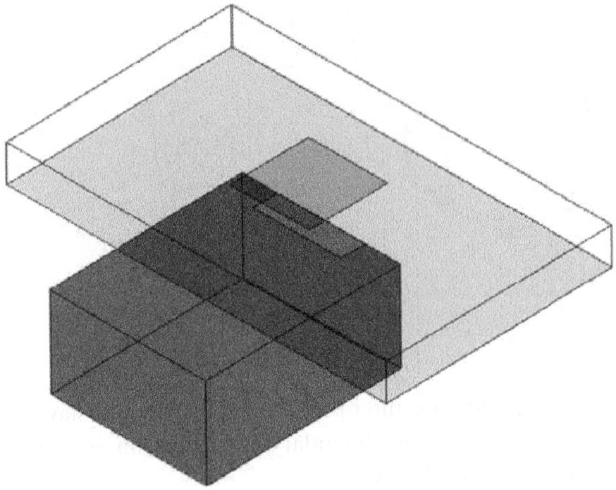

Figure 13.14: The HFSS model of a waveguide-fed slot coupled patch antenna.

element in the infinite array environment. This input impedance can be dramatically different from the case for the single element radiating in free space with no neighboring antenna elements, and significantly different from that of elements radiating in a finite array size.

In the design of the element, the element pitch is determined and controlled by the scan angle and grating lobe conditions for the frequency of operation and desired return loss across the frequency of operation and scan angles. Under these conditions, the desired dimensions and spacing of the elements can be found through an iterative approach. Most full-wave tools have optimizer routines that can make this quite automatic. The optimizer is normally bound by a set of physical constraints imposed by system requirements and a set of performance specifications, which may or may not be possible to meet inside the prescribed design space. Another optimizer constraint is the material design rules for manufacturability. This is a critical constraint for practical phased array antenna designs.

To demonstrate the effects of mutual coupling on the input impedance of an antenna element radiating in an infinite array environment, the slot coupled patch antenna of Figure 13.14 is used for an infinite array analysis. The slot and patch are optimized to operate at 94.175 GHz. The array is scanned from 0 to 60° in the E-, H-, and diagonal plane.

The impedance as a function of scan angles at 94.175 GHz in the H-plane is visualized on the Smith chart as shown in Figure 13.15(a). Note that the impedance of the single radiating element has significantly changed from the case where the array is radiating at broadside. At broadside the input impedance of the array element matches the generator impedance. As the array scans in the H-plane the input impedance becomes more capacitive, with a decreasing real part. In Figure 13.15(b), we illustrate the effects of scan angle in the H-plane on the return loss across 90 GHz to 98 GHz. The resonant design point degrades and moves to a lower frequency, albeit in a complicated manner, since as can be seen at a scan angle of 30°, the return loss is quite good, being −30 dB at 93 GHz.

As the array scans in the E-plane at a constant 94.175 GHz, the input impedance becomes more inductive with a decreasing real part as shown in Figure 13.16(a). Figure 13.16(b) shows

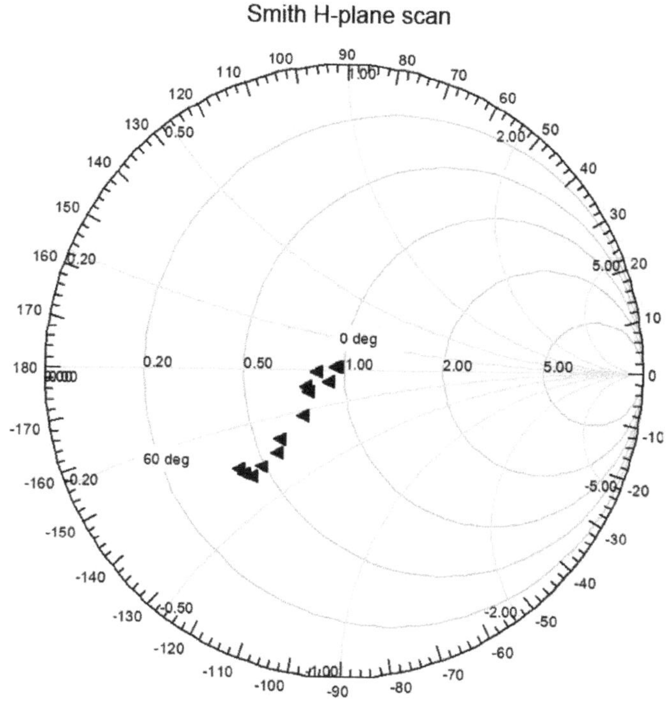

(a)

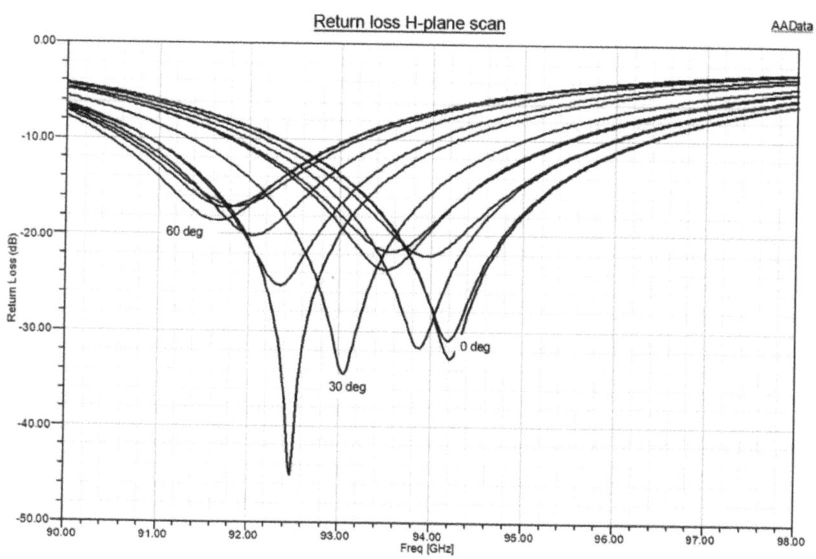

(b)

Figure 13.15: Impedance of the antenna as a function of H-plane scan angle: (a) Smith chart of the input impedance as a function of the H-plane scan angle; (b) return loss of the antenna as a function of the H-plane scan angle.

the change in return loss as a function of frequency. For the scan in the E-plane, the changing input impedance leads to degradation in performance at the desired frequency; however, the bandwidth as a function of scan angle can be seen to be greater than that of the bandwidth in the H-plane scan case. Also, the shift in array resonance is towards higher frequencies, as can be seen in Figure 13.16(b).

For the case of scanning the array in the diagonal plane, Figure 13.17(a) shows that the drive point impedance at 94.175 GHz behaves similarly to that of the E-plane scan in that the input impedance becomes increasingly capacitive with a lower real part. Figure 13.17(b) shows the return loss as a function of frequency.

It is clear from the infinite array results that consideration of mutual coupling between elements is a critical aspect in phased array antenna design. The active input impedance is seen to be a strong function of scan angle in the H-plane scan of a slot-coupled patch infinite array. In the E-plane scan, the array input impedance dependence on scan angle is seen to be relatively not as strong, but significant nonetheless. Although periodic cell analysis available to analyze an infinite array allows in effect the simulation of a single element to take into account the coupling effects of an infinite array, the application of infinite array results to design the element in a physical large array is subject to how well the large array model approximates the characteristics of a physically realizable smaller array. In the next section we discuss this approximation in detail through an example case study.

13.2.3 Finite Array Design Methodology

To design a finite array we must consider if it is sufficiently large to be considered 'large'. Holter [11] has developed a rule of thumb that states that for the narrow band case the finite array should be at least $5\lambda \times 5\lambda$ in size. For the frequency of 94 GHz, this translates to a required antenna size of about 628 mils by 628 mils. Depending on the computer resources available, this can pose a challenging if not prohibitive antenna size for direct computational analysis. In an extreme case, we consider the infinite array approach to designing a relatively small array which was able to be solved by HFSS on a 64-bit dual processor computer. We simply take the optimized element discussed in the previous section and form a 64-element array as shown in Figure 13.18. This antenna array is a 64-element 8×8 slot-coupled patch radiator array having a pitch of 2 mm in the H-plane and 1.21 mm pitch in the E-plane. The pitch dimensions for the elements of the array are those for achieving minimum reflection at the design frequency of 94.175 GHz and broadside radiation.

The antenna pattern for broadside radiation is shown in Figure 13.19(a). The array elements are uniform in amplitude and are all in phase. As can be seen, the antenna pattern is well behaved. In Figure 13.19(b), we show the antenna element return loss for each element. It is clear that the relative position of the element in the array has a significant impact on the return loss. However, as can be noted, the usable 10 dB return loss is still available across a 2 GHz band from 93.4 GHz to 95.4 GHz, so even though this finite array size falls far short of being considered large, the infinite array approach, at least in the case of a broadside radiation, provides a quick and accurate design methodology for even much smaller arrays than might otherwise be thought not to be amenable to this approach.

In Figures 13.20(a)–(d), we examine the impact of scanning in the E-plane on pattern performance and return loss from zero to 30°. The antenna pattern gain can be seen to be reduced as a function of the element factor, and the sidelobe levels outside the first sidelobes

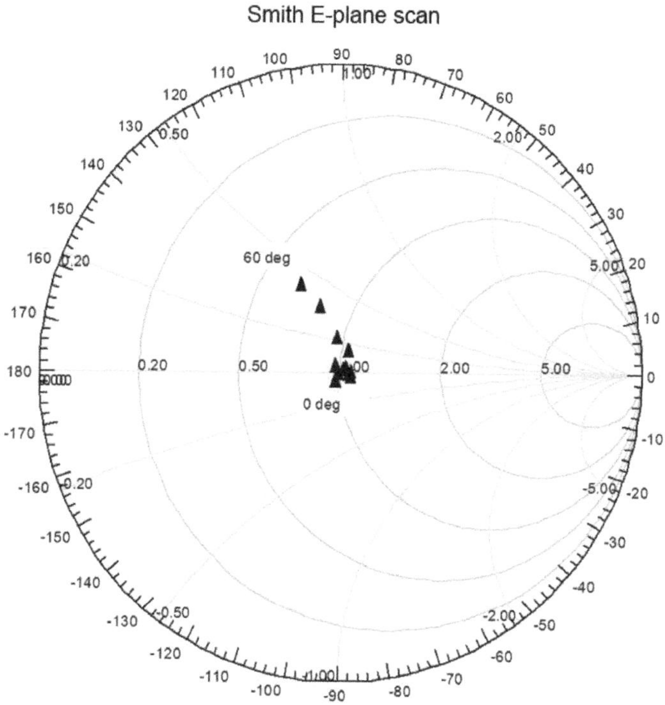

(a)

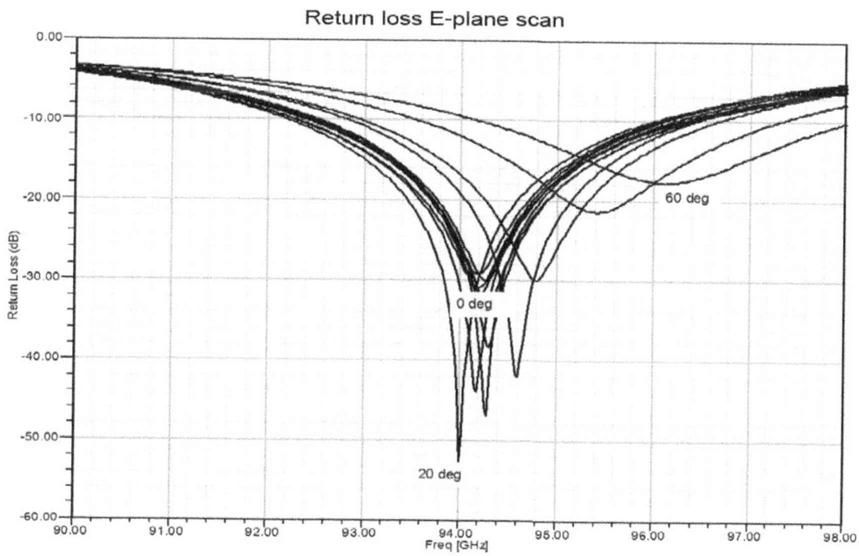

(b)

Figure 13.16: Impedance of the antenna as a function of E-plane scan angle: (a) Smith chart of input impedance as a function of E-plane scan angle; (b) return loss of antenna as a function of E-plane scan angle.

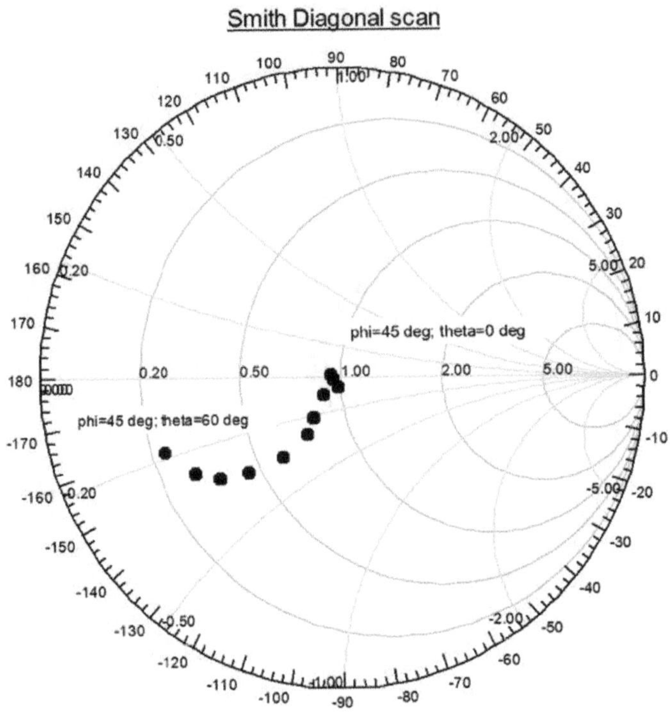

(a)

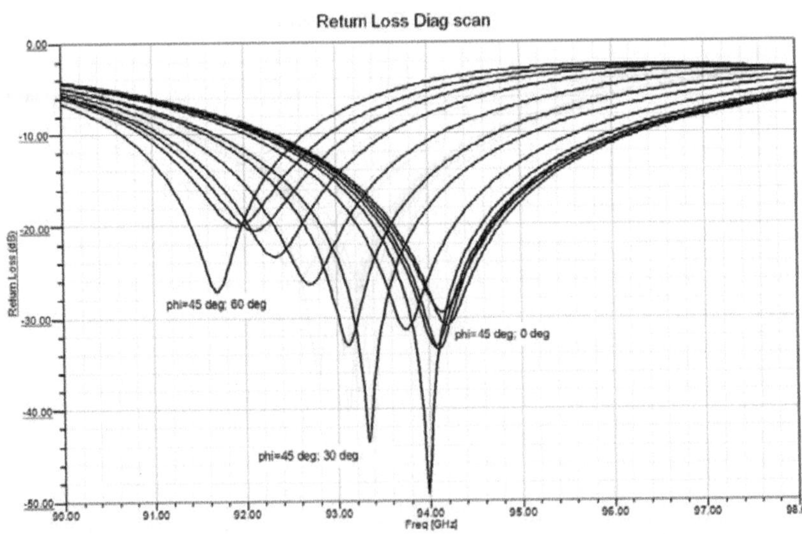

(b)

Figure 13.17: Impedance of the antenna as a function of diagonal-plane scan angle: (a) Smith chart of input impedance as a function of diagonal-plane scan angle; (b) return loss of antenna as a function of diagonal-plane scan angle.

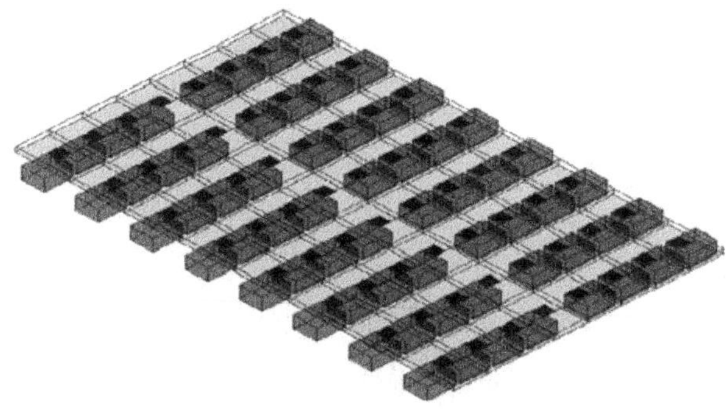

Figure 13.18: A 64-element finite antenna array of slot-coupled radiating.

are significantly raised for the 10° and 30° scan angle. The return loss indicates that we maintain a 1 GHz 10 dB return loss bandwidth even at a scan angle of 30° in the E-plane. At a scan angle of 10°, the 10 dB return loss bandwidth is 1.5 GHz from 93.5 GHz to 95 GHz. At a scan angle of 20°, the 10 dB return loss bandwidth is still about 1.5 GHz from 93.5 GHz to 95 GHz.

In Figures 13.21(a)–(e), we examine the impact on pattern performance and return loss as the 64-element array is scanned from 0 to 40° in the H-plane. We include 40°, which is beyond the scan angle limit dictated by the avoidance of a grating lobe limit at 34°, to illustrate the effects of scanning beyond this grating lobe limit. As can be seen in Figure 13.21(a), the antenna performance has significant sidelobe levels as we approach the grating lobe scan limit (30°) and exceed it (40° scan). By examining the return loss for each element in the array, it is noteworthy to point out that the H-plane scan bandwidth maintains a 2 GHz −10 dB return loss bandwidth even up to the scan angle of 30°. At a 10° scan, the usable bandwidth occurs from 93 to 95 GHz. At a 20° scan, the usable bandwidth occurs from 92.8 GHz to 94.8 GHz. At a 30° scan, the usable band width occurs from 92.4 to 94.4 GHz. Of course we must note that the center frequency is shifting to a lower frequency and the guard band of performance of our original design frequency of 94.175 GHz has been degraded. For the case of a 40° scan, the usable bandwidth is still almost 1.8 GHz; however, its span is well below our frequency of interest. These simulations illustrate the utility of using the infinite array design approach. A small array can be designed to perform sufficiently well even for scan angles up to 30°. The impact of edge elements is significant and in theory could be tuned for optimum performance in a finite array; however, the computation requirements can be very time-consuming and possibly prohibitive. Of course the level of effort is dictated by the requirements of the system specifications, and given adequate computer resources a rather brute force approach to designing small to mid-size arrays could be taken, each element being tuned and optimized for the entire performance space of pattern and impedance requirements.

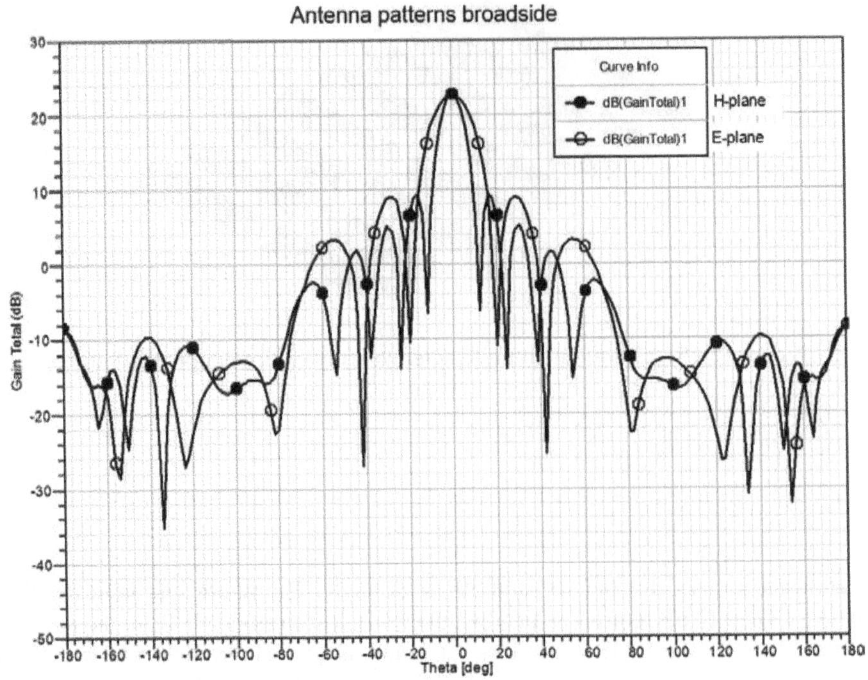

(a)

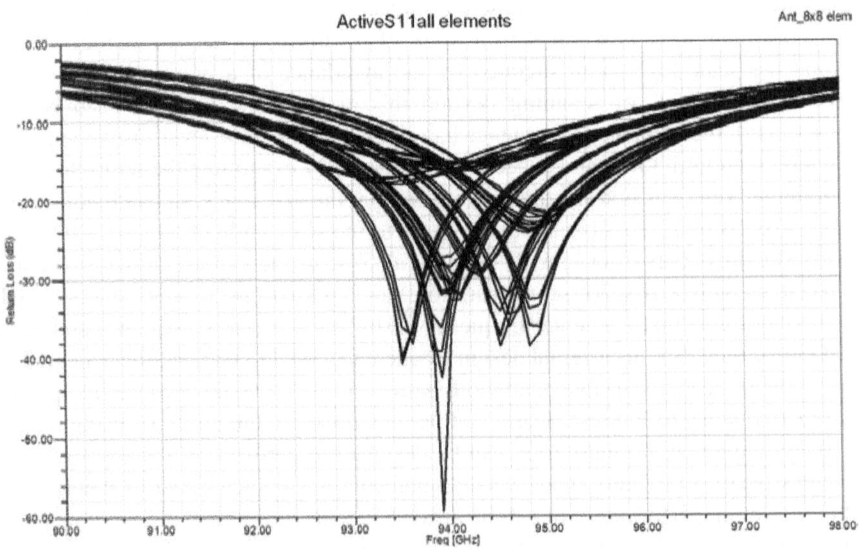

(b)

Figure 13.19: A 64-element finite antenna array performance (broadside): (a) antenna pattern in E- and H-plane for 64-element broadside array; (b) return loss for 64-elements in array.

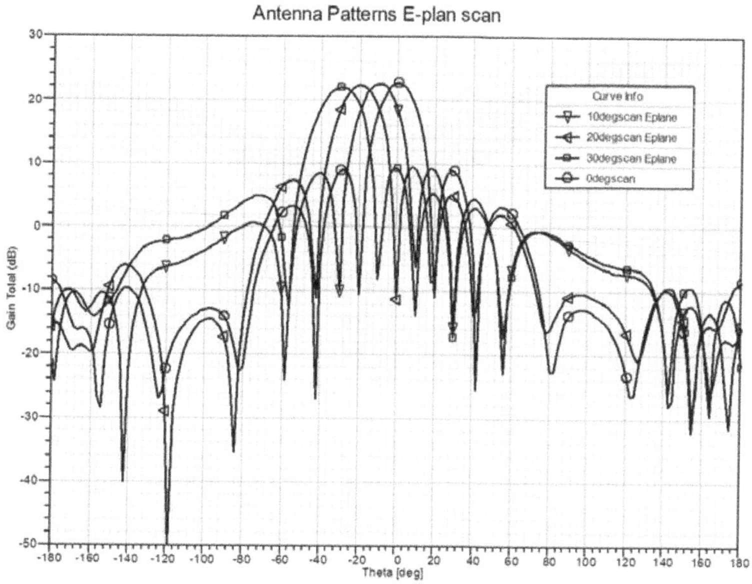

(a)

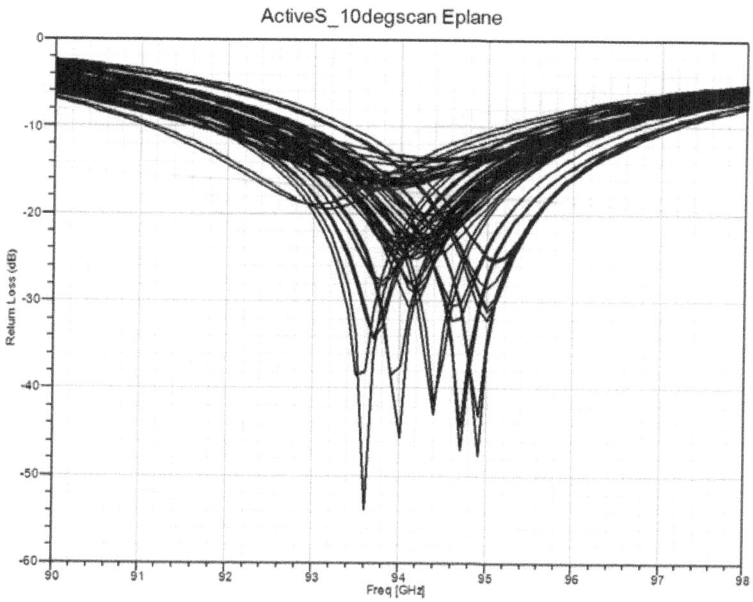

(b)

Figure 13.20: A 64-element finite antenna array performance (E-plane scan): (a) antenna pattern in E-plane for 64-element array with different scan angles; (b) return loss for 64 elements in array with 10° E-plane scan; (c) return loss for 64 elements in array with 20° E-plane scan; (d) return loss for 64 elements in array with 30° E-plane scan.

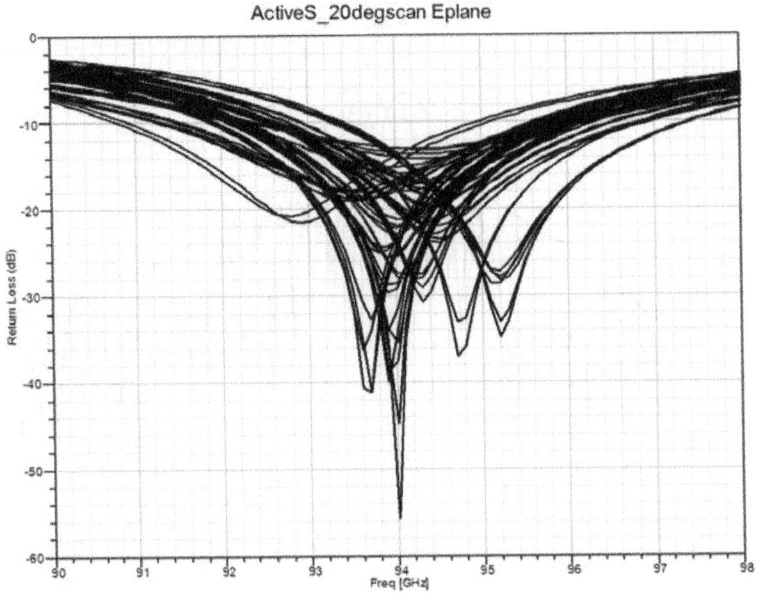

(c)

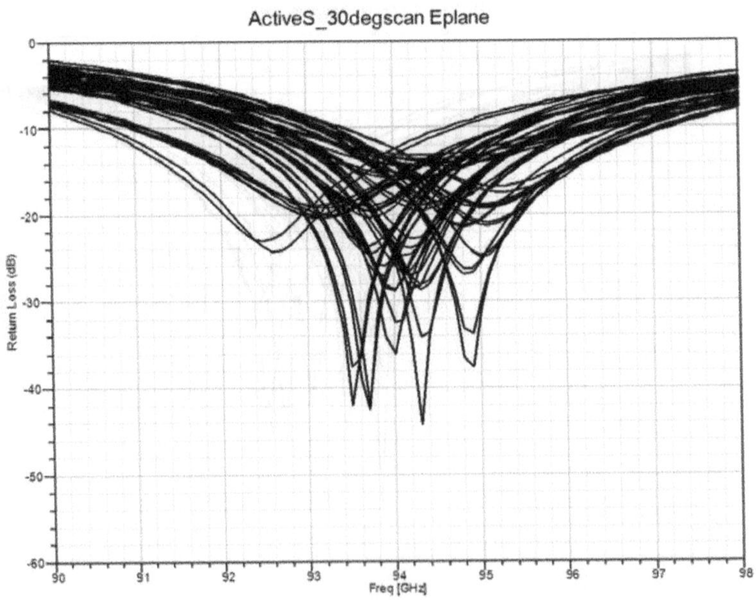

(d)

Figure 13.20: Continued.

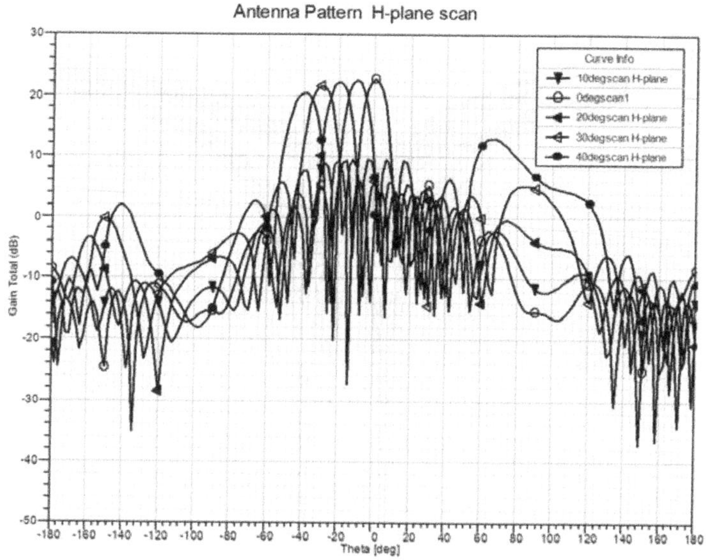

(a)

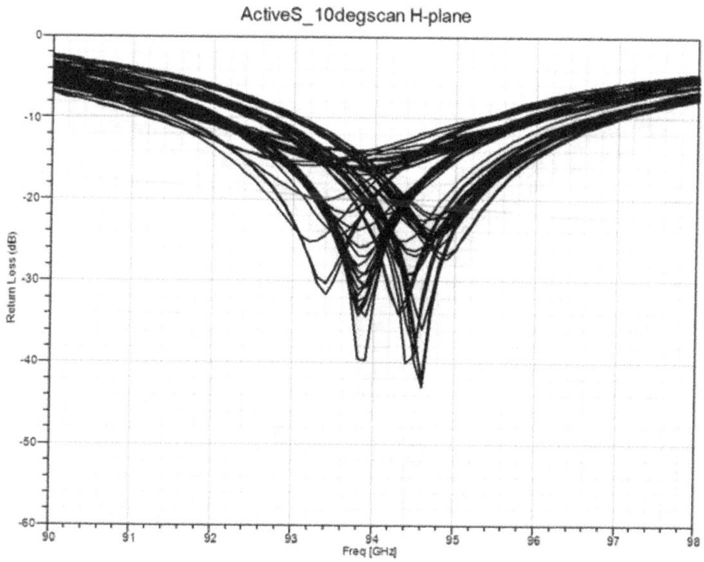

(b)

Figure 13.21: 64-element finite antenna array performance (H-plane scan): (a) antenna pattern in H-plane for 64-element array with different scan angles; (b) return loss for 64 elements in array with 10° H-plane scan; (c) return loss for 64 elements in array with 20° H-plane scan; (d) return loss for 64 elements in array with 30° H-plane scan; (e) return loss for 64 elements in array with 40° H-plane scan.

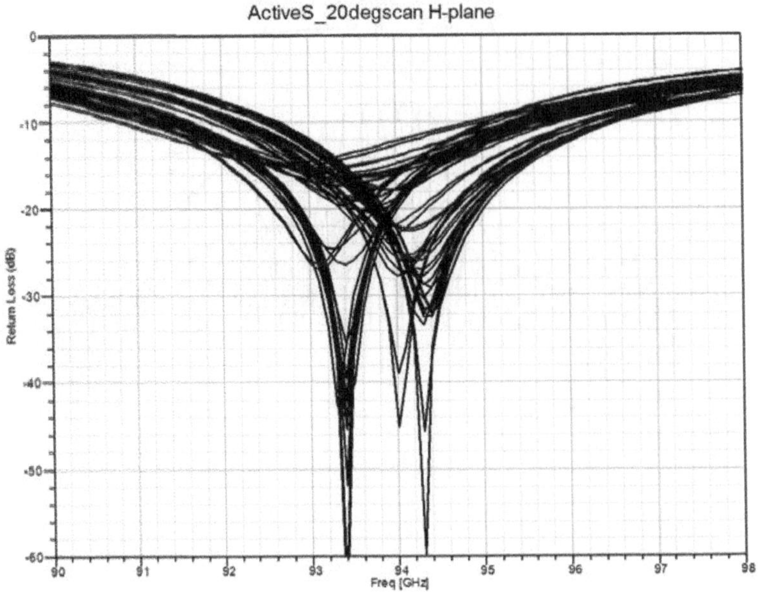

(c)

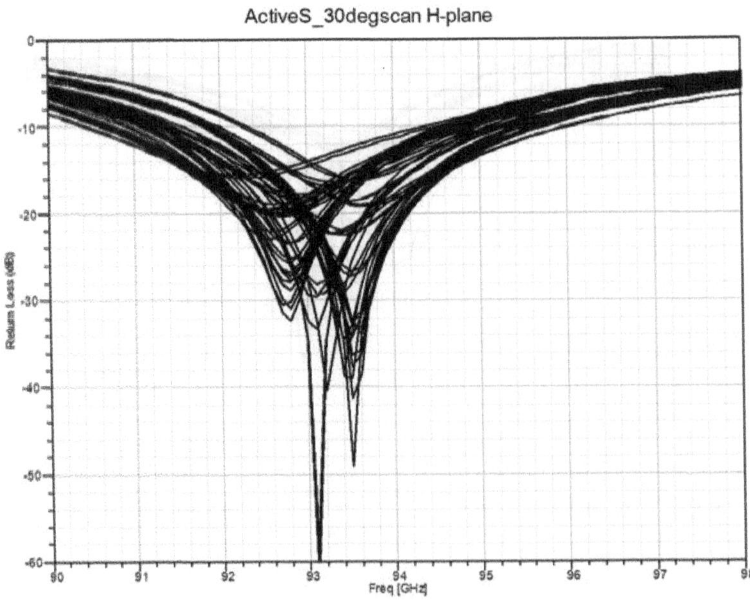

(d)

Figure 13.21: Continued.

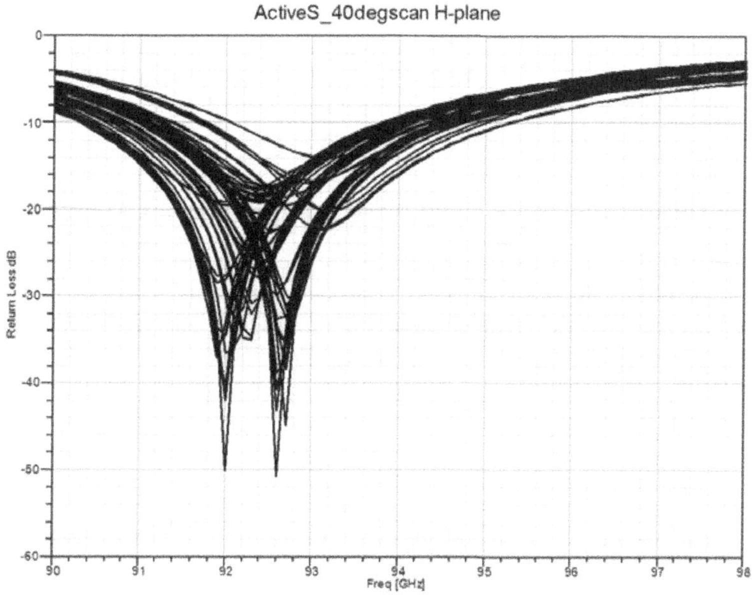

(e)

Figure 13.21: Continued.

13.3 Beam-forming Network

13.3.1 Introduction

In the previous sections we discussed a few fundamentals of the phased array antennas, in addition to discussing the effective design of the radiation elements, which are the important components of the phase array antenna, since they are the physical components that send and receive the electromagnetic energy relaying information between two points in space.

In this section we introduce and briefly discuss another aspect of phased array antenna design: beam-forming network design, also known as array factor design. For the purposes of discussion in this section, we make the assumption that there is a signal, or there are multiple signals in some region of the space-time field. In addition, there is noise and/or interference in some region of a space-time field. In the applications of interest these regions have some overlap. Furthermore, without loss of generality, we assume that the array elements have an isotropic pattern (i.e. uniform in all directions).

The primary goal of the phase array antenna is to exploit the spatial characteristics of the space-time field by filtering the signal in that field. It is well known that the solutions of the wave equation carry both spatial and temporal information; therefore, they represent the space-time fields. Since the foundation of antenna theory is the wave equation, we can proceed by expressing the filtering of the signal in terms of an angle, or a wave number. In the frequency domain, this filtering is done by combining the outputs of the phase array antenna elements with complex gains that enhance or reject a particular signal. We want this spatial filtering process such that a signal from a particular angle, or set of angles, is enhanced by a constructive beam forming, and noise or interference from different angles is rejected by destructive interference. This spatial filter is called a beam-forming network [12, 13]. The radiation elements of the phased array antenna collect spatial samples of propagating wave fields, and the beam-forming network processes the data.

13.3.2 Different Beam-forming Network of Complex Weightings

In the previous sections we have demonstrated that the beam can be scanned by appropriate introduction of phase shifts to the phased array antenna elements. In this section we will further demonstrate different element weightings and analyze their impact on the performance of the phased array antenna. In Section 13.1.3 we have discussed uniform weightings. In this section we first consider cosine weightings and use the results obtained from uniform weightings as the base line for making comparisons.

13.3.2.1 Cosine Weightings

We consider a linear array with an odd number of elements N. The cosine weighting is

$$a_n = \sin\left(\frac{\pi}{2N}\right)\cos\left(\pi\frac{n}{N}\right), \quad -\frac{N-1}{2} \le n \le \frac{N-1}{2} \tag{13.73}$$

where the $\sin(\pi/2N)$ term is a constant for normalization. Writing the cosine in exponential form

$$a_n = \sin\left(\frac{\pi}{2N}\right)\left(\frac{e^{i(n\pi/N)} + e^{-i(n\pi/N)}}{2}\right), \quad -\frac{N-1}{2} \le n \le \frac{N-1}{2} \tag{13.74}$$

The array factor of the cosine weighting in v space is

$$\begin{aligned} A.F. &= \frac{1}{2}\sin\left(\frac{\pi}{2N}\right)\left(\sum_{n=-(N-1)/2}^{(N-1)/2} e^{i(n\pi)/N} e^{inv} + \sum_{n=-(N-1)/2}^{(N-1)/2} e^{-i(n\pi/N)} e^{inv}\right) \\ &= \frac{1}{2}\sin\left(\frac{\pi}{2N}\right)\left[\sum_{n=-(N-1)/2}^{(N-1)/2} e^{in\pi(v/\pi+1/N)} + \sum_{n=-(N-1)/2}^{(N-1)/2} e^{in\pi(v/\pi-1/N)}\right] \end{aligned} \tag{13.75}$$

The first term of (13.75) represents a conventional array factor pattern steered to $v_s = -\pi/N$, and the second term of (13.75) represents a conventional array factor pattern steered to $v_s = \pi/N$. It is straightforward to find the summations of (13.75) by recognizing that the sums are geometric series. The array factor can thus be written

$$A.F. = \frac{1}{2}\sin\left(\frac{\pi}{2N}\right)\left[\frac{\sin(N\pi/2)(v/\pi + 1/N)}{\sin(\pi/2)(v/\pi + 1/N)} + \frac{\sin(N\pi/2)(v/\pi - 1/N)}{\sin(\pi/2)(v/\pi - 1/N)}\right] \tag{13.76}$$

We show an 11-element array factor (13.76) with the uniform weighting array factor pattern in Figure 13.22. We find that the sidelobes of the cosine weighting are lower than the referenced uniform weighting array factor, but the beamwidth is wider since the cosine weighting array factor is the sum of two beams which are shifted π/N in opposite directions. The parameters for the two array factor patterns are given in Table 13.1.

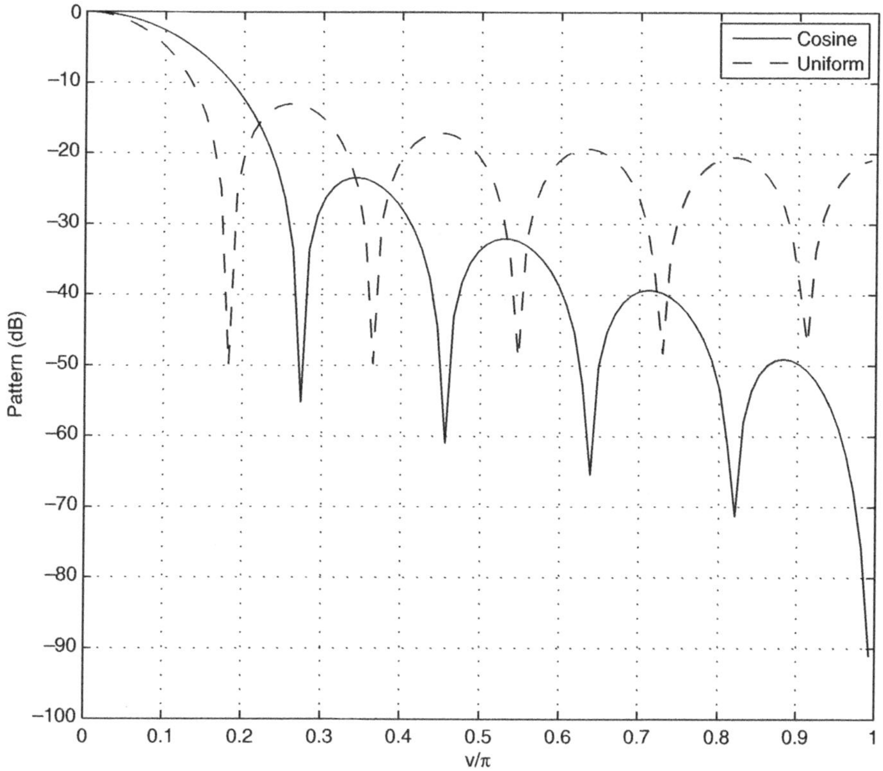

Figure 13.22: Array factor pattern for cosine and uniform weighting: $N = 11$.

Table 13.1: The characteristics of the cosine weighting compared with uniform weighting.

Weighting	HPBW	BW	First sidelobe
Cosine	$1.18\frac{2}{N}$	$3.0\frac{2}{N}$	−23.5 dB
Uniform	$0.89\frac{2}{N}$	$2.0\frac{2}{N}$	−13.0 dB

13.3.2.2 Dolph–Chebyshev Weightings

If it is desired to have antenna sidelobes that are lower than the 13 dB sidelobes of the uniform excited array, it is necessary to taper the aperture distribution across the array. The phased array antennas we present in this section are equispaced excited in phase. This assumption gives an array factor with one main beam plus sidelobes. Furthermore we assume that the feeding coefficients are symmetrical and real, thus the coefficients of (13.45) are

$$a_{-n} = a_n \tag{13.77}$$

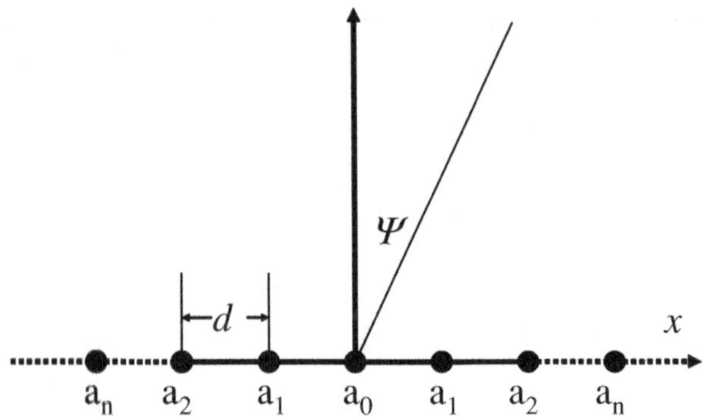

Figure 13.23: An odd element linear broadside array with uniform spacing.

The array factor (13.45) then takes the polynomial form

$$A.F. = \sum_{n=0}^{(N-1)/2} b_n \cos(nv) \tag{13.78}$$

To find b_n let us first consider an odd number of elements of linear array as shown in Figure 13.23. The main beam is at broadside. The coefficients of the array factor are symmetric about the origin. The coefficients of the array factor are

$$b_n = \begin{cases} a_0, & n = 0 \\ 2a_n, & n = 1, 2, \ldots, \dfrac{N-1}{2} \end{cases} \tag{13.79}$$

We can rewrite the normalized odd number element linear array factor (13.78) as

$$\begin{aligned} A.F. &= 1 + \sum_{n=1}^{(N-1)/2} \frac{b_n}{b_0} \cos(nv) \\ &= 1 + \sum_{n=1}^{(N-1)/2} \frac{b_n}{b_0} \cos(nkd \sin\varphi) = 1 + \sum_{n=1}^{(N-1)/2} \frac{b_n}{b_0} \cos(nkd \cos\theta) \end{aligned} \tag{13.80}$$

In Figure 13.24 we show an array where the element number is even. The array factor coefficients (13.78) are

$$b_n = 2a_n, \quad n = 1, 2, \ldots, \frac{N}{2} \tag{13.81}$$

The even number element linear array factor (13.78) is

$$\begin{aligned} A.F. &= \sum_{n=1}^{N/2} b_n \cos\left(\frac{2n-1}{2} v\right) = \sum_{n=1}^{N/2} b_n \cos\left(\frac{2n-1}{2} kd \sin\varphi\right) \\ &= \sum_{n=1}^{N/2} b_n \cos\left(\frac{2n-1}{2} kd \cos\theta\right) \end{aligned} \tag{13.82}$$

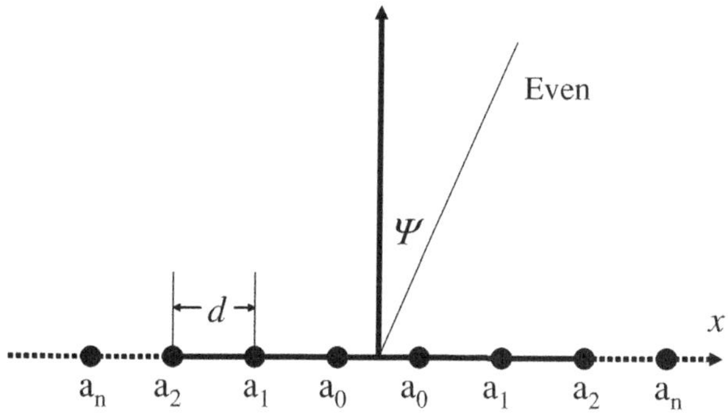

Figure 13.24: An even element linear broadside array with uniform spacing.

Our goal is to design a phased array antenna for which the radiation pattern has a minimum null-to-null beamwidth for a desired sidelobe level. An array antenna that has a far-field pattern for which all the sidelobes have the same amplitude will meet this criterion. The effect of maintaining the sidelobes at a constant value is to narrow the main beam resulting in the narrowest possible beamwidth for a given sidelobe level. In order to achieve this goal we must determine the array element current distributions, i.e. the coefficients in array factor (13.80) and (13.82). C. L. Dolph in his classic paper [14] presented the solutions of the proper array excitations to give a radiation pattern with uniform sidelobes at a specified height. He developed a procedure to utilize the properties of the Chebychev polynomials to design an optimum radiation pattern.

The Chebyshev polynomials are the solutions of the differential equation

$$(1-x^2)\frac{\mathrm{d}^2 T_m}{\mathrm{d}x^2} - x\frac{\mathrm{d}T_m}{\mathrm{d}x} + m^2 T_m = 0 \tag{13.83}$$

To find a solution we assume that $T_m(x)$ takes a power series form

$$T_m(x) = \sum_{k=0}^{\infty} a_k x^k \tag{13.84}$$

We can therefore write

$$\frac{\mathrm{d}T_m}{\mathrm{d}x} = \sum_{k=1}^{\infty} k a_k x^{k-1} \tag{13.85}$$

and

$$\frac{\mathrm{d}^2 T_m}{\mathrm{d}x^2} = \sum_{k=2}^{\infty} k(k-1) a_k x^{k-2} \tag{13.86}$$

Substituting (13.85) and (13.86) in (13.83), we find that

$$\sum_{k=2}^{\infty} k(k-1)a_k x^{k-2} - \sum_{k=2}^{\infty} k(k-1)a_k x^k - \sum_{k=1}^{\infty} k a_k x^k + m^2 \sum_{k=0}^{\infty} a_k x^k = 0 \tag{13.87}$$

We can find a recursive relationship between the coefficients by considering different values of k in each summation of (13.87) such that the sum of all coefficients of the polynomial with the same power of x is equal to zero. For example, the coefficients of x to the power 0 are found by letting $k = 2$ in the first summation and $k = 0$ in the last summation. For this case we can write

$$2a_2 + m^2 a_0 = 0, \quad k = 0 \tag{13.88}$$

Similarly, following the same procedure $k = 1$ and $k \geq 2$, we can write

$$6a_3 + (m^2 - 1)a_1 = 0, \quad k = 1 \tag{13.89}$$

$$(k+2)(k+1)a_{k+2} + (m^2 - k^2)a_k = 0, \quad k \geq 2 \tag{13.90}$$

Therefore, from (13.90) we find the recursive relationship between coefficients to be

$$a_{k+2} = -a_k \frac{(m+k)(m-k)}{(k+2)(k+1)} \tag{13.91}$$

It is clear that the highest value k could be is m if m is a positive integer.

By careful inspection of (13.88) and (13.91), we can conclude that all the even coefficients can be expressed by a_0 if m is an even number. Furthermore, all the odd coefficients are zero by (13.91) if a_1 is zero. The solution of (13.83) then is a degree m polynomial with only one arbitrary constant a_0 and even powers of x.

Similarly by inspecting (13.89) and (13.91) together, we find that all the odd coefficients can be expressed in terms of a_1. In the same manner all the even coefficients are zero by (13.91) if a_0 is zero. Thus the solution of (13.83) is a degree m polynomial with only one arbitrary constant a_1 and odd powers of x.

For an even positive integer m (13.88) and (13.91) give

$$\left.\begin{aligned} a_2 &= -a_0 \frac{m \cdot m}{2 \cdot 1} \\ a_4 &= a_0 \frac{(m+2)m \cdot m(m-2)}{4 \cdot 3 \cdot 2 \cdot 1} \\ &\vdots \\ a_k &= (-1)^{-k/2} a_0 \frac{m(m+2)\cdots(m+k-2)m(m-2)\cdots[m-(k-2)]}{k!} \end{aligned}\right\} \tag{13.92}$$

After some manipulation, (13.92) can be written as

$$a_k = (-1)^{-k/2} \times a_0 \frac{2^k (m/2)(m/2+1)\cdots(m/2+k/2-1)(m/2)(m/2-1)\cdots(m/2-k/2+1)}{k!} \tag{13.93}$$

If $m = 2N$ and $k = 2n$, we have

$$\begin{aligned} a_{2n} &= (-1)^{-n} a_0 \frac{2^{2n}}{(2n)!}[N(N+1)\cdots(N+n-1)N(N-1)\cdots(N-n+1)] \\ &= (-1)^{-n} a_0 \frac{2^{2n}}{(2n)!} \cdot \frac{(N+n-1)!}{(N-1)!} \cdot \frac{N!}{(N-n)!} \end{aligned} \tag{13.94}$$

From (13.84) we can write

$$
\begin{aligned}
T_{2n}(x) &= \sum_{n=0}^{\infty} a_{2n} x^{2n} \\
&= a_0 \sum_{n=0}^{\infty} (-1)^{-n} 2^{2n} \frac{N(N+n-1)!}{(2n)!(N-n)!} x^{2n} \\
&= a_0 \sum_{n=0}^{\infty} (-1)^{-n} \frac{N}{N+n} \cdot \frac{(N+n)!}{(2n)!(N-n)!} (2x)^{2n} \\
&= a_0 \sum_{n=0}^{\infty} (-1)^{-n} \frac{N}{N+n} \cdot \binom{N+n}{2n} (2x)^{2n}
\end{aligned} \tag{13.95}
$$

where $\binom{N}{n} = N!/(n!(N-n))$.

One approach to determine the constant a_0 is set $T_{2N}(0) = (-1)^N$,

$$
T_{2N}(0) = (-1)^N = a_0 (2)^{-1} \frac{2N}{N} \cdot \frac{N!}{0!N!} = a_0 \tag{13.96}
$$

By inserting (13.96) into (13.95), we finally have the general expression for an even-degree Chebyshev polynomial

$$
T_{2N}(x) = \sum_{n=0}^{N} (-1)^{N-n} \frac{N}{N+n} \cdot \binom{N+n}{2n} (2x)^{2n} \tag{13.97}
$$

By following the same procedure, we find out the odd number coefficients can be expressed as

$$
\left.
\begin{aligned}
a_3 &= -a_1 \frac{(m+1)\cdot(m-1)}{3\cdot 2\cdot 1} \\
a_5 &= a_1 \frac{(m+3)(m+1)(m-1)(m-3)}{5\cdot 4\cdot 3\cdot 2\cdot 1} \\
&\vdots \\
a_k &= (-1)^{-(k-1)/2} a_1 \frac{(m+k+2)\cdots(m+3)(m+1)(m-1)(m-3)\cdots[m-(k-2)]}{k!}
\end{aligned}
\right\} \tag{13.98}
$$

If $m = 2N - 1$ and $n = 2k - 1$, (13.98) becomes

$$
a_{2n-1} = (-1)^{-(n-1)} a_1 \frac{2^{2n-2}}{N+n-1} \cdot \binom{N+n-1}{2n-1} \tag{13.99}
$$

The solution of the odd-degree Chebyshev polynomial is

$$
T_{2N-1}(x) = a_1 \sum_{n=1}^{N} (-1)^{-(n-1)} \frac{2^{2n-2}}{N+n-1} \cdot \binom{N+n-1}{2n-1} x^{2n-1} \tag{13.100}
$$

We can choose a_1 by setting $T_{2N-1}(0) = (-1)^{N-1}(2N-1)$. We obtain the general expression for an odd-degree Chebyshev polynomial as

$$T_{2N-1}(x) = \sum_{n=1}^{N}(-1)^{N-n}\frac{2^{2n-2}(2N-1)}{N+n-1}\cdot\binom{N+n-1}{2n-1}x^{2n-1} \tag{13.101}$$

We list the first eight Chebyshev polynomials

$$\left.\begin{aligned} T_0(x) &= 1 \\ T_1(x) &= x \\ T_2(x) &= 2x^2 - 1 \\ T_3(x) &= 4x^3 - 3x \\ T_4(x) &= 8x^4 - 8x^2 + 1 \\ T_5(x) &= 16x^5 - 20x^3 + 5x \\ T_6(x) &= 32x^6 - 48x^4 + 18x^2 - 1 \\ T_7(x) &= 64x^7 - 112x^5 + 56x^3 - 7x \end{aligned}\right\} \tag{13.102}$$

We note that the degree of the polynomial in (13.102) is the same as the value of m.

The general expressions of the Chebyshev polynomial (13.97) and (13.101) can also be written in the alternative form

$$T_m(x) = \begin{cases} \cos(m\cos^{-1}x), & |x| \le 1 \\ \cosh(m\cosh^{-1}x), & x > 1 \\ (-1)^m\cosh(m\cosh^{-1}x), & x < -1 \end{cases} \tag{13.103}$$

The Chebyshev polynomial is orthogonal over the interval $-1 \le x \le 1$ with respect to the weighting function $w(x) = 1/\sqrt{1-x^2}$ [15]

$$\int_{-1}^{1}\frac{1}{\sqrt{1-x^2}}T_m(x)T_n(x)\,\mathrm{d}x = \int_0^{\pi}T_m(\cos\theta)T_n(\cos\theta)\,\mathrm{d}\theta = c_m\delta_{mn} \tag{13.104}$$

where the constant is

$$c_{mn} = \begin{cases} \pi, & m = 0, \\ \dfrac{\pi}{2}, & m \neq 0 \end{cases} \tag{13.105}$$

Up to this point, we have introduced the necessary mathematical background to develop the Dolph–Chebyshev phased array antenna. We can now use the mathematical tools developed thus far in consideration with the array factors (13.80) and (13.82).

If we examine both array factors (13.80) and (13.82), we find that they are accurately represented by the Fourier Cosine series. The Cosine function can be expressed by the Euler identity as

$$\cos\frac{n\phi}{2} = \Re(e^{in\phi/2}) = \Re\left(\cos\frac{\phi}{2} + i\sin\frac{\phi}{2}\right)^n \tag{13.106}$$

Using the binomial series expansion, we obtain

$$\begin{aligned} \cos\frac{n\phi}{2} &= \cos^n\frac{\phi}{2} - \frac{n(n-1)}{2!}\cos^{n-2}\frac{\phi}{2}\sin^2\frac{\phi}{2} \\ &+ \frac{n(n-1)(n-2)(n-3)}{4!}\cos^{n-4}\frac{\phi}{2}\sin^4\frac{\phi}{2} - \cdots \end{aligned} \tag{13.107}$$

Substituting trigonometry relation $\sin^2(\phi/2) = 1 - \cos^2(\phi/2)$ in (13.107) results in

$$\left.\begin{aligned} n &= 0, & \cos n\frac{\phi}{2} &= 1 \\ n &= 1, & \cos n\frac{\phi}{2} &= \cos\frac{\phi}{2} \\ n &= 2, & \cos n\frac{\phi}{2} &= 2\cos^2\frac{\phi}{2} - 1 \\ n &= 3, & \cos n\frac{\phi}{2} &= 4\cos^3\frac{\phi}{2} - 3\cos\frac{\phi}{2} \\ n &= 4, & \cos n\frac{\phi}{2} &= 8\cos^4\frac{\phi}{2} - 8\cos^2\frac{\phi}{2} + 1 \end{aligned}\right\} \tag{13.108}$$

If we let

$$x = \cos\frac{\phi}{2} \tag{13.109}$$

then (13.108) is the same as (13.102). We can therefore conclude that the Chebyshev polynomials in general are designated by

$$T_n(x) = \cos\left(n\frac{\phi}{2}\right)\bigg|_{\cos(\phi/2)=x} \tag{13.110}$$

We conclude that for a broadside, N-element symmetric equispaced isotropic phased array antenna the far-field pattern is an $N - 1$ degree Chebyshev polynomial.

The Chebyshev polynomial $T_m(x)$ has m real roots in the interval $|x| \leq 1$. These roots take place as $\cos(m\phi_r/2) = 0$ or

$$m\frac{\phi_r}{2} = \frac{(2n-1)\pi}{2}, \qquad n = 1, 2, \ldots, m \tag{13.111}$$

The roots of x are

$$x_r = \cos\left[\frac{(2n-1)\pi}{2m}\right] \tag{13.112}$$

Figure 13.25 shows the first three polynomials.

The polynomial demonstrates the oscillation with the same unity maxima and minima. Therefore, the Chebyshev polynomials have an equal ripple in $|x| \leq 1$ region. The polynomials increase monotonically for the range $|x| > 1$. Figure 13.26 plots the seventh degree Chebyshev polynomial.

In the procedure developed by Dolph, the main beam magnitude is the value of $T_m(x_0)$ where $T_m(x_0) > 1$, and $x_0 > 1$. The magnitude of the sidelobes remains the same and is unity. We define the main beam maximum to the sidelobe ratio R

$$R = \frac{\text{main beam maximum}}{\text{sidelobe level}} \tag{13.113}$$

We can determine x_0 by the second formula in (13.103) if we know R

$$T_{N-1}(x_0) = \cosh[(N-1)\cosh^{-1} x_0] = R, \qquad |x_0| > 1 \tag{13.114}$$

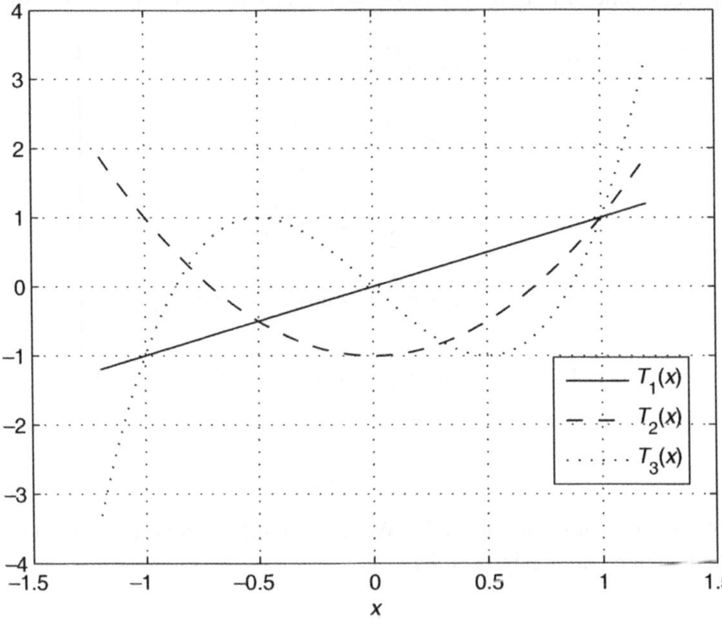

Figure 13.25: The first three Chebyshev polynomials.

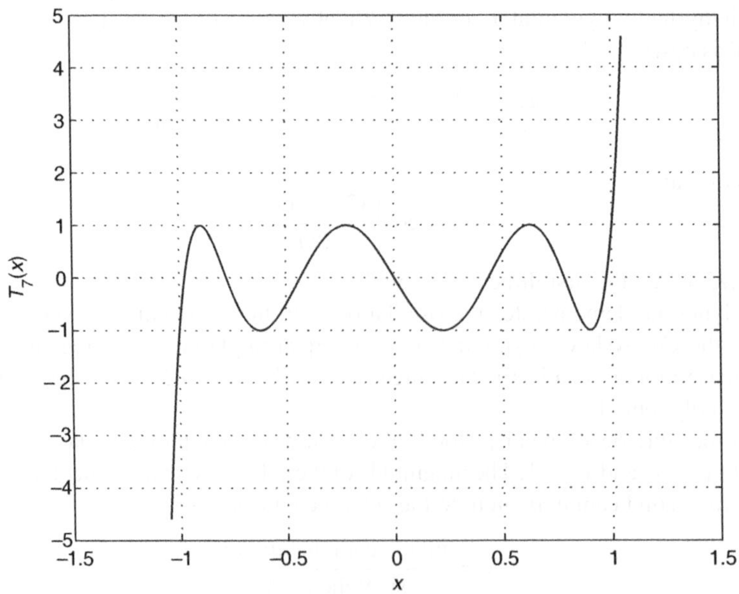

Figure 13.26: The seventh-degree Chebyshev polynomial.

or

$$x_0 = \cosh\left(\frac{1}{N-1}\cosh^{-1} R\right), \quad |x_0| > 1 \tag{13.115}$$

The design procedure presented by Dolph begins with the selection of the Chebyshev polynomial $T_m(x)$ as the array polynomial. For an N-element array we make the following variable change

$$m = N - 1 \tag{13.116}$$

The next step is to find x_0 from the main beam maximum to sidelobe ratio (13.115). After a change in scale we have

$$x = x_0 \cos\frac{\phi}{2} \tag{13.117}$$

the normalized far-field pattern of the array factor is

$$A.F. = \frac{1}{R} T_{N-1}\left[x_0 \cos\left(\frac{\phi}{2}\right)\right] \tag{13.118}$$

In order to determine the coefficients of the array factor, we need to find out the roots x_r of the Chebyshev polynomial $T_m(x_r)$ from (13.112). We find the roots ϕ_r by combining (13.111), (13.112) and (13.117) to find

$$\phi_r = 2\cos^{-1}\left(\frac{x_r}{x_0}\right) \tag{13.119}$$

We make a variable change as follows

$$z = e^{iv} \tag{13.120}$$

therefore, we can rewrite the array factor (13.44) in an explicit polynomial form

$$A.F. = f(z) = a_0 + a_1 z + a_2 z^2 + \cdots + a_{N-1} z^{N-1} \tag{13.121}$$

We can determine the coefficients by considering the fundamental theorem of algebra

$$f(z) = (z - z_1)(z - z_2)\cdots(z - z_N) \tag{13.122}$$

where z_n is the nth root. We equal (13.121) and (13.122) to find the desired sidelobe level array factor coefficients.

As an example, we consider a five-element array, $m = N - 1 = 4$. The far-field pattern is desired -20 dB sidelobe level. In this case, $R = 10$. From (13.115) we can write

$$x_0 = \cosh\left(\frac{1}{4}\cosh^{-1} 10\right) = 1.2933 \tag{13.123}$$

From (13.112) the roots of $T_4(x)$ are ± 0.9239 and ± 0.3827. These roots correspond to $\pm 88.82°$ and $\pm 145.16°$ in ϕ space by (13.119), or ± 1.55 radians and ± 2.54 radians. Using (13.122) we have

$$\begin{aligned} f(z) &= (z - e^{i1.55})(z - e^{-i1.55})(z - e^{i2.54})(z - e^{-i2.54}) \\ &= [z^2 + 2\cos(1.55)z + 1][z^2 + 2\cos(2.54)z + 1] \\ &= z^4 + 1.60z^3 + 1.93z^2 + 1.60z + 1 \end{aligned} \tag{13.124}$$

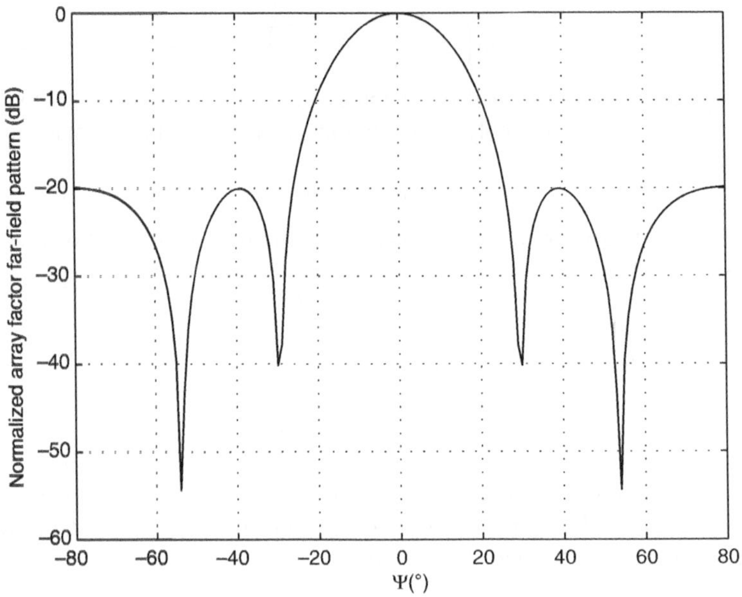

Figure 13.27: The normalized array factor far-field pattern of the five-element array.

We determine the relative magnitudes for each element as

$$a_0 = 1,$$

$$a_1 = 1.60,$$

$$a_2 = 1.93,$$

$$a_3 = 1.60,$$

$$a_4 = 1$$

In Figure 13.27 we plot the normalized array factor far-field pattern of the five element array with the half-wavelength inter-element space.

We learn from this example that the procedure to carry out the necessary multiplications of (13.122) after finding the root ϕ_r and z_m will be tedious for a large element number N. With the help of

$$\cos^{2m}\frac{\phi}{2} = \frac{1}{2^{2m}}\sum_{l=0}^{m}\epsilon_l\binom{2m}{m-l}\cos\frac{2l\phi}{2} \tag{13.125}$$

where

$$\epsilon_l = \begin{cases} 1 & l = 0, \\ 2 & \text{otherwise} \end{cases} \tag{13.126}$$

and

$$\cos^{2m-1}\frac{\phi}{2} = \frac{1}{2^{2m-2}}\sum_{l=0}^{m}\binom{2m-1}{m-l}\cos(2l-1)\frac{\phi}{2} \tag{13.127}$$

we present an alternate procedure for the determination of the array factor coefficients for large element numbers.

If we insert (13.125) and (13.127) into (13.97) and (13.101), respectively, we can rewrite the even and odd degree Chebyshev polynomials as

$$T_{2N}\left(\frac{\phi}{2}\right) = \sum_{m=0}^{N}\sum_{l=0}^{m} \epsilon_l(-1)^{N-m}\frac{N}{N+m}\binom{N+m}{2m}\binom{2m}{m-l}x_0^{2m}\cos\frac{2l\phi}{2} \tag{13.128}$$

and

$$T_{2N-1}\left(\frac{\phi}{2}\right) = \sum_{m=1}^{N}\sum_{l=1}^{m}(-1)^{N-m}\frac{2N-1}{N+m-1}\binom{N+m-1}{2m-1} \times \binom{2m-1}{m-l}x_0^{2m-1}\cos(2l-1)\frac{\phi}{2} \tag{13.129}$$

Here we used (13.109).

The coefficients of $\cos 2m\phi/2$ are in the same relation in (13.97) and (13.128) for all m. Also, the coefficients of $\cos(2m-1)\phi/2$ are in the same relation in (13.101) and (13.129). The relative current distribution for an odd-element array, therefore, is written as

$$a_m = \sum_{l=m}^{N}(-1)^{N-l}\frac{N}{N+l}\binom{N+l}{2l}\binom{2l}{l-m}x_0^{2l} \tag{13.130}$$

and for an even array the coefficients are written as

$$a_m = \sum_{l=m}^{N}(-1)^{N-l}\frac{2N-1}{N+l-1}\binom{N+l-1}{2l-1}\binom{2l-1}{l-m}x_0^{2l-1} \tag{13.131}$$

To note, there is a function CHEBWIN in Matlab which calculates the Chebyshev polynomial coefficients; w = CHEBWIN(L,R) returns the Lth order Chebyshev polynomial coefficients w with the sidelobe level R dB below the main lobe.

13.3.2.3 Array Null Steering

In the beginning of this section we stated that the primary function of a phased array antenna is spatial filtering. So far we have introduced one aspect of this filtering procedure: enhancement of the signal we desire. Now we will briefly describe another aspect of this filtering: rejection of noise, or interference; this is called array null steering.

Let us consider a uniformly spaced linear phased array. Assume that there are N noise sources at the direction $\theta_1, \theta_2, \ldots, \theta_N$. We set these angles as the nulls. The desired array factor is then

$$A.F. = c\prod_{n=1}^{N}(z-z_n) = c(I_0 + I_1 z + I_2 z^2 + \cdots + I_N z^N) \tag{13.132}$$

where c is a normalized constant, $z_n = e^{\Psi_n}$, $\Psi_n = kd\cos\theta_n$, and I_n is a complex coefficient for the Nth element.

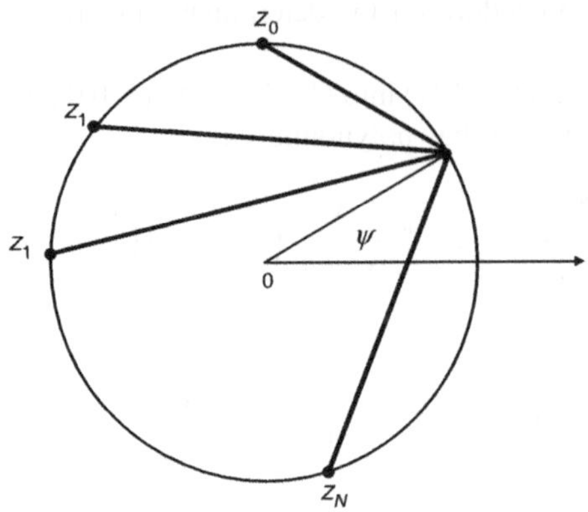

Figure 13.28: Zeros of an array polynomial on a unit circle.

As shown in Figure 13.28, $|z - z_n|$ is the length between a position $z(\theta)$ on the unit circle and z_n. If we want a deep null at θ_n, multiplicity of the root z_n can be imposed. In fact $(z - z_n)^p$, with p an integer, implies that all derivatives up to the $(p-1)$th order vanished at $z = z_n$. Let us consider a binomial array. If all roots $\{z_n\}$ coincide and are equal to $z_1 = e^{jk\cos\theta_1}$, the array factor is

$$A.F. = c(z - z_1)^N = c\left[z^N + \binom{N}{1} z^{N-1}(-z_1) + \cdots + (-z_1)^N\right]$$
$$= c\sum_{k=0}^{N} \binom{N}{k} z^{N-k}(-z_1)^k \tag{13.133}$$

where c is a constant and θ_1 is a null angle. The magnitude of the array factor in decibels is

$$|A.F.| = |c||z - z_1|^N = 20(\log|c| + N\log|z - z_1|) \tag{13.134}$$

This array has no sidelobe if the inter-element separation is less than, or equal to half of a wavelength.

If the null steering is desired, we can change the root locations on the unit circle.

13.4 Design and Manufacture Issues

13.4.1 Design Considerations

One of the first and major considerations in mmWave applications is the selection of a material and process technology from the myriad of available choices. A down select for a material can be quite complicated, involving many trade-off factors including cost, performance and typical delivery times for manufactured parts. Because of the small feature sizes necessary, in addition to performance requirements, at the operation frequencies in mmWave applications, the design and manufacturing practices are well beyond conventional microwave techniques. It is critical to understand both the high-frequency material properties and manufacturing tolerances as a foundation to any mmWave antenna array design. For mmWave applications, there is little to no reliable material data available in the frequency range above 60 GHz because measurements of the dielectric properties of materials at this frequency band are extremely difficult to measure accurately. This want of accurate dielectric data makes it extremely difficult if not impossible for a precision engineering antenna design in the mmWave bands. Thus, the design and manufacturing of the phased array antennas in mmWave bands is much more difficult than the design and manufacturing of phased array antennas in microwave frequencies.

In this section we will discuss the material selection approach useful for a successful mmWave antenna design presenting both material property measurements and process developments needed to execute a multi-element array applicable to W-band applications. We begin with a brief survey of material and process choices which might be considered, and focus on a few salient characteristics that can be used to make a quick material selection decision based on the critical parameters of a material appropriate for a W-band application. Based on these key characteristics we highlight the selection of low temperature co-fired ceramic (LTCC) as the material of choice in mmWave applications. We compare different transmission line structures in LTCC to illustrate and compare the best-case transmission line losses for microstrip and stripline. The main component in a W-band antenna is what is known as a laminate waveguide (LWG) [16, 17]. We discuss and describe the manufacturing details for a W-band antenna array using utilizing laminated waveguides. Finally we demonstrate the measurement issues also.

13.4.1.1 Materials

We begin our discussion in this section by considering some key characteristics of materials currently being used in W-band applications. Fused quartz is one such material. It has attractive qualities such as being low loss with a typical low dielectric constant of about 3.5 and loss tangent of about 0.002. However, it does have serious drawbacks for incorporating embedded components for small footprint designs. Fused quartz is a single-layer substrate with metalization on either side limiting it to surface structures, which can add to conductor losses and make mechanical alignment challenging. In addition, adherence of metals to the quartz can be challenging; and typical W-band radiation elements are frequently fragile, very thin structures, ranging from 0.004 to 0.006 inches in thickness. These thicknesses lead to serious handling and mounting problems.

Another material which could conceivably be considered because of reported low losses are certain plastic laminates. These substrates have low-loss, low-dielectric constants and overcome the single-layer limitation of the quartz material by having a multi-layer processing capability. The single most serious limitation to using organic multilayer materials is the dimensional tolerances in the layer-to-layer alignment. Typically, layer-to-layer alignment in organic multilayer processes is ± 0.005 inches. This drawback makes it unable to meet the much tighter alignment requirements for multi-layer slot-coupled structures at W-band frequencies. Also, the mechanical stability of the material is temperature-dependent. The material cannot be assumed to be stable over temperature, and hence, again, the required stringent and tight tolerances required at W-band applications cannot be met, especially on multi-layer designs, with organic materials. Also, best microwave practices for via pitch placement at about a tenth of a wavelength cannot in general be met for mmWave frequencies by organic technology, which typically has as a standard 0.010 inch diameter vias on a 0.030 inch center-to-center pitch.

To meet the very stringent and mechanical requirements at mmWave frequencies it becomes clear that we must have a material that is temperature-stable, low loss, and having process capabilities for extremely tight via spacing. These requirements lead one to consider LTCC as a first option for high-performance low-loss mmWave applications. It is stable over temperature, has a low-loss tangent, and a dielectric constant suitable for miniaturization of antenna components.

There are many flavors and variations in LTCC processes on the market. To reduce the risk on any project, material data in the band of interest must be available of course; however, as previously indicated, data at these high mmWave frequencies is lacking for most LTCCs. Among the most common LTCCs for volume production, the most likely first candidate is Ferro A6M owing to its having published data for permittivity and a loss tangent at 95 GHz, which for the purposes of this discussion is our frequency of interest. The permittivity of A6M is nominally 6.2 and the loss tangent is nominally 0.002. The metalization conductivity for A6M is typically 1.97×10^7 Mho/m.

13.4.1.2 Transmission Line Structures

Before we discuss the use of a laminate waveguide in LTCC, we first consider typical transmission lines used in mmWave applications: the stripline and the microstrip. A loss in any waveguide or transmission line is a combination of conductor and dielectric loss. For example, in Figure 13.29 we examine the losses per inch of a stripline fabricated in A6M as a function of frequency ranging from near DC to 40 GHz. The substrate thickness is 0.0296 inches and the width of the corresponding trace width is 9 mils for a 50 Ω impedance structure. As can be seen and expected for a low-loss ceramic, the conductor losses dominate for an A6M stripline structure. To avoid multi-moding possibilities at higher frequencies, the stripline would need to decrease from a height of 0.0296 inches to a value much smaller than a wavelength at 95 GHz. Using a commonly distributed and free impedance calculator (TxLine ref: http://web.awrcorp.com/), the loss of the stripline is approximately 4.7 dB/inch for a stripline structure having 11.1 mils of thickness and a 3 mil-wide line at 95 GHz. The 3 mil line is the minimum standard line width, and it is also the constraining dimension determining the substrate height of the stripline. The permittivity was chosen as 6.25, and the loss tangent is 0.003, which is, as we will discuss later, considered more accurate than the nominal permittivity of 6.2 and loss tangent of 0.002.

Another common transmission line that can be used at 95 GHz is the microstrip line shown in Figure 13.30. The microstrip line is a single trace patterned on a conductor backed substrate. A critical factor to making a fair comparison of losses between different transmission lines is to account for all loss mechanisms. Instead of using TxLine, which does not account for surface roughness, we instead use a tool from Eagleware, TLine. For the input to this tool, we assume a dielectric constant of 6.25, a loss tangent of 0.003, and a surface roughness of 0.5 μm, which is a typical surface roughness number for A6M. For a 50 Ω impedance system, the loss for a transmission line with typical dimensions is tabulated in Table 13.2. For the thinnest of substrates, the line loss is greater than that of the thinnest manufacturable stripline in A6M: 6.3 dB/in versus 4.7 dB/in. Of course,

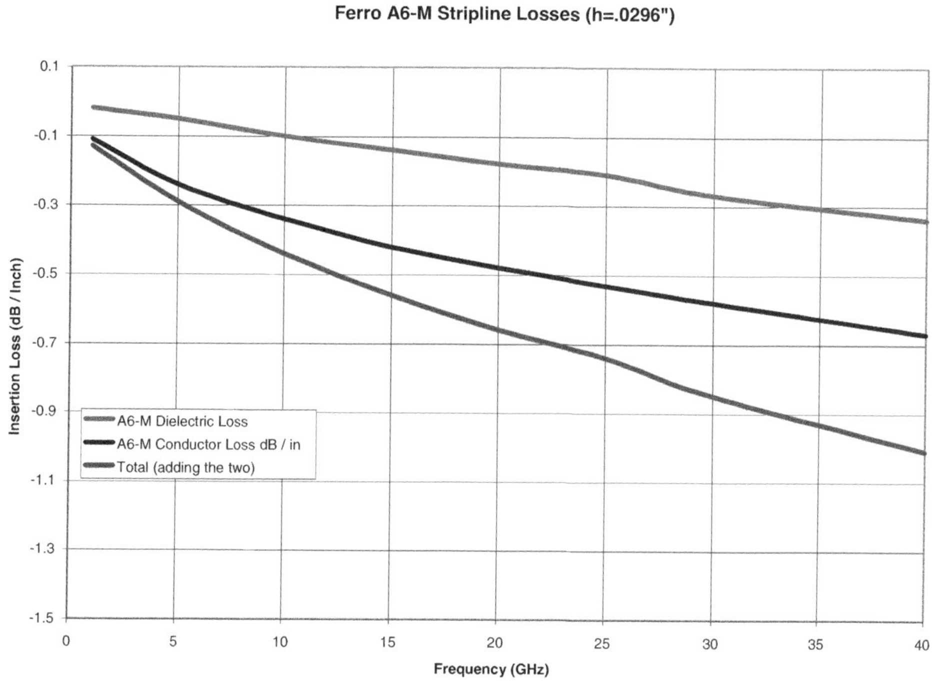

Figure 13.29: The losses per inch of a stripline fabricated in A6M as a function of frequency ranging from near DC to 40 GHz.

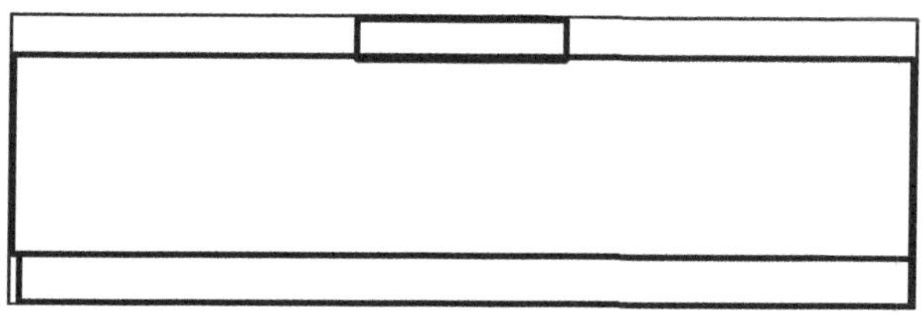

Figure 13.30: The microstrip transmission line: A6M.

the microstrip structure as well as the stripline structure can be dimensioned to have lower losses by increasing the substrate thickness, and hence the signal conductor width, but the risk becomes great in exciting other modes as the thickness for lower losses becomes on the order of a quarter wavelength at 95 GHz.

The waveguide structure of choice at mmWave frequencies is a LWG first described by reference [16]. The main advantage that LWGs have over stripline transmission lines is the low loss at W-band frequency applications. Waveguide structures can be built in LTCC by printing ground planes connected through layers by tightly spaced via fences. As an example, Figure 13.31 shows a T-junction splitter fabricated by two intersecting LWGs in A6M.

Table 13.2: The insertion loss of the microstrip transmission line at 95 GHz.

Z_0	h (mil)	w (mil)	Loss (dB/in)
50	3.7	5.6	6.3
50	7.4	12.1	4.2

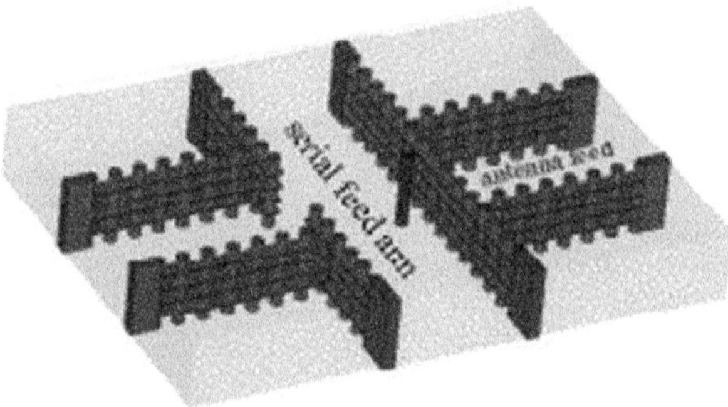

Figure 13.31: T-junction formed with laminate waveguides.

For this particular T-junction, the via fences are connected by ground strips for increased isolation between LWGs. The vias are 3 mils in diameter and the via pitch is 6 mils. A structure modeled to compute the isolation afforded between two well-separated LWGs is shown in Figure 13.32(a). Figure 13.32(b), is a field strength plot across a line perpendicular to the LWG. The isolation is seen to be approximately 30 dB between the inside via fence wall of the LWG and the outside of the via fence wall as shown in Figure 13.33. Hence the isolation from one LWG to an adjacent is of the order of about 60 dB.

The mmWave frequency band is between microwave frequency band and optical frequency band. It is well known that the primary transmission line structures for microwaves are constructed by conductor (metal waveguide) while the primary transmission lines for optic are constructed by dielectric (optical fiber). What is the best transmission line structure for the upper mmWave band (90 GHz and above)? Conductor or dielectric? Or a combination of both types of transmission line structure.

There is almost no reliable material data available in the frequency above 60 GHz range because measurements of the dielectric properties of materials at this frequency band are extremely difficult to carry out accurately. The lack of the accurate dielectric data makes it impossible for the precision engineering in mmWave band applications. We do not know if a material is 'opaque' or 'transparent' at mmWave bands.

We attribute the difficulty in acquiring accurate data to the extrapolated microwave methods, or extrapolated optical methods, which have many serious limitations and uncertainties. The most frequently used instruments in microwave, such as a waveguide interferometer, or a cavity resonator, or a Fabry–Perot open resonator, are based on the frequency resonant principle. These principles are beyond the limit of their classical capabilities as used in millimetre wavelengths. For example, the millimetre wavelengths are too short for the

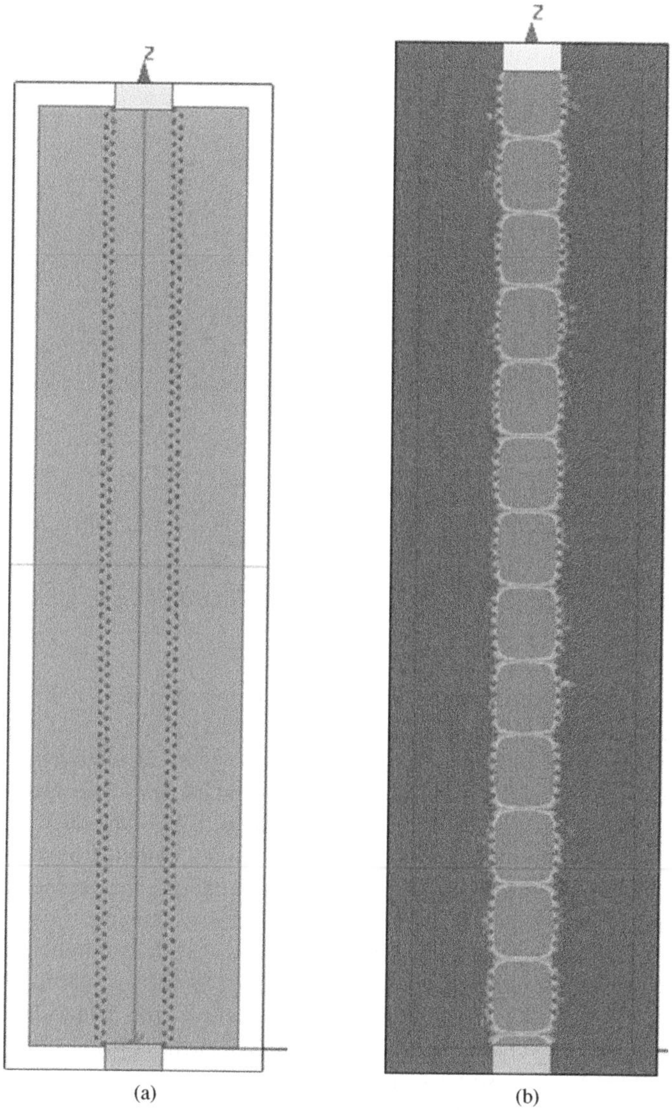

Figure 13.32: The construction of LWG in A6M and its electrical field strength: (a) top view of a LWG using two staggered rows of vias; (b) field strength inside and outside laminate waveguides.

practical use of a microwave single mode resonant cavity. On the other hand, the millimetre wavelengths are too long at the optical spectrum for the familiar blackbody source radiation originating, for example, from mercury-vapor lamps. Normally these types of source provide too little energy for mmWave measurements with a Fourier spectrometer. The use of

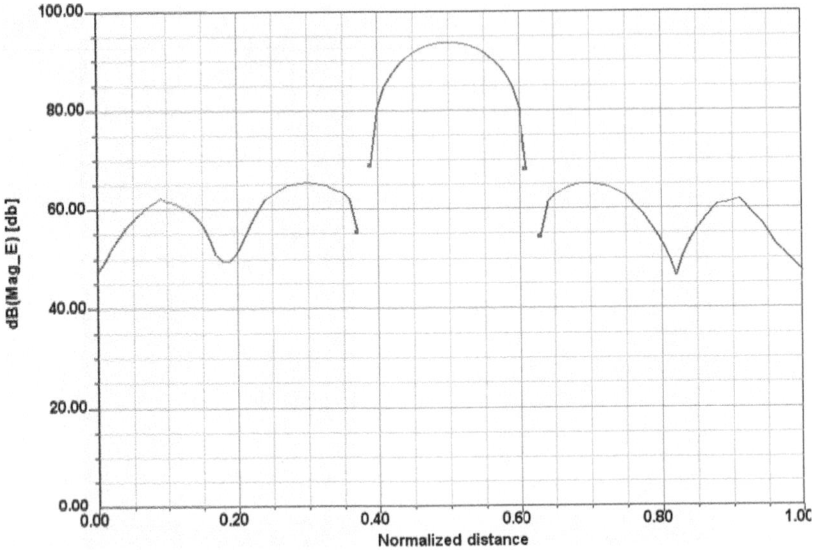

Figure 13.33: The electrical field distribution in a cross section of LWG.

a conventional plane wave interference technique employing a mercury lamp to obtain mmWave dielectric data is almost impossible.

The numbers used for the return loss and insertion loss calculations using impedance calculators was based upon the measurement results and analysis presented in this section. Figure 13.34(a) shows the measured insertion loss for an LWG fabricated in A6M material technology from 90 to 98 GHz. The insertion loss curve computed by full-wave analysis tool shown in Figure 13.34(a) was computed by assuming that the geometry features were precisely manufactured, but that the permittivity and loss tangent values were slightly different at W-band from those of the published values. It is a fair assumption that the nominal number published by a vendor might be slightly different as one considers the same material processed under slightly different conditions. For example, the actual dielectric constant is slightly dependent on various processing variations, such as firing time, and the firing temperature profile. As can also be seen from the Figures 13.34(a) and (b), a value of 6.25 for permittivity and a loss tangent value of 0.003 provides the closest match to return and insertion loss measurements. An antenna built using these parameter values confirmed the performance prediction by simulation using HFSS, and successfully met the specifications at the first time. No more iterations required.

13.4.2 Fabrication

As we have alluded to, the fabrication challenges necessitated by the requirement for small feature variation and control of small feature sizes based on wavelength in the substrate is critical. One critical fabrication challenge is maintaining the relative alignment of tightly spaced vias to slot structures in antenna and antenna feed networks (a typical challenging process limit is a 0.004 inches in diameter via on a 0.010 inches pitch, whereas W-band

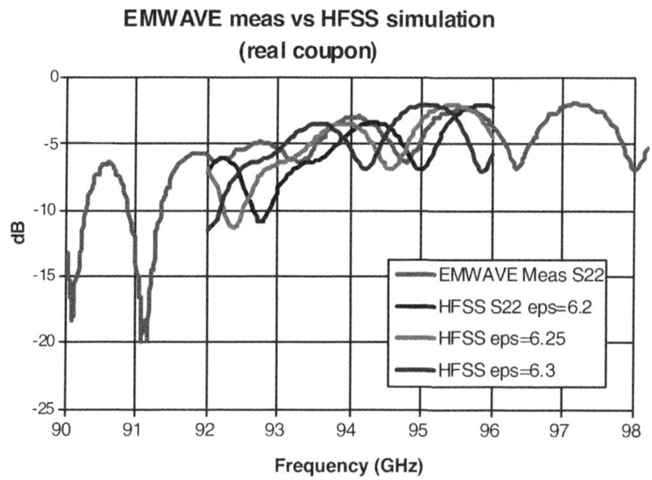

(a)

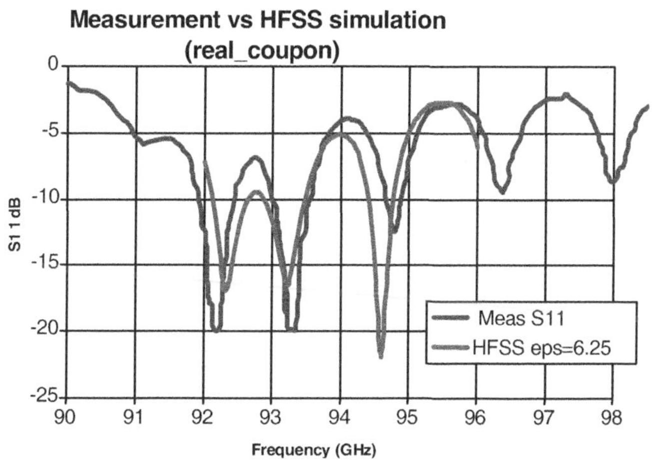

(b)

Figure 13.34: The insertion loss and return loss simulation versus measurement for LWG in A6M at W-band: (a) insertion loss for an LWG in A6M: measurement, and simulation for LWG using three values of permittivity; (b) return loss simulation versus measurement for LWG in A6M for dielectric constant of 6.25.

designs require a 0.003 inches in diameter via on a 0.006 inches pitch). Although the pitch can be loosened based on isolation requirements as well as return loss and insertion loss requirements, the via-fence wall typically requires an isolation between transmission structures of at least 60 dB. This is the requirement that forces the extremely tight via pitch. This represents not only a challenge to punching the vias, but also to via filling [18]. The via filling material for A6M is a gold paste that needs careful control of both the alignment of

Figure 13.35: Tightly spaced (0.006 inch centers) via test.

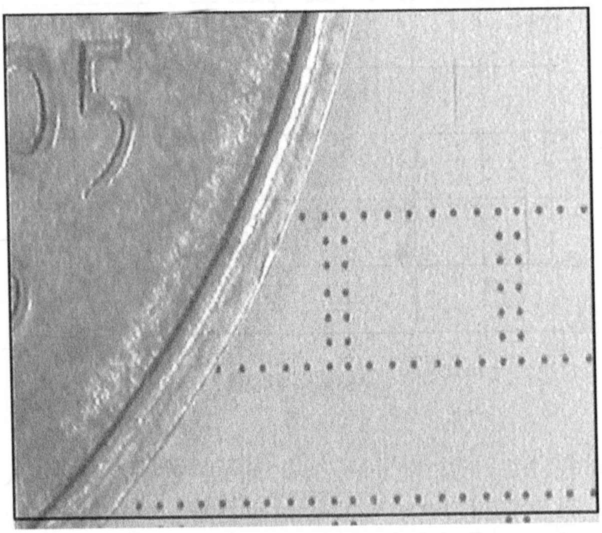

Figure 13.36: Stencil of one layer for showing the via pitch required.

the stencil and ceramic substrate in addition to close monitoring of the viscosity of the paste. Figures 13.35 and 13.36 show an initial test punching result and the actual via stencil (used to fill the vias with metal paste) for a specific layer of the resultant design.

As is clear from the previous discussions, the material and the process selected is critical for design success. Although many materials can have attractive features, any one critical feature required for mmWave applications that is not met automatically eliminates

Figure 13.37: A rectangular transmission line with two cross-section dimension uncertainties at position A and position B.

it from consideration. It is the authors' experience that LTCC affords the best performance solution, especially with regards to miniaturization, although we recognize that there are many considerations in the design space that allow for other material and process selections.

13.4.3 Assembly

13.4.3.1 Tolerance

It is evident that any discontinuities, or the dimension uncertainties in a transmission line will cause a transmission line impedance mismatch. Figure 13.37 shows a rectangular transmission line with two cross-section dimension uncertainties at position A and position B.

For a perfect lossless transmission line, the voltage is

$$V = V_1 e^{\gamma l} + V_2 e^{-\gamma l} \tag{13.135}$$

where V_1 and V_2 are sinusoidal voltage waves, γ is the propagation constant and l is the position along the line [19]. If a signal is applied to the transmission line which is terminated with a load impedance Z_L not equal to the characteristic impedance Z_0, where Z_0 is given as

$$Z_0 = \sqrt{\frac{R + j\omega L}{G + j\omega C}} \tag{13.136}$$

then the load impedance, Z_L, will not be able to absorb all the energy. Part of the energy will be reflected from this mismatch in impedance. R, L, G and C in (13.136) are unit resistance, unit inductance, unit conductance and unit capacitance of the uniform lossless transmission line, respectively. When the transmission line dimensions change as indicated in Figure 13.37, the associated R, L, G and C will also change for that small section; ultimately the transmission line characteristic impedance changes. These changes in characteristic impedances lead to reflected waves in the transmission line. The larger these impedance changes, the larger the impedance mismatch, and hence the larger reflection the signal will undergo as it propagates along the transmission line. Therefore, the control of the tolerance is absolutely essential for the design and fabrication of the phased array antenna developments in mmWave frequency.

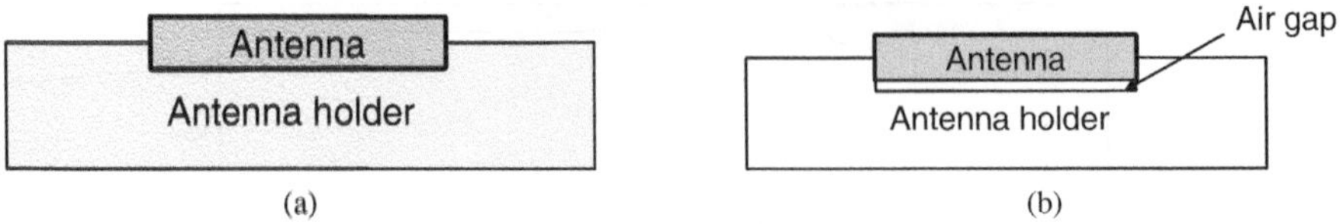

Figure 13.38: A planar array antenna held by a fixture with perfect holding and defect holding: (a) planar array antenna on the antenna holder; (b) planar array antenna on the antenna holder with a air gap.

Over the years, efforts to establish, modify and improve the tolerance standards have been made involving both industry and research institutions. These efforts led to the creation of the tolerance standards for design and development of the components and systems in the mmWave band. The widely adopted tolerance standard for W-band is ranged from one-tenth mil to one mil [20,21]. These standards impose a very difficult manufacturing task.

This is difficult because

- the CNC machine usually has only a 0.5 mil accuracy after being well calibrated and using good machining procedures;
- the possibility of bad programming in the machining programming can easily introduce a 5 mil difference;
- any manual alignment could introduce a 4 mil tolerance.

There appears to be a need to explore manufacturing techniques having greater precision in addition to the ability to hold material properties to closer specifications.

13.4.3.2 Fixture

Another important aspect of the design and manufacturing issues for mmWave phased array antenna that deserves mention is the test fixture. The fixture is a necessary part or subsystem of the phased array antenna in mmWave length. We use a W-band example shown in Figure 13.38 to explain.

Figure 13.38(a) shows the concept for how the test fixture operates. A planar array antenna sits in the fixture so that the back of the antenna has good electric contact with the antenna holder. However, from direct experience, we have found that the return loss in a test antenna is sensitive to the flatness of the holder surface. In the W-band antenna to which we refer, measurements had several frequency points with return losses that were out of specification. A review of all antenna test parts and test procedures did not reveal an obvious problem.

By going back to simulation studies using full-wave solvers, we were able to identify the source of the problem. In our simulation models we inserted an air gap between the antenna and the fixture as shown in Figure 13.38(b) and reran the simulations. As the thickness of the air gap approached 3 mils, an interesting phenomenon was observed: two peaks in return loss appeared at around 94 GHz and 96 GHz (Figure 13.39(b)). We compared this simulation result with the expected benchmark return loss of Figure 13.39(a), simulated earlier, and concluded that these two peaks suggested the existence of an air gap between the antenna

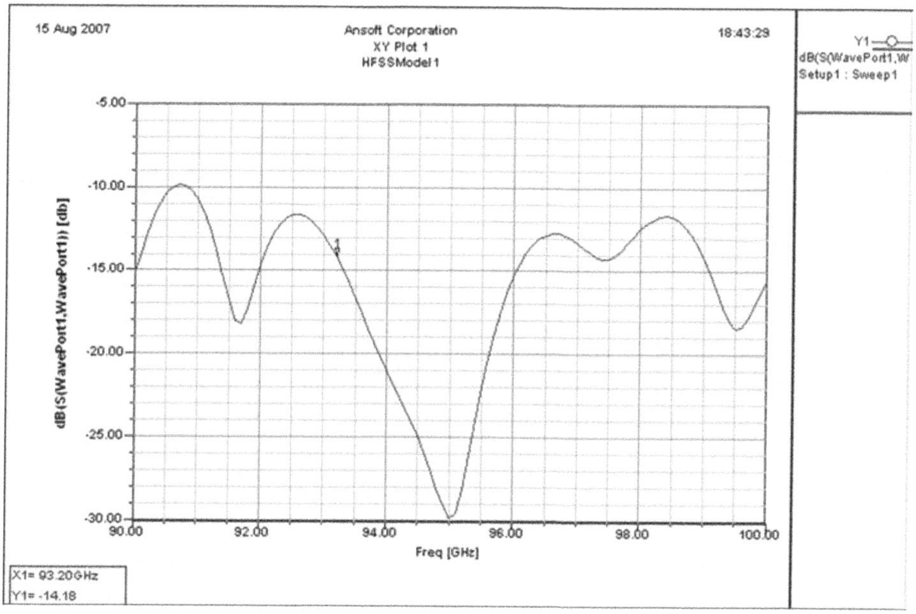

(a)

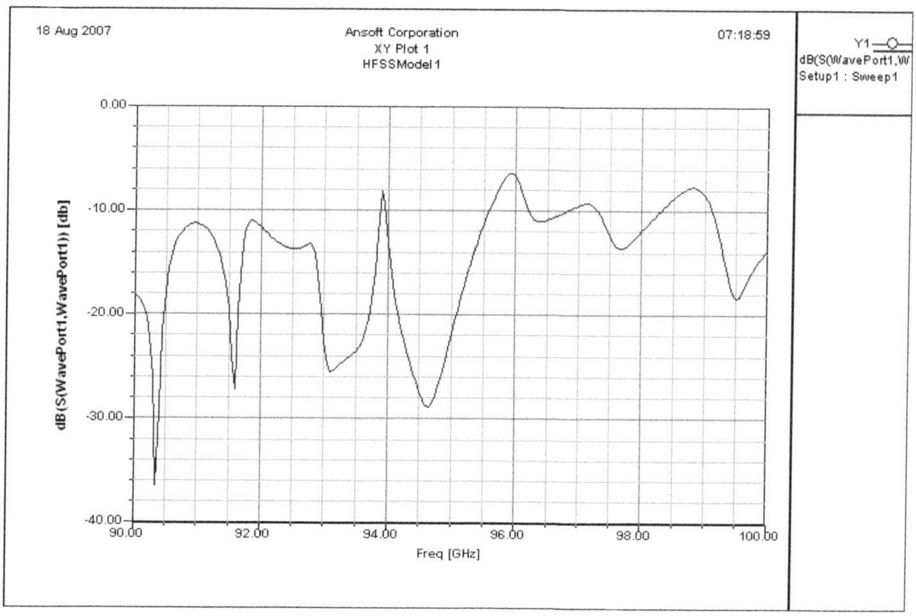

(b)

Figure 13.39: The return loss of a planar array antenna held by a fixture with perfect holding and defect holding: (a) expected benchmark return loss; (b) return loss with a 3 mil air gap between the antenna and fixture.

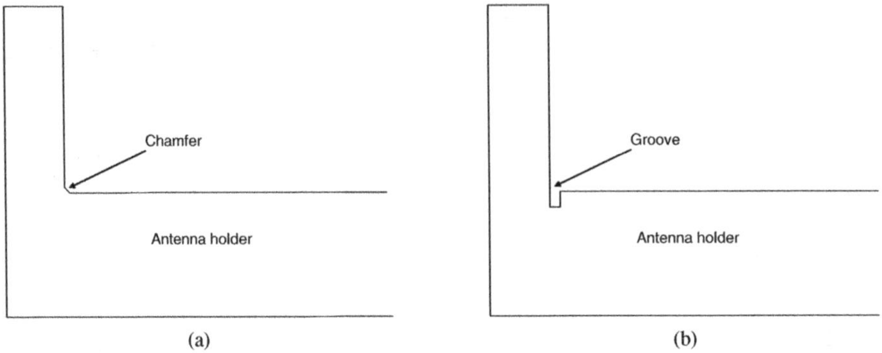

Figure 13.40: The machined antenna holders before and after the improvement: (a) antenna holder with a chamfer at the bottom corner; (b) antenna holder with a groove at the bottom corner.

and the test fixture. After careful inspection, we found that the antenna holder had about a 3 mil chamfer at the corner of the antenna holder as shown in Figure 13.40(a). It is this chamfer that prevented the antenna from being flat inside the fixture. The resulting air gap formed a resonant cavity, which caused the antenna performance to be out of specification.

Although we identified the problem, it remained a difficult one to solve. A milling machine will not remove a 3 mil chamfer without damaging the fixture.

Since the antenna fixture was fabricated from aluminum, we took advantage of the relative softness of this metal. We used a thin wood dowel to scribe the corners while intermittently connecting the test part to a network analyzer to monitor the return loss. Through this iterative method, the resonant problem was fixed.

The lesson we learned from this example revealed that the machining process will not build a sufficiently perpendicular wall at the corner of the fixture because of the inherent tool limitations. We therefore propose another fixture. In the bottom corner of the holder surface we propose cutting a groove along the fixture wall as shown in Figure 13.40(b). This fixture configuration will eliminate the chamfer at the bottom corner and thus remove the resonances at a certain frequency associated with any air gaps between the antenna and fixture resulting from this chamfer.

Acknowledgments

One of the authors would like to express his deepest gratitude to a special colleague and personal friend, Dr. Joe Anderson. Having worked with Dr. Anderson in what was once the Hughes Missile System and now part of Raytheon, this author recognizes him as one of the best phased array antenna designers in the world. Dr. Anderson's expertise in phased array antenna design is observed via this author's professional interaction with him over the past 15 years as well as casual conversation regarding this topic. His thoughts and ideas have permeated this chapter, as he has greatly affected this author's beliefs and knowledge in this field. This author also thanks Dr. James Candy, Chief Scientist of the Lawrence Livermore National Laboratory, for his helpful comments and suggestions in the preparation of this manuscript of the beam-forming network section.

References

[1] J. A. Stratton and L. J. Chu, 'Steady-state solutions of electromagnetic field problems, I. Forced oscillations of a cylindrical conductor', *Journal of Applied Physics* **12**(3) (1941), pp. 230–235.

[2] R. S. Elliott, *Antenna Theory and Design* (Englewood Cliffs, Prentice-Hall, 1981).

[3] P. W. Hannan, 'The element-gain paradox for a phased-array antenna', *IEEE Transactions on Antennas and Propagation* **12**(4) (1964), pp. 423–433.

[4] H. A. Wheeler, 'The radiation resistance of an antenna in an infinite array or waveguide', *Proceedings of the IRE* 36, pp. 478–487, April 1948.

[5] P. W. Hannan and M. A. Balfour, 'Simulation of a phased-array antenna in waveguide', *IEEE Antennas and Propagation* **13** (1965), pp. 342–353.

[6] L. C. Shen and J. A. Kong, *Applied Electromagnetism* (Monterey, Brooks/Cole Engineering Division, 1983).

[7] R. F Harrington, *Time-Harmonic Electromagnetic Fields* (New York, John Wiley & Sons Ltd/Inc., 2001).

[8] A. A. Oliner and R. G. Malech, 'Mutual coupling in infinite scanning arrays', in *Microwave Scanning Antennas (Array Theory & Practice)*, Ed. by R. C. Hansen, vol. 2, Ch. 3 (New York, Academic Press, 1966).

[9] S. Edelberg and A. A. Oliner, 'Mutual coupling effect in large antenna arrays: Part I – Slot arrays', *IRE Transactions on Antennas and Propagation* **8**(5) (1960), pp. 286–297.

[10] R. L. Eisenhart and P. K. Park, 'Phased array scanning performance simulation', *IEEE Antennas and Propagation Society International Symposium*, Newport Beach, CA, pp. 2002–2005, June 1995.

[11] H. Holter, H. Steyskal, 'On the size requirement for finite phase-array models', *IEEE Transactions on Antennas and Propagation* **50**(6) (2002), pp. 836–840.

[12] B. D. Van Veen and K. M. Buckley, 'Beamforming: A versatile approach to spatial filtering', *IEEE ASSP Magazine* **5**(2) (1988), pp. 4–24.

[13] H. Krim and M. Viberg, 'Two decades of array signal processing research', *IEEE Signal Processing Magazine* **13**(4) (1996), pp. 67–94.

[14] C. L. Dolph, 'A current distribution for broadside arrays which optimizes the relationship between beamwidth and side lobe level', *Proceedings of the IRE*, vol. 34, pp. 335–348, June, 1946.

[15] M. Abramowitz and I. A. Stegun, *Handbook of Mathematical Functions* (New York, Dover Publications, 1965).

[16] H. Uchimara, T. Takenoshita and M. Fujii, 'Development of a 'laminated waveguide', *IEEE Transactions Microwave Theory Tech.* **46**(12) (1998), pp. 2438–2443.

[17] F. Xu and K. Wu, 'Guide-wave and leakage characteristics of substrate integrated waveguide', *IEEE Transactions Microwave Theory Tech.* **53**(1) (2005), pp. 66–73.

[18] P. Garland, H.-Y. Pao, H.-S. Lin, J. Aguirre, D. O'Neill and K. Horton, 'Manufacturing challenges for a W-band laminated waveguide phased array', *2008 IEEE Antennas and Propagation Society International Symposium*, San Diego, CA, July 2008.

[19] S. F. Adam, *Microwave Theory and Applications* (Englewood Cliffs, Prentice-Hall, 1969).

[20] H. Eskelinen and P. Eskelinen, *Microwave Component Mechanics* (Boston, MA, Artech House, 2003).

[21] J. L. Hesler, A. R. Kerr, W. Grammer and E. Wollack, 'Recommendations for waveguide interfaces to 1 THz', National Radio Astronomy Observatory, Electronics Division, Internal Report no. 319, July 26, 2007.

14

Integrated Phased Arrays

Sanggeun Jeon, Aydin Babakhani and Ali Hajimiri

14.1 Introduction

Phased arrays have been widely employed in radar and communication systems in the area of military, space, and radio astronomy since their advent in the 1950s [1–3]. Basically, phased arrays consist of multiple antennas followed by phase control units, which enables multiple receive/transmit signals to be combined coherently at a certain direction(s). By steering a beam direction electronically, phased arrays provide several benefits including high directivity, high sensitivity, fast beam scanning, multiple beamforming, and spatial filtering. In addition, the parallel configuration of array elements relieves the requirements of power handling and noise characteristics for each individual array component. This also makes the entire system performance less sensitive to the failure of individual components. Recently, due to these advantages, there is also substantial interest in civil applications of phased arrays at microwave and millimeter-wave (mmWave) frequencies. These include point-to-point high-speed communications at 24.0–24.25 GHz and 57–64 GHz, and automotive radars at 23.12–29.0 GHz and 76–77 GHz, which have been recently approved by the Federal Communications Commission (FCC) [4].

Traditionally, phased arrays have been built using module-based approaches [2, 5, 6]. Most array components including antennas, filters, front-end active blocks, phase shifters, and LO sources are implemented as separate modules and then interconnected by external cables, waveguides, or microstrip lines. Particularly, active RF circuit blocks are implemented usually in compound semiconductors such as gallium arsenide (GaAs). Although these approaches can be and have been employed for various applications at different frequencies, they are not efficient nor reliable for phased arrays operating at mmWave frequencies for several reasons. First, the interconnection between array components is very challenging and costly at mmWave frequencies. This is because the interconnection performance such as

Advanced Millimeter-wave Technologies: Antennas, Packaging and Circuits
Duixian Liu, Ulrich Pfeiffer, Janusz Grzyb and Brian Gaucher © 2009 John Wiley & Sons, Ltd

loss and phase distortion at those high frequencies is significantly sensitive to the material and geometric characteristics of interconnection, which demands special handling and high cost accordingly. The challenging interconnection issue will be exacerbated as one wants to combine a larger number of array elements in order to take more advantage of phased-array operation. This leads to a rapid increase in interconnection complexity, resulting not only in higher cost but also in less reliable operation of the array system. The second reason is that monolithic microwave integrated circuits (MMICs) based on compound semiconductors are expensive and take a considerable portion of the total array system cost, for instance, about 25% in an X-band radar [7]. The MMIC cost is further raised for mmWave circuits owing to a need for substrate materials with a higher electron mobility, which thereby limits the maximum achievable number of array elements depending on the system budget.

Integrated phased arrays in silicon offer a promising and unique solution to address the issues of conventional module-based approaches at mmWave frequencies [8–14]. The capability to integrate a complete system including antennas, RF front-ends, and digital circuitry on a single silicon chip opens a new opportunity for achieving unprecedented system performance with substantial reduction in cost, size, and complexity. The driving force enabling a complete mmWave phased array to be fully integrated on a silicon chip is given as follows.

First, the advancement of silicon processing technology provides faster transistors and higher-Q passive components e.g. inductors, capacitors, transmission lines), which allow for the design of high-performance circuits at mmWave frequencies. Continuous transistor scaling keeps pushing its operating frequency higher and higher, taking over the area where compound semiconductors have been used historically. For instance, a unity current-gain cutoff frequency (f_T) is reported as high as 485 GHz in 45-nm silicon on insulator (SOI) complementary metal–oxide–semiconductor (CMOS) technology [15]. Furthermore, the high yield and repeatability of silicon integrated circuits (ICs) offers dramatic improvement in cost, size, and reliability of such high-frequency systems, compared with compound-semiconductor counterparts.

Second, the ability to add a virtually unlimited number of transistors on a silicon chip with little incremental cost maximizes the system integration level and flexibility, which makes it possible to implement several innovative architectures. Particularly, even digital-signal processing (DSP) units for baseband processing and/or beamforming can be integrated together with RF and analog circuit blocks on the same die with little cost increase. This integration eliminates the need to employ and interconnect off-chip DSP components, leading to additional reduction in cost and power consumption for phased-array systems.

Finally, the effective wavelength at mmWave frequencies on silicon becomes small in the order of a millimeter. For instance, the wavelength at 77 GHz becomes 3.9 mm and 1.1 mm in free space ($\varepsilon_r = 1$) and silicon substrate ($\varepsilon_r = 11.7$), respectively. This short wavelength enables antennas to be integrated on-chip in silicon for phased arrays if special attention is paid to remove undesirable antenna modes which will be discussed in Section 14.2.4 [10, 13]. The on-chip antennas eliminate, otherwise, the demanding interconnection between off-chip antennas and RF components at mmWave frequencies. Hence, the system reliability and repeatability is improved significantly, while the system cost and complexity is reduced accordingly.

Integrated phased arrays in silicon reforms existing applications of phased arrays in a revolutionized way that would not be possible by traditional module-based approaches in compound semiconductors. In particular, military and space applications such as high-resolution radars and tracking systems can be highly benefited by the integrated solution in silicon. The unmatched level of integration and the corresponding reduction in cost, size, and complexity of silicon integrated systems make it possible to realize very large-scale phased arrays with millions of elements for those strategic applications in a cost-effective manner [12]. On the other hand, several new applications of integrated phased arrays are recently emerging at microwave and mmWave frequencies. One application is high-speed wireless communications including giga-bit point-to-point communications and wireless local area networks (LANs). The wide bandwidth available in those high frequencies and the improvement of sensitivity and directivity in phased-array systems allow for communications with high data throughput and quality of service [8, 9]. Another commercial application of integrated phased arrays is the ranging and sensing systems such as automotive radars, short-range surveillance systems, and microwave/mmWave imaging systems [10, 11, 14]. Particularly, automotive radars, now actively developed in industry at 23.12–29.0 GHz and 76–77 GHz, provide features such as autonomous cruise control, collision avoidance, low-visibility driving aid, early warning and brake priming, self-parking, and global traffic control.

The integration ability of a complete system including on-chip antennas, RF front-ends, and digital control units in silicon generalizes the concept of phased arrays and introduces a new modulation technique based on multiple antenna elements, called direct antenna modulation (DAM) [13]. In DAM, by combining multiple radiating waves with different modes, desired data is transmitted in a direction-dependent fashion, which is similar to phased arrays. The radiated far-field signal is modulated by time-varying changes in the antenna near-field electromagnetic (EM) boundary conditions. However, unlike phased arrays, DAM-based transmit systems radiate desired data only in a single narrow beam without leaking the data at sidelobes, thereby enabling a secure communication link. Moreover, the system is capable of transmitting independent data concurrently in multiple directions at a full rate by using a single transmitter, which is practically impossible by conventional radio architectures.

This chapter is organized as follows. In Section 14.2, an overview of phased arrays is given, followed by the design and implementation issues associated with integrated phased arrays in silicon. In Section 14.3, a fully-integrated mmWave phased-array transceiver is presented with detailed discussion on the system architecture, building blocks, and experimental results. The concept of a novel modulation technique using multiple antennas, i.e. DAM, is introduced and demonstrated in an mmWave integrated transmitter in Section 14.4. Finally, Section 14.5 presents an integrated solution to implement very large-scale phased arrays (with millions of array elements) with low cost, small size, and high reliability.

14.2 Integrated Phased Arrays

A phased array imitates a directional antenna whose bearing is controlled electronically. To achieve the electronic beamsteering, it combines radiating or incoming waves spatially using multiple antenna elements followed by RF front-ends, beamforming networks, and digital control units. In the past, such phased-array systems have been implemented using a large number of microwave or mmWave modules, increasing their cost and manufacturing complexity [2, 5, 6]. However, the continuous transistor scaling and the high yield and repeatability in silicon technology make it possible to fully integrate multi-antenna phased-array systems in a single chip [14, 16, 17]. This section briefly reviews the principle and associated benefits of phased arrays. Then, the challenges involved with silicon integration of phased arrays including on-chip antennas are discussed. Based on the design challenges and constraints in silicon, several architectures are discussed and compared for implementing high-performance integrated phased arrays.

14.2.1 Principles of Phased Arrays

As illustrated in Figure 14.1, phased arrays consist of multiple antenna elements spaced with a certain distance (d), followed by time-delay elements for electronic beamforming at a given incident angle (θ) in space [2,6]. The operation principle of phased arrays is similar for both receivers and transmitters. In a phased-array receiver, the radiated signal arrives at each of the spatially separated antenna elements at different times. The arrival time difference between two adjacent elements is given by

$$\Delta t = \frac{d \sin \theta}{c} \tag{14.1}$$

where c is the speed of light. As shown in Figure 14.1(a), an ideal phased-array receiver compensates for the different time delays from multiple antennas in such a way that the resultant output signals from multiple elements are combined all in phase to enhance the reception from the desired direction(s).

On the other hand, incoming waves at other directions do not add up coherently and thus are significantly attenuated at the array output. Similarly, in a phased-array transmitter (Figure 14.1(b)), the signals in different elements are delayed by different amounts so that the radiated signals from multiple antenna elements add up coherently in space only at the desired direction(s). Again, incoherent addition of the signals in other directions results in lower radiated power in those directions.

With ideal time-delay elements used after or before each antenna, phased arrays can achieve the ideal beamforming regardless of the frequency and bandwidth of the signal. However, there are practical challenges to implement such broadband time-delay elements at microwave and mmWave frequencies, such as signal attenuation, noise and linearity degradation, as well as signal dispersion. Fortunately, in many practical applications, particularly in wireless communications, the bandwidth of interest is a small fraction of the center frequency. If the bandwidth is sufficiently smaller than the center frequency, the time delay can be effectively approximated by a constant phase shift at the center frequency. With this narrow-band approximation, the arrival time delay given in Equation (14.1) is translated to a constant phase difference between two adjacent elements

$$\Delta \varphi = \frac{2\pi d \sin \theta}{\lambda} \tag{14.2}$$

where λ is the wavelength at the center frequency. Thus, the coherent signal generation from each element in phased arrays can be achieved by a phase-shifting component that is relatively easier to implement than a time-delay component at microwave and mmWave frequencies. On the other hand, in wideband systems with very large instantaneous bandwidths, the constant-phase approximation does not work appropriately, resulting in non-constant group delay and non-negligible signal dispersion. This gives rise to an increased bit

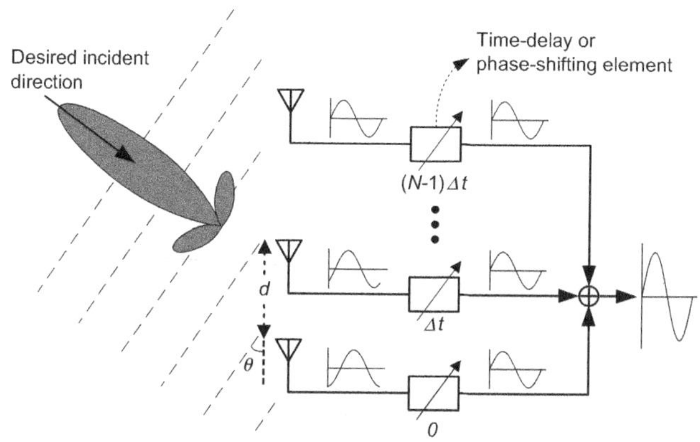

(a)

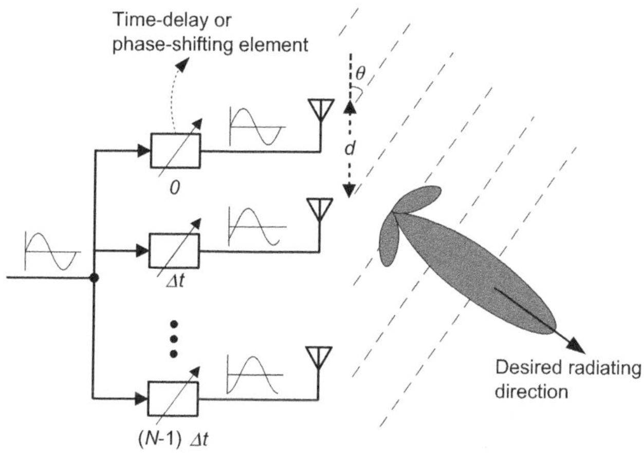

(b)

Figure 14.1: Basic configuration of phased arrays: (a) receiver; (b) transmitter.

error rate (BER) in wireless communications and beam squinting in radar applications [16]. To avoid the error for such systems, true time-delay components must be employed, where, for instance, the lengths of transmission lines are adjusted to generate coherent signals from each element [2].

14.2.2 Benefits of Phased Arrays

Due to the electronic beamforming capability based on multiple antennas, phased arrays have several benefits compared with single-directive-antenna systems driven mechanically [2,3,6, 16,17].

Array Gain: In phased-array receivers, several in-phase signals are combined coherently at the array output, resulting in an effectively higher gain than a single-element receiver. When multiple element signals are combined coherently in the amplitude domain (current or voltage) with a same output load, the array gain is given by

$$G_{\text{array}} = G_{\text{single}} + 20\log_{10}(N)\ (\text{dB}) \tag{14.3}$$

where G_{single} is the gain of each single element and N is the number of array elements. Similarly, in a phased-array transmitter with N elements, if each element radiates P watts isotropically, the effective isotropic radiated power (EIRP) in the desired direction is N^2P W. This means that the total radiated power that will be seen at the target receiver increases by a factor of N^2. The improvement comes from the coherent addition of the electromagnetic fields in the desired direction. For example, the total power radiated by a four-element array is 12 dB [$20\log_{10}(4)$] higher than the power radiated by each element. This improvement in signal power at the target receiver is particularly useful for integrated phased arrays at high frequencies, where the efficiency and output power of silicon-based power amplifiers are low, path loss is high, and the receiver sensitivity is low.

Interference Rejection: While the signals at the desired direction are combined coherently in phased arrays, undesired signals at other directions do not add up coherently and thus are attenuated. This spatial filtering property of phased arrays enables the receiver to filter out interference signals even at the same RF frequency as the desired one as long as they have different directions of arrival at receive antennas. The interference signals may originate from intentional sources such as jammers in military radar applications or from unintentional different access points that reuse the same channel frequency in wireless communications. In either case, the improvement of the signal-to-interference ratio (SIR) alleviates the nonlinearity requirement of receiver blocks and enhance the system reliability and capacity. It is noteworthy that the interference rejection at the same channel frequency cannot be easily achieved by any physical filters. In principle, $(N-1)$ different interference signals can be simultaneously and perfectly rejected with N antenna elements by the spatial filtering property. The passband of the spatial filtering also becomes narrower as N increases. As shown in Figure 14.2, as N increases from 4 to 32, the beamwidth between first nulls (BWFN) decreases from 60° to 7.2° with the number of null positions increased from 3 to 31.

By reciprocity, in phased-array transmitters, less radiated power is generated at different directions from the desired beam. This delivers fewer interference signals to receivers that are not targeted, thereby reducing the nonlinearity requirement of the receivers.

Sensitivity Improvement: Phased arrays improve a receiver sensitivity, which allows for channel capacity increase for a given bandwidth. This enhances the data throughput of communication receiver systems without degrading the BER. The sensitivity is determined by the noise figure (NF) which, in turn, is dependent on the output signal-to-noise ratio (SNR) for a given input noise. NF is defined as the ratio of the total output noise power to the output noise power caused only by the source [18]. Consider an n-path phased-array receiver shown in Figure 14.3. Since the input desired signals (S_{in}) add up coherently, the combined output is given by

$$S_{\text{out}} = n^2 G_1 G_2 S_{\text{in}} \tag{14.4}$$

where n is the number of elements and G_1 and G_2 correspond to the gains before and after the signal combining. The antenna's noise temperature is primarily determined by the temperature of the object(s) it is pointed at. In general, the amount of SNR improvement in a phased-array receiver depends on the nature and location of the objects in the environment

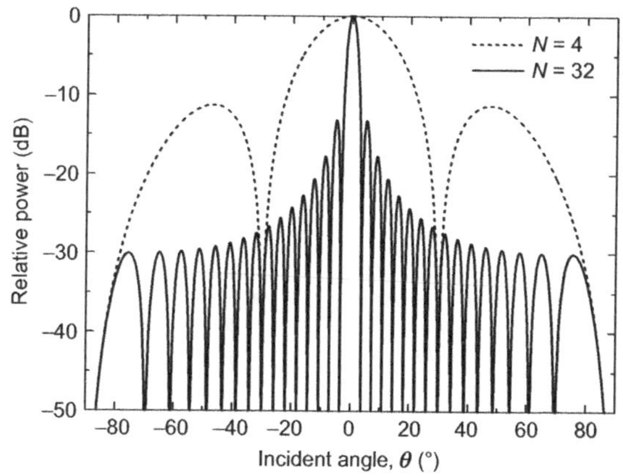

Figure 14.2: Theoretical array patterns with different array sizes: $N = 4$ (dashed) and $N = 32$ (solid). A uniformly excited and equally spaced linear array with a half wavelength spacing is assumed.

that generate noise, correlation between such noise generators, multi-path effects, coupling between antenna elements, input impedance mismatch, angle of incidence, and the antenna beam pattern. Assuming that the antenna noise contributions in different elements are uncorrelated, the output total noise power is given by

$$N_{out} = n(N_{in} + N_1)G_1G_2 + N_2G_2 \tag{14.5}$$

where N_1 and N_2 are the input-referred noise contributions of the stages corresponding to gains G_1 and G_2, and N_{in} is the noise at the input of each antenna. Thus, compared to the output SNR of a single-path receiver, the output SNR of the array can be improved by up to a factor of n depending on the noise and gain contribution of different stages. The array noise factor can be expressed as

$$F = \frac{n(N_{in} + N_1)G_1G_2 + N_2G_2}{nN_{in}G_1G_2} \tag{14.6}$$

$$= n\frac{\mathrm{SNR_{in}}}{\mathrm{SNR_{out}}} \tag{14.7}$$

which shows that the SNR at the phased-array output can be even smaller than the SNR at the input if $n > F$. For a given NF, an n-element receiver can improve the sensitivity by $10\log_{10}(n)$ in decibels compared to a single-path receiver. For instance, if noise from the antennas is uncorrelated, an eight-path phased array can improve the receiver sensitivity by 9 dB.

Fast Multi-beam Scanning: Phased arrays are able to not only form a beam at the desired direction, but also steer it electronically through the dynamic control of each element. Hence, fast beam scanning at arbitrary angles can be achieved relatively easily without the mechanical steering of antenna directions which takes substantial time and power. For this reason, phased arrays have been widely used for radars in military applications which often operate in harsh environment such as high-gravity situations. The electronic beam control

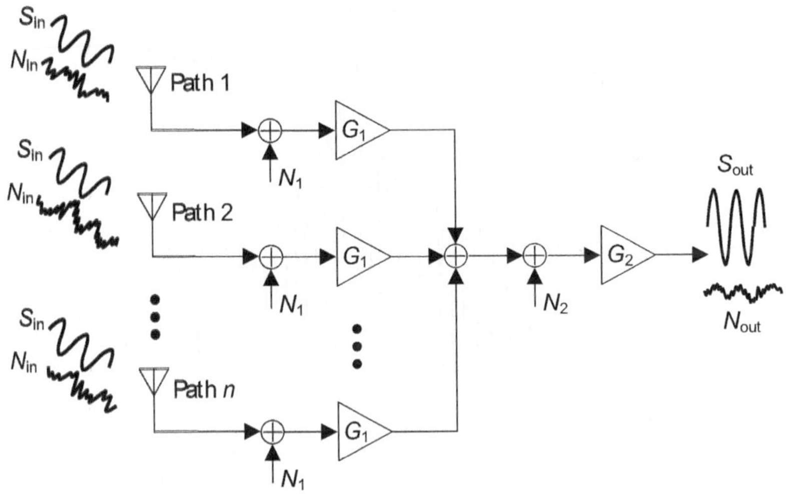

Figure 14.3: Output SNR improvement in phased-array receivers.

also allows multiple beams to be formed at different directions at the same time by sharing the same set of antennas. The multiple beamforming makes it possible to track multiple targets simultaneously in radars and to communicate with multiple points concurrently in wireless communications.

14.2.3 Silicon Integration Challenges

Integration of phased arrays in silicon provides further advantages such as low cost, low complexity, and high reliability, while taking full advantage of the benefits of phased arrays themselves. Nonetheless, there are several challenges to integrating mmWave phased-array systems on a silicon substrate, which should be addressed appropriately [10, 11, 14].

Compared to high resistivities of compound semiconductors, silicon integrated circuits are implemented on a conductive substrate with a typical substrate resistivity lower than 10 $\Omega \cdot$ cm. This high conductivity causes energy loss due to magnetically induced eddy currents. Accordingly, on-chip passive devices such as inductors, capacitors, and transmission lines suffer from high loss and low Q-factor, resulting in degradation of the system gain and noise performance. Antennas, if implemented on chip, can also couple energy into the substrate, leading to substantial additional loss, as is discussed in Section 14.2.4.

The finite conductivity of metal structures causes further energy loss in the integrated system. Since the skin depth becomes very small at mmWave frequencies (e.g. the skin depth of copper at 60 GHz is approximately 300 nm), the ohmic loss in metal structures significantly increases, degrading the performance of passive devices further. This is particularly problematic in chemical-mechanical-polishing (CMP) processes because the metal surface becomes rougher on which most of the current flows. The additional 'fill' and 'cheese' rules for wide metal strips also degrade the current handling capabilities of the metal lines.

The high dielectric constant of the silicon substrate ($\epsilon_r = 11.7$) poses another challenge to mmWave integrated systems. The rectangular silicon substrate forms a dielectric waveguide that can sustain propagating modes depending on its physical dimensions. Unfortunately, many of the natural modes for a typical silicon die size fall right in the middle of the mmWave bands of interest. This in turn creates an alternative mechanism for energy loss in on-chip components. Energy could easily leak into the substrate modes and be dissipated by ohmic loss within the substrate, or even worse, be radiated in undesirable directions [10, 14].

The coupling and crosstalk among different on-chip components can be easily generated through a silicon substrate. The coupling is parasitic and unavoidable, but may be strong enough to be detrimental to the system performance. For instance, a strong RF signal generated by an on-chip power amplifier may pull the frequency of a voltage-controlled oscillator (VCO) implemented on the same chip, or may give rise to a spurious oscillation in other unrelated on-chip blocks. The crosstalk between RF and digital blocks on the same chip may also be critical to perform proper digital functionalities. Hence, special care must be paid to the overall design to avoid such undesirable coupling and crosstalk. Fully differential design is also helpful to improve immunity to the crosstalk or noise.

The relatively low breakdown voltage of transistors in silicon presents a serious challenge to power generation in integrated systems. Current technology trends such as the scaling process and the shrinking of transistor depletion regions lowers the breakdown voltage further, which demands the use of a lower supply voltage. The low supply voltage, therefore, makes it very challenging to implement power amplifiers in silicon that can generate a sufficient power level for transmitters. Rather than implementing a single amplifying block requiring a high voltage swing, it is more promising to use a parallel structure of multiple amplifying elements with a smaller swing and combine the output power. This approach necessitates novel techniques for efficient power combining as described in references [19, 20].

Finally, the accurate modeling of transistors and passive devices is more challenging as the operating frequency increases. Even very small value of parasitic components has a substantial effect on system performance at mmWave frequencies. For instance, an inductance of 100 pH translates to 48 Ω at 77 GHz. Hence, special attention must be paid to model the devices including all parasitics precisely. The modeling becomes even more challenging in the presence of process variations and environmental changes.

14.2.4 Integrated Antennas in Silicon

On-chip antennas integrated with RF and digital circuit blocks on the same die enable a true one-chip radio system providing several benefits in a system perspective. By eliminating the challenging interface between an RF front-end chip and an off-chip antenna, the system cost and complexity is significantly reduced while the reliability increases due to the excellent repeatability of silicon processes. Also, unavoidable mismatch between different chip-antenna interfaces can be minimized by the on-chip antenna implementation, which is very important in large arrays. Note that these benefits are more obvious for mmWave systems because the chip-antenna interface present more difficulties and mismatches at mmWave frequencies. One promising fact in on-chip antennas is that the physical dimension of antennas is inversely proportional to the operating frequency and thus becomes small enough to be implemented on chip at mmWave frequencies. On the other hand, the integration

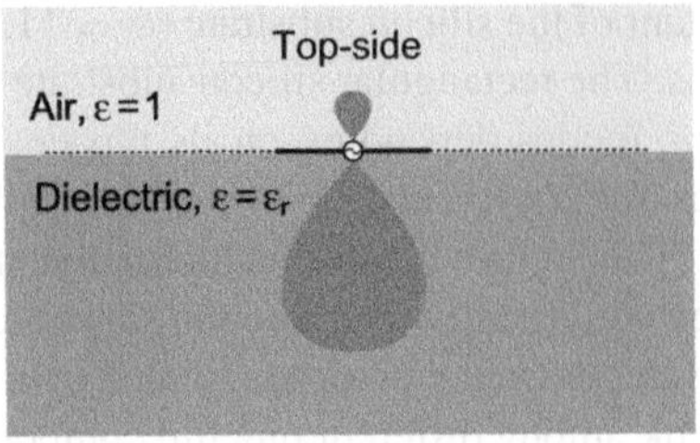

Figure 14.4: Dipole antenna at the interface of air and a semi-infinite dielectric substrate.

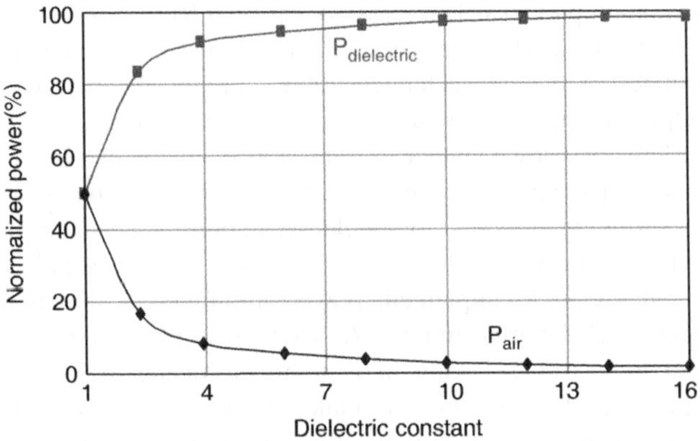

Figure 14.5: Ratio of the radiated power into air and a dielectric substrate as a function of the dielectric constant.

of antennas on a silicon chip has several technical challenges. Due to the high dielectric constant and conductivity of the silicon substrate, on-chip antennas with a reasonable gain and efficiency are not easy to implement in silicon, as discussed in the following [10].

The seemingly simplest way to implement on-chip antennas in silicon is to put metal lines on top of a substrate. Let us consider a dipole antenna placed at the interface of air and a semi-infinite dielectric substrate, as depicted in Figure 14.4. It is desired that most power is radiated into the air to communicate with the outside world. However, the radiated power of a dipole placed at the interface is more coupled into a dielectric material because it presents a lower electromagnetic impedance than the air. Figure 14.5 shows the ratio of radiated power from a dipole placed at the interface of air and a dielectric material as a function of the relative permittivity [21]. As can be seen, for a silicon substrate with $\epsilon_r = 11.7$, more than 95% of power is radiated into the silicon rather than the air, suggesting that this antenna configuration is very inefficient.

To avoid the radiating problem into a silicon substrate, it appears that a ground plane can be placed to reflect the wave back into the air. There are two ways to implement this approach. One way is to make an on-chip ground plane using a lowest metal layer (e.g. M_1) while placing an on-chip antenna on the topmost metal layer, as shown in Figure 14.6.

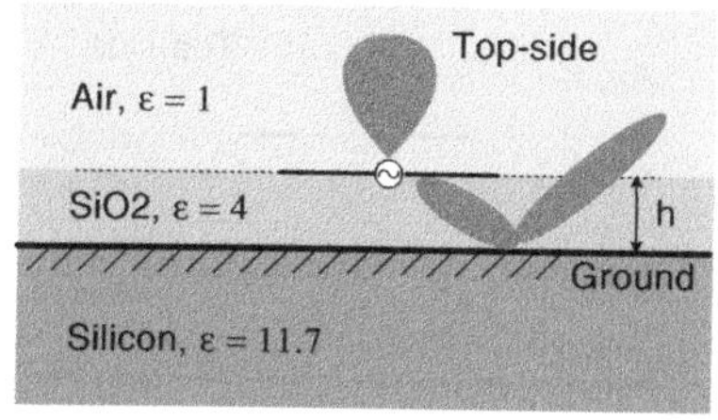

Figure 14.6: On-chip antenna with a ground plane on top of the silicon substrate.

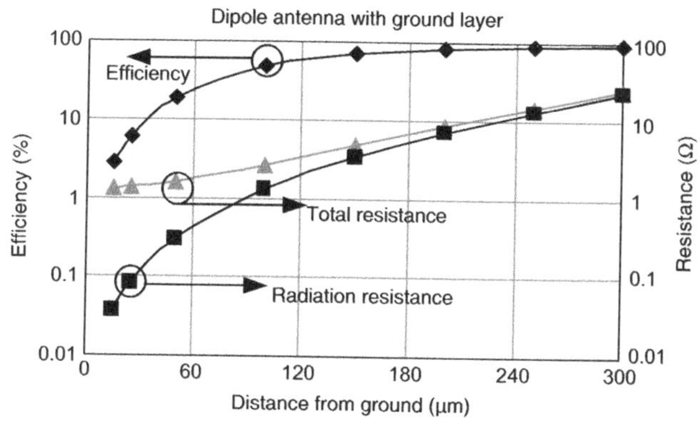

Figure 14.7: Radiation resistance and efficiency of a on-chip dipole antenna as a function of distance to the ground plane.

In this configuration, the distance between the antenna and the ground plane determines the radiation resistance and efficiency. For example, Figure 14.7 shows the electromagnetic simulation result of a copper dipole antenna that is placed over a metal ground plane with a SiO_2 dielectric sandwiched in between. The dipole length is equal to a length of a resonant dipole at 77 GHz. Then, the radiation resistance and efficiency is simulated as a function of the dielectric thickness, i.e. the distance between the antenna and the metal ground plane. According to the simulation result, as the distance increases, the antenna efficiency and resistance improves monotonically. Unfortunately, the distance between the topmost and the lowest metal layers is as small as 10–20 μm in today's process technologies. This means that the maximum achievable radiation efficiency in this configuration is only around 5% according to Figure 14.7.

The other way to implement a ground shield is to place an off-chip metal plane onto the backside of a silicon substrate as shown in Figure 14.8. In this configuration, the distance between the antenna and the ground plane is much larger than the SiO_2 layer because the substrate thickness is usually more than 100 μm. Although the increased distance can improve the radiation resistance, it gives rise to another problem that limits the antenna efficiency. Due to the high dielectric constant and the large thickness of the substrate, the entire substrate behaves as a dielectric waveguide generating multiple propagation waves [21, 22].

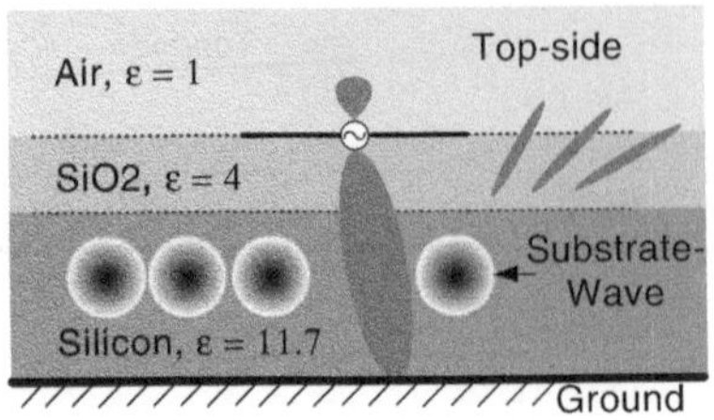

Figure 14.8: On-chip antenna with a ground plane on the backside of the silicon substrate.

Unfortunately, several modes of the propagation waves fall into the frequency of interest in the mmWave band. For example, at 77 GHz, the coupled power into the substrate propagating modes in the silicon substrate with a thickness of 290 μm is 2.7 times greater than the useful radiated power [10]. The substrate-mode power will be eventually wasted because it is either dissipated into heat in the substrate or radiated in undesired directions at the edge of the chip.

Although none of the approaches given so far provide a promising solution to on-chip antennas, they suggest a completely different approach: a wave can be radiated from the backside of a silicon chip by taking advantage of the coupled power into the substrate. In order to make this concept realized, the power dissipated by the substrate modes must be suppressed and converted to a desired radiative mode. One way to achieve the mode conversion is to use a hemispherical silicon lens on the backside of the chip [21], as shown in Figure 14.9. The lens changes the shape of the silicon substrate in such a way that the coupled EM energy is converted into a dominant radiation mode. Using this approach, an on-chip antenna has been fabricated in silicon at 77 GHz [10]. In this structure, antennas are fabricated by several low metal layers connected in parallel to minimize the associated ohmic loss and the distance to the substrate. The silicon substrate is further thinned down to 100 μm to reduce ohmic loss during the travel of radiated waves inside the conductive substrate. Instead, an undoped silicon wafer with a thickness of 500 μm is placed underneath the chip to increase the mechanical stability. Then, a silicon lens is mounted on the backside of the undoped wafer. A two-axis spherical far-field measurement is performed to measure the radiation pattern while a W-band horn antenna is used to irradiate the integrated dipoles. The measured E-plane patterns are shown in Figure 14.10. The maximum peak gain of 8 dBi is achieved in the measurement. The peaks of two antennas occur at different directions due to the off-axis properties [23].

14.2.5 Architectural Considerations

Phased arrays can be categorized into two different architectures, passive and active, depending on the location of RF front-end blocks and phase shifters [2,6].

In passive phased arrays shown in Figure 14.11(a), phase shifters are located right behind the antenna while active RF front-end blocks such as low-noise amplifiers (LNAs) and power amplifiers (PAs) are placed back after the feed network. Although it is simple and thus cost effective, the passive phased array has several drawbacks. Since phase shifters receive (and transmit) signals directly from (and to) antennas, the loss of phase shifters degrade the efficiency and noise figure of the system significantly. Therefore, the phase shifters and the

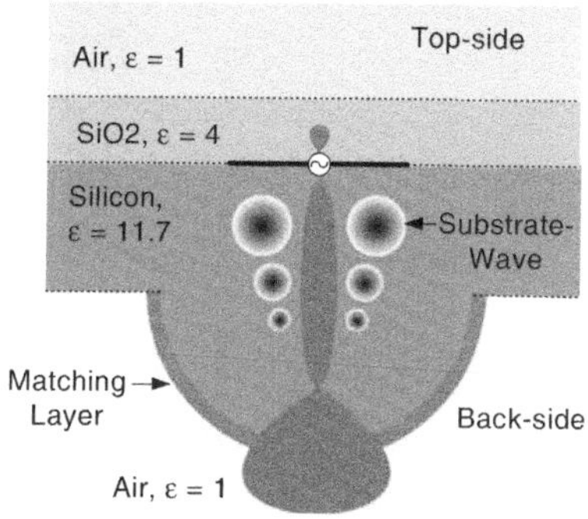

Figure 14.9: On-chip antenna with a silicon lens on the backside of the silicon substrate.

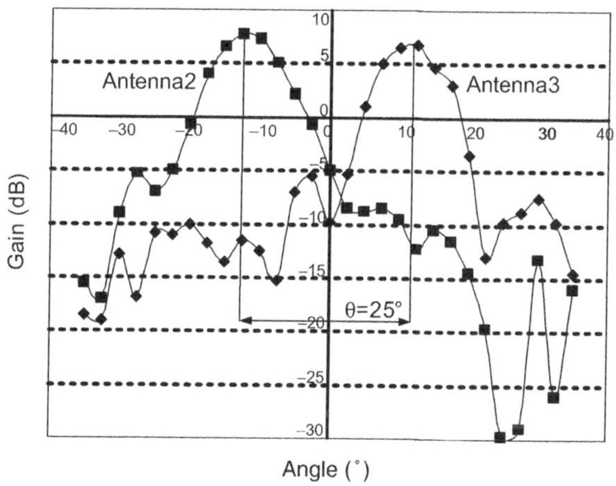

Figure 14.10: Measured E-plane patterns of two adjacent antennas on a silicon chip with a backside silicon lens.

following feed network must be designed with a low insertion loss, which is challenging, particularly at mmWave frequencies.

On the other hand, active phased arrays shown in Figure 14.11(b) use an active RF front-end block in each antenna element, followed by a phase shifter. With this configuration, the system efficiency and noise figure is not primarily affected by the loss of phase shifters and feed network, but is set by the active circuit blocks. Hence, the system performance improves significantly while the design requirements for phase shifters and feed network are relieved. Note that these advantages of active phased arrays are achieved at the expense of increased array complexity and cost. Active blocks at mmWave frequencies added to each antenna element are expensive by themselves. Furthermore, in traditional module-based arrays, the additional blocks must be interconnected externally in the system, which becomes complicated and challenging at such high frequencies, thereby increasing the system cost further. Fortunately, these issues can be well addressed by integrating the required blocks

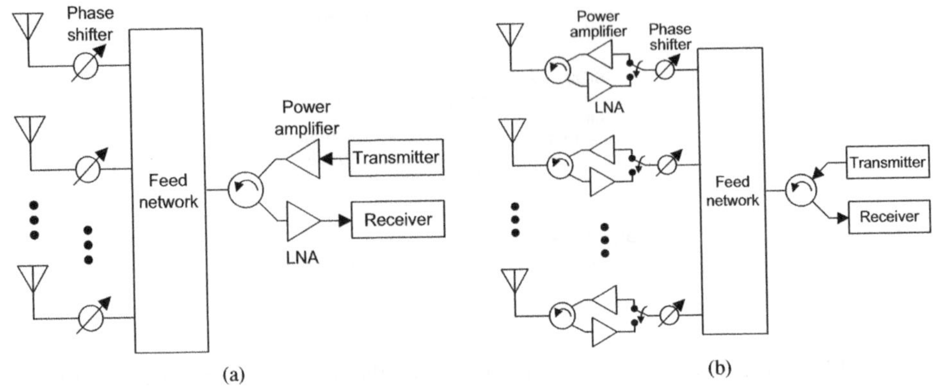

Figure 14.11: Basic phased-array architectures: (a) passive phased array; (b) active phased array.

and components in silicon. The silicon technology provides a practically unlimited number of transistors available on a chip with low cost, which enables the integration of all or most of the active circuit blocks. In addition, the high-level integration including on-chip antennas can avoid the challenging high-frequency interconnections, leading to dramatic reduction in complexity and cost. For this reason, an active phased-array configuration is preferred for integrated phased arrays.

In traditional active phased arrays, phase shifters are placed in the RF signal path before down-conversion, as shown in Figure 14.11(b). However, in principle, the phase shifting necessary for each element can be achieved at RF, intermediate frequency (IF), baseband, or in the local oscillator (LO) path [16, 24], as depicted in Figure 14.12. Each architecture has its unique properties, advantages, and shortcomings.

A phased-array architecture with phase shifting performed in each RF path is shown in Figure 14.12(a). Since RF signals are combined after the phase shifting, other RF, IF, and baseband blocks are shared for all array elements, resulting in low power and small chip-area consumption. Additionally, since all the interference signals are cancelled out at RF after signal combining, the linearity requirements of the IF and baseband blocks are relieved. However, a serious drawback of RF phase-shifting architecture is that it requires low-loss phase shifters operating at the RF frequency. Also, the insertion loss (or gain) of the phase shifters should be constant across all phase-shifting angles, otherwise, variable-gain amplifiers with a fine resolution would be necessary to offset the amplitude imbalance among the elements. At high RF frequencies including the mmWave band, such a low-loss flat-gain phase shifter is a challenging circuit block to implement in integrated phased-array systems.

Phase shifting can be performed in the IF stage, as shown in Figure 14.12(b). In this architecture, the design requirement of phase shifters is somewhat relieved due to the lower operating frequency. However, the power consumption in the array system increases because each path must have a RF mixer before the phase shifter. Also, the aforementioned advantage of the RF phase shifting, i.e. the lower linearity requirement for RF mixers become less effective in this architecture. Since the values of passive components such as inductors,

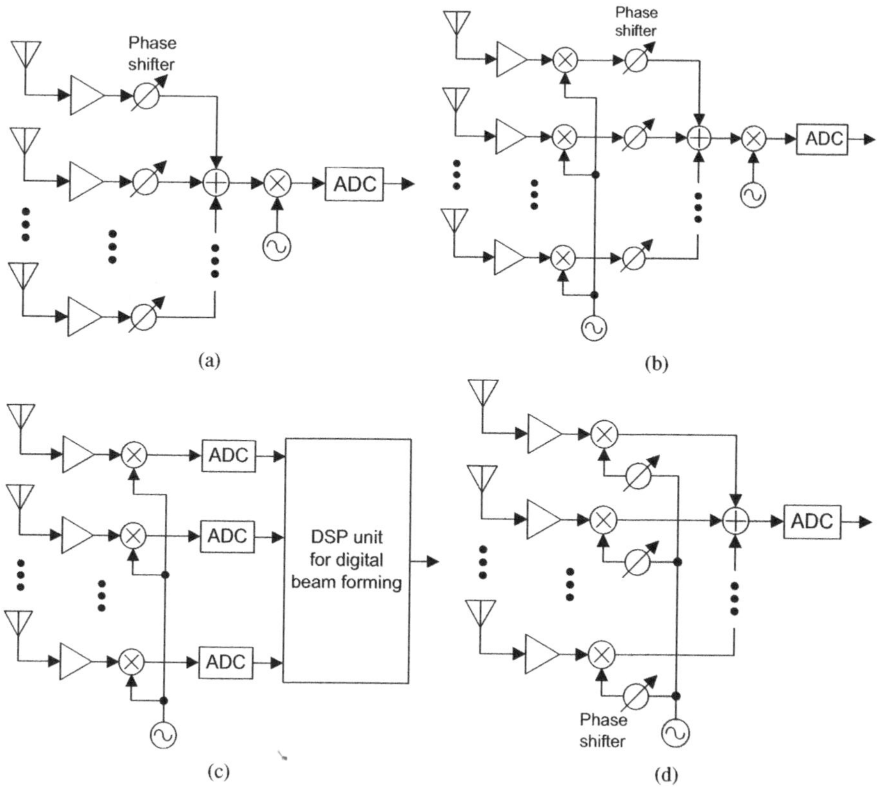

Figure 14.12: Different phase-shifting architectures: (a) RF phase shifting; (b) IF phase shifting; (c) digital phase shifting; (d) LO phase shifting.

capacitors, and transmission lines are inversely proportional to the operating frequency, the phase shifters implemented at IF will increase the chip area in the integrated system.

Another possible architecture is to perform the phase shifting in the digital domain after converting each baseband signal to a digital one, as shown in Figure 14.12(c). This architecture is very flexible, such that it can be reconfigured as a MIMO or a smart-antenna system relatively easily. However, the power consumption required for this system is severely high because the system is essentially equivalent to N receivers (or transmitters) operating in parallel without sharing any RF blocks except for frequency synthesizers. In addition, the 'raw' signal input without any interference cancellation requires analog-to-digital converters (ADCs) to have a large dynamic range, resulting in further high power consumption. The high processing speed of ADCs and DSP units required for high-resolution beam steering is another bottleneck to implement this architecture. For instance, suppose a digital array of eight receivers where each has an 8-bit ADC that samples the signal with a 100 MHz channel bandwidth at twice the Nyquist rate. Then, the baseband data rate in the whole system becomes as high as 76.8 Gb/s. This requires a high-speed interface, and a power-hungry and expensive signal processing core.

Finally, the phase shifting can be performed in the LO signal that drives down-conversion (or up-conversion) mixers, as shown in Figure 14.12(c). A down-converted signal after mixing the RF and LO signals is written by

$$V_{IF}(t) = \cos(\omega_{RF}t + \varphi_{RF}) \times \cos(\omega_{LO}t + \varphi_{LO}) \tag{14.8}$$

$$= \frac{1}{2}\cos[(\omega_{RF} - \omega_{LO})t + (\varphi_{RF} - \varphi_{LO})] \tag{14.9}$$

where ω_{RF} and φ_{RF} are the RF frequency and phase, respectively, and ω_{LO} and φ_{LO} are the LO frequency and phase, respectively. Hence, the LO phase shifting with an RF phase fixed is effectively same as the RF phase shifting, both resulting in the same amount of phase shift in the IF output signal. The LO phase-shifting architecture is relatively advantageous for integrated phased arrays, compared to the RF, IF, and digital phase-shifting counterparts. Since phase shifters are located in the LO path, their loss and nonlinearity have less impact on the RF signal-path performance of the system. Therefore, the requirements for the linearity, noise figure and bandwidth of the phase shifters are substantially reduced. Additionally, with switching-type of mixers (e.g. current-commutating Gilbert mixer), the down-converted (or up-converted) output amplitude does not depend primarily on the LO amplitude as long as the mixers are hard driven. Thus, the gain of each element path is less sensitive to the output amplitude variation of the phase shifters. Furthermore, the loss in the LO phase-shifting networks can be easily compensated by high-gain amplifiers (e.g. limiters) without signal-path linearity degradation and/or the need for any amplitude tuning. These factors make the LO phase-shifting architecture attractive to silicon integrated systems where a large number of transistors and active circuit blocks are implemented on the same die with a low cost.

14.3 Fully Integrated mmWave Phased-array Transceiver

To demonstrate the advantages of integrated phased arrays in the mmWave band, a fully integrated phased-array transceiver at 77 GHz is implemented in a 0.13 μm SiGe bipolar CMOS (BiCMOS) process [10, 11]. The transceiver chip integrates on-chip antennas, phase-shifting blocks located in the LO path, LO generators, as well as RF front-end blocks including power amplifiers, LNAs, and mixers. In this section, a detailed description on the system architecture and circuit block design is provided, followed by experimental results including the array and on-chip antenna performance.

14.3.1 Architecture

The 77 GHz phased-array transceiver fully integrates four array elements on a single chip, including a receiver block, a transmitter block, a phase-shifting block, a frequency generator, and on-chip antennas. Figure 14.13 shows the transceiver architecture.

The receiver block adopts a double down-conversion scheme that converts an RF frequency of 76–81 GHz to an IF in the 25–27 GHz. This frequency plan makes it possible to generate both the first and second LO signals by a single frequency synthesizer because the second LO frequency is a half of the first one. The receiver front-end in each element path consists of an on-chip dipole antenna, a LNA, an RF mixer, and a dual-mode IF amplifier.

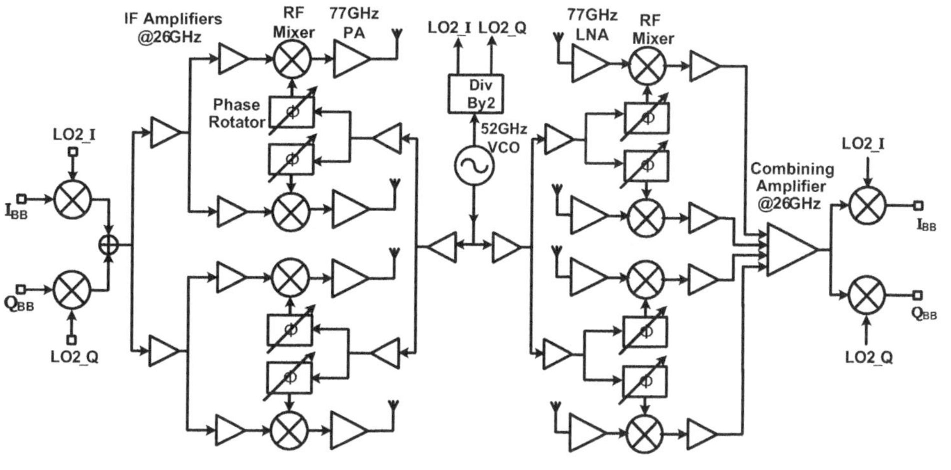

Figure 14.13: 77 GHz phased-array transceiver architecture.

The gain of the IF amplifier is varied by 15 dB using a single digital control bit. The four-path IF signals are combined using a symmetric active combining amplifier. The combined IF signal is further down-converted to the quadrature baseband signals by IF mixers.

The transmitter block also utilizes a double up-conversion scheme with the same frequency plan as the receiver block, presenting an IF frequency of 26 GHz. Therefore, the LO signal for the first up-conversion at 26 GHz can be obtained by dividing the 52 GHz LO signal for the second up-conversion generated from an on-chip frequency generator. The quadrature baseband signals are up-converted to 26 GHz by a pair of quadrature up-conversion mixers. The up-converted IF signal is distributed to each of the four paths through a network of distribution amplifiers. The RF mixer in each path up-converts the 26 GHz IF signal to 77 GHz, providing the input that feeds a driver and a on-chip 77 GHz PA subsequently. The adopted frequency plan leads to the image frequency of the second up-conversion falling at 26 GHz while the RF is at 77 GHz. Since the tuned mixer, driver, and PA stages provide sufficient attenuation at 26 GHz, the second up-conversion does not employ quadrature configuration for image rejection.

The LO generation block is shared between the receiver and the transmitter. A differential, cross-coupled VCO generates the LO signal at 52 GHz. A quadrature injection-locked divide-by-two circuit is following the VCO to produce the LO signal at 26 GHz. Subsequently, a cascade of divide-by-two frequency divider blocks are followed to provide the 50 MHz signal that is used by an off-chip phase-frequency-detector (PFD) to lock the VCO.

An LO phase-shifting architecture is adopted in this transceiver, as shown in Figure 14.14, owing to its advantages discussed in Section 14.2.5. The 52 GHz LO signal generated by the VCO is symmetrically distributed by differential transmission lines to each phase rotator driving an RF mixer in each path of the transmitter and receiver. A network of LO distribution buffers is used to compensate for the transmission-line insertion loss, thereby ensuring a sufficient amplitude at the phase-rotator input. The phase rotator generates a quadrature component of the LO signal locally and then interpolates between the in-phase (I) and

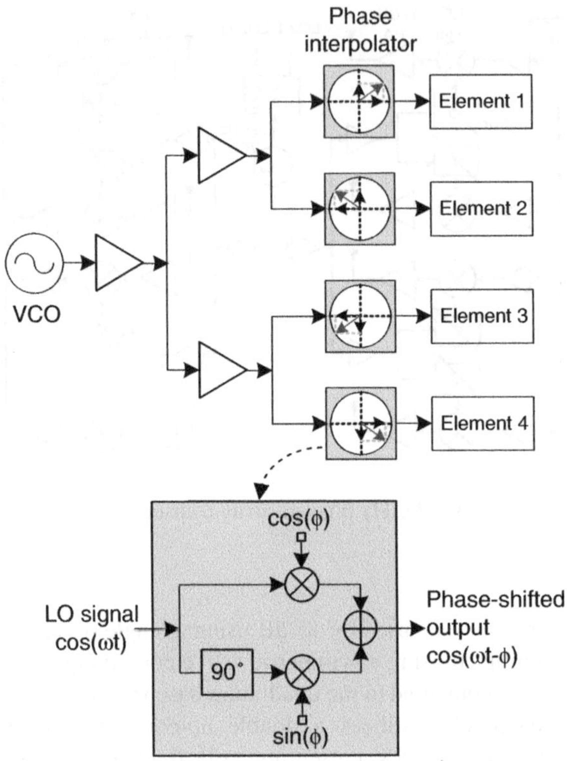

Figure 14.14: Local LO phase-shifting architecture.

quadrature (Q) components of LO signals to achieve the desired amount of phase shifting in each path. The phase-shifting resolution in this approach depends primarily upon the resolution of interpolating weights (cos ϕ and sin ϕ). By using a high-performance off-chip digital-to-analog converter (DAC), the interpolating weights can be generated with a high resolution, resulting in a fine step of phase shifting in each path. This enhances the accuracy of calibration procedures, thereby bringing the array performance improvement such as high-resolution beam steering and a low sidelobe level. It should also be noted that each transmit and receive path has a separate phase rotator, so that the transmit and receive beamforming can be achieved at different directions independently.

On-chip antennas are integrated at LNA inputs and PA outputs, following the backside radiating approach with a silicon lens discussed in Section 14.2.4. Due to layout limitations in the design, the antennas are placed at the edge of the chip and a slab of undoped silicon is abutted to the substrate to maintain a uniform dielectric-constant substrate underneath the antenna, as shown in Figure 14.15. For mechanical stability, a 500 μm-thick undoped silicon wafer is placed underneath the chip and the silicon lens is mounted on the backside. All of the low-frequency connections are brought to the chip by board metal traces and wire-bond connections. As this setup is highly compatible with flip-chip technology, all of these low-frequency signals can be carried by flip-chip connections as well.

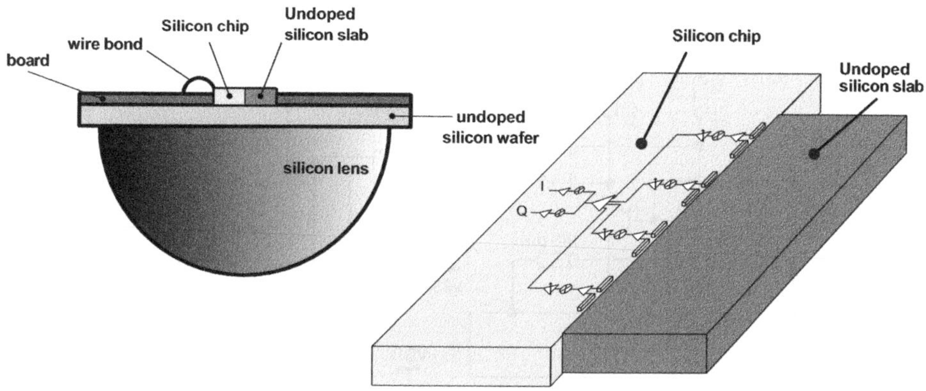

Figure 14.15: Radiating component configuration for on-chip antennas.

14.3.2 Circuit Blocks

In this section, detailed description on the circuit-level design is presented for the 77 GHz phased-array transceiver. Each circuit block employed in the receive chain is described first, followed by the circuit blocks in the transmitter chain. Subsequently, the design of the on-chip frequency generator and phase rotator is presented.

14.3.2.1 77 GHz LNA in Receiver

Since it is fed by an on-chip dipole antenna, the LNA is designed differentially using a two-stage cascode configuration, as shown in Figure 14.16. The differential input and output impedances are matched to 50 Ω and 100 Ω, respectively, at 77 GHz, by employing shunt and series transmission lines.

14.3.2.2 77–26 GHz RF Mixer in Receiver

A double-balanced current-commutating mixer is designed to down-convert the 77 GHz RF signal to a 25.5 GHz IF, as illustrated in Figure 14.17. Both the RF and LO ports of the mixer are matched to 100 Ω differentially by L-section transmission lines [25]. The IF bandwidth of the RF mixer is primarily determined by a Q-factor of the resonant load at the IF port

$$\mathrm{BW} = \frac{\omega_0}{Q} \tag{14.10}$$

where ω_0 is the load resonant frequency, i.e. 26 GHz in this mixer. On the other hand, the load impedance of a parallel inductor and capacitor (LC) tank at the resonance frequency is given by [25]

$$Z_\mathrm{L} = \omega_0 L Q \tag{14.11}$$

which is proportional to the Q-factor. Therefore, there is a trade-off in choosing the Q-factor between the operating bandwidth and gain. To achieve the desired bandwidth, a Q of 3.5 is chosen, which leads to 0.4 nH inductance and 250 Ω de-Q resistance to form a resonant

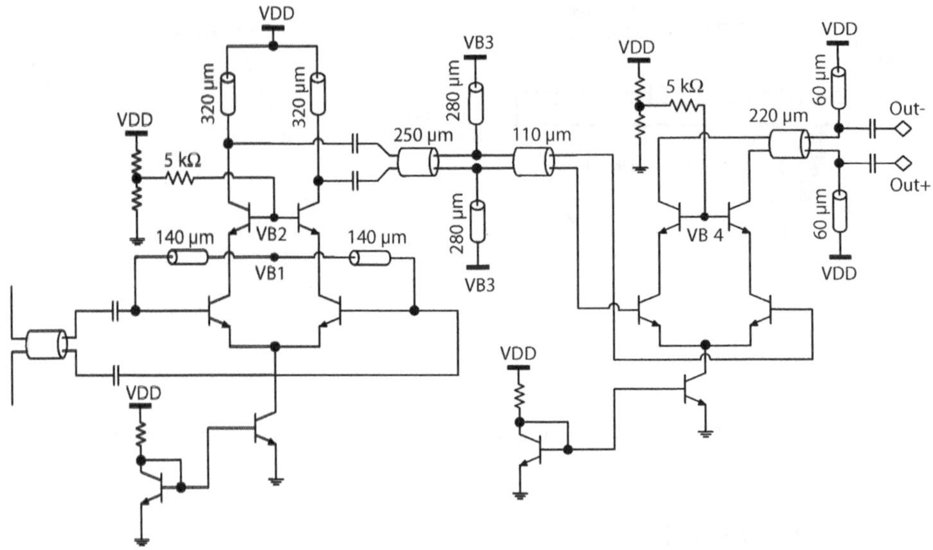

Figure 14.16: Schematic of the 77 GHz LNA.

load with the transistor parasitics capacitance and the input impedance of the subsequent stage. Simulation results indicate that the RF mixer achieves 5 dB voltage conversion gain, a 10 GHz bandwidth, and an 11 dB noise figure. The input return loss is 20 dB at the RF port and 11 dB at the LO port. Resistive emitter degeneration is used to enhance the linearity, which saves chip area compared with inductive degeneration. The common node of the degeneration resistors is connected to the ground instead of a tail current source further to improve the linearity.

14.3.2.3 26 GHz Dual-mode Amplifier in Receiver

The dual-mode amplifier is designed using a resistively degenerated differential cascode (Q_0–Q_3) with a differential current-bleeding network (Q_4 and Q_5), as shown in Figure 14.18. The amplifier is operated in either a high-gain or a low-gain mode by toggling a digital switch that changes the base voltage of Q_4 and Q_5. In a high-gain mode, Q_4 and Q_5 are off, so that all bias current flow into cascode devices (Q_2 and Q_3). On the other hand, in a low-gain mode, a bias voltage is applied to the base of Q_4 and Q_5 so that a part of bias current leaks into the current-bleeding network. The resulting low gain normalized to its high gain is approximately given by

$$\frac{A_{v,\text{low}}}{A_{v,\text{high}}} = \left[1 + \frac{A_5}{A_2}\exp\left(\frac{V_{b,\text{low}} - V_{\text{dd}}}{V_{\text{T}}}\right)\right]^{-1} \tag{14.12}$$

where A_2 and A_5 are the emitter area of Q_2 and Q_5, respectively, and $V_{b,\text{low}}$ is the base voltage of Q_4 and Q_5 in a low-gain mode. In this design, $V_{b,\text{low}}$ is set to V_{dd}, and A_5/A_2 is fixed at 11/3. The simulation result shows 15 dB gain variation between the two modes while Equation (14.12) predicts 13.5 dB gain variation.

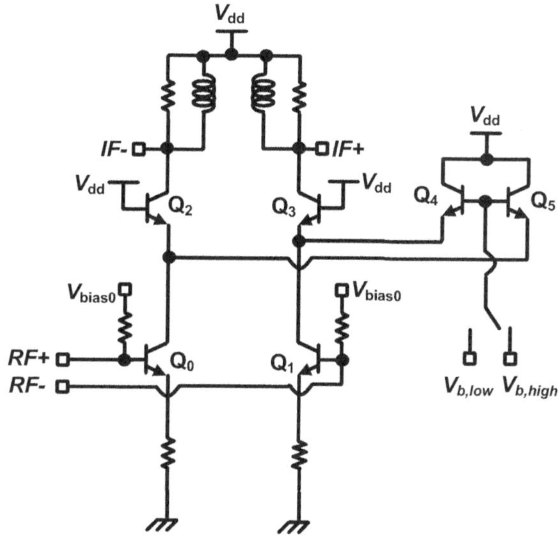

Figure 14.17: Schematic of the 77–26 GHz RF mixer in the receiver.

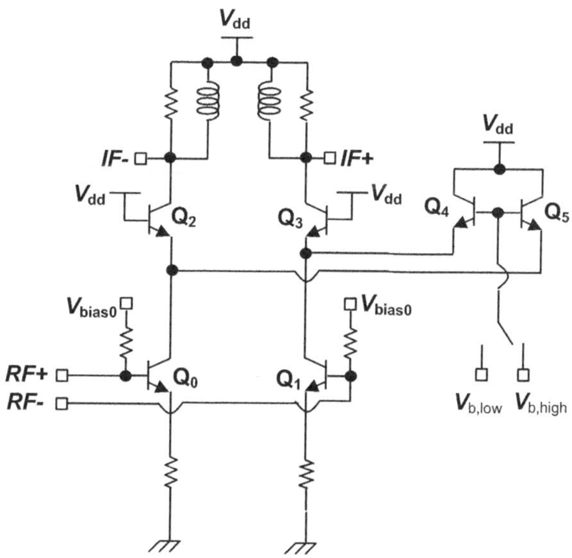

Figure 14.18: Schematic of the 26 GHz dual-mode amplifier in the receiver.

14.3.2.4 26 GHz Distributed Active Combiner in Receiver

After the first down-conversion and the phase shifting by RF mixers, the IF signals at 26 GHz in the four paths are combined through a distributed active combiner, as shown in Figure 14.19.

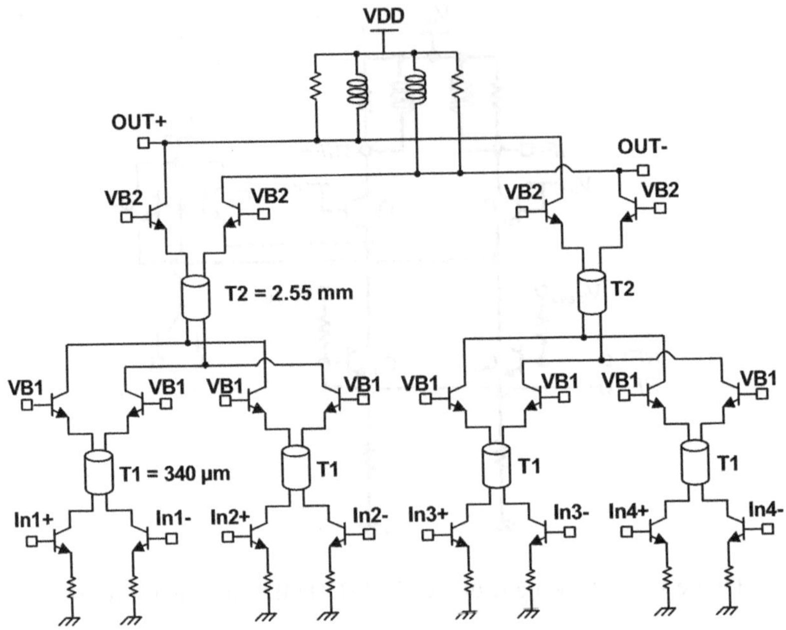

Figure 14.19: Schematic of the 26 GHz distributed active combiner in the receiver.

The input transistors are degenerated by resistors to improve the linearity and accordingly the dynamic range of the system. The output current from each of the four transconductance stage is combined through a symmetric two-stage binary structure. A pair of cascode transistors is inserted at each combining junction to isolate the input and output ports, thereby improving the overall stability of the amplifier.

The total length of each routing transmission line, T_1 and T_2, is 340 μm and 2.55 mm. The transmission lines are implemented by a differential line structure with ground and side metal shields to minimize the substrate loss and cross coupling. The bias current of cascode transistors is adjusted, such that its input conductance (g_m) is matched to the characteristic impedance of the transmission lines. For the operating frequency of ω_0, the imaginary part of the emitter-base admittance, $j\omega_0 c_\pi$, is much smaller than g_m if the transistor transition frequency ω_T is much higher than ω_0. Therefore, impedance matching can be achieved even without additional passive tuning. To match the odd-mode characteristic impedance of 64 Ω and 32 Ω for T_1 and T_2, respectively, DC bias current of 1 mA is applied to each branch. Simulations show that the return loss is better than 10 dB at each termination of T_1 and T_2. The differential output of the active combiner is loaded with a LC tank in parallel with a resistor to improve the bandwidth.

14.3.2.5 26 GHz-to-baseband IF Mixer in Receiver

A pair of double-balanced current-commutating mixers down-converts the 26 GHz IF signal to the baseband. As shown in Figure 14.20, each mixer is driven by either an in-phase or quadrature LO signal through an LO buffer that compensates for loss in the LO distribution

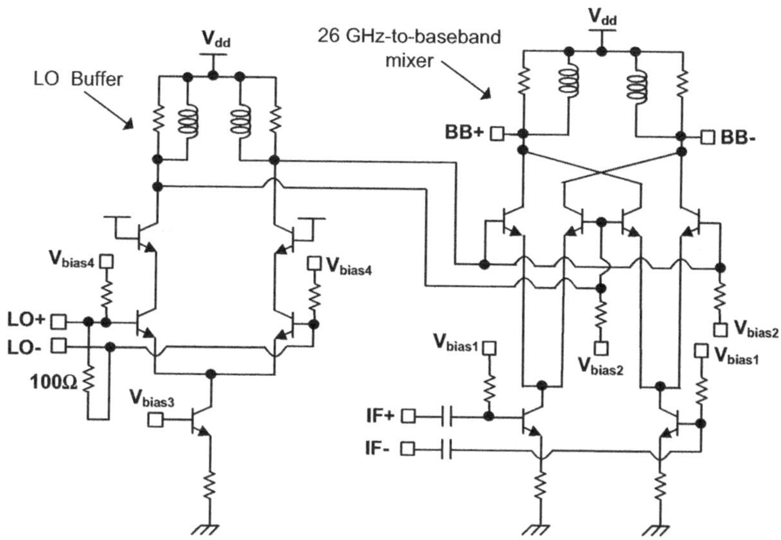

Figure 14.20: Schematic of the 26 GHz-to-baseband IF mixer in the receiver.

network, thereby ensuring the saturated mixer operation. A 0.9 pF metal–insulator–metal (MIM) capacitor couples the 26 GHz IF signal to the mixer input. The input differential pair of the mixer is degenerated by 30 Ω resistors at the emitter to improve linearity. The input matching of the LO buffer is achieved by a 100 Ω resistor directly connected between the differential inputs in order to reduce the chip area. The LO buffer is loaded with 0.6 nH spiral inductors and 320 Ω de-Q resistors, providing a gain of 15 dB. With 280 Ω load resistance, the IF mixer achieves a 6 dB conversion gain and a 8 GHz IF-referred bandwidth with 4 mA DC current consumption.

14.3.2.6 Baseband-to-26 GHz IF Mixer in Transmitter

The baseband-to-IF up-conversion is achieved by double-balanced current-commutating quadrature mixers with a shorted transmission line as a load. As shown in Figure 14.21, each mixer output is matched to a 50 Ω differential impedance because the output signal is divided into two IF paths driving IF distribution buffers with a 100 Ω input impedance. The mixers and buffers draw 46 mA from a 2.5 V supply. Since four array elements are integrated on the same die, the routing length of IF and LO signals is as long as 1.5 mm that is translated as 0.52 λ at the LO frequency of 52 GHz. Therefore, it is beneficial to use well-modeled transmission lines for the signal routing and to ensure the impedance matching between the interconnecting blocks and transmission line. This makes the system performance more reliable and less dependent on the floorplan.

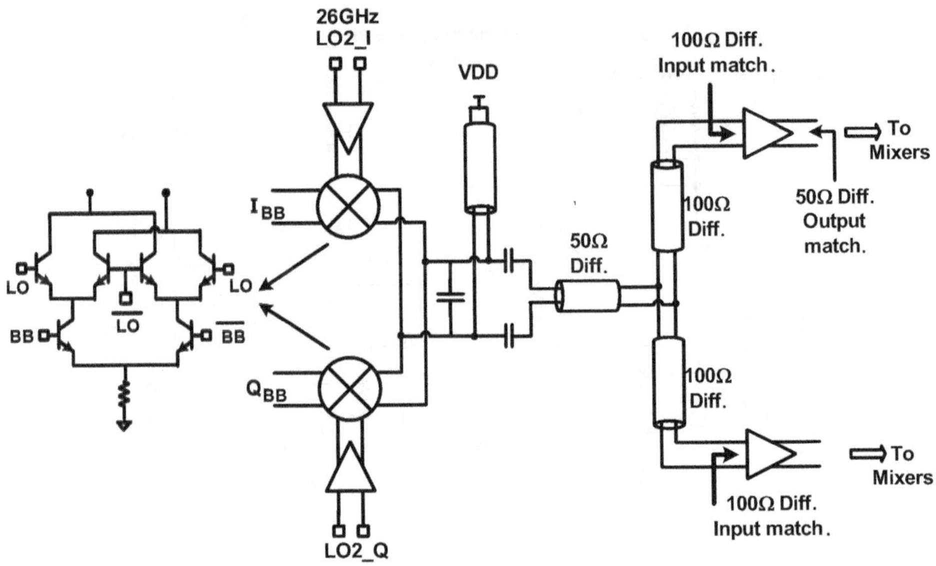

Figure 14.21: Baseband-to-26 GHz IF mixer and IF distribution in the transmitter.

14.3.2.7 26–77 GHz RF Mixer in Transmitter

As shown in Figure 14.22, the 26 GHz IF signal is up-converted to 77 GHz by a double-balanced current-commutating mixer in each of the four transmit paths. Since the mixer is driven by a large LO amplitude for its saturated operation, the RF output also presents a large-swing signal. Therefore, the load at the RF port should not be matched to its small-signal port impedance. Instead, the lengths of the shunt and series transmission lines at the RF output port are optimized by a methodology similar to the load-pull technique used in PA design, to maximize the output power into a 50 Ω load under large signal conditions.

The differential output of the RF mixer feeds a single-ended PA driver and a PA subsequently, which requires a balun. Although an on-chip mmWave balun can be implemented, similar to [26], it takes substantial chip area. Therefore, in this transceiver, one of the differential mixer outputs is terminated to 50 Ω through a capacitor while the other is fed to the PA driver that is input matched to 50 Ω.

14.3.2.8 77 GHz Power Amplifier in Transmitter

Four transmitter outputs are generated by on-chip PAs, one for each of the four transmit paths. As shown in Figure 14.23, the PA consist of four stages with the transistor size doubled in each stage to ensure that the output stage saturates first provided each stage has at least 3 dB gain reference [27]. While the first three stages of each PA are designed for the maximum gain, the final stage is designed for the maximum efficiency. Each of the PAs is connected to an on-chip dipole antenna that can optionally be trimmed out using a laser for direct electrical measurements via bonding pads.

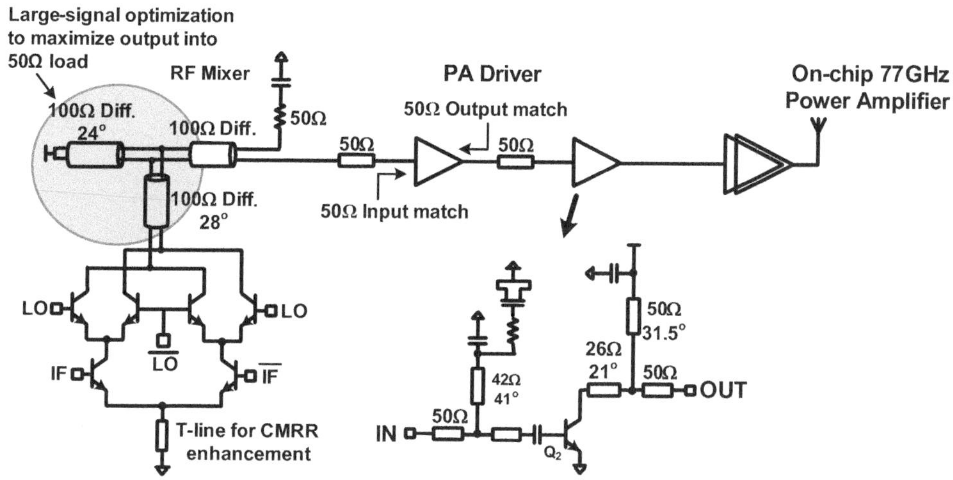

Figure 14.22: 26–77 GHz RF mixer and PA driver in the transmitter.

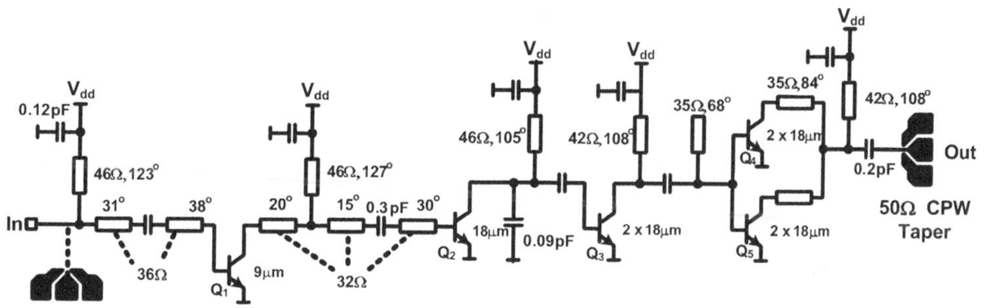

Figure 14.23: Schematic of the 77 GHz power amplifier.

14.3.2.9 52 GHz Voltage-controlled Oscillator

A cross-coupled LC VCO is designed at 52 GHz, as shown in Figure 14.24. It is noteworthy that a single VCO generates all LO signals required for RF and IF mixers in receive and transmit paths. The inductance in the LC tank is implemented by a shorted differential transmission line that provides approximately 95 pH at 52 GHz. The proximity of a return-path in the chosen transmission line reduces the inductance per unit length and increases current crowding which leads to an inductor Q of 24 at 53 GHz. However, the well-defined path for return current ensures accurate modeling of the transmission line. When accompanied by careful extraction of interconnect parasitics, it guarantees that the VCO operates at the desired frequency of 52.5 GHz with a tuning range of 10%. The VCO tuning is achieved through varactors that have a simulated Q of 40 at 52.5 GHz.

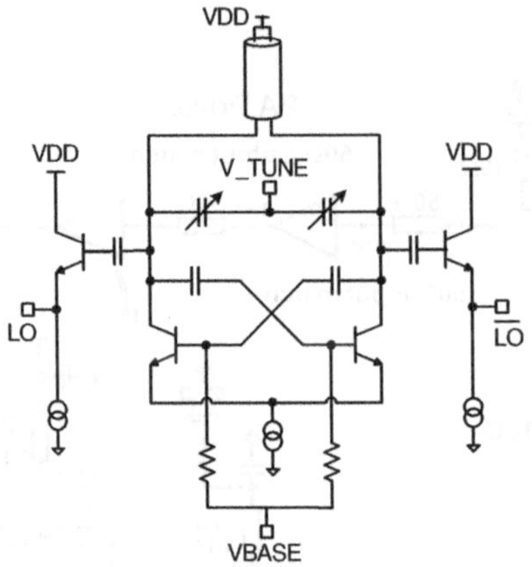

Figure 14.24: Schematic of the 52 GHz VCO.

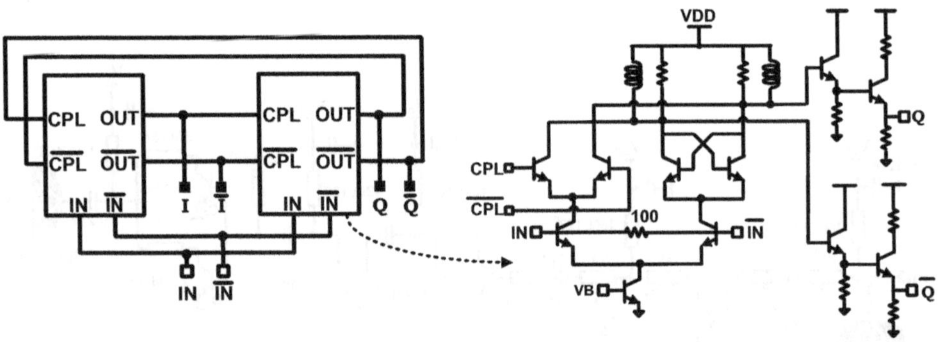

Figure 14.25: 52 GHz quadrature injection-locked frequency divider.

14.3.2.10 52 GHz Injection-locked Divider

The frequency plan of the transceiver requires a divide-by-two circuit to generate from the 52 GHz VCO a 26 GHz LO signal driving IF mixers. Since a digital divider with emitter-coupled logic (ECL)-based D-flip flops consumes substantial power at high frequencies, an injection-locked divider is implemented in this mmWave transceiver, as shown in Figure 14.25. While an injection-locked divider consumes much less power, it operates over a narrow frequency range and thus the parasitics have to be carefully modeled to ensure the proper operation at the frequencies of interest. The 52 GHz VCO input signal is provided at the tail current. The quadrature phases of the divided signal are generated by the cross-coupling of two injection-locked dividers. The divider core draws 3.1 mA from a 2.5 V supply.

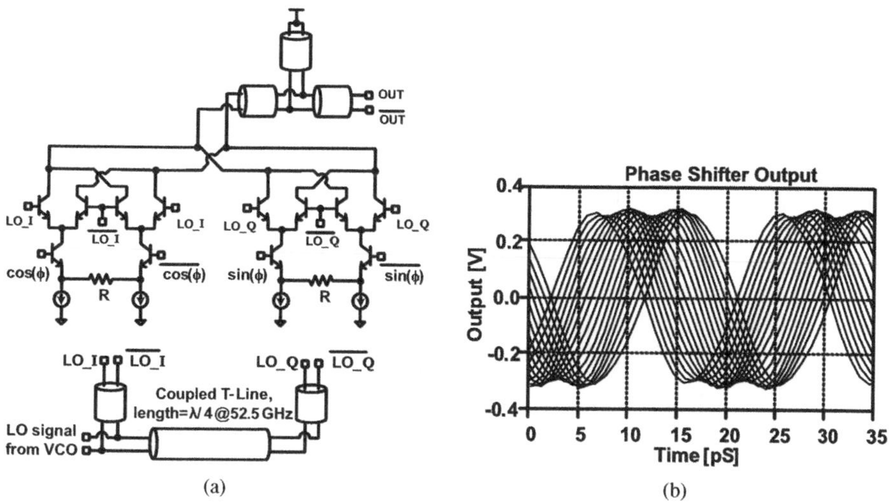

Figure 14.26: 52 GHz phase rotator: (a) schematic; (b) simulated performance.

14.3.2.11 52 GHz Phase Rotator

A phase rotator needs both in-phase and quadrature components of the 52 GHz LO signal, base on which it interpolates the required phase shifting at the output. Since only an in-phase component generated by the VCO is distributed to each phase rotator, a quadrature component must be produced locally at the phase rotator input. A quarter-wavelength transmission line is used to generate the quadrature LO component, as shown in Figure 14.26(a). The phase rotator consists of two double-balanced differential pairs driven by in-phase and quadrature LO signals, respectively. The output signals of the two differential pairs are combined in the current domain. Therefore, the output phase can be interpolated by controlling the tail current weights ($\cos\phi$ and $\sin\phi$) of the two differential pairs relatively, which are provided by an off-chip DAC. The emitter degeneration of each differential pair increases the voltage range of the weights in the rotator, thereby relaxing the DAC requirements. Figure 14.26(b) shows the simulated phase-shifting performance, which indicates that the amplitude variation for different phase-shift settings is around 1.5 dB. This variation is further reduced in the entire system because the mixer is designed to operate in a switching mode, so that the conversion gain is not sensitive to the LO amplitude variation as long as it is driven by a sufficiently large amplitude. This is one of the advantages of the LO phase-shifting architecture adopted in this phased-array transceiver.

14.3.3 Experimental Results

The 77 GHz transceiver with four array elements is implemented in a 0.13 μm SiGe BiCMOS process that provides SiGe heterojunction bipolar transistors (HBTs) with an f_T of 200 GHz. Figure 14.27 shows a die micrograph of the transceiver, which occupies a die area of 6.8 × 3.8 mm^2.

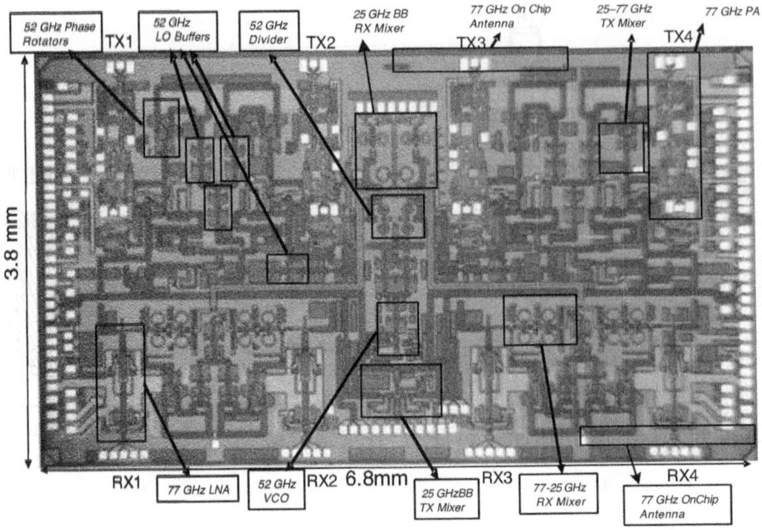

Figure 14.27: Die micrograph of the 77 GHz four-element phased-array transceiver with on-chip antennas.

14.3.3.1 Receiver Performance

Before characterizing the performance of the entire receiver chain, a test structure consisting of a stand-alone LNA and an on-chip balun is measured. The gain and noise figure of the LNA only is characterized by de-embedding the on-chip balun loss, which is shown in Figure 14.28. The stand-alone LNA exhibits a peak gain of 23.8 dB at 77 GHz and a 3 dB bandwidth of more than 6 GHz, while the lowest noise figure of 5.7 dB is measured at 75.7 GHz. The LNA consumes 17.5 mA from a 3.5 V supply.

The electrical performance of the entire receiver chain is measured by laser-trimming the on-chip antennas and feeding a single-ended RF signal through a WR-12 planar wafer probe. The measured conversion gain and noise figure is shown in Figure 14.29. A 37 dB single-path receiver gain is obtained at 79.8 GHz with a 2 GHz bandwidth, which corresponds to an inferred array gain of 49 dB after combining four-path signals. The minimum receiver noise figure is measured to be 8 dB at 78.8 GHz.

The radiation patterns of the on-chip antennas are measured using the setup shown in Figure 14.15, where a printed circuit board (PCB) provides DC supplies and digital control signals and takes baseband output signals through wirebond connections. A W-band standard horn antenna is used to irradiate the receiver. Photos of the radiation measurement setup and the close-up view of the receiver with a silicon lens are provided in Figure 14.30. The measurement radiation patterns of two adjacent on-chip antennas are shown in Figure 14.10, where the maximum peak gain of 8 dBi is achieved.

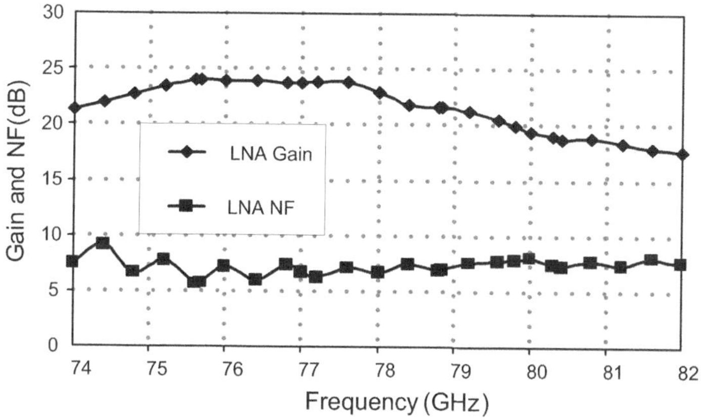

Figure 14.28: Measured LNA performance.

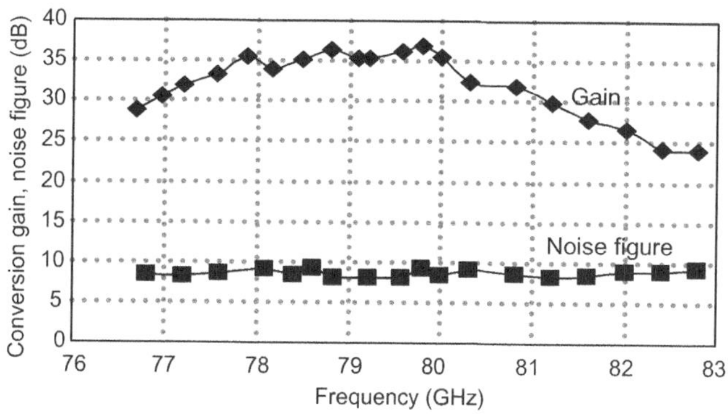

Figure 14.29: Measured conversion gain and noise figure of the receiver (single path).

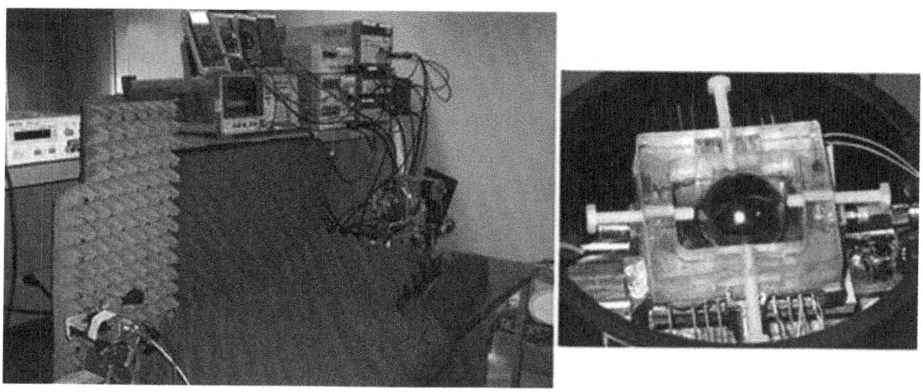

Figure 14.30: On-chip antenna measurement setup.

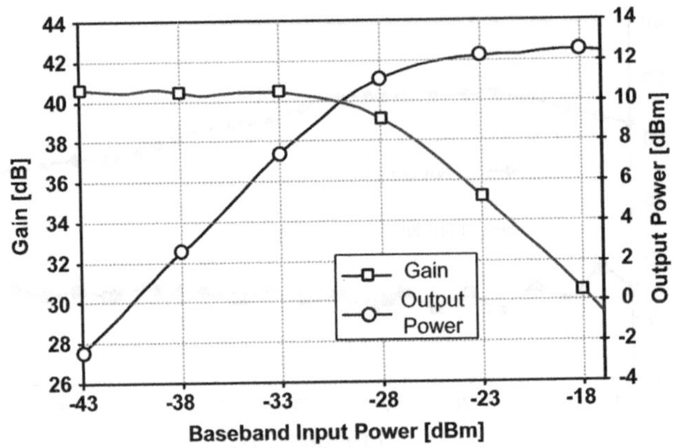

Figure 14.31: Measured conversion gain and output power of the transmitter (single path).

14.3.3.2 Transmitter Performance

The electrical transmitter performance is characterized by using a WR-12 wafer probe in a similar way to the receiver measurement. Figure 14.31 shows the measured conversion gain and output power of the transmitter as a function of input power at 77 GHz. Each array element in the transmitter generates up to 12.5 dBm with a 1 dB compression point of 10.2 dBm. The transmitter 3 dB bandwidth is measured to be 2.5 GHz. A test structure of a stand-alone PA is also measured, exhibiting a maximum power of 17.5 dBm with a power-added efficiency (PAE) of 12.8%.

14.3.3.3 LO Generation Performance

The 52 GHz VCO and the injection-locked frequency divider are characterized by probing internal test points. The measured VCO tuning range is shown in Figure 14.32(a) with the maximum divider locking rage superimposed. The VCO frequency can be tuned from 50.35 GHz to 55.49 GHz, which corresponds to a tuning range of 9.7%. The divider is locked to the VCO input from 51.4 GHz to 54.5 GHz. The VCO phase noise is measured as −95 dBc/Hz at a 1 MHz offset from the 54 GHz carrier frequency. The divider's input sensitivity, which is the input power necessary for the divider to achieve locking, is plotted in Figure 14.32(b). The VCO tuning range and the divider locking range are sufficiently large to down-convert or up-convert the frequencies of interest in this transceiver system.

14.3.3.4 Array Performance

An internal loopback option is used to characterize the array performance of the four-element transceiver. To implement the loopback option, a RF up-conversion mixer output at 77 GHz in the transmit element is connected directly to the input port of a RF down-conversion mixer in the receive element on the same chip. It should be noted that during stand-alone receiver and transmitter characterization, the internal loopback connections between the transmit and receive elements are laser trimmed. However, in the loopback mode, the internal loopback

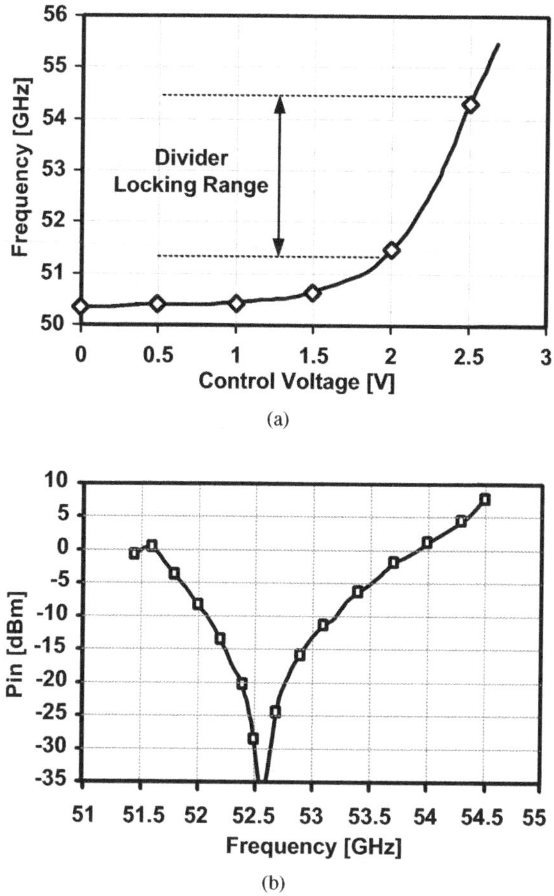

Figure 14.32: Measured frequency-generation performance: (a) VCO tuning rage; (b) divider sensitivity.

connections are preserved while the PA in the transmitter element and the LNA in the receiver element are bypassed by laser-trimming its output and its input lines, respectively. Figure 14.33 shows the on-chip configuration for the internal loopback measurement.

Electrical array patterns can be measured without using antennas by feeding different-phase signals to each receiver path with the phase linearly progressed, which emulates an incoming wavefront at a given direction [8, 24]. In the loopback mode, the linearly progressed phase shifting of the transmitting signal can be achieved by the on-chip phase shifting of each transmit element. Therefore, it is possible to measure the transmitter pattern for a particular transmitter phase-shift setting by varying the phase-shift settings in each of the receive elements. This emulates the situation that the receiver antennas receive the transmitted signal at different directions. Thus, the loopback option allows for transmitter and receiver array patterns to be measured using baseband input and output signals, with no off-chip mmWave connection. Figure 14.34 shows the measured array patterns with two

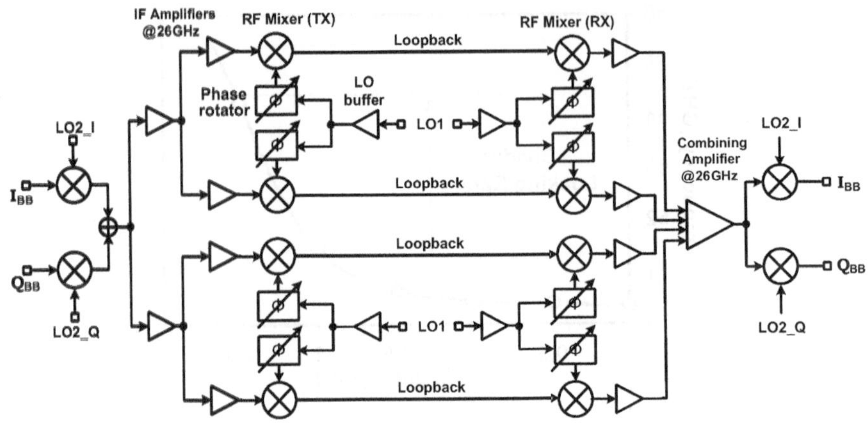

Figure 14.33: Internal loopback configuration for array pattern measurement.

transmit-receive pairs active in the loopback mode. The good match between expected and measured beam direction demonstrates the beamforming capabilities of the transmitter. The peak-to-null ratio is measured to be more than 12 dB.

Power consumption of the transceiver is measured as follows. In the receiver, each LNA and each down-conversion chain draw 17.5 mA at 3.5 V and 40 mA at 2.5 V, respectively. In the transmitter, the PA and PA driver per element draws 200 mA at 1.5 V and the up-conversion circuitry draws 30 mA at 2.5 V. The VCO and injection-locked divider core and the divider buffers draw 13 mA and 28 mA, respectively, at 2.5 V. Each phase rotator draws 14 mA from a 2.5 V supply.

14.4 Direct Antenna Modulation (DAM)

The availability of a large number of transistors with little incremental cost and the high reproducibility in silicon enables a designer to take into account multiple levels of abstraction in a single chip: from the system-, architecture-, circuit-, device-level to the electromagnetic structures. This interdisciplinary integration in multiple levels allows for implementing novel concepts and architectures that would not be feasible for conventional module-based approaches.

The DAM is a novel concept that is established by taking full advantage of such interdisciplinary integration capability [13]. By combining antenna electromagnetic boundary conditions with data-transmitting architectures and circuits all together in a single integrated system, an efficient and secured communication scheme is possible, which is fundamentally different from conventional modulation schemes. The DAM-based system allows for direction-dependent data transmission with higher efficiency and security than the conventional modulation and phased-system systems. In this section, a detailed concept of the DAM technique is described along with the associated advantages. Subsequently, the implementation of a DAM-based 60 GHz integrated transmitter is described as a proof of the concept, followed by the measurement results.

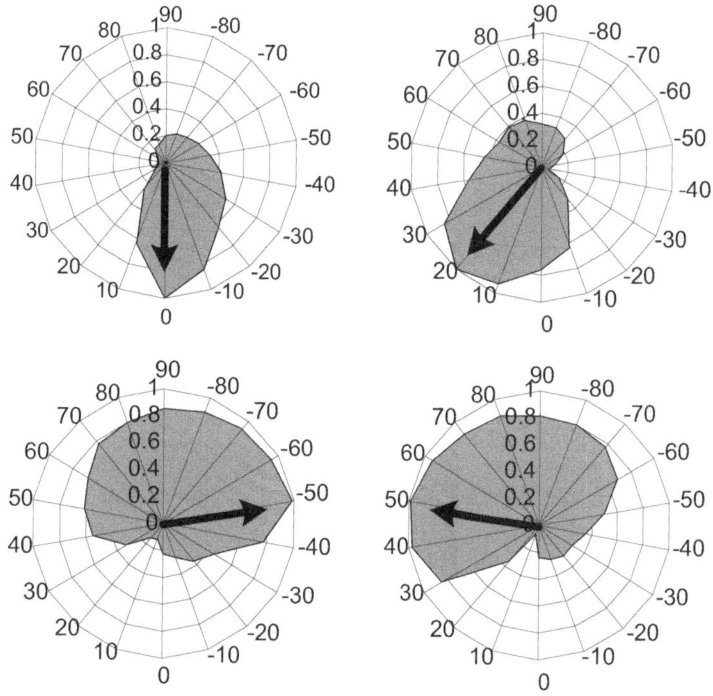

Figure 14.34: Measured loopback array patterns with two array elements activated.

14.4.1 Concept

In conventional radio transmitters, the desired information is modulated at the baseband and up-converted to the RF frequency by quadrature mixers, as shown in Figure 14.35. The modulated signal feeds a PA which, subsequently, drives an antenna to radiate the signal into air. Since the modulation is performed at the baseband before the antenna, it radiates the same data information (the same modulated phase and amplitude) in all directions, albeit with different gain factors and time delays. Even if a directional antenna or a phased array is used for radiation, practically it should present a sidelobe(s) at a different direction(s) from the desired one. Therefore, the same data constellation and information is sent in other directions that can be intercepted by an unintended receiver with good sensitivity.

The DAM technique is a fundamentally different approach to modulating and transmitting data that combines digital circuitry and electromagnetics to modulate the signal directly at the antenna through digital manipulation of the electromagnetic boundary conditions [13]. Since the modulation is performed by changing the near field of the radiation characteristics at the antenna structure, the DAM does not need up-conversion mixers and high-performance linear PAs that are required in conventional transmitters. Instead, only a locked RF source and a possibly nonlinear PA are used to drive the antenna, reducing the complexity in transmitter systems. The near-field modulation in the DAM technique is dependent on the radiating direction. Therefore, it can be used to create a secure communication channel by sending

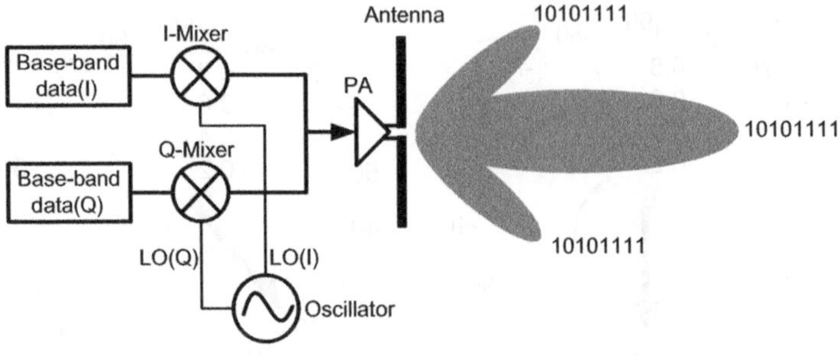

Figure 14.35: Conventional radio transmitter architecture.

the desired data only in the intended direction, while a scrambled data stream is transmitted in other directions.

The basic principle behind the DAM technique is illustrated in Figure 14.36, where a continuous wave (CW) signal of a constant amplitude and phase drives an antenna with an adjacent reflector. The reflector is a two-piece metal line with comparable dimensions to the wavelength which can be shorted or opened by a ideal switch. The amplitude and phase of the reflected signal depends on the electromagnetic boundary conditions that the reflector imposes and can be varied by turning the digital switch on or off. Hence, the far-field signal in the z-direction presents two different phases and amplitudes, which correspond to two distinct points in the I–Q plane, respectively. This scheme enables a simple one-bit digital modulation by controlling the reflector switch digitally.

The basic scheme in Figure 14.36 can be extended generally for a practical communication system that requires more than two constellation points to achieve high data throughput. The number of constellation points can be increased, by introducing multiple reflectors, each with multiple switches, as shown in Figure 14.37. In principle, 2^N different constellation points are generated for N switches in the reflectors. Hence, by using a sufficiently large number of reflectors and switches, it is possible to generate all required constellation points for a given digital modulation scheme at a desired direction (Figure 14.37(a)). On the other hand, the same set of switch combinations generates different constellation points at a different direction (Figure 14.37(b)) because the scattering properties of the reflectors and the resulting phase and amplitude vary with radiating directions. Accordingly, the constellation points are completely scrambled with sufficiently a large difference in the radiation angle, which prevents an undesired eavesdropper from demodulating and recovering the signal, thus enabling a secure communication link. Figure 14.38 shows the simulated error rate versus a radiation angle offset from the desired direction on the E-plane and the H-plane of an on-chip dipole antenna. The error rate is calculated by comparing a set of 210 equally spaced constellation points at the bore-sight (desired direction) with a scrambled version at different radiation angles on the E- and H-planes. The error rate increases rapidly and reaches 50% at an offset angle of 2–3° and 6–7° on the H- and E-planes, respectively. Therefore, target receivers located at an angle within $\pm 1°$ can recover the modulated signal completely without any error in the absence of noise and other channel non-idealities, while unintended receivers

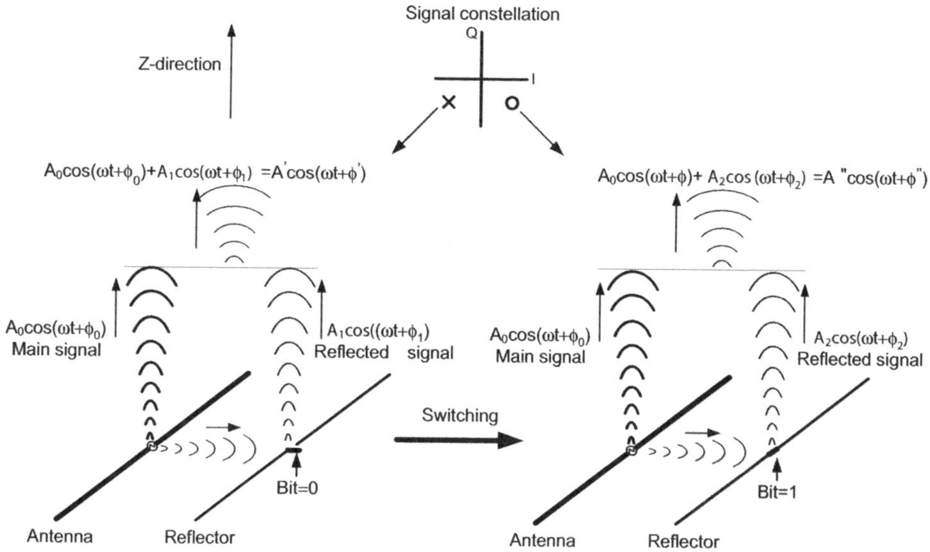

Figure 14.36: Single-bit modulation of the transmitted signal by switching a reflector in the near field.

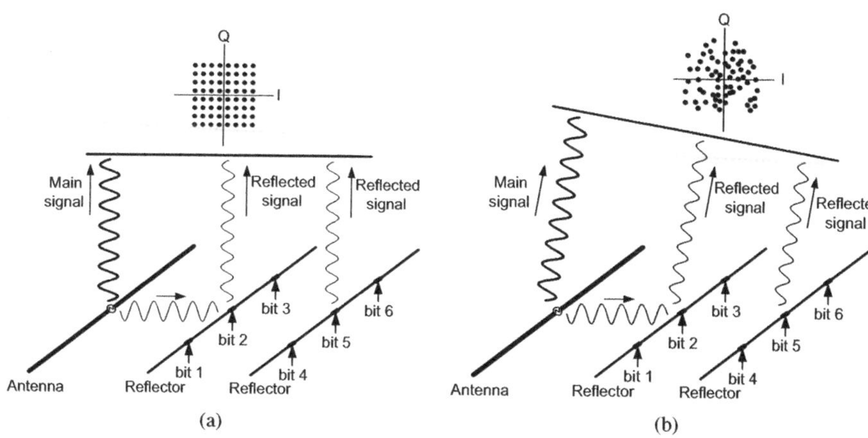

Figure 14.37: Multiple reflectors with multiple switches for more constellation points: (a) desired direction; (b) undesired direction.

at other directions are practically impossible to demodulate the signal with a reasonable error rate.

In integrated systems including on-chip antennas, it is relatively less difficult to implement a large number of antenna reflectors and switches on a single chip. With the large number of reflectors and switches, a huge number of constellation points can be generated from a

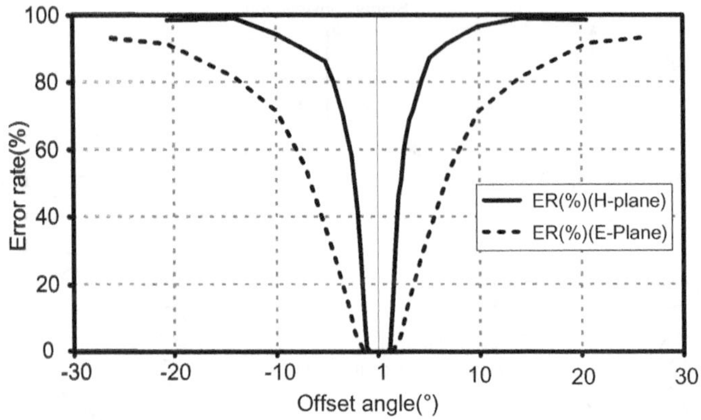

Figure 14.38: Simulated error rate in the DAM-based transmission.

single transmitter system. For instance, a total number of 90 switches in multiple reflectors provide $2^{90} \approx 10^{27}$ different constellation points. Although not all of them are used for a given modulation scheme, the huge number of constellation points provide high redundancy which can be exploited to transmit multiple independent data streams in multiple directions simultaneously at a full rate. Due to the high redundancy, it is possible to find a set of switch combinations that simultaneously generates two arbitrarily selected symbols of a given modulation scheme(s) at two different directions. These switch combinations allow a single transmitter system to perform concurrent transmission of multiple independent data in different directions, as depicted in Figure 14.39 where 16-QAM and 16-QPSK modulated signals are concurrently transmitted from a single DAM-based system into two different directions.

It is noteworthy that the modulation is performed not in the transmitting active-circuit chain but in the antenna level for the DAM-based system, so that it requires neither a broadband nor a linear PA to drive the antenna. Only a narrow-band unmodulated signal needs to be amplified and generated by a PA in the DAM technique. Moreover, since the PA is fed by a constant-envelope signal, a nonlinear PA can be used to amplify the signal, which increases the system efficiency significantly.

14.4.2 Implementation

A 60 GHz transmitter based on the DAM technique is implemented in a 130-nm SiGe BiCMOS process to demonstrate the viability of the technique [13]. The transmitter block diagram is shown in Figure 14.40. An on-chip dipole antenna with a length of 835 μm and a width of 20 μm is implemented by stacking three low metal layers (M_1, M_2, and M_3) in parallel. Five reflectors are located at each side of the antenna, each reflector with nine switches, resulting in a total number of 90 switches. This configuration provides $2^{90} \approx 10^{27}$ switching combinations. The location and spacing of the reflectors are optimized in simulation to maximize the coverage of the constellation points which are determined

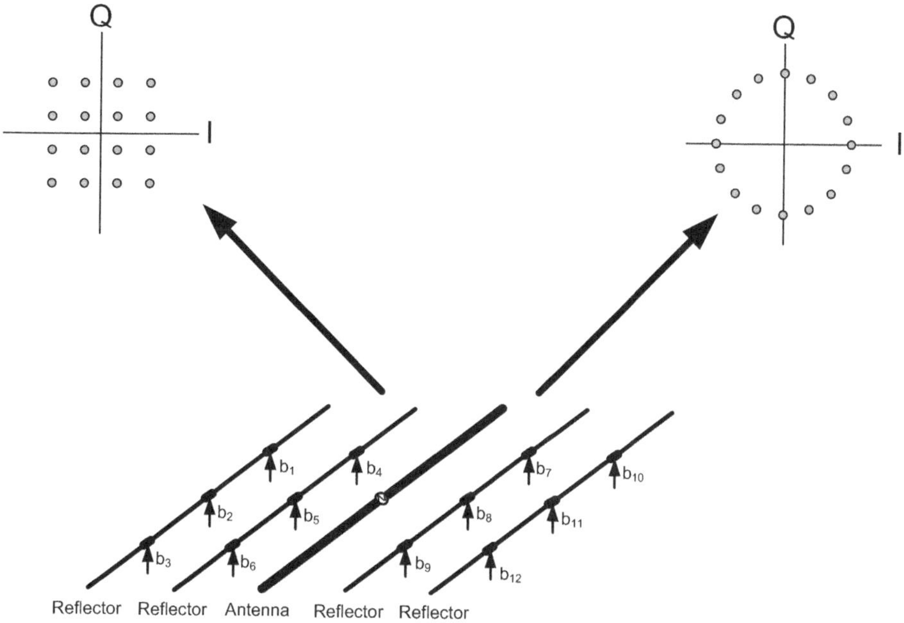

Figure 14.39: Concurrent multi-beam transmission in a DAM-based system.

by different radiating modes. The switches are controlled by a digital control unit with a baseband data sequence as an input.

The on-chip antenna is fed by a PA through a shielded differential transmission line. The input signal for the PA is provided by a locked oscillator with or without an optional quadrant-selection unit. The optional unit improves the constellation coverage further by applying ± 1 to the control signals A and B. The control signals preselect one of the four quadrants in the I–Q plane and then the reflector switches generate constellation points further in the selected quadrant.

The reflector switch is implemented by an n-type metal-oxide-semiconductor (NMOS) with a resonator connected between the drain and source terminals, as shown in Figure 14.41. The resonator is implemented by a circular shielded transmission line which resonates out the switch capacitance when it is off, mainly the parasitic transistor capacitance (C_{gd} and C_{gs}), at 60 GHz. With an NMOS of a 150 μm width and a circular transmission line of a 60 μm diameter, the switch achieves an on impedance of 1–5 Ω and an off impedance of 70 Ω.

The 60 GHz PA is implemented in three differential stages in cascade, as shown in Figure 14.42. The impedance matching for the input, output, and interstage is performed by differential transmission lines which have a grounded coplanar-waveguide (CPW) structure [27]. A 225 μm transmission line along with a 40 μm series line provides conjugate impedance matching between the output of the un-conversion mixer and the input of the PA. The 225 μm transmission line also feeds a DC bias voltage to each transistor base. AC coupling between the adjacent stages is fulfilled by 150 fF MIM capacitors.

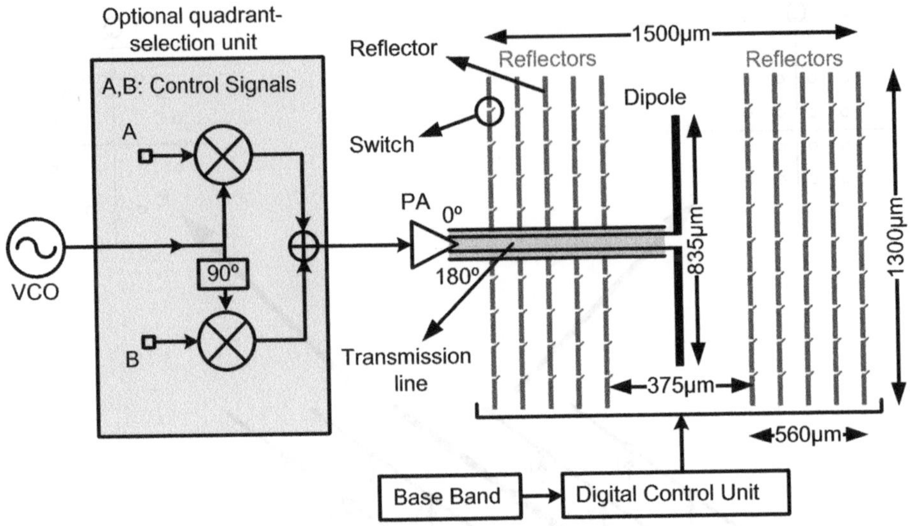

Figure 14.40: Implementation of a 60 GHz DAM-based transmitter.

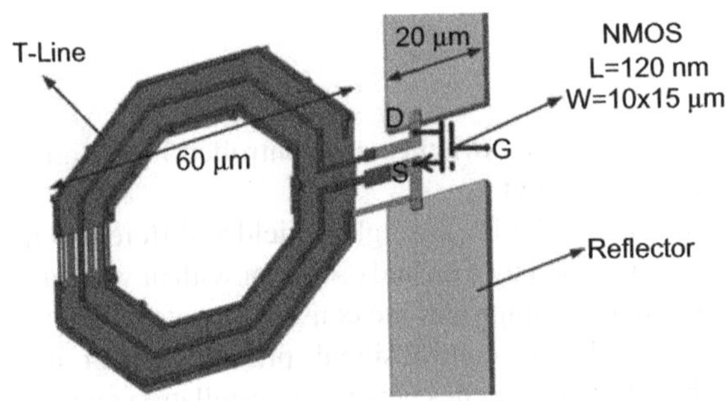

Figure 14.41: 60 GHz NMOS switch with a transmission-line resonator.

A cross-coupled on-chip VCO generates a 60 GHz LO signal to drive the PA or the optional up-conversion mixer. The oscillation frequency is tuned by NMOS varactors with a width of 15 μm. An inductor for the LC tank is implemented by a differential coupled-wire transmission line. The VCO output is divided down by a factor of 1024 using an injection-locked divider and a subsequent divider chain.

The chip micrograph is shown in Figure 14.43. The on-chip antenna and reflectors occupy an area of 1.3×1.5 mm^2. In the chip, a 60 GHz receiver with an on-chip dipole antenna is also implemented, which consists of a 60 GHz LNA, down-conversion mixers, and baseband amplifiers.

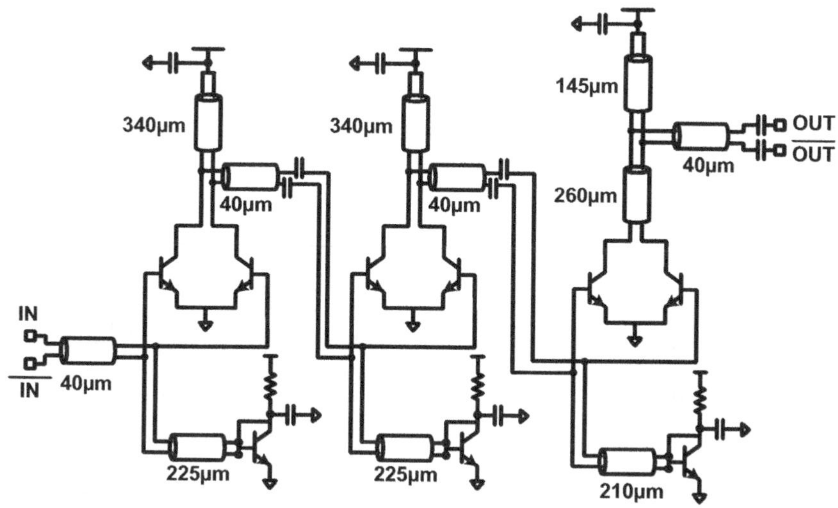

Figure 14.42: Schematic of the 60 GHz power amplifier.

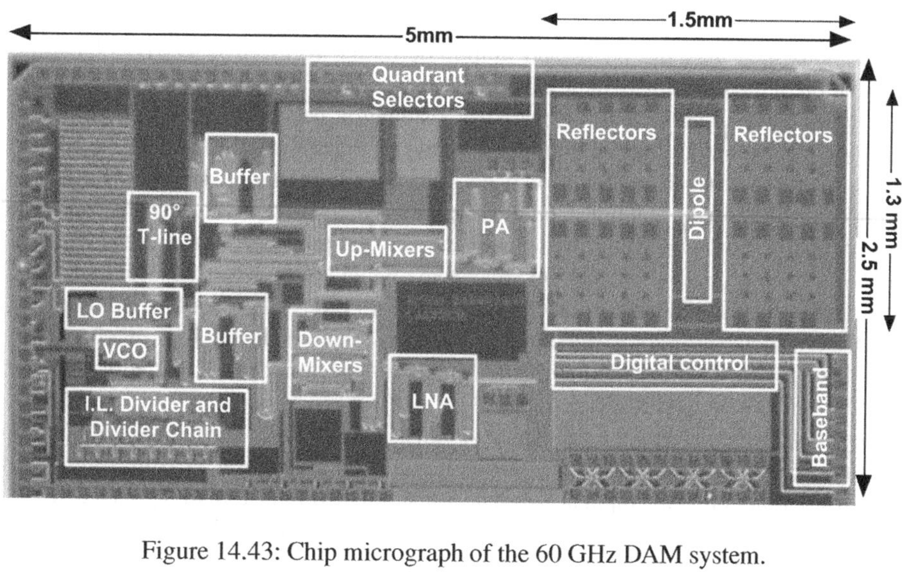

Figure 14.43: Chip micrograph of the 60 GHz DAM system.

14.4.3 Experimental Results

For efficient radiation from an on-chip dipole antenna and reflectors, a hemispherical silicon lens is attached on the backside of the silicon chip, as shown in Figure 14.9. The radiated power from the backside of the chip is received by a horn antenna. A network analyzer connected to the input of the on-chip antenna-reflector structure and the output of the

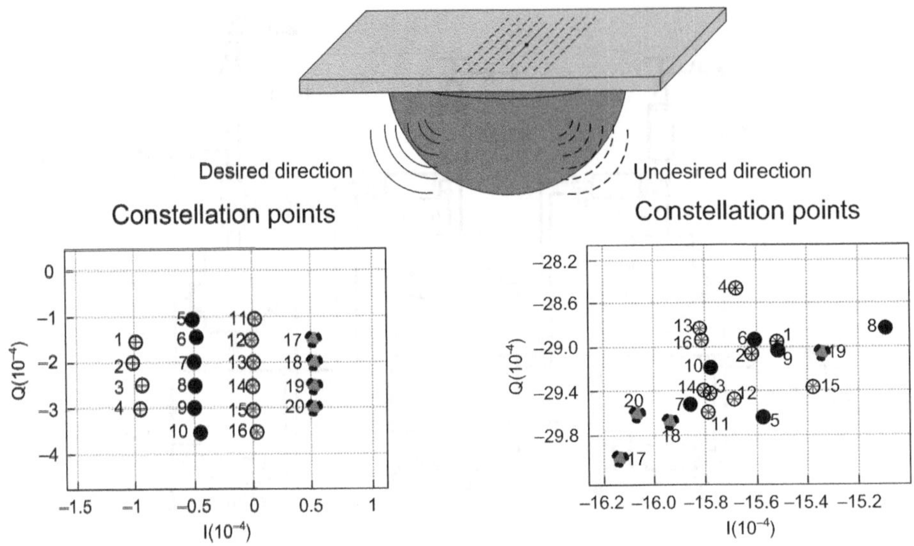

Figure 14.44: Measured constellation points in two different directions with an angular separation of 90°.

horn antenna measures the phase and amplitude of the received wave. By changing the combination of the reflector switches, the variations of the phase and amplitude are measured in the I–Q plane. Figure 14.44 shows 20 constellation points measured in two different directions with an angular separation of 90° while the combination of the reflector switches is kept the same to each other. As can be seen, the points are completely scrambled in the undesired direction, making it practically impossible for a receiver at the undesirable angle to recover this signal, no matter how sensitive it may be. This confirms the secure communication capability of the DAM system, which is hard to achieve in traditional transmitter systems. While the constellation points in Figure 14.44 is obtained by changing the reflector switches only, a full coverage of constellation in the four quadrants of the I–Q plane is achieved by using the optional quadrant-selection control unit with the reflector switches, which is shown in Figure 14.45.

The transmit output power and linearity is measured by disconnecting the antenna from the PA through optional laser-trimming work. Figure 14.46 shows the measured output power and gain as a function of the input power. The transmitter exhibits a small-signal gain of 33 dB with a saturated output power of 7 dBm and a saturated gain of 25 dB.

14.5 Large-scale Integrated Phased Arrays

The advantages of using phased arrays over a single-element transceiver become more obvious as the number of combined array elements (N) increases, as described in Section 14.2.2. The array gain and the transmitter EIRP increase by a factor of $10 \log_{10} N$ while the receiver sensitivity improves by a factor of $10 \log_{10} N$. The spatial filtering property of phased array also improves because the main beamwidth is narrowed with a larger number of null positions

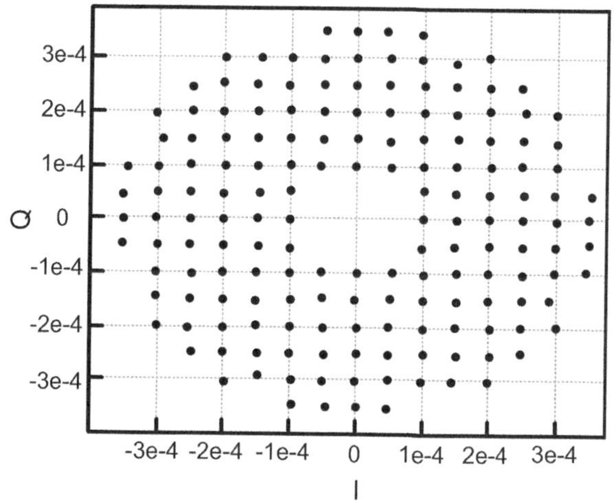

Figure 14.45: Measured constellation points fully covering the four quadrants by using the quadrant-selection unit with the reflector switches.

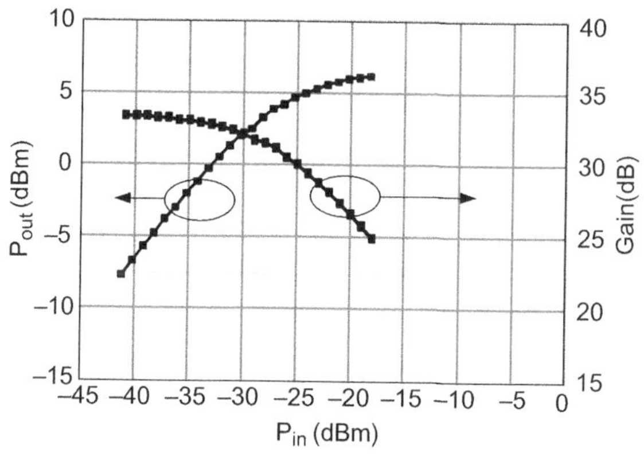

Figure 14.46: Measured output power and gain of the transmitter.

as N increases. Although these benefits raise the desire of combining a huge number of elements for high array performance, large-scale phased arrays have been implemented only for limited applications such as military and space radars, and even in those applications, the element numbers are limited up to 10^4 or 10^5 at most. One of the major reasons for the limitation is high complexity and cost associated with the large-scale array system. In conventional module-based array systems, most transmitter/receiver components, such as LNAs, PAs, phase shifters, attenuators, filters, mixers, and LO sources, are implemented in separate modules and then interconnected to each other externally [5, 6]. This approach not only increases the assembly size and cost, but also degrades the system reliability due to the complicated configuration. Furthermore, several transmit/receive module components have been implemented using expensive compound semiconductors such as GaAs, which takes a substantial portion of the overall system cost [5–7].

The integrated approach in silicon provides a promising solution to deal with those difficulties in large-scale phased arrays. The high yield and repeatability of silicon ICs allows the entire or most of the receiver/transmitter modules to be integrated on a single chip. Accordingly, this leads to dramatic reduction in cost and complexity of large-scale phased arrays, compared to the conventional module-based approaches. Moreover, the cost of silicon chips is substantially lower than that of GaAs chips, which further reduce the system cost. The silicon integrated solution is also appropriate for implementing multi-beam wideband phased arrays demanded for multi-target tracking or multi-point high-speed communications. The recent development of high-frequency silicon transistors and the practically unlimited number of transistors available on a single chip make it possible to integrate the required components for multi-beam phased arrays operating in a wide range of frequencies with low cost and complexity.

In this section, the first CMOS dual-band quad-beam phased-array receiver operating in a tritave bandwidth, from 6 to 18 GHz [12], is presented to demonstrate the integration capability for large-scale arrays. Since all components required for a wideband multi-beam array are integrated on a single chip only except for the antenna and front-end LNA, the receiver is easily scalable to implement a very large-scale array (up to 10^6) by adding extra receiver chips with low cost. First, the architecture of a large-scale phased-array system based on the receiver chips will be given, followed by a detailed description of the architecture and circuit building blocks of the receiver chip. The experimental results is then provided, including the performance of a four-element phased array that is implemented based on the receiver chips for demonstration purposes.

14.5.1 Large-scale Phased-array Architecture

The system architecture of the 6–18 GHz dual-band quad-beam phased-array receiver is shown in Figure 14.47. The CMOS receiver chip (a shaded block in Figure 14.47) is the key building block where most of the receiver components are integrated on the same die, including tunable concurrent amplifiers (TCAs), down-conversion mixers, phase shifters, frequency synthesizers, and baseband buffers [12]. With this single-chip solution, the need for a costly large number of separate component modules and their complicated interconnection can be avoided to build a large-scale array, resulting in a dramatic cost reduction. The only feed signal which needs to be distributed among the elements other than DC supplies is a 50 MHz reference signal for on-chip frequency synthesizers. Therefore, this system is easily scalable to build very large arrays by combining a large number of the receiver chips without severe increase of complexity and cost.

The phased-array system is programmable to concurrently receive two different frequencies between 6 and 18 GHz (a tritave) while forming four independently controlled beams. As can be seen in Figure 14.47, each of the horizontal and vertical polarizations (HP and VP) of the received signal, separated and amplified by the active antenna module, is fed to a 6–18 GHz CMOS receiver chip. The CMOS receiver is tunable to receive two different frequencies concurrently, such that each incoming signal is split into two separate frequency bands on chip; a low-band (LB) from 6 to 10.4 GHz and a high-band (HB) from

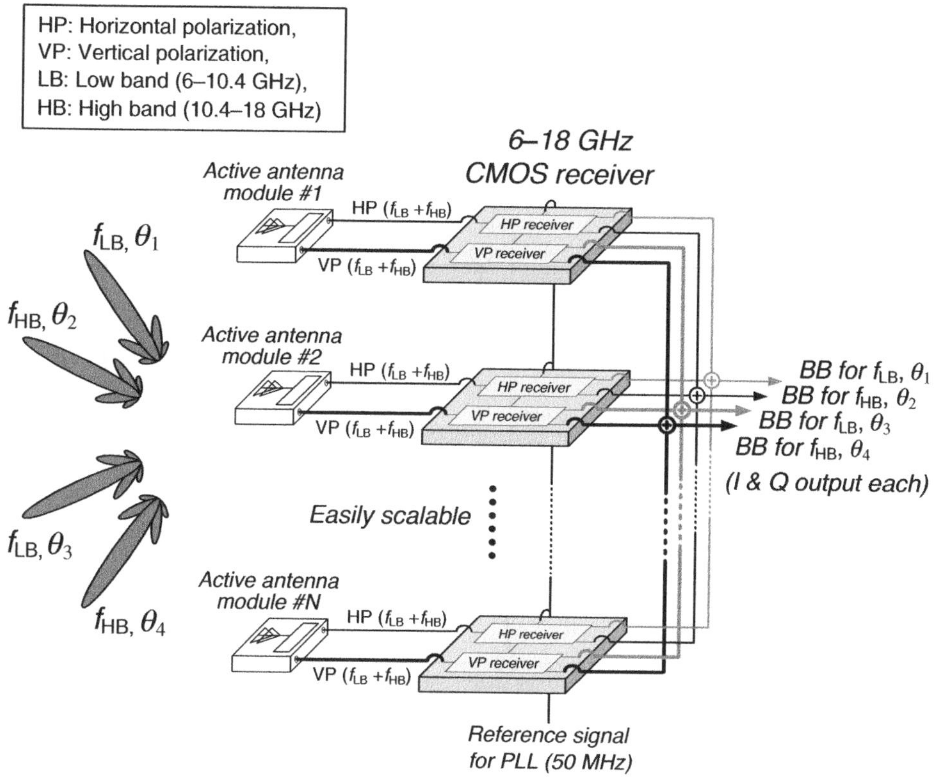

Figure 14.47: Architecture of the 6–18 GHz dual-band quad-beam phased-array receiver system.

10.4 to 18 GHz. The LB and HB signals are then down-converted by two separate down-conversion chains, respectively. The down-conversion is performed in two steps by RF and IF mixers. This phased-array system adopts the LO phase-shifting scheme due to the advantages discussed in Section 14.2.5. Independent phase shifting is performed to each of the LO signals driving four pairs of in-phase and quadrature IF mixers (two pairs per polarization and two pairs per band). Hence, the array system can receive and steer four different beams at two different frequencies concurrently. The down-converted signals from each array element are then combined at the baseband in a hierarchical fashion allowing for full scalability to very large arrays.

The LO signals required for the down-conversion and phase shifting are generated by on-chip frequency synthesizers. Therefore, the phase-noise sources are uncorrelated among frequency synthesizers in different elements except for the ones associated with a common reference signal. This leads to effective improvement in phase noise of the array output signal outside the synthesizer loop bandwidth, by a factor of $10 \log_{10} N$ (where N is the number of elements combined in the array). The phase-noise improvement makes it possible to use on-chip frequency synthesizers for high-precision beamforming applications which require low phase noise.

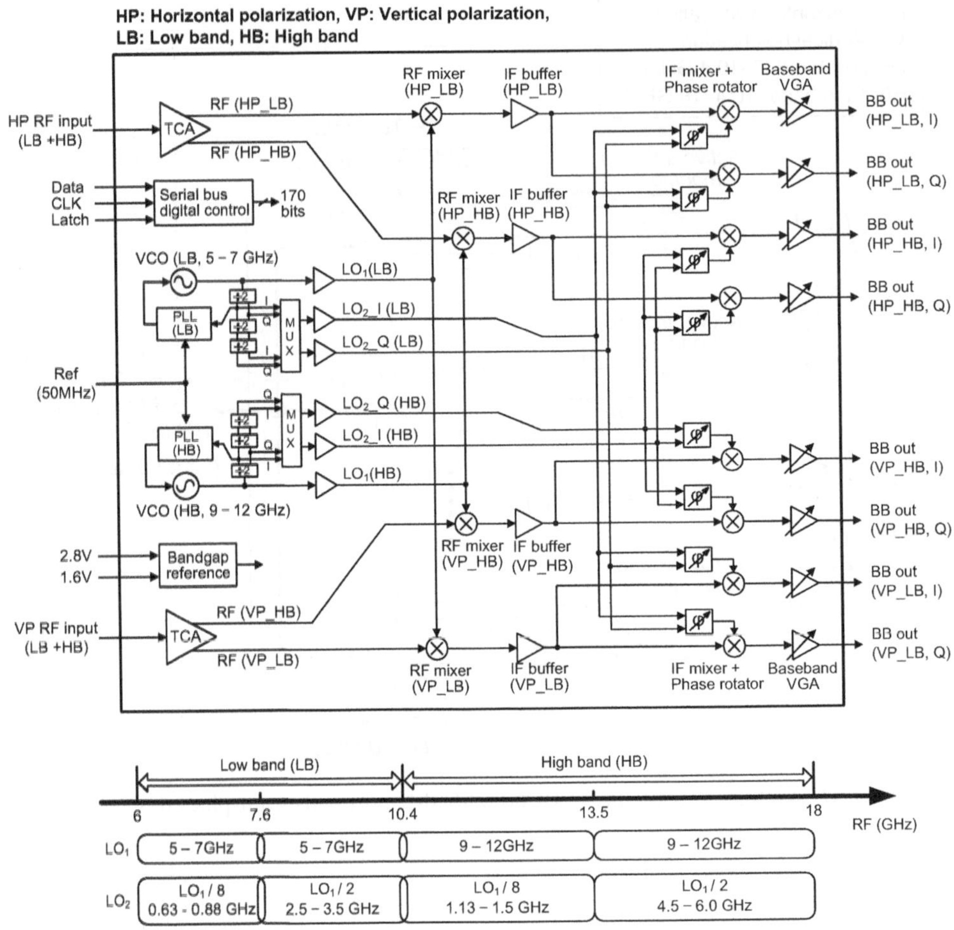

Figure 14.48: Architecture and frequency plan of the CMOS phased-array receiver element.

14.5.2 CMOS Phased-array Element

Figure 14.48 shows the architecture of the CMOS phased-array receiver element with the frequency plan. The incoming RF signal, containing two frequencies at LB and HB respectively, feeds a front-end TCA. The TCA amplifies, filters, and finally splits the RF signal into two separate outputs; one at LB and the other at HB, as shown in Figure 14.49. The wideband input matching to 50 Ω is accomplished by an active termination with resistive feedback and an impedance transformation network. The active termination contributes less noise to the subsequent blocks than a simple shunt resistive termination [28]. The RF signals are then selectively amplified by two separate cascode amplifiers that have tunable LC output loads. A 3 bit switched capacitor bank at each output load is tuned to cover the entire LB and HB frequencies. This allows for the digital tuning of the amplifier so that it can provide the maximum gain at the desired frequency while attenuating out-of-band signals prior to the first down-conversion.

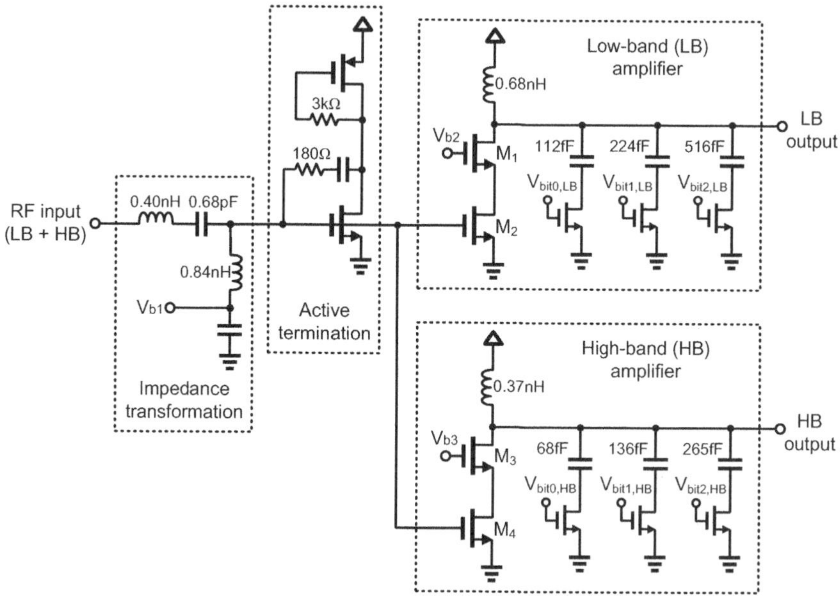

Figure 14.49: Schematic of the 6–18 GHz tunable concurrent amplifier.

Each of the LB and HB signals is down-converted in two steps via current-commutating double balanced mixers. While a shunt-peaking inductor is employed in the LB RF mixers to extend the operating bandwidth, the HB RF mixers have a tunable LC load with a 3 bit switched capacitor bank, as shown in Figure 14.50. The quadrature IF mixers perform not only the second down-conversion but also the phase shifting with the digitally controlled phases of LO signals. The baseband variable-gain amplifiers (VGAs) provide 11 dB of gain variation in five steps. Since there are two identical RF signal paths for the HP and VP signals as shown in Figure 14.48, the receiver chip presents eight differential baseband outputs in total, one for each combination of two frequency bands, two polarizations, and I and Q.

Two on-chip frequency synthesizers generate LO signals for LB and HB, respectively. As can be seen in the frequency plan shown in Figure 14.48, the receiver relies upon a dual frequency scheme where the LO frequency for IF mixers (LO_2) switches between 1/2 and 1/8 of the LO frequency for the first down-conversion (LO_1). This dual scheme reduces the required VCO tuning range from 1:1.7 to 1:1.3 while the entire tritave frequencies are covered with no blind spot. The on-chip VCOs are implemented by a cross-coupled differential PMOS pair with a two-step frequency tuning method, as shown in Figure 14.51. The first coarse tuning is achieved by a 2 bit switched-capacitor bank, and then MOS varactors tune the frequency further in a fine manner. The VCO frequency is tuned from 5 to 7 GHz for LB and from 9 to 12 GHz for HB. Each VCO is locked to a 50 MHz reference signal by a fully-programmable phased-locked loop (PLL), which synthesize a required LO frequency with a 200 MHz step.

A 10 bit digital phase rotator shifts the phase of LO signal driving each IF mixer. As shown in Figure 14.52, the phase rotator takes and amplifies the quadrature (I and Q) components

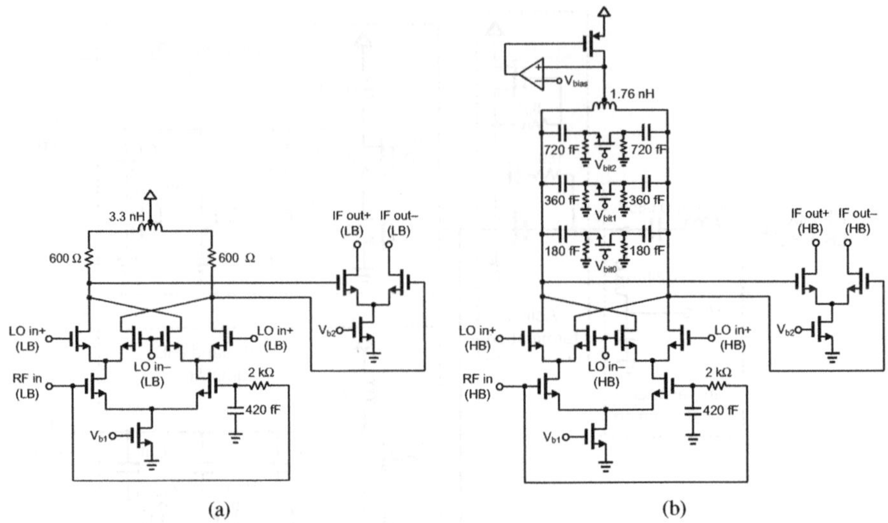

Figure 14.50: Schematic of the RF mixers: (a) LB RF mixer; (b) HB RF mixer.

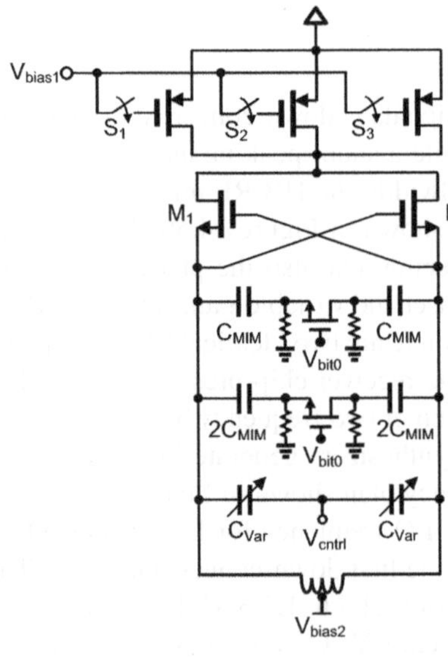

Figure 14.51: Schematic of the on-chip VCOs.

of the LO signal with different weighting coefficients by two VGAs [29]. By combining the VGA outputs, the required phase and amplitude is interpolated in the Cartesian coordinates of the I and Q outputs. Since each VGA is controlled by five bits, this scheme independently generates 2^{10}(= 1024) distinct interpolation points for the LO phase and amplitude. This

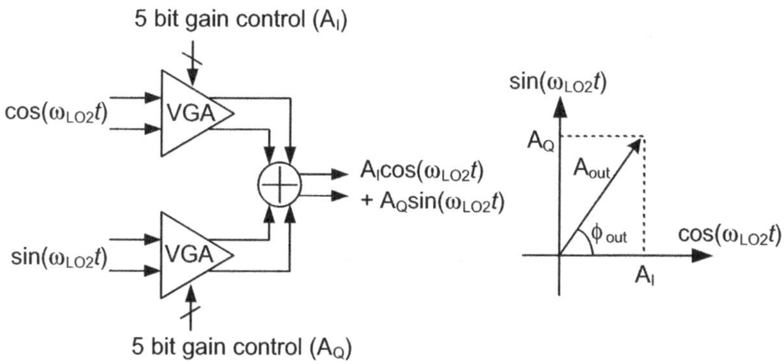

Figure 14.52: Digital phase rotator.

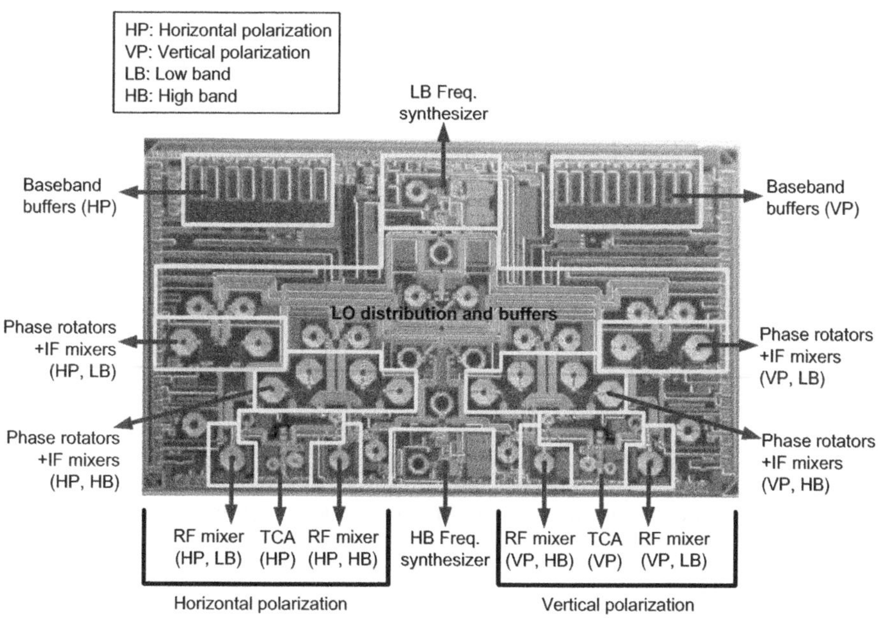

Figure 14.53: Die micrograph of the 6–18 GHz dual-band quad-beam phased-array receiver.

dense phase configuration of the receiver makes it possible to accurately compensate for any phase errors occurred between the array elements. The phase errors originate from the I/Q mismatches and excessive harmonic contents of the LO signals and/or the systematic skews in the distribution of the off-chip reference signal, which results in degradation of the array performance such as increase of the beam-pointing error and sidelobe level.

The receiver functionalities including the RF receiving frequency, on-chip LO frequency, phase shifting, and receiver gain are controlled by a built-in digital serial bus. This digital

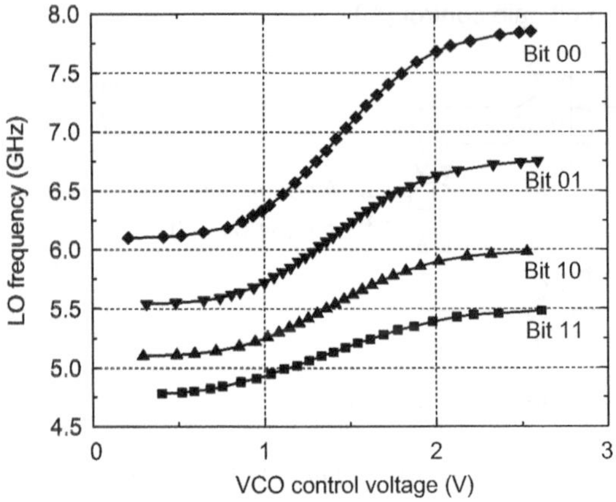

(a)

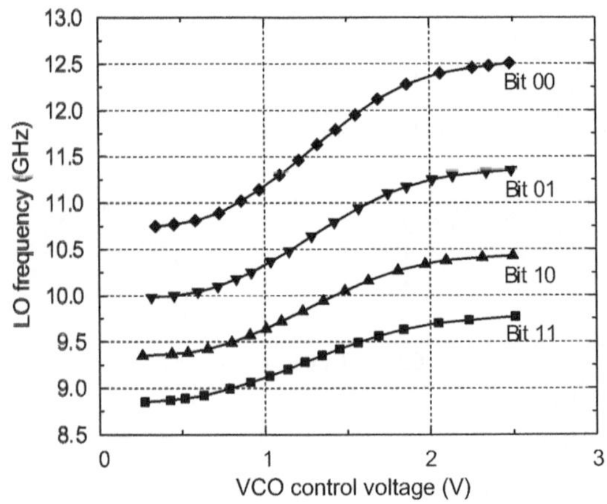

(b)

Figure 14.54: Measured performance of the on-chip frequency synthesizers: (a) LB frequency synthesizer; (b) HB frequency synthesizer.

controllability maximizes the flexibility of the receiver for extended applications such as smart-antenna systems.

14.5.3 Experimental Results

The 6–18 GHz dual-band quad-beam phased-array receiver is implemented in a 130 nm CMOS process with eight metal layers. Figure 14.53 shows a micrograph of the receiver chip that occupies 3.0×5.2 mm^2.

The measured performance of the on-chip frequency synthesizers is presented in Figure 14.54, where the LO frequency is plotted as a function of the VCO control voltage for LB and HB, respectively. Each curve represents one of the four different bit settings for the VCO

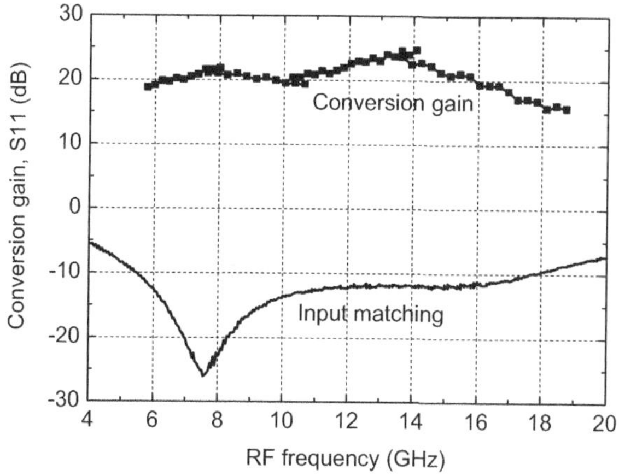

Figure 14.55: Measured conversion gain and input matching performance.

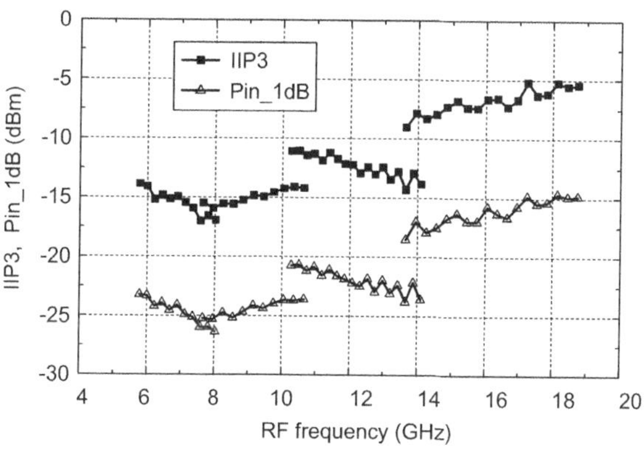

Figure 14.56: Measured nonlinearity performance.

switched capacitors. The LB and HB synthesizers generate 4.8–7.8 GHz and 8.8–12.5 GHz of LO signals, respectively, without any blind spot.

Figure 14.55 shows the measured conversion gain and input matching performance over the tritave. The conversion gain ranges from 16 to 24 dB with nominal baseband VGA settings. The discontinuities at 7.6, 10.4 and 13.5 GHz are due to the switching of either the frequency band or the IF frequency scheme. The input return loss is more than 9.8 dB across the entire band from 6 to 18 GHz. The measured nonlinearity performance is shown in Figure 14.56, where the input-referred values of 1 dB gain compression points (Pin_1dB) and third-order intercept points (IIP3) are presented.

The measured noise figure of the CMOS receiver ranges from 8 to 14 dB, as shown in Figure 14.57. However, when incorporating a 2.5 dB noise figure and a 20 dB gain of the active antenna module (Figure 14.47) with the CMOS chip, the entire system

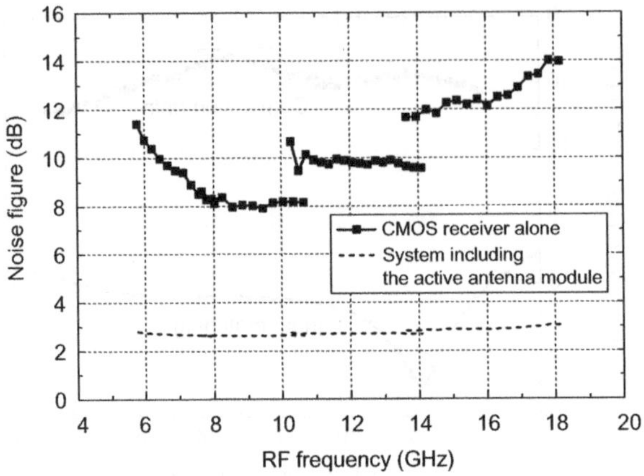

Figure 14.57: Measured noise figure.

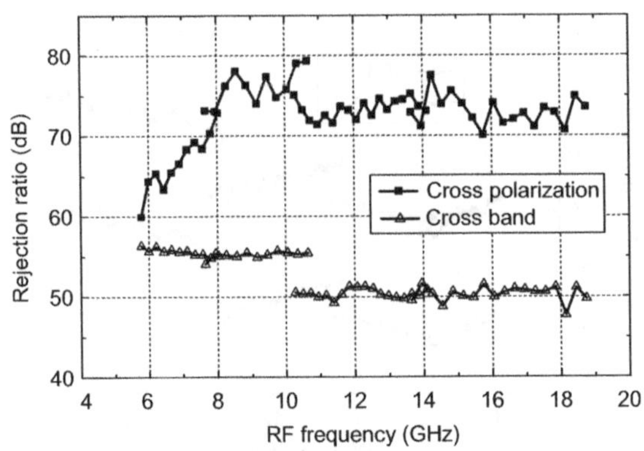

Figure 14.58: Measured isolation performance.

noise figure reduces significantly to 2.6–3.0 dB. The isolation performance between two different frequency bands and between two different polarizations are measured as shown in Figure 14.58. The worst-case cross-band and cross-polarization rejections are 48.4 dB and 63.4 dB, respectively.

The on-chip phase-shifting performance is characterized by measuring a relative delay of the down-converted baseband signal while changing the LO phase by 1024 different interpolating settings in the phase rotator. For the entire operating frequency, the worst-case RMS phase error is measured as 0.5° within a RMS amplitude variation of 0.4 dB.

To demonstrate the array performance, a four-element phased array is implemented by combining four CMOS receiver chips. The on-chip LO phase of each element is synchronized by symmetrically distributing a 50 MHz reference signal generated by an off-chip low-noise source. The array pattern is measured by an electrical way in which four external

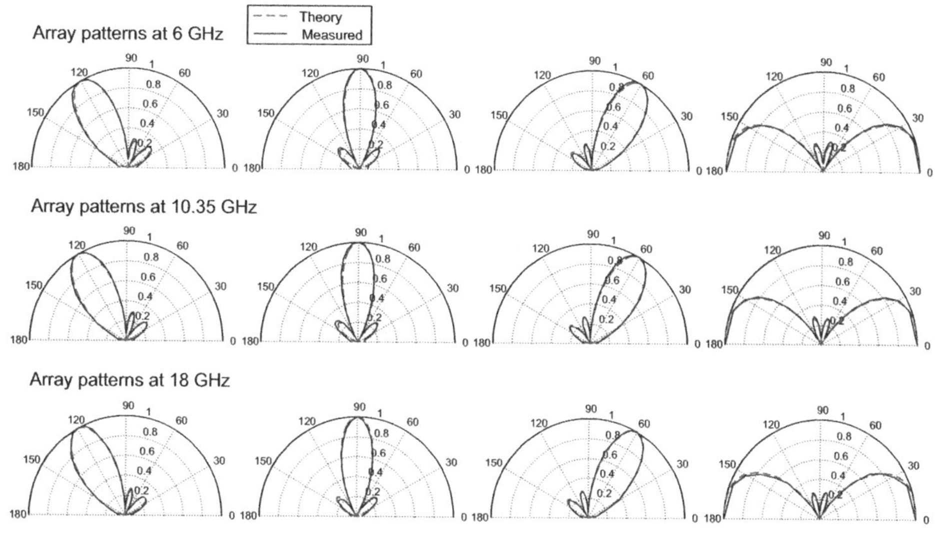

Figure 14.59: Measured electrical array patterns at 6, 10.35 and 18 GHz.

variable phase shifters are used to emulate an incoming wavefront at a given incident angle. Figure 14.59 shows the measured electrical array patterns with four different beam-pointing angles set at 6, 10.35, and 18 GHz, respectively. Theoretical array patterns are superimposed on the measured ones. As can be seen, the measured beam patterns are well steered in good agreement with the theoretical ones. The worst-case peak-to-null ratio is 21.5 dB. This good array performance is due to the high resolution of the on-chip phase shifting, which allows accurate array calibration to offset any unavoidable skews and mismatches presented among the array elements.

Each array element draws 658 mA and 217 mA from DC supplies of 2.7 V and 1.6 V, respectively. In addition, each baseband buffer draws 34 mA from a 1.5 V DC supply.

14.6 Conclusions

As the application of wireless technologies expands rapidly from its historical military and space areas to commercial markets, the system requirements become more and more stringent. A reliable system supporting a fast data rate needs to be implemented in low power consumption and low cost. This recent trend strongly drives the motivation to integrate as many components as possible on a single chip, particularly, in silicon. Phased arrays are one of the exemplary systems that need the high-level integration due to the large number of components associated with the system. Compared with conventional module-based ways of building phased arrays, the integrated solution provides several benefits such as low cost, low complexity, small volume, and high reliability. Fortunately, the integrated phased arrays at mmWave frequencies become feasible in today's technologies because of two major reasons. First, the advancement in silicon process technology makes it possible to integrate mmWave active and passive circuit blocks on the same die with low cost and high repeatability. The scaling of transistors keeps pushing the maximum operating frequency up to several hundreds of GHz and the improved silicon process provides passive elements (e.g. inductors, capacitors, transmission lines) with good performance at high frequencies. The second reason is that the wavelength is reduced to a reasonably small value at mmWave frequencies, such that wavelength-dependent blocks such as antennas are able to be integrated on the same die. By exploiting this favorable environment and matured technology, a fully-integrated phased-array transceiver is implemented in silicon at 77 GHz. The transceiver also includes on-chip antennas that can radiate from the backside of the chip with a silicon lens attached to improve the radiation efficiency. In addition, the flexibility of silicon ICs to incorporate multiple fields from electromagnetics to a system level enables the implementation of a novel concept for directional data transmission – the direct antenna modulation. By modulating the antenna near-field characteristics, an efficient and secured communication link can be established in a fundamentally different way from conventional radio techniques. Finally, an efficient architecture suitable for very large-scale phased arrays is presented and implemented by taking full advantage of the high-level integration capability in silicon. The integrated approach dramatically reduces the cost and complexity of wideband multi-beam phased-array systems.

Acknowledgments

The authors would like to thank H. Wang, Y.-J. Wang, F. Bohn of the California Institute of Technology (Caltech) and Dr A. Natarajan, Dr X. Guan and Dr A. Komijani, formerly of Caltech for their contributions to microwave and mmWave integrated circuits presented in this chapter. The authors also would like to thank Prof. D. Rutledge, Dr S. Weinreb of Caltech, J. DeFalco, R. Healy, and J. Holley of Raytheon for helpful technical discussions.

References

[1] J. Spradley, 'A volumetric electrically scanned two-dimensional microwave antenna array', *IRE Int. Conv. Rec.* **6** (1958), pp. 204–212.

[2] D. Parker and D. C. Zimmermann, 'Phased arrays – Part I: Theory and architectures', *IEEE Trans. Microw. Theory Tech.* **50**(3) (2002), pp. 678–687.

[3] W. L. Stutzman and G. A. Thiele, *Antenna Theory and Design*, 2nd edn (New York: John Wiley & Sons Ltd/Inc., 1998).

[4] Federal Communications Commission, FCC Part 15, sec. 15.249, 15.252, 15.253, 15.255.

[5] D. Parker and D. C. Zimmermann, 'Phased arrays – Part II: Implementations, applications, and future trends', *IEEE Trans. Microw. Theory Tech.* **50**(3) (2002), pp. 688–698.

[6] R. J. Mailloux, *Phased Array Antenna Handbook*, 2nd edn (Norwood, MA: Artech House, 2005).

[7] P. Lacomme, 'New trends in airborne phased array radars', in *IEEE Int. Symp. on Phased Array Syst. and Technol.*, October 2003, pp. 17–22.

[8] X. Guan, H. Hashemi and A. Hajimiri, 'A fully integrated 24 GHz eight-element phased-array receiver in silicon', *IEEE J. Solid-State Circuits* **39**(12) (2004), pp. 2311–2320.

[9] A. Natarajan, A. Komijani, and A. Hajimiri, 'A fully integrated 24 GHz phased-array transmitter in CMOS', *IEEE J. Solid-State Circuits* **40**(12) (2005), pp. 2502–2514.

[10] A. Babakhani, X. Guan, A. Komijani, A. Natarajan and A. Hajimiri, 'A 77 GHz phased-array transceiver with on-chip antennas in silicon: Receiver and antennas', *IEEE J. Solid-State Circuits* **41**(12) (2006), pp. 2795–2806.

[11] A. Natarajan, A. Komijani, X. Guan, A. Babakhani and A. Hajimiri, 'A 77 GHz phased-array transceiver with on-chip antennas in silicon: transmitter and local LO-path phase shifting', *IEEE J. Solid-State Circuits* **41**(12) (2006), pp. 2807–2819.

[12] S. Jeon, Y.-J. Wang, H. Wang, F. Bohn, A. Natarajan, A. Babakhani and A. Hajimiri, 'A scalable 6-to-18 GHz concurrent dual-band quad-beam phased-array receiver in CMOS', in *IEEE Int. Solid-State Circuits Conf. (ISSCC) Dig. Tech. Papers*, San Francisco, CA, February 2008, pp. 186–187.

[13] A. Babakhani, D. B. Rutledge and A. Hajimiri, 'A near-field modulation technique using antenna reflector switching', in *IEEE Int. Solid-State Circuits Conf. (ISSCC) Dig. Tech. Papers*, San Francisco, CA, February 2008, pp. 188–189.

[14] A. Hajimiri, 'mm-Wave silicon ICs: Challenges and opportunities', in *IEEE Custom Integrated Circuits Conf.*, San Jose, CA, September 2007, pp. 741–747.

[15] S. Lee *et al.* 'Record RF performance of 45 nm SOI CMOS technology', in *IEEE Int. Electron Devices Meeting*, Washington, DC, December 2007, pp. 255–258.

[16] A. Hajimiri, A. Komijani, A. Natarajan, R. Chunara, X. Guan and H. Hashemi, 'Phased array systems in silicon', *IEEE Commun. Mag.* (2004), **42**(8), pp. 122–130.

[17] A. Hajimiri, H. Hashemi, A. Natarajan, X. Guan and A. Komijani, 'Integrated phased array system in silicon', *Proc. IEEE* **93**(3) (2005), pp. 1637–1655.

[18] 'IRE standards on electron tubes: Definition of terms', *Proc. IRE*, vol. 45, pp. 983–1010, July 1957.

[19] I. Aoki, S. D. Kee, D. B. Rutledge and A. Hajimiri, 'Fully integrated CMOS power amplifier design using the distributed active-transformer architecture', *IEEE J. Solid-State Circuits* **37**(3) (2002), pp. 371–383.

[20] E. Afshari, H. Bhat, X. Li and A. Hajimiri, 'Electrical funnel: A broadband signal combining method', in *IEEE Int. Solid-State Circuits Conf. (ISSCC) Dig. Tech. Papers*, San Francisco, CA, February 2006, pp. 751–760.

[21] D. B. Rutledge, D. P. Neikirk and D. Kasilingam, 'Integrated-circuit antennas', *Infrared and Millimeter-waves*, ed. K. J. Button (New York: Academic, 1983), vol. 10, pp. 1–90.

[22] H. Kogelnik, 'Theory of dielectric waveguides', *Integrated Optics*, ed. T. Tamir (New York: Springer, 1975) Chapter 2.

[23] D. F. Filipovic, G. P. Gauthier, S. Raman and G. M. Rebeiz, 'Off axis properties of silicon and quartz dielectric lens antennas', *IEEE Trans. Antennas Propagat.* **45**(5) (1997), pp. 760–766.

[24] H. Hashemi, X. Guan, A. Komijani and A. Hajimiri, 'A 24 GHz SiGe phased-array receiver—LO phase shifting approach', *IEEE Trans. Microw. Theory Tech.* **53**(2) (2005), pp. 614–626.

[25] D. M. Pozar, *Microwave Engineering*, 3rd edn (New York: John Wiley & Sons Ltd/Inc., 2005).

[26] M. K. Chirala and B. A. Floyd, 'Millimeter-wave lange and ring-hybrid couplers in a silicon technology for E-band applications', in *IEEE MTT-S Microwave Symp. Dig.*, San Francisco, CA, June 2006, pp. 1547–1550.

[27] A. Komijani and A. Hajimiri, 'A wideband 77 GHz, 17.5 dBm power amplifier in silicon', *IEEE J. Solid-State Circuits* **41**(8) (2006), pp. 1749–1756.

[28] P. Ikalainen, 'Low-noise distributed amplifier with active load', *IEEE Microwave Guided Wave Lett.* **6**(1) (1996), pp. 7–9.

[29] H. Wang and A. Hajimiri, 'A wideband CMOS linear digital phase rotator', in *IEEE Custom Integrated Circuits Conf.*, San Jose, CA, September 2007, pp. 671–674.

15

Millimeter-wave Imaging

Zuowei Shen and Neville C. Luhmann, Jr

15.1 Introduction to mmWave and THz Imaging

Millimeter-wave (mmWave) and THz imaging are proving to be valuable adjuncts to visible, infrared (IR), and X-ray imaging systems. The region of the electromagnetic spectrum where humans can see is the visible or optical spectrum which is approximately 380 to 750 nm, corresponding to sunlight. In that region, humans see different colored lights with different wavelengths. In situations without sunlight, appropriate sensors can detect emission from objects in the IR regime (governed by Planck's radiation law). The IR sensor technology advances in the last 30 years have thus made true night vision possible. To further expand the realm of our 'vision', researchers have focused on explorations of the mmWave regime and the further THz regime.

The mmWave (THz) regime roughly corresponds to electromagnetic waves with frequency between 30 and 300 GHz (0.3–3 THz), with corresponding free space wavelength between 10 and 1 mm (1000 and 100 μm). The advantage of mmWave radiation is that, in addition to clear weather day and night operation, it can also be used in low visibility conditions such as in smoke, fog, clouds, and even sandstorms [1, 2]. In this way, mmWave imaging expands our vision by letting us 'see' things under poor visibility conditions. With this extended vision ability, a wide range of military imaging missions can benefit, such as surveillance, precision targeting, navigation, search, and rescue. In the commercial realm, aircraft landing aids, airport operations, and highway traffic monitoring in fog can also benefit from mmWave imaging. For security concerns, concealed weapon and explosives detection in airports and other locations can be aided with several passive as well as active mmWave imaging techniques [1]. Compared with X-ray imaging, active mmWave imaging does not present a health hazard to people under surveillance or to operating personnel because of the non-ionizing nature of the radiation. In addition, mmWave imaging is also applied in

Advanced Millimeter-wave Technologies: Antennas, Packaging and Circuits
Duixian Liu, Ulrich Pfeiffer, Janusz Grzyb and Brian Gaucher © 2009 John Wiley & Sons, Ltd

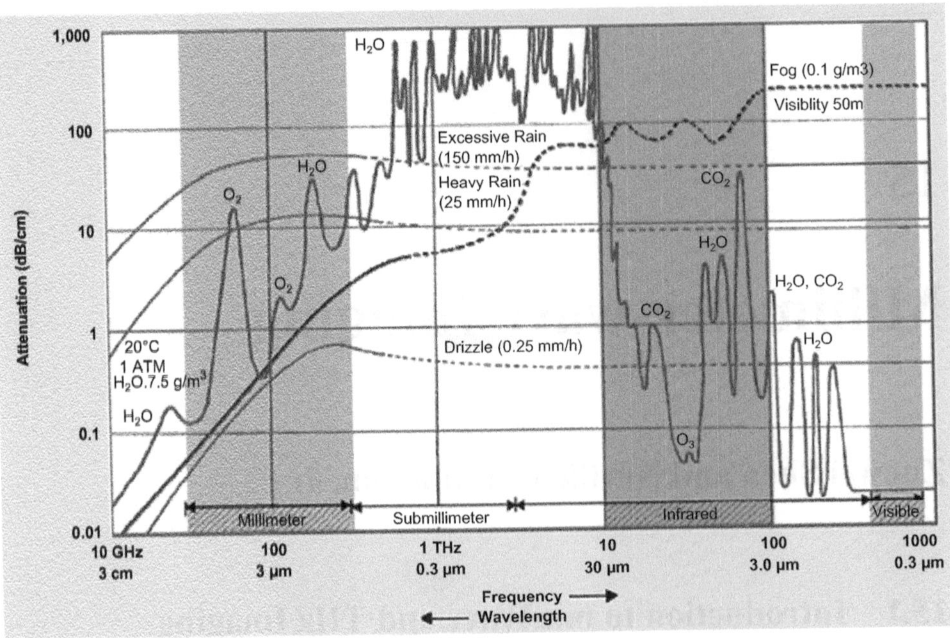

Figure 15.1: Attenuation of mmWaves by atmospheric gases, rain and fog. Reproduced by permission of © 2003 IEEE [1].

remote sensing, radio astronomy, and plasma diagnostics to visualize objects that cannot be seen directly [3]. Compared to microwaves, mmWave/THz imaging also offers an advantage in either reduced antenna aperture size or higher resolution for a given aperture since the beam divergence $\theta_{3\text{ dB}}$ (beamwidth) is given by: $\theta_{3\text{ dB}} = (\kappa\lambda/d)$. Here, κ is a constant (70 for degrees and 1.22 for radians), λ is the wavelength (m), and d is the aperture diameter (m).

The development of mmWave and THz imaging systems has also been strongly motivated by early modeling that suggested a considerable advantage over infrared and visible systems, which are adversely affected by atmospheric conditions such as dust and fog. Figure 15.1 contains an often used compilation of plots that have provided impetus for the developments. mmWave propagation is affected by the resonant absorption from various atmospheric constituents such as water, oxygen, and carbon dioxide. IR and visible radiation can propagate with little attenuation in clear, dry weather; however, they both suffer significant attenuation and scattering when water vapor appears in the form of fog, clouds, and rain. Fortunately, in the mmWave regime, there exist so-called propagation 'windows' at 35, 94, 140, and 220 GHz, where the attenuation is relatively modest in both clear air and fog. Even taking into account the significantly higher blackbody radiation at IR and visible frequencies for normal temperatures, mmWaves still provide the strongest signals in fog since mmWave radiation has significantly less attenuation under low-visibility conditions than visible or IR radiation [1, 2]. These mmWave windows make possible both visualization under low visibility conditions and high-speed wireless communications (i.e. the new E-band).

Table 15.1: Weather conditions employed for atmospheric propagation estimates.

Condition	Temperature (C)	Pressure (hPa)	Relative humidity(%)	Rain (mm/hr)	Fog (m)	Dust (m)
STD	20	1013	44	0	0	0
HHH	35	1013	90	0	0	0
Winter	−35	1013	30	0	0	0
Fog	20	1013	44	0	100	0
Dust	20	1013	44	0	0	10
Rain	20	1013	44	4	0	0

Reproduced by permission of © 2007 IEEE [5].

In recent years, there has been considerable effort devoted to developing both mmWave and THz imagers and this has led to careful consideration of the effects of atmospheric conditions on the range of systems using the latest models, since propagation directly impacts the applicability for various applications [4,5]. These recent studies have modeled conditions such as humidity, dust, rainfall, and snow which can all lead to absorption and thus range degradation. References [4] and [5] contain interesting discussions of the genesis/parentage of the data plots contained in Figure 15.1 together with the discrepancies with modeling predictions based on the current understanding of atmospheric effects as well as references to the models used in their predictions. A general observation is that above 100 GHz the atmosphere is more opaque than in the optical region for clear atmospheres and that only under adverse conditions such as fog does the mmWave region experience less attenuation, with this advantage disappearing above 400 GHz.

To provide quantitative guidance for the assessment of the potential of mmWave and THz imaging systems, references [4, 5] have calculated the attenuation for six different atmospheric conditions, all at sea level, representing the extremes from very low atmospheric attenuation in the winter, to very high levels in tropical climates (see Table 15.1 for the specific atmospheric conditions). The results of such studies are indicated in Figure 15.3 and, as noted in reference [4], the atmospheric windows are seen to be rather broad below 300 GHz, becoming less well defined as attenuation increases and with more absorption lines appearing for higher frequencies. Using these data together with knowledge of the current state-of-the-art of source and detector technology (see Section 15.6) as well as considerations of reasonable antenna size, one may assess the practicality of 'remote' imaging. For example, assuming a distance of $\sim$10 m as a minimum range, attenuation in excess of $\sim$1 dB m^{-1} (10^3 dB km^{-1}) appears impractically large, thereby limiting one to operation below $\sim$400 GHz except for frequencies of particular interest which include: 340 GHz, 650 GHz, 850 GHz, 1.05 THz, and 1.5 THz.

The images in Figure 15.2 demonstrate the advantages of mmWave imaging over visible imaging under low visibility conditions. The visible images (Figure 15.2(c)) are totally obscured, while the PMMW images (Figure 15.2(d)) are nearly unaffected by the fog. As discussed in detail in Section 15.2, in PMMW imaging, naturally occurring radiation is collected and images are formed through contrasts between both thermally warmer and colder objects as well as differences in material emissivities. Conventional cameras detect

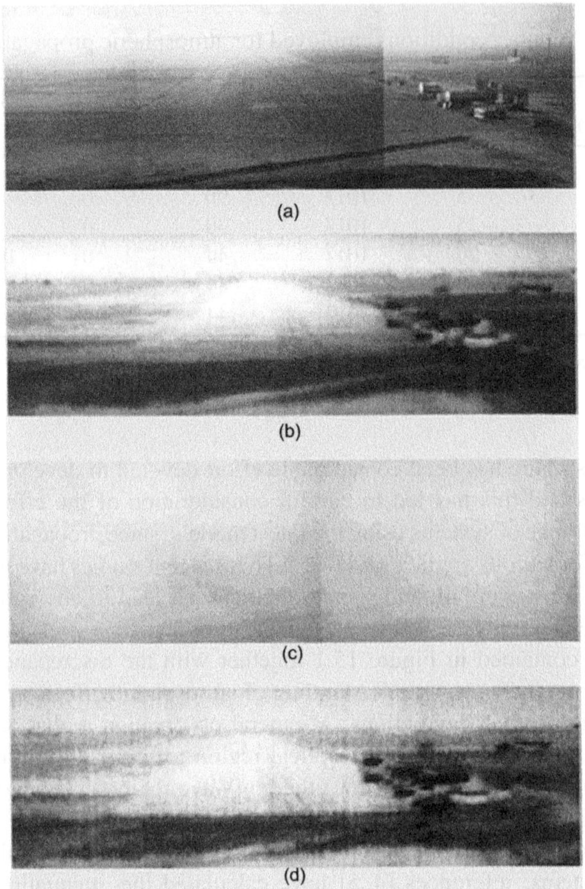

Figure 15.2: Images of an airport runway viewed from the glide slope before touchdown: (a) and (c) show visible images in clear and foggy weather; (b) and (d) show the corresponding passive mmWave (PMMW) images. Reproduced by permission of © 2003 IEEE [1].

visible energy when there is light while infrared cameras can detect infrared thermal energy in the night; mmWave imaging 'cameras' provide capability for detecting energy under low visibility conditions.

The combination of the above advantages and recent advances in the state of the associated technologies has resulted in the application of mmWave/THz imaging in a variety of areas including advanced plasma diagnostics, radio astronomy, atmospheric radiometry, concealed weapon detection, all-weather aircraft landing, contraband goods detection, helicopter and automotive collision avoidance in fog, harbor navigation/ surveillance and highway traffic monitoring in fog, and environmental remote sensing data associated with weather, pollution, soil moisture, and oil spill detection. They have also been applied to measure surface moisture, snow cover, vegetation type and extent, mineral type and extent, and surface

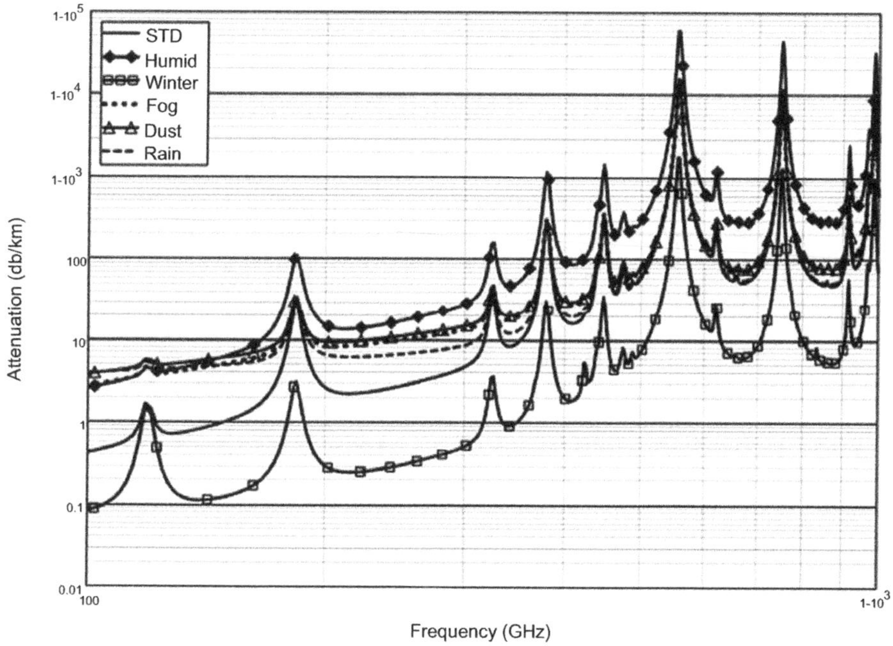

Figure 15.3: Calculated atmospheric absorption as a function of frequency for a number of different atmospheric conditions as described in Table 15.1. Reproduced by permission of © 2007 IEEE [5].

temperature and thermal inertia and also map fires and volcanic lava flows through obscuring clouds and smoke as well as provide cloud and satellite images. Since a number of these applications are aimed at imaging through materials such as clothing, it is important to have quantitative measurements of transmission as a function of frequency. Figure 15.4 displays the measured transmission as a function of frequency for a number of common materials such as wool, denim, silk, and nylon [6]. It is seen that the transmission is quite good in the mmWave region (< 2 dB at 200 GHz) and ~2–10 dB at 800 GHz.

mmWave imaging can basically be divided into two separate approaches: passive mmWave imaging and active mmWave imaging. PMMW systems directly detect the natural radiation from the objects or the reflection from the environment; the concept is analogous to radiometry. Active mmWave imaging systems first illuminate the objects and subsequently detect the reflected mmWaves. Many active imaging systems are essentially radar-based. In the following sections, passive and active mmWave imaging concepts and technologies, as well as some representative examples will be briefly described.

15.2 Passive mmWave Imaging Systems

PMMW imaging is a method of forming images through the passive detection of naturally occurring mmWave radiation from a scene and making use of differences in temperature

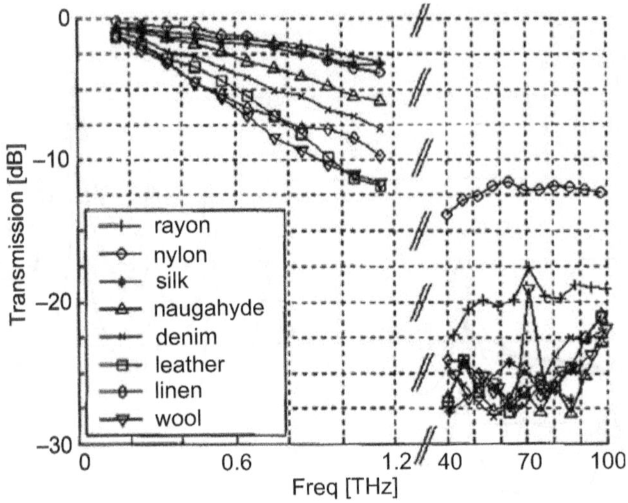

Figure 15.4: THz and mid-IR transmission through eight clothing samples with a common vertical axis (and scale) but broken horizontal axis (and different frequency scales). Reproduced by permission of © 2004 American Institute of Physics [6].

and emissivity [1, 6, 7]. Such imaging has been performed for decades; however, the new mmWave technologies, particularly new sensor technologies such as mmWave antennas and the Schottky diode mixers and hot electron bolometers described in Section 15.6, have enabled PMMW imaging at video rates [8, 9], thus offering the promise of additional applications. For the military, the enhanced vision systems afford advantages under low-visibility conditions. In the commercial arena, aircraft can be operated under all-weather conditions when fog-bound airports are eliminated as a cause for flight delays or diversions. In the security realm, as mentioned previously, with the ability to non-intrusively detect concealed weapons, it is predicted that PMMW imaging will form the core of a gateway security system and be used at entry points in transportation terminals and in buildings such as court houses, offices, and secure facilities [1, 6, 10].

Objects reflect and emit radiation naturally in the mmWave regime. The 'emissivity (ϵ)' parameter is employed to quantify how much an object reflects and emits. A blackbody has $\epsilon = 1$ as a perfect radiator, and a perfect reflector has $\epsilon = 0$. The emissivity of an object is a function of its dielectric properties, surface roughness, and angle of observation. Since metal, dielectric material, and the human body have different emissivity and reflectivity values (as shown in Table 15.2), they can be discriminated with mmWaves.

An object's radiometric temperature T_s is defined as the product of the object's physical thermodynamic temperature T_0 and its emissivity.

$$T_s = \epsilon T_0 \tag{15.1}$$

However, imaging is also critically dependent upon the way in which the object is illuminated. For example, a perfect reflector with $\epsilon = 0$ will have $T_s = 0$, but it still appears to possess finite radiometric temperature through the reflected energy. This effect is accounted

Table 15.2: Effective emissivity of common materials at various frequencies.

Surface	Effective emissivity		
	44 GHz	94 GHz	140 GHz
Bare metal	0.01	0.04	0.06
Painted metal	0.03	0.10	0.12
Painted metal under canvas	0.18	0.24	0.30
Painted metal under camouflage	0.22	0.39	0.46
Dry gravel	0.88	0.92	0.96
Dry asphalt	0.89	0.91	0.95
Dry concrete	0.86	0.91	0.95
Smooth water	0.47	0.59	0.66
Rough or hard-packed dirt	1.00	1.00	1.00

Reproduced by permission of © 2007 IEEE [1].

for through the surface-scattered radiometric temperature T_{SC}, which is the product of the reflectivity and the radiometric temperature of the illuminator, $T_{ILLUMINATOR}$:

$$T_{SC} = \rho T_{ILLUMINATOR} \tag{15.2}$$

Thus, the effective radiometric temperature T_E, which mmWave systems measure, is the sum of these two terms.

$$T_E = T_S + T_{SC} = \epsilon T_0 + \rho T_{ILLUMINATOR} \tag{15.3}$$

When a mmWave imaging system observes a ground scene from above, as depicted in Figure 15.5, the observed object temperature is dependent upon the emission from objects, reflected sky emission, and atmospheric emission between the scene and the observer [1]. The received effective temperature by the radiometers is also determined by the beamwidth of the main antennas. There are also other factors such as viewing angle, surface orientation, and polarization effects. Those factors have been discussed in more detail and accurately modeled in simulations. The received signal temperatures under various atmospheric conditions for a reflective object with low emissivity ($\epsilon = 0.1$) and for an absorbing object with higher emissivity ($\epsilon = 0.8$) are shown in Table 15.3 [11]. Figure 15.6 shows the images of a harbor scene obtained with a single 94 GHz radiometer with a 48 inch aperture dish antenna. Note that black represents cold, so that the metal domed structure behind the Queen Mary, reflecting the cold sky, appears black. Also note that the chimney in the middle appears brighter than the other two. That is because the one in the middle was repaired and different material was used. In order to assess the applicability of exploiting sky temperature as an illuminator, references [4] and [5] performed calculations of the frequency dependence of the sky temperature for the various atmospheric conditions enumerated in Table 15.1 (see Figure 15.7). The authors conclude that sky illumination is practical for 94 GHz PMMW imaging systems, and that the available contrast is significantly reduced for increasing operating frequency.

PMMW imaging systems are quite similar to visible light video camera imaging systems. mmWave imaging systems usually consist of an array ofmmWave detectors, collecting optics,

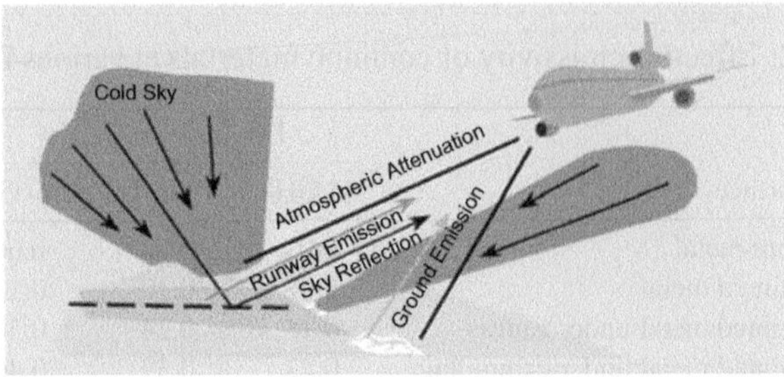

Figure 15.5: The observed radiometric image is composed of various effects. Reproduced by permission of © 2003 IEEE [1].

Table 15.3: Received signal temperatures (K) for various atmospheric conditions when observing an object which fills the antenna beam from a range of 750 m.

	$T_{received}$		
Atmospheric conditions	T_{sky}	$\epsilon = 0.1$	$\epsilon = 0.8$
Clear, smoke, light fog ($L_{atm} = 0.6$ dB km^{-1})	60	103	249
Thick fog, overcast ($L_{atm} = 1.0$ dB km^{-1})	120	161	261
Thick clouds ($L_{atm} = 2.0$ dB km^{-1})	180	220	274

Reproduced by permission of © 1986 IEEE [11].

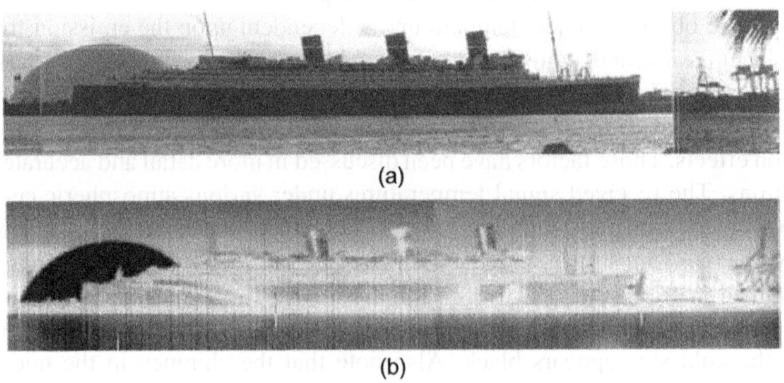

Figure 15.6: (a) Visible and (b) PMMW images of the Queen Mary and former Spruce Goose dome in Long Beach, California, harbor. The ship is 1,700 ft from the imaging system; black represents cold. Reproduced by permission of © 2003 IEEE [1].

intermediate frequency (IF) electronics, and a data acquisition system together with imaging processing algorithms. All cameras are comprised of a means to measure the radiation level from the scene and a means for forming the image (directing the radiation onto

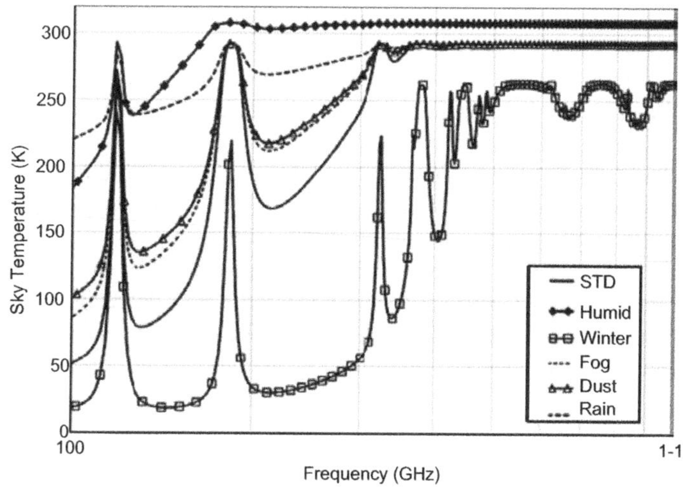

Figure 15.7: Calculated apparent radiometric sky temperature as a function of frequency for a number of different atmospheric conditions as described in Table 15.1. Reproduced by permission of © 2007 IEEE [5].

the measurement devices). Correspondingly, mmWave imaging systems are comprised of focusing optics and an energy detector array. The mmWave radiation from the objects within a scene first passes through the focusing optics, and is subsequently collected by afocal-plane array (FPA) of mmWave detectors. By measuring and recording the variation in the detected power or intensity levels, sequences of scene images can be obtained and a video-rate image stream can be formed [1]. Several existing passive millimeter imaging systems are described in Section 15.4 to illustrate representative configurations of PMMW systems.

15.3 Active mmWave Imaging

Active mmWave imaging [12] is considered to be an extremely promising approach for the development of practical and versatile systems to address the needs of indoor security applications where the range is limited. In an indoor environment, in order to achieve a sufficient difference in the radiation temperature between the objects to be imaged and the background noise level of a typical indoor environment, an active imaging approach with higher sensitivity is desirable. Active mmWave imaging systems illuminate the scene with a beam of mmWave energy and subsequently image the reflected energy by recording the origin and strength of the received signal from objects within the system's field of view. This concept is similar to that used in radar. Compared to PMMW imaging, active mmWave imaging has increased capabilities such as higher sensitivity as well as the ability to 'probe', thereby enabling additional applications. For example, active mmWave imaging systems are capable of 'seeing' through many building materials to detect the situation on the other side. This enables remote scanning of potentially dangerous areas prior to entry [1]. However, active mmWave imaging systems are limited by systematic uncertainties such as glint effects. In active mmWave imaging systems, truly incoherent sources are difficult to realize. Speckle occurs when scattered radiation from objects or rough surfaces randomly destructively interferes and degrades image smoothness.

Similar to PMMW imaging systems, active mmWave imaging systems commonly contain a focal plane antenna array with a detector/mixer combination. The antenna array can be simply a mechanically scanned one-dimensional array or a two-dimensional matrix for direct real-time imaging. For PMMW imaging, staring arrays are always superior to scanned arrays [10]. However, for active systems, scanning systems are more sensitive because the illumination can be focused on just those pixels being observed at each instant of the scan. The illumination geometry is critical to the appearance and detectability of targets [13]. The simplest illumination source can be a horn antenna. Other scanning and beam-forming technologies can be implemented for illumination and more details are provided in the next section.

15.4 Representative Examples of Passive and Active mmWave Imaging Systems

mmWave/THz imaging systems are a promising approach to screen people forconcealed weapons or contraband goods. As noted above, the radiation does not present a health hazard to people under surveillance or operating personnel (in contrast to ionizing radiation) and the radiation is transmitted through many optically opaque materials such as clothing fabrics, allowing for the identification of concealed objects. Even if active mmWave imaging systems are used, they can be operated at extremely low power levels, thereby reducing the risk of harm to people. mmWave imaging systems are also capable of detecting plastic and liquid explosives, or plastic or ceramic guns and knives which metal detectors cannot detect. Thus, mmWave imaging systems are being applied for concealed weapon detection. For example, following the suicide bomb attacks on London, the UK government and police are beginning to employ scanning technology that can see through clothes and walls and detect concealed bombs and weapons on people.

Figures 15.8 and 15.9 illustrate two imaging systems for this purpose designed to detect not just metal but other threats, such as ceramic knives and hidden drugs. Both detection systems are PMMW imaging systems which measure the naturally emitted energy by the human body, exposing cold objects such as metal, plastics, or ceramics concealed under clothing. It is reported that a single scanning portal is capable of scanning moving images of up to 500 people an hour [14]. Representative examples of mmWave imaging systems are described briefly in the following. The first one employs active mmWave illumination to form high-resolution three-dimensional images of the person being screened and it is being developed by Pacific Northwest National Laboratory (PNNL) to inspect persons rapidly for concealed explosives, hand guns, or other threats [16]. The second one is the Northrop Grumman Space Technology (NGST) PMMW video camera, which was the first FPA PMMW video camera [1, 8, 9]. The third is a 94 GHz passive imaging camera developed for security scanning purposes. The final example is a combined passive imaging and active imaging system referred to as an ECEI (electron cyclotron emission imaging)/MIR (microwave imaging reflectrometry) system, which has been developed for plasma diagnostics [3].

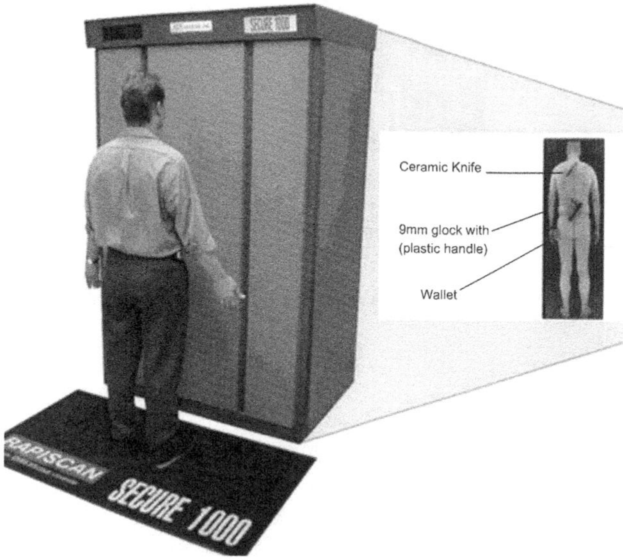

Figure 15.8: Rapiscan Security Products [14].

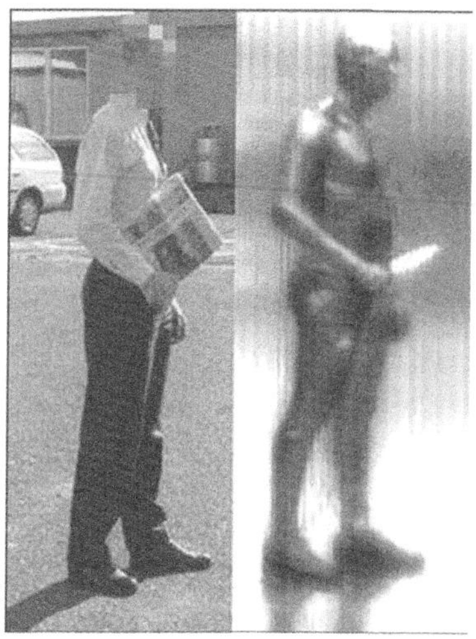

Figure 15.9: mmWave technology from security specialist Qinetiq [15].

15.4.1 Three-dimensional Active mmWave Video Camera

The active imaging system operates at 27–33 GHz and employs coherent illumination and detection (magnitude and phase) of the scattered wavefront [12, 16]. This wavefront is then mathematically reconstructed to form a full three-dimensional image. Three-dimensional

Figure 15.10: Prototype wideband mmWave imaging system. Reproduced by permission of © 1996 SPIE [16].

Figure 15.11: A 128 element, 27–33 GHz mmWave linear array. Reproduced by permission of © 1996 SPIE [16].

image construction is necessary because the human body is not flat, and it is impossible to obtain images in which all of the body is in focus at the same time without wideband three-dimensional operation. The approach and reconstruction algorithm are described in references [12] and [16].

In this active PNNL Ka-band system, the 128-element linear switched antenna array is supported within a 2 m vertical high-speed scanner. A transceiver creates the wideband illumination and measures the magnitude and phase of the scattered wavefront. The collected data from the transceiver are subsequently digitized with an analog-to-digital (A/D) converter and transferred to the computer. After a full aperture of data is collected, a computer reconstruction algorithm developed at PNNL is then used to convert the data into a three-dimensional image. Figure 15.10 shows the prototype imaging system and Figure 15.11 provides a close-up photograph of the 128-element mmWave array with transceiver and interface electronics [16].

The 128-element mmWave array is comprised of two 64-element arrays placed back-to-back. They are sequentially switched with control from the interface board. One 64-element array is employed as the mmWave signal transmitter and is composed of nine single-pole, eight-throw (SP8T) switch modules. The SP8T switch modules are fabricated in a split-block waveguide structure using a fin-line pin-diode configuration. The horizontal aperture is 0.75 mm and an interleaving scheme is used to obtain 127 independent samples of the scattered wavefront across the aperture. The 64-element receiver array is identical [16].

Two voltage controlled oscillators (VCOs) are employed in the transceiver, and their frequency difference is the IF of the system. The VCO for the transmitter is frequency doubled to 27–33 GHz, while the other VCO is frequency doubled to provide the local-oscillator (LO) signal for the receiver electronics. The received mmWave signal is down-converted to the IF frequency, which is subsequently down-converted to baseband to yield the in-phase signal (I). This signal contains the amplitude and phase of the scattered wavefront, which is the desired measurement.

Figures 15.12 and 15.13 are two imaging results obtained in real-time (1–2 s scans) using the active PNNL Ka-band system described above. Figure 15.12 shows six wideband three-dimensional images of a man carrying two concealed handguns as well as several harmless items. The first image shows a Glock-17 handgun tucked at the belt line under the man's shirt. The second image shows a small handgun in the man's left pants pocket, and it is close to his left hand. The third image shows a checkbook in the left back pocket. The fifth image shows a leather wallet in the man's right back pocket. The fourth and sixth images only show the human body. All those imaging results were obtained in near real-time scanning (1–2 s) [16].

Recent news reports have underscored the critical need for explosive detection in personnel screening applications, since relatively small amounts of explosives can result in considerable loss of life. This has become a particularly serious problem with the advent of numerous incidents of improvised explosive devices concealed on persons engaged in suicide attacks worldwide dictating that detection systems should protect operators through standoff. This three-dimensional active mmWave imaging system can also be used to detect explosives and affords the possibility of operation where the operator is at a safe distance from the threat. Figure 15.13 shows a thin sheet of simulated plastic explosive (duct putty) concealed between the man's shoulder blades. The optical image and active mmWave image are compared.

15.4.2 PMMW Cameras

The first mmWave monolithic integrated circuit (MMIC)-based direct amplification and detection (DAD) FPA video camera was developed by TRW, now NGST [1, 8, 9, 17]. The camera is capable of generating a real-time display of the imaged scene, similar to video cameras. The camera operates at 89 GHz, and it acquires images at a frame rate of 17 Hz. Its typical minimum resolvable temperature is 2 K at the lens input. It has an instantaneous field of view (IFOV) of $15°H \times 10°V$. The center frequency of the camera was chosen at 89 GHz to avoid interference from radars operating at 94 GHz while providing a center frequency range where it was possible to have 10 GHz bandwidth as well as acceptable atmospheric absorption. The PMMW video camera was successfully tested on the ground, in flight, and at sea [18].

Figure 15.14 shows a block diagram of the camera and its subsystems. The camera employs quasi-optical imaging through plastic lenses, focusing the incoming mmWave

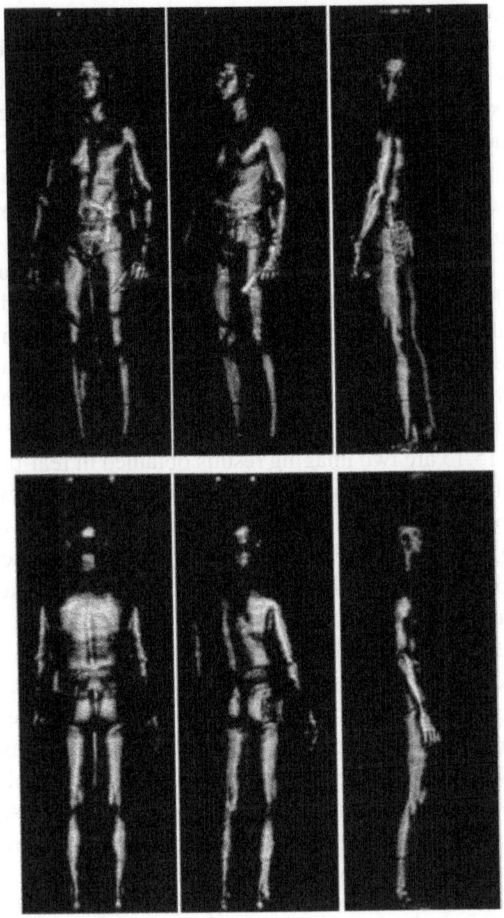

Figure 15.12: Wideband (27–33 GHz) images of a man carrying two concealed handguns along with several harmless items. Reproduced by permission of © 1996 SPIE [16].

radiation onto the 1040 (40 horizontal $\times 26$ vertical) receiving antennas of the FPA within a diffraction-limited 0.5° angular resolution. A high-pass filter is inserted directly after the lenses to block out unwanted electromagnetic interference (EMI)/radio frequency interference (RFI) from the environment. After the filter, a 45° image oversampling mirror with a factor of 2 in the vertical and 2 in the horizontal planes follows. The 45° turning mirror directs the focused radiation onto the FPA and it also moves the image to four locations at the corners of a 0.5×0.5 pixel square, thereby quadrupling the number of points and permitting oversampling of the scene [1]. The mmWave radiation is then amplified and rectified in each of the 1040 MMIC chips, and the resultant 4160 video signals then pass through various electronic and image processing steps which condition the image for final display. These processing steps include image calibration and super-resolution (suggested for future implementation) processing. Here, it is noted that super resolution is a technique that provides an enhanced resolution image by fusing together several low resolution images.

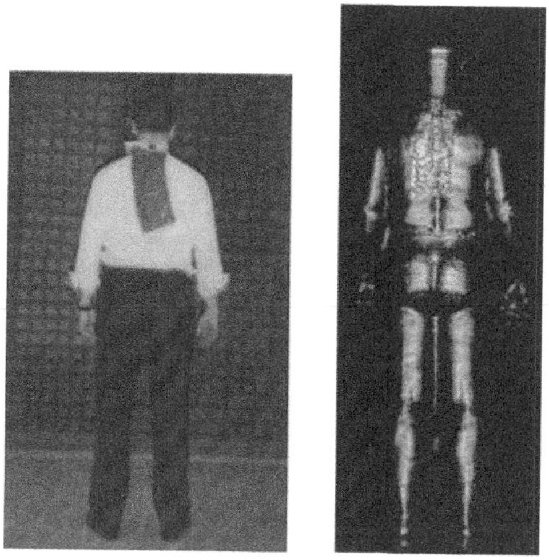

Figure 15.13: Wideband image of a man with a thin sheet of simulated plastic explosive between his shoulder blades: (a) optical picture; (b) mmWave image. Reproduced by permission of © 1996 SPIE [16].

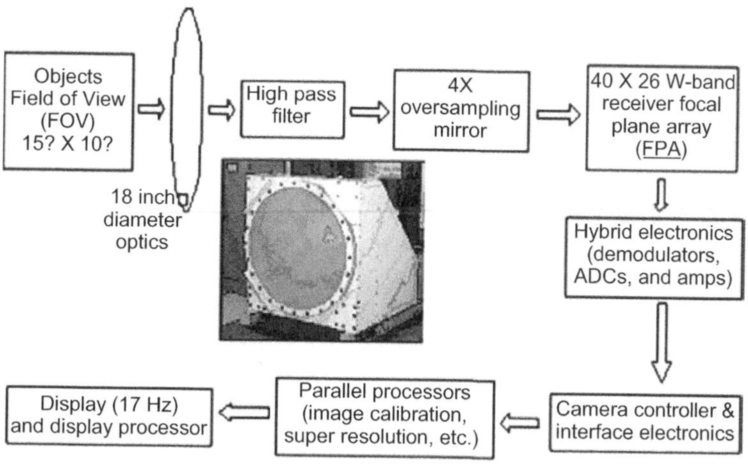

Figure 15.14: PMMW Camera block diagram, after Figure 1 in reference [18]. Reproduced by permission of © 1998 SPIE.

The combination of high-sensitivity, low-cost, and low-power operation provided by this MMIC device was the enabling technology for developing the world's first FPA PMMW video camera. Low-noise amplifiers utilizing pseudomorphic high electron mobility transistor (PHEMT) technology on gallium arsenide (GaAs) substrates were first developed, and then later combined with on-chip switches and a sensitive detector diode to form a complete MMIC-based, 89 GHz Dicke-like receiver on a single 2×7 mm chip which has

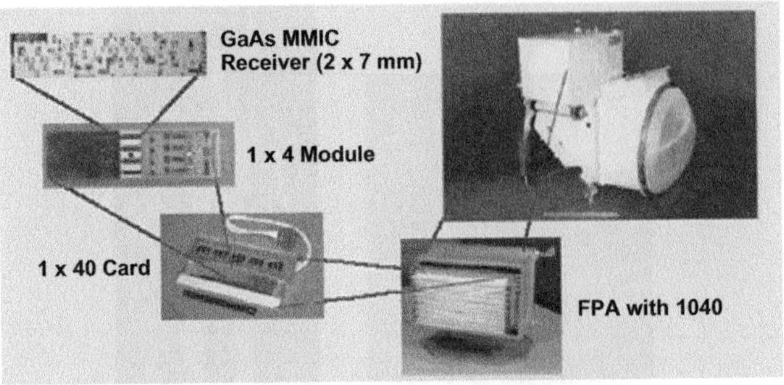

Figure 15.15: The NGST PMMW video camera. Reproduced by permission of © 2003 IEEE [1].

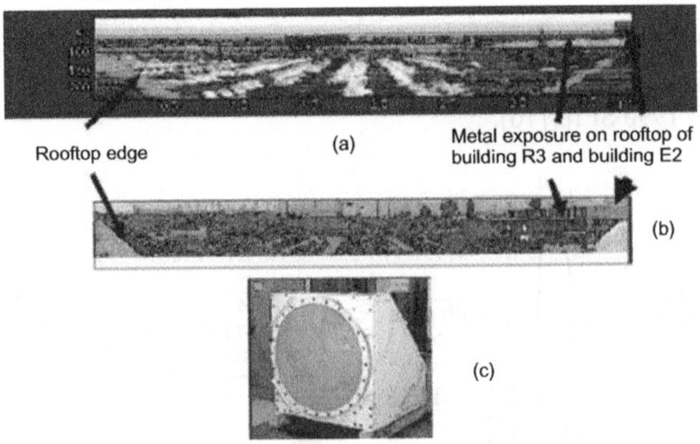

Figure 15.16: Panoramic view from the rooftop of a building, overlooking a parking lot and other structures. The PMMW image (a) was taken with (c) and covers a 180° horizontal and 30° vertical view. White is cold in this image. The visible-light image (b) is a collage of video camera image frames. Reproduced by permission of © 2003 IEEE [1].

a noise figure of ~5.5 dB and bandwidth of 10 GHz. As shown in Figure 15.15, the FPA is configured in a modular form, beginning with modules that contain four MMIC receiver chips, a four-element antenna array, current regulator, and video amplifiers. There are 10 such modules placed on a '1 × 40 card', along with electronics for handling the signals from the 40 receivers. The FPA consists of 26 '1 × 40' cards (1040 receivers). This electronics module includes hybrid components containing circuitry for further signal amplification, integration, filtering, and digitization. Figure 15.16 contains a panoramic view obtained with this camera.

A 94 GHz passive imaging camera developed for security scanning purposes has been reported in references [5] and [17]. The mechanically scanned system employs a 5 mm spacing linear array of InP MMIC amplifiers with a thermal sensitivity of ~1 K [19]. The optical configuration consists of a Schmidt camera arrangement with polarizers employed to fold the camera. A Cassegrain subreflector maximizes spatial sampling and provides a diffraction spot in the image plane with a half-power width of 6.1 mm, allowing the image to be sampled once per beamwidth.

15.4.3 ECEI/MIR

In addition to concealed weapon and contraband detection, mmWave imaging systems are currently also being developed for a wide range of applications, including remote sensing, radio astronomy, environmental measurements, and plasma diagnostics. The examples presented here concern mmWave imaging systems [3] which have made possible advanced imaging and visualization of magnetohydrodynamic (MHD) fluctuations and microturbulence in fusion plasmas [20]. As discussed in detail in reference [20], mmWave-based diagnostics (most of which are based directly on the dispersive properties of the plasma medium) are widely used in magnetic fusion plasma diagnostics. Here, recent advances in mmWave technology have also made possible the use of imaging techniques which have dramatically extended the range of measurements and insights possible through the provision of imaging and visualization capability.

The temperature of magnetic fusion-relevant plasmas is extremely high ($\sim 10^8$ K); consequently, direct measurements are impossible in this extremely hot environment. Numerous studies have stressed the need for high-resolution imaging diagnostics, which will ultimately permit the visualization of the complicated two-dimensional and three-dimensional structures of both electron temperature and density profiles and fluctuations which determine particle and energy confinement. Full three-dimensional simulations produced by the latest massively parallel computers have shed much light on the physics of hot plasmas. However, the largest scale computers today can only provide such simulations for periods of a few microseconds, whereas hundreds of milliseconds or even seconds are required to understand the phenomena. In addition, it is essential to benchmark these simulations with actual experimental data. Consequently, it is essential to visualize the plasma via experimental measurements. Fortunately, the microwave/mmWave (MMW) portion of the electromagnetic spectrum is ideally suited for performing a variety of measurements of magnetic fusion plasma equilibrium parameters as well as their fluctuations. Here, we will restrict our attention to imaging examples involving two techniques: (a) electron cyclotron emission, which provides a measurement of electron temperature and its fluctuations; and (b) the radar-based microwave reflectometry technique which measures the electron density profile and its fluctuations by means of the reflection of electromagnetic waves at the plasma cutoff layer.

In magnetic fusion plasmas, the conventional technique to measure electron temperature is via a one-dimensional electron cyclotron emission (ECE) radiometer, and the conventional technique to measure electron density is microwave (non-imaging) radar reflectometry. ECE occurs in magnetized plasmas because the gyromotion of electrons results in radiation at the electron cyclotron frequency $\omega_{ce} = eB/\gamma m_e$ and its harmonics [20]. Here, B is the magnetic field strength, e is the electron charge, m_e is the kinetic electron mass, and γ is the familiar Lorentz relativistic factor. When the plasma density and temperature are sufficiently high, the plasma becomes optically thick to some harmonics of the ECE and, since the emission from plasmas of magnetic fusion interest is in the Rayleigh–Jeans limit ($\hbar\omega \ll kT_e$), the radiation intensity of optically thick ECE harmonics reaches that of black body radiation, i.e. $I(\omega) = \omega^2 kT_e/(8\pi^3 c^2)$. In magnetic fusion devices such as tokamaks, the ECE frequency is inversely proportional to the major radius R because the magnetic field strength B is inversely proportional to R. Thus, in a conventional ECE radiometer, a horn antenna receives the ECE radiation at the outboard side, which is separated into different frequency bands, each corresponding to a different horizontal location in the plasma (see Figure 15.18). Through these frequency-resolved radiometric measurements, time-resolved one-dimensional electron temperature T_e profiles can be obtained. To obtain multi-dimensional temperature profile and fluctuation data, a PMMW imaging technique, ECEI, has been developed [3, 20] and is described in Section 15.4.3.1

Microwave reflectometry first saw use in probing the height of ionospheric plasmas where it was called ionosonde. Reflectometry is a form of microwave radar that uses the plasma as a reflector and has been widely employed to determine the equilibrium electron density (n_e) profile [3,20]. Here, electromagnetic waves are reflected at the plasma cutoff layer, when the refraction index goes to zero for the particular wave frequency. The solution to the problem of determining the phase delay of a wave which propagates through the inhomogeneous plasma and reflects at the plasma cutoff layer was investigated by Ginzburg. He found that

$$\phi(\omega) = \frac{2\omega}{c} \int_{r_c(\omega)}^{r_i} \eta(r) \cdot \mathrm{d}r - \frac{\pi}{2} \tag{15.4}$$

where r_i is the radius at the plasma edge, $r_c(\omega)$ the position of the cutoff layer (also called the 'critical layer') for the wave at frequency ω, and η is the refractive index. This basically yields the delay along the two-way propagation path plus the negative shift $\pi/2$, which accounts for the reflection at the cutoff layer. Electron density profile determination is based on the measurement of the time delay of the reflected wave τ. If one measures $\tau(\omega)$ for a series of ω values ranging from those reflecting at the plasma edge ω_e to the ones reflecting at a given depth, the corresponding density profile can be derived by an inversion of Equation (15.4).

From the outset, microwave reflectometry has also been seen as a tool to aid an understanding of the relationship between fluctuations and transport by providing high-resolution localized measurements of density turbulence in fusion plasmas. Unfortunately, this technique has limited capability in the presence of two-dimensional fluctuations. Thus, to capture multi-dimensional images of plasma density fluctuations, the MIR concept was developed [3, 20] and is described in detail in Section 15.4.3.2

To resolve the relation between anomalous transport and microturbulence, there is a need for simultaneous n_e and T_e fluctuation measurements. Consequently, mmWave imaging systems, comprised of a passive mmWave imaging system ECEI and an active mmWave imaging system MIR, are being developed to measure both fluctuations simultaneously. These new mmWave diagnostics technologies are currently being applied in (or developed for) a number of toroidal devices including TEXTOR, LHD, NSTX, KSTAR, and EAST to help understand turbulence physics by visualizing plasma temperature and density fluctuations [3, 20].

15.4.3.1 Electron Cyclotron Emission Imaging System

As discussed in the preceding subsection, in a conventional ECE radiometer, time-resolved one-dimensional T_e profiles can be obtained from the mmWave signals collected by a single horn antenna as shown in Figure 15.17. In the two-dimensional ECEI system, the single antenna in the conventional ECE radiometer is replaced by an array of vertically aligned broad bandwidth mixers. It is similar to the many vertical layers of a conventional

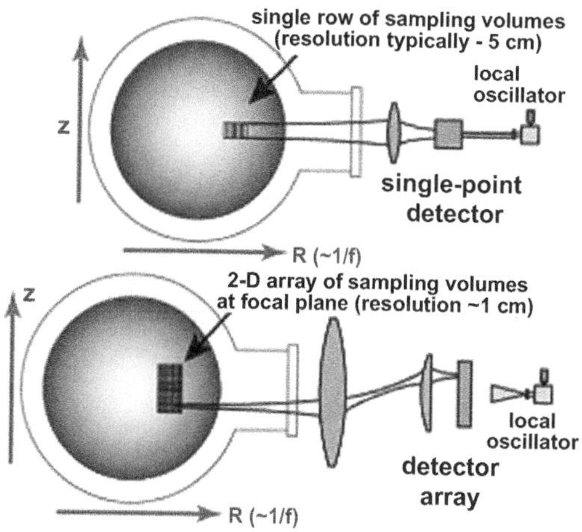

Figure 15.17: Comparison of the principles of a two-dimensional ECEI imaging system and a conventional one-dimensional ECE system.

ECE system. ECE radiation is collected and imaged onto a mixer/receiver array comprised of broad bandwidth planar antennas with integrated Schottky diodes. The vertically arranged mixer array elements are aligned along the E-field (vertical) direction. Large diameter optics image the plasma emission onto the array, enabling each array element to view a distinct plasma chord. The horizontal positions of the sample volumes are determined by the receiver frequencies. Using the one-to-one mapping of ECE frequency to major radius mentioned above in tokamak plasmas allows the ECEI focal plane to be swept through the plasma by sweeping the receiver frequency of the array, thereby forming two-dimensional images of the T_e profile [3, 20]. In addition to two-dimensional measurement capability, ECEI diagnostic systems have excellent spatial resolution ($\sim$1 cm) in both poloidal and toroidal directions.

Figure 15.18 shows the configuration of an ECEI system. No illumination is needed for this PMMW imaging system. Large aperture optics focus ECE radiation on to the Schottky diode mixer array. A quasi-optical notch filter has been developed to shield the mixer array from stray radiation from the high power gyrotron electron heating sources which are narrowband and typically in the 84–170 GHz range [21]. Detected RF signals are down-converted in the mixer array and go to the IF detection system where the IF signals are converted into video signals and subsequently digitized. The resultant data are then processed with a computer. Figure 15.19 shows the receiver architecture as well as photographs of its components.

mmWave radiation (emission: passive system ECEI; or reflection: active system MIR) from the plasma exits from a window following which it passes through the optical system and then is separately imaged onto the MIR and ECEI mixers arrays. For ECEI, the goal of the optical design is to achieve the highest possible spatial resolution given constraints such as the port access, need for reasonable dimension optics, and affordable fabrication cost. For MIR, there is an added complication related to the probing beam. Optics are required to match (both toroidally and poloidally) the probing beam with the curvature of the cutoff surface. The illumination beam and reflected signal are both focused with the optics. Ray tracing and

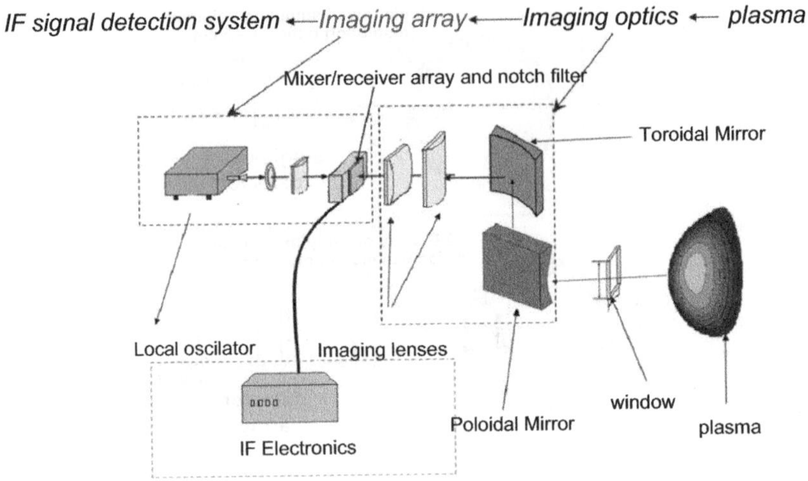

Figure 15.18: ECEI system configuration.

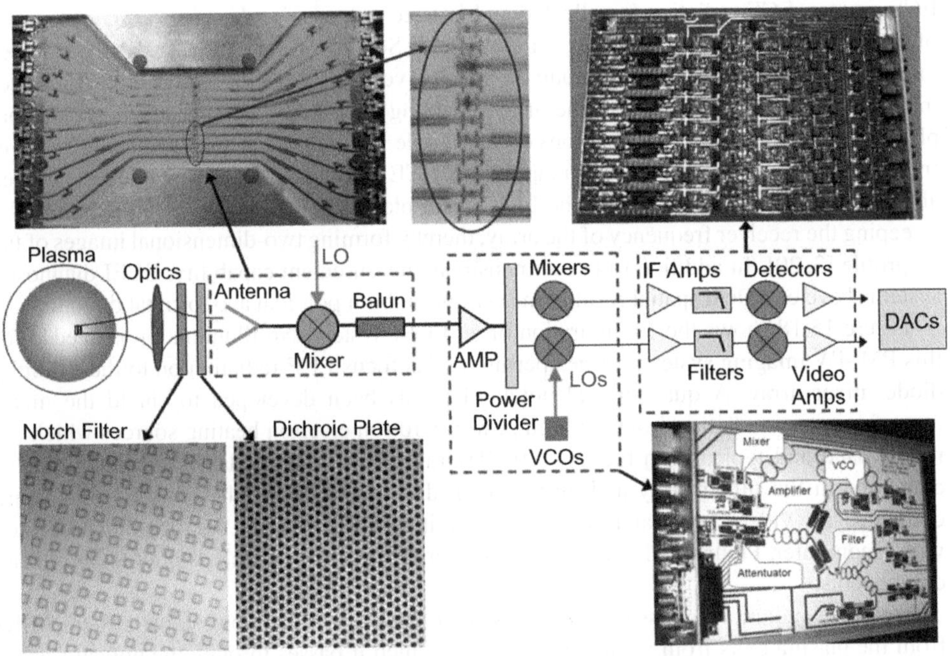

Figure 15.19: Schematic diagram of the ECEI detection scheme, illustrating the steps required to convert wideband ECE radiation into 2.0–8.4 GHz IF signals and then to generate 16 × 8 channel ECEI output signals.

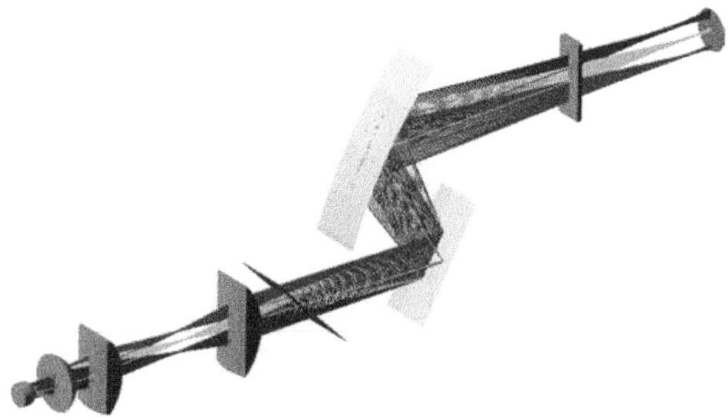

Figure 15.20: ECEI ray tracing figure for the TEXTOR tokamak system.

Gaussian propagation analysis are used to provide a one-to-one mapping between the array elements and images and to calculate the focal plane spot size.

For the TEXTOR tokamak plasma device (located in Julich, Germany), the ECEI and MIR systems share a 42×42 cm^2 vacuum window and two large primary focusing vertical aligned cylindrical mirrors as shown in Figure 15.20. The mirrors are first designed to tailor the illumination beam wave front to match the cutoff surface. The reflected MIR beam passes through the same window and mirrors, but is separated from the illumination beam by a beam splitter. After passing through the imaging lenses, the reflected signal is collected by the mixer array. The higher frequency (>110 GHz) ECEI signal is separated from the lower frequency (<90 GHz) MIR signal with a dichroic plate as shown and then focused onto the ECEI mixer array using separate imaging optics. Plasma facing cylindrical mirrors are employed instead of lenses because lens surfaces can cause internal reflections which could lead to interference in the case of MIR. The mirrors are comprised of a polystyrene-backed aluminum film bonded to a machined substrate and all lenses are made of high-density polyethylene. The spatial resolution is about 1 cm at the center channel. Off-axis channels have slightly degraded performance including larger spot size and lower power. Figure 15.20 illustrates ray-tracing of the optical design in TEXTOR.

The function of the quasi-optical frequency selective surface (FSS) notch filter in the optical system is to protect the mixer array from spurious heating power from the 800 kW, 140 GHz gyrotron heating source [21]. Because the filter must be mounted between the optical lenses in the imaging system, a FSS is suitable as a thin planar filter and is easy to implement. FSSs consist of an array of periodic metallic patches on a dielectric substrate or a conducting sheet periodically perforated with apertures. Figure 15.21 shows unit cell structure parameters and a photograph of the 140 GHz FSS notch filter which exhibits total reflection at the resonant frequency. Unlike waveguide filters, the 140 GHz band stop filter must retain its filtering characteristics over a wide range of incident angles, which corresponds to $\sim 15^\circ$ for the TEXTOR application. The $8'' \times 8''$ square loop frequency selective surface filter shown in Figure 15.20 has a measured 35 dB rejection at normal incidence, and retains this high performance at incidence angles as high as $\pm 15^\circ$ at which point the rejection has only fallen to 25 dB [21]. The FSS filter can be easily implemented

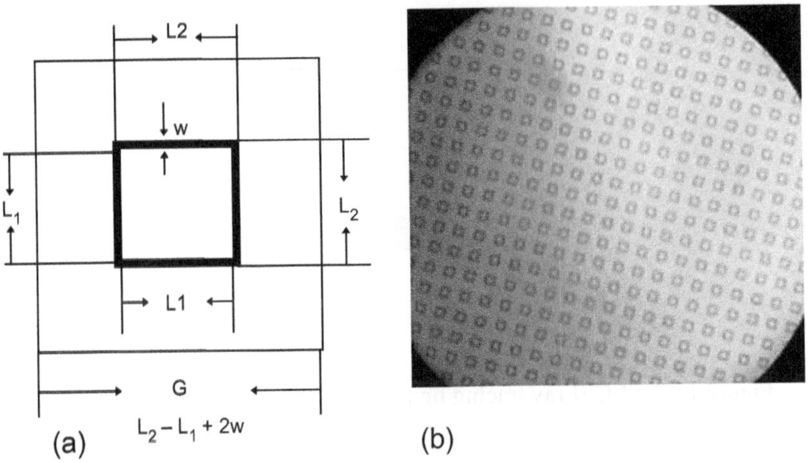

Figure 15.21: (a) Square loop unit cell parameters $G = 951$ μm, $L_1 = 320$ μm, $L_2 = 410$ μm, $w = 45$ μm; (b) photographs of 140 GHz FSS notch filter with square loop structure.

between the lenses and its excellent angle insensitivity allow it to reject most of the spurious heating power even if it is mounted on the lens whose incident beam is uncollimated.

Quasi-optical planar antenna mixer arrays are used as shown in Figure 15.19. They have advantages over waveguide mixers at mmWave frequencies. They are smaller and lighter than waveguide systems and can be easily produced in large numbers for low-cost applications such as mmWave imaging systems. Technological challenges for the mixer include (a) achieving wide bandwidth, (b) minimizing the intra-element spacing while maintaining low crosstalk and sidelobe levels, and (c) supplying sufficient LO power to each antenna element, all the while achieving sufficiently low noise for hot–cold load calibration [22]. The mixer arrays in both the ECEI and MIR systems employ vertically aligned coplanar strip (CPS)-fed dual dipole antennas with beam-lead Schottky diodes. The antennas are closely spaced in the vertical or E-plane, and fabricated on a low dielectric constant elliptical substrate lens. The substrate lens is used to eliminate substrate modes. The ECEI signals down-converted in the Schottky diode mixers are subsequently coupled out on the CPS lines. Wideband microwave baluns are employed to convert the IF signals from the balanced CPS lines to unbalanced transmission lines leading to end launch subminiature version A (SMA) connectors. The applied balun has an extremely wide IF bandwidth whose 3 dB bandwidth extends past 30 GHz.

Another recent development thrust has been to move from the use of one-dimensional detector arrays to the use of two-dimensional receiver arrays. In this case, the use of multi-frequency illumination together with wideband two-dimensional MIR antenna arrays makes it possible spatially to resolve density fluctuations over an extended three-dimensional plasma volume. The schematic of a two-dimensional dual dipole antenna array is shown in Figure 15.22 [23].

Subsequently, the signals are amplified by ~35 dB using low-noise preamplifiers, and transmitted to the wideband IF electronics modules (see Figure 15.20), where they are converted to baseband IF signals. The RF circuits equally split each 2.0–8.4 GHz input signal

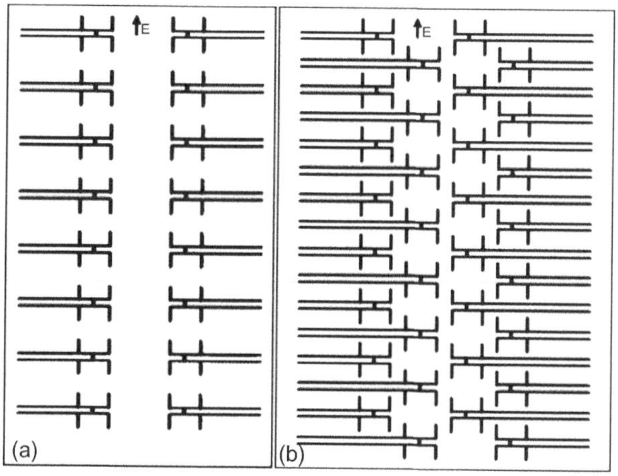

Figure 15.22: Schematic layout of two-dimensional (a) 8 × 2 and (b) 8 × 4 dual dipole antenna arrays.

into eight portions with an RF power divider, and each portion is down-converted again using a surface mount mixer with a 2.0–8.4 GHz LO. The second circuit, 'IF Electronics', rectifies the IF signals from the first circuit to video signals whose amplitude is proportional to the ECE signal strength.

The TEXTOR ECEI system employs a 16 element array with eight frequency bands spaced 800 MHz apart, generating time-resolved 16 × 8 images of T_e profiles and fluctuations of the TEXTOR plasma. Despite a limited image size, the system has greatly contributed to the understanding of the fundamental MHD physics such as the magnetic reconnection process of the so-called sawtooth oscillation [24].

The ECEI measurements provide two-dimensional movies of the magnetic field perturbation, and make it possible directly to compare the measurements with predictions from various theoretical models. Figure 15.23 shows a two-dimensional image obtained by this PMMW imaging system obtained in the study of magnetic reconnection processes during a so-called sawtooth crash. This common periodic instability in the core of the tokamak has important consequences for a fusion reactor: reducing the efficiency or triggering even more violent instabilities. However, despite intensive study ever since its discovery 30 years ago, no final explanation for the crash event had been found. Now the ECEI results have revealed the details of this phenomenon as a result of the ability to image the region where the crash occurs. Understandings of such phenomena are crucial to the successful achievement of magnetic fusion energy since they repeatedly expel large amounts of energy and particles from the plasma. Such physics understanding will also contribute to other areas of science since the behavior resembles in many aspects that observed in solar flares and astrophysical plasmas.

An ECEI system was first developed and implemented on the TEXT-U tokamak. This was followed by an ECEI system on the RTP tokamak and more recently on the TEXTOR tokamak. ECEI systems have now also been developed and widely applied on mirror machines and stellarators [3, 20].

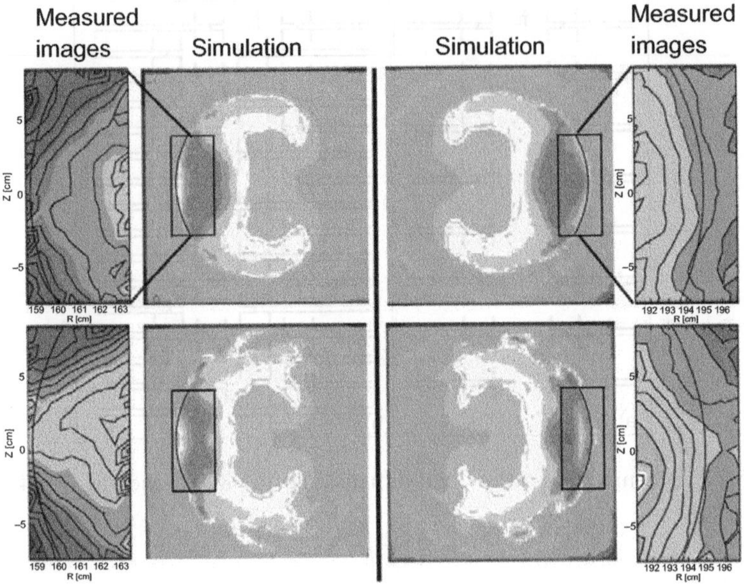

Figure 15.23: Experimental two-dimensional images of hot spot and island formation are directly overlaid for comparison with 'ballooning' mode model images, after reference [24]. Reproduced by permission of © 2006 The American Physical Society.

15.4.3.2 Microwave Imaging Reflectometry System

As noted above, microwave reflectometry is a radar technique for the detection of plasma fluctuations from the reflection of microwaves from plasma cutoff surfaces. It has proven to be an extremely useful and sensitive tool for measuring low-level density fluctuations in some circumstances; however, this technique has been shown to have limited viability for two-dimensional turbulence in tokamaks. For one-dimensional fluctuations, the reflection layer will move back and forth in the radial direction, resulting in simple phase changes in the reflected wave. The one-dimensional fluctuation is analogous to a moving mirror which changes the phase of the reflected waves (see Figure 15.24). In the presence of two-dimensional turbulent fluctuations, the interpretation of reflectometry becomes considerably more complex. Unfortunately, this is precisely the point of interest for tokomak plasmas, which exhibit both radial and poloidal fluctuations. The difficulty arises from the fact that when the plasma permittivity fluctuates perpendicularly to the direction of propagation of the probing wave, the spectral components of the reflected field propagate in different directions. This can result in a complicated interference making it difficult to extract any information about the plasma fluctuations. In essence, the measurement of the fluctuations can be limited by the fluctuations themselves. Imaging optics is then required to recover the phase information from waves reflected from localized region of cutoff layer, to restore the phase front. Thus, to capture multi-dimensional images of plasma density fluctuations, the MIR concept was developed [3, 20].

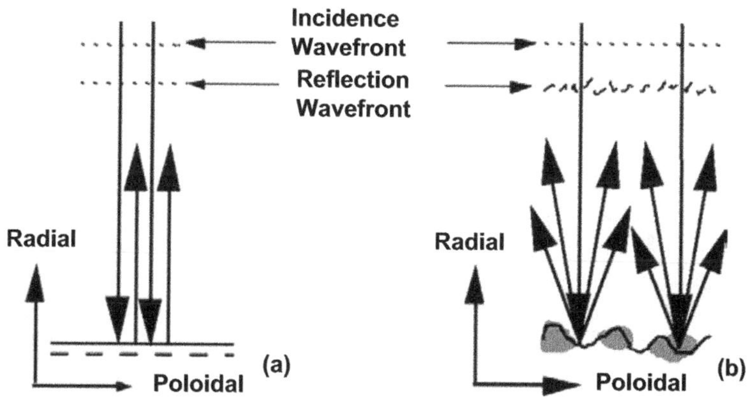

Figure 15.24: Comparison of (a) one-dimensional (with phase modulation) (b) two-dimensional reflectometry (with both phase and amplitude modulations). Reproduced by permission of © 2003 American Institute of Physics [3].

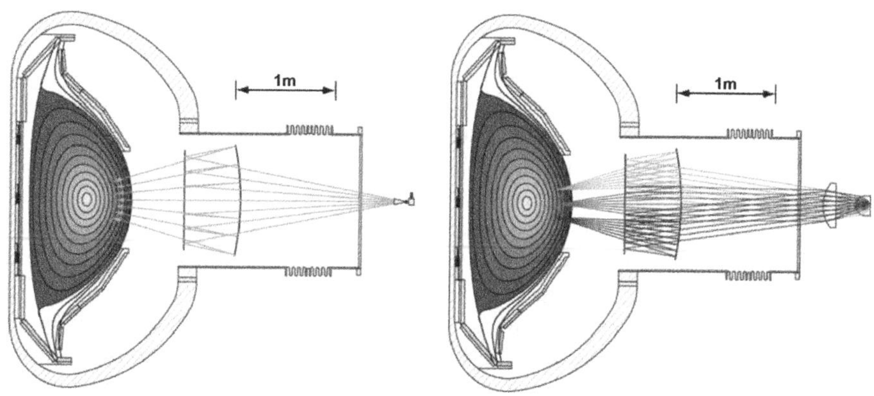

Figure 15.25: Illumination and detection paths for the MIR system.

MIR is a technique with both illumination and detection paths, as shown in Figure 15.25. The plasma is illuminated with mmWaves (~30–140 GHz), and the focusing mirrors match the curvature of the illuminating beam to that of the cutoff surfaces (toroidal and poloidal). The reflected mmWaves are then collected with the large aperture optics at the plasma edge which optically focus an image of the cutoff surface onto an array of detectors, thus restoring the integrity of the phase measurement as described above. As density fluctuations near the plasma cutoff layer modulate the reflected beam, MIR effectively spatially resolves n_e fluctuations on the extended plasma cutoff surface; the use of multiple frequency illumination results in the formation of two-dimensional n_e fluctuation images.

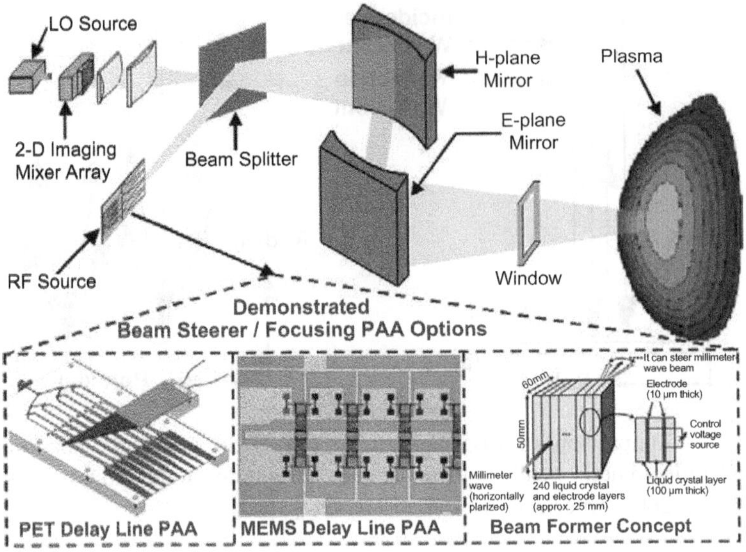

Figure 15.26: Schematic diagram of the MIR diagnostic employed on TEXTOR, with insets showing PAA options for a beam steering/focusing system.

Figure 15.26 shows the schematic diagram of the MIR diagnostic employed on the TEXTOR tokamak, together with phased antenna array (PAA) options for a beam steering/focusing system required electronically to shape the incident beam to match the plasma cutoff surfaces. The detection system uses technologies similar to those described for the ECEI system. A beam splitter is used to separate the illumination and reflected signals. Alignment of the probing beam with respect to the cutoff surface is a key technological challenge in an MIR system, especially for MIR applied in highly shaped plasma machines [3, 25]. Thus, steering and beam shaping PAA technology specifically for this application is under development. By controlling the illumination signal bandwidth electronically, the PAA system can function like an 'artificial lens' and the focal properties of the launching beam can be changed at high speed without mechanically moving the launching antenna [25, 26].

Three technologies under investigation for this application are shown in Figure 15.26. The first of three PAA technologies under development for reflectometry beam steering and shaping is a Ka-band piezoelectric transducer (PET) controlled PAA system using an antipodal elliptically-tapered slot antenna [25]. The PET-controlled delay line incorporates a dielectric perturber whose height above a transmission line is electronically controlled [27, 28]. Recent experiments with the Ka-band eight-element PAA using antipodal elliptically tapered slot antennae have demonstrated wide angle scanning from $-21°$ to $+24°$ [25, 26]. The second true time delay technology is the microelectromechanical systems (MEMS) nonlinear delay line [29], employing voltage-controlled varactors whose varying capacitance results in a change in propagation velocity along the transmission line. The third is a liquid crystal 60 GHz mmWave beam former developed by NHK Labs in Japan [30].

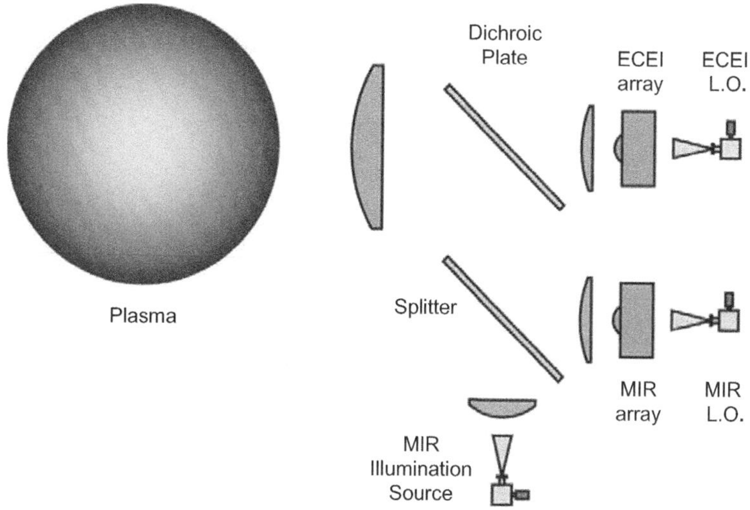

Figure 15.27: Concept of combined ECEI and MIR System. Reproduced by permission of © 2003 American Institute of Physics [3].

15.4.3.3 Combined ECEI/MIR System

To resolve the relation between anomalous transport and microturbulence, there is a need for simultaneous n_e and T_e fluctuation measurements. Fortunately, it has been demonstrated on the TEXTOR tokamak that it is possible to implement a combined ECEI/MIR system to measure both fluctuations simultaneously. Since both the ECEI and MIR systems require similar collection optics, and the two systems operate in contiguous, but non-overlapping, frequency ranges, it is feasible to combine these two systems by sharing the optical system and window. Consequently, a combined ECEI/MIR system is being developed simultaneously to measure core plasma temperature and density fluctuations in the toroidal machines. The primary difference between the two systems is that the MIR system has the added complication of the probing beam. A dichroic plate is used to separate the ECEI and MIR beams because they have different frequencies, as shown in Figure 15.27. Currently, a combined ECEI and MIR system is being operated on the TEXTOR tokamak. In this system, the primary components of a microwave optical system are shared between the ECE and reflectometer subsystems, with each subsystem employing single/multiple dedicated high-resolution multi-channel detector arrays [3, 20].

15.4.4 mmWave Imaging System Applications in Astronomy

The submmWave region of the spectrum ($\sim$30–500 μm) is of great interest to the radio astronomy community [31]. Through spectrally resolved radiometric observations in the submmWave region, astronomers are able to gain insights into the formation and evolution of galaxies, the creation of stars and their interaction with the interstellar medium, and the composition of cometary, planetary, and satellite atmospheres. Some lines of interest are listed below in Table 15.4 illustrating the spectral range required for submmWave astronomy

Table 15.4: Representative molecular emission and atomic fine structure lines [32].

Species	Frequency (THz)
C	0.809
N^+	1.461
CO	1.726
C^+	1.901
N^+	2.495
H_2O	2.640
H_2	2.670
OH	3.545
O	4.746

measurements [32]. As noted in Section 15.1, sea-level atmospheric absorption is extremely high; consequently, submmWave astronomy measurements are conducted from mountains, airborne observatories, and from space.

At the extreme limit of mmWave/submmWave imaging systems, the mmWave and submmWave astronomy community is developing impressive telescope arrays. A noteworthy example is the Atacama large millimeter array (ALMA) which is a large international effort to build a mmWave radio telescope array comprised of 64 12-m antennas located at a 5000 m altitude site in the Atacama desert of Chile [33]. The surface accuracy of the antennas is ±25 μm, with 0.6″ tracking, and 2″ absolute pointing over the entire sky. The antennas are located on tracks permitting a range of array configurations extending from 150 m to ~15 km. The system covers the spectral region from 31–950 GHz in 10 bands with the details including noise temperatures displayed in Table 15.5. Correction for atmospheric water vapor fluctuations is provided through observations using a 183 GHz water vapor radiometer. The IF amplifiers cover either a 4–12 GHz or a 4–8 GHz range and provide the correlator with 16 GHz of data. The correlator is a major development and digitizes the signals from the receivers on 64 antennas. There are eight spectral windows, each spanning 2 GHz which are to be digitized and transmitted at a rate of over 100 Gb/s from each antenna. The signals travel over fiber optic cables to digital filters, and then are introduced to the correlator which must process 1.6 × 1016 multiplication and addition operations per second. It cross-correlates signals from 32 × 63 = 2016 pairs of antennas on a timescale of 16 ms, and also autocorrelates signals from 64 antennas on 1 ms timescales, producing 32 Gb/s output.

Another interesting example concerns the Antarctic Submillimeter Telescope and Remote Observatory (AST/RO) which is located at the South Pole where the cold, dry conditions, coupled with its relatively high altitude (10 500 ft) make it an excellent location for observations at submillimeter wavelengths [34]. This 1.7 m diameter off-axis telescope is employed for research in astronomy and aeronomy at wavelengths between 200 and 2000 μm and features five heterodyne receivers mounted on an optical table suspended on the telescope. One of those receivers is the 'PoleSTAR' array which consists of four 800–820 GHz fixed-tuned SIS waveguide mixer receivers with double-sideband (DSB) noise temperatures of 850–1500 K. Over the range 806–809 GHz, the noise temperature is 900–950 K DSB. In this arrangement, a pair of parabolic mirrors and two optical flats are

Table 15.5: ALMA receiver details.

Frequency range (GHz)	Maximum noise temperature (K)	Receiver technology
31–45	26	HEMT
67–90	47	HEMT
84–116	60	SIS
125–163	82	SIS
163–211	105	SIS
211–275	136	SIS
275–373	219	SIS
385–500	292	SIS
602–720	261	SIS
787–950	344	SIS

HEMT - high electron mobility transistor. SIS - superconductor–insulator–superconductor

employed to re-image the AST/RO focal plane onto a 2×2 array of lenses located in the array cryostat.

Currently under development is 'SuperCam', which will be a 64-pixel heterodyne imaging array comprised of eight stacks of 1×8 rows of fixed tuned, SIS mixers optimized for operation in the 320–360 GHz atmospheric window [35]. IF outputs from each mixer are to be connected to low-noise, broadband MMIC amplifiers integrated into the mixer blocks. The instantaneous IF bandwidth of each pixel will be $\sim$2 GHz, with a center frequency of 5 GHz.

As a final submmWave astronomy imaging example, we consider the Stratospheric Observatory for Infrared Astronomy (SOFIA), which features a 2.5 m airborne telescope carried on a modified Boeing 747SP [36]. Flying at an altitude of 12.5 to 13.7 km (41 000–45 000 ft), water vapor absorption becomes small to negligible, thereby permitting observations in the 30–1000 μm region from this platform not possible from ground-based observations. Under development is the SOFIA Terahertz Array Receiver (STAR) consisting of a 4×4 element heterodyne niobium hot electron bolometer (HEB) mixer array covering the frequency range from 1.6 to 1.9 THz (187 to 158 μm). Its main scientific goal is large scale mapping of the 158 μm fine structure transition of singly ionized carbon.

15.4.5 mmWave and THz Radars

There have been a number of mmWave radars designed with an emphasis on imaging. A particularly noteworthy example is the 94 GHz WARLOC (W-band Advanced Radar for Low Observable Control) radar which features scanning capability. This employs a gyro-klystron amplifier producing 80 kW peak power with 3–10 kW average power. This has considerably advanced cloud research. The operation at the W-band (in common with other mmWave radars) provides significant improvement over microwave radars as well as lidars since mmWave radars produce strong signals from cloud water droplets (scattered power inversely proportional to wavelength) and are able to penetrate the clouds and provide images of the internal structure. The advantage over previous mmWave radars is that the WARLOC transmitter has about three orders-of-magnitude more average power than the W-band radars used in previous cloud studies and greatly improved resolution and scanning capability [37]. Using the WARLOC radar, high-resolution images of the internal structure of clouds were obtained over regions many square kilometers in area can be scanned with scans obtained at 15 m resolution in about a minute even through intervening cloud layers. The scanned cloud reflectivity yields two-dimensional cloud turbulence correlation functions and power spectra directly from spatial measurements. This W-band technology is being employed to upgrade the MIT Haystack X-Band Observatory Radar in Westford, Massachusetts [38] with the new system designated the Haystack Ultra-wideband Satellite Imaging Radar (HUSIR). The combination of increased operating frequency and replacement of the previous 37 m (120 ft) antenna with a new dish, accurate to 0.1 mm (0.004 in) over its entire surface is predicted to provide an order of magnitude increase in resolution.

There have been a number of mmWave radar system demonstrations with the emphasis on their suitability for tracking and imaging applications. Long-term applications envisioned for the technology are for autonomous guidance and navigation. These have tended to operate at 77 GHz or 94 GHz and typically operate in frequency modulated continuous wave (FMCW) mode since it facilitates range gate limited imaging. Using mirror scanners, three-dimensional terrain images are obtained. Using a transmit power of 10 mW and a sweep bandwidth of 600 MHz and a 250 mm diameter antenna (beam divergence of $\sim 0.8°$, a range resolution of 50 cm is achieved at a range of 500 m [39].

Very recently, a high-resolution imaging radar operating at 580 GHz was reported [40]. The FMCW radar employed a solid-state Schottky diode frequency multiplier chain to provide coherent illumination in the 576–589 GHz range at the 0.3–0.4 mW output level and utilized phase-sensitive detection. The system employed a plano-convex Teflon lens with a diameter of 20 cm which focused the THz beam to a spot size of 1 cm at a standoff range of 4 m. The system was mechanically scanned using a flat mirror on a two-axis rotational stage to deflect the beam in the desired direction. The typical image acquisition time was 100 ms per pixel (50 ms ramp up and down). They obtained centimeter-scale range resolution while utilizing fractional bandwidths of less than 3% and also achieved centimeter-scale cross-range resolution at several-meter standoff distances despite the relatively small apertures.

With the advent of affordable mmWave technology, there has also been increasing activity in developing imaging radars for automotive applications. While earlier systems tended to employ GUNN device oscillators as the transmitter source, the recent advances in SiGe technology have changed this and are resulting in cheaper, higher performance systems. For example, reference [41] describes 77 GHz radar technology developments featuring a 77 GHz SiGe VCO MMIC integrated into an automotive long-range radar system. They report both the achievement of higher signal power (compared to a GUNN) with the SiGe VCO at identical noise level, together with an improvement in signal-to-noise (S/N) ratio. Additional information is contained in Chapter 18.

15.5 THz Imaging Technology

In recent years, there has been explosive growth in the THz field which continues unabated due to the numerous potential application areas [42–45]. Many of these applications involve THz imaging ranging from security screening systems (e.g. recognition of explosives and hazardous materials), genetic engineering, pharmaceutical quality control, and medical imaging. In addition to taking advantage of the moderate transmission of THz waves through clothing, many of these applications directly depend upon spectroscopic signatures such as characteristic phonon vibrational modes in solids which can be used for material identification and which possess frequencies above 300 GHz [46–48]. For example, in the case of explosive detection studies, reference [46] showed that the explosive 1,3,5-trinitro-1,3,5-triazacyclohexane (RDX) possesses seven absorption features centered at 27.2, 34.9, 46.0, 51.8, 65.5, 74.2, and 105.5 cm^{-1} in the spectral range, of 5–120 cm^{-1}.

As discussed in Section 15.1, atmospheric absorption primarily limits THz imaging to relatively short-range applications with the exception of high altitude, airborne, and space-based applications such as submillimeter astronomy. Here, we will briefly discuss some of the supporting technologies for THz imaging together with some examples of their applications.

Pioneering far-infrared (THz) imaging studies were conducted by Hartwick *et al.* [49]. Here, reflection and transmission images were obtained using both an HCN laser (337 μm) and an optically pumped far infrared (FIR) waveguide laser (at both 554 μm and 1020 μm). A fixed reflecting Cassegrain optical system comprised of dual 6″ diameter parabolic mirrors was employed to collect the transmitted or reflected radiation and focus it on the liquid helium-cooled GaAs detector. At the other focus of the $f/14$ optical system was a scanning stage, which mechanically provided a raster scan by translating the object to be imaged. A variety of images were obtained which provided a glimpse into the potential of THz imaging. However, widespread application of THz imaging awaited the work of Hu and Nuss [50] which made use of advancements in ultra-fast femtosecond pulsed lasers and electro-optics. Here, THz radiation was generated by illuminating a photoconducting transmitter consisting of two strip lines lithographically defined on a semi-insulating GaAs substrate, with 100 fs duration, 800 nm pulses from a Ti:Sapphire laser. Figure 15.28 illustrates their experimental arrangement where the imaging optics employed two pairs of off-axis parabolic mirrors. The incident THz radiation was focused down to a diffraction-limited spot on the object by the first pair of mirrors with the second pair used to collect and focus the THz radiation from the sample onto the optically gated silicon-on-sapphire photoconducting dipole antenna detector. The scanning delay line temporally downconverts the THz waveforms into the kHz range, following which the downconverted signals are amplified, digitized, and processed.

Many of these technology advances and applications have occurred through the so-called ultrafast optics or femtosecond electronics areas. These 'laser'-based temporal techniques are of importance to a number of imaging applications and the reader is referred to several excellent reviews for further information [51,52]. The recent review article by reference [43] covers both applications as well as the status of the technology. Comprehensive details about THz optoelectronics technology including imaging may be found in the book edited by Sakai [42].

A major limitation of the THz imaging system reported by Hu and Nuss was obviously the need for mechanical raster scanning. A major advance permitting real-time THz imaging was the approach employing a two-dimensional electro-optic (EO) sampling technique where the THz images were acquired with a CCD camera. Usami *et al.* [53] have described a two-dimensional imaging system intended for real-time THz imaging aimed at applications in industrial non-destructive inspection for semiconductors, plastics, dry foods, and pharmaceutical products. Their system operated at a video rate of up to 30 frames/s and achieved a spatial resolution of 1.4 mm.

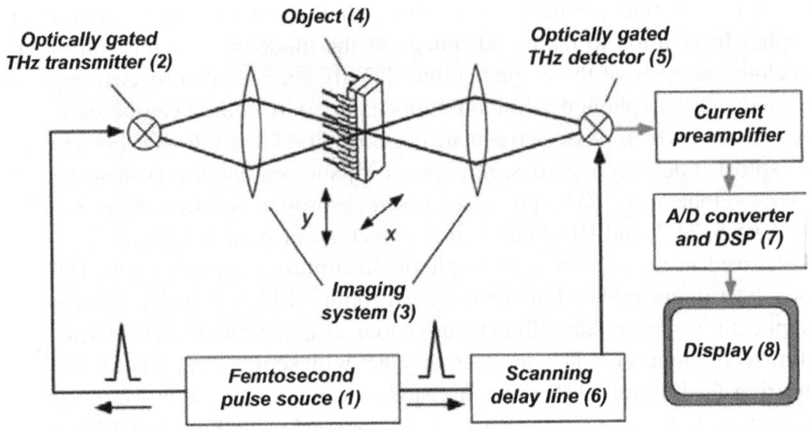

Figure 15.28: Schematic of THz imaging system. Reproduced by permission of © 1995 Optical Society of America [50].

Many imaging applications including those in medicine and biology require spatial resolution significantly better than that possible using far-field imaging. One method of overcoming this limitation is near-field imaging. Here, one makes use of the principle that the distribution of radiation directly behind a small aperture is a function of the aperture diameter rather than a function of wavelength or diffraction. Consequently, for sufficiently short distance between the object to be imaged and the aperture (less than a wavelength separation), spatial resolutions of a fraction of a wavelength (λ) can be obtained as first shown by Ash and Nicholls in the microwave region who reported resolutions of $\sim\lambda/60$ [54].

At 1 THz, the Rayleigh criterion limits the far-field resolution to approximately 300 μm. Using the near-field aperture approach with aperture sizes d as small as $\lambda/300$, Mitrofanov *et al.* [55] reported resolutions of 7 μm. A slightly different approach is the so-called 'dynamic' aperture where an optical gating beam is employed to modulate the THz transmission of a small area on a semiconductor surface near the object [56]. Yet another near-field imaging technique employs sharp metal tips to locally enhance the field [57]. By using submicrometer dimension tips, sub-wavelength resolution can be obtained. An interesting application of near-field imaging has been the use of a sample consisting of multiple, sub-wavelength slits in a metal foil where two-dimensional measurements of the full THz electric field behind the sample has made possible the reconstruction of the magnetic field and thus the spatially and temporally resolved Poynting vector which permits visualization of the flow of energy behind the sample [58]. Here, it should be noted that near-field imaging techniques have also been found useful in the mmWave region [59].

The above have primarily described THz imaging systems employing ultra-fast femtosecond pulsed lasers. However, there have been a number of other systems featuring other transmitter technologies. For example, Dobroiu *et al.* [60] have reported a system based on a backward-wave oscillator (BWO) tunable from 520 to 710 GHz. The output power of the BWO is $\sim$15 mW which they suggest is a major reason for their high observed signal-to-noise ratios (10,000:1). The measured THz spot size in the sample plane was 550 μm which was nearly the diffraction limit for the BWO operating conditions. They compared the performance of their BWO-based systems to two other THz imagers (one that employed a THz parametric oscillator THz-wave parametric oscillator (TPO) source and another based on femtosecond time-domain spectroscopy (TDS)).

Recently, there has been great interest in the use of broadband, high brightness coherent synchrotron radiation (CSR) from electron storage rings. An example of its use in THz near-field imaging of biological tissues is described in reference [61].

Another major technological advance has been the development of the quantum cascade laser (QCL) (see Section 15.6 for details). Using a 50 mW peak power QCL operating at the 4.3 THz 'window' as the illumination source, researchers have developed a real-time imaging system operating over moderate distances [62]. Images in both transmission and reflection mode are obtained with a 320 × 240-element room-temperature vanadium oxide microbolometer focal-plane array detector. Differential imaging is accomplished by synchronously modulating the QCL with the focal-plane array. At 20-frame/s acquisition rate, the signal-to-noise- ratio is 340 and the optical noise equivalent power of the detector array at 4.3 THz is estimated to be 320 pW/$\sqrt{\text{Hz}}$. Using this system, real-time imaging was demonstrated in transmission mode at a stand-off distance >25 m [63].

15.6 Technologies in mmWave/THz Imaging

The preceding sections of this chapter have provided detailed examples of the usefulness of the mmWave and THz regions of the electromagnetic spectrum for passive and active imaging. It is therefore useful briefly to survey the current state-of-the art of the technologies employed in these systems as well as discuss some of the techniques involved in imaging systems, particularly beam shaping and formation. More details on the underlying technologies may be found in the other chapters. Readers should regard this section only as an attempt to highlight some of the current areas of technological development relevant to mmWave/THz imaging.

There are several methods to detect the emitted/reflected radiation. The traditional method is to use doped germanium bolometers cooled to liquid helium temperatures to improve noise performance over a large bandwidth. Direct detection receivers are also widely used in mmWave imaging systems. Superheterodyne receivers using a tuned preamplifier before the detector have been developed. Several examples of the FPA approach are described here, representing FPA architectures using different methods.

15.6.1 Mixers

The heterodyne mixer is the 'work horse' of mmWave receiver architecture and is the most widely used receiver type for mmWave imaging. The mixer can be a monolithic circuit composed of a planar antenna integrated with a matching network and a mixer. They are easier to fabricate, more reliable, smaller, lighter, and much lower cost than waveguide receivers. The integration also allows the use of linear or two-dimensional arrays for imaging applications without a dramatic increase in cost and weight of the system. Heterodyne mixer arrays offer much faster imaging of a scene than a scanned single-element system, and avoid mechanical scanning problems. They also allow a long integration time in radio-astronomy and remote-sensing applications where signals are very weak, and can provide short integration time for imaging the rapidly changing scene in military or plasma diagnostic applications.

The heart of most mmWave receivers has been a low noise mixer. Traditionally, waveguide-mounted, whisker-contacted GaAs Schottky diodes have been employed as the mixing element. As interest in operation above 300 GHz increased, quasi-optical structures such as the biconical and corner-cube mixers were developed. The problems with the whisker contact were eliminated by the development of planar Schottky diodes which have also led to the realization of integrated mixers. A significant advance took place via the use of micromachining techniques [64]. Here, using standard semiconductor processing techniques, the authors fabricated a 585 GHz fundamentally pumped Schottky mixer comprised of an etched silicon horn, a diced waveguide, and a lithographically formed microstrip channel for the diode circuit. The measured double sideband mixer temperature was 1200 K. The authors suggest that their approach is suitable for cryogenic operation and note that a single 8 in silicon wafer would yield over eighty 600 GHz mixer housings. The use of micromachining has also been used together with so-called electromagnetic bandgap (EBG) technology which eliminates substrate modes to realize a variety of mixers. Another development of interest is the fabrication of silicon Schottky diodes with THz cutoff frequencies [65,66]. Very recently, cutoff frequencies of 1.5 THz have even been achieved using a standard CMOS process which offers the promise of diagnostic instruments of increased complexity and performance in the future. Additional information regarding silicon activity may be found in Chapter 18.

For extremely low noise, sensitive heterodyne receiver applications in the mmWave region, a commonly employed device is the superconductor–insulator–superconductor (SIS) tunnel junction which makes use of the sharp nonlinearity produced by single-electron quasi-particle tunneling between the two superconductors (typically niobium) approximately 200 nm thick separated by an oxide layer a few atoms thick [67,68]. Here, in the mmWave region, the quantum energy $\hbar\omega/e$ may exceed the voltage width for onset of quasi-particle tunneling so that the absorption of a single photon can cause one additional electron to tunnel through the barrier. The required LO power is much lower than for Schottky diodes, which partially makes up for the need to cool the junctions to 4 K to become superconducting and sufficiently nonlinear. The upper frequency limit is reached when the signal frequency exceeds the superconductor energy gap, above which the radiation has sufficient energy to destroy the superconducting properties of the material (i.e. sufficiently high to break Cooper pairs, leading to losses and degraded sensitivity). For NbN with a gap of 5 meV (approximately twice that of Nb), this corresponds to an upper frequency for lossless coupling of 1.3 THz. In theory, the sensitivity of SIS mixers can approach the quantum limit, where the quantum limited noise temperature $T_N = h\nu/k_B$, which is ~48 K at 1 THz. However, typically, below the gap frequency, the sensitivity limit is 2–5 times the quantum limit.

At frequencies above the limit of SIS mixers, radio astronomers have concentrated on the development of hot electron bolometer junctions which are comprised of a thin short superconducting strip sandwiched between two thick normal metal pads [69]. The required detector nonlinearity is a result of the fact that the incident radiation causes the electron temperature in the strip to be raised to the critical temperature T_c of the superconductor with the resistance varying in a highly nonlinear fashion. The small heat capacity, due to the submicrometer size, and the high thermal conductance, due to cooling by electron diffusion, leads to very fast response times (~ tens of picoseconds). Another advantage of this device (compared to the SIS junction) is that the intrinsic capacitance is negligible,

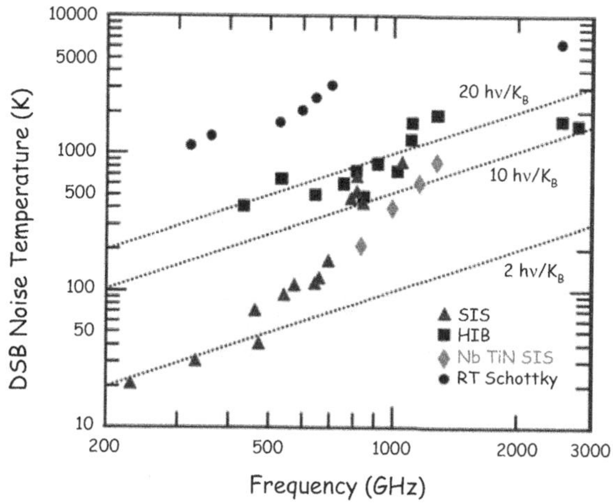

Figure 15.29: Comparison of Schottky, SIS, and HEB mixer technology status (courtesy of Imran Medhi).

which greatly facilitates matching of the incoming radiation to the junction. In addition, the required LO power is quite low which is of paramount importance in the THz region. HEBs have been employed as heterodyne mixers up to 5 THz and possess IF bandwidths up to 6–10 GHz.

A view of the state-of-the-art of mixer technology is provided in Figure 15.29. This figure contains a comparison of Schottky, SIS, and HEB mixer technology status. Plotted are contours of multiples of $h\nu/k_B$ providing a measure of how close the performance approaches the quantum limit. The choice of mixer technology is dependent upon the particular application as briefly discussed below.

For many applications, an imaging mixer array is essential and some loss in overall noise performance can be tolerated as long as array operation is possible. An early development involved an array of planar Schottky diodes which was monolithically integrated with bow-tie antennas to form a one-dimensional 94 GHz array.

As discussed previously in Section 15.4.3, there is increasing emphasis on imaging and visualization in fusion plasma diagnostics. Here, even in the case of radiometry, one measures emission from extremely hot plasmas; consequently, receiver noise temperatures of 3000 to 10 000 K suffice since this permits the instruments to be absolutely calibrated using standard hot–cold load techniques. Both hybrid and monolithic Schottky diode imaging arrays have been employed in ECEI and MIR plasma imaging applications [3, 23]. The hybrid arrays have been comprised of printed circuit dual-dipole antennas coupled with beam lead diodes that are attached to the feed of the individual antennas (see Figure 15.30).

In imaging applications requiring the lowest possible noise temperature, the radio astronomy community has initiated developments of focal plane arrays following the availability of SIS and HEB mixers with their relatively low power LO requirements compared with Schottky mixers [70,71]. The example of the design and successful of a three-element prototype, 1.63 THz FPA based on HEB mixers and wide-band MMIC IF amplifiers is provided in reference [70].

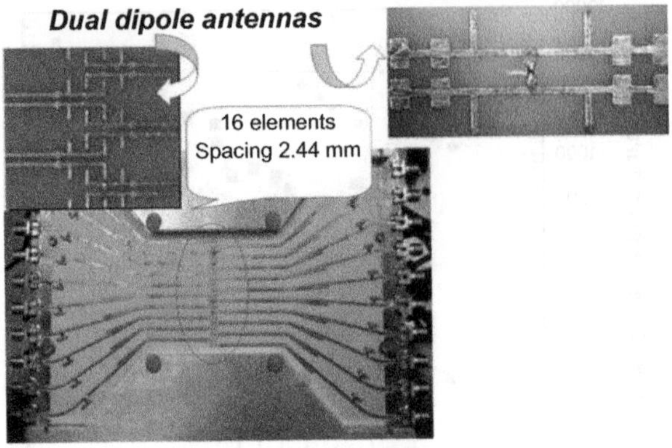

Figure 15.30: Photograph of dual dipole antenna/mixer array.

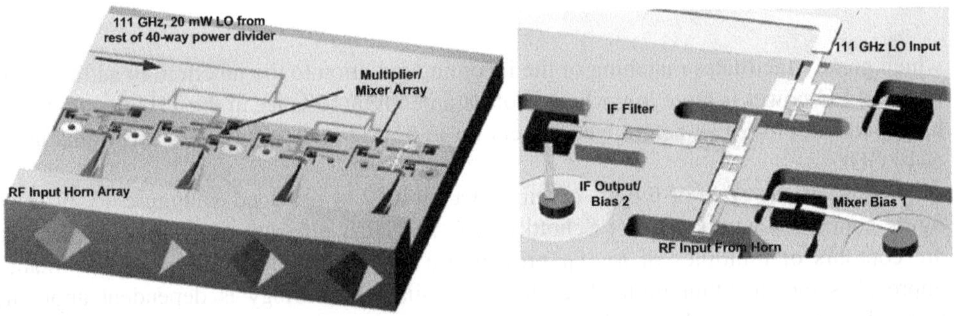

Figure 15.31: Waveguide focal plane technologies (courtesy of Imran Medhi).

Figure 15.31 shows an approach to mmWave/THz focal plane technology using waveguide technology. In this approach, most functionality is incorporated directly into each channel, so that the design can be easily replicated to a large array. The advantages claimed for this waveguide design are that: it eliminates most RF and IF crosstalk between channels; it possesses high beam quality and polarization properties; LO injection does not require complicated optics; it affords easy boresighting on all channels. The difficulties are said to be micromachining circuits above 1 THz and providing uniform LO distribution.

15.6.2 Direct Detection Receiver

The direct detection receivers are comprised of detectors and high-gain low-noise amplifiers (LNAs). They have advantages such as low noise temperature, no need of an LO, low power consumption, and low cost.

15.6.2.1 MMIC LNA

The development and availability of low-noise MMIC amplifiers in the mmWave region has made possible direct amplification and detection, with subsequent filtering and video detection. This is facilitated by the developments in InP and InAs/AlSb HEMT-based MMICs which include 20 dB gain at 220 GHz with simulated noise levels of 6.7 dB and ~30 K noise temperature over 75–110 GHz. An example of the direct amplification approach is the passive 89 GHz mmWave camera developed by Northrop–Grumman (TRW), which was described earlier. A camera operating at 140 GHz was also developed [9]. Two-dimensional tapered slot antennas are utilized in the focal plane of the imaging optics of these cameras.

Recently, there has been considerable activity directed toward the development of InAs-channel HEMTs for which it is claimed that there is the potential to enable revolutionary low-noise, low-power, and high-speed applications [72, 73]. InAs electronic properties, such as electron mobility and peak velocity, are seen to be nearly twice as large those of state-of-the-art $In_XGa_{1-X}As$-channels. Modeling by reference [73] suggests that an HBT could achieve f_T and f_{max} approaching 1 THz for an emitter-base and collector-base dimension of $0.5 \times 4.0\ \mu m^2$ ($2\ \mu m^2$) and that a simple metal-oxide-semiconductor field-effect transistor (MOSFET) model predicts an f_T of about 3 THz for a gate length of 0.1 μm. Recently, an ultra-low power InAs/AlSb HEMT W-band LNA was reported with a 3.9 dB and an associated gain of 20.5 dB together with a dc power dissipation less than one-tenth of a typical equivalent InGaAs/AlGaAs/GaAs HEMT LNA [74]. The envisioned applications include handheld mmWave (94–220 GHz) passive imagers and active-array space-based radar.

15.6.2.2 Video Detectors

An interesting development has been the Hughes Research Laboratory InAs/AlGaSb detectors which operate at zero bias and provide direct detection capability [75]. These exhibit voltage responsivities of up to 8000 mV/mW from 75 to 93 GHz, with input powers from −50 to −30 dBm (0.01–1 μW). These Sb-heterostructure diodes have been employed as detectors in W-band mmWave imaging cameras [76]. Reference [77] reports the development of a 2×2 array of these radiometers which exhibited a noise equivalent temperature difference of 10 K, calculated assuming a 30 Hz frame rate.

Another interesting technology development involves Semimetal Semiconductor Schottky (S3) diodes which are intended for direct detector applications from 94 GHz to 30 THz [78]. These are based on the InAlGaAs/InP material system and employ ErAs for the Schottky contact. The measured responsivity was 800–1000 V/W at 77 GHz and 500–600 V/W at 94 GHz. The authors predict that the detector capacitance and resistance can be reduced, thereby increasing the cutoff frequency of the detectors to the THz range.

15.6.2.3 Direct Detection Receiver Examples

Reference [79] describes an imaging array with direct detection technology. Each pixel in the focal plane array is composed of a Q-band Vivaldi antenna and a beam lead Schottky diode detector preamplified by an InGaAs pseudomorphic HEMT MMIC LNA. LO power is not required in this approach, so the array complexity and manufacturing cost are both reduced. As described earlier, the Schottky diode detector has advantages such as wide bandwidth, operational simplicity, and MMIC compatibility. In order to achieve high sensitivity, an LNA is used in front of the detector. Another example is the TRW W-band camera discussed earlier; the responsivity of a single Schottky diode detector is measured to be 7.2 V/mW at 94 GHz and the tangential sensitivity is −45 dBm. When a 15 dB LNA is added before the detector, the responsivity increases to 300 V/mW at 94 GHz, and the sensitivity becomes −62 dBm. The increase in receiver sensitivity enables the array to image faster. Following the detector, a video amplifier is used to increase the output signal to a level suitable for signal processing. W-band (94 GHz) MMIC direct direction receivers were also demonstrated using InGaAs HEMT devices based on W-band LNAs and preamplified detector MMICs. The receiver uses a preamplified Schottky barrier diode detector with a 50 dB gain, 6 dB noise figure amplifier from 90 to 95 GHz. The detector circuit consists of a two-stage W-band LNA in front of the diodes. The waveguide mounted W-band direct direction receiver subassembly was then connected to the feed of a large dish antenna. An azimuth-elevation drive system controls the vertical and horizontal scanning of the dish antenna. The passive mmWave camera described in Section 15.4.2 uses a direct detection receiver array.

This technology is being extended to considerably higher frequencies. Reference [80] describes the development of 220 GHz LNA MMICs for use in high-resolution active and PMMW imaging systems using InAlAs/InGaAs HEMT technology. Gain in excess of 10 dB was obtained over the bandwidth from 180 to 225 GHz. reference [81] has reported an InP HEMT MMIC amplifier with 15 dB of gain over the range 150–205 GHz.

15.6.3 Microbolometer Focal Plane Arrays

Another approach to mmWave imaging is the use of microbolometer arrays which were first fabricated by integrating the microbolometers with bow-tie antennas. More recently, uncooled mmWave microbolometer-based, 120-element focal plane imaging arrays have been developed employing Nb films, coupled to the incident field by annular slot-ring antennas. Using these microbolometer arrays, a 95 GHz active mmWave imaging system for concealed weapons detection has been demonstrated where pulsed illuminators are employed for illumination [13]. The authors contend that their results show that active imaging with antenna-coupled microbolometers can yield images comparable to that obtained with systems using MMIC amplifiers, but with a cost per pixel, that is, orders of magnitude lower.

For indoor applications with low-level signals, active mmWave imaging is favored for providing higher sensitivity than PMMW imaging. The widely used high-frequency MMIC amplifier-based imaging array will be extremely expensive when a large number of pixels are required to achieve video rate images and the operating frequency is still limited by the availability of the high-frequency amplifier, although a 260 GHz MMIC amplifier has already been demonstrated [82, 83]. In addition, those imaging arrays based on total power detection cannot benefit from the coherent nature of the return signal in active imaging systems either. Consequently, an antenna-coupled microbolometer focal plane array (as shown in Figure 15.32(a)) has been proposed in reference [13]. They are simple to fabricate, even in large arrays, and they have more than adequate sensitivity when used in an active imaging system.

The FPA described in reference [13] is fabricated on a 75 mm diameter, high-resistivity silicon wafer, and incorporates 120 microbolometers coupled to slot-ring antennas. The antennas collect the incident mmWave radiation and a voltage difference proportional to the incident power is produced across the bolometer. In the active mmWave imaging

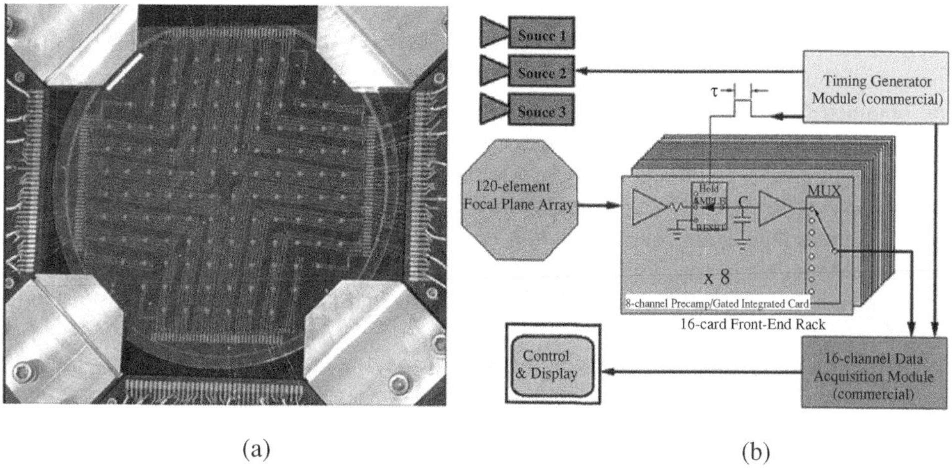

Figure 15.32: (a) The 120-element antenna-coupled microbolometer focal plane array. (b) System architecture. Reproduced by permission of © 2004 SPIE [13].

system using this antenna-coupled microbolometer FPA, three 95 GHz IMPATT pulsed noise oscillators with 1 W peak power each are configured to illuminate the scene from three different directions. The illumination from multiple angles helps signal detection regardless of the orientation of typical targets. A 30 cm diameter off-axis ellipsoid main aperture located 1.5 m from the object plane focuses the reflected power to the FPA. Each pixel is connected via coaxial cable to the 16 × 8 IF readout electronics, which is similar to other mmWave imaging systems. The digitized data are then displayed in real time at 30 frames/s on the PC display [13].

Another type of indirect FPA is composed of ferroelectric elements. The pyroelectric effect occurs when radiant energy causes an increase in the temperature of a ferroelectric material and causes a charge accumulation in the material to be detected. Those charges are sensed by the intermediate frequency signal processing circuits in the readout devices and a signal indicating the image of the scene will be provided. In reference [84], an approach is described where the array of thin ferroelectric film, the readout electronics and the semiconductor layer can form an integrated circuit based on one substrate, and which can be manufactured in a cost efficient manner. This technique can increase the signal-to-noise ratio and reduce the size of the system. Each of the ferroelectric elements represents a pixel of the imaging system. A lens is required to focus the radiation onto the ferroelectric element array [84].

15.6.4 LO and Probe Sources

mmWave imaging systems instruments require LO sources for their heterodyne receiver implementations as well as to serve as probe sources in the case of active imagers. Here, we briefly review the state-of-the-art and provide references for further investigation.

A commonly utilized two-terminal solid-state oscillator is the GUNN diode because of its low noise properties. This has virtually eliminated reflex klystrons since it operates up to

~200 GHz (in InP devices) and produces ~100 mW with 45 μW reported at 409 GHz as well as 6.1 mW at 285 GHz and 2.7 mW at 316 GHz from two devices each in an in-line power combining circuit [85–87]. Higher power (as well as higher frequency operation) may be obtained from the Impact Ionization Avalanche Transit Time or IMPATT diode where the negative differential resistance arises from the combination of avalanche breakdown delays and transit time effects which lead to the phase relation between voltage and current. Typical power levels at W-band are 150 mW CW and 10 W pulsed, making them higher power sources than Gunn devices, but also considerably noisier due to the carrier generation process. More details concerning these and related devices may be found in reference [87]. The potential for two such terminal devices to operate in the submmWave region is discussed in reference [87].

A traditional method of generating high-frequency radiation is via frequency multiplication using device nonlinearities; both the current–voltage (I–V) and capacitance–voltage (C–V) nonlinearities have been employed, but the former suffers from reduced efficiency, particularly at higher harmonic numbers, due to dissipative losses. The most commonly employed frequency multiplier is the GaAs varactor diode which has produced output power to beyond 2 THz [88–90]. Power outputs of 30 mW have been obtained between 184 and 212 GHz with a substrateless Schottky doubler with peaks in excess of 50 mW [90]. A broadband doubler produced 1.1 mW output at 765 GHz with 10% efficiency [88]. Although tripler design is difficult for Schottky varactors (owing to the need to provide a reactive idler circuit), recent results are still quite impressive. A planar Schottky tripler operating at room temperature has produced 0.6–1.6 mW at 540–640 GHz while other triplers produced 25 μW at 1130–1260 GHz. Another varactor, the so-called quantum barrier varactor (QBV) or heterostructure barrier varactor (HBV), has attracted interest because of its symmetric C–V characteristic, which means that it does not require dc bias and that it produces only odd harmonics when pumped with power at the fundamental frequency, thereby eliminating the need for an idler circuit to suppress even harmonics. Another feature which has attracted interest is the ease in which junctions can be epitaxially stacked to increase the breakdown voltage. It is predicted that a MMIC realization will yield 100 mW at W-band. At 250 GHz, 9 mW of output power was obtained at 12% efficiency from an InP-based heterostructure barrier varactor waveguide tripler [91].

MIMIC chip amplifier technology has also made significant advances, thereby permitting them to be used as both active imager probe sources and LOs for large Schottky diode heterodyne receiver arrays at frequencies up to ~200 GHz [92]. In addition, MMIC amplifiers are essential constituents of THz frequency multiplier chains. As examples of the state of the technology, single chip MMIC amplifiers have demonstrated 427 mW output at 95 GHz [93] and 2.4 W in a waveguide combined system [94]. More recently, the performance of InP HBT PAs has increased dramatically. For example, at 176 GHz a two-stage common-base amplifier exhibited 8.1 dBm output power with 6.3 dB associated power gain and exhibited 9.1 dBm saturated output power [92].

Traditionally, plasma diagnostics systems operating in the submmWave region (~1.5 mm to ~50 μm) have employed molecular lasers which produce mW to W level power at discrete wavelengths [95]. These continue to be used in fusion plasma laboratories throughout the world, particularly in interferometer and polarimeter systems. Considerable effort has been devoted by the submmWave astronomy community into developing space qualified, THz optically pumped lasers [96]. These employ sealed-off, rf excited, folded-cavity, waveguide

CO_2 laser pumps together with folded cavity THz laser cells with ultra-high vacuum designs. Using a 50 W pump laser, cw output powers up to 100 mW are obtained in the 50 μm to 1 mm region with the output power exceeding 10 mW over the range 70 μm to 600 μm and above 1 mW out to 1.22 mm.

In recent years, there has been increasing interest in the THz QCL where the lasing mechanism is based on intersubband transitions in the conduction band of GaAs/AlGaAs heterostructures [97]. This device technology has demonstrated single mode, narrowline lasing between 1.9 THz and 4.8 THz with output powers up $\sim$90 mW and is being actively pursued as a local oscillator for submmWave heterodyne receivers where noise temperatures of 2700 K were obtained at 2.5 THz from an HEB pumped by a QCL [98]. A complete heterodyne receiver using an NbN superconducting HEB as mixer and a quantum cascade laser operating at 2.8 THz as local oscillator has been developed for high-resolution spectroscopy above 2 THz on space-based observatories [99]. They report a double sideband receiver noise temperature of 1400 K at 2.8 THz and 4.2 K, and claim that the free-running QCL has sufficient power stability for a practical receiver.

Another development trend which may prove of great use to passive and active mmWave/THz imaging systems is the recent application of microfabrication techniques to the manufacture of miniature vacuum 'tubes' producing mW to 100 W outputs and beyond [100]. These include reflex klystrons for use as THz LOs as well as travelling wave tubes (TWTs), multi-cavity klystron amplifiers and BWOs.

15.6.5 Quasi-optical Power Combining

An obvious approach to increasing the available output power from solid-state sources is to combine their outputs. However, when this is accomplished via techniques such as waveguide power combining, Ohmic losses limit the number of devices which can be practically combined. This limitation led to interest in the so-called quasi-optical or spatial power combining techniques. The first quasi-optical spatial power combining investigations extended the traditional approach to obtaining power in the mmWave region which is to employ waveguide mounted Schottky diodes as frequency multipliers, utilizing either the diode nonlinear I–V or C–V characteristics to provide the desired harmonic multiplication. Here, the output power levels can be dramatically increased by means of quasi-optical spatial power combining of the output of large planar grid arrays of solid state devices which can be employed to provide high output power levels, as well as to avoid the Ohmic losses and limitations associated with conventional power combining techniques. This approach has yielded 5 W in a quasi-optical frequency tripler grid array [101] at 99 GHz in a pulsed proof-of-principle experiment. At Rockwell Scientific Co. (now Teledyne Scientific Company), CW output power of 684 mW with a peak conversion efficiency of 11.3% was obtained at 93 GHz using a W-band quasi-optical array consisting of 196 HBV diodes mounted in a waveguide where the 32 mm^2 array was fabricated using 2×2 barriers per diode on a lattice-matched InP substrate [102]. Frequency doubler grid arrays have also demonstrated the potential for high output power levels with 2.1 W reported at 66 GHz and efficiencies of 11.5% obtained.

The power levels available from MMIC amplifiers can also be dramatically increased through the use of spatial power combining. Two reviews provide an excellent overview of the current state-of-the-art in spatial power combining [103, 104] which has advanced dramatically. Particularly noteworthy is the work by researchers at Sanders who reported an output power of 35 W at 61 GHz. This amplifier employed 272 individual MMICs arranged in a 'tray' architecture and provided 60 dB of small-signal gain. At Ka-band, a grid amplifier module has produced over 16 W of saturated power while a W-band waveguide-packaged InP HEMT reflection grid amplifier has produced 264 mW at 79 GHz.

These spatial power combining techniques are also applicable to oscillator configurations; both GUNN device and MESFET grid oscillators were developed at the end of the 1980s. In addition, CW power levels of 1.5 W at 61.4 GHz and 0.45 W at 98.8 GHz have been reported using a 3 × 3 waveguide Gunn diode array [105].

15.6.6 Beam Formation and Shaping

Key to both active and passive mmWave imaging is beam formation and shaping. A considerable number of imaging systems make use of phased antenna arrays which are the subject of Chapter 13 so the reader is referred there for additional details on PAAs and their use in beam formation. In some of the early imaging systems, a single Schottky diode heterodyne mixer was employed using mechanical scanning for the laboratory experiments. Airborne imaging experiments made use of a combination of mechanical scanning and the use of the forward velocity to realize a 'push-broom raster' scan [11, 106]. A number of more modern embodiments arc described by reference [107]. Here, an eight-channel 94 GHz imager is described which had a 1.2 m square Cassegrain antenna mounted on an azimuth and elevation gimbal, which in turn was operated by dc servos under computer control. Although the image quality was excellent, the system was limited by the acquisition time of approximately 10 min. A later 94 GHz imager intended for poor-weather navigation [107,108] featured 150 direct-detection receiver channels and had a 50 cm aperture.

Reference [109] has described a single channel 94 GHz opto-mechanically scanned imager for monitoring the ground movement of aircraft in adverse weather conditions. This employs an InP LNA followed by a down-converter and covers a field of view of 60° horizontally and 20° vertically. The system employs two counter-rotating mirrors that are tilted about their axes of rotation and which simulate the linear scan of a single high speed, large aperture flapping mirror. The two counter-rotating mirrors were driven from a single servo motor, using timing belts and toothed pulleys. The slower flapping mirror was slaved to the motion of the rotating discs using an electronic camera. The video output from the receiver was displayed as a two-dimensional image on a conventional PC.

Another interesting approach of relevance for passive imagers concerns aperture synthesis which permits high resolution using a number of small, distributed apertures, rather than a single, large aperture [110]. Here, one interferometrically combines the individual mmWave signals to form the image. The paper describes the technique of coherent optical beam forming which performs optical transport of the up-converted mmWave signal and direct image formation at the optical frequency.

Traditionally, dish or slotted array antennas, have been employed and use physical shape and direction to form and steer beams. However, the demand for PAAs has increased significantly during the past decades. PAAs utilize the interference between multiple radiating elements to achieve beam forming and beam steering. In radar and communication systems, they are typically employed to transmit/receive a signal or search/trace a target. Existing mechanical scanning methods are inherently slow and require large amounts of power in order to respond sufficiently rapidly to deal with large numbers of high-speed maneuvering targets. With mechanically scanned systems, antenna inertia and inflexibility prevent employment of optimum radar beam positioning patterns that can reduce reaction times and increase target capacity. With electronic scanning, the radar beams are positioned almost instantaneously and completely without the inertia, time lags, and vibration of mechanical systems. These same attributes have attracted equal interest from the imaging community.

The fundamental principles underlying the concept of electronic beam steering are readily understood: The electromagnetic energy received at a point in space from two or more closely spaced radiating elements is a maximum when the energy from each radiating element arrives at the point in phase. Controlling the phase through the many segments of the antenna system allows the beam to be rapidly directed in different directions.

The scan angle θ_0 from a PAA depends on the operating frequency, spacing between antenna elements, and the phase offset between signals in the individual elements. The far-field pattern of a phased antenna array is controlled by the relative phase and the amplitude distribution of the microwave signals emitted by regularly spaced radiating elements. Since antenna characteristics are reciprocal, the phased antenna array can also be used as a receiving system, as well as a transmitter. Often one is concerned with maintaining a fixed steering angle θ_0 as the operation frequency f varies. However, it is readily seen that if the phase shift angle ϕ is fixed, changing the operation frequency f results in a change in the steering angle θ_0. Therefore, beam squinting arises from this distortion. This problem can be ameliorated by a true time delay approach in which a physical time delay element is introduced to effect the inter-element phase differential necessary to steer the beam:

$$\tau = \left.\frac{\mathrm{d}\phi}{\mathrm{d}\omega}\right|_{\omega=\omega_c} = \frac{\phi_0}{2\pi f_c} = \frac{nd}{c}\sin(\theta_0).$$

Here, n is the number of elements and d is the element spacing. Note that the steering angle θ_0 remains unchanged when the operation frequency f is varied. Consequently, there is considerable interest in providing true time delay excitation. The radiation pattern is given by

$$G_a(\theta) = \frac{|E_a|^2}{n^2} = \frac{\sin^2\{(n\pi d/\lambda)(\sin\theta - \sin\theta_0)\}}{n^2\sin^2\{(\pi d/\lambda)(\sin\theta - \sin\theta_0)\}} \tag{15.5}$$

Grating lobes occur at $\pi d/\lambda(\sin\theta_g - \sin\theta_0) = \pm n$ so that for a scan over $\pm 90°$, the element spacing should be $d = \lambda/2$.

Although scanning capability is the most common function, a phased antenna array can also provide beam-shaping capability by appropriate arrangement of the feed signals. This topic has been investigated by many researchers. Recently, the so-called flat parabolic surface (FLAPS) reflector [111], which consists of an array of dipole scatters and is spatially fed using a feed assembly as in a conventional reflector system, has been employed to realize beam shaping and beam scanning functions by adjusting the length of the dipoles so that the phase shift between the incident and reflected wave can be controlled.

To illustrate both beam steering and shaping, it is convenient to examine the example of MIR. As noted in Section 15.4.3, to ensure that the MIR system is easily able to accommodate changes of curvature of cutoff layer, it is desirable to develop an electronically controlled

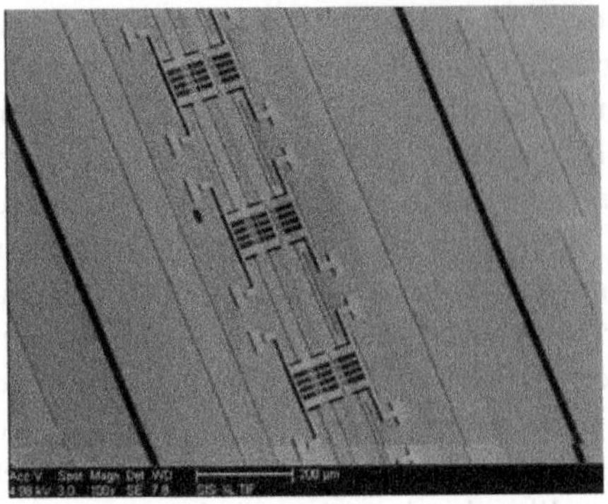

Figure 15.33: SEM picture of a MEMS varactor-based true time delay line. Reproduced by permission of © 2007 IEEE [29].

'lens' whose focal properties can be adjusted to the particular plasma of interest. A realization of such a 'lens' involves the use of PAAs where a quadratic phase shift is programmed across the array using true time delay [112]. The controlled time delay can be generated by a variety of techniques including nonlinear delay lines (NDLs) [112], MEMS switched delay lines [113], piezoelectric controlled phase shifters [27,28], ferroelectric phase shifters [114], liquid crystal beam formers [115], tunable electromagnetic crystal (EMXT) surfaces [116], integrated antenna and phase management chip [117], and optical delay lines [118].

The NDL principle is readily understood; here, the propagation velocity on a Schottky or MEMS varactor diode loaded transmission line is varied by changing the varactor diode bias voltage since $v_{\text{prop}} = 1/\sqrt{LC(V)}$. For frequencies well below the Bragg cutoff, this provides frequency independent true time delay. When used in conjunction with PAAs, beam steering is effected by programming a linear delay across the array while beam focusing/defocusing is accomplished by employing a quadratic delay.

As discussed in Chapter 12, MEMS are miniature, microfabricated devices using mechanical movement to achieve open or short circuit conditions on a transmission line. Although the first microwave MEMS switches and varactors were only developed in the early 1990s, the technology has progressed rapidly so that high-performance devices are available at 50 GHz and even higher. A major focus of this work has been aimed at the development of phase shifters where the application is phased antenna arrays. This interest has been driven by their low insertion loss and high Q. The average on-wafer loss for RF MEMS phase shifters is ~0.6 dB/bit at Ka-Band. An excellent review of RF MEMS technology may be found in [119]. MEMS varactors are also receiving increasing interest since the extended tuning varactor concept has significantly increased the capacitance ratios [29, 120]. Figure 15.33 displays a scanning electron microscope (SEM) photograph of a MEMS extended tuning range varactor-based true time delay line where the diodes are loaded on a 70 Ω CPW transmission line operated at 28 GHz [29]. Here, a distributed LC ladder structure is realized by parallel loading the MEMS varactors on high impedance coplanar waveguide (CPW) transmission lines.

In contrast, the distributed type delay lines take advantage of the possibility of controlling the wave velocity, while keeping the length of the propagation path constant. The liquid

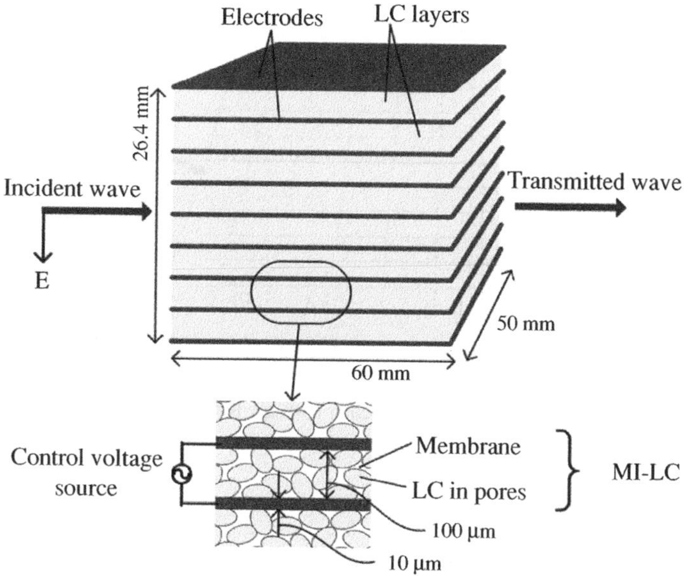

Figure 15.34: Structure of a liquid crystal beam former. Reproduced by permission of © 2004 IEEE [115].

crystal phase shifter is one of the distributed types of delay lines which operate by controlling the wave velocity, while keeping the length of the propagation path constant. In liquid crystal, the dielectric constant is anisotropic, which means it is different in the direction along the director (molecule) from the one in the direction perpendicular to it. This occurs when the charge distribution along the molecule responds differently to the parallel component of the local electric field than the distribution perpendicular to the length does to the perpendicular component, thereby yielding a difference in dielectric constants. When an electric field is applied, the molecule tends to rotate and align parallel or perpendicular to the electric field which depends on whether the LC has positive or negative anisotropy. Consequently, the dielectric constant can be changed and therefore the propagation constant can be changed. Figure 15.34 below schematically illustrates the LC beam former structure. As shown in Figure 15.35, beam-steering (or shaping) is provided by appropriate biasing of the electrodes (0 to 120 V). The LC phase shifter is an easy, relatively inexpensive true time delay technology; however, the modulation speed is low, which is on the order of several Hz. Reference [115] demonstrated a V-band liquid crystal beam former (LCBF) which could can steer the beam by approximately $\pm 13°$ by changing the applied control voltage and also operate as a lens that can adjust the wavefront of an incident mmWave.

More recent work has focused on reducing the conductor losses, thereby helping to make the LC beam former more practical for mmWave applications. By taking advantage of the fact that the surface currents flow on both sides of the conductors, but in opposite directions, the losses are reducing by thinning of the electrodes and cancellation of the surface currents [121]. EM simulations predict that attenuation constants of ~ 2 dB m^{-1} can be

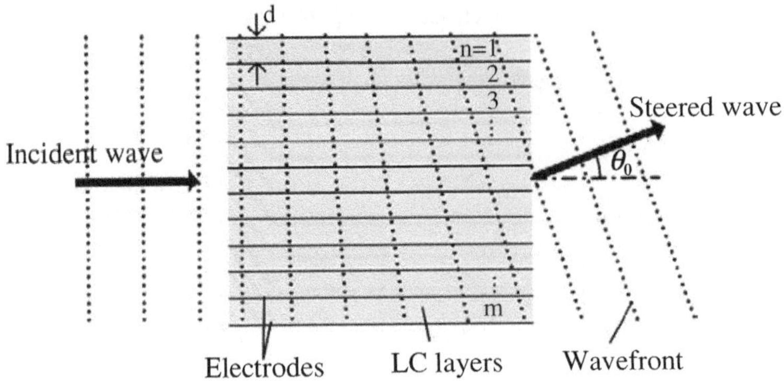

Figure 15.35: Structure of a liquid crystal beam former. Reproduced by permission of © 2004 IEEE [115].

obtained at 60 GHz using this approach for copper electrode thicknesses values of 0.2 μm while keeping the leakage to the adjacent LC layer to −19 dB.

Another interesting true time delay line is the PET controlled delay line [25, 26]. A dielectric material used as a perturber is attached with a piezoelectric bending element, which is placed above a microwave transmission line (ex. microstrip line). A voltage applied to the piezoelectric bender causes a vertical deflection. Therefore, the air gap between the transmission line and perturber changes when different dc bias levels are applied, resulting in different values of effective dielectric constant, which in turn results in a change in the propagation velocity on the transmission line. A schematic of this multilayer structure is shown in Figure 15.36. Using an appropriately shaped perturber above an array feed network, quadratic and linear phase functions can be applied to the array to broaden and scan the beams. Reference [26] has reported a 20–40 GHz beam-shaping/steering phased antenna array system using Fermi tapered slot antennas and PET perturbers.

Figures 15.37 and 15.38 shows both schematics and photographs of the configurations. CPW is chosen over microstrip line because larger phase delays can be generated. An unequal power divider provides an appropriate amplitude taper (Taylor N-Bar distribution) to lower the sidelobe ratio (SLR) in beam shaping. In the beam-steering demonstration, an eight-way T-junction power divider with equal output powers is designed with an output port spacing of $0.66\lambda_0$. In the beam-shaping demonstration, a 16-way T-junction unequal power divider is employed. A Taylor N-bar distribution with N-bar = 6 and SLR = 30 dB is used for sidelobe suppression. The spacing of the output ports is $0.8\lambda_0$ to decrease the mutual coupling effect. The measured beam-scanning radiation pattern at 30 GHz shows a −17° to +19° beam-scanning range. The beam-shaping array shows a 12° 3 dB beamwidth broadening range keeping SLR above 15 dB from 20 to 40 GHz.

As an example of the use of time delay technology, we briefly consider the beam shaping technology being developed for a proposed three-dimensional MIR system to be installed on a number of magnetic fusion plasma devices. In this approach, the three-dimensional MIR receiver is comprised of a two-dimensional array of dual dipole antennas (Figure 15.21) where the third (radial) dimension is obtained through the use of multiple illumination

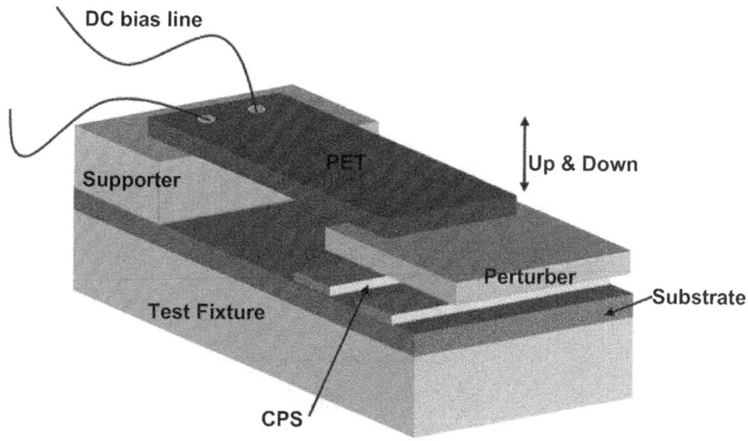

Figure 15.36: Schematic of PET controlled phase shifter. Reproduced by permission of © 2007 IEEE [26].

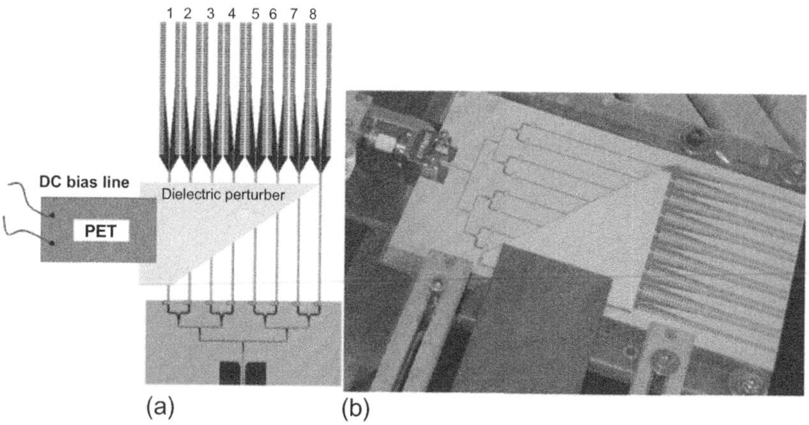

Figure 15.37: (a) Schematics and (b) photographs of beam steering with amplitude taper PAA systems using PET-controlled phase shifter. Reproduced by permission of © 2007 IEEE [26].

frequencies over the frequency range (see Figure 15.39), associated with reflecting from the right-hand X-mode plasma cutoff layer (f_R). Although the eventual frequencies of interest range from about 40 GHz to 140 GHz depending upon the particular device, the initial developments have been carried out at Ka-band.

15.6.7 Imaging Optics

In mmWave imaging systems, optical components or lenses are used to collect the radiation from objects and diffract them to the focal plane. The goal of the optical design is to provide

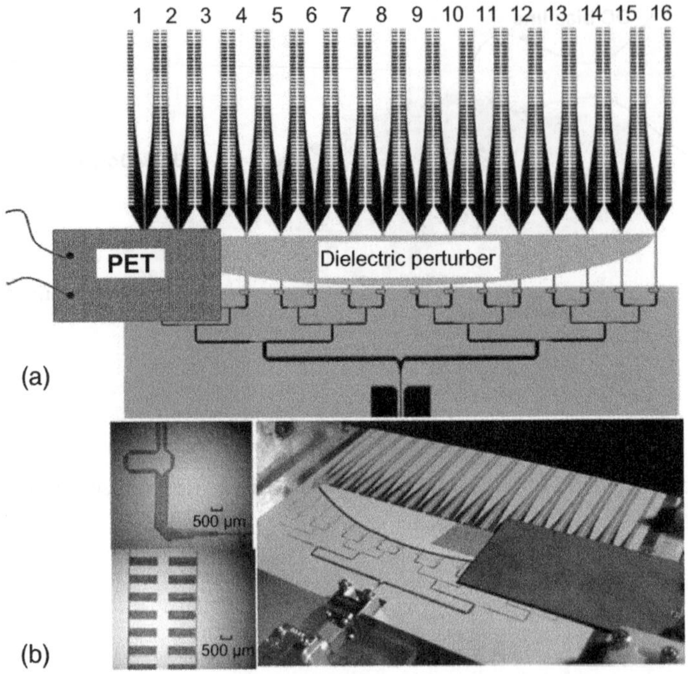

Figure 15.38: (a) Schematics and (b) photographs of beam shaping with amplitude taper PAA systems using PET-controlled phase shifter. Reproduced by permission of © 2007 IEEE [26].

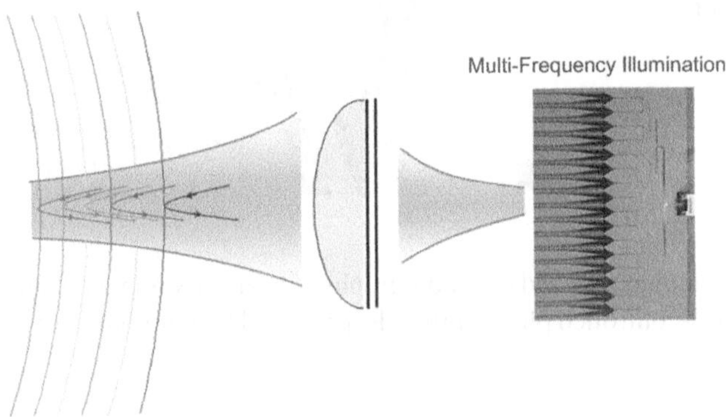

Figure 15.39: Schematic illustrating how a simultaneous 'comb' of illumination frequencies can probe multiple cutoff layers, as each distinct frequency reflects from a distinct cutoff layer.

high resolution with minimal aberration over large fields of view. The imaging quality can be established if the lens could give a diffraction limited pattern on the focal plane in a wide field of view. Compared to visible and IR imaging, the lens and aperture sizes are much larger due to the longer wavelength of the mmWave radiation. The lens and aperture dimensions can be 0.5 m or larger. Large optics is not convenient for many applications, but compact structures will cause serious aberrations, thereby restricting imaging capability. Geometric ray tracing and Gaussian optics can be used to design the optics to reduce the aberrations and expand fields of view. Some microwave components such as filters and beam splitters are also implemented among the imaging optics.

For systems requiring scanning, there are additional requirements for the optics. The transmitted and received beams must track one another on the target to within a diffraction spot. The aberration should be minimized because aberrations or other optical imperfections not only degrade the spatial resolution by diffusing the receive beam, but also degrade the sensitivity by diffusing precious source power in the transmit beam over unobserved target pixels.

15.7 Conclusion and Outlook

Given that microwave, mmWave, and THz technology frontiers are rapidly advancing due to the numerous scientific, commercial, and military applications which demand ever increasing levels of performance while also requiring reduced cost, increased level of performance can be expected from imaging systems. Besides the advanced SiGe technologies [122], a particularly important driver will be advances in CMOS technology which have been extended to the mmWave region [123]. For example, VCOs have operated at 99 GHz with a tuning range of 2.5 GHz and 192 GHz with ~ -20 dBm output power using 0.13 μm CMOS technology [124, 125]. At 140 GHz, −22 to −19 dBm output power with a frequency tuning range of 1.2 GHz has been reported using a 90 nm logic CMOS process [126]. Recently, a complete 60-GHz-band wireless direct-conversion transceiver with an output power of 7 mW in standard 1V 90 nm CMOS technology was reported [127]. Also reported using 90 nm CMOS technology are power amplifiers producing 10.6 dBm (60 GHz) and 6.3 dBm (77 GHz) from a 1.2 V supply [128]. Given the above advances as well as recent progress in developing fully functional ICs using the 45 nm CMOS process, it can be anticipated that imaging applications will dramatically increase. Already, a 410 GHz push–push oscillator with an on-chip patch antenna fabricated using low leakage transistors of a 6 M 45 nm CMOS process has been reported [129].

Acknowledgments

The contributions of the authors' colleagues and collaborators are gratefully acknowledged as are Drs I. Mehdi and H. Bruce Wallace for kindly providing figures. Support is also acknowledged from the US Department of Energy.

References

[1] L. Yujiri, M. Shoucri and P. Moffa, 'Passive millimeter wave imaging', in *Microwave Magazine, IEEE* **4**(3) (2003), pp. 39–50.

[2] V. E., Lyubchenko, 'millimeter wave imaging systems', in *13th International Conference on Microwaves, Radar and Wireless Communications, 2000*, Wroclaw, Poland, *MIKON-2000,* vol. 3, pp. 63–69, 2000.

[3] H. Park, *et al.*, 'Recent advancements in microwave imaging plasma diagnostics', *Review of Scientific Instruments* **74**(10) (2003), pp. 4239–4262.

[4] M. J. Rosker and H. B. Wallace, 'Imaging through the atmosphere at terahertz frequencies', in *IEEE/MTT-S International Microwave Symposium, 2007*, Honolulu, HI, pp. 773–776, June 3–8, 2007.

[5] R. Appleby and H. B. Wallace, 'Standoff detection of weapons and contraband in the 100 GHz to 1 THz region', *IEEE Transactions on Antennas and Propagation* **55**(11) (2007), pp. 2944–2956.

[6] J. E. Bjarnason, T. L. J. Chan, A. W. M. Lee, M. A. Celis and E. R. Brown, 'Millimeter-wave, terahertz, and mid-infrared transmission through common clothing', in *Applied Physics Letters* **85**(4) (2004), pp. 519–521.

[7] R. Olsen, S. Clark, J. Galliano, Chung-Ta Hseih, J. Lovberg, C. Martin and C. Phillips, 'Video-rate passive millimeter-wave imager', in *Proceedings of the 18th Conference on Digital Avionics Systems*, 1999. vol. 1, pp. 3.C.1-1–3.C.1-8, November 1999.

[8] R. Kuroda, R. Kuroda, G. S. Dow, Y. Guo, R. Johnson, M. Biedenbender, A. Marashi, L. Yujiri, and M. Shoucri, 'Direct detection MMIC FPAs for MMW imaging', in *Proceedings of SPIE Conference, Passive Millimeter-Wave Imaging Technology*, Orlando, FL, vol. 3064, pp. 90–97, 1997.

[9] R. T. Kuroda, G. S. Dow, D. Moriarty, R. Johnson, A. Quil, S. D. Tran, V. Pajo, S. Fornaca and L. Yujiri, 'Large scale W-band focal plane array developments for passive millimeter-wave imaging', in *Proceedings of SPIE Conference on Passive Millimeter-Wave Imaging Technology II*, Orlando, FL, vol. 3378, pp. 57–62, 1998.

[10] G. R. Huguenin, 'millimeter-wave concealed weapons detection and through-the-wall imaging systems', in *Proceedings of SPIE, Command, Control, Communications, and Intelligence Systems for Law Enforcement*, vol. 2938, no. 1, pp. 152, 1997.

[11] W. J. Wilson, R. J. Howard, A. C. Ibbott, G. Parks and W. B. Ricketts, 'Millimeter-waveimaging sensor', in *IEEE Transactions on Microwave Theory and Techniques* **34**(10) (1986), pp. 1026–1035.

[12] D. M. Sheen and D. L. McMakin, 'Three-dimensional millimeter-wave imaging for concealed weapon detection', *IEEE Transactions on Microwave Theory and Techniques* **49**(9) (2001), pp. 1581–1592.

[13] A. Luukanen, A. J. Miller and E. N. Grossman, 'Active millimeter-wave video rate imaging with a staring 120-element microbolometer array', in *Proceedings of SPIE – The International Society for Optical Engineering*, vol. 5410, no. 1, pp. 195–201, 2004.

[14] 'UK looks at Qinetiq's millimeter wave imaging for bomb detection', July 2005, http://www.engadget. com/2005/07/23/uk-looks-at-qinetiqs-mmWave-imaging-for-bomb/ (last accessed in April 2008).

[15] 'Qinetiq's millimeter wave sees through walls and your clothes', Mar. 2005, http://www.engadget.com/2005/03/17/qinetiqs-millimeter-wave-sees-through-walls-and-your-clothes/ (last accessed in April 2008).

[16] D. M. Sheen, D. L. McMakin, H. D. Collins, T. E. Hall and R. H. Severtsen, 'Concealed explosive detection on personnel using a wideband holographic millimeter wave imaging system', in *Proceedings of SPIE, Signal Processing, Sensor Fusion, and Target Recognition V*, vol. 2755, no. 1, pp. 503–513, 1996.

[17] R. N. Anderton, R. Appleby, J. E. Beale, P. R. Coward and S. Price, 'Security scanning at 94 GHz', in *Proceedings of SPIE Passive Millimeter-Wave Imaging Technology IX,* vol. 6211, pp. 62–110C, 2006.

[18] L. Yujiri, *et al.*, Passive millimeter wave video camera', in *Proceedings of SPIE Conference on Passive Millimeter-Wave Imaging Technology II*, vol. 3378, pp. 14–19, 1998.

[19] R. Appleby, R. N. Anderton, R. G. Humphreys, P. Munday and S. Price, 'Components for passive millimetre wave imaging at 94 GHz', *4th ESA Workshop on Millimetre-Wave Technology and Applications*, February 2006,

[20] N. C. Luhmann, Jr., H. Bindslev, H. Park, J. Sánchez, G. Taylor and C. X. Yu, (invited) Ch. 3: Microwave diagnostics technical paper 'Plasma Diagnostics for Magnetic Fusion Research', *Special Issue of Fusion Science and Technology on MFE Diagnostics* **53**(2) (2008), pp. 335–396.

[21] Z. Shen, N. Ito, E. Sakata, C. Domier, Y. Liang, N. C. Luhmann, Jr. and A. Mase, 'Frequency selective surface notch filter for use in a millimeter wave imaging system', *Antennas and Propagation Society IEEE International Symposium 2006*, pp. 4191–4194, 2006.

[22] Z. Shen, N. Ito, E. Sakata, C. Domier, N. C. Luhmann, Jr. and A. Mase, 'D-band double dipole antenna for use in millimeter wave imaging systems', *Antennas and Propagation IEEE International Symposium, 2007*, pp. 2658–2661, June 9–15, 2007.

[23] L. Yang, C. Domier and N. C. Luhmann, Jr., 'Q-band to V-Band 1-D and 2-D elliptical lens antenna arrays', *Microwave and Technology Letters* **49**(8) (2007), pp. 1798–1801.

[24] H. K. Park, *et al.*, 'Comparison study of 2D images of temperature fluctuations during sawtooth oscillation with theoretical models', *Physical Review Letters* **96** (2006), pp. 195004.

[25] L. Yang, C. W. Domier and N. C. Luhmann, Jr., 'Ka-band E-plane beam steering/shaping phased array system using antipodal elliptically-tapered slot antenna', *International Journal of Infrared Millimeter Waves* **28** (2007), pp. 283–289.

[26] L. Yang, N. Ito, C. Domier, N. C. Luhmann, Jr. and A. Mase, '20 to 40 GHz beam shaping/steering phased antenna array system using Fermi tapered slot antennas', *IEEE/MTT-S International Microwave Symposium, 2007*, pp. 1887–1890, June 3–8, 2007.

[27] T.-Y. Yun and K. Chang, 'A low loss time-delay phase shifter controlled by piezoelectric transducer to perturb microstrip line', *IEEE Microwave and Guided Wave Letters* **10** (2000), pp. 96–98.

[28] T.-Y. Yun and K. Chang, 'A low-cost 8 to 26.5 GHz phased array antenna using a piezoelectric transducer controlled phase shifter', *IEEE Transactions on Antennas and Propagation* **48**(9) (2001), pp. 1290–1298.

[29] Y. Liang, C. W. Domier and N.C. Luhmann, Jr., 'MEMS based true time delay technology for phased antenna array systems', *Asian Pacific Microwave Conference*, 2007, Asia-Pacific, pp. 1–4, 11–14 December 2007.

[30] H. Kamoda, T. Kuki, H. Fujikake and T. Nomoto, 'Millimeter-wave beam former using liquid crystal', *Electronics and Communications in Japan*, Part 2 (Electronics) **88** (2005), pp. 10–18.

[31] T. De Graauw and F. P. Helmich, 'Herschel-HIFI: the heterodyne instrument for the far-infrared,' in *Proceedings of the Promise of the Herschel Space Observatory Symposium*, pp. 45–51, 2001.

[32] H. M. Pickett, R. L. Poynter, E. A. Cohen, M. L. Delitsky, J. C. Pearson and H. S. P. Muller, 'Submillimeter, millimeter, and microwave spectral line catalog,' *Journal of Quantitative Spectroscopy and Radiative Transfer* **60** (1998), pp. 883–890.

[33] R. L. Brown, W. Wild and C. Cunningham, 'ALMA – the Atacama large millimeter array, Advances in Space Research', *Astronomy at IR/Submm and the Microwave Background* **34**(3) (2004), pp. 555–559.

[34] A. A. Stark, *et al.*, 'The Antarctic Submillimeter Telescope and Remote Observatory (AST/RO)', *Publications of the Astronomical Society of the Pacific* **113** (2001), pp. 567–585.

[35] C. E. Groppi, *et al.*, 'SuperCam, a 64-pixel heterodyne imaging array for the 870 micron atmospheric window, millimeter and submillimeter detectors and instrumentation for astronomy III', *Proceedings of SPIE*, vol. 6275, pp. 62750O, 2006.

[36] Stratospheric Observatory for Infrared Astronomy (SOFIA), *Proceedings of SPIE on Infrared Spaceborne Remote Sensing and Instrumentation XV*, 2007.

[37] W. M. Manheimer, A. W. Fliflet, G. J. Linde, W. J. Cheung, V. Gregers-Hansen, M. T. Ngo and B. G. Danly, 'The structure of turbulence in clouds measured by a high power 94 GHz radar', *Journal of Physics of Plasma* **11**(5) (2004), pp. 2852–2856.

[38] M. A. Kempkes, T. J. Hawkey, M. P. J. Gaudreau and R. A. Phillips, 'W-band transmitter upgrade for the Haystack UltraWideband Satellite. Imaging Radar (HUSIR)', *Vacuum Electronics Conference, 2006 Held Jointly with 2006 IEEE International Vacuum Electron Sources*, IEEE International, pp. 551–552, 25–27 April 2006.

[39] S. Scheding, G. Brooker, R. Hennessy, M. Bishop and A. Maclean, 'Terrain imaging and perception using millimetre wave radar', in *Proceedings 2002 Australasian Conference on Robotics and Automation*, pp. 13–18, Auckland, November 27–29, 2002.

[40] K. B. Cooper, R. J. Dengler, G. Chattopadhyay, E. Schlecht, J. Gill, A. Skalare, I. Mehdi and P. H. Siegel, 'A high-resolution imaging radar at 580 GHz', *IEEE Microwave and Wireless Components Letters* **18**(1) (2008), pp. 64–66.

[41] R. Schneider, H.-L. Blocher and K. M. Strohm, 'KOKON – automotive high frequency technology at 77/79 GHz', in *European Radar Conference, 2007. EuRAD 2007*, pp. 247–250, October 10–12, 2007.

[42] K. Sakai (Ed.), 'Terahertz optoelectronics', *Topics in Applied Physics, Springer/National Institute of Information and Communications Technology*, vol. 97, 2005.

[43] M. Tonouchi, 'Cutting-edge terahertz technology', *Nature Photonics* **1**(2) (2007), pp. 97–105.

[44] P. H. Siegel, 'THz technology', *IEEE transactions on Microwave Theory and Techniques, 50th Anniversary Issue* **50**(3) (2002), pp. 910–928.

[45] P. H. Siegel, 'THz technology in biology and medicine', *IEEE Transactions on Microwave Theory and Techniques* **52**(10) (2004), pp. 2438–2447.

[46] Y. C. Shen, T. Lo, P. F. Taday, B. E. Cole, W. R. Tribe and M. C. Kemp, 'Detection and identification of explosives using terahertz pulsed spectroscopic imaging', *Applied Physics Letters* **86**(24) (2005), pp. 241116.

[47] K. Yamamoto, 'Noninvasive inspection of C-4 explosive in mails by terahertz time-domain spectroscopy', *Japanese Journal of Applied Physics* **43**(3B) (2004), pp. L414–L417.

[48] H.-B. Liu, Y. Chen, G. J. Bastiaans and X.-C. Zhang, 'Detection and identification of explosive RDX by THz diffuse reflection spectroscopy', *Optics Express* **14**(1) (2006), pp. 415–423.

[49] T. S. Hartwick, D. T. Hodges, D. H. Barker and F. B. Foote, 'Far infrared imagery', *Applied Optics* **15**(8) (1976), pp. 1919–1922.

[50] B. B. Hu and M. C. Nuss, 'Imaging with terahertz waves', *Optics Letters* **20**(16) (1995), pp. 1716–1718.

[51] D. M. Mittelman, M. Gupta, R. Neelamani, R. G. Baraniuk, J. V. Rudd and M. Koch, 'Recent advances in terahertz imaging', *Applied Physics* **68**(6) (1999), pp. 1085–1094.

[52] D. L. Woodlard, W. R. Loerop and M. S. Shur, Editors, 'Special issue on Terahertz sensing technology, vol 1: Electronic devices & advanced systems technology', *International Journal of High Speed Electronics and Systems* **13**(2) (2003), pp. 351–699.

[53] M. Usami, T. Iwamoto, R. Fukasawa, M. Tani, M. Watanabe and K Sakai, 'Development of a THz spectroscopic imaging system', *Physics in Medicine and Biology* **47**(21) (2002), pp. 3749–3753.

[54] E. A. Ash and G. Nicholls, 'Super-resolution aperture scanning microscope', *Nature* **237** (1972), pp. 510–512.

[55] O. Mitrofanov, M. Lee, J. W. P. Hsu, L. N. Pfeiffer, K. W. West, J. D. Wynn and J. F. Federici, 'Terahertz pulse propagation through small apertures', *Applied Physics Letters* (2001), **79**(7), pp. 907–909.

[56] Q. Chen, Z. Jiang, G. X. Xu and X.-C. Zhang, 'Near-field terahertz imaging with a dynamic aperture', *Optics Letters* **25**(15) (2000), pp. 1122–1124.

[57] J. Aurèle, L. Adam, Nick C. J. van der Valk and Paul C. M. Planken, 'Measurement and calculation of the near field of a terahertz apertureless scanning optical microscope', *Journal of the Optical Society of America B* **24** (2007), pp. 1080–1090.

[58] M. A. Seo, A. J. L. Adam, J. H. Kang, J. W. Lee, S. C. Jeoung, Q. H. Park, P. C. M. Planken and D. S. Kim, 'Fourier-transform terahertz near-field imaging of one-dimensional slit arrays: mapping of electric-field-, magnetic-field-, and Poynting vectors', *Optics Express* **15**(5) (2007), pp. 11781–11789.

[59] T. Nozokido, R. Iibushi, J. Bae, K. Mizuno and H. Kudo, 'Millimeter-wave scanning near-field anisotropy microscopy', *Review of Scientific Instruments* **76**(3) (2005), pp. 033702-1–033702-6.

[60] A. Dobroiu, M. Yamashita, Y. N. Ohshima, Y. Morita, C. Otani and K. Kawase, 'Terahertz imaging system based on a backward-wave oscillator', *Applied Optics* **43**(30) (2004), pp. 5637–5645.

[61] U. Schade, K. Holldack, P. Kuske, G. Wüstefeld and H.-W. Hübers, 'THz near-field imaging of biological tissues employing synchrotron radiation', *Applied Physics Letters* **84** (2004), 1422–1424.

[62] A. W. M. Lee, B. S. Williams, S. Kumar, Q. Hu and J. L. Reno, 'Real-time imaging using a 4.3-THz quantum cascade laser and a 320 × 240 microbolometer focal-plane array', *IEEE Photonics Technology Letters* **18**(13) (2006), pp. 1415–1417.

[63] A. W. M. Lee, *et al.*, 'Real-time terahertz imaging over a standoff distance (25 meters)', *Applied Physics Letters* **89** (2006), pp. 141125.

[64] K. Hui, J. L. Hesler, D. S. Kurtz, W. L. Bishop and T. W. Crowe, 'A micromachined 585 GHz Schottky mixer', *IEEE Microwave and Guided Wave Letters [see also IEEE Microwave and Wireless Components Letters]* **10**(9) (2000), pp. 374–376.

[65] M. Morschbach, A. Müller, C. Schöllhorn, M. Oehme, T. Buck and E. Kasper, 'Integrated silicon Schottky mixer diodes with cutoff frequencies above 1 THz', *IEEE Transactions on Microwave Theory and Techniques* **53**(6) (2005), pp. 2013–2018.

[66] U. R. Pfeiffer, C. Mishra, R. M. Rassel, S. Pinkett and S. K. Reynolds, 'Schottky barrier diode circuits in silicon for future millimeter-wave and terahertz applications', *IEEE Transactions on Microwave Theory and Techniques* **56**(2) (2008), pp. 364–371.

[67] S.-K. Pan, *et al.*, 'A fixed-tuned SIS mixer with ultra-wide-band IF and quantum-limited sensitivity for ALMA band 3 (84–116 GHz) receivers', in *Proceedings for the 15th International Symposium on Space Terahertz Technology*, 2004.

[68] A. Karpov, D. Miller, F. Rice, J. A. Stern, B. Bumble, H. G. LeDuc and J. Zmuidzinas, 'Low noise 1.2 THz SIS mixer for Herschel radio observatory', in *Proceedings for the 15th International Symposium on Space Terahertz Technology*, 2004.

[69] Y. Uzawa, S. Miki, Z. Wang, A. Kawakami, M. Kroug, P. Yagoubov and E. Kollberg, 'Performance of a quasi-optical NbN hot-electron bolometric mixer at terahertz frequencies', *Superconductor Science and Technology* **15** (2002), pp. 141–145.

[70] F. Rodriguez-Morales, K. S. Yngvesson, E. Gerecht, N. Wadefalk, J. Nicholson, D. Gu, X. Zhao, T. Goyette and J. Waldman, 'A terahertz focal plane array using HEB superconducting mixers and MMIC IF amplifiers', *IEEE Microwave and Wireless Components Letters* **15**(4) (2005), pp. 199–201.

[71] C. K. Walker, *et al.*, 'Integrated heterodyne array receivers for submillimeter astronomy' in *Proceedings of SPIE 4855 of Millimeter and Submillimeter Detectors for Astronomy*, ed. T. G. Phillips and J. Zmuidzinas, pp. 349–354, 2002.

[72] R. Tsai, *et al.*, 'Manufacturable AlSb/InAs HEMT technology for ultra-low power millimeter-wave integrated circuits', *GaAs MANTECH*, vol. 4, pp. 4, 2004.

[73] J. M. Woodall, 'High bandwidth devices: faster materials versus nanoscaled Si and SiGe', *GaAs MANTECH*, vol 1, pp. 4, 2004.

[74] J. B. Hacker, J. Bergman, G. Nagy, G. Sullivan, C. Kadow, H.-K. Lin, A. C. Gossard, M. Rodwell and B. Brar, 'An ultra-low powerAlSb/InAs HEMT W-band low-noise amplifier', *IEEE MTT-S International Microwave Symposium Digest,* vol. 2, pp. 1029–1032, 2005.

[75] P. Fay, J. N. Schulman, S. Thomas III, D. H. Chow, Y. K. Boegeman and K. S. Holabird, 'High-performance antimonide-based heterostructure backward diodes for millimeter-wave detection', *IEEE Electron Device Letters* **23**(10) (2002), pp. 585–587.

[76] H. P. Moyer, R. L. Bowen, J. N. Schulman, D. H. Chow, S. Thomas III, T. Y. Hsu, J. J. Lynch and K. S. Holabird, 'Sb-heterostructure diode detector W-band NEP and NEDT optimization on passive millimeter-wave imaging technology IX', in *Proceedings of the SPIE,* R. Appleby, and D. A. Wikner, eds, vol. 6211, pp. 62110J, 2006.

[77] J. Lynch, H. Moyer, P. L. Schulman, P. Lawyer, R. Bowen, J. Schaffner, D. Choudhury, J. Foschaar and D. Chow, 'Unamplified direct detection sensor for passive millimeter wave imaging', in *Proceedings of the SPIE,* vol. 6211, pp. 621101-1–621101-7, 2006.

[78] H. Kazemi, J. D. Zimmerman, E. R. Brown, A. C. Gossard, G. D.Boreman, J. B. Hacker, B. Lail and C. Middleton, 'First MMW characterization of ErAs/InAlGaAs/InP semimetal semiconductor-Schottky diode (S3) detectors for passive millimeterwave and infrared imaging', in *Proceedings of SPIE, International Society Optical Engineers* **5789** (2005), pp. 80–83.

[79] G. R. Huguenin, 'Millivison millimeter wave Imagers', Final Report to Department of Justice, March 28, 2000.

[80] A. Tessmann, '220 GHz metamorphic HEMTs amplifier MMICs for high resolution imaging applications', *IEEE Journal of Solid-State Circuits* **40**(10) (2005), pp. 2070–2076.

[81] C. Pobanz, *et al.*, 'High performance MMICs with submillimeter wave InP HEMTs', in *Proceedings of 2000 International Conference on Indium Phosphide and Related Materials,* pp. 67–70, 2000.

[82] W. R. Deal, *et al.*, 'A 245-GHz MMIC amplifier with 80-μm output periphery and 12-dBGain', *IEEE/MTT-S International Microwave Symposium, 2007*, pp. 329–332, June 3–8, 2007.

[83] A. Tessmann, A. Leuther, H. Massler, W. Bronner, M. Schlechtweg and G. Weimann, 'Metamorphic H-band low-noise amplifier MMICs', *IEEE/MTT-S International Microwave Symposium, 2007*, pp. 353–356, June 3–8, 2007.

[84] G. W. McIver and S. R. Sashital, 'Millimeter wave imager device using pyroelectric effect', US Patent 5,530,247, June 25, 1996.

[85] H. Eisele and R. Kamoua, 'High-performance oscillators and power combiners with InP Gunn devices at 260–330 GHz', *IEEE Microwave and Wireless Components Letters* **16**(5) (2006), pp. 284–286.

[86] H. Eisele and R. Kamoua 'Submillimeter-wave InP Gunn devices', *IEEE Transactions on Microwave Theory and Techniques* **52**(10) (2004), pp. 2371–2378.

[87] G. I. Haddad and R. Trew, 'Microwave solid-state active devices', *IEEE Transactions on Microwave Theory and Techniques* **50** (2002), pp. 760–779.

[88] G. Chattopadhyay, E. Schlecht, J. Gill, S. Martin, A. Maestrini, D. Pukala, F. Maiwald and I. Mehdi 'A broadband 800 GHz Schottky balanced doubler', *IEEE Microwave and Wireless Components Letters* **12**(4) (2002), pp. 117–118.

[89] Alain Maestrini, J. Ward, J. Gill, H. Javadi, D. Pukala, F. Maiwald and I. Mehdi, 'A planar-diode frequency tripler at 1.9THz', *2003 IEEE MTT-S International Microwave Symposium Digest*, vol. 2, pp. 747–750, 2003.

[90] E. Schlecht, G. Chattopadhyay, A. Maestrini, D. Pukala, J. Gill, S. Martin, F. Maiwald and I. Mehdi, 'A high–power wideband cryogenic 200 GHz Schottky 'substrateless' multiplier: modeling, design, and results', in *Proceedings of European THz*, 2001.

[91] H. Xu, Y. Duan, J. L. Hesler, T. W. Crowe and R. M. Weikle II, 'Subharmonically pumped millimeter-wave upconverters based on heterostructure barrier varactors', *IEEE Transactions on Microwave Theory and Techniques* **54**(10) (2006), pp. 3648–3653.

[92] V. K. Paidi, *et al.*, 'G-band (140–220 GHz) and W-band (75–110 GHz) InP DHBT medium power amplifiers', *IEEE Transactions on Microwave Theory and Techniques* **53**(2) (2005), pp. 598–605.

[93] Y. C. Chen, D. L. Ingram, R. Lai, M. Barsky, R. Grunbacher, T. Block, H. C. Yen and D. C. Streit, 'A 95-GHz InP HEMT MMIC amplifier with 427-mW power output', *IEEE Microwave and Guided Wave Letters* **8**(11) (1998), pp. 399–401.

[94] D. L. Ingram, *et al.*, 'Compact W-band solid-state MMIC high power sources', *2000 IEEE MTT-S International Microwave Symposium Digest,* vol. 2, pp. 955–958.

[95] T. Lehecka, N. C. Luhmann, Jr., W. A. Peebles, J. Goldhar and S. P. Obenschain, 'Gas Lasers', in *The Handbook of Microwave and Optical Components*, ed. K. Chang (John Wiley and Sons, Inc., New York, 1990), vol. 3, Ch. 9, pp. 451–596.

[96] E. R. Mueller, J. Fontanella and R. Henschke, 'Far-infrared laser system for NASA/Goddard space flight center', in *Proceedings of the 11th International Symposium on Space Terahertz Technology*, Ann Arbor, MI, 2000.

[97] R. Köhler, A. Tredicucci, F. Beltram, H. E. Beere, E. H. Linfield, A. G. Davies, D. A. Ritchie, R. C. Iotti and F. Rossi, 'Terahertz semiconductor-heterostructure laser', *Nature* **417** (2002), pp. 156–159.

[98] H. Hübers, S. G. Pavlov, A. D. Semenov, R. Köhler, L. Mahler, A. Tredicucci, H. E. Beere, D. A. Ritchie and E. H. Linfield 'Terahertz quantum cascade laser as local oscillator in a heterodyne receiver', *Optics Express* **13**, (2005), pp. 5890–5896.

[99] J. R. Gao, *et al.*, 'Terahertz heterodyne receiver based on a quantum cascade laser and a superconducting bolometer', *Applied Physics Letters* **86**(24) (2005), pp. 244104.

[100] G. P. Scheitrum, 'Microfabricated MVEDS, in modern microwave and millimeter-wave power electronics', in *Modern Microwave and Millimeter-Wave Power Electronics*, ed. R. J. Barker, J. H. Booske, N. C. Luhmann, Jr and G. S. Nusinovich (Wiley-VCH, April 2005) pp. 343–392.

[101] H-X. L. Liu, L. B. Sjogren, C. W. Domier, N. C. Luhmann, Jr., D. L. Sivco and A.Y. Cho, 'Monolithic quasi-optical frequency tripler array with 5-W output power at 99 GHz', *IEEE Electron. Device Letters* **14**(7) (1993), pp. 329–331.

[102] J. B. Hacker, A. L. Sailer, B. Brar, G. Nagy, R. L. Pierson and J. A. Higgins, 'A high-power W-band quasi-optical frequency tripler', *IEEE MTT-S International Microwave Symposium Digest, 2003*, vol. 3, pp. 1859–1862, 2003.

[103] J. Harvey, E. R. Brown, D. B. Rutledge and R. A. York, 'Spatial power combining for high-power transmitters', *IEEE Microwave Magazine* **1**(4) (2000), pp. 48–59.

[104] M. P. DeLisio and R. A. York, 'Quasi-optical and spatial power combining', *IEEE Transactions on Microwave Theory and Techniques* **50**(3) (2002), pp. 929–936.

[105] J. Bae, T. Unou, T. Fujii and K. Mizuno, 'Spatial power combining of Gunn diodes using an overmoded waveguide resonator at millimeter wavelengths', *IEEE Transactions on Microwave Theory and Techniques* **46**(12) (1998), pp. 2289–2294.

[106] Hollinger, J. E. Kenney and B. E. Troy, Jr., 'A versatile millimeter-wave imaging system', *IEEE Transactions on Microwave Theory and Techniques* **24**(11) (1976), pp. 786–793.

[107] R. Appleby, 'Passive millimetre-wave imaging and how it differs from terahertz imaging', *Philosophical Transactions of the Royal Society of London* A , **362** (2004) pp. 379–394.

[108] R. Appleby, *et al.*, 'Mechanically scanned real time passive millimetre wave imaging at 94 GHz', in *Proceedings of SPIE*, vol. 5077, pp. 1–6, 2003.

[109] A. H. Lettington, D. Dunn, N. E. Alexander, A. Wabby, B. N. Lyons, R. Doyle, J. Walshe, M. Attia and I. Blankson, 'Design and development of a high performance passive millimeter-wave imager for aeronautical applications', in *Proceedings of SPIE*, vol. 5410, pp. 210–218, 2004.

[110] P. M. Blanchard, A. H. Greenaway, A. R. Harvey and K. Webster, 'Coherent optical beam forming with passive millimeter wave arrays', *Journal of Lightwave Technology* **17**(3) (1999), pp. 418–425.

[111] Malibu Research, http://www.maliburesearch.com/technology.htm (last accessed in April 2008).

[112] C. C. Chang, C. Liang and N. C. Luhmann, Jr., 'Phase antenna array based on nonlinear delay line technology', in *Novel Technologies for Microwave and Millimeter Wave Applications*, ed. Jean-Fu Kiang (Kluwer Academic Publishers, 2003).

[113] J. Rizk, G.-L. Tan, J. B. Muldavin and G. M. Rebeiz, 'High-isolation W-band MEMS switches', *IEEE Microwave and Wireless Components Letters* **11**(1) (2001), pp. 10–12.

[114] B. Noren, 'Thin film barium strontium titanate (BST) for a new class of tunable RF components', *Microwave Journal* **47**(5) (2004), pp. 210–220.

[115] H. Kamoda, *et al.*, 'Millimeter-wave beam former using liquid crystal', *34th European Microwave Conference, 2004*, vol. 3, pp. 1141–1144, October 11–15, 2004.

[116] H. Xin, J. B. West, J. C. Mather, J. P. Doane, J. A. Higgins, H. Kazemi and M. J. Rosker, 'A two-dimensional millimeter-wave phase scanned lens utilizing analog electromagnetic crystal (EMXT) waveguide phase shifters', *IEEE Transactions on Antennas and Propagation* **53**(1) (2005), pp. 151–159.

[117] F. Mohamadi, 'Si integration with millimeter wave phased array antenna', *RF Design* February (2004), pp. 40–48.

[118] W. D. Jemison, 'Analysis of the AO-FDPC optical heterodyne technique for microwave time delay and phased array beamsteering applications', *IEEE Transactions on Microwave Theory and Techniques* **50**(7) (2002), pp. 1832–1843.

[119] S. Lucyszyn, 'Review of radio frequency microelectromechanical systems technology', *IEE Proceedings Science, Measurement and Technology,* vol. 151, no. 2, pp. 93–103, March 3, 2004.

[120] Jun Zou; *et al.*, 'Development of a wide tuning range MEMS tunable capacitor for wireless communication systems', *IEDM International Technical Digest on Electron Devices Meeting, 2000*, pp. 403–406, 2000.

[121] H. Kamoda, T. Kuki and T. Nomoto, Conductor loss reduction for liquid crystal millimeter-wave beam former, *IEICE Electronic Express* **2**(18) (2005), pp. 471–476.

[122] B. Floyd, B. Gaucher, S. Reynolds, A. Valdes-Garcia, U. Pfeiffer, D. Liu, J. Grzyb, N. Hoivik and B. Jagannathan, 'SiGe vs. CMOS for 60–100 GHz: technology, circuits, packages, and systems', *Government Microcircuit Applications Conference Digest Papers*, March 2007.

[123] L. M. Franca-Neto, R. J. Eline and B. Bisla, 'Fully integrated CMOS radios from RF to millimeter wave frequencies', *Intel®Technology Journal* **8**(3) (2004), pp. 241–258.

[124] C. Cao and Kenneth K. O, 'Millimeter-wave voltage-controlled oscillators in 0.13 μm CMOS technology', *IEEE Journal of Solid-State Circuits* **41**(6) (2006), pp. 1297–1304.

[125] C. Cao and Kenneth K. O, '192 GHz push-push VCO in 0.13 μm CMOS', *Electronics Letters* **42**(4) (2006), pp. 208–210.

[126] C. Cao and Kenneth K. O, 'A 140-GHz fundamental mode voltage-controlled oscillator in 90-nm CMOS technology', *IEEE Microwave and Wireless Components Letters* **16**(10) (2006), pp. 555–557.

[127] M. Tanomura, Y. Hamada, S. Kisimoto, M. Ito, N. Orihashi, K. Maruhashi and H. Shimawaki, 'TX and RX front-ends for the 60 GHz band in 90 nm standard bulk CMOS', in *IEEE Int. Solid-State Circuits Conf. (ISSCC) Dig. Tech. Papers*, 2008, pp. 556–557.

[128] T. Suzuki, Y. Kawano, M. Sato, T. Hirose and K. Joshin, '60 and 77 GHz Power Amplifiers in Standard 90nm CMOS', in *IEEE Int. Solid-State Circuits Conf. (ISSCC) Dig. Tech. Papers*, 2008, pp. 560–561.

[129] E. Seok, K. O, 'A 410 GHz CMOS push-push oscillator with an on-chip patch antenna', in *IEEE Int. Solid-State Circuits Conf. (ISSCC) Dig. Tech. Papers*, 2008, pp. 472–473.

16

Millimeter-wave System Overview

Scott K. Reynolds, Alberto Valdes-Garcia, Brian A. Floyd, Yasunao Katayama and Arun Natarajan

16.1 Outlook for Low-cost, High-volume mmWave Systems

A flood of recent publications has made it clear that silicon technology is now capable of addressing millimeter-wave (mmWave) applications in the 60–100 GHz region [1–6]. The rising f_T and f_{MAX} which have come with shrinking integrated circuit (IC) line widths have pushed the frequency capabilities of silicon transistors into a region previously occupied only by compound semiconductors such as GaAs. Recently, highly integrated mmWave systems and high performance components have been reported in SiGe [1, 6], SiGe:C [5], silicon-on-insulator (SOI) complementary metal–oxide–semiconductor (CMOS) [3] and bulk CMOS [2, 4, 7]. The superior level of integration allowed by silicon technologies allows the implementation of mmWave wireless links with complex and efficient modulation schemes; the 60 GHz systems so far demonstrated with III/V semiconductors have been limited to amplitude-shift keying (ASK) modulation [8, 9]. In addition, silicon IC technology's economy of scale promises dramatically to reduce the form factor, power, and cost of existing mmWave applications while also opening up new mmWave market opportunities. These include 60 GHz wireless communications, 77 GHz automotive radar, and 94 GHz imagers and radiometers.

The principal advantage of mmWave frequencies such as 60 GHz for wireless communications is the wide bandwidth available. Around 60 GHz, there are international unlicensed bands in the range 57–66 GHz, with 4–6 GHz of contiguous bandwidth available in most countries. This makes possible extremely high data rates, and wireless transmissions of data at rates from 2 to 15 Gb/s have already been reported. These data rate capabilities make

Advanced Millimeter-wave Technologies: Antennas, Packaging and Circuits
Duixian Liu, Ulrich Pfeiffer, Janusz Grzyb and Brian Gaucher © 2009 John Wiley & Sons, Ltd

possible streaming of uncompressed 1080i or 1080p format video data at approximately 2 or 4 Gb/s. However, other characteristics of mmWave frequencies make this application challenging. The line-of-sight (LOS) nature of mmWave signals combined with their rapid attenuation when passing through most materials means that a video stream transmitted from a video-disc player could be blocked by a person or a piece of furniture. A simple system could be placed where it is not easily blocked, but widespread consumer acceptance probably depends on the development of a robust system which cannot be easily interrupted, even when there are several people in a room. To achieve this, the signals can be bounced off walls, or the ceiling, or other hard, smooth surfaces – but since there is loss with each bounce, a single reflection is probably the most that can be tolerated. The success of this application will depend on the development of antenna arrays and transceiver systems capable of achieving high antenna gains and dynamically steering the beam around obstructions.

For short-range file transfer applications at 60 GHz, the system design is much simpler. One could easily imagine a hand-held device with a fixed antenna (such as a camera) that the user could point at another device (such as a laptop computer or kiosk). A large volume of data could be transmitted in burst mode, without the need for beam steering or antenna arrays; but in this application, mmWave solutions would compete with other established technologies operating at lower frequencies, such as ultra-wide band (UWB).

The directionality of mmWave signals can be used to advantage in other applications. Automotive radars at 77 GHz depend on the ability to accurately focus and steer a beam at a range of $\approx$100 m. It then has to be reflected off an object with relatively low loss and detected back at the source, either with the same antenna or an adjacent one. Steerable, high-gain antennas with narrow beam widths are needed, achieved either by a phased array or a switched array of antennas, or by a single high-gain antenna which can be mechanically steered.

Although these mmWave systems share many of the same building blocks (such as low-noise amplifiers, power amplifiers, mixers, digital signal processors, etc.) and could all be implemented in silicon, they are very different system designs, with different antennas, packaging requirements, and cost points. Short-range file transfer applications could have enormous volumes if incorporated into multi-function cell phones or cameras, but the entire system, including packaged IC(s) and antenna, should cost no more than other local-area network (LAN)/personal-area network (PAN) solutions currently found in handheld devices (802.11b transceiver ICs and modules cost less than US $10 from catalog electronic parts vendors). The relatively more relaxed performance requirements could certainly be met by mainstream silicon processes. Video streaming and radar applications have lower volumes (because there are fewer televisions and cars sold than cellphones), but still possibly in the tens of millions with ten percent market penetration. They have more complex and demanding performance requirements, but can support slightly higher prices because the consumer end products are more expensive. The current IEEE standardization efforts for the 60 GHz band include the above-mentioned commercial usage scenarios [10].

It is important to understand that systems will consist of more than the RF 'front-end' ICs, package, and antennas. Systems will also require digital ICs to perform digital baseband and media-access functions. For instance, a digital baseband IC may perform analog-to-digital conversion, digital filtering, digital demodulation, decoding, and error correction. A media-access controller (MAC) controls the interaction between multiple wireless devices within communication range of each other, and it also controls the transfer of data between the host system and wireless-link ICs. The complexity of these ICs may vary widely, depending on the application. Short-range file transfer applications may use very simple modulation schemes such as ASK or FSK (frequency-shift keying). In this case, modulation and demodulation can be handled entirely in the analog domain, and no analog-to-digital converter (ADC) or digital baseband IC may be required. On the other hand, advanced digital modulations such as orthogonal frequency-division multiplexing (OFDM) running at multiple gigabits per second place enormous computational burdens on the digital baseband IC, such that the digital baseband may be more of a design challenge than the RF ICs. The role of the transceiver ICs in a larger digital system is discussed in Section 16.4.

Detailed discussion of system design issues for mmWave applications is beyond the scope of this chapter, but many of the system design issues can be discussed with the aid of an example, as follows.

16.2 Example: 60 GHz SiGe Transceiver

Over the past three years, the authors and many colleagues at IBM have been developing a prototype chipset for high-speed wireless communications at 60 GHz. It is intended for LOS (i.e. non-beam steered) applications such as fixed point-to-point data links, video streaming where the beam is not easily blocked, and file transfer. It has also been a good learning vehicle for development of a beam-steered chipset. The chipset has now been through several iterations; the first-generation SiGe 60 GHz chipset featured separate receiver (Rx) and transmitter (Tx) ICs and achieved a very high level of integration together with excellent RF performance [11]. The two chips were implemented in 0.13 μm SiGe BiCMOS technology with $f_T = 200$ GHz and $f_{MAX} = 250$ GHz. The Rx IC achieved an average noise figure (NF) of 6 dB across the 59–64 GHz band, and the Tx IC had an output power of 10 dBm at the 1 dB compression point. In addition to the silicon chips, first-generation 60 GHz dipole antennas were developed [12] along with low-cost packaging techniques [13]. The result was a fully-packaged 60 GHz transceiver chipset, with antennas, in an area of $13 \times 13\text{mm}^2$.

The second-generation 60 GHz transceiver ICs follow the same general architecture as the first-generation parts and have been developed for enhanced functionality and simplified usability [14]. The discussion in this chapter will focus on the second-generation chipset, a block diagram of which is shown in Figure 16.1. Both Rx and Tx chips operate from a 2.7 V supply, with an additional 4 V supply for the Tx PA, and are implemented in the same SiGe BiCMOS technology as the first generation. A dual-conversion, sliding-intermediate frequency (IF) superheterodyne radio architecture is used, which requires only a single frequency synthesizer. Referring to the receiver, the input 57–64 GHz signal is amplified by an image-rejecting low-noise amplifier (LNA). The signal is then mixed down to approximately a 9 GHz IF, at which point it is filtered and amplified by the IF amplifier. The signal is then split and fed to a pair of double-balanced mixers which convert the IF signal to in-phase and quadrature-phase (I and Q) baseband signals.

The frequency synthesizer is the same for both the Rx and Tx ICs, and it operates in the 16.3–18.3 GHz range, or exactly two-sevenths of the RF input frequency. In the receiver, the local-oscillator (LO) signal for the first down-conversion mixer is derived from a frequency tripler, so that the LO frequency is exactly six-sevenths of the RF input frequency. That places the IF in the 8.1–9.1 GHz IF range, or one-seventh the RF. The quadrature LO signals for

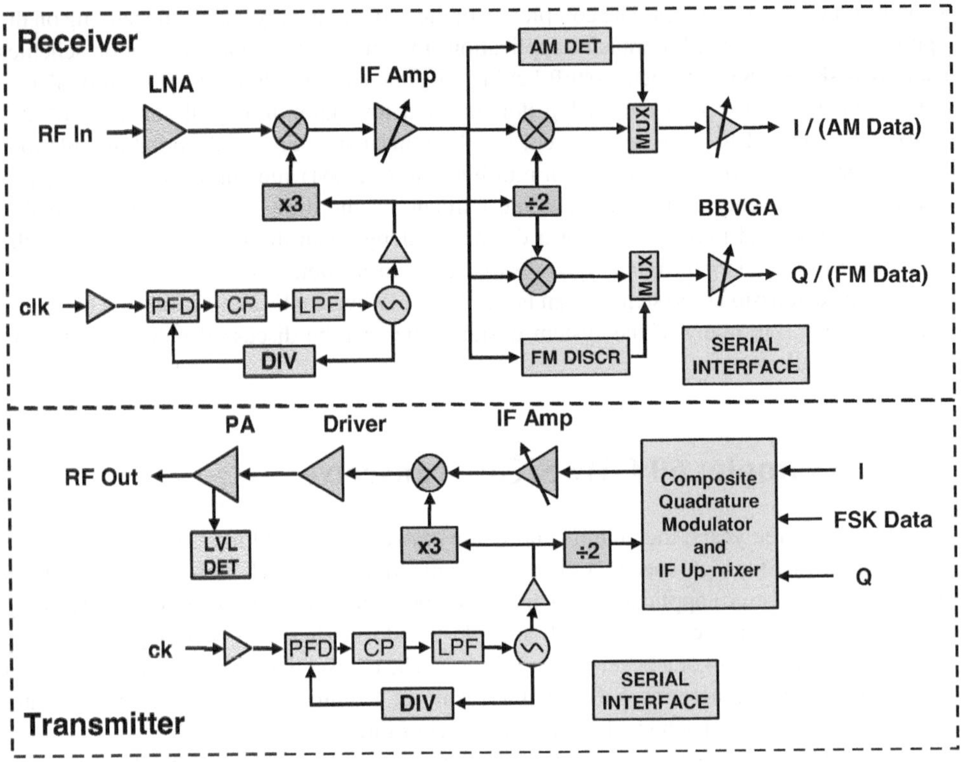

Figure 16.1: Block diagram of the 60 GHz transceiver chipset.

the IF-to-baseband mixers are obtained from a divide-by-two operating from the synthesizer output. This frequency plan results in an image frequency at the RF input of five-sevenths the input frequency, or 40.7–45.7 GHz. Signals at the undesired image frequency are removed by notch filters in the LNA.

The LNA provides about 20 dB of gain, with an additional 9 dB of gain in the first mixer, and adjustable gain of +10 to −10 dB in the IF amplifier. The IF-to-baseband mixers are unity (0 dB) gain, and the baseband variable-gain amplifier (BBVGA) provides approximately −5 to +33 dB of gain, in 1 dB steps. Thus, the receiver gain can be varied in the range 14–72 dB, in 1 dB steps.

The transmitter is essentially a mirror-image of the receiver, taking baseband I and Q signals, up-converting to the ≈9 GHz IF, filtering and amplifying, and up-converting to the 57–64 GHz RF band. An image-rejecting pre-driver (with notch filters such as those in the LNA) provides additional amplification before the final power amplifier (PA) stage, which provides 10 dBm into the antenna. The PA has a differential output and is designed to drive an antenna with 100 Ω differential input impedance. This allows the use of a push–pull output stage in the PA without having to perform on-chip power combining. Total power gain of the transmitter from each of the baseband inputs to the PA differential output is approximately 32 dB.

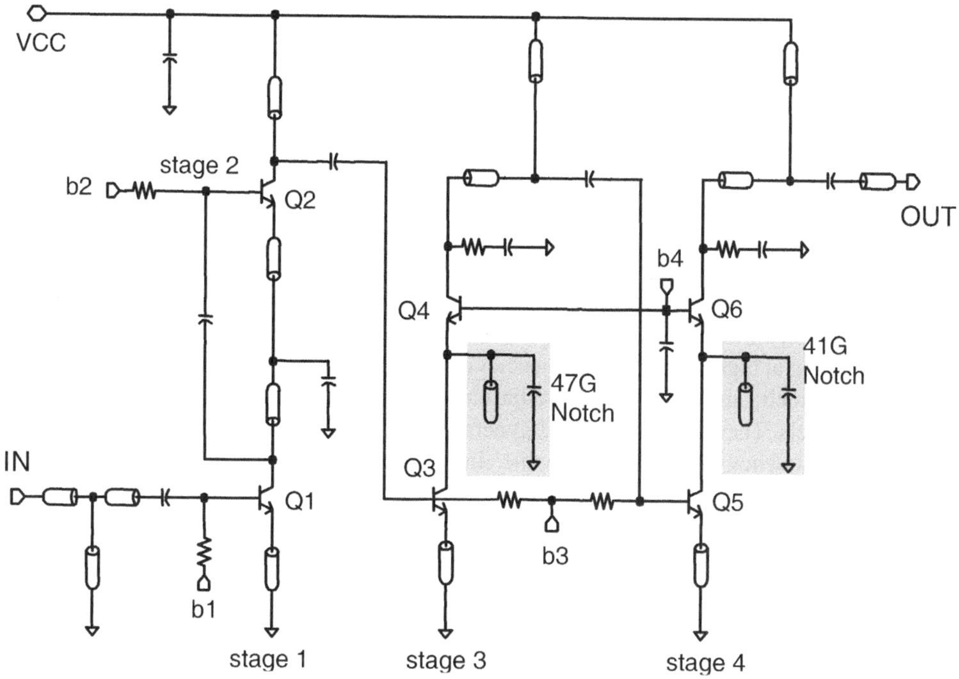

Figure 16.2: Schematic of the 60 GHz image-reject LNA (bias details not shown).

Detailed, circuit-level description of the individual blocks within the Rx and Tx ICs can be found in references [11, 14], and [15]. However, it is interesting to examine how certain system level issues are influenced by the performance of the circuit blocks.

The LNA is a key circuit block in the receiver because it primarily determines the NF. Figure 16.2 shows a schematic of the four-stage LNA. The first two stages (transistors Q1 and Q2) are inductively-degenerated common-emitter amplifiers which are stacked with Q2 above Q1 for current reuse. The inductors are realized by short lengths of microstrip transmission lines formed on the top metal level and running over a ground plane on one of the lower metal levels. Stages 3 and 4 of the LNA are cascode amplifiers, and image rejection is provided in these two stages by notch filters at 47 and 41 GHz. Each notch filter, placed at the junction between the two cascode devices (Q3–Q4 and Q5–Q6), contains a small capacitor in parallel with an open-circuited microstrip stub, $\lambda/4$ in length at the notch frequency. The result is a series resonance to ground at the image and a parallel resonance at RF. The noise figure of each cascode amplifier is approximately 6 dB; hence, stages 1 and 2, each with 5–6 dB gain and approximately 4 dB NF, were added to reduce the cascaded NF and increase gain. Measurements made on a stand-alone LNA show 5–6.2 dB NF, 20 dB gain, >12 dB input/output return losses, and < -40 dB S_{12}. More than 26 dB of image rejection is achieved, and power consumption is 10 mA from 2.7 V.

The first down-conversion mixer also influences the receiver NF, and the mixer combined with the IF amplifier determines the input compression point. The mixer is a single-balanced Gilbert cell with a common-base input, similar to one described in reference [16]; and provides about 9 dB of conversion gain. The cascaded upper single-sideband (SSB) NF of the mixer and subsequent stages is 13 dB (measured) at 25°C. The first stage of IF filtering is provided by passive R-L-C loads at the differential mixer outputs, centered around 8.8 GHz. The inductances at IF frequencies are on the order of 500 pH to 1.2 nH and are formed by spiral inductors on the top metal level.

In the first-generation receiver, the IF amplifier consisted of an inductively-degenerated differential pair with a quad of current-steering devices stacked above the lower pair, steering the output current either to tuned R-L-C loads or to the supply. One of the system considerations in using a variable-IF architecture with fixed-frequency filters is that the IF-filter response can only be perfectly centered at one frequency. Thus, the IF filters are symmetrical only when the Rx (or Tx) is tuned to one RF input frequency. The asymmetry when operating at higher or lower frequencies can become an issue for high data-rate signals, and the asymmetry may also degrade adjacent-channel rejection in a system with closely-spaced channels. Therefore, the second-generation chipset used tunable IF filters, with both the center frequency and filter Q tunable. In the receiver, the R-L-C loads for both the first mixer and the IF amplifier were made tunable by digitally switched MOS-capacitors, and digitally switched resistors were used to de-Q the loads for wider bandwidth. In the transmitter, the R-L-C loads for the first baseband-to-IF mixer and the intermediate-frequency variable-gain amplifier (IFVGA) are similarly tuned.

The compression point for the receiver is set at the input to the IFVGA, which compresses around −1 dBm, so the 29 dB of gain between this point and the Rx input yields an input-referred 1 dB Rx compression point of around −30 dBm. This compression point could be increased by moving the first signal attenuation point from the IFVGA output loads to the mixer output, or by designing the LNA for adjustable gain. Either of these steps would be at the expense of slightly increased system NF.

The baseband VGA maximum bandwidth is around 1.5 GHz, which when cascaded with the bandwidth of the two IF filters, results in an effective signal bandwidth of about 1 GHz. Low-pass filters in the BBVGA allow the signal bandwidth to be reduced to 400, 300, or 250 MHz. The BBVGA filters can be optimized to precisely set the channel bandwidth defined for a particular standard.

Another important element in the system link margin is the transmitter power output. Using a single PA in silicon technology, it is practical to develop power at 60 GHz in the range of 5 to 20 dBm. The saturated power for a CMOS PA is about 9 dBm for a single 65 nm device [17] and up to 11 dB m [18] and 12 dBm [19] when using power-combining techniques. On the other hand, SiGe PAs have reported up to 23 dBm of saturated power [20]. It is important to keep in mind that a stand-alone PA may have a higher output 1 dB compression point than an integrated transmitter incorporating the same PA, since the stages preceding the PA may also begin to compress.

In the present example, the transmitter PA is made of two single-stage cascode amplifiers operating in a push-pull configuration from a 4 V supply, as shown in Figure 16.3. The bases of the upper devices in the cascode (Q3–Q4) are connected to a low-output-impedance on-chip regulator supplying approximately 2 V. Under these conditions, the output voltage at the collectors of Q3–Q4 is not limited by BV_{CEO} (1.7 V), but can swing up to ±2.5 V around the 4 V supply, limited instead by BV_{CBO} (5.5 V). The PA differential output is designed to drive a folded-dipole antenna directly through flip-chip interconnect, so no balun is required. The linear, class-AB amplifier has 10 to 15 dB gain from 59 to 64 GHz. Measurements on

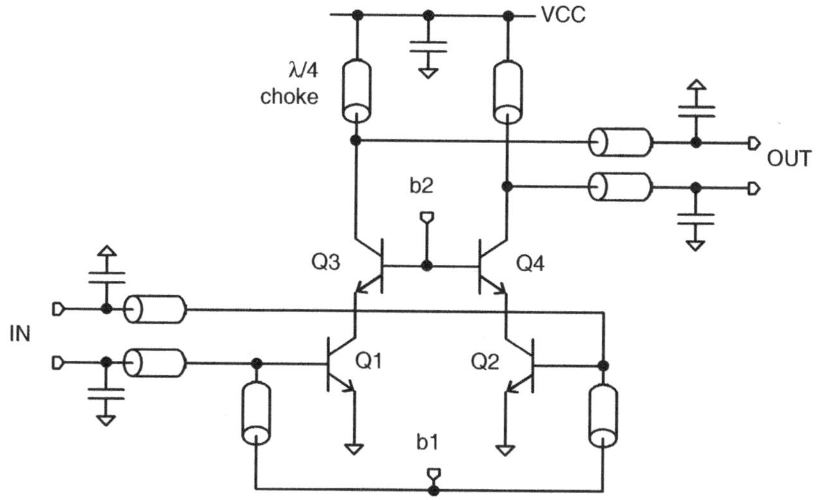

Figure 16.3: Transmitter power amplifier simplified schematic.

the entire Tx show a 10 to 12 dBm output P_{1dB} and 14 to 16 dBm P_{sat} over the band, and the circuit consumes 72 mA from 4 V at P_{1dB}. Across the ISM band, the PA peak power-added efficiency (PAE) is 10% at 59 GHz and drops gradually to 6% at 64 GHz.

The programmable frequency synthesizer is new to both transmitter and receiver for the second-generation chipset [15]. It was designed prior to the existence of any communication standard for dividing the 57–66 GHz band into channels, so it was designed to allow the RFICs to tune the 56.5–64 GHz band in 500 MHz steps, although the 500 MHz steps should not be confused with channels; the channel bandwidth is set by the cascaded baseband bandwidth as discussed above. The synthesizer features a phase-rotating multi-modulus divider capable of sub-integer division, and operates from a 285.7 MHz reference. The output frequency range from the synthesizer is 15.3–18 GHz, which feeds a frequency tripler as shown in Figure 16.1. The measured RMS phase noise of the synthesizer is 0.9° (1 MHz to 1 GHz integration), while phase noise at 100 kHz and 10 MHz offsets are −90 and −124 dBc/Hz, respectively. This noise is suitable for OFDM-16QAM (quadrature-amplitude modulation) modulation. Note that the phase-noise requirement for this radio is more difficult to meet compared to lower frequency OFDM systems owing to the higher carrier frequency. Reference spurs are −69 dBc; sub-integer spurs are −65 dBc.

In addition to the described synthesizer, the Gen-2 receiver included several new features. First, the receiver included an integrated frequency modulation (FM) discriminator for the demodulation of high data-rate signals using frequency and phase shifting modulation techniques such as FSK or minimum-shift keying (MSK) [21]. An amplitude modulation (AM) detector for high data-rate ASK demodulation at IF was also included. These circuits can be used for very high data rate (∼2 Gb/s) directional point-to-point links, without requiring analog-to-digital converters (ADCs) or digital demodulation. The Gen-2 transmitter included a new feature to complement the FM discriminator; a multi-mode modulator was embedded in the IF upconversion mixer. It provides a simple way to generate a radio

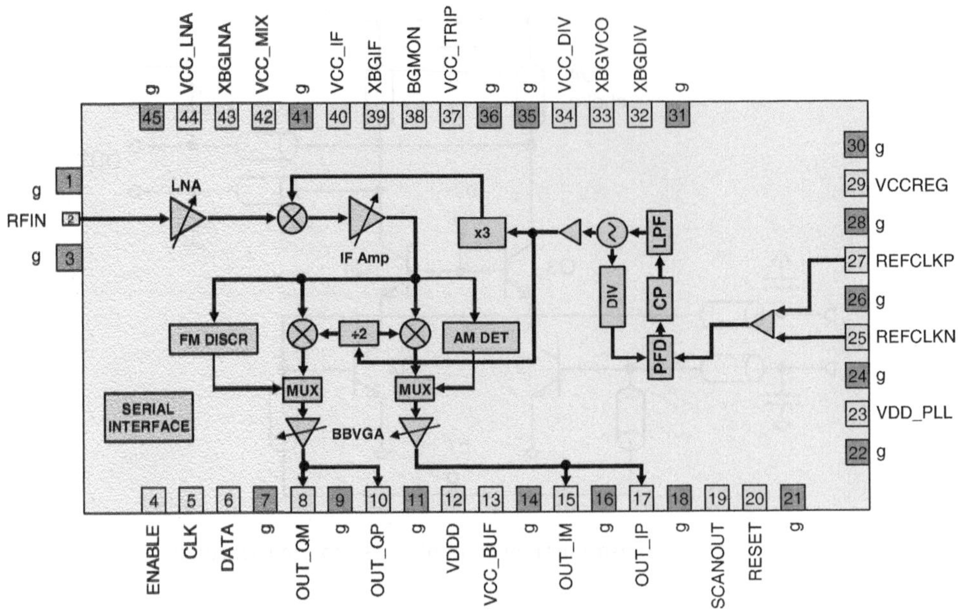

Figure 16.4: More detailed block diagram of the 60 GHz receiver, including IC pinout details.

frequency (RF) modulated waveform directly from a digital bit-stream in FSK, BPSK (binary-phase shift keying) MSK and other similar signaling schemes [21].

16.3 Demonstration Board for 60 GHz SiGe Transceiver

Some of our early 60 GHz wireless link experiments were done with a demonstration printed cicuit board (PCB) which we assembled for the Gen-2 receiver and transmitter ICs. This demo board nicely illustrates the simple, low-cost packaging possible for 60 GHz radios, and the data and video links which were set up provide a nice introduction to some of the system issues.

Figures 16.4 and 16.5 show more detailed block diagrams of the receiver and transmitter ICs, including details of the chip pinouts. All Rx and Tx IC functions and internal bias points are controlled by a 128-bit on-chip register array which is programmed through a three-wire serial digital interface, which allows configuration of the ICs by a system microcontroller. The ENABLE, CLOCK, DATA, SCANOUT, and RESET pins are 1.2 V CMOS logic signals used to write or read the register array.

The evaluation boards can be assembled with a receiver IC, transmitter IC, or both. Photographs of the evaluation board with either a receiver IC or a transmitter IC mounted on it are shown in the left and right portions of Figure 16.6, respectively. The encapsulated ICs with attached 7 dBi gain folded-dipole antennas are located in the center of the boards, with the antennas positioned above a gold-plated ground plane. The Rx and Tx ICs are wirebonded to the FR-4 circuit board. Planar antennas (formally, cavity-backed superstrate

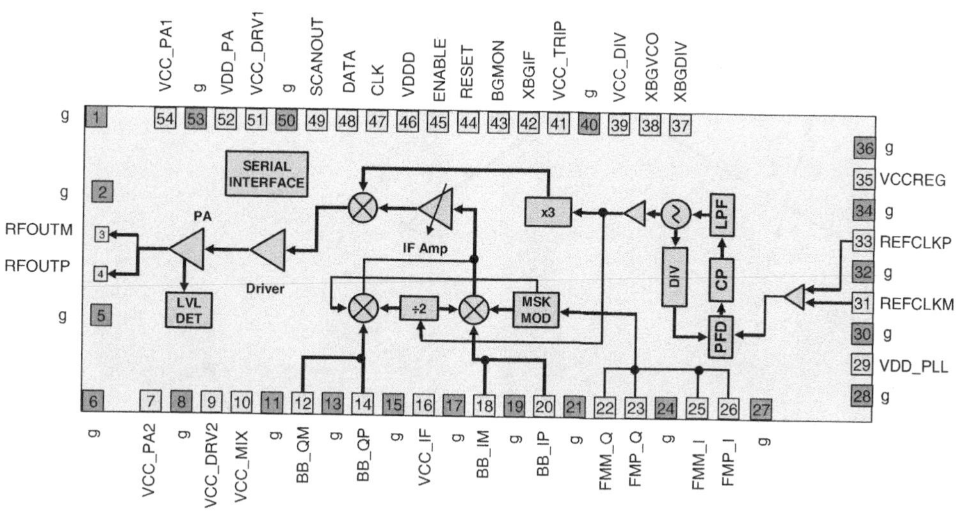

Figure 16.5: Detailed block diagram of the 60 GHz transmitter, including IC pinout details.

antennas) fabricated on fused-silica are flip-chip attached to the ICs using gold studbumps. Frames support the antennas at the same height as the chips and provide mechanical stability and a controlled electromagnetic environment. Note that all of the wire-bond connections are for signals at relatively low frequencies, less than 1 GHz.

The Rx IC has two pairs of baseband differential outputs which are routed to four coaxial connectors along the bottom of the board. Each pair carries either the I or Q portion of the quadrature signal and is designed to be loaded with 100 Ω differential (or 50 Ω each side). The maximum output signal levels are approximately 1 V peak-to-peak differential (ppd) into 100 Ω at the 1 dB compression point. Depending on how the Rx ICs are used, these baseband outputs carry either analog signals to an ADC located in the digital baseband circuitry, or they carry demodulated digital signals to the baseband.

The transmitter has four pairs of differential inputs which are routed to eight coaxial connectors along the top of the board. There are two pairs of differential inputs which carry either the I or Q portion of the baseband signal to a conventional quadrature upmixer. There are also two pairs of differential inputs for the multi-mode modulator which carry digital signals when the FSK/MSK modulator is used. The modulator inputs can be left disconnected if the frequency modulation feature is not used.

The four pairs of differential inputs to the Tx IC are terminated on-chip with 50 Ω to AC ground, and DC blocking capacitors are included on the board. Each pair of quadrature inputs (I-channel and Q-channel) is intended to be driven with a 100 Ω differential source. To generate the differential inputs, either a balun or complementary signals can be used. A 100 Ω differential signal level of $\approx$100 mV ppd will drive the final PA at approximately 1 dB compression, when the Tx IC is in maximum gain mode.

The demo boards have a 285.7 MHz crystal and ECL clock buffers installed as a frequency reference for the synthesizers on the Rx and Tx ICs. With this crystal the ICs tune from 56.5 GHz to 64 GHz in 500 MHz digitally controlled steps. In addition to the on-board

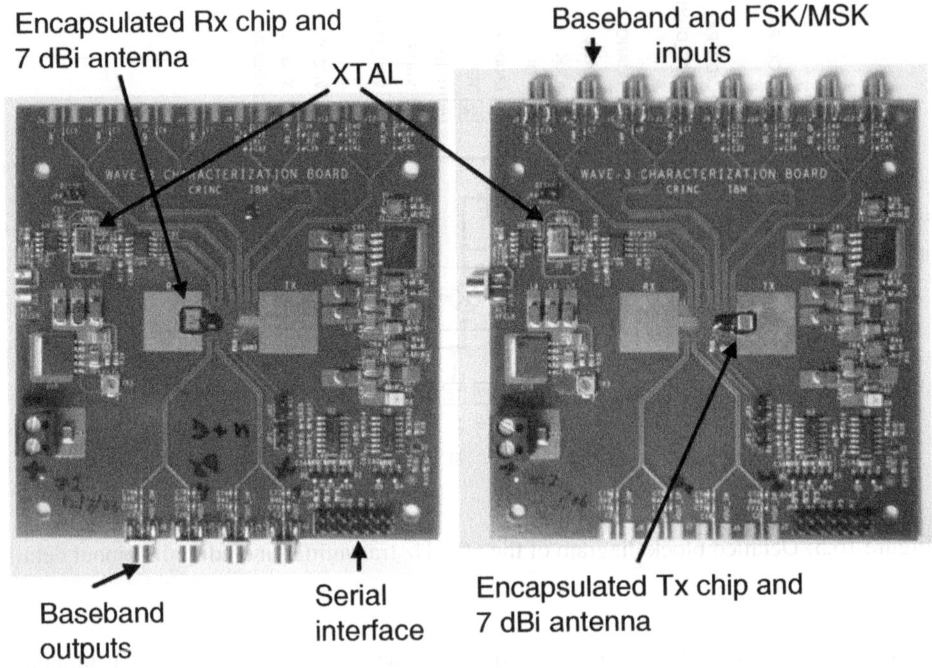

Figure 16.6: Photographs of the demonstration board used for many of our link experiments, populated with an Rx IC (left) and a Tx IC (right).

crystal, there is the option of using an external synthesizer as a frequency reference. This option might be useful in order to phase-lock the synthesizer to an external ADC sample clock, or to tune in steps other than 500 MHz.

16.4 Transceiver ICs as Part of Larger Digital System

The Tx and Rx modules have been used to perform wireless link experiments with a variety of modulation formats, summarized in Table 16.1. The first two rows of Table 16.1 describe systems which involve an external ADC and a digital demodulator. The first, using QPSK (quadrature-phase shift keying)-OFDM modulation, employed an arbitrary waveform generator and software demodulator to demonstrate a system involving an advanced digital modulation, but the high latency of the software prevented the transmission of real-time video signals. The algorithm to demodulate the QPSK-OFDM signal was too complicated to run in real time on the FPGA we used.

The system in the second row used a simpler differential-QPSK modulation which could be handled in real time at both the Tx and Rx ends of the link by an FPGA, but retained an external ADC at the Rx end of the link. The systems in the third and fourth rows made use of the internal FM limiter-discriminator in the Rx IC to eliminate the need for an external ADC, and for these systems modulations were chosen which could be handled by the FM detector. The systems described in the first and fourth rows will be discussed in more detail.

Table 16.1: Table summarizing the link experiments performed with the evaluation boards. The first two rows, in gray, describe experiments done with an external ADC and demodulator. The second two rows used the Rx IC's internal FM limiter-discriminator for demodulation.

Modulation	Rate and distance	Real time?	ADC requirement	Demodulator
QPSK-OFDM	630 Mb/s @ 10 m (ADC limited)	No, data	6-b 700 MS/s	software
DQPSK	2 Gb/s @ 1–6 m	Yes, Video	6–b 2 GS/s	FPGA
QFSK	**2 Gb/s @ 3 m**	**Yes, video**	**None**	**FM limiter-discriminator**
MSK	**2 Gb/s @ 3.5 m**	**Yes, Video**	**None**	**FM limiter-discriminator**

A block diagram of the equipment used to set up the QPSK-OFDM links is shown in Figure 16.7. The modulation was implemented using a 700 Ms/s arbitrary waveform generator for baseband IQ modulation and a 700-Ms/s 8-bit PCI analog-to-digital converter and software demodulator. Bit-error rate (BER) testing of the link has been completed using an IEEE 802.11a-based OFDM-QPSK modulation at 630 Mb/s. The equipment on either end of the link was mounted on carts as shown in Figure 16.8, and the link was tested in different environments and with different antenna separations. The OFDM modulation format was effective at handling multi-path interference due to reflections off the walls, floor, and ceiling. The setup achieved $\leq$10% packet-error rate at a 10 m separation, which defines the sensitivity point by the 802.11a standard convention. These measurements agree well with some simple link budget calculations, which predict a LOS 10 m range with receiver sensitivity of -68 dBm at 630 Mb/s. These same link budget calculations predict an LOS 8 m range at 1 Gb/s. The maximum tested data rate in these experiments was limited by the sampling rate and bandwidth of the ADC, not the Rx and Tx ICs. To extend the coverage range, higher-gain antennas can be used at the transmitter and/or receiver. For example, our link budget calculations predict that increasing the receiver antenna gain by 12 dBi would increase the range by a factor of four, assuming an inverse-square-law (free space) propagation channel with negligible loss from oxygen absorption for $<$100 m ranges.

To transmit digital video signals at a rate of approximately 2 Gb/s, a modulation scheme with lower computational requirements for demodulation was chosen. Although there is no theoretical reason why a QPSK-OFDM system could not operate at the required rate, it requires an field-programmable gate array (FPGA) board with very high computational throughput, and for our initial experiments we opted for a simpler system. Figures 16.9 to 16.12 show a setup that we have used to transmit uncompressed digital video images over a 60 GHz link using MSK. A high-definition video camera produces a 1080i video signal with a data rate of 1.485 Gb/s, which is the input to the Tx FPGA board. The FPGA performs video encoding, grouping the digital data into packets, and then performs error-correction

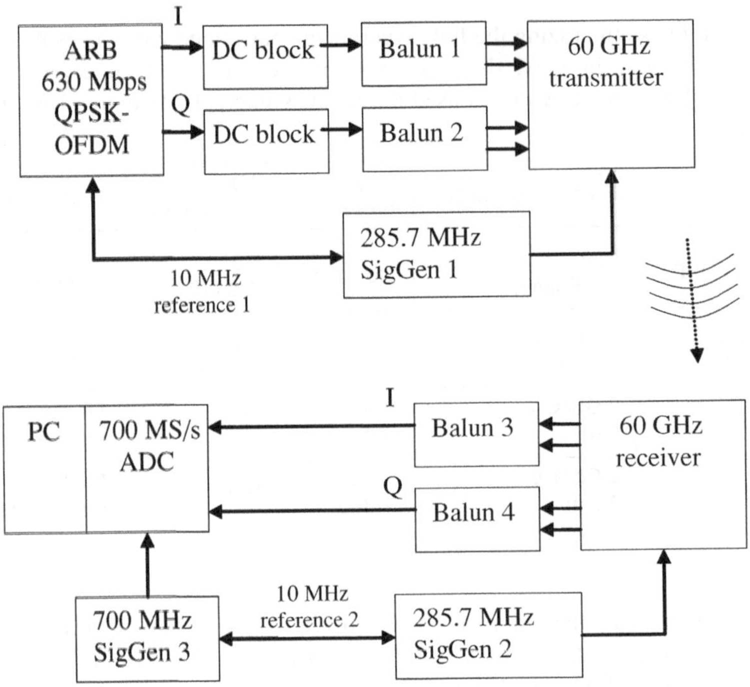

Figure 16.7: Block diagram of measurement setup used to demonstrate QPSK-OFDM 60 GHz link.

coding (ECC) and MSK modulation. With packet overhead and Reed–Solomon coding, the output data rate of the FPGA is exactly 2 Gb/s. After passing through the RF channel, the MSK modulated signal is demodulated in the receiver using the FM limiter-discriminator. The FM limiter-discriminator output is amplified and applied to the digital inputs of the Rx FPGA board, where it undergoes MSK and ECC decoding, is reformatted out of packets, and undergoes video decoding for the high-definition video display.

In this configuration, MSK modulation is performed partly by the Tx FPGA board and partly by a modulator on the Tx IC. Figure 16.10 depicts a conceptual block diagram of the modulator implemented in the Tx IC. Transconductors convert quadrature baseband voltage signals (which in the case of MSK are fixed sinusoids with a 90° phase difference) into current. The digital data comes into the modulator in two branches (I and Q) and changes the polarity of the current. Both current signals are multiplied by an LO signal and then combined. This operation completes the modulation operation and up-converts the resultant signal to a desired IF frequency. Finally, a band-pass filter is incorporated at the output to reduce undesired side-lobe emissions.

To create the needed baseband signals, the Tx FPGA generates two offset digital sequences with the pattern '. . . 0011001100110011 . . .', which are then low-pass filtered to produce the quadrature sinusoids for the modulator's baseband I and Q inputs. The Tx FPGA also produces appropriately coded and synchronized binary polarity control signals on the Data I and Data Q lines. Through the action of the modulator in the Tx IC, the polarity

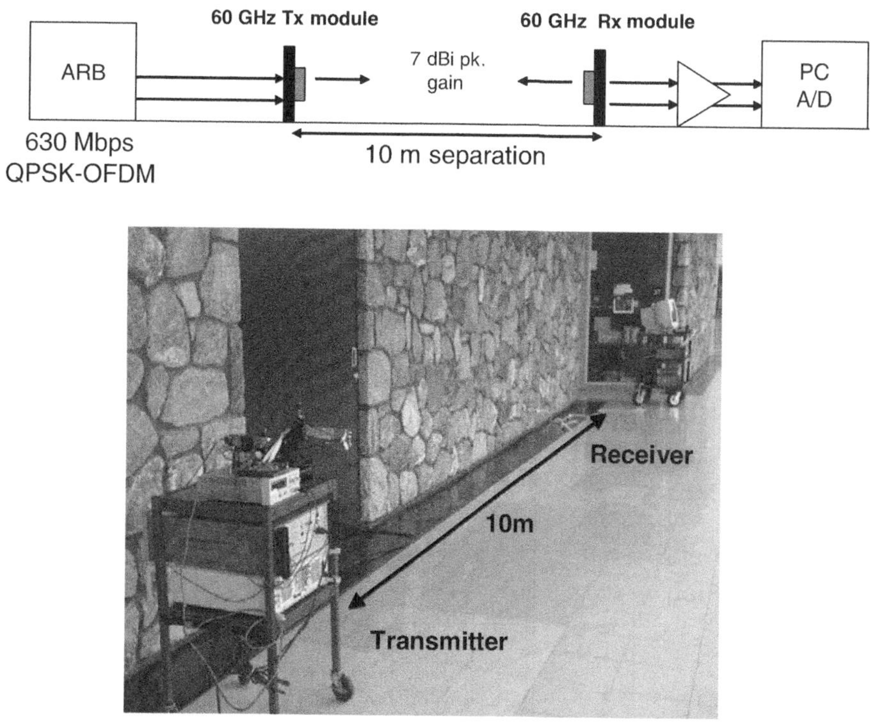

Figure 16.8: QPSK-OFDM link demonstration at 630 Mb/s.

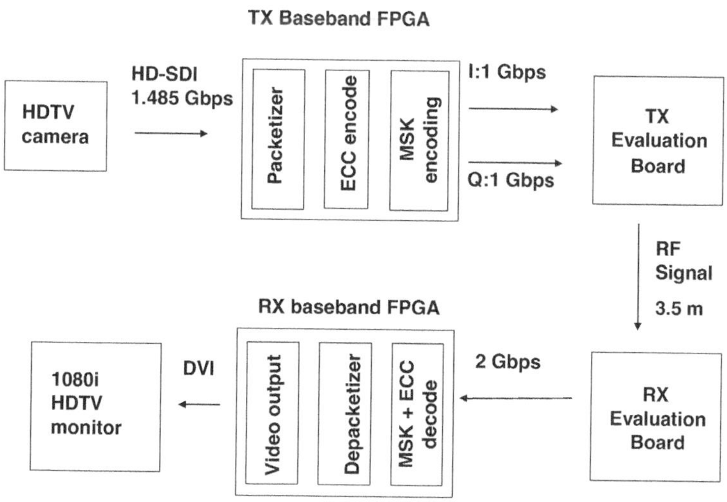

Figurè 16.9: Block diagram of the setup used to demonstrate MSK 60 GHz link.

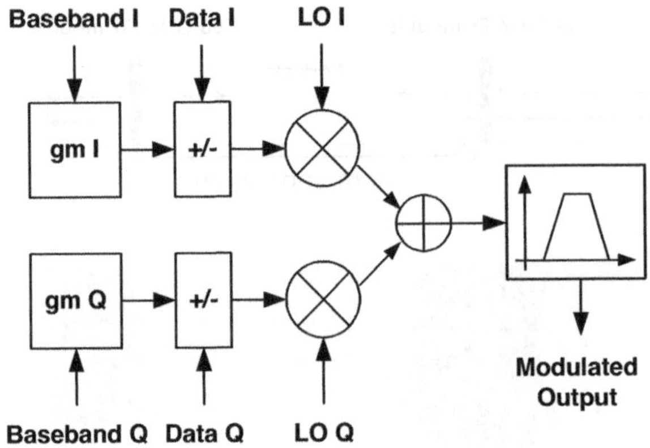

Figure 16.10: Conceptual block diagram of the quadrature modulator in the Tx IC (IBM patent pending).

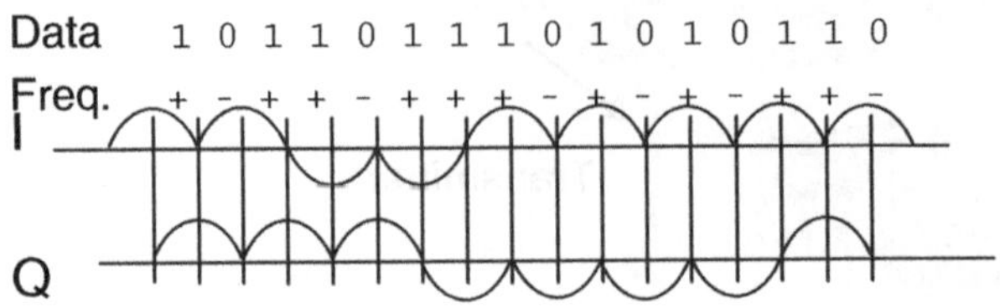

Figure 16.11: MSK modulation mechanism and data encoding.

of the cosine and sine pulses change at their crossover points to encode a bit stream onto the frequency of the output signal, as shown in Figure 16.11. In this example encoding, an information data bit of 1 generates a positive frequency and a data bit of 0 generates a negative frequency. Other frequency modulations, such as FSK, BPSK, QPSK, and offset-QPSK can also be performed using this modulator.

Figure 16.12 is a photograph of the MSK link demonstration in our lab. The HD camera is focused on the front of the Rx board, the resulting image of which is displayed on the HD monitor. Although quite simple and elegant, the video link depicted in Figures 16.9 to 16.12 has two shortcomings: the RF signal cannot pass through an obstruction in the LOS between transmitter and receiver, and the modulation scheme employed cannot compensate for multipath interference. To overcome these problems, it would be advantageous to have higher gain antennas which could be steered automatically in response to a changing environment. Higher gain antennas would improve the link margin while simultaneously reducing multipath reflections, but the narrower beam width makes alignment more difficult and increases the need for beam steering.

A phased-array transceiver system can simultaneously increase antenna gain and facilitate beam steering, as illustrated in Figures 16.13 and 16.14. Figure 16.13 is a conceptual block diagram of a phased-array receiver system and illustrates several principles of signal combining from multiple antennas. The phased-array receiver is capable of receiving signals

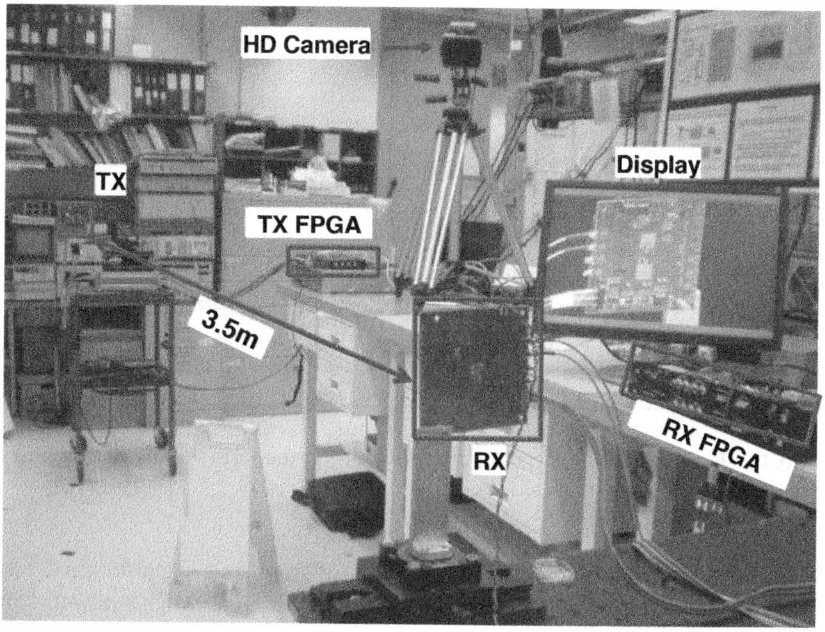

Figure 16.12: MSK Link demonstration photograph.

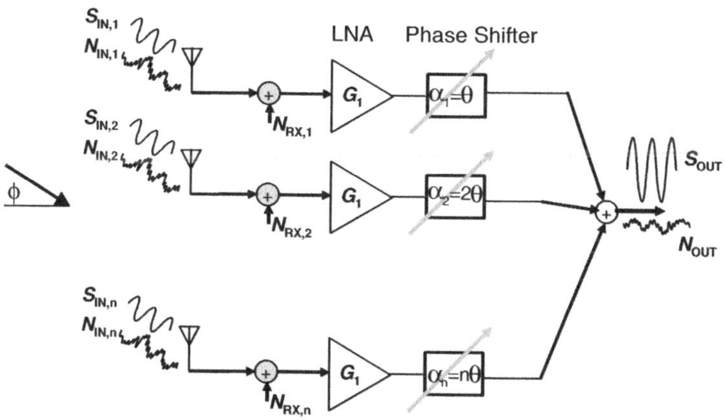

Figure 16.13: Conceptual block diagram of a phased-array receiver system.

from a particular direction and also exhibits improved signal-to-noise ratio. RF signals from n antennas each pass through a phase shifter and are combined in the receiver. By applying a phase shift of $n\theta$ to each of the paths ($-\pi \leq \theta \leq \pi$), the array can be pointed to receive signals from direction ϕ. Theoretically, $\phi = \arcsin(\theta/\pi)$ and can range from $\pi/2$ to $\pi/2$ radians, but non-idealities in the antennas limit the practical range to approximately $\pm 60°$. In Figure 16.13, the RF signals are shown as being combined at RF, but in a practical receiver might be combined after conversion to IF or baseband frequencies.

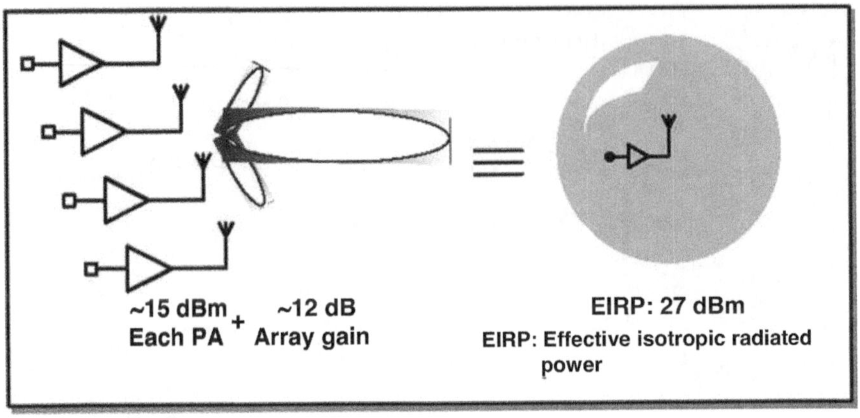

Figure 16.14: Conceptual block diagram illustrating the array gain in a phased-array transmitter.

Also illustrated in Figure 16.13 is the concept of improved signal-to-noise ratio in a phased-array receiver. Since the RF signals are combined in-phase, signal powers from each of the paths are summed and the total power scales as the square of the number of paths, n^2. Noise from each of the paths is not correlated, and total noise power scales linearly with the number of paths, n. Thus, the overall signal-to-noise ratio increases linearly with n.

Figure 16.14 illustrates the concept of array gain in a phased-array transmitter. Each of the transmit paths incorporates a phase shifter, as in the receiver. Once again, by applying a phase shift of $n\theta$ to each of the paths ($-\pi \leq \theta \leq \pi$), the array can be pointed to transmit signals in direction ϕ. If each antenna element radiates power P, the total effective isotropic radiated power (EIRP) is $n^2 P$. For example, if each element in a four-element array radiates 15 dBm, the EIRP is 27 dBm in the direction to which the array is pointed. This is the same power which would be radiated to that point by an element which radiated 27 dBm in an omni-directional fashion. Beamforming in this way allows the signal power to be focused in a particular direction, not only improving the link margin, but also reducing interference for receivers which are not targeted.

In this manner, phased arrays offer the possibility of significantly improved system performance, at the cost of dramatically increased complexity. For an n-element array, a transmitter or receiver with n outputs or inputs and n phase shifters is required, which implies increased die area and power consumption. An n-element phased-array antenna is larger and more complicated than a single-element antenna, and it also drives more complicated packaging for the antenna and mmWave IC. There are also increases in digital system complexity. Figure 16.15 shows a block diagram for a steered-beam, 60 GHz transceiver system in which the high-data-rate channel is unidirectional. Such a system might be used for video streaming. In addition to the high-data-rate channel, there is a need for a low-rate, bidirectional channel to carry control signals between the two sides of the link. This is needed to support the beam-steering algorithm, and also to assist the MAC in controlling the interaction between multiple 60 GHz devices within communication range of each other. The low-rate channel needs to have ample link margin so that it cannot be broken by people or objects which may interrupt the direct LOS. This might be achieved by transmitting at very

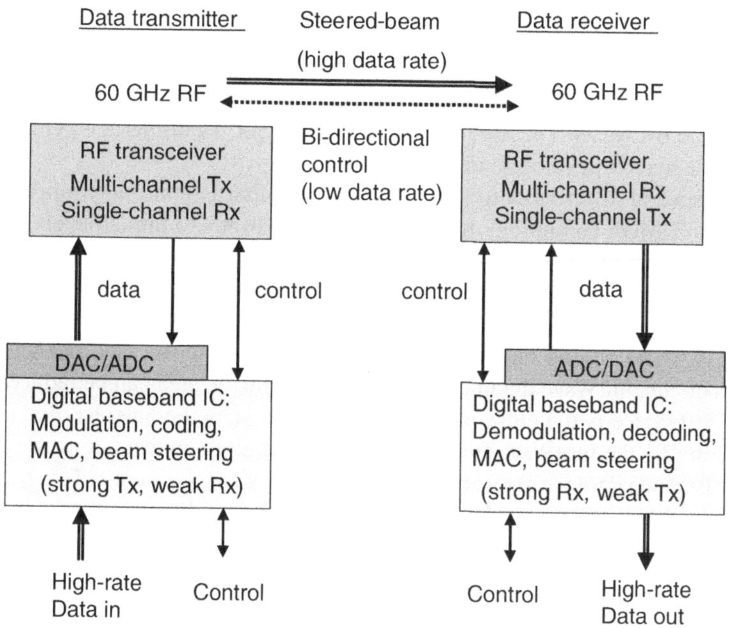

Figure 16.15: Block diagram for a 60 GHz steered-beam transceiver system. Even if the high-data-rate channel is unidirectional, as shown here, a bidirectional, low-rate, non-steered channel is required for control functions.

low data rates (or very high redundancy) with an approximately omni-directional antenna, at a different frequency than the main high-rate channel.

The beam-steering algorithm needs to locate the best direct or reflected path from transmitter to receiver, and it must be capable of switching to the next-best secondary path in near real time, since it will be impractical to cache more than 100 ms or so of data. The digital baseband ICs are thus very busy parts; they must handle data conversion, modulation, demodulation, coding and decoding at 4 Gb/s for 1080p video. They must also handle MAC functions, support the beam-steering algorithm, and serve as a controller for the mmWave ICs. Referring to the block diagram in Figure 16.15, it appears that early systems will have at least four unique ICs to support a beam-steered link, or more if the data converters are not initially integrated on the digital baseband ICs.

16.5 Future Evolution

It is reasonable to assume that 60 GHz wireless systems will evolve in much the same way that the ubiquitous 802.11a/b/g/n wireless local-area network (WLAN) systems have done. Wireless network systems following the 802.11a and 802.11b standards began more than a decade ago as a collection of Si BiCMOS radio frequency integrated circuits (RFICs), discrete RF transistors and passive components, bipolar CMOS (BiCMOS) data converters, and CMOS digital baseband components. Increasing product volumes have driven the market toward ever higher levels of integration, until today when the systems consist of a single, highly-integrated SoC (system-on-chip) in 130-nm CMOS [22]. In a similar way, commercial 60 GHz products are likely to start out as multi-chip systems, with the digital baseband and data converter ICs in 45- or 65-nm CMOS, and the mmWave ICs in either 65- or 90-nm CMOS, or SiGe BiCMOS. The time it takes to develop SoC solutions at 60 GHz will depend on how fast the market develops and how large it is.

There are several differences to keep in mind. The first is that many mmWave ICs will already start in CMOS technology, so the technology barriers to higher levels of integration are lower in that regard; but multi-element, beam-steered systems will probably be the norm from the beginning, so the mmWave ICs will start out very large and with relatively high power dissipation (certainly >1 W). The economics may not be present to develop an even larger IC by including the digital baseband on the same die, and the size of passive components on the mmWave die may limit how much the die area can be reduced.

Another difference is the close inter-relationship between the antenna, die, and package made necessary by the mmWave frequencies. It is difficult to envision how the antenna and die can be easily separated as they are in 2.4 and 5 GHz WLAN systems. This means that the antenna, die, and package assembly will remain a non-commodity item for the foreseeable (2 to 5 year) future, given the difficult technological problems that have to be solved to make it.

Ultimately, however, we should expect mmWave systems to become as inexpensive and ubiquitous as 2.4 and 5 GHz WLAN systems are today. Some of the early companies developing products in the mmWave space will succeed and become profitable, and some will fail; but the end result will be 'millimeter-waves for the masses'.

Acknowledgments

The authors would like to acknowledge Troy Beukema, Chuck Haymes, Ullrich Pfeiffer, and Thomas Zwick for their early contributions to the work described here. The authors would also like to thank Brian Gaucher, Sudhir Gowda, and Mehmet Soyuer for ongoing management support at IBM. Early parts of this work received financial support under DARPA contracts N66001-05-C-8013 and N66001-02-C-8014.

References

[1] B. Floyd, U. Pfeiffer,S. Reynolds, A. Valdes-Garcia, C. Haymes, Y. Katayama, D. Nakano, T. Beukema, B. Gaucher and M. Soyuer, 'Silicon millimeter wave radio circuits at 60–100 GHz', in *Proceedings IEEE Topical Meeting on Silicon Monolithic Integrated Circuits in RF Systems*, pp. 213–218, Long Beach, CA, January 2007.

[2] C.-H. Wang, H.-Y. Chang, P.-S. Wu, K.-Y. Lin, T.-W. Huang, H. Wang and C.-H. Chen, 'A 60 GHz low-power six-port transceiver for Gigabit software defined transceiver applications', in *IEEE Int. Solid State Circuits Conf. Dig. Tech. Papers*, pp. 192–193, San Francisco, CA, February 2007.

[3] D. Lim, J. Kim, J.-O. Plouchart, D. C. Cho and D. S. Boning, 'Performance and yield optimization of mm-Wave PLL front-end in 65 nm SOI CMOS', in *IEEE RFIC Dig. Tech. Papers*, pp. 525–529, Honolulu, HI, June 2007.

[4] T. Mitomo, R. Fujimoto, N. Ono, R. Tachibana, H. Hoshino, Y. Yoshihara, Y. Tsutsumi and I. Seto, 'A 60 GHz CMOS receiver with frequency synthesizer', in *IEEE Symp. on VLSI Circuits Dig. Tech. Papers*, pp. 172–173, Kyoto, Japan, June 2007.

[5] M. Hartmann, C. Wagner, K. Seemann, J. Platz, H. Jager and R. Weigel, 'A low-power low-noise single-chip receiver front-end for automotive radar at 77GHz in silicon-germanium bipolar technology', in *IEEE RFIC Dig. Tech. Papers*, pp. 149–152, Honolulu, HI, June 2007.

[6] E. Laskin, P. Chevalier, A. Chantre, B. Sautreuil and S. P. Voinigescu, '80/160 GHz transceiver and 140 GHz amplifier in SiGe technology', in *IEEE RFIC Dig. Tech. Papers*, pp. 153–156, Honolulu, HI, June 2007.

[7] S. Pinel, S. Sarkar, P. Sen, B. Perumana, D. Yeh, D. Dawn and J. Laskar, 'A 90nm CMOS 60 GHz radio', in *IEEE Int. Solid State Circuits Conf. Dig. Tech. Papers*, pp. 130–131, San Francsisco, CA, February 2008.

[8] D. Y. Jung, W. I. Chang, K. C. Eun and C. S. Park, '60 GHz system-on-package transmitter integrating sub-harmonic frequency amplitude shift-keying modulator', *IEEE Trans. Microwave Theory Tech.* **55**(8) (2007), pp. 1786–1793.

[9] S. E. Gunnarsson, C. Kärnfelt, H. Zirath, R. Kozhuharov, D. Kuylenstierna, C. Fager, M. Ferndahl, B. Hansson, A. Alping and P. Hallbjörner, '60 GHz single-chip front-end MMICs and systems for multi-Gb/s wireless communication', *IEEE J. Solid State Circuits* **42**(5) (2007), pp. 1143–1157.

[10] A. Sadri, '802.15.3c usage model document', *IEEE 802.15 Working Group for WPAN*, no. IEEE15-06-0055-06-003c, January 2006.

[11] S. Reynolds, B. Floyd, U. Pfeiffer, T. Beukema, J. Grzyb, C. Haymes, B. Gaucher and M. Soyuer 'A silicon 60 GHz receiver and transmitter chipset for broadband communications', *IEEE J. Solid State Circuits* **41**(12) (2006), pp. 2820–2831.

[12] J. Grzyb, D. Liu, U. Pfeiffer and B. Gaucher, 'Wideband cavity-backed folded dipole superstrate antenna for 60 GHz applications', in *Proceedings IEEE Antennas and Propagation Soc. Int. Symp.*, pp. 3939–3942, Albuquerque, NM, June 2006.

[13] U. R. Pfeiffer, J. Grzyb, D. Liu, B. Gaucher, T. Beukema, B. A. Floyd and S. K. Reynolds, 'A chip-scale packaging technology for 60 GHz wireless chipsets', *IEEE Trans. Microwave Theory Tech.* **54**(8) (2006), pp. 3387–3397.

[14] S. A. Valdes-Garcia, B. Floyd, B. Gaucher, D. Liu and N. Hoivik, 'Second generation transceiver chipset supporting multiple modulations at Gb/s data rates', invited to *IEEE Bipolar/BiCMOS Circuits and Technol. Meeting*, pp. 192–197, Boston, MA, October 2007.

[15] B. Floyd, 'A 15 to 18 GHz programmable sub-integer frequency synthesizer for a 60 GHz transceiver', in *IEEE RFIC Dig. Tech. Papers*, pp. 529–532, Honolulu, HI, June 2007.

[16] S. Reynolds, 'A 60 GHz superheterodyne downconversion mixer in silicon-germanium bipolar technology', *IEEE J. Solid State Circuits* **39**(11) (2004), pp. 2065–8.

[17] A. Valdes-Garcia, S. Reynolds and J.-O. Plouchart, '60 GHz transmitter circuits in 65 nm CMOS', *IEEE RFIC*, pp. 641–644, Atlanta, GA, June 2008.

[18] M. Tanomura, Y. Hanada, S. Kishmoto, M. Ito, N. Orihashi, K. Maruhashi and H. Shimawaki, 'TX and RX front-ends for 60 GHz band in 90 nm standard bulk CMOS', in *IEEE Int. Solid State Circuits Conf. Dig. Tech. Papers*, pp. 558–559, San Francisco, CA, February 2008.

[19] D. Chowdhury, P. Reynaert and A. M. Niknejad, 'A 60 GHz 1 V + 12.3 dBm transformer-coupled wideband PA in 90nm CMOS', in *IEEE Int. Solid State Circuits Conf. Dig. Tech. Papers*, pp. 560–561, San Francisco, CA, February 2008.

[20] U. Pfeiffer and D. Goren, 'A 23-dBm 60 GHz distributed active transformer in a silicon process technology', *IEEE Trans. Microwave Theory Tech.* **55**(5) (2007), pp. 857–865.

[21] A. Valdes-Garcia, T. Beukema and S. Reynolds, 'Multi-mode modulator and frequency demodulator circuits for Gb/s data rate 60 GHz wireless transceivers', *IEEE Custom Integrated Circuits Conf.*, pp. 639–642, San Jose, CA, September 2007.

[22] L. Nathawad *et al.*, 'A dual-band CMOS MIMO radio SoC for IEEE 802.11n wireless LAN', *ISSCC Dig. Tech. Papers*, pp. 358–359, San Francisco, CA, February 2008.

17

Special Millimeter-wave Measurement Techniques

Thomas Zwick and Ullrich Pfeiffer

17.1 Introduction

At millimeter-wave (mmWave) frequencies measurements are especially challenging. Any cable, connector or probe used to contact the device under test (DUT) has a non-neglectable effect on the measurement result. Therefore the frequent use of special calibration techniques is essential, especially after any change in the setup. In addition, structures on semiconductor die or mmWave substrates are often much smaller than probe pitches or connector dimensions. This requires launching structures and probe pads which have a substantial effect on the measured signals. After a brief overview of modern vector error calibration methods, special mmWave measurement techniques are presented where the effect of the launching structures are de-embedded based on lumped element models of the launching structure. In addition, the determination of transmission line parameters from S-parameter measurements is presented.

In the second part of this chapter, two very special measurement techniques for highly integrated mmWave systems are described. The first special measurement technique allows measurement of the complex impedance and the radiation pattern in an anechoic chamber while contacting the antenna with a coplanar probe. This technique has the advantage of measuring exactly at the reference plane of interest and avoids the additional effort of mounting a connector.

Finally a special non-destructive integrated circuit (IC) package characterization method is described. At mmWave frequencies the IC package has a large influence on signal integrity. Therefore measurement methods for the characterization of interconnects to the semiconductor die in a package are essential. The new mmWave measurement technique

Advanced Millimeter-wave Technologies: Antennas, Packaging and Circuits
Duixian Liu, Ulrich Pfeiffer, Janusz Grzyb and Brian Gaucher © 2009 John Wiley & Sons, Ltd

uses a recursive un-termination method for non-destructive *in situ* S-parameter measurements of multi-port packages. The methodology is based on programmable terminations fully integrated in a standard silicon process technology and a recursive unterminating method implemented in software.

17.2 Overview of Modern Vector Error Calibration Methods

In this section an overview about vector network analyzer (VNA) calibration methods is given. More information about vector network analyzers in general can be found in references [1, 2]. By far the most commonly used two-port method is the short-open-load-thru (SOLT) calibration [3, 4] which is based on a 12-term error model as shown in references [2, 4]. Normally the isolation terms are neglected since they are very small. Additionally, the case of on-wafer testing, will not be the same for the calibration standards and the DUT as they will also change with probe separation. Modern VNAs are able to calibrate the switching terms internally [2], which results in a reduced 8-term error model. This improvement enabled a series of additional calibration methods which have better performance than SOLT in some applications (mainly towards higher frequencies). In reference [5] the formulation is given to convert the error models mathematically. The most important calibration methods used today are listed here:

- SOL (short-open-load): This is the only available one-port calibration method. Three calibration standards (short, open and load), which have to be perfectly known, are connected in order to compute the three error terms.

- SOLT (short-open-load-thru): This calibration is the combination of two one-port short-open-load (SOL) calibrations with additional measurements of a thru connection to obtain the complete set of error terms. The SOLT calibration method is based on a 12-term error model as shown in references [2, 4] and uses different error terms for forward and backward stimulation. It should be noted that the SOLT calibration method provides an over-determined set of equations which can cause problems [2].

- SOLR (short-open-load-reciprocal): The SOLR calibration method is an alternative to the SOLT calibration where the perfect transmission line thru is replaced by a non-ideal thru which just has to be the reciprocal. More details can be found in reference [2].

- TRL (thru-reflect-line): The TRL method was first presented in reference [6] based on an 8-term error model. The characteristic impedance of a transmission line is used as reference impedance instead of a resistor. An extremely accurate version of TRL calibration using multiple line standards can be found in references [7, 8].

- LRM (line-reflect-match): The LRM calibration method is similar to the TRL method [9]. A match standard is used instead of the transmission line as reference impedance, so no change of probe spacing is required during calibration.

- LRM+ (line-reflect-match+): A special LRM calibration technique was presented in reference [10]. The approach relies on a non-perfect thru standard, a reflection standard

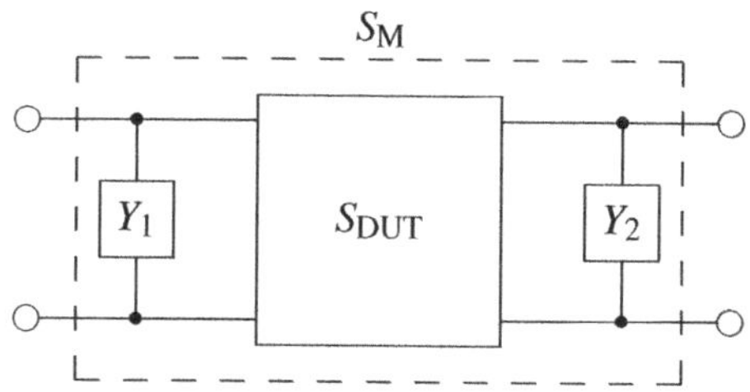

Figure 17.1: Lumped element model using just one parallel admittance.

which is symmetrical for both ports and a complex match (Load) standard which can be different for both ports.

- LRRM (line-reflect-reflect-match): The LRRM calibration is a variation of the LRM calibration which requires two reflect standards (open and load) and a match standard at only one port [11, 12]. The mathematical solution is presented in reference [13]. The LRRM method requires the same standards as SOLT but less knowledge about the standards.

More literature about calibration techniques in general can be found in references [14–17]. In recent years multi-port VNAs have become available and are growing in importance due to the differential mmWave circuits becoming more and more important. Background on multi-port solutions and calibration techniques can be found in references [18–21].

17.3 Lumped Element De-embedding

Lumped element de-embedding techniques are widely used to de-embed microwave fixtures after one-port or two-port VNA measurements. A model based on a few lumped elements (resistors, capacitors, inductors or even transmission lines) which describe the fixture has to be assumed and sufficient calibration structures have to be measured prior to the DUT measurements to be able to determine the values of the lumped elements. Here, the most commonly used lumped element de-embedding techniques are explained.

The most simplest de-embedding technique uses one parallel admittance (mostly a capacitance) to model the probe pad [22–24]. The model is also shown in Figure 17.1. First a separate structure ('open') with just the probe pads is measured as $\mathbf{S_O}$. Then the DUT together with the probe pads is measured as $\mathbf{S_M}$. After converting both S-parameter matrices into Y-matrices (see reference [25]) the pad effects can be de-embedded by

$$\mathbf{Y_{DUT}} = \mathbf{Y_M} - \mathbf{Y_O} \tag{17.1}$$

which yields $\mathbf{S_{DUT}}$ by converting $\mathbf{Y_{DUT}}$ back to S-parameters.

In references [26, 27] a more sophisticated de-embedding technique is suggested which adds a second calibration structure and therefore can also use a more complex Π-model as

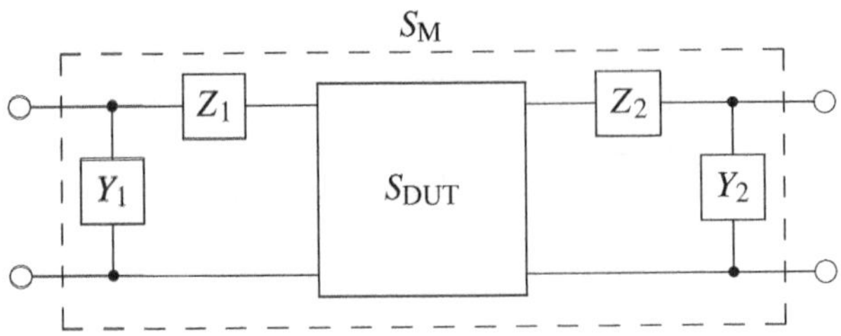

Figure 17.2: Lumped element model using a parallel admittance and a series impedance.

shown in Figure 17.2. As a second calibration structure, the end of the fixture is connected to a short instead of the DUT. After measuring the S-parameters of the two calibration standards, open ($\mathbf{S}_O$) and short ($\mathbf{S}_S$), the DUT ($\mathbf{S}_M$) is measured. These S-matrices are all converted into Y-matrices (see [25]) to $\mathbf{Y}_O$, $\mathbf{Y}_S$ and $\mathbf{Y}_M$. The de-embedded Y-matrix of the DUT yields

$$\mathbf{Y}_{DUT} = [(\mathbf{Y}_M - \mathbf{Y}_O)^{-1} - (\mathbf{Y}_S - \mathbf{Y}_O)^{-1}]^{-1} \tag{17.2}$$

which then can be converted to the S-parameter matrix for the DUT $\mathbf{S}_{DUT}$.

Instead of the above described open-short technique which uses a Π-model (see Figure 17.2) in some cases it makes more sense to flip the order of parallel admittance and series impedance. This T-model has been suggested in reference [27] using the same two calibration structures. Therefore it will be denoted as short-open method in the following. In this case Z-matrices are used instead of the Y-matrices. The de-embedded Z-matrix of the DUT yields

$$\mathbf{Z}_{DUT} = [(\mathbf{Z}_M - \mathbf{Z}_S)^{-1} - (\mathbf{Z}_O - \mathbf{Z}_S)^{-1}]^{-1} \tag{17.3}$$

which then can be converted to the S-parameter matrix for the DUT $\mathbf{S}_{DUT}$.

The same lumped element model as used for the open-short technique (see Figure 17.2) was also used in reference [28] but with a thru calibration standard instead of the two standards open and short. The circuit for the thru connection is given together with the Π-model in Figure 17.3. In this case both fixtures are assumed to be identical. First the measured S-parameter matrix of the thru $\mathbf{S}_T$ is converted to an ABCD-matrix $\mathbf{ABCD}_T$ (see reference [25]) which can be represented by Y and Z:

$$\begin{aligned}\begin{pmatrix} A_T & B_T \\ C_T & D_T \end{pmatrix} &= \begin{pmatrix} 1 & 0 \\ Y & 1 \end{pmatrix} \cdot \begin{pmatrix} 1 & Z \\ 0 & 1 \end{pmatrix} \cdot \begin{pmatrix} 1 & Z \\ 0 & 1 \end{pmatrix} \cdot \begin{pmatrix} 1 & 0 \\ Y & 1 \end{pmatrix} \\ &= \begin{pmatrix} 1 & Z \\ Y & YZ+1 \end{pmatrix} \cdot \begin{pmatrix} YZ+1 & Z \\ Y & 1 \end{pmatrix} \\ &= \begin{pmatrix} 2YZ+1 & 2Z \\ 2Y(1+YZ) & 2YZ+1 \end{pmatrix} \end{aligned} \tag{17.4}$$

Y and Z can now be determined from (17.4) as

$$Z = B_T/2 \tag{17.5}$$

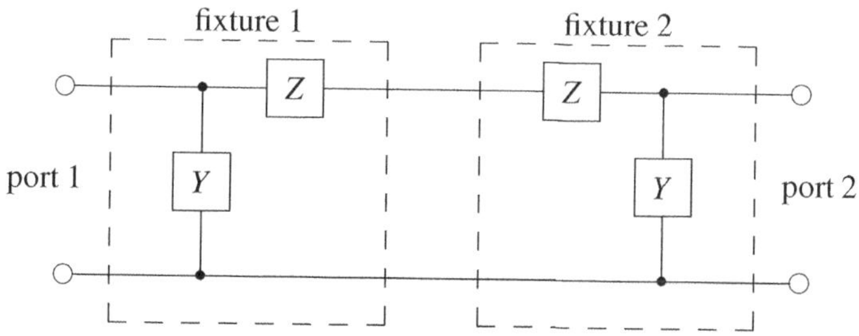

Figure 17.3: Lumped element representation (Π-model) of microwave fixture using thru calibration structure.

and

$$Y = \frac{2C_{\mathrm{T}}}{A_{\mathrm{T}} + D_{\mathrm{T}} + 2} \tag{17.6}$$

The measured ABCD-matrix of an unknown DUT can be expressed by the ABCD-matrix of the DUT together with Y and Z as

$$\begin{pmatrix} A_{\mathrm{M}} & B_{\mathrm{M}} \\ C_{\mathrm{M}} & D_{\mathrm{M}} \end{pmatrix} = \begin{pmatrix} 1 & Z \\ Y & YZ+1 \end{pmatrix} \cdot \begin{pmatrix} A_{\mathrm{DUT}} & B_{\mathrm{DUT}} \\ C_{\mathrm{DUT}} & D_{\mathrm{DUT}} \end{pmatrix} \cdot \begin{pmatrix} YZ+1 & Z \\ Y & 1 \end{pmatrix} \tag{17.7}$$

which can be solved after the ABCD-matrix for the DUT as

$$\begin{pmatrix} A_{\mathrm{DUT}} & B_{\mathrm{DUT}} \\ C_{\mathrm{DUT}} & D_{\mathrm{DUT}} \end{pmatrix} = \begin{pmatrix} 1 & Z \\ Y & YZ+1 \end{pmatrix}^{-1} \cdot \begin{pmatrix} A_{\mathrm{M}} & B_{\mathrm{M}} \\ C_{\mathrm{M}} & D_{\mathrm{M}} \end{pmatrix} \cdot \begin{pmatrix} YZ+1 & Z \\ Y & 1 \end{pmatrix}^{-1} \tag{17.8}$$

It is obvious that the thru standard can also be used for a T-model instead of the Π-model.

In reference [28] a de-embedding technique has been suggested which uses the same lumped element models for the fixture as the techniques described above but derives the equivalent of a thru structure from measuring two lines with different lengths ℓ_1 and ℓ_2, where $\ell_2 = 2\ell_1$. In reference [29] this method has been extended to the more general case $\ell_2 = N\ell_1$ ($N = 2, 3, 4 \ldots$). Here transport matrices are used in the mathematical analysis. The transport matrices of the measured transmission lines with length $\ell_1 = \ell$ and $\ell_2 = N\ell$ are

$$\mathbf{T}_{\ell\mathbf{M}} = \mathbf{T}_{\mathrm{L}}\mathbf{T}_{\ell}\mathbf{T}_{\mathbf{R}} \tag{17.9}$$

and

$$\mathbf{T}_{\mathbf{N}\ell\mathbf{M}} = \mathbf{T}_{\mathrm{L}}\mathbf{T}_{\mathbf{N}\ell}\mathbf{T}_{\mathbf{R}} \tag{17.10}$$

respectively, with $\mathbf{T}_{\mathrm{L}}$ and $\mathbf{T}_{\mathbf{R}}$ describing the two fixtures and $\mathbf{T}_{\ell}$ and $\mathbf{T}_{\mathbf{N}\ell}$ denoting the transmission lines after fixture de-embedding. The thru equivalent transport matrix can be given by

$$\mathbf{T}_{\mathrm{L}}\mathbf{T}_{\mathbf{R}} = (\mathbf{T}_{\mathbf{N}\ell\mathbf{M}} \cdot \mathbf{T}_{\ell\mathbf{M}}{}^{-1})^{1/(1-N)} \cdot \mathbf{T}_{\ell\mathbf{M}} \tag{17.11}$$

From now on the same methodology as for the thru de-embedding technique given above can be used to derive the de-embedded transmission line S-parameters. In case the condition

$\ell_2 = N\ell_1$ is undesirable, three lines of lengths ℓ_1, ℓ_2 and $\Delta\ell = \ell_2 - \ell_1$ can be used [29]. The thru equivalent transport matrix can be derived by

$$\mathbf{T_L T_R} = \mathbf{T_{\Delta\ell M}} \cdot \mathbf{T_{\ell_2 M}}^{-1} \cdot \mathbf{T_{\ell_1 M}} \tag{17.12}$$

More de-embedding techniques using different variants of fixture models can be found in references [23, 30–38]. In reference [39] a very interesting method is presented for the measurement of on-chip transistors. This method uses a kind of de-embedding technique which does not require any model for the fixture.

17.4 Determination of Transmission Line Parameters from S-Parameter Measurements

A very common measurement problem is the determination of the transmission line characteristics from S-parameter measurements. In case the transmission line can directly be contacted by the probes (which already have been calibrated before), the transmission line parameters Z_L and γ can directly be obtained from the measured two-port S-parameters using the formulas given in the following. Towards higher frequencies the accuracy of this method decreases due to probe placement accuracy and repeatability problems as well as the effect of the transmission line stub, which is left outside the probe contact. In most cases though the transmission line cannot be directly probed and a launching structure has to be used. One could principally use the same techniques as in Section 17.3 to first de-embed the launching structures and then extract the transmission line characteristics using the formulas given below, but more accurate methods are described in Section 17.4.1 and 17.4.2.

A lossy transmission line can be described by its impedance Z_L and its propagation constant

$$\gamma = \alpha + j\beta = \alpha + j\frac{2\pi}{\lambda_z} = j2\pi f\sqrt{\mu\varepsilon_{\mathrm{eff}}} \tag{17.13}$$

as well as by a series impedance $Z_s = R + j\omega L$ and a parallel admittance $Y_p = G + j\omega C$ with $\omega = 2\pi f$. With

$$\gamma = \sqrt{Z_s \cdot Y_p} = \sqrt{(R + j\omega L)(G + j\omega C)} \tag{17.14}$$

and

$$Z_L^2 = \frac{R + j\omega L}{G + j\omega C} \tag{17.15}$$

the parameters R, L, C and G for the transmission line (see Figure 17.4) are given by

$$R = \mathrm{Re}\{\gamma Z_L\} \tag{17.16}$$

$$L = \mathrm{Im}\{\gamma Z_L\}/\omega \tag{17.17}$$

$$G = \mathrm{Re}\{\gamma / Z_L\} \tag{17.18}$$

$$C = \mathrm{Im}\{\gamma / Z_L\}/\omega \tag{17.19}$$

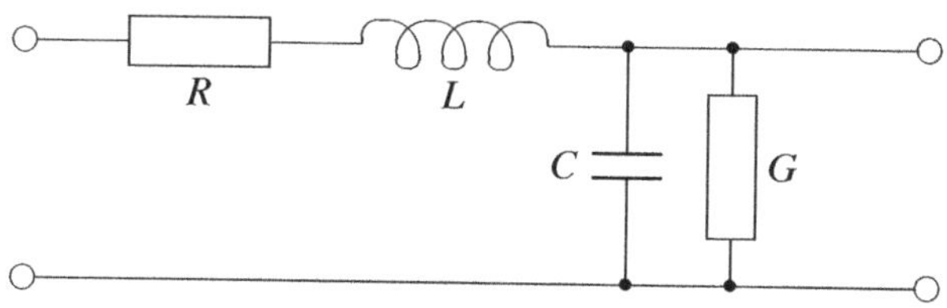

Figure 17.4: Telegrapher's equation model representation for single conductor transmission line.

The lossy transmission line can be explicitly described by its ABCD parameters

$$\mathbf{ABCD_{TL}} = \begin{pmatrix} \cosh \gamma\ell & Z_L \sinh \gamma\ell \\ \dfrac{\sinh \gamma\ell}{Z_L} & \cosh \gamma\ell \end{pmatrix} \tag{17.20}$$

Using the formulations given in reference [25] the matrix $\mathbf{ABCD_{TL}}$ can be converted to S-parameters as

$$\mathbf{S_{TL}} = \frac{1}{D_S} \begin{pmatrix} (Z_L^2 + Z_0^2) \sinh \gamma\ell & 2Z_0 Z_L \\ 2Z_0 Z_L & (Z_L^2 + Z_0^2) \sinh \gamma\ell \end{pmatrix} \tag{17.21}$$

with

$$D_S = 2Z_L Z_0 \cosh \gamma\ell + (Z_L^2 + Z_0^2) \sinh \gamma\ell \tag{17.22}$$

where Z_0 is the reference impedance of the S-parameters.

After measuring the complex two-port S-parameters of a transmission line of length ℓ its parameters can be obtained as follows [23]

$$e^{-\gamma\ell} = \left[\frac{1 - S_{11}^2 + S_{21}^2}{2S_{21}} \pm K \right]^{-1} \tag{17.23}$$

where

$$K = \sqrt{\frac{(S_{11}^2 - S_{21}^2 + 1)^2 - 4S_{11}^2}{4S_{21}^2}} \tag{17.24}$$

$$Z_L^2 = Z_0^2 \frac{(1 + S_{11})^2 - S_{21}^2}{(1 - S_{11})^2 - S_{21}^2} \tag{17.25}$$

The correct sign in (17.23) can be detected by enforcing $\alpha \geq 0$.

17.4.1 Propagation Constant Determination from Measurement of Two Transmission Lines of Different Length

The propagation constant γ of a transmission line can be accurately determined from measurements of two transmission lines with identical geometry but different lengths without de-embedding any launching structures (pads, probes etc.) [38, 40]. First the measured S-parameters of both lines $\mathbf{S}_1$ and $\mathbf{S}_2$ have to be transformed to transport matrices $\mathbf{M}_1$ and $\mathbf{M}_2$ (see reference [25]). Assuming the two error matrices for the unknown launching structures are **A** and **B** the two measurements yield

$$\mathbf{M}_1 = \mathbf{A} \cdot \mathbf{T}_1 \cdot \mathbf{B} \tag{17.26}$$

and

$$\mathbf{M}_2 = \mathbf{A} \cdot \mathbf{T}_2 \cdot \mathbf{B} \tag{17.27}$$

where $\mathbf{T}_1$ and $\mathbf{T}_2$ are the transport matrices of the lines itself

$$\mathbf{T}_1 = \begin{pmatrix} e^{-\gamma \ell_1} & 0 \\ 0 & e^{\gamma \ell_1} \end{pmatrix} \tag{17.28}$$

$$\mathbf{T}_2 = \begin{pmatrix} e^{-\gamma \ell_2} & 0 \\ 0 & e^{\gamma \ell_2} \end{pmatrix} \tag{17.29}$$

The error network **B** can be eliminated by the multiplication of the matrix $\mathbf{M}_1$ with the inverse matrix $\mathbf{M}_2$

$$\mathbf{M}_1 \cdot \mathbf{M}_2^{-1} = \mathbf{A} \cdot \mathbf{T}_{12} \cdot \mathbf{A}^{-1} \tag{17.30}$$

with

$$\mathbf{T}_{12} = \begin{pmatrix} e^{-\gamma(\ell_1 - \ell_2)} & 0 \\ 0 & e^{\gamma(\ell_1 - \ell_2)} \end{pmatrix} \tag{17.31}$$

The propagation constant γ can now be derived from an eigenvalue calculation, since it can be shown that the eigenvalues of two matrices **T** and $\mathbf{T}'$ with

$$\mathbf{T}' = \mathbf{C} \cdot \mathbf{T} \cdot \mathbf{C}^{-1} \tag{17.32}$$

are identical. The eigenvalues λ_1 and λ_2 of

$$\mathbf{Q} = \mathbf{M}_1 \cdot \mathbf{M}_2^{-1} \tag{17.33}$$

are calculated by solving

$$(Q_{11} - \lambda)(Q_{22} - \lambda) = Q_{12} \cdot Q_{21} \tag{17.34}$$

The eigenvalues λ_1 and λ_2 can now be used to derive γ by

$$\gamma = \frac{1}{(\ell_1 - \ell_2)} \operatorname{acosh} \frac{\lambda^2 + 1}{2\lambda} \tag{17.35}$$

since they are also the eigenvalues of $\mathbf{T}_{12}$ and therefore have to satisfy

$$(e^{-\gamma(\ell_1 - \ell_2)} - \lambda)(e^{\gamma(\ell_1 - \ell_2)} - \lambda) = 0 \tag{17.36}$$

The sign ambiguity of the acosh function can be solved since the real part of γ has to be negative. Also the imaginary part of γ should be decreasing over frequency.

Important: The length difference $\Delta\ell = |\ell_1 - \ell_2|$ between the two lines should be as large as possible to minimize the effect of probe placement differences in the two measurements.

17.4.2 Accurate Impedance Determination of Transmission Lines

As shown in references [32–34, 38, 41] the methods from Section 17.3 can be used together with the formulation in Section 17.4 to measure and extract transmission line characteristic, but the accuracy of these methods for transmission line parameter extraction is not sufficient at very high frequencies. In Section 17.4.1 a method for the accurate determination of γ from measurements of two transmission lines with different lengths is shown. The accurate impedance determination of transmission lines is a much more difficult task than finding the gamma. In reference [42] it has been shown that for transmission lines on low loss substrates G is negligible ($G \ll \omega C$). In addition C has been found independent of frequency and conductivity in the case of very low substrate losses [43, 44], so C may be well approximated using C_0, the dc-capacitance assuming perfect conductors. These assumptions allow one to derive the characteristic impedance of the transmission line Z_{L} directly from the propagation constant γ by

$$Z_{\mathrm{L}} \approx \frac{\gamma}{j\omega C_0} \tag{17.37}$$

The dc-capacitance C_0 can be obtained from measurements using one of the de-embedding techniques from Section 17.3 at very low frequencies, where these techniques are quite accurate. In reference [45] results of on-chip transmission line measurements using this technique can be found.

In references [44, 46, 47] more sophisticated methods are given which try to solve the problem of characteristic impedance measurement on lossy transmission lines. Another effect which might complicate the understanding of transmission line measurement results is the conductor surface roughness which has been investigated in reference [48].

17.5 Probe-based Antenna Measurement

The tremendous improvements in high-speed semiconductor technologies (e.g. SiGe) have made highly integrated mmWave systems possible. Full radios can be implemented on a single semiconductor die and then integrated in a standard IC package together with the antennas. Thereby no standard connectors (e.g. coax or waveguide) are used to connect the antenna to the transceiver. Instead the antenna is directly connected to the semiconductor (e.g. via solder balls) or integrated on the semiconductor die itself. Therefore a probe based antenna measurement has the advantages of first, measuring exactly at the reference plane of interest and second, avoiding the effort of mounting a connector. This new antenna measurement setup has the following three major challenges:

- a special calibration method is required since the system has two different ports (probe on DUT side and coax or waveguide on second port);

- rotating the DUT (as is usually done in antenna chambers) is not possible (without very high effort) since motor vibration could disrupt the connection between probe and DUT;

- the shielding of probe connections is worse than the shielding of connectors limiting the dynamic range.

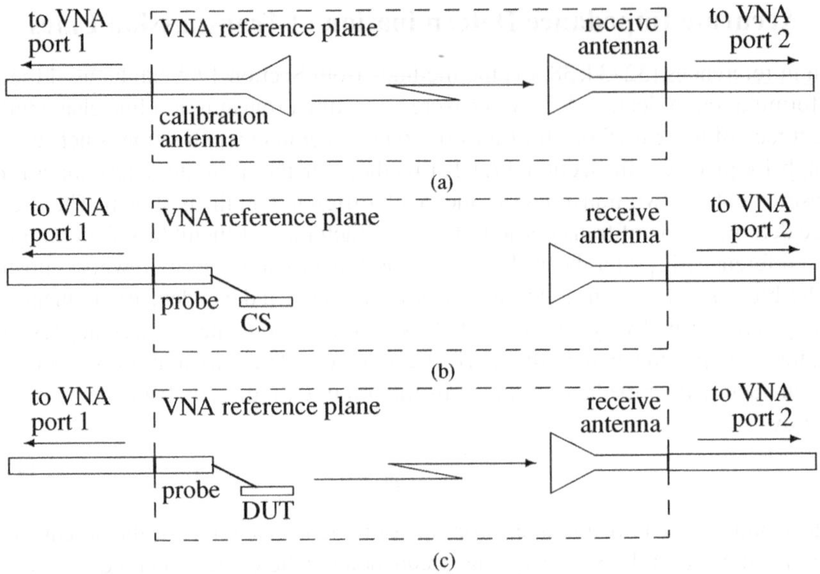

Figure 17.5: Calibration concept for probe based mmWave antenna measurement setup: (a) gain calibration; (b) SOL calibration with calibration substrate (CS); (c) DUT/antenna measurement.

In the following a calibration method for the probe-based antenna measurement setup and a realization example are presented.

17.5.1 Calibration Method

The probe-based mmWave antenna measurement setup can be calibrated in three steps as follows. First, the network analyzer is calibrated at the reference planes given in Figure 17.5. Now only the imperfections of the components in the dashed box are left to be calibrated. In the second step a scalar gain calibration is required. Therefore the probe is replaced by a standard gain horn antenna with well known gain G_{cal} as shown in Figure 17.5(a). This results in

$$|S_{21\mathrm{c}}| = k \cdot G_{\mathrm{cal}} \tag{17.38}$$

with k representing the free space loss and the gain of the receive antenna. After measuring antenna samples contacted by the probe, the gain of those DUTs G_{DUT} can be obtained by

$$G_{\mathrm{DUT}} = \frac{|S_{21\mathrm{m}}|G_{\mathrm{cal}}}{|S_{21\mathrm{c}}|G_{\mathrm{probe}}} \tag{17.39}$$

In (17.39) $S_{21\mathrm{m}}$ is the backward transmission measured with the DUT. G_{probe} is the gain of the probe which is obtained from the third calibration step as shown in the following.

In the third calibration step, the reference plane of port 1 must be moved to the probe tip to enable the determination of the complex input impedance. Since only one probe at port 1 is used the SOL calibration is the only viable method. Instead of the DUT, a calibration

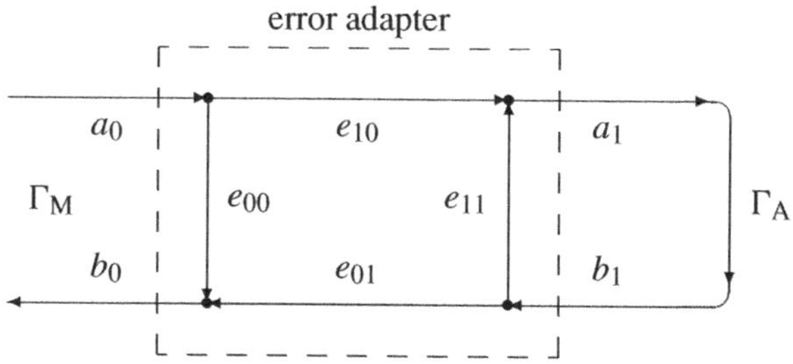

Figure 17.6: One-port SOL error flow chart.

substrate is connected with the probe as shown in Figure 17.5(b). In Figure 17.6 the error adapter definition for the probe is given. Based on the error model in Figure 17.6 the following relations result between the measured reflection coefficient Γ_M and the actual 'correct' reflection coefficient Γ_A at the device itself

$$\Gamma_M = \frac{b_0}{a_0} = e_{00} + \frac{e_{10}e_{01}\Gamma_A}{1 - e_{11}\Gamma_A} \tag{17.40}$$

$$\Gamma_A = \frac{\Gamma_M - e_{00}}{e_{11}(\Gamma_M - e_{00}) + e_{10}e_{01}} \tag{17.41}$$

To enable the calculation of Γ_A from Γ_M three calibration measurements with known on-wafer reflection standards are required. The calibration standards for on-wafer calibrations usually are manufactured on very well controlled alumina substrates. The models used to describe their behavior are more dependent on the probe than on the substrate. In the following, the three calibration standards are given with the usual functions to accurately describe their reflection behavior:

- SHORT: The reflection factor of the on-wafer short is fully described by a known inductance L_S as

$$\Gamma_S = \frac{j\omega L_S - Z_0}{j\omega L_S + Z_0} \tag{17.42}$$

with $\omega = 2\pi f$ and Z_0 being the reference impedance (usually 50 Ω).

- OPEN: The reflection factor of the open includes its capacitance C_O

$$\Gamma_O = \frac{1 - j\omega C_O Z_0}{1 + j\omega C_O Z_0} \tag{17.43}$$

- LOAD: The on-wafer load is usually laser trimmed to reach a very high accuracy. Its susceptance is given by

$$Y_L = \frac{1}{R_L} + j\omega C_L \tag{17.44}$$

with R_L being the real impedance of the load (usually 50 Ω) and C_L being the additional, parallel load capacitance. The reflection factor of the load can then be given by

$$\Gamma_L = \frac{1 - Y_L Z_0}{1 + Y_L Z_0} \tag{17.45}$$

Based on the three calibration measurements, Γ_A and Γ_M in (17.40) and (17.41) can be replaced by the reflection coefficients of the three different but known calibration standards of the short Γ_{SA}, the open Γ_{OA} and the load Γ_{LA} as well as their measured counterparts of the short Γ_{SM}, the open Γ_{OM} and the load Γ_{LM} which yields three equations. Solving these three equations (a detailed derivation is given in Section 17.5.2) results in

$$e_{11} = -\frac{a \cdot (\Gamma_{LA} - \Gamma_{SA}) - (\Gamma_{LM} - \Gamma_{SM})}{(\Gamma_{LA}\Gamma_{LM} - \Gamma_{SA}\Gamma_{SM})} \tag{17.46}$$

$$e_{00} = \Gamma_{SM} - a \cdot \Gamma_{SA} - e_{11} \cdot \Gamma_{SA}\Gamma_{SM} \tag{17.47}$$

and

$$e_{10}e_{01} = a + e_{11}e_{00} \tag{17.48}$$

with

$$a = \frac{(\Gamma_{LM} - \Gamma_{SM})(\Gamma_{LA}\Gamma_{LM} - \Gamma_{OA}\Gamma_{OM}) - (\Gamma_{LM} + \Gamma_{OM})(\Gamma_{LA}\Gamma_{LM} - \Gamma_{SA}\Gamma_{SM})}{(\Gamma_{LA} - \Gamma_{SA})(\Gamma_{LA}\Gamma_{LM} - \Gamma_{OA}\Gamma_{OM}) - (\Gamma_{LA} + \Gamma_{OA})(\Gamma_{LA}\Gamma_{LM} - \Gamma_{SA}\Gamma_{SM})} \tag{17.49}$$

Using the results of (17.46), (17.47) and (17.48) in (17.41) yields corrected S_{11} reflection coefficients for a reference impedance of Z_0.

Unfortunately only the product of e_{10} and e_{01} can be determined, which is sufficient for normal SOL calibration procedures. Here, however, the gain of the probe is required for a proper gain calibration (17.39). The only solution, therefore, is to assume that the error adapter is perfectly reciprocal ($e_{10} = e_{01}$), which should be true for passive structures. The gain of the probe can be expressed as

$$G_{\mathrm{probe}} = \sqrt{|e_{10}e_{01}|} \tag{17.50}$$

In references [49, 50] measurement results of folded dipole antennas performed in the probe-based mmWave antenna measurement setup are presented. Since the folded dipole antennas have been designed for a characteristic impedance of $Z_{\mathrm{ref}} = 100\ \Omega$ the measured results must be re-normalized to Z_{ref}. The input impedance Z_{in1} of port 1 can be obtained from S_{11,Z_0} by

$$Z_{\mathrm{in1}} = Z_0 \frac{1 + S_{11,Z_0}}{1 - S_{11,Z_0}} \tag{17.51}$$

Now $S_{11,Z_{\mathrm{ref}}}$ normalized to Z_{ref} can be obtained as

$$S_{11,Z_{\mathrm{ref}}} = \frac{Z_{\mathrm{in1}} - Z_{\mathrm{ref}}}{Z_{\mathrm{in1}} + Z_{\mathrm{ref}}} \tag{17.52}$$

The gain error caused by the mismatch can be corrected by

$$G_{\mathrm{DUT},Z_0,Z_{\mathrm{ref}}} = G_{\mathrm{DUT},Z_0,Z_0} \left| \frac{Z_{\mathrm{in1}} + Z_0}{Z_{\mathrm{in1}} + Z_{\mathrm{ref}}} \right| \sqrt{\frac{Z_{\mathrm{ref}}}{Z_0}} \tag{17.53}$$

17.5.2 Derivation of Error Terms for SOL Calibration

Referring to Figure 17.6, using

$$c = -e_{11} \tag{17.54}$$

$$b = e_{00} \tag{17.55}$$

and

$$a = b \cdot c + e_{10}e_{01} \tag{17.56}$$

(17.40) and (17.41) can be rewritten to

$$\Gamma_{\mathrm{M}} = \frac{a \cdot \Gamma_{\mathrm{A}} + b}{c \cdot \Gamma_{\mathrm{A}} + 1} \tag{17.57}$$

and

$$\Gamma_{\mathrm{A}} = \frac{b - \Gamma_{\mathrm{M}}}{c \cdot \Gamma_{\mathrm{M}} - a} \tag{17.58}$$

Based on the three calibration measurements Γ_{A} and Γ_{M} can be replaced in (17.57) by the reflection coefficients of the three different but known calibration standards of the short Γ_{SA}, the open Γ_{OA} and the load Γ_{LA} as well as their measured counterparts of the short Γ_{SM}, the open Γ_{OM} and the load Γ_{LM} which yield the following three equations

$$a \cdot \Gamma_{\mathrm{SA}} + b - c \cdot \Gamma_{\mathrm{SA}}\Gamma_{\mathrm{SM}} = \Gamma_{\mathrm{SM}} \tag{17.59}$$

$$a \cdot \Gamma_{\mathrm{OA}} + b - c \cdot \Gamma_{\mathrm{OA}}\Gamma_{\mathrm{OM}} = \Gamma_{\mathrm{OM}} \tag{17.60}$$

$$a \cdot \Gamma_{\mathrm{LA}} + b - c \cdot \Gamma_{\mathrm{LA}}\Gamma_{\mathrm{LM}} = \Gamma_{\mathrm{LM}} \tag{17.61}$$

(17.61) – (17.59) yields

$$a \cdot (\Gamma_{\mathrm{LA}} - \Gamma_{\mathrm{SA}}) - \Gamma_{\mathrm{LM}} + \Gamma_{\mathrm{SM}} = c \cdot (\Gamma_{\mathrm{LA}}\Gamma_{\mathrm{LM}} - \Gamma_{\mathrm{SA}}\Gamma_{\mathrm{SM}}) \tag{17.62}$$

which can be rewritten as

$$a \cdot O_{\mathrm{LS}} - P_{\mathrm{LS}} = c \cdot Q_{\mathrm{LS}} \tag{17.63}$$

(17.61) – (17.60) yields

$$a \cdot (\Gamma_{\mathrm{LA}} - \Gamma_{\mathrm{OA}}) - \Gamma_{\mathrm{LM}} + \Gamma_{\mathrm{OM}} = c \cdot (\Gamma_{\mathrm{LA}}\Gamma_{\mathrm{LM}} - \Gamma_{\mathrm{OA}}\Gamma_{\mathrm{OM}}) \tag{17.64}$$

which can be rewritten as

$$a \cdot O_{\mathrm{LO}} - P_{\mathrm{LO}} = c \cdot Q_{\mathrm{LO}} \tag{17.65}$$

(17.63)/(17.65) yields

$$\frac{a \cdot O_{\mathrm{LS}} - P_{\mathrm{LS}}}{a \cdot O_{\mathrm{LO}} - P_{\mathrm{LO}}} = \frac{Q_{\mathrm{LS}}}{Q_{\mathrm{LO}}} \tag{17.66}$$

which can be solved after a as

$$a = \frac{P_{\mathrm{LS}}Q_{\mathrm{LO}} - P_{\mathrm{LO}}Q_{\mathrm{LS}}}{O_{\mathrm{LS}}Q_{\mathrm{LO}} - O_{\mathrm{LO}}Q_{\mathrm{LS}}} \tag{17.67}$$

Using (17.63) c can be obtained to

$$c = \frac{a \cdot O_{\mathrm{LS}} - P_{\mathrm{LS}}}{Q_{\mathrm{LS}}} \tag{17.68}$$

Now b can be obtained from (17.59) as

$$b = \Gamma_{\mathrm{SM}} - a \cdot \Gamma_{\mathrm{SA}} + c \cdot \Gamma_{\mathrm{SA}}\Gamma_{\mathrm{SM}} \tag{17.69}$$

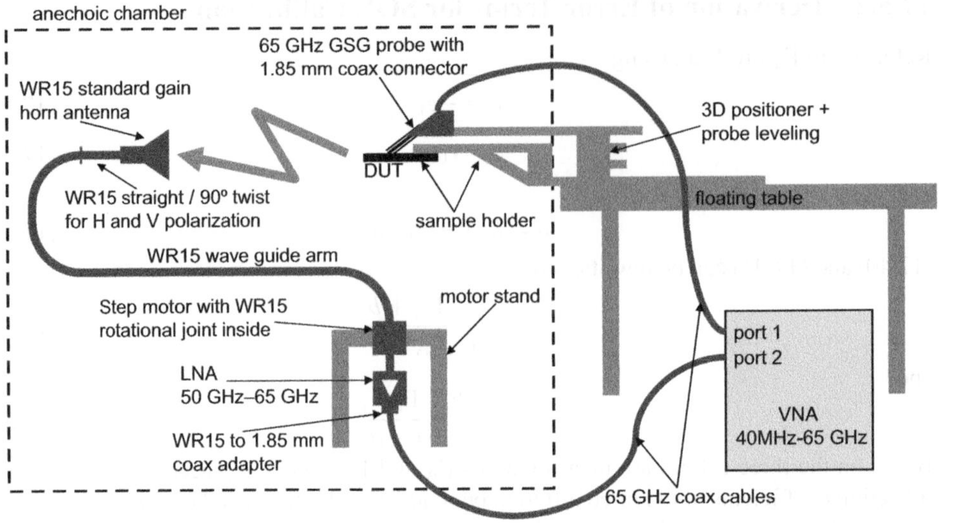

Figure 17.7: mmWave antenna measurement setup for 50 GHz to 65 GHz (from [51], reproduced by permission of © 2004 IEEE).

17.5.3 Example of Setup for the Frequency Range of 50 GHz to 65 GHz

In this section an example for a probe-based antenna measurement setup for the frequency range of 50 GHz to 65 GHz is presented. Details and further results can be found in reference [51]. Figure 17.7 shows a diagram of the test setup. The system is based on a two-port VNA. The DUT is held by a custom sample holder which reaches into the anechoic chamber (size: $140 \times 120 \times 120$ cm^3, no roof) from an isolation table. A second arm, with the microwave probe, is mounted on a three-dimensional positioner with additional probe leveling, which also sits on the isolation table. A special waveguide arm (all WR15, see Figure 17.7) with a standard gain horn can be rotated around the DUT at a distance of 38 cm to ensure a far field condition. A WR15 rotational joint with a low-noise amplifier is mounted in the center of the stepper motor to avoid any cable bending due to motor movement and to enhance the sensitivity for the gain measurement. By changing all band-limited components, the setup can be built for the different waveguide bands (e.g. WR15, WR12, WR10. . .). At frequencies above 65 GHz, the mmWave modules of the VNA should be mounted close to the probe and on the rotational arm to provide a high dynamic range and repeatability and to avoid the fragile waveguide rotational joint. In Figure 17.8 the whole setup can be seen with the motor mounted for measurement of the horizontal radiation pattern. The axis of the motor can be mounted in three different ways with respect to the DUT (vertically below, as shown in Figure 17.8, horizontally in front and horizontally on the side) so radiation patterns with a $\pm 90°$ range can be measured in three different planes. The last straight piece of WR15 waveguide in the antenna arm before the standard gain horn can be replaced by a 90° twisted piece to switch polarizations (see Figure 17.8). Figure 17.9 shows the DUT in the sample holder together with the probe. Special dielectric sample holders have been designed for

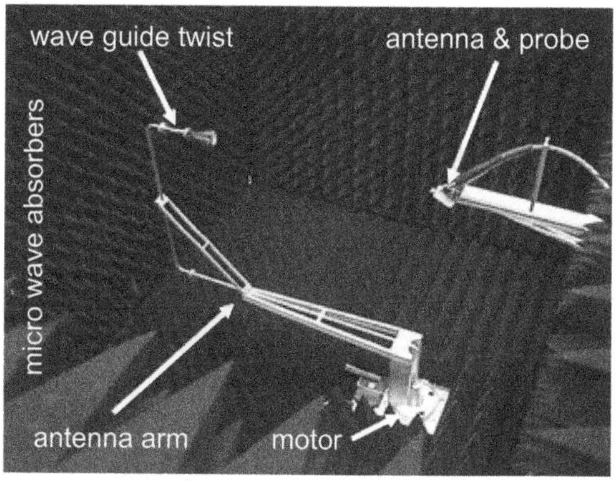

Figure 17.8: Anechoic chamber with waveguide arms and antennas (from reference [51], reproduced by permission of © 2004 IEEE).

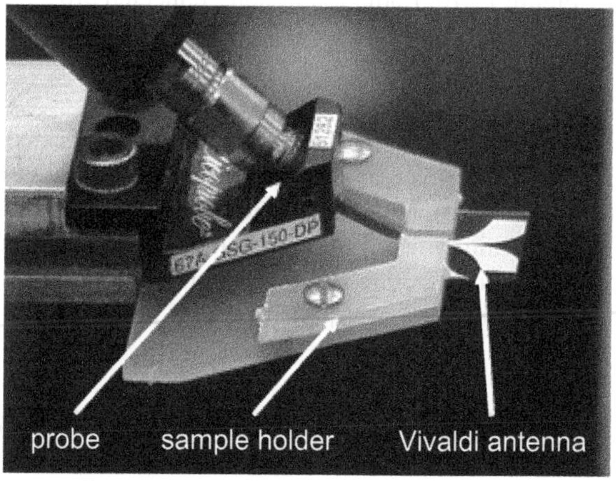

Figure 17.9: DUT (here Vivaldi antenna) in sample holder contacted by probe (from reference [51], reproduced by permission of © 2004 IEEE).

the different antennas to minimize interference with their near field or obstruction of the radiation pattern. Since no other probe configurations besides coplanar probes are available at these frequencies yet, only antennas with three contact pads in ground–signal–ground (GSG) configuration can be measured.

Since the microwave probe is not shielded, it also radiates some energy that can limit the setup's dynamic range. The sensitivity/minimum measurable gain of the setup without probe radiation is between −40 dBi and −30 dBi. To quantify the effect of the probe, the radiation of the probe without any DUT has been measured over frequency for both polarizations in all three measurement planes. The result for the horizontal plane is shown in Figure 17.10.

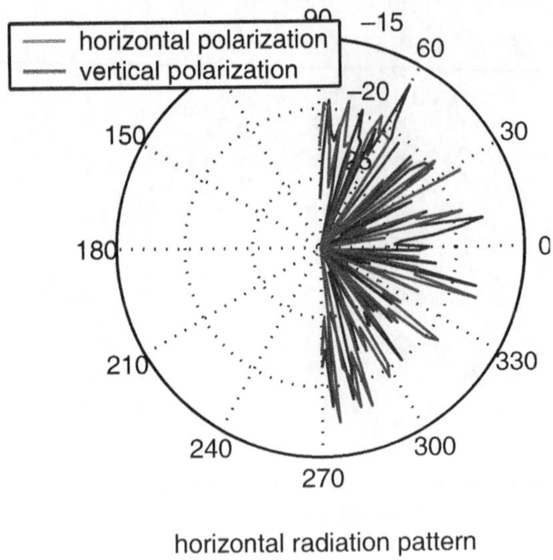

Figure 17.10: Radiation of non-connected probe.

A maximum radiation of −15.5 dBi has been found, but it is expected to be lowered when the probe is loaded by the DUT in a real measurement. The sensitivity of the setup without probe radiation could easily be improved by 20–30 dB using more amplifiers, but since the probe was found to be the limiting factor for sensitivity and no measurements without probe are anticipated, no effort in this direction has been taken yet. Measurement results using the presented setup can be found in references [49, 50].

17.6 Non-destructive IC Package Characterization

The electrical characterization of electronic packages is usually destructive. This means that internal signals are not available from the outside, since the purpose of the package is hermetically to protect the chip from its surrounding environment. To make internal ports available to external measurement equipment, one either has to drill a hole into the package exactly where the chip connects to a wire bond or flip-chip interconnect, or one has to at least reroute the signal internally to another pin to measure it from outside. Drilling a hole into an electronic package, however, has a considerable effect on the electrical properties of the component. Some of the package dielectric is removed and electric fields that usually would extend into the package substrate or the encapsulation material are now extending into air. Fields are perturbed and the air replaces the otherwise higher dielectric constant (ϵ_r) material. Furthermore, it is extremely difficult to get exactly to the location of the internal port and to launch a signal into it. This is because an electrical probe requires a reference ground that is simultaneously connected to a probe, at a pitch on the order of 100–300 μm. The electrical properties of the ground return path are equally important to the signal path. The location and

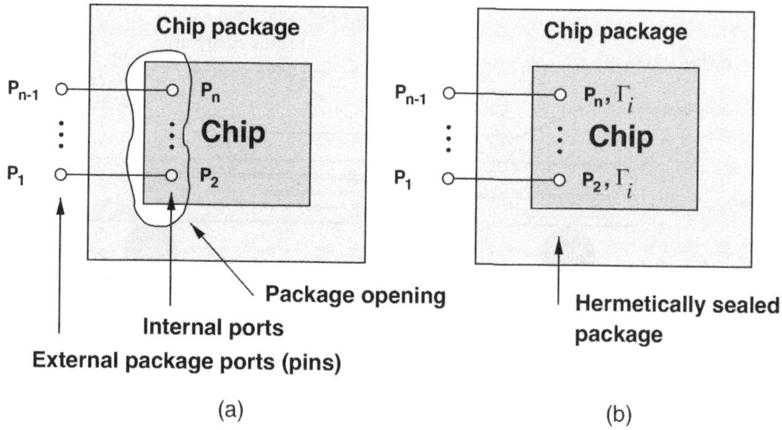

Figure 17.11: (a) Standard chip package characterization; (b) *in situ* non-destructive characterization, where additional terminations Γ_i are integrated on chip [52], reproduced by permission of © 2005 IEEE.

number of ground down-bonds to the die paddle and the number of package ground pins all affect the characteristics of a signal transition to a printed circuit board.

A solution to this problem is a non-destructive measurement technique as described in this chapter [52–54]. Unlike conventional destructive characterization techniques, the package remains virtually unaffected, as shown conceptually in Figure 17.11(a) and (b). A programmable termination network is implemented on a test chip. The test chip replaces the real chip inside of a package and enables *in situ* characterization without internal probing by means of a recursive un-termination method. The knowledge of at least three on-chip programmable terminations with reflection coefficients (Γ_i, $i = 1, 2, 3$) are needed at each internal port. The test chip may be designed with a size and layout typical for many applications. The terminations may also be part of a commercial product with only minor modifications, e.g. in high-speed serial link applications, programmable active terminations are already part of the I/O circuits. Due to the non-destructive feature of the presented methodology, a variety of different package technologies can be measured directly in the frequency domain. Figure 17.12 shows a cross-section of a possible ball grid array (BGA) package characterization scenario.

Alternative *in situ* package characterization and calibration techniques are based on a single bias-dependent active standard, embedded within a package, to accurately characterize the package from external reflection coefficient measurements. Early work by R. Bauer and P. Penfield, for instance, describes the process of un-terminating known impedance standards from a two-port network [55]. They show first *in situ* characterization of a diode package by means of three different impedances of the diode itself (active, avalanche, and drift region). Similar work was done by K. Phillips and D. Williams for monolithic microwave integrated circuit (MMIC) microwave package characterization [56]. However, the use of diodes in characterization of multi-port packages requires gluing/soldering of the standards into the package which makes it impossible to measure high-density multi-port integrated circuit packages. Moreover, setting the correct DC-bias current for many diodes remains problematic

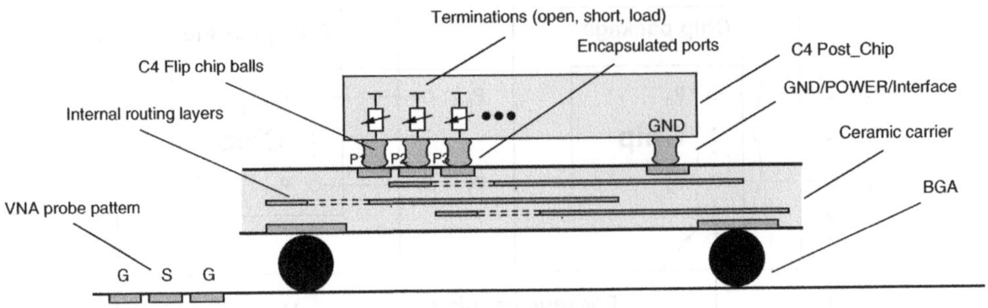

Figure 17.12: Illustration of a test configuration for *in situ* non-destructive characterization of a BGA package.

for multi-port packages. A mathematical extension for multi-port networks has recently been introduced by H. Lu and T. Chu in references [57,58]. They propose different port reduction methods to solve multi-port networks with a minimum number of reflection coefficients recursively. However, for *in situ* package characterization, integrating more than the required termination states is desirable since the frequency range can be divided into bands with optimum performance. In recent years, much advancement has also been made in the field of electronic calibrations. Recently, programmable terminations have been used to simplify the calibration of fixtures and vector network analyzers. For example, Agilent's series of Electronic Calibration Modules uses known reflective impedance standards for automated one-port error correction [59–61]. These standards are used for calibration purposes only and are too large to be used for IC package characterization.

17.6.1 Formulation of the Algorithm

In non-destructive *in situ* package characterization one is interested in measuring the scattering matrix **S** for an n-port network normalized to a characteristic impedance Z_0 (e.g. 50 Ω), if only an x-port VNA is available with $x < n$. This section describes a conventional probing technique for multi-port package characterization first (see also reference [62]), and then derives the formulation for non-destructive *in situ* characterization. The mathematical nomenclature is summarized in Section 17.6.7 for reference.

17.6.1.1 Conventional Multi-port Characterization

To recall, the simplest way to measure the matrix elements S_{ij} with $i, j = 1, \ldots, n$ is using a conventional VNA directly. This requires reconnecting the VNA ports since several partial measurements are needed. In the following, scattering parameter measurements are represented by a matrix **M** of size $x \times x$, where x is the VNA port number (e.g. $x = 2$ for a two-port VNA). In each of these partial measurements, the $n - x$ unused ports of the network should ideally be terminated with perfectly matched loads with values equal to the characteristic impedance Z_0. This way, each VNA measurement will contain some of the n-port S-parameter elements ($S_{ij} = M_{ij}$). For an n-port network the minimum required number of multi-port scattering matrix measurements can be expressed by the use of the binomial notation $\binom{n}{x}$, i.e. a two-port VNA requires $\binom{n}{2} = n(n-1)/2$ matrix measurements.

This direct method, however, has a few drawbacks. First, some ports might be inaccessible during measurement and cannot be probed directly. Second, terminations are unlikely to be perfect and cannot be easily connected and removed in small integrated circuit packages. Considerations on how to eliminate the error induced due to the imperfection of the terminations exist and are described in references [63, 64].

17.6.1.2 Non-destructive In-situ Package Characterization

A typical integrated circuit package will usually protect half of its ports from the environment, at the internal chip-to-package transition. It is therefore desirable to have on-chip terminations. For *in situ* package characterization these terminations are most likely imperfect due to manufacturing tolerances associated with a typical silicon process technology. In general, if a load at a port k is imperfect and its value differs from the required termination impedance Z_0 a reflection amounts to an incident wave defined by the reflection coefficient $\Gamma^{(k)}$ at port k. In such a case, the elements S_{ij} of the scattering matrix $\mathbf{S}$ will be transformed to the new $S_{ij}^{(k)}$ elements as given by [65]

$$S_{ij}^{(k)} = S_{ij} + \frac{S_{ik} S_{kj} \Gamma^{(k)}}{1 - S_{kk} \Gamma^{(k)}}, \quad i, j \neq k \tag{17.70}$$

If three different but known terminations $\Gamma_t^{(k)}$ with $t = 1, 2, 3$ are available at port k, one can obtain three equations like (17.70) for each $i, j \neq k$ combination. All the unknowns S_{ij}, S_{kk} and the product $S_{ik}S_{kj}$ can be derived from these equations. Knowing these values, one has removed the influence from the miss-terminated port k. Based on induction one can now go to the next miss-terminated port $k' = k + 1$ and it can be easily shown that

$$S_{ij}^{(k')} = S_{ij}^{(k)} + \frac{S_{ik'}^{(k)} S_{k'j}^{(k)} \Gamma^{(k')}}{1 - S_{k'k'}^{(k)} \Gamma^{(k')}}, \quad i, j \neq k' \tag{17.71}$$

is true since reflections from the kth and the $(k + 1)$-port are superimposed on each other.

If all ports not connected to a VNA have three terminations available, one can recursively apply this algorithm until the full n-port matrix $\mathbf{S}$ has been reached. Note, that this is only true except for a sign ambiguity in the $S_{ik}S_{kj}$ product. It is, however, possible to set the S_{ik} to $\rho\sqrt{S_{ik}S_{kj}}$ for reciprocal networks ($S_{ik} = S_{ki}$), where the factor ρ is ± 1. A sign ambiguity will cause a 180° phase change for the S_{ik} elements in the complex S-parameter plane. However, if frequency dependent S-parameters are measured starting near DC upwards, an initial boundary condition for ρ can be set to solve the sign ambiguity. For example, in a simple two-port network case, the insertion loss (S_{21}) will start with a perfect through going capacitive in the Smith chart. Therefore, the insertion loss has a negative phase $\rho = -1$ near DC, which can be tracked along consecutive frequency points. This way, a sudden phase change can easily be discovered up to four ports. For an electrically small package the phase changes slowly along frequency and a 180° phase change due to a sign ambiguity might not even occur up to the maximum frequency of interest. In the case of $n > 4$, one could also use data from terminated ports ($\neq k$) of a previously calculated step to resolve the sign ambiguity and to generalize the formulation for non-reciprocal networks, similar to what H. Lu and T. Chu propose in references [57, 58]. However, chip packages are usually reciprocal networks (passive and contain only isotropic materials). The initial boundary condition allows a simple matrix implementation of the algorithm in the form of

$$\text{RUM} : \mathbf{M}_t^{(k)}, \Gamma_t^{(k+1)} \rightarrow \mathbf{M}^{(k+1)} \quad t = 1, 2, 3 \tag{17.72}$$

where the three matrices $\mathbf{M}_t^{(k)}$ and the three reflections coefficients $\Gamma_t^{(k+1)}$ with $t = 1, 2, 3$ are used to calculate the intermediate matrix $\mathbf{M}^{(k+1)}$. This is done for all the possible combinations of the termination states $\Gamma_t^{(k')}$ with higher port orders ($k' > k + 1$) and $k = x, \ldots, n$. The intermediate matrices are input to the algorithm in the next iterative step.

17.6.1.3 Recursive Implementation of the Algorithm

The implementation of the recursive un-termination algorithm is described in the following. First, the x-port VNA has to be connected to arbitrary but fixed ports of the n-port network. At these fixed port locations one has to measure the S-parameters $\mathbf{M}_t^{(x)}$ with $\dim(\mathbf{M}_t^{(x)}) = x$ for all possible $\Gamma_t^{(k)}$ combinations with $k = x + 1, \ldots, n$ and $t = 1, 2, 3$. These are (3^{n-x}) initial measurements to start with. In the case of a one-port VNA ($x = 1$) and a four-port network ($n = 4$) these are $3^{4-1} = 27$ $\mathbf{S}_{11}$ measurements at port one, for example. Note, that all x ports of the VNA are fixed during these steps and do not need to be reconnected to other network ports. In the first step, the recursive un-termination algorithm will be applied to the port $k = x$ for all possible combinations of the termination states with higher port orders ($k' > k + 1$). In each of the following steps the recursive un-termination method (rum) will be recursively applied with

$$\mathbf{M}^{(k+1)} = \text{rum}(\mathbf{M}_t^{(k)}, \Gamma_t^{(k+1)}) \quad t = 1, 2, 3 \tag{17.73}$$

where

$$\dim(\mathbf{M}^{(k+1)}) = \dim(\mathbf{M}_t^{(k)}) + 1 \tag{17.74}$$

In the final step, where $k = n - 1$ the result is the scattering matrix of the n-port network

$$\mathbf{S} = \mathbf{M}^{(n)} \tag{17.75}$$

The rum function in (17.73) solves a linear system of equations of the form $\mathbf{Ax} = \mathbf{b}$, where the matrix $\mathbf{A}$ and the vectors $\mathbf{b}$ and $\mathbf{x}$ are as follows

$$\mathbf{A} = \begin{bmatrix} \mathbf{I} \cdot (1/\Gamma_3^{(k+1)} - 1/\Gamma_1^{(k+1)}) & [\mathbf{M}_3^{(k)} - \mathbf{M}_1^{(k)}]_{(k^2 \times 1)} \\ \mathbf{I} \cdot (1/\Gamma_2^{(k+1)} - 1/\Gamma_1^{(k+1)}) & [\mathbf{M}_2^{(k)} - \mathbf{M}_1^{(k)}]_{(k^2 \times 1)} \end{bmatrix} \tag{17.76}$$

$$\mathbf{b} = \begin{bmatrix} [\mathbf{M}_3^{(k)}/\Gamma_3^{(k+1)} - \mathbf{M}_1^{(k)}/\Gamma_1^{(k+1)}]_{(k^2 \times 1)} \\ [\mathbf{M}_2^{(k)}/\Gamma_2^{(k+1)} - \mathbf{M}_1^{(k)}/\Gamma_1^{(k+1)}]_{(k^2 \times 1)} \end{bmatrix} \tag{17.77}$$

$$\mathbf{x} = \begin{bmatrix} [\mathbf{M}^{(k+1)}]_{(k^2 \times 1)} \\ M_{k+1,k+1} \end{bmatrix}_{((k^2+1) \times 1)} \tag{17.78}$$

Note, the matrix $\mathbf{I}$ is the identity with size $k^2 \times k^2$ and the matrix subscript notation $[\ldots]_{(k^2 \times 1)}$ indicates a k^2-by-1 matrix whose elements are taken column-wise from the square matrix in brackets. Reordering the matrices and vectors simplifies the given linear equation system for implementation purposes. In each step k, the number of equations obtained by this will be $2k^2$. For $k > 1$ the system will be over-determined and a singular value decomposition (SVD) may be used to solve an equivalent linear equation system of the form $\mathbf{x} = \mathbf{V} \cdot \mathrm{diag}(1/\mathrm{diag}(\mathbf{W})) \cdot \mathbf{U}^{\mathrm{T}}\mathbf{b}$, where $\mathbf{A} = \mathbf{UWV}^{\mathrm{T}}$ is the SVD of matrix $\mathbf{A}$. It solves the linear system of equations in a least-square sense and has shown better numerical stability [66]. This is one important difference compared to the work done by Lu and Chu in references [57, 58], where the multi-port network equations are solved with a minimum number of reflection coefficients. Finally, with the definition of the vector

$$\mathbf{r} = \rho \cdot \sqrt{\mathrm{diag}\left(\left(\mathbf{M}_t^{(k)} - [\mathbf{M}^{(k+1)}]_{(k\times k)}\right)\frac{1 - M_{k+1,k+1}\Gamma_1^{(k+1)}}{\Gamma_1^{(k+1)}}\right)} \tag{17.79}$$

follows the matrix

$$\mathbf{M}^{(k+1)} = \begin{bmatrix} [\mathbf{M}^{(k+1)}]_{(k\times k)} & \mathbf{r} \\ \mathbf{r}^{tr} & M_{k+1,k+1} \end{bmatrix} \tag{17.80}$$

where $\mathbf{r}^{tr}$ is the array transpose of $\mathbf{r}$ and the matrix $[\mathbf{M}^{(k+1)}]_{(k\times k)}$ is the square sub-matrix of $\mathbf{M}^{(k+1)}$ including the elements already solved by the SVD method. In (17.79) there are three possibilities to choose the $\mathbf{M}_t$ where $t = 1$. As mentioned before, the sign of ρ has been set by the DC-boundary conditions and the phase has been tracked along consecutive frequency points in the Smith chart to avoid a sign ambiguity.

The computational effort involved is illustrated in Figure 17.13. The figure shows how the recursive un-termination method can be applied to a four-port network based on measurements acquired only at one single port (P1). In this case ($x = 1, n = 4$) the algorithm requires $3^{4-1} = 27$ initial S_{11}-measurements listed as $\mathbf{M}_t^{(1)}$ in the top first line of Figure 17.13. Although the measurements are shown only in repeated groups of $\mathbf{M}_1^{(1)}\mathbf{M}_2^{(1)}\mathbf{M}_3^{(1)}$ their values differ since other terminations have been used at port P3 and P4 while the measurement was performed.[1] Remember, none of these measurements will show up in the final **S**-matrix directly. To solve for the **S**-matrix the algorithm has to generate intermediate matrices with a growing dimension until it reaches the full **S**-matrix in step 4. The boxed elements in the figure represent the input to the algorithm going from one step to another and 'rum' stands for the recursive un-termination method used. For each call to the rum function the linear system of equations is solved as described earlier.

17.6.2 Test Chips for Non-destructive Package Characterization

The recursive untermination method requires at least three programmable on-chip terminations for each wire bond or flip-chip bond pad. The complex reflection coefficients should be maximally separated in a Smith chart and frequency independent if possible for optimum sensitivity (see in Section 17.6.2.1). In an integrated circuit, however, it is difficult to

[1] To uniquely identify all initial measurements as well as the generated intermediate matrices $\mathbf{M}_t^{(k)}$ with $k > 1$ the index t could be expanded in this example to include all the permutations of the termination states of the entire network at port P2, P3 and P4 (e.g. $t = [123]$ for $\Gamma_1^{(2)}$ used at port P2, $\Gamma_2^{(3)}$ used at port P3, and $\Gamma_3^{(4)}$ used at port P4). Due to the limited space available in the figure the index t is only shown for terminations which change (e.g. $t = 1$ in the first line corresponds to the use of $\Gamma_1^{(2)}$ at port P2. Those who remain fixed for each intermediate matrix calculation are omitted for the index.

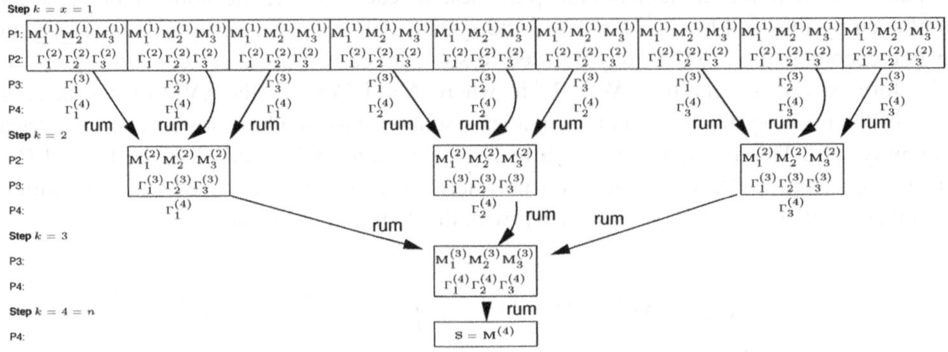

Figure 17.13: Visualization of the algorithm for a four-port network and a one-port VNA ($x = 1$) used for measurements. The boxed elements represent the input to the algorithm. Note, 'rum' in the figure stands for recursive un-termination method, which has to run for all possible termination states in each step [52], reproduced by permission of © 2005 IEEE.

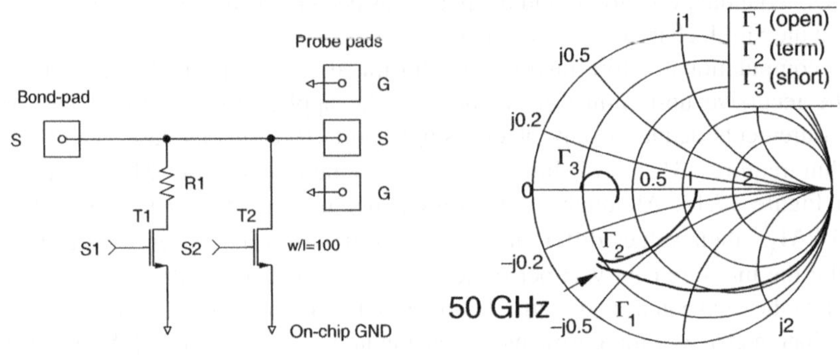

(a) (b)

Figure 17.14: (a) Simplified schematics of a programmable FET-based termination cell; (b) associated S-parameters up to 50 GHz for three states: open, short, and load.

implement all three states equally well and trade-offs have to be made for their values. Finally, unavoidable parasitics of bond-pads and of the termination circuit itself set a limit on the high-frequency performance. One possibility, however, is to implement more than three terminations and divide the frequency into bands where at least three distinguishable terminations can be found. Possible termination networks are shown in Figures 17.14 and 17.15.

The simple termination cell in Figure 17.14(a) consists of two field-effect (FET) transistors (T1 and T2). The transistors are identical ($w/l = 100$) and require an additional series 50 Ω resistor R1 to realize a load termination. Additional on-chip probe structures are included for calibration purposes before packaging. The four different termination states

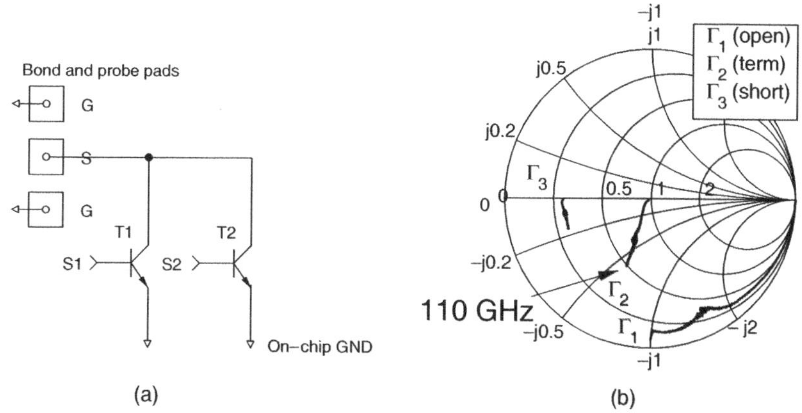

Figure 17.15: (a) Alternative schematics of a bipolar termination cell; (b) associated S-parameters up to 110 GHz. The S-parameters are clearly separated and allow non-destructive measurements with good sensitivity up to 110 GHz.

can be programmed with the S1 and S2 signals: open (S1 = off, S2 = off), short (S1 = off, S2 = on), load (S1 = on, S2 = off), and short+term (S1 = on, S2 = on). The inductive and capacitive parasitics of the pads, interconnects, transistors, and the resistor limit the high-frequency performance of the termination cells. Figure 17.14(b) shows the measured complex reflection coefficients of the three FET states up to 50 GHz. With all the switches open, the termination just looks like an open-circuit transistor with the pad parasitics in parallel. The data moves around the periphery of the Smith chart and accumulates phase shift. With S1 switched on, the termination looks like a short-circuited transistor with an impedance defined by the transistors channel resistance. The switching of the terminations may be automated and controllable via software. A silicon process exhibits excellent thermal stability and no temperature cycling and burn-in is required. The reflection coefficients are stable and do not change with the incident power of a VNA. In the following, the effect of the impedance convergence at higher frequencies is analyzed by means of statistical methods, including the effect of lossy DUT networks.

An alternative termination circuit is shown in Figure 17.15(a). The circuit is based on NPN transistors and is optimized for lower parasitics and higher frequencies. The cell omits additional on-chip probe pads and differently sized NPN transistors accommodate the load impedance without the need of an additional resistor. The measured complex reflection coefficients of the three NPN states are shown in Figure 17.15(b). The cell clearly shows a better separation of the termination states and can be used for non-destructive package characterization up to 110 GHz with good sensitivity.

An example test chip is shown in Figure 17.16. The chip has 14-port programmable terminations located at each bond pad. The transistor base contacts S1 and S2 are connected to a 28-bit register, which holds the termination state for each of the 14 pads. The content of the register can be downloaded via a 6 bit-wide serial interface. The chip was designed in an IBM SiGe 6HP process. It has a 300 μm bonding pitch for most of the pads on the outer ring, while the inner-pad ring uses a reduced 150 μm pitch. All ground pads are connected with a

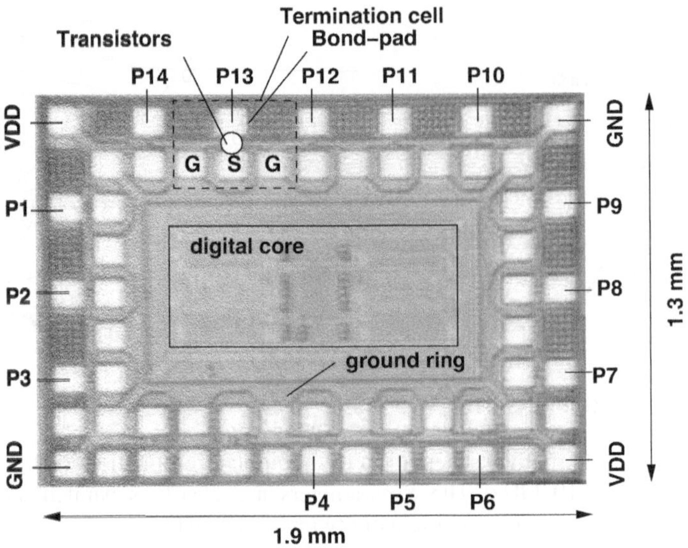

Figure 17.16: The chip micrograph indicates one termination cell with its bond-pad and GSG probe structure is indicated. The inner pad-ring consists of GSG probe pattern only, which share adjacent grounds. The serial interface and scan chain signals are located between the bottom pads on the outer pad ring [52], reproduced by permission of © 2005 IEEE.

wide top-metal ground ring to achieve a low on-chip ground impedance. The chip has a size of $1350 \times 1950\ \mu m^2$ and a height of 300 μm.

17.6.2.1 Achievable Measurement Sensitivity

Any frequency domain S-parameter measurement is limited to the first degree by the accuracy of the available measurement equipment and its calibration (e.g. VNA calibration accuracy, VNA noise floor, probe contact resistance, environmental temperature drifts). However, the recursive nature of the algorithm and the fact that the termination states inherently have some degree of uncertainty need to be considered as well. The insertion loss of the package interconnect and the separation of the termination impedance over frequency will affect the accuracy of the algorithm.

The performance of the algorithm is shown in Figure 17.17 for low insertion loss interconnects. The figure shows the degraded sensitivity $\langle \delta_{ij} \rangle$, after the recursive untermination method has been applied, versus frequency from DC up to 40 GHz. The analysis is based on Monte Carlo simulations which take the actual on-chip termination impedance separation over frequency into account (see reference [52]). The analysis assumes random measurement errors for all one-port S-parameters, e.g. for the on-chip termination reflection coefficients and the measurements performed at the coplanar wave (CPW) launch structure. A white noise error distribution with a noise power equivalent to the dynamic range ($D =$ 73 dB) of a typical VNA has been assumed. See reference [52] for more details on how the error terms have been calculated.

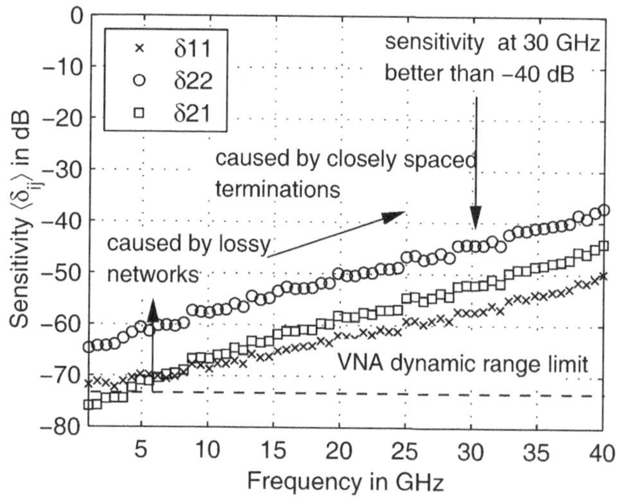

Figure 17.17: Uncorrelated experimental errors. Each reflection coefficient state was perturbed by adding an *independent* white noise error term. The sensitivity $\langle\delta_{ij}\rangle$ is shown versus frequency for random networks with a typical insertion loss of better than −0.9 dB up to 30 GHz [54], reproduced by permission of © 2005 IEEE.

The simulation result shows uncorrelated experimental errors only, where each on-chip reflection coefficient state was perturbed by adding *uncorrelated* white noise. White noise has also been added to the three measurements at the CPW launch structure. Note, the sensitivity $\langle\delta_{ij}\rangle$ is the mean value of the equivalent noise power expressed in dB, where the equivalent noise power is calculated from the squared magnitude deviation $\delta_{ij} = |S_{ij} - S'_{ij}|^2$ with $(i, j = 1, 2)$. The S'_{ij} are the elements of the reconstructed S-parameters and the S_{ij} the elements of the known network. At each data point the mean value was taken from 100 simulations. The sensitivity is plotted for random networks with an insertion loss of −0.9 dB and better. The figure shows, that at near DC, the sensitivity of the return loss S_{11} is of the same order as for a direct two-port measurement. Surprisingly, the offset of the insertion loss (S_{21}) is only slightly higher. However, the accuracy of the return-loss magnitude of the hidden port (S_{22} at P2) is degraded as if a VNA two-port measurement with a reduced dynamic range of 40 dB had been available. The degradation over frequency is due to the convergence of the e.g. 'open' and 'load' reflection coefficients of the implemented terminations. The method does *not* require ideal calibration standards, but it will not be applicable at crossovers, where the termination reflection coefficients are equal, since the sensitivity will be substantially degraded. The greater the spacing between the termination S-parameters, the better the quality and the higher the upper frequency limit of the method gets. Higher frequencies will have to be addressed by a new chip design where less frequency dependence exists or where more than three terminations are available to ensure that at least three widely separated terminations can be selected. However, the following measurements will show that the de-embedding of the CPW launch structure has a much larger effect on the results.

17.6.3 Non-destructive COB and QFN Package Characterization

The non-destructive approach has been applied to encapsulated quad flat non-leaded (QFN) package in reference [67] and [68]. Such packaging technologies are of particular interest since all internal ports are fully encapsulated and inherently inaccessible [67, 68]. The measurements were based on FET terminations integrated in a test chip as shown in Figure 17.14. The results have been verified with EM simulations and a direct measurement of an uncovered chip-on-board technology up to 4 GHz. Beyond 4 GHz the EM simulations start to deviate from the direct measurement and the extracted results. There are several possible reasons for this, including inaccuracies in the material parameters. EM simulations had to rely on reported material parameter values and are limited due to inaccuracies in the three-dimensional physical computer aided design (CAD) model of the bonding wires and on-chip ground structure. The bonding wires had to be modeled after close-up photos had been taken. Several sensitivity studies were performed and it was finally concluded that inaccuracies in the three-dimensional CAD model can explain most of the deviations found. Hence, it can be stated that accurate and detailed three-dimensional CAD models are needed beyond 4 GHz for good model-to-hardware correlation. This shows the value of the proposed methodology, where detailed information about the physical structure is not required.

17.6.4 Non-destructive FC-PBGA Package Characterization

Figure 17.18 shows a non-destructive measurement setup for a flip-chip plastic ball grid array (FC-PBGA) package from IBM. The method was used to investigate the differential cross-talk from external ports on a printed circuit board (PCB) through a ball-grid array up to flip-chip (C4) ball interconnects on the chip surface. A differential port with a reference impedance of 100 Ω was reconstructed from two 50 Ω single-ended measurements. Cut-outs were used for the probes on the PCB side to improve the signal excitation [69]. Note, that such a measurement would require eight ports, where four ports are located on the chip side and another set of four ports is required on the board side. Today's vector network analyzers, however, are typically limited to four ports and such differential cross-talk measurements are not possible without this non-destructive technique. The results demonstrated very good package performance up to 10 GHz and sufficiently low cross-talk for many applications.

17.6.5 Non-destructive Flip-chip Ball Interconnect Characterization

A great deal of research has been done on flip-chip bonding for mmWave applications [70,71] and excellent mechanical and thermal reliability has been reported [72]. However, CPW-to-CPW transitions made of CPW structures as shown in Figure 17.19(a) are predominantly of interest. The performance of flipped CPW transition lines has been thoroughly investigated, and the capability for mmWave applications has been verified [73, 74]. Despite the great advantages in combining CPW and flip-chip technologies for good measurement and three-dimensional simulation purposes, it remains problematic to isolate the influence of the flip-chip interconnect from the CPW lines. For instance, on-chip CPW lines may detune due to their proximity to the substrate. Although standard TRL-like calibrations as described in Section 17.2 could be employed to resolve the characteristics of detuned on-chip CPW lines, it is, however, difficult to manufacture a set of identically flipped calibration structures to provide the required calibration accuracy.

Figure 17.18: Non-destructive measurement setup of a FC-PBGA package. The setup is used to measure differential cross-talk through a BGA package including the influence of flip-chip ball (C4) interconnects. The figure shows the location of one differential port on the PCB side, while the second differential port is located on the side and is not visible in the picture. Reproduced by permission of © IBM.

This section describes an alternative measurement technique to isolate the electrical performance of the raw flip-chip interconnect based on the *in situ* non-destructive measurement approach. Instead of CPW-to-CPW transitions, a test chip has been designed to provide programmable terminations on chip at the location where the flip-chip balls are attached to the chip. This may be done in two ways: (i) the measurement structure consists of only one CPW launching structure including three flip-chip balls in a CPW configuration as shown in Figure 17.19(b) [54], or (ii) the balls are probed directly as shown in Figure 17.19(c) [53]. This way, the measurement setup does not alter the chip package and the package can be characterized *in situ*. There is neither an on-chip CPW line nor a second CPW-to-CPW transition involved, to provide the ability for a two-port through measurement in a back-to-back mounted configuration.

In Figure 17.19(b) the hidden second port (P2) is located at the flip-chip ball to bond-pad transition. It is covered by the silicon chip and cannot be probed directly without influencing the measurement or without partially destroying the flip-chip mount. This port has been reconstructed by means of a recursive un-termination method. The limitations and implementation of this method are described in detail in reference [52]. The method requires knowledge of at least three on-chip programmable terminations with reflection coefficients (Γ_i, $i = 1, 2, 3$) located at each internal port. These on-chip terminations have to be known accurately and therefore they need to be measured on-wafer in an initial step before flip-chip mounting. While the method can be used to extract packages with higher port orders, it was used here to reconstruct the two-port information from one-port measurements only. In principle, the method can be extended to extract the near-end and far-end crosstalk of a four-port network where only two ports are accessible from the outside. However, there are some

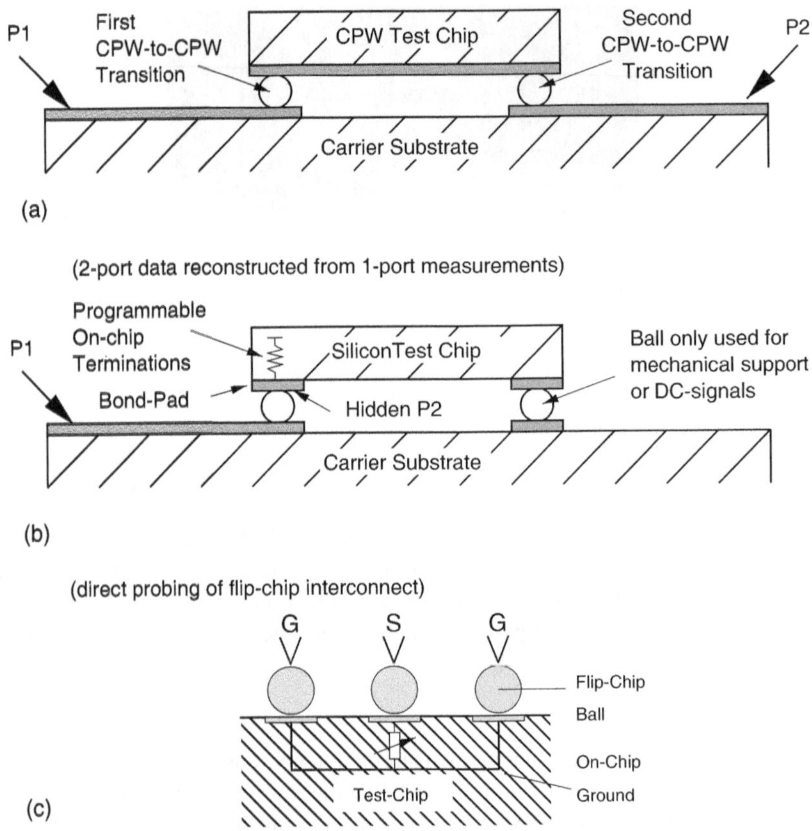

Figure 17.19: (a) Shows a standard flip-chip characterization setup based on two CPW-to-CPW transitions. (b) and (c) Show the alternative non-destructive measurement approach. Note, in (a) and (b) the full three-ball (GSG) interconnect is hidden due to the figures perspective.

limitations to that owing to the presence of the CPW launching structures which inherently introduce additional crosstalk to the measurement, as will be shown later. The sensitivity of the reconstructed scattering parameters depends strongly on the impedance separation of the integrated on-chip terminations and the insertion loss of the interconnect as it will be described in the following.

The investigated flip-chip technology is a thermal-compression gold stud-bumping technology. This single-chip bumping technique has the advantage that the on-chip terminations can be measured first in the absence of gold stud bumps and then re-measured after the bump as been applied. After thermal-compression bonding a final ball height and width are 30 μm and 90 μm, respectively.

In the following section, the non-destructive *in situ* approach to flip-chip characterization has been compared with standard back-to-back mounted CPW-to-CPW transitions and a direct ball probing approach. Experimental results are presented and compared.

17.6.5.1 Case (a): Regular Back-to-back Mounted CPW-to-CPW Transition

The standard flip-chip characterization of two 'back-to-back' mounted CPW-to-CPW transitions has been evaluated first. As mentioned before, it remains problematic to provide accurate two-port S-parameters and to isolate the influence of the flip-chip interconnect from such configurations. Note, that this measurement technique requires a two-port S-parameter measurement. The side-view of this setup is shown previously in Figure 17.19(a), which indicates the close proximity of the flipped CPW line to the bottom substrate. A common procedure is to compare the results with measurements of a straight through-line to extract the additional insertion loss caused by the interconnect.

A straight, 11.5 mm-long CPW line, has only 0.7 dB (0.06 dB mm^{-1}) insertion loss at 30 GHz whereas the flipped sample doubles the insertion loss to 1.4 dB (0.12 dB mm^{-1}). Consequently, the assumption is often made that the additional insertion loss can be attributed to the two flip-chip interconnects (e.g. 0.35 dB/interconnect). A conceptual drawing of this setup was previously shown in Figure 17.19(b) and a photo of the setup is shown in Figure 17.20. The *in situ* measurement technique requires only one-port measurements from which the two-port data is reconstructed. From this reconstructed data one may de-embed the CPW launch to only show the influence of the flip-chip interconnect alone. This greatly relaxes the complexity since only one CPW line has to be de-embedded and on-chip parasitics are naturally excluded by the method. However, it remains a moot question whether or not the problem can be electrically separated into two, fully isolated, series-connected networks which is the underlying idea of any de-embedding technique.

A small alumina sample has been flipped onto two CPW launching structures for test purposes. The investigated low-loss ceramic substrate was 125 μm thick with back-side metalization and it was made of a 99.6% alumina with a polished finish. The relative dielectric constant (ϵ_r) was 9.6 and the loss-tangent ($\tan\delta$) was better than 0.001 up to 60 GHz. These material properties have been measured with an open resonator technique [75]. The CPW line has a 50 μm-wide center conductor with a 40 μm gap between its 170 μm-wide finite-ground structure (50/40/170). The probe-pitch as well as the flip-chip pitch is 150 μm.

17.6.5.2 Case (b): Non-destructive Approach for a Single CPW Launch

A conceptual drawing of this setup was previously shown in Figure 17.19(b) and a photo of it is shown in Figure 17.20 for better clarity. The *in situ* measurement technique requires only one-port measurements from which the two-port data is reconstructed. From this reconstructed data one may de-embed the CPW launch to only show the influence of the flip-chip interconnect alone. This greatly relaxes the complexity since only one CPW line has to be de-embedded and on-chip parasitics are naturally excluded by the method.

Extracted two-port results for a 100/50/100 μm launch are shown in Figure 17.21 for all five flip-chip interconnects on the carrier up to 40 GHz. The return loss (S_{11}) in air is better than −15 dB at 30 GHz. Sensitivity studies from Section 17.6.2.1 have shown a sensitivity of −55 dB for S_{11} data and −45 dB for S_{22} at 30 GHz, which is well below the measured return-loss and therefore has a negligible effect on the data. Figure 17.21 shows that there is a variation of the return-loss from port to port. The difference is due to the in-line configuration of the five ports, where the sides do not have two adjacent neighbors. The results for the 50/40/170 μm launch in air with underfill have been previously published in reference [76]

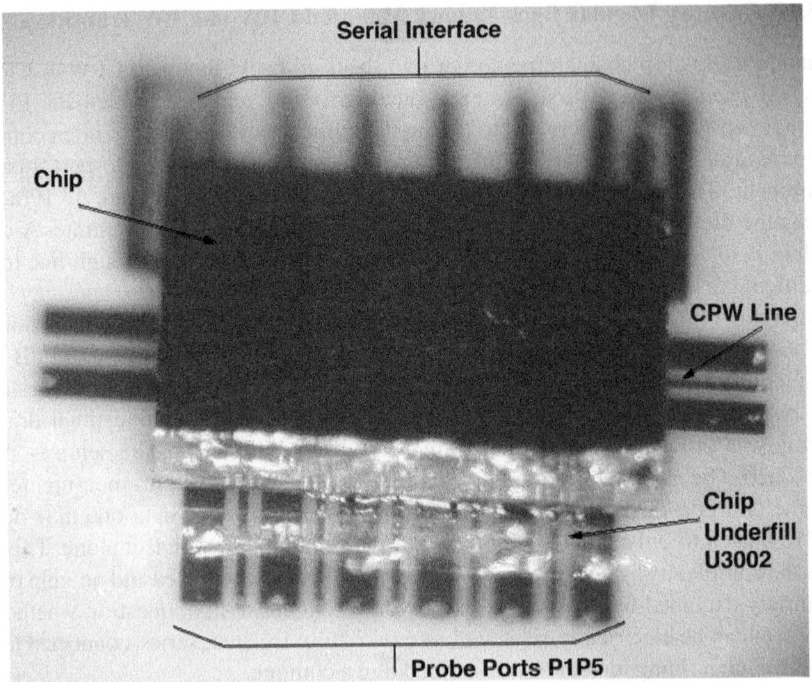

Figure 17.20: Device under test. The figure shows the mounted test-chip after underfilling [54], reproduced by permission of © 2005 IEEE.

and are not repeated here. However, the insertion loss for both launches shows comparable results up to 30 GHz where the insertion loss is between 0.8 dB and 1 dB.

These results demonstrate that the *in situ* approach provides results which are independent from the geometry or impedance of the launch up to 30 GHz. Above 30 GHz however, the method starts to include residual contributions from the CPW launch de-embedding technique. Above 30 GHz the launch and the chip are 'electrically bound' together and the previously measured launch characteristic is not a sufficient representation at higher frequencies. The reason being, that both launches are susceptible to the presence of the chip. The amount of cross-talk between the five CPW lines has changed with the bonded chip in place. Adjacent lines on the CPW launch structure have been left unterminated on both sides while each two-port CPW line data has been measured for later de-embedding purposes. During the actual *in situ* measurement, however, these lines are terminated by the test chip at F1–F5 which will influence the amount of cross-talk. Consequently, the characteristic impedance of the CPW lines has now been changed due to a modified coupling. The presence of cross-talk within the launch has been verified with three-dimensional EM-simulations. Above 30 GHz the simulations showed significant far-end cross-talk along the 1.7 mm long diagonal axis of the launch structure, which supports the resonant observations made above. In the end, it comes down to the fact that the CPW launch is a multi-port (10-port) network which has to be characterized and de-embedded as such. Considering the launch as a two-port network only is not sufficient at frequencies above 30 GHz. To achieve

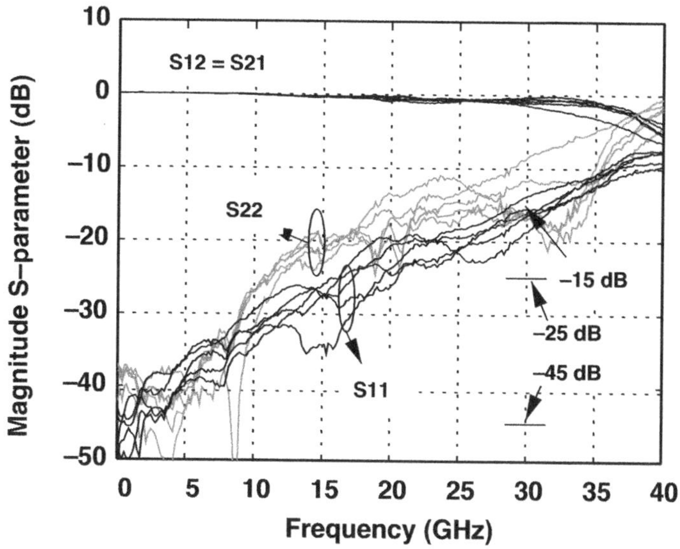

Figure 17.21: S-parameters of the five raw gold stud bump interconnects without under-fill mounted on the 100/50/100 μm sample. The return-loss is about −15 dB at 30 GHz. The sensitivity on S_{22} data is −45 dB for a simulated interconnect insertion loss of better than −0.9 dB [54], reproduced by permission of © 2005 IEEE.

better performance of the interconnect measurement at higher frequencies requires either a lower interconnect density on the substrate with a better isolation from its neighbors or the full n-port information of the launch needs to be considered. It is noteworthy, that while under-filling the chip the characteristic of the de-embedded launch structure will also change slightly. Seen from that perspective, the presented measurement technique unfolds the performance of the raw flip-chip interconnect in the presence of a given chip package.

17.6.5.3 Case (c): Non-destructive Approach for Direct Ball Probing

A direct ball probing requires two steps. First, the on-chip terminations have to be measured prior to bumping. Second, the tips of the wafer probe have to be placed on the flat top-side of the balls where a good electrical contact can be made. In order to achieve a flat ball top and a ball shape similar to a real mounted configuration the bumped chip has been pressed onto a planar silica substrate. Figure 17.22(a) shows the ball interconnect after its shape compression and Figure 17.22(b) shows the probe set-up during measurement. A thermal-compression gold stud-bumping technology has been used for these measurements. The DC probe-wedge required to program the terminations is not shown in the figure. Note, due to the programability of the chip, the probe remains fixed and does not need to be re-connected while the VNA measures the three one-port responses from the terminations.

Figure 17.23 shows the equivalent circuit model that has been found and which led to very good model-to-hardware correlation. It should be noted, that the difficulty in all modeling work is to find simple models that can represent all the aspects of the measured

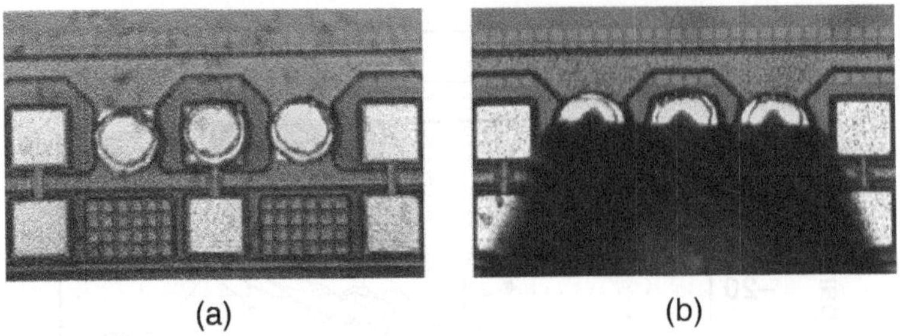

Figure 17.22: Flip-chip interconnect probe set-up: (a) ball after shape compression; (b) while the interconnect is being probed [53], reproduced by permission of © 2005 IEEE.

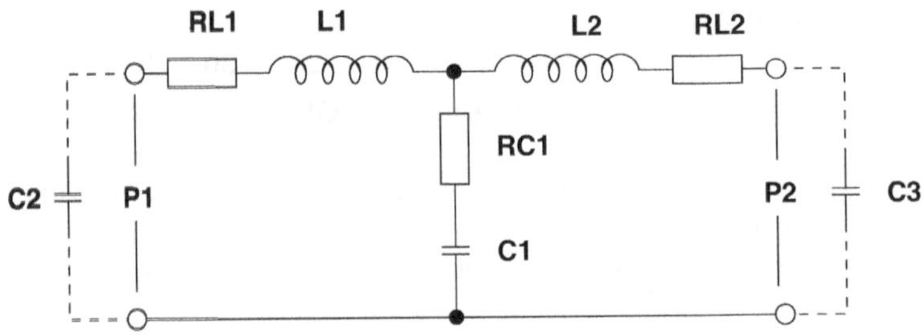

Figure 17.23: Equivalent circuit model representation including the location of negative capacitances associated with calibration errors. Port P1 is located at the ball top side whereas port P2 is located on-chip [53], reproduced by permission of © 2005 IEEE.

S-parameter data over a very wide frequency range. Often models can only predict the frequency dependence of the magnitude (e.g. of the insertion loss) but not its phase. Needless to say, the extraction of a simple model that can represent magnitude and phase for all the three S-parameters (S_{11}, $S_{21} = S_{12}$, S_{22}) from DC up to 40 GHz is a challenge.

Although the presented method eliminates any probe test-fixture and subsequent deembedding, attention has to be paid to the off-wafer probe calibration process. The Infinity probe from Cascade Microtech uses an alumina LRRM calibration substrate with a different dielectric constant than the silicon chip. The different substrate material causes a slight calibration error which can be described as a negative capacitive offset. Usually, small calibration errors can be tolerated for electrically large structures, however a flip-chip interconnect is approaching electrical transparency and calibration errors will affect the data. The negative capacitance was measured to be only -4 fF on-chip and about -8 fF in air at a height equivalent to the ball top side. The latter one is more negative since an elevated probe has a slightly modified electrical field around it, which degrades the calibration further. The capacitance C3 is associated with the on-chip port P2 and C2 with port P1 located at the ball top side. These capacitances change little over frequency (DC to 40 GHz) and were assumed to be constant in the following. After de-embedding of these capacitances, the model was derived empirically with the help of supporting analytic model extraction based on π, T, or

Table 17.1: Model parameter of GSG flip-chip interconnect (DC-40 GHz).

Parameter	Value	Units
Total loop inductance[1]		
L1	<1	pH
L2	25	pH
Series resistance[2]		
RL1	<0.1	Ω
RL2	<0.1	Ω
Coupling[3]		
C1	$17(1+f)$	fF
RC1	19	Ω
Calibration capacitances[4]		
C2	−8	fF
C3	−4	fF

[1]Main contribution is coming from L2, L1 has little effect. [2]Within limit no contribution found. [3]C1 is frequency dependent with f given in GHz. [4]Negative capacitance is caused by calibration errors. Table from reference [53], reproduced by permission of © 2005 IEEE.

L-network topologies. This was used to gain first order estimates on the total loop inductance or shunt capacitance. The frequency characteristic of the shunt capacitance has been found to be $17(1+f)$ fF where f is the frequency of operation in GHz. The shunt capacitance is located between the center ball and the adjacent ground balls. A summary of all model parameters is given in Table 17.1 for which the model-to-hardware correlation has shown best results. Note, the inductance L1, the series resistance RL1 and RL2 are very small and can be neglected from the model. The inductance L1 $\ll$ L2 cause an asymmetry in the model which can be understood in parts by the shape of the ball and the fact that the top ball side is not connected to a pad. Hence, a chip-package co-simulation requires an additional bondpad model for the package side.

The model-to-hardware correlation of the flip-chip interconnect has been verified up to 40 GHz. This is done in two ways to stress the effect of calibration artifacts on the model extraction process: first, the raw measured data is compared with the model that includes a negative capacitance at each port as shown previously in Figure 17.23. Second, the capacitances were omitted from the model, but instead de-embedded from the measured data. The first case is shown in Figure 17.24(a) for the magnitude and in (b) for the phase. The latter case is shown right below with the magnitude plot in (c) and the phase plot in (d). For the sake of clarity, only S_{11} and S_{21} data is shown here. Note, measured data is always shown in gray color where the model is shown in black color.

The first thing to note is that the model is clearly able to predict the trend of the measured data in all aspects. Nevertheless, slight deviations are noticeable, for example, the phase of the insertion loss in Figure 17.24(b) and (d) is increasing slightly faster over frequency then the measured data. However, this happens on a very small scale since the height of the interconnect is so short (about 30 μm) and the phase plot scale is only 5° per division. It is

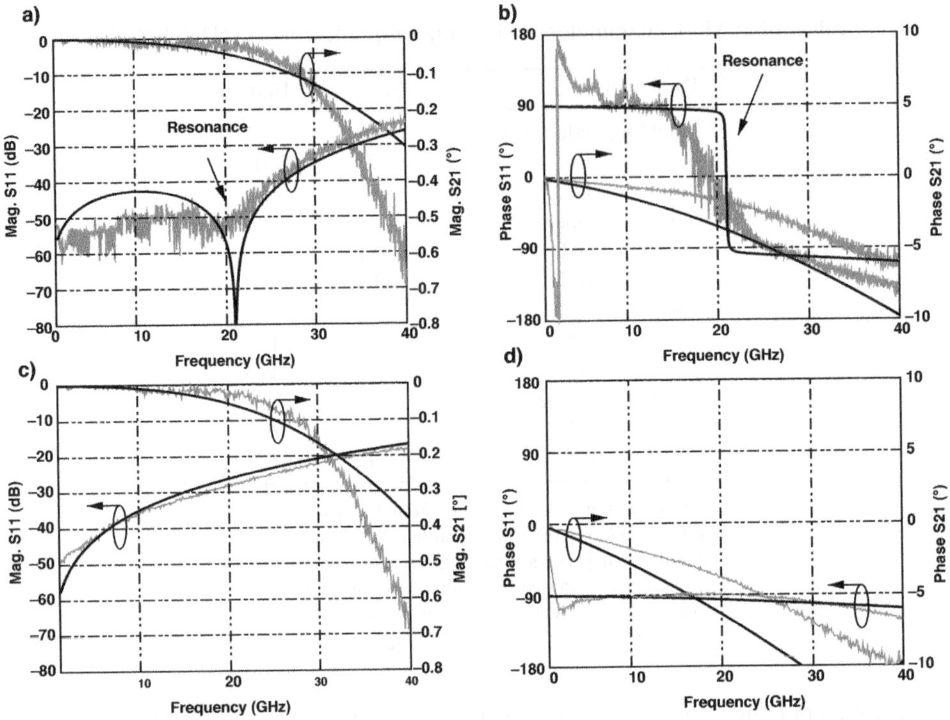

Figure 17.24: Correlated S-parameters of the flip-chip interconnect. Figure (a) and (b) compare the raw measurement with the model having the negative capacitances included. Figure (c) and (d) show the same plots where the negative capacitances have instead been de-embedded from the measurement and where they have been omitted in the model. Measured data is always shown in gray color where the model is shown in black [53], reproduced by permission of © 2005 IEEE.

interesting to see the pronounced effect of the negative port capacitances in figures (a) and (b). Although the port capacitances are small (−4 fF and −8 fF) they cause a resonance with the interconnect inductance at about 20 GHz (see S_{11}). The measured S_{11} is therefore almost flat (−50 dB) through 20 GHz and then increases up to −25 dB at 40 GHz, closely followed by the model which indicates a clear notch at 20 GHz. The measurement is not fully able to resolve the notch due to sensitivity limits of the VNA and the used reconstruction method. The same 20 GHz resonance is visible in the phase plot of the return-loss in (b). The return-loss looks first inductive (90°) as one would expect for a negative capacitance but then changes at 20 GHz to a more capacitive characteristic (−90°) when the frequency dependent capacitive coupling between the signal ball and its adjacent ground neighbors overtakes. If the capacitances are omitted in the model but instead deembedded from the measured data, it is becoming obvious that the resonance is not caused by the interconnect alone as can be seen in Figure 17.24(c) and (d).

17.6.6 Discussion and Outlook

A novel methodology was presented for non-destructive S-parameter measurements of hermetically encapsulated packages. A programmable termination network was implemented on a chip. The chip replaces the real chip inside of a package and enables *in situ* characterization by means of a recursive un-termination method. The detailed implementation of the algorithm was presented, including a statistical analysis of its performance limitations. Due to the non-destructive feature of the presented methodology, a variety of different package technologies can be measured directly in the frequency domain using this methodology.

The measurement technique has been used for the characterization of flip-chip interconnects up to mmWave frequencies. The results have been compared with standard two-port CPW-to-CPW transition measurements and a direct ball probing attempt. Similar results for the interconnect insertion loss have been found for a standard two-port CPW-to-CPW transition measurement and the direct ball probing (0.35 dB and 0.2 dB), however, the *in situ* approach taking the mounted chip into account has shown an insertion loss of 0.8 dB at 30 GHz. A non-destructive measurement technique was used to extract the two-port S-parameters of a raw flip-chip interconnect excluding any launch structures. This new concept combines the advantages of *in situ* characterization, by means of a real silicon chip, with improved models for chip-package co-design. Measured results have demonstrated the concept and have indicated key issues for improvement of high-density mmWave packaging concepts. The described non-destructive measurement approach can also be used for high pin-count packages. In such applications, internal ports can always be added by integrating more programmable termination on chip, however, the number of external ports are limited by the available measurement equipment since external switchable terminations are bulky and can not be considered as an option. Typically, VNAs are limited to four ports, which would enable the measurement of a eight-port package network. Of course, one might use additional external probe configurations on the board side in conjunction with broadband 50 Ω loads to increase the external port number further. The sign ambiguity problem at higher port orders can be resolved by known boundary conditions at DC or by sharing termination data between two consecutive iterative steps of the algorithm as described in references [57, 58]. The recursive un-termination algorithm and the software, which was used for post processing the measured data, will have to be extended in the future to accommodate for that. However, it goes without saying, that the complexity of the problem should be limited to the main aggressors and package symmetries should be considered where possible to minimize the number of ports.

An equivalent circuit model for a flip-chip ball interconnect up to 40 GHz has been presented. The model was derived based on data taken directly from a flip-chip interconnect. It has been found that the calibration substrate has a substantial effect on the measured data at mmWave frequencies and cannot be neglected for simple and yet accurate equivalent circuit models.

The sign ambiguity problem can be resolved due to the known boundary condition at DC. In principle, these sign ambiguities could be resolved if termination data is shared between two consecutive iterative steps of the algorithm as described in references [57, 58]. Also, one might use additional external probe configurations on the board side to resolve the ambiguity. This would of course increase the number of external probe configurations needed. The recursive un-termination algorithm and the software, which was used for post processing the measured data, will have to be extended to accommodate for that.

Limitations exist for the maximum frequency up to which the test-chip can be successfully used. The method does *not* require ideal calibration standards, but it will not be applicable at crossovers, where the different port termination reflection coefficients will be equal. The larger the S-parameter separation is, the better the quality and the higher the upper frequency limit of the method. In principle, it should be possible to measure electrically small packages with low insertion loss up to about 50 GHz with the current implementation. Higher frequencies will have to be addressed by a new chip design, where more than three terminations are available to ensure that at least three widely separated terminations can be selected. Note that the current test chip design is only one possible implementation of a termination network. Other versions are easily fabricated to accommodate different package geometries and pin layouts.

Overall, the measurements showed excellent agreement. The minor degradation of the sensitivity did not have a noticeable effect on the accuracy of the reconstructed on-chip port. For the measured 2 dB insertion loss, the Monte Carlo analysis showed a sensitivity on S22 of better than −45 dB up to 10 GHz which is well below the measured −9 dB for this package. The fact that the algorithm is insensitive to correlated errors greatly relaxes any contact resistance related issues as compared to the direct measurement approach. Since the on-chip terminations are programmable it can be ensured that the contact resistance remains the same and correlates during measurement.

Future activities will include the design of representative bond-pad structures to study the bond-pad to ball interaction by the same technique. Finally, three-dimensional-modeling of the physical structure and its correlation with the obtained data will follow.

17.6.7 Nomenclature

The mathematical nomenclature for the non-destructive measurement approach presented in this section is summarized as follows:

- *Matrix:* The matrix notation is *capital bold letters*, e.g. $\mathbf{S}$ for the scattering parameters of an n-port network, or $\mathbf{M}$ for measured or intermediate matrices with $\dim(\mathbf{M}) \leq n$.

- *Vectors:* The vector notation is *small bold letters*, e.g. $\mathbf{b}$ or $\mathbf{x}$ in matrix equations like $\mathbf{Ax} = \mathbf{b}$.

- *Components:* The component notation is *capital cursive letters*, e.g. S_{ij} with $i, j = 1, \ldots, n$. Component indices are *small subscript cursive letters*.

- *Terminated Ports:* The notation to indicate the port location of terminations is with a *parenthesize superscript index*, e.g. $S_{ij}^{(k)}$ are the elements of an $\mathbf{S}$ matrix terminated at port k with a reflection coefficient $\Gamma^{(k)}$ at port k.

- *Reshaped Matrices:* The reshape notation is *squared bracket matrix with subscript size indices*, e.g. the matrix $[\mathbf{M}]_{(n^2 \times 1)}$ is a n^2-by-1 matrix whose elements are taken column-wise from the square matrix $\mathbf{M}$ of dimension n.

Acknowledgments

The authors would like to thank their former team at the IBM communication technology group (IBM T. J. Watson Research Center, Yorktown Heights, New York, USA) and the IBM SiGe technology group (IBM Burlington, Essex Jct., VT 05452, USA). Much appreciation goes to C. Baks for his help in building the probe based antenna measurement setup and R. John for preparing the antenna samples and sample holders. Special thanks also go to M. Soyuer, M. Oprysko, B. Gaucher for the mmWave projects leadership and management support. This project was partially funded by DARPA (contracts N66001-02-C-8014 and N66001-05-C-8013) and NASA (contract NAS3-03070).

References

[1] D. Ballo, 'Network analyzer basics', Hewlett-Packard Back to Basics Seminar, Santa Rosa, CA, Tech. Rep., 1998.

[2] 'On-wafer vector network analyzer calibration and measurements', Cascade Microtech Application Note, Tech. Rep., March 27, 2002.

[3] 'Applying error correction to network analyzer measurements', Agilent Application Note, Tech. Rep. AN 1287-3, March 1, 2002.

[4] D. Rytting, 'An analysis of vector measurement accuracy enhancement techniques', Hewlett-Packard Presentation, Tech. Rep., March, 1982.

[5] R. B. Marks, 'Formulations of the basic vector network analyzer error model including switching terms', in *IEEE Automatic RF Techniques Group (ARFTG) Conference*, Portland, OR, December 1997, pp. 115–126.

[6] G. F. Engen and C. A. Hofer, 'Thru-reflect-line: An improved technique for calibrating the dual six-port automatic network analyzer', *IEEE Trans. Microwave Theory Tech.* **27**(12) (1979), pp. 987–993.

[7] R. B. Marks, 'A multiline method of network analyzer calibration', *IEEE Trans. Microwave Theory Tech.* **39**(7) (1991), pp. 1205–1215.

[8] D. F. Williams, J. C. M. Wang and U. Arz, 'An optimal vector-network-analyzer calibration algorithm', *IEEE Trans. Microwave Theory Tech.* **51**(12) (2003), pp. 2391–2401.

[9] H.-J. Eul and B. Schiek, 'Thru-match-reflect: One result of a rigorous theory for de-embedding and network analyzer calibration', in *European Microwave Conference*, Stockholm, Sweden, September 1988, pp. 909–914.

[10] R. F. Scholz, F. Korndörfer, B. Senapati and A. Rumiantsev, 'Advanced technique for broadband on-wafer RF device characterization', *IEEE Automatic RF Techniques Group (ARFTG) Conference*, Fort Worth, TX, June 2004, pp. 83–90.

[11] A. Davidson, K. Jones and E. Strid, 'LRM and LRRM calibrations with automatic determination of load inductance', in *IEEE Automatic RF Techniques Group (ARFTG) Conference*, Monterey, CA, November 1990, pp. 57–63.

[12] J. Pence, 'Verification of LRRM calibrations with load inductance compensation for CPW measurements on GaAs substrates', in *IEEE Automatic RF Techniques Group (ARFTG) Conference*, San Jose, CA, December 1993, pp. 45–47.

[13] F. Purroy and L. Pradell, 'New theoretical analysis of the LRRM calibration technique for vector network analyzers', *IEEE Trans. Instrum. Meas.* **50**(5) (2001), pp. 1307–1314.

[14] R. A. Speciale, 'A generalization of the TSD network-analyzer calibration procedure, covering n-port scattering-parameter measurements, affected by leakage errors', *IEEE Trans. Microwave Theory Tech.* **25**(12) (1977), pp. 1100–1115.

[15] D. Williams, R. B. Marks and A. Davidson, 'Comparisons of on-wafer calibrations', in *IEEE Automatic RF Techniques Group (ARFTG) Conference*, San Diego, CA, Fall 1991, pp. 68–81.

[16] D. K. Walker and D. F. Williams, 'Comparison of SOLR and TRL calibrations', in *IEEE Automatic RF Techniques Group (ARFTG) Conference*, Baltimore, MD, June 1998, pp. 83–87.

[17] J. A. Jargon, R. B. Marks and D. K. Rytting, 'Robust SOLT and alternative calibrations for four-sampler vector network analyzers', *IEEE Trans. Microwave Theory Tech.* **47**(10) (1999), pp. 2008–2013.

[18] D. E. Bockelman and W. R. Eisenstadt, 'Calibration and verification of the pure-mode vector network analyzer', *IEEE Trans. Microwave Theory Tech.* **46**(7) (1998), pp. 1009–1012.

[19] T. Zwick and U. R. Pfeiffer, 'Pure-mode network analyzer concept for on-wafer measurements of differential circuits at millimeter-wave frequencies', *IEEE Trans. Microwave Theory Tech.* **53**(3) (2005), pp. 934–937.

[20] H. Heuermann, A. Rumiantsev and S. Schott, 'Advanced on-wafer multiport calibration methods for mono- and mixed-mode device characterization', *IEEE Automatic RF Techniques Group (ARFTG) Conference*, Fort Worth, TX, June 2004, pp. 91–96.

[21] L. Hayden, 'A hybrid probe-tip calibration for multiport vector network analyzers', *IEEE Automatic RF Techniques Group (ARFTG) Conference*, Broomfield, CO, December 2006.

[22] P. J. van Wijnen, H. R. Claessen, and E. A. Wolsheimer, 'A new straightforward calibration and correction procedure for on wafer high frequency S-parameter measurements (45 MHz–18 GHz)', in *IEEE Bipolar Circuits and Technology Meeting*, Minneapolis, MN, September 1987, pp. 70–73.

[23] W. R. Eisenstadt and Y. Eo, 'S-parameter-based IC interconnect transmission line characterization', *IEEE Trans. Comp., Hybrids, Manufact. Technol.* **15**(4) (1992), pp. 483–490.

[24] P. Heymann, H. Prinzler and F. Schnieder, 'De-embedding of MMIC transmission-line measurements', in *IEEE MTT-S International Microwave Symposium*, San Diego, CA, June 1994, pp. 1045–1048.

[25] D. M. Pozar, *Microwave Engineering* (New York: John Wiley & Sons Ltd/Inc., 1998).

[26] M. C. A. M. Koolen, J. A. M. Geelen and M. P. J. G. Versleijen, 'An improved de-embedding technique for on-wafer high frequency characterization', in *IEEE Bipolar/BiCMOS Circuits and Technology Meeting (BCTM)*, Minneapolis, MN, October 1991, pp. 188–191.

[27] L. F. Tiemeijer and R. J. Havens, 'A calibrated lumped element de-embedding technique for on-wafer RF characterization of high-quality inductors and high-speed transistors', *IEEE Trans. Electron Devices* **50**(3) (2003), pp. 822–829.

[28] J. Song, F. Ling, G. Flynn, W. Blood and E. Demircan, 'A de-embedding technique for interconnects', in *IEEE Meeting on Electrical Performance of Electronic Packaging*, Boston, MA, October 2001, pp. 129–132.

[29] Y. Tretiakov, J. Rascoe, K. Vaed, W. Woods, S. Venkatadri and T. Zwick, 'A new on-wafer de-embedding technique for on-chip RF transmission line interconnect characterization', in *IEEE Automatic RF Techniques Group (ARFTG) Conference*, Ft. Worth, TX, June 2004, pp. 69–72.

[30] H. Cho and D. E. Burk, 'A three-step method for the de-embedding of high frequency S-parameter measurements', *IEEE Trans. Electron Devices* **38**(6) (1991), pp. 1371–1375.

[31] Y. Eo and W. R. Eisenstadt, 'High-speed VLSI interconnect modeling based on S-parameter measurements', *IEEE Trans. Comp., Hybrids, Mfg. Technol.* **16**(5) (1993), pp. 555–562.

[32] T.-M. Winkel, L. S. Dutta and H. Grabinski, 'An accurate determination of the characteristic impedance of lossy lines on chips based on high frequency S-parameter measurements', in *IEEE Multi-Chip Module Conference*, Santa Cruz, CA, February 1996, pp. 190–195.

[33] T.-M. Winkel, L. S. Dutta and H. Grabinski, 'An accurate determination of the characteristic impedance matrix of coupled symmetrical lines on chips based on high frequency S-parameter measurements', in *IEEE MTT-S International Microwave Symposium*, Denver, CO, June 1997, pp. 1769–1772.

[34] T.-M. Winkel, 'An accurate and complete frequency dependent transmission line characterization using S-parameter measurements', in *IEEE Topical Meeting on Electrical Performance of Electronic Packaging*, San Diego, CA, October 1999, pp. 133–136.

[35] T. E. Kolding, O. K. Jensen and T. Larsen, 'Ground-shielded measuring technique for accurate on-wafer characterization of RF CMOS devices', in *IEEE International Conference on Microelectronic Test Structures*, Monterey, CA, March 2000, pp. 246–251.

[36] T. E. Kolding, 'A four-step method for de-embedding gigahertz on-wafer CMOS measurements', *IEEE Trans. Electron Devices* **47**(4) (2000), pp. 734–740.

[37] E. P. Vandamme, D. M. M.-P. Schreurs and C. van Dinther, 'Improved three-step de-embedding method to accurately account for the influence of pad parasitics in silicon on-wafer RF test-structures', *IEEE Trans. Electron Devices* **48**(4) (2001), pp. 737–742.

[38] T.-M. Winkel, M. F. Ktata, T. Ludwig, H. Schettler, H. Grabinski and E. Klink, 'Determination of frequency dependent transmission line parameters on product related on chip test line structures using S-parameter measurements', in *IEEE Topical Meeting on Electrical Performance of Electronic Packaging*, Princeton, NJ, October 2003, pp. 97–100.

[39] Q. Liang, J. D. Cressler, G. Niu, Y. Lu, G. Freeman, D. C. Ahlgren, R. M. Malladi, K. Newton and D. L. Harame, 'A simple four-port parasitic de-embedding methodology for high-frequency scattering parameter and noise characterization of SiGe HBTs', *IEEE Trans. Microwave Theory Tech.* **51**(11) (2003), pp. 2165–2174.

[40] J. P. Mondal and T.-H. Chen, 'Propagation constant determination in microwave fixture de-embedding procedure', *IEEE Trans. Microwave Theory Tech.* **36**(4) (1988), pp. 706–714.

[41] J. Kim and D. P. Neikirk, 'Experimental characterization of copper/low-k transmission line interconnects through microwave measurements', in *IEEE Topical Meeting on Electrical Performance of Electronic Packaging*, Princeton, NJ, October 2003, pp. 93–96.

[42] R. B. Marks and D. F. Williams, 'Characteristic impedance determination using propagation constant measurements', *IEEE Microwave Guided Wave Lett.* **1**(6) (1991), pp. 141–143.

[43] D. F. Williams and R. B. Marks, 'Transmission line capacitance measurement', *IEEE Microwave Guided Wave Lett.* **1**(9) (1991), pp. 243–245.

[44] D. F. Williams and R. B. Marks, 'Accurate transmission line characterization', *IEEE Microwave Guided Wave Lett.* **3**(8) (1993), pp. 247–249.

[45] T. Zwick, Y. Tretiakov, and D. Goren, 'On-chip SiGe transmission line measurements and model verification up to 110 GHz', in *IEEE Microwave and Wireless Components Letters* **15**(2) (2005), pp. 65–67.

[46] D. F. Williams, 'Accurate characteristic impedance measurement on silicon', in *IEEE MTT-S International Microwave Symposium*, Baltimore, MD, June 1998, pp. 1917–1920.

[47] M. D. Janezic, D. F. Williams, V. Blaschke, A. Karamcheti, and C. S. Chang, 'Permittivity characterization of low-k thin films from transmission-line measurements', *IEEE Trans. Microwave Theory Tech.* **51**(1) (2003), pp. 132–136.

[48] L. Proekt and A. C. Cangellaris, 'Investigation of the impact of conductor surface roughness on interconnect frequency-dependent ohmic loss', in *IEEE Electronic Components and Technology Conference*, Princeton, NJ, November 2003, pp. 1004–1010.

[49] T. Zwick, D. Liu, and B. Gaucher, 'Broadband planar superstrate antenna for integrated millimeter wave transceivers', in *IEEE Trans. Antennas and Propagation* **54**(10) (2006), pp. 2790–2796.

[50] T. Zwick, D. Liu and B. Gaucher, 'Broadband planar millimeter wave dipole with flip-chip interconnect', in *IEEE International Symposium on Antennas and Propagation*, Honolulu, HI, June 2007, pp. 5047–5050.

[51] T. Zwick, C. Baks, U. R. Pfeiffer, D. Liu and B. P. Gaucher, 'Probe based MMW antenna measurement setup', in *IEEE International Symposium on Antennas and Propagation*, Monterey, CA, June 2004, vol. 1, pp. 747–750.

[52] U. Pfeiffer and C. Schuster, 'A recursive un-termination method for non-destructive in-situ s-parameter measurement of hermetically encapsulated packages', *IEEE Trans. Microw. Theory and Tech.* **53** (2005), pp. 1845–1855.

[53] U. Pfeiffer and B. Welch, 'Equivalent circuit model extraction of flip-chip ball interconnects based on direct probing techniques', *IEEE Microw. Wireless Compon. Lett.* **15** (2005), pp. 594–596.

[54] U. Pfeiffer and A. Chandrasekhar, 'Characterization of flip-chip interconnects up to millimeter wave frequencies based on a non-destructive in-situ approach', *IEEE Trans. Advd. Packag.* **28** (2005), pp. 160–167.

[55] R. F. Bauer and P. Penfield, 'De-embedding and unterminating', *IEEE Trans. Microw. Theory and Tech.* **22** (1974), pp. 282–288.

[56] K. Phillips and D. Williams, 'MMIC package characterization with active loads', in *Digest of the 36th ARFTG Conference*, pp. 64–72, Dallas, TX, November 1990.

[57] H. Lu and T. Chu, 'Port reduction method for scattering matrix measurement of an n-port network', *IEEE Trans. Microw. Theory and Tech.* **48** (2000), pp. 959–968.

[58] H. Lu and T. Chu, 'Multiport scattering matrix measurement using a reduced-port network analyzer', *IEEE Trans. Microw. Theory and Tech.* **51** (2003), pp. 1525–1533.

[59] A. Corporation, 'Agilent user characterization: New electronic calibration feature allows users to customize to specific needs'. White Paper, 2003.

[60] V. Adamian, 'A novel procedure for network analyzer calibration and verification', in *Digest of the 41st ARFTG Conference*, pp. 8–17, Atlanta, GA, 1993.

[61] V. Adamian, 'Electronic calibration of a vector network analyzer for non-insertable devices', in *Digest of the 43rd ARFTG Conference*, pp. 1–10, Boulder, CO, 1994.

[62] L. Martens, *High-Frequency Characterisation of Electronic Packaging*, 2nd edn (Kluwer Academic Publishers, 1998).

[63] E. V. Lil, 'A rigorous technique for measuring the scattering matrix of a multiport device with a two-port network analyzer', *IEEE Trans. Microw. Theory and Tech.* **33** (1985), pp. 286–287.

[64] J. C. Tippet and R. A. Speciale, 'A rigorous technique for measuring the scattering matrix of a multiport device with a 2-port network analyzer', *IEEE Trans. Microw. Theory and Tech.* **82** (1982), pp. 661–666.

[65] W. Lin and C. Ruan, 'Measurement and calibration of a universal six-port network analyzer', *IEEE Trans. Microw. Theory and Tech.* **37** (1989), pp. 734–742.

[66] W. H. Press, S. A. Teukolsky, W. T. Vetterling and B. P. Flannery, *Numerical Recipes in C*, 2nd edn (Cambridge University Press, New York, 1996), ch. 2.6.

[67] U. R. Pfeiffer and C. Schuster, 'Non-destructive S-parameter measurement of a hermetically encapsulated package with comparison to high-frequency simulation', in *Proceedings of the IEEE Topical Meeting on Electrical Performance of Electronic Packaging*, pp. 323–326, California, USA, October 2002.

[68] U. R. Pfeiffer and A. Chandrasekhar, 'Statistical analysis and modelling of low-cost leadless packages for wireless applications based on non-destructive measurements', in *Proceedings of the IEEE Topical Meeting on Electrical Performance of Electronic Packaging*, pp. 79–82, October 2003.

[69] C. Schuster, Y. Kwark, R. Frech, E. Klink, J. Diepenbrock, G. Edlund, T. Gneiting and R. Modinger, 'Issues and challenges of Gbps backplane connector characterization', *Proceedings of the 8th IEEE Workshop on Signal Propagation on Interconnects, 2004*, pp. 63–66, Heidelberg, Germany, May 9–12, 2004.

[70] A. Jentzsch and W. Heinrich, 'Theory and measurements of flip-chip interconnects for frequencies up to 100 GHz', *IEEE Trans. Microw. Theory and Tech.* **49** (2001), pp. 871–878.

[71] W. Heinrich, A. Jentzsch and G. Baumann, 'Millimeter-wave characteristics of flip-chip interconnects for multichip modules', *IEEE Trans. Microw. Theory and Tech.* **46** (1998), pp. 2264–2268.

[72] F. Zhiping, Z. Wemge, S. Bingzhi, K. C. Gupta and Y. C. Lee, 'RF and mechanical characterization of flip-chip interconnects in CPW circuits with underfill', *IEEE Trans. Microw. Theory and Tech.* **46** (1998), pp. 2269–2275.

[73] T. Hirose, K. Makiyama, K. Ono, T. M. Shimura, S. Aoki, Y. Ohashi, S. Yokokawa and Y. Watanabe, 'A flip-chip MMIC design with coplanar waveguide transmission line in the W-band', *IEEE Trans. Microw. Theory and Tech.* **46** (1998), pp. 2276–2282.

[74] T. Tokumitsu, 'An over-110-GHz InP HEMT flip-chip distributed baseband amplifier with inverted microstrip line structure for optical transmission system', *IEEE J. Solid-State Circuits*, **38** (2003), pp. 1479–1484.

[75] N. J. Damaskos and B. J. Kelsall, 'Cavity techniques for substrate properties at microwave/millimeter-wave bands', *Microwave Journal* **46** (2003), pp. 112–124.

[76] U. R. Pfeiffer, A. Chandrasekhar and T. Zwick, 'A new in-situ approach to flip-chip interconnect characterization up to millimeter wave frequencies', in *Proceedings of the IEEE Topical Meeting on Electrical Performance of Electronic Packaging*, May 2004.

18

Silicon-based Packaging and Silicon Micromachining

Cornelia K. Tsang, Paul S. Andry and Michelle L. Steen

18.1 Introduction to mmWave Packaging

18.1.1 Review Existing Packaging Technology

All electronic packages perform a variety of basic but vital electronic and physical functions. They house and mechanically protect active electronic integrated circuits (ICs), hermetically sealing them off from moisture, airborne contaminants, corrosive environments, electromagnetic radiation and physical contact of any kind. Moreover, they perform all these functions while reliably supplying electrical signals, power and ground to active ICs contained within, fanning those connections out from a relatively small on-chip pitch to the much larger contact structure existing at board level. Increasingly, as is the case with multi-chip-modules (MCMs) and systems-in-package/systems-on-package (SiPs/SOPs), more and more functionality has migrated from the board to the first level package; including a mix of analog and digital ICs, or flash memory and DRAM/SRAM, along with a variety of passives, all within a single package. The move to higher levels of integration and miniaturization of packaged components in the RF space has been driven primarily by the huge market for wireless communication devices. As the frequency of operation for a given application increases up into the GHz range, the package design, physical size, materials, construction and integrated components all have to keep pace with the active IC in order that a specified level of performance be attained. A simple example would be the decision to remap a wire-bonded IC to a peripheral solder grid array to limit the parasitic inductive effects of the wire bonds. As the frequencies increase up into the millimeter-wave (mmWave) regime, every element of the package has to be effectively 'tuned'. Characteristic length, spacing and thickness of wiring

Advanced Millimeter-wave Technologies: Antennas, Packaging and Circuits
Duixian Liu, Ulrich Pfeiffer, Janusz Grzyb and Brian Gaucher © 2009 John Wiley & Sons, Ltd

traces have to be reproduced with high fidelity, dielectric thicknesses must be maintained, and every possible discontinuity must be taken into account, whether it is a metallurgical connection from a stripline to the monolithic microwave integrated circuit (MMIC) or a section of metal trace over which a bead of encapsulating material runs to join a hermetic lid.

As discussed in Chapter 4, typical interconnect technologies used for mmWave applications include MCM-L (laminated), low-temperature co-fired ceramic (LTCC) and MCM-D (deposited) package types. Traditional MCM-L technology derived from printed wiring board (PWB) fabrication techniques using laminated metal foils suffer from poor dimensional and electrical stability of the materials as a function of temperature and humidity, and 25 μm line and via sizes are considered fairly aggressive. LTCC packages are also limited in the dimensional tolerance they can achieve since the ceramic sheets shrink when fired, and minimum linewidths are fairly coarse, on the order of 100 μm. The MCM-D technology based on polyimide (PI) or benzocyclobutene (BCB) build-up layers have a number of advantages over the first two types. The spin-on layers can be from tens of micrometers thick down to a few micrometers thick. Photo-patternable materials allow precise definition of features down to a few micrometers with excellent fidelity. Further, the lower dielectric constants of the PI or BCB over dielectrics used in LTCC give larger physical dimensions for a given electrical length, thus effectively tightening the tolerance that can be achieved. Regardless of which of these technologies is used, the requirement of adding components such as filters and an antenna further complicates the choice. This is because expensive customized solutions must be used to tune the off and on chip connections at mmWave frequencies, and to ensure reliability of the multi-component system regardless of the number of added metal connections, mismatch of material constants of the added components, or disparate encapsulation requirements of the chip and antenna.

18.1.2 Advantages and Limitations

Set against the standard backdrop of packaging technologies, silicon has a number of obvious advantages for integrated mmWave applications. While antennas for microwave frequencies are relatively large, mmWave antennas are smaller and it is not only possible but highly advantageous to integrate them 'on-chip'. As an example, a quarter wavelength on-chip is around 400 μm at 60 GHz. Silicon technology allows the patterning and etching of precise features at dimensions and tolerances which exceed those of standard packaging technologies by many orders of magnitude. Metal and dielectric film thicknesses can be similarly well-controlled. Precision micromachining techniques can be used to produce simple cavity structures right up to far more complex and intricate micro-electro-mechanical systems (MEMS) switches, filters and resonators [22]. With the advent of robust through-silicon via (TSV) technology and high-precision packaging concepts such as silicon carriers [2], it is now possible to produce highly integrated multi-component systems or sub-systems entirely out of silicon using fairly standard semiconductor processing equipment. Such all-silicon systems allow a common set of design tools to lay out the chip, the package, the antenna and even the passive devices, leading to higher performance and reliability at reduced cost. A few caveats of using silicon arise because, unlike most other starting substrate packaging materials, it is a semiconductor rather than an insulator. For this reason, even 'un-doped' bulk wafers cut directly from Czochralski ingots will contain enough impurities to be somewhat conductive. Highly purified float zone silicon wafers must be used to obtain the highest resistivity possible, and even here it is difficult to push the resistivity too far beyond ~10 000 Ω cm. A fixed dielectric constant of 11.9 is also higher than most LTCC substrates, and substantially higher than laminate core materials. Nevertheless, undesirable inductive losses and capacitive parasitics can be somewhat controlled by keeping the total package thickness very small. Wafer thinning has already become fairly commonplace for producing low-profile chip stacks, and its importance is only expected to increase as three-dimensional applications using TSVs begin to appear.

18.2 Introduction to Silicon-based Packaging

18.2.1 Key Silicon-based Packaging Technology Elements and Application Examples

Silicon-based packaging, and in particular silicon carrier technology, is ideally suited as a platform for dense, low-profile, high-frequency applications. The key technology elements of the silicon carrier which allow for these include fine-pitch interconnects, fine-pitch wiring and TSVs [2, 3]. Each of these elements have been fabricated and characterized, and their applicability in the high-frequency regime is expected to be due largely to the fact that undesirable electrical parasitics associated with much larger interconnects and/or much taller vias of standard packages are vastly reduced in the corresponding silicon carrier technology.

Fine-pitch interconnects used to join die or other surface-mounted elements to the silicon carrier are also known as micro-joins. In an extension of standard packaging terminology for solder bump area arrays, fine-pitched versions of these are called micro-bumps. The salient feature of a micro-bump array is that the bumps are of significantly finer pitch than the industry standard first-level solder C4 array pitch today, which is typically 200 μm. They are also much smaller in diameter, height and hence total volume, than a typical C4 solder ball. Area array micro-bumps with nominal diameters of 50 μm and 25 μm on pitches of 100 μm and 50 μm, respectively, have been fabricated and electrically characterized [4, 5]. Contact resistances equaling 14 mΩ and 17 mΩ were obtained for 50 μm and 25 μm micro-bumps, respectively, which is only around three times the value for a typical 100 μm diameter C4 bump. This has dramatic implications for power delivery for all-silicon three-dimensional systems because the density of interconnects grows as the inverse square of the pitch, while the resistance per bump clearly does not increase at a commensurate rate. The net result is a much lower total parallel resistance of the power delivery grid from silicon carrier to the active die. For the purpose of high-speed signaling, the slight increase in resistance of the individual microjoin connection is far outweighed by the decrease of the pad capacitance which does in fact scale with pad area. Micro-bumps with 50 μm diameter were characterized by time and frequency domain measurements up to 40 GHz using transmission line test macros [6] and the net result was 6 ps of additional delay and less than 0.5 dB of transmission loss per bump. The delay and the losses are expected to be further reduced as the micro-bump diameter and its respective pad is scaled down to smaller dimensions.

The wiring on the silicon carrier can help to bridge the large gap that exists between standard complementary metal–oxide–semiconductor (CMOS) copper back-end-of-line (BEOL) wiring ground-rules and those for organic and ceramic packages. In general, the wiring levels on the silicon carrier are similar to the uppermost on-chip BEOL levels, the main difference being that silicon carrier wires can be built to be two to three times thicker than BEOL wires. As a result, silicon carrier wires can have lower transmission and power losses than most silicon ICs, although usually not as low as similar length nets on standard packages. Nevertheless, high-fidelity signaling and low crosstalk can be achieved by careful shielding of the signal lines in-plane and/or out-of-plane through various signaling schemes.The high-speed signal performance of silicon carrier wiring has been investigated using multiple test sites [2–7]. Hundreds of test macros were designed to cover a wide range of operating frequencies, net lengths and signaling schemes, with measurements being made in both the time and frequency domain. One such test site explored high-speed differential microstrip lines designed to operate in the 10–20 GHz frequency range with less than 5 dB/cm transmission losses [7]. Three levels of wiring (one each for power, ground and signal) were used to fabricate differential microstrip transmission lines having line width of ~4.5 μm, space of 15 μm and a channel-to-channel spacing of approximately 100 μm. As shown in Figure 18.1, signal attenuation on these silicon carrier lines was measured to be ~4.3 dB/cm at 20 GHz and relatively wide eye openings were measured at a data rate of 20 Gbps.

TSVs bring the power and signals from the 'outside' world and distribute them to surface-mounted die by way of pads and wiring on the silicon carrier. The inductance and the current-carrying capacity of a through-via are important parameters that aid in efficient power distribution. Low inductance is highly desirable to minimize LdI/dt voltage drops in power delivery networks for high-performance application specific integrated circuits (ASICs) or microprocessors. Moreover, because silicon carriers are intended to be platforms for multiple active components, TSVs are also required to supply significant currents. A number of TSV process flows have been developed, with structures ranging from simple cylinders, to annular shapes as well as multi-bar or slotted vias [8–10]. Annular tungsten TSVs have been shown to be particularly robust and well-suited for silicon carrier applications requiring low resistance, with typical DC values ranging from about 10 to 20 mΩ, with no increase in resistance or yield loss after 1000+ deep thermal cycles [8]. Preliminary measurements on current-carrying capacity indicate that 0.8 to 1.0 A can be passed through these structures, and they are expected to be at least as reliable as standard solder C4 connections, if not more so [10]. Depending on the type of application, the width of the annulus ranges from about 2 to 8 μm whereas its height can range from less than 50 to over 150 μm. Because they are designed to mate with typical first-level packages, the annular diameter is usually between 80 μm and 50 μm, while the pitch is typically 200 μm. Cylindrical vias of similar diameter can be etched to even greater depths, and hence thicker silicon carriers are possible; however, practical metal filling of such large structures can be quite challenging.

For high-frequency applications, TSV inductance should be minimized as much as possible. Measurement of inductance for two different heights of silicon carrier TSV [11] yielded a per unit length inductance of approximately 0.15 pH/μm for Cu-lined, composite-filled cylindrical TSVs – at present the smallest reported value for through-vias in silicon. This value was observed to scale linearly with the height of the through-vias, yielding 41 pH for a 185 μm-tall TSV and 63 pH for a 271 μm-tall TSV. To ascertain the relative delay incurred as a result of sending a signal through a TSV, time delay transmission (TDT) measurements were made on sample macros with and without through-vias. For the case of tungsten-filled annular TSVs of ~60 μm height, the TDT results for two separate macros revealed that the presence of the TSV introduced a 3 dB delay of 0.75 ps.

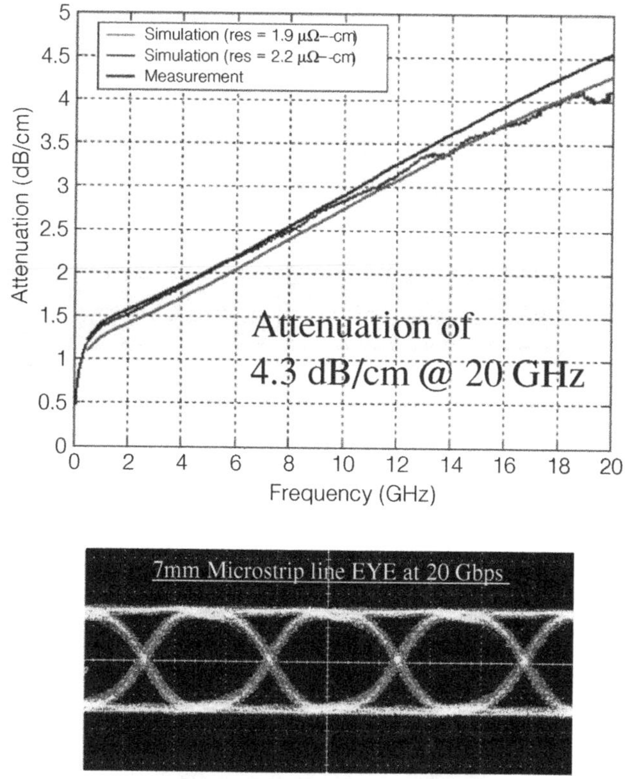

Figure 18.1: Measured and simulated attenuation per unit length for differential microstrips and corresponding eye diagram (from reference [7], reproduced by permission of © 2005 IEEE).

The silicon carrier's fine-pitched interconnects, high-speed transmission line wiring, and low inductance TSVs have been effectively combined with precision silicon micromachining to enable high-frequency applications such as the parallel optical transceiver shown in Figure 18.2 [6]. The components of the parallel optical transceiver include an 'optochip' silicon carrier and an 'optocard'. The 'optochip' consisted of a laser diode driver (LDD) or a transimpedance amplifier (TIA) IC, a vertical cavity surface emitting laser (VCSEL) or photodiode (PD) optoelectronic (OE) integrated using a 300 μm thick silicon carrier which included a rectangular through-cavity as shown in Figure 18.3. The 'optocard' was equipped with integrated parallel optical waveguides that optically connect the transmitter and receiver modules on an organic 'surface laminar circuit' card (SLC). Designed to support Tb/s bandwidth for future high-performance computing systems, the results demonstrated that each component could be effectively fabricated and assembled, while characterization showed that the silicon carrier 'optochip' could easily support the required electrical transmission data rates of 20 Gbps per channel with an acceptable attenuation.

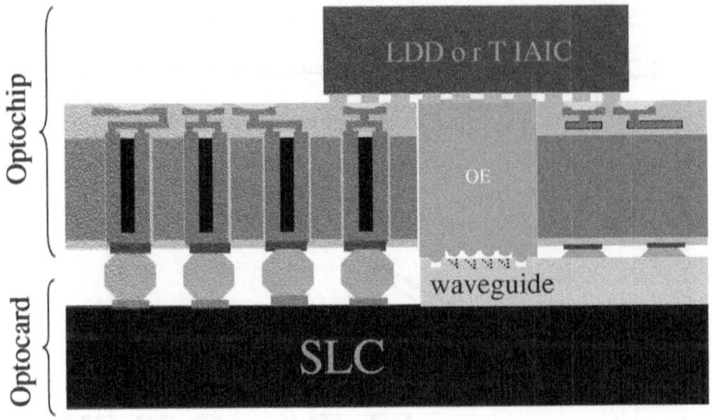

Figure 18.2: Parallel optical transceiver employing a silicon carrier comprising of high-speed transmission lines, tsvs, microjoins and a cavity containing an OE device (from reference [6], reproduced by permission of © 2005 IEEE).

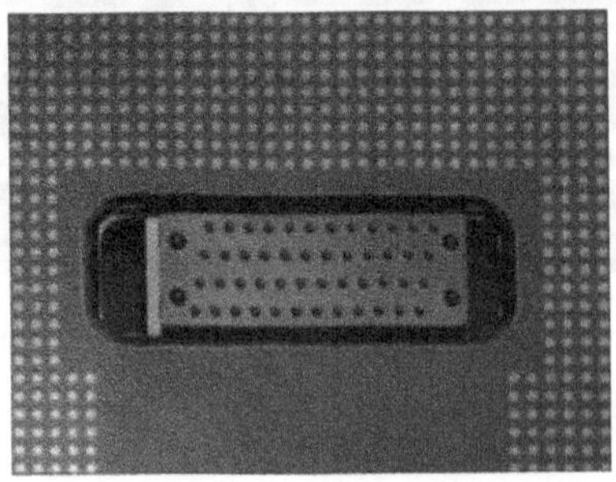

Figure 18.3: Bottom view of 'optochip' silicon carrier showing OE in cavity and under-bump metallurgy (UBM) pads (from reference [6], reproduced by permission of © 2005 IEEE).

18.3 Silicon-based Packaging: Process Options

18.3.1 Introduction to Semiconductor Processing

As previously mentioned, the processes that are useful for silicon-based packaging are primarily those found in the BEOL of the integrated circuit fabrication flow. This designation encompasses all processes that involve building of metal wiring lines, especially copper damascene lines, which have become the standard in semiconductor device fabrication. After the transistors have been formed in the 'front-end-of-line' (FEOL), a series of process steps as shown in Figure 18.4 are repeated until all metal wiring levels have been completed. Lithography, copper electroplating and chemical-mechanical planarization are specific processes that can be adapted to manufacture silicon-based packages.

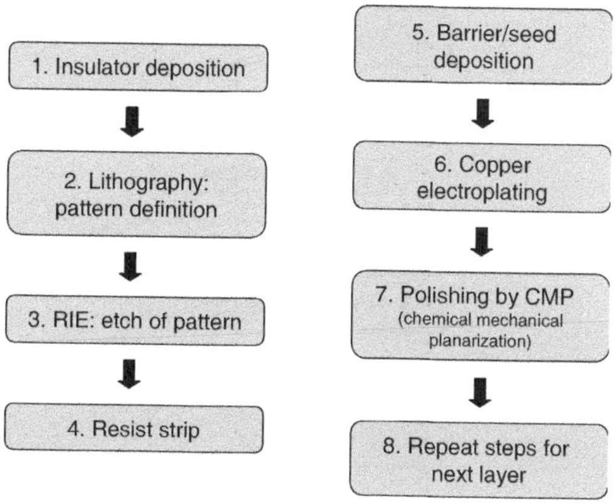

Figure 18.4: Flowchart showing typical back-end-of-line Cu damascene processes.

18.3.2 Lithography

18.3.2.1 Spin-on Photoresist

Standard lithography consists of spin-coating wafers with photoresist in a range of thicknesses, for silicon packaging this could span from 1 to 10 μm or beyond. The film is then soft-baked (temperature and time-dependent on specific material) to drive off solvents and exposed with ultraviolet light, mostly in the mid-ultraviolet (MUV) range, 356 nm and 436 nm wavelengths. Post-exposure bake takes place at resist vendor determined conditions and then the resist is developed and potentially hard-baked. The hard-bake step is especially important with thicker resists as this allows more stability in the film during subsequent etch processes. A schematic of the lithography process is shown in Figure 18.5.

With standard BEOL metal wiring, the final step of each wiring level is a planarization step which results in very small topographical variations. This is often not the case in silicon packaging fabrication where large cavities, with sloped walls possibly up to hundreds of micrometers deep, or via holes may be etched first in order to facilitate a certain process integration flow. These large cavities or vias would result in significant topographical changes and breaks in the wafer surface. If spin-on photoresists were used with standard process parameters, a non-uniform coating with ‘holes’, ‘comets’ or ‘rings’ could form in the film, see Figure 18.6, along with a tendency for the resist to pull from the edges of the three-dimensional features in the silicon as shown in Figure 18.7. Sometimes this can be overcome. For example, a higher viscosity resist could be used with a slower spin speed, therefore allowing the resist to flow over topographical changes in the wafer and still maintain adhesion to the surface. This is dependent upon topography on the wafer that is not greater than the thickness of the resist coating. If the surface topography is greater, another method would be to dispense the resist onto the entire surface of the wafer and allow the resist to flow into the features before slowly spinning the excess off [12]. This is followed with a high-speed spin to dry the resist and stop further flowing of the material. Such a method has been shown to give films with 10–20% uniformity and is useful when the desired pattern to be exposed is at the bottom of a large cavity [12]. Since spin-on resist application is a standard

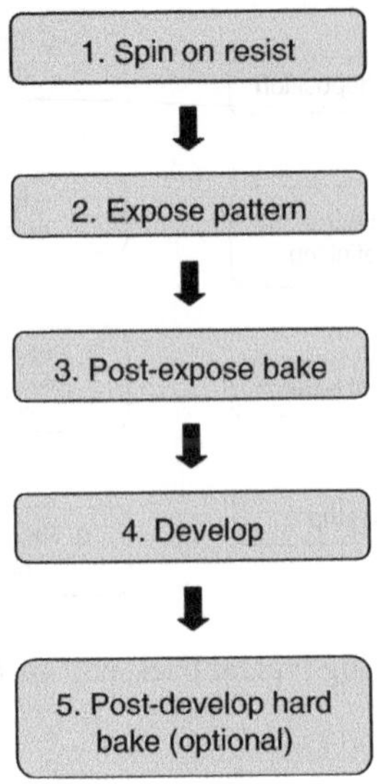

Figure 18.5: Flowchart depicting the typical steps in a photolithography process.

semiconductor process, the tool sets and materials needed are all readily available to users. Another advantage in choosing spin-on resist is that there are few process parameters that can be modified and thus optimization is easier. However, if one considers the process method mentioned above, it is easily observed that much excess material is used in order to achieve good film coverage over large step changes. If the desire for less cost and materials waste is important, then other lithography methods may be considered. Lastly, there will be features that simply cannot be coated due to extreme step heights, large cavities, or high densities and therefore other coating techniques must be investigated.

18.3.2.2 Spray-coat Photoresist

The spray coat method of photoresist application was designed specifically to resolve difficulties in applying films to topographically challenging wafer surfaces. In contrast to the spin method, which is governed mainly by flow dynamics, spray coating distributes an even thickness of individual droplets of resist onto the wafer from an ultrasonic nozzle above the surface of the wafer. This allows the sprayed resist to coat surfaces with a variety of three-dimensional structures and surfaces.

There are four main parameters that control the quality of the film coating: (1) viscosity of the resist, (2) angle of the spray nozzle, (3) scanning speed of the nozzle, and (4) rotation/position of the wafer. Since these resist droplets are microscopic, it is very important to control the solids content of the resist. Some vendors recommend droplet sizes on the order of 20 μm and viscosity below 20 centistokes (cSt) [14]. Compatible solvents, such as propylenglycol methyl ethyl acetate (PGMEA) or methyl ethyl ketone (MEK), are generally used to dilute commercially available photoresists to the right consistency. The photoresist is then propelled through the nozzle with either nitrogen or very clean dry air. There are slight

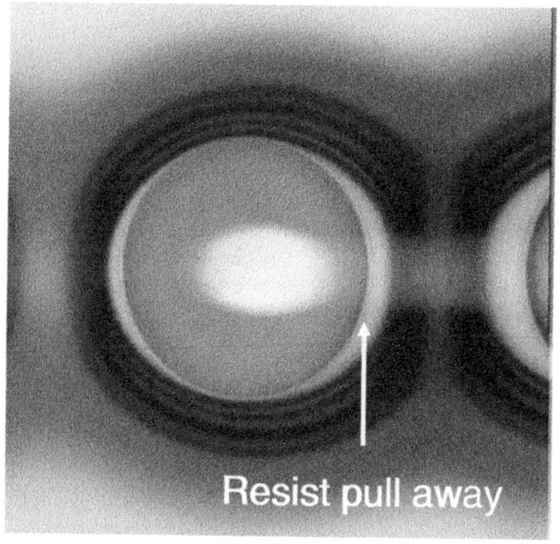

Figure 18.6: Optical image showing spin-on resist that has pulled away from a topographical feature.

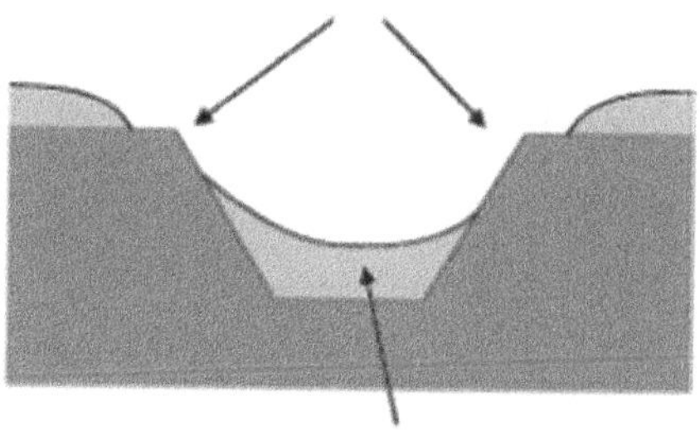

Figure 18.7: Schematic of resist with insufficient coating uniformity (from reference [13], reproduced by permission from K. A. Cooper, *et al.*, 'Conformal photoresist coatings for high aspect ratio features', *Proceedings of 2007 International Wafer-level Packaging Conference*, San Jose, CA, June 2007).

differences in the tool configurations of spray coaters, but each has the goal of uniformly distributing the spray coating over the surface of the wafer. Two spray system layouts are shown where Figure 18.8 depicts a swivel arm spray system and Figure 18.9 shows a spray system mounted on an x–y stage [13, 14].

In Figure 18.8, the swivel arm system is shown to have a nozzle that oscillates at a frequency of 100 Hz back and forth over the surface of a wafer that is slowly rotating at speeds of 50–100 rpm, which will reduce the centrifugal forces on the outer edge of the wafer [14]. The speed of the oscillation is carefully controlled so that when the nozzle is at

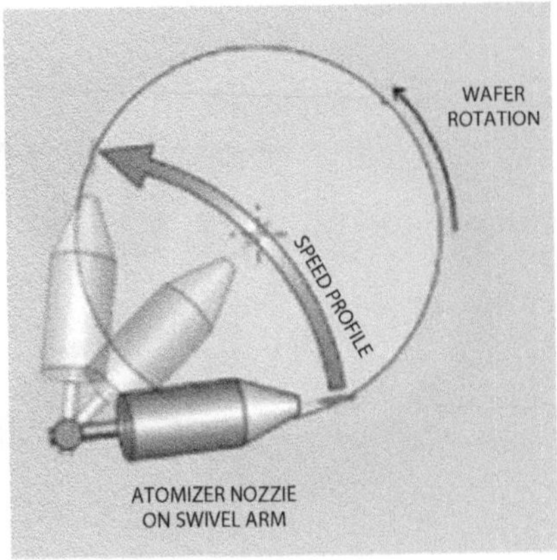

Figure 18.8: Schematic of a swivel arm spray resist tool (reproduced by permission of © EV Group E. Thallner GmbH).

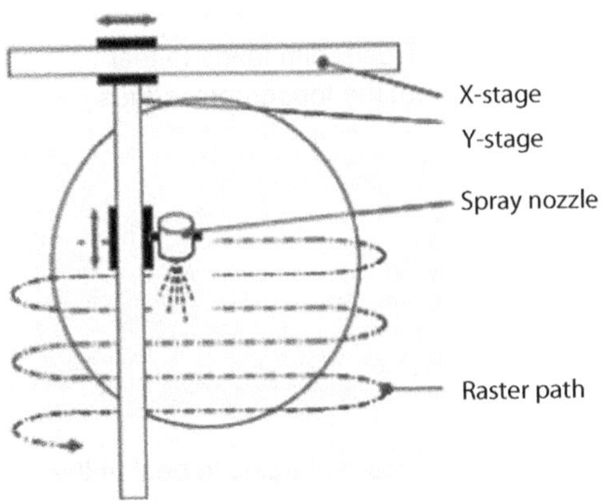

Figure 18.9: Schematic of a spray resist system with x–y stage (from reference [13], reproduced by permission from K. A. Cooper, *et al.*, 'Conformal photoresist coatings for high aspect ratio features', *Proceedings of 2007 International Wafer-level Packaging Conference*, San Jose, CA, June 2007).

the center of the wafer, with the smallest surface area requiring coating, the film thickness deposited is not greater than on the outer edges of the wafer. This is achieved by increasing the velocity of the nozzle sweep at the center of the wafer. In contrast, Figure 18.9 shows a spray nozzle mounted onto an arm on an x–y stage. The nozzle is moved from one end of the wafer to the other according to a programmed slew rate, following the raster path. After a single pass over the entire wafer, the wafer is rotated at a predetermined angle and the process is repeated until the desired thickness of resist has been achieved [13]. Sample coatings over v-grooves are shown in Figure 18.10. In both tool configurations, multiple types of photoresist have been deposited at single digit film uniformities. In addition to the

Figure 18.10: Scanning electron microscope (SEM) image from a spray-coated 150 μm deep v-groove performed on an EVG 150 Series Coating System (reproduced by permission of © EV Group E. Thallner GmbH).

capability of coating three-dimensional structures that spin-on resist systems are not able to accommodate, there are also other advantages to the spray system. A further advantage of spray resist is that silicon need not be the only substrate used; other materials such as glass, gallium arsenide, metals, etc., can also be coated [13, 14]. Finally, a key advantage is the reduction in materials waste. When compared to comparable resist thickness coatings, the spray system has been calculated to save 70% more material [14], with the primary reduction occurring with the elimination of the spin-off step. However, there are still some drawbacks to the spray-coat system. When a wafer surface includes cavities of varying sizes, the amount of resist coated inside the cavity and especially at the bottom of the cavity will vary. This may present a challenge to selecting exposure dosages that will clear thinner resist on some surfaces versus the thicker resist at the bottom of the well. Furthermore, over sharp, angled corners, the resist thickness will still be thinner compared to the flat surface of the wafer. Therefore, while spray-coat resist is able to overcome resist pullback from edges (as shown in Figure 18.11), it may not provide enough resist coverage for some applications, depending on the subsequent process steps post-lithography.

18.3.2.3 Electrodeposited Photoresist

Electrodeposited photoresist results in a very uniform film as one would expect from an electroplated material, hence the reason why it is particularly useful for addressing the challenges of three-dimensional structures. Aqueous baths that contain resist emulsion with high solids content, 10% with particles around 100 nm in size [15], can be commercially purchased in a negative or positive form. The wafer must have a conductive seed layer over all areas that should be coated. The wafer is then placed into the solution, contained within special plating equipment, and a voltage is applied. The charged resist particles are attracted to the surface of the conductive substrate. The entire process only lasts for a few seconds

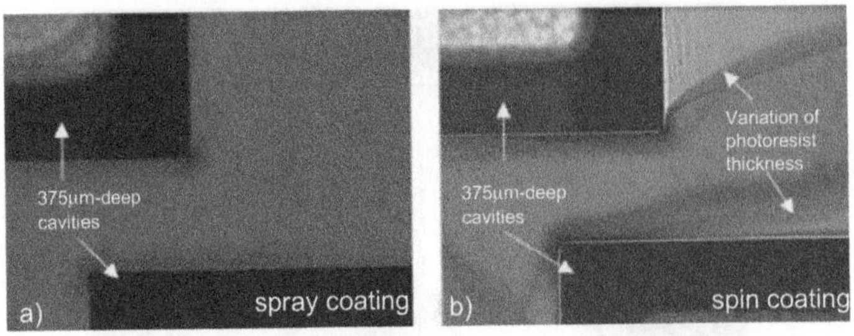

Figure 18.11: SEM images showing the comparison between coverage of spray and spin resist systems around corners (from reference [12], reproduced by permission of © 2004 IEEE).

and is self-limiting because the resist layer that builds up is insulating. When the current drops to zero, the process is completed. The thickness of the resist deposited is controlled by the following parameters: the bath composition, temperature and voltage applied. It is very important in this process to maintain a bath and wafer surface that is free of gas bubbles as this would result in defective resist coatings.

Electrodeposited resist is especially good at coating corners at the bottom of a large cavity and at exposing patterns that run over surfaces with a sloped path. However, after a post-expose bake is conducted, there is the potential for resist pullback from the top surfaces surrounding a large cavity. If this occurs, then it is necessary to deposit a thicker resist in order to overcome the surface pullback of the resist. Another consideration when using electrodeposited resist is the necessity of the conductive seed layer for plating. This must be evaluated early on in the process integration scheme so as to ensure that the seed layer can be deposited and successfully removed without causing any problems with the rest of the process steps. Furthermore, since this process involves maintenance of an electrochemical bath, it can be more expensive and complicated compared to the spray and spin methods of resist coating. However, the benefits of uniform coatings over difficult topographies may be worth the effort.

18.3.2.4 Dry Film Resist

Finally, one other method of resist coating that could be useful to silicon-based packaging is laminated or dry-film photoresist. The method of using laminated resists is frequently utilized in semiconductor processing to form patterns for large features such as solder bumps and is also used in printed circuit board (PCB) fabrication technologies, for which it was originally introduced by Dupont in 1970 [16]. Various types of dry-film are commercially available under names such as Dupont Riston, Sunfort DRF, Eterec, Ordyl, and others.

Dry-film photoresist is a very viscous liquid which appears 'dry' and is sandwiched between a cover film and a carrier film, frequently polyolefin or polyester. The material is rolled onto a support core for easy handling. Before application of the resist, the cover film is removed. The wafer is laminated by insertion into a hot roll laminator tool which applies heat and pressure, resulting in conformal resist application. The resist can then be patterned under UV exposure and finally the carrier or support film is removed.

The advantages of dry film resist include excellent thickness uniformity, good conformality and adhesion to surface, ease of handling and potentially lower costs and process times. It is useful in covering over topographical changes on the wafer surface which would traditionally result in resist film pullback or thickness variability. However, due to the air that could be trapped in a cavity between the resist and the wafer surface, care must be taken in considering whether or not vacuum conditions will be encountered.

In summary, four different methods of lithography have been briefly introduced and the advantages and disadvantages of each have been evaluated. These methods enable fabrication of unique and challenging topographies found in silicon-based packaging processes.

18.3.3 Silicon Micromachining

This section describes the removal of bulk silicon to create cavities, vias, trenches and other silicon structures. This entails etching silicon selective to a masking layer such as photoresist or silicon dioxide that has been patterned to define the desired features. The patterning method, known as photolithography, was described above. The type of masking layer used is dependent on the silicon etching method. There are two basic categories for etching processes: (1) wet etching; and (2) dry etching. Wet and dry etching methods may be used in combination or exclusively, depending on the size, depth and desired sidewall morphology of the silicon structure. Etching methods are generally categorized as: (1) isotropic, etching equally in all directions; and (2) anisotropic, etching preferentially in one direction. Etching methods are selected based on the process requirements of the application.

18.3.3.1 Wet Etching

Isotropic etching. Many materials such as silicon (including polycrystalline silicon), silicon dioxide, silicon nitride, aluminum and gold can be isotropically etched [17]. For silicon etching, the most common chemistry used is nitric acid (HNO_3) mixed with hydrofluoric acid (HF) and diluted with water or acetic acid, commonly referred to as the HNA etch system [18]. Isotropic etching can be done at room temperature and proceed at faster rates than anisotropic etching. Silicon can be etched selectively to photoresist [18]. Hard mask materials such as silicon dioxide and silicon nitride can be used when selectivity to the masking material is a process requirement. Figure 18.12 shows the isotropic etch rate of silicon with both water and acetic acid diluents. The variation of etch concentrations and agitation affect the shape of the etched feature. Agitation tends to form more hemispherical shapes [18]. By definition, isotropic etching does not have the lateral dimensional control to form cavities and vias desired for silicon-based packages. However, isotropic etching can be used in combination with anisotropic methods (both wet and dry etching) to change the angle of the sidewall. Specifically, the tops of straight, cylindrical vias or re-entrant vias ($< 90°$ sidewall) etched in silicon can be flared out to produce a wine-glass shape, see Figure 18.13. This sidewall morphology is useful in certain applications (e.g. ease of fill) and the isotropic etch can be done as a large batch process. Furthermore, sharp edges can be rounded or smoothed out with an isotropic etch.

Anisotropic etching. Anisotropic etching for silicon bulk micro-machining was first investigated by Bell Telephone Laboratories in the mid-1960s. The most widely known

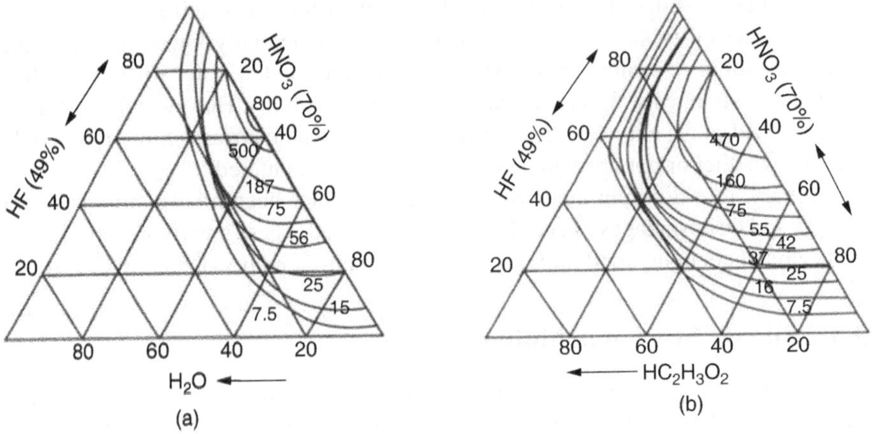

Figure 18.12: Silicon isotropic etch rate in μm/min: (a) consists of a $HF{-}HNO_3{-}H_2O$ system, and (b) consists of $HF{-}HNO_3{-}CH_3COOH$.

Figure 18.13: SEM image a flared via defined through isotropic etch.

etchant is potassium hydroxide (KOH) mixed with water, and sometimes with the addition of alcohol. Anisotropy is achieved by preferential etching of silicon in the (110) and (100) crystal plane directions compared to (111). More specifically, the (110) plane etches faster than the (111) plane at a rate of 400:1 and the ratio of the respective etch rates of the (100) and (111) planes is 200:1. Anisotropic etching in KOH can produce a variety of sidewall morphologies depending on the crystalline orientation of the bulk silicon wafer, the mask pattern, and the rotation of the wafer during exposure. For a (100) silicon substrate, etch of the (111) plane produces a sidewall angle of 54.74°. Representative drawings of vias etched by anisotropic and isotropic methods are shown in Figures 18.14–18.16.

Anisotropic etching can be achieved by a variety of alkaline chemistries including sodium hydroxide cesium hydroxide and ammonium hydroxide. Alkaline organic etches include ethylenediamine pyrocatechol (EDP) and tetramethylammonium hydroxide (TMAH). For EDP and TMAH etch, hard mask materials such as silicon nitride, silicon dioxide and some metallic films (e.g. gold and chromium) are commonly used. Table 18.1 shows

Figure 18.14: Drawing showing an example of anisotropic etching (from reference [18], reproduced by permission of © 1998 Cambridge University Press).

Figure 18.15: Drawing showing an alternate example of anisotropic etching (from reference [18], reproduced by permission of © 1998 Cambridge University Press).

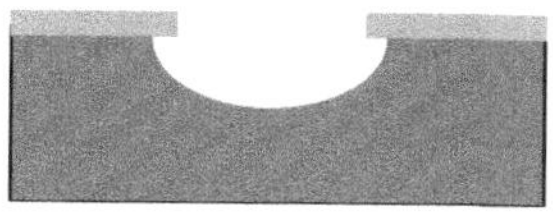

Figure 18.16: Drawing of isotropic etching (from reference [18], reproduced by permission of © 1998 Cambridge University Press).

the mask selectivity and etch rates for common preparations of KOH, EDP and TMAH. The process temperature can substantially influence the silicon etch rate as shown graphically in Figure 18.17.

Anisotropic etch is frequently used to create: (1) v-grooves trenches; (2) large cavities for accommodating MEMS devices; and (3) large vias in silicon interposers for chip-stacking. Due to the ability to process wafers in parallel, it is an attractive low-cost process compared to some other options for silicon micro-machining. Some applications using this process are pictured in Figure 18.18.

18.3.3.2 Reactive Ion Etching

Semiconductor processing relies heavily on plasma etching for patterning a wide range of materials. Plasma etching relies on source gases being broken down in a high-density plasma region to form highly reactive species, such as electrons, reactive neutrals, positive ions and photons. These species can simultaneously interact with a substrate to chemically and/or physically alter the material. For example, in plasma etching, radicals react with surface atoms to form stable compounds. Volatile byproducts are pumped away, thereby exposing a fresh, reactive surface to impinging plasma species. Simultaneously, other species in the plasma assist in the removal of etch products from the surface. Specifically, ions can have a major effect on the chemical reactions occurring at the plasma-surface interface and impart enough energy to physically remove material, ion-assisted etching.

Table 18.1: Sample of anisotropic etchants, etch rates and mask materials.

Etchant	Etch rate (100) (μm/min)	Etch rate ratio ratio (100)/(111)	Mask	Remarks
KOH, H_2O, 20%, 80°C	1.5	400	Si_3N_4 negligible, SiO_2 (46 Å/min)	IC incompatible, H_2 evolution
EDP, 115°C	0.75	100	Si_3N_4 (10 Å/min) SiO_2 (20 Å/min)	Carcinogenic, silicates may precipitate, ages in O_2
TMAH, H_2O 20%, 80°C	0.7	25	Si_3N_4 negligible, Si(100) etch selectivity to SiO_2 10^4:1	IC compatible, easy to handle, smooth surface

From references [19–21].

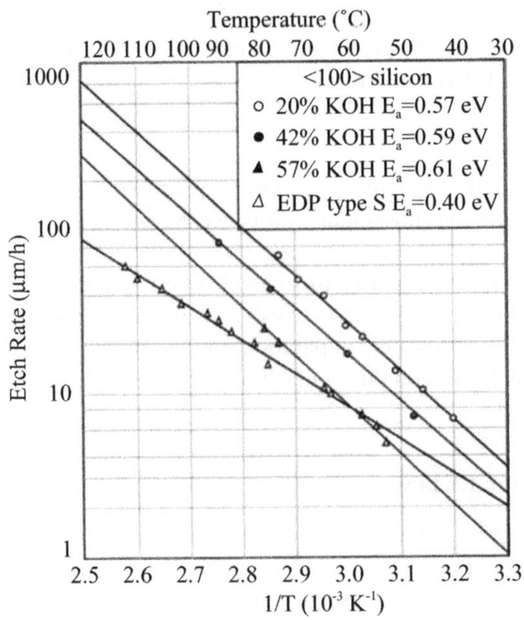

Figure 18.17: Arrhenius diagram of silicon etch rates in EDP and KOH (from reference [22], reproduced by permission of © 1990 The Electrochemical Society).

Reactive ion etching (RIE) is used routinely to fabricate high aspect ratio structures (HARS) in silicon. Figure 18.19 shows a SEM of highly anisotropic deep trench capacitors.

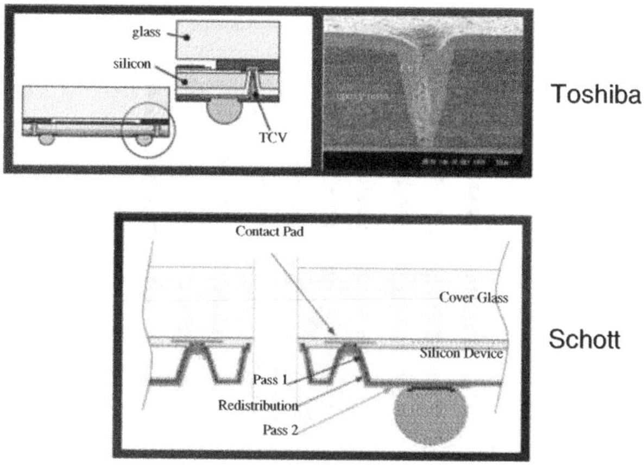

Figure 18.18: Examples of silicon-based packaging applications utilizing anisotropic etch. Toshiba – CMOS image sensor with TSV. Schott OPTO-WLP technology (from references [23, 24], reproduced by permission of © 2005, 2006 IEEE).

Anisotropy can be achieved by careful selection of the plasma parameters and gas precursors. For example, plasma parameters such source power, substrate bias, and system pressure can be tuned to increase the relative concentrations and energies of reactive species, thereby shifting the balance from isotropic chemical etching to predominantly anisotropic ion-assisted etching. In addition, selecting certain plasma precursors result in the formation of a protecting layer of material over the silicon substrate. This phenomenon known as passivation occurs *in situ* as a result of etch-competing processes: (1) recombination of radical species at the surface and (2) plasma deposition. For example, oxygen is often added as a plasma precursor to hydrogen bromide (HBr) plasmas used to etch silicon. Recombination at the silicon surface results in the formation of a low-volatility etch product (e.g. $Si_xBr_yO_z$). Alternately, plasma deposition occurs when precursors to the deposited film are generated by fragmentation of the parent gas and react at the surface to form a stable solid film. For example, octofluorocyclobutane (C_4F_8) can be broken apart in the plasma to generate deposition precursors such as CF_x radicals. These species polymerize in the plasma and on the surfaces of the wafer to form a stable fluorocarbon film. Thin films formed or deposited *in situ* by these methods act as sacrificial layers in plasma etching, preventing etch of the underlying material. The passivating layer is continually removed by ion bombardment and regenerated during plasma etching. This process occurs at a slower rate on the sidewall where the impact of line-of-sight species such as ions is much less, thereby slowing the progress of etch in the lateral direction (sidewall passivation). The balance between etch and passivation can be optimized by varying the gas precursors and plasma parameters (e.g. substrate bias) to achieve the desired etch rate, selectivity and sidewall morphology.

Deep silicon etching. Although RIE can achieve silicon HARS with aspect ratios as high as 80:1, it is a low-rate/throughput method and therefore confined to shallow structures typically on the order of 1–5 μm in depth. Depending on the application, deep silicon etching produces structures ranging from a few micrometers deep to through-wafer etching. In addition to high

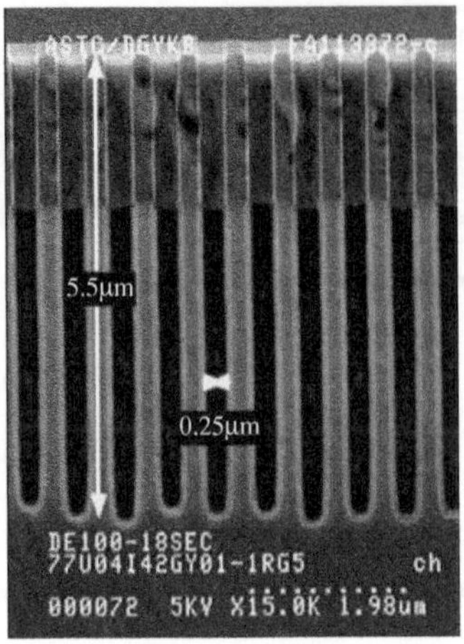

Figure 18.19: SEM of highly anisotropic deep trench capacitors.

throughput, many applications have added demands on profile shape and surface morphology. Supporting such applications hinges on rational control of sidewall passivation.

The most widely used method for deep silicon etching is time multiplexed deep etching (TMDE), patented by Robert Bosch Gmbh (US 5,501,892; March 26, 1996). Etching and passivating cycles are performed sequentially. Sidewall passivation is accomplished via plasma deposition at room temperature. Figure 18.20 shows a SEM of 250 μm deep vias in silicon. Silicon etch rates as high as 9 μm/min can be achieved with high SF_6 flow rates (750 sccm) and 6:1 SF_6:C_4F_8 cycle times. However, sidewall profile is compromised under these conditions by: (1) severe bowing (Figure 18.21); (2) significant aspect-ratio-dependent etching (ARDE) (Figure 18.22); (3) sidewall and bottom roughness (Figure 18.23); and (4) mask undercut and pronounced scalloping inherent to the TMDE process (Figure 18.24). Smooth, straight profiles with minimal mask undercut and scalloping can be achieved at the expense of etch rate (Figure 18.25). Some ARDE is inherent to the TMDE regardless of conditions as shown in Figure 18.26.

A second deep silicon etching method, known as cryogenic etching, consists of cooling the wafer to approximately −110°C. Unlike TDME, SF_6 is flowed continually throughout the process. The method is based on the reduced volatility of etch byproducts at these temperatures to minimize lateral etching of the sidewalls by spontaneous chemical etching. To enhance anisotropy, a passivating gas such as oxygen is commonly used, forming a SiO_xF_y blocking layer. Silicon etch rates as high as 5 μm/min can be achieved. In cryogenic etching, the SiO_xF_y layer formed on the sidewalls is very thin, ranging from 5 to 20 Å. The thickness of this layer is believed to increase with etch depth typically resulting in retrograde profiles as shown in Figure 18.27.

18.3.4 Metallization

Any silicon-based packaging must enable signal propagation off the package. This can take place in the form of electronic, optical or rf signals as required by the devices. In particular,

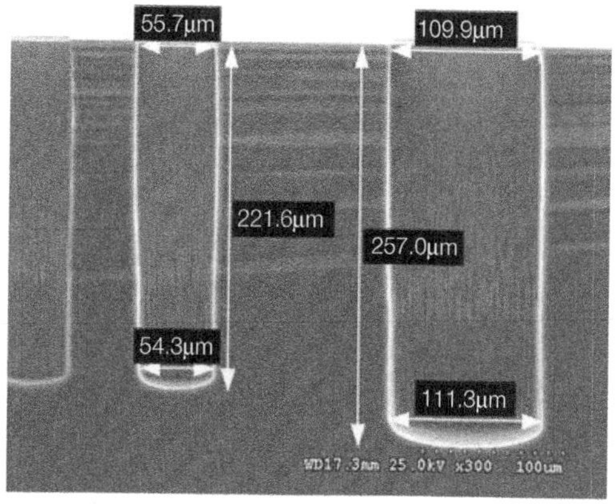

Figure 18.20: SEM of 250 μm deep vias etched in silicon.

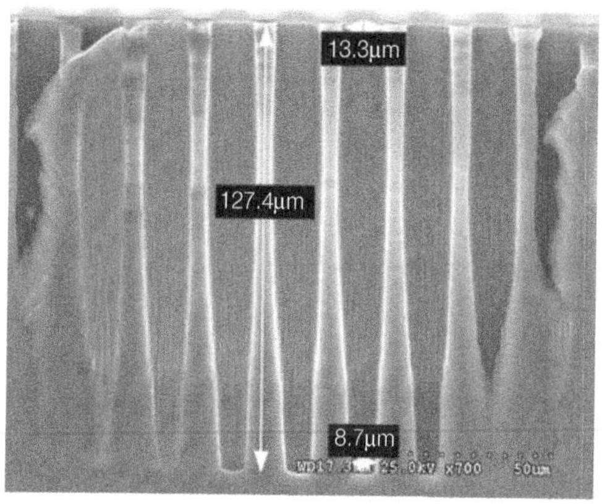

Figure 18.21: SEM demonstrating bowing in deep silicon etched vias.

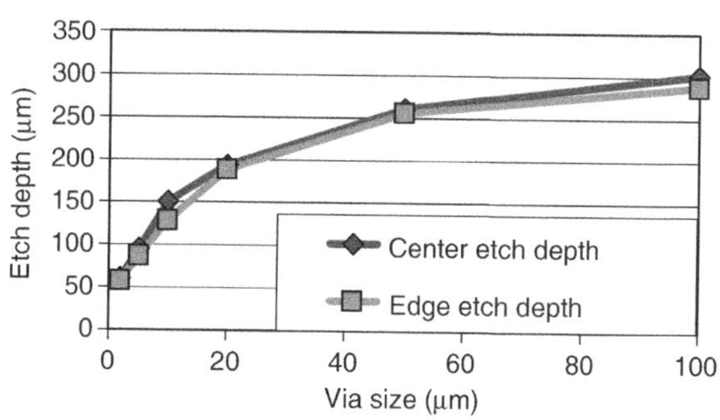

Figure 18.22: Graph showing the variation of etch depth as a function of via dimension.

(a) (b)

Figure 18.23: SEM showing (a) bottom and (b) sidewall roughness during deep silicon etch.

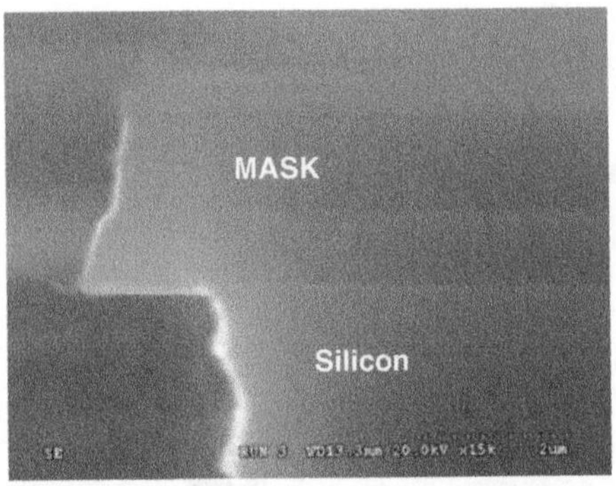

Figure 18.24: SEM showing mask undercut and pronounced scalloping inherent to the time multiplexed deep etching process.

electronic signals can be transmitted primarily through wire bonds and solder bumps interconnected with electrical through-vias; the focus here will be on electrical through-vias. Much research has taken place in recent years to determine the possible dimensions and materials available for fabricating through-vias in silicon. (The main impetus has been to create a three-dimensional interconnection which will help ease interconnect delay issues found in traditional integrated circuit devices and packaging.) Materials such as polysilicon, chemical vapor deposition (CVD) tungsten, electroplated copper, conductive adhesives, silver paste and others have been investigated. Some are limited to certain applications by the current they can carry (polysilicon), the thickness that can be deposited (tungsten), and the manufacturability (conductive adhesives/paste) in a semiconductor fab. Therefore, only copper electroplating, which is one of the most commonly used metallization processes, will be discussed.

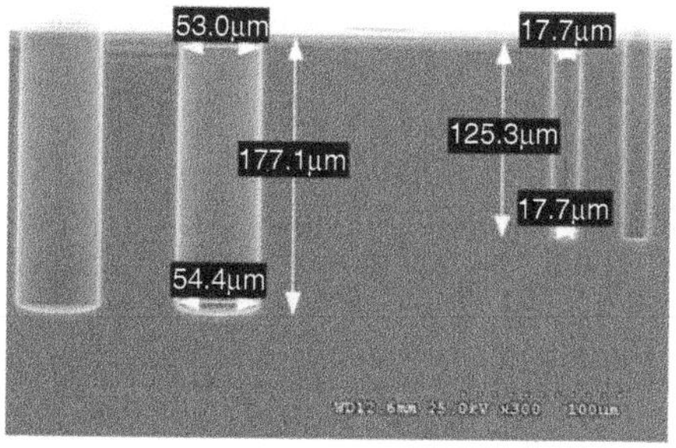

Figure 18.25: SEM showing silicon vias with smooth, straight profiles and minimal mask undercut and scalloping.

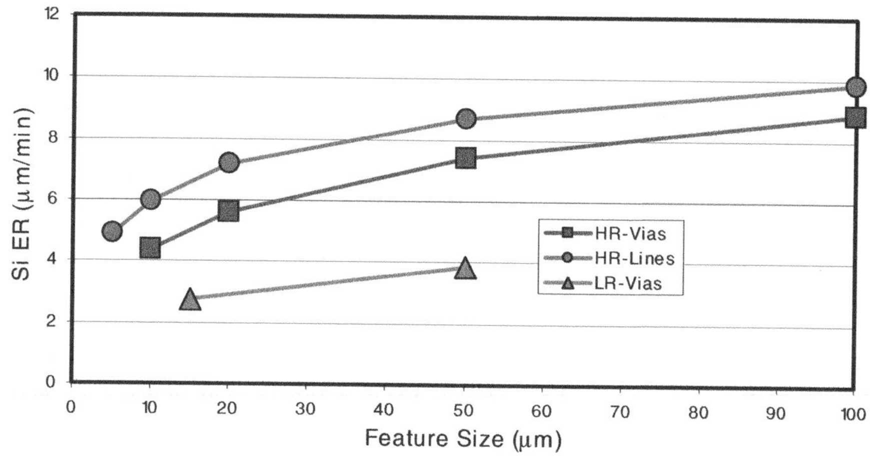

Figure 18.26: Graph showing silicon etch rate (ER) as a function of feature size.

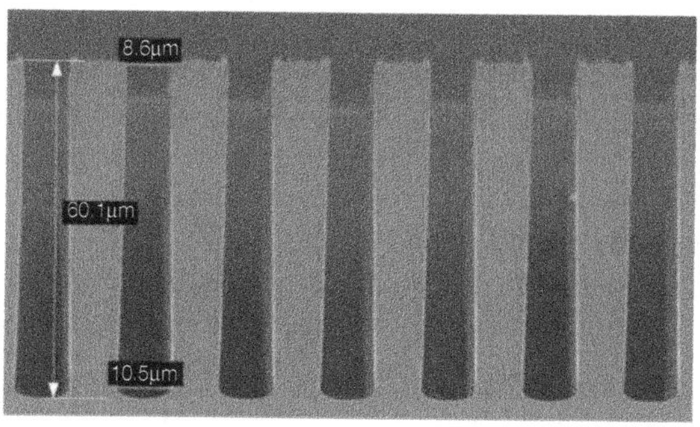

Figure 18.27: SEM showing retrograde profiles in cryogenically etched deep silicon vias.

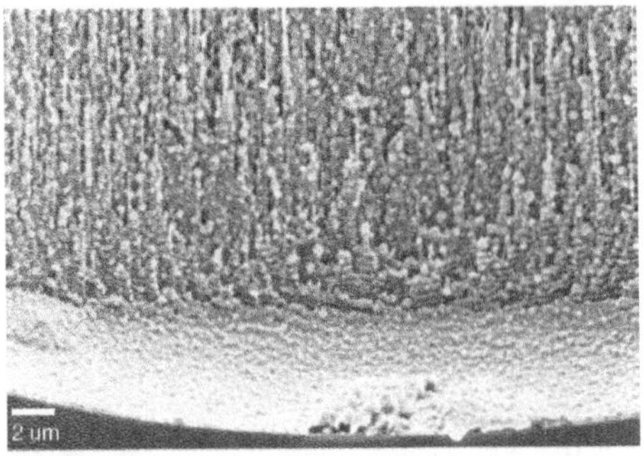

Figure 18.28: SEM showing PVD copper seed agglomeration.

18.3.4.1 Physical Vapor Deposition Copper Seed

All electroplating must be enabled by an electrically conducting seed layer for nucleation and growth of the desired plated metal. In standard semiconductor BEOL fabrication, this consists most commonly of physical vapour deposition (PVD) tantalum nitride/tantalum (TaNTa) and copper (Cu). However, other materials have also been pursued, especially when metallizing TSVs. TSV dimensions are often larger in scale and aspect ratio than standard copper damascene features found in the BEOL wiring build-up. Thus, the need for improved materials and processes has been pursued. Some of the other metal liners that have been evaluated include: metal organic-chemical vapor deposited titanium nitride (MO-CVD TiN), copper (MO-CVD Cu), PVD titanium (Ti), chromium/gold (Cr/Au) and others [25–27].

It is critical that the seed layer deposited has the optimal properties for promoting void-free copper electroplating. First, the coverage of the film inside the TSV must be continuous and uniform. A PVD source is often utilized because of its highly ionized (directional) deposition which results in seed at the bottom and lower sidewalls of the vias. Seed re-sputtering can be utilized to compensate for areas of thinner film deposition [28]. A related feature is the morphology of the film. If the seed layer is discontinuous, then the plated copper will form voids or agglomerations in the areas where the nucleation fails. This is shown in Figure 18.28. The underlying TSV feature also influences the quality of the film and its ability to initiate the desired via metallization. If the top of the via has an overhang, then the seed layer deposited will duplicate this overhang, resulting in a pinching off of the via before complete copper filling has taken place; see Figure 18.29. Lastly, oxidation (in the range 50–100 Å) of the copper seed layer will produce larger voids at the bottom of the vias being plated [28]. It is critical to reduce the copper oxides on the seed and additional steps are taken at the beginning of the copper electroplating process to address this.

If the seed layer deposited is not ideal to electroplating, changes to the process can be taken to address discontinuous film morphology or inadequate coverage. One of the most common solutions is to increase the thickness of the PVD Cu seed [29]. However, if this cannot be accomplished without creating a copper overhang at the top of the via, then another method that has been developed is electrochemical repair of the discontinuous copper seed. During this process, an electrolyte operating in a neutral to basic pH is used to plate a thin layer of Cu on the exposed liner and existing copper seed, resulting in a repaired, continuous seed

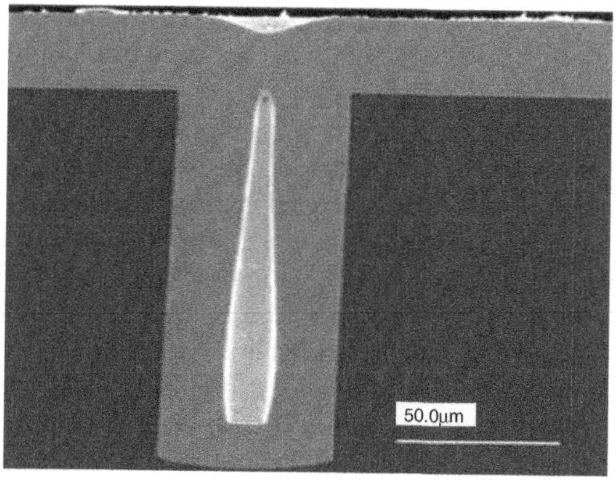

Figure 18.29: SEM showing pinched-off copper plated deep silicon via (from reference [9], reproduced by permission from C. K. Tsang, *et al.*, 'CMOS-Compatible silicon through-vias for 3D process integration', ed. C.A. Bower, P. Garrou, P. Ramm and K. Takahashi, *Mater. Res. Soc. Symp. Proc.*, vol. 970, Warrendale, PA, 2006, pp. 145–153).

Figure 18.30: SEMs taken at the bottoms of deep etched vias demonstrating the progression of Cu seed deposition and seed repair to conformally plated copper (from left to right).

layer [30]. This restoration process has been successfully used to enable Cu metallization of high aspect ratio through-vias [31]. Once a high-quality Cu seed layer has been achieved, the via is now finally ready for electroplating. Figure 18.30 demonstrates the progression of seed deposition, repair and plating within a via.

18.3.4.2 Copper Electroplating

Electrodeposition of copper, otherwise known as copper plating, has been used in the microelectronic industry for a variety of applications. Copper plating has long been used for filling through-holes and blind vias with conducting material. Most recently, copper plating has been developed for filling very small interconnects; copper for this application replaced aluminum and improved the resistivity and reliability of these ever-shrinking features. The development of these processes is a starting point for metallizing TSVs.

Electrodeposition of a metal from solution requires a conductive substrate. The substrate is brought in contact with an electrolyte, in this case an aqueous solution containing metal ions. A potential is applied between two electrodes; one being a cathode which is the conductive substrate of interest (i.e. wafer) and the other being an anode. This potential difference drives a current through the circuit between the anode and cathode. The current supplies electrons so that copper ions in solution are reduced to copper metal on the surface of the cathode. Copper metal is usually used as anode; copper from this electrode is dissolved into the solution. In this way, copper is being replenished in the solution as it is plated. As long as the conductive seed is continuous on the surface of the substrate, recesses (lines or vias) on the substrate will be plated with copper along with the blanket/flat areas of the sample.

Lack of wetting inside tall vias is a prevalent problem in this particular application of copper plating. A variety of techniques are used to overcome the wetting issue. One technique uses a vacuum to evacuate air pockets in features while the substrate is submerged in a solution (water, acid, etc.). Traditional spin-rinse-dry processes prior to plating also had been found to wet tall vias adequately. The success of these techniques is dependent on the shape and size and aspect ratio of the feature. One caveat of wetting the features before plating is subsequent mixing of the plating solution inside features that are wet with water (or another solution) when the substrate is immersed in the plating solution. Significant dwell times at open circuit are necessary in some cases to achieve good mixing before plating starts.

The goal for defect-free copper fill is superconformally to fill the features with copper from the bottom-up. Superconformal fill, or superfilling, mitigates the possibility of having seams, keyholes or voids inside the feature. Plating parameters such as plating bath constituents, applied waveform, cell hydrodynamics and bath temperature determine the degree of superfilling.

The plating bath constituents, both organic and inorganic, are very important. The main components of copper electrolyte for this application are copper sulfate, sulfuric acid, trace amounts of chloride ions (usually added as hydrochloric acid) and a number of organic additives. Mass transport of copper is important when plating in tall vias and large cavities. The copper plating baths used for this application usually contain higher copper concentrations than those used for plating smaller vias (i.e. interconnect sized features). The higher concentration of copper ensures that the solution is not depleted of copper ions deep inside the feature where there is no convection and little or no transport from the bulk solution. The acid concentration of the electrolyte is also important because it determines the conductivity of the solution (higher acid concentration allows higher current with less driving force/applied potential) and also affects the strength and interactions between the organic additives. Chloride ions are also included in copper plating baths; these halide ions impact dissolution of the anode and interact with the additive components of the bath.

With respect to organic additives used in the plating solution, there are usually three of interest: accelerators, suppressors and levelers. As their names suggest, accelerators accelerate or increase the rate of copper deposition, while suppressors suppress the rate or deposition. Levelers can be thought to actually 'level' or flatten the surface of the copper. The relative diffusion and adsorption of these molecules on the surface determine the quality of the copper fill inside vias and cavities. Superfilling of features relies on strong suppression in the field, outside the feature, and on the corners at the mouth of the feature while there is a highly accelerated rate of plating on the inside of the feature. The strong suppression can be realized with larger, bulkier molecules that have very fast adsorption kinetics (and

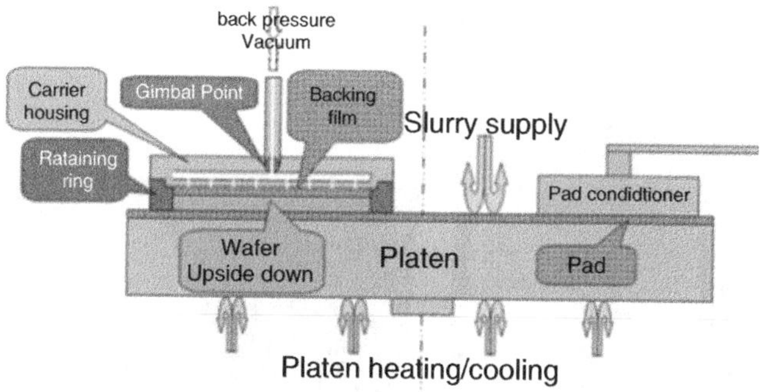

Figure 18.31: Schematic showing a typical CMP tool [33].

possibly slower diffusion inside the feature). Faster acceleration inside the feature relative to the outside is possible when the surface is pre-derivatized with the accelerator before plating starts [32]. Other ideas have been explored to 'shut down' acceleration in the field outside the via with the introduction of oxygen in solution during plating with results showing fully metallized vias [25].

The waveform is also a critical parameter in plating void-free copper. The deposits plated for this application are usually on the order of micrometers and faster plating is needed. Higher applied currents can compromise the fill capability of the bath. It is sometimes helpful to use stepped waveforms or pulse-plating. With pulsed waveforms, the average current is higher and plating is faster. Pulse-reverse is also an option because it allows concentration gradients to relax and copper to be redistributed inside the feature where it is being depleted.

18.3.4.3 Copper Chemical-mechanical Planarization

After copper electroplating of through-vias or of other Cu damascene wiring levels, the over-plated copper must be removed and polished flat. This is accomplished through chemical-mechanical planarization (CMP), sometimes called polishing. During the CMP process, overburden material is removed not only through a mechanically abrasive interaction between a polishing pad and the silicon wafer surface, but also through a chemical reaction of a corrosive slurry (a colloid solution) and the material being removed. The polish pad also aids in wiping away the reacted materials. This combination of chemical and mechanical interaction results in faster material removal and a uniform, topographically flat surface. A typical CMP setup is shown in Figure 18.31. The platen is covered with the abrasive pad and the wafer is placed faced down onto the pad. Pressure on the wafer is applied through vacuum on the wafer carrier which also contains a backing film. The wafer is confined within the retaining ring and cooled through either back spray technology (shown in schematic) or with contact on a water-cooled surface. It is important that the wafer temperature be controlled as this can influence the rate of the polishing process.

While CMP is routine in BEOL Cu wiring fabrication, some special considerations may be necessary when addressing unusually large features and cavities, increased material thicknesses, and larger open areas of metallurgy sometimes found in silicon package

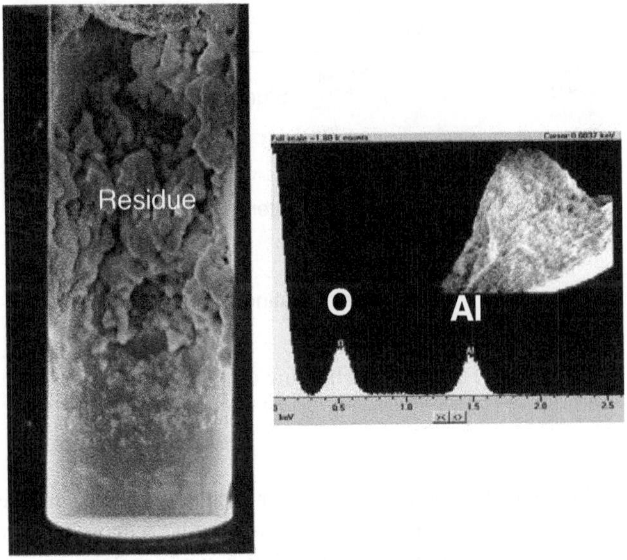

(a) (b)

Figure 18.32: (a) Cross-sectional SEM image showing CMP residue on the sidewalls of a partially plated Cu via and (b) EDS spectrum identifying residue of alumina oxide from CuCMP slurry (from reference [34], reproduced by permission from C. K. Tsang, *et al.*, 'Advanced CMP for enabling silicon carrier integration with through-vias and fine pitch wiring', *Proceedings of 11th International CMP Planarization for ULSI Multilevel Interconnection Conference (CMP-MIC)*, Fremont, CA, Feb. 21–23, 2006, pp. 63–72).

fabrication. One such example is the CMP process when incorporating large blind through-vias (in the range of tens of micrometers of vias in diameter and hundreds of micrometers in depth) or large cavities. Some applications may not require a completely metallized via or cavity. Instead, a sidewall of copper may be opted for as it is able to provide the needed electrical signal, but the whole via may not be filled with the purpose of reducing the CTE (coefficient of thermal expansion) mismatch between silicon and copper (see reference [6] for more detail). In such an example, the CMP process would occur in the presence of large open vias which could result in trapped slurry solution during processing, Figure 18.32(a). The initial approach to CMP of such a structure could be based on conventional Cu CMP and use existing commercial slurries for CMOS BEOL. In the first step, the overburden copper would be removed with the polish stopping on the underlying liner layer. Then, the liner is removed with a final topography planarization step. Such a process ends with a roller brush clean of the wafer. However, as Figure 18.32(a) demonstrates, the alumina-based slurry material could become trapped by the open via top and this was verified through energy dispersive x-ray spectroscopy (EDS), Figure 18.32(b). These large blind vias create an area of slurry entrapment and is a stagnant and inaccessible area for the final brush cleaning step.

In order to address this issue, alternative slurries, such as silica-based versions, and additional cleaning steps have been investigated and reported [34]. A cleaning solution

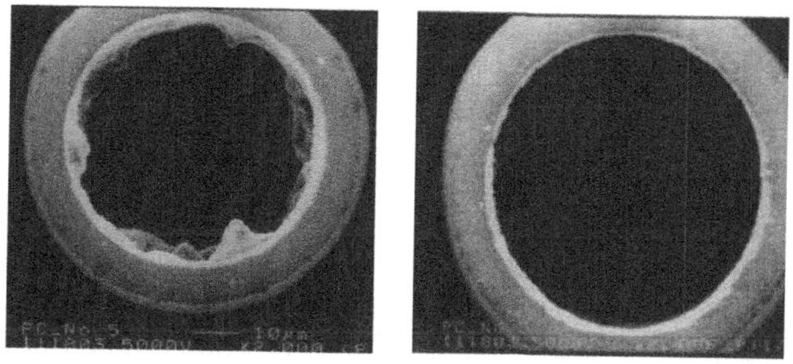

(a) (b)

Figure 18.33: (a) Top-down SEM image of post-CMP deep via without megasonics cleaning and (b) top-down SEM image of post-CMP megasonically-cleaned deep via (from reference [34], reproduced by permission from C. K. Tsang, *et al.*, 'Advanced CMP for enabling silicon carrier integration with through-vias and fine pitch wiring', *Proceedings of 11th International CMP Planarization for ULSI Multilevel Interconnection Conference (CMP-MIC)*, Fremont, CA, Feb. 21–23, 2006, pp. 63–72).

that could be implemented includes pre-wetting the wafers to introduce water into the vias, keeping the silicon wafers wet throughout the CMP process steps, and utilizing a megasonics clean after all the overburden Cu and liner layers have been removed [34]. This method is able successfully to prevent entrapment of slurry as shown in Figure 18.33.

Another challenge that could be encountered when fabricating silicon-based packages is an increase in the total metal open area of a wiring level or redistribution level. Sometimes silicon-based packaging will feature open pads, on the order of hundreds of micrometers, which are larger than used in conventional CMOS BEOL. These features tend to dish during CMP, resulting in greater than desired topography. In order to address this difference, additional CMP steps may be added to correct planarity. In one example that has been observed [34], a wiring level of greater than 85% open Cu area was fabricated. A consequence of this open area was topography of 200 nm over a 5 mm profilometry scan, with copper recessed below the surrounding dielectric surface. An additional polish with silica-based slurry planarized the raised areas of silicon dioxide in the field. Therefore, these examples show that some changes in conventional CMP processes may be needed in order to accommodate more challenging features found on some silicon packages.

18.3.5 Wafer Thinning

Another process that is heavily utilized in fabricating silicon packages is wafer thinning. Wafer thinning has become particularly important as there is an increasing interest in applications requiring stacked die packages, reduced die thicknesses, and other novel silicon-based packaging applications. These could then be used in popular consumer products to reduce size, weight and increase functionality. A standard semiconductor wafer is produced in the range 730–750 μm; however, dies can now be reduced to even 50 μm thickness, although the standard minimum thickness achieved with backgrinding is 150 μm [35, 36]. In order to meet these demands, cost-effective and reliable wafer thinning technology must be made available. There are three main methods of wafer thinning: (1) mechanical backgrinding and polishing, (2) wet/chemical etching, and (3) plasma etching. These may be used individually or in conjunction with each other, as needed.

18.3.5.1 Wafer Backgrinding

Wafer backgrinding is the most popular method of thinning because it is a mature, fast process and less costly compared to the others listed above. The steps involved in backgrinding are as follows: (1) taping, (2) coarse grinding, (3) fine grinding, (4) polishing (optional), and (5) detaping. The multi-step grinding process is utilized for improved productivity with 90% of the thinning achieved by the coarse grind [37]. Since the coarse grind leaves residual scratches in the surface of the wafer, the fine grind step is then applied to remove the scratches. Figures 18.34(a) and (b) show scratches left by backgrinding at 2000 and 1200 grit, respectively, [37]. If a mirror-finish surface is desired, an additional CMP-like polish with a SiO_2+H_2O slurry is utilized to remove around 2 μm of silicon before final detaping. The silica particle size ranges from 5 to 100 nm and is around 40–50 weight% [36]. The additional polish step could also reduce wafer warpage and strengthen the silicon wafer by removing any residual micro-cracks. Throughout the process, deionized water is used to keep the wafer clean. Process parameters that can be varied to influence the final wafer finish characteristics include initial and final wafer thickness desired, grit of grinder, speed of grind, and water temperature and flow rate.

18.3.5.2 Chemical Wet Etching

If the final die thickness is less than 150 μm, then additional processes are often combined with wafer backgrinding. Silicon strength is inversely related to the depth of scratches [37], and consequently chemical wet etching can be used to remove residual damage and strengthen the final product. This process is generally conducted in special, single-wafer tools designed to protect the front-side of the wafer during etch [38]. A thin stream of etchant is applied over the surface of the rotating wafer. Different etchant chemistries can be applied with some targeted for bulk silicon wafer etch, wafer polishing, and texture etch for uniform roughening. Most of these chemistries are still based on a mixture of HF (hydrofluoric acid) and HNO_3 (nitric acid) as previously mentioned in the section on isotropic wet etch. An average silicon etch rate is 10 μm per minute [38] and some suggest that at least 20 μm of silicon must be removed before good thickness variation can be achieved [36]. Figure 18.35 contains a schematic of a chemical wet etch tool and Figure 18.36 shows the increase in wafer strength following the addition of a wet etch after wafer backgrinding.

18.3.5.3 Plasma Etching

Plasma etching can be used not only to define vias and cavities on the front-side, but it is also a useful process in wafer thinning. In particular, plasma etch can be used to relieve stress built up in the silicon as a result of wafer backgrinding [35]. The application of plasma etching to wafer thinning uses gases including Ar/CF_4 and SF_6. A high wafer throughput with etch rates such as 20 μm/min has been demonstrated for the Ar/CF_4 plasma etch system [38]. A deep silicon RIE tool used for front-side wafer etch could also be modified to accept wafers for

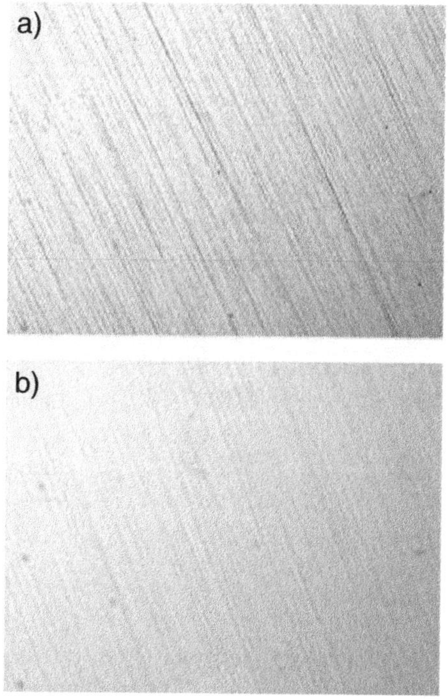

Figure 18.34: Silicon wafers background with (a) 2000 grit and (b) 1200 grit grinding wheels (30X) (from reference [37], reprinted with permission from E. G. Combs, 'The back-end process: step 3 wafer backgrinding', *Advanced Packaging*, vol. 11, no. 3, Mar. 2002, cited Jan. 11, 2008).

backside etching. The capital investment and process gas expense involved must be weighed when evaluating this process in comparison to the others described.

18.4 Assembly Options for Silicon-based Packaging

An important aspect of silicon-based packaging that must be considered is the test and assembly procedure. This is usually considered early in the design stage of the process flow as special test features, die dimensions and necessary bonding structures may need to be taken into account. In this section, a brief overview of wafer-to-wafer and die-to-die or die-to-wafer assembly will be given.

18.4.1 Wafer-level Processes

Wafer bonding is a broad and versatile process technology that has been utilized in many different areas such as silicon-on-insulator (SOI) technology, SOI-epi technology, photonics, optoelectronics, MEMS, three-dimensional integrated circuits, to name a few. This section will cover a few types of wafer bond technology that can be utilized for

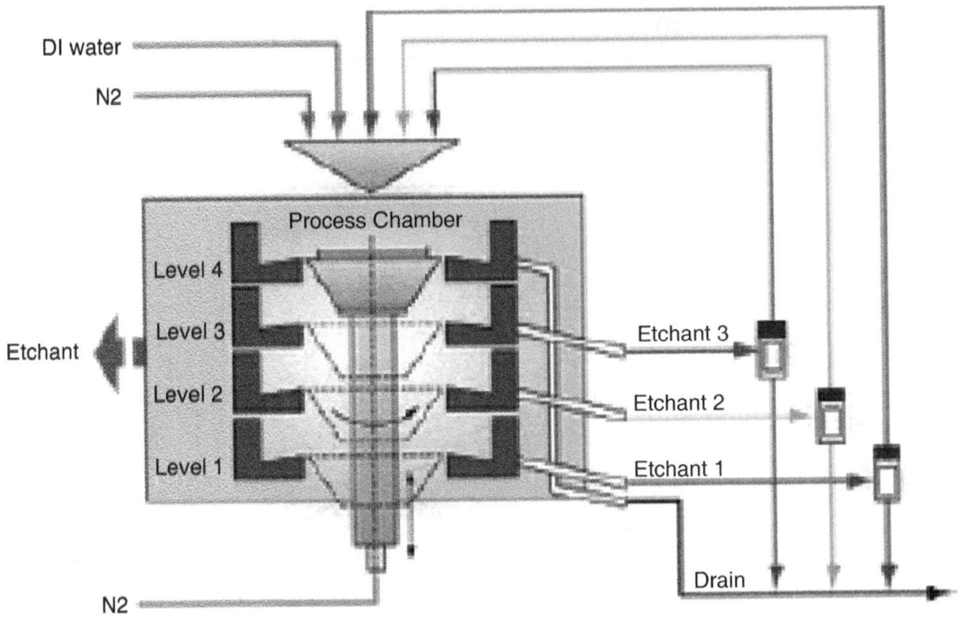

Figure 18.35: Schematic of a spin wet etch tool (from reference [38], reproduced with permission from M. Reiche and G. Wagner, 'Wafer thinning: techniques for ultra-thin wafers', *Advanced Packaging*, vol. 12, no. 3, Mar. 2003, pp. 29).

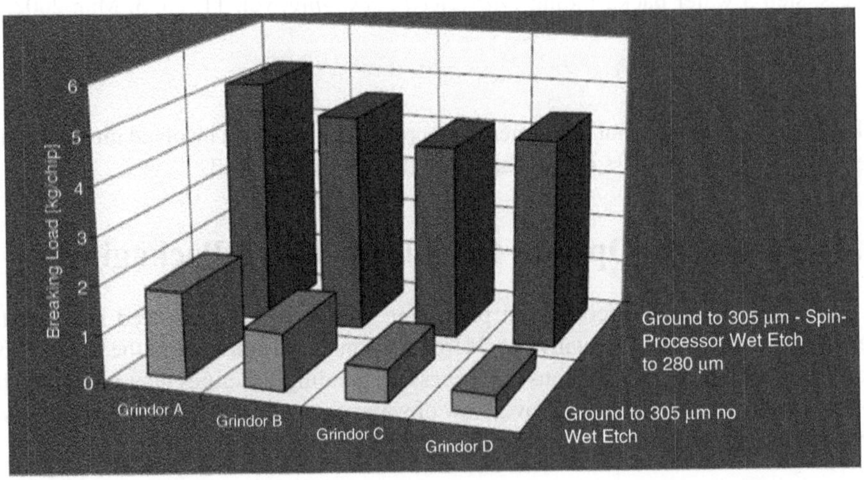

Figure 18.36: Comparison of wafer strength following wafer backgrinding process and backgrinding combined with chemical wet etch (from reference [39], courtesy of © SEZ Group).

wafer-level processing: oxide fusion bonding, copper thermo-compression bonding, and bonding with adhesives. Each of these bonding methods have their own specific process parameters, but the critical elements include wafer surface topography, surface cleanliness, and temperature/pressure conditions [40].

18.4.1.1 Oxide Fusion Bonding

Oxide fusion bonding is enabled by intermolecular van der Waals attraction forces between the two extremely smooth oxide surfaces. The oxide can be activated to form Si–O–H bonds through a wet dip in acid or a plasma treatment. Two prepared surfaces can then be brought together where the hydrogen bonds from hydroxyl groups and van der Waals attraction work in conjunction to bond the wafers at room temperature without the use of external adhesives or pressure [41]. The bond strength is determined in part by the bonding temperature and thus further anneal is often implemented after the initial bond. The advantage of this type of bonding is that very fine alignment can be achieved and preserved since there is no CTE mismatch experienced by the two wafers during the bonding process and no liquefying or movement of the bond material. Other dielectric materials beyond oxide can also be used, including SiO_2–Si, Si–O_2–Si_3N_4, Si_3N_4–Si_3N_4, and Si_3N_4–Si [41], providing many options to the fabrication process.

With all of these materials, careful surface preparation is necessary to ensure good bond results. Firstly, the smoothness of the surface (for a variety of materials considered) must be below 10 nm rms [42] with some using oxide surfaces of only 5 Å rms [43]. Surface roughness can be removed with CMP. The surface cleanliness is another concern because contaminants can reduce the surface reactivity and therefore reduce bond strength. Particles of just 1 μm diameter will result in a bubble between the bonded surfaces of 5000 times larger on a 4 in wafer [44], so other wafer cleans may be necessary post CMP. Another source of voiding or bubbles within the bonded surface could come from the dielectric material itself due to out-gassing. As such, sometimes it is necessary to implement a pre-bond anneal to densify the dielectric, generally done at temperatures similar to or above the film deposition temperature. The oxide wafers are then activated in common solutions such as a 1:1:5 solution of $NH_4OH + H_2O_2 + H_2O$ at 75 or 80° followed by drying. Bonding takes place and studies have shown that post-bond anneals at less than 300°C are sufficient for achieving bond strengths in excess of 1000 mJ/m^2 [41–43, 45]. Figure 18.37 shows an example of a desired oxide–oxide bonded interface.

18.4.1.2 Copper Thermo-compression Bonding

Another wafer-level adhesion method that is utilized is copper thermo-compression bonding. As the name suggests, this takes advantage of temperature and pressure to form strong bonds between copper surfaces. In contrast to the method described above, this process would only be useful once the wafers have entered the BEOL where copper is an accepted metal, however an advantage is that the metal bond could also be used as a conductive pathway for signals or even heat dissipation. With copper bonding, bond pad structures must be designed for the final top surface on each wafer. The wafers can then be aligned and bonded at temperatures in the range 300–400°C for 30–60 min and at down forces around 10 kN. Frequently, a post-bond anneal is also used to ensure stronger bond strength. Figure 18.38 shows a variety of

Figure 18.37: Oxide bonding interface shown in a TEM image of two surfaces brought together through oxide fusion bonding (from reference [46], reproduced with permission from A. W. Topol, *et al.*, 'Three-dimensional integrated circuits', *IBM J. Res. and Dev.*, vol. 50, no. 4/5, July/Sep. 2006, pp. 491–506).

process parameters with the types of bonding interfaces achieved [45]. Wafer level alignment of approximately 1 μm has been claimed by tool vendors [47]; however, the copper material will shift during the high temperature and pressure process, resulting in some misalignment post-bond. The tolerance post-bond is generally in the range of 5 μm. Similarly to oxide bonding, the surface preparation of the copper bond pads prior to thermo-compression will greatly affect the final bond interface and strength.

Copper bond pads for silicon-based packaging and other MEMS and three-dimensional IC applications are often prepared by standard Cu damascene processes as previously described. Thus, the CMP process is used to remove overburdened copper leaving a very planar surface between the dielectric in the field and the copper pad. If the copper pad is very large, there could be dishing of the pad up to 50 nm. If this dishing remains, then the two bonded pads will have a significant gap between the copper surfaces prior to bonding as shown in Figure 18.39. Various methods of reducing this gap and ensuring a proper contact surface have been previously described [48] utilizing oxide CMP and field etching with weak acids. Once the desired topography for bonding has been achieved (in some studies this includes copper that is above the surface of the oxide), then the copper must be cleaned to remove all copper oxides. Commonly used etchants include dilute sulfuric acid or hydrochloric acid [49]. Commercially available wafer align and bond tools are then used to complete the process. Figure 18.40 shows a FIB (focused ion beam) cut into a copper-copper bond indicating no voids, trapped particulates or bond interface.

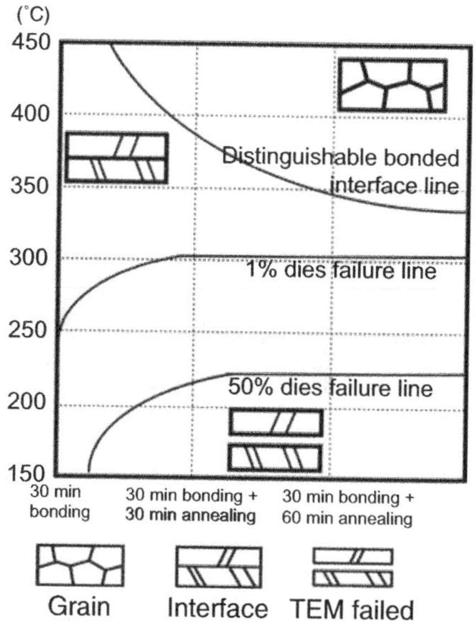

Figure 18.38: Graphical representation of resulting bond structures as a function of bonding conditions (from reference [45], reproduced by permission of © 2004 The Electrochemical Society).

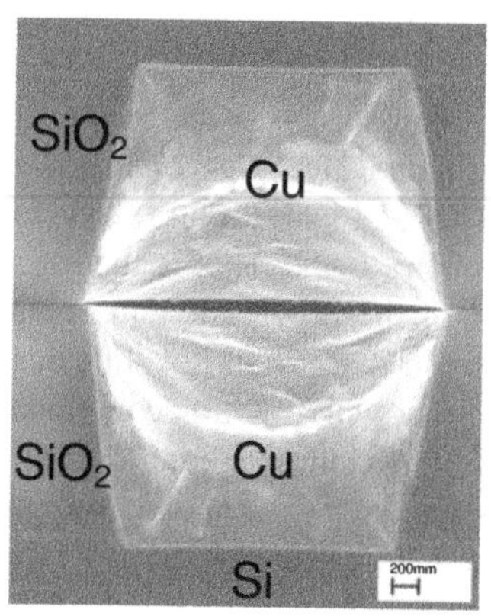

Figure 18.39: Representation using SEM images to show the resultant gap which would occur if two copper pads that display post-CMP dishing are brought in contact.

18.4.1.3 Adhesive Bonding

Adhesive bonding spans many different forms from temporary to permanent, low-temperature to high-temperature, non-conductive to conductive. They can consist of epoxies, polyimides, waxes, BCB, glues and many others. The selection of such adhesives would be driven by the specific application of the wafer bond and the type of device wafers involved. The thermal

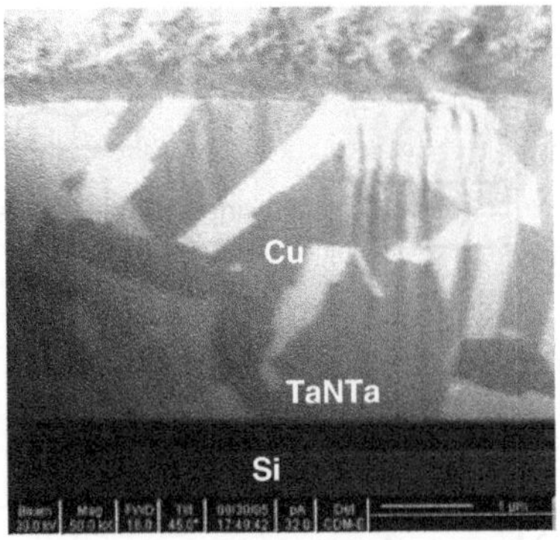

Figure 18.40: SEM image of two copper bond pads which were bonded through thermocompression. FIB was used to cut into the bonding interface, showing good cross-boundary grain structure.

budget of semiconductor wafers after BEOL processing is nominally 400°C. If the bond is to be permanent, then another consideration is whether or not electrical bonds must also be made with the other wafer. Electrical bonds could be achieved through some conductive adhesives or also through a combination of adhesive and copper bonding. An example of a BCB and copper bonding method has been previously detailed [50].

There are also applications which may require a temporary adhesive bonding during the fabrication and assembly process at the wafer-level, followed by dicing and die-level attach in the end. This is particularly useful in situations where a support wafer may be needed for the target wafer which is to be thinned. One example of this describes the attachment of a glass handler wafer to a target silicon wafer [42–52]. In this assembly, polyimide was applied and cured on the target wafer. Then, a PFTE (Teflon) layer is laid on the target wafer and the glass wafer is aligned on top of the stack. This is then laminated under isostatic pressure and a temperature in the range of the polyimide cure temperature [51]. The glass used is thermally-matched to the silicon so that it will not shift during the lamination process. The PTFE foil acts as a compliant layer which will accommodate the presence of particles and imperfections in flatness. In this way, adhesive bonding could have less stringent parameters compared to oxide bonding. When the glass handler is no longer needed, it can be removed through laser ablation and the remaining residue is easily cleaned with an oxygen ash.

18.4.2 Die-level Processing

One of the advantages to wafer-level assembly is certainly the expected throughput compared to die-to-die assembly. However, there are also some drawbacks that can be better addressed with die-to-die or die-to-wafer joining alternatives. As mentioned previously, high wafer co-planarity is needed in order to assure complete joining across the wafer. Furthermore, yield of the wafers before and after assembly is an important issue. High yield is required of each individual wafer and of the process itself, otherwise the defects would multiply. Lastly, wafer-level assembly is limited to corresponding chip and wafer sizes, which would limit the chip design and flexibility of assembly.

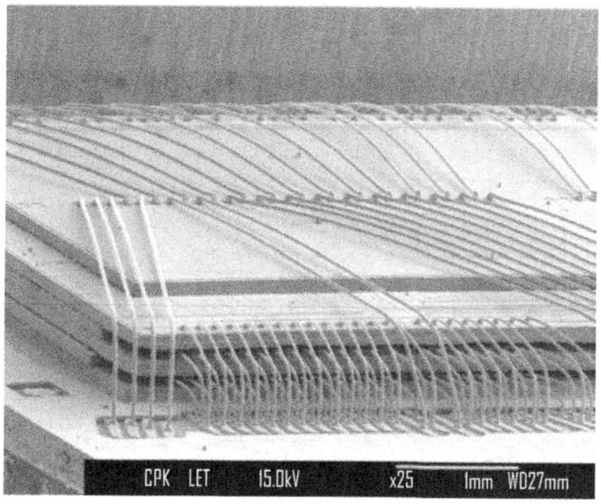

Figure 18.41: SEM image of a stacked-die package consisting of four die (from reference [53], reproduced by permission from EVG SmartView Wafer-to-Wafer Bond Aligner, cited on Jan. 11, 2008 *Advanced Packaging*, vol. 13, no. 8, Aug. 2004).

If die-level assembly were utilized, then known-good-dies (KGDs) of heterogeneous devices consisting of varying dimensions, fabricated in different manufacturing lines could be brought together. Many different styles and materials for bonding are available. Wire-bonding is commonly selected as a cost-effective method of die-to-die assembly. Figure 18.41 shows an example of a wire-bonded assembly, which involves a pad on the die side, a pad on the package/substrate and the wire connection. The types of wire used are gold, aluminum, copper and palladium, through either a ball-bond (90% of all electronics) or a wedge-bond (Figure 18.42). Table 18.2 shows the types of pad used for wire-bonding along with the method of bonding: thermo-compression, ultrasonic or thermosonic. Another option for die-level assembly is flip-chip interconnection which might be advantageous in higher density, area array configurations. Flip-chip is a die-attach structure where the active face of the device faces down on the package. It can also be used with silicon-based packages where assemblies may include bonding a device into a cavity followed by flip-chip join onto another substrate. The connections here are both mechanical and electrical and could be made with solder connections, including high-lead (97Pb3Sn), eutectic, and lead-free could be considered. Other metal alloys that are also chosen for bonding include Cu–Cu, Sn–capped Cu, Au–Au, Au–Sn, to name a few. The physical process of attachment takes place through melting or inter-diffusion. After bonding, under-fill materials may be used to act as a stress-reliever and/or environmental seal. The bonding metallurgy must be chosen to allow for correct thermal hierarchy if multiply joins are needed and another consideration could include hermeticity of the assembly.

18.5 Example of mmWave System on Silicon Package

Thus far, various CMOS BEOL processes have been examined for their suitability in fabricating silicon-based packages. Here the pieces are connected and an example of a

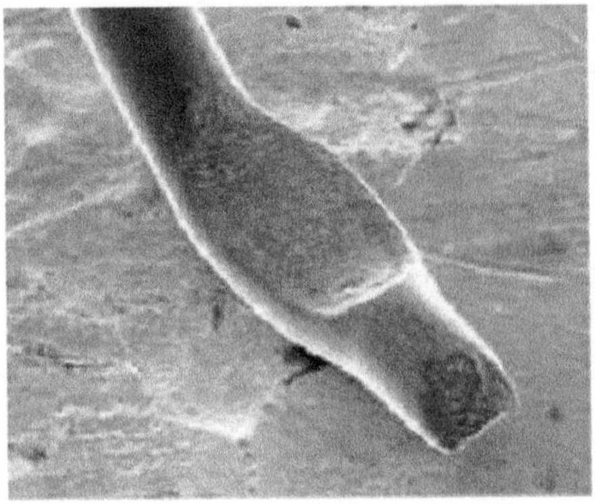

Figure 18.42: SEM image of a wedge bond (courtesy of NASA Goddard Space Flight Center).

Table 18.2: Wirebonding methods and materials.

Wirebond method	Temperature (°C)	Wire materials	Pad materials	Remarks
Thermocompression	300–500	Au	Al, Au	High pressure, no ultrasonic energy
Ultrasonic	25	Al, Au	Al, Au	Low pressure in ultrasonic energy
Thermosonic	100–240	Au, Cu	Al, Au	Low pressure in ultrasonic energy

Reproduced from 'The Nordic Electronics Packaging Guideline', Chapter 'W Wirebonding', Jan. 30, 2008.

mmWave system, fabricated and assembled with silicon-packaging technology, is presented. Figure 18.43 shows a schematic of this silicon package concept containing through-silicon vias, RF IC chip, on-package antennas and MEMS antenna switch. In order to determine the feasibility of this mmWave Si-based system, a 60 GHz antenna was designed with the desired bandwidth, efficiency and performance. The antenna itself was fabricated on high resistivity silicon (1000 Ω cm, shown in Figure 18.44(a) and was assembled on top of the metallized cavity (Figure 18.44(b)) as conceived in Figure 18.43.

The top antenna was fabricated through Cu damascene methods as previously described in earlier sections. Dielectric layers of 70 nm thick silicon nitride and 1.0 μm thick oxide separated the Cu level from the silicon substrate. The dielectric was then patterned, etched and Cu electroplated to metallize the 1.2 μm thick Cu antennas. The special considerations for planarization, mentioned earlier in the section discussing CMP, was applied to result in planar Cu antennas. Once BEOL process steps were completed, the full thickness wafer of 725 μm was background to 150 μm.

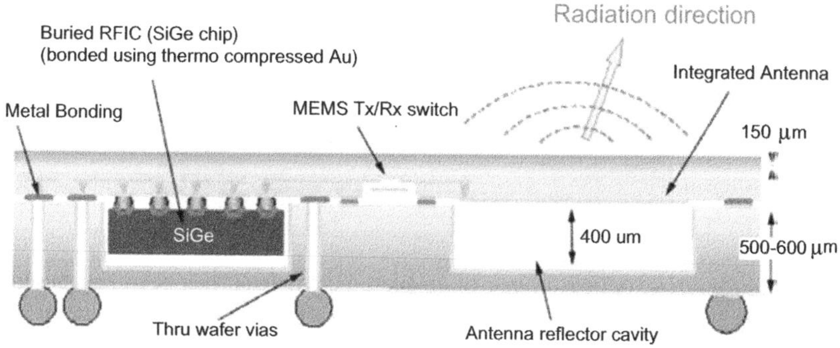

Figure 18.43: Schematic of a mmWave package fabricated with silicon-based packaging, (from reference [1], reproduced by permission of © 2007 IEEE).

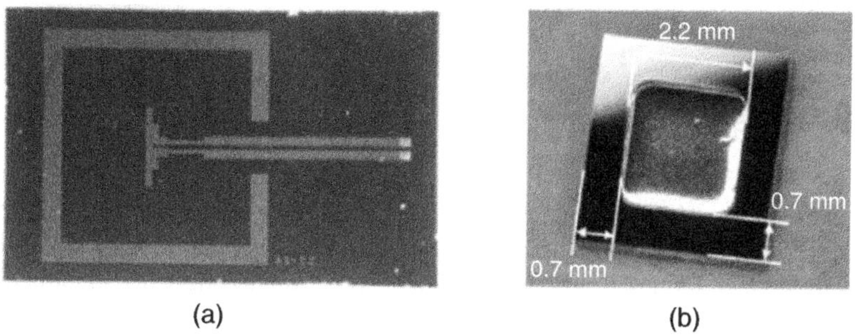

Figure 18.44: (a) Optical image of an antenna structure fabricated with a 1.2 μm Cu damascene process and (b) a 400 μm deep cavity etched in silicon and plated with 8 μm of Cu, (from reference [1], reproduced by permission of © 2007 IEEE).

The bottom piece of the package contained TSVs and large cavities. Both of these were etched with deep reactive ion etch through the Bosch process. The vias etched were 500 μm deep and the cavities were 2×2 mm^2 and 400 μm deep. Following etch definition, the features were lined with tantalum nitride and tantalum liner and a copper seed layer. Some of the processes described in the earlier sections were applied in order to electroplate 8 μm of copper on the walls of the vias and the cavities. Then, the overburdened Cu and liner was removed through CMP (see Figure 18.45). The next process steps were to apply needed metallurgy for assembly and to thin the wafer to the desired package thickness. In this example, the wafer was not thinned because the tests were focused on the antenna. The antenna and cavity wafers were subsequently diced and the two parts were assembled by hand with an adhesive. Alignment of the top and bottom dies was achieved between 50 and 100 μm accuracy and verified through IR imaging. An antenna-on-cavity package joined to a FR4 test fixture which connected to a coplanar probe. The test allowed for measurement of the antenna resonant frequency which was very close to simulated numbers. A gain between 6 and 8 dBi in the frequency range 60–65 GHz was measured showing a well-defined uniform

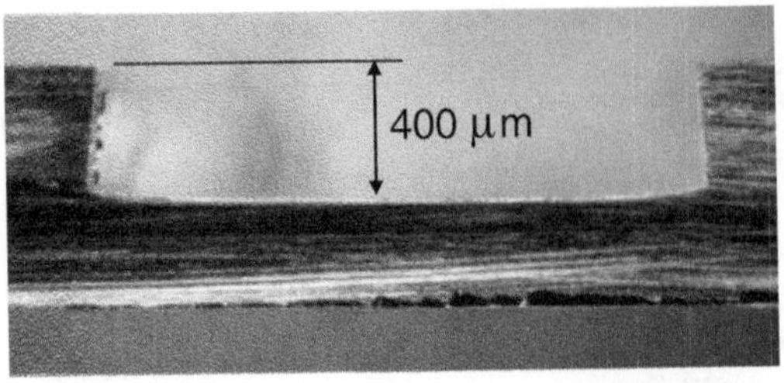

Figure 18.45: Side-view, optical image of cu-plated cavity that has been processed through CMP and dicing steps.

radiation pattern. This example showed that indeed silicon-based packaging technology and semiconductor processes can be used to form the elements of a mmWave system and can further be investigated for novel applications.

Acknowledgments

The authors wish to thank Brett Baker-O'Neal and James Kelly for their assistance in writing this chapter along with the following IBM personnel for their contributions to process development: Edmund Sprogis, Chirag Patel, Bing Dang, Steven Wright, Nils Hoivik, Christopher Jahnes, John Cotte, Anna Topol, Kuan-neng Chen, Leathen Shi, Roy Yu, Fuad Doany, Niranjana Ruiz, John Wilson, Daniel Lisounenko, William Graham, James Vichiconti, Keith Kwietniak, James Tornello, Michael Lofaro, Donald Canaperi, Igor Konovalov, Matthew Farinelli, Timothy Krywanczyk, Steve Codding, David Dmilia, and Ranjani Sirdeshmukh. They also appreciate the support from IBM management including John Knickerbocker, Timothy Chainer, David Seeger and T. C. Chen.

Some of the examples presented were supported through funding by DARPA contract N66001-02-C-8014, the Chip-to-Chip Optical Interconnects (C2OI) Program Agreement MDA 972-03-3-0004, the PERCS program Agreement NBCH30390004, Maryland Procurement Office (MPO) Contract H98230-04-C-0920, the SSC SAN DIEGO Contract N66001-04-C-8032 and N66001-00-C-8003.

References

[1] N. Hoivik, *et al.*, 'High-efficiency 60 GHz antenna fabricated using low-cost silicon micromachining techniques', *IEEE International Symposium on Antennas and Propagation*, Honolulu, Hawaii, June 9–15, 2007, pp. 5043–5046.

[2] J. U. Knickerbocker, *et al.*, 'Development of next-generation system-on-package (SOP) technology based on silicon carriers with fine-pitch chip interconnection', *IBM Journal of Research and Development* **49** (2005), pp. 725–753.

[3] J. U. Knickerbocker, *et al.*, 'Three-dimensional silicon integration and silicon packaging technology using silicon through-vias', *Journal of Solid State Circuits* **41** (2006), pp. 1718–1785.

[4] H. Gan, *et al.*, 'Pb-free micro-joints for the next generation micro-systems: the fabrication, assembly and characterization', *Proc. of 56th Electronic Components and Technology Conference*, San Diego, CA, May 30–June 2, 2006, pp. 1210–1215.

[5] S. L. Wright, *et al.*, 'Characterization of micro-bump C4 interconnects for Si-Carrier SOP applications', *Proc. of 56th Electronic Components and Technology Conference*, San Diego, CA, May 30–June 2, 2006, pp. 633–640.

[6] C. S. Patel, P. S. Andry, K. A. Jenkins, B. Dang, R. Horton, R. J. Polastre and C. K. Tsang, 'Characterization of flip chip microjoins up to 40 GHz using silicon carrier', *Proc. of 2005 International Interconnect Technology Conference*, June 2005, pp. 129–131.

[7] C. S. Patel, *et al.*, 'Silicon carrier with deep through vias, fine pitch wiring and through cavity for parallel optical transceiver', *Proc. of the 55th Electronic Components and Technology Conference*, Lake Buena Vista, FL, May 31–June 3, 2005, pp. 1318–1324.

[8] P. S. Andry, *et al.*, 'A CMOS-compatible process for fabricating electrical through-vias in silicon', *Proc. of 56th Electronic Components and Technology Conference*, San Diego, CA, May 30–June 2, 2006, pp. 831–837.

[9] C. K. Tsang, P. S. Andry, E. J. Sprogis, C. S. Patel, B. C. Webb, D. G. Manzer and J. U. Knickerbocker, 'CMOS-Compatible silicon through-vias for three-dimensional process integration', ed. C. A. Bower, *et al.*, *Materials Research Society Symposium Proceedings*, vol. 970, Warrendale, PA, 2006, pp. 145–153.

[10] P. S. Andry, C. K. Tsang, B. C. Webb, E. J. Sprogis, S. L. Wright, B. Dang and D. G. Manzer, 'Fabrication and characterization of robust through-silicon vias for silicon carrier applications', *IBM Journal of Research and Development* **52** (2008), pp. 571–582.

[11] K. A. Jenkins and C. S. Patel, 'Copper-filled through wafer vias with very low inductance', *Proc. of 2005 International Interconnect Technology Conference*, San Francisco, CA, June 2005, pp. 144–146.

[12] N. P. Pham, E. Boellaard, J. N. Burghartz and P. M. Sarro, 'Photoresist coating methods for the integration of novel 3-D RF picrostructures', *Journal of Microelectromechanical Systems* **13**(3) (2004), pp. 491–499.

[13] K. A. Cooper, C. Hamel and B. Whitney, 'Conformal photoresist coatings for high aspect ratio features', *Proc. of 2007 International Wafer-level Packaging Conference*, San Jose, CA, June 2007.

[14] B. Wieder, C. Brubaker, T. Glinsner, P. Kettner and N. Nodes, 'Spray coating for MEMS, NEMS and Micro systems', *Proc. of Pacific Rim Workshop on Transducers and Micro/Nano Technologies*, Xiamen, China, 2002.

[15] N. P. Pham, P. M Sarro, J. Bertens and L. Van den Brekel, 'Comparing the costs of photoresist coating using spin, spray or electrodeposition systems', *Micromagazine*, April 2005, pp. 45–49, available from: http://www.micromagazine.com/archive/05/04/pham.html.

[16] Available from: http://heritage.dupont.com/floater/fl_riston/floater.shtml [cited on April 7, 2008].

[17] R. R. Tummala, eds., *Fundamentals of Microsystems Packaging* (New York: McGraw-Hill, 2001).

[18] M. Elwenspoek and H. V. Jansen, *Silicon Micromachining*, no. 7 of *Cambridge Studies in Semiconductor Physics and Microelectronic Engineering* (New York: Cambridge University Press, 1998).

[19] 'Silicon VLSI technology fundamentals, practice and modeling' [cited on April 7, 2008], available from http://www.iaa.ncku.edu/˜ggau/Lab/HTML/C_CourseData/SILICON012.pdf

[20] H. Ni, H.-J. Lee and A. G. Ramirez, 'A robust two-step etching process for larger-scale microfabricated SiO_2 and Si_3N_4 MEMS membranes', *Sensors and Actuators A* **119** (2005), pp. 553–558.

[21] 'KOH Etching of silicon wafers, silicon dioxide, and silicon nitride', cited on April 7, 2008, available from: http://www.ee.byu.edu/cleanroom/KOH.phtml.

[22] H. Seidel, L. Csepregi, A. Heuberger and H. Baumgartel, 'Anisotropic etching of crystalline silicon in alkaline solutions', *Journal of the Electrochemical Society* **137**(11) (1990), pp. 3612–3626.

[23] J. Leib, D. Mund and M. Topper, 'Novel hermetic wafer-level-packaging technology using low-temperature passivation', *Proc. of 55th Electronic Components and Technology Conference*, Lake Buena Vista, FL, May 31–June 3, 2005, pp. 562–565.

[24] M. Sekiguchi, H. Numata, N. Sato, T. Shirakawa, M. Matsuo, H. Yoshikawa, M. Yanagida, H. Nakayoshi and K. Takahashi, 'Novel low cost integration of through chip interconnection and application to CMOS image sensor', *Proc. of 56th Electronic Components and Technology Conference*, San Diego, CA, May 30–June 2, 2006, pp. 1367–1374.

[25] K. Takahashi, *et al.*,'Process integration of three-dimensional chip stack with vertical interconnection', *Proc. of 54th Electronic Components and Technology Conference*, Las Vegas, NV, 2004, pp. 601–609.

[26] S. W. Ricky Lee, R. Hon, S. X. D. Zhang and C. K. Wong, 'Three-dimensional stacked flip chip packaging with through silicon vias and copper plating or conductive adhesive filling', *Proc. of 55th Electronic Components and Technology Conference*, Lake Buena Vista, FL, May 31–June 3, 2005, pp. 795–801.

[27] C. K. Tsang and A. W. Topol, 'Three-dimensional integrated circuits and silicon carrier packaging realization', *Proc. of 23rd International VLSI Multilevel Interconnection Conference (VMIC)*, Fremont, CA, September 26–28, 2006, pp. 61–69.

[28] J. Reid, S. Mayer, E. Broadbent, E. Klawuhn and K. Ashtiani, 'Factors influencing damascene feature fill using copper PVD and electroplating', *Journal of Solid State Technology* **43**(7) (2000), pp. 86–103.

[29] J. J. Kelly, M. Zheng, H. Deligianni, B. C. Baker-O'Neal and C. Cabral, Jr., 'Electrodeposition of Cu on Ta-based layer: Cu electrodeposition on Cu/Ta bilayers', *Journal of the Electrochemical Society* **154**(8) (2007), pp. D406–D410.

[30] M. Zheng, J. J. Kelly and H. Deligianni, 'Electrodeposition of Cu on Ta-based layers: Electrodeposition on Ta', *Journal of the Electrochemical Society* **154**(8) (2007), pp. D400–D405.

[31] B. Kim, C. Sharbono, T. Ritzdorf and D. Schmauch, 'Factors affecting copper filling process within high aspect ratio deep vias for three-dimensional chip stacking', *Proc. of 56th Electronic Components and Technology Conference*, San Diego, CA, May 30–June 2, 2006, pp. 838–843.

[32] T. P. Moffat, D. Wheeler, W. H. Huber and D. Josell, 'Superconformal electrodeposition using derivitized substrates', *Electrochemical Solid-State Letters* **5**(12) (2002), pp. C110–C112.

[33] 'CMP Process', cited on January 11, 2008, available at http://crystec.com/alpovere.htm.

[34] C. K. Tsang, D. F. Canaperi, J. A. Tornello, L. P. Buchwalter and D. L. Rath, 'Advanced CMP for enabling silicon carrier integration with through-vias and fine pitch wiring', *Proc. of 11th International CMP Planarization for ULSI Multilevel Interconnection Conference (CMP-MIC)*, Fremont, CA, February 21–23, 2006, pp. 63–72.

[35] W.C. Lo, Y. H. Chen, J. D. Ko, T. Y. Kuo, C. W. Chien, Y. C. Chen, W. Y. Chen, F. J. Leu and H. T. Hu, 'Development and characterization of low cost ultrathin three-dimensional interconnect', *Proc. of 55th Electronic Components and Technology Conference*, Lake Buena Vista, FL, May 31–June 3, 2005, pp. 337–342.

[36] L. Wu, J. Chan and C. S. Hsiao, 'Cost-performance wafer thinning technology', *Proc. of 53rd Electronic Components and Technology Conference*, New Orleans, LA, May 27–30, 2003, pp. 1463–1467.

[37] E. G. Combs, 'The back-end process: step 3 wafer backgrinding', *Advanced Packaging*, vol. 11, no. 3, Mar. 2002, cited Jan. 11, 2008, available at http://ap.pennnet.com/display_article/138561/36/ARTCL/none/none/1/Thebackendprocess:Step-3Wafer-backgrinding/.

[38] M. Reiche and G. Wagner,'Wafer thinning: techniques for ultra-thin wafers', *Advanced Packaging* **12**(3) (2003), pp. 29.

[39] 'Benefits of wet chemical spin-processing for ultra-thin wafer applications' [cited Jan. 11, 2008], available at http://www.apialliance.com/pdf/Archive_02/SEZ.pdf.

[40] J. Haisma, 'Direct bonding, fusion bonding, anodic bonding, wafer bonding: A historical patent picture of the worldwide moving front of the state-of-the-art of contact bonding', ch. 1 in *Wafer Bonding: Applications and Technology*, vol. 75 of *Springer Series in Materials Science*, ed. R. Hull, *et al.* (New York: Springer, 2004).

[41] A. W. Topol, *et al.*, 'Enabling SOI-based assembly technology for three-dimensional (3-D) integrated circuits (ICs)', *IEDM Technical Digest*, December 5–7, 2005, pp. 352–355.

[42] A. W. Topol, S. J. Koester, D. C. La Tulipe and A. M. Young, '3-D fabrication options for high performance CMOS technology' in '*Wafer Level 3-D ICs Process Technology*', ed. C. S. Tan, R. Gutmann, and R. Reif *(Wafer Level 3-D ICs Process Technology Series: Series on Integrated Circuits and Systems, XII)*, 410 pp., 2008.

[43] C. K. Chen, K. Warner, D. R. W. Yost, J. M. Knecht, V. Sutharalingam, C. L. Chen, J. A. Burns and C. L. Keast, 'Scaling three-dimensional SOI integrated-circuit technology', *Proc. of 2007 IEEE International SOI Conference*, Indian Wells, CA, October 14, 2007, pp. 87–88.

[44] Q.-Y. Tong, 'Principles of Wafer Bonding', chap. 1 in *Silicon Wafer Bonding Technology for VLSI and MEMS Applications*, vol. 1 of *EMIS (Electronic Materials Information Service) Processing Series*, ed. S. S. Iyer and A. J. AubertonHerve (London: INSPEC, The Institution of Electrical Engineers, 2002).

[45] K. N. Chen, C. S. Tan, A. Fan and R. Reif, 'Morphology and bond strength of copper wafer bonding', *Electrochemical and Solid-State Letters* **7**(1) (2004), pp. G14–G16.

[46] A. W. Topol, *et al.*, 'Three-dimensional integrated circuits', *IBM Journal of Research and Development* **50**(4/5) (2006), pp. 491–506.

[47] 'EVG SmartView Wafer-to-Wafer Bond Aligner' [cited on Jan. 11, 2008], available at http://www.evgroup.com/market_to_product.asp?MTP_id18.

[48] K. N. Chen, *et al.*, 'Improved manufacturability of Cu bond pads and implementation of seal design in three-dimensional integrated circuits and packages', *Proc. of 23rd International VLSI Multilevel Interconnection Conference (VMIC)*, Fremont, CA, September 26–28, 2006, pp. 195–202.

[49] K. N. Chen, A. Fan, C. S. Tan, R. Reif and C. Y. Wen, 'Microstructure evolution and abnormal grain growth during copper wafer bonding', *Applied Physics Letters* **81**(20) (2002), pp. 3774–3776.

[50] J. J. McMahon, J.-Q. Lu and R. J. Gutmann, 'Wafer bonding of damascene-patterned metal/adhesive redistribution layers for via-first three-dimensional (three-dimensional) interconnect', *Proc. of 55th Electronic Components and Technology Conference*, Lake Buena Vista, FL, May 31–June 3, 2005, pp. 331–336.

[51] M. Despont, 'Wafer-scale microdevice transfer/interconnect: its application in an AFM-based data-storage system', *Journal of Microelectromechanical Systems* **13**(6) (2004), pp. 895–901.

[52] C. Narayan, S. Purushothaman, F. Doany and A. Deutsch, 'Thin film transfer process for low cost MCM-D fabrication', *IEEE Trans. on Components, Packaging, and Manufacturing Technology Part B* (1995), **18**(1), pp. 42–46.

[53] M. Karnezos, 'Stacked-die Packaging: Technology Toolbox, Step 8', *Advanced Packaging* **13**(8) (2004). Available at http://ap.pennnet.com/Articles/Article_Display.cfm?Section=Archiv... Display&ARTICLE_ID=209839 [cited January 11, 2008].

[54] 'The Nordic Electronics Packaging Guideline, Chapter W Wirebonding' [cited on January 30, 2008], available at http://extra.ivf.se/ngl/A-WireBonding/ChapterA1.htm#A1.1.

Index

 © 2009 John Wiley & Sons, Ltd

Printed and bound by CPI Group (UK) Ltd, Croydon, CR0 4YY

16/09/2026

14963475-0007